SPECTRAL METHODS IN INFINITE-DIMENSIONAL ANALYSIS

Mathematical Physics and Applied Mathematics

Editor:

The titles published in this series are listed at the end of this volume.

Volume 12/1

SPECTRAL METHODS IN INFINITE-DIMENSIONAL ANALYSIS

Volume 1

by

Y. M. Berezansky

and

Y. G. Kondratiev

Institute of Mathematics,
Ukrainian Academy of Science,
Kiev, Ukraine

SPRINGER-SCIENCE+BUSINESS MEDIA, B.V.

Library of Congress Cataloging-in-Publication Data

Berezanskiĭ, Ĭ͡U. M. (Ĭ͡Uriĭ Makarovich)
[Spektral'nye metody v beskonechnomernom analize. English]
Spectral methods in infinite-dimensional analysis / by Yurij M. Berezansky and Yurij G. Kondratiev.
p. cm. -- (Mathematical physics and applied mathematics ; v. 12)
Includes bibliographical references and index.
ISBN 978-94-010-4227-7 ISBN 978-94-011-0509-5 (eBook)
DOI 10.1007/978-94-011-0509-5
1. Quantum field theory--Mathematics. 2. Degrees of freedom. 3. Statistical physics. 4. Spectral theory (Mathematics) 5. Infinite groups. 6. Mathematical physics. I. Kondrat'ev, Ĭ͡U. G. (Ĭ͡Uriĭ Grigor'evich) II. Title. III. Series.
QC174.52.D43B47 1994
530.1'55782--dc20 94-14885

ISBN 978-94-010-4227-7

This is the updated and revised translation of the original Russian work,
Spectral Methods in Infinite-Dimensional Analysis.
Translated by P.V. Malyshev and D. V. Malyshev.
Published by Naukova Dumka, Kiev, © 1988.

Printed on acid-free paper

Contents

Volume 1

Preface to the English Edition

The Russian edition of this book appeared 5 years ago. Since that time, many results have been improved upon and new approaches to the problems investigated in the book have appeared. But the greatest surprise for us was to discover that there exists a large group of mathematicians working in the area of the so-called White Noise Analysis which is closely connected with the essential part of our book, namely, with the theory of generalized functions of infinitely many variables. The first papers dealing with White Noise Analysis were written by T. Hida in Japan in 1975. Later, this analysis was developed intensively in Japan, Germany, U.S.A., Taipei, and in other places.

The related problems of infinite-dimensional analysis have been studied in Kiev since 1967, and the theory of generalized functions of infinitely many variables has been investigated since 1973. However, due to the political system in the U.S.S.R., contact between Ukrainian and foreign mathematicians was impossible for a long period of time. This is why, to our great regret, only at the end of 1988 did one of the authors meet L. Streit who told him about the existence of White Noise Analysis. And it become clear that many results in these two theories coincide and that, in fact, there exists a single theory and not two distinct ones.

Certainly, the authors were unable to revise the English translation of the book in order to discuss this situation in any detail — for this purpose, one should write a completely new book. This is why we have only added several most important papers on White Noise Analysis to our list of references, and added some remarks to the Bibliographical Notes.

For technical reasons, we have found it reasonable to divide the English edition into two volumes. This first volume contains Chapters 1, 2, 3, and 4. In Volume 2, we study the application of the spectral projection theorem to harmonic analysis (the infinite-dimensional moment problem, positive definite functions of infinitely many variables, etc.) and the application of the spectral theory of infinite-dimensional differential operators to models of quantum statistical physics and field theory. Note that the general problems of the spectral theory of infinite-dimensional differential operators are presented in the books of Berezansky [18, 26]. At the end of this volume, we also give the contents of Volume 2. For convenience, we decided to give the full list of references in both volumes. Certainly, we have eliminated the misprints and errors which were found in the Russian edition.

Finally, we are happy to express our deep gratitude to Kluwer Academic Publishers for publishing the English edition of our book.

September, 19, 1993 — Yu. M. Berezansky, Yu. G. Kondratyev

Preface

This book appeared as a result of our interest in different areas of infinite-dimensional analysis. This part of functional analysis has no strictly outlined boundaries and, possibly, the only feature that unites it is using (in various aspects) the analysis of (nonlinear) functions of infinitely many variables.

We have no intention of covering all the problems of infinite-dimensional analysis (this task seems to be almost unrealizable because of the great variety of problems to be considered), but we give a sufficiently complete presentation of the spectral approaches to the corresponding problems. To some extent, this monograph can be considered as a textbook. Anyway, this was one of our aims, and we have tried to make our presentation as self-consistent as possible and such that any mathematician with a university education can read it.

This book is connected with the previous monograph by Berezansky [18], and its English translation [26] which was considerably revised and complemented. Here we develop many ideas of these two books, paying special attention to the applications. Perhaps, the first two sections are written in a somewhat concise manner, since we did not want to repeat the well-known definitions of a space with negative norm, of a tensor product, etc., because they can be easily found on the first pages of the above-mentioned books by Berezansky, in the books by Maurin [2], by Reed and Simon [1], and in many other publications. It should be also noted that here we do not consider the problems of the spectral theory of differential operators admitting the separation of variables and their potential perturbations, the problem of self-adjointness of finite-dimensional differential operators, and some other problems which were investigated in Berezansky's books.

Some parts of this book were included to the special lecture courses which we gave at the Mechanics and Mathematics Department of Kiev State University. Many results presented here are due to discussions with our colleagues and students at the seminars held at the Kiev Institute of Mathematics (Ukrainian Academy of Science). We are very grateful to everybody who took part in these discussions.

July 1987 Yu. M. Berezansky, Yu. G. Kondratyev

Introduction

In this monograph, we present a part of infinite-dimensional analysis. Certainly, this area of mathematics cannot be considered as an already terminated list of ideas and methods, or even of spheres of application. On the contrary, the last two decades were years of essential changes in infinite-dimensional analysis which were influenced mainly by modern mathematical physics on the one hand and by the theories of random processes and of representations of infinite-dimensional groups on the other hand.

As in each developing area of mathematics, the choice of problems that are investigated is determined to a large extent by the accentuation of certain original problems.

In the case where this relation becomes commonplace, it can be easily replaced by references to the internal logic of the development of a subject. Although this formulation seems to be rather aphoristic, it cannot serve as a criterion governing progress in the investigations (if such criteria are necessary at all). And even a superficial examination of the development of mathematical physics in recent years has revealed not only the necessity for perfection of existing methods of infinite-dimensional analysis, but also the necessity for the investigation of absolutely new subjects for infinite-dimensional analysis which cannot be derived from already existing ones without using "external" arguments. One may also try to understand the problems of infinite-dimensional analysis by using only analogy with the finite-dimensional case, but this approach very quickly proves to be ineligible, since in a series of concrete situations, this analogy does not reflect the essence of the problem or fails completely.

When writing this book, we paid special attention to the investigation of those classes of problems which are frequently met with in applications of infinite-dimensional analysis to the mathematical description of physical systems with infinitely many degrees of freedom. To be more specific, we mainly investigated the problems of quantum field theory and statistical physics. Our other purpose was to present the different aspects of infinite-dimensional harmonic analysis. These two sets of problems are closely connected. One of the authors realized this for the first time in 1967, when he revealed the connection between Wightman functionals and the infinite-dimensional moment problem. To study the latter, he had to develop the spectral theory of infinite families of commuting selfadjoint operators.

The problem considered here are united by the spectral theory of operators which has determined the title of this book. On the one hand, this is a technical tool for investigation of various subjects in infinite-dimensional analysis. By using this theory we study the representations of different algebraic structures and the infinite-dimensional moment problem, we obtain integral representations of positive definite functions of infinitely many variables, etc. On the other hand, examination of the spectral properties of operators acting in spaces of functions on infinitely many variables is of independent interest. In particular, infinite-dimensional analysis plays an important role in the constructive mathematical physics of systems with infinite number of particles. As an example, we can mention the investigation of the spectral characteristics of the field theory models (and of the physical structure of these models which is closely connected with the spectral characteristics). In our book, this side of the spectral theory is presented in the chapters dealing with the theory of infinite-dimensional differential operators and their applications.

Let us dwell upon the contents of the book (a more detailed presentation of the contents of each chapter can be found in the introduction preceding it).

The first chapter is auxiliary.

In Section 1, we give a rather concise exposition of the theory of rigged Hilbert spaces that serves as an abstract version of the theory of generalized functions.

The technique of finite and infinite tensor products of Hilbert spaces is considered in Section 2.

In Section 3, by using the methods of rigged spaces, we construct the theory of unbounded bilinear forms. The material of this chapter forms the language frequently used in subsequent parts of the book.

In the second chapter, we study some parts of the theory of functions of infinitely many variables.

Thus, Section 1 contains the presentation of the measure theory on infinite-dimensional spaces which is adjusted for our purposes. These measures appear in almost all parts of the book.

The structure of the space of functions square-integrable with respect to Gaussian measure is analyzed in Section 2. Here, we also introduce a series of important concepts and constructions, e.g., the Fock space, its functional realization, and the Fourier-Wiener transform.

The necessary information about differentiable functions on linear spaces is collected in Section 3. In this section, we also describe a general construction of the spaces of smooth functions with infinite-dimensional argument and study the approach to the theory of generalized functions of infinitely many arguments proposed by the authors.

In Section 4, we present the coordinate version of this theory based on the technique of infinite tensor products.

Section 5 deals with the invariant case in which no fixed coordinate system is chosen in the space of arguments. The spaces of test and generalized functions that appear in this chapter are frequently used in our further considerations.

In the third chapter, we investigate arbitrary families of commuting selfadjoint (or normal) operators that are, generally speaking, unbounded.

First, in Section 1, we present some auxiliary facts and constructions on the joint resolution of the identity of a family of operators.

Then, in Section 2, we prove the central result of this chapter, namely, the spectral projection theorem which shows that the identity operator can be represented as an integral of the generalized projectors onto generalized eigenvectors of the corresponding family of operators.

The results which complement this theorem are gathered in Section 3. Thus, in particular, here we define the Fourier transform which corresponds to a given family of operators and construct the expansion of the initial Hilbert space into a direct integral of eigensubspaces (in the generalized sense), the spectral theory of Carleman operators, etc.

The fourth chapter deals with some applications of the spectral projection theorem.

In Section 1 of this chapter, we consider a situation when a given family of commuting operators is a representation of an algebraic structure (group, semi-group, linear space, etc.). In this case, the spectral projection theorem allows us to obtain spectral representations for the operators of this family. As a typical example, we can mention the analogue of Stone's theorem for the unitary representation of the real Hilbert space which is proved in this Section.

In Section 2, we study the families of commuting operators with more complicated relations between operators. We assume that the family of operators realizes a representation of some hypercomplex system with a locally compact basis and obtain a representation for the operators of this family in terms of the spectral integral. This section also contains the presentation of the necessary elements of the theory of hypercomplex systems with locally compact basis.

In Section 3, we present a general scheme for the construction of representations of a family of noncommuting operators by means of transition to the Fourier transform in the expansions into generalized joint eigenvectors of the family of commuting operators related to the given one.

The fifth chapter is devoted to applications of the spectral projection theorem to infinite-dimensional harmonic analysis.

In Section 1 of this chapter, we give several criteria of the selfadjointness of general operators. These criteria are grouped around the evolution criteria, i.e., criteria which enable us to establish selfadjointness by using the properties of the corresponding evolution equations, and the quasi-analytic criterion which follows from the evolution equations. The results of Section 1 are used both in Chapter 5 and in our further discussions.

In Section 2, we consider the infinite-dimensional moment problem, i.e., the problem of representability of the functionals with an infinitely increasing number of variables as a moment of some measure on an infinite-dimensional space (for example, one can take Schwinger functions (in Euclidean field theory) as these functionals).

The positive definite functions given on the layer in the space $\mathbb{R}^\infty$ (or on the layer in the Hilbert space) are examined in Sections 3 and 4, respectively. We prove the theorem on the extendability of these functions onto the whole space and establish their spectral representation (the generalization of the Minlos–Sazonov theorem for the case of a layer).

In Section 5, we present a general scheme for the construction of spectral representations of positive definite kernels in terms of generalized joint eigenvectors of the families of commuting selfadjoint operators. Sections 2–4 deal with the special realizations of this scheme. It should be noted that this scheme is a generalization of M.Krein's approach (employing the method of directing functionals) to the investigation of a single operator.

The sixth chapter deals with the theory of infinite-dimensional differential operators.

In Section 1, we study the second quantization operators in the Schrödinger representation and the functional integrals related to these operators.

The potential perturbations of the second quantization operators and the spectral properties of the operators that appeared as a result of these perturbations are discussed in Section 2. In this section, we also prove the analogue of the Feynman–Kac formula for potential perturbations of the second quantization operators.

In Section 3, we establish the relation between the potential perturbations of the second quantization operators and the infinite-dimensional elliptic differential operators of the second order generated by the Dirichlet forms for measures on linear spaces (so-called Dirichlet operators). In the case of measures on Hilbert spaces, we obtain the conditions of the selfadjointness of Dirichlet operators. The last section of this chapter is devoted to the investigation of the infinite-dimensional differential equations that correspond to Dirichlet operators. In particular, we prove that the velocity of propagation of perturbations is finite for the corresponding hyperbolic equations, and this enables us to apply the hyperbolic criteria of selfadjointness of the potential perturbations of Dirichlet operators. In addition, we investigate the elliptic equations generated by the Dirichlet operators.

In the seventh chapter, we examine some models of quantum field theory and quantum statistical physics by using the methods of infinite-dimensional analysis.

First of all, in Section 1, we construct a general scheme of an operator realization of formal Hamiltonians which determine these models. This scheme is based on the results of Section 3 in Chapter 6 and leads to renormalized Hamiltonians that are Dirichlet operators. In this section, we also present an interpretation of the procedure for the construction of renormalized Hamiltonians in terms of functional integrals.

In Section 2, we illustrate the general scheme of operator realization established above, considering examples of linear and quadratic field theory models.

In Section 3, this scheme is applied to the construction of the dynamics of so-called harmonic systems which describe interacting quantum oscillators on a lattice. We also discuss a modification of the renormalization scheme for quantum lattice systems with anharmonicity.

The study of quantum lattice systems is continued in Section 4, where, by using the methods of nonstationary scattering theory, we investigate their spectral properties. Since the usual statement of the scattering problem fails here, we modify it in order to include the potential perturbations of harmonic systems in our consideration. This statement of the problem is based on the realization of the Hamiltonians of the initial harmonic and perturbed systems as a pair of Dirichlet operators. This realization combines the results of Section 3 in Chapter 6 and the general scheme of operator realization. Finally, in Section 4, we examine the examples of potentials for which wave operators can be found explicitly. These examples show the limits of the applicability of the method for the investigation of the spectrum of potential perturbations of harmonic systems which is developed in this book.

We now present some notes about the structure of the book. We were trying to make our exposition selfcontained and based on the standard university courses in functional analysis. In a few cases, when we need some facts lying outside the standard courses, we refer to the corresponding literature. But for the most part, references are contained in the

"Bibliographical Notes". The enumeration of formulas, theorems, and lemmas is independent in each chapter.

It is necessary to add few words about the examples given in the book. As a rule, they are used for several purposes. Thus, they illustrate general schemes and show the relations with those applications of the infinite-dimensional analysis which motivated the creation of those schemes. Moreover, in many cases, the contents of these examples are of independent interest and are used later on as elements of our considerations. Therefore, one should treat the examples as an important part of our presentation.

The concepts and notations used in this book are standard. We also give a list of notations with necessary explanations.

1. RIGGED SPACES

Throughout this book, we use the theory of generalized functions, and it is convenient to present this theory within the framework of the abstract scheme of rigged spaces. In this case, Hilbert riggings are initial, and the role of the classical Sobolev–Schwartz chain $\mathcal{D}'(\mathbb{R}^d) \supset L_2(\mathbb{R}^d) \supset \mathcal{D}(\mathbb{R}^d)$ is played by the chain of Hilbert spaces $H_- \supset H_0 \supset H_+$ which consists of the space H_+ of "test" (positive) vectors, the space H_- (dual to H_+) of "generalized" (negative) vectors, and the "zero" space H_0 setting the duality. These constructions are now well known and frequently used. Nevertheless, in Section 1, we give the necessary information with proofs (sometimes concise).

In Section 2, we present the theory of tensor products of separable Hilbert spaces and their riggings which is also widely used in the book. We consider both finite tensor products and infinite (countable) tensor products which are separable subspaces of the complete von Neumann tensor product (this tensor product is not used in the book and, thus, we do not give the corresponding theory). These products are introduced by use of a somewhat nontraditional coordinate method. Thus, if H_1 and H_2 are two Hilbert spaces, and $(e^{(1)}_{\alpha_1})^{\infty}_{\alpha_1=0}$ and $(e^{(2)}_{\alpha_2})^{\infty}_{\alpha_2=0}$ are orthonormal bases in H_1 and H_2, respectively, then the Hilbert space $H_1 \otimes H_2$ spanned by the formal vectors $e^{(1)}_{\alpha_1} \otimes e^{(2)}_{\alpha_2}$ $(\alpha_1, \alpha_2 = 0, 1, \ldots)$ (considered as mutually orthogonal unit vectors) is called a tensor product of these Hilbert spaces.

In the case of the infinite sequence of Hilbert spaces $H_1, H_2, \ldots$ we act analogously, namely, as $\otimes^{\infty}_{k=1} H_k = H_1 \otimes H_2 \otimes \ldots$, we understand the Hilbert space spanned by the formal mutually orthogonal unit vectors $e^{(1)}_{\alpha_1} \otimes e^{(2)}_{\alpha_2} \otimes \ldots$, where $\alpha_1, \alpha_2, \ldots = 0, 1, \ldots$ and the sequence $(\alpha_1, \alpha_2, \ldots)$ is finite. The sequence $e = \left(e^{(1)}_0, e^{(2)}_0, \ldots\right)$ is "stabilizing" in the following sense: Any considered unit vector differs from the unit vector $e^{(1)}_0 \otimes e^{(2)}_0 \otimes \ldots$ only by a finite number of factors. This construction of the infinite tensor product is more useful for our purposes than the traditional one. In particular, we can now easily construct infinite products of the chains

$$H_{-,k} \supset H_{0,k} \supset H_{+,k} \quad (k = 1 \cdot 2 \cdot \ldots),$$

i.e., the chain

$$\bigotimes_{k=1}^{\infty} H_{-,k} \supset \bigotimes_{k=1}^{\infty} H_{0,k} \supset \bigotimes_{k=1}^{\infty} H_{+,k}$$

is the main construction, with the help of which we introduce the generalized functions of infinitely many variables in Chapter 2.

In Section 2, we also present a version of the Schwartz kernel theorem that is closely connected with the concept of tensor products. In the simplest situation, this result can be formulated as follows. Assume that the continuous operator A acts in the space $L_2(\mathbb{R}^1)$ with respect to the Lebesgue measure dx. Then it allows the following representation for smooth finite φ and ψ:

$$(A\varphi, \psi)_{L_2(\mathbb{R}^1)} = \int_{\mathbb{R}^1}\int_{\mathbb{R}^1} T(x, y)\, \varphi'(y)\, \overline{\psi'(x)}\, dx\, dy, \tag{0.1}$$

where T is a continuous kernel. If this kernel is sufficiently smooth, then by integrating by parts, the derivatives can be transferred from φ and ψ onto T. As a result, the relation (0.1) turns into a usual integral representation for A with the kernel $A(x, y) = \left(\left(\partial^2 / \partial x \partial y\right) T\right)(x, y)$. In the general case, it is natural to understand the relation (0.1) as an "integral" representation for A with the generalized kernel $A = \left(\partial^2 / \partial x \partial y\right) T$. The proof of equality (0.1) itself is quite simple. Indeed, if for $x \in \mathbb{R}^1$ we denote $\omega(x, \xi) = -\operatorname{sgn} x\, \kappa_{(0,x)}(\xi)$, where $\xi \in \mathbb{R}^1$ and κ_α is a characteristic function of the set α, and put

$$T(x, y) = (A\,\omega(y, \cdot), \omega(x, \cdot))_{L_2(\mathbb{R}^1)}, \quad \text{where } x, y \in \mathbb{R}^1,$$

then (0.1) can be verified by integrating by parts. In fact, in Section 2, we generalize representation (0.1) for continuous multilinear forms in Hilbert spaces.

Below, we shall use the theory of semi-bounded bilinear forms. In Section 3, we present this theory on the basis of Hilbert riggings. We show how simple results about positive and negative spaces may be interpreted as theorems of the theory of these forms. In our presentation, we essentially use the following remark: Setting a positive closed form $a(f, g)$ (with the domain of definition $\mathfrak{D}(a) \subset H_0$) in the Hilbert space H_0 is equivalent to setting the positive space $H_+ = \mathfrak{D}(a)$ with the scalar product

$$(f, g)_{H_+} = a(f, g) \quad (f, g \in \mathfrak{D}(a)).$$

§1. Rigged Hilbert Spaces

The procedure for the construction of Hilbert space riggings may be considered as an abstract method of introducing generalized functions. The objects, which appear as a result, prove to be useful in different areas of functional analysis and its applications. Below, we shall systematically use the rigged spaces technique to which the present section is devoted. Since this information can be found in many books, we give only a concise exposition (for more detailed information, see Berezansky [18, Chapter 1, Section 1]).

1.1. Hilbert Riggings

First, let us consider a notion of the space with negative norm. Let H_0 be a complex Hilbert space with scalar product $(\cdot, \cdot)_{H_0}$ and norm $\|\cdot\|_{H_0}$; and let $f, g, \ldots$ be its elements. Assume that the linear set H_+ is dense in H_0 (H_+ itself is a Hilbert space with respect to a new scalar product $(\cdot, \cdot)_{H_+}$) and, moreover, the inequality

$$\|\varphi\|_{H_0} \leq \|\varphi\|_{H_+} \quad (\varphi \in H_+) \tag{1.1}$$

holds for the corresponding norm (the more general case when $\|\cdot\|_{H_0} \leq c\|\cdot\|_{H_+}$ can be reduced to (1.1) by introducing the equivalent norm in H_+). Elements of H_+ play the role of test functions, we denote them by $\varphi, \psi, \ldots$; and H_+ itself is called a space with positive norm.

Each element $f \in H_0$ generates the antilinear continuous functional l_f on H_+, according to the formula

$$l_f(\varphi) = (f, \varphi)_{H_0} \quad (\varphi \in H_+), \tag{1.2}$$

and if $l_f = 0$, then $f = 0$. Let us introduce a norm in H_0 by

$$\|f\|_{H_-} = \|l_\varphi\| = \sup_{\varphi \in H_+} \frac{|(f, \varphi)_{H_0}|}{\|\varphi\|_{H_+}} \tag{1.3}$$

and denote by H_- the completion of H_0 with respect to this norm. Extending $(f, g)_{H_0}$ onto $H_- \times H_+$ by continuity, we define the bilinear form

$$H_- \times H_+ \ni (\xi, \varphi) \mapsto (\xi, \varphi)_{H_0} \in \mathbb{C}^1. \tag{1.4}$$

According to the definition, we have

$$(\varphi, \xi)_{H_0} = \overline{(\xi, \varphi)_{H_0}} \quad (\xi \in H_-, \varphi \in H_+).$$

Clearly, the following generalization of the Cauchy–Bunyakovsky inequality holds

$$|(\xi, \varphi)_{H_0}| \leq \|\xi\|_{H_-} \|\varphi\|_{H_+} \quad (\xi \in H_-, \ \varphi \in H_+). \tag{1.5}$$

The space H_- is called a space with negative norm, and its elements $\xi, \eta, \ldots$ play the role of generalized functions in the abstract situation considered.

In fact, H_- is a Hilbert space with respect to the scalar product $(\cdot, \cdot)_{H_-}$ which is constructed as follows. Let $O: H_+ \to H_0$ be a continuous imbedding operator. For the operator $I = O^*: H_0 \to H_+$, we have

$$(f, \varphi)_{H_0} = (f, O\varphi)_{H_0} = (If, \varphi)_{H_+} \quad (f \in H_0, \varphi \in H_+) \tag{1.6}$$

and, thus, $\| If \|_{H_+} = \| l_f \| = \| f \|_{H_-}$. Therefore, the bilinear form

$$H_0 \times H_0 \ni (f, g) \mapsto (f, g)_{H_-} = (If, Ig)_{H_+} \tag{1.7}$$

is a scalar product in H_0 consistent with the negative norm $\| \cdot \|_{H_-}$. The space H_- is a completion of H_0 with respect to this scalar product.

If we consider I as an operator from H_- to H_+ defined on the set H_0 dense in H_-, and extend it by continuity, then we obtain the isometry $\boldsymbol{I}$ which maps the whole H_- onto the whole H_+ and, moreover,

$$(\xi, \varphi)_{H_0} = (\boldsymbol{I}\xi, \varphi)_{H_+} \quad (\xi \in H_-, \varphi \in H_+). \tag{1.8}$$

Equality (1.8) shows that $(\cdot, \cdot)_{H_0}$ defines a pairing between H_+ and its dual space $(H_+)'$ of antilinear functionals. Since each functional $l \in (H_+)'$ can be written in the form

$$l(\varphi) = (a, \varphi)_{H_+} = (\boldsymbol{I}^{-1} a, \varphi)_{H_0} = (\xi, \varphi)_{H_0} \quad (\varphi \in H_+),$$

where $a \in H_+$ corresponds to l, according to the Riesz theorem, and $\xi = \boldsymbol{I}^{-1} a \in H_-$, we can write $(H_+)' = H_-$. This enables us to say that the space H_- is also dual to H_+ with respect to H_0 (this means that duality of H_+ and H_- is written in terms of H_0). The following theorem gives a useful decomposition of the operator $\boldsymbol{I}$.

Theorem 1.1. *The isometry* $\boldsymbol{I}$: $H_- \to H_+$ *admits a decomposition into the product of two isometries*

$$\boldsymbol{I} = JJ; \; J: H_- \to H_0, \; J: H_0 \to H_+; \; OJ = \boldsymbol{J} \restriction H_0. \tag{1.9}$$

Proof. We set $A = O: H_0 \to H_0$. This operator is bounded and nonnegative (according to (1.6), we have

$$(Af, f)_{H_0} = (OIf, f)_{H_0} = (If, If)_{H_+} \geq 0, \quad \text{for } f \in H_0).$$

Let $B = \sqrt{A} : H_0 \to H_0$; consider this operator as acting from the dense subset of the space H_- into H_0. Then it is isometric

$$(Bf, Bg)_{H_0} = (B^2 f, g)_{H_0} = (If, If)_{H_+} = (f, g)_{H_-} \quad (f, g \in H_0)$$

and its closure by continuity also gives the isometric operator $\boldsymbol{J}$: $H_- \to H_0$.

Let us prove the equality $\mathfrak{R}(\boldsymbol{J}) = H_0$. It suffices to show that if $f \perp \mathfrak{R}(\boldsymbol{J})$ in H_0, then $f = 0$. For arbitrary $g \in H_0$, we have

$$0 = (\boldsymbol{J}g, f)_{H_0} = (Bg, f)_{H_0} = (g, Bf)_{H_0}.$$

This yields $Bf = 0$ and, therefore, $OIf = B^2 f = 0$, i.e., $If = 0$ and, hence, $f = 0$. We now show that $\mathfrak{R}(B) \subset H_+$. It suffices to establish the equality

$$B\boldsymbol{J} = O\boldsymbol{I}. \tag{1.10}$$

For $f \in H_0$, we have $B\boldsymbol{J}f = B^2 f = OIf = O\boldsymbol{I}f$. Since H_0 is dense in H_-, and the operators $\boldsymbol{J}$: $H_- \to H_0$, $B : H_0 \to H_0$, and $O\boldsymbol{I} : H_- \to H_0$ are continuous, this immediately yields (1.10).

Denote by J the operator B, regarded as an operator from H_0 to H_+. Then (1.10) implies that $J\boldsymbol{J} = \boldsymbol{I}$. It remains to prove that J is an isometric operator and that $\mathfrak{R}(J) = H_+$. But this follows from the equality $J = \boldsymbol{I}\boldsymbol{J}^{-1}$, because $\boldsymbol{J}^{-1}$: $H_0 \to H_-$ and $\boldsymbol{I} : H_- \to H_+$ are isometric operators.

■

Let us write the main relations for the constructed rigging of the Hilbert space H_0 by the spaces H_+ and H_- (of "smooth" and "generalized" vectors, respectively) or, in other words, by the constructed chain $H_- \supset H_0 \supset H_+$:

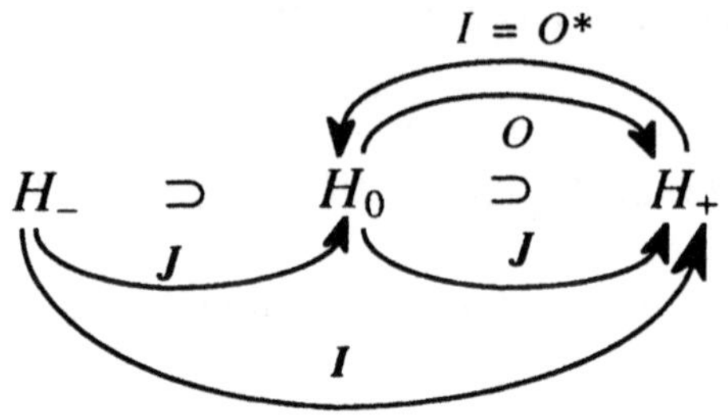

(1.11)

$$\|\cdot\|_{H_-} \le \|\cdot\|_{H_0} \le \|\cdot\|_{H_+},$$

$$OJ = J \restriction H_0, \; I = J\,J, \tag{1.12}$$

$$(\xi, \varphi)_{H_0} = (I\,\xi, \varphi)_{H_+} = (\xi, I^{-1}\varphi)_{H_-},$$

$$(I\,\xi, \eta)_{H_0} = (\xi, J\,\eta)_{H_0}, \; (J\xi, f)_{H_0} = (\xi, Jf)_{H_0}$$

$$(\xi, \eta \in H_-; \quad f \in H_0; \quad \varphi \in H_+)$$

(in order to obtain the last two inequalities of (1.12), one must set $\varphi = I\eta$ or $\varphi = Jf$, respectively).

Example 1.1. Let

$$H_0 = L_2(R, d\mu(x)) = L_2(R, \mathfrak{R}, \mu),$$

where R is a space with a measure μ given on a certain σ-algebra $\mathfrak{R}$ of subsets of R and such that $\mu(R) \le \infty$. We set $H_+ = L_2(R, p(x)\,d\mu(x))$, where $p(x) \ge 1$ $(x \in R)$ is some measurable weight finite almost everywhere. One can easily see that the chain (1.11) now takes the form

$$H_- = L_2(R, p^{-1}(x), d\mu(x)) \supset L_2(R, d\mu(x)). \tag{1.13}$$

Therefore, the "generalized nature" of the vector $\xi = \xi(\cdot) \in H_-$ becomes apparent in the fact that this function can grow sufficiently fast near the points x at which $p(x)$ grows, and the "smoothness" of $\varphi = \varphi(\cdot)$ means that this function decreases near these points. We now have

$$(I\xi)(x) = p^{-1}(x)\,\xi(x), \quad (J\xi)(x) = p^{-1/2}(x)\,\xi(x) \;\; (\xi \in H_-).$$

Example 1.2. Let G be a region (bounded or not) in the space $\mathbb{R}^d$ $(d \in \mathbb{N})$ with a piecewise continuously differentiable boundary ∂G. The Sobolev space $W_2^l(G)$ $(l \in \mathbb{Z}_+)$ is defined as a completion of $C^\infty(\tilde{G})$ with respect to the scalar product

$$(\varphi, \psi)_{W_2^l(G)} = \sum_{|\alpha| \le l} (D^\alpha \varphi, D^\alpha \psi)_{L_2(G)} \quad (\varphi, \psi \in C^\infty(\tilde{G})), \tag{1.14}$$

where $L_2(G) = L_2(G, \mathfrak{B}(G), dx)$ is constructed with respect to the Lebesgue measure dx. We set $H_0 = L_2(G)$ and $H_+ = W_2^l(G)$, then H_- is called a Sobolev space with negative index $-l$ and is denoted by $W_2^{-l}(G)$.

In the case $l > d/2$, we have $W_2^l(G) \subset C(\tilde{G})$ by virtue of the imbedding theorems and, thus, the δ-function δ_x concentrated at the point $x \in \tilde{G}$ lies in $W_2^{-l}(G)$. This function is defined by

$$(\delta_x, \varphi)_{H_0} = \overline{\varphi(x)} \quad (\varphi \in H_+).$$

The elements $\xi \in H_-$ are now called generalized functions of a finite (lth) order; the operators $\boldsymbol{I}$ and $\boldsymbol{J}$ can be expressed in terms of Green's function for the boundary value problem generated by (1.14). For additional information about the spaces $W_2^{-l}(G)$, see Berezansky [18, Chapter 1, Section 3, Subsections 2, 3, and 7].

Note also that in the case $G = \mathbb{R}^d$, by using the Fourier transforms, one can pass from the chain

$$W_2^{-l}(\mathbb{R}^d) \supset L_2(\mathbb{R}^d) \supset W_2^l(\mathbb{R}^d)$$

to the isomorphic chain of Example 1.1 with $R = \mathbb{R}^d$, $d\mu(x) = dx$ and $p(x) = \sum_{|\alpha| \le l} x^{2\alpha}$.

Above, we have constructed the chain (1.11) by using the given pair of spaces H_0 and H_+. Let us show that this chain can be also constructed by use of the pair H_0 and H_-.

Lemma 1.1. *Suppose that a dense linear set H_0 is chosen in the given Hilbert space H_- and that this set is itself a Hilbert space with respect to its scalar product and, moreover, $\| f \|_{H_-} \le \| f \|_{H_0}$ $(f \in H_0)$. Then one can construct a positive space $H_+ \in H_0$ such that the corresponding negative space coincides with H_-.*

Proof. The inequality $\| f \|_{H_-} \le \| f \|_{H_0}$ $(f \in H_0)$ shows the continuity of the bilinear form

$$H_0 \times H_0 \ni (f, g) \mapsto (f, g)_{H_-} \in \mathbb{C}^1$$

and, therefore, ensures the representation

$$(f, g)_{H_-} = (Kf, g)_{H_0} \quad (f, g \in H_0)$$

in which $K \in \mathfrak{L}(H_0, H_0)$, $\|K\| \le 1$, and $K > 0$. By virtue of the denseness of H_0 in H_-, the range of K is dense in H_0. On $\mathfrak{R}(K)$, there exists the inverse operator K^{-1}: if $Kf = 0$, then $(f, g)_{H_-} = (Kf, g)_{H_0} = 0$, where $g \in H_0$, i.e., $f = 0$.

Let us set

$$(\varphi, \psi)_{H_+} = (K^{-1}\varphi, \psi)_{H_0} \quad (\varphi, \psi \in \mathfrak{R}(K)). \tag{1.15}$$

Since $\|K\| \le 1$, we have $\|\varphi\|_{H_+} \le \|\varphi\|_{H_0}$ $(\varphi \in \mathfrak{R}(K))$. This implies that the space H_+ obtained by the completion of $\mathfrak{R}(K)$ in the scalar product $(\cdot, \cdot)_{H_+}$ satisfies all the conditions imposed on a space positive with respect to H_0. By using H_0 and H_+, let us construct a negative space G_- and show that $G_- = H_-$. To this end, we note that $K = OI$, where the last two operators are constructed from the chain $G_- \supset H_0 \supset H_+$. This follows from the relation

$$(If, \varphi)_{H_+} = (f, \varphi)_{H_0} \quad (f \in H_0, \varphi \in H_+),$$

the equality

$$(Kf, \varphi)_{H_+} = (f, \varphi)_{H_-} \quad (f \in H_0, \varphi \in \mathfrak{R}(K))$$

(see (1.15)), and the fact that $\mathfrak{R}(K)$ is dense in H_+. Hence,

$$(f, g)_{H_-} = (Kf, g)_{H_0} = (\boldsymbol{I}f, g) = (f, g)_{G_-}$$

for $f, g \in H_0$. By virtue of the fact that H_0 is dense in H_- and in G_-, this yields the equality $H_- = G_-$.

■

Example 1.3. Just as in Example 1.1, assume that $H_0 = L_2(R, d\mu(x))$, and let

$$R \times R \ni (x, y) \mapsto K(x, y) \in \mathbb{C}^1$$

be some bounded (e.g., by one) kernel measurable with respect to $\mathfrak{R} \times \mathfrak{R}$. We set

$$(f, g)_{H_K} = \int_R \int_R K(x, y) f(y) \overline{g(x)} \, d\mu(x) \, d\mu(y) \quad (f, g \in H_0) \tag{1.16}$$

and demand that the kernel of K is positive definite, i.e., $\forall f \in H_0$ we have $(f, f)_{H_K} > 0$. Then (1.16) defines the scalar product in H_0 if, in addition, we assume that K is nondegenerate, i.e., that the equality $(f,f)_{H_K} = 0$ implies that $f = 0$ in H_0. The completion of H_0 with respect to (1.16) is denoted by H_K. According to Lemma 1.1, one can interpret the space H_K as negative with respect to H_0 and corresponding H_+. In the case of topological R, continuous K, and sufficiently regular measure μ, this space contains δ-functions. The Sobolev spaces $W_2^{-l}(G)$ are examples of the spaces H_K. For more details, see Berezansky [18, Chapter 1, Section 1, Subsection 7; Section 3, Subsection 12] and Chapter 5, Section 5, Subsection 1.

Formulas (1.12) yield

$$(\varphi, \psi)_{H_+} = (J^{-1}\varphi, J^{-1}\psi)_{H_0} \quad (\varphi, \psi \in H_+),$$

$$(\xi, \eta)_{H_-} = (J\varphi, J\psi)_{H_0} \quad (\xi, \eta \in H_-). \tag{1.17}$$

These relations show that one can construct the riggings of H_0 starting from some operator in this space similar to J or to J^{-1} (the latter is more convenient).

Assume that a closed operator D with a dense domain of definition $\mathfrak{D}(D)$ is given in H_0, and that

$$\| D\varphi \|_{H_0} \geq \| \varphi \|_{H_0} \quad (\varphi \in \mathfrak{D}(D)). \tag{1.18}$$

Since D is closed, the set $\mathfrak{D}(D)$ is a Hilbert space with the scalar product

$$(\varphi, \psi)_{H_+} = (D\varphi, D\psi)_{H_0}. \tag{1.19}$$

Let us take it as a positive space H_+ and construct the corresponding Hilbert space H_- dual to H_+ with respect to H_0. Let J be an isometric operator that corresponds to the chain $H_- \supset H_0 \supset H_+$. Comparing (1.17) with (1.19), we find that $J_{H_0}^{-1} = \sqrt{D^* D}$, where $J_{H_0}^{-1}$ is the operator J^{-1} understood as an operator in H_0 with the domain of definition H_+ (we have used the fact that $J_{H_0}^{-1}$ is positive and the theorem on metrically equal operators). If, in addition, D is a positive selfadjoint operator, then $J_{H_0}^{-1} = D$. In this case $OJ = D^{-1}$, and (1.17) yields

$$(f, g)_{H_-} = (D^{-1}f, D^{-1}g)_{H_0},$$

where $f, g \in H_0$. The negative space H_- is obtained as a completion of H_0 with respect to the last scalar product.

The concept of adjointness with respect to H_0 can be introduced naturally for the operators acting continuously between the spaces of the chain (1.11). Let $A \in \mathcal{L}(H_+, H_-)$, then the operator $A^+ \in \mathcal{L}(H_+, H_-)$ adjoint to A with respect to H_0 is given by

$$(A\varphi, \psi)_{H_0} = (\varphi, A^+\psi)_{H_0} \quad (\varphi, \psi \in H_+). \tag{1.20}$$

Clearly, $A^+ = \boldsymbol{I}^{-1}A^*\boldsymbol{I}$, where $A^* \in \mathcal{L}(H_-, H_+)$ is an operator adjoint to A in the usual sense. For $A \in \mathcal{L}(H_+, H_-)$, the notion of selfadjointness can be generalized: We say that A is selfadjoint if

$$(A\varphi, \psi)_{H_0} = (\varphi, A\psi)_{H_0}$$

for all $\varphi, \psi \in H_+$, i.e., $A^+ = A$. Similarly, we say that the operator $A \in \mathcal{L}(H_+, H_-)$ is nonnegative if $(A\varphi, \varphi)_{H_0} \geq 0$ for all $\varphi \in H_+$.

The bounded operator A in H_0, selfadjoint in the usual sense, is clearly selfadjoint in the generalized sense introduced above if it is regarded as an operator acting from H_+ to H_-. Generally speaking, the operator $A \in \mathcal{L}(H_+, H_-)$ is selfadjoint in this new sense if and only if the operator $\boldsymbol{I}A \in \mathcal{L}(H_+, H_+)$ is selfadjoint in H_+ (or if $A\boldsymbol{I}^{-1} \in \mathcal{L}(H_-, H_-)$ is selfadjoint in H_-). This follows from the equality (see (1.12))

$$(\boldsymbol{I}A\varphi, \psi)_{H_+} = (A\varphi, \psi)_{H_0} = (\varphi, A\psi)_{H_0} = (\varphi, \boldsymbol{I}A\psi)_{H_+} \quad (\varphi, \psi \in H_+).$$

Example 1.4. Consider the chain (1.11) from Example 1.2 with a bounded domain G : $W_2^{-l}(G) \supset L_2(G) \supset W_2^l(G)$ and the operator $A \in \mathcal{L}(H_+, H_-)$ defined by $H_+ = W_2^l(G) \ni \varphi \mapsto \mathcal{L}\varphi$, where $\mathcal{L}$ is a linear differential expression of the lth order with sufficiently smooth coefficients. Then A^+ is constructed by using the formally adjoint expression $\mathcal{L}^+$ and the boundary expression defined by the form

$$\Gamma(\varphi, \psi) = (\mathcal{L}\varphi, \psi)_{H_0} - (\varphi, \mathcal{L}^+\psi)_{H_0},$$

where $\varphi, \psi \in H_+$. Thus, for instance, for $l = 2$ and $\mathcal{L} = \Delta$, we have $A^+ = A + B$, where $B \in \mathcal{L}(H_+, H_-)$ is given by

$$(\varphi, B\psi)_{H_0} = \int_{\partial G}\left[\left(\frac{\partial\varphi}{\partial n}\right)(x)\overline{\psi(x)} - \varphi(x)\left(\frac{\overline{\partial\varphi}}{\partial n}\right)(x)\right]dx.$$

Here, $n(x)$ is an outward normal to ∂G at the point x; $\varphi, \psi \in H_+$.

Assume that the space H_+ is separable. The nonnegative operator $A \in \mathcal{L}(H_+, H_-)$ is called an operator with finite trace $\mathrm{Tr}_{H_0} A$ if for some orthonormal basis $(e_j)_{j=1}^{\infty}$ in H_+, we have

$$\mathrm{Tr}_{H_0} A = \sum_{j=1}^{\infty} (Ae_j, e_j)_{H_0} < \infty$$

(this is equivalent to the finiteness of the trace of the operator $\boldsymbol{I}A$ in H_+, moreover $\mathrm{Tr}_{H_0} A = \mathrm{Tr}_{H_+} \boldsymbol{I}A$). One can easily check that this definition is independent of the choice of a basis.

For nonnegative $A \in \mathcal{L}(H_+, H_-)$ the inequality $|A| \leq \mathrm{Tr}_{H_0} A$ holds, where $|A|$ is a Hilbert norm of the operator A.

In fact, for the orthonormal basis $(e_j)_{j=1}^{\infty}$ in H_+, the system of vectors $(\boldsymbol{I}^{-1} e_j)_{j=1}^{\infty}$ is an orthonormal basis in H_- and, thus,

$$|A|^2 = \sum_{j=1}^{\infty} \|Ae_j\|_{H_-}^2 = \sum_{j,k=1}^{\infty} |(Ae_j, \boldsymbol{I}^{-1} e_k)_{H_-}|^2 \sum_{j,k=1}^{\infty} |(Ae_j, e_k)_{H_0}|^2$$

$$\leq \sum_{j,k=1}^{\infty} (Ae_j, e_j)_{H_0} (Ae_k, e_k)_{H_0} = (\mathrm{Tr}_{H_0} A)^2$$

(we have used the inequality

$$|(A\varphi, \varphi)|^2 \leq (A\varphi, \varphi)_{H_0} (A\psi, \psi)_{H_0}$$

for $\varphi, \psi \in H_+$, which is valid, because A is nonnegative).

■

Remark 1. The adjointness with respect to H_0 can be introduced similarly for the operators acting between other spaces in the chain (1.11). Then it follows from (1.12) that $\boldsymbol{I}^+ = \boldsymbol{I} \geq 0$ and $\boldsymbol{J}^+ = J$. The operator $O^+ \in \mathcal{L}(H_0, H_-)$ is just the imbedding of H_0 into H_-, i.e.,

$$(O\varphi, f)_{H_0} = (\varphi, O^+ f)_{H_0} \quad (\varphi \in H_+, f \in H_0). \tag{1.21}$$

■

Suppose that the chain (1.11) (or the Hilbert rigging of the space H_0) is constructed.

We shall often encounter the situation when the operator of imbedding $O\colon H_+ \to H_0$ is a quasinuclear (i.e., Hilbert–Schmidt) operator. In this case, we shall say that the space H_+ is imbedded into H_0 quasinuclearly. The corresponding rigging (or the chain (1.11)) will be also called quasinuclear.

Note that the quasinuclearity of O is equivalent to the quasinuclearity of $OJ\colon H_0 \to H_0$ because J is an isometry between H_0 and H_+. It is convenient to use this remark when we establish the quasinuclearity of the chain, since the operator OJ acts in the single space H_0.

Example 1.5. Let us present some examples of quasinuclear chains. First, we note that the chain from Example 1.1 will never be quasinuclear (clearly, with exception of a finite-dimensional $L_2(R, d\mu(x))$). This is a consequence of the form of the operator $OJ = J \restriction H_0$.

The chain $W_2^{-l}(G) \supset L_2(G) \supset W_2^l(G)$ *in Example 1.2 is quasinuclear for* $l > d/2$ *if* $G \subset \mathbb{R}^d$ *is bounded.*

This is proved as follows. Since $\delta_x \in W_2^{-l}(G)$ for $x \in \tilde{G}$, we have

$$(OJf)(x) = (Jf)(x) = (f, J^+\delta_x)_{L_2(G)} = (f, J\delta_x)_{L_2(G)} \int_G K(x,y) f(y)\,dy,$$

$$K(x,y) = \overline{(J\delta_x)(y)}.$$

But the imbedding $W_2^l(G) \subset C(\tilde{G})$ is topologic and, therefore, $W_2^{-l}(G) \supset C'(\tilde{G})$ and $\delta_x \in C'(\tilde{G}) \subset W_2^{-l}(G)$; $\exists c \in (0, \infty)$:

$$\|\delta_x\|_{W_2^{-l}(G)} \le c\|\delta_x\|_{C'(\tilde{G})} = c \quad (x \in \tilde{G}).$$

Hence,

$$\int_G\int_G |K(x,y)|^2\,dx\,dy = \int_G \|J\delta_x\|^2_{L_2(G)}\,dx = \int_G \|\delta_x\|^2_{W_2^{-l}(G)}\,dx \le cm(G).$$

Therefore, OJ is a Hilbert–Schmidt operator, i.e., the imbedding $W_2^l(G) \subset L_2(G)$ is quasinuclear.

■

By using this fact, we can prove a more general statement, namely, that for bounded $G \subset \mathbb{R}^d$, the imbedding $W_2^{l''}(G) \to W_2^{l'}(G)$ is quasinuclear if $l'' - l' > d/2$.

For unbounded G, this is not true (as a rule). Thus, if $G \subset \mathbb{R}^d$, the chain $W_2^{-l}(\mathbb{R}^d) \supset L_2(\mathbb{R}^d) \supset W_2^l(\mathbb{R}^d)$ is isomorphic to the nonquasinuclear chain from Example 1.1. The chains of Sobolev-type spaces with quasinuclear imbedding are usually obtained by passing to the Sobolev spaces $W_2^l(G, p(x)\,dx)$ with weight $p(x) > 0$. This space is defined just as $W_2^l(G)$, but in (1.14) $L_2(G)$ should be replaced by $L_2(G, p(x)\,dx)$. The following result can be proved:

Suppose that $G \subset \mathbb{R}^d$ is regular in the sense that it does not contain "spikes" reaching the infinity (e.g., $G = \mathbb{R}^d$) and that $l'', l' \in \mathbb{Z}_+$ are such that $l'' - l' > d/2$. Consider two weights $q_1, q_2 \in C^{l''-l'}(\tilde{G})$ such that $0 < q_1(x) \le q_2(x)$ $(x \in \tilde{G})$ and

$$\int_G q_1^2(x)\, q_2^{-2}\, x\, dx < \infty.$$

Then the imbedding

$$W_2^{l''}(G, q_{2,l''-l'}^2(x)\,dx) \subset W_2^{l'}(G, q_1^2(x)\,dx),$$

where

$$q_{2,l''-l'}(x) = \max_{|\alpha| \le l''-l'} \left|(D^\alpha q_2)(x)\right| \quad (x \in \tilde{G}),$$

is quasinuclear.

For more details concerning the results of this example, see Berezansky [18, Chapter 1, Section 3, Subsections 4, 7, 8].

1.2. Rigging of a Hilbert Space by Linear Topologic Spaces

Assume that a topologic space Φ with elements $\varphi, \psi, \ldots$ is topologically imbedded into a Hilbert space H_0. Just as before, each element $f \in H_0$ generates the antilinear continuous functional on Φ according to the formula $l_f(\varphi) = (f, \varphi)_{H_0}$ for $\varphi \in \Phi$. Identifying f with l_f, we obtain the imbedding of H_0 into the space Φ' of antilinear continuous functionals on Φ. If we equip Φ' with a weak topology, then, clearly, the imbedding $H_0 \to \Phi'$ becomes continuous. Hence, we have constructed the chain

$$\Phi' \supset H_0 \supset \Phi. \tag{1.22}$$

We also say that (1.22) is a rigging of H_0 by the spaces Φ and Φ'. Henceforth, unless otherwise stated, we shall always consider Φ' to be equipped with a weak topology.

Below, we shall mainly encounter the situation when Φ is a projective limit of Hilbert spaces. Let us recall the corresponding definitions. Suppose that a family of Hilbert spaces $(H_\tau)_{\tau \in \mathrm{T}}$ parametrized by the elements of an arbitrary set of indices T is given.

Assume that the set $\Phi = \bigcap_{\tau \in T} H_\tau$ is dense in each H_τ, and that the family $(H_\tau)_{\tau \in T}$ is directed by imbedding, i.e.,

$$\forall\ \tau', \tau'' \in T\ \exists\ \tau''' \in T: H_{\tau'''} \subset H_{\tau'},\ H_{\tau'''} \subset H_{\tau''}, \tag{1.23}$$

where the imbeddings are topologic. On Φ, we introduce a projective topology with respect to the families of Hilbert spaces $(H_\tau)_{\tau \in T}$ and natural imbeddings $O_\tau: \Phi \to H_\tau$. According to the definition, this means that we consider the weakest topology on Φ for which all the mappings O_τ $(\tau \in T)$ are continuous. One can easily see that the collection of all possible balls

$$U(\varphi_1; \tau; \varepsilon) = \{\varphi \in \Phi \mid \|\varphi - \varphi_1\|_{H_\tau} < \varepsilon\} \quad (\varphi_1 \in \Phi, \tau \in T, \varepsilon > 0)$$

may be taken as a system of base neighborhoods of this topology. The space Φ constructed above is called a projective limit of the family $(H_\tau)_{\tau \in T}$ and is denoted by $\operatorname{pr\,lim}_{\tau \in T} H_\tau$. Sometimes this projective limit is called reduced: This stresses the denseness of Φ in each[1] $(H_\tau)_{\tau \in T}$.

The space $\Phi = \operatorname{pr\,lim}_{\tau \in T} H_\tau$ is called nuclear if, for any $\tau \in T$, one can find $\tau' \in T$ such that $H_{\tau'}$ is quasinuclearly imbedded into H_τ. In the case of a nuclear Φ, the chain (rigging) (1.22) is also called nuclear.

Assume that the rigging (1.22) is given with $\Phi = \operatorname{pr\,lim}_{\tau \in T} H_\tau$.

Without loss of generality, one can regard each H_τ as imbedded topologically into H_0 such that $\|\varphi\|_{H_0} \le \|\varphi\|_{H_\tau}$ $(\varphi \in H_\tau, \tau \in T)$, where T *contains the index* 0.

Indeed, instead of the spaces H_τ, let us consider Hilbert spaces $F_\tau = H_0 \cap H_\tau$ with the scalar product

$$(\varphi, \psi)_{F_\tau} = (\varphi, \psi)_{H_0} + (\varphi, \psi)_{H_\tau} \quad (\varphi, \psi \in F_\tau)$$

and $F_0 = H_0$. Since the mapping $\Phi \ni \varphi \mapsto \varphi \in H_0$ is continuous at 0, there exist $\tau_0 \in T$ and $c_{\tau_0} \in (0, \infty)$ such that $\|\varphi\|_{H_0} \le c_{\tau_0} \|\varphi\|_{H_{\tau_0}}$ for $\varphi \in \Phi$ and, hence, for $\varphi \in F_{\tau_0}$. This implies that norms in H_τ $(\tau \in T)$ and $F_\tau (\tau \in T \cup \{0\})$ define the same topology in Φ. The spaces F_τ possess all the required properties.

■

---

[1] Note that the projective limit for Banach spaces can be constructed similarly. One should only replace H_τ by a Banach space B_τ everywhere in the above argument.

Hence, for every $\tau \in \mathrm{T}$, one can construct the chain of type (1.11)

$$H_{-\tau} \supset H_0 \supset H_\tau, \tag{1.24}$$

where $H_{-\tau}$ is a space dual to H_τ with respect to H_0, i.e., the negative space which corresponds to the positive space[2] H_τ.

Lemma 1.2. *We have*

$$\Phi' = \bigcup_{\tau \in \mathrm{T}} H_{-\tau}. \tag{1.25}$$

Proof. The imbedding $H_{-\tau} = (H_\tau)' \subset \Phi'$ is evident. Conversely, let $l \in \Phi'$. It follows from the continuity of l at the origin that one can find a base neighborhood $U(0, \tau, \varepsilon)$ in Φ such that $|l(\varphi)| < 1$ for $\varphi \in U(0, \tau, \varepsilon)$. This means that the functional l, defined on the dense subset Φ of the space H_τ, is continuous on H_τ. Extending this functional by continuity, we get that $l \in H_{-\tau}$.

■

As in the case of the chain (1.11), the scalar product in H_0 determines the pairing between Φ and Φ' in (1.22), i.e., if $\xi \in \Phi'$, then $\xi(\varphi) = (\xi, \varphi)_{H_0}$ ($\varphi \in \Phi$), where

$$\Phi' \times \Phi \ni (\xi, \varphi) \mapsto (\xi, \varphi)_{H_0} \in \mathbb{C}^1$$

is an extension of $(f, \varphi)_{H_0}$ by continuity for $H_0 \ni f \to \xi$ This clearly follows from (1.25) and the pairing between H_τ and $H_{-\tau}$.

As a space of antilinear functionals dual to Φ with respect to H_0, the set Φ' can be topologized in many ways leading to nonequivalent topologies for infinite-dimensional Φ. First, we examine the case when $\Phi = \operatorname{pr} \lim_{\tau \in \mathrm{T}} H_\tau$. The decomposition (1.25) enables us to equip Φ' with the topology of an inductive limit of the space $H_{-\tau}$. The relation (1.23) involves the following property of directedness for the family $(H_{-\tau})_{\tau \in \mathrm{T}}$:

$$\forall\ \tau', \tau'' \in \mathrm{T}\ \exists\ \tau''' \in \mathrm{T}: H_{-\tau'''} \subset H_{-\tau'},\ H_{-\tau'''} \subset H_{-\tau''}, \tag{1.26}$$

and, in addition, the imbeddings are topologic. By using (1.26), one can easily find that all possible sets of the form

[2] Below, it will be convenient to denote the chains of type (1.24) by $H_{-,\tau} \supset H_0 \supset H_{+,\tau}$, where $H_{+,\tau} = H_\tau$ and $H_{-,\tau} = H_{-\tau}$. This emphasizes the fact that H_τ plays a role of a positive space.

$$U(\xi_1; \varepsilon(\cdot)) = \bigcup_{\tau \in T} \{\xi \in \Phi' \mid \xi - \xi_1 \in H_{-\tau}, \ \|\xi - \xi_1\|_{H_{-\tau}} < \varepsilon(\tau)\}$$

$$(\xi_1 \in \Phi', T \ni \tau \mapsto \varepsilon(\tau) > 0). \tag{1.27}$$

form a system of base neighborhoods in Φ' (this system converts Φ' into a linear topologic space). In other words, for $\xi_1 = 0$, all the possible unions of open balls with positive radii centered at the origin form a system of base neighborhoods in the spaces $H_{-\tau}$ ($\tau \in T$) (at the origin). Then these neighborhoods are shifted by arbitrary $\xi_1 \in \Phi'$. The space Φ', topologized by using (1.27), is called an inductive limit of the family $(H_{-\tau})_{\tau \in T}$ and is denoted as follows: $\Phi' = \text{ind lim}_{\tau \in T} H_{-\tau}$. One can easily find that this topology may be also described in the following way: Consider natural imbeddings $O'_\tau : H_{-\tau} \to \Phi'$. We provide Φ' with the strongest locally convex topology under which all the imbeddings O'_τ ($\tau \in T$) are continuous. Then (1.27) is a systemof base neighborhoods generating this topology.

Let us mention some other topologies in the space Φ' of (1.22) which will be encountered below (generally speaking, here we do not assume that Φ is a projective limit of Hilbert spaces). First of all, we recall the already mentioned weak topology $\sigma(\Phi', \Phi)$ given by the system of base neighborhoods

$$U(\xi_1; \varphi_1, \dots, \varphi_n) = \{\xi \in \Phi' \mid |(\xi - \xi_1, \varphi_k)_{H_0}| < 1; \ k = 1, \dots, n\}$$

$$(\xi_1 \in \Phi', \varphi_k \in \Phi; \ \ k = 1, \dots, n; \ \ n \in \mathbb{N}). \tag{1.28}$$

We shall also consider the strong topology $\beta(\Phi', \Phi)$ whose system of base neighborhoods is given by the same formula (1.28), but with an arbitrary bounded set $\delta \subset \Phi$:

$$U(\xi_1; \delta) = \{\xi \in \Phi' \mid |(\xi - \xi_1, \varphi_k)_{H_0}| < 1; \ \varphi \in \delta\} \quad (\xi_1 \in \Phi')$$

instead of finite collections of[3] $\varphi_1, \dots, \varphi_n$.

The weak topology $\sigma(\Phi', \Phi)$ is consistent with the given duality in the following sense: A set of linear continuous functionals over Φ' coincides with Φ in this topology, i.e., an arbitrary functional L of this kind admits a representation $L(\xi) = (\xi, \varphi)_{H_0}$ with some $\varphi \in \Phi$, and an arbitrary vector $\varphi \in \Phi$ generates (according to this formula) a linear $\sigma(\Phi', \Phi)$-continuous functional on Φ'.

For the topology $\beta(\Phi', \Phi)$, more rich in neighborhoods, this is, generally speaking, not true, as the $\beta(\Phi', \Phi)$-continuous linear functionals may appear which cannot be

[3] Recall that a set $\delta \subset \Phi$ is called bounded if for an arbitrary neighborhood v of the origin in Φ, there exists $\lambda \in (0, \infty)$ such that $\delta \subset \lambda v$.

given by the elements of Φ. It is useful therefore to consider one more topology, namely, the Mackey topology $\tau(\Phi', \Phi)$ which is defined as the strongest topology on Φ' consistent with the given duality between Φ and Φ' in the above-mentioned sense. One can show that for $\Phi = \operatorname{pr}\lim_{\tau \in \mathrm{T}} H_\tau$ the Mackey topology $\tau(\Phi', \Phi)$ admits a constructive description which coincides with the topology of the inductive limit $\operatorname{ind}\lim_{\tau \in \mathrm{T}} H_{-\tau}$.

The space $\Phi = \operatorname{pr}\lim_{\tau \in \mathrm{T}} H_\tau$ is called countably Hilbert or countably normed; in the case of Banach spaces $H_\tau = B_\tau$, if T is countable, then $\mathrm{T} = \mathbb{N}$. Projective and inductive limits are now denoted by $\operatorname{pr}\lim_{\tau \in \mathrm{T}} H_\tau$ and $\operatorname{ind}\lim_{\tau \in \mathrm{T}} H_{-\tau}$, as well. Finally, let us note that for a nuclear countable Hilbert space Φ, the Mackey topology $\tau(\Phi', \Phi)$ coincides with the strong topology $\beta(\Phi', \Phi)$, and in this topology the dual space Φ' is nuclear (in the sense that Φ' can be represented as a projective (nuclear) limit of some Hilbert spaces). This, in particular, yields the nuclearity of $\operatorname{ind}\lim_{\tau \in \mathrm{T}} H_{-\tau}$ if $\operatorname{pr}\lim_{\tau \in \mathrm{T}} H_\tau$ is nuclear. The proofs of the above statements on the consistency of topologies are given, e.g., in Schaefer [1, Chapters 2, 4].

Note that it is sometimes useful to topologize the set Φ' of the type (1.25) by means of a system of base neighborhoods consisting of all possible intersections of finitely many balls in the spaces $H_{-\tau}$:

$$U(\xi_1, \dots, \xi_n; \tau_1, \dots, \tau_n; \varepsilon_1, \dots, \varepsilon_n) =$$

$$\Big\{ \xi \in \Phi' \mid \xi \in \bigcap_{k=1}^{n} H_{-\tau_k}, \ \| \xi - \xi_1 \|_{H_{-\tau_1}} < \varepsilon_1, \operatorname{pr}\lim_{\tau \in \mathrm{T}} H_\tau \dots, \| \xi - \xi_n \|_{H_{-\tau_n}} < \varepsilon_n \Big\}$$

$$(\tau_k \in \mathrm{T}, \ \xi_k \in H_{-\tau_k}, \ \varepsilon_k > 0; \ \ k = 1, \dots, n; \ \ n \in \mathbb{N}).$$

Using (1.26), one can easily find that under this topologization Φ' turns into a topologic space in which the operations

$$\Phi' \times \Phi' \ni (\xi, \eta) \mapsto \xi + \eta \in \Phi', \ \ (\mathbb{C}^1 \setminus \{0\}) \times \Phi' \ni (\lambda, \xi) \mapsto \lambda\xi \in \Phi'$$

are continuous. The convergence $\Phi' \ni \xi_n \to \xi \in \Phi'$ in this topology means that

$$\exists\, \tau \in \mathrm{T}, \ m \in \mathbb{N} : \ \xi, \xi_n \in H_{-\tau} \ \ (n = m+1, m+2, \dots)$$

and $\| \xi - \xi_n \|_{H_{-\tau}} \to 0$ as $n \to \infty$. This is why, the considered convergence resembles the convergence in an elementary inductive limit, i.e., in the space $\mathbb{C}_0^\infty$ of finite sequences $x = (x_k)_{k=-\infty}^{\infty}$ $(x_k \in \mathbb{C}^1)$ which is understood as a union of finite-dimensional spaces H_τ ($\tau \in \mathbb{N}$) consisting of finite sequences $x = (x_k)_{k=-\tau}^{\tau}$ that are continued by zeros onto $k : |k| > \tau$ (the convergence in $\mathbb{C}_0^\infty$ is understood as the uniform finiteness plus the coordinatewise convergence). This enables us to call Φ' topologized in this way "an inductive limit of the spaces $H_{-\tau}$" too.

Clearly, all the above considerations are also valid in the case of a real Hilbert space H_0. We shall often encounter this situation in what follows.

Example 1.6. Let $H_0 = l_2 = l_2(\mathbb{R}^1)$ be a space of square summable real sequences $l_2 \ni x = (x_k)_{k=1}^\infty$ with a scalar product $(x, y)_{H_0} = \sum_{k=1}^\infty x_k y_k$. Denote by T a set of all possible weights $\tau = (\tau_k)_{k=1}^\infty$, $\tau_k \geq 1$, and put in correspondence with each $\tau \in \mathrm{T}$ a Hilbert space

$$H_\tau = l_2(\tau) = \left\{ x \in l_2 \;\middle|\; \sum_{k=1}^\infty x_k^2 \tau_k = \|x\|_{H_\tau}^2 < \infty \right\},$$

$$(x, y)_{H_\tau} = \sum_{k=1}^\infty x_k y_k \tau_k; \quad H_1 = H_0. \tag{1.29}$$

Evidently, $H_\tau \subset H_0$ topologically and $\|\cdot\|_{H_\tau} \geq \|\cdot\|_{H_0}$. The family of Hilbert spaces $(H_\tau)_{\tau \in \mathrm{T}}$ is directed by imbedding, i.e., if for given $\tau' = (\tau_k')_{k=1}^\infty \in \mathrm{T}$ and $\tau'' = (\tau_k'')_{k=1}^\infty \in \mathrm{T}$, we choose, for example, $\tau''' = (\tau''' = \tau' + \tau_k'')_{k=1}^\infty \in \mathrm{T}$, then $H_{\tau'''} \subset H_{\tau'}$ and $H_{\tau'''} \subset H_{\tau''}$ topologically. Consider a space $\Phi = \mathrm{pr\,lim}_{\tau \in \mathrm{T}} H_\tau$. This space is nuclear since, for every $\tau \in \mathrm{T}$, one can take $\tau' = (2^k \tau_k)_{k=1}^\infty$ such that the imbedding $O_{\tau',\tau}\colon H_{\tau'} \to H_\tau$ is quasinuclear. Indeed, let $(e_k)_{k=1}^\infty$ be a natural basis in l_2. Then the vectors $(\tau_k^{-1/2} e_k)_{k=1}^\infty$ form a basis in H_τ and, therefore, for the Hilbert norm of the imbedding operator $O_{\tau',\tau}$, we have

$$|O_{\tau',\tau}|^2 = \sum_{k=1}^\infty \left\| (\tau_k')^{-1/2} e_k \right\|_{H_\tau}^2 = \sum_{k=1}^\infty 2^{-k} < \infty.$$

Obviously, the set Φ coincides with a collection of finite real sequences $\mathbb{R}_0^\infty$, i.e., $\mathbb{R}_0^\infty \ni \varphi = (\varphi_1, \ldots, \varphi_n, 0, 0, \ldots)$, where $n = n(\varphi)$ depends on a given sequence. This follows from the equality $\Phi = \bigcap_{\tau \in \mathrm{T}} l_2(\tau)$ and the fact that for a given sequence $\varphi = (\varphi_k)_{k=1}^\infty \in \Phi$, one can always take a weight $\tau \in \mathrm{T}$ such that $\tau_k = |\varphi_k|^{-2} + 1$ provided that $\varphi_k \neq 0$, and $\tau_k = 1$ otherwise. Then the vector $\varphi \in H_\tau$ only in the case when it has a finite number of nonzero coordinates.

For every $\tau \in \mathrm{T}$, the Hilbert space $H_{-\tau} = l_2(\tau^{-1})$ is dual to $H_\tau = l_2(\tau)$ with respect to $H_\tau = l_2$. Here, $l_2(\tau^{-1})$ is constructed just as (1.29) by using the weight $\tau^{-1} = (\tau_k^{-1})_{k=1}^\infty$ (see Example 1.1 in the case $R = \mathbb{N}$ and $\mu(\{x\}) = 1$). According to the above argument, the space Φ' coincides with $\bigcup_{\tau \in \mathrm{T}} H_{-\tau}$ and, hence, $\Phi' = \mathbb{R}^\infty$ ($\mathbb{R}^\infty$ is a set of all real sequences). In fact, for every vector $\xi = (\xi_k)_{k=1}^\infty \in \mathbb{R}^\infty$, let us take $\tau \in \mathrm{T}$: $\tau_k = (|\xi_k| + 1)^2 2^k$ $(k \in \mathbb{N})$. Then

$$\| \xi \|_{H_{-\tau}}^2 = \sum_{k=1}^\infty |\xi_k|^2 (1 + |\xi_k|)^{-2} 2^{-k} < \infty,$$

i.e., $\xi \in H_{-\tau}$. The scalar product in $H_0 = l_2$ defines a natural pairing of the elements of $\mathbb{R}_0^\infty$ and $\mathbb{R}^\infty$, namely,

$$(\xi, \varphi)_{H_0} = \sum_{k=1}^\infty \xi_k \varphi_k \quad (\xi \in \mathbb{R}^\infty, \xi \in \mathbb{R}_0^\infty).$$

Hence, we have constructed the nuclear rigging

$$\mathbb{R}^\infty \supset l_2(\tau^{-1}) \supset l_2 \supset l_2(\tau) \supset \mathbb{R}_0^\infty \tag{1.30}$$

(here, $\tau \in \mathrm{T}$ is arbitrary) which will be frequently used below.

Example 1.7. We now give the simplest example of a nuclear space of functions. Let G be a bounded region in $\mathbb{R}^d$ $(d \in \mathbb{N})$ with a piecewise continuously differentiable boundary, and let $\mathrm{T} = \mathbb{Z}_+$. We set $B_\tau = C^\tau(\tilde{G})$ with the norm

$$\| \varphi \|_{B_\tau} = \max_{x \in \tilde{G}, |\alpha| \le \tau} |D^\alpha \varphi(x)|.$$

Clearly, $C^\infty(\tilde{G}) = \operatorname{pr\,lim}_{\tau \in \mathrm{T}} C^\tau(\tilde{G})$. The convergence in $C^\infty(\tilde{G})$ thus topologized is uniform with each derivative.

This space $C^\infty(\tilde{G})$ can be transformed into a countably Hilbert space if we note that by virtue of the imbedding theorems, $W_2^{l''}(G) \subset C^{l'}(G)$ topologically for $l'' - l' > d/2$ $(l', l'' \in \mathbb{Z}_+)$. This imbedding and the trivial topologic imbedding $C^l(\tilde{G}) \subset W_2^l(G)$ $(l \in \mathbb{Z}_+)$ show that the space $C^\infty(\tilde{G})$ introduced above coincides with $\operatorname{pr\,lim}_{\tau \in \mathrm{T}} W_2^\tau(G)$ both as a set and topologically.

The space $C^\infty(\tilde{G})$ is nuclear. This follows from the facts that $\forall \tau \in \mathrm{T}\ \exists\ \ \tau' = \tau + [d/2] + 1$ and that the imbedding $W_2^{\tau'}(G) \subset W_2^{\tau}(G)$ is quasinuclear (see Example 1.5).

Example 1.8. The classical Schwartz space of rapidly decreasing infinitely many times differentiable test functions $\mathcal{S}(\mathbb{R}^d)$ is usually introduced as a countably normed space $\mathcal{S}(\mathbb{R}^d) = \operatorname{pr\,lim}_{\tau \in \mathrm{T}} \mathcal{S}_\tau(\mathbb{R}^d)$, where $\mathcal{S}_\tau(\mathbb{R}^d)$ is a completion of $C_0^\infty(\mathbb{R}^d)$ with respect to the norm

$$\|\varphi\|_{\mathcal{S}_\tau(\mathbb{R}^d)} = \max_{x \in \mathbb{R}^d} \left[(1+|x|^2)^{\tau/2} \sum_{|\alpha| \le \tau} \left| (D^\alpha \varphi)(x) \right| \right] \quad (\tau \in \mathrm{T} = \mathbb{Z}_+). \tag{1.31}$$

It can be transformed into a countably Hilbert space if we set $\mathrm{T} = \mathbb{Z}_+$ (as before) and take the Sobolev space with the weight $p(x) = \left(1+|x|^2\right)^\tau$ as H_τ (see Example 1.5), i.e.,

$$H_\tau = S_\tau(\mathbb{R}^d) = W_2^\tau(\mathbb{R}^d, (1+|x|^2)^\tau dx).$$

As in Example 1.7, the coincidence of the two projective limits $\operatorname{pr\,lim}_{\tau \in \mathrm{T}} \mathcal{S}_\tau(\mathbb{R}^d)$ and $\operatorname{pr\,lim}_{\tau \in \mathrm{T}} S_\tau(\mathbb{R}^d)$ is proved by using the imbedding theorems in the bounded domain and the corresponding estimates (more exactly, we establish the topologic imbeddings $S_\tau(\mathbb{R}^d) \subset \mathcal{S}_{\tau+[d/2]+1}(\mathbb{R}^d)$ and $\mathcal{S}_\tau(\mathbb{R}^d) \subset S_{\tau+[d/2]+1}(\mathbb{R}^d)\ (\tau \in \mathrm{T})$). The nuclearity of $\mathcal{S}(\mathbb{R}^d)$ follows from the fact that the imbedding $S_{l''}(\mathbb{R}^d) \subset S_{l'}(\mathbb{R}^d)$ is quasinuclear for $l', l'' \in \mathbb{Z}_+$ and $l'' - l' > d/2$. The latter follows, in turn, from the condition of quasinuclearity of the imbedding of Sobolev spaces with weight in the unbounded domain which was presented in Example 1.5.

For a detailed presentation, see, for instance, Berezansky [18, Chapter 1, Section 3, Subsections 8, 9].

Example 1.9. We shall always understand the classical space $\mathcal{D}(\mathbb{R}^d)$ of finite infinitely differentiable test functions as the set of functions $C_0^\infty(\mathbb{R}^d)$ equipped with the topology of the projective limit (but not of the inductive limit as is usual). Just as in Example 1.8, the space $\mathcal{D}(\mathbb{R}^d)$ is initially constructed as the projective limit of the Banach spaces $\mathcal{D}_\tau(\mathbb{R}^d)$. The indexing set T now consists of pairs $\tau = (\tau_1, \tau_2(x))$, where $\tau_1 \in \mathbb{Z}_+$, $\tau_2 \in C_0^\infty(\mathbb{R}^d)$, and $\tau_2(x) \ge 1\ (x \in \mathbb{R}^d)$. The space $\mathcal{D}_\tau(\mathbb{R}^d)$ is defined as a completion of $C_0^\infty(\mathbb{R}^d)$ with respect to the norm

$$\|\varphi\|_{\mathcal{D}_\tau(\mathbb{R}^d)} = \max_{x \in \mathbb{R}^d} \left(\tau_2(x) \sum_{|\alpha| \le \tau_1} \left| (D_\alpha^\varphi)(x) \right| \right) \quad (\tau \in \mathrm{T}). \tag{1.32}$$

It is assumed that $\mathcal{D}(\mathbb{R}^d) = \operatorname{pr}\lim_{\tau\in T}\mathcal{D}(\mathbb{R}^d)$. One can easily show that $\mathcal{D}(\mathbb{R}^d)$ coincides (as a set) with the class $C_0^\infty(\mathbb{R}^d)$, and that the convergence of the sequence $\varphi_n \to \varphi$ (in this space) is the usual convergence in the space of test functions, i.e., the finiteness of functions (uniform in n) and their convergence to the limit uniform on $\mathbb{R}^d$ with all the derivatives. The proof of this fact is quite similar to the proof of the equality $\Phi = \mathbb{R}_0^\infty$ in Example 1.6. Here, we essentially use the arbitrariness of $\tau_2(\cdot)$.

By analogy with Example 1.8, the space $\mathcal{D}(\mathbb{R}^d)$ can be represented as a projective limit of Hilbert spaces $H_\tau = W_2^{\tau_1}(\mathbb{R}^d, \tau_2(x)dx)$ (the set T is the same as in (1.32)). The scheme of the proof remains the same, but the results of Example 1.5 are used. The same results lead to the nuclearity of $\mathcal{D}(\mathbb{R}^d)$.

For the set of elements of the dual spaces $\mathcal{S}'(\mathbb{R}^d)$ and $\mathcal{D}'(\mathbb{R}^d)$, i.e., for slowly decreasing generalized functions and for generalized functions of the general form, the decomposition (1.25) holds. This decomposition shows that each generalized function belongs to some (corresponding) space $H_{-\tau}$, i.e., that it is (by definition) a generalized function of the finite order.

§2. Tensor Products and the Kernel Theorem

For a detailed presentation, see, e.g., Berezansky [18, Chapter 1, Section 3, Subsection 10].

Below, we give the definitions of tensor product of both finitely and infinitely many Hilbert spaces and prove some facts concerning these products which are necessary for further discussion. We also construct the tensor products of Hilbert and nuclear riggings and give the version of the Schwartz kernel theorem for the case of Hilbert riggings.

2.1. Finite Tensor Products

Suppose that $(H_k)_{k=1}^n$ is a finite sequence of separable Hilbert spaces, and that $(e_j^{(k)})_{j=0}^\infty$ is some orthonormal basis in H_k. Let us construct a formal product

$$e_\alpha = e_{\alpha_1}^{(1)} \otimes \cdots \otimes e_{\alpha_n}^{(n)}, \tag{2.1}$$

where $\alpha = (\alpha_1, \ldots, \alpha_n) \in \mathbb{Z}_+^n = \mathbb{Z}_+^1 \times \ldots \times \mathbb{Z}_+^1$ (n times), i.e., let us consider the ordered sequence $\left(e^{(1)}_{\alpha_1}, \ldots, e^{(n)}_{\alpha_n}\right)$ and span the Hilbert space by the formal vectors (2.1), assuming that they form the orthonormal basis of this space. The separable Hilbert space obtained as a result is called a tensor product of the spaces $H_1, \ldots, H_n$ and is denoted by $H_1 \otimes \cdots \otimes H_n = \otimes_{k=1}^{n} H_k$. Its vectors have the form

$$f = \sum_{\alpha \in \mathbb{Z}_+^n} f_\alpha e_\alpha \quad (f_\alpha \in \mathbb{C}^1), \quad \|f\|^2_{\overset{n}{\underset{k=1}{\otimes}} H_k} = \sum_{\alpha \in \mathbb{Z}_+^n} |f_\alpha|^2 < \infty, \tag{2.2}$$

$$(f, g)_{\overset{n}{\underset{k=1}{\otimes}} H_k} = \sum_{\alpha \in \mathbb{Z}_+^n} f_\alpha \overline{g_\alpha}, \quad g = \sum_{\alpha \in \mathbb{Z}_+^n} g_\alpha e_\alpha \in \overset{n}{\underset{k=1}{\otimes}} H_k.$$

Let $f^{(k)} = \sum_{j=0}^{\infty} f_j^{(k)} e_j^{(k)} \in H_k \ (k = 1, \ldots, n)$ be some vectors. According to the definition,

$$f = f^{(1)} \otimes \cdots \otimes f^{(n)} = \sum_{\alpha \in \mathbb{Z}_+^n} f^{(1)}_{\alpha_1} \ldots f^{(n)}_{\alpha_n} e_\alpha. \tag{2.3}$$

The coefficients $f_\alpha = f^{(1)}_{\alpha_1} \ldots f^{(n)}_{\alpha_n}$ of decomposition (2.3) satisfy (2.2). Therefore, the vector (2.3) belongs to $\otimes_{k=1}^{n} H_k$ and, moreover,

$$\|f\|_{\overset{n}{\underset{k=1}{\otimes}} H_k} = \prod_{k=1}^{n} \|f_k\|_{H_k}. \tag{2.4}$$

Clearly, the function

$$\overset{n}{\underset{k=1}{\otimes}} H_k \ni \left(f^{(1)}, \ldots, f^{(n)}\right) \mapsto f^{(1)} \otimes \ldots \otimes f^{(n)} \in \overset{n}{\underset{k=1}{\otimes}} H_k$$

is linear in each argument, and the linear span of the vectors (2.3) is dense in $\otimes_{k=1}^{n} H_k$. This linear span is called the algebraic (noncompleted) tensor product of the spaces $H_1, \ldots, H_n$ and is denoted by $\text{a.}\otimes_{k=1}^{n} H_k$. If L_k is a linear set in $H_k \ (k = 1, \ldots, n)$, then similarly,

$$\text{a.} \overset{n}{\underset{k=1}{\otimes}} L_k = \text{l.s.} \{f^{(1)} \otimes \cdots \otimes f^{(n)} \mid f^k \in L_k, \ k = 1, \ldots, n\},$$

$$\overset{n}{\underset{k=1}{\otimes}} L_k = \text{c.l.s.} \{f^{(1)} \otimes \cdots \otimes f^{(n)} \mid f^k \in L_k, \ k = 1, \ldots, n\}.$$

Clearly, this definition of a tensor product depends on the choice of the orthonormal basis $(e_j^{(k)})_{j=0}^{\infty}$ in each factor H_k. However, one can easily understand that after changing the bases, we obtain a tensor product isomorphic to the initial one and preserving its structure.

We now proceed to the definition of the tensor product of bounded operators.

Theorem 2.1. *Let* $(H_k)_{k=1}^n$ *and* $(G_k)_{k=1}^n$ *be two sequences of Hilbert spaces, and let* $(A_k)_{k=1}^n$ *be a sequence of operators* $A_k \in \mathcal{L}(H_k, G_k)$. *We define the tensor product* $A_1 \otimes \cdots \otimes A_n = \otimes_{k=1}^n A_k$ *by*

$$\left(\bigotimes_{k=1}^{n} A_k\right) f = \left(\bigotimes_{k=1}^{n} A_k\right)\left(\sum_{\alpha \in \mathbb{Z}_+^n} f_\alpha e_\alpha\right) = \sum_{\alpha \in \mathbb{Z}_+^n} f_\alpha (A_1 e_{\alpha_1}^{(1)} \otimes \cdots \otimes A_n e_{\alpha_n}^{(n)})$$

$$\left(f \in \bigotimes_{k=1}^{n} H_k\right). \tag{2.5}$$

Then the series on the right-hand side of (2.5) *converges weakly in* $\otimes_{k=1}^n G_k$ *and defines the operator* $\otimes_{k=1}^n A_k \in \mathcal{L}(\otimes_{k=1}^n H_k, \otimes_{k=1}^n G_k)$; *moreover,*

$$\left\| \bigotimes_{k=1}^{n} A_k \right\| = \prod_{k=1}^{n} \| A_k \|. \tag{2.6}$$

Proof. It is sufficient to consider the $n = 2$ case since, by virtue of the equality $H_1 \otimes \cdots \otimes H_n = (H_1 \otimes \cdots \otimes H_{n-1}) \otimes H_n$ (the associativity of the tensor product), the general case can be then obtained by induction.

Thus, we set $n = 2$. Denote by $(l_j^{(k)})_{j=0}^{\infty}$ some orthonormal basis in G_k $(k = 1, 2)$. Assume that

$$g = \sum_{\beta \in \mathbb{Z}_+^2} g_\beta l_{\beta_1}^{(1)} \otimes l_{\beta_2}^{(2)} \in G_1 \otimes G_2.$$

Let f be a vector from $H_1 \otimes H_2$ with finitely many nonzero coordinates f_α. We fix $\alpha_2, \beta_1 \in \mathbb{Z}_+^1$ and denote the vector $\sum_{\alpha_1=0}^{\infty} f_\alpha e_{\alpha_1}^{(1)}$ by $f(\alpha_2) \in H_1$ and the vector $\sum_{\beta_2=0}^{\infty} g_\beta l_{\beta_2}^{(2)}$ by $g(\beta_1) \in G_2$. We obtain

$$\left| \left(\sum_{\alpha \in \mathbb{Z}_+^2} f_\alpha A_1 e_{\alpha_1}^{(1)} \otimes A_2 e_{\alpha_2}^{(2)}, g \right)_{G_1 \otimes G_2} \right|^2$$

$$= \left| \sum_{\alpha, \beta \in \mathbb{Z}_+^2} f_\alpha \bar{g}_\beta \left(A_1 e_{\alpha_1}^{(1)}, l_{\beta_1}^{(1)} \right)_{G_1} \left(A_2 e_{\alpha_2}^{(2)}, l_{\beta_2}^{(2)} \right)_{G_2} \right|^2$$

$$= \left| \sum_{\alpha_2=0}^{\infty} \sum_{\beta_1=0}^{\infty} \left(A_1 f(\alpha_2), l_{\beta_1}^{(1)} \right)_{G_1} \overline{\left(A_2^* g(\beta_1), e_{\alpha_2}^{(2)} \right)_{H_2}} \right|^2$$

$$\leq \sum_{\alpha_2=0}^{\infty} \sum_{\beta_1=0}^{\infty} \left| \left(A_1 f(\alpha_2), l_{\beta_1}^{(1)} \right)_{G_1} \right|^2 \sum_{\alpha_2=0}^{\infty} \sum_{\beta_1=0}^{\infty} \left| \left(A_2^* g(\beta_1), e_{\alpha_2}^{(2)} \right)_{H_2} \right|^2$$

$$= \sum_{\alpha_2=0}^{\infty} \| A_1 f(\alpha_2) \|_{G_1}^2 \sum_{\beta_1=0}^{\infty} \| A_2^* g(\beta_1) \|_{H_2}^2$$

$$\leq \| A_1 \|^2 \| A_2^* \|^2 \sum_{\alpha_2=0}^{\infty} \| f(\alpha_2) \|_{H_1}^2 \sum_{\beta_1=0}^{\infty} \| g(\beta_1) \|_{G_2}^2$$

$$= \| A_1 \|^2 \| A_2 \|^2 \sum_{\alpha \in \mathbb{Z}_+^2} | f_\alpha |^2 \sum_{\beta \in \mathbb{Z}_+^2} | g_\beta |^2.$$

This inequality obviously implies weak convergence of the series

$$\sum_{\alpha \in \mathbb{Z}_+^2} f_\alpha A_1 e_{\alpha_1}^{(1)} \otimes A_2 e_{\alpha_2}^{(2)}$$

(in $G_1 \otimes G_2$) for arbitrary $f \in H_1 \otimes H_2$ and the upper estimate for its norm (in $G_1 \otimes G_2$) in terms of $\| A_1 \| \| A_2 \| \| f \|_{H_1 \otimes H_2}$. Therefore, the operator $A_1 \otimes A_2 : H_1 \otimes H_2 \to G_1 \otimes G_2$ is correctly defined by (2.5), it is bounded and its norm does not exceed $\| A_1 \| \; \| A_2 \|$.

On the other hand, taking into account (2.4) and (2.5), we get

$$\| (A_1 \otimes A_2)(f_1 \otimes f_2) \|_{G_1 \otimes G_2} = \| A_1 f_1 \|_{G_1} \| A_2 f_2 \|_{G_2} \quad (f_k \in H_k, \; k = 1, 2).$$

By a proper choice of the unit vectors f_1 and f_2, we can make the last product as close to $\| A_1 \| \; \| A_2 \|$ as desired. Hence, the inequality $\| A_1 \otimes A_2 \| < \| A_1 \| \; \| A_2 \|$ cannot hold, i.e., for $n = 2$ the relation (2.6) is proved.

■

Remark 1. The following equality can be easily obtained from the definition (2.5):

$$\left(\bigotimes_{k=1}^{n} A_k\right)(f^{(1)} \otimes \cdots \otimes f^{(n)}) = A_1 f^{(1)} \otimes \cdots \otimes A_n f^{(n)} \tag{2.7}$$

$$(f_k \in H_k,\ k = 1, \ldots, n).$$

This equality determines the operator $\otimes_{k=1}^{n} A_k$ uniquely. The mapping

$$\mathop{\times}_{k=1}^{n} \mathfrak{L}(H_k, G_k) \ni (A_1, \ldots, A_n) \mapsto \bigotimes_{k=1}^{n} A_k \in \mathfrak{L}\left(\bigotimes_{k=1}^{n} H_k, \bigotimes_{k=1}^{n} G_k\right)$$

is linear in each variable. Note that by using (2.7) one can easily obtain the following relations

$$\left(\bigotimes_{k=1}^{n} B_k\right)\left(\bigotimes_{k=1}^{n} A_k\right) = \bigotimes_{k=1}^{n} (B_k A_k); \quad \left(\bigotimes_{k=1}^{n} A_k\right)^* = \bigotimes_{k=1}^{n} A_k^* \tag{2.8}$$

for $A_k \in \mathfrak{L}(H_k, G_k), B_k \in \mathfrak{L}(G_k, F_k)$ $(k = 1, \ldots, n)$.

■

Remark 2. *Suppose that each* A_k *in Theorem* 2.1 *is a Hilbert–Schmidt operator. Then* $\otimes_{k=1}^{n} A_k$ *is also a Hilbert–Schmidt operator, and*

$$\left| \bigotimes_{k=1}^{n} A_k \right| = \prod_{k=1}^{n} |A_k|. \tag{2.9}$$

Indeed, according to (2.5) and (2.4),

$$\left| \bigotimes_{k=1}^{n} A_k \right|^2 = \sum_{\alpha \in \mathbb{Z}_+^n} \left\| \left(\bigotimes_{k=1}^{n} A_k\right) e_\alpha \right\|^2_{\bigotimes_{k=1}^{n} G_k} = \sum_{\alpha \in \mathbb{Z}_+^n} \left\| A_1 e_{\alpha_1}^{(1)} \otimes \cdots \otimes A_n e_{\alpha_n}^{(n)} \right\|^2_{\bigotimes_{k=1}^{n} G_k}$$

$$= \prod_{k=1}^{n} \left(\sum_{\alpha_k = 0}^{\infty} \left\| A_k e_{\alpha_k}^{(k)} \right\|^2_{G_k} \right) = \prod_{k=1}^{n} |A_k|^2.$$

■

Remark 3. In what follows, we shall also consider the tensor products $\otimes_{k=1}^{n} A_k$ of unbounded operators $A_k : H_k \supset \mathfrak{D}(A_k) \to G_k$ $(k = 1, \dots, n)$. Here, we only note that they are defined correctly by (2.5) on the domain of definition $\mathfrak{D}(\otimes_{k=1}^{n} A_k) = \text{a.} \otimes_{k=1}^{n} \mathfrak{D}(A_k)$.

■

Corollary 1. *Assume that the Hilbert spaces $H_k \subset G_k$ $(k = 1, \dots, n)$ are given such that the imbedding operators $O_k: H_k \to G_k$ are continuous. Then $\otimes_{k=1}^{n} H_k \subset \otimes_{k=1}^{n} G_k$, and $O = \otimes_{k=1}^{n} O_k$ is the corresponding imbedding operator. If O_k are quasinuclear for any k, then the same is true for the operator O.*

This statement follows immediately from Theorem 2.1 and Remark 2 if we consider the mapping

$$\bigotimes_{k=1}^{n} H_k \ni \sum_{\alpha \in \mathbb{Z}_+^n} f_\alpha e_\alpha \mapsto \sum_{\alpha \in \mathbb{Z}_+^n} f_\alpha O_1 e_{\alpha_1}^{(1)} \otimes \dots \otimes O_n e_{\alpha_n}^{(n)} \in \bigotimes_{k=1}^{n} G_k.$$

■

Consider a collection of chains of the form (1.11)

$$H_{-,k} \supset H_{0,k} \supset H_{+,k} \quad (k = 1, \dots, n). \tag{2.10}$$

According to Corollary 1 of Theorem 2.1, we have

$$\bigotimes_{k=1}^{n} H_{-,k} \supset \bigotimes_{k=1}^{n} H_{0,k} \supset \bigotimes_{k=1}^{n} H_{+,k}. \tag{2.11}$$

Lemma 2.1. *The Hilbert space $\otimes_{k=1}^{n} H_{-,k}$ can be understood as a negative space with respect to the zero space $\otimes_{k=1}^{n} H_{0,k}$ and the positive space $\otimes_{k=1}^{n} H_{+,k}$, i.e.,* (2.11) *is a chain.*

Proof. Denote by G_- a negative space with respect to the zero space $\otimes_{k=1}^{n} H_{0,k}$ and the positive space $\otimes_{k=1}^{n} H_{+,k}$. Let O and I be the operators connected with the chain

$$G_- \supset \bigotimes_{k=1}^{n} H_{0,k} \supset \bigotimes_{k=1}^{n} H_{+,k},$$

and let O_k and I_k be the same operators connected with (2.10). Then $O = \otimes_{k=1}^{n} O_k$, and according to (2.8), $I = O^* = \otimes_{k=1}^{n} I_k$. But G_- is the completion of $\otimes_{k=1}^{n} H_{0,k}$ with respect to the scalar product

$$(f, g)_{G_-} = (If, g)_{\underset{k=1}{\overset{n}{\otimes}} H_{0,k}}.$$

By virtue of the equality $I = \otimes_{k=1}^{n} I_k$, on the dense set $\text{a.}\otimes_{k=1}^{n} H_{0,k}$ this scalar product coincides with the scalar product in $\otimes_{k=1}^{n} H_{-,k}$.

■

The above proof yields the following equalities for the operators connected with the chains (2.10) and (2.11):

$$I = \underset{k=1}{\overset{n}{\otimes}} I_k, \quad \boldsymbol{I} = \underset{k=1}{\overset{n}{\otimes}} \boldsymbol{I}_k, \quad J = \underset{k=1}{\overset{n}{\otimes}} J_k, \quad \boldsymbol{J} = \underset{k=1}{\overset{n}{\otimes}} \boldsymbol{J}_k. \tag{2.12}$$

Employing the already proved statements concerning the tensor products of Hilbert spaces, we can investigate the similar problem for the riggings by linear topologic spaces.

Consider the collection of riggings

$$\Phi'_k \supset H_{0,k} \supset \Phi_k \quad (k = 1, \dots, n), \tag{2.13}$$

where $\Phi_k = \text{pr lim}_{\tau_k \in \mathrm{T}_k} H_{+,\tau_k}$ is a projective limit of the family of Hilbert spaces $(H_{+,\tau_k})_{\tau_k \in \mathrm{T}_k}$ $(k = 1, \dots, n)$ directed by inclusion and satisfying the conditions of Subsection 1.2. For every multi-index $\tau = (\tau_1, \dots, \tau_n) \in \mathrm{T} = \times_{k=1}^{n} T_k$, consider the family of Hilbert riggings

$$H_{-,\tau_k} \supset H_{0,k} \supset H_{+,\tau_k} \quad (k = 1, \dots, n), \tag{2.14}$$

where H_{-,τ_k} is the Hilbert space dual to H_{+,τ_k} with respect to $H_{0,k}$. According to Lemma 2.1, for fixed $\tau \in \mathrm{T}$ the tensor product of the chains (2.14) is also the chain

$$\underset{k=1}{\overset{n}{\otimes}} H_{-,\tau_k} \supset \underset{k=1}{\overset{n}{\otimes}} H_{0,k} \supset \underset{k=1}{\overset{n}{\otimes}} H_{+,\tau_k}. \tag{2.15}$$

Since each family $(H_{+,\tau_k})_{\tau_k \in \mathrm{T}_k}$ $(k = 1, \dots, n)$ is directed, the family of Hilbert spaces $(\otimes_{k=1}^{n} H_{+,\tau_k})_{\tau \in \mathrm{T}}$ is also directed by inclusion, moreover, the set $\bigcap_{\tau \in \mathrm{T}} \otimes_{k=1}^{n} H_{+,\tau_k}$ is dense in each space of this family. According to our assumption, we have $\|\cdot\|_{H_{0,k}} \leq \|\cdot\|_{H_{+},\tau_k}$ for any $\tau_k \in \mathrm{T}_k$ and, hence, by virtue of Corollary 1 to Theorem 2.1, we conclude that

$$\|\cdot\|_{\bigotimes_{k=1}^{n} H_{0,k}} \leq \|\cdot\|_{\bigotimes_{k=1}^{n} H_{+},\tau_k}$$

for any $\tau \in \mathrm{T}$.

We define the tensor product $\otimes_{k=1}^{n}\Phi_k$ of the spaces Φ_k $(k = 1, \ldots, n)$ as a projective limit

$$\bigotimes_{k=1}^{n} \Phi_k = \mathrm{pr} \lim_{\tau=(\tau_1,\ldots,\tau_n)\in\mathrm{T}} \bigotimes_{k=1}^{n} H_{+,\tau_k} \quad \left(\mathrm{T} = \bigtimes_{k=1}^{n} \mathrm{T}_k\right). \tag{2.16}$$

Therefore, if we vary the multi-index $\tau = (\tau_1, \ldots, \tau_n)$ in (2.15) over the indexing set T, we obtain the family of chains of the form (1.24) with $H_0 = \otimes_{k=1}^{n} H_{0,k}$. This enables us to apply the scheme of Subsection 1.2 and construct the chain

$$\Phi' \supset H_0 = \bigotimes_{k=1}^{n} H_{0,k} \supset \mathrm{pr} \lim_{\tau\in\mathrm{T}} \bigotimes_{k=1}^{n} H_{+,\tau_k} = \bigotimes_{k=1}^{n} \Phi_k = \Phi.$$

The space Φ' can be topologized by the topology of the inductive limit of the negative spaces $\otimes_{k=1}^{n} H_{-,\tau_k}$ of the chain (2.15). According to the definition,

$$\bigotimes_{k=1}^{n} \Phi'_k = \mathrm{ind} \lim_{\tau=(\tau_1,\ldots,\tau_n)\in\mathrm{T}} \bigotimes_{k=1}^{n} H_{-,\tau_k}. \tag{2.17}$$

Finally, we obtain the chain

$$\bigotimes_{k=1}^{n} \Phi'_k \supset \bigotimes_{k=1}^{n} H_{0,\tau_k} \supset \bigotimes_{k=1}^{n} \Phi_k. \tag{2.18}$$

Corollary 1 implies that the space $\otimes_{k=1}^{n}\Phi_k$ is nuclear provided that each Φ_k $(k = 1, \ldots, n)$ is nuclear.

Let us recapitulate the above statements.

Lemma 2.2. *The tensor product of the chains*

$$\Phi'_k \supset H_{0,k} \supset \Phi_k \quad (k = 1, \ldots, n),$$

where $\Phi_k = \operatorname{pr}\lim_{\tau_k \in T_k} H_{+,\tau_k}$, *is also a chain*

$$\bigotimes_{k=1}^{n} \Phi'_k \supset \bigotimes_{k=1}^{n} H_{0,\tau_k} \supset \bigotimes_{k=1}^{n} \Phi_k,$$

where the spaces $\otimes_{k=1}^{n}\Phi_k$ *and* $\otimes_{k=1}^{n}\Phi'_k$ *are defined by* (2.16) *and* (2.17). *The rigging* (2.18) *(thus constructed) is nuclear provided that each rigging* (2.14) *is also nuclear.*

■

In many cases, we shall also need a definition of the tensor product for Banach spaces. We do not consider the general definition and study only the example of the specific situation which shall be used below.

Example 2.1. Let X_k be a space with a measure μ_k given on some σ-algebra $\mathfrak{R}_k$ $(k = 1, \ldots, n)$. Consider the spaces $L_p(X_k, \mu_k) = L_p(X_k, \mathfrak{R}_k, \mu_k)$ and define, as before, the algebraic tensor product

$$\text{a.}\bigotimes_{k=1}^{n} L_p(X_k, \mu_k) = \text{l.s.}\left\{ \bigotimes_{k=1}^{n} f_k(\cdot) \mid f_k \in L_p(X_k, \mu_k),\ k = 1, \ldots, n) \right\}.$$

There exists the natural imbedding

$$\text{a.}\bigotimes_{k=1}^{n} L_p(X_k, \mu_k) \ni \bigotimes_{k=1}^{n} f_k \mapsto \prod_{k=1}^{n} f_k(x) \in L_p\left(\mathop{\times}_{k=1}^{n} X_k, \mathop{\times}_{k=1}^{n} \mathfrak{R}_k, \mathop{\times}_{k=1}^{n} \mu_k \right)$$

$$\left(x = (x_1, \ldots, x_n) \in \mathop{\times}_{k=1}^{n} X_k \right),$$

which maps $\text{a.}\otimes_{k=1}^{n} L_p(X_k, \mu_k)$ into a dense subset of the space $L_p(\times_{k=1}^{n} X_k, \times_{k=1}^{n}\mu_k)$. It is natural to introduce the norm on $\text{a.}\otimes_{k=1}^{n} L_p(X_k, \mu_k)$ (which is induced by this imbedding from $L_p(\times_{k=1}^{n} X_k, \times_{k=1}^{n}\mu_k))$ and to define $\otimes_{k=1}^{n} L_p(X_k, \mu_k)$ as a completion of $\text{a.}\otimes_{k=1}^{n} L_p(X_k, \mu_k)$ with respect to this norm. Clearly, the space $\otimes_{k=1}^{n} L_p(X_k, \mu_k)$ is isomorphic to $L_p(\times_{k=1}^{n} X_k, \times_{k=1}^{n}\mu_k)$. It can be easily shown that for $p = 2$ the tensor product introduced above coincides with the tensor product introduced according to the general definition.

Now, let $A_k \in \mathcal{L}(L_p(X_k, \mu_k), L_p(X_k, \mu_k))$ $(k = 1, \ldots, n)$. We define the map

$$\bigotimes_{k=1}^{n} L_p(X_k, \mu_k) \ni f_1 \otimes \ldots \otimes f_n \mapsto \left(\bigotimes_{k=1}^{n} A_k \right)(f_1 \otimes \ldots \otimes f_n)$$

$$= A_1 f_1 \otimes \ldots \otimes A_n f_n \in \bigotimes_{k=1}^{n} L_p(X_k, \mu_k),$$

which can be extended by linearity to the map

$$\bigotimes_{k=1}^{n} A_k\text{: a. } \bigotimes_{k=1}^{n} L_p(X_k, \mu_k) \to \text{a. } \bigotimes_{k=1}^{n} L_p(X_k, \mu_k).$$

The operator $\otimes_{k=1}^{n} A_k$ defined by this relation is called the algebraic tensor product of the operators A_k.

Lemma 2.3. *Let $A_k \in \mathfrak{L}(L_p(X_k, \mu_k), L_p(X_k, \mu_k))$ $(k = 1, \ldots, n)$ be given integral operators with nonnegative kernels $R_k(\cdot, \cdot)$, i.e.,*

$$(A_k f)(x_k) = \int_{X_k} R_k(x_k, y_k) f(y_k)\, d\mu_k(y_k) \quad (f \in L_p(X_k, \mu_k)), \tag{2.19}$$

and $R_k(\cdot, \cdot) \geq 0$ almost everywhere with respect to $\mu_k \times \mu_k$. Assume that $\| A_k \| \leq 1$ $(k = 1, \ldots, n)$, i.e., that each operator A_k is contracting. Then the algebraic tensor product $\otimes_{k=1}^{n} A_k$ can be extended to a contracting operator from $L_p(\times_{k=1}^{n} X_k, \times_{k=1}^{n} \mu_k)$ to $L_q(\times_{k=1}^{n} X_k, \times_{k=1}^{n} \mu_k)$.

Proof. By virtue of associativity, it suffices to investigate the case $n = 2$. Consider the Banach space $L_p(X_1 \to L_p(X_2, \mu_2), \mu_1)$ consisting of the mappings

$$X_1 \ni x_1 \mapsto F(x_1) \in L_p(X_2, \mu_2),$$

such that the norm

$$\| F \|_{p,p} = \left[\int_{X_1} \| F(x_1) \|_{L_p(X_2, \mu_2)}^{p} \, d\mu_1(x_1) \right]^{1/p} \tag{2.20}$$

is bounded. It follows from (2.20) and the Fubini theorem that this space is isometrically isomorphic ($\cong$) to $L_p(X_1 \times X_2, \mu_1 \times \mu_2)$. By analogy, we define the space $L_q(X_1 \to L_p(X_2, \mu_2), \mu_1)$ with the norm

$$\| F \|_{q,p} = \left[\int_{X_1} \| F(x_1) \|_{L_p(X_2, \mu_2)}^{q} d\mu_1(x_1) \right]^{1/q}.$$

$$(\tilde{A}_1 F)(x_1) = \int_{X_1} R_1(x_1, y_1)\, F(y_1)\, d\mu_1(y_1). \tag{2.21}$$

Relation (2.21) and the inequality $R_1 \geq 0$ yield

$$\| (\tilde{A}_1 F)(x_1) \|_{L_p(X_2, \mu_2)} \leq \int_{X_1} R_1(x_1, y_1) \, \| F(y_1) \|_{L_p(X_2, \mu_2)} \, d\mu_1(y_1)$$

$$= (A_1 \| F(\cdot) \|_{L_p(X_2, \mu_2)})(x_1),$$

whence

$$\| \tilde{A}_1 F \|_{q,p} \leq \| A_1 \| F(\cdot) \|_{L_p(X_2, \mu_2)} \|_{L_q(X_1, \mu_1)} \leq \| \| F \|_{L_p(X_2, \mu_2)} \|_{L_q(X_1, \mu_1)} = \| F \|_{p,p}$$

by virtue of the contractibility of A_1. Therefore, $\tilde{A}_1$ acts as a contracting operator to the space $L_q(X_1 \to L_p(X_2, \mu_2), \mu_1)$.

Let us introduce the operator

$$\tilde{A}_2 \colon L_q(X_1 \to L_p(X_2, \mu_2), \mu_1) \to L_q(X_1 \to L_q(X_2, \mu_2), \mu_1),$$

given by the equality $(\tilde{A}_2 F)(x_1) = A_2 F(x_1)$, which is a contracting operator since the operator $A_2 \colon L_p(X_2, \mu_2) \to L_q(X_2, \mu_2)$ is contracting. The operator $\tilde{A} = \tilde{A}_2 \tilde{A}_1$ is defined as a contracting operator from

$$L_p(X_1 \to L_p(X_2, \mu_2), \mu_1) \cong L_p(X_1 \times X_2, \mu_1 \times \mu_2)$$

to

$$L_q(X_1 \to L_q(X_2, \mu_2), \mu_1) \cong L_q(X_1 \times X_2, \mu_1 \times \mu_2).$$

The Fubini theorem shows that

$$\tilde{A} \restriction \text{a.}\, L_p(X_1, \mu_1) \otimes L_p(X_2, \mu_2) = A_1 \otimes A_2.$$

■

2.2. Infinite Tensor Products

We now pass on to the definition of the tensor product of infinitely many Hilbert spaces (or, more exactly, to the definition of a separable subspace of the complete von Neumann tensor product).

Let $(H_k)_{k=1}^{\infty}$ be a sequence of Hilbert spaces, and let $e = (e^{(k)})_{k=1}^{\infty}$ $(e^{(k)} \in H_k)$ be a fixed "stabilizing" sequence of unit vectors. In each H_k consider an orthonormal basis $(e_j^{(k)})_{j=0}^{\infty}$ such that $e_0^{(k)} = e^{(k)}$. Denote by $\mathbb{Z}_{+,0}^{\infty} \subset \mathbb{Z}_+^{\infty} = \mathbb{Z}_+ \times \mathbb{Z}_+ \times \ldots$ the set of finite multi-indices with integer nonnegative coordinates. In other words, $\mathbb{Z}_{+,0}^{\infty}$ consists of multi-indices $\alpha = (\alpha_1, \alpha_2, \ldots)$ $(\alpha_k \in \mathbb{Z}_+)$ such that $\alpha_{k+1} = \alpha_{k+2} = \ldots = 0$ beginning with a certain number $k \in \mathbb{N}$ depending on α. Denote the minimal such number by $\nu(\alpha)$ and call it the length of the multi-index $\alpha \in \mathbb{Z}_{+,0}^{\infty}$. The set $\mathbb{Z}_{+,0}^{\infty}$ is clearly countable.

For every $\alpha \in \mathbb{Z}_{+,0}^{\infty}$, we construct the formal product

$$e_\alpha = e_{\alpha_1}^{(1)} \otimes e_{\alpha_2}^{(2)} \otimes \cdots. \tag{2.22}$$

Here $\alpha_k = 0$ for $k > \nu(\alpha)$, i.e., for these k we have $e_{\alpha_k}^{(k)} = e^{(k)}$. Therefore, this product coincides with the corresponding part of the product $e^{(1)} \otimes e^{(2)} \otimes \ldots$ beginning with the $(\nu(\alpha) + 1)$ th place (it is getting "stabilized").

The infinite tensor product $\mathcal{H}_e = \otimes_{k=1;e}^{\infty} H_k$ of the Hilbert spaces H_k with the stabilizing sequence e is defined to be the Hilbert space spanned by the basis $(e_\alpha)_{\alpha \in \mathbb{Z}_{+,0}^{\infty}}$, which is regarded as orthonormal. Consequently, the vectors from $\mathcal{H}_e$ have the form

$$f = \sum_{\alpha \in \mathbb{Z}_{+,0}^{\infty}} f_\alpha e_\alpha \quad (f_\alpha \in \mathbb{C}^1), \tag{2.23}$$

where

$$\|f\|_{\mathcal{H}_e}^2 = \sum_{\alpha \in \mathbb{Z}_{+,0}^{\infty}} |f_\alpha|^2 < \infty, \quad (f, g)_{\mathcal{H}_e} = \sum_{\alpha \in \mathbb{Z}_{+,0}^{\infty}} f_\alpha \overline{g_\alpha}. \tag{2.24}$$

Of course, should we take another basis in H_k but preserve e, we obtain the same tensor product.

Let $f^{(k)} = \sum_{j=0}^{\infty} f_j^{(k)} e_j^{(k)} \in H_k$ $(k = 1, \ldots, n)$. Just as in (2.3), we set

$$\begin{aligned} f &= f^{(1)} \otimes \cdots \otimes f^{(n)} \otimes e^{(n+1)} \otimes e^{(n+2)} \otimes \cdots \\ &= f_{\alpha_1}^{(1)} \cdots f_{\alpha_n}^{(n)} e_{\alpha_1}^{(1)} \otimes \cdots \otimes e_{\alpha_n}^{(n)} \otimes e^{(n+1)} \otimes \cdots. \end{aligned} \tag{2.25}$$

It is easy to see that the coefficients f_α of the expansion (2.25) satisfy condition (2.24); therefore, $f \in \mathcal{H}_e$ and for its norm we have

$$\|f\|_{\mathcal{H}_e} = \prod_{k=1}^{n} \|f^{(k)}\|_{H_k}.$$

As in the case of finite tensor products, we associate the linear sets $L_k \subset H_k$ with

$$(\text{a.} \bigotimes_{k=1}^{n} \mathfrak{L}_k) \otimes e^{(n+1)} \otimes e^{(n+2)} \otimes \cdots$$

$$= \text{l.s.} \{ f^{(1)} \otimes \cdots \otimes f^{(n)} \otimes e^{(n+1)} \otimes e^{(n+2)} \otimes \cdots | f^{(k)} \in L_k;\ k = 1, \ldots, n \},$$

$$\text{a.} \bigotimes_{k=1}^{n} L_k = \text{l.s.} \{ f^{(1)} \otimes \cdots \otimes f^{(n)} \otimes e^{(n+1)} \otimes e^{(n+2)} \otimes \cdots | f^{(k)} \in L_k; k = 1, \ldots, n, n \in \mathbb{N} \}$$

$$= \text{l.s.} \bigcup_{n=1}^{\infty} (\text{a.} \bigotimes_{k=1}^{n} L_k) \otimes e^{(n+1)} \otimes e^{(n+2)} \otimes \cdots.$$

The closures of these sets in $\mathcal{H}_e$ are denoted by $(\otimes_{k=1}^{n} L_k) \otimes e^{(n+1)} \otimes e^{(n+2)} \otimes \ldots$ and $\otimes_{k=1;e}^{\infty} L_k$, respectively.

Let us associate the vectors (2.3) and (2.25) with each other. Then extending this correspondence by linearity and continuity, we get a natural isomorphism between the Hilbert spaces $\otimes_{k=1}^{n} H_k$ and $(\otimes_{k=1}^{n} H_k) \otimes e^{(n+1)} \otimes e^{(n+2)} \otimes \ldots$. The image of a vector $f_n \in \otimes_{k=1}^{n} H_k$ under this isomorphism is denoted by $f_n \otimes e^{(n+1)} \otimes e^{(n+2)} \otimes \cdots$.

Corollary 1 to Theorem 2.1 implies that for finite tensor products, the continuous imbedding of cofactors involves the imbedding of tensor products. The following theorem establishes a similar result for the case of infinite tensor products.

Theorem 2.2. *Let $(H_k)_{k=1}^{\infty}$ and $(G_k)_{k=1}^{\infty}$ be two sequences of Hilbert spaces such that $H_k \subset G_k$, and let the imbedding operator O_k: $H_k \to G_k$ be continuous and $\prod_{k=1}^{\infty} \| O_k \| = c < \infty$. If the sequence of vectors $e = \left(e^{(k)}\right)_{k=1}^{\infty}$, $e^{(k)} \in H_k$, is such that $\| e^{(k)} \|_{H_k} = \| e^{(k)} \|_{G_k} = 1$, then $\otimes_{k=1;e}^{\infty} H_k \subset \otimes_{k=1;e}^{\infty} G_k$; furthermore, this imbedding is continuous and its norm does not exceed c.*

Proof. Let $(e_j^{(k)})_{j=0}^{\infty}$, $e_0^{(k)} = e^{(k)}$, be an orthonormal basis in H_k ($k \in \mathbb{N}$). By using this basis, we form the basis vectors e_α (see (2.22)) and consider an arbitrary vector

$$f = \sum_{\alpha \in \mathbb{Z}^{\infty}_{+,0}} f_\alpha e_\alpha \in \bigotimes_{k=1;e}^{\infty} H_k.$$

For $n \in \mathbb{N}$, we define

$$f_n = \sum_{\nu(\alpha)\le n} f_\alpha e_\alpha = h_n \otimes e^{(n+1)} \otimes e^{(n+2)} \otimes \cdots,$$

where $h_n \in \bigotimes_{k=1}^{n} H_k$. According to Corollary 1 of Theorem 2.1, we have $\otimes_{k=1}^{n} H_k \subset \otimes_{k=1}^{n} G_k$ and, hence, $h_n \in \otimes_{k=1}^{n} G_k$. Consequently, $f_n \in \otimes_{k=1;e}^{\infty} G_k$ and $f_n = \sum_{\nu(\alpha)\le n} g_\alpha l_\alpha$, where l_α are constructed (just as e_α) by using the bases in the spaces G_k, and $g_\alpha \in \mathbb{C}^1$.

Clearly,

$$\| f_n \|^2_{\bigotimes_{k=1,e}^{\infty} G_k} = \sum_{\nu(\alpha)\le n} | g_\alpha |^2 = \| h_n \|^2_{\bigotimes_{k=1}^{\infty} G_k}.$$

Since the equality $\| e^{(k)} \|_{H_k} = \| e^{(k)} \|_{G_k}$ involves $\| O_k \| \ge 1$, we have, by virtue of (2.6),

$$\| f_n \|_{\bigotimes_{k=1,e}^{\infty} G_k} = \| h_n \|_{\bigotimes_{k=1}^{\infty} G_k} \le \left(\prod_{k=1}^{n} \| O_n \| \right) \| h_n \|_{\bigotimes_{k=1}^{\infty} H_k}$$

$$\le \left(\prod_{k=1}^{n} \| O_n \| \right) \| f_n \|_{\bigotimes_{k=1,e}^{\infty} H_k}. \tag{2.26}$$

It remains to note that the sequence $(f_n)_{n=1}^{\infty}$ is fundamental in $\otimes_{k=1;e}^{\infty} G_k$. This follows from the inequality (2.26) applied to $f_n - f_m$

■

Remark 1. It is easy to show that the norm of the imbedding operator $\mathbb{O}$: $\otimes_{k=1;e}^{\infty} H_k \to \otimes_{k=1;e}^{\infty} G_k$ equals $\prod_{k=1}^{\infty} \| O_k \|$.

■

Lemma 2.4. *Assume that all assumptions of Theorem 2.2 are satisfied and, in addition, that the imbeddings O_k are quasinuclear. In order that $\otimes_{k=1;e}^{\infty} H_k$ be quasinuclearly imbedded into $\otimes_{k=1;e}^{\infty} G_k$, it is necessary and sufficient that the following condition hold*

$$\sum_{k=1}^{\infty} (\boldsymbol{|} O_k \boldsymbol{|} - 1) < \infty, \tag{2.27}$$

i.e., that the infinite product $\prod_{k=1}^{\infty} \boldsymbol{|} O_k \boldsymbol{|}$ *converge. Moreover,* $\boldsymbol{|} \mathbb{O} \boldsymbol{|} = \prod_{k=1}^{\infty} \boldsymbol{|} O_k \boldsymbol{|}$. *Note that it is not necessary to presuppose the convergence of* $\prod_{k=1}^{\infty} \| O_k \|$. *This follows from the inequality* $\boldsymbol{|} O_k \boldsymbol{|} \geq \| O_k \| \geq 1$. *The same inequality implies that the terms of series* (2.27) *are nonnegative.*

Proof. Clearly, the set $\mathbb{Z}^{\infty}_{+,0}$ can be decomposed into disjoint subsets in the following manner

$$\mathbb{Z}^{\infty}_{+,0} = \bigcup_{n=1}^{\infty} \mathrm{A}_n, \qquad \mathrm{A}_n = \{\alpha \in \mathbb{Z}^{\infty}_{+,0} \,|\, \nu(\alpha) = n\}.$$

This means that we can write

$$\sum_{\alpha \in \mathbb{Z}^{\infty}_{+,0}} c_\alpha = \sum_{n=1}^{\infty} \sum_{\alpha \in \mathrm{A}_n} c_\alpha = c_{(0,0,\ldots)} + \sum_{n=1}^{\infty} \left\{ \sum_{\alpha_1,\ldots,\alpha_{n-1}=0}^{\infty} \sum_{\alpha_n=1}^{\infty} c_\alpha \right\}. \tag{2.28}$$

Assembling the terms of the series

$$|\mathbb{O}|^2 = \sum_{\alpha \in \mathbb{Z}^{\infty}_{+,0}} \| e_\alpha \|^2_{\overset{\infty}{\underset{k=1,e}{\otimes}} G_k} = \sum_{\alpha \in \mathbb{Z}^{\infty}_{+,0}} \| e^{(1)}_{\alpha_1} \|^2_{G_1} \cdots \| e^{(\nu(\alpha))}_{\alpha_{\nu(\alpha)}} \|^2_{G_{\nu(\alpha)}}$$

in correspondence with (2.28), we can easily obtain the equality $\boldsymbol{|} \mathbb{O} \boldsymbol{|} = \prod_{k=1}^{\infty} \boldsymbol{|} O_k \boldsymbol{|}$ by use of the relation

$$1 + \sum_{n=1}^{\infty} \left(\prod_{k=1}^{n-1} c_k \right) (c_n - 1) = c_k \quad (c_k \geq 1). \tag{2.29}$$

The convergence of the so-obtained tensor product is equivalent to the convergence of the series (2.27).

∎

We shall often meet the situation when each imbedding $O_k \colon H_k \to G_k$ is quasinuclear but the series (2.27), generally speaking, diverges. Note that one can always make this series convergent (and, thus, make the imbedding $\otimes^{\infty}_{k=1;e} H_k \subset \otimes^{\infty}_{k=1;e} G_k$ quasinuclear) by proper renormalization of the spaces H_k $(k \in \mathbb{N})$. We have the following lemma.

Lemma 2.5. *Let H and G be two Hilbert spaces such that $H \subset G$ and the imbedding operator $O: H \to G$ is quasinuclear. Suppose that there exists a vector $e \in H: \|e\|_G = \|e\|_H = 1$. Then, for any $\varepsilon > 0$, one can choose a scalar product $(\cdot, \cdot)_{H,q}$ on H equivalent to the original one and such that $\|e\|_{H,q} = 1$, as before, whereas $1 \le |O|^2 \le 1 + \varepsilon$.*

Proof. Let $(e_j)_{j=0}^{\infty}$ be an orthonormal basis in H such that $e_0 = e$. Then

$$|O|^2 = 1 + \sum_{j=1}^{\infty} \|e_j\|_G^2 .$$

We introduce a new scalar product in H by setting

$$(f, g)_{H,q} = (f, e)_H \overline{(g, e)_H} + c\,(Pf, Pg)_H \qquad (f, g \in H), \tag{2.30}$$

where P is the projector onto the orthogonal complement of e in H, and $c > 0$ is a constant. This product is equivalent to the original product $(\cdot, \cdot)_H$, and the vectors $l_0 = e_0$ and $l_j = c^{-1/2} e_j$ $(j \in \mathbb{N})$ form an orthonormal basis in H. Let us choose a sufficiently large c such that the inequality $\sum_{j=1}^{\infty} \| l_j \|_G^2 \le \varepsilon$ is satisfied. Now $1 \le |O_k|^2 \le 1 + \varepsilon$.

■

The procedure of renormalization can be realized automatically if we introduce the concept of a weighted infinite tensor product of Hilbert spaces H_k $(k \in \mathbb{N})$ with stabilizing sequence e. In fact, assume, in addition, that a weight sequence $\delta = (\delta_n)_{n=1}^{\infty}$ of numbers $\delta_n > 0$ is given. The weighted infinite tensor product of the spaces H_k $(k \in \mathbb{N})$ with stabilizing sequence e and weight δ is defined as a Hilbert space spanned by the basis $(\delta_{\nu(\alpha)}^{-1/2} e_\alpha)_{\alpha \in \mathbb{Z}_{+,0}^{\infty}}$, which is regarded as orthonormal by definition (here e_α has the form (2.22) as before). We denote this product as $\otimes_{k=1; e; \delta}^{\infty} H_k = \mathcal{H}_{e,\delta}$. Therefore, the norm and the scalar product of the vectors (2.23) are now calculated by use of the following modification of formulae (2.24) :

$$\|f\|_{\mathcal{H}_{e,\delta}}^2 = \sum_{n=1}^{\infty} \delta_n \Big(\sum_{\alpha \in A_n} c_\alpha |f_\alpha|^2 \Big), \quad (f, g)_{\mathcal{H}_{e,\delta}} = \sum_{n=1}^{\infty} \delta_n \Big(\sum_{\alpha \in A_n} f_\alpha \overline{g_\alpha} \Big). \tag{2.31}$$

It is easy to see that if we renormalize each H_k $(k \in \mathbb{N})$ by (2.30) with $e = e^{(k)}$ and constants $c = c_k$ and construct the tensor product of these renormalized spaces with stabilizing sequence e, then we obtain the weighted tensor product $\mathcal{H}_{e,\delta}$, where $\delta_n = (c_1 \ldots c_n)^{-1}$. Conversely, each weighted product can be understood as the usual one if we first perform the corresponding renormalizations with $c_1 = \delta_1^{-1}$ and $c_n = \delta_{n-1}\delta_n^{-1}$ $(n = 2, 3, \ldots)$.

It follows from the arguments presented above that Theorem 2.2 and the successive facts remain valid for weighted tensor products (one must use the renormalized spaces when calculating $\| O_k \|$). The imbedding $\mathbb{O}\colon \otimes_{k=1; e; \delta}^{\infty} H_k \to \otimes_{k=1; e}^{\infty} G_k$ is quasinuclear if and only if

$$1 + \sum_{n=1}^{\infty} \frac{1}{\delta_n} \left(\prod_{k=1}^{n-1} | O_k |^2 \right) (| O_n |^2 - 1) < \infty, \tag{2.32}$$

where $| O_k |$ is the Hilbert norm of the imbedding operator $O_k\colon H_k \to G_k$ for the original spaces (the left-hand side of this inequality equals $| \mathbb{O}_n |^2$). As in Lemma 2.4, the proof reduces to the direct calculation of $| \mathbb{O}_n |^2$ with the use of the equality (2.28).

If $\delta = 1$ (i.e., if $\delta_n = 1$ for any n), then $\mathcal{H}_{e,1} = \mathcal{H}_e$, and by virtue of (2.29), (2.32) turns into $(\prod_{k=1}^{\infty} | O_k |)^2$.

Remark 1. *Assume that the imbedding $H_n \subset G_n$ is quasinuclear for $n \in \mathbb{N}$. Then for each weight ε, one can find a weight δ such that the imbedding $\mathcal{H}_{e,\delta} \supset \mathcal{H}_{e,\varepsilon}$ is quasinuclear.*

This immediately follows from the criterion (2.32) if we first renormalize H_n such that $\mathcal{H}_{e,\varepsilon}$ turns into the usual infinite tensor product with stabilizing sequence e.

■

Finally, consider the infinite product of chains. First, we note that *for the rigging $H_- \supset H_0 \supset H_+$, for which there exists a vector $e \in H_+$ such that $\| e \|_{H_+} = \| e \|_{H_0} = 1$, the equalities $Ie = e$ and $\| e \|_{H_-} = 1$ hold.*

Indeed, let P be the projector onto the orthogonal complement of e in H_+. Then

$$\| \varphi \|_{H_+}^2 = |(\varphi, e)_{H_+}|^2 + \| P\varphi \|_{H_+}^2$$

for $\varphi \in H_+$. In particular,

$$\| Ie \|_{H_+}^2 = |(Ie, e)_{H_+}|^2 + \| PIe \|_{H_+}^2 = |(e, e)_{H_0}|^2 + \| PIe \|_{H_+}^2 = 1 + \| PIe \|_{H_+}^2 .$$

On the other hand, $\|I\| = 1$ and, thus, $\|Ie\|_{H_+} \le \|e\|_{H_0} = 1$. Consequently, $PIe = 0$, i.e., $Ie = (Ie, e)_{H_+} e = e$, and

$$\|e\|^2_{H_-} = (Ie, e)_{H_0} = (e, e)_{H_0} = 1.$$

■

Let the sequence of chains

$$H_{-,k} \supset H_{0,k} \supset H_{+,k} \qquad (k \in \mathbb{N}) \tag{2.33}$$

be given. Assume that there exists a sequence $e = (e^{(k)})_{k=1}^{\infty}$ of vectors $e^{(k)} \in H_{+,k}$ such that $\|e^{(k)}\|_{H_{+,k}} = \|e^{(k)}\|_{H_{0,k}}$ for any $k \in \mathbb{N}$. Then $\|e^{(k)}\|_{H_{-,k}} = 1$, as has been shown above. Denote $\mathcal{H}_{0,e} = \otimes_{k=1;e}^{\infty} H_{0,k}$ and $\mathcal{H}_{\pm,e} = \otimes_{k=1;e}^{\infty} H_{\pm,k}$. For the norm of O_k: $H_{+,k} \to H_{0,k}$, we now have $\|O_k\| = 1$. Therefore, Theorem 2.2 yield the inclusions

$$\mathcal{H}_{-,e} \supset \mathcal{H}_{0,e} \supset \mathcal{H}_{+,e}. \tag{2.34}$$

The space $\mathcal{H}_{+,e}$ is dense in each space of (2.34), and $\|\varphi\|_{\mathcal{H}_{0,e}} \le \|\varphi\|_{\mathcal{H}_{+,e}}$ $(\varphi \in \mathcal{H}_{+,e})$.

Lemma 2.6. *The sequence of spaces* (2.34) *is a chain.*

In other words, in (2.33) the space $\mathcal{H}_{-,e}$ coincides with the negative space G_- constructed by use of the positive space $\mathcal{H}_{+,e}$ and the zero space $\mathcal{H}_{0,e}$. The proof of this fact is similar to the proof of Lemma 2.1. Namely, we verify the coincidence of the scalar products $(\cdot,\cdot)_{G_-}$ and $(\cdot,\cdot)_{\mathcal{H}_{-,e}}$ on the vectors of the form (2.25), where $f^{(k)} \in H_{0,k}$, then take the linear span of these vectors and complete it. The proof of the coincidence of scalar products is based mainly on the fact that $Ie^{(k)} = e^{(k)}$ for any k, and this has been proved above.

■

When constructing the chains (2.34) from the chains (2.33) for concrete spaces, we essentially use one more renormalization. The problem can be formulated as follows: If $H_{0,k} \supset H_{+,k}$ with $\|\cdot\|_{H_{0,k}} \le \|\cdot\|_{H_{+,k}}$, and $e = (e^{(k)})_{k=1}^{\infty}$ $(e^{(k)} \in H_{+,k})$ is some stabilizing sequence for the spaces $H_{0,k}$, then, generally speaking, $\|e^{(k)}\|_{H_{+,k}} \ge 1$, and this sequence is not suitable to be a stabilizing sequence for $H_{+,k}$. The following general lemma improves the situation.

Lemma 2.7. *Let H and G be two Hilbert spaces such that $H \subset G$ topologically, $\|\cdot\|_G \le \|\cdot\|_H$, and there exists $e \in H$ for which $\|e\|_G = 1$. Then one can choose a scalar product $(\cdot,\cdot)_{H,\mathrm{st}}$ in H equivalent to the original one and such that $\|\cdot\|_G \le \|\cdot\|_{H,\mathrm{st}}$ and $\|e\|_{H,\mathrm{st}} = 1$, as before.*

Proof. Let P be the projector onto the orthogonal complement to e in G. Define the scalar product in H by

$$(f, g)_{H,\mathrm{st}} = (f, e)_G \overline{(g, e)_G} + (Pf, Pg)_H \quad (f, g \in H). \tag{2.35}$$

Since $H \ni f = (f, e)_G e + Pf$ and $e \in H$, we have $Pf \in H$, and (2.35) is meaningful. The relations for $\|\cdot\|_{H,\mathrm{st}}$ clearly hold and, therefore, it remains to prove that the norms $\|\cdot\|_{H,\mathrm{st}}$ and $\|\cdot\|_H$ are equivalent. For $f \in H$, we have

$$\|f\|_H = \|(f, e)_G e + Pf\|_H \le |(f, e)_G| \|e\|_H + \|Pf\|_H \le \sqrt{2}\,\|e\|_H \|f\|_{H,\mathrm{st}}.$$

Conversely, let $f_n \in H$ be such that $\|f_n\|_H \to 0$ as $n \to \infty$. Then

$$|(f_n, e)_G| \le \|f_n\|_G \le \|f_n\|_H \to 0,$$

and thus

$$\|Pf_n\|_H = \|f_n - (f_n, e)_G e\|_H \le \|f_n\|_H + |(f_n, e)_G| \|e\|_H \to 0.$$

This yields that $\|f_n\|_{H,\mathrm{st}} \to 0$.

∎

This lemma concludes our presentation of the general constructions related to infinite tensor products. In our further discussions, we shall often encounter examples of these products. The infinite product of probability measures should be mentioned as the first classical example of this sort (see Chapter 2, Section 1, Subsection 1.2). The infinite tensor products of chains constructed using projective and inductive limits of Hilbert spaces (i.e., the results such as Lemma 2.2) will be studied in detail in Chapter 2, Section 4, Subsection 4.1.

2.3. The Kernel Theorem

The well-known Schwartz kernel theorem states that every continuous bilinear form

$a(\varphi, \psi)$ ($\varphi, \psi \in \mathcal{S}(\mathbb{R}^d)$) can be represented in the form

$$a(\varphi, \psi) = \langle \xi, \varphi \otimes \overline{\psi} \rangle, \tag{2.36}$$

where $\xi \in \mathcal{S}'(\mathbb{R}^{2d})$ is a generalized kernel, and

$$(\varphi \otimes \overline{\psi})(x_1, \dots, x_{2d}) = \varphi(x_1, \dots, x_d)\, \overline{\psi}(x_{d+1}, \dots, x_{2d}).$$

A similar representation holds for multilinear forms as well. This theorem is generalized below for the case of abstract riggings.

First, let us introduce the concept of a generalized kernel. Consider n chains of the form

$$H_{-,k} \supset H_{0,k} \supset H_{+,k} \quad (k = 1, \dots, n) \tag{2.37}$$

and their tensor product

$$\bigotimes_{k=1}^{n} H_{-,k} \supset \bigotimes_{k=1}^{n} H_{0,k} \supset \bigotimes_{k=1}^{n} H_{+,k}\,. \tag{2.38}$$

The elements

$$F, G, \dots \in \otimes_{k=1}^{n} H_{0,k}, \quad \Phi, \Psi, \dots \in \otimes_{k=1}^{n} H_{+,k}, \quad \text{and } \Theta, \Xi, \dots \in \otimes_{k=1}^{n} H_{-,k}$$

are called ordinary, smooth, and generalized kernels, respectively.

Consider a continuous n-linear form $a(f^{(1)}, \dots, f^{(n)})$, i.e., a continuous function

$$\bigoplus_{k=1}^{n} H_{0,k} \ni (f^{(1)}, \dots, f^{(n)}) \mapsto a(f^{(1)}, \dots, f^{(n)}) \in \mathbb{C}^1, \tag{2.39}$$

linear in each variable $f^{(k)}$ (all other variables are fixed). It is a well-known fact that the continuity of (2.39) is equivalent to the existence of the estimate

$$|a(f^{(1)}, \dots, f^{(n)})| \le c \prod_{k=1}^{n} \|f^{(k)}\|_{H_{0,k}} \quad (f^{(k)} \in H_{0,k};\ k = 1, \dots, n) \tag{2.40}$$

for some $c > 0$ (this is proved just as in the case of $n = 1$). Fix an orthonormal basis $(e_j^{(k)})_{j=0}^{\infty}$ in each space $H_{0,k}$. Let $f^{(k)} = \sum_{\alpha_k=0}^{\infty} f_{\alpha_k} e_{\alpha_k}^{(k)}$ be the decomposition of the vector $f^{(k)} \in H_{0,k}$ in this basis. We set $\alpha = (\alpha_1, \dots, \alpha_n) \in \mathbb{Z}_+^n$. The continuity and multilinearity of a imply the following representation of this form in terms of its coordinates a_α and the coordinates of the vectors $f^{(k)}$:

$$a(f^{(1)}, \dots, f^{(n)}) = \sum_{\alpha \in \mathbb{Z}_n^+} a_\alpha f_{\alpha_1}^{(1)} \dots f_{\alpha_n}^{(n)}, \tag{2.41}$$

$$a_\alpha = a(e_{\alpha_1}^{(1)}, \dots, e_{\alpha_n}^{(n)}).$$

The proof of the kernel theorem is based on the following two lemmas:

Lemma 2.8. *Let a be a continuous n-linear form* (2.39), *and let $A_k \in \mathcal{L}(H_{0,k}, H_{0,k})$ ($k = 2, \dots, n$) be Hilbert–Schmidt operators. Consider the continuous n-linear form*

$$\bigoplus_{k=1}^{n} H_{0,k} \ni (f^{(1)}, \dots, f^{(n)}) \mapsto b(f^{(1)}, \dots, f^{(n)}) = a(f^{(1)}, A_2 f^{(2)}, \dots, A_n f^{(n)})$$

Then the coordinates $(b_\alpha)_{\alpha \in \mathbb{Z}_+^n}$ of the form b are such that

$$\sum_{\alpha \in \mathbb{Z}_+^\infty} |b_\alpha|^2 < \infty. \tag{2.42}$$

Conversely, if $A_k \in \mathcal{L}(H_{0,k}, H_{0,k})$, $A_k \neq 0$ ($k = 2, \dots, n$), and for any continuous form a, the coordinates of the form b satisfy the condition (2.42), *then all the operators A_k are Hilbert–Schmidt operators.*

Proof. Fix $f^{(k)} \in H_{0,k}$ ($k = 2, \dots, n$). Then $H_{0,1} \ni f^{(1)} \mapsto l(f^{(1)}) = a(f^{(1)}, f^{(2)}, \dots, f^{(n)})$ is a continuous linear functional on $H_{0,1}$ whose norm is at most $\prod_{k=2}^n \|f^{(k)}\|_{H_{0,k}}$ (by virtue of (2.40)). Since $l(f^{(1)}) = (f^{(1)}, h_{H_{0,1}})$, where the coordinates $(h_{\alpha_1})_{\alpha_1=0}^\infty$ of the vector $h \in H_{0,1}$ in the basis $(e_{\alpha_1}^{(1)})_{\alpha_1=0}^\infty$ have the form $\overline{a\left(e_{\alpha_1}^{(1)}, f^{(2)}, \dots, f^{(n)}\right)}$, we obtain

$$\sum_{\alpha_1=0}^{\infty} |a(e_{\alpha_1}^{(1)}, f^{(2)}, \dots, f^{(n)})|^2 = \|l\|^2 \leq c^2 \prod_{k=2}^{n} \|f^{(k)}\|_{H_{0,k}}$$

$$(f^{(k)} \in H_{0,k};\ k = 2, \dots, n).$$

This estimate yields

$$\sum_{\alpha \in \mathbb{Z}_+^\infty} |b_\alpha|^2 = \sum_{\alpha \in \mathbb{Z}_+^\infty} |b(e_{\alpha_1}^{(1)}, \dots, e_{\alpha_n}^{(n)})|^2 = \sum_{\alpha \in \mathbb{Z}_+^\infty} |a(e_{\alpha_1}^{(1)}, A_2 e_{\alpha_2}^{(2)}, \dots, A_n e_{\alpha_n}^{(n)})|^2$$

$$\leq c^2 \sum_{\alpha_2,\ldots,\alpha_n}^{\infty} \prod_{k=2}^{n} \| A_k e^{(k)}_{\alpha_k} \|^2_{H_{0,k}} = c^2 \prod_{k=2}^{n} | A_k |^2 < \infty.$$

The converse statement can be obtained if, for given $k = 2, \ldots, n$, we construct a continuous form a such that $\sum_{\alpha \in \mathbb{Z}^n_+} |b_\alpha|^2 = c_k |A_k|^2$ with some $c_k > 0$ (b is constructed by use of the form a). Then the convergence of the last series implies $|A_k| < \infty$. The form a is given by its coordinates a_α having the form $\delta_{\alpha_1 \alpha_k} \delta_{\alpha_2 \beta_2} \ldots \delta_{\alpha_{k-1} \beta_{k-1}} \times \delta_{\alpha_{k+1} \beta_{k+1}} \ldots \delta_{\alpha_n \beta_n}$ ($\alpha \in \mathbb{Z}^n_+$), where δ_{jk} is the Kronecker symbol, and $\beta_2, \ldots, \beta_{k-1}, \beta_{k+1}, \ldots, \beta_n = 0, 1, \ldots$ are fixed. By calculating the sum $\sum_{\alpha \in \mathbb{Z}^n_+} |b_\alpha|^2$ for this form, we easily get the required equality with the constant

$$c_k = \prod_{j \neq k, j=2}^{n} \| A_j^* e^{(j)}_{\beta_j} \|_{H_{0,j}}.$$

Since $A_j \neq 0$ ($j = 2, \ldots, n$), one can always find β_j such that $c_k > 0$.

■

Lemma 2.9. *Let*

$$\bigoplus_{k=1}^{n} H_{0,k} \ni (f^{(1)}, \ldots, f^{(n)}) \mapsto b(f^{(1)}, \ldots, f^{(n)})$$

be a continuous n-linear form. It can be represented as

$$b(f^{(1)}, \ldots, f^{(n)}) = (f^{(1)} \otimes \ldots \otimes f^{(n)}, K)_{\bigotimes_{k=1}^{n} H_{0,k}},$$

where $K \in \otimes_{k=1}^{n} H_{0,k}$ *if and only if for its coordinates* $(b_\alpha)_{\alpha \in \mathbb{Z}^n_+}$, *condition* (2.42) *holds.*

Proof. Suppose that (2.42) holds. We set

$$K = \sum_{\alpha \in \mathbb{Z}^\infty_+} \overline{b_\alpha} e_\alpha \in \bigotimes_{k=1}^{n} H_{0,k},$$

where $(e_\alpha)_{\alpha \in \mathbb{Z}^n_+}$ is the basis of type (2.1) in $\bigotimes_{k=1}^{n} H_{0,k}$. Then, clearly,

$$(f^{(1)} \otimes \dots \otimes f^{(n)}, K)_{\bigotimes_{k=1}^{n} H_{0,k}} = \sum_{\alpha \in \mathbb{Z}_+^\infty} f_{\alpha_1}^{(1)} \dots f_{\alpha_n}^{(n)} b_\alpha = b(f^{(1)}, \dots, f^{(n)})$$

$$(f^{(k)} \in H_{0,k};\ k = 1, \dots, n).$$

Conversely, if the required representation of the form b exists, then

$$b(e_{\alpha_1}^{(1)}, \dots, e_{\alpha_n}^{(n)}) = (e_\alpha, K)_{\bigotimes_{k=1}^{n} H_{0,k}} = \overline{K_\alpha} \ (\alpha \in \mathbb{Z}_+^n)$$

and condition (2.42) is satisfied due to the inclusion $K \in \bigotimes_{k=1}^{n} H_{0,k}$.

■

Theorem 2.3. *Let the chains* (2.37) *be such that the imbeddings* O_k: $H_{+,k} \to H_{0,k}$, $k = 2, \dots, n$ *are quasinuclear. Then a generalized kernel* $\Theta \in H_{0,1} \otimes H_{-,2} \otimes \dots \otimes H_{-,n}$ *can be uniquely associated with each continuous n-linear form*

$$\bigoplus_{k=1}^{n} H_{0,k} \ni (f^{(1)}, \dots, f^{(n)}) \mapsto a(f^{(1)}, \dots, f^{(n)}) \in \mathbb{C}^1.$$

This kernel is such that

$$a(f^{(1)}, \varphi^{(2)}, \dots, \varphi^{(n)}) = (f^{(1)} \otimes \varphi^{(2)} \otimes \dots \otimes \varphi^{(n)}, \Theta)_{\bigotimes_{k=1}^{n} H_{0,k}}$$

$$(f^{(1)} \in H_{0,1},\ \varphi^{(k)} \in H_{+,k};\ k = 2, \dots, n). \tag{2.43}$$

Conversely, if the representation (2.43) *is valid for each form a of the indicated type with* $\Theta \in H_{0,1} \otimes H_{-,2} \otimes \dots \otimes H_{-,n}$, *then the imbeddings* O_k $(k = 2, \dots, n)$ *are quasinuclear.*

Proof. Let O_k and J_k $(k = 2, \dots, n)$ be the operators connected with the chain (2.37); for $k = 1$, we assume that $H_{+,1} = H_{0,1} = H_{-,1}$. For $f^{(1)} \in H_{0,1}$ and $\varphi^{(k)} \in H_{+,k}$ $(k = 2, \dots, n)$, we have

$$a(f^{(1)}, \varphi^{(2)}, \dots, \varphi^{(n)}) = a(f^{(1)}, O_2 J_2 J_2^{-1} \varphi^{(2)}, \dots, O_n J_n J_n^{-1} \varphi^{(n)})$$

$$= b(f^{(1)}, J_2^{-1} \varphi^{(2)}, \dots, J_n^{-1} \varphi^{(n)}), \tag{2.44}$$

where

$$b(f^{(1)}, \dots, f^{(n)}) = a(f^{(1)}, O_2 J_2 f^{(2)}, \dots, O_n J_n f^{(n)}) \tag{2.45}$$

$$(f^{(k)} \in H_{0,k};\ k = 1, \dots, n).$$

The operator O_k is the Hilbert–Schmidt operator for $k = 2, \dots, n$. The same is true for $A^{(k)} = O_k J_k \colon H_{0,k} \to H_{0,k}$. Therefore, from Lemma 2.8, the coordinates b_α of the form b satisfy condition (2.42) and, consequently, the following representation holds

$$b(f^{(1)}, \dots, f^{(n)}) = (f^{(1)} \otimes \cdots \otimes f^{(n)}, K)_{\otimes_{k=1}^{n} H_{0,k}}, \qquad K \in \bigotimes_{k=1}^{n} H_{0,k}$$

by virtue of Lemma 2.9. Hence, (2.44) can be continued as follows

$$\begin{aligned} a(f^{(1)}, \varphi^{(2)}, \dots, \varphi^{(n)}) &= a(f^{(1)} \otimes J_2^{-1}\varphi^{(2)} \otimes \cdots \otimes J_n^{-1}\varphi^{(n)}, K)_{\otimes_{k=1}^{n} H_{0,k}} \\ &= ((1 \otimes J_2^{-1} \otimes \cdots \otimes J_n^{-1})(f^{(1)} \otimes \varphi^{(2)} \otimes \cdots \otimes \varphi^{(n)}), K)_{\otimes_{k=1}^{n} H_{0,k}} \\ &= (f^{(1)} \otimes \varphi^{(2)} \otimes \cdots \otimes \varphi^{(n)}, (1 \otimes J_2^{-1} \otimes \cdots \otimes J_n^{-1})^{+} K)_{\otimes_{k=1}^{n} H_{0,k}} \\ &= (f^{(1)} \otimes \varphi^{(2)} \otimes \cdots \otimes \varphi^{(n)}, (1 \otimes \boldsymbol{J}_2^{-1} \otimes \cdots \otimes \boldsymbol{J}_n^{-1}) K)_{\otimes_{k=1}^{n} H_{0,k}}. \end{aligned} \tag{2.46}$$

Here, we have used the relation $(\otimes_{k=1}^{n} J_k^{-1})^{+} = \otimes_{k=1}^{n} \boldsymbol{J}_k^{-1}$, where the conjugation + is taken with respect to the chain (2.38). Setting

$$\Theta = (1 \otimes \boldsymbol{J}_2^{-1} \otimes \dots \otimes \boldsymbol{J}_n^{-1}) K \in H_{0,1} \otimes H_{-,2} \otimes \cdots \otimes H_{-,n}$$

in (2.46), we get the required representation (2.43). The fact that the kernel Θ can be determined uniquely in terms of the form a follows from the denseness of the linear span of the vectors $f^{(1)} \otimes \varphi^{(2)} \otimes \cdots \otimes \varphi^{(n)}$ $(f^{(1)} \in H_{0,1}, \varphi^{(k)} \in H_{+,k}; k = 2, \dots, n)$ in $H_{0,1} \otimes H_{-,2} \otimes \cdots \otimes H_{-,n}$.

Let us prove the last statement of the theorem. Assume that the representation (2.43) holds for the form a. Then the following representation for the form b defined by (2.45) can be obtained just as above

$$b(f^{(1)}, \dots, f^{(n)}) = (f^{(1)} \otimes J_2 f^{(2)} \otimes \cdots \otimes J_n f^{(n)}, \Theta)_{\otimes_{k=1}^{n} H_{0,k}}$$

$$= (f^{(1)} \otimes \cdots \otimes f^{(n)}, (1 \otimes J_2 \otimes \cdots \otimes J_n)\,\Theta)_{\bigotimes_{k=1}^{n} H_{0,k}}$$

$$(f^{(k)} \in H_{0,k};\ k = 1, \dots, n).$$

But $(1 \otimes J_2 \otimes \cdots \otimes J_n)\,\Theta \in \otimes_{k=1}^{n} H_{0,k}$ and, therefore, using Lemma 2.9 we find that condition (2.42) holds for the coordinates b_α of the form b. Since the form a is arbitrary, and $A_k = O_k J_k \neq 0$, Lemma 2.8 implies that $\| A_k \| < \infty$, i.e., that $\| O_k \| < \infty$ $(k = 2, \dots, n)$.

■

Corollary 1. *The statement of Theorem 2.3 can be formulated in a more symmetric way (however, the result becomes somewhat rougher):*

Assume that each imbedding $H_{+,k} \to H_{0,k}$ $(k = 1, \dots, n)$ in (2.37) is quasinuclear. Then any continuous n-linear form (2.39) allows the representation

$$a(\varphi^{(1)}, \dots, \varphi^{(n)}) = (\varphi^{(1)} \otimes \cdots \otimes \varphi^{(n)}, \Theta)_{\bigotimes_{k=1}^{n} H_{0,k}} \tag{2.47}$$

$$(f^{(k)} \in H_{0,k};\ k = 1, \dots, n)$$

with the kernel $\Theta \in \bigotimes_{k=1}^{n} H_{-,k}$ which is determined uniquely.

Theorem 2.3 will be employed mainly in this form.

■

We now present one more modification of Theorem 2.3 related to the case of nuclear riggings and forms on nuclear spaces. Assume that the following collection of n nuclear riggings of the form (1.22) is given

$$\operatorname{ind} \lim_{\tau_k \in T_k} H_{-,\tau_k} = \Phi'_k \supset H_{0,k} \supset \Phi_k = \operatorname{pr} \lim_{\tau_k \in T_k} H_{+,\tau_k} \quad (k = 1, \dots, n). \tag{2.48}$$

According to the construction given in Subsection 2.1 (see (2.18)), we form the nuclear chain

$$\bigotimes_{k=1}^{n} \Phi'_k \supset \bigotimes_{k=1}^{n} H_{0,k} \supset \bigotimes_{k=1}^{n} \Phi_k \tag{2.49}$$

and consider the n-linear forms (cf. (2.39))

$$\bigoplus_{k=1}^{n} \Phi_k \ni (\varphi^{(1)}, \dots, \varphi^{(n)}) \mapsto a(\varphi^{(1)}, \dots, \varphi^{(n)}) \in \mathbb{C}^1, \tag{2.50}$$

continuous on the direct product $\times_{k=1}^{n} \Phi_k$ of linear topologic spaces Φ_k. Since

$$\bigotimes_{k=1}^{n} \Phi_k = \mathrm{pr} \lim_{(\tau_1, \dots, \tau_n) \in \mathrm{T}} \bigotimes_{k=1}^{n} H_{+, \tau_k} \quad (\mathrm{T} = \times_{k=1}^{n} \mathrm{T}_k),$$

the continuity of the form (2.50) is equivalent to the continuity of this form in the norm of the space $\otimes_{k=1}^{n} H_{+, \tau_k}$ with a certain collection $\tau = (\tau_1, \dots, \tau_n)$, and, consequently, to the fact that

$$|a(\varphi^{(1)}, \dots, \varphi^{(n)})| \le c_\tau \prod_{k=1}^{n} \|\varphi^{(k)}\|_{H_{+, \tau_k}} \tag{2.51}$$

$$(c_\tau > 0, \quad \varphi^{(k)} \in H_{+, \tau_k}; \ k = 1, \dots, n).$$

Theorem 2.4. *Suppose that the nuclear riggings* (2.48) *are given. Then any continuous n-linear form* (2.50) *can be represented as*

$$a(\varphi^{(1)}, \dots, \varphi^{(n)}) = (\varphi^{(1)} \otimes \cdots \otimes \varphi^{(n)}, \Theta)_{\bigotimes_{k=1}^{n} H_{0,k}} \tag{2.52}$$

$$(\varphi^{(k)} \in \Phi_k; \ k = 1, \dots, n),$$

where the generalized kernel $\Theta \in \bigotimes_{k=1}^{n} \Phi_k'$ *is determined uniquely in terms of a.*

Proof. Due to the continuity of a, we can choose $\tau = (\tau_1, \dots, \tau_n) \in \mathrm{T}$ so that (2.51) holds. Using the nuclearity of Φ_k, for every τ_k, one can find $\tau_k' \in \mathrm{T}$ such that $H_{+,\tau_k'} \subset H_{+,\tau_k}$, and the imbedding operator $O_{\tau_k', \tau_k} : H_{+,\tau_k'} \to H_{+,\tau_k}$ is quasinuclear. Consider the collection of chains

$$H_{-, \tau_k'} \supset H_{-, \tau_k} \supset H_{0, k} \supset H_{+, \tau_k} \supset H_{+, \tau_k'} \quad (k = 1, \dots, n)$$

and their tensor product

$$\bigotimes_{k=1}^{n} H_{-, \tau_k'} \supset \bigotimes_{k=1}^{n} H_{-, \tau_k} \supset \bigotimes_{k=1}^{n} H_{0, k} \supset \bigotimes_{k=1}^{n} H_{+, \tau_k} \supset \bigotimes_{k=1}^{n} H_{+, \tau_k'}.$$

The form $a(\varphi^{(1)}, \dots, \varphi^{(n)})$ $(\varphi^{(k)} \in \Phi_k)$ is continuous on $\otimes_{k=1}^{n} H_{+, \tau_k}$, and the imbedding $H_{+,\tau_k'} \to H_{+,\tau_k}$ is quasinuclear ($k = 1, \dots, n$). Therefore, by virtue of Corollary 1 of Theorem 2.3, we have the representation

$$a(\varphi^{(1)}, \dots, \varphi^{(n)}) = (\varphi^{(1)} \otimes \cdots \otimes \varphi^{(n)}, \hat{\Theta})_{\bigotimes_{k=1}^{n} H_{+,\tau_k}} \qquad (\varphi^{(k)} \in \Phi_k;\ k = 1, \dots, n). \quad (2.53)$$

Here the kernel $\hat{\Theta} \in \otimes_{k=1}^{n} H_{-,\tau_k'}^{(\tau_k)}$, where the space $H_{-,\tau_k'}^{(\tau_k)}$ is dual to the space $H_{+,\tau_k'}$ with respect to H_{+,τ_k}, i.e., it is the element of the chain

$$H_{-,\tau_k'}^{(\tau_k)} \supset H_{+,\tau_k} \supset H_{+,\tau_k'} \quad (k = 1, \dots, n).$$

Denote by $I_{\tau_k'}^{(\tau_k)} : H_{-,\tau_k'}^{(\tau_k)} \to H_{+,\tau_k'}$ the isometries connected with the last chains. Then (2.53) can be rewritten as

$$a(\varphi^{(1)}, \dots, \varphi^{(n)}) = (\varphi^{(1)} \otimes \cdots \otimes \varphi^{(n)}, \hat{\Xi})_{\bigotimes_{k=1}^{n} H_{+,\tau_k'}} \quad (2.54)$$

where

$$\Xi = \left(\bigotimes_{k=1}^{n} I_{\tau_k'}^{(\tau_k)} \right) \hat{\Theta} \in \bigotimes_{k=1}^{n} H_{+,\tau_k'} .$$

Finally, by using the isometries $I_{\tau_k'} : H_{-,\tau_k'} \to H_{+,\tau_k'}$, we can transform (2.54) into the required representation (2.52) with the kernel

$$\Theta = \left(\bigotimes_{k=1}^{n} I_{\tau_k'}^{-1} \right) \hat{\Xi} \in \bigotimes_{k=1}^{n} H_{-,\tau_k'} \subset \bigotimes_{k=1}^{n} \Phi_k' .$$

As above, the fact that the kernel Θ is determined uniquely for a given a follows from the denseness of the linear span of the vectors

$$\varphi^{(1)} \otimes \cdots \otimes \varphi^{(n)} \quad (\varphi^{(k)} \in \Phi_k) \quad \text{in} \quad \bigotimes_{k=1}^{n} \Phi_k .$$

■

Remark 1. In the case where each Φ_k in (2.50) is a countable Hilbert space, any separately continuous (this means that it is continuous with respect to each variable when all the others are fixed) multilinear form (2.50) is continuous with respect to the collection of its variables, too, i.e., it is continuous in $\times_{k=1}^{n} \Phi_k$ (see, e.g., Gelfand and Vilenkin [1, Chapter 1, Section 1]). Hence, in this case, Theorem 2.4 also holds for separately continuous forms. Similar situations take place for some other Φ_k, for example, for the space $\mathcal{D}(\mathbb{R}^d)$ from Example 1.9. For details, see Maurin [2, Chapter 3, Section 10].

■

2.4. The Case of Bilinear Forms

The kernel theorem for bilinear forms can be also obtained by use of Theorem 2.3. Consider the chain

$$H_- \supset H_0 \supset H_+. \tag{2.55}$$

Assume that there exists an involution in H_+ which is at the same time an involution in H_0. This means that an antilinear mapping $H_+ \ni \varphi \mapsto \varphi^* \in H_+$ is given in H_+ such that $(\varphi^*)^*$, $(\varphi^*, \psi^*)_{H_+} = \overline{(\varphi, \psi)_{H_+}}$, and $(\varphi^*, \psi^*)_{H_0} = \overline{(\varphi, \psi)_{H_0}}$ $(\varphi, \psi \in H_+)$.

It is easy to see that $(\varphi^*, \psi^*)_{H_-} = \overline{(\varphi, \psi)_{H_-}}$, i.e., the operation * is also an involution in H_- and, consequently, can be extended by continuity to the involution $H_- \ni \xi \mapsto \xi^* \in H_-$ defined on the whole of H_-. The restriction of this mapping $H_- \supset H_0 \ni f \mapsto f^* \in H_0$ is the involution in H_0. Clearly, $(I\xi)^* = I\xi^*$ $(\xi \in H_-)$. If the involution described above is introduced into the spaces of the chain (2.55), then we say that (2.55) is a chain with an involution *.

Let $a(f, g)$ be some bilinear form on H_0, i.e., a continuous function $H_0 \otimes H_0 \ni (f, g) \mapsto a(f, g) \in \mathbb{C}^1$ linear in the first variable and antilinear in the second one. We have the following theorem.

Theorem 2.5. *Suppose that the imbedding* $H_+ \to H_0$ *in the chain* (2.55) *is quasinuclear. Then for every continuous bilinear form* $a(f, g)$ $(f, g \in H_0)$, *a generalized kernel* $\Theta_a \in H_0 \otimes H_-$ *can be constructed uniquely to satisfy*

$$a(\varphi, g) = (\Theta_a, g \otimes \varphi^*)_{H_0 \otimes H_0} \quad (\varphi \in H_+, g \in H_0). \tag{2.56}$$

Conversely, if each form a of the indicated type can be represented by (2.56) *with* $\Theta_a \in H_0 \otimes H_-$, *then the imbedding* $H_+ \to H_0$ *is quasinuclear.*

Proof. Let us construct a continuous two-linear form b by setting $b(f, g) = \overline{a(g^*, f)}$ $(f, g \in H_0)$. According to Theorem 2.3, there exists a generalized kernel $\Theta \in H_0 \otimes H_-$ such that

$$b(f, \varphi) = (f \otimes \varphi, \Theta)_{H_0 \otimes H_0} \quad (f \in H_0, \varphi \in H_+).$$

Therefore,

$$a(\varphi, g) = \overline{b(g, \varphi^*)} = \overline{(g \otimes \varphi^*, \Theta)_{H_0 \otimes H_0}} = (\Theta, g \otimes \varphi^*)_{H_0 \otimes H_0},$$

for $\varphi \in H_+$ and $g \in H_0$, i.e., the representation (2.56) with $\Theta_a = \Theta$ holds. It is now obvious that Θ_a is determined uniquely for the given a.

The converse statement immediately follows from the analogous statement of Theorem 2.3: The representation (2.56) involves the representation (2.43) for corresponding b, which is arbitrary due to the arbitrariness of a.

■

Remark 1. Clearly, in Theorem 2.3 we could have fixed any other variable instead of the first one. Hence, the following representation holds for a along with (2.56)

$$a(\varphi, g) = (\Theta'_a, \psi \otimes f^*) \quad (f \in H_0, \psi \in H_+), \tag{2.57}$$

where the kernel $\Theta'_a \in H_- \otimes H_0$, and is determined uniquely for given a. It follows from (2.56) and (2.57) that

$$a(\psi^*, \varphi) = (\Theta_a, \varphi \otimes \psi)_{H_0 \otimes H_0} = (\Theta'_a, \varphi \otimes \psi)_{H_0 \otimes H_0}$$

for $\varphi, \psi \in H_+$. This means that the restrictions of the functionals Θ_a and Θ'_a to $H_+ \otimes H_+$ coincide. Denoting this common restriction by Θ and assuming that conditions of Theorem 2.5 are satisfied, we obtain the representation

$$a(\varphi, \psi) = (\Theta, \psi \otimes \varphi^*)_{H_0 \otimes H_0} \quad (\varphi, \psi \in H_+), \tag{2.58}$$

with the kernel $\Theta \in H_- \otimes H_-$.

■

Remark 2. The results of Theorem 2.4 for nuclear riggings can be easily extended to the case of bilinear forms. Thus, let Φ be a nuclear space with the involution $*$ defined in a natural way, namely, $\Phi \to \Phi$. We consider a bilinear form on Φ to be a continuous function $\Phi \times \Phi \ni (\varphi, \psi) \mapsto a(\varphi, \psi) \in \mathbb{C}^1$ linear in the first variable and antilinear in the second one. Then $a(\varphi, \psi) = (\Theta, \psi \otimes \varphi^*)_{H_0 \otimes H_0}$, where $\Theta \in \Phi' \otimes \Phi'$.

■

Remark 3. The results of this section can be easily extended to the cases of real Hilbert spaces and nuclear spaces.

■

Let $A \in \mathcal{L}(H_0, H_0)$. By using A, we introduce a continuous bilinear form $a(f, g) = (Af, g)_{H_0}$ $(f, g \in H_0)$. Given this form, we construct, in turn, the kernel $\Theta_A = \Theta \in H_- \otimes H_-$ according to (2.58). This kernel is called a kernel of the operator A, and we have

$$(A\varphi, \psi)_{H_0} = (\Theta_A, \psi \otimes \varphi^*)_{H_0 \otimes H_0} \quad (\varphi, \psi \in H_+). \tag{2.59}$$

Example 2.2. Let $H_0 = L_2(X, \mu) = L_2(X, \mathfrak{R}, \mu)$, where X is a space with a measure μ given on a certain σ-algebra $\mathfrak{R}$; and let A be a Hilbert–Schmidt operator in H_0, i.e.,

$$(Af)(x) = \int_X K(x, y) f(y)\, d\mu(y), \tag{2.60}$$

$$K \in L_2(X \times X, \mathfrak{R} \times \mathfrak{R}, \mu \times \mu) = H_0 \otimes H_0.$$

Then

$$(Af, g)_{H_0} = \int_{X \times X} K(x, y) f(y) \overline{g(x)}\, d\mu(x) d\mu(y) = (K, g \otimes f^*)_{H_0 \otimes H_0}, \tag{2.61}$$

where $f^*(\cdot) = \overline{f(\cdot)}$. Comparing (2.61) with (2.59), we note that in this case $\Theta_A = K$. Relation (2.59) shows that an arbitrary bounded operator A in $L_2(X, \mu)$ admits (in a certain sense) the representation (2.60) with $K = \Theta_A$. This means that A is an "integral operator with a generalized kernel". On the other hand, (2.59) clarifies what arrangement of the variables on the right-hand sides of (2.56)–(2.58) is appropriate.

It is now convenient to introduce some definitions for the chain (2.55) with the involution $*$ (they will be exploited in the next chapters).

The generalized kernel $K \in H_- \otimes H_-$ is called Hermitian if

$$(K, \psi \otimes \varphi^*)_{H_0 \otimes H_0} = \overline{(K, \varphi \otimes \psi^*)}_{H_0 \otimes H_0} \quad (\varphi, \psi \in H_+). \tag{2.62}$$

The kernel $K \in H_- \otimes H_-$ is called positive definite if

$$(K, \varphi \otimes \varphi^*)_{H_0 \otimes H_0} \geq 0 \quad (\varphi \in H_+). \tag{2.63}$$

In other words, the continuous bilinear form

$$H_+ \oplus H_+ \ni (\varphi, \psi) \mapsto a(\varphi, \psi) = (K, \psi \otimes \varphi^*)_{H_0 \otimes H_0}$$

should be either Hermitian or positive definite. Clearly, a positive definite kernel is always Hermitian. If Θ_A is a kernel of the operator $A \in \mathcal{L}(H_0, H_0)$, then it is Hermitian or positive definite if and only if A is selfadjoint or nonnegative.

Example 2.3. In order to formulate the kernel theorem for the case of multilinear and bilinear forms on the spaces $L_2(G)$ $(G \subset \mathbb{R}^d, d \in \mathbb{N})$, one can use the Sobolev spaces quasinuclearly imbedded into $L_2(G)$ (see Example 1.5) or (in the case of nuclear riggings) the spaces $C^\infty(G)$, $\mathcal{S}(\mathbb{R}^d)$, and $\mathcal{D}(\mathbb{R}^d)$ (see Examples 1.7-1.9) which prove to be suitable.

The following "elementary" kernel theorem holds:

Suppose that

$$L_2(G) \oplus L_2(G) \ni (f, g) \mapsto a(f, g) \in \mathbb{C}^1$$

is a continuous bilinear form. Then there exists a kernel $T \in C(\mathbb{R}^d \times \mathbb{R}^d)$ *such that*

$$a(\varphi, \psi) = \int_{G\times G} T(x,y)(\mathcal{D}\varphi)(y)\overline{(\mathcal{D}\psi)(x)}\, dx\, dy \tag{2.64}$$

$$(\mathcal{D} = D_1 \cdots D_d, \quad \varphi, \psi \in C_0^d(G)).$$

To prove this, let us consider the characteristic function of the open parallelepiped in $\mathbb{R}^d$ formed by the coordinate hyperplanes and the hyperplanes which pass through some point $x = (x_1, \dots, x_d) \in \mathbb{R}^d$ and are parallel to the coordinate ones. Denote by $\omega(x, \cdot)$ the product of this function, the characteristic function of the region G, and $(-1)^d \operatorname{sgn} x_1 \dots \operatorname{sgn} x_d$. Shifting the derivatives D_k by integrating by parts, we can easily establish the equality

$$\int_G (f, \omega(x,\cdot))_{L_2(G)} \overline{(\mathcal{D}\varphi)(x)}\, dx \quad (f \in L_2(G),\ \varphi \in C_0^d(G)). \tag{2.65}$$

By using (2.65), we can prove (2.64). Indeed, let $A \in \mathcal{L}(L_2(G), L_2(G))$ be an operator associated with a, i.e., $a(f, g) = (Af, g)_{L_2(G)}$. We set $T(x, y) = (A\,\omega(y, \cdot), \omega(x, \cdot))_{L_2(G)}$, insert this expression into the right-hand side of (2.64), use (2.65), and then, after applying A to the second factor in the expression obtained, we use (2.65) once again. As a result, we get $(A\varphi, \psi)_{L_2(G)}$.

Note that if the kernel T is smooth, then (2.64) has the form (2.58) with $\Theta = \mathcal{D}_x\mathcal{D}_y T$. In the general case these derivatives should be regarded as generalized functions, and the last kernel is also generalized. One can find a quasinuclear rigging (2.55) of the space $H_0 = L_2(G)$ for which (2.58) turns into (2.64). Note finally that the formula of type (2.64) is valid for multilinear forms, too. This example is considered in more detail in Berezansky [18, Chapter 1, Section 3, Subsection 11].

§3. Bilinear Forms

It is well known that a continuous bilinear form a in a Hilbert space H can be represented in the form $a(f, g) = (Af, g)_H$ $(f, g \in H)$, where A is a bounded operator in H. For the forms which are not continuous but possess certain additional properties, similar theorems on representation also play an important role. In this Section, we present the corresponding theory and clarify its connection with Hilbert riggings. More exactly, we show how the simple results concerning Hilbert riggings are interpreted as theorems of the theory of bilinear forms.

3.1. Some Remarks on the Completion of Spaces

Let L be a linear space, $L \ni f \mapsto \|f\|_E \geq 0$ be a norm in this space, and E be the completion of L in this norm. Recall that E consists of the classes f_E of mutually equivalent fundamental sequences $(f_n)_{n=1}^\infty$ $(f_n \in L)$. The equivalence $(f_n)_{n=1}^\infty \sim (g_n)_{n=1}^\infty$ means that $\|f_n - g_n\|_E \to 0$ as $n \to \infty$. Linear operations over these sequences induce the linear structure in E. If $\|\cdot\|_E$ is a Hilbert norm (this means that a scalar product $(f, g)_E$ $(f, g \in L)$ is introduced in L, and $\|\cdot\|_E = \sqrt{(\cdot,\cdot)_E}$), then E is a Hilbert space. The space L is imbedded into E by the identification of each $f \in L$ with the class containing the stationary sequence $(f, f, \ldots)$.

Now assume that L is a linear set in which the following two norms are given: $L \ni f \mapsto \|f\|_{E_1} \geq 0$ and $L \ni f \mapsto \|f\|_{E_2} \geq 0$, where E_1 and E_2 are the corresponding completions of L. Suppose that these norms are comparable in the following sense

$$\|f\|_{E_1} \leq \|f\|_{E_2} \quad (f \in L). \tag{3.1}$$

At first sight, it seems that the inequality (3.1) involves the inclusion $E_1 \supset E_2$ for the completions and the inequality $\|f\|_{E_1} \leq \|f\|_{E_2}$ $(f \in E_2)$. In fact, this is not true. Let us clarify the situation rigorously.

Assume that $(f_n)_{n=1}^\infty$ $(f_n \in L)$ is a fundamental sequence with respect to the norm $\|\cdot\|_{E_2}$. By virtue of (3.1), it is fundamental with respect to $\|\cdot\|_{E_1}$, too. Let $(f_n)_{n=1}^\infty \in f_{E_2} \in E_2$, and $(f_n)_{n=1}^\infty \in f_{E_1} \in E_1$. We associate the vector f_{E_2} with the vector f_{E_1}. This mapping is defined correctly. Indeed, if we also have $(g_n)_{n=1}^{\infty} \in f_{E_2}$, i.e., $(f_n)_{n=1}^{\infty} \sim (g_n)_{n=1}^{\infty}$ with respect to $\|\cdot\|_{E_2}$, then, according to (3.1), the same equivalence relation holds with respect to $\|\cdot\|_{E_1}$.

Thus, we have constructed the mapping $E_2 \ni f_{E_2} \mapsto Qf_{E_2} = f_{E_1} \in E_1$. It follows from the method by which the linear structure is inserted into the completion, that Q is linear. By virtue of (3.1), it is also continuous. In fact,

$$\| Qf_{E_2} \|_{E_1} = \| f_{E_1} \|_{E_1} = \lim_{n\to\infty} \| f_n \|_{E_1} \le \lim_{n\to\infty} \| f_n \|_{E_2} = \| f_{E_2} \|_{E_2}$$

$$((f_n)_{n=1}^{\infty} \in f_{E_2} \in E_2).$$

Note that the restriction $Q \upharpoonright L$ is an imbedding operator which imbeds $L \subset E_2$ into the set L regarded as a set from E_1.

Consider the subspace

$$\operatorname{Ker} Q = \{ f \in E_2 \mid Qf = 0 \} \subset E_2. \tag{3.2}$$

If $\operatorname{Ker} Q = \{0\}$, then we can identify E_2 with the range of values $\mathfrak{R}(Q)$ and consider that $E_2 \subset E_1$ and $\|f\|_{E_1} \le \|f\|_{E_2}$ $(f \in E_2)$. In the general case, a similar inclusion and inequality hold for the factor space $E_2 / \operatorname{Ker} Q \subset E_1$. In the case where E_2 is a Hilbert space, one may take the orthogonal complement $E_2 \ominus \operatorname{Ker} Q$ instead of the factor space. Note that the second extreme case ($\operatorname{Ker} Q = E_2$) is impossible. Furthermore, $L \cap \operatorname{Ker} Q = \{0\}$. This follows from the fact that $Q \upharpoonright L$ is the above-mentioned imbedding.

Summarizing all that has been said, we obtain the following assertion:

Assume that L is a linear set on which two norms $\|\cdot\|_{E_1}$ *and* $\|\cdot\|_{E_2}$ *comparable in the sense of* (3.1) *are given (here,* E_1 *and* E_2 *are the corresponding completions of L), and that Q is the operator introduced above. Then*

$$E_1 \supset E_2 / \operatorname{Ker} Q, \quad \|f\|_{E_1} \le \|f\|_{E_2 / \operatorname{Ker} Q} \quad (f \in E_2 / \operatorname{Ker} Q). \tag{3.3}$$

If E_2 *is a Hilbert space, then the role of a factor space in* (3.3) *is played by* $E_2 \ominus \operatorname{Ker} Q$. *If* $\operatorname{Ker} Q = \{0\}$, *then* $E_1 \supset E_2$ *and* $\|f\|_{E_1} \le \|f\|_{E_2}$ $(f \in E_2)$.

The fact that factorization is absent (or present) in (3.3) is important in the study of bilinear forms. Therefore, this will be always specified in what follows. The construction of the operator Q and (3.2) imply that

$\operatorname{Ker} Q = \{0\}$ *if and only if any sequence* $(f_n)_{n=1}^{\infty}$ $(f_n \in L)$, *fundamental in norm*

$\|\cdot\|_{E_2}$ *and convergent (to 0) in norm* $\|\cdot\|_{E_1}$, *converges to zero in norm* $\|\cdot\|_{E_2}$, *too.*

Example 3.1. Let $G \subset \mathbb{R}^d$ $(d \in \mathbb{N})$ be a bounded region, $L = C^1(\tilde{G})$, $\|f\|_{E_1} = \|f\|_{L_2(G)}$, and let $\|f\|_{E_2} = \|f\|_{W_2^1(G)}$. Here, $L_2(G) = L_2(G, \mathfrak{B}(G), dx)$ is constructed by using the Lebesgue measure dx, and $W_2^1(G)$ is a Sobolev space.

In this case $\operatorname{Ker} Q = \{0\}$.

In fact, let $(f_n)_{n=1}^{\infty}$ $(f_n \in L)$ be fundamental with respect to the norm of $E_2 = W_2^1(G)$, and let $f_n \to \infty$ in $E_1 = L_2(G)$. Then for every derivative D_j, the sequence $(D_j f_n)_{n=1}^{\infty}$ is fundamental in $L_2(G)$. Let $h_j \in L_2(G)$ be the corresponding limit in $L_2(G)$. For a function $g \in C_0^{\infty}(G)$ finite with respect to G, we have

$$(h_j, g)_{L_2(G)} = \lim_{n\to\infty} (D_j f_n, g)_{L_2(G)} = -\lim_{n\to\infty} (f_n, D_j g)_{L_2(G)} = 0,$$

and this implies that $h_j = 0$ $(j = 1, \ldots, d)$. By passing to the limit $n \to \infty$ in the expression

$$\|f_n\|^2_{W_2^1(G)} = \|f_n\|^2_{L_2(G)} + \sum_{j=1}^{d} \|D_j f_n\|^2_{L_2(G)},$$

we get that $f_n \underset{n\to\infty}{\to} 0$ in $W_2^1(G)$.

■

Hence, we now have $E_1 = L_2(G)$ and $E_2 = W_2^1(G)$. Therefore, we can write the imbedding $W_2^1(G) \subset L_2(G)$ (and this is the usual practice). Clearly, the situation is similar in the case where $E_1 = L_2(G)$ and $E_2 = W_2^m(G)$, where $m \geq l$.

Example 3.2. Let $G \subset \mathbb{R}^d$ be a bounded domain, $L = C(\tilde{G})$,

$$\|f\|_{E_1} = \|f\|_{L_2(G)}, \quad (f, g)_{E_2} = (f, g)_{L_2(G)} + f(x_0)\overline{g(x_0)} \quad (f, g \in L), \tag{3.4}$$

where $x_0 \in \tilde{G}$ is a fixed point. Consider the Hilbert space $L_2(G) \oplus \mathbb{C}^1$ consisting of pairs (f, p) $(f \in L_2(G), p \in \mathbb{C}^1)$. The space $C(\tilde{G})$ can be imbedded into $L_2(G) \oplus \mathbb{C}^1$ by identifying each $f \in C(\tilde{G})$ with the pair $(f, f(x_0))$; obviously, $C(\tilde{G})$ is dense in $L_2(G) \oplus \mathbb{C}^1$. It follows from the foregoing that $E_2 = L_2(G) \oplus \mathbb{C}^1$ and $E_1 = L_2(G)$. The operator Q has the form

$$Q(f, p) = f((f, p) \in E_2), \quad \operatorname{Ker} Q = \{(0, p) \mid p \in \mathbb{C}^1\} \subset E_2,$$

i.e., $\operatorname{Ker} Q = \mathbb{C}^1$ regarded as a subspace of E_2.

Example 3.3. Let $G \subset \mathbb{R}^d$ be a bounded domain, $L = C^1(\tilde{G})$, $\|f\|_{E_1} = \|f\|_{L_2(G)}$, and

$$(f, g)_{E_2} = (f, g)_{W_2^1(G)} + f(x_0)\overline{g(x_0)} \quad (f, g \in L) \tag{3.5}$$

($x_0 \in \tilde{G}$ being fixed). By virtue of the imbedding theorems, in the case where $d = 1$, we have $W_2^1(G) \subset L_2(\tilde{G})$ and, moreover, this imbedding is continuous. Therefore, the scalar product (3.5) is equivalent to the scalar product in $W_2^1(G)$ and, consequently, $E_2 = W_2^1(G)$, and $E_1 = L_2(G)$. According to Example 3.1, we have Ker $Q = \{0\}$.

Let $d \geq 2$. Consider the Hilbert space $W_2^1(G) \oplus \mathbb{C}^1$ of pairs (φ, p) ($\varphi \in W_2^1(G)$, $p \in \mathbb{C}^1$). The space $C^1(\tilde{G})$ can be imbedded into this space if we identify each $\varphi \in C^1(\tilde{G})$ with $(\varphi, \varphi(x_0))$. Since $W_2^1(G)$ cannot be imbedded into $C(\tilde{G})$, it is easy to see that the set $C^1(\tilde{G})$ is dense in $W_2^1(G) \oplus \mathbb{C}^1$. Hence, $E_2 = W_2^1(G) \oplus \mathbb{C}^1$ and $E_1 = L_2(G)$. The operator Q has the form

$$Q(\varphi, p) = \varphi((\varphi, p) \in E_2); \quad \text{Ker } Q = \mathbb{C}^1 \subset E_2.$$

Along with the auxiliary facts stated above, we shall need the following fact from the theory of Hilbert riggings. Consider the chain (1.11) and the operator $\boldsymbol{I}$ associated with it which maps the whole of H_- onto the whole of H_+ isometrically.

Lemma 3.1. *Let* $\mathfrak{D}(A) = \{\varphi \in H_+ \mid \boldsymbol{I}^{-1} \in H_0\}$, *and consider the operator* $A = \boldsymbol{I}^{-1} \restriction \mathfrak{D}(A)$ *in* H_0. *It is stated that A is selfadjoint and satisfies the relations* $A \geq 1$, $\|A\varphi\|_{H_0} \geq \|\varphi\|_{H_+}$ $(\varphi \in \mathfrak{D}(A))$, *and*

$$(\varphi, \psi)_{H_+} = (A\varphi, \psi)_{H_0} \quad (\varphi \in \mathfrak{D}(A),\ \psi \in H_+),$$

$$(\varphi, \psi)_{H_+} = (\sqrt{A}\,\varphi, \sqrt{A}\,\psi)_{H_0} \quad (\varphi, \psi \in H_+ = \mathfrak{D}(\sqrt{A}\,)). \tag{3.6}$$

Proof. Since H_0 is dense in H_-, the set $\mathfrak{D}(A)$ is dense in H_+ and, consequently, in H_0. Further, according to (1.12), we have

$$(\xi, \psi)_{H_0} = (\boldsymbol{I}\,\xi, \psi)_{H_+} \quad (\xi \in H_-, \psi \in H_+)$$

and thus,

$$(\varphi, \psi)_{H_+} = (\boldsymbol{I}^{-1}\varphi, \psi)_{H_0} \quad (\varphi, \psi \in H_+). \tag{3.7}$$

It follows from (3.7) and the definition of A that

$$(\varphi, \psi)_{H_+} = (A\varphi, \psi)_{H_0} \quad (\varphi \in \mathfrak{D}(A), \psi \in H_+)$$

and this is just the first relation (3.6). By setting $\psi = \varphi$ in it, we obtain that

$$(A\varphi, \varphi)_{H_0} = (\varphi, \varphi)_{H_+} \geq 0 \quad (\varphi \in \mathfrak{D}(A)),$$

but this means that A is Hermitian and the inequality $A \geq 1$ holds. Further, for any $\varphi \in \mathfrak{D}(A)$ we have

$$\| A\varphi \|_{H_0} \geq \| \boldsymbol{I}^{-1}\varphi \|_{H_0} \geq \| \boldsymbol{I}^{-1}\varphi \|_{H_-} \geq \| \varphi \|_{H_+}.$$

Let us establish the second relation (3.6). We shall use the operators shown on (1.11). According to the proof of Theorem 1.1, we have $J = \sqrt{OI}$ if the operator to the right is regarded as acting from H_0 to H_+. Hence,

$$(\sqrt{OI}\, f, \sqrt{OI}\, g)_{H_+} = (f, g)_{H_0} \quad (\varphi, g \in H_0)$$

and $\mathfrak{R}(\sqrt{OI}) = H_+$. In other words,

$$(\varphi, \psi)_{H_+} = ((OI)^{-1/2}\varphi, (OI)^{-1/2}\psi)_{H_0} \quad (\varphi, \psi \in H_+). \tag{3.8}$$

But $I = \boldsymbol{I} \restriction H_0$ and, therefore, $(OI)^{-1} = A$. This means that (3.8) turns into the required second relation (3.6).

■

3.2. Two Equivalent Languages: Pre-Chains and Positive Forms

Let H_0 be a Hilbert space, and let L be a linear set dense in this space on which a scalar product $(f, g)_{L_+}$ $(f, g \in L)$ is given such that

$$\| f \|_{H_0} \leq \| f \|_{L_+} \quad (\| f \|_{L_+} = ((f, f)_{L_+})^{1/2}, \ f \in L).$$

In this case, we say that the pre-chain

$$H_0 \supset L \tag{3.9}$$

is given.

Denote by L_+ the completion of L with respect to the norm $\| \cdot \|_{L_+}$. The premises of the scheme presented in Subsection 3.1 now hold for $E_2 = H_0$ and $E_2 = L_+$. Let Q: $L_+ \to H_0$ be the corresponding operator. According to (3.3), for the given pre-chain (3.9), one can construct the chain

$$H_- \supset H_0 \supset H_+ = L_+ \ominus \operatorname{Ker} Q. \tag{3.10}$$

We shall say that the pre-chain (3.9) is closed if L is complete with respect to $\|\cdot\|_{L_+}$. In the case where $\operatorname{Ker} Q = \{0\}$, we shall say that (3.9) is closable (or, in other words, that it admits a closure). Obviously, if a chain is closed, then it admits a closure. According to Subsection 3.1, we can formulate the following criterion:

The pre-chain (3.9) *is closable if and only if any sequence* $(f_n)_{n=1}^{\infty}$ $(f_n \in L)$ *fundamental in norm* $\|\cdot\|_{L_+}$ *and convergent* (*to* 0) *in norm* $\|\cdot\|_{H_+}$, *converges to zero also in norm* $\|\cdot\|_{L_+}$.

The pre-chain $H_0 \supset L_+$, constructed by the completion of L with respect to $\|\cdot\|_{L_+}$ from the pre-chain (3.9) which is closable, is called a closure of this pre-chain. Everywhere in this Section (except Subsection 3.5), we shall exploit either closed pre-chains or pre-chains that admit a closure $H_0 \supset L_+$ (in the case of closeness $L_+ = L$). Each pre-chain of this sort can be extended to a chain

$$H_- \supset H_0 \supset H_+ = L_+ \tag{3.11}$$

by constructing a negative space. Then one may employ the general facts from Section 1.

We now turn to the language of forms. A function $\mathfrak{D}(a) \times \mathfrak{D}(a) \ni (f, g) \mapsto a(f, g) \in \mathbb{C}^1$ linear in the first variable and antilinear in the second one, is called a bilinear (sesquilinear) form a in the Hilbert space H_0 (here $\mathfrak{D}(a)$ is a linear set dense in H_0 which is called the domain of definition of the form a). The diagonal values of a form the quadratic form $a[\cdot]$ associated with the bilinear one, $\mathfrak{D}(a) \ni f \mapsto a[f] = a(f, f) \in \mathbb{C}^1$. A bilinear form can be uniquely defined in terms of the associated quadratic form by using the polarization formula

$$a(f, g) = \frac{1}{4}(a[f+g] - a[f-g] + ia[f+ig] - ia[f-ig]) \quad (f, g \in \mathfrak{D}(a)). \tag{3.12}$$

The linear operations on the set of bilinear forms are introduced naturally. Thus, if a and b are two bilinear forms and $\mathfrak{D}(a) \cap \mathfrak{D}(b)$ is dense in H_0, then the bilinear form $a + b$ is defined by

$$(a+b)(f, g) = a(f, g) + b(f, g) \quad (f, g \in \mathfrak{D}(a+b) = \mathfrak{D}(a) \cap \mathfrak{D}(b)).$$

For the product λa, where $\lambda \in \mathbb{C}^1$, is always defined, we have

$$\lambda a(f, g) = \lambda a(f, g) \quad (f, g \in \mathfrak{D}(\lambda a) = \mathfrak{D}(a)).$$

For a given bilinear form a, let us construct an adjoint bilinear form a^* by the equality

$$a^*(f, g) = \overline{a(g, f)} \quad (f, g \in \mathfrak{D}(a^*) = \mathfrak{D}(a)).$$

A bilinear form a is called Hermitian if $a^* = a$. It follows from (3.12) that a is Hermitian if and only if the quadratic form $a[\cdot]$ is real-valued. Every bilinear form a can be expressed as a linear combination of the Hermitian forms $\operatorname{Re} a$ and $\operatorname{Im} a$ as follows

$$a = \operatorname{Re} a + i \operatorname{Im} a, \quad \operatorname{Re} a = \frac{1}{2}(a + a^*), \quad \operatorname{Im} a = \frac{1}{2i}(a - a^*).$$

A bilinear form a is called positive bilinear form with a vertex $\alpha > 0$ provided that

$$a(f, f) \geq \alpha \|f\|_{H_0}^2 \quad (f \in \mathfrak{D}(a)). \tag{3.13}$$

In this subsection, it is convenient to assume that $\alpha = 1$. A positive form is always Hermitian, since $a[\cdot]$ is real-valued.

Having a positive form a in H_0, we can construct the pre-chain (3.9) naturally by setting $L = \mathfrak{D}(a)$ and

$$(f, g)_{L_+} = a(f, g) \quad (f, g \in L).$$

Conversely, the pre-chain (3.9) determines the positive form

$$a(f, g) = (f, g)_{L_+} \quad (f, g \in \mathfrak{D}(a) = L).$$

The definitions which have been introduced above for pre-chains can be easily reformulated for positive forms. Thus, a positive form a is called closed if the corresponding pre-chain is closed. The closure $\tilde{a}$ of the closable form given by $\tilde{a}(f, g) = f, g)_{L_+}$, where $f, g \in \mathfrak{D}(\tilde{a}) = L_+$; L and $(\cdot, \cdot)_{L_+}$ are constructed in terms of a. Therefore, to calculate $\tilde{a}(f, g)$ for $f, g \in \mathfrak{D}(\tilde{a}) \subset H_0$ one should construct the sequences $(f_n)_{n=1}^{\infty}$ and $(g_m)_{m=1}^{\infty}$ $(f_n, g_m \in \mathfrak{D}(\tilde{a}))$ which are fundamental with respect to the norm $(a[\cdot])^{1/2}$ and converge in H_0 to f and g, respectively. Then

$$\tilde{a}(f, g) = \lim_{n,m \to \infty} a(f_n, g_m).$$

Taking into account what has been said in this subsection, we can formulate the following theorem on the representation of positive form.

Theorem 3.1. *Let a be a closed positive bilinear form with the vertex $\alpha = 1$. Then there exists a selfadjoint operator $A \geq 1$ acting in the space H_0 such that*

$$a(f, g) = (Af, g)_{H_0} \quad (f \in \mathfrak{D}(A) = \mathfrak{D}(a), g \in \mathfrak{D}(a)). \tag{3.14}$$

The domain of definition $\mathfrak{D}(A)$ is dense in $\mathfrak{D}(a)$ with respect to the norm $(a[\cdot])^{1/2}$, and, moreover, $\|Af\|_{H_0} \geq (a[f])^{1/2}$ $(f \in \mathfrak{D}(a))$. Along with (3.14), we have the following representation in terms of the operator $\sqrt{A}$:

$$a(f, g) = (\sqrt{A}f, \sqrt{A}g)_{H_0} \quad (f, g \in \mathfrak{D}(\sqrt{A}) = \mathfrak{D}(a)). \tag{3.15}$$

Proof. By using the method described above, we proceed from the form a to the closed pre-chain (3.9) and then to the chain (3.10). Applying Lemma 3.1 to (3.10), we obtain the statement of the theorem.

■

If the form a is not closed but admits the closure $\tilde{a}$, then (3.14) and (3.15) can be rewritten for $\tilde{a}$. Setting in these formulas $f \in \mathfrak{D}(a) \cap \mathfrak{D}(A)$, $g \in \mathfrak{D}(a)$ or $f, g \in \mathfrak{D}(\sqrt{A}) \cap \mathfrak{D}(a)$, respectively, we get the representation of the original form.

Let us also note the following. Let a and b be two positive forms such that $\mathfrak{D}(a) \cap \mathfrak{D}(A)$ is dense in H_0. Then $a + b$ is also a positive form. It is easy to see that if a and b are closed (or can be closed), then $a + b$ is also closed (can be closed). This fact immediately follows from the definitions and the criterion presented at the beginning of this Subsection. One should only note that the fact that some sequence is fundamental with respect to the norm $(a[\cdot] + b[\cdot])^{1/2}$ is equivalent to the fact that it is fundamental with respect to both $(a[\cdot])^{1/2}$ and $(b[\cdot])^{1/2}$.

3.3. Semi-Bounded Forms and Operators

A bilinear form a in the space H_0 is called semi-bounded (from below) with a vertex $\alpha \in \mathbb{R}^1$ if

$$a(f, f) \geq \alpha \|f\|^2_{H_0} \quad (f \in \mathfrak{D}(a)). \tag{3.16}$$

If $\alpha = 0$, the form a is called nonnegative, and if $\alpha > 0$, we get the positive form introduced above. It is clear that semi-bounded forms are Hermitian.

By setting

$$a_p(f, g) = a(f, g) + (1-\alpha)(f, g)_{H_0} \quad (f, g \in \mathfrak{D}(a_p) = \mathfrak{D}(a)), \tag{3.17}$$

we associate a positive form a_p (with a vertex equal to unity) with a semi-bounded form a. The definitions introduced for the form a can be now expressed in terms of the form a_p. Thus, a is closed (closable) if a_p is closed (closable). According to (3.17), the closure $\tilde{a}$ of the closable form a is defined by

$$\tilde{a}(f, g) = \tilde{a}_p(f, g) - (1-\alpha)(f, g)_{H_0} \quad (f, g \in \mathfrak{D}(\tilde{a}) = \mathfrak{D}(\tilde{a}_p)). \tag{3.18}$$

It is a semi-bounded form with the same vertex α.

In the case of positive forms, we can proceed in a different way. If a is a positive form with vertex α, then $\alpha^{-1}a$ is a positive form with the vertex equal to unity. Therefore, the definitions in Subsection 3. 2 applied to $\alpha^{-1}a$ lead to the definitions for a (since the norms $(a_p[\cdot])^{1/2}$ and $(\alpha^{-1}a[\cdot])^{1/2}$ are equivalent). Further, this implies that in the case of semi-bounded forms, one may also define a_p by formula (3.17) with the only difference being that the unity is replaced by $\varepsilon > 0$.

For semi-bounded forms, Theorem 3.1 can be formulated as follows.

Theorem 3.2. *Let a be a closed semi-bounded bilinear form with a vertex $\alpha \in \mathbb{R}^1$. Then there exists a selfadjoint operator $A \geq \alpha 1$ acting in the space H_0 such that the representation* (3.14) *holds. Its domain of definition $\mathfrak{D}(A)$ is dense in $\mathfrak{D}(a)$ with respect to the norm $(a_p[\cdot])^{1/2}$. In the case when a is nonnegative, formula* (3.15) *holds, as well.*

Proof. Let us write representation (3.14) for the form a_p. Let $A_p \geq 1$ be the corresponding operator. Taking into account relation (3.17), we obtain (3.14) for a with the operator $A = A_p - (1-\alpha)1 \geq \alpha 1$. To obtain representation (3.15) in the case of a positive form a, it suffices to write it for $\alpha^{-1}a$. Now let the form a be nonnegative, and A be a nonnegative selfadjoint operator associated with this form. For every $\varepsilon > 0$, the form

$$a(f, g) + \varepsilon(f, g)_{H_0} \quad (f, g \in \mathfrak{D}(a))$$

is positive and, thus,

$$a(f, g) + \varepsilon(f, g)_{H_0} = (\sqrt{A+\varepsilon 1}\, f, \sqrt{A+\varepsilon 1}\, g)_{H_0} \quad (f, g \in \mathfrak{D}(a)).$$

It follows from the spectral decomposition of the operator A that for every $f \in \mathfrak{D}(\sqrt{A}) = \mathfrak{D}(\sqrt{A+\varepsilon 1})$ in H_0, we have $\sqrt{A+\varepsilon 1}f \to \sqrt{A}f$ as $\varepsilon \to 0$. Hence, by taking the limit, we find that (3.15) holds also for a.

■

If the form a admits a closure $\tilde{a}$, we can write (3.14) and (3.15) for this closure. Then the representations for the form a itself are obtained as a consequence.

Let us consider an important procedure for the extension of semi-bounded to selfadjoint operators which is due to Friedrichs. Assume that an Hermitian semi-bounded operator $A \geq \alpha 1$ $(\alpha \in \mathbb{R}^1)$ with dense domain of definition is given in H_0. It generates a semi-bounded form a with vertex α in a standard way

$$a(f, g) + (Af, g)_{H_0} \quad (f, g \in \mathfrak{D}(a) = \mathfrak{D}(A)). \tag{3.19}$$

The following statement is valid for this form.

Theorem 3.3. *The bilinear form* (3.19) *admits a closure $\tilde{a}$ for which the representation*

$$\tilde{a}(f, g) = (A_F f, g)_{H_0} \quad (f \in \mathfrak{D}(A_F) \subset \mathfrak{D}(\tilde{a}),\ g \in \mathfrak{D}(\tilde{a}))$$

holds, where $A_F \geq \alpha 1$ is a selfadjoint operator in H_0 being an extension of the operator A (the so-called Friedrichs extension). The operator A_F is the only selfadjoint extension of A whose domain of definition belongs to $\mathfrak{D}(\tilde{a})$.

Proof. We set $A_p = A + (1-\alpha)1 \geq 1$. Then

$$a_p(f, g) = (A_p f, g)_{H_0} \quad (f, g \in \mathfrak{D}(A_p) = \mathfrak{D}(A)).$$

The fact that a is closable means that a_p is closable. We show the latter. Let $(f_n)_{n=1}^{\infty}$ $(f_n \in \mathfrak{D}(A_p))$ be such that

$$(A_p(f_n - f_m), f_n - f_m)_{H_0} = a_p[f_n - f_m] \to 0, \quad n, m \to \infty,$$

and $\|f_n\|_{H_0} \to 0$ as $n \to \infty$. It is necessary to prove that

$$(A_p f_n, f_n)_{H_0} = a_p[f_n] \xrightarrow[n\to\infty]{} 0.$$

For an arbitrary $\varepsilon > 0$, one can find $N(\varepsilon)$ such that

$$(A_p(f_n - f_m), f_n - f_m)_{H_0} < \varepsilon$$

for $n, m > N(\varepsilon)$. The fundamental sequence is bounded and, therefore, with the help of the Cauchy–Bunyakovsky inequality, we obtain

$$(A_p f_n, f_n)_{H_0} \le |(A_p f_n, f_n - f_m)_{H_0}| + |(A_p f_n, f_m)_{H_0}|$$

$$\le (A_p f_n, f_n)_{H_0}^{1/2} (A_p(f_n - f_m), f_n - f_m)_{H_0}^{1/2} + |(A_p f_n, f_m)_{H_0}|$$

$$\le c\varepsilon^{1/2} + |(A_p f_n, f_m)_0|$$

for $n, m > N(\varepsilon)$. By passing to the limit as $m \to \infty$ in the last inequality, we get $(A_p f_n, f_n)_{H_0} \le c\varepsilon^{1/2}$ for $n > N(\varepsilon)$, and this proves the required result.

Applying Theorem 3.2 to a closed semi-bounded form a with vertex α, we obtain the representation

$$\tilde{a}(f, g) = (A_F f, g)_{H_0} \quad (f \in \mathfrak{D}(A_F) \subset \mathfrak{D}(\tilde{a}),\ g \in \mathfrak{D}(\tilde{a})),$$

where $A_F \ge \alpha 1$ is selfadjoint.

We now show that if an Hermitian operator $B \supset A$ is such that $\mathfrak{D}(B) \subset \mathfrak{D}(\tilde{a})$, then $B \subset A_F$. To do this, it is necessary to establish the equality

$$(f, Bg)_{H_0} = \tilde{a}(f, g) \quad (f \in \mathfrak{D}(\tilde{a}),\ g \in \mathfrak{D}(B)). \tag{3.20}$$

First, let $f \in \mathfrak{D}(A) \subset \mathfrak{D}(B)$, then

$$(f, Bg)_{H_0} = (Bf, g)_{H_0} = (Af, g)_{H_0}.$$

Since $\mathfrak{D}(B) \subset \mathfrak{D}(\tilde{a})$, there exists a sequence $(g_n)_{n=1}^{\infty}$ $g_n \in \mathfrak{D}(a) = \mathfrak{D}(A)$ such that $g_n \to g$ as $n \to \infty$ in norm $(a_p[\cdot])^{1/2}$. This implies that

$$(f, Bg)_{H_0} = (Af, g)_{H_0} = \lim_{n\to\infty} (Af, g_n)_{H_0} = \lim_{n\to\infty} a(f, g_n) = \tilde{a}(f, g).$$

Thus, we have established equality (3.20) for $f \in \mathfrak{D}(A)$. Now let $f \in \mathfrak{D}(\tilde{a})$ and $(f_n)_{n=1}^{\infty}$ $(f_n \in \mathfrak{D}(a) = \mathfrak{D}(A)$ be a sequence such that $f_n \to f$ as $n \to \infty$ in the norm $(a_p[\cdot])^{1/2}$. If we now write (3.20) for $f = f_n \in \mathfrak{D}(A)$ and pass to the limit, then we obtain (3.20) in the general case.

By employing the definition of A_F and (3.20), we get

$$(A_F f, g)_{H_0} = \tilde{a}(f, g) = (f, Bg)_{H_0} \quad (f \in \mathfrak{D}(A_F) \subset \mathfrak{D}(\tilde{a}),\ g \in \mathfrak{D}(B)).$$

This and the selfadjointness of A_F together imply that $g \in \mathfrak{D}(A_F)$ and $A_F g = Bg$, i.e., $A_F \supset B$.

The proof of the theorem easily follows from the proved inclusion. In fact, by setting $B = A$, we get $A_F \supset A$. Further, if, in addition, B is selfadjoint, then the inclusion $A_F \supset B$ implies that $B = A_F$.

■

If A is selfadjoint, then clearly $A_F = A$. Assume that A is selfadjoint and positive. Then it follows from the spectral decomposition of A that $\mathfrak{D}(\tilde{a}) = \mathfrak{D}(\sqrt{A})$, and

$$\tilde{a}(f, g) = (\sqrt{A} f, \sqrt{A} g)_{H_0} \quad (f, g \in \mathfrak{D}(\tilde{a})).$$

This enables us to conclude that in the case of unbounded A, the form (3.19) is always nonclosed.

At the end of this subsection, we consider, in brief, sectorial forms (the concept of sectorial forms is a generalization of the concept of semi-bounded forms). A bilinear form a in the space H_0 is called a sectorial form with a vertex $\alpha \in \mathbb{R}^1$ if, in the complex plane, there exists an angle $S(\alpha, k)$ $(k \in [0, \infty))$ smaller than π (with a vertex at the point α) which has the form

$$S(\alpha, k) = \{z \in \mathbb{C}^1 \mid \operatorname{Re} z \geq \alpha,\ |\operatorname{Im} z| \leq k(\operatorname{Re} z - \alpha)\}$$

and is such that

$$\frac{a(f, f)}{\|f\|_{H_0}^2} \in S(\alpha, k) \quad (f \in \mathfrak{D}(a)).$$

In the case $k = 0$, we obtain a semi-bounded form with a vertex α. If a is sectorial with the vertex α, then $\operatorname{Re} a$ is semi-bounded and has the same vertex.

Let us define, for a given sectorial form a, a sectorial form a_p by (3.17) (a_p is connected with the angle $S(1, k)$). The form $\operatorname{Re} a_p$ is a positive form with $\alpha = 1$. By using this form, we can introduce the scalar product $(f, g)_{L_+} = (\operatorname{Re} a_p)(f, g)$ $(f, g \in L)$ on $L = \mathfrak{D}(\operatorname{Re} a_p) = \mathfrak{D}(a_p) = \mathfrak{D}(a)$ employing the method described above.

The study of sectorial forms is based on the following simple remark:

Assume that the form a is sectorial, then the following inequality holds for the form a_p

$$|a_p(f,g)| \le (k+1)\|f\|_{L_+}\|g\|_{L_+} \quad (f,g \in L). \tag{3.21}$$

Indeed, since $a_p(f,f)\|f\|_{H_0}^{-2} \in S(1,k)$, we have

$$|\operatorname{Im}(a_p(f,f)\|f\|_{H_0}^{-2})| \le k(\operatorname{Re}(a_p(f,f)\|f\|_{H_0}^{-2}-1))$$

$$\le k(\operatorname{Re}(a_p(f,f)\|f\|_{H_0}^{-2})) \quad (f \in L).$$

In other words,

$$|(\operatorname{Im} a_p)(f,f)| \le k\|f\|_{L_+}^2 \quad (f \in L).$$

Since the form $\operatorname{Im} a_p$ is Hermitian, the last inequality implies that

$$|(\operatorname{Im} a_p)(f,g)| \le k\|f\|_{L_+}\|g\|_{L_+} \quad (f,g \in L).$$

Hence, taking into account that $a_p = \operatorname{Re} a_p + i\operatorname{Im} a_p$ and the form of $(\cdot,\cdot)_{L_+}$, we obtain (3.21).

■

A sectorial form a is called closed (closable) if $\operatorname{Re} a_p$ is closed (closable). The closure $\tilde{a}$ of the closable form a is defined as follows: We complete L to the space L_+; then, due to (3.21), a_p can be extended by continuity to the form $\tilde{a}_p$ defined on $\mathfrak{D}(\tilde{a}_p) = L_+$; and finally, we define $\tilde{a}$ according to (3.18).

Formula (3.14) can be generalized as follows:

Let a be a closed sectorial form. Then it admits the representation (3.14) *with a certain closed operator A acting in H_0 whose domain of definition is dense in $\mathfrak{D}(a)$ with respect to the norm* $((\operatorname{Re} a_p)[\cdot])^{1/2}$.

In fact, the space L is now dense, and the inequality (3.21) yields the representation

$$a_p(f,g) = (Bf,g)_{L_+} = (\operatorname{Re} a_p)(Bf,g) \quad (f,g \in L), \tag{3.22}$$

where B is a operator bounded in L. Since

$$\|Bf\|_{L_+}\|f\|_{L_+} \ge |(Bf,f)_{L_+}| = |a_p(f,f)| \ge (\operatorname{Re} a_p)(f,f) = \|f\|_{L_+}^2,$$

we have $\|Bf\|_{L_+} \ge \|f\|_{L_+}$ $(f \in L)$. This inequality guarantees the existence of the inverse operator B^{-1} in L.

Let us apply Theorem 3.1 to the form $\operatorname{Re} a_p$. Then (3.14) yields

$$(\operatorname{Re} a_p)(f, g) = (Cf, g)_{H_0} \quad (f \in \mathfrak{D}(C) \subset L, g \in L),$$

where $C \geq 1$ is selfadjoint. The domain $\mathfrak{D}(C)$ is dense in L, and $\| Cf \|_{H_0} \geq \| f \|_{L_+}$ $(f \in L)$. It follows from this representation and (3.22) that

$$a(f, g) = (\operatorname{Re} a_p)(Bf, g) = (CBf, g)_{H_0} \quad (f \in \{f \in L \mid Bf \in \mathfrak{D}(C)\}, g \in L). \quad (3.23)$$

The set $\mathfrak{D}(A_p) = \{f \in L \mid Bf \in \mathfrak{D}(C)\}$ is dense in L (and, hence, in H_0) since $\mathfrak{D}(C)$ is dense in L, and B is an operator invertible in L. Let us define the operator A_p in H_0 by setting $A_p f = CBf$ $(f \in \mathfrak{D}(A_p))$. Therefore, by virtue of (3.23), we have

$$a(f, g) = (A_p f, g)_{H_0} \quad (f \in \mathfrak{D}(A_p), \ g \in L).$$

It follows from the invertibility of the operators C and B in the corresponding spaces that $\mathfrak{R}(A_p) = H_0$. Further,

$$\| A_p f \|_{H_0} = \| CBf \|_{H_0}$$

$$\geq \| Bf \|_{L_+} \geq \| B^{-1} \|^{-1} \| f \|_{L_+} \geq \| B^{-1} \|^{-1} \| f \|_{H_0} \quad (f \in \mathfrak{D}(A)).$$

Thus, the inverse operator A_p^{-1} exists in the space H_0. This implies, in particular, that A_p is closed.

To obtain representation (3.14), one should take the operator $A = A_p - (1 - \alpha)\, 1$.

■

The operator A acting in H_0 is called sectorial (with a vertex $\alpha \in \mathbb{R}^1$) if the corresponding bilinear form (3.19) is sectorial (this is the generalization of the concept of the Hermitian semi-bounded operator). Formula (3.14) proved above shows that every closed sectorial form is generated by some sectorial operator. Moreover, unlike an arbitrary sectorial operator, this one possesses a property of maximality obtained from the representation $A = CB - (1 - \alpha)\, 1$, where B is a bounded invertible operator in L, and C is a selfadjoint invertible operator in H_0 (the maximal sectorial operator generalizing the concept of a selfadjoint semi-bounded operator).

We shall not investigate the properties of the maximal sectorial operator and describe these operators by themselves. These properties can be obtained quite easily from the formula for A given above. Note also that it is not difficult to prove the generalization of

Theorem 3.3 for the case of the extension of an arbitrary sectorial operator to the maximal sectorial operator.

3.4. Form-Sums of Operators

Consider an often-encountered problem: Let A be a selfadjoint operator in the space H_0, and B be an Hermitian operator in the same space. We want to clarify under what conditions the operator $A + B$ ($\mathfrak{D}(A + B) = \mathfrak{D}(A) \cap \mathfrak{D}(B)$) is selfadjoint or essentially selfadjoint. Here, we present several approaches to this problem which make use of forms.

First, let us consider the well-known result for forms which sometimes enables us to give a reasonable meaning to the operator $A + B$, even in the case where $\mathfrak{D}(A) \cap \mathfrak{D}(B) = \{0\}$ (this is the KLMN Theorem, see Reed and Simon [2, Chapter 10, Theorem X.17]).

Theorem 3.4. *Let a be a closed positive form, and b be an Hermitian form on $\mathfrak{D}(b) = \mathfrak{D}(a)$, such that*

$$|b(f,f)| \leq pa(f,f) + q(f,f)_{H_0} \quad (f \in \mathfrak{D}(A)). \tag{3.24}$$

for some $p \in [0, 1)$ and $q \in \mathbb{R}^1$. Then the form $a + b$ ($\mathfrak{D}(a + b) = \mathfrak{D}(a)$) is semibounded and closed.

Proof. By virtue of the positivity of a, we have $a(f,f) \geq \varepsilon \|f\|^2_{H_0}$ ($f \in \mathfrak{D}(A)$) for some $\varepsilon \in (0, 1)$. It follows from (3.24) and this estimate that

$$\alpha \|f\|^2_{H_0} \leq (1-p)\,a(f,f) - q(f,f)_{H_0} \leq a(f,f) + b(f,f)$$

$$\leq (1+p)\,a(f,f) - q(f,f)_{H_0} \quad (\alpha = (1-p)\varepsilon - q,\ f \in \mathfrak{D}(A)). \tag{3.25}$$

This inequality ensures, in particular, that the form $a + b$ is semi-bounded. Taking into account (3.17), we conclude from (3.25) that

$$(1-p)\,a(f,f) \leq (1-p)\,a(f,f) + (1-\alpha-q)(f,f)_{H_0} \leq (a+b)_p(f,f)$$

$$\leq (1+p)\,a(f,f) + (1-\alpha+q)(f,f)_{H_0} \quad (f \in \mathfrak{D}(A)).$$

This and the inequality $a(f,f) \geq \varepsilon \| f \|^2_{H_0}$ imply that the norms $(a[\cdot])^{1/2}$ and

$((a+b)_p[\cdot])^{1/2}$ are equivalent on $\mathfrak{D}(a)$. By virtue of the fact that a is closed and positive, $\mathfrak{D}(a)$ is complete with respect to $(a[\cdot])^{1/2}$, and, hence, with respect to $((a+b)_p[\cdot])^{1/2}$. The latter means that $a+b$ is closed.

■

Let us present a scheme which enables us, in particular, to give an operatorial meaning to this theorem. Let $A \geq 1$ be a selfadjoint operator in the space H_0. We set $D = A^{1/2}$ and construct the chain

$$H_- \supset H_0 \supset H_+ = \mathfrak{D}(A^{1/2}), \tag{3.26}$$

$$(\varphi, \psi)_{H_+} = (A^{1/2}\varphi, A^{1/2}\psi)_{H_0} \quad (\varphi, \psi \in H_+),$$

according to the procedure (1.19). We now have

$$(f, g)_{H_-} = (A^{-1/2}f, A^{-1/2}g)_{H_0} \quad (f, g \in H_+),$$

and H_- coincides with the completion of H_0 with respect to this scalar product. The operator A can be regarded as an isometric operator acting from $\mathfrak{D}(A) \subset H_+$ into H_-, i.e.,

$$\|A\varphi\|_{H_-} = \|A^{1/2}\varphi\|_{H_0} = \|\varphi\|_{H_+} \quad (\varphi \in \mathfrak{D}(A)).$$

Therefore, it can be extended by continuity to the operator $\mathcal{A}: H_+ \to H_-$ realizing the isometry between H_+ and H_- (and coinciding with I^{-1}). Clearly, $A = \mathcal{A} \restriction \mathfrak{D}(A)$, where $\mathfrak{D}(A) = \{\varphi \in H_+ \mid \mathcal{A}\varphi \in H_0\}$. The operator $\mathcal{A}$ is selfadjoint with respect to H_0, i.e.,

$$(\mathcal{A}\varphi, \psi)_{H_0} = (\varphi, \mathcal{A}\psi)_{H_0}.$$

The following form can be defined for given A (according to (3.14))

$$a(f, g) = (Af, g)_{H_0} = \int_1^\infty \lambda \, d(E(\lambda)f, g)_{H_0} \quad (f, g \in \mathfrak{D}(A)),$$

where E is a resolution of identity that corresponds to A. It follows from the spectral representation written above that, for the closure of this form $\tilde{a}\left(\mathfrak{D}(\tilde{a}) = H_+\right)$, we have

$$\tilde{a}(f, g) = (A^{1/2}f, A^{1/2}g)_{H_0} = (f, g)_{H_+} = (\mathcal{A}f, g)_{H_0} \quad (f, g \in H_+).$$

Let us introduce a perturbation of the operator A as follows: Assume that there exists an operator $\mathfrak{B}$ acting continuously from the whole of H_+ into H_- and selfadjoint with respect to the zero space of the chain (3.26). Therefore, the operator $B = \mathfrak{B} \upharpoonright \mathfrak{D}(B)$, where $\mathfrak{D}(B) = \{\varphi \in H_+ \mid \mathfrak{B}\varphi \in H_0\}$, is Hermitian, although it may happen that it is nondensely defined or equal to zero on $\mathfrak{D}(B)$. The sum $A + B$, which is directly defined, is given on $\mathfrak{D}(A) \cap \mathfrak{D}(B)$ and may coincide with A or be defined only at 0. Thus, in the latter case, it is, generally speaking, impossible to indicate whether this sum is selfadjoint. However, we can employ the following construction.

By using $\mathfrak{B}$, we introduce a bilinear Hermitian form

$$b(f, g) = (\mathfrak{B}f, g)_{H_0} \quad (f, g \in \mathfrak{D}(b) = H_+ = \mathfrak{D}(\tilde{a}))$$

in H_0. Assume that condition (3.24) holds for it. This condition now has the form

$$|(\mathfrak{B}f, f)_{H_0}| \leq p(A^{1/2}f, A^{1/2}f)_{H_0} + q(f, f)_{H_0} \quad (f \in H_+;\ p \in [0, 1), q \in \mathbb{R}^1). \quad (3.27)$$

Then, according to Theorem 3.4, the form $\tilde{a} + b$ $(\mathfrak{D}(\tilde{a} + b) = H_+)$ is semi-bounded and closed. According to Theorem 3.2 on representation, there exists a selfadjoint semi-bounded operator C in H_0 such that

$$\tilde{a}(f, g) + b(f, g) = (Cf, g)_{H_0} \quad (f \in \mathfrak{D}(C) \subset H_+,\ g \in H_+).$$

Therefore, the relation between A, B, and C is given by

$$(A^{1/2}f, A^{1/2}g)_{H_0} + (\mathfrak{B}f, g)_{H_0} = (Cf, g)_{H_0} \quad (3.28)$$

$$(f \in \mathfrak{D}(C) \subset H_+ = \mathfrak{D}(A^{1/2}),\ g \in H_+).$$

The operator C so defined is called a form-sum of the operators A and B and is denoted by $A \dotplus B$ (the notation $A \dotplus \mathfrak{B}$ is more precise).

The form-sum can be also defined in the following natural way

$$A \dotplus B = (\mathcal{A} + \mathfrak{B}) \upharpoonright \mathfrak{D}(A \dotplus B), \quad \mathfrak{D}(A \dotplus B) = \{\varphi \in H_+ \mid (\mathcal{A} + \mathfrak{B})\varphi \in H_0\} \quad (3.29)$$

(*the restriction is understood as an operator in* H_0).

In fact, denote the operator defined by (3.29) by F. It is easy to see that $C \subset F$. Indeed, we have

$$(A^{1/2}f, A^{1/2}g)_{H_0} = (\mathcal{A}f, g)_{H_0} \quad (f, g \in H_+).$$

Hence, by taking into account (3.28), we obtain the equality $H_0 \ni Cf = (\mathcal{A} + \mathfrak{B})f = Ff$ for $f \in \mathfrak{D}(C)$. Further, F is obviously an Hermitian operator in H_0. Consequently, by virtue of the selfadjointness of C, we have $F = C$.

■

Thus, the operator $A \dotplus B$, defined by (3.28) or (3.29), is selfadjoint in H_0 provided that condition (3.27) holds. Note that the form b has been introduced in terms of the operator B. Clearly, for the form b satisfying the conditions of Theorem 3.4, this operator always exists since condition (3.24) guarantees the continuity of b on $H_+ = \mathfrak{D}(\tilde{a})$. Therefore, we can say that "the operator A is perturbed by the form b".

Remark 1. The form $\tilde{a} + b$ may be closed, even if the assertion (3.27) is violated; it is obvious that the form-sum $A \dotplus B$ is defined in this case, as well.

■

Remark 2. The concept of a form-sum can be generalized as follows: Assume that the operator

$$H_+ \supset \mathfrak{D}(\mathfrak{B}) \ni f \mapsto \mathfrak{B}f \in H_-$$

is given, with the domain of definition which is dense in H_+, and that it is nonnegative with respect to H_0, i.e., $(\mathfrak{B}f, f)_{H_0} \geq 0$ $(f \in \mathfrak{D}(\mathfrak{B}))$ (one may also consider the case when the operator is semi-bounded, i.e., when

$$(\mathfrak{B}f, f)_{H_0} \geq \alpha (f, f)_{H_0} \quad (f \in \mathfrak{D}(\mathfrak{B}))$$

for any $\alpha \in \mathbb{R}^1$). In the space H_0, we construct the form

$$b(f, g) = (\mathfrak{B}f, g)_{H_0} \quad (f, g \in \mathfrak{D}(b) = \mathfrak{D}(\mathfrak{B}) \subset H_+ \subset H_0).$$

It is easy to prove that the form $\tilde{a} + b\,(\mathfrak{D}(\tilde{a} + b) = \mathfrak{D}(b))$ admits a closure. The selfadjoint nonnegative operator C in H_0, which corresponds to this closure according to (3.14), is just the required generalization of the form-sum $A \dotplus B$.

Let us show that the form $\tilde{a} + b$ admits a closure. In fact, it can be represented as

$$(\tilde{a} + b)(f, g) = ((\mathcal{A} + \mathfrak{B})f, g)_{H_0} = (\mathcal{A}^{-1}(\mathcal{A} + \mathfrak{B})f, g)_{H_+} \quad (f, g \in \mathfrak{D}(\mathfrak{B})),$$

and, therefore, according to the beginning of the proof of Theorem 3.3, it can be closed as a form in H_+. This means that the corresponding completion of $\mathfrak{D}(\mathfrak{B})$ belongs to H_+ and, hence, to $H_0 \supset H_+$.

■

The examples in which the form-sums of operators are employed will be studied in Chapters 5 and 6. Here, we only present one useful illustrative example.

Example 3.4. (Schrödinger operator with the δ-potential). In the space $L_2(\mathbb{R}^1) = L_2(\mathbb{R}^1, \mathfrak{B}(\mathbb{R}^1), dx)$ with the Lebesgue measure dx, consider the operator

$$L_2(\mathbb{R}^1) \supset C_0^\infty(\mathbb{R}^1) \ni u \mapsto -u'' + u \in L_2(\mathbb{R}^1).$$

Let $A \geq 1$ be its closure which is a selfadjoint operator in $L_2(\mathbb{R}^1)$. For the corresponding form $\tilde{a}$, we now have

$$H_+ = \mathfrak{D}(\tilde{a}) = \mathfrak{D}(A^{1/2}) = W_1^2(\mathbb{R}^1) \supset C(\mathbb{R}^1)$$

(by virtue of the imbedding theorems, $1 > d/2$, $d = 1$). The following operator is defined

$$H_+ = W_1^2(\mathbb{R}^1) \ni \varphi \mapsto \mathfrak{B}\varphi = \varphi(0)\,\delta_0 \in W_1^2(\mathbb{R}^1) = H_-$$

where δ_0 is a δ-function concentrated at the origin, and $\mathfrak{B} \in \mathcal{L}(H_+, H_-)$ being nonnegative with respect to H_0.

It can be easily shown that the corresponding form b satisfies the estimate (3.27) with an arbitrarily small $p \in [0, 1)$ and some $q = q(p) \in [0, \infty)$. Indeed, by virtue of the imbedding theorem, there exists $c \in (0, \infty)$ such that

$$|u(0)| \leq c\,\|u\|_{W_2^1((-1,1))} \leq c\,\|u\|_{W_2^1(\mathbb{R}^1)} = c\,(\|u'\|^2_{L_2(\mathbb{R}^1)} + \|u\|^2_{L_2(\mathbb{R}^1)})^{1/2}$$

for any $u \in C_0^\infty(\mathbb{R}^1)$. In this inequality, we substitute $u_\varepsilon(x) = u(\varepsilon x) \in C_0^\infty(\mathbb{R}^1)$ $(\varepsilon \in (0, \infty))$ for $u(x)$. Then, by differentiating and changing the variables ($\varepsilon x = y$), we obtain

$$|u(0)| = |u_\varepsilon(0)| \leq c\,(\varepsilon\|u'\|^2_{L_2(\mathbb{R}^1)} + \varepsilon^{-1}\|u\|^2_{L_2(\mathbb{R}^1)})^{1/2}.$$

This inequality leads to (3.27) with an arbitrarily small $p \in [0,1)$ and, therefore, the form-sum $A \dotplus B$ is defined. For this form-sum, we can write relation (3.28) showing that the singularity of the operator $\mathfrak{B}$ at the origin is compensated by the corresponding singularity of $\mathcal{A}$.

The operator $A \dotplus B$ constructed above may be regarded as an operator realization of the differential expression $-u'' + u + \delta_0 u$. It was essential here that the Schrödinger operator is one-dimensional. In the case where dimensionality is greater than one, the corresponding imbedding theorem does not hold. We also stress that $B = 0$ on

$$\mathfrak{D}(\mathfrak{B}) = \{\varphi \in W_1^2(\mathbb{R}^1) \mid \varphi(0) = 0\}.$$

3.5. Nonclosable Forms

Here, we present some facts concerning general (nonclosable) forms. We restrict ourselves to the case of positive forms with a vertex equal to unit. As in Subsection 3.2, it is convenient to start with the study of the pre-chains (3.9). Recall that $Q: L_+ \to H_0$. In the general case, the relation $H_0 \supset L_+ \ominus \operatorname{Ker} Q$ holds. This yields $H_0 \oplus \operatorname{Ker} Q \supset L_+$. Therefore, if we set $L_0 = H_0 \oplus \operatorname{Ker} Q$, we can expect that the theory, which we are planning to construct, will be analogous to the theory developed above but with H_0 replaced by L_0. Consider these questions in detail.

As mentioned in Subsection 3.1, the restriction $Q \upharpoonright L$ is the imbedding operator which imbeds L regarded as a part of L_+ into L regarded as a part of H_0. We also have $L \cap \operatorname{Ker} Q = \{0\}$. The restriction $Q \upharpoonright (L_+ \ominus \operatorname{Ker} Q)$ is the operator which imbeds $L_+ \ominus \operatorname{Ker} Q$ into H_0. This is so because we identify the vector $f \in L_+ \ominus \operatorname{Ker} Q$ with the vector $Qf \in H_0$. Thus, the operator Q (as a whole) can be regarded as the imbedding of L_+ into $H_0 \oplus \operatorname{Ker} Q = L_0$.

Let us illustrate this by Example 3.3 in the case $d \geq 2$ where $H_0 = L_2(G)$, $L = C^1(\tilde{G})$, and $(\cdot, \cdot)_{L_+}$ is given by (3.5)[4]. We have $L_+ = W_1^2(G) \oplus \mathbb{C}^1$ and, therefore, L can be identified with the collection of pairs $(\varphi, \varphi(x_0))$ $(\varphi \in C^1(\tilde{G}) = L)$. The operator Q has the form

$$L_+ = W_1^2(G) \oplus \mathbb{C}^1 \ni (\varphi, p) \mapsto Q(\varphi, p) = \varphi \in L_2(G) = H_0$$

and it imbeds $L_+ = W_1^2(G) \oplus \mathbb{C}^1$ into $L_0 = L_2(G) \oplus \mathbb{C}^1$.

As the space $L_+ \ominus \operatorname{Ker} Q$ is dense in H_0, L_+ is dense in L_0, and one can construct the chain

$$L_- \supset L_0 \supset L_+. \tag{3.30}$$

By applying Lemma 3.1 to (3.30), we obtain the representation with the proper operator A

$$(\varphi, \psi)_{L_+} = (A\varphi, \psi)_{L_0} \quad (\varphi \in \mathfrak{D}(A), \psi \in L_+). \tag{3.31}$$

Let us now proceed from the positive form $a(f, g)$ $(f, g \in \mathfrak{D}(a) \subset H_0$, $\mathfrak{D}(a)$ is dense in $H_0)$ with the vertex 1. As above, we assume that $L = \mathfrak{D}(a)$ and

$$(f, g)_{L_+} = a(f, g) \quad (f, g \in L).$$

By passing to the completion of L, we obtain the space L_+ whose vectors can be inter-

[4] One may also consider Example 3.2. In this case, $L_+ = L_2(G) \oplus \mathbb{C}^1 = L_0$.

preted as vectors from $L_0 = H_0 \oplus \operatorname{Ker} Q$, where the operator Q is, in fact, generated by the form a. We set $\hat{a}(\varphi, \psi) = (\varphi, \psi)_{L_+}$ $(\varphi, \psi \in L_+)$. This expression is a positive bilinear form (with vertex 1) given in the space L_0 whose domain of definition $\mathfrak{D}(\hat{a}) = L_+$ is dense in L_0. Equality (3.31) implies the theorem on the representation of this form

$$\hat{a}(\varphi, \psi) = (A\varphi, \psi)_{L_0} \quad (\varphi \in \mathfrak{D}(A), \psi \in \mathfrak{D}(\hat{a})), \tag{3.32}$$

where $A \geq 1$ is a selfadjoint operator in the space L_0 whose domain of definition is dense in $\mathfrak{D}(\hat{a})$ with respect to $(\hat{a}[\cdot])^{1/2} = \|\cdot\|_{L_+}$, and

$$\|A\varphi\|_{L_0} \geq (\hat{a}[\varphi])^{1/2} \quad (\varphi \in \mathfrak{D}(A)).$$

Clearly, we can also write the representation of the type (3.15).

The form $\hat{a}$ which always exists will be called a generalized closure of a. Therefore, we have obtained the representation for $\hat{a}$ as in Theorem 3.1, but in the bigger space $L_0 = H_0 \oplus \operatorname{Ker} Q$ instead of the space H_0. In order to return to the representation of a in (3.32), one should take

$$\varphi \in \mathfrak{D}(A) \cap \mathfrak{D}(A) \subset H_0 \quad \text{and} \quad \varphi \in \mathfrak{D}(a) \subset H_0,$$

then

$$a(\varphi, \psi) = \hat{a}(\varphi, \psi) = (A\varphi, \psi)_{L_0} = = (PA\varphi, \psi)_{H_0},$$

where P is the projector onto H_0 given in L_0. However, in this representation, $\mathfrak{D}(a) \cap \mathfrak{D}(A)$ is, generally speaking, nothing more than the proper part of $\mathfrak{D}(a)$ and, hence, the representation obtained is not a representation of the original form a.

In addition, let us investigate several useful concepts and facts concerning nonclosable forms. Denote $H_+ = L_+ \ominus \operatorname{Ker} Q$. Consequently, $L_+ = H_+ \oplus \operatorname{Ker} Q$. Let P_r and P_s be the projectors in L_+ onto H_+ and $\operatorname{Ker} Q$, respectively. Consider a positive bilinear form a (with vertex 1) which is, generally speaking, nonclosable or, in other words, consider a pre-chain $H_0 \supset L$, where $L = \mathfrak{D}(a)$ and

$$(f, g)_{L_+} = a(f, g) \quad (f, g \in L).$$

Let us define two bilinear forms a_r and a_s such that the form a can be decomposed into their sum. We have

$$a_r(f, g) = (P_r f, g)_{L_+}, \quad a_s(f, g) = (P_s f, g)_{L_+} \tag{3.33a}$$

$$(f, g \in L = \mathfrak{D}(a) = \mathfrak{D}(a_r) = \mathfrak{D}(a_s)),$$

$$a(f, g) = a_r(f, g) + a_s(f, g) \qquad (f, g \in L). \tag{3.33b}$$

This definition implies that $a_r = 0$ and $a_s = 0$ provided that a is closable.

The form a_r is positive (with vertex 1) and closable.

Indeed, taking into account that $H_0 \supset H_+$ is the right-hand side of the chain, and Q is the imbedding operator from H_+ into H_0, we obtain

$$a_r(f, f) = (P_r f, f)_{L_+} = (P_r f, P_r f)_{L_+} = \| P_r f \|^2_{L_+} \geq \| Q P_r f \|^2_{H_0} \quad (f \in L). \tag{3.34}$$

But

$$f = Qf = Q(P_r f + P_s f) = QP_r f$$

for $f \in L$, since $P_s f \in \operatorname{Ker} Q$. Therefore, the right-hand side of (3.34) is equal to $\| f \|^2_{H_0}$, i.e., the form a_r is positive with vertex 1. This form is closable. In fact, assume that the sequence $(f_n)_{n=1}^{\infty}$ $(f_n \in L)$ is fundamental in the norm $(\hat{a}[\cdot])^{1/2}$ and converges to zero in the norm $\| \cdot \|_{H_0}$; or, in other words, that $(P_r f_n)_{n=1}^{\infty}$ is fundamental in the norm $\| \cdot \|_{H_+}$, and $QP_r f_n \underset{n\to\infty}{\to} 0$ in H_0. By virtue of the fact that $H_0 \supset H_+$ is the right-hand side of the chain, we conclude that $P_r f_n \underset{n\to\infty}{\to} 0$ in H_+, i.e., that $f_n \to 0$ in the norm $(\hat{a}[\cdot])^{1/2}$.

■

Comparing the definition of the generalized closure $\hat{a}$ of the form a with (3.33) and (3.34), we find that

$$a_r(f, g) = \hat{a}(P_r f, P_r g) \quad (f, g \in \mathfrak{D}(a_r)). \tag{3.35}$$

The form a_r is called a regular part of a, whereas the form a_s (which is obviously nonnegative) is called its singular part. If a_s differs from the zero form, then, clearly, it is nonclosable. Indeed, the inequality $a_s \neq 0$ means that $\operatorname{Ker} Q \neq \{0\}$, i.e., that a is nonclosable. On the other hand, the equality $a = a_r + a_s$, the statement just proved above, and the fact that a_s is closable, together imply that a is closable (see Subsection 3.2).

Let us illustrate the definitions introduced above by use of Example 3.2. In terms of forms, this example can be interpreted as follows: We take $H_0 = L_2(G)$,

$$a(f, g) = (f, g)_{L_+(G)} + f(x_0)\overline{g(x_0)} \quad (f, g \in \mathfrak{D}(a) = C(\tilde{G})),$$

and the form a is nonclosable. Its generalized closure $\hat{a}$ is a form in the space $L_0 =$

$H_0 \oplus \mathbb{C}^1 = L_+$ and can be written as

$$\hat{a}(\varphi, \psi) = \hat{a}((f, p), (g, q)) = (f, g)_{L_2(G)} + p\,\bar{q},$$

where $\varphi = (f, p)$ and

$$\psi = (g, q) \in L_+ = \mathfrak{D}(\hat{a}).$$

In representation (3.32), we have $A = 1: L_0 \to L_0$ ($\mathfrak{D}(A) = L_0$). We turn to the form a. The set $\mathfrak{D}(a) \cap \mathfrak{D}(A)$ consists of the vectors $\varphi = (f, 0)$, where $f \in C(\tilde{G})$ and, thus,

$$\hat{a}(\varphi, \psi) = (f, g)_{L_2(G)} \quad (f, g \in C(\tilde{G})).$$

This means that we have obtained the form without the term $f(x_0)\overline{g(x_0)}$ instead of the form a. For the latter, we have

$$a_r(f, g) = (f, g)_{L_2(G)} \qquad (f, g \in \mathfrak{D}(a_r) = C(\tilde{G})),$$

$$a_r(f, g) = f(x_0)\overline{g(x_0)} \qquad (f, g \in \mathfrak{D}(a_s) = C(\tilde{G})).$$

This example admits the following natural generalization which determines the names to the forms a_r and a_s. Assume that R is a compact set and μ and ρ are finite measures on the σ-algebra $\mathfrak{B}(R)$ of Borel sets from R such that $p(\alpha) \geq \mu(\alpha)$ ($\alpha \in \mathfrak{B}(R)$). In the space

$$H_0 = L_2(R, d\mu(x)) = L_2(R, \mathfrak{B}(R), d\mu),$$

we consider the positive bilinear form

$$a(f, g) = \int_R f(x)\,\overline{g(x)}\,d\rho(x) \quad (f, g \in \mathfrak{D}(a) = C(R))$$

with vertex 1. Then the forms a_r and a_s are given by the integrals of the same type where integration is carried over $d\rho_a(x)$ and $d\rho_s(x)$, respectively (here, ρ_a and ρ_s are, respectively, the absolutely continuous and singular components of the measure ρ with respect to μ. To prove this fact, one must verify the following two simple assertions:

(i) $L_2(R, d\rho(x)) = L_2(R, d\rho_a(x)) \oplus L_2(R, d\rho_s(x))$;

(ii) $b(f, g) = \int_R f(x)\,\overline{g(x)}\,d\rho_a(x)$ $(f, g \in C(R))$ is closable in H_0.

Let us establish several more facts concerning regular and singular parts of the form.

Suppose that B *is a nonnegative bounded operator in* H_0 *and*

$$b(f, g) = (Bf, g)_{H_0} \quad (f, g \in \mathfrak{D}(b) = H_0)$$

is a nonnegative form associated with it (clearly, this form is closable). Let a *be an arbitrary positive form with vertex* 1. *Then*

$$(a+b)_r = a_r + b. \tag{3.36}$$

In fact, the completions of $L = \mathfrak{D}(a)$ with respect to the scalar products which correspond to a and $a+b$, respectively, coincide, since

$$a[f] \le a[f] + b[f] \le a[f] + \| B \| \| f \|^2_{H_0} \le (1 + \| B \|) a[f].$$

The construction of the operator Q is determined only by the topology of the space L_+ and does not depend on the form of the scalar product in it. Therefore, by virtue of the above estimate, the operators Q are identical for a and $a+b$ and, consequently, the decompositions $L_+ = H_+ \oplus \operatorname{Ker} Q$ are the same in both cases. This yields (3.36).

■

Theorem 3.5. *Let* a *and* b *be two arbitrary forms with vertex* 1. *We write* $a \le b$ *if* $\mathfrak{D}(a) \supset \mathfrak{D}(b)$ *and*

$$a(f,f) \le b(f,f) \quad (f \in \mathfrak{D}(b)).$$

It is stated that if $a \le b$, *then* $a_r \le b_r$.

Proof. First suppose that a is closable. Then $a_r = a$, and the situation is as follows. Without loss of generality, we can assume that $\mathfrak{D}(a) = \mathfrak{D}(b)$, because otherwise one should replace a by its restriction a' to $\mathfrak{D}(b)$; the form a' is also closable, and the inequality $a' \le b_r$. obviously involves the required inequality $a \le b_r$. Therefore, we now consider the space H_0 and the dense linear set $L = \mathfrak{D}(a) = \mathfrak{D}(b)$ in it; H_1 is the completion of L with respect to the scalar product

$$(f, g)_{H_1} = a(f, g) \quad (f, g \in L).$$

According to the assumption, $H_1 \subset H_0$ and $\| f \|_{H_0} \le \| f \|_{H_1}$ $(f \in H_1)$.

Consider two pre-chains $H_0 \supset L$ and $H_1 \supset L$ such that

$$(f, g)_{L_+} = b(f, g) \quad (f, g \in L)$$

in both cases. Let $Q: L_+ \to H_0$ and $Q': L_+ \to H_1$ be the corresponding operators and let $L_+ = H_+ \oplus \operatorname{Ker} Q$ and $L_+ = H'_+ \oplus \operatorname{Ker} Q'$ be the decompositions associated with these. It is easy to see that

$$\operatorname{Ker} Q = \operatorname{Ker} Q'. \tag{3.37}$$

In fact, Ker Q consists of the classes of equivalent fundamental sequences $(f_n)_{n=1}^{\infty}$ ($f_n \in L$) such that $\| f_n - f_m \|_{L_+} \to 0$ as $n, m \to \infty$, and $\| f \|_{H_0} \to 0$ as $n \to \infty$, whereas Ker Q' consists of the analogous classes with the only difference that the last condition should be replaced by $\| f \|_{H_1} \to 0$ as $n \to \infty$. Clearly, Ker $Q' \subset$ Ker Q. On the other hand, consider a class from Ker Q. Let $(f_n)_{n=1}^{\infty}$ ($f_n \in L$) be a representative of this class. Then $\| f_n - f_m \|_{H_1} \to 0$ as $n, m \to \infty$, and $\| f_n \|_{H_0} \to 0$ as $n \to \infty$. Therefore, $\| f_n \|_{H_1} \to 0$ as $n \to \infty$ by virtue of the fact that a is closable. Hence, this class from Ker Q belongs to Ker Q', but this means that (3.37) is valid.

It follows from (3.37) that the projector P_r onto $L_+ \ominus \operatorname{Ker} Q$ (from the definition of b_r) coincides with the projector P'_r associated with the pre-chain $H_1 \supset L$. Consequently, taking (3.33) and (3.34) into account, we obtain

$$b_r(f, f) = \| P_r f \|^2_{L_+} = \| P'_r f \|_{L_+} = \| P'_r f \|_{H'_+}$$

$$\geq \| Q' P'_r f \|^2_{H_1} = \| f \|^2_{H_1} = a(f, f) \quad (f \in L),$$

i.e., $b_r \geq a$.

Now assume that the form a is arbitrary. Then $a_r \leq a \leq b$ and, therefore, since a_r is closable, $a_r \leq b_r$, by virtue of what has been proved.

■

Remark 1. The form a_r can be singled out by the following maximality property: a_r is the maximal closable form which does not exceed a, i.e., in other words, if the form $b \leq a$ is positive with vertex 1, closable, and such that $b \geq a_r$, then $b = a_r$. Indeed, by virtue of the inequality $b \leq a$, Theorem 3.5 implies that $b = b_r \leq a_r$.

■

Remark 2. Generally speaking, the inequality $a \leq b$ does not imply the inequality $a_s \leq b_s$. In fact, let $G \subset \mathbb{R}^1$ be a bounded region, $H_0 = L_2(G, dx)$, $\mathfrak{D}(a) = \mathfrak{D}(b) = C(\tilde{G})$, and let

$$a(f, g) = \varepsilon f(x_0) \overline{g(x_0)} \quad \text{and} \quad b(f, g) = (f, g)_{W_2^1(G)}$$

(here, $x_0 \in \tilde{G}$ is fixed, and $f, g \in C(\tilde{G})$). Then, by virtue of the imbedding theorems, for sufficiently small $\varepsilon > 0$ we have that $a \le b$, and $a_s = a$, $b_s = 0$ at the same time.

■

Remark 3. The constructions presented in this Subsection can be easily generalized for the case of semi-bounded forms. According to (3.17) and (3.18), one can set

$$a_r(f, g) = (a_p)_r(f, g) = (1 - \alpha)(f, g)_{H_0}$$

and

$$a_s(f, g) = (a_p)_s(f, g) \quad (f, g \in \mathfrak{D}(a)).$$

Clearly, a_r is closable, a_s is nonclosable, and $a = a_r + a_s$.

■

2. GENERALIZED FUNCTIONS OF INFINITELY MANY VARIABLES. GAUSSIAN MEASURES

In this chapter, we construct test and generalized functions of infinitely many variables and study the properties of Gaussian measures that are necessary to construct these functions and are necessary in our further discussions. If we understand a function of infinitely many variables as a function of a point $x = (x_1, x_2, \ldots) \in \mathbb{R}^1 \times \mathbb{R}^1 \times \ldots = \mathbb{R}^\infty$, then it is reasonable to start constructing the theory of generalized functions of this sort by replacing $\mathbb{R}^d$ by $\mathbb{R}^\infty$ in the classical Sobolev–Schwartz chain $\mathcal{D}'(\mathbb{R}^d) \supset L_2(\mathbb{R}^d) \supset \mathcal{D}(\mathbb{R}^d)$ and finding a proper analogue for $\mathcal{D}(\mathbb{R}^d)$. However, the standard measure of the Lebesgue type is absent in $\mathbb{R}^\infty$ and, thus, the choice of a space $L_2(\mathbb{R}^\infty)$ is ambiguous. Therefore, there is no natural way of identifying an ordinary function with a generalized one, and this yields difficulties.

Different ways exist to overcome these difficulties. The construction presented below has the following peculiarities distinguishing it from other approaches: This construction is substantially based on employing the classical finite-dimensional analysis, and there is no need, for example, to establish the theory of infinitely differentiable measures; our spaces of test functions are nuclear; these spaces are convenient for some applications (e.g., for the spectral theory of operators and for quantum field theory).

In the simplest situation, this construction is as follows. Consider a sequence of Hilbert riggings of the "spaces of functions of one variable"

$$H_{-,k} \supset H_{0,k} \supset H_{+,k} \quad (k = 1, 2, \ldots). \tag{0.1}$$

As a Hilbert chain for "infinitely many variables", we take the infinite tensor product of these riggings with a stabilization e

$$\bigotimes_{k=1;e}^{\infty} H_{-,k} \supset \bigotimes_{k=1;e}^{\infty} H_{0,k} \supset \bigotimes_{k=1;e}^{\infty} H_{+,k} \tag{0.2}$$

(we assume that (0.1) and e are related properly so that the products in (0.2) are meaningful). Under certain conditions, the quasinuclearity of the chains (0.1) implies the quasinuclearity of (0.2). By renormalizing the spaces or (this is the same) by taking a weighted tensor product, one can always ensure that these conditions are satisfied.

To obtain a nuclear chain instead of (0.2), we realize the following modification. Instead of (0.1), we consider a sequence of nuclear riggings

$$\Phi_k' \supset H_{0,k} \supset \Phi_k \ (k = 1, 2, \dots), \tag{0.3}$$

where each Φ_k is a projective limit of Hilbert spaces, i.e., $\Phi_k = \mathrm{pr}\ \lim_{\tau_k \in T_k} H_{+,\tau_k}$. Then, in according to (0.1) and (0.2), we construct a tensor product

$$\bigotimes_{k=1;e}^{\infty} H_{-,\tau_k} \supset \bigotimes_{k=1;e}^{\infty} H_{0,k} \supset \bigotimes_{k=1;e}^{\infty} H_{+,\tau_k}$$

by using the chains $H_{-,\tau_k} \supset H_{0,k} \supset H_{+,\tau_k}$ (the sequence $\tau = (\tau_k)_{k=1}^{\infty}$, $\tau_k \in T_k$, is fixed). Finally, we construct the required chain

$$\Phi' \supset \bigotimes_{k=1;e}^{\infty} H_{0,k} \supset \Phi, \tag{0.4}$$

where $\Phi = \mathrm{pr}\lim_{\tau} \otimes_{k=1;e}^{\infty} H_{+,\tau_k}$ (the projective limit with respect to all possible $\tau \in \times_{k=1}^{\infty} T_k$ or with respect to a certain part of these).

This construction is presented in Section 4. By employing it, we construct important examples of spaces of the test functions of infinitely many variables. One of these, namely the space $\mathcal{A}(\mathbb{R}^{\infty})$, plays a role quite similar to the role of the space $\mathcal{S}(\mathbb{R}^d)$ in the classical theory.

The space $\mathcal{A}(\mathbb{R}^{\infty})$ and its generalizations are studied in detail in Section 5. The elements of $\mathcal{A}(\mathbb{R}^{\infty})$ (these are the complex-valued functions $\varphi(x)$ of the point $x \in \mathbb{R}^{\infty}$) are determined by their values on a certain Hilbert space from $\mathbb{R}^{\infty}$. This is why, it is reasonable to work without choosing any basis in this space, i.e., in the invariant form.

The relation between these problems and the measure theory becomes clear from the above presentation: As usual, the spaces $H_{0,k}$ from the chains (0.1) or (0.3) are the spaces of the functions on the line $\mathbb{R}^1$ that are L_2 (square summable) with respect to a certain probability measure. Then the zero space in the chains (0.2) or (0.4) is the space of the functions on $\mathbb{R}^{\infty}$ that are L_2 with respect to the infinite product of these measures. Since we must study such products, it is convenient not to take the space $\mathcal{S}(\mathbb{R}^1)$ as $H_{+,k}$, but to use its isomorphic image (obtained by a proper change of variable) which rigs L_2 with respect to the Lebesgue measure.

For our purposes, we always assume that probability measures are Gaussian. This is why, in Sections 1 and 2, we give a detailed exposition of the theory of Gaussian measures and present some facts from analysis necessary for studying these measures. Unfortunately, we failed to find the complete presentation of the required facts in the literature, and therefore, we found it reasonable to make our presentation of this theory self-contained.

In the case of Gaussian measures, the role of the zero space $\otimes_{k=1;e}^{\infty} H_{0,k}$ in the rigging (0.4) is played by the space of functions on $\mathbb{R}^{\infty}$ that are L_2 with respect to a Gaussian product measure. However, it is often necessary to construct the rigging of L_2 functions on a linear space with respect to a Gaussian measure that is not a product measure. It is sometimes inconvenient and unnatural to diagonalize this measure reducing the problem to product measures (this is another reason for the invariant form of construction). Since the presentation of these constructions is necessary, we also give the required facts on the differentiability of functions given on Hilbert linear topologic spaces (Section 3).

In this chapter, an important role is played by three objects, namely, the Fock space, the Segal isomorphism, and the Wiener–Itô decomposition. They are related with the space of functions on the space dual to the nuclear space that are square summable with respect to the Gaussian measure γ. It is useful to clarify that if one passes from the complicated space L_2 to the space $L_2(\mathbb{R}^1, d\gamma(x))$ of the functions on the $\mathbb{R}^1$ axis, then these objects become very simple: the role of the Fock space is played by the usual space l_2 of sequences of the form $(f_n)_{n=0}^{\infty}$; the Segal isomorphism turns into the reconstruction of the function $f \in L_2(\mathbb{R}^1, d\gamma(x))$ from the sequence $(f_n)_{n=0}^{\infty}$ of its Fourier coefficients for the expansion of $L_2(\mathbb{R}^1, d\gamma(x))$ in the Hermitian polynomials; and the Wiener–Itô decomposition turns into the latter expansion itself.

§1. Measures on Linear Spaces

In this section, we present the main facts from the measure theory on infinite-dimensional spaces that are necessary for further discussion. This approach to the contents of the section stipulates the style of our presentation. We briefly consider some important theorems of general theory and not in the most general cases. At the same time, we pay much more attention to some more specific questions. The case of Gaussian measures is analyzed with maximal completeness. Here, the exposition is selfcontained, and we present a collection of statements (sometimes of a technical character) which are widely used in different parts of our book.

1.1. Cylindrical Sets and Product Measures

First, we give some information about product measures, i.e., about measures that are constructed as a product of an arbitrary collection of probability measures.

Let X be a set of indices x of arbitrary cardinality, and let $(R_x)_{x\in X}$ be a family of abstract spaces with σ-algebras $\mathfrak{R}_x$ of subsets defined in each space. Assume that the measures on the spaces R_x are given. We now want to construct a measure on the direct product $R_X = \times_{x\in X} R_x$ of these spaces which, according to definition, consists of all possible mappings of the form $X \ni x \mapsto \lambda(x) \in R_x$. For this purpose, let us introduce a suitable σ-algebra in R_X. For arbitrary *different* points $x_1, \dots, x_n \in X$, let us denote $R_{x_1,\dots,x_n} = R_{x_1} \times \dots \times R_{x_n}$ $n \in N$). In $R_{x_1,\dots,x_n}$, we consider the σ-algebra $\mathfrak{R}_{x_1,\dots,x_n} = \mathfrak{R}_{x_1} \times \dots \times \mathfrak{R}_{x_n}$ of its subsets[5]. The set $\mathfrak{C} \in R_x$ is called cylindrical if it is determined by these points $x_1, \dots, x_n$ (its coordinates) and the base $\delta \in R_{x_1,\dots,x_n}$ according to the relation

$$\mathfrak{C} = \mathfrak{C}(x_1, \dots, x_n; \delta) = \{\lambda(\cdot) \in R_X \mid (\lambda(x_1), \dots, \lambda(x_n)) \in \delta\}. \tag{1.1}$$

Note that if δ is a rectangle, i.e., $\delta = \alpha_{x_1} \times \dots \times \alpha_{x_n}$, then $\mathfrak{C}$ consists of all the mappings passing through the "gates" α_{x_k} over the point x_k, where $k = 1, \dots, n$ (in the simplest situations $\mathfrak{C}$ consists of functions).

Being written in the form of (1.1), the given cylindrical set $\mathfrak{C}$ does not determine the sequence of points $x_1, \dots, x_n$ and the base δ uniquely. Thus, one can change the order of these points (e.g.,

$$\mathfrak{C}(x_1, x_2; \delta_1 \times \delta_2) = \mathfrak{C}(x_2, x_1; \delta_2 \times \delta_1) \quad (\delta_1 \in \mathfrak{R}_{x_1}; \delta_2 \in \mathfrak{R}_{x_2}))$$

or insert an extra one $y \in X$ multiplying the base by R_y: $\mathfrak{C}(x_1, \dots, x_n; \delta) = \mathfrak{C}(x_1, \dots, x_n, y; \delta \times R_y)$, or exclude y and the corresponding factor R_y if this is possible.

The collection of all cylindrical sets forms an algebra $\mathfrak{C}(R_X)$. In fact, to prove the inclusions $\mathfrak{C}' \cup \mathfrak{C}'' \in \mathfrak{C}(R_X)$ and $\mathfrak{C}' \setminus \mathfrak{C}'' \in \mathfrak{C}(R_X)$ if $\mathfrak{C}', \mathfrak{C}'' \in \mathfrak{C}(R_X)$, it is sufficient to pass to the common coordinates $x_1, \dots, x_n$ in the relation (1.1) for $\mathfrak{C}'$ and $\mathfrak{C}''$ adding extra points where necessary (by using the above-mentioned procedure). Then our statement follows from the formula

$$\mathfrak{C}' \cup \mathfrak{C}'' = \mathfrak{C}(x_1, \dots, x_n; \delta') \cup \mathfrak{C}(x_1, \dots, x_n; \delta'') = \mathfrak{C}(x_1, \dots, x_n; \delta' \cup \delta'')$$

and from a similar formula for the difference.

■

Denote by $\mathfrak{C}_\sigma(R_X)$ a σ-span of the algebra $\mathfrak{C}(R_X)$. This is the required σ-algebra.

[5] Recall that this σ-algebra consists of the σ-span of all possible rectangles $\alpha_{x_1} \times \dots \times \alpha_{x_n}$, $\alpha_{x_k} \in \mathfrak{R}_{x_k}$ $(k = 1, \dots, n)$.

We shall study the structure of its sets in detail in Subsection 1.1 of Chapter 3. Here, we note only that if X is more than countable, then the supply of sets in $\mathfrak{C}_\delta(R_X)$ is small. For example, the single point set does not belong to $\mathfrak{C}_\delta(R_X)$ (if R_X are nontrivial).

Assume that, for any $y \in X$, some probability measure μ_x is given on the σ-algebra $\mathfrak{R}_x$, i.e., $\mu_x(R_x) = 1$. There exists the standard Kolmogorov procedure which enables us to construct a measure μ_X on the σ-algebra $\mathfrak{C}_\delta(R_X)$ from the family of measures $(\mu_x)_{x\in X}$; the measure μ_X is called a product of measures μ_x and is denoted by $\mu_X = \times_{x\in X}\mu_x$. Let us employ this procedure. Denote by $\mu_{x_1,\ldots,x_n}$ $(n \in N)$ the measure on $\mathfrak{R}_{x_1,\ldots,x_n}$ obtained by the usual procedure of multiplying out a finite number of measures $\mu_{x_1,\ldots,x_n} = \times_{k=1}^n \mu_{x_k}$[6]. On cylindrical sets $\mathfrak{C} \in \mathfrak{C}(R_X)$, the measure μ_X is defined by

$$\mu_X(\mathfrak{C}) = \mu_X(\mathfrak{C}(x_1, \ldots, x_n; \delta)) = \mu_{x_1,\ldots,x_n}(\delta). \tag{1.2}$$

This definition is correct: We put the number $\mu_X(\mathfrak{C}) \in [0, 1]$ in correspondence with each cylindrical set $\mathfrak{C}$. Indeed, any change of the form of $\mathfrak{C}$ is connected either with a permutation of coordinates, or with multiplying δ by R_y (directly), or with removing this factor from δ. Since $\mu_y(R_y) = 1$, it follows from the construction of $\mu_{x_1,\ldots,x_n}(\delta)$ that $\mu(\mathfrak{C})$ is independent of the form of $\mathfrak{C}$.

The function of sets (1.2) satisfies the equalities $\mu_X(\varnothing) = 0$ and $\mu_X(R_X) = 1$ and is finitely additive. To prove this, one must write the considered mutually disjoint sets $\mathfrak{C}_1, \ldots, \mathfrak{C}_m \in \mathfrak{C}(R_X)$ using the same set of coordinates $x_1, \ldots, x_n$ and then taking the additivity of the measure $\mu_{x_1,\ldots,x_n}$ into account. According to the classical theory of extension of measures (e.g., see Halmos [1, Chapter 3]), the finitely additive measure μ_X can be uniquely extended to the measure on the σ-span $\mathfrak{C}_\delta(R_X)$ of the algebra $\mathfrak{C}(R_X)$, provided that it satisfies the condition of absolute additivity on $\mathfrak{C}(R_X)$.

Theorem 1.1. *The finitely additive measure (1.2) on the algebra $\mathfrak{C}(R_X)$ possesses the property of absolute additivity and, hence, can be extended to the probability product measure μ_X on $\mathfrak{C}_\delta(R_X)$, i.e., to the product $\times_{x\in X}\mu_x$ of the measures μ_x.*

Proof. Absolute additivity is equivalent to the upper continuity of the measure μ_X on the empty set. Therefore, it is necessary to prove that if $\mathfrak{C}_1, \mathfrak{C}_2, \ldots \in \mathfrak{C}(R_X)$ are such that $\mathfrak{C}_1 \supset \mathfrak{C}_2 \supset \ldots$ and $\bigcap_{j=1}^\infty \mathfrak{C}_j$, then $\mu_X(\mathfrak{C}_j) \underset{j\to\infty}{\to} 0$. By virtue of finite additivity, μ_X is monotonic. Hence, $\mu_X(\mathfrak{C}_1) \geq \mu_X(\mathfrak{C}_2) \geq \ldots$ and, consequently, there exists $\lim_{j\to\infty}\mu_X(\mathfrak{C}_j) = c \geq 0$. We now prove that $c = 0$. Assume the opposite: Let $c > 0$.

[6] Recall that $\mu_{x_1,\ldots,x_n}(\alpha_{x_1} \times \ldots \times \alpha_{x_n}) = \mu_{x_1}(\alpha_{x_1}) \times \ldots \times \mu_{x_n}(\alpha_{x_n})$, where $\alpha_{x_k} \in \mathfrak{R}_{x_k}$. Then this function of sets is extended to the measure on $\mathfrak{R}_{x_1,\ldots,x_n}$.

Consider a sequence $(x_k)_{k=1}^{\infty}$ of different points $x_k \in X$ composed of the coordinates of all the sets $\mathfrak{C}_j$ (the case of a finite sequence is trivial). Denote $X_1 = X \setminus \{x_1\}$, and construct R_{X_1}, $\mathfrak{S}(R_{X_1})$ and μ_{X_1} as above. Clearly, $R_X = R_{x_1} \times R_{X_1}$, $\mathfrak{S}(R_X) = \mathfrak{S}(R_{x_1}) \times \mathfrak{S}(R_{X_1})$, and $\mu_X = \mu_{x_1} \times \mu_{X_1}$, where the last product is to be understood as a product of finitely additive measures. Let $r_1 \in R_{x_1}$ and $\mathfrak{C} \in \mathfrak{S}(R_X)$. The set $\mathfrak{C}[r_1]$ of all mappings $\{\lambda(\cdot) \in \mathfrak{C} \mid \lambda(x_1) = r_1\}$ (considered only on X_1) is called a cross-section of the set $\mathfrak{C}$ by the point r_1. It can be easily shown [7] that $\mathfrak{C}[r_1] \in \mathfrak{S}(R_X)$, $\mu_{X_1}([r_1])$ is measurable with respect to $\mathfrak{R}_{x_1}$ and that

$$\mu_X(\mathfrak{C}) = \int_{R_{X_1}} \mu_{X_1}(\mathfrak{C}[r_1])\, d\mu_{x_1}(r_1). \tag{1.3}$$

Let us write (1.3) for $\mathfrak{C} = \mathfrak{C}_j$. We have

$$\mu_X(\mathfrak{C}_j) = \int_{R_{X_1}} \mu_{X_1}(\mathfrak{C}_j[r_1])\, d\mu_{x_1}(r_1) \qquad (j \in \mathbb{N}). \tag{1.4}$$

For the cross-sections, we have $\mathfrak{C}[r_1] \supset \mathfrak{C}[r_2] \supset \ldots$ for every $r_1 \in R_{x_1}$, therefore, $\mu_{X_1}(\mathfrak{C}[r_1]) \geq \mu_{X_1}(\mathfrak{C}[r_2]) \geq \ldots$ and, consequently, there exists

$$\lim_{j\to\infty} \mu_{X_1}(\mathfrak{C}_j[r_1]) = c_1(r_1) \geq 0.$$

The inequality $c(r_1) > 0$ holds at least for one $r_1^0 \in R_{x_1}$. To prove this, we must take the limit as $j \to \infty$ under the integral sign in (1.4). Then

$$\int_{R_{X_1}} c_1(r_1)\, d\mu_{x_1}(r_1) = c > 0,$$

and this yields the required result.

Let us replace the sequence $\mu_X(\mathfrak{C}_1) \geq \mu_X(\mathfrak{C}_2) \geq \ldots$, $\lim_{j\to\infty} \mu_X(\mathfrak{C}_j) = c > 0$, by the sequence $\mu_{X_1}(\mathfrak{C}_1[r_1^0]) \geq \mu_{X_1}(\mathfrak{C}_2[r_1^0]) \geq \ldots$, and X and $X \setminus \{x_1\}$ by X_1 and $X_1 \setminus \{x_2\} = X \setminus \{x_1, x_2\}$, respectively, and then repeat our arguments. If we continue this process, we construct the sequence of points $(r_m^0)_{m=1}^{\infty}$ ($r_m^0 \in R_{x_m}$) such that

[7] One should consider the cases where x_1 is one of the coordinates of the set $\mathfrak{C}$ and where this is not true, and employ the well-known formula for cross-section when multiplying two measures (e.g., see Halmos [1, Chapter 7, Section 35]).

$$\mu_{X_m}(\mathfrak{C}_j[r_1^0, \ldots, r_m^0]) > 0 \quad (j \in \mathbb{N}), \tag{1.5}$$

for any $m \in \mathbb{N}$. Here, $\mathfrak{C}_j[r_1^0, \ldots, r_m^0]$ denotes the cross-section of the set $\mathfrak{C}_j$ by the points $r_1^0, \ldots, r_m^0$, i.e., the set of all the considered mappings $\lambda(\cdot) \in \mathfrak{C}_j$ on $X \setminus \{x_1, \ldots, x_m\}$ for which $\lambda(x_1) = r_1^0, \ldots, \lambda(x_m) = r_m^0$. Of course, the additive measure μ_{X_m} is constructed just as μ_X but with $X \setminus \{x_1, \ldots, x_m\}$ substituted for X.

Consider the mapping $\lambda^0(\cdot) \in R_X$ such that $\lambda^0(x_1) = r_1^0$, $\lambda^0(x_2) = r_0^2, \ldots$ and taking certain values in the other points of X. We state that $\lambda^0(\cdot)$ belongs to each $\mathfrak{C}_j$, $j \in \mathbb{N}$. In fact, $\mathfrak{C}_j = \mathfrak{C}(x_{k_1}, \ldots, x_{k_n}; \delta)$, where the coordinates and the base clearly depend on j. Assume that m is sufficiently large so that $\{x_{k_1}, \ldots, x_{k_n}\} \subset \{x_1, \ldots, x_m\}$. According to (1.5), we have $\mu_{X_m}(\mathfrak{C}_j[r_1^0, \ldots, r_m^0]) > 0$ and, therefore, the cross-section $\mathfrak{C}_j[r_1^0, \ldots, r_m^0]$ of the set $\mathfrak{C}$ is nonempty, i.e., in $\mathfrak{C}_j$ there exists a mapping $\lambda(\cdot)$ such that $\lambda(x_1) = r_1^0, \ldots, \lambda(x_m) = r_m^0$ and, in particular, $\lambda(x_{k_1}) = r_{k_1}^0, \ldots, \lambda(x_{k_n}) = r_{k_n}^0$. The set $\mathfrak{C}_j$ is defined only by the values of the mappings at the points $x_{k_1}, \ldots, x_{k_n}$, and their values at the other points of X are arbitrary. Hence, $\lambda^0(\cdot)$ belongs to $\mathfrak{C}_j$ along with $\lambda(\cdot)$. Consequently, $\lambda^0(\cdot) \in \bigcap_{j=1}^{\infty} \mathfrak{C}_j$, but this contradicts the condition $\bigcap_{j=1}^{\infty} \mathfrak{C}_j = \varnothing$.

■

The products of more than countably many measures will appear only in Chapter 3. In this Chapter, we encounter the simpler situation where X is countable. Unless stated otherwise, we shall always understand the infinite product of measures as a product of, at most, a countable number of measures. So, let $X = \mathbb{N}$ and let the sequence of spaces $(R_k)_{k=1}^{\infty}$, σ-algebras $(\mathfrak{R}_k)_{k=1}^{\infty}$, and probability measures $(\mu_k)_{k=1}^{\infty}$ be given. In accordance with the above, the product $R_X = \times_{k=1}^{\infty} R_k$, $\mathfrak{C}(R_X)$ and the probability measure $\mathfrak{C}_\delta(R_X) \ni \alpha \mapsto \mu_X(\alpha) = (\times_{k=1}^{\infty} \mu_k)(\alpha)$ are defined. Henceforth, we shall also use the notations $R_\infty = R_X$ and $\mu_\infty = \mu_X$. In particular, if $R_1 = R_2 = \ldots = R$, then we write $R^\infty = R \times R \times \ldots = R_\infty$ (hence, $\mathbb{R}^\infty$ and $\mathbb{C}^\infty$ have the same sense as before).

The infinite product $\mu_\infty = \times_{k=1}^{\infty} \mu_k$ is calculated most simply on the cylindrical set $\mathfrak{C}_n$ whose base is the rectangle $\alpha_1 \times \ldots \times \alpha_n$ $(\alpha_k \in \mathfrak{R}_k)$. According to (1.1), one can write

$$\mathfrak{C}_n = \alpha_1 \times \alpha_2 \times \ldots \times \alpha_n \times R_{n+1} \times R_{n+2} \times \ldots,$$

and, by virtue of (1.2), we have

$$\mu_\infty(\mathfrak{C}_n) = \mu_1(\alpha_1) \ldots \mu_n(\alpha_n).$$

The "infinite-dimensional" rectangles $\times_{k=1}^{\infty} \alpha_k$ ($\alpha_k \in \mathfrak{R}_k$) also belong to our σ-algebra $\mathfrak{C}_\delta(R_\infty)$. They coincide with $\bigcap_{n=1}^{\infty} \mathbf{C}_n$. By passing to the limit, we find that $\mu_\infty(\times_{k=1}^{\infty}\alpha_k) = \prod_{k=1}^{\infty}\mu_k(\alpha_k)$; the infinite product always converges (possibly to zero) because $\mu_k(\alpha_k) \in [0, 1]$.

We now somewhat change the character of our notations. The "spectral" (due to λ) notations $((R_x)_{x\in X}, R_X, \lambda(\cdot) \in R_X$ and others) which were used above, are reserved for the case where a product of more than countably many measures is considered. For the case of countably many measures, we now introduce a more natural system of notations: R_k and R_∞ are replaced by $X_k (k \in \mathbb{N})$ and X_∞, respectively, where $x = (x_k)_{k=1}^{\infty} \in X_\infty$, $x_k \in X_k$. According to these changes, $\mathfrak{C}(X_\infty)$ and $\mathfrak{C}_\delta(X_\infty)$ denote the previous algebra of cylindrical sets and the σ-algebra spanned by this algebra. We also denote

$$X_{(n)} = X_1 \times \ldots \times X_n, \quad \mathfrak{R}_{(n)} = \mathfrak{R}_1 \times \ldots \times \mathfrak{R}_n, \quad X^{(n+1)} = X_{n+1} \times X_{n+2} \times \ldots,$$

$$X_\infty = X_{(n)} \times X^{(n+1)} \ni x = ((x_1, \ldots, x_n), x^{(n+1)}) = (x_{(n)}, x^{(n+1)}) \quad (n \in \mathbb{N}).$$

1.2. Functions Summable with Respect to Infinite Products of Measures

Let us pass to the investigation of the spaces of functions on $X_\infty = X_1 \times X_2 \times \ldots$ summable with respect to the measure μ_∞ given on the σ-algebra $\mathfrak{C}_\sigma(X_\infty)$. Along with this σ-algebra, we consider a σ-algebra $\mathfrak{C}_{\sigma,n}(X_\infty)$ $(n \in \mathbb{N})$ consisting of all cylindrical sets of the form $\alpha \times X_{n+1} \times X_{n+2} \times \ldots$, where $\alpha \in \mathfrak{R}_{(n)} = \mathfrak{R}_1 \times \mathfrak{R}_2 \times \ldots \times \mathfrak{R}_n$. Clearly,

$$\mathfrak{C}_{\sigma,1}(X_\infty) \subset \mathfrak{C}_{\sigma,2}(X_\infty) \subset \ldots, \tag{1.6}$$

i.e., we have a "flow" of expanding σ-algebras; their union coincides with $\mathfrak{C}(X_\infty)$ and is only an algebra. Each $\mathfrak{C}_{\sigma,n}(X_\infty)$ is isomorphic to the σ-algebra $\mathfrak{R}_{(n)}$ in the natural sense.

Denote by $L_0(X_\infty, \mathfrak{C}_\sigma(X_\infty))$ a set of complex-valued functions f on X_∞ which are measurable with respect to $\mathfrak{C}_\sigma(X_\infty)$. A function f of this sort is called cylindrical if it is measurable with respect to the σ-algebra $\mathfrak{C}_{\sigma,n}(X_\infty)$ for some $n \in \mathbb{N}$, i.e., $f \in L_0(X_\infty, \mathfrak{C}_{\sigma,n}(X_\infty))$. In other words, in this case $f(x)$ depends only on a finite number n of coordinates of the point x, i.e., $f(x) = u_f(x_1, \ldots, x_n) = u_f(x_{(n)})$, where u_f is a function on $X_{(n)}$ measurable with respect to $\mathfrak{R}_{(n)}$.

The space $L_p = L_p(X_\infty, \mathfrak{C}_\sigma(X_\infty), \mu_\infty)$ $(p \in [1, \infty))$ is defined as usual; $L_p(X_\infty, \mathfrak{C}_{\sigma,n}(X_\infty), \mu_\infty)$ is considered to be its subspace consisting of cylindrical functions with fixed n that are summable with respect to μ_∞ with the p-th power. On the other hand, $L_p = L_p(X_\infty, \mathfrak{C}_{\sigma,n}(X_\infty), \mu_\infty)$ can be identified with the space $L_p(X_{(n)}, \mathfrak{R}_{(n)}, \times_{k=1}^n \mu_k)$.

For every $n \in \mathbb{N}$, the measure μ_∞ can be represented in the form $\mu_\infty = (\times_{k=1}^n \mu_k) \times \mu^{(n+1)}$ according to the expansion $X_\infty = X_{(n)} \times X^{(n+1)}$. By virtue of the Fubini theorem, the function

$$f_n(x_1, \dots, x_n) = f_n(x_{(n)}) = \int_{X^{(n+1)}} f(x)\, d\mu^{(n+1)}(x^{(n+1)}) \tag{1.7}$$

belongs to $L_1(X_{(n)}, \mathfrak{R}_{(n)}, \times_{k=1}^n \mu_k) = L_1(X_\infty, \mathfrak{C}_{\sigma,n}(X_\infty), \mu_\infty)$ for $f \in L_1$ and, moreover,

$$\int_{X_\infty} f(x)\, d\mu_\infty(x) = \int_{X_{(n)}} f_n(x_{(n)})\, d\mu_1(x) \dots d\mu_n(x_n).$$

We introduce a linear operator E_n in L_1 by setting $E_n f = f_n$, where f_n is defined by (1.7) $(n \in \mathbb{N})$. The properties of the operator f_n are described by the following theorem.

Theorem 1.2. 1. *Let* $f \in L_p$ $(p \in [1, \infty))$, *then* $E_n f \underset{n\to\infty}{\to} f$ *in* L_p.

2. *The restriction* $E_n \upharpoonright L_p$ *is a projector of* L_p *onto its subspace* $L_p(X_\infty, \mathfrak{C}_{\sigma,n}(X_\infty), \mu_\infty)$. *In particular, this projector is orthogonal for* $p = 2$.

Proof. For $f \geq 0$, we have $E_n f = f_n \geq 0$ and, hence, $\|f\|_{L_1} = \|f_n\|_{L_1}$. For arbitrary $f = f^+ - f^-$ $(f^\pm \geq 0)$, the equality $f_n = (f^+)_n - (f^-)_n$ holds, and thus $|f_n| \leq (f^+)_n - (f^-)_n$ and

$$\|E_n f\|_{L_1} \leq \int_{X_\infty} (f^+)_n\, d\mu_\infty + \int_{X_\infty} (f^-)_n\, d\mu_\infty = \int_{X_\infty} f^+\, d\mu_\infty + \int_{X_\infty} f^-\, d\mu_\infty = \|f\|_{L_1}.$$

Consequently,

$$\|E_n f\|_{L_1} \leq \|f\|_{L_1} \quad (f \in L_1,\ n \in \mathbb{N}). \tag{1.8}$$

Cylindrical functions are dense in the space L_1 (the same is true for all L_p). This follows, e.g., from the fact that an arbitrary set from $\mathfrak{C}_\sigma(X_\infty)$ can be approximated by the same cylindrical set (in the sense of the measure μ_∞). Therefore, its characteristic function can be approximated (in the metric of L_1) by the characteristic function of the cylindrical set which is a cylindrical function. But finite linear combinations of characteristic functions are dense in all L_p, and this completes the proof of the required result. By using this fact, for given $\varepsilon > 0$, we choose a cylindrical function $u \in L_1(X_\infty, \mathfrak{C}_{\sigma,n}(X_\infty), \mu_\infty)$ such that $\|f - u\|_{L_1} < \varepsilon$. Then by virtue of the equality $E_n u = u$ (see (1.7)) and (1.8), we have for this n

$$\| E_n f - f\|_{L_1} \leq \| E_n (f - u)\|_{L_1} + \| u - f\|_{L_1} \leq 2 \|f - u\|_{L_1} < 2\varepsilon.$$

This proves the first statement of the theorem for $p = 1$.

Now let $f \in L_p$, $p > 1$. Since the functions t^p $(t \in [0, \infty))$ are convex, the Jensen inequality yields

$$E_n |f|^p = \int\limits_{X^{(n+1)}} |f|^p \, d\mu^{(n+1)} \geq \Big(\int\limits_{X^{(n+1)}} |f| \, d\mu^{(n+1)} \Big)^p = (E_n |f|^p).$$

Therefore,

$$\| E_n f\|_{L_p}^p \leq \int\limits_{X_\infty} | E_n f|^p \, d\mu_\infty \leq \int\limits_{X_\infty} (E_n f)^p \, d\mu_\infty$$

$$\leq \int\limits_{X_\infty} E_n |f|^p \, d\mu_\infty = \int\limits_{X_\infty} |f|^p \, d\mu_\infty = \|f\|_{L_p}^p. \qquad (1.9)$$

This means that the sequence $(| E_n f|^p)_{n=1}^\infty$ is uniformly integrable. Taking into account that $E_n f \underset{n\to\infty}{\to} f$ in the measure μ_∞ (since $E_n f \underset{n\to\infty}{\to} f$ in L_1), we get $E_n f \underset{n\to\infty}{\to} f$ in L_p.

2. The statement that E_n is a projector of L_p onto $L_p(X_\infty, \mathfrak{C}_{\sigma,n}(X_\infty), \mu_\infty)$ follows from the inclusion

$$E_n f = f_n \in L_1(X_\infty, \mathfrak{C}_{\sigma,n}(X_\infty), \mu_\infty),$$

the estimate (1.9), and the evident idempotency of E_n. The equality $\|E_n\| = 1$ is a consequence of (1.9) and the equality $E_n 1 = 1$ (1 is a function identically equal to one).

■

Remark 1. In the language of probability theory, the operator E_n is called an operator of conditional mathematical expectation with respect to the σ-algebra $\mathfrak{C}_{\sigma,n}(X_\infty)$. It can be directly determined by the equality (1.7), because μ_∞ is a product measure consistent with the flow (1.6)

■

Corollary 1. *Assume that* $f \in L_1(X_\infty, \mathfrak{C}_\sigma(X_\infty))$ *and remains unchanged under arbitrary changes of any finite number of coordinates* x_k. *Then* f *is constant almost everywhere with respect to* μ_∞.

In fact, suppose that $f \in L_1$ (otherwise, we shall consider $f_1 = \arctan f$). The cylindrical function $E_n f$ does not change under arbitrary perturbations of each coordinate $x_1, \ldots, x_n$, i.e., this function is constant almost everywhere with respect to μ_∞. Thus, the same is true for $f = \lim_{n \to \infty} E_n f$.

■

Corollary 2. *Assume that* $\alpha \in \mathfrak{C}_\sigma(X_\infty)$, $x \in \alpha$, *and that* α *is such that an arbitrary point* $y \in X_\infty$, *for which* $y_k = x_k$, $k \geq n(y)$, *belongs to* α *along with* x. *Then* $\mu_\infty(\alpha) = 0$ *or* $\mu_\infty(\alpha) = 1$.

Indeed, let us apply Corollary 1 to the characteristic function $\kappa_\alpha = f$. By virtue of the condition imposed on α, the function κ_α is constant almost everywhere. But this just means that $\mu_\infty(\alpha) = 0$ or $\mu_\infty(\alpha) = 1$.

■

The product measures are the simplest class of measures on infinite-dimensional spaces, and their properties can be investigated more profoundly by analyzing the properties of the factor measures. This can be done, for example, when we study the question about absolute continuity of one product measure with respect to another. Consider the following example:

Example 1.1. (the Kakutani theorem). *Assume that* $\mu = \times_{k=1}^{\infty} \mu_k$, $\nu = \times_{k=1}^{\infty} \nu_k$ *and that the measure* μ_k *is absolutely continuous with respect to* ν_k *for any* k. *Let also* $\rho_k(x_k) = (d\mu_k / d\nu_k)(x_k)$. *We show that the measure* μ *is either absolutely continuous with respect to* ν *or singular if the infinite product*

$$\prod_{k=1}^{\infty} \int_{X_k} \sqrt{\rho_k(x_k)}\, d\nu_k(x_k).$$

is, respectively, convergent or divergent.

In fact, suppose that this product converges. This fact immediately implies that the sequence of cylindrical functions $r_n(x_{(n)}) = \prod_{k=1}^{n} \rho_k(x_k)$ $(n \in \mathbb{N})$ is fundamental in L_2. Therefore, the sequence r_n^2 converges (in L_1) to the function $\rho(x) = \prod_{k=1}^{\infty} \rho_k(x_k) \in L_1$. For the bounded cylindrical function $u(x) = u(x_{(n)})$, we have

$$\int_{X_\infty} u(x)\, d\mu(x) = \int_{X_{(m)}} u(x)\, d\mu_1(x_1) \dots d\mu_m(x_m)$$

$$= \int_{X_{(m)}} u(x)\, r_m^2(x_{(m)})\, d\nu_1(x_1) \dots d\nu_m(x_m)$$

$$= \int_{X_\infty} u(x)\, r_m^2(x_{(m)})\, d\nu(x) \xrightarrow[m\to\infty]{} \int_{X_\infty} u(x)\, \rho(x)\, d\nu(x)$$

for $m \geq n$. Approximating this equality, we can obtain the same formulas for an arbitrary bounded $u \in L_0(X_\infty, \mathfrak{C}_\sigma(X_\infty))$. This proves that μ is absolutely continuous with respect to ν and $(d\mu / d\nu)(x) = \rho(x)$. We leave it to the reader to examine the case where the product is divergent.

Let us establish the relation between

$$L_2(X_\infty, \mu_\infty) = L_2(X_\infty, \mathfrak{C}_\sigma(X_\infty), \mu_\infty)$$

and the infinite tensor product of Hilbert spaces

$$L_2(X_k, \mu_k) = L_2(X_k, \mathfrak{R}_k, \mu_k) \qquad (k \in \mathbb{N})$$

(see Chapter 1, Subsection 2.2). Suppose that each space $L_2(X_k, \mu_k)$ is separable and consider the infinite tensor product $\otimes_{k=1;e}^{\infty} L_2(X_k, \mu_k)$, where the sequence of functions $e^{(k)}(x_k) = 1$ is taken as a stabilizing sequence $e = (e^{(k)})_{k=1}^{\infty}$.

Let us put some cylindrical function $e_\alpha(x) = e_{\alpha_1}^{(1)}(x_1) \dots e_{\alpha_\nu}^{(\nu)}(x_\nu)$ in correspondence with each basis vector from the tensor product of the form $e_\alpha = e_{\alpha_1}^{(1)} \otimes \dots \otimes e_{\alpha_\nu}^{(\nu)} \otimes e^{(\nu+1)} \otimes e^{(\nu+2)} \otimes \cdots$. The functions $(e_\alpha(\cdot))_{\alpha \in \mathbb{Z}_{+,0}^{\infty}}$ form an orthonormal basis in $L_2(X_\infty, \mu_\infty)$. Indeed, one should only verify the completeness of this system which follows from the fact that the orthonormal basis in $L_2(X_{(n)}, \mathfrak{R}_{(n)}, \times_{k=1}^{n} \mu_k)$ is formed by the functions $(e_\alpha)_{\alpha \in \mathbb{Z}_+^n}$, and from the above-mentioned fact that cylindrical functions are dense in $L_2(X_\infty, \mu_\infty)$ (statement 1 of Theorem 1.2). This correspondence of bases enables us to identify $\otimes_{k=1;e}^{\infty} L_2(X_k, \mu_k)$ with $L_2(X_\infty, \mu_\infty)$.

Comparing the established equality

$$\bigotimes_{k=1;e}^{\infty} L_2(X_k, \mathfrak{R}_k, \mu_k) = L_2(X_\infty, \mathfrak{C}_\sigma(X_\infty), \mu_\infty)$$

with the construction of Subsection 1.2, we can define the "continual" tensor product $\otimes_{x\in X;e} H_x$ with the stabilization sequence $e = (e_x)_{x\in X}$ for the family $(H_x)_{x\in X}$ of separable Hilbert spaces H_x (the cardinality of this family may be arbitrary). Thus, let e_x be the orthonormal basis in H_x such that $e_0^{(x)} = e^{(x)}$. We take the formal vectors $e_{\alpha(\cdot)}$, where $\alpha(\cdot)$ is a mapping $X \ni x \mapsto \alpha(x) \in \mathbb{Z}_+^1$ whose values are zero everywhere except finitely many points x, and consider these vectors as an orthonormal basis. Then we span a Hilbert space by this basis, and according to the definition, this is just $\otimes_{x\in X;e} H_x$. The other (sometimes more reasonable) definitions that are connected with the topologization of the set X of indices of multiplied spaces can be found, e.g., in Guichardet [1] and Streater, Wulfsohn [1].

1.3. Products of Gaussian Measures

Consider the important special case of an infinite-dimensional (countable) product of measures, namely, the Gaussian product measures. The study of these measures is useful from different points of view: On the one hand, the properties of Gaussian product measures reflect (in the simplest case) specific peculiarities of the measure theory on infinite-dimensional spaces. On the other hand, measures of this sort often appear in applications, for example, in the description of various models of physical systems with infinitely many degrees of freedom.

In the situation which we are interested in, the factor measures are given on the spaces $X_k = \mathbb{R}^1$ with the Borel σ-algebra $\mathfrak{R}_k = \mathfrak{B}(\mathbb{R}^1)$ $(k \in \mathbb{N})$. The space $\mathbb{R}^\infty = \times_{k=1}^\infty \mathbb{R}^1$ is equipped with the σ-algebra $\mathfrak{C}_\sigma(\mathbb{R}^\infty)$ generated by the cylindrical sets which now have the form

$$\mathfrak{C} = \mathfrak{C}(1, \dots, n; \delta) = \{x \in \mathbb{R}^\infty \mid (x_1, \dots, x_n) \in \delta \in \mathfrak{B}(\mathbb{R}^n)\}.$$

Note that if $\mathbb{R}^\infty$ is considered as a topologic space with the Tikhonov topology, then $\mathfrak{C}_\sigma(\mathbb{R}^\infty) = \mathfrak{B}(\mathbb{R}^\infty)$ (this simple fact is proved in Chapter 3, Subsection 1.4).

Given an arbitrary sequence $\kappa = (\kappa_k)_{k=1}^\infty$, we introduce a Gaussian product measure $\gamma_\kappa = \times_{k=1}^\infty \gamma_{\kappa_k}$ in $\mathbb{R}^\infty$, where

$$d\gamma_{\kappa_k}(x_k) = \frac{1}{\sqrt{\pi\,\kappa_k}} \exp\left(-\frac{x_k^2}{\kappa_k}\right) dx_k \quad (k \in \mathbb{N}) \tag{1.10}$$

is a one-dimensional Gaussian measure on the axis $\mathbb{R}^1$. The existence of the measure γ_κ is ensured by the statement of Theorem 1.1.

The product measures on $\mathbb{R}^\infty$ possess one essential peculiarity. This space is dual to the nuclear space $\mathbb{R}_0^\infty$ in the sense of duality established with the help of the space $l_2 = l_2(\mathbb{R}^1)$, namely

$$\mathbb{R}^\infty \supset l_2 \supset \mathbb{R}_0^\infty \tag{1.11}$$

(see Chapter 1, Section 1, Example 1.6).

The presence of the linear structure in the space of arguments creates additional interesting problems concerning the measures on $\mathbb{R}^\infty$. First, we mention the problem of finding linear manifolds of full measure in $\mathbb{R}^\infty$. It follows from Corollary 2 of Theorem 1.2 that, in the case of product measures, each measurable linear subset of $\mathbb{R}^\infty$ that contains $\mathbb{R}_0^\infty$ has either measure 0 or measure 1. To formulate a statement describing some sets of a full measure for Gaussian product measures, we establish the following lemma.

Lemma 1.1. *For each weight $p = (p_k)_{k=1}^\infty$ $(p_k > 0)$, the Hilbert space*

$$l_2(p) = \left\{ x \in \mathbb{R}^\infty \;\middle|\; \sum_{k=1}^\infty x_k^2 p_k = \|x\|_{l_2(p)}^2 < \infty \right\}$$

belongs to $\mathfrak{C}_\sigma(\mathbb{R}^\infty)$.

Proof. Let us represent $l_2(p)$ as a union of closed balls, namely, $l_2 = \bigcup_{R=1}^\infty \tilde{B}_R(0)$, where

$$\tilde{B}_R(0) = \{x \in l_2(p) \mid \|x\|_{l_2,(p)} \le R\}$$

We also introduce the set

$$\beta_{n,R} = \left\{ x \in \mathbb{R}^\infty \;\middle|\; \sum_{k=1}^n x_k^2 p_k \le R \right\} \in \mathfrak{C}_\sigma(\mathbb{R}^\infty),$$

where $n, R \in \mathbb{N}$.

Then, clearly, $\tilde{B}_R(0) = \bigcap_{n=1}^\infty \beta_{n,R}$. Therefore, $\tilde{B}_R(0)$ and, hence, $l_2(p)$ belong to $\mathfrak{C}_\sigma(\mathbb{R}^\infty)$.

■

The next theorem is known as the Kolmogorov–Khinchin criterion.

Theorem 1.3. *If the series $\sum_{k=1}^\infty p_k \kappa_k$ converges, then $\gamma_\kappa(l_2(p)) = 1$, otherwise, $\gamma_\kappa(l_2(p)) = 0$.*

Proof. The function

$$\mathbb{R}^\infty \ni x \mapsto f_\varepsilon(x) = \exp\left(-\varepsilon\sum_{k=1}^{\infty} p_k x_k^2\right) \qquad (\varepsilon > 0)$$

possesses the following property: $f_\varepsilon(x) > 0$ if $x \in l_2(p)$, and $f_\varepsilon(x) = 0$ if $x \notin l_2(p)$. The sequence of cylindrical functions

$$f_{\varepsilon,n}(x) = \exp\left(-\varepsilon\sum_{k=1}^{n} p_k x_k^2\right) \qquad (n \in \mathbb{N})$$

is monotonically decreasing and converges (as $n \to \infty$) to $f_\varepsilon(x)$ pointwise. According to the Levy theorem, we have

$$\int_{\mathbb{R}^\infty} f_\varepsilon(x)\, d\gamma_\kappa(x) = \lim_{n\to\infty} \int_{\mathbb{R}^\infty} f_{\varepsilon,n}(x)\, d\gamma_\kappa(x). \tag{1.12}$$

But

$$\int_{\mathbb{R}^\infty} f_{\varepsilon,n}(x)\, d\gamma_\kappa(x) = \prod_{k=1}^{n} \frac{1}{\sqrt{\pi\kappa_k}} \int_{\mathbb{R}^1} \exp\left[-x_k^2\ (\varepsilon p_k + \kappa_k^{-1})\right] dx_k$$

$$= \prod_{k=1}^{n} \frac{1}{\sqrt{(\varepsilon p_k + \kappa_k^{-1})\ \kappa_k}} = \left[\prod_{k=1}^{n}(1+\varepsilon p_k\kappa_k)\right]^{-1/2}.$$

Therefore, (1.12) yields

$$\int_{\mathbb{R}^\infty} f_\varepsilon(x)\, d\gamma_\kappa(x) = \left[\prod_{k=1}^{n}(1+\varepsilon p_k\kappa_k)\right]^{-1/2}. \tag{1.13}$$

If $\sum_{k=1}^{\infty} p_k\kappa_k = \infty$, then by virtue of (1.13), we obtain

$$\int_{\mathbb{R}^\infty} f_\varepsilon(x)\, d\gamma_\kappa(x) = 0.$$

Consequently, $\gamma_\kappa(l_2(p)) = 0$.

If $\sum_{k=1}^{\infty} p_k\kappa_k < \infty$, then $\prod_{k=1}^{\infty}(1 + \varepsilon p_k\kappa_k)$ converges and, hence, there exists

$$\int_{\mathbb{R}^\infty} f_\varepsilon(x) d\gamma_\kappa(x) > 0.$$

We have

$$[0, 1] \ni f_\varepsilon(x) \to \kappa_{l_2(p)}(x)$$

as $\varepsilon \to 0$ pointwise. Here, $\kappa_{l_2(p)}(x)$ is the characteristic function of the set $l_2(p)$. Therefore,

$$\int_{\mathbb{R}^\infty} f_\varepsilon(x)\, d\gamma_\kappa(x) \underset{\varepsilon\to\infty}{\to} \gamma_\kappa(l_2(p)).$$

On the other side, $\prod_{k=1}^\infty (1 + \varepsilon p_k \kappa_k) \to 0$ as $\varepsilon \to 0$, and this proves that $\gamma_\kappa(l_2(p)) = 1$ in the considered case.

■

Each vector $a = (a_1, \ldots, a_n, 0, 0, \ldots) \in \mathbb{R}_0^\infty$ is associated with some linear function $l_a(x) = \sum_{k=1}^n x_k a_k \; (x \in \mathbb{R}^\infty)$. Clearly,

$$l_a(\cdot) \in L_2(\mathbb{R}^\infty, \gamma_\kappa) = L_2(\mathbb{R}^\infty, \mathfrak{C}_\sigma(\mathbb{R}^\infty), \gamma_\kappa).$$

We show that the functions l_a can be also defined as elements of $L_2(\mathbb{R}^\infty, \gamma_\kappa)$ for a wider collection of vectors a. Let us introduce a correspondence between a sequence $\kappa = (\kappa_k)_{k=1}^\infty$ and a Hilbert space

$$l_2(\kappa) = \left\{ x \in \mathbb{R}^\infty \mid \|x\|^2_{l_2(\kappa)} = \sum_{k=1}^\infty x_k^2 \kappa_k < \infty \right\}.$$

The following theorem holds:

Theorem 1.4. *For all* $a \in l_2(\kappa)$, *the series* $\sum_{k=1}^\infty a_k x_k \; (x \in \mathbb{R}^\infty)$ *converges in* $L_2(\mathbb{R}^\infty, \gamma_\kappa)$ *and, hence, for* $a \in l_2(\kappa)$ *the functions* $l_a(\cdot)$ *can be defined as measurable linear functions on* $\mathbb{R}^\infty$ *belonging to* $L_2(\mathbb{R}^\infty, \gamma_\kappa)$.

Proof. We show that the sequence $l_{a(n)}(x) = \sum_{k=1}^n a_k x_k \; (n \in \mathbb{N})$ is fundamental in $L_2(\mathbb{R}^\infty, \gamma_\kappa)$. In fact, we have

$$\| l_{a_{(n+p)}} - l_{a_{(n)}} \|^2_{L_2(\mathbb{R}^\infty, \gamma_\kappa)} = \int_{\mathbb{R}^\infty} \Big(\sum_{k=n+1}^{n+p} (a_k x_k) \Big)^2 d\gamma_\kappa(x)$$

$$= \sum_{k=n+1}^{n+p} a_k^2 \int_{\mathbb{R}^\infty} x_k^2 \, d\gamma_\kappa(x) + \sum_{k,j=n+1;\, k\neq j}^{n+p} a_k a_j \int_{\mathbb{R}^\infty} x_k x_j \, d\gamma_\kappa(x)$$

$$= \sum_{k=n+1}^{n+p} a_k^2 \int_{\mathbb{R}^1} x_k^2 \frac{1}{\sqrt{\pi \kappa_k}} \exp(-x_k^2 \kappa_k^{-1}) \, dx_k = \frac{1}{2} \sum_{k=n+1}^{n+p} a_k^2 \kappa_k, \quad (1.14)$$

for $n, p \in \mathbb{N}$, and this yields the required result. The next equality also follows from (1.14)

$$\| l_a \|^2_{L_2(\mathbb{R}^\infty, \gamma_\kappa)} = \sum_{k=1}^{\infty} a_k^2 \kappa_k = \| a \|^2_{l_2(\kappa)}. \quad (1.15)$$

Note that the condition of linearity holds for the function l_a only γ_κ-almost everywhere on $\mathbb{R}^\infty$. This follows from the facts that the functions $l_{a_{(n)}}$ are linear and that l_a is the limit of a certain subsequence $(l_{a_{(n)}})_{j=1}^{\infty}$ which converges γ_κ-almost everywhere.

■

Let us define the Fourier transform (the characteristic functional) of the measure γ_κ. By direct calculation, we obtain

$$\tilde{\gamma}_\kappa(y) = \int_{\mathbb{R}^\infty} e^{i(y,x)} \, d\gamma_\kappa(x) = \prod_{k=1}^{n} \int_{\mathbb{R}^1} e^{i(y_k, x_k)} \frac{1}{\sqrt{\pi \kappa_k}} \exp(-x_k^2 \kappa_k^{-1}) \, dx_k$$

$$= \prod_{k=1}^{n} \exp\Big(-\frac{1}{4} \kappa_k y_k^2 \Big) = \exp\Big(-\frac{1}{4} \| y \|^2_{l_2(\kappa)} \Big) \quad (1.16)$$

for $y = (y_1, \ldots, y_n, 0, 0, \ldots) \in \mathbb{R}_0^\infty$. According to Theorem 1.4, for a given $y \in l_2(\kappa)$, we can choose a sequence $(y^{(k)})_{k=1}^{\infty} \subset \mathbb{R}_0^\infty$ such that $l_{y^{(k)}} \to l_y$ as $k \to \infty$ almost everywhere with respect to γ_κ. Thus, using the Lebesgue theorem, equality (1.16) can be extended onto all $y \in l_2(\kappa)$. Therefore, the Fourier transform of the measure γ_κ has the form

$$\tilde{\gamma}_\kappa(y) = \int_{\mathbb{R}^\infty} e^{i(y,x)_{l_2}} \, d\gamma_\kappa(x) = \exp\Big(-\frac{1}{4} \| y \|^2_{l_2(\kappa)} \Big) \quad (y \in l_2(\kappa)) \quad (1.17)$$

and is thus a continuous function on $l_2(\kappa)$.

Let us examine the already established facts concerning the measures γ_κ assuming that the condition $\kappa_k \geq \varepsilon > 0$ $(k \in \mathbb{N})$ holds. In this case, two Hilbert spaces correspond to the measure γ_κ, namely, $l_2(\kappa) \subset l_2$ and the space $l_2(\kappa^{-1})$ dual to $l_2(\kappa)$ with respect to l_2. This means that the following chain corresponds to γ_κ

$$l_2(\kappa^{-1}) \supset l_2 \supset l_2(\kappa). \tag{1.18}$$

The Fourier transform $\tilde{\gamma}_\kappa$ may be expressed in terms of the norm in $l_2(\kappa)$ by (1.17) and determines the measure γ_κ uniquely. In this sense, we can say that γ_κ is a Gaussian measure canonically generated by the Hilbert space $l_2(\kappa)$.

By virtue of Theorem 1.3, the space $l_2(\kappa^{-1})$ has a measure γ_κ equal to zero. Let $l_2(p^{-1}) \supset l_2(\kappa^{-1})$. Then this theorem can be reformulated as follows:

$\gamma_\kappa(l_2(p^{-1})) = 1$ *if and only if the imbedding operator* $O_{\kappa^{-1},p^{-1}} : l_2(\kappa^{-1}) \supset l_2(p^{-1})$ *is quasinuclear.*

Indeed, if $(e_k)_{k=1}^\infty$ is a natural orthonormal basis in l_2, then $(\kappa_k^{1/2} e_k)_{k=1}^\infty$ is an orthonormal basis in $l_2(\kappa^{-1})$, and thus,

$$| O_{\kappa^{-1},p^{-1}} |^2 = \sum_{k=1}^{\infty} \| \kappa_k^{1/2} e_k \|^2_{l_2(p^{-1})} = \sum_{k=1}^{\infty} \kappa_k / p_k .$$

According to Theorem 1.3, the condition under which the last series converges, is equivalent to the requirement that $\gamma_\kappa(l_2(p^{-1})) = 1$.

■

Later we shall see that the situation described here is typical for the Gaussian measures on linear spaces. A rigging of type (1.18) can be attached to each measure of this sort, and this rigging possesses the same properties with respect to this measure as the rigging (1.18) does with respect to γ_κ.

1.4. Measures in Hilbert Spaces

Let H be a real Hilbert space. In order to introduce and study probability measures in H, it is convenient to choose a class of subsets of H having the simplest possible structure (just as in the case of measures on the products of spaces).

For the *finite-dimensional* subspace $K \subset H$ and the Borel set $\delta \in \mathfrak{B}(K)$, we introduce the set

$$\mathfrak{C} = \mathfrak{C}(K; \delta) = \{ f \in H \mid P_K f \in \delta \}, \tag{1.19}$$

where P_K is an orthogonal projector (in H) onto K. As in Subsection 1.1, this set is called cylindrical (K is its "coordinate", and δ is its base). Evidently, the coordinate K and the base δ are ambiguously determined by $\mathfrak{C}$: If $L \subset H$ is finite-dimensional and orthogonal to K, then the set (1.19) can be also written in the form $\mathfrak{C} = \mathfrak{C}(K \oplus L, \delta \times L)$. Clearly, it is possible to include the direct factors coinciding with some $L = \mathbb{R}^d$ from δ if these factors are present.

For our purposes, it is not necessary to consider arbitrary cylindrical sets of the form (1.19). We must study only a certain class of these sets determined by a given class $\mathcal{K}$ of their coordinates which satisfies certain conditions. We define $\mathcal{K}$ as a set of finite-dimensional subsets of H for which

(i) if $K \in \mathcal{K}$, then an arbitrary subspace $L \subset H$ belongs to $\mathcal{K}$;

(ii) if $K, L \in \mathcal{K}$, then l.s. $(K, L \in \mathcal{K})$;

(iii) the linear set $\bigcup_{K \in \mathcal{K}} K$ is dense in H.

Denote $\mathfrak{T}(\mathcal{K}, H) = \{\mathfrak{C}(K, \delta) \mid \delta \in \mathfrak{B}(K), K \in \mathcal{K}\}$. It can be easily shown that this set is defined correctly.

The collection of $\mathfrak{T}(\mathcal{K}, H)$ *is an algebra of sets.*

In fact, let $\mathfrak{C}' = \mathfrak{C}(K'; \delta')$ and $\mathfrak{C}'' = \mathfrak{C}(K''; \delta'') \in \mathfrak{T}(\mathcal{K}, H)$. We set $K =$ l.s. (K', K''), represent K in the form $K = K' \oplus L' = K'' \oplus L''$, and pass to the common coordinate K in the notation of $\mathfrak{C}'$ and $\mathfrak{C}''$. Then $\mathfrak{C}' = \mathfrak{C}(K; \gamma')$ and $\mathfrak{C}'' = \mathfrak{C}(K; \gamma'')$ with some bases $\gamma', \gamma'' \in \mathfrak{B}(K)$ and $\mathfrak{C}' \cup \mathfrak{C}'' = \mathfrak{C}(K; \gamma' \cup \gamma'') \in \mathfrak{T}(\mathcal{K}, H)$. The set $\mathfrak{C}' \setminus \mathfrak{C}''$ can be examined similarly.

■

Denote the σ-span of the algebra $\mathfrak{T}(\mathcal{K}, H)$ by $\mathfrak{T}_\sigma(\mathcal{K}, H)$. On the one hand, it is convenient to construct measures on this σ-algebra, and on the other hand, this σ-algebra is rigidly connected with H. Thus, the following lemma is valid:

Lemma 1.2. *For an arbitrary class* $\mathcal{K}$, *we have* $\mathfrak{T}_\sigma(\mathcal{K}, H) = \mathfrak{B}(H)$.

Proof. Since $\bigcup_{K \in \mathcal{K}} K$ is dense in H, we can choose a sequence of vectors dense in H such that each vector belongs to a certain $K \in \mathcal{K}$. By orthogonalizing this sequence, we construct the orthonormal basis $(e_j)_{j=1}^\infty$ in H for which $\forall n \in \mathbb{N}\ \exists K_n \in \mathcal{K}: e_1, \ldots, e_n \in K_n$. By $(f_j)_{j=1}^\infty$, we denote a sequence of coordinates of a vector $f \in H$ in this basis.

The following relation is obviously true for the closed ball $\tilde{B}_r(a) \subset H$ with radius $r > 0$ centered at $a \in H$

$$\tilde{B}_r(a) = \left\{ f \in H \,\middle|\, \sum_{k=1}^{\infty} (f_k - a_k)^2 \le r^2 \right\} = \beta_{n,r},$$

$$\beta_{n,r} = \left\{ f \in H \,\Big|\, \sum_{k=1}^{n} (f_k - a_k)^2 \le r^2 \right\}. \tag{1.20}$$

However, $\beta_{n,r} = \mathfrak{C}(K_n; \delta)$, where $\delta \subset K_n$ is a closed ball with the radius r centered at $P_{K_n} a$. Therefore, $\beta_{n,r} \in \mathfrak{C}(\mathcal{K}, H)$. By virtue of (1.20), we have $\tilde{B}_r(a) \in \mathfrak{C}_\sigma(\mathcal{K}, H)$. The open ball $B_r(a) = \bigcup_{n=1}^{\infty} \tilde{B}_{r-1/n}(a)$ also belongs to $\mathfrak{C}_\sigma(\mathcal{K}, H)$. Finally, any open set in H can be obtained as a union of countably many open spheres and, hence, also belongs to $\mathfrak{C}_\sigma(\mathcal{K}, H)$. So, $\mathfrak{B}(H) \subset \mathfrak{C}_\sigma(\mathcal{K}, H)$. The inverse inclusion follows from the fact that $\mathfrak{C}(K, \delta) \in \mathfrak{B}(H)$ for any $K \in \mathcal{K}$.

■

Example 1.2. Let $H = l_2(\mathbb{R}^1)$, and let $\mathcal{K}$ be a collection of all finite-dimensional subspaces of $l_2(\mathbb{R}^1)$ consisting of finite vectors from $l_2(\mathbb{R}^1)$. Along with the cylindrical sets defined by (1.19), we consider the cylindrical sets of the form (1.1), where $R_x = \mathbb{R}^1$, $x \in X = \mathbb{N}$. Clearly, all the possible intersections $\mathfrak{C}(1, \ldots, n; \delta) \cap l_2(\mathbb{R}^1)$ form the whole algebra $\mathfrak{C}(\mathcal{K}, H)$, whereas the intersections $\alpha \cap l_2(\mathbb{R}^1)$, where $\alpha \in \mathfrak{C}_\sigma(\mathbb{R}^\infty)$, form the whole σ-algebra $\mathfrak{C}_\sigma(\mathcal{K}, H)$.

Remark 1. It is sometimes convenient to introduce cylindrical sets (1.19) by employing the "coordinate" method: Choose an orthonormal basis $(e_k)_{k=1}^{n}$ in K, then

$$\mathfrak{C} = \mathfrak{C}(K; \delta) = \{ f \in H \mid ((f, e_1)_H, \ldots, (f, e_n)_H) \in \delta \}. \tag{1.21}$$

Furthermore, if we replace the vectors e_k in (1.21) by arbitrary vectors $h_k \in K$, then the set

$$\{ f \in H \mid ((f, h_1)_H, \ldots, (f, h_n)_H) \in \gamma \} \quad (\gamma \in \mathfrak{B}(\mathbb{R}^n)) \tag{1.22}$$

obtained as a result, will be cylindrical, too. One must only choose an orthonormal basis $(e_k)_{k=1}^{m}$ $(m \le n)$ in the l.s. $(h_1, \ldots, h_n)$ and rewrite the condition $((f, h_1)_H, \ldots, (f, h_n)_H) \in \gamma$ in the form $((f, e_1)_H, \ldots, (f, e_m)_H) \in \delta$ with a properly chosen $\delta \in \mathfrak{B}(\mathbb{R}^n)$ (it should be noted that (1.22) may be empty even if γ is nonempty). Certainly, h_j and e_k are determined by the definition of $\mathfrak{C}$ ambiguously (cf. Subsection 1.1).

■

Remark 2. It follows from the definition of a cylindrical set written in the form (1.22) that $\mathfrak{C}_\sigma(\mathcal{K}, H) = \mathfrak{B}(H)$ can be also defined as a minimal σ-algebra with respect to which all the functionals $l_h(f) = (f, h)_H$ $(f \in H)$, where $h \in H$, are measurable[8].

■

[8] Here, we can obviously restrict ourselves to a sufficiently large collection of $h \in H$, e.g., to take h from a set that is dense in H.

Remark 3. Consider the chain of real spaces $H_- \supset H \supset H_+$ (see Chapter 1, (1.11)). When constructing $\mathfrak{B}(H)$ ($\mathfrak{B}(H_-)$), one can start, for example, from the cylindrical sets (1.19), where $\mathcal{K}$ is a set of all the finite-dimensional subspaces of H_+ (H or H_+).

■

We now proceed to constructing the measures on $\mathfrak{B}(H)$. Fix some class $\mathcal{K}$. The function of sets $\mathfrak{C}(\mathcal{K}, H) \ni \mathbf{C} \mapsto \gamma(\mathbf{C}) \in [0, 1]$ is called a cylindrical measure if it possesses the property of absolute additivity on the sets with fixed coordinate and $\gamma(H) = 1$. This absolute additivity means that

$$\gamma\Big(\bigcup_{j=1}^{\infty} \mathbf{C}(K; \delta_j)\Big) = \sum_{j=1}^{\infty} \gamma(\mathbf{C}(K; \delta_j)) \tag{1.23}$$

for any $K \in \mathcal{K}$ and mutually disjoint sets $\mathbf{C}(K; \delta_j)$ (or equivalently δ_j). The property of absolute additivity can be also formulated as follows: Denote the σ-algebra of cylindrical sets with the fixed coordinate $K \in \mathcal{K}$ by $\mathfrak{C}(K, H)$. Then, for any $K \in \mathcal{K}$, the restriction $\gamma \upharpoonright \mathfrak{C}(K, H)$ is absolutely additive.

If $\mathfrak{B}(H) \ni \alpha \mapsto \mu(\alpha) \geq 0$ is a probability measure, then its restriction to $\mathfrak{C}(\mathcal{K}, H)$ is evidently a cylindrical measure. However, the inverse statement, generally speaking, is not true because the cylindrical measure γ is not always absolutely additive on $\mathfrak{C}(\mathcal{K}, H)$, and therefore, it is sometimes impossible to extend this measure to the measure on the σ-span $\mathfrak{C}_\sigma(\mathcal{K}, H) = \mathfrak{B}(H)$ (see Example 1.6). Certainly, γ is always additive: Assume that there exist finitely many mutually disjoint sets from $\mathfrak{C}(\mathcal{K}, H)$. The required result is obtained if we write the definitions of these sets by using the common coordinate K and (1.23). Let us formulate the general criterion of the extendability of a cylindrical measure to a measure on $\mathfrak{B}(H)$.

Lemma 1.3. *In order that the cylindrical measure γ be extendable to the probability measure on $\mathfrak{B}(H)$, it is necessary and sufficient that the following condition hold: For any $\varepsilon > 0$ one can find $r > 0$ such that $\gamma(\mathbf{C}) < \varepsilon$ for every $\mathbf{C} \in \mathfrak{C}(\mathcal{K}, H)$ lying outside the ball $\tilde{B}_r(0) = \{f \in H \mid \|f\|_H \leq r\}$.*

Proof. The necessity of this condition is obvious. In fact, let γ be extendable to the probability measure μ. Then, for given $\varepsilon > 0$, one can find $r > 0$ such that

$$\mu(H \setminus \tilde{B}_r(0)) < \varepsilon \quad (\lim_{r \to \infty} \mu(\tilde{B}_r(0)) = 1).$$

Consequently,

$$\gamma(\mathfrak{C}) = \mu(\mathfrak{C}) \leq \mu(H \setminus \tilde{B}_r(0)) < \varepsilon,$$

for $\mathfrak{C} \subset H \setminus \tilde{B}_r(0)$.

Sufficiency. We must prove that γ is absolutely additive on $\mathfrak{C}(\mathcal{K}, H)$. For this purpose, it is necessary to verify the following statement: Let $\mathfrak{C}_1 \supset \mathfrak{C}_2 \supset \ldots$ be a sequence of sets $\mathfrak{C}_n \in \mathfrak{C}(\mathcal{K}, H)$ with an empty intersection $\bigcap_{n=1}^{\infty} \mathfrak{C}_n$. Then $\lim_{n\to\infty} \gamma(\mathfrak{C}_n) = 0$.

We now prove that if the last statement is valid for the sets $\mathfrak{C}_n = \mathfrak{C}(K_n; \delta_n)$ with the bases δ_n closed in K_n, then it is valid in the general case, too.

First we note that for every $\mathfrak{C}(K; \delta) \in \mathfrak{C}(\mathcal{K}, H)$ and $\varepsilon > 0$ one can find $\varphi \subset \delta$ (closed in K) such that $\gamma(\mathfrak{C}(K; \delta) \setminus \mathfrak{C}(K; \varphi)) < \varepsilon$. Hence, for fixed K, the function of sets $\mathfrak{B}(K) \ni \alpha \mapsto \nu(\alpha) = \gamma(\mathfrak{C}(K; \alpha))$ is a nonnegative finite measure on the σ-algebra of Borel sets of a finite-dimensional Euclidean space. Therefore, according to the well known facts (see, e.g., Halmos [1, Chapter 10, Sections 50–52]), it is regular, i.e., $\nu(\alpha) = \sup \nu(\varphi)$, where supremum is taken over the set of all closed $\varphi \subset \alpha$ $(\varphi \in \mathfrak{B}(K))$.

Let $\mathfrak{C}_n = \mathfrak{C}(K_n; \delta_n) \in \mathfrak{C}(\mathcal{K}, H)$, $\bigcap_{n=1}^{\infty} \mathfrak{C}_n = \varnothing$, and $\varepsilon > 0$. For every $n \in \mathbb{N}$, we choose a set $\varphi_n \subset \delta_n$ closed in K_n so that $\gamma(\mathfrak{C}_n \setminus \mathfrak{C}(K_n; \varphi_n)) < \varepsilon 2^{-n}$. We set

$$\mathfrak{C}'_n = \bigcap_{k=1}^{\infty} \mathfrak{C}(K_k; \varphi_k) \subset \mathfrak{C}(K_n; \varphi_n) \subset \mathfrak{C}_n \quad (n \in \mathbb{N}).$$

Clearly, $\mathfrak{C}'_1 \supset \mathfrak{C}'_2 \supset \ldots$, $\bigcap_{n=1}^{\infty} \mathfrak{C}'_n = \varnothing$. Further, let us write the definitions of the sets $\mathfrak{C}(K_1; \delta_1), \ldots, \mathfrak{C}(K_n; \varphi_n)$ using the common coordinate K'_n: $\mathfrak{C}(K_j; \varphi_j) = \mathfrak{C}(K'_n; \psi_j)$. If φ_j is closed, then ψ_j is also closed $(j = 1, \ldots, n)$. Therefore, $\mathfrak{C}'_n = \mathfrak{C}\left(K'_n, \bigcap_{j=1}^{\infty} \psi_j\right)$, and $\mathfrak{C}'_n$ has a closed base in K'_n. By virtue of our assumption, $\gamma(\mathfrak{C}'_n) \to 0$ as $n \to 0$. On the other hand,

$$\mathfrak{C}_n \setminus \mathfrak{C}'_n = \mathfrak{C}_n \setminus \bigcap_{j=1}^{n} \mathfrak{C}(K_j; \varphi_j) = \bigcup_{j=1}^{n} (\mathfrak{C}_n \setminus \mathfrak{C}(K_j; \varphi_j)) \subset \bigcup_{j=1}^{n} (\mathfrak{C}_j \setminus \mathfrak{C}(K_j; \varphi_j))$$

for any $n \in \mathbb{N}$. Consequently,

$$\gamma(\mathfrak{C}_n) - \gamma(\mathfrak{C}'_n) = \gamma(\mathfrak{C}_n \setminus \mathfrak{C}'_n) \leq \sum_{j=1}^{\infty} \gamma(\mathfrak{C}_j \setminus \mathfrak{C}(K_j; \varphi_j)) < \varepsilon.$$

This implies that $\gamma(\mathfrak{C}_n) \to 0$ if $n \to \infty$.

Thus, to complete the proof of Lemma 1.3, it remains to show that if $\mathfrak{C}_1 \supset \mathfrak{C}_2 \supset \ldots$, $\bigcap_{n=1}^{\infty} \mathfrak{C}_n = \varnothing$, and $\mathfrak{C}_n = \mathfrak{C}(K_n; \delta_n)$, where δ_n are closed in K_n, then $\lim_{n\to\infty} \gamma(\mathfrak{C}_n) = 0$. Consider the weak topology in H. In this topology, the sets $\mathfrak{C}_n$ are closed as cylindrical sets with closed bases, and each ball $\tilde{B}_r(0)$ is compact. For given $\varepsilon > 0$, let us find a number $r > 0$ as prescribed in the condition of the lemma. Then

$$(\mathfrak{C}_1 \cap \tilde{B}_r(0)) \supset (\mathfrak{C}_2 \cap \tilde{B}_r(0)) \supset \ldots, \quad \bigcap_{n=1}^{\infty} (\mathfrak{C}_n \cap \tilde{B}_r(0)) = \varnothing,$$

and the sets $\mathfrak{C}_n \cap \tilde{B}_r(0)$ are compact. Hence, $\mathfrak{C}_n \cap \tilde{B}_r(0) = \varnothing$ starting with sufficiently large n and, consequently, $\gamma(\mathfrak{C}_n) < \varepsilon$.

■

We shall use Lemma 1.3 in Subsection 1.6 when constructing a Gaussian measure in a Hilbert space.

Let us remark that instead of the notion of a cylindrical measure, the equivalent notion of a weak distribution is often employed. We say that a weak distribution is given if we have a collection of measures γ_K of sets δ belonging to all possible finite-dimensional subspaces $K \in \mathcal{K}$ and these measures are constructed from the cylindrical measure γ (in H) as follows: $\mathfrak{B}(K) \ni \delta \mapsto \gamma_K(\delta) \geq \gamma(\mathfrak{C}(K; \delta))$. Conversely, the measure γ of the cylindrical sets can be constructed by use of the measures γ_K consistent in the natural sense.

1.5. Gaussian Measures in Finite-Dimensional Spaces

Consider Gaussian measures in the Euclidean space $\mathbb{R}^n$ $(n \in \mathbb{N})$. The definition of these measures and some of their properties are necessary to construct Gaussian measures in Hilbert spaces. Fix some positive operator S acting in $\mathbb{R}^n$ and a vector $a \in \mathbb{R}^n$. The following probability measure is called a Gaussian measure in $\mathbb{R}^n$ with the correlation operator S and the mean value a:

$$\mathfrak{B}(\mathbb{R}^n) \ni \delta \mapsto \gamma_{S,a}(\delta) = \pi^{-n/2} (\operatorname{Det} S)^{-1/2} \int_{\delta} \exp[-(S^{-1}(x-a), x-a)_{\mathbb{R}^n}] dx, \quad (1.24)$$

where dx denotes the integration in the Lebesgue measure in $\mathbb{R}^n$.

Therefore, the measures γ_{κ_k}[9] of the type (1.10) given in $\mathbb{R}^1$ are Gaussian measures with the correlation operator κ_k and the mean value zero. Henceforth, we shall

[9] These measures have been considered in Subsection 1.3.

often meet measures with mean value zero; they are denoted by $\gamma_S = \gamma_{S,0}$. The measure $\gamma_1 = \gamma_{1,0}$ is called a canonical Gaussian measure. Clearly,

$$d\gamma_1(x) = \pi^{-n/2} \exp(-\|x\|^2_{\mathbb{R}^n})\, dx = d\gamma_1(x_1) \times \dots \times d\gamma_n(x_n). \tag{1.25}$$

Let us change the variables in the integral (1.24) $\mathbb{R}^n \ni x \mapsto y \in \mathbb{R}^n$ by setting $x = a + C(y-b)$, where C is a linear invertible operator in $\mathbb{R}^n$, and $b \in \mathbb{R}^n$ is a fixed vector. Since the Jacobian $dx/dy = \operatorname{Det} C$, we have

$$\gamma_{S,a}(\delta) = \pi^{-n/2} (\operatorname{Det} S)^{-1/2} \int\limits_{C^{-1}(\delta - a)+b} \exp[-(S^{-1}C(y-b), C(y-b))_{\mathbb{R}^n}]\, dy. \tag{1.26}$$

Here, $(C^{-1}(\delta - a) + b)$ is the image of the set δ under the mapping

$$\mathbb{R}^n \ni x \mapsto y = b + C^{-1}(x-a) \in \mathbb{R}^n.$$

Shifting C to the first factor in (1.25) and comparing (1.26) with (1.24), we conclude that

$$\gamma_{S,a}(\delta) = \gamma_{C^{-1}SC^{*-1},b}(C^{-1}(\delta - a) + b) \quad (\delta \in \mathcal{B}(\mathbb{R}^n)). \tag{1.27}$$

For fixed S and arbitrary C, the operators $C^{-1}SC^{*-1}$ run over the whole set of positive operators in $\mathbb{R}^n$. Hence, the equality (1.27) shows that if we have two arbitrary Gaussian measures in $\mathbb{R}^n$, then one of these can be expressed in terms of the other (and, in particular, in terms of the canonical measure, namely,

$$\gamma_{S,a}(\delta) = \gamma_1(S^{-1/2}(\delta - a))$$

for any $\delta \in \mathcal{B}(\mathbb{R}^n)$.

We now present some necessary formulas as the following remarks.

Remark 1. As in the case of the canonical Gaussian measure, an arbitrary measure γ_S with mean value zero can be represented as a product measure of the form $\times_{k=1}^n \gamma_{\kappa_k}$. The following generalization of formula (1.25) holds: Suppose that $e_1, \dots, e_n > 0$ and $e_1, \dots, e_n$ are the eigenvalues of the operator S and the corresponding eigenvectors, respectively. Then

$$d\gamma_S(x) = d\gamma_{s_1}(x_1) \times \dots \times d\gamma_{s_n}(x_n), \tag{1.28}$$

where $x_1, \dots, x_n$ are the coordinates of the vector $x \in \mathbb{R}^n$ in the orthonormal basis $(e_k)_{k=1}^n$ (but not in the original standard basis). Relation (1.28) immediately follows from (1.24) if we write the Lebesgue measure in the form $dx = dx_1 \ldots dx_n$. It is clear that a formula similar to (1.28) holds in the case of a nonzero mean value a, as well. In this case, the right-hand side of (1.28) is a product of one-dimensional measures of type $d\gamma_{s_k}$ with mean values a_k.

■

Remark 2. The Fourier transform of the measure $\gamma_{S,a}$ has the form

$$\tilde{\gamma}_{S,a}(y) = \int_{\mathbb{R}^n} \exp\,[i\,(x, y)_{\mathbb{R}^n}]\,d\gamma_{S,a}(x)$$

$$= \exp\,[\,i\,(y, a)_{\mathbb{R}^n} - \tfrac{1}{4}(Sy, y)_{\mathbb{R}^n}\,] \quad (y \in \mathbb{R}^n). \tag{1.29}$$

Indeed, changing the variables in the integral of (1.29) by setting $y = x - a$ and taking (1.27) into account, we obtain

$$\tilde{\gamma}_{S,a}(y) = \exp\,[i\,(x, y)_{\mathbb{R}^n}]\,\tilde{\gamma}_S(y) \quad (y \in \mathbb{R}^n).$$

To calculate $\tilde{\gamma}_S(y)$, we pass to the variables $x_1, \ldots, x_n$ (see (1.28)) in the corresponding integral. Just as in (1.16), we find

$$\tilde{\gamma}_S\,(y) = \exp\left(-\tfrac{1}{4}\sum_{k=1}^{n} s_k\, y_k^2\right) = \exp\left(-\tfrac{1}{4}\,(Sy, y)_{\mathbb{R}^n}\right) \quad (y \in \mathbb{R}^n).$$

■

Remark 3. Consider the decomposition $\mathbb{R}^n = K \oplus \mathbb{R}^{n-m}$, where $K = \mathbb{R}^m$ ($m = 1, \ldots, n-1$ is fixed). We state that the measure $\mathfrak{B}(\mathbb{R}^m) \ni \delta \mapsto \gamma_{S,a}(\delta \times \mathbb{R}^{n-m})$ is Gaussian in $\mathbb{R}^m$ with the correlation operator $P_K S \upharpoonright K$ and mean value $P_K a$ (P_K is the projector onto K in $\mathbb{R}^n$). In other words, we have

$$\mathfrak{B}(\mathbb{R}^m) \ni \delta \mapsto \gamma_{S,a}(\delta \times \mathbb{R}^{n-m}) = \gamma_{S_K,\,P_K a}(\delta), \quad S_K = P_K S \upharpoonright K. \tag{1.30}$$

To prove this, we calculate the Fourier transform of the measure μ on the left-hand side of (1.30). According to (1.29), we have

$$\tilde{\mu}(y) = \gamma_{S,a}(\cdot \times \mathbb{R}^{n-m})^{\sim}(y) = \int_{\mathbb{R}^m} \exp\,[i\,(y, x)_{\mathbb{R}^m}]\,d\gamma_{S,a}(x \times \mathbb{R}^{n-m})$$

$$= \int_{\mathbb{R}^n} \exp\,[i\,(y, P_K x)_{\mathbb{R}^n}]\,d\gamma_{S,a}(x) = \int_{\mathbb{R}^n} \exp\,[i\,(P_K y, x)_{\mathbb{R}^n}]\,d\gamma_{S,a}(x)$$

$$= \exp\left[i(P_k y, a)_{\mathbb{R}^n} - \tfrac{1}{4}(SP_k y, P_k y)_{\mathbb{R}^n}\right] = \exp\left[i(y, P_k a)_{\mathbb{R}^m} - \tfrac{1}{4}(S_K y, y)_{\mathbb{R}^m}\right]$$

$$= \tilde{\gamma}_{S_K, P_K a}(y),$$

for $y = P_k y \in \mathbb{R}^m \subset \mathbb{R}^n$. Since the Fourier transforms are equal, we can conclude that $\mu = \gamma_{S_K, P_K a}$, i.e., that (1.30) holds.

Remark 4. For any vector $b \in \mathbb{R}^n$ and operator C in $\mathbb{R}^n$, the following formulae are valid

$$\int_{\mathbb{R}^n} (x, b)_{\mathbb{R}^n} d\gamma_{S,a}(x) = (b, a)_{\mathbb{R}^n},$$

$$\int_{\mathbb{R}^n} (Cx, x)_{\mathbb{R}^n} d\gamma_{S,a}(x) = \tfrac{1}{2}\mathrm{Tr}\,(SC) + (Ca, a)_{\mathbb{R}^n}. \tag{1.31}$$

Integrals (1.31) obviously converge, since the measure $\gamma_{S,a}$ rapidly decreases at infinity. To calculate the first integral, we differentiate (1.29) with respect to y_k, set $y = 0$, multiply the resulting equality by b_k, and sum over k. To calculate the second one, we take the second derivatives of both sides of (1.29), namely $\partial^2 / \partial y_j \partial y_k$, and set $y = 0$. As a result, we get

$$\int_{\mathbb{R}^n} x_j x_k d\gamma_{S,a}(x) = \tfrac{1}{2}s_{jk} + a_j a_k \quad (j, k = 1, \dots, n). \tag{1.32}$$

Multiplying this equality by $c_{jk}((c_{jk})_{j,k=1}^n$ is the matrix of the operator C) and summing over j and k, we obtain the second formula of (1.31). The first formula of (1.31) shows that a is indeed the mean value of the measure $\gamma_{S,a}$. It follows from (1.32) that

$$\int_{\mathbb{R}^n} (x-a, f)_{\mathbb{R}^n}(x-a, g)_{\mathbb{R}^n} d\gamma_{S,a}(x) = \int_{\mathbb{R}^n} (y, f)_{\mathbb{R}^n}(y, g)_{\mathbb{R}^n} d\gamma_{S,0}(x) = \tfrac{1}{2}(Sf, g)_{\mathbb{R}^n}.$$

for any $f, g \in \mathbb{R}^n$. Hence, according to the terminology of probability theory, the correlation operator of this measure is $\frac{1}{2}S$ but not S. Nevertheless, we leave our terminology unchanged, because it is in common use in mathematical physics.

Example 1.3. It is now clear that formula (1.29)[10] is very convenient for the investigation of Gaussian measures, as the calculations become rather simple. Let us demonstrate, for example, how relation (1.30) can be established by direct calculation. We must prove that

[10] This formula is often used as a basis of the definition of a Gaussian measure: A probability measure μ on $\mathfrak{B}(\mathbb{R}^n)$ is called Gaussian if $\tilde{\mu}(y)$ is given by the right-hand side of (1.29).

$$\pi^{-n/2}(\mathrm{Det}\, S)^{-1/2} \int_{\delta\times\mathbb{R}^{n-m}} \exp\,[S^{-1}(x-a), x-a)_{\mathbb{R}^n}]\,dx$$

$$= \pi^{-m/2}(\mathrm{Det}\, S_K)^{-1/2} \int_{\delta} \exp\,[S_K^{-1}(y-P_k a),\ y-P_k a)_{\mathbb{R}^m}]\,dy \quad (\delta \in \mathfrak{B}(\mathbb{R}^m)). \quad (1.33)$$

We represent the operator S^{-1} as $S^{-1} = T^*T$, where T is an invertible operator with a triangular matrix having zeros above the main diagonal. Let us change the variables $\mathbb{R}^n \ni x \mapsto \xi \in \mathbb{R}^n$ in the integral on the left-hand side of (1.33) by setting $x = a + T^{-1}\xi$. According to (1.27), we obtain

$$\gamma_{S,a}(\delta \times \mathbb{R}^{n-m}) = \gamma_1(T(\delta \times \mathbb{R}^{n-m} - a)).$$

Since T is triangular, we have

$$T(\delta \times \mathbb{R}^{n-m} - a) = (T_K(\delta - P_k a)) \times \mathbb{R}^{n-m} \quad (T_K = P_K T \restriction K).$$

Calculating the measure γ_1 of this set and using (1.25), we find that the $(n-m)$-fold multiple integral over $\mathbb{R}^{n-m}$ vanishes. Finally, we get

$$\gamma_{S,a}(\delta \times \mathbb{R}^{n-m}) = \int_{T_K(\delta - P_K a)} \exp\,(-\|\xi\|^2_{\mathbb{R}^m})\, d\xi.$$

The same integral is obtained if we change the variables $\mathbb{R}^m \ni y \mapsto \eta \in \mathbb{R}^m$ in the integral on the right-hand side of (1.33) by setting $y = P_K a + (T_K)^{-1}\eta$. To prove this fact, it suffices to note that if $S^{-1} = T * T$, then $(S_K)^{-1} = (T_K) * T_K$, and to use formula (1.27) (for the space $\mathbb{R}^m$).

1.6. Gaussian Measures in Hilbert Spaces

A Gaussian measure in a Hilbert space will be defined as an extension of a cylindrical measure given by Gaussian measures on finite-dimensional spaces. Let H be a real Hilbert space, and let S be a positive nuclear operator given in H. Let also $(e_j)_{j=1}^{\infty}$ be an orthonormal basis in H consisting of its eigenvectors, and let $a \in H$ be fixed. Denote by $\mathcal{K}$ a collection of all finite-dimensional subspaces of H spanned by e_j $(j \in \mathbb{N})$. The class $\mathcal{K}$ clearly satisfies conditions (i)–(iii) of Subsection 1.4. We define a cylindrical measure γ on the algebra $\mathfrak{C}(\mathcal{K}, H)$ by setting

$$\mathfrak{C}(\mathcal{K}, H) \ni \mathfrak{C}(K; \delta) \mapsto \gamma(\mathfrak{C}(K; \delta))$$

$$= \pi^{-1/2 \dim K}(\operatorname{Det} S_K)^{-1/2} \int_\delta \exp[-((S_K)^{-1}(f - P_K a), f - P_K a)_H]\, dm_K(f), \quad (1.34)$$

$$S_K = P_K S \upharpoonright K: K \to K,$$

where P_K is the projector onto K, and m_K is the Lebesgue measure in K induced by the metric of H.

Theorem 1.5. *Formula* (1.34) *defines a cylindrical measure in H which can be extended to the measure on $\mathfrak{C}_\sigma(\mathcal{K}, H) = \mathfrak{B}(H)$. The measure $\gamma_{S,a}$ obtained as a result is called the Gaussian measure with correlation operator S and mean value a.*

Proof. Let us first show that this definition of γ is correct, i.e., that it does not depend on the way a cylindrical set is denoted. Indeed, let $\mathfrak{C}(K; \delta) = \mathfrak{C}(L; \beta)$ be another notation of this set, then one of the subspaces K or L belongs to another, for example, $K \subset L$. Introducing the coordinates properly, we can write $L = \mathbb{R}^n = K \oplus \mathbb{R}^{n-m}$ and $\beta = \delta \times \mathbb{R}^{n-m}$ for some n and m ($m = 1, \ldots, n-1$). According to (1.34) and (1.30), we have

$$\gamma(\mathfrak{C}(L; \beta)) = \gamma_{S_L, P_L a}(\beta) = \gamma_{S_L, P_L a}(\delta \times \mathbb{R}^{n-m}) = \gamma_{(S_L)_K, P_L a}(\delta)$$

$$= \gamma_{S_K, P_L a}(\delta) = \gamma(\mathfrak{C}(K; \delta)) \qquad (\delta \in \mathfrak{B}(K)).$$

The evident remark that $(S_L)_K = S_K$ completes the proof of this statement.

Clearly, the function of sets (1.34) is absolutely additive for fixed K, i.e., (1.23) holds. Therefore, the measure γ defined above is cylindrical. To prove the theorem, one must show that the condition of Lemma 1.3 is satisfied.

Consider the ball $\tilde{B}_r(0) = \{f \in H \mid \|f\|_H \leq r\}$ in H and the cylindrical set $\mathfrak{C}(K; \delta)$ ($K \in \mathcal{K}$) lying outside this ball. Then the base $\delta \subset \mathfrak{C}(K; \delta)$ is also situated outside this ball and, thus, $\|f\|_H > r$ for $f \in \delta$. Therefore, by virtue of (1.34), (1.24), and (1.31), we get

$$\gamma(\mathfrak{C}(K; \delta)) \leq \pi^{-n/2}(\operatorname{Det} S_K)^{-1/2} \int_\delta \|f\|_H^2\, r^{-2} \exp[-((S_K)^{-1}(f - P_K a), f - P_K a)_H]\, dm_K(f)$$

$$= r^{-2} \int_\delta (f, f)_{\mathbb{R}^n}\, d\gamma_{S_K, P_L a}(f)$$

$$\leq r^{-2}\left(\tfrac{1}{2} \operatorname{Tr} S_K + \|P_K a\|_H^2\right) \leq r^{-2}\left(\tfrac{1}{2} \operatorname{Tr} S + \|a\|_H^2\right) \quad (n = \dim K) \quad (1.35)$$

(we have used the evident inequality $\operatorname{Tr} S_K \le \operatorname{Tr} S$). It follows from (1.35) that the measure $\gamma(\mathfrak{C}(K;\delta))$ can be made arbitrarily small for large r. Thus, the conditions of Lemma 1.3 are satisfied.

■

Let us present another "coordinate" method of constructing Gaussian measures in H which proves to be useful. We shall first construct a Gaussian product measure in $\mathbb{R}^\infty$ and then "land" it onto $l_2 = l_2(\mathbb{R}^1) = H$ — this means that, in fact, we shall try to obtain formula (1.28) in the infinite-dimensional case. Let us restrict ourselves to the case of the mean value zero.

As above, we assume that S is a positive nuclear operator in H and that $s = (s_k)_{k=1}^\infty$ is a sequence of its eigenvalues such that $s_k > 0$ and $\sum_{k=1}^\infty s_k < \infty$.

According to Subsection 1.3, we construct a Gaussian product measure in $\mathbb{R}^\infty$

$$\gamma_s = \mathop{\times}_{k=1}^{\infty} \gamma_{s_k}, \quad \gamma_{s_k}(x_k) = \frac{1}{\sqrt{\pi s_k}} \exp(-x_k^2/s_k^{-1})\, dx_k \qquad (k \in \mathbb{N}) \tag{1.36}$$

defined on the σ-algebra $\mathfrak{C}_\sigma(\mathbb{R}^\infty)$. By virtue of Lemma 1.1 and Theorem 1.3, $l_2 \in \mathfrak{C}_\sigma(\mathbb{R}^\infty)$ and $\gamma_S(l_2)$, i.e., l_2 is a set of full measures γ_S.

Consider an orthonormal basis $(e_k)_{k=1}^\infty$ in H composed of the eigenvectors of the operator S, i.e., $Se_k = s_k e_k$ $(k \in \mathbb{N})$. Let us realize the space H as a space l_2 in this basis.

Theorem 1.6. *For given positive nuclear operator S, let us construct a Gaussian measure γ_S in H and (using (1.36)) a Gaussian product measure γ_s in $\mathbb{R}^\infty$. Let us also realize H as $l_2 \subset \mathbb{R}^\infty$, as described above. It is stated that $\gamma_S = \gamma_s \restriction l_2$.*

Proof. The measures γ_S and $\gamma_s \restriction l_2$ are defined on the σ-algebras $\mathfrak{B}(H) = \mathfrak{B}(l_2)$ and $\{\alpha \in \mathfrak{C}_\sigma(\mathbb{R}^\infty) \mid \alpha \subset l_2\}$, respectively. It is easy to see that these two σ-algebras coincide. Thus, the latter consists of all sets $\beta \cap l_2$ $(\beta \in \mathfrak{C}_\sigma(\mathbb{R}^\infty))$ and, thereby, it is the σ-span of all sets $\beta \cap l_2$, where $\beta \in \mathfrak{C}_\sigma(\mathbb{R}^\infty)$. But these sets are cylindrical sets $\mathfrak{C}(K;\delta)$ from l_2, where K is a finite-dimensional subspace of l_2 belonging to the l.s. of the vectors $\delta_k = (\delta_{kj})_{j=1}^\infty$ $(k \in \mathbb{N})$. Let $\mathcal{K}$ be a class of all the subspaces of this sort. The σ-algebra under consideration coincides with $\mathfrak{C}_\sigma(\mathcal{K}, l_2)$ and, according to Lemma 1.2, with $\mathfrak{B}(l_2)$.

Hence, the measures γ_S and $\gamma_s \restriction l_2$ are defined on the same σ-algebra $\mathfrak{B}(l_2) = \mathfrak{C}_\sigma(\mathcal{K}, l_2)$. It suffices to check their coincidence only on $\mathfrak{C}_\sigma(\mathcal{K}, l_2)$. Suppose that $\mathfrak{C}(K;\delta) \in \mathfrak{C}_\sigma(\mathcal{K}, l_2)$. Consequently, we can assume that $K = \text{l.s.}((\delta_k)_{k=1}^n)$ with some $k \in \mathbb{N}$. Now $S_K = (s_j \delta_{jk})_{j,\,k=1}^n$. Therefore, according to (1.34), we obtain

$$\gamma(\mathfrak{C}(K;\delta)) = \pi^{-n/2}(s_1 \ldots s_n)^{-1/2} \int_\delta \exp[-(s_1^{-1}x_1^2 + \ldots + s_n^{-1}x_n^2)]\, dx_1 \ldots dx_n. \quad (1.37)$$

On the other hand, $\mathfrak{C}(K;\delta) = \mathfrak{C}(1, \ldots, n; \delta) \cap l_2$, where $\mathfrak{C}(1, \ldots, n; \delta) \in \mathfrak{C}(\mathbb{R}^\infty)$. Thus, taking the equality $\gamma_s(l_2) = 1$ into account, we find

$$(\gamma_s \upharpoonright l_2)(\mathfrak{C}(K;\delta)) = (\gamma_s \upharpoonright l_2)(\mathfrak{C}(1, \ldots, n; \delta) \cap l_2) = \gamma_s(\mathfrak{C}(1, \ldots, n; \delta)).$$

Computing this measure according to (1.36), we obtain the same expression (1.37). ■

The Fourier transform of an arbitrary finite measure $\mathfrak{B}(H) \ni \alpha \mapsto \mu(\alpha)$ in the Hilbert space H is defined by the following natural formula

$$\tilde{\mu}(y) = \int_H \exp[i(y, x)_H]\, d\mu(x). \quad (1.38)$$

It is well known that this transformation determines the measure μ uniquely (the proof of this fact is given in Chapter 5, Subsection 4.1).

Let us calculate $\tilde{\gamma}_S$. The simplest way to do that is to employ the coordinate setting of γ_S: We realize H as l_2 in the basis of eigenvectors, construct γ_s according to (1.36), and set $\gamma_S = \gamma_s \upharpoonright l_2$. Taking into account (1.38), (1.36), and (1.17), we find that

$$\tilde{\gamma}_S(y) = \int_{l_2} \exp[i(y, x)_{l_2}]\, d(\gamma_s \upharpoonright l_2)(x) = \int_{\mathbb{R}^\infty} \exp[i(y, x)_{l_2}]\, d\gamma_s(x)$$

$$= \exp(-\tfrac{1}{4}\|y\|^2_{l_2(s)}) = \exp\Big(-\tfrac{1}{4}\sum_{k=1}^{\infty} s_k y_k^2\Big) = \exp(-\tfrac{1}{4}(Sy, y)_{l_2})$$

for $y \in \mathbb{R}_0^\infty$. If we approximate an arbitrary vector in l_2 by vectors from $\mathbb{R}_0^\infty$, pass to the limit in the equality obtained, and return to the space H, then we get

$$\tilde{\gamma}_S(y) = \exp(-\tfrac{1}{4}(Sy, y)_H) \quad (y \in H). \quad (1.39)$$

Now we introduce a Hilbert space H_S as a completion of H with respect to the scalar product $(x, y)_{H_S} = (Sx, y)_H$ $(x, y \in H)$. Clearly, $H \subset H_S$ topologically, and this imbedding is quasinuclear. The Fourier transform $\tilde{\gamma}_S(y) = \exp(-\tfrac{1}{4}\|y\|_{H_S})$ $(y \in H)$ can be extended onto H_S by continuity.

The last statement concerning the Fourier transform of the measure γ_S proves to be of a much more general nature and does not depend on the Gaussian character of this measure in any way. Indeed, let us fix a certain class $\mathcal{K}$ and consider a cylindrical measure γ on $\mathfrak{C}(\mathcal{K}, H)$. For this measure, the Fourier transform is defined by

$$\tilde{\gamma}(y) = \int_H \exp[i(y, x)_H]\, d\gamma(x) \qquad (y \in \cup_{K \in \mathcal{K}} K \subset H) \tag{1.40}$$

(recall that $\cup_{K\in\mathcal{K}} K$ is a linear set dense in H). The integral in (1.40) is correctly defined for every fixed y: Denote by $K_y \in \mathcal{K}$ the one-dimensional space spanned by y. Then the function of sets $\mathfrak{B}(\mathbb{R}^1) \ni \delta \mapsto \gamma(K_y; \delta)$ is a measure and, thus, (1.40) can be rewritten in the form $\int_{\mathbb{R}^1} e^{it}\, d\gamma(K_y; t)$. In terms of $\tilde{\gamma}$, one can prove the following convenient criterion of the extendability of γ to a measure on $\mathfrak{B}(H)$ (the Minlos–Sazonov theorem).

Theorem 1.7. *A cylindrical measure γ in H can be extended to a measure on $\mathfrak{B}(H)$ if and only if one can find a Hilbert space $F \supset H$ with dense quasinuclear imbedding such that the Fourier transform $\tilde{\gamma}$ is continuous in the topology of F.*

The proof of this theorem can be actually reduced to the verification of the validity of the conditions of Lemma 1.3. We shall return to this theorem and present another proof in Chapter 5, Subsection 4.3 (while studying positive definite functions on H).

1.7. Gaussian Measures in Hilbert Riggings

Let us examine an important point of view on the construction of Gaussian measures in Hilbert spaces. We note first that formula (1.34) also determines the cylindrical measure γ (i.e., the weak distribution) in the case of arbitrary positive operators S. The condition $\operatorname{Tr} S < \infty$ appears when we extend this measure onto $\mathfrak{B}(H)$. If we compare this condition with the fact that we can construct an arbitrary Gaussian product measure γ_k on $\mathbb{R}^\infty$ and with Theorem 1.6, then it is natural to expect that this condition may be weakened when we try to extend γ onto $\mathfrak{B}(G)$ (but not onto $\mathfrak{B}(H)$). Here, $G \supset H$ is some broader Hilbert space[11]. It is convenient to take the negative space from the chain of real spaces to be G (in what follows, H is denoted by H_0)

$$H_- \supset H_0 \supset H_+. \tag{1.41}$$

We now proceed to a rigorous presentation. Consider the chain (1.41), the vector $a \in H_0$, and the positive operator S acting in H_0 which is unbounded, in general. As-

[11] Since γ can be interpreted as a weak distribution in H and $H \subset G$, we can consider γ as a weak distribution in G, as well.

sume that its domain of definition $\mathfrak{D}(S)$ and range of values $\mathfrak{R}(S)$ are dense in H_0 and that S^{-1} exists as an operator in H_0 acting from $\mathfrak{R}(S)$ to $\mathfrak{D}(S)$. Consider the generalized cylindrical sets $\mathfrak{C}(K;\delta)$ in the space H_-, where the class $\mathcal{K}$ is understood as a certain collection of finite-dimensional subspaces $K \subset \mathfrak{D}(S)$ satisfying conditions (i)–(iii) of Subsection 1.4 for $H = H_0$ (and, hence, for $H = H_-$). The set $\mathfrak{C}(K;\delta)$ itself is defined by a formula of type (1.19), namely

$$\mathfrak{C}(K;\delta) = \{\xi \in H_- \mid P_k\xi \in \delta\},$$

where P_k is an extension by continuity of the projector P_k given in H_0 to H_-. Since K is finite-dimensional, and $K \subset H_+$, this extension exists (cf. Subsection 1.9). One can easily see that everything that has been said in Subsection 1.4 remains valid for the collection $\mathfrak{C}(\mathcal{K}, H_-)$ of the sets $\mathfrak{C}(K;\delta)$ $(K \in \mathcal{K},\ \delta \in \mathfrak{B}(K))$. Here, the role of H is played by H_-. A cylindrical measure γ on the algebra $\mathfrak{C}(\mathcal{K}, H_-)$ of these sets is defined by

$$\mathfrak{C}(\mathcal{K}, H_-) \ni \mathfrak{C}(K;\delta) \mapsto \gamma(\mathfrak{C}(K;\delta))$$

$$= \pi^{-1/2 \dim K}(\operatorname{Det} S_K)^{-1/2} \int_\delta \exp[-((S_K)^{-1}(f - P_K a), f - P_K a)_{H_0}]\, dm_K(f), \quad (1.42)$$

$$(S_K = P_K S \restriction K:\ K \to K).$$

Let us describe the difference between (1.34) for $H = H_-$ and (1.42): Here, P_K is the projector in H_0 but not in H_-; the scalar product under the integral sign is taken in the sense of H_0 but not of H_-, and the Lebesgue measure m_K is also induced by the metric of H_0. In other words, the measure γ of the cylindrical set $\mathfrak{C}(K;\delta)$ from H_- is understood as the measure (1.34) of the corresponding cylindrical set $\mathfrak{C}^0(K;\delta)$ from H_0 with the same coordinate and base.

Comparing (1.42) and (1.34), we can make the following conclusion: Define the cylindrical measure γ in $H_0 = H$ by (1.34), where the class $\mathcal{K}$ is the same as above. The weak distribution γ_K is introduced as usual: We set

$$\mathfrak{B}(K) \ni \delta \mapsto \gamma_K(\delta) = \gamma(\mathfrak{C}(K;\delta)) \qquad (K \in \mathcal{K}).$$

Taking into account that $H_0 \subset H_-$, we can consider γ_K as a weak distribution in H_-, too. If we construct the cylindrical measure in H_- that corresponds to γ_K, then we obtain just formula (1.42). This means that (1.42) actually realizes the idea formulated above. It is also clear that this construction generalizes the construction presented earlier (see Subsection 1.6) for $H_+ = H_0 = H_-$.

Theorem 1.8. *Formula* (1.42) *defines the cylindrical measure on* H_- *which can be extended to the measure on* $\mathfrak{C}_\sigma(\mathcal{K}, H_-) = \mathfrak{B}(H_-)$ *provided that the estimate*

$$\operatorname{Tr}(P_K S P_K OI \restriction K) \le c < \infty \tag{1.43}$$

holds uniformly with respect to all the projectors of the form P_K $(K \in \mathcal{K})$ *in the space* H_0 *(*O *and* I *are standard operators connected with the chain* (1.41)*). This measure is called a Gaussian measure in the negative space* H_- *with correlation operator* S *and mean value* a *and is denoted by* $\gamma_{S,a}$ *as before.*

Let us show that $P_K S P_K OI \restriction K = (P_K S \restriction K)(P_K OI \restriction K)$, where both operators in parentheses are positive and act in K. By virtue of the properties of traces of operators in a finite-dimensional space, the trace in (1.43) is equal to $\operatorname{Tr}((P_K S \restriction K)^{1/2}(P_K OI \restriction K)(P_K S \restriction K)^{1/2})$ and, therefore, is nonnegative.

Proof. It has been already stated that comparing (1.42) and (1.34), one can establish the following relation $\gamma(\mathfrak{C}(K;\delta)) = \gamma^0(\mathfrak{C}^0(K;\delta))$, where γ^0 is the corresponding cylindrical measure in H_0. Hence, Theorem 1.5 implies that (1.42) defines the cylindrical measure on $\mathfrak{C}(\mathcal{K}, H_-)$ correctly.

Let us show that the condition of Lemma 1.3 is satisfied. In the space H_-, we consider the ball $\tilde{B}_r(0) = \{\xi \in H_- \mid \|\xi\|_{H_-} \le r\}$ and the cylindrical set $\mathfrak{C}(K;\delta)$ $(K \in \mathcal{K})$ which lies outside this ball. Then the base δ also lies outside $\tilde{B}_r(0)$ and, therefore, $\|f\|_{H_-} > r$ for $f \in \delta$. By using (1.42), the definition of the negative norm, (1.24), and (1.31), we obtain (cf. (1.35))

$$\gamma(\mathfrak{C}(K;\delta)) \le \pi^{-n/2}(\operatorname{Det} S_K)^{-1/2}\int_\delta r^{-2}\|f\|^2_{H_-}\exp[-((S_K)^{-1}(f-P_K a), f-P_K a)_{H_0}]\,dm_K(f)$$

$$= r^{-2}\int_\delta (OIf,f)_{H_0}\,d\gamma_{S_K,P_K a}(f) = r^{-2}\int_\delta (P_K OI \restriction Kf,f)_{\mathbb{R}^n}\,d\gamma_{S_K,P_K a}(f)$$

$$= r^{-2}\left[\tfrac{1}{2}\operatorname{Tr}(S_K(P_K OI \restriction K)) + ((P_K OI \restriction K)P_K a, P_K a)_{\mathbb{R}^n}\right]$$

$$\le r^{-2}\left[\tfrac{1}{2}\operatorname{Tr}(S_K(P_K OI \restriction K)) + \|(OI \restriction K)P_K a\|_{H_0}\|a\|_{H_0}\right]$$

$$\le r^{-2}(c/2 + \|a\|^2_{H_0}) \qquad (n = \dim K). \tag{1.44}$$

Here, we have used the estimate (1.43) and the fact that $\|OI\| \le 1$. The inequality (1.44) implies that the condition of Lemma 1.3 is satisfied.

■

Clearly, this theorem generalizes Theorem 1.5. Thus, for $H_+ = H_0 = H_-$ and positive nuclear S (restricted to the l.s. $((e_k)_{k=1}^{\infty})$), we have $P_K S P_K OI \restriction K = (P_K S \restriction K) = S_K$, and condition (1.43) is satisfied by virtue of the inequality $\operatorname{Tr} S_K \le \operatorname{Tr} S$. Below, we shall note two other important cases when condition (1.43) is satisfied. But first of all we stress one essential fact:

The measure $\gamma_{S,a}$ is completely determined by the space H_0, the positive operator S acting in this space, and the mean value $a \in H_0$ (the class $\mathcal{K}$ plays an auxiliary role).

As above, S is called a correlation operator. The chain (1.41) is chosen (ambiguously) with the only purpose being to "land" the cylindrical measure correctly as a measure on Borel subsets of the space H_-. This situation is quite similar to the case of the Gaussian product measure $\gamma_\kappa = \times_{k=1}^{\infty} \gamma_{\kappa_k}$ which is determined by the sequence $\kappa = (\kappa_k)_{k=1}^{\infty}$ ($\kappa_k > 0$) (more exactly, it generalizes this case). This measure can "land" either on $\mathbb{R}^\infty$ or on an arbitrary Hilbert space $l_2(p^{-1}) \subset \mathbb{R}^\infty$, where $\sum_{k=1}^{\infty} \kappa_k / p_k < \infty$ (according to Theorem 1.3).

Corollary 1. *Suppose, in addition, that the operator S is bounded in H_0. Then the measure $\gamma_{S,a}$ is defined on $\mathfrak{B}(H_-)$ for an arbitrary chain (1.41) in which the imbedding $H_+ \to H_0$ is quasinuclear. This means that an arbitrary quasinuclear chain is suitable for "landing" of this measure on the negative space.*

In fact,

$$\begin{aligned} \operatorname{Tr}(P_K S P_K OI \restriction K) &= \operatorname{Tr}((P_K S \restriction K)^{1/2} (P_K OI \restriction K)(P_K S \restriction K)^{1/2} \\ &\le \| (P_K S \restriction K)^{1/2} \|^2 \operatorname{Tr}(P_K OI \restriction K) \le (1 + \| S \|)^2 \operatorname{Tr}(OI) \\ &= (1 + \| S \|)^2 \operatorname{Tr}(OJ)^2 = (1 + \| S \|)^2 | OJ |^2 < \infty. \end{aligned}$$

Here, we have used the inequality for the selfadjoint operator A, namely, $\|A\| \le 1 + \|A^2\|$ and the relation $OI = (OJ)^2$ which follows from (1.9) of Chapter 1.

■

As in the finite-dimensional case, the measure $\gamma_1 = \gamma_{1,0}$ is called a canonical Gaussian measure (generated by the chain (1.41)). According to the corollary established above, it is defined on $\mathfrak{B}(H_-)$ for any quasinuclear chain (1.41).

Corollary 2. *For any positive operator S in H_0 with both dense domain of defi-*

nition and range of values and for any $a \in H_0$, *there exists a rigging* (1.41) *such that the measure* $\gamma_{S,a}$ *is defined on* $\mathfrak{B}(H_-)$.

Indeed, let us choose an orthonormal basis $(e_j)_{j=1}^{\infty} \subset \bigcup_{K \in \mathcal{K}} K$ in H_0 and consider an operator D acting in H_0 with a diagonal matrix $(d_j \delta_{kj})_{k,j=1}^{\infty}$, where $d_j \geq 0$. As mentioned in Chapter 1, Section 1, (1.19), by using the operator D, one can construct chain (1.41) so that $OJ = D^{-1}$. Then $OI = (OJ) = D^{-2}$. Assume that $\mathcal{K}$ is a collection of finite-dimensional subspaces from l.s. $((e_j)_{j=1}^{\infty})$, and that $K \in \mathcal{K}$. Choose $n \in \mathbb{N}$ sufficiently large so that $K \subset K_n = \text{l.s.}((e_j)_{j=1}^{n})$. Then

$$\operatorname{Tr}(P_K S P_K OI \upharpoonright K) = \operatorname{Tr}(P_K (P_{K_n} S P_K OI \upharpoonright K_n) \upharpoonright K) \leq \operatorname{Tr}(P_{K_n} S P_K OI \upharpoonright K_n)$$

$$= \sum_{j=1}^{n} ((P_{K_n} S P_K D^{-2} \upharpoonright K_n) e_j, e_j)_{H_0} = \sum_{j=1}^{n} d_j^{-2} (P_K e_j, S e_j)_{H_0}$$

$$\leq \sum_{j=1}^{n} d_j^{-2} \| S e_j \|_{H_0} \leq \sum_{j=1}^{\infty} d_j^{-2} \| S e_j \|_{H_0} = c.$$

It remains to take d_j rapidly increasing towards $+\infty$ so that the last series is convergent. The condition of Theorem 1.7 is satisfied.

■

Remark 1. It follows from the proof that the choice of rigging (1.41) is ambiguous and is determined by the behavior of $\| S e_j \|_{H_0}$ as $j \to \infty$.

■

Generally speaking, constructions such as Theorem 1.6 are now incorrect, because the operator S may possess a continuous spectrum. However, in the case when the chain (1.41) is quasinuclear, and the operator S commutes with OI, the situation can be diagonalized by decomposing H_0 in the eigenvectors of the operator OI (but not of S). The following two examples clarify this.

Example 1.4. Let us construct the canonical Gaussian measure γ_1 (with the original H_0) as a product measure. To do this, we take an arbitrary quasinuclear chain (1.41). We have

$$\operatorname{Tr}(OI) = \operatorname{Tr}((OJ^2)) = | OJ |^2, \quad \| OI \| \leq 1.$$

Let $(e_k)_{k=1}^{\infty}$ be an orthonormal basis in H_0 composed from the eigenvectors of the operator OI, i.e., $Ole_k = p_k^{-1} e_k$ ($p_k \geq 1$, $k \in \mathbb{N}$). Denote $p^{-1} = (p_k^{-1})_{k=1}^{\infty}$, $\sum_{k=1}^{\infty} p_k^{-1} < 1$. Consider the Gaussian product measure γ_1 (see (1.10)), where $1 = (1, 1, \dots)$. It is defined in $\mathbb{R}^\infty$ on $\mathcal{C}_\sigma(\mathbb{R}^\infty)$. However, according to Theorem 1.3, the space $l_2(p^{-1}) \in \mathcal{C}_\sigma(\mathbb{R}^\infty)$ has the full measure γ_1. With such a choice of the basis $(e_k)_{k=1}^{\infty}$, the chain (1.41) turns into the chain $l_2(p^{-1}) \supset l_2 \supset l_2(p)$ and, therefore, the measure γ_1, considered as a measure in H_-, should be a measure in $l_2(p^{-1})$ defined on $\mathfrak{B}(l_2(p^{-1}))$. Just as in the proof of Theorem 1.6, we easily get that $\gamma_1 = \gamma_1 \upharpoonright l_2(p^{-1})$ (and clearly,

$$\mathfrak{B}(l_2(p^{-1})) = \{\alpha \in \mathcal{C}_\sigma(\mathbb{R}^\infty) \mid \alpha \subset l_2(p^{-1})\}).$$

This example also clarifies the statement made at the end of Subsection 1.3.

Example 1.5. Let us compare two measures given in the coordinate form , namely the measure in H_0 with nuclear S (see Theorem 1.6) and the measure γ_1 from Example 1.4. In the first case, a Gaussian product measure γ_s on $\mathbb{R}^\infty$ is given by (1.36) with $\sum_{k=1}^{\infty} s_k < \infty$. We study the restriction of γ_s onto l_2, which is defined on $\mathfrak{B}(l_2)$ and equals γ_S with S induced in l_2 by the matrix $(s_j \delta_{jk})_{j,k=1}^{\infty}$. In the second case, we consider the Gaussian product measure γ_1 given on $\mathbb{R}^\infty$ (here, $1 = (1, 1, \dots)$), and its restriction to $l_2(p^{-1})$. As we have shown above, this restriction is equal to the canonical Gaussian measure γ_1 on $\mathfrak{B}(l_2(p^{-1}))$ (the chain (1.41) now has the form $l_2(p^{-1}) \supset l_2 \supset l_2(p)$).

It is easy to see that these two measures coincide for $s_k = p_k^{-1}$. In fact, let us map the space $l_2(p^{-1})$ onto the whole l_2 by introducing the coordinates $y_k = p_k^{-1/2} x_k$ of the vector $x = (x_k)_{k=1}^{\infty} \in l_2(p^{-1})$. As x runs through the set $\alpha \in \mathfrak{B}(l_2(p^{-1}))$, the vector $y = (y_k)_{k=1}^{\infty} \in l_2$ runs through the set $\alpha' \in \mathfrak{B}(l_2)$. This procedure results in a one-to-one correspondence between $\mathfrak{B}(l_2(p^{-1}))$ and $\mathfrak{B}(l_2)$. Changing the variables when computing the measure on cylindrical sets, we finally show that the measure $\gamma_1 \upharpoonright l_2(p^{-1})$ coincides with the measure $\gamma_s \upharpoonright l_2$ in the coordinates y_k.

We now formulate the coincidence of measures established in Example 1.5 in the form of the following simple but useful theorem.

Theorem 1.9. *Assume that the quasinuclear chain* (1.41) *and the canonical Gaussian measure* γ_1 *induced by this chain in* H_- *are given. Then this measure coincides with the measure* γ_S *in the Hilbert space* H_- *constructed by use of the positive nuclear operator* $S = O^+OI \in \mathcal{L}(H_-, H_-)$. *Conversely, each measure* γ_S *in* H_- *constructed using the positive nuclear operator* $S \in \mathcal{L}(H_-, H_-)$ *coincides with the canonical measure* γ_1 *induced by the properly chosen quasinuclear chain* (1.41).

Proof. The complete proof of this theorem can be found in Examples 1.4 and 1.5. One should only return from the chain $l_2(p^{-1}) \supset l_2 \supset l_2(p)$ to its coordinate free form (1.41), and note that, in fact, the mapping $l_2(p^{-1}) \ni x \to y \in l_2$ in Example 1.5 has the form $y = \boldsymbol{J}x$ (since the matrix $((p_j^{-1}\delta_{jk})_{j,k=1}^{\infty})$ corresponds to OI). This yields the formula $S = \boldsymbol{J}^{-1}OI\boldsymbol{J}$.

It remains to prove the equality $\boldsymbol{J}^{-1}OI\boldsymbol{J} = O^+OI$, i.e., $OI = \boldsymbol{J}O^+O\boldsymbol{J}$. But $(OIf, g)_{H_0} = (If, Ig)_{H_+}$ for any $f, g \in H_0$. On the other hand,

$$(\boldsymbol{J}O^+O\boldsymbol{J}f, g)_{H_0} = \left(O^+O\boldsymbol{J}f, \boldsymbol{J}g\right)_{H_0} = (O\boldsymbol{J}f, O\boldsymbol{J}g)_{H_0}$$

$$= (\boldsymbol{J}f, \boldsymbol{J}g)_{H_0} = (\boldsymbol{J}\boldsymbol{J}f, \boldsymbol{J}\boldsymbol{J}g)_{H_+} = (\boldsymbol{I}f, \boldsymbol{I}g)_{H_+} = (\boldsymbol{J}f, \boldsymbol{J}g)_{H_+},$$

and this gives the required result (above we have used relations (1.11) of Chapter 1). ■

Remark. The connection between the operator S and the chain (1.41) (in Theorem 1.9) can be also expressed by

$$(\xi, S^{-1}\varphi)_{H_-} = (\xi, \varphi)_{H_0}, \quad \mathfrak{D}(S^{-1}) = H_+ \quad (\xi \in H_-, \varphi \in H_+). \tag{1.45}$$

To prove this, we set $S^{-1}\varphi = \eta$. Then

$$(\xi, \varphi)_{H_0} = (\xi, S\eta)_{H_0} = \left(\xi, O^+OI\eta\right)_{H_0} = (\xi, I\eta)_{H_0} = (\xi, \eta)_{H_-}.$$

■

1.8. Gaussian Measures on Hilbert Riggings. The Language of the Fourier Transform

In Subsection 1.7, we have constructed the measure $\gamma_{S,a}$ in H_- (defined on $\mathfrak{B}(H_-)$) by using an arbitrary positive operator S acting in H_0 (with dense domain of definition and range of values) and $a \in H_0$. The chain $H_- \supset H_0 \supset H_+$ is chosen for the correlation operator S according to Corollary 2 of Theorem 1.8. Let us examine this problem with the help of the Fourier transform and employing Theorem 1.7. For the sake of simplicity, we consider the case when the mean value a is zero.

Let us first establish how to construct the cylindrical measure (1.42) from (1.34). We use the following scheme. Consider the class $\mathfrak{K}$ of coordinates from H_0 satisfying conditions (i)–(iii) of Subsection 1.4 with $H = H_0$. By using this class, we construct both cylindrical sets $\mathfrak{C}^0(K, \delta)$ from H_0 and $\mathfrak{C}(K, \delta)$ from H_- (this class evidently suits the constructions in $H = H_-$). We obtain two algebras of sets, namely $\mathfrak{C}(\mathcal{K}, H_0)$ from H_0 and $\mathfrak{C}(\mathcal{K}, H_-)$ from H_- with the same set of coordinates (and bases); there exists a one-to-one correspondence between these algebras $\mathfrak{C}(\mathcal{K}, H_0) \ni \mathfrak{C}^0(K, \delta) \mapsto \mathfrak{C}(K, \delta) \in \mathfrak{C}(\mathcal{K}, H_-)$[12].

In terms of the cylindrical measure γ^0 on $\mathfrak{C}(\mathcal{K}, H_0)$, we construct a cylindrical measure γ on $\mathfrak{C}(\mathcal{K}, H_-)$ by

$$\gamma(\mathfrak{C}(K, \delta)) = \gamma^0(\mathfrak{C}^0(K, \delta)) \qquad (\mathfrak{C}^0(K, \delta) \in \mathfrak{C}(\mathcal{K}, H_0))$$

(clearly, the inverse statement is also true: γ determines γ^0). In this sense, we can say that the cylindrical measure γ^0 can be extended to the cylindrical measure γ, and moreover, if γ is absolutely additive, then γ^0 can be extended to the measure γ_S. This is just the construction of γ_S from γ^0 that has been employed in Subsection 1.7 (the role of γ^0 was played by the measure (1.34) with $H = H_0$).

For cylindrical measures γ^0 and γ, the Fourier transform is introduced by (1.40) with $H = H_0$ and H_- on the same linear set of vectors $y \in L = \bigcup_{K \in \mathcal{K}} K$ dense in H_0. Theorem 1.7 implies the following result.

Theorem 1.10. *Assume that the Fourier transform* $\tilde{\gamma}^0(y)$ $(y \in L)$ *is continuous in the topology of* H_0. *Then for every quasinuclear chain* (1.41), *the cylindrical measure* γ *can be extended to the measure on* $\mathfrak{B}(H_-)$ *(or in other words, the original cylindrical measure* γ^0 *can be extended to the measure on* $\mathfrak{B}(H_-)$).

Proof. According to Theorem 1.7, it suffices to show that the Fourier transform $\tilde{\gamma}(y)$ $(y \in L \subset H_0 \subset H_-)$ is continuous in the topology of some Hilbert space $F \supset H_-$ with dense quasinuclear imbedding. Using (1.40) and the connection between γ^0 and γ, we find

$$\tilde{\gamma}(y) = \int_{H_-} \exp\big(i(y, x)_{H_-}\big)\, d\gamma(x) = \int_{H_-} \exp\big(i(\boldsymbol{I}y, x)_{H_0}\big)\, d\gamma(x)$$

$$= \int_{H_0} \exp\big(i(\boldsymbol{I}y, x)_{H_0}\big)\, d\gamma^0(x) = \tilde{\gamma}^0(\boldsymbol{I}y) \qquad (y \in H_-). \tag{1.46}$$

Let us introduce a Hilbert space F as a completion of H_- with respect to the scalar product $(\xi, \eta)_F = (\boldsymbol{I}\xi, \boldsymbol{I}\eta)_{H_0}$. It is easy to see that the imbedding $H_- \subset F$ is quasi-

[12] It is more logical to write $\mathfrak{C}(K; \delta) \subset H_-$ with index "–", but we shall not do this for the sake of simplicity.

nuclear: if $(e_j)_{j=1}^{\infty}$ is an orthonormal basis in H_-, then $(\boldsymbol{I}e_j)_{j=1}^{\infty}$ is an orthonormal basis in H_+ and, thus,

$$\sum_{j=1}^{\infty} \|e_j\|_F^2 = \sum_{j=1}^{\infty} \|\boldsymbol{I}e_j\|_{H_0}^2 < \infty,$$

because the imbedding $H_+ \subset H_0$ is quasinuclear. At the same time, (1.46) and the continuity of $\tilde{\gamma}^0$ in the topology of H_0 together imply that $\tilde{\gamma}$ is continuous in the topology of F.

■

Remark 1. The theorem proved enables us, in particular, to construct a Gaussian measure with bounded correlation operator S. In the case of unbounded S, we shall use the following generalization of this theorem:

Assume that the chain of Hilbert spaces $H_- \supset G_- \supset H_0 \supset G_+ \supset H_+$ can be constructed from (1.41) *so that $L \supset G_+$, $\tilde{\gamma}^0(y)$ $(y \in L)$ is continuous in the topology of G_+, and the imbedding $H_+ \subset G_+$ is quasinuclear; then γ can be extended to a measure on $\mathfrak{B}(H_-)$ (or the original measure γ^0 can be extended to a measure on $\mathfrak{B}(H_-)$).*

In fact, we can repeat the proof of Theorem 1.10, introducing F as the completion of H_- with respect to $(\xi, \eta)_F = (\boldsymbol{I}\xi, \boldsymbol{I}\eta)_{G_+}$ $(\xi, \eta \in H_-)$ and employing the quasinuclearity of the imbedding $H_+ \subset G_+$.

■

We now return to Gaussian measures. Consider an arbitrary positive operator S in H_0 with dense domain of definition and range of values. On $\mathfrak{C}(\mathcal{K}, H_0)$, we define a Gaussian cylindrical measure γ^0 by (1.34) with $H = H_0$ and $a = 0$; here, the class $\mathcal{K}$ satisfies conditions (i)–(iii) of Subsection 1.4 and consists of subspaces $K \in \mathfrak{D}(S)$. *Then the Fourier transform* (1.40) *of this measure has the form*

$$\tilde{\gamma}^0(y) = \int_{H_0} \exp\left[i\,(y, x)_{H_0}\right] d\gamma^0(y) = \exp\left(-\tfrac{1}{4}(Sy, y)_{H_0}\right) \quad (y \in L). \qquad (1.47)$$

Indeed, let us fix $y \in L$ and denote the one-dimensional space spanned by this vector by $K = K_y$. Let e_1 be the unit vector from K. Then according to (1.34), we have

$$\gamma^0(\mathbb{C}(K; \delta)) = \pi^{-1/2} s_K^{-1/2} \int_{\delta} \exp(-s_K^{-1} x_1^2)\, dx_1 \quad (\gamma \in \mathfrak{B}(\mathbb{R}^1)), \qquad (1.48)$$

where $s_K = (Se_1, e_1)_{H_0}$ (the action of the operator $S_K = P_K S \upharpoonright K$ is reduced to the multiplication by s_K in $K = \mathbb{R}^1$). Inserting (1.48) into (1.47), we obtain

$$\tilde{\gamma}^0(y) = \int_{\mathbb{R}^1} \exp\{i y_1 x_1\}\, d\gamma^0(\mathfrak{C}(K; x_1)) = \tilde{\gamma}_{s_K}(y_1) = \exp(-\tfrac{1}{4}(Sy, y)_{H_0}).$$

■

Assume that $S \in \mathfrak{L}(H_0, H_0)$. Then, by virtue of (1.47), $\tilde{\gamma}^0(y)$ $(y \in L)$ is continuous in the topology of H_0 and, therefore, according to Theorem 1.10, for every quasinuclear chain (1.41), the corresponding cylindrical measure γ can be extended to the measure on $\mathfrak{B}(H_-)$. Hence, we have established the existence of the Gaussian measure γ_S on $\mathfrak{B}(H_-)$ (this is a consequence of Corollary 1 of Theorem 1.8 for $a = 0$).

In the case of unbounded S, according to Remark 1, we proceed as follows. Denote by G_+ the completion of $\mathfrak{D}(S)$ in the norm of the graph of operator S, i.e., with respect to

$$(f, g)_{G_+} = (f, g)_{H_0} + (Sf, Sg)_{H_0} \quad (f, g \in \mathfrak{D}(S) \subset H_0).$$

Clearly, $L \subset G_+$, $G_+ \subset H_0$ topologically, and $\tilde{\gamma}^0(y)$ $(y \in L)$ is continuous in the topology of G_+ by virtue of (1.47). Thus, for any chain (1.41) such that the imbedding $H_+ \subset G_+$ is quasinuclear, the cylindrical measure γ can be extended to the measure on $\mathfrak{B}(H_-)$. This means that we have established the existence of the measure γ_S once again (this is a consequence of Corollary 2 of Theorem 1.8 for $a = 0$).

This approach to construction of the Gaussian measure γ_S with arbitrary S is less efficient than the approach of Subsection 1.7. However, one should keep in mind that it is often important to establish only the fact of the existence of a Gaussian measure with a given correlation operator, whereas the information about the space in which this measure is defined is unnecessary (this space is ambiguously defined for given H_0 and S).

■

Remark 2. In Subsection 1.6, we have shown that if $S \in \mathfrak{L}(H, H)$ is nuclear, then there exists the measure γ_S on $\mathfrak{B}(H)$. It is easy to see that the condition of nuclearity is also a necessary condition for the existence of γ_S. Indeed, it follows from Theorem 1.7 and (1.47) that it is necessary that a Hilbert space $F \supset H$ with dense quasinuclear imbedding exists such that the function $F \supset L \ni y \mapsto \exp(\tfrac{1}{4}(Sy, y)_H)$ is continuous. This is also evident in the case when the Gaussian measure γ_K is given in the coordinate form (1.10): If $\sum_{k=1}^{\infty} = \infty$, then $\gamma_K(l_2) = 0$, according to Theorem 1.3.

■

Example 1.6. The canonical Gaussian cylindrical measure H_0 cannot be extended to a measure on $\mathfrak{B}(H_0)$. We show this by direct calculation. Let $H_0 = H_+ = H_- = l_2$. For every $n \in \mathbb{N}$, we consider the set

$$\mathfrak{C}_n = \{f = (f_k)_{k=1}^{\infty} \in l_2 \mid |f_1| \leq n, \ldots, |f_{p_n}| \leq n\},$$

where $p_n \in \mathbb{N}$ is a fixed number. Evidently, $\mathfrak{C}_n \in \mathfrak{C}(\mathcal{K}, H_0)$, where $\mathcal{K}$ is a collection of all possible finite-dimensional subspaces from H_0. According to (1.42), we have

$$\gamma(\mathfrak{C}_n) = \left(\pi^{-1/2} \int_{-n}^{n} e^{-t^2} dt \right)^{p_n}.$$

For every n, we choose the number p_n to be sufficiently large so that $\gamma(\mathfrak{C}_n) \leq \frac{1}{2}$ and fix all these p_n. The sets $\mathfrak{C}'_n = l_2 \setminus \mathfrak{C}_n$ are also cylindrical, $\gamma(\mathfrak{C}'_n) \geq \frac{1}{2}$, and $\|f\|_{l_2} > n$ $(n \in \mathbb{N})$ for every $f \in \mathfrak{C}'_n$. Therefore, for each ball $\tilde{B}_r(0) = \{f \in l_2 \mid \|f\|_{l_2} \leq r\}$ one can find $\mathfrak{C}'_n$ such that $\mathfrak{C}'_n \subset l_2 \setminus \tilde{B}_r(0)$ and $\gamma(\mathfrak{C}'_n) \geq \frac{1}{2}$. Conditions of Lemma 1.3 do not hold and, consequently, the cylindrical measure γ under consideration cannot be extended to $\mathfrak{B}(l_2)$.

Remark 3. We now mention the following natural generalization of the problem of extension of the cylindrical measure in H_0 to $\mathfrak{B}(H_-)$.

Let B be a Banach space containing H_0. In what cases can we guarantee the extendability of this cylindrical measure (considered as a weak distribution) to $\mathfrak{B}(B)$? Here, the answer can be given with the use of more complicated terms as compared to (1.43) (see Kuo [1, Chapter 1, Section 4]).

■

Remark 4. Gaussian measures in the finite-dimensional space $\mathbb{R}^n$ are always absolutely continuous with respect to each other. This follows from (1.27), because each measure of this sort is equivalent to the Lebesgue measure in $\mathbb{R}^n$. In the case of a Hilbert space, the absolute continuity of Gaussian measures with respect to each other does not always occur. We shall return to this problem in Section 2, Subsection 2.4.

■

1.9. Gaussian Measures in Nuclear Riggings

In many cases, the formalism concerning the extension of measures in Hilbert riggings proves to be inadequate. First, we present the following example.

Example 1.7. Assume that the family $(S_\tau)_{\tau \in \mathrm{T}}$ of correlation operators in H_0 is given such that $\mathfrak{D}(S_\tau) \supset \mathfrak{D}$ for any $\tau \in \mathrm{T}$, where $\mathfrak{D}$ is a dense linear set in H_0. By using this family, we construct a family of cylindrical Gaussian measures $(\gamma^\tau)_{\tau\in \mathrm{T}}$. According to Subsections 1.7 and 1.8, for every measure γ^τ, we introduce the positive space H_τ and extend γ^τ to the measure γ_{S_τ} in $H_{-\tau}$. If we want to study the family of measures $(\gamma_{S_\tau})_{\tau \in \mathrm{T}}$ on a common Hilbert space, we must demand the existence of a Hilbert space H_+ such that $H_\tau \supset H_-$ $(\tau \in \mathrm{T})$ topologically. This space exists, e.g., in the case when $S_\tau \in \mathfrak{L}(H_+, H_0)$ $(\tau \in \mathrm{T})$, where $H_+ \subset H_0$ topologically. It is clear, however, that for an arbitrary family of correlation operators, this space may not exist. In this case, it is sometimes possible to realize the family $(\gamma_{S_\tau})_{\tau \in \mathrm{T}}$ as a family of measures on a linear topologic space more general than a Hilbert space. Thus, let $H_0 = l_2 = l_2(\mathbb{R}^1)$, and $\mathfrak{D} = \mathbb{R}^\infty_0$. Then every cylindrical measure γ^τ can be extended to a measure on a certain Hilbert space $H_{-\tau} \subset \mathbb{R}^\infty$. This means that $(\gamma_{S_\tau})_{\tau \in \mathrm{T}}$ can be regarded as a family of Gaussian measures on $\mathbb{R}^\infty$. Under additional assumptions, we can pass from $\mathbb{R}^\infty$ to a narrower linear manifold in $\mathbb{R}^\infty$. As will be shown below, the latter is directly determined by the properties of the family of correlation operators.

We now proceed to the realization of these concepts. Assume that the nuclear rigging

$$\Phi' \supset H_0 \supset \Phi \tag{1.49}$$

of the real space H_0 is given (see Chapter 1, Subsection 1.2). Generally speaking, the space $\Phi = \mathrm{pr}\lim_{\tau \to \mathrm{T}} H_\tau$ is not assumed to be separable.

Let us define the cylindrical sets in Φ'. Since Φ' is not a Hilbert space, we use a definition different from that presented in Subsection 1.4; it is, in fact, an advanced version of the definition given in Subsection 1.7. Let K be a finite-dimensional subspace of Φ, and let P_K be the orthogonal projector onto K defined in H_0. The operator P_K can be extended by continuity to a "projector" acting from Φ' to Φ. In fact, let us choose in K an orthonormal (in H_0) basis $(e_k)_{k=1}^n$ $(n = \dim K)$. Then

$$P_K f = \sum_{k=1}^{n} (f, e_k)_{H_0} e_k \qquad (f \in H_0), \tag{1.50}$$

and each term in (1.50) can be extended by continuity to $\xi \in \Phi'$ due to the fact that $e_k \in \Phi$. Hence, we obtain the extension of P_K to the operator $P_K \in \mathfrak{L}(\Phi', \Phi)$ with $\mathfrak{R}(P_K) = K \subset \Phi$.

For given finite-dimensional $K \subset \Phi$ and $\delta \in \mathfrak{B}(K)$, we introduce the set

$$\mathbb{C} = \mathbb{C}(K; \delta) = \{\xi \in \Phi' \mid P_K \xi \in \delta\}, \tag{1.51}$$

which will be called the cylindrical set from Φ' with the coordinate K and base δ.

Let us compare this definition for the chain (1.41) taken instead of (1.49) (see Subsection 1.7) with the definition (1.19) in the case of the space $H = H_-$. The vectors $e_k \in H_+$ from (1.50) now form an orthonormal basis in H_0, while the operator P_K takes the form

$$P_K \xi = \sum_{k=1}^{n} (\xi, \boldsymbol{I}^{-1} e_k)_{H_-} e_k \quad (\xi \in H_-).$$

Thus, realizing K as $\mathbb{R}^n$ in the basis $(e_k)_{k=1}^n$, we can rewrite the definition (1.51) as

$$\mathfrak{C}(K; \delta) = \{\xi \in H_- \mid (\xi, \eta_1)_{H_-}, \dots, (\xi, \eta_n)_{H_-} \in \delta\},$$

where $\eta_k = \boldsymbol{I}^{-1} e_k \in H_-$ and $\delta \in \mathfrak{B}(\mathbb{R}^n)$. This means that we have written (1.51) as a cylindrical set (1.19) but in the form (1.22).

We can now repeat all that was said in Subsection 1.4 concerning the ambiguity of the definition of a coordinate and a base for a given $\mathfrak{C}$. It is also clear that the collection of all cylindrical sets $\mathfrak{C}(K; \delta)$ $(\delta \in \mathfrak{B}(K))$ forms the algebra $\mathfrak{C}(\Phi')$, $\mathfrak{C}(K, \Phi')$ is its σ-subalgebra consisting of all sets of this sort with a fixed coordinate. Denote by $\mathfrak{C}_\sigma(\Phi')$ the σ-span of the algebra $\mathfrak{C}(\Phi')$. We stress that the role of the coordinates K is now played by arbitrary finite-dimensional subspaces of Φ. Therefore, the notion of a class $\mathcal{K}$ is absent from the notations connected with the measures on Φ'.

It is easy to see that *in the case of a countable Hilbert space* Φ, *we have* $\mathfrak{C}_\sigma(\Phi') = \mathfrak{B}(\Phi')$, *where* Φ' *is topologized by the weak topology*[13].

In fact, (1.51) can be rewritten as

$$\mathfrak{C}(K; \delta) = \{\xi \in \Phi' \mid (\xi, e_1)_{H_0}, \dots, (\xi, e_n)_{H_0} \in \delta\} \quad (e_k \in \Phi; \delta \in \mathfrak{B}(\mathbb{R}^n)). \tag{1.52}$$

Comparing (1.28) of Chapter 1 with (1.52), we conclude that $\mathfrak{C}(K; \delta) \in \mathfrak{B}(\Phi')$, i.e., $\mathfrak{C}_\sigma(\Phi') \subset \mathfrak{B}(\Phi')$. Let us prove that in the case $\Phi = \operatorname{pr}\lim_{\tau \to T} H_\tau$, where T is countable, this inclusion turns into the equality. It is easy to show (see Lemma 1.2) that every closed ball in $H_{-\tau}$ centered at 0 belongs to $\mathfrak{C}_\sigma(\Phi') \cap \mathfrak{B}(\Phi')$ and, hence, $H_{-\tau} \in \mathfrak{C}_\sigma(\Phi') \cap \mathfrak{B}(\Phi')$. Using this inclusion, Lemma 1.2 of Chapter 1, and the fact that T is countable, one can easily prove that $\mathfrak{C}_\sigma(\Phi') = \bigcup_{\tau \in T} \mathfrak{C}_\sigma(\mathcal{K}, H_{-\tau})$ and $\mathfrak{B}(\Phi') = \bigcup_{\tau \in T} \mathfrak{B}(H_{-\tau})$. However, according to Lemma 1.2, $\mathfrak{C}_\sigma(\mathcal{K}, H_{-\tau}) = \mathfrak{B}(H_{-\tau})$ and, consequently, $\mathfrak{C}_\sigma(\Phi') = \mathfrak{B}(\Phi')$.

■

[13] Note that, in this case, the classes of Borel sets from Φ' coincide for all topologizations of Φ' mentioned in Chapter 1, Subsection 1.2.

In the case of an arbitrary Φ, the σ-algebra $\mathfrak{C}_\sigma(\Phi')$ is, generally speaking, a proper part of $\mathfrak{B}(\Phi')$. In the case of a nuclear rigging, measures are constructed on $\mathfrak{C}_\sigma(\Phi')$.

The following statement is a consequence of (1.52) (cf. Subsection 1.4):

Remark 1. $\mathfrak{C}_\sigma(\Phi')$ can be also defined as the minimal σ-algebra with respect to which all the functionals $l_\varphi(\xi) = (\xi, \varphi)_{H_0}$ $(\xi \in \Phi')$, where $\varphi \in \Phi'$ (or φ belongs to a certain set dense in Φ) are measurable.

■

As in Subsection 1.4, the set function $\mathfrak{C}(\Phi') \ni \mathbb{C} \to \gamma(\mathbb{C}) \in [0, 1]$ is called a cylindrical measure in Φ' provided that for any $K \subset \Phi$ such that $\dim K < \infty$, the restriction $\gamma \restriction \mathfrak{C}(K, \Phi')$ is absolutely additive.

The Fourier transform of the cylindrical measure γ is defined by the equality

$$\tilde{\gamma}(\varphi) = \int_{\Phi'} \exp\left[i(\varphi, \xi)_{H_0}\right] d\gamma(\xi) \qquad (\varphi \in \Phi), \tag{1.53}$$

which makes sense by virtue of the facts that $\exp\left[i(\varphi, \cdot)_{H_0}\right]$ is measurable with respect to $\mathfrak{C}(K_\varphi, \Phi')$ and that $\gamma \restriction \mathfrak{C}(K_\varphi, \Phi')$, where K_φ is spanned by the given φ, is absolutely additive. This definition somewhat differs from definition (1.40), because here we have used another definition of a cylindrical set. Conditions of absolute additivity of a cylindrical measure on $\mathfrak{C}(\Phi')$ are given by the following Minlos theorem.

Theorem 1.11. *If the Fourier transform* $\tilde{\gamma}(\varphi)$ $(\varphi \in \Phi)$ *of the cylindrical measure* γ *is continuous in the topology of* Φ, *then this measure is absolutely additive on* $\mathfrak{C}(\Phi')$ *and, consequently, can be extended to the measure on* $\mathfrak{C}_\sigma(\Phi')$.

The proof of this theorem is presented in Gelfand, Vilenkin [1, Chapter 4, Section 4] and Daletsky, Fomin [1, Chapter 3, Theorem 1.3]. Nevertheless, we shall now establish the validity of Theorem 1.11 for the case of Gaussian measures γ_S with a certain correlation operator S, since in this case Theorem 1.11 is frequently used below. Let us first present some preliminary facts.

Consider an operator $S\colon \Phi \to \Phi'$ for which there exist the indices $\tau', \tau'' \in \mathrm{T}$ (which depend on S) and $c_{\tau'\tau''} \in (0, \infty)$ such that

$$\| S\varphi \|_{H_{-\tau''}} \le c_{\tau'\tau''} \| \varphi \|_{H_{\tau'}} \quad (\varphi \in \Phi).$$

Since the set $(H_\tau)_{\tau \in \mathrm{T}}$ is directed by inclusion, and the set $(H_{-\tau})_{\tau \in \mathrm{T}}$ is directed accordingly, the collection $\mathfrak{L}_0(\Phi, \Phi')$ of all the operators of this sort is a linear set.

Each operator $S \in \mathcal{L}_0(\Phi, \Phi')$ acts continuously from Φ to Φ' (equipped with a weak topology), i.e.,

$$|(S\varphi, \psi)_{H_0}| \le \|S\varphi\|_{H_{-\tau''}} \|\psi\|_{H_{\tau''}} \le c_{\tau'\tau''} \|\varphi\|_{H_{\tau'}} \|\psi\|_{H_{\tau''}} \quad (\varphi \in \Phi)$$

for any $\psi \in \Phi$. In other words, $\mathcal{L}_0(\Phi, \Phi') \subset \mathcal{L}(\Phi, \Phi')$. Generally speaking, the inverse inclusion is not true. However, *if* Φ *is a countable Hilbert nuclear space, then* $\mathcal{L}_0(\Phi, \Phi') = \mathcal{L}(\Phi, \Phi')$. In fact, one can always assume that $0 \in T$, and $H_\tau \subset H_0$ $(\tau \in T)$ topologically (Chapter 1, Subsection 1.2). Let $S \in \mathcal{L}(\Phi, \Phi')$. Consider a two-linear form

$$\Phi \times \Phi \ni (\varphi, \psi) \mapsto a(\varphi, \psi) = (S\varphi, \psi)_{H_0} \in \mathbb{R}^1.$$

Obviously, it is separately continuous. According to Remark 1 in Chapter 1, Subsection 2.3, the kernel theorem (Theorem 2.4) holds for this form, namely,

$$(S\varphi, \psi)_{H_0} = a(\varphi, \psi) = (\varphi \otimes \psi, \Theta)_{H_0 \otimes H_0} \quad (\varphi, \psi \in \Phi),$$

where $\Theta \in \Phi' \otimes \Phi'$. By virtue of (2.17) of Chapter 1, one can find $\tau \in T$ such that $\Theta \in H_{-\tau} \otimes H_{-\tau}$. But then

$$|(S\varphi, \psi)_{H_0}| = |(\varphi \otimes \psi, \Theta)_{H_0 \otimes H_0}| \le \|\Theta\|_{H_{-\tau} \otimes H_{-\tau}} \|\varphi\|_{H_\tau} \|\psi\|_{H_\tau} \quad (\varphi, \psi \in \Phi).$$

■

Consider an operator $S \in \mathcal{L}_0(\Phi, \Phi')$ positive in the sense of the chain $\Phi' \supset H_0 \supset \Phi$, i.e., $(S\varphi, \varphi)_{H_0} > 0$ for any nonzero $\varphi \in \Phi$. Let us construct a Gaussian measure γ_S with the correlation operator S. First, we note that, by virtue of positivity, the form $\Phi \otimes \Phi \ni (\varphi, \psi) \mapsto (S\varphi, \psi)_{H_0} = (\varphi, \psi)_{H_S}$ can be regarded as a scalar product in Φ.

Let H_S *be the completion of* Φ *with respect to this scalar product. Then* $\Phi \subset H_S$ *topologically.*

In fact, one should only prove that the imbedding $\Phi \to H_S$ is continuous, but this is an immediate consequence of the estimate

$$|(\varphi, \psi)_{H_S}| = |(S\varphi, \psi)_{H_0}| \le c_{\tau'\tau''} \|\varphi\|_{H_{\tau'}} \|\psi\|_{H_{\tau''}} \quad (\varphi, \psi \in \Phi),$$

where $\tau', \tau'' \in T$ are connected with $S \in \mathcal{L}_0(\Phi, \Phi')$.

■

According to the definition, a Gaussian cylindrical measure γ on Φ' is given on the sets $\mathfrak{C}(K;\delta)\in\mathfrak{C}(\Phi')$ by the formula similar to (1.42)

$$\mathfrak{C}(\Phi') \ni \mathfrak{C}(K;\delta) \mapsto \gamma(\mathfrak{C}(K;\delta))$$

$$= \pi^{-1/2\dim K}(\det S_K)^{-1/2}\int_\delta \exp[-((S_K)^{-1}\varphi,\varphi)_{H_0}]\,dm_K(\varphi) = \gamma_{S_K}(\delta) \quad (1.54)$$

$$(S_K = P_K S \upharpoonright K: K\to K).$$

Here, as before, P_K is a projector in H_0 extended to Φ', and the Lebesgue measure m_K is induced by the metric of H_0 and by the correlation operator S formally regarded as an operator in H_0. In other words, $\gamma(\mathfrak{C}(K;\delta)) = \gamma^0(\mathfrak{C}^0(K;\delta))$, where γ^0 is the original cylindrical measure on $\mathfrak{C}(\mathfrak{K}, H_0)$ with a chosen class $\mathfrak{K}$. The invertibility of the operator S_K in (1.54) follows from the fact that the equality $S_K\varphi = 0$ yields $(S\varphi,\varphi)_{H_0} = 0$ for some nonzero $\varphi\in K\subset\Phi$, but this contradicts the positivity of S.

The Fourier transform of the cylindrical measure γ is calculated just as in (1.47)

$$\tilde{\gamma}(\varphi) = \int_{\Phi'} \exp(i(\varphi,\xi)_{H_0})\,d\gamma(\xi) = \exp(-\tfrac{1}{4}(S\varphi,\varphi)_{H_0}) \quad (\varphi\in\Phi). \qquad (1.55)$$

It follows from (1.55) that $\tilde{\gamma}(\varphi)$ $(\varphi\in\Phi)$ is continuous in the topology of Φ. According to the first part of Theorem 1.11, γ can be extended to a measure on $\mathfrak{C}_\sigma(\Phi')$. We now shall prove this fact by direct calculation.

Let $\tau',\tau''\in T$ be indices dependent on the operator $S\in\mathfrak{L}_0(\Phi,\Phi')$. Then, for any $\varphi\in\Phi$, we have

$$|(S\varphi,\varphi)_{H_0}| \le c_{\tau',\tau''}\|\varphi\|_{H_{\tau'}}\|\varphi\|_{H_{\tau''}} \le \|\varphi\|^2_{H_\tau},$$

where $\tau\in T$ is such that $H_{\tau'}, H_{\tau''}\subset H_\tau$ topologically. Therefore, the quadratic form $(S\varphi,\varphi)_{H_0}$ is continuous in the topology of the space H_τ and, hence, by virtue of (1.55), this is also true for the measure $\tilde{\gamma}=\tilde{\gamma}^0$. By using the nuclearity of Φ, for this $\tau\in T$, we can choose $\tau'''\in T$ such that $H_{\tau'''}\subset H_\tau$ densely, and the imbedding operator is quasinuclear. By virtue of Remark 1, Subsection 1.8, the original cylindrical measure γ^0 can be extended to the measure μ_1 on $\mathfrak{B}(H_{-\tau'''})$. Furthermore, γ^0 can be extended to the measure μ_2 on Φ'. Indeed, taking into account that $H_{-\tau'''}\subset\Phi'$, we obtain (by definition) $\mu_2(\alpha)=\mu_1(\alpha\cap H_{-\tau'''})$ $(\alpha\in\mathfrak{C}_\sigma(\Phi')$ and $\alpha\cap H_{-\tau'''}\in\mathfrak{B}(H_{-\tau'''})$, see Lemma 1.2). We set $\gamma=\mu_2$.

■

Let us summarize the facts concerning Gaussian measures that were presented above.

Assume that the nuclear rigging

$$\Phi' \supset H_- \supset H_0 \supset H_+ \supset \Phi \tag{1.56}$$

is given, and that conditions imposed on the imbedding operator and on the positive operator $S \in \mathfrak{L}_0(\Phi, \Phi')$ *are usual. Then*

(i) *a Gaussian measure* γ_S *with a correlation operator* S *is defined on the* σ-*algebra* $\mathfrak{C}_\sigma(\Phi')$;

(ii) *if* $S \in \mathfrak{L}(H_+, H_-)$, *then the measure* γ_S *can be regarded as a measure in an arbitrary extension* $F \supset H_-$ *with a quasinuclear imbedding on the* σ-*algebra* $\mathfrak{B}(F)$;

(iii) *if* $S \in \mathfrak{L}(H_0, H_0)$ *and the imbedding of* H_+ *into* H_0 *is quasinuclear, then* γ_S *is a measure on* $\mathfrak{B}(H_-)$;

(iv) *if* S *is a nuclear operator in* H_0, *then* γ_S *is a measure on* $\mathfrak{B}(H_0)$.

In fact, statement (i) has been already established. Statement (ii) follows from Remark 1 of Subsection 1.8, because the continuity of $S: H_+ \to H_-$ and (1.47) together imply that $\tilde{\gamma}^0(y)$ $(y \in L \subset H_+)$ is continuous in the topology of H_+. Statement (iii) follows from statement (ii) or, for example, from Corollary 1 of Theorem 1.8. Finally, statement (iv) results from Theorem 1.5.

■

As a rule, having given operator S from H_0, in what follows we choose the chain (1.56) possessing the required properties.

Concluding this subsection, we stress that below the Fourier transform of a probability measure $\mathfrak{C}_\sigma(\Phi') \ni \alpha \mapsto \mu(\alpha) \geq 0$ will be always defined as a function

$$\tilde{\mu}(\varphi) = \int_{\Phi'} \exp\left(i(\varphi, \xi)_{H_0}\right) d\mu(\xi) \qquad (\varphi \in \Phi). \tag{1.57}$$

In the case of a probability measure $\mathfrak{B}(H_-) \ni \alpha \mapsto \mu(\alpha) \geq 0$ (for the Hilbert rigging (1.41)) the Fourier transform $\tilde{\mu}$ is defined analogously. One should only replace Φ' by H_- and take $\varphi \in H_+$ in (1.57). Therefore, for the Gaussian measure γ_S, we have

$$\tilde{\gamma}_S(\varphi) = \exp\left(-\tfrac{1}{4}(S\varphi, \varphi)_{H_0}\right) \tag{1.58}$$

$$(S \in \mathfrak{L}_0(\Phi, \Phi'),\ \varphi \in \Phi,\ S \in \mathfrak{L}(H_+, H_-),\ \varphi \in H_+).$$

The Fourier transform (1.38) corresponds to the case $H = H_- = H_0 = H_+$.

1.10. Cylindrical Functions

Suppose that the assumption of Subsection 1.9 holds. Let us study the concept of a cylindrical function on Φ' in detail. The function $\Phi' \ni \xi \mapsto f(\xi) \in \mathbb{C}^1$, measurable with respect to $\mathfrak{C}_\sigma(\Phi')$, is called cylindrical if and only if there exists a finite-dimensional subspace $K \subset \Phi$ such that f is measurable with respect to the σ-algebra $\mathfrak{C}(K, \Phi')$.

Lemma 1.4. *The function $\Phi' \ni \xi \mapsto f(\xi) \in \mathbb{C}^1$ is measurable with respect to the σ-algebra $\mathfrak{C}(K, \Phi')$ if and only if there exists a Borel function $K \ni \xi \mapsto f_K(\xi) \in \mathbb{C}^1$ such that $f(\xi) = f_K(P_K \xi)$, where P_K is the extension of the orthogonal projector* (1.50) *(onto the subspace K) defined in H_0 to the space Φ'.*

Proof. Let $K \subset \Phi$ ($\dim K < \infty$) be given. We associate the Borel function $K \ni \varphi \mapsto f_K(\varphi) \in \mathbb{C}^1$ with the function $f(\xi) = f_K(P_K \xi)$ ($\xi \in \Phi'$) on Φ'. Then

$$\{\xi \in \Phi' \mid f(\xi) \in \alpha\} = \{\xi \in \Phi' \mid P_K \xi \in \delta_{f_K}\},$$

for $\alpha \in \mathfrak{B}(\mathbb{C}^1)$, where

$$\delta_{f_K} = \{\varphi \in K \mid f_K(\varphi) \in \alpha\} \in \mathfrak{B}(K),$$

but this means that this function is measurable with respect to the σ-algebra $\mathfrak{C}(K, \Phi')$. Conversely, assume that f is measurable with respect to $\mathfrak{C}(K, \Phi')$. It is easy to see that $\mathfrak{C}(K, \Phi')$ can be defined as the minimal σ-algebra with respect to which all linear functions $l_\varphi(\xi) = (\xi, \varphi)_{H_0}$ (here, $\xi \in \Phi'$, $\varphi \in K$) are measurable. As a sufficiently wide collection of $\varphi \in K$, one can take any total set from K. Let us choose a basis $(e_k)_{k=1}^n$ ($n = \dim K$) in K which is orthonormal in H_0. Then the σ-algebra $\mathfrak{C}(K, \Phi')$ is generated by the functions $l_{e_K}(\cdot)$ ($k = 1, \ldots, n$) and, consequently, any $\mathfrak{C}(K, \Phi')$-measurable function f can be represented as

$$f(\xi) = g((\xi, e_1)_{H_0}, \ldots, (\xi, e_n)_{H_0}),$$

where g is a Borel function on $\mathbb{R}^n$. We introduce the function f_K by setting

$$f_K(\varphi) = g((\varphi, e_1)_{H_0}, \dots, (\varphi, e_n)_{H_0}) \quad (\varphi \in K).$$

The measurability of this function with respect to $\mathfrak{B}(K)$ follows from the measurability of the linear function and from the fact that g is a Borel function. Taking the form of P_K into account, we obtain the required representation

$$f(\xi) = f_K(P_K \xi) \qquad (\xi \in \Phi').$$

■

The next statement follows immediately from the proof of Lemma 1.4.

Corollary 1. *Each cylindrical function f admits a representation*

$$f(\xi) = g((\xi, \varphi_1)_{H_0}, \dots, (\xi, \varphi_n)_{H_0}),$$

where $\varphi_k \in \Phi$ $(k = 1, \dots, n)$, and g is a Borel function on $\mathbb{R}^n$. Also note that the equality

$$f(\xi) = f_K(P_K \xi).$$

holds for this function.

■

The concept of a cylindrical function on H_- is introduced quite analogously, and the evident analogue of Lemma 1.4 is valid in this case.

Cylindrical functions possess one important property: They are dense in functional spaces connected with the measures in linear spaces. This makes it possible to approximate the functions of infinitely many variables by the functions of a finite-dimensional argument.

Lemma 1.5. *Let μ be a measure on $\mathfrak{C}_\sigma(\Phi')$. The set of bounded cylindrical functions is dense in each space $L_p = L_p(\Phi', \mathfrak{C}_\sigma(\Phi'), \mu)$, $p \geq 1$.*

Proof. It suffices to show that the characteristic function κ_α of an arbitrary set $\alpha \in \mathfrak{C}_\sigma(\Phi')$ can be approximated by bounded cylindrical functions in the sense of the metric of L_p.

Taking into account that $\mathfrak{C}_\sigma(\Phi')$ is generated by the algebra $\mathfrak{C}(\Phi')$, for given $\varepsilon > 0$ one can find a cylindrical set $\alpha_\varepsilon \in \mathfrak{C}(\Phi')$ such that $\mu((\alpha \setminus \alpha_\varepsilon) \cup (\alpha_\varepsilon \setminus \alpha)) < \varepsilon$. It remains to note that $\kappa_{\alpha_\varepsilon}$ is a cylindrical function and

$$\int_{\Phi'} \left|\kappa_\alpha - \kappa_{\alpha_\varepsilon}\right|^p d\mu \leq \mu((\alpha \setminus \alpha_\varepsilon) \cup (\alpha_\varepsilon \setminus \alpha)) < \varepsilon.$$

■

1.11. The Case of Abstract Duality

It is sometimes necessary to construct measures in the dual space without using the duality given by a certain Hilbert space. In this case, there is no imbedding of the original space onto the dual one, and the chain (1.49) breaks. This leads to some technical modifications. Let us briefly examine them.

Let Φ be a real nuclear space and Φ' be the space dual to Φ. The duality between Φ and Φ' is denoted by $\langle \cdot, \cdot \rangle$, and this means that each $\varphi \in \Phi$ is associated with the linear functional $l_\varphi(\xi) = \langle \xi, \varphi \rangle$ $(\xi \in \Phi')$ continuous in the weak topology on Φ'. A finite-dimensional subspace $K \subset \Phi$ is associated with a σ-algebra $\mathfrak{C}(K, \Phi')$ of sets from Φ' generated by the sets of the form

$$\{\xi \in \Phi' \mid (\langle \xi, \varphi_1 \rangle, \dots, \langle \xi, \varphi_n \rangle) \in \delta\}, \tag{1.59}$$

where $\varphi_1, \dots, \varphi_n \in \Phi$ is a certain basis in K, and $\delta \in \mathfrak{B}(\mathbb{R}^n)$. In other words, $\mathfrak{C}(K, \Phi')$ is the minimal σ-algebra with respect to which the functionals $l_\varphi(\cdot)$, $\varphi \in K$ are measurable. The algebra of cylindrical sets $\mathfrak{C}(\Phi')$ is introduced as a union over all $K \subset \Phi$. Finally, the σ-algebra generated by $\mathfrak{C}(\Phi')$ is denoted by $\mathfrak{C}_\sigma(\Phi')$. Note that in the case where the duality between Φ and Φ' is established (as before) by the chain $\Phi' \supset H_0 \supset \Phi$, we obtain the same objects as in Subsection 1.9. This is why we keep the same notations for the system of sets introduced above.

A set function γ $(\mathfrak{C}(\Phi') \ni \mathbb{C} \mapsto \gamma(\mathbb{C}) \in [0, 1])$, such that the restriction $\gamma \restriction \mathfrak{C}(K, \Phi')$ is a measure on $\mathfrak{C}(K, \Phi')$ for any $K \subset \Phi$ $(\dim K < \infty)$, is called a cylindrical measure in Φ'.

For any $\varphi \in \Phi$, the function $\exp(i\langle \xi, \varphi \rangle)$ is measurable with respect to the σ-algebra $\mathfrak{C}(K_\varphi, \Phi')$, where K_φ is spanned by φ. Therefore, the function

$$\tilde{\gamma}(\varphi) = \int_{\Phi'} \exp(i\langle \xi, \varphi \rangle)\, d\gamma(\xi) \quad (\varphi \in \Phi),$$

called the Fourier transform of the cylindrical measure γ, is well-defined. The following theorem now serves as an analogue of Theorem 1.11 (for the proof, see the references to Theorem 1.11).

Theorem 1.12. *In order that the cylindrical measure γ on $\mathfrak{C}(\Phi')$ be absolutely additive, it is sufficient that $\tilde{\gamma}(\varphi)$ $(\varphi \in \Phi)$ be continuous in the topology of Φ.*

Let us show how Gaussian measures on Φ' can be constructed with the help of Theorem 1.12.

Let $\Phi \ni \varphi \mapsto a(\varphi, \psi)$ be a positive continuous quadratic form on Φ (i.e., $a(\varphi, \varphi) > 0$ for any nonzero $\varphi \in \Phi$). Then a probability measure γ_a such that

$$\tilde{\gamma}_a(\varphi) = \exp(-\tfrac{1}{4}\, a(\varphi, \varphi)) \quad (\varphi \in \Phi)$$

is defined on the σ-algebra $\mathfrak{C}_\sigma(\Phi')$.

Indeed, for a finite-dimensional subspace $K \subset \Phi$ and an orthonormal basis $e = (e_k)_{k=1}^n$ in K, we can construct the function

$$f_e(x) = \exp\left(-\frac{1}{4} a\left(\sum_{k=1}^n x_k e_k, \sum_{k=1}^n x_k e_k\right)\right) \quad (x = (x_1, \dots, x_n) \in \mathbb{R}^n).$$

By virtue of the fact that a is positive definite, this function is the Fourier transform of a certain Gaussian measure γ_S in $\mathbb{R}^n$ (S can be easily expressed in terms of a and e). For a cylindrical set

$$\mathbb{C} = \{\xi \in \Phi' \mid (\langle \xi, e_1 \rangle, \dots, \langle \xi, e_n \rangle) \in \delta\} \quad (\delta \in \mathfrak{B}(\mathbb{R}^n)), \tag{1.60}$$

we set $\gamma(\mathbb{C}) = \gamma_S(\delta)$. It is easy to verify that this definition is correct, i.e., that it does not depend on the representation of the cylindrical set in the form (1.60). The set function γ is absolutely additive on $\mathfrak{C}(K, \Phi')$ due to the fact that γ_S is a measure. Moreover, the definition of γ yields

$$\tilde{\gamma}(\varphi) = \int_{\Phi'} \exp(i\langle \xi, \varphi \rangle)\, d\gamma(\xi) = \int_{\mathbb{R}^1} e^{it}\, d\gamma_S(t) = \exp(-\tfrac{1}{4}\, a(\varphi, \varphi)) \quad (\varphi \in \Phi).$$

Hence, we have constructed the cylindrical measure γ on $\mathfrak{C}(\Phi')$ with the Fourier transform

$$\tilde{\gamma}(\varphi) = \exp(-\tfrac{1}{4}\, a(\varphi, \varphi)) \quad (\varphi \in \Phi).$$

The continuity of a on Φ involves the continuity of $\tilde{\gamma}(\varphi)$ in the topology of Φ. Therefore, Theorem 1.12 implies the absolute additivity of γ. This means that we have constructed the measure γ_a on $\mathfrak{C}_\sigma(\Phi')$.

■

§2. Some Problems of Analysis in the Case of Gaussian Measures

In the previous section, we have mainly discussed the methods of setting measures on linear spaces. There the central problem was the extension of a cylindrical measure to the measure on a properly chosen space. For some classes of measures on linear spaces, one can obtain additional information both about the measures and about the functional spaces connected with them. This is particularly true for Gaussian measures. The importance of this special case can be explained not only by the fact that it allows the most advanced investigation, but also by the particular role played by Gaussian measures in applications: various model systems in quantum field theory and statistical physics are described with the help of these measures.

2.1. Measurable Linear and Multilinear Functionals

We shall consider the most general case of measures in nuclear riggings. However, in our presentation, we also discuss the special cases when the additional properties of the correlation operator allow us to make the corresponding statements more exact.

Consider a nuclear riggings of a real space H_0

$$\Phi' \supset H_0 \supset \Phi. \tag{2.1}$$

In what follows, it is convenient to denote the vectors from Φ', H_0, and Φ by $x, y, \dots$; $f, g, \dots$; and $\varphi, \psi, \dots$, respectively. Assume that the operator $S \in \mathfrak{L}_0(\Phi, \Phi')$ is given which is positive in the sense that $(S\varphi, \varphi)_{H_0} > 0$ for any nonzero $\varphi \in \Phi$. Denote by γ_S the Gaussian measure on $\mathfrak{C}_\sigma(\Phi')$ with the correlation operator S and mean value zero, constructed in Subsection 1.9.

Consider the Hilbert space H_S that has already appeared in Subsection 1.9 and was obtained as a result of the completion of Φ with respect to the scalar product

$$(\varphi, \psi)_H = (S\varphi, \psi)_{H_0} \quad (\varphi, \psi \in \Phi). \tag{2.2}$$

Since $S \in \mathfrak{L}_0(\Phi, \Phi')$, the imbedding $\Phi \subset H_S$ is continuous. The space H_S defines the Gaussian measure γ_S canonically in the following sense.

Let H be a Hilbert space such that $\Phi \subset H$ topologically. The function $\exp(-\frac{1}{4}\|\varphi\|_H^2)$ $(\varphi \in \Phi)$ is a Fourier transform of some Gaussian measure γ_H on Φ'. In fact, let us take the operator $S \in \mathfrak{L}_0(\Phi, \Phi')$ which realizes the representation $(\varphi, \psi)_H = (S\varphi, \psi)_{H_0}$ (we assume that this operator exists), and consider the measure γ_S on Φ'. Then $\tilde{\gamma}(\varphi) = \exp(-\frac{1}{4}\|\varphi\|_H^2)$ $(\varphi \in \Phi)$. Therefore, we can set $\gamma_H = \gamma_S$.

By virtue of this correspondence, below we shall use equivalently (depending on the situation) both the Gaussian measure γ_S which corresponds to the given correlation operator $S \in \mathfrak{L}_0(\Phi, \Phi')$ and the Gaussian measure γ_H generated by the Hilbert space H, where $H \supset S$ topologically. Thus, $\gamma_{H_0} = \gamma_1$.

In this section, we always use the spaces $L_p(\Phi', \gamma_S) = L_p(\Phi', \mathfrak{C}_\sigma(\Phi'), \gamma_S)$ constructed by using the measure γ_S (or γ_H). Recall also that $L_0(\Phi', \mathfrak{C}_\sigma(\Phi'))$ denotes the collection of functions $\Phi' \ni x \mapsto f(x) \in \mathbb{C}^1$ measurable with respect to $\mathfrak{C}_\sigma(\Phi')$.

Many characteristics of the measure γ_S can be expressed in terms of the space H_S. First, let us show how to calculate the moments of the measure γ_S. For given $\varphi_1, \dots, \varphi_n \in \Phi$, the n-th moment of the measure μ in the space Φ' given on the σ-algebra $\mathfrak{C}_\sigma(\Phi')$ is defined as follows:

$$s_n^\mu(\varphi_1, \dots, \varphi_n) = \int_{\Phi'} (\varphi_1, x)_{H_0} \cdots (\varphi_n, x)_{H_0} \, d\mu(x), \tag{2.3}$$

provided that the last integral exists. In the case of the Gaussian measure all the moments exist, and one can easily obtain an explicit expression for them.

Lemma 2.1. *The following equalities hold*

$$s_{2n+1}^{\gamma_s}(\varphi_1, \dots, \varphi_{2n+1}) = 0, \tag{2.4}$$

$$s_{2n}^{\gamma_s}(\varphi_1, \dots, \varphi_{2n}) = 2^{-n} \sum (\varphi_{k_1}, \varphi_{j_1})_{H_S} \cdots (\varphi_{k_n}, \varphi_{j_n})_{H_S} \tag{2.5}$$

$$(\varphi_k \in \Phi;\ k = 1, \dots, 2n+1;\ n \in \mathbb{Z}_+),$$

where summation is carried out over all $(2n)!/2^n n!$ *pairings of the numbers* $(1, \dots,$

2n), i.e., over all possible decompositions of this collection of numbers into n pairs $(k_1, j_1), \dots, (k_n, j_n)$.

Proof. For fixed nonzero $\varphi \in \Phi$, the function $(\varphi, x)_{H_0}$ $(x \in \Phi')$ is measurable with respect to the σ-algebra $\mathcal{C}_\sigma(K_\varphi, \Phi')$ (K_φ is spanned by φ) and, hence, by virtue of (1.54), we get

$$\int_{\Phi'} f((\varphi, x)_{H_0})\, d\gamma_S(x) = \pi^{-1/2} S_{K_\varphi}^{-1/2} \int_{\mathbb{R}^1} f(t) \exp(-S_{K_\varphi}^{-1} t^2)\, dm_{K_\varphi}(t)$$

$$= \pi^{-1/2} (S\varphi, \varphi)_{H_0}^{-1/2} \int_{\mathbb{R}^1} f(t) \exp[-(S\varphi, \varphi)_{H_0}^{-1} t^2]\, dt$$

for every Borel function $f \in L_0(\mathbb{R}^1, \mathcal{B}(\mathbb{R}^1))$. In particular,

$$\int_{\Phi'} (\varphi, x)_{H_0}^k\, d\gamma_S(x) = \pi^{-1/2} (S\varphi, \varphi)_{H_0}^{-1/2} \int_{\mathbb{R}^1} t^k \exp[-(S\varphi, \varphi)_{H_0}^{-1} t^2]\, dt. \qquad (2.6)$$

Taking into account the well-known formula

$$\frac{1}{\sigma\sqrt{\pi}} \int_{\mathbb{R}^1} t^k \exp(-t^2/\sigma^2)\, dt = \begin{cases} 0, & k = 2n+1, \\ 2^{-n}(2n-1)!!\, \sigma^n = \dfrac{(2n)!\, 2^{-2n}}{n!} \sigma^n, & k = 2n, \end{cases}$$

we obtain the statement of the lemma for the case of $\varphi_1 = \dots = \varphi_{2n+1} = \varphi \in \Phi$. To pass to the general case, we employ the polarization identity

$$\prod_{k=1}^{m} (\varphi_k, x)_{H_0} = \frac{1}{m!} \prod_{k=1}^{m} (-1)^{m-p} \sum_{k_1 < \dots < k_p} (\varphi_{k_1} + \dots + \varphi_{k_p}, x)_{H_0}^m \qquad (2.7)$$

$$(\varphi_k \in \Phi;\ x \in \Phi';\ m \in \mathbb{N}),$$

which immediately yields formulas (2.4) and (2.5).

■

Remark 1. Along with the chain $\Phi' \supset H_0 \supset \Phi$, we consider its complexification $\Phi'_c \supset H_{0,c} \supset \Phi_c$. Denote by $(\cdot, \cdot)_{H_{0,c}}$ the scalar product in $H_{0,c}$. In addition, let us introduce a two-linear extension of the scalar product from H_0 into $H_{0,c}$ preserving the old notation $(\cdot, \cdot)_{H_0}$. Then $(\varphi, \psi)_{H_{0,c}} = (\varphi, \overline{\psi})_{H_0}$, where $\varphi, \psi \in H_{0,c}$ (by definition, if $\varphi = u + i v$, $u, v \in H_0$, then $\overline{\varphi} = u - i v$). Due to multilinearity, the equalities (2.4) and (2.5) can be extended to the case of $\varphi_k \in \Phi_c$.

■

Assume that the kernel $a_n \in \Phi_c^{\otimes n} = \Phi_c \otimes \ldots \otimes \Phi_c$ (n times; $n \in \mathbb{N}$) is given. Introduce a continuous n-linear monomial $m_{a_n}(x)$ ($x \in \Phi'$) by the equality

$$m_{a_n}(x) = (a_n, x^{\otimes n})_{H_0^{\otimes n}},$$

where $x^{\otimes n} = x \otimes \ldots \otimes x$ (n times). For a given collection $(a_k)_{k=0}^n$, where $a_0 \in \mathbb{C}^1$, $a_k \in \Phi_c^{\otimes k}$ $(k = 1,\ldots,n)$, let us introduce a continuous polynomial "with complex coefficients of a real variable x"

$$p(x) = a_0 + \sum_{k=1}^{n} (a_k, x^{\otimes k})_{H_0^{\otimes k}} \quad (x \in \Phi';\ n \in \mathbb{N}) \tag{2.8}$$

(for $n = 0$, we have $p(x) = a_0$). Denote the collection of all continuous polynomials on Φ' by $\mathcal{P}(\Phi')$.

Let us select a subset of cylindrical polynomials $\mathcal{P}_{\text{cyl}}(\Phi')$ from the set of continuous polynomials. This subset consists of those polynomials of the type (2.8) which are cylindrical functions on Φ'. Let us show that $\mathcal{P}_{\text{cyl}}(\Phi')$ *coincides with the linear span of the set of cylindrical functions of the form*

$$l_{\varphi_1, \ldots, \varphi_n}(x) = \prod_{k=1}^{n} (\varphi_k, x)_{H_0} \quad (x \in \Phi';\ n \in \mathbb{N})$$

and constants $c \in \mathbb{C}^1$. Indeed, it is easy to see that the fact that a polynomial p of the type (2.8) is cylindrical means that for any $k = 1, \ldots, n$ and some finite-dimensional $K \subset \Phi$ we have $a_k = a_k' + i a_k''$, where $a_k', a_k'' \in K^{\otimes k}$. To prove the last statement, we employ the equality $p(x) = p(P_K x)$ (which follows from the definition of a cylindrical function on Φ') and the relation

$$(a_k, (P_K x)^{\otimes k})_{H_0^{\otimes k}} = ((P_K \otimes \ldots \otimes P_K) a_k, x^{\otimes k})_{H_0^{\otimes k}} \quad (k = 1, \ldots, n)$$

in each term of (2.8). But then

$$a_k = \sum_{\alpha_1, \ldots, \alpha_k}^{m} a_{\alpha_1, \ldots, \alpha_k} e_{\alpha_1} \otimes \ldots \otimes e_{\alpha_k},$$

where $m = \dim K$, $(e_j)_{j=1}^m$ is a basis in K, and $a_{\alpha_1, \ldots, \alpha_k} \in \mathbb{C}^1$. Therefore,

$$(a_k, x^{\otimes k})_{H_0^{\otimes k}} = \sum_{\alpha_1,\ldots,\alpha_k}^{m} a_{\alpha_1,\ldots,\alpha_k} l_{e_1,\ldots,e_k}(x).$$

■

For given $n \in \mathbb{Z}_+$, we denote by $\mathcal{P}_n(\Phi')$ the collection of continuous polynomials on Φ' whose degree is at most n, i.e., the set of functions of the form (2.8) with fixed n given on Φ'. The closure of $\mathcal{P}_n(\Phi')$ in $L_2(\Phi', \gamma_S)$ is denoted by $\mathcal{P}_{\gamma_S, n}(\Phi')$. We shall call it the subspace of measurable polynomials of a degree not exceeding n. The elements of the set $\mathcal{P}_{\gamma_S}(\Phi') = \bigcup_{n=0}^{\infty} \mathcal{P}_{\gamma_S, n}(\Phi')$ will be called measurable polynomials. It follows from the argument presented in Lemma 2.2 that $\mathcal{P}_{\gamma_S, n}(\Phi')$ coincides with the closure of the linear span of monomials $\prod_{k=1}^{m} (\varphi_k, x)_{H_0}$ $(\varphi_k \in \Phi;\ k = 1, \ldots, m;\ m \le n)$ and constants in $L_2(\Phi', \gamma_S)$.

We now present the essential consequence of Lemma 2.1.

Corollary 1. *For all $h \in H_S$, the measurable linear function $l_h(x) = (h, x)_{H_0}$ $(x \in \Phi')$ is defined, moreover, $l_h \in L_p(\Phi', \gamma_S)$ for all $p \ge 1$.*

Indeed, for the sequence $(\varphi_j)_{j=1}^{\infty} \subset \Phi$, $\varphi_j \underset{j\to\infty}{\to} h$ in H_S, the equality (2.5) implies that $(l_{\varphi_j})_{j=1}^{\infty}$ is fundamental in $L_2(\Phi', \gamma_S)$. We set $l_h = \lim_{j\to\infty} l_{\varphi_j}$. The equality

$$\int_{\Phi'} (\varphi_j, x)_{H_0}^{2n}\, d\gamma_S(x) = c_n \|\varphi_j\|_{H_S}^{2n} \qquad (n \in \mathbb{N}) \tag{2.9}$$

(see (2.5)) shows that, for all $p \ge 1$, we have $l_{\varphi_j} \underset{j\to\infty}{\to} l_h$ in $L_p(\Phi', \gamma_S)$.

■

By virtue of the above argument, the measurable monomial

$$l_{h_1,\ldots,h_n}(x) = (h_1 \otimes \ldots \otimes h_n, x^{\otimes n})_{H_0^{\otimes n}} = \prod_{k=1}^{n} (h_k, x)_{H_0} \tag{2.10}$$

can be defined for $h_1, \ldots, h_n \in H_S$ as a function from the space $L_p(\Phi', \gamma_S)$ for any $p \ge 1$. Note that in the case of a general kernel $a_n \in H_S^{\otimes n}$, it is impossible to define the monomial $m_{a_n}(x) = (a_n, x^{\otimes n})_{H_0^{\otimes n}}$ as an element of $L_2(\Phi', \gamma_S)$ (and, in general, as a measurable function). We clarify this by the following example related to the case $n = 2$, i.e., to the case of a quadratic monomial.

Example 2.1. Assume that $S = 1$, i.e., γ_S is a canonical Gaussian measure γ_{H_0} corresponding to the Hilbert space H_0. For a basis $(e_k)_{k=1}^{\infty}$ which is orthonormal in H_0 and a sequence of coefficients $(\lambda_k)_{k=1}^{\infty}$, $\lambda_k \geq 0$ $(k \in \mathbb{N})$ such that $\sum_{k=1}^{\infty} \lambda_k^2 < \infty$, the kernel $a = \sum_{k=1}^{\infty} \lambda_k e_k \otimes e_k \in H_0^{\otimes 2}$ since $\| a \|_{H_0^{\otimes 2}}^2 = \sum_{k=1}^{\infty} \lambda_k^2 < \infty$. If, in addition, we assume that $\sum_{k=1}^{\infty} \lambda_k^2 < \infty$, then, according to the Kolmogorov-Khinchin criterion (see Section 1, Theorem 1.3), the function

$$(a, x^{\otimes 2})_{H_0^{\otimes 2}} = \sum_{k=1}^{\infty} \lambda_k (x, e_k)_{H_0}^2 \quad (x \in \Phi')$$

is given by a series convergent γ_{H_0}-almost everywhere on Φ'. Moreover, by direct calculation, we obtain

$$\|(a, \cdot)_{H_0^{\otimes 2}}\|_{L_2(\Phi', \gamma_{H_0})}^2 = c_1 \sum_{k=1}^{\infty} \lambda_k^2 + c_2 \sum_{\substack{k,j=1 \\ k \neq j}}^{\infty} \lambda_k \lambda_j,$$

where

$$c_1 = \int_{\Phi'} (e_k, x)_{H_0}^4 \, d\gamma_{H_0}(x) = \frac{3}{4}, \qquad c_1 = \left(\int_{\Phi'} (e_k, x)_{H_0}^2 \, d\gamma_{H_0}(x) \right)^2 = \frac{1}{4}$$

(we have used Lemma 2.1 when calculating these integrals). Thus, the monomial under consideration belongs to $L_2(\Phi', \gamma_{H_0})$, provided that $\sum_{k=1}^{\infty} \lambda_k < \infty$.

However, in the case when $\sum_{k=1}^{\infty} \lambda_k = \infty$ the same Kolmogorov-Khinchin criterion gives $(a, x^{\otimes 2})_{H_0^{\otimes 2}} = \infty$ for γ_{H_0}-almost all $x \in \Phi'$. Hence, the fact that the inclusion $a \in H_0^{\otimes 2}$ holds is not sufficient in order to define the monomial $(a, x^{\otimes 2})_{H_0^{\otimes 2}} = \infty$ $(x \in \Phi')$ as a function from $L_2(\Phi', \gamma_{H_0})$.

We now leave the question about the internal description of a set of measurable polynomials open for a moment, since we need an auxiliary construction to give a complete answer (this answer will be given in Corollary 2 of Theorem 2.2; see also Remark 2).

Thus, we have constructed the following systems of polynomials

$$\mathcal{P}_{\mathrm{cyl}}(\Phi') \subset \mathcal{P}(\Phi') \subset \mathcal{P}_{\gamma_S}(\Phi').$$

Some properties of these systems (necessary in what follows) will be established below. The detailed analysis of these properties is given in the next subsection. Here, we only note the fact that the polynomials are dense in $L_p(\Phi', \gamma_S)$.

Lemma 2.2. *For all* $p \geq 1$, *the set of cylindrical polynomials* $\mathcal{P}_{\text{cyl}}(\Phi')$ *is dense in* $L_p(\Phi', \gamma_S)$.

Proof. Since cylindrical functions are dense in each $L_p(\Phi', \gamma_S)$ (Lemma 1.5), it suffices to show that each cylindrical function f can be approximated by the elements of $\mathcal{P}_{\text{cyl}}(\Phi')$. Therefore, let $f \in L_p(\Phi', \gamma_S)$ be measurable with respect to the σ-algebra $\mathcal{C}(K, \Phi')$, $K \subset \Phi$, $\dim K < \infty$. The restriction $\gamma_S \upharpoonright \mathcal{C}(K, \Phi')$ given by the formula (1.54) induces a Gaussian measure on $\mathcal{B}(K)$ with a correlation operator $S_K = P_K S \upharpoonright K$, where P_K is an extension to Φ' of an orthogonal projector onto the subspace K acting in H_0. The function f can be represented as $f(x) = f_K(P_K x)$ $(x \in \Phi')$, where f_K is a Borel function on K, and the inclusion $f \in L_p(\Phi', \gamma_S)$ means that

$$f_K \in L_p(K, \gamma_{S_K}) = L_p(K, \mathcal{B}(K), \gamma_{S_K}). \tag{2.11}$$

In K, we choose a basis $(e_k)_{k=1}^n$ $(n = \dim K)$ orthonormal in H_0. Every polynomial $q \in \mathcal{P}(K)$ on the finite-dimensional space K can be represented as follows

$$q(x) = \sum_{|\alpha| \leq m} b_\alpha x^\alpha \qquad (x \in K), \tag{2.12}$$

where $b_\alpha \in \mathbb{C}^1$, $x^\alpha = (x, e_1)_{H_0}^{\alpha_1} \dots (x, e_n)_{H_0}^{\alpha_n}$, and $\alpha = (\alpha_1, \dots, \alpha_n)$ is a multiindex with nonnegative integer components. Clearly, for a given $\varepsilon > 0$, one can find a polynomial q of the form (2.12) for which

$$\int_K |f_K(x) - q(x)|^p d\gamma_{S_K}(x) < \varepsilon. \tag{2.13}$$

Let us define the polynomial $p \in \mathcal{P}_{\text{cyl}}(\Phi')$ by setting

$$p(x) = q(P_K x) = \sum_{|\alpha| \leq m} b_\alpha (P_K x, e_1)_{H_0}^{\alpha_1} \dots (P_K x, e_n)_{H_0}^{\alpha_n}$$

$$= \sum_{|\alpha| \leq m} b_\alpha (x, e_1)_{H_0}^{\alpha_1} \dots (x, e_n)_{H_0}^{\alpha_n} \qquad (x \in \Phi').$$

Then, taking (2.13) into account, we obtain

$$\int_{\Phi'} |f(x) - p(x)|^p d\gamma_S(x) = \int_{\Phi'} |f_K(P_K x) - q(P_K x)|^p d\gamma_S(x)$$

$$= \int_K |f_K(x) - q(x)|^p \, d\gamma_{S_K}(x) < \varepsilon.$$

■

Here, we have also proved that, for a cylindrical function measurable with respect to $\mathfrak{C}(K, \Phi')$, one can find an approximating sequence of polynomials measurable with respect to the same σ-algebra.

Lemma 2.3. *Let K_1 and K_2 be finite-dimensional subspaces of Φ that are orthogonal in H_S. Then for all $f_1, f_2 \in L_2(\Phi', \gamma_S)$ and $f_j \in L_0(\Phi', \mathfrak{C}(K_j, \Phi'))$ $(j = 1,$ 2), the following relation holds:*

$$\int_{\Phi'} f_1(x) f_2(x) \, d\gamma_S(x) = \int_{\Phi'} f_1(x) \, d\gamma_S(x) \int_{\Phi'} f_2(x) \, d\gamma_S(x). \tag{2.14}$$

Proof. First, let $f_1, f_2 \in \mathcal{P}_{\mathrm{cy}\,1}(\Phi')$. Then each of these functions is a sum of monomials of the form

$$(\varphi_1^{(j)}, x)_{H_0} \cdots (\varphi_{m_j}^{(j)}, x)_{H_0} \qquad (\varphi_k^{(j)} \in K_j;\ j = 1, 2;\ x \in \Phi').$$

To establish the desired equality in this case, we must show that

$$\int_{\Phi'} \prod_{k=1}^{m_1} (\varphi_k^{(1)}, x)_{H_0} \prod_{l=1}^{m_2} (\varphi_l^{(2)}, x)_{H_0} \, d\gamma_S(x)$$

$$= \int_{\Phi'} \prod_{k=1}^{m_1} (\varphi_k^{(1)}, x)_{H_0} \, d\gamma_S(x) \int_{\Phi'} \prod_{l=1}^{m_2} (\varphi_l^{(2)}, x)_{H_0} \, d\gamma_S(x). \tag{2.15}$$

Taking (2.5) and the orthogonality of $\varphi_k^{(1)}$ and $\varphi_l^{(2)}$ in H_S into account, we obtain the following expression on the right-hand side of (2.15):

$$2^{-(n_1+n_2)} \sum (\varphi_{s_1}^{(1)}, \varphi_{t_1}^{(1)})_{H_S} \cdots (\varphi_{s_{n_1}}^{(1)}, \varphi_{t_{n_1}}^{(1)})_{H_S} (\varphi_{p_1}^{(2)}, \varphi_{q_1}^{(2)})_{H_S} \cdots (\varphi_{p_{n_2}}^{(2)}, \varphi_{q_{n_2}}^{(2)})_{H_S}$$

for $m_1 = 2n_1$ and $m_2 = 2n_2$. Here, the sum is taken only over the pairings inside each subset $(1, \ldots, 2n_1)$ and $(1, \ldots, 2n_2)$. But the same expression can be obtained if we apply (2.5) to the right-hand side twice. In the case where $m_1 + m_2$ is odd, both the right-hand and left-hand sides vanish by virtue of (2.4). Finally, if $m_1 + m_2$ is even but m_1 and m_2 are odd, then the right-hand side is zero because each pairing of numbers from the set $(1, \ldots, m_1, m_1 + 1, \ldots, m_1 + m_2)$ necessarily contains the pair (p, q) such that $p \le m_1$, $q > m_1$. Thus, (2.15) is proved.

For arbitrary f_1 and f_2 of the above-mentioned type we choose, according to Lemma 2.2, the sequence $(f_j^{(n)})_{n=1}^{\infty} \subset \mathcal{P}_{\mathrm{cyl}}(\Phi')$, $(j = 1, 2)$ such that $f_j^{(n)}$ is measurable with respect to $\mathfrak{C}(K_j, \Phi')$ and $f_j^{(n)} \underset{n\to\infty}{\to} f_j$ in $L_2(\Phi', \gamma_S)$. Then, according to the result already proved, we have

$$\int_{\Phi'} f_1^{(n)}(x) f_2^{(n)}(x)\, d\gamma_S(x) = \int_{\Phi'} f_1^{(n)}(x)\, d\gamma_S(x) \int_{\Phi'} f_2^{(n)}(x)\, d\gamma_S(x),$$

and this equality turns into (2.14) as $n \to \infty$.

■

2.2. The Fock Space and the Wiener–Itô Decomposition

The Hilbert space $L_2(\Phi', \gamma_S)$ admits a decomposition which is very useful from the technical point of view, namely the decomposition into the orthogonal sum of subspaces which consist of polynomials of a special type (the so-called Wick monomials). Before we proceed with the description of this decomposition in the general case, let us consider the simplest example related to the case $H_0 = \mathbb{R}^1$.

Example 2.2. Let γ_1 be a canonical Gaussian measure on the one-dimensional Hilbert space $\mathbb{R}^1$ (we exploit this high-sounding terminology in such a trivial case in order to regard this situation as the most simplified version of the general case). Clearly, we now have $d\gamma_1(x) = \pi^{-1/2} \exp(-x^2)\,dx$ and $\Phi' = \Phi = \mathbb{R}^1$. Consider the Hermite polynomials given by

$$H_n(x) = (-1)^n e^{x^2} \frac{d^n}{dx^n}(e^{-x^2}) \quad (n \in \mathbb{Z}_+,\ x \in \mathbb{R}^1). \tag{2.16}$$

The orthogonality relation for Hermite polynomials

$$\int_{\mathbb{R}^1} H_n(x)\, H_m(x)\, d\gamma_1(x) = 2^n\, n!\, \delta_{nm} \quad (n, m \in \mathbb{Z}_+) \tag{2.17}$$

and the well-known formulas

$$x^{2m} = \frac{(2m)!}{2^{2m}} \sum_{n=0}^{m} \frac{1}{(2n)!\,(m-n)!} H_{2n}(x),$$

$$x^{2m+1} = \frac{(2m+1)!}{2^{2m+1}} \sum_{n=0}^{m} \frac{1}{(2n+1)!\,(m-n)!} H_{2n+1}(x) \quad (m \in \mathbb{Z}_+) \tag{2.18}$$

imply that the normalized polynomials

$$h_n(x) = \left(2^n n!\right)^{-1/2} H_n(x) \qquad (n \in \mathbb{Z}_+)$$

form an orthonormal basis in the Hilbert space $L_2(\mathbb{R}^1, \gamma_1)$. This basis is constructed by use of the Schmidt orthogonalization of the powers $1, x, x^2, \dots$, so that $h_n \perp (\text{l.s.}(1, x, \dots, x^{n-1}))$. Consequently, every $f \in L_2(\mathbb{R}^1, \gamma_1)$ can be expanded into the Fourier–Hermite series

$$f(x) = \sum_{n=0}^{\infty} f_n h_n(x), \quad f_n = (f, h_n)_{L_2(\mathbb{R}^1, \gamma_1)} \quad (n \in \mathbb{Z}_+). \tag{2.19}$$

This enables us to set the unitary isomorphism between the space $L_2(\mathbb{R}^1, \gamma_1)$ and $l_2(\mathbb{C}^1)$ given by the relation

$$L_2(\mathbb{R}^1, \gamma_1) \ni f \leftrightarrow \vec{f} = (f_0, f_1, \dots) \in l_2(\mathbb{C}^1). \tag{2.20}$$

Below, we obtain a generalization of the expansion (2.19) for the infinite-dimensional case, which allows us to introduce an analogue of the isomorphism (2.20) with the so-called Fock space instead of the space $l_2(\mathbb{C}^1)$. We now proceed to describe the Fock space.

Let H be a real Hilbert space, and let H_c be its complexification. For every $n \in \mathbb{N}$ we denote by $H_c^{\otimes n} = H_c \otimes \dots \otimes H_c$ the n-th tensor power of H_c. The object of our interest is a subspace of symmetric elements from $H_c^{\otimes n}$. To define this subspace, we associate the unitary operator $U_{\sigma, n}$ in $H_c^{\otimes n}$, introduced by the formula

$$U_{\sigma, n}(h_1 \otimes \dots \otimes h_n) = h_{\sigma(1)} \otimes \dots \otimes h_{\sigma(n)}$$

on a total set of elements of the form $h_1 \otimes \dots \otimes h_n \in H_c^{\otimes n}$, with each permutation $\sigma = (\sigma(1), \dots, \sigma(n))$ taken from the group of n-dimensional permutations $\mathfrak{S}_n$. It is easy to verify directly that, for the operator

$$P_n = \frac{1}{n!} \sum_{\gamma \in \mathfrak{S}_n} U_{\sigma, n},$$

we have $P_n^2 = P_n$, $P_n^* = P_n$ and, therefore, it is an orthogonal projector in $H_c^{\otimes n}$. The closed subspace of $H_c^{\otimes n}$ onto which P_n projects is denoted by $\mathfrak{F}_n(H)$. It is called the n-th symmetric tensor power of H_c. In addition, we set $\mathfrak{F}_0(H) = \mathbb{C}^1$. The Fock space (symmetric or Bose) $\mathfrak{F}(H)$ is defined as a Hilbert orthogonal sum

$$\mathfrak{F}(H) = \bigoplus_{n=0}^{\infty} \mathfrak{F}_n(H) \tag{2.21}$$

and, hence, consists of the sequences $f = (f_n)_{n=0}^{\infty}$ $(f_n \in \mathfrak{F}_n(H))$ for which

$$\|f\|_2^{\mathfrak{F}(H)} = \sum_{n=0}^{\infty} \|f_n\|^2_{H_c^{\otimes n}} < \infty.$$

The subspaces $\mathfrak{F}_n(H)$ are often called n-particle subspaces, and $\mathfrak{F}_0(H)$ is called the vacuum subspace.

Let us introduce a series of objects in the Fock space which are relevant in what follows. Denote by $\mathfrak{F}_{\text{fin}}(H)$ the subset of finite vectors from $\mathfrak{F}(H)$, i.e., vectors of the form $f = (f_0, \dots, f_n, 0, 0, \dots)$. Obviously, $\mathfrak{F}_{\text{fin}}(H)$ is dense in $\mathfrak{F}(H)$. For a given linear manifold $L \subset H$, denote by $\overset{\circ}{\mathfrak{F}}_n(L) \subset \mathfrak{F}_n(H)$ the n-th algebraic symmetric tensor power of $L_c = L \oplus iL \subset H_c$ and by $\overset{\circ}{\mathfrak{F}}_{\text{fin}}(L) \subset \mathfrak{F}_{\text{fin}}(H)$ the subset of finite vectors with components from $\overset{\circ}{\mathfrak{F}}(L)$[14].

For $h \in H_c$, we set

$$h^{\otimes n} = h \otimes \dots \otimes h \in \mathfrak{F}_n(H).$$

The vectors

$$\Gamma(h) = (1, (1!)^{-1/2} h, \dots, (n!)^{-1/2} h^{\otimes n}, \dots)$$

are called coherent states. It is easy to show that, for different $h \in H_c$, they are linearly independent and form a total set in $\mathfrak{F}(H)$ (even in the case of the real vectors $h \in H$). According to the definition of $\mathfrak{F}(H)$, we have

$$(\Gamma(h_1), \Gamma(h_2))_{\mathfrak{F}(H)} = \exp\,(h_1, h_2)_{H_c},$$

$$\|\Gamma(h)\|_{F(H)} = \exp\,(\tfrac{1}{2}\,\|h\|^2_{H_c}) \qquad (h_1, h_2, h \in H_c).$$

The vector $\Gamma(0) = (1, 0, 0, \dots)$ is called a vacuum.

Let us construct a special basis in the Fock space which is often called the occupation numbers basis. We start from the orthonormal basis $(e_j)_{j=1}^{\infty}$ in H. Let $\mathbb{Z}_{+,0}^{\infty} \subset \mathbb{Z}_{+}^{\infty} = \mathbb{Z}_+ \times \mathbb{Z}_+ \times \dots$ be a set of finite multi-indices $\alpha = (\alpha_1, \dots, \alpha_\nu, 0, 0, \dots)$ with integer nonnegative coordinates; here $\nu = \nu(\alpha)$ is the length of the multi-index α, i.e., the minimal $k \in \mathbb{N}$ for which $\alpha_{k+1} = \alpha_{k+2} = \dots = 0$. For each $\alpha \in \mathbb{Z}_{+,0}^{\infty}$ such that $|\alpha| = \alpha_1 + \dots + \alpha_\nu = n \in \mathbb{N}$, the vector

[14] Since the operator P_n transforms algebraic tensor powers into the sums of algebraic powers, the definition of $\overset{\circ}{\mathfrak{F}}_n(L)$ is correct.

$$e_\alpha^{(n)} = \left(\frac{n!}{\alpha_1!\ldots\alpha_\nu!}\right)^{1/2} P_n(e_1^{\otimes\alpha_1}\otimes\ldots\otimes e_\nu^{\otimes\alpha_\nu})$$

belongs to $\mathfrak{F}_n(H)$ (in the case where $\alpha_k = 0$, the vector e_k is absent in the tensor product, and $0! = 1$). The vectors $e_\alpha^{(n)}$, where $\alpha \in \mathbb{Z}^\infty_{+,0}$, $|\alpha| = n$, form an orthonormal basis in $\mathfrak{F}_n(H)$. This follows from the fact that vectors of the type $e_{j_1} \otimes \ldots \otimes e_{j_n}$ ($j_k \in \mathbb{N}$) form an orthonormal basis in $H_c^{\otimes n}$, and the numerical coefficient in the expression for $e_\alpha^{(n)}$ is chosen so that the equality $\| e_\alpha^{(n)} \|_{H_c^{\otimes n}} = 1$ holds. Therefore, the vectors

$$e_\alpha = (\underbrace{0,\ldots,0}_{|\alpha|-1}, e_\alpha^{(n)}, 0, 0, \ldots) \qquad (\alpha \in \mathbb{Z}^\infty_{+,0},\ |\alpha| > 0),$$

$$e_{(0,0,\ldots)} = (1, 0, 0, \ldots) = \Gamma(0)$$

form an orthonormal basis in $\mathfrak{F}_n(H)$. This is just the occupation numbers basis.

Thus, we have constructed an analogue of the space $l_2(\mathbb{C}^1)$ (strictly speaking, its generalization). The role of the space $L_2(\mathbb{R}^1, \gamma_1)$ is now played by $L_2(\Phi', \gamma_S)$. Let us generalize the expansion (2.19) of $L_2(\mathbb{R}^1, \gamma_1)$ into the series of Hermite polynomials for the case of this space. This expansion is constructed by use of a procedure somewhat different from that in Example 2.1. It is based on the fact that Hermite polynomials are constructed by the orthogonalization of powers in the space $L_2(\mathbb{R}^1, \gamma_1)$.

As before, we denote by $\mathcal{P}_{\gamma_S,n}(\Phi')$ $(n \in \mathbb{Z}_+)$ the closed subspace of $L_2(\Phi', \gamma_S)$ which consists of measurable polynomials whose degree is at most n. As mentioned above, $\mathcal{P}_{\gamma_S,n}(\Phi')$ can be also defined as a closed linear span of monomials $\prod_{k=1}^m (\varphi_k, x)_{H_0}$ $(\varphi_k \in \Phi;\ m = 1, \ldots, n)$ and constants. Introduce a subspace

$$\Gamma_n(\Phi', \gamma_S) \subset L_2(\Phi', \gamma_S) \qquad (n \in \mathbb{Z}_+)$$

by setting

$$\Gamma_n(\Phi', \gamma_S) = \mathcal{P}_{\gamma_S,n}(\Phi') \ominus \mathcal{P}_{\gamma_S,n-1}(\Phi') \qquad (n \in \mathbb{N})$$

and $\Gamma_0(\Phi', \gamma_S) = \mathbb{C}^1$. The subspaces $\Gamma_n(\Phi', \gamma_S)$ $(n \in \mathbb{Z}_+)$ are mutually orthogonal by virtue of their definition. Moreover, the equality

$$\mathcal{P}_{\gamma_S,n}(\Phi') = \bigoplus_{k=0}^{n} \Gamma_k(\Phi', \gamma_S)$$

and the denseness of polynomials in $L_2(\Phi', \gamma_S)$ yield

$$L_2(\Phi', \gamma_S) = \bigoplus_{n=0}^{\infty} \Gamma_n(\Phi', \gamma_S). \tag{2.22}$$

This decomposition of the space $L_2(\Phi', \gamma_S)$ is sometimes called the Wiener–Itô expansion.

Let $\alpha_n \in \Phi_c^{\otimes n}$, $n \in \mathbb{N}$. With each continuous monomial $(a_n, x^{\otimes n})_{H_0^{\otimes n}}$ $(x \in \Phi')$, we associate the Wick monomial

$$:(a_n, x^{\otimes n})_{H_0^{\otimes n}}: \; = P_{\Gamma_n}((a_n, x^{\otimes n})_{H_0^{\otimes n}}),$$

where P_{Γ_n} is the projector onto the subspace $\Gamma_n(\Phi', \gamma_S)$ defined in $L_2(\Phi', \gamma_S)$. For the polynomial

$$p(x) = \sum_{k=0}^{n} (a_k, x^{\otimes k})_{H_0^{\otimes k}}$$

the Wick polynomial is defined by

$$:p(x): \; = \sum_{k=0}^{n} :(a_k, x^{\otimes k})_{H_0^{\otimes k}}: \quad (:a_0: \; = a_0). \tag{2.23}$$

We set $a_n = \varphi^{\otimes n}$, where $\varphi \in \Phi$. Then $(a_n, x^{\otimes n})_{H_0^{\otimes n}} = (\varphi, x)_{H_0}^n$ and, therefore, the Wick monomial $:(\varphi, x)_{H_0}^n: \, = P_{\Gamma_n}((\varphi, x)_{H_0}^n)$ is defined. Moreover, it has been already shown that the linear function $l_\varphi(x) = (\varphi, x)_{H_0}$ $(x \in \Phi')$, regarded as a function of $\varphi \in \Phi$ with values in $L_p(\Phi', \gamma_S)$ (for any $p \geq 1$), can be extended to a measurable linear function $l_h(x) = (h, x)_{H_0}$ by continuity; here, $h \in H_S$ $(\varphi \to h$ in $H_S)$. Hence, for any $n \in \mathbb{N}$, the Wick monomial

$$:(h, x)_{H_0}^n: \, = P_{\Gamma_n}((h, x)_{H_0}^n) \qquad (h \in H_S)$$

is also defined. The main relations for Wick monomials are given by the following theorem:

Theorem 2.1. *(i) Suppose that $h \in H_S$ and $n \in \mathbb{N}$. Then the following representation in terms of Hermite polynomials holds:*

$$:(h, x)_{H_0}^n: \; = 2^{-n} \|h\|_{H_S}^n H_n(\|h\|_{H_S}^{-1} (h, x)_{H_0}). \tag{2.24}$$

(ii) For the vectors $h_1, \dots, h_m \in H_S$ mutually orthogonal in H_S and the collection of powers $n_1, \dots, n_m \in \mathbb{N}$, we have in $L_2(\Phi', \gamma_S)$

$$:\prod_{k=1}^{m} (h_k, x)_{H_0}^{n_k} : = \prod_{k=1}^{m} :(h_k, x)_{H_0}^{n_k}: . \tag{2.25}$$

(iii) For $h_k, g_k \in H_S$, we have

$$\int_{\Phi'} :\prod_{k=1}^{m} (h_k, x)_{H_0}: :\prod_{k=1}^{m} (g_k, x)_{H_0}: d\gamma_S(x)$$

$$= 2^{-n} \sum_{\sigma \in \mathfrak{S}_n} (h_{\sigma(1)}, g_1)_{H_S} \dots (h_{\sigma(n)}, g_n)_{H_S}. \tag{2.26}$$

Proof. (i) Since the leading coefficient of the Hermite polynomial H_n equals 2^n, the difference

$$(h, x)_{H_0}^n - 2^{-n} \|h\|_{H_S}^n H_n(\|h\|_{H_S}^{-1} (h, x)_{H_0})$$

belongs to $\mathcal{P}_{\gamma_S, n-1}(\Phi')$. Therefore, to prove the required statement it suffices to show that $H_n(\|h\|_{H_S}^{-1}(h, x)_{H_0})$ is orthogonal to the subspace $\mathcal{P}_{\gamma_S, n-1}(\Phi')$. Let us show that $H_n(\|h\|_{H_S}^{-1}(h, x)_{H_0})$ is orthogonal to each element of the set $\{(g, x)_{H_0}^q \mid g \in \Phi;\ q = 0, \dots, n-1\}$ which is total in $\mathcal{P}_{\gamma_S, n-1}(\Phi')$. Denote $h_1 = \|h\|_{H_S}^{-1} h$ and decompose H_S as $H_S = H_S^{(1)} \oplus H_S^{(2)}$, where $H_S^{(1)}$ is spanned by h_1. Then $g = ch_1 + g_1$, $g_1 \perp h_1$, and

$$(g, x)_{H_0}^q = \sum_{j=0}^{q} C_q^j c^j (h_1, x)_{H_0}^j (g, x)_{H_0}^{q-j}.$$

By virtue of the definition of the measure γ_S and orthogonality $g_1 \perp h_1$, for $j < n$, we have

$$\int_{\Phi'} (h_1, x)_{H_0}^j H_n((h_1, x)_{H_0}) (g_1, x)_{H_0}^{q-j} d\gamma_S(x)$$

$$= \int_{\Phi'} (h_1, x)_{H_0}^j H_n((h_1, x)_{H_0}) d\gamma_S(x) \int_{\Phi'} (g_1, x)_{H_0}^{q-j} d\gamma_S(x)$$

$$= \int_{\mathbb{R}^1} t^j H_n(t) \pi^{-1/2} e^{-t^2} \int_{\Phi'} (g_1, x)_{H_0}^{q-j} d\gamma_S(x) = 0,$$

since $H_n(t)$ is orthogonal to t^j in $L_2(\mathbb{R}^1, \gamma_1)$ for $j = 0, \dots, n-1$ (see (2.17) and (2.18)).

(ii) Similarly, the difference

$$\prod_{k=1}^{m} (h_k, x)_{H_0}^{n_k} - \prod_{k=1}^{m} (h_k, x)_{H_0}^{n_k}$$

belongs to $\mathcal{P}_{\gamma_S, l}(\Phi')$, where $l = \sum_{k=1}^{m} n_k - 1$. Therefore, it suffices to show that $\prod_{k=1}^{m} :(h_k, x)_{H_0}^{n_k}:$ is orthogonal to each monomial $(g, x)_{H_0}^{q}$ $(g \in \Phi;\ q = 0, \dots, l)$. By expanding

$$g = \sum_{k=1}^{m} c_k h_k + g_0, \quad g_0 \perp h_k \quad (k = 1, \dots, m),$$

we obtain

$$(g, x)_{H_0}^{q} = \sum_{\alpha_0 + \ldots + \alpha_m = q} c_1^{\alpha_1} \dots c_m^{\alpha_m} (h_1, x)_{H_0}^{\alpha_1} \dots (h_m, x)_{H_0}^{\alpha_m} (g_0, x)_{H_0}^{\alpha_0}.$$

By virtue of the fact that the vectors $h_1, \dots, h_m, g_0$ are mutually orthogonal, we get

$$\int_{\Phi'} (g, x)_{H_0}^{q} \prod_{k=1}^{m} :(h_k, x)_{H_0}^{n_k}: d\gamma_S(x)$$

$$= \sum_{\alpha_0 + \ldots + \alpha_m = q} \prod_{k=1}^{m} \int_{\Phi'} (h_k, x)_{H_0}^{\alpha_k} :(h_k, x)_{H_0}^{n_k}: c_k^{\alpha_k} d\gamma_S(x) \int_{\Phi'} (g_0, x)_{H_0}^{\alpha_0} d\gamma_S(x) = 0$$

according to what has been proved in (i). Here, we have taken into account the fact that $\sum_{k=0}^{m} \alpha_k = q \le l = \sum_{k=1}^{m} n_k - 1$, and thus, one can find a component $\alpha_k < n_k$ in each term of the sum.

(iii) First, let $g_k = h_k = h \in H_S$ $(k = 0, \dots, n)$. Then

$$\int_{\Phi'} :(h, x)_{H_0}^{n}:^2 d\gamma_S(x) = \int_{\mathbb{R}^1} \|h\|_{H_s}^{2n} 2^{-2n} H_n^2(t)\, d\gamma_1(t) = \|h\|_{H_0}^{2n} 2^{-n} n!\,.$$

Here, we have employed the relation (2.17). Therefore, (2.26) is proved for the case considered. The validity of the corresponding assertion in the general case follows from the result proved above and from the following consequence of the polarization identity.

In the space $L_2(\Phi', \gamma_S)$, we have

$$:\prod_{k=1}^{n} (h_k, x)_{H_0}: \;=\; \frac{1}{n!}\sum_{k=1}^{n}(-1)^{n-k} \sum_{j_1<\ldots<j_k} :(h_{j_1} + \ldots + h_{j_k}, x)^n_{H_0}:$$

$$(h_1, \ldots , h_n \in H_S;\; x \in \Phi';\; n \in \mathbb{N}).$$

■

Remark 1. The equalities (2.24)–(2.26) can be obviously extended into the complex case when, instead of H_S, the space $H_{S,c}$ is constructed by use of a similar procedure, i.e., by natural extension of S onto Φ_c and $H_{0,c}$. In this case, (2.26) transforms into

$$\left(:\prod_{k=1}^{n} (h_k, \cdot)_{H_0}:,\; :\prod_{k=1}^{n} (g_k, \cdot)_{H_0}: \right)_{L_2(\Phi', \gamma_S)} = 2^{-n} (h_{\sigma(1)}, g_1)_{H_{S,c}} (h_{\sigma(n)}, g_n)_{H_{S,c}}$$

$$(h_k, g_k \in H_{S,c};\; k = 1, \ldots , n).$$

■

2.3. The Segal Isomorphism

Let us fix $S \in \mathfrak{L}_0(\Phi, \Phi')$ and construct the space H_S and the corresponding Fock space $\mathfrak{F}(H_S)$, on the one hand, and the space $L_2(\Phi', \gamma_S)$, on the other. As in the simplest case of Example 2.1, these spaces are unitary isomorphic, and this isomorphism is realized by an operator I_S. Let us construct this operator. Introduce the mapping $I_S: \mathfrak{F}(H_S) \to L_2(\Phi', \gamma_S)$, given on the set of vectors of the form $P_n(h_1 \otimes \ldots \otimes h_n)$ $(h_k \in H_S)$ which is total in $\mathfrak{F}(H_S)$, by the formula

$$I_S P_n (h_1 \otimes \ldots \otimes h_n) = 2^{n/2} (n!)^{-1/2} :\prod_{k=1}^{n} (h_k, x)_{H_0}: \quad (n \in \mathbb{N}),$$

$$I_S \Gamma(0) = 1. \tag{2.27}$$

According to (2.26) and the definition of the projector P_n, this mapping is isometric and, consequently, can be continued onto $\mathfrak{F}(H_S)$.

Theorem 2.2. *The image of the Fock space $\mathfrak{F}(H_S)$ under the mapping I_S coincides with $L_2(\Phi', \gamma_S)$ and, hence, I_S realizes a unitary isomorphism of $\mathfrak{F}(H_S)$ and $L_2(\Phi', \gamma_S)$.*

Proof. Since the operator I_S is isometric, it suffices to show that the image $I_S(\mathfrak{F}(H_S))$ is dense in $L_2(\Phi', \gamma_S)$. However, by virtue of formula (2.27) with $h_1 = \ldots =$

$h_n = h$ and (2.24), this set includes functions of the form $H_n((h, x)_{H_0})$ ($h \in H_S$; $n \in \mathbb{Z}_+$) and, hence (by virtue of linearity), the monomials $(h, x)^n_{H_0}$ and constants. Further, according to the polarization identity (2.7), we have $\mathcal{P}_{\text{cyl}}(\Phi') = I_S(\mathfrak{F}(H_S))$, and $\mathcal{P}_{\text{cyl}}(\Phi')$ is dense in $L_2(\Phi', \gamma_S)$ by virtue of Lemma 2.2. Note that we have simultaneously proved the equality

$$I_S(\mathring{\mathfrak{F}}_{\text{fin}}(\Phi)) = \mathcal{P}_{\text{cyl}}(\Phi').$$

■

Corollary 1. *Equality* (2.23) *defines the Wick polynomial* $:p:$ *which corresponds to the given continuous polynomial* $p \in \mathcal{P}(\Phi')$.

By Theorem 2.2, we can also define the Wick monomial $:(a_n, x^{\otimes n})_{H_0^{\otimes n}}:$ for the case of the kernel $a_n \in H_S^{\otimes n}$ by setting (cf. (2.27))

$$:(a_n, x^{\otimes n})_{H_0^{\otimes n}}: = 2^{-n/2}\sqrt{n!}\, I_S(P_n a_n) \in \Gamma_n(\Phi', \gamma_S)$$

(the value of the monomial is determined by the symmetric part $P_n a_n$ of the kernel a_n). In the case when $a_n \in H_{S,c}^{\otimes n}$, the Wick monomial is defined analogously (see Remark 1).

■

Corollary 2. *Every measurable polynomial* $p \in \mathcal{P}_{\gamma_S}(\Phi')$ *has the form*

$$p = \sum_{k=0}^{n} p_k, \quad p_k \in \Gamma_k(\Phi', \gamma_S)$$

(see (2.22)*). Therefore, by virtue of Corollary* 1, *we have the following representation*

$$p(x) = \sum_{k=0}^{n} :(a_k, x^{\otimes k})_{H_0^{\otimes k}}: \quad (a_k \in H_{S,c}^{\otimes k}).$$

To prove this, one must decompose the vector $I_S^{-1}p \in \mathfrak{F}(H_S)$ into its k-particle components a_k ($a_{n+1} = a_{n+2} = \ldots = 0$, since the same is true for continuous polynomials). If we restrict ourselves to the case of symmetric kernels, i.e., $a_k \in \mathfrak{F}(H_S)$, then this representation is univalent. Thus, we have obtained the internal description of the set of measurable polynomials $\mathcal{P}_{\gamma_S}(\Phi')$.

■

Remark 2. In Section 1 of Chapter 6, by exploiting the so-called hyper-contractivity estimates, we shall show that $\mathcal{P}_{\gamma_S}(\Phi') \subset L_q(\Phi', \gamma_S)$ for all $q \geq 2$ and, moreover,

$$\|p\|_{L_q(\Phi', \gamma_S)} \leq (q-1)^{n/2} \|p\|_{L_2(\Phi', \gamma_S)}$$

for $p \in \mathcal{P}_{\gamma_S}(\Phi')$. Thus, the set $\mathcal{P}_{\gamma_S}(\Phi')$ can be defined as a closure of $\mathcal{P}_{\text{cyl}}(\Phi')$ not only in $L_2(\Phi', \gamma_S)$, but also in each space $L_q(\Phi', \gamma_S)$, $q \geq 1$ (equivalently). This fact clarifies the meaning of the concept "a measurable polynomial".

∎

The isomorphism of the Fock space $\mathcal{F}(H_S)$ and $L_2(\Phi', \gamma_S)$ realized by the unitary operator I_S is called the Segal isomorphism. In this case, each Fock vector $f = (f_0, f_1, \dots) \in \mathcal{F}(H_S)$ is associated with the function

$$(I_S f)(x) = \sum_{n=0}^{\infty} \frac{2^{n/2}}{\sqrt{n!}} :(f_n, x^{\otimes n})_{H_0^{\otimes n}}: \in L_2(\Phi', \gamma_S), \tag{2.28}$$

and, vice versa, each function from $L_2(\Phi', \gamma_S)$ admits expansion of the form (2.28). Thus, we have obtained the desired analogue of expansion (2.19) for the infinite-dimensional case.

The idea of introducing the Segal isomorphism by formula (2.27) may seem to be artificial. This isomorphism is, in fact, a Fourier transform with respect to generalized common eigenvectors of some family of commuting selfadjoint operators in the Fock space. This will be shown in Chapter 3, Section 3, Subsection 3.8.

In the following examples, we shall find the explicit form of the functions in $L_2(\Phi', \gamma_S)$ which correspond to the coherent states and to the occupation numbers basis under the Segal isomorphism.

Example 2.3. Let $h \in H_S$. According to the definition of the Segal isomorphism, the Fock vector

$$\Gamma(h) = (1, h, \dots, (n!)^{-1/2} h^{\otimes n}, \dots)$$

is associated with the following function from $L_2(\Phi', \gamma_S)$:

$$I_S(\Gamma(h))(x) = \sum_{n=0}^{\infty} \frac{2^{n/2}}{n!} :(h, x)_{H_0}^n: = \sum_{n=0}^{\infty} \frac{2^{-n/2}}{n!} \|h\|_{H_S}^n H_n(\|h\|_{H_S}^{-1} (h, x)_{H_0})$$

$$= \exp(\sqrt{2}(h, x)_{H_0} - \tfrac{1}{2}\|h\|_{H_S}^2) \quad (x \in \Phi'). \tag{2.29}$$

Here, we have used the generating function for Hermite polynomials

$$\sum_{n=0}^{\infty} \frac{z^n}{n!} H_n(t) = \exp(2tz - z^2) \qquad (z \in \mathbb{C}^1,\ t \in \mathbb{R}^1). \tag{2.30}$$

Using the expansion

$$\exp(h, x)_{H_0} = \sum_{n=0}^{\infty} \frac{1}{n!} (h, x)_{H_0}^n$$

for $h \in H_S$, we define a "Wick exponential" by the equality

$$:\exp(h, x)_{H_0}: = \sum_{n=0}^{\infty} \frac{1}{n!} :(h, x)_{H_0}^n: = \exp\left((h, x)_{H_0} - \tfrac{1}{4}\|h\|_{H_S}^2\right) \tag{2.31}$$

(we have used the relations (2.24) and (2.30)). Note that the definition of the Wick monomials is often based on formula (2.31). Equality (2.31) implies, in particular, that

$$\int_{\Phi''} :\exp(h, x)_{H_0}: d\gamma_S(x) = 1.$$

Therefore, it can be rewritten as follows

$$:\exp(h, x)_{H_0}: = \exp(h, x)_{H_0} \left(\int_{\Phi'} \exp(h, x)_{H_0}\, d\gamma_S(x) \right)^{-1}. \tag{2.32}$$

Relations (2.29) and (2.31) give the following expression for $I_S(\Gamma(h))$ in terms of the Wick exponential

$$I_S(\Gamma(h)) = :\exp(\sqrt{2}\,h, \cdot)_{H_0}: \qquad (h \in H_S).$$

Taking into account the properties of coherent states and (2.32), we conclude that the functions $\exp(h, \cdot)_{H_0}$ ($h \in H_S$, and even $h \in \Phi$) form a set total in $L_2(\Phi', \gamma_S)$. In what follows, we shall use this consequence.

Example 2.4. Let us describe the basis in the space $L_2(\Phi', \gamma_S)$ which is obtained from the occupation numbers basis under the Segal isomorphism. Let $(e_j)_{j=1}^{\infty}$ be an orthonormal basis in H_S, and let $(e_\alpha)_{\alpha \in \mathbb{Z}_{+,0}^{\infty}}$ be the corresponding occupation numbers basis in $\mathcal{F}(H_S)$. It follows from (2.27) that

$$I_S e_\alpha = I_S\left[\left(\frac{|\alpha|!}{\alpha_1! \ldots \alpha_\nu!} \right)^{1/2} P_{|\alpha|} \left(e_1^{\otimes \alpha_1} \otimes \ldots \otimes e_\nu^{\otimes \alpha_\nu}\right) \right]$$

$$= 2^{|\alpha|/2} (\alpha_1! \ldots \alpha_\nu!) :(e_1, x)_{H_0}^{\alpha_1} \ldots (e_\nu, x)_{H_0}^{\alpha_\nu}: .$$

Relations (2.24) and (2.25) yield

$$I_S e_\alpha = h_{\alpha_1}((e_1, x)_{H_0}) \dots h_{\alpha_\nu}((e_\nu, x)_{H_0}) \quad (\alpha \in \mathbb{Z}^\infty_{+,0}) \tag{2.33}$$

($h_n(t) = 2^{-n/2}(n!)^{-1/2} H_n(t)$) is a normalized Hermite polynomial). Recall that $(e_j, x)_{H_0}$ is a measurable linear function on Φ' by virtue of the inclusion $e_j \in H_S$. Hence, $I_S e_\alpha$ is a measurable polynomial. If we proceed from the basis $(e_j)_{j=1}^\infty \subset \Phi$, then $I_S e_\alpha \in \mathcal{P}_{\text{cyl}}(\Phi')$.

The basis $(I_S e_\alpha)_{\alpha \in \mathbb{Z}^\infty_{+,0}}$ constructed above allows us to obtain a convenient realization of $L_2(\Phi', \gamma_S)$ as a space of functions on $\mathbb{R}^\infty$. In the space $\mathbb{R}^\infty$, we consider the canonical Gaussian measure γ_1 on the σ-algebra $\mathfrak{C}_\sigma(\mathbb{R}^\infty)$ which corresponds to the rigging $\mathbb{R}^\infty \supset l_2(\mathbb{R}^1) \supset \mathbb{R}^\infty_0$ (see Subsection 1.9). In other words, we have $\gamma_1 = \gamma_\kappa$ in terms of (1.10), where $\kappa = (1, 1, \dots)$, and thus, $\gamma_1(x) = \otimes_{k=1}^\infty \gamma_1(x_k)$, $d\gamma_1(x_k) = \pi^{-1/2} e^{-x_k^2} dx_k$. In this case

$$L_2(\mathbb{R}^\infty, \gamma_1) = \bigotimes_{k=1;e}^\infty L_2(\mathbb{R}^1, \gamma_1), \qquad e = (1, 1, \dots)$$

(see Subsection 1.2), and an orthonormal basis in $L_2(\mathbb{R}^\infty, \gamma_1)$ is formed by the polynomials

$$h_\alpha(x) = h_{\alpha_1}(x_1) \dots h_{\alpha_\nu}(x_\nu) \quad (x = (x_k)_{k=1}^\infty;\ \alpha \in \mathbb{Z}^\infty_{+,0}), \tag{2.34}$$

since $(h_k)_{k=0}^\infty$ is an orthonormal basis in $L_2(\mathbb{R}^1, \gamma_1)$. Below, the basis $(h_\alpha)_{\alpha \in \mathbb{Z}^\infty_{+,0}}$ in the space $L_2(\mathbb{R}^\infty, \gamma_1)$ is referred to as the basis of Hermite polynomials by analogy with the one-dimensional case.

Let $(e_k)_{k=1}^\infty \subset \Phi$ be an orthonormal basis in H_S. We introduce the mapping

$$\Phi' \ni x \mapsto j_e x = ((e_k, x)_{H_0})_{k=1}^\infty \in \mathbb{R}^\infty.$$

The mapping j_e induces the mapping J_e (which maps cylindrical functions given on $\mathbb{R}^\infty$ onto cylindrical functions given on Φ') by the equality

$$(J_e f)(x) = f(j_e x) \quad (x \in \Phi').$$

Comparing (2.33) with (2.34), we conclude that $J_e h_\alpha = J_S e_\alpha$ ($\alpha \in \mathbb{Z}^\infty_{+,0}$). This establishes the correspondence between the orthonormal bases in $L_2(\Phi', \gamma_S)$ and $L_2(\mathbb{R}^\infty, \gamma_1)$ and, hence, the isomorphism of these Hilbert spaces.

We have described the Segal isomorphism, starting from the given chain (2.1) and the operator $S \in \mathfrak{L}_0(\Phi, \Phi')$. This means that, on the one hand, we have constructed the space H_S (as a completion of Φ with respect to $(\varphi, \psi)_{H_S} = (S\varphi, \psi)_{H_0}$ $(\varphi, \psi \in \Phi))$ and the corresponding Fock space $\mathfrak{F}(H_S)$ and, on the other hand, we have constructed the Gaussian measure γ_S and the space $L_2(\Phi', \gamma_S)$.

The mapping $I_S: \mathfrak{F}(H_S) \to L_2(\Phi', \gamma_S)$ gives the functional (as functions from L_2) realization, of the Fock space. However, one can also start from a given real Hilbert space H and the Fock space $\mathfrak{F}(H)$ constructed in terms of H. To obtain its functional realization, it is necessary to construct a nuclear chain $\Phi' \supset H \supset \Phi$ (e.g., by passing to coordinates and setting $\mathbb{R}^\infty \supset l_2(\mathbb{R}^1) \supset \mathbb{R}_0^\infty$), to take $S = 1$, and apply the above scheme.

Example 2.5. (The space of states for a free field in the Schrödinger representation). In field theory, the space of states of the free boson field with mass $m_0 > 0$ in $d+1$-dimensional space-time is defined as a Fock space $\mathfrak{F}(\mathfrak{H}_{m_0})$. Here, $\mathfrak{H}_{m_0}$ is constructed as follows. As (2.1), we take the nuclear chain of the real spaces

$$\mathfrak{S}'_{\mathrm{Re}}(\mathbb{R}^d) \supset L_{2,\mathrm{Re}}(\mathbb{R}^d) \supset \mathfrak{S}_{\mathrm{Re}}(\mathbb{R}^d),$$

where $\mathfrak{S}_{\mathrm{Re}}(\mathbb{R}^d)$ is the classical Schwartz space (see Chapter 1, Section 1, Example 1.8). The operator $S = (-\Delta + m_0^2 1)^{-1/2}$, where $-\Delta + m_0^2 1$, is a selfadjoint positive operator in $L_2(\mathbb{R}^d)$ equal to the closure of the operator

$$C_0^\infty(\mathbb{R}^d) \ni \varphi \mapsto (-\Delta + m_0^2 1)\varphi.$$

Obviously, $S \in \mathfrak{L}_0(\mathfrak{S}_{\mathrm{Re}}(\mathbb{R}^d), \mathfrak{S}'_{\mathrm{Re}}(\mathbb{R}^d))$. In this case, $\mathfrak{H}_{m_0} = H_S$, i.e., $\mathfrak{H}_{m_0}$ is equal to the completion of $\mathfrak{S}_{\mathrm{Re}}(\mathbb{R}^d)$ with respect to the scalar product

$$(\varphi, \psi)_{\mathfrak{H}_{m_0}} = ((-\Delta + m_0^2 1)^{-1/2} \varphi, \psi)_{L_2(\mathbb{R}^d)} \quad (\varphi, \psi \in \mathfrak{S}_{\mathrm{Re}}(\mathbb{R}^d)). \tag{2.35}$$

The functional realization of $\mathfrak{F}(\mathfrak{H}_{m_0})$ is the space $L_2(\mathfrak{S}'_{\mathrm{Re}}(\mathbb{R}^d), \mu_{m_0})$, where $\mu_{m_0} = \gamma_S$ is the Gaussian measure on $\mathfrak{S}'_{\mathrm{Re}}(\mathbb{R}^d)$ with the Fourier transform

$$\tilde{\mu}_{m_0}(\varphi) = \int_{\mathfrak{S}'_{\mathrm{Re}}(\mathbb{R}^d)} \exp(i(\varphi, \xi)_{L_2(\mathbb{R}^d)})\, d\mu_{m_0}(\xi) = \exp(-\tfrac{1}{4}(\varphi, \varphi)_{\mathfrak{H}_{m_0}}) \tag{2.36}$$

$$(\varphi \in \mathfrak{S}_{\mathrm{Re}}(\mathbb{R}^d)).$$

The transition from $\mathfrak{F}(\mathfrak{H}_{m_0})$ to $L_2(\mathfrak{S}'_{\mathrm{Re}}(\mathbb{R}^d), \mu_{m_0})$ is called the transition to the Schrödinger representation. Under this transition many field-like objects obtain new useful properties. For instance, the operator of the free field energy becomes an infinite-dimensional elliptic operator of second order and its potential perturbations are associated with typical interactions of the constructive field theory (see Chapter 6, Sections 1 and 2).

2.4. Shifts of Gaussian Measures and Absolute Continuity

All previous considerations in this section were connected with Gaussian measures with zero mean. In what follows, we shall use the measures obtained from these by shifting the arguments. The main problem in this case is to clarify the conditions under which the shifted measure is absolutely continuous with respect to the original one. For Gaussian measures, this problem admits a comprehensive solution in terms of a correlation operator which determines a measure.

Let S be a positive operator in the real Hilbert space H_0 with a dense domain of definition $\mathfrak{D}(S)$ and a dense range of values $\mathfrak{R}(S)$. Denote by H_S the Hilbert space obtained as a result of the completion of $\mathfrak{D}(S) \subset H_0$ with respect to the scalar product $(x, y)_{H_0} = (Sx, y)_{H_0}$ $(x, y \in \mathfrak{D}(S))$; the positiveness of S guarantees that this definition is correct. Clearly, the operator S is invertible on $\mathfrak{R}(S) = \mathfrak{D}(S^{-1})$, and S^{-1} is positive. Introduce the space $H_{S^{-1}}$ constructed analogously by exploiting the operator S^{-1}. The relations

$$\| Sx \|_{H_{S^{-1}}} = \| x \|_{H_S} \quad \text{and} \quad \| S^{-1} y \|_{H_S} = \| y \|_{H_{S^{-1}}} \quad (x \in \mathfrak{D}(S),\ y \in \mathfrak{D}(S^{-1})) \quad (2.37)$$

involve the isometricity of the mappings

$$H_S \supset \mathfrak{D}(S) \ni x \mapsto Sx \in H_{S^{-1}}$$

and

$$H_{S^{-1}} \supset \mathfrak{D}(S^{-1}) \ni y \mapsto S^{-1}y \in H_S$$

and, hence, these mappings can be continued to the isometric operators

$$S \in \mathfrak{L}(H_S, H_{S^{-1}}) \quad \text{and} \quad S^{1-} \in \mathfrak{L}(H_{S^{-1}}, H_S). \quad (2.38)$$

Introduce a Gaussian measure with the correlation operator S. To do this, we choose a nuclear rigging $\Phi' \supset H_0 \supset \Phi$ such that $\Phi \subset \mathfrak{D}(S)$ and $S \in \mathfrak{L}_0(\Phi, \Phi')$ (this choice is always possible, see Maurin [2, Chapter 17, Section 5]). According to Subsection 1.9, the Gaussian measure γ_S is defined on $\mathfrak{C}_\sigma(\Phi')$. Recall that the only condition imposed on the choice of a nuclear rigging , which allows us to introduce γ_S as a measure on Φ', is $S \in \mathfrak{L}_0(\Phi, \Phi')$. One can easily verify the validity of the topological imbeddings

$$\Phi' \supset H_{S^{-1}} \quad \text{and} \quad \mathfrak{H} \supset \Phi. \tag{2.39}$$

In fact, the second imbedding (2.39) follows from the inclusion $S \in \mathfrak{L}_0(\Phi, \Phi')$ and from Subsection 1.9. Let us prove the first imbedding. According to (2.37), for $y \in \mathfrak{D}(S^{-1})$ and $\varphi \in \Phi$ we have

$$|(y, \varphi)_{H_0}| = |(S^{-1}y, \varphi)_{H_S}| \leq \|S^{-1}y\|_{H_S} \|\varphi\|_{H_S} = \|y\|_{H_{S^{-1}}} \|\varphi\|_{H_S}.$$

This and the imbedding $\Phi \in H_S$ imply that $l_y(\varphi) = (y, \varphi)_{H_0}$ can be extended (in $y \in \mathfrak{D}(S^{-1})$) by continuity to an antilinear continuous functional l_z ($z \in H_{S^{-1}}$) on Φ. Since H_0 is dense in Φ', and $\mathfrak{D}(S^{-1})$ is dense in H_0, we find that $H_{S^{-1}} \supset \mathfrak{D}(S^{-1}))$ is dense in Φ'. It is also clear that if $H_{S^{-1}} \ni y_n \xrightarrow[n\to\infty]{} y \in H_{S^{-1}}$ in $H_{S^{-1}}$, then $l_{y_n} \xrightarrow[n\to\infty]{} l_y$ weakly in Φ', i.e. the imbedding $H_{S^{-1}} \subset \Phi'$ is continuous.

∎

If, in addition, $S \geq 1$, then, clearly, (2.39) transforms into the single chain $\Phi' \supset H_{S^{-1}} \supset H_S \supset \Phi$.

Let us define a shifted measure which corresponds to some measure μ on Φ' defined on the σ-algebra $\mathfrak{C}_\sigma(\Phi')$. For every $y \in \Phi'$, we introduce the shift $T_y : \Phi' \to \Phi'$ given by the equality $T_y x = x + y$ ($x \in \Phi'$). Obviously, T_y is a bijective mapping. Moreover, for $\alpha \in \mathfrak{C}_\sigma(\Phi')$, we have

$$T_y \alpha = \{x + y \mid x \in \alpha\} \in \mathfrak{C}_\sigma(\Phi').$$

In fact, in the case of cylindrical sets of the form

$$\mathfrak{C} = \{x \in \Phi' \mid ((\varphi_1, x)_{H_0}, \ldots, (\varphi_n, x)_{H_0}) \in \delta\} \quad (\varphi_k \in \Phi', \delta \in \mathfrak{B}(\mathbb{R}^n)),$$

we have

$$T_y \mathfrak{C} = \{z \in \Phi' \mid (z - y \in \mathfrak{C}\} = \{z \in \Phi' \mid ((z - y, \varphi_1)_{H_0}, \ldots, (z - y, \varphi_n)_{H_0}) \in \delta\}$$

$$= \{z \in \Phi' \mid ((z, \varphi_1)_{H_0}, \ldots, (z, \varphi_n)_{H_0}) \in \delta_y\},$$

where

$$\delta_y = \delta + a_y, \quad a_y = ((y, \varphi_1)_{H_0}, \ldots, (y, \varphi_n)_{H_0}) \in \mathbb{R}^n, \quad \delta_y \in \mathfrak{B}(\mathbb{R}^n).$$

Therefore, $T_y \mathfrak{C} \in \mathfrak{C}(\Phi')$. But the σ-span of the sets $\mathfrak{C}$ gives the whole of the σ-algebra $\mathfrak{C}_\sigma(\Phi')$. This and the bijectiveness of T_y together imply the required statement.

Define the measure μ_y by the equality

$$\mu_y(\alpha) = \mu(T_y\alpha) \quad (\alpha \in \mathfrak{C}_\sigma(\Phi');\ y \in \Phi'). \tag{2.40}$$

The measure μ_y is called the shift of the measure μ by the vector $y \in \Phi'$. For the characteristic function κ_α, we obtain (taking (2.40) into account)

$$\int_{\Phi'} \kappa_\alpha(x)\, d\mu_y(x) = \int_{\Phi'} \kappa_\alpha(x-y)\, d\mu(x) \quad (y \in \Phi').$$

Hence, for every measurable bounded function f on Φ', we have

$$\int_{\Phi'} f(x)\, d\mu_y(x) = \int_{\Phi'} f(x-y)\, d\mu(x) \quad (y \in \Phi'), \tag{2.41}$$

since f is the uniform limit of finite linear combinations of characteristic functions. Formula (2.41) involves, in particular, the relation between the Fourier transforms of the measures μ and μ_y

$$\tilde{\mu}_y(\varphi) = \int_{\Phi'} \exp(i(\varphi, x)_{H_0})\, d\mu_y(x) = \exp(-i(\varphi, y)_{H_0})\, \tilde{\mu}(\varphi) \quad (\varphi \in \Phi). \tag{2.42}$$

The absolute continuity of the measure μ_y with respect to μ is essential in many problems of finite-dimensional analysis. The shifts $y \in \Phi'$ for which μ_y is absolutely continuous with respect to μ, are called admissible. The set of admissible shifts is an important characteristic of a measure. However, in the infinite-dimensional case, it may have a rather complicated structure and, thus, the investigation of this set may be very difficult. Anyhow, on the infinite-dimensional space, there are no measures for which all the shifts are admissible. The Gaussian measures belong to a class of measures for which a complete description of the set of admissible shifts is possible. Below, we shall obtain the necessary and sufficient conditions for the admissibility of shifts in the case of the Gaussian measure γ_S in Φ' given on $\mathfrak{C}_\sigma(\Phi')$.

From the definition of the measure γ_S on cylindrical sets given by (1.54), one should expect that the Radon–Nikodym derivative $\rho_{\gamma_S}(y, \cdot)$ of the measure $\gamma_{S,y}$ with respect to γ_S has the form

$$\rho_{\gamma_S}(y, \cdot) = \exp(-(S^{-1}y, y)_{H_0} - 2(S^{-1}y, \cdot)_{H_0}). \tag{2.43}$$

(if this derivative exists). This follows from the formal representation (written by analogy with (1.54))

$$d\gamma_S = N^{-1} \exp(-(S^{-1}x, x)_{H_0})\, dm(x), \tag{2.44}$$

which has a rigorous meaning only in the case $\dim H_0 < \infty$ (N^{-1} is a normalizing vector). Nevertheless, while studying the measure γ_S, it proves useful to exploit representation (2.44) as a source of heuristic arguments.

Lemma 2.4. *Suppose that* $y \in H_{S^{-1}}$. *Then* $\rho_{\gamma_S}(y, \cdot) \in \mathfrak{L}(\Phi', \gamma_S)$ *for every* $p \geq 1$.

Proof. Taking into account the definition of the Hilbert space $H_{S^{-1}}$, we can represent $\rho_{\gamma_S}(y, \cdot)$ as follows

$$\rho_{\gamma_S}(y, x) = \exp(-\|y\|^2_{H_{S^{-1}}}) \exp(2(S^{-1}y, x)_{H_0}) \quad (x \in \Phi'). \tag{2.45}$$

By virtue of the isometricity of the operator $S^{-1} \in \mathfrak{L}(H_{S^{-1}}, H_S)$, we have $S^{-1}y \in H_S$. Therefore, $(S^{-1}y, x)_{H_0}$ is defined as a measurable linear function on Φ'.

Note that (2.31) gives the representation

$$\rho_{\gamma_S}(y, x) = :\exp(2(S^{-1}y, x)_{H_0}): \quad (x \in \Phi') \tag{2.46}$$

which is important for our further discussion.

By equating the right-hand sides of (2.31) and (2.32), we obtain

$$\int_{\Phi'} \exp((h, x)_{H_0})\, d\gamma_S(x) = \exp\left(\frac{1}{4}(h, h)_{H_S}\right) \quad (h \in H_S). \tag{2.47}$$

It follows from (2.45) and (2.47) that

$$\|\rho_{\gamma_S}(y, \cdot)\|_{L_p(\Phi', \gamma_S)} = \exp((p-1)\|y\|^2_{H_{S^{-1}}}) < \infty$$

for all $p \geq 1$.

■

Theorem 2.3. *If* $y \in H_{S^{-1}}$, *the measures* γ_S *and* $\gamma_{S,y}$ *are mutually absolutely continuous and, moreover,*

$$(d\gamma_{S,y}/d\gamma_S)(\cdot) = \rho_{\gamma_S}(y, \cdot).$$

Otherwise, the measures γ_S *and* $\gamma_{S,y}$ *are orthogonal.*

Proof. By virtue of Lemma 2.2, it suffices to show that if $y \in H_{S^{-1}}$, then the equality

$$\int_{\Phi'} f(x)\, d\gamma_{S,y}(x) = \int_{\Phi'} f(x)\, \rho_{\gamma_S}(y,x)\, d\gamma_S(x)$$

holds for $f \in \mathcal{P}_{\mathrm{cyl}}(\Phi')$. Moreover, we can confine ourselves to the case when

$$f(x) = :(\varphi, x)^n_{H_0}: \qquad (n \in \mathbb{Z}_+,\ \varphi \in \Phi),$$

since the polarization identity employed in the proof of Theorem 2.1 extends this equality onto the whole of $\mathcal{P}_{\mathrm{cyl}}(\Phi')$. Taking the definition of Wick powers and (2.11) into account, we get

$$\int_{\Phi'} :(\varphi, x)^n_{H_0}: d\gamma_{S,y}(x) = \int_{\Phi'} :(\varphi, x-y)^n_{H_0}: d\gamma_S(x)$$

$$= \sum_{k=0}^{n} (-1)^{n-k} C_n^k\, (\varphi, y)^{n-k}_{H_0} \int_{\Phi'} :(\varphi, x)^k_{H_0}: d\gamma_S(x) = (-1)^n (\varphi, y)^n_{H_0}.$$

On the other hand, the expansion

$$\rho_{\gamma_S}(y, x) = :\exp\left(-2\,(S^{-1}y, x)_{H_0}\right): = \sum_{k=0}^{\infty} (-1)^k \frac{2^k}{k!} :(S^{-1}y, x)^k_{H_0}: \tag{2.48}$$

(see (2.29)) and (2.26) together yield

$$\int_{\Phi'} :(\varphi, x)^n_{H_0}: \rho_{\gamma_S}(y,x)\, d\gamma_S(x) = \frac{(-1)^n \cdot 2^n}{n!} \int_{\Phi'} :(\varphi, x)^n_{H_0}::(S^{-1}y, x)^n_{H_0}: d\gamma_S(x)$$

$$= (-1)^n (\varphi, S^{-1}y)^n_{H_S} = (-1)^n (\varphi, y)^n_{H_0}.$$

Hence, (2.48) is proved and, thus, the absolute continuity of $\gamma_{S,y}$ with respect to γ_S is proved, too.

In order to ascertain the absolute continuity of γ_S with respect to $\gamma_{S,y}$, we note that these measures are obtained from the measures $\gamma_{S,-y}$ and γ_S by means of the mapping $T_y\colon \Phi' \to \Phi'$. Taking into account that $-y \in H_{S^{-1}}$, we find, by virtue of what has been proved, that $\gamma_{S,-y}$ is absolutely continuous with respect to γ_S. Consequently, γ_S is absolutely continuous with respect to $\gamma_{S,y}$. Here, we have used the fact that images of absolutely continuous measures (obtained as a result of mappings of measurable spaces) are also absolutely continuous.

To prove the last statement of the theorem, we consider the mapping $j_e\colon \Phi' \to \mathbb{R}^\infty$ introduced in Example 2.3 by use of the orthonormal basis $e = (e_k)_{k=1}^\infty \subset \Phi$ in H_S; we choose the basis e so that $e_k \in \mathfrak{D}(S^{-1})$ for every $k \in \mathbb{N}$. As has been shown in this example, the image of the measure γ_S under the mapping j_e is the Gaussian measure $\gamma_1(x) = \times_{k=1}^\infty \gamma_1(x_k)$ on $\mathbb{R}^\infty$, i.e., for any $\alpha \in \mathfrak{C}_\sigma(\mathbb{R}^\infty)$ we have $\gamma_1(\alpha) = \gamma_S(j_e^{-1}(\alpha))$.

Let us find the image of the shifted measure $\gamma_{S,y}$ under the mapping j_e. For this purpose, we determine the image of its Fourier transform $\tilde{\gamma}_{S,y}$. According to (2.42), we have

$$\tilde{\gamma}_{S,y}(\varphi) = \exp\left[-i(\varphi, y)_{H_0}\right]\exp\left[-\frac{1}{4}\|\varphi\|_{H_S}^2\right]$$

$$= \exp\left[-i\sum_{k=1}^n \varphi_k z_k\right]\exp\left[-\frac{1}{4}\sum_{k=1}^n \varphi_k^2\right]$$

for

$$\varphi = \sum_{k=1}^n \varphi_k e_k \qquad (\varphi_k \in \mathbb{R}^1),$$

where $z = (z_k)_{k=1}^\infty \in \mathbb{R}^\infty$, $z_k = (y, S^{-1}y)_{H_0}$. But the last expression coincides with the Fourier transform of the measure $\gamma_{1,z} = \gamma_1(\cdot + z)$ on $\mathbb{R}^\infty$:

$$\int_{\mathbb{R}^\infty} \exp\left(i\sum_{k=1}^n x_k\varphi_k\right) d\gamma_{1,z}(x) = \int_{\mathbb{R}^\infty} \exp\left[i\sum_{k=1}^n (x_k - z_k)\varphi_k\right] d\gamma_1(x)$$

$$= \exp\left[-i\sum_{k=1}^n \varphi_k z_k\right]\exp\left[-\frac{1}{4}\sum_{k=1}^n \varphi_k^2\right] \quad ((\varphi_1, \ldots, \varphi_n, 0, 0, \ldots) \in \mathbb{R}_0^\infty).$$

Thus, under the mapping j_e, the measure $\gamma_{S,y}$ turns into the measure $\gamma_{1,z} = \gamma_1(\cdot + z)$. It can be easily found from the assertion of Example 1.1 (Section 1) that the measures γ_1 and $\gamma_{1,z}$ are orthogonal for $z \notin l_2(\mathbb{R}^1)$ and, hence, the measures γ_S and $\gamma_{S,y}$ also are orthogonal in this case. It remains to note that the inclusion $z \in l_2(\mathbb{R}^1)$ is equivalent to $S^{-1}y \in H_S$, or $y \in H_{S^{-1}}$ (this is just the same).

■

2.5. The Fourier–Wiener Transform

Before we proceed to study the infinite-dimensional case, let us show how the Fourier–Wiener transform appears in the one-dimensional case. For this purpose, we consider the following example.

Example 2.6. As mentioned in Example 2.1, the orthonormal basis in the space $L_2(\mathbb{R}^1, \gamma_1)$, $d\gamma_1(x) = \pi^{-1/2}\exp(-x^2)dx$, is formed by the normalized Hermite polynomials $(h_n)_{n=0}^{\infty}$. The well-known relation for Hermite polynomials

$$\int_{\mathbb{R}^1} e^{iyx} h_n(x) \frac{e^{-x^2/2}}{\sqrt{2\pi}}\, dx = i^n e^{-y^2/2} h_n(y) \quad (y \in \mathbb{R}^1;\ z \in \mathbb{Z}_+) \tag{2.49}$$

allows us to define the unitary operator W acting in $L_2(\mathbb{R}^1, \gamma_1)$ which is given on the dense set of polynomials $\mathcal{P}(\mathbb{R}^1) \subset L_2(\mathbb{R}^1, \gamma_1)$ by

$$(Wf)(y) = \int_{\mathbb{R}^1} f(x) e^{iyx + y^2/2} \frac{1}{\sqrt{2\pi}} e^{-x^2/2}\, dx \quad (f \in \mathcal{P}(\mathbb{R}^1);\ y \in \mathbb{R}^1). \tag{2.50}$$

The operator W possesses the property $W h_n = i^n h_n$ which involves the unitarity of W and the fact that the equality $W\mathcal{P}(\mathbb{R}^1) = \mathcal{P}(\mathbb{R}^1)$ holds. Consider the representation (2.50) in a somewhat different way: Introduce the measure

$$d\gamma_2(x) = \frac{1}{\sqrt{2\pi}} e^{-x^2/2}\, dx$$

on $\mathbb{R}^1$. Then $\tilde{\gamma}_2(y) = e^{-y^2/2}$, and (2.50) can be rewritten as follows

$$(Wf)(y) = (\tilde{\gamma}_2(y))^{-1} \int_{\mathbb{R}^1} e^{iyx} f(x)\, d\gamma_2(x)$$

or, if the equality $iyx + y^2/2 - x^2/2 = -1/2\,(x - iy)^2$ is taken into account, as follows

$$(Wf)(y) = \int_{\mathbb{R}^1} f(x + iy)\, d\gamma_2(x) \quad (f \in \mathcal{P}(\mathbb{R}^1);\ y \in \mathbb{R}^1). \tag{2.51}$$

Note that $W^2 h_n(y) = (-1)^n h_n(y) = h_n(-y)$ by virtue of the fact that each Hermite polynomial includes the powers of the variable having the same parity as the index of this polynomial. This proves the equalities

$$(W^2 f)(y) = f(-y),$$

$$(W^{-1}f)(y) = \int_{\mathbb{R}^1} f(x - iy)\,d\gamma_2(x) \quad (f \in \mathcal{P}(\mathbb{R}^1);\ y \in \mathbb{R}^1). \tag{2.52}$$

The unitary transform of the functions from $L_2(\mathbb{R}^1, \gamma_1)$ realized by the operator W is called the Fourier–Wiener transform. Its representation by (2.50) admits the natural extension to the infinite-dimensional case which is given below.

As in Subsection 2.4, consider a Gaussian measure γ_S given on $\mathcal{C}_\sigma(\Phi')$ and assume that the nuclear rigging $\Phi' \supset H_0 \supset \Phi$ is chosen such that the condition imposed on the operator $S \in \mathcal{L}_0(\Phi, \Phi')$ in the previous subsection remains valid. For a cylindrical polynomial $f \in \mathcal{P}_{\mathrm{cyl}}(\Phi')$ and fixed $y \in \Phi'$, the function $f(\cdot + iy) \in \mathcal{P}_{\mathrm{cyl}}(\Phi')$. This follows from the representation

$$(\varphi, x + iy)_{H_0}^n = \sum_{k=0}^{n} C_n^k (\varphi, x)_{H_0}^k (\varphi, x)_{H_0}^{n-k} \quad (\varphi \in \Phi,\ x \in \Phi',\ n \in \mathbb{Z}_+) \tag{2.53}$$

and from the fact that, according to the polarization identity (2.7), the monomial $(\varphi, x)_{H_0}^n$ $(\varphi \in \Phi,\ n \in \mathbb{Z}_+)$ generates the whole of $\mathcal{P}_{\mathrm{cyl}}(\Phi')$. Note also that $f(x + iy)$ belongs to $\mathcal{P}_{\mathrm{cyl}}(\Phi')$ as a function of $y \in \Phi'$ for fixed $x \in \Phi'$.

Along with γ_S, we introduce the measure γ_{2S} on $\mathcal{C}_\sigma(\Phi')$ with the correlation operator $2S$. This measure will play the same role as the measure γ_2 in Example 2.6, when we proceed to the definition of the Fourier–Wiener transform in the space $L_2(\Phi', \gamma_S)$. By analogy with (2.51), let us define the integral representation

$$(W_S f)(y) = \int_{\Phi'} f(x + iy)\,d\gamma_{2S}(x) \quad (y \in \Phi') \tag{2.54}$$

for $f \in \mathcal{P}_{\mathrm{cyl}}(\Phi')$.

Lemma 2.5. *For every polynomial $f \in \mathcal{P}_{\mathrm{cyl}}(\Phi')$, we have $W_S f \in \mathcal{P}_{\mathrm{cyl}}(\Phi')$. For all $\varphi_1, \dots, \varphi_n \in \Phi$, $n \in \mathbb{Z}_+$, we have*

$$W_S :\prod_{k=1}^{n} (\varphi_k, x)_{H_0}: \; = i^n :\prod_{k=1}^{n} (\varphi_k, x)_{H_0}:. \tag{2.55}$$

Proof. It suffices to prove the first statement of the lemma for the monomial of the form $(\varphi, x)_{H_0}^n$ $(\varphi \in \Phi,\ n \in \mathbb{Z}_+)$, whose linear span, according to (2.7), coincides with $\mathcal{P}_{\mathrm{cyl}}(\Phi')$. By integrating (2.52) with respect to $x \in \Phi'$, we obtain the desired assertion.

First, we prove (2.55) for the case $\varphi_1 = \ldots = \varphi_n = \varphi \in \Phi$. The representation (2.24) yields

$$:(\varphi, x+iy)_{H_0}^n: \; = \sum_{k=0}^{n} C_n^k i^{n-k} (\varphi, y)_{H_0}^{n-k} :(\varphi, x)_{H_0}^k:$$

$$= \sum_{k=0}^{n} C_n^k i^{n-k} 2^{-k} \|\varphi\|_{H_S}^k (\varphi, y)_{H_0}^{n-k} H_k(\|\varphi\|_{H_S}^{-1} (\varphi, x)_{H_0}). \quad (2.56)$$

But (2.6) and (2.49) imply

$$\int_{\Phi'} H_k(\|\varphi\|_{H_S}^{-1} (\varphi, x) H_0)\, d\gamma_{2S}(x) = \int_{\mathbb{R}^1} H_k(t) \frac{1}{\sqrt{2\pi}} e^{-t^2/2} dt = i^k H_k(0).$$

By integrating (2.56) with respect to $x \in \Phi'$, we get

$$\int_{\Phi'} :(\varphi, x+iy)_{H_0}^n: d\gamma_{2S}(x) = \sum_{k=0}^{n} C_n^k i^k (\varphi, x)_{H_0}^k \|\varphi\|_{H_S}^k 2^{-k} H_k(0)$$

$$= 2^{-n} \|\varphi\|_{H_S}^n i^n \sum_{k=0}^{n} C_n^k 2^{n-k} (\varphi / \|\varphi\|_{H_S}, y)_{H_0}^{n-k} H_k(0).$$

Taking into account the relations

$$H_{2m+1}(0) = 0, \quad H_{2m}(0) = (-1)^m \frac{(2m)!}{m!},$$

and (2.18), we find that the last expression is equal to

$$2^{-n} i^n \|\varphi\|_{H_S}^n H_n(\|\varphi\|_{H_S}^{-1} (\varphi, x)_{H_0}) = i^n :(\varphi, y)_{H_0}^n:.$$

The general case follows from the polarization identity which has been already used

$$:\prod_{k=1}^{n} (\varphi_k, x)_{H_0}: \; = \frac{1}{n!} \sum_{k=1}^{n} (-1)^{n-k} \sum_{j_1 < \ldots < j_k} :(\varphi_{j_1} + \ldots + \varphi_{j_k}, x)_{H_0}^n:. \quad (2.57)$$

■

Theorem 2.4. *The mapping* W_S *can be extended to the unitary operator in* $L_2(\Phi', \gamma_S)$. *Moreover, for* $f \in \mathcal{P}_{\mathrm{cyl}}(\Phi')$, *we have*

$$(W_S^{-1} f)(y) = \int_{\Phi'} f(x - iy)\, d\gamma_{2S}(x) \qquad (y \in \Phi'). \quad (2.58)$$

Proof. By virtue of Lemma 2.5, the mapping W_S is isometric on the dense set $\mathcal{P}_{\mathrm{cyl}}(\Phi')$. In fact, every cylindrical polynomial $f \in \mathcal{P}_{\mathrm{cyl}}(\Phi')$ admits the expansion $f = \sum_{k=0}^{n} f_k$, where $f_k \in \Gamma_k(\Phi', \gamma_S)$, and f_k is a finite sum of Wick monomials of the form

$$:\prod_{j=1}^{n} (\varphi_j, x)_{H_0}: \qquad (\varphi_j \in \Phi_c,\ j = 1, \dots, k).$$

Therefore, $W_S f = \sum_{k=0}^{n} i^k f_k$ and, due to the mutual orthogonality of the subspaces $\Gamma_k(\Phi', \gamma_S)$ $(k = 1, \dots, n)$, we have

$$\| W_S f \|_{L_2(\Phi', \gamma_S)} = \| f \|_{L_2(\Phi', \gamma_S)}.$$

Further, the image $W_S(\mathcal{P}_{\mathrm{cyl}}(\Phi'))$ contains the Wick monomials $:(\varphi, x)_{H_0}^n:$ for all $\varphi \in \Phi$, $n \in \mathbb{Z}_+$. Consequently, by virtue of (2.24), it contains the functions $H_n((\varphi, x)_{H_0}^n)$ $(x \in \Phi')$ with arbitrary $\varphi \in \Phi$, $\|\varphi\|_{H_S} = 1$, and $n \in \mathbb{Z}_+$. Since every polynomial can be expressed in terms of the Hermite polynomials (see (2.18)), this implies that $W_S(\mathcal{P}_{\mathrm{cyl}}(\Phi'))$ contains the monomials $(\varphi, x)_{H_0}^n$ $(x \in \Phi')$ for all $\varphi \in \Phi$. Further, the polarization identity (2.7) allows us to obtain from this fact the equality

$$W_S(\mathcal{P}_{\mathrm{cyl}}(\Phi')) = \mathcal{P}_{\mathrm{cyl}}(\Phi').$$

Therefore, the isometric operator W_S can be extended from the dense set $\mathcal{P}_{\mathrm{cyl}}(\Phi')$ to $L_2(\Phi', \gamma_S)$ as a unitary operator.

For all $f \in \mathcal{P}_{\mathrm{cyl}}(\Phi')$, the equality (2.55) involves the relation $(W_S^2 f)(y) = f(-y)$ $(y \in \Phi')$ analogous to (2.52) ; and this proves the validity of (2.58).

∎

The unitary operator which extends W_S onto the whole of space $L_2(\Phi', \gamma_S)$ will be denoted, as before, by W_S and called the Fourier–Wiener transform corresponding to the measure γ_S.

Corollary 1. *For the restriction* $W_S \upharpoonright \Gamma_n(\Phi', \gamma_S)$, *we have*

$$W_S \upharpoonright \Gamma_n(\Phi', \gamma_S) = i^n 1.$$

Hence, in terms of the Wiener–Itô expansion, we have

$$W_S f = \sum_{n=0}^{\infty} i^n f_n, \tag{2.59}$$

$f_n \in \Gamma_n(\Phi', \gamma_S)$, *for*

$$f = \sum_{n=0}^{\infty} f_n \in L_2(\Phi', \gamma_S).$$

Indeed, in the case where

$$f_n = :\prod_{k=1}^{n} (\varphi_k, x)_{H_0}: \qquad (\varphi_k \in \Phi)$$

the equality (2.59) (that is, $W_S f = i^n f_n$) is established by Lemma 2.5. But the linear span of these functions (for fixed $n \in \mathbb{Z}_+$) is dense in $\Gamma_n(\Phi', \gamma_S)$ (see Subsection 2.2), and the operator W_S is unitary.

It is well known that the Fourier transform of the functions $f(x)$ $(x = (x_1, \ldots, x_n) \in \mathbb{R}^n)$ can be constructed as an expansion in common generalized eigenfunctions of n commuting selfadjoint operators which are generated by the derivatives $i\,\partial/\partial x_k$ and act in the space L_2 with the Lebesgue measure. This scheme cannot be extended directly to the case of $n = \infty$, because there is no Lebesgue measure in this case. However, one can pass to the Gaussian measure in $\mathbb{R}^\infty$ by changing the operators $i\,\partial/\partial x_k$ properly (in order to make them Hermitian in the corresponding space L_2). The expansion in common generalized eigenfunctions of the corresponding countable family of commuting selfadjoint operators obtained, in this case coincides with the Fourier–Wiener transform introduced above. This point of view will be presented in Chapter 3, Subsection 3.7.

§3. Differentiable Functions on Infinite-Dimensional Spaces

When studying the differential operators and constructing the generalized functions in the infinite-dimensional case, we shall need to exploit the ideas of the differential calculus on linear spaces. Below, we present some elements of this theory necessary for our discussion and construct some spaces of smooth functions of infinitely many variables useful in what follows.

3.1. Derivatives of Functions on a Hilbert Space

Let X, Y be linear normed spaces and, moreover, let X be real. As above, $\mathcal{L}(X, Y)$ denotes a space of linear continuous operators acting from X into Y with an operator norm. A mapping $X \supset U \ni x \mapsto f(x) \in Y$ defined in a certain neighborhood $U \subset X$ is called differentiable at the point $x_0 \in U$ if there exists an operator $f'(x_0) \in \mathcal{L}(X, Y)$ for which

$$f(x_0 + h) = f(x_0) + f'(x_0)\, h + \varepsilon_1(x_0, h),$$

$$\varepsilon_1(x_0, h) = o(\| h \|_X), \qquad \| h \|_X \to 0. \tag{3.1}$$

The operator $f'(x_0)$ is called a derivative of the mapping f at the point x_0.

By definition, f is (one time) differentiable in U if it is differentiable at all the points $x \in U$. For the given mapping f, one can construct a new mapping

$$X \supset U \ni x \mapsto f'(x) \in \mathcal{L}(X, Y).$$

If, in turn, this map is differentiable at the point $x_0 \in U$, then the second derivative is defined

$$f''(x_0) = (f'(x_0))' \in \mathcal{L}(X, \mathcal{L}(X, Y)).$$

By continuing this process, we can define a derivative of an arbitrary order

$$f_n(x_0) \in \mathcal{L}(X, (\mathcal{L}(X, \ldots, \mathcal{L}(X, Y), \ldots)$$

and an n times differentiable mapping in U. If the mapping

$$X \supset U \ni x \mapsto f_n(x) \in \mathcal{L}(X, (\mathcal{L}(X, \ldots, \mathcal{L}(X, Y), \ldots)$$

is continuous, then we say that f is n times continuously differentiable in U. Clearly, as in the finite-dimensional case, an n times differentiable mapping is $n-1$ times continuously differentiable.

For the mapping f which is n times continuously differentiable in U, the Taylor formula holds and has the form

$$f(x + h) = f(x) + \sum_{k=1}^{n} \frac{1}{k!}\, d^k f(x)(h) + \varepsilon_n(x, h), \tag{3.2}$$

where $\varepsilon_n(x, h) = o(\| h \|_X^n)$, $\| h \|_X \to 0$, the value of $\| h \|_X$ is sufficiently small, and

$d^k f(x)(h) = (f^{(k)}(x)h)h\dots)h$ is the k-th differential of the mapping f at the point $x \in U$ in the direction of $h \in U$.

The derivatives $f^{(n)}(x)$ introduced above may be interpreted as vector functions of a point $x \in U$ with values in the space of multilinear mappings of X into Y. Let us describe this approach in the case of complex-valued functions f, i.e., for $Y = \mathbb{C}^1$. Denote

$$\mathfrak{L}^{(1)}(X, \mathbb{C}^1) = \mathfrak{L}(X, \mathbb{C}^1)$$

and

$$\mathfrak{L}^{(n)}(X, \mathbb{C}^1) = \mathfrak{L}(X, \mathfrak{L}^{(n-1)}(X, \mathbb{C}^1)) \qquad (n = 2, 3, \dots).$$

Therefore, $\mathfrak{L}^{(1)}(X, \mathbb{C}^1)$ coincides with the space X^* of all complex-valued linear continuous functionals on the real space X, whereas each $\mathfrak{L}^{(n)}(X, \mathbb{C}^1)$ with $n > 1$ is the corresponding space of operators from X into $\mathfrak{L}^{(n-1)}(X, \mathbb{C}^1)$. According to what has been said above, the derivative $f^{(n)}(\cdot)$ defines the mapping

$$X \supset U \ni x \mapsto f^{(n)}(x) \in \mathfrak{L}^{(n)}(X, \mathbb{C}^1).$$

Consider a collection of all n-linear continuous forms

$$X^n = \mathop{\times}_{k=1}^{n} X \ni (h_1, \dots, h_n) \mapsto a(h_1, \dots, h_n) \in \mathbb{C}^1$$

and denote it by $\mathfrak{F}(X^n, \mathbb{C}^1)$. The space $\mathfrak{F}(X^n, \mathbb{C}^1)$ is obviously a linear normed space with a norm

$$\|a\| = \sup_{h_1, \dots, h_n \in X} \frac{|a(h_1, \dots, h_n)|}{\|h_1\|_X \dots \|h_n\|_X}.$$

It is easy to show that there exists a natural isomorphism between Banach spaces $\mathfrak{L}^{(n)}(X, \mathbb{C}^1)$ and $\mathfrak{F}(X^n, \mathbb{C}^1)$, namely, an operator $A \in \mathfrak{L}^{(n)}(X, \mathbb{C}^1)$ is associated with the form

$$a_A(h_1, \dots, h_n) = (\dots(Ah_n)h_{n-1})\dots)h_1 \qquad (h_1, \dots, h_n \in X).$$

Due to this correspondence, $f^{(n)}$ can be regarded as a vector-function

$$X \supset U \ni x \mapsto f^{(n)}(x) \in \mathfrak{F}(X^n, \mathbb{C}^1). \tag{3.3}$$

One can prove that the derivative $f^{(n)}(x)$ is a symmetric form in the variables $h_1, \dots,$

$h_n \in X$. The k-th differential of the function f can be now written as

$$d^k f(x)(h) = f^{(k)}(x)\underbrace{(h, \ldots, h)}_{k},$$

and the Taylor formula (3.2) turns into

$$f(x+h) = f(x) + \sum_{k=1}^{n} \frac{1}{k!} f^{(k)}(x)\underbrace{(h, \ldots, h)}_{k} + \varepsilon_n(x, h),$$

$$\varepsilon_n(x, h) = o(\|h\|_X^n), \quad \|h\|_X \to 0 \quad (x \in U). \tag{3.4}$$

In the case when X is a real Hilbert space, it is convenient to pass to the interpretation of derivatives in terms of the kernels associated with the forms (3.3). For this purpose we fix a certain rigging of the space X

$$X_- \supset X \supset X_+, \tag{3.5}$$

with the quasinuclear imbedding of X_- into X. Consider the complexification of the chain (3.5), namely

$$X_{-,c} \supset X_c \supset X_{+,c}$$

(here the imbedding of $X_{+,c}$ in X_c is also quasinuclear, it is now convenient to omit the index 0 of the space X). Then

$$X_{-,c}^{\otimes n} \supset X_c^{\otimes n} \supset X_{+,c}^{\otimes n} \tag{3.6}$$

is a chain for all $n \in \mathbb{N}$. According to the kernel theorem (Chapter 1, Section 2, Theorem 2.3, Corollary 1), every form $a \in \mathfrak{F}(X^n, \mathbb{C}^1)$ admits the representation

$$a(\varphi_1, \ldots, \varphi_n) = (\Theta_a, \varphi_1 \otimes \ldots \otimes \varphi_n)_{X_{-,c}^{\otimes n}} \quad (\varphi_k \in X_+), \tag{3.7}$$

where $\Theta_a \in X_{-,c}^{\otimes n}$ is the kernel of this form uniquely determined by a. When a changes in $\mathfrak{F}(X^n, \mathbb{C}^1)$, the kernels Θ_a run through a dense linear manifold F_n in $X_{-,c}^{\otimes n}$ which does not coincide with the whole of $X_{-,c}^{\otimes n}$ (for $\dim X = \infty$). The norm $\|\Theta_a\|_{F_n} = \|a\|$ transforms F_n into a normed space isomorphic to $\mathfrak{F}(X^n, \mathbb{C}^1)$. Clearly, $\|\Theta\|_{X_{-,c}^{\otimes n}} \le \|\Theta\|_{F_n}$ $(\Theta \in F_n)$.

Let us apply the representation (3.7) to the form $f^{(n)}(x)$ ($x \in U$ is fixed). We have

$$f^{(n)}(x)(\varphi_1, \ldots, \varphi_n) = (f^{(n)}(x), \varphi_1 \otimes \ldots \otimes \varphi_n)_{X^{\otimes n}_{-,c}} \qquad (\varphi_k \in X_+), \tag{3.8}$$

where $f^{(n)}(x)$ on the right-hand side of (3.8) denotes the kernel of the form $f^{(n)}(x)$. We do not introduce a separate notation for this kernel, since henceforth, as a rule, we will use the notation of the n-th derivative to just denote a kernel of this sort. Thus, we proceed to the vector functions

$$X \supset U \ni x \mapsto f^{(n)}(x) \in X^{\otimes n}_{-,c} \quad (n \in \mathbb{N}). \tag{3.9}$$

Let us also set $f^{(0)}(x) = f(x)$ $(x \in U)$. The symmetry of the kernel $f^{(n)}(x)$ follows from the symmetry of the forms $f^{(n)}(x)$; this means that the function $(f^{(n)}(x), \varphi_1 \otimes \ldots \otimes \varphi_n)_{X^{\otimes n}_{-,c}}$ is symmetric with respect to the variables $h_k \in X_+$.

The Taylor formula (3.4) can be rewritten in the form

$$f(x+h) = f(x) + \sum_{k=0}^{n} \frac{1}{k!} (f^{(k)}(x), h^{\otimes k})_{X^{\otimes k}_{c}} + \varepsilon_n(x, h),$$

$$\varepsilon_n(x, h) = o(\| h \|^n_X), \quad \| h \|_X \to 0 \quad (x \in U, h \in X_+). \tag{3.10}$$

Let us mention one purely infinite-dimensional effect connected with the possibility of varying the range of values of the derivatives in the sense of "smoothness" (more exactly, in the sense of the regularity in the number of variables). Namely, for some functions their derivatives may belong not to $X^{\otimes n}_{-,c}$, but to a narrower linear topologic space $V_n \subset X^{\otimes n}_{-,c}$, and this imbedding is assumed to be topological. We can select different classes of functions by choosing proper spaces V_n to which the derivatives of the functions under consideration should belong (i.e., we demand that the inclusion $f^{(n)}(x) \in V_n$ $(x \in X)$ holds). It is convenient to employ the spaces V_n possessing some additional properties, e.g., Hilbert spaces or spaces smooth in the sense of the rigging (3.6), etc.[15]

Consider the example in which linear differential expressions given on the functions $X \ni x \mapsto f(x) \in \mathbb{C}^1$ are defined. In this example the opportunity of choosing different spaces V_n enables us to extend the class of coefficients of differential expressions.

Example 3.1. For $\xi \in X^{\otimes n}_{-,c}$ and $\varphi \in X^{\otimes n}_{-,c}$, we denote by $(\xi, \varphi)_{X^{\otimes n}}$ a two-linear extension of the scalar product in $X^{\otimes n}$ onto $X^{\otimes n}_c$ or, in other words, $(\xi, \varphi)_{X^{\otimes n}_c} = (\xi, \overline{\varphi})_{X^{\otimes n}}$, where the line denotes the n-th tensor power of the natural involution in the complexification $X_{+,c}$ $(n \in \mathbb{N})$. In order to define the expression of the r-th order, it is necessary to set its coefficients

[15] Note that $f^{(n)}(x)$ always belongs not to $X^{\otimes n}_{-,c}$ but to a narrower space $V_n = F_n$.

$$X \supset U \ni x \mapsto a_k(x) \in X_{+,c}^{\otimes k} \quad (k = 0, \dots, r;\ X_{+,c}^{\otimes 0} = \mathbb{C}^1). \tag{3.11}$$

We define a linear differential expression L by

$$Lf(x) = \sum_{k=0}^{r} (a_k(x), f^{(k)}(x))_{X^{\otimes k}} \quad (x \in U). \tag{3.12}$$

This expression is always defined correctly on the r times differentiable functions $U \ni x \mapsto f(x) \in \mathbb{C}^1$. Furthermore, the function Lf is continuous for the continuous mapping (3.11) and r times continuously differentiable function f. The values of coefficients $a_k(x)$ now lie in positive spaces, i.e., the expression (3.12) is highly regular.

At the same time, we can act as follows. Assume that the triple of spaces V, X_c, V' is given, where the Hilbert space V and its dual space V' are in duality with respect to X_c, and $V \subset X_{c,-}$ (the imbedding is topological). Then it is easy to show that $V^{\otimes n}$, $X_c^{\otimes n}$, and $V'^{\otimes n}$ also form the triple in the sense that $V^{\otimes n}$ and $V'^{\otimes n}$ are in duality with respect to $X_c^{\otimes n}$ (for the proof and detailed presentation, see Berezansky [18, Chapter 1, Section 2, Subsection 11]). Therefore, the form $(\xi, \varphi)_{X_c^{\otimes n}} = (\xi, \overline{\varphi})_{X^{\otimes n}}$, where $\xi \in V'^{\otimes n}$, $\varphi \in V^{\otimes n}$, is meaningful.

Let us fix the triple V, X_c, V' and consider the coefficients of L having the form

$$X \supset U \ni x \mapsto a_k(x) \in V'^{\otimes k} \quad (k = 0, \dots, r). \tag{3.13}$$

For the functions $U \ni x \mapsto f(x) \in \mathbb{C}^1$ such that $f^{(k)}(x) \in V^{\otimes k}$ $(x \in U, k = 0, \dots, r)$, the differential expression (3.12) is defined correctly. If the mapping (3.13) and $U \ni x \mapsto f^{(k)}(x) \in V^{\otimes k}$ $(k = 0, \dots, r)$ are continuous, then the function Lf is continuous on U. In particular, for $V = X_{+,c}$ the values of the coefficients (3.13) may be generalized vectors since, in this case, $V' = X_{+,c}$. However, in this case, functions f should be highly regular, i.e., $f^{(k)}(x) \in X_{+,c}^{\otimes k}$.

Let us consider the coordinate form of the derivatives of the function $X \supset U \ni x \mapsto f(x) \in \mathbb{C}^1$. We fix a certain orthonormal basis $(e_j)_{j=1}^{\infty}$ in X and identify $x \in X$ with the sequence $(x_j)_{j=1}^{\infty}$ of its coordinates in this basis. Let us introduce the coordinates of the multilinear form $f^{(n)}(x)$ $(n \in \mathbb{N})$

$$f^{(n)}(x)(e_{\alpha_1}, \dots, e_{\alpha_n}) = \left(\frac{\partial^n f}{\partial x_{\alpha_1} \dots \partial x_{\alpha_n}} \right)(x). \tag{3.14}$$

The expression $\partial^n f / \partial x_{\alpha_1} \dots \partial x_{\alpha_n}$ on the right-hand side of (3.14) is only a symbol denoting the coordinates of the form, but it is quite easy to understand that it may be interpreted as a "partial derivative" of the function $f(x) = f(x_1, x_2, \dots)$. Let us introduce the rigging (3.5) and choose $e_j \in X_+$, then we can write the partial derivatives (3.14) in the form $(f^{(n)}(x), e_{\alpha_1} \otimes \dots \otimes e_{\alpha_n})_{X_c^{\otimes n}}$ ($f^{(n)}(x)$ is now regarded as a kernel).

Note also that, for any $x \in U$ and $n = 1$, we have

$$((\partial^n f / \partial x_{\alpha_1})(x))_{\alpha_1=1}^{\infty} \in l_2(\mathbb{C}^1).$$

This follows from the definition of the derivative. For $n > 1$, the matrices (3.14) do not belong to such simple classes.

Let us rewrite Example 3.1 in the coordinate form.

Example 3.2. Let $X = l_2(\mathbb{R}^1)$. We choose the rigging (3.5) in the form

$$l_2(\mathbb{R}^1, p^{-1}) \supset l_2(\mathbb{R}^1) \supset l_2(\mathbb{R}^1, p), \tag{3.15}$$

where $p = (p_j)_{j=1}^{\infty}$ $(p_j \geq 1)$ is a weight for which $\sum_{j=1}^{\infty} p_j^{-1} < \infty$ (see Chapter 1, Section 1, Example 1.6). The complexification of this chain is reduced to the substitution of $\mathbb{C}^1$ for $\mathbb{R}^1$ in (3.15). Tensor powers from (3.6) consist of n-dimensional sequences of complex numbers of the form $z = (z_{\alpha_1,\dots,\alpha_n})_{\alpha_1,\dots,\alpha_n=1}^{\infty}$ for which the following series

$$\sum_{\alpha_1,\dots,\alpha_n=1}^{\infty} |\xi_{\alpha_1,\dots,\alpha_n}|^2 p_{\alpha_1}^{-1} \cdots p_{\alpha_n}^{-1} = \|\xi\|_{X_{-,c}^{\otimes n}}^2 < \infty, \tag{3.16}$$

$$\sum_{\alpha_1,\dots,\alpha_n=1}^{\infty} |x_{\alpha_1,\dots,\alpha_n}|^2 = \|x\|_{X_c^{\otimes n}}^2 < \infty,$$

$$\sum_{\alpha_1,\dots,\alpha_n=1}^{\infty} |\varphi_{\alpha_1,\dots,\alpha_n}|^2 p_{\alpha_1} \cdots p_{\alpha_n} = \|\varphi\|_{X_{+,c}^{\otimes n}}^2 < \infty$$

converge, respectively.

Consider a function $l_2 \ni x \mapsto f(x) \in \mathbb{C}^1$ n times differentiable in l_2. Taking into account that we regard derivatives as kernels from $X_{-,c}^{\otimes n}$, by virtue of (3.16) and (3.14), we can say that

$$f^{(n)}(x) = \left(\left(\frac{\partial^n f}{\partial x_{\alpha_1} \cdots \partial x_{\alpha_n}} \right)(x) \right)_{\alpha_1,\dots,\alpha_k=1},$$

$$\sum_{\alpha_1,\dots,\alpha_n=1}^{\infty} \left| \left(\frac{\partial^n f}{\partial x_{\alpha_1} \cdots \partial x_{\alpha_n}} \right)(x) \right|^2 p_{\alpha_1}^{-1} \cdots p_{\alpha_n}^{-1} = \| f^{(n)}(x) \|_{X_{-,c}^{\otimes n}}^2 < \infty$$

for every $x \in l_2(\mathbb{R}^1)$. Note that we can now compute the derivatives using the standard rule, i.e., as a limit of the difference quotient of f and x.

Setting of the coefficients (3.12) and the differential expression L is now equivalent to setting of the mappings of $l_2(\mathbb{R}^1)$ into the space of sequences

$$l_2(\mathbb{R}^1) \ni x \mapsto a_k(x) = (a_{\alpha_1,\ldots,\alpha_k}(x))_{\alpha_1,\ldots,\alpha_k=1}^{\infty},$$

$$\sum_{\alpha_1,\ldots,\alpha_k=1}^{\infty} |a_{\alpha_1,\ldots,\alpha_k}|^2 p_{\alpha_1} \cdots p_{\alpha_k} = \| a_k \|^2_{X_{+,c}^{\otimes k}} < \infty. \tag{3.17}$$

The differential expression L now takes the form

$$Lf(x) = \sum_{k=0}^{r} \sum_{\alpha_1,\ldots,\alpha_k=1}^{\infty} a_{\alpha_1,\ldots,\alpha_k}(x) \left(\frac{\partial^k f}{\partial x_{\alpha_1} \ldots \partial x_{\alpha_k}} \right)(x) \quad (x \in l_2(\mathbb{R}^1)). \tag{3.18}$$

Conditions (3.17) imposed on the coefficients of L are, surely, rather restrictive. However, if we define expression (3.18) only on cylindrical functions f, then it is defined at every point x without any restrictions such as (3.17)[16].

The situation described in this example is general. In fact, assume that we have the chain (3.5) with the quasinuclear imbedding $X_+ \to X$. The operator $OJ : X \to X$ is positive and quasinuclear. Moreover, $\| OJ \| \le 1$. If we take the sequence of unit eigenvectors of the operator $OJ : X \to X$ as an orthonormal basis $(e_j)_{j=1}^{\infty}$ in X, then we find that the chain (3.5) is isomorphic to (3.14). The weight p is constructed by use of the eigenvalues λ_j of the operator OJ as follows: $p_j = \lambda_j^{-2}$, $OJ e_j = \lambda_j e_j$ $(j \in \mathbb{N})$.

Example 3.3. Variational derivatives are included into this scheme as follows. Consider the chain

$$S_{-l,\,\mathrm{Re}}(\mathbb{R}^d) \supset L_{2,\,\mathrm{Re}}(\mathbb{R}^d) \supset S_{l,\,\mathrm{Re}}(\mathbb{R}^d) \quad (l > d/2) \tag{3.19}$$

consisting of real spaces of Example 1.8 (Chapter 1, Section 1) as a chain (3.5). The derivative $f^{(n)}(x)$ of the function $L_{2,\,\mathrm{Re}}(\mathbb{R}^d) \ni x \mapsto f(x) \in \mathbb{C}^1$ is understood as a generalized kernel from $(S_{-l,\mathrm{Re}}(\mathbb{R}^d))^{\otimes n}$, that is, as a generalized function of the variables $t_1, \ldots, t_n$ (t is a point in $\mathbb{R}^d$).

For this generalized kernel, we use the following notation

$$\left(\frac{\delta^{(n)} f}{\delta(t_1) \ldots \delta(t_n)} \right)(x). \tag{3.20}$$

---

[16] In order to include this case to the general scheme of Example 3.1, it suffices to extend this scheme onto the spaces V, which are projective limits of Hilbert spaces, and set $V = \mathbb{R}_0^{\infty}$.

Expression (3.20) is also called a variation derivative of the n-th order. It may happen that (3.20) is an ordinary function of the point $(t_1, \dots, t_n)$ (this is always true for $n = 1$); but in the general case, (3.20) should be regarded only as a notation of the corresponding generalized kernel.

Let us make three remarks concerning variational derivatives:

(i) It is sometimes convenient to consider other riggings of the space $X = L_{2,\mathrm{Re}}(\mathbb{R}^d)$ (with a quasinuclear imbedding of the positive space into the zero space) instead of (3.19). Thus, we can employ the rigging from Example 2.3 (Chapter 1, Section 2). If we write the second derivative $f''(x)$, regarded as a two-linear form using (2.64) of Chapter 1, then we obtain

$$(f''(x))(\varphi_1, \varphi_2) = \left(\left(\frac{\delta^2 f}{\delta(t_1)\,\delta(t_2)}\right)(x), \varphi_1(t_1) \otimes \varphi_2(t_2)\right)_{X^{\otimes 2}}$$

$$= \int_{\mathbb{R}^d}\int_{\mathbb{R}^d} T''(x))(t_1, t_2)(\mathcal{D}\varphi_1)(t_1)(\mathcal{D}\varphi_2)(t_2)\,dt_1\,dt_2 \quad (\varphi_1, \varphi_2 \in C^\infty_{0,\mathrm{Re}}(\mathbb{R}^d)) \tag{3.21}$$

for every $x \in L_{2,\mathrm{Re}}(\mathbb{R}^d)$. Here,

$$T''(x))(\cdot,\cdot) \in C(\mathbb{R}^{2d}), \quad \left(\frac{\delta^2 f}{\delta(t_1)\,\delta(t_2)}\right)(x) = (\mathcal{D}_{t_1}\mathcal{D}_{t_2}T''(x))(t_1, t_2). \tag{3.22}$$

Clearly, the representations similar to (3.21) and (3.22) hold for variational derivatives of an arbitrary order n.

(ii) Assume that a function $L_{2,\mathrm{Re}}(\mathbb{R}^d) \ni x \mapsto f(x) \in \mathbb{C}^1$ satisfies $(\delta f/\delta(t_1))(x_0) \in S_l(\mathbb{R}^d)$ for some x_0. The expression

$$f(x_0) + \left(\frac{\delta f}{\delta(t_1)}(x_0), \delta_{t_1}\right)_X$$

approximately extending this function to the vectors $x_0 + \delta_{t_1} \in S_{-l,\mathrm{Re}}(\mathbb{R}^d)$, now makes sense. Conversely, if the function f can be extended to $S_{-l,\mathrm{Re}}(\mathbb{R}^d)$ and is differentiable, then, according to (3.13), one can regard $(\delta f/\delta(t_1))(x)$ as $(f'(x)(\delta_{t_1}))$.

(iii) Let us examine the relation between variational derivatives and derivatives given by (3.14). First, assume that $f^{(n)}(x) \in X^{\otimes n}_{-,c}$ for given f and fixed x. Denote by $(e_j(t_j))_{j=1}^\infty$ ($t \in \mathbb{R}^d$) an orthonormal basis chosen in the way discussed at the end of Example 3.2. Then definitions (3.20) and (3.13) imply that

$$\left(\frac{\delta^{(n)} f}{\delta(t_1)\dots\delta(t_n)}\right)(x) \tag{3.23}$$

in the sense of convergence in $L_{2,\mathrm{Re}}(\mathbb{R}^d)$. In the general case, where $f^{(n)}(x) \in X^{\otimes n}_{-,c}$, this series converges in $X_{-,c}^{\otimes n}$. This and Example 3.1 also clarify how to write differential expressions in terms of variational derivatives.

3.2. Derivatives of Functions Given in Riggings

In the case where the role of a space of arguments is played by the negative space of some given chain, it is natural to interpret the derivatives by use of the scalar product in the zero space. Let us describe the corresponding construction. Assume that the quasi-nuclear rigging

$$H_- \supset H_0 \supset H_+ \tag{3.24}$$

of the real Hilbert space H_0 is given. For the function $H_- \supset U \ni x \mapsto f(x) \in \mathbb{C}^1$, the derivative $f^{(n)}(x)$ can be regarded as a multilinear form, i.e., $f^{(n)}(x) \in \mathfrak{F}(H_-^n, \mathbb{C}^1)$, or as a corresponding generalized kernel.

Introduce a space $\hat{H}_0$ dual to H_0 with respect to H_-, i.e., $\hat{H}_0 \supset H_- \supset H_0$. Then the representation (3.8) has the form

$$f^{(n)}(x)\,(h_1, \ldots, h_n) = (f^{(n)}(x), h_1 \otimes \ldots \otimes h_n)_{H_{-,c}^{\otimes n}} \quad (h_k \in H_0), \tag{3.25}$$

where $f^{(n)}(x)$ on the right-hand side is the generalized kernel from $\hat{H}_{0,c}^{\otimes n}$. By using the isometric operator $I : \hat{H}_0 \to H_0$ connected with the chain $\hat{H}_0 \supset H_- \supset H_0$, we can rewrite (3.25) as follows

$$f^{(n)}(x)\,(h_1, \ldots, h_n) = (D^n f(x), h_1 \otimes \ldots \otimes h_n)_{H_{0,c}^{\otimes n}} \quad (h_j \in H_0). \tag{3.26}$$

Here,

$$D^n f(x) = (I^{\otimes n}) f^{(n)}(x) \in H_{0,c}^{\otimes n}$$

is the image of the kernel $f^{(n)}(x)$ under the isometry $I^{\otimes n} : \hat{H}_{0,c}^{\otimes n} \to H_{0,c}^{\otimes n}$.

For the functions given in riggings, we shall successively apply the representation (3.26); taking this into account, we can identify the notations $f^{(n)}$ and $D^n f$. If it is necessary to distinguish the derivatives defined by (3.25) and (3.26), we shall call them the derivatives in the sense of H_- and H_0, respectively. The interpretation of derivatives as kernels from $H_{0,c}^{\otimes n}$ is particularly convenient, due to the fact that they can be defined in terms of the fixed Hilbert space H_0. The latter enables us to study functions defined on H_- for different riggings of this Hilbert space simultaneously. In addition, this interpretation allows us to clarify the direct dependence of "smoothness" of the values of derivatives on the topological properties of riggings in which the functions are defined. In somewhat simplified terms, we can formulate this dependence as follows: The broader is the space H_- on which the function f is differentiable, the smoother is the kernel $f^{(n)}(x)$. The exact statement is given in the following lemma.

Lemma 3.1. *Assume that the rigging*

$$H_{-2} \supset H_{-1} \supset H_0 \supset H_1 \supset H_2 \tag{3.27}$$

is given in which the spaces H_j and H_{-j} are dual with respect to H_0, and the imbedding of H_2 into H_1 is quasinuclear. Then, for the derivatives of a function f n times differentiable in the region $U \supset H_{-2}$, we have $f^{(k)}(x) \in H_{1,c}^{\otimes k}$ $(x \in U;$ $k = 1, \ldots, n)$.

Proof. Let us introduce a space $\hat{H}_{-1}$ dual to H_{-1} with respect to H_{-2} and the corresponding rigging $\hat{H}_{-1} \supset H_{-2} \supset H_{-1}$. Since f is differentiable, we obtain the following representation:

$$f^{(k)}(x)\,(h_1, \ldots, h_k) = ((I^{\otimes k})f^{(k)}(x), h_1 \otimes \ldots \otimes h_k)_{H_{-1,c}^{\otimes k}} \quad (x \in U;\, h_j \in H_{-1}) \tag{3.28}$$

with the kernel $f^{(k)}(x) \in \hat{H}_{-1,c}^{\otimes k}$. Let $\hat{I}\colon \hat{H}_{-1} \to H_{-1}$ be the isometry associated with the chain $\hat{H}_{-1} \supset H_{-2} \supset H_{-1}$. Then we can rewrite (3.28) in the form

$$f^{(k)}(x)\,(h_1, \ldots, h_k) = (f^{(k)}(x), h_1 \otimes \ldots \otimes h_k)_{H_{-2,c}^{\otimes k}} \quad (x \in U;\, h_j \in H_{-1}) \tag{3.29}$$

with the kernel $(\hat{I}^{\otimes k})f^{(k)}(x) \in \hat{H}_{-1,c}^{\otimes k}$ and, using the isometry $I\colon H_{-1} \to H_1$, pass to the representation in terms of the scalar product in H_0

$$f^{(k)}(x)\,(h_1, \ldots, h_n) = (D^k f(x), h_1 \otimes \ldots \otimes h_k)_{H_{0,c}^{\otimes k}} \quad (x \in U;\, h_j \in H_{-1}). \tag{3.30}$$

Here, the kernel $D^k f(x) \in H_{1,c}^{\otimes k}$ $(k = 1, \ldots, n)$.

■

Remark 1. It is often convenient to apply Lemma 3.1 in the following situation. Having given the necessary smoothness of values of the derivatives, i.e., having given the space H_1, we want to construct a functional space on quasinuclear extensions of the space H_{-1}.

■

When studying infinite-dimensional differential operators in Chapter 6, we shall employ not only the representation of the first and second derivatives in terms of kernels, but also their operator representation. It proves now to be more convenient not to exploit the original operator interpretation in the space H_-, but to pass to the representation in terms of H_0. This transition corresponds to the transition from the derivatives defined above as kernels from tensor products to the operators associated with these kernels.

Consider the rigging (3.24) and a twice differentiable function $H_- \supset U \ni x \mapsto f(x) \in \mathbb{C}^1$. The Taylor formula (3.4) holds for this function

$$f(x+h) = f(x) + df(x)(h) + \tfrac{1}{2} d^2 f(x)[h] + o(\| h \|^2_{H_-}), \quad \| h \|_{H_-} \to 0,$$

where $df(x)(h)$ and $d^2 f(x)[h] = d^2 f(x)(h, h)$ are linear and quadratic continuous complex-valued forms on H_-, respectively. Accordingly, we have

$$df(x)(h) = (f'(x), h)_{H_{0,c}},$$

$$d^2 f(x)[h] = (f''(x)h, h)_{H_{0,c}} \qquad (h \in H_-), \tag{3.31}$$

where $f'(x) \in H_{+,c}$ and $f''(x) \in \mathcal{L}(H_-, H_{+,c})$. We do not introduce any special notations to write the derivatives — it is clear that this notation should be understood as the operator $f''(x)$ or the kernel associated with it.

Note that in the case where the imbedding $H_+ \subset H_0$ is quasinuclear, the operator $f''(x) \in \mathcal{L}(H_-, H_{+,c})$ is nuclear as an operator considered in $H_{0,c}$, i.e., the operator $O(f''(x) \restriction (O^+ H_{0,c}))$ is nuclear. This fact is exploited when we define differential expressions of the second order. It is a consequence of the fact that the product of two quasinuclear operators and a bounded operator is a nuclear operator.

Example 3.4. Assume that the following three mappings are given: $H_- \supset U \ni x \mapsto a(x) \in \mathcal{L}(H_{0,c}, H_{0,c})$, $b(x) \in H_{-,c}$, and $c(x) \in \mathbb{C}^1$. The differential expression

$$(Lf)(x) = \mathrm{Tr}_0 (a(x) f''(x)) + (b(x), f'(x))_{H_0} + c(x) f(x) \tag{3.32}$$

is defined on twice differentiable functions $H_- \supset U \ni x \mapsto f(x) \in \mathbb{C}^1$. Here, $\mathrm{Tr}_0 (a(x) \times f''(x))$ is a trace of the operator $a(x) f''(x)$ in $H_{0,c}$. This trace is finite due to the nuclearity of $f''(x))$ in $H_{0,c}$ and the boundedness of $a(x)$. Using differential expressions of the type (3.32), we can define a number of infinite-dimensional elliptic differential operators which are important in applications (see Chapter 6). The coordinate form of the expression (3.32) (see Example 3.2) is given by

$$(Lf)(x) = \sum_{j,k=1}^{\infty} a_{jk}(x)\left(\frac{\partial^2 f}{\partial x_j \partial x_k}\right)(x) + \sum_{j}^{\infty} b_j(x)\left(\frac{\partial f}{\partial x_j}\right)(x) + c(x)f(x). \qquad (3.33)$$

For the mappings acting between the spaces of the chain, we can introduce a representation of derivatives similar to (3.31). Consider the case of the mapping $H_- \supset U \ni x \mapsto f(x) \in H_{-,c}$ which will be exploited below. Assume that f is twice differentiable. Let us associate the scalar function $f_\varphi(x) = (f(x), \varphi)_{H_{0,c}}$ $(x \in U)$ with each $\varphi \in H_+$. This function is obviously twice differentiable, too. Therefore, according to (3.31), the derivatives $f'_\varphi(x) \in H_{+,c}$ and $f''_\varphi(x) \in \mathcal{L}(H_-, H_{+,c})$ are defined. For every $x \in U$, consider the mapping

$$H_+ \ni \varphi \mapsto f'_\varphi(x) \in H_{+,c}, \quad H_+ \ni \varphi \mapsto f''_\varphi(x) \in \mathcal{L}(H_-, H_{+,c}).$$

According to the construction of f_φ, this mapping is linear and continuous on H_+. Hence, we can introduce the operators

$$f'(x) \in \mathcal{L}(H_+, H_{+,c}), \quad f''(x) \in \mathcal{L}(H_+, \mathcal{L}(H_-, H_{+,c})),$$

$$((f')\varphi)(x) = (f'_\varphi)(x), \quad ((f'')\varphi)(x) = (f''_\varphi)(x) \quad (x \in U). \qquad (3.34)$$

We shall call these operators the derivatives of the mappings f in the sense of the space H_0. By employing these operators, we can rewrite the Taylor formula in terms of the scalar product in $H_{0,c}$

$$(f(x+h), \varphi)_{H_{0,c}} = (f(x), \varphi)_{H_{0,c}} + (f'(x), \varphi)_{H_{0,c}} + \tfrac{1}{2}(f''(x)\varphi)h, h)_{H_{0,c}} + o(\|h\|^2_{H_-}),$$

$$\|h\|_{H_-} \to 0 \quad (\varphi \in H_+;\ h \in H_-). \qquad (3.35)$$

3.3. Smooth Functions in Nuclear Riggings

As in the finite-dimensional case, in order to construct the analysis of functions of infinitely many variables, it is necessary to use various functional spaces including the spaces of smooth functions. These spaces should possess good topological properties and, besides, satisfy a series of additional requirements depending on the class of problems under consideration. Thus, when studying differential operators, we need spaces of functions, whose derivatives satisfy certain conditions, possessing a sufficiently large collection of multiplicators. It turns out that, when passing to the infinite-dimensional case, most of spaces of smooth functions either lose their good topological properties or do not allow such transition at all. Another essential infinite-dimensional effect consists in the following: The properties of functional spaces depend on class of linear topological spaces to which the arguments of functions belong. Below, we consider some spaces that are dual to nuclear spaces. This is just the case in which we can construct a sufficiently wide collection of spaces of smooth functions possessing the required properties.

Assume that the nuclear rigging

$$\Phi' \supset H_0 \supset \Phi \tag{3.36}$$

of the real Hilbert space H_0 is given such that, in addition, Φ is a countably Hilbert space. Without loss of generality, we can suppose that Φ is the projective limit of the sequence $(H_j)_{j=1}^{\infty}$ of Hilbert spaces satisfying the following conditions

$$H_0 \supset H_1 \supset H_2 \supset \ldots,$$

$$\|\varphi\|_{H_0} \le \|\varphi\|_{H_1} \le \|\varphi\|_{H_2} \le \ldots \quad \Big(\varphi \in \Phi = \bigcap_{j=1}^{\infty} H_j\Big). \tag{3.37}$$

For each $j \in \mathbb{N}$, we introduce the Hilbert space H_{-j} dual to H_j with respect to H_0. In Chapter 1, Subsection 1.2, we mentioned that the following topologies on Φ' are equivalent in this case: the strong topology $\beta(\Phi, \Phi')$, the Mackey topology $\tau(\Phi, \Phi')$, and the topology of inductive limit $\Phi' = \operatorname{ind}\lim_{j\to\infty} H_{-j}$. Moreover, the space Φ' is nuclear, i.e., it can be represented as a nuclear projective limit of some Hilbert spaces. Below, we consider Φ' topologized by this common topology. For $k > j$, we denote by O_{kj} the operator of imbedding of H_k into H_j and by $O_{-j,-k}$ the operator which imbeds H_{-j} into H_{-k}. We shall also exploit the isometric operators $I_j : H_{-j} \to H_j$ associated with the chains $H_{-j} \supset H_0 \supset H_j$ $(j \in \mathbb{N})$.

Lemma 3.2. *(i) The closed ball*

$$\tilde{B}_{R,-j}(x_0) = \{x \in H_{-j} \mid \|x - x_0\|_{H_{-j}} \le R\} \subset \Phi' \quad (x_0 \in H_{-j},\ j \in \mathbb{N};\ R > 0)$$

is compact in Φ'.

(ii) Let $k > j$ be such that the imbedding $H_k \subset H_j$ is quasinuclear, let $\alpha \subset H_{-j}$ be some set bounded in H_{-j}, and let $(\tilde{\alpha})_{-k}$ be its closure in H_{-k}. Then the restriction of the topology $\beta(\Phi, \Phi')$ to $(\tilde{\alpha})_{-k}$ coincides with the topology induced by H_{-k} on $(\tilde{\alpha})_{-k}$.

Proof. (i) Since the set $\tilde{B}_{R,-j}(x_0)$ is bounded in H_{-j}, it is also bounded in the strong topology $\beta(\Phi, \Phi')$. Hence, it is precompact as a bounded set in the nuclear space. We show that it is closed in the topology $\beta(\Phi, \Phi')$. Let as assume the opposite. Let $y \in \Phi'$ be a limiting point of $\tilde{B}_{R,-j}(x_0)$ and let $y \notin \tilde{B}_{R,-j}(x_0)$. We choose the basis $(e_n)_{n=1}^{\infty} \subset \Phi$ as orthonormal in H_j (this is possible by virtue of the fact that Φ is dense in the space H_j). By using our assumption, we fix $N \in \mathbb{N}$ and $\delta > 0$ such that

$$\sum_{n=1}^{N} (y - x_0, e_n)_{H_0}^2 = R^2 + \delta$$

(otherwise, $y \in \tilde{B}_{R,-j}(x_0)$). We set

$$c = \max_{n=1,\ldots,N} |(y - x_0, e_n)_{H_0}| > 0.$$

For $x \in \tilde{B}_{R,-j}(x_0)$, we have

$$\begin{aligned} \|x - x_0\|_{H_{-j}}^2 &\geq \sum_{n=1}^{N} (x - x_0, e_n)_{H_0}^2 = \sum_{n=1}^{N} (y - x_0, e_n)_{H_0}^2 \\ &\quad + \sum_{n=1}^{N} (x - y, e_n)_{H_0}^2 + 2\sum_{n=1}^{N} (x - y, e_n)_{H_0} (y - x_0, e_n)_{H_0} \\ &\geq R^2 + \delta + 2\sum_{n=1}^{N} (x - y, e_n)_{H_0} (y - x_0, e_n)_{H_0} \\ &\geq R^2 + \delta - 2c\sum_{n=1}^{N} |(x - y, e_n)_{H_0}|. \end{aligned} \tag{3.38}$$

Let us consider the neighborhood of the point $y \in \Phi'$ having the form $U = \{x \in \Phi' \mid |(x - y, e_n)_{H_0}| < \delta / 2cN,\ n = 1, \ldots, N\}$. This is not only a neighborhood in the weak topology $\sigma(\Phi, \Phi')$ but also in the strong topology $\beta(\Phi, \Phi')$. Since y is a limiting point of $\tilde{B}_{R,-j}(x_0)$, one can find a point x such that $x \in U \cap \tilde{B}_{R,-j}(x_0)$. But then (3.38) yields

$$\|x - x_0\|_{H_{-j}}^2 > R^2 + \delta - \delta = R^2,$$

i.e., $x \notin \tilde{B}_{R,-j}(x_0)$. Thus, we get a contradiction.

(ii) The set $(\tilde{\alpha})_{-k}$ is compact in H_{-k} by virtue of the facts that α is bounded in H_{-j} and the imbedding of H_{-j} into H_{-k} is quasinuclear. The operator of the imbedding of H_{-k} into Φ' is continuous. Therefore, due to the separability of the topology β (Φ,

Φ'), the set $(\tilde{\alpha})_{-k}$ is compact in the strong topology on Φ', being a continuous image of the compact set. The statement of the lemma now follows from the fact that the imbedding

$$H_{-k} \supset (\tilde{\alpha})_{-k} \to (\tilde{\alpha})_{-k} \subset \Phi'$$

is a continuous bijection of compact Hausdorff spaces and, therefore, a homeomorphism (see, e.g., Bourbaki [1, Chapter 1, Section 9, Subsection 4].

■

We define $C(\Phi')$ to be a set of complex-valued functions on Φ' continuous on every Hilbert space H_{-j} ($j \in \mathbb{N}$). Lemma 3.2 shows that $C(\Phi')$ coincides with the set of functions on Φ' continuous in the topology $\beta(\Phi, \Phi')$ on every bounded set in Φ'. Let us topologize $C(\Phi')$ by a system of seminorms which separate the points

$$p_{j,R}(u) = \sup_{\tilde{B}_{R,-j}(0)} |u(\cdot)| \quad (u \in C(\Phi');\ j \in \mathbb{N};\ R > 0). \tag{3.39}$$

Denote by $C(\Phi')$ the space of linear continuous functionals on $C(\Phi')$. The next theorem shows that if Φ' is chosen as a space of arguments, then the situation considered is similar to the finite-dimensional case.

Theorem 3.1. *The space $C'(\Phi')$ can be identified with a set of finite complex measures in Φ' given on $\mathfrak{B}(\Phi')$ and concentrated on the sets bounded with respect to the topology $\beta(\Phi, \Phi')$.*

Proof. We present only a sketch of the proof. First, note that the continuity of the functional $l \in C'(\Phi')$ in the topology given by the system of seminorms (3.39) implies that for some $j \in \mathbb{N}$ and $R > 0$, the value $l(u)$ depends only on the behavior of $l \in C(\Phi')$ on the set $\tilde{B}_{R,-j}(0)$. Therefore, l can be regarded as a linear continuous functional on $C(\tilde{B}_{R,-j}(0))$, and $\tilde{B}_{R,-j}(0)$ is compact in the topology $\beta(\Phi, \Phi')$ by Lemma 3.2. The Riesz–Markov theorem yields the representation

$$l(u) = \int u(x)\,d\mu(x) \qquad (u \in C(\tilde{B}_{R,-j}(0)),$$

where the complex measure μ is concentrated on the set $\tilde{B}_{R,-j}(0) \in \mathfrak{B}(\Phi')$. The measure μ can be naturally extended to $\mathfrak{B}(\Phi')$ by the formula

$$\mu(\alpha) = \mu(\alpha \cap \tilde{B}_{R,-j}(0)) \qquad (\alpha \in \mathfrak{B}(\Phi')).$$

It realizes the required representation of the functional l, namely,

$$l(u) = \int_{\Phi'} u(x)\, d\mu(x) \qquad (u \in C(\Phi')). \tag{3.40}$$

It is obvious that (3.40) associates some functional with each measure (possessing the required properties).

■

It is interesting to note that in the infinite-dimensional case, the strict inclusion $C_\beta(\Phi') \subset C(\Phi')$ is valid, where $C_\beta(\Phi')$ is a set of functions on Φ' continuous in the topology $\beta(\Phi, \Phi')$. Another effect, connected with employing Φ' as a space of arguments, can be formulated as follows.

There are no functions with bounded support in $C(\Phi')$.

Indeed, assume that there exists $u \in C(\Phi')$ such that for any $x \in \Phi' \setminus \tilde{B}_{R,-j}(0)$ and some $j \in \mathbb{N}$ and $R > 0$, we have $u(x) = 0$. We show that in this case, $u(x) = 0$ for every $x \in \Phi'$. Let us choose $x_0 \in \tilde{B}_{R,-j}(0)$ and an orthonormal basis $(e_n)_{n=1}^{\infty}$ in H_{-j}. For $k \in \mathbb{N}$, such that $\| O_{kj} \| < \infty$, the sequence $(y_n = x_0 + 3\mathrm{R}e_n)_{n=1}^{\infty}$ converges to x_0 strongly in H_{-k} and, moreover, $y_n \in \Phi' \setminus \tilde{B}_{R,-j}(0)$. By virtue of the continuity of u on H_{-k}, we have $0 = u(y_n) \xrightarrow[n\to\infty]{} u(x_0)$, i.e., $u(x_0) = 0$.

■

We now proceed to the study of differentiable functions on Φ'. Let us define $C^\infty(\Phi')$ to be a set of complex-valued functions on Φ' differentiable infinitely many times on each Hilbert space H_{-j} $(j \in \mathbb{N})$. We denote by Φ_c the complexification of the space Φ. For $n \in \mathbb{N}$, we introduce the n-th tensor power $\Phi_c^{\otimes n} = \mathrm{pr\,lim}_{j\to\infty} H_{j,c}^{\otimes n}$ (see Chapter 1, Section 2).

Lemma 3.3. *For each function* $u \in C^\infty(\Phi')$ *and every* $k \in \mathbb{N}$, *we have* $u^{(k)}(x) \in \Phi_c^{\otimes k}$ $(x \in \Phi')$.

Proof. Assume that $k \in \mathbb{N}$ is fixed. According to the definition of $\Phi_c^{\otimes k}$, it suffices to show that $u^{(k)}(x) \in H_{j,c}^{\otimes k}$ $(x \in \Phi')$ for all $j \in \mathbb{N}$. Using the nuclearity of Φ, for given $j \in \mathbb{N}$, we can choose $j_1 > j$ such that $| O_{j_1 j} | < \infty$. Taking into account that u is infinitely differentiable on H_{-j_1}, and that the imbedding of H_{j_1} into H_j is quasinuclear, we get that $u^{(k)}(x) \in H_{j,c}^{\otimes k}$ for any $x \in H_{-j_1} \supset H_{-j}$ (by virtue of Lemma 3.1). Consequently,

$$u^{(k)}(x) \in \bigcap_{j=1}^{\infty} H_{j,c}^{\otimes k} = \Phi_c^{\otimes k}$$

for $x \in \bigcup_{j=1}^{\infty} H_{-j} = \Phi'$.

■

Therefore, the derivatives of the functions from $C^\infty(\Phi')$ belong to the nuclear spaces $\Phi_c^{\otimes k}$ $(k \in \mathbb{N})$ and this, in turn, gives us the possibility of constructing spaces of differentiable functions with good topological properties. In order to construct the spaces of smooth functions with some conditions imposed on the growth of these functions, we fix the sequence of weight functions $M = (m_i(t))_{i=1}^\infty$ $([0, \infty) \ni \mapsto m_i(t) \in (0, \infty))$ such that the following conditions hold:

(i) $0 < m_i(t) \le m_{i+1}(t) \le 1$, for any $i \in \mathbb{N}$ and $t \in [0, \infty)$;
(ii) for any $i \in \mathbb{N}$, the function $m_i(t)$ decreases monotonically to zero as $t \to \infty$;
(iii) for any $i_1 \in \mathbb{N}$, there exists $i_2 \in \mathbb{N}$ such that $\lim_{t\to\infty} m_{i_1}(t)/m_{i_2}(t) = 0$.

We introduce a system of norms $|u|_{i,k,j}$ $((i, k, j) \in \mathbb{N}^3)$ on $C^\infty(\Phi')$ by

$$|u|_{i,k,j} = \sup_{\substack{x \in H_{-j} \\ l=0,\dots,k}} \{\| u^{(l)}(x)\|_{H_{j,c}^{\otimes l}} m_i(\|x\|_{H_-})\} \tag{3.42}$$

(for the functions $u \in C^\infty(\Phi')$ for which these norms are finite). The axioms of norm can be verified directly if we take into account that the weight functions are positive and that $u \restriction H_{-j} = 0$ implies that $u = 0$ on Φ' (by virtue of the continuity of $u \restriction H_{-j}$ $(j \in \mathbb{N})$ and the denseness of H_{-j} in H_{-k} for $k > j$). It follows from (3.41) and (3.42) that

$$|\cdot|_{i,k,j} \le |\cdot|_{i_1,k_1,j_1} \tag{3.43}$$

for all $i \le i_1$, $j \le j_1$, and $k \le k_1$.

Denote by $\mathcal{K}_M(\Phi')$ the collection of all functions from $C^\infty(\Phi')$ for which all the norms $|\cdot|_{i,k,j}$ $((i, k, j) \in \mathbb{N}^3)$ are finite. On $\mathcal{K}_M(\Phi')$, we introduce the topology generated by the system of norms (3.42). Define the Banach space $\mathcal{K}_{i,k,j}(\Phi')$ as the completion of $\mathcal{K}_M(\Phi')$ with respect to the fixed norm $|\cdot|_{i,k,j}$. It follows from (3.43) that the family of Banach spaces $(\mathcal{K}_{i,k,j}(\Phi'))_{(i,k,j) \in \mathbb{N}^3}$ is directed by imbedding and, hence, the projective limit $\operatorname{pr\,lim}_{(i,k,j) \in \mathbb{N}^3} K_{i,k,j}(\Phi')$ is defined (see Chapter 1, Section 1).

Lemma 3.4. *The space $\mathcal{K}_M(\Phi')$ coincides with the projective limit* $\operatorname{pr\,lim}_{(i,k,j) \in \mathbb{N}^3} K_{i,k,j}(\Phi')$.

Proof. It is obvious that $\mathcal{K}_M(\Phi') \subset \bigcap_{(i,k,j) \in \mathbb{N}^3} K_{i,k,j}(\Phi')$ and that the topology of $\mathcal{K}_M(\Phi')$ coincides with the topology of projective limit. It remains to prove the equality

$$\mathcal{K}_M(\Phi') = \bigcap_{(i,k,j)\in\mathbb{N}^3} K_{i,k,j}(\Phi').$$

Let

$$u \in \bigcap_{(i,k,j)\in\mathbb{N}^3} K_{i,k,j}(\Phi').$$

We prove that $u \in \mathcal{K}_M(\Phi')$. Assume that $j > j_0$ is such that H_j is imbedded into H_{j_0} quasinuclearly. Using Lemma 3.1 and the definition of the norm in $K_{i,k,j}(\Phi')$, one can show that every function from $K_{i,k,j}(\Phi')$ is differentiable on H_{-j_0} up to the order k inclusively, and that the norm $|\cdot|_{i,k,j_0}$ is finite for this function. But this yields $u \in C^\infty(\Phi')$ for $u \in \bigcap_{(i,k,j)\in\mathbb{N}^3} K_{i,k,j}(\Phi')$, and the finiteness of all the norms $|\cdot|_{i,k,j}$ $((i,k,j)\in\mathbb{N}^3)$ means that $u \in \mathcal{K}_M(\Phi')$.

■

Before we proceed to the main theorem on topologic properties of the space $\mathcal{K}_M(\Phi')$, recall the definition of perfect spaces (see Gelfand, Shilov [4, Chapter 1, Section 6]). These are countably normed spaces in which every bounded subset is compact. Perfect spaces possess a series of remarkable properties which do not hold in the case of infinite-dimensional normed spaces. First of all, we should mention that in perfect spaces (and in spaces dual to them), strong convergence coincides with weak convergence. This property is of particular importance if we employ perfect spaces as spaces of test functions when constructing the theory of generalized functions. In addition, each perfect space is reflexive and, thus, perfect spaces belong to the class of Montel spaces[17].

Theorem 3.2. *The space $\mathcal{K}_M(\Phi')$ is perfect.*

Proof. Lemma 3.4 establishes that $\mathcal{K}_M(\Phi')$ is countably normed. So, it remains to prove that closed bounded sets in $\mathcal{K}_M(\Phi')$ are compact. First, note the following characteristic of the convergence in $\mathcal{K}_M(\Phi')$.

Lemma 3.5. *The sequence $(u_n)_{n=1}^\infty \subset \mathcal{K}_M(\Phi')$ converges to zero in $\mathcal{K}_M(\Phi')$ if and only if it is bounded in $\mathcal{K}_M(\Phi')$, and the sequence of derivatives $\Phi' \ni x \mapsto u_n^{(l)}(x) \in \Phi_c^{\otimes l}$ converges to zero uniformly on bounded sets in Φ' for all $l \in \mathbb{Z}_+$.*

Proof. The necessity is obvious. Let us show that the conditions are sufficient. Using condition (iii) of (3.41), we can choose $i_2 \in \mathbb{N}$ such that $\lim_{t\to\infty}(m_{i_1}(t)/m_{i_2}(t)) = 0$ for arbitrary $(i_1, k, j) \in \mathbb{N}^3$. It follows from the boundedness of $(u_n)_{n=1}^\infty$ in $\mathcal{K}_M(\Phi')$ that

[17] These are reflexive locally convex spaces in which every closed bounded set is compact.

$$\exists\, c_{i_2,k,j}\colon\ |u|_{i_2,k,j} \le c_{i_2,k,j}. \tag{3.44}$$

Given $\varepsilon > 0$, we choose $R > 0$ such that

$$\|x\|_{H_{-j}} < R\colon\ \frac{m_{i_1}(\|x\|_{H_{-j}})}{m_{i_2}(\|x\|_{H_{-j}})} < \frac{\varepsilon}{c_{i_2,k,j}}, \tag{3.45}$$

for any $x \in H_{-j}$. Then

$$\sup_{\|x\|_{H_{-j}} > R;\ l=0,\ldots,k} \{\| u_n^{(l)}(x)\|_{H_{j,c}^{\otimes l}}\, m_{i_1}(\|x\|_{H_{-j}})\}$$

$$= \sup_{\|x\|_{H_{-j}} > R;\ l=0,\ldots,k} \left\{ \| u_n^{(l)}(x)\|_{H_{j,c}^{\otimes l}}\, m_{i_2}(\|x\|_{H_{-j}}) \frac{m_{i_1}(\|x\|_{H_{-j}})}{m_{i_2}(\|x\|_{H_{-j}})} \right\}$$

$$\le c_{i_2,k,j} \frac{\varepsilon}{c_{i_2,k,j}} = \varepsilon \quad (n \in \mathbb{N}).$$

On the other hand, the conditions of the lemma imply that for any $\varepsilon > 0$ there exists $n(\varepsilon) \in \mathbb{N}$ such that

$$\sup_{\|x\|_{H_{-j}} > R;\ l=0,\ldots,k} \left\{ \| u_n^{(l)}(x)\|_{H_{j,c}^{\otimes l}}\, m_{i_2}(\|x\|_{H_{-j}} \frac{m_{i_1}(\|x\|_{H_{-j}})}{m_{i_2}(\|x\|_{H_{-j}})} \right\}$$

$$\le \sup_{\|x\|_{H_{-j}} > R;\ l=0,\ldots,k} \{\| u_n^{(l)}(x)\|_{H_{j,c}^{\otimes l}}\} < \varepsilon$$

for all $n \ge n(\varepsilon)$.

Therefore, for any $(i_1, k, j) \in \mathbb{N}^3$ and $\varepsilon > 0$ there exists $n_0 \in \mathbb{N}$ such that $|u|_{i,k,j} < \varepsilon$ for all $n \ge n_0$. This means that $(u_n)_{n=1}^{\infty}$ converges to zero in the topology of $\mathcal{K}_M(\Phi')$.

■

The space $\mathcal{K}_M(\Phi')$ possesses the first axiom of countability. Therefore, to prove the theorem, it suffices to show that a subsequence, satisfying the conditions of Lemma 3.5, can be extracted from each sequence bounded in $\mathcal{K}_M(\Phi')$. For this purpose, we employ the following version of the Arzella–Ascoli theorem (see Bourbaki [1, Chapter 10, Section 2, Subsection 5]):

Let X be a separable topological space, σ be its covering by compact sets, and let Q be a family of mapping from X into Y, where Y is a metrizable space. Assume that for every mapping $u \in Q$ and each set $\alpha \in \sigma$, the restriction $u \restriction \alpha$ is continuous.

In order that Q be precompact with respect to the topology of uniform convergence on the sets from the system σ, it is necessary and sufficient that the following two conditions hold:

(a) for every $\alpha \in \sigma$, the set $\{u \restriction \alpha, u \in Q\} \in C(\alpha, Y)$ is equicontinuous;

(b) for every $x \in X$, the set $\{u(x), x \in Q\} \in Y$ is precompact in Y.

In our case, $Q = (u_n)_{n=1}^{\infty}$ is a sequence bounded in $\mathcal{K}_M(\Phi')$. Let us take σ as the covering of Φ' by the sets $\tilde{B}_{j,-j}(0)$ $(j \in \mathbb{N})$ (by virtue of Lemma 3.2, each closed ball $\tilde{B}_{j,-j}(0)$ is compact in the topology $\beta(\Phi, \Phi')$). We consider a fixed $l \in \mathbb{Z}_+$ and show that the family of mappings $X = \Phi' \ni x \mapsto u_n^{(l)}(x) \in \Phi_c^{\otimes l}$ $(n \in \mathbb{N})$ satisfies the conditions of the Arzella–Ascoli theorem (note that $\Phi_c^{\otimes l}$ is metrizable as a countably normed space).

It follows from the boundedness of $(u_n)_{n=1}^{\infty}$ in $\mathcal{K}_M(\Phi')$ that

$$\sup_{n \in \mathbb{N}} \sup_{\|x\|_{H_{-j}} \leq j} \{\| u_n^{(l+1)}(x) \|_{H_{j,c}^{\otimes(l+1)}} m_j(\| x \|_{H_{-j}})\} \leq c_{ij}$$

$$(c_{ij} > 0;\ j \in \mathbb{N})$$

(l being fixed). Condition (ii) for the family of weights M implies that

$$\sup_{n \in \mathbb{N}} \sup_{\|x\|_{H_{-j}} \leq j} \{\| u_n^{(l+1)}(x) \|_{H_{j,c}^{\otimes(l+1)}}\} \leq \frac{c_{ij}}{m_i(j)} = c_j \qquad (j \in \mathbb{N})$$

for fixed $i \in \mathbb{N}$. According to the mean-value theorem, for $x_1, x_2 \in \tilde{B}_{j,-j}(0)$ the last statement yields

$$\| u_n^{(l)}(x_1) - u_n^{(l)}(x_2) \|_{H_{j,c}^{\otimes l}} \leq c_j \| x_1 - x_2 \|_{H_{-j}} \qquad (n \in \mathbb{N}).$$

This inequality shows that the family $(u_n^{(l)})_{n=1}^{\infty}$ is equicontinuous on each ball $\tilde{B}_{j,-j}(0)$ as a family of mappings in $H_{j,c}^{\otimes l}$. But since $j \in \mathbb{N}$ is arbitrary, this involves the equicontinuity of $(u_n^{(l)})_{n=1}^{\infty}$ as mappings in $\Phi_c^{\otimes l} = \operatorname{pr}\lim_{j \to \infty} H_{j,c}^{\otimes l}$. The space $\Phi_c^{\otimes l}$ is nuclear, and, therefore, every bounded set from $\Phi_c^{\otimes l}$ is precompact. Hence, to verify the second condition of the Arzella–Ascoli theorem, it suffices to show that the set

$(u_n^{(l)}(x))_{n=1}^{\infty} \subset \Phi_c^{\otimes l}$ is bounded in $\Phi_c^{\otimes l}$ for every $x \in \Phi'$. But $\Phi' = \bigcup_{j=1}^{\infty} H_{-j}$. Consequently, there exists $j_0(x) \in \mathbb{N}$ such that $x \in H_{-j}$ for all $j \geq j_0(x)$. Taking into account the estimate

$$\| u_n^{(l)}(x) \|_{H_{j,c}^{\otimes l}} m_j(\| x \|_{H_{-j}}) \leq c_{i,l,j} \quad (n \in \mathbb{N}),$$

and the monotonicity of $\| \cdot \|_{H_j}$ in $j \in \mathbb{N}$, we calculate that for any $j \in \mathbb{N}$ there exists $d(j) > 0$ such that

$$\| u_n^{(l)}(x) \|_{H_{j,c}^{\otimes l}} \leq d(j)$$

for all $n \in \mathbb{N}$, and this means that the set under consideration is bounded in $\Phi_c^{\otimes l}$.

Thus, we have shown that the conditions of the Arzella–Ascoli theorem hold for every sequence $(u_n^{(l)})_{n=1}^{\infty} \subset \Phi_c^{\otimes l}$ $(l \in \mathbb{Z}_+)$. Employing this theorem, we extract from the sequence $(u_n)_{n=1}^{\infty}$ a subsequence $(u_{n_1})_{n_1=1}^{\infty}$ uniformly convergent on the balls $\tilde{B}_{j,-j}(0)$ $(j \in \mathbb{N})$. From this subsequence we, in turn, extract another subsequence $(u_{n_2})_{n_2=1}^{\infty}$ such that $(u_{n_2})_{n_2=1}^{\infty}$ converges uniformly on $\tilde{B}_{j,-j}(0)$ $(j \in \mathbb{N})$, etc. The diagonal sequence $(u_{n_n})_{n=1}^{\infty}$ satisfies the conditions of Lemma 3.5 and is thus convergent in $\mathcal{K}_M(\Phi')$.

■

Consider an important example of the space $\mathcal{K}_M(\Phi')$ which corresponds to the case of a specific choice of the family of weights M.

Example 3.5. Introduce the weight functions by $m_i(t) = \exp(-t^2 / i)$ $(i \in \mathbb{N},\ t \in [0, +\infty))$. It is easy to see that all conditions (3.41) for the family of weights are satisfied. Here, the corresponding space $\mathcal{K}_M(\Phi')$ is denoted by $\mathcal{K}(\Phi')$. Below, we shall see that this space is very well suited for the construction of riggings of the space $L_2(\Phi', \mathcal{C}_\sigma(\Phi'), \gamma_s)$. Lemma 3.4 shows that the space $\mathcal{K}(\Phi')$ can be described internally as a space of functions u on Φ' which are infinitely many times differentiable on each H_{-j} $(j \in \mathbb{N})$ and admit the estimate

$$\| u^{(l)}(x) \|_{H_{j,c}^{\otimes l}} \leq c_{l,j,\varepsilon}\, e^{\varepsilon \| x \|_{H_{-j}}^2} \quad (l, j \in \mathbb{N},\ \varepsilon > 0,\ x \in H_{-j}). \tag{3.46}$$

We stress that the space $\mathcal{K}(\Phi')$ consists of smooth functions which may be increasing (to a certain extent). The last statement clarifies the fact that in the infinite-dimensional case, it is hard to combine the following two requirements: the functional space should be perfect and the functions belonging to it should be bounded.

§4. Spaces of Test and Generalized Functions of Infinitely Many Variables as Infinite Tensor Products

The Hilbert space $L_2(X_\infty, \mu_\infty)$, consisting of functions of an infinite-dimensional argument $x = (x_k)_{k=1}^\infty \in X_\infty = \otimes_{k=1}^\infty X_k$ square summable with respect to the product measure $\mu_\infty = \otimes_{k=1}^\infty \mu_k$, $\mu_k(X_k) = 1$ $(k \in \mathbb{N})$, can be identified with the infinite tensor product $\otimes_{k;e}^\infty L_2(X_k, \mu_k)$ with a stabilizing sequence $e = (e^{(k)})_{k=1}^\infty$, $e^{(k)}(\cdot) = 1$ $(k \in \mathbb{N})$. Therefore, it is natural to construct spaces of test and generalized functions on X_∞ (connected with the rigging of $L_2(X_\infty, \mu_\infty)$) as infinite tensor products of the corresponding riggings of each tensor factor $L_2(X_k, \mu_k)$ $(k \in \mathbb{N})$. We shall describe the corresponding construction in the general situation and study its realizations related to the construction of spaces of test and generalized functions in the case of the Gaussian product measure.

4.1. Infinite Tensor Products of Nuclear Riggings

Let the sequence of Hilbert spaces $(H_{0,k})_{k=1}^\infty$ and the stabilizing sequence

$$e = (e^{(k)})_{k=1}^\infty, \quad e^{(k)} \in H_{0,k}, \quad \| e^{(k)} \|_{H_{0,k}} = 1 \quad (k \in \mathbb{N})$$

be given. Assume that for every $k \in \mathbb{N}$ a linear topologic space $\Phi_k \subset H_{0,k}$ is chosen which satisfies the following properties:

(i) $\Phi_k = \operatorname{pr\,lim}_{\tau_k \in \mathrm{T}_k} H_{\tau_k}$ is the projective limit of the family of Hilbert spaces $(H_{\tau_k})_{\tau_k \in \mathrm{T}_k}$ directed by imbedding. Moreover, $H_{\tau_k} \subset H_{0,k}$ is topologically dense for any $\tau_k \in \mathrm{T}_k$, and $\| \cdot \|_{H_{0,k}} \le \| \cdot \|_{H_{\tau_k}}$. The directedness of the family $(H_{\tau_k})_{\tau_k \in \mathrm{T}_k}$ by imbedding induces a partial ordering of the set T_k, i.e., we say that $\tau_k' \le \tau_k''$ $(\tau_k', \tau_k'' \in T_k)$ if $H_{\tau_k''}$ is topologically imbedded into $H_{\tau_k'}$;

(ii) for any $\tau_k' \le \tau_k''$, we have $\| O_{\tau_k'' \tau_k'} \| \le 1$ for the imbedding operator $O_{\tau_k'' \tau_k'}$: $H_{\tau_k''} \to H_{\tau_k'}$;

(iii) for any $k \in \mathbb{N}$ and $\tau_k \in \mathrm{T}_k$, $e^{(k)} \in \Phi$, we have $\| e^{(k)} \|_{H_{\tau_k}} = 1$.

Note that properties (ii) and (iii) yield the equality $\| O_{\tau_k'' \tau_k'} \| = 1$ for $\tau_k'' \leq \tau_k'$.

Let us introduce a family of multi-indices $\mathrm{T}_\infty = \times_{k=1}^{\infty} \mathrm{T}_k \ni \tau = (\tau_k)_{k=1}^{\infty}$ and construct a Hilbert space $\mathcal{H}_\tau = \otimes_{k=1;e}^{\infty} H_{\tau_k}$ for each $\tau \in \mathrm{T}_\infty$. The set T_∞ is naturally directed by the induced directedness of each T_k, $k \in \mathbb{N}$, namely

$$\tau' = (\tau_k')_{k=1}^{\infty} \leq \tau'' = (\tau_k'')_{k=1}^{\infty} \Leftrightarrow \tau_k' \leq \tau_k'' \quad (k \in \mathbb{N}).$$

The family of Hilbert spaces $(\mathcal{H}_\tau)_{\tau \in \mathrm{T}_\infty}$ is directed by imbedding with respect to the directedness of the set T_∞. This follows from the directedness of each family $(H_{\tau_k})_{\tau_k \in \mathrm{T}_k}$ $(k \in \mathbb{N})$ and condition (ii) for the norms of imbedding operators. In fact, according to the equality $\| O_{\tau_k'' \tau_k'} \| = 1$ $(k \in \mathbb{N})$ and Theorem 2.2 of Chapter 1, for $\tau' \leq \tau''$ there exists a continuous imbedding $\otimes_{k=1;e}^{\infty} H_{\tau_k''} \subset \otimes_{k=1;e}^{\infty} H_{\tau_k'}$. An arbitrary directed set $\mathrm{T} \subset \mathrm{T}_\infty$ enables us to define the projective limit $\mathrm{pr}\, \mathcal{H}_\tau$, which will be called an infinite tensor product of the spaces $(\Phi_k)_{k=1}^{\infty}$ with a given stabilizing sequence e and set of multi-indices $\mathrm{T} \subset \mathrm{T}_\infty$ and denoted by $\Phi_{\mathrm{T}} = \otimes_{k;e;T}^{\infty} \Phi_k$. Note that, generally speaking, the space Φ_{T} depends on the chosen representation of each Φ_k as a projective limit of Hilbert spaces.

Among the spaces just constructed, the space Φ_{T_∞} has a minimal collection of elements. As we shall see below, for concrete functional spaces Φ_k $(k \in \mathbb{N})$, the possibility of choosing different sets $\mathrm{T} \subset \mathrm{T}_\infty$ enables us to vary (within certain limits) not only the collection of elements in Φ_{T} but also their properties (smoothness, growth, etc.).

Since $\| \cdot \|_{H_{0,k}} \geq \| \cdot \|_{H_{\tau_k}}$ $(k \in \mathbb{N}, \tau_k \in \mathrm{T}_k)$, the imbedding of each space $\mathcal{H}_\tau$ into $\mathcal{H}_0 = \otimes_{k;e}^{\infty} H_{0,k}$ is topologically dense. This gives the imbedding $\Phi_{\mathrm{T}} \subset \mathcal{H}_0$ which is also topologically dense. Let us describe the space Φ_{T}' dual to Φ_{T} with respect to $\mathcal{H}_0$. Denote by $H_{-\tau_k}$ the Hilbert space dual to H_{τ_k} with respect to $H_{0,k}$ $(k \in \mathbb{N}, \tau_k \in \mathrm{T}_k)$. According to Lemma 2.6 of Chapter 1, the space $\mathcal{H}_{-\tau} = \otimes_{k=1;e}^{\infty} \mathcal{H}_{-\tau_k}$ is dual to $\mathcal{H}_\tau$ with respect to $\mathcal{H}_0$. The family of Hilbert spaces $(\mathcal{H}_{-\tau})_{\tau \in \mathrm{T}_\infty}$ is clearly directed by increasing with respect to the directedness of T. Therefore, one can define the inductive limit $\mathrm{ind}\lim_{\tau \in \mathrm{T}} \mathcal{H}_{-\tau}$.

Lemma 4.1. *The space* Φ_{T}' *equipped with the Mackey topology* $\tau(\Phi_{\mathrm{T}}', \Phi_{\mathrm{T}})$ *coincides with* $\mathrm{ind}\lim_{\tau \in \mathrm{T}} \mathcal{H}_{-\tau}$.

Proof. The statement of the lemma follows from the duality of projective and inductive limits described in Section 1 of Chapter 1 and from the fact that the Mackey topology $\tau(\Phi'_{\mathrm{T}}, \Phi_{\mathrm{T}})$ coincides with the strong topology of the Hilbert space $\mathcal{H}_{-\tau}$.

∎

Thus, we have constructed the chain

$$\Phi'_{\mathrm{T}} \supset \mathcal{H}_0 \supset \Phi_{\mathrm{T}}, \tag{4.1}$$

which can be regarded as one of the versions of an infinite tensor product of riggings $\Phi'_k \supset H_{0,k} \supset \Phi_k$ $(k \in \mathbb{N})$. If the spaces Φ_k $(k \in \mathbb{N})$ are nuclear, it is natural to ask whether the space Φ_{T} is nuclear.

Lemma 4.2. *The space* Φ_{T} *is nuclear if and only if for any* $\tau' \in \mathrm{T}$ *there exists* $\tau'' \in \mathrm{T}$ *such that*

$$\sum_{k=1}^{\infty} \left(| O_{\tau''_k \tau'_k} |^2 - 1 \right) < \infty, \tag{4.2}$$

where $| O_{\tau''_k \tau'_k} |$ *is a Hilbert norm of the imbedding operator*

$$O_{\tau''_k \tau'_k} : H_{\tau''_k} \to H_{\tau'_k} \qquad (k \in \mathbb{N}).$$

Proof. Fix arbitrary $\tau, \eta \in \mathrm{T}$. Since T is directed, one can find $\tau' \in \mathrm{T}$ such that $\tau \le \tau'$ and $\eta \le \tau'$. As mentioned above, in this case, the imbeddings $\mathcal{H}_{\tau'} \subset \mathcal{H}_{\tau}$ and $\mathcal{H}_{\tau'} \subset \mathcal{H}_{\eta}$ are continuous. Let us choose a multi-index $\tau'' \in \mathrm{T}$ satisfying condition (4.2). Just as in Section 2 of Chapter 1, one can show that the imbedding operator $\mathcal{O}_{\tau''\tau'} : H_{\tau''} \to H_{\tau'}$ has the form

$$\mathcal{O}_{\tau''\tau'} = \bigotimes_{k=1}^{\infty} O_{\tau''_k \tau'_k}; \qquad | \mathcal{O}_{\tau''\tau'} | = \prod_{k=1}^{\infty} | O_{\tau''_k \tau'_k} |,$$

and that the convergence of this infinite product is ensured by the convergence of the series (4.2). We have thus shown that the imbeddings $\mathcal{H}_{\tau''} \subset \mathcal{H}_{\tau}$ and $\mathcal{H}_{\tau''} \subset \mathcal{H}_{\eta}$ are quasinuclear as compositions of continuous and quasinuclear imbeddings. This proves that Φ_{T} is nuclear, since $\tau, \eta \in \mathrm{T}$ are arbitrary.

The necessity of condition (4.2) follows from the definition of a nuclear space and from the equality

$$| \mathcal{O}_{\tau''\tau'} | = \prod_{k=1}^{\infty} | O_{\tau''_k \tau'_k} |.$$

∎

This method of constructing infinite tensor products of nuclear spaces is very convenient in applications, especially in the case when the spaces Φ_k $(k \in \mathbb{N})$ are Köthe spaces (or ladder spaces, according to other terminology). This means that the Hilbert spaces H_{τ_k} $(\tau_k \in T_k)$ which generate the spaces under consideration, determine (in terms of Fourier coefficients) the expansions of the elements of these spaces in a certain fixed basis in $H_{0,k}$. First, let us present an example in which the representation of the Schwartz space of test functions $\mathcal{S}(\mathbb{R}^1)$ as a Köthe space is given.

Example 4.1. In $L_2(\mathbb{R}^1)$ an orthonormal basis is formed by the Hermite functions

$$\psi_k(x) = \pi^{-1/4} h_k(x) e^{-x^2/2} \quad (k \in \mathbb{Z}_+, \; x \in \mathbb{R}^1),$$

where $(h_k)_{k=1}^{\infty}$ are the normed Hermite polynomials introduced in Section 2. For every $l \in \mathbb{Z}_+$, we define the Hilbert space

$$H_l(\mathbb{R}^1) = \left\{ f = \sum_{k=0}^{\infty} f_k \psi_k \in L_2(\mathbb{R}^1) \mid f_k \in \mathbb{C}^1, \; \sum_{k=0}^{\infty} |f_k|^2 (1+k)^l < \infty \right\} \tag{4.3}$$

with the scalar product

$$(f, g)_{H_l(\mathbb{R}^1)} = \sum_{k=0}^{\infty} f_k \overline{g}_k (1+k)^l.$$

It is clear that $H_0(\mathbb{R}^1) = L_2(\mathbb{R}^1)$ and the imbedding $H_{l_1}(\mathbb{R}^1) \supset H_{l_2}(\mathbb{R}^1)$, $l_1 \le l_2$, is topologically dense. Consequently, the projective limit $\mathrm{pr}\lim_{l \to \infty} H_l(\mathbb{R}^1)$ is defined. One can show that

$$\mathcal{S}(\mathbb{R}^1) = \mathrm{pr}\lim_{l \to \infty} H_l(\mathbb{R}^1).$$

In fact, it is well known that the basis $(\psi_k)_{k=0}^{\infty}$ consists of the eigenfunctions of the operator of a harmonic oscillator N given for $f \in \mathcal{S}(\mathbb{R}^1)$ by the equality

$$(Nf)(x) = -1/2 f''(x) + 1/2 (x^2 - 1) f(x)$$

and, thus, $N\psi_k = k\psi_k$ $(k \in \mathbb{Z}_+)$.

It follows from (4.3) that

$$H_l(\mathbb{R}^1) = \mathfrak{D}((N+1)^{l/2}) \subset L_2(\mathbb{R}^1)$$

and, therefore,

$$\text{pr}\lim_{l\to\infty} H_l(\mathbb{R}^1) = \bigcap_{l=0}^{\infty} \mathfrak{D}((N+1)^{l/2}) = C^\infty(N).$$

The last equality means that the functions from $\text{pr}\ \lim_{l\to\infty} H_l(\mathbb{R}^1)$ are infinitely differentiable and such that, being multiplied by x^{2p}, $p \in \mathbb{N}$, they remain in $L_2(\mathbb{R}^1)$. Hence, they belong to $\mathcal{S}(\mathbb{R}^1)$. The inclusion $\mathcal{S}(\mathbb{R}^1) \subset H_l(\mathbb{R}^1)$ is obvious, since $\mathcal{S}(\mathbb{R}^1) \subset \mathfrak{D}((N+1)^{l/2})$ $(l \in \mathbb{N})$. It is also easy to show that the convergences in $\mathcal{S}(\mathbb{R}^1)$ and $\text{pr}\ \lim_{l\to\infty} H_l(\mathbb{R}^1)$ coincide. But this means that these spaces coincide as linear topological spaces (by virtue of the first axiom of countability which holds for both spaces).

Let us choose an orthonormal basis $(e_j^{(k)})_{j=0}^{\infty}$, $e_0^{(k)} = e^{(k)}$, $k \in \mathbb{N}$, in each space $H_{0,k}$. In order to construct the space Φ_k, we set the family of weights $\mathrm{T}_k \ni \tau_k = (\tau_k^{(j)})_{j=0}^{\infty}$ for which the following conditions hold:

(i) $\tau_k^{(j)} \geq 1$ $(j \in \mathbb{N})$, $\tau_k^{(0)} = 1$;

(ii) for any $\tau_k \in \mathrm{T}_k$ and $\eta_k \in \mathrm{T}_k$, there exists $\xi_k \in \mathrm{T}_k$ such that

$$\tau_k^{(j)},\ \eta_k^{(j)} \leq \xi_k^{(j)} \quad (j \in \mathbb{Z}_+); \tag{4.4}$$

(iii) for any $\tau_k \in \mathrm{T}_k$, there exists $\eta_k \in \mathrm{T}_k$ such that $\sum_{j=0}^{\infty} \tau_k^{(j)} \eta_k^{(j)} < \infty$.

For each weight $\tau_k \in \mathrm{T}_k$, we construct a Hilbert space

$$H_{\tau_k} = \left\{ f = \sum_{j=0}^{\infty} f_j e_j^{(k)} \in H_{0,k} \,\Big|\, \sum_{j=0}^{\infty} |f_j|^2 \tau_k^{(j)} < \infty \right\} \tag{4.5}$$

with the scalar product

$$(f, g)_{H_{\tau_k}} = \sum_{j=0}^{\infty} f_j \bar{g}_k \tau_k^{(j)} \quad (f, g \in H_{\tau_k}). \tag{4.6}$$

Conditions (4.4), imposed on the family of weights T_k, guarantee that the family $(H_{\tau_k})_{\tau_k \in \mathrm{T}_k}$ is directed by imbedding. Consider the linear topologic space $\Phi_k = \text{pr}\lim_{\tau_k \in \mathrm{T}_k} H_{\tau_k}$ $(k \in \mathbb{N})$. Conditions (4.4) imply that Φ_k satisfies conditions (i)–(iii) necessary to define the infinite tensor product of the spaces $(\Phi_k)_{k=1}^{\infty}$ with the stabilizing sequence $e = (e^{(k)})_{k=1}^{\infty}$. Furthermore, each Φ_k is nuclear. This follows from the fact that for $\tau_k, \eta_k \in \mathrm{T}_k$, such that $\tau_k^{(j)} \leq \eta_k^{(j)}$ $(j \in \mathbb{Z}_+)$, the Hilbert norm of the imbedding operator $O_{\eta_k \tau_k}: H_{\eta_k} \to H_{\tau_k}$ equals

$$\| O_{\eta_k \tau_k} \| = \left(\sum_{j=0}^{\infty} \tau_k^{(j)} \eta_k^{(j)} \right)^{1/2}.$$

Using representation (4.5), we can easily obtain a similar description of the spaces $\mathcal{H}_\tau$, $\tau \in \mathrm{T}_\infty$, and Φ_{T} in terms of expansions in some basis in the space $\mathcal{H}_0$. We have no intention of doing this in the general situation. Let us investigate the concrete examples of infinite tensor products of nuclear spaces instead.

4.2 The space $\mathcal{A}(\mathbb{R}^1)$

Define the nuclear space of functions of one variable which will be used when constructing the spaces of test functions of infinitely many variables. In the space $L_2(\mathbb{R}^1, \gamma_1)$, $d\gamma_1(t) = \pi^{-1/2} \exp(-t^2)\, dt$, an orthonormal basis is formed by the Hermite polynomials $(h_k(t))_{k=0}^{\infty}$ normed in the measure γ_1 (see Section 2). With each function $f \in L_2(\mathbb{R}^1, \gamma_1)$, we associate the sequence $(f_k)_{k=0}^{\infty}$ of its Fourier coefficients obtained as a result of the expansion of this function in the given basis, namely, $f_k = (f, h_k)_{L_2(\mathbb{R}^1, \gamma_1)}$ ($k \in \mathbb{Z}_+$). For $p \in [1, \infty)$, we define the Hilbert space

$$A_p(\mathbb{R}^1) = \left\{ f \in L_2(\mathbb{R}^1, \gamma_1) \,\middle|\, \sum_{j=0}^{\infty} |f_j|^2 p^j < \infty \right\} \tag{4.7}$$

with the scalar product

$$(f^{(1)}, f^{(2)})_{A_p(\mathbb{R}^1)} = \sum_{j=0}^{\infty} f_j^{(1)} \overline{f_j^{(2)}} p^j < \infty \quad (f^{(1)}, f^{(2)} \in A_p(\mathbb{R}^1)). \tag{4.8}$$

Clearly, $A_1(\mathbb{R}^1) = L_2(\mathbb{R}^1, \gamma_1)$ and $A_p \subset L_2(\mathbb{R}^1, \gamma_1)$ for $p \in [1, \infty)$ (the imbedding is topological). The family of Hilbert spaces $(A_p(\mathbb{R}^1))_{p \in [1, \infty)}$ thus constructed is ordered by the imbedding

$$A_{p'}(\mathbb{R}^1) \supset A_{p''}(\mathbb{R}^1), \quad p' \le p'' \quad (p', p'' \in [1, \infty)),$$

$\bigcap_{p \in [1, \infty)} A_p(\mathbb{R}^1)$ contains finite linear combinations of the Hermite polynomials and, consequently, is dense in $L_2(\mathbb{R}^1, \gamma_1)$. Let us define the space

$$\mathcal{A}(\mathbb{R}^1) = \text{pr} \lim_{p\in[1,\infty)} A_p(\mathbb{R}^1).$$

Let us introduce the Hilbert spaces of generalized functions $A_{-p}(\mathbb{R}^1)$ to be dual to $A_p(\mathbb{R}^1)$ $(p \in [1, \infty))$ with respect to $L_2(\mathbb{R}^1, \gamma_1)$. The family of Hilbert spaces $(A_{-p}(\mathbb{R}^1))_{p\in[1,\infty)}$ is ordered by increasing, and this enables us to define their inductive limit $\text{ind}\lim_{p\in[1,\infty)} A_{-p}(\mathbb{R}^1)$.

Lemma 4.3. *The space* $\mathcal{A}(\mathbb{R}^1)$ *is nuclear. Its dual space* $\mathcal{A}'(\mathbb{R}^1)$ *coincides with* $\text{ind}\lim_{p\in[1,\infty)} A_{-p}(\mathbb{R}^1)$ *in the Mackey topology* $\tau(\mathcal{A}'(\mathbb{R}^1), \mathcal{A}(\mathbb{R}^1))$. *Furthermore,*

$$A_{-p}(\mathbb{R}^1) = \left\{ \xi = \sum_{j=0}^{\infty} \xi_j h_j \,\middle|\, \xi_j \in \mathbb{C}^1;\ \sum_{j=0}^{\infty} |\xi_j|^2 p^{-j} < \infty \right\} \tag{4.9}$$

and the scalar product of generalized functions $\xi^{(1)}, \xi^{(2)} \in A_{-p}(\mathbb{R}^1)$ *in* $A_{-p}(\mathbb{R}^1)$ *is equal to*

$$(\xi^{(1)}, \xi^{(2)})_{A_{-p}(\mathbb{R}^1)} = \sum_{j=0}^{\infty} \xi_j^{(1)} \overline{\xi_j^{(2)}} p^j, \tag{4.10}$$

where $\xi_j^{(k)} = (\xi^k, h_j)_{L_2(\mathbb{R}^1,\gamma_1)}$ $(j \in \mathbb{Z}_+)$ *is an action of the generalized function* $\xi^{(k)}$ *on the test function* $h_j \in \mathcal{A}(\mathbb{R}^1)$ $(k = 1, 2)$. *The functional* $\xi = \sum_{j=0}^{\infty} u_j h_j \in A_{-p}(\mathbb{R}^1)$ *acts on the test function according to the formula*

$$(\xi, u)_{L_2(\mathbb{R}^1,\gamma_1)} = \sum_{j=0}^{\infty} \xi_j \bar{u}_j.$$

Proof. We show that the imbedding operator $O_{p+1,p} : A_{p+1}(\mathbb{R}^1) \to A_p(\mathbb{R}^1)$ is quasinuclear for all $p \in [1, \infty)$. In fact, according to (4.8), the orthonormal basis in $A_{p+1}(\mathbb{R}^1)$ is formed by the polynomials $((1+p)^{-j/2} h_j)_{j=0}^{\infty}$. By definition,

$$| O_{p+1,p} |^2 = \sum_{j=0}^{\infty} \| (1+p)^{-j/2} h_j \|^2_{A_p(\mathbb{R}^1)} = \sum_{j=0}^{\infty} (p/p+1)^j = 1 + p < \infty.$$

Therefore, the arbitrariness of $p \in [1, \infty)$ involves the nuclearity of $\mathcal{A}(\mathbb{R}^1)$.

The coincidence of $\mathcal{A}'(\mathbb{R}^1)$ with $\text{ind}\lim_{p\in[1,\infty)} A_{-p}(\mathbb{R}^1)$ in the Mackey topology $\tau(\mathcal{A}'(\mathbb{R}^1), \mathcal{A}(\mathbb{R}^1))$ is a consequence of the general facts concerning the duality between the projective and inductive limits which were studied in Section 1 of Chapter 1.

The remaining statements of the lemma follow from the natural isomorphism of the spaces $L_2(\mathbb{R}^1, \gamma_1)$ and $l_2(\mathbb{C}^1)$ under which $A_p(\mathbb{R}^1)$ is isomorphic to the space of complex sequences absolutely square summable with the weight $(p^j)_{j=0}^{\infty}$.

■

According to its construction, the space $\mathcal{A}(\mathbb{R}^1)$ can be described as a set of functions $u \in L_2(\mathbb{R}^1, \gamma_1)$ whose Fourier coefficients $(u_j)_{j=0}^{\infty}$ satisfy the following conditions: $\sum_{j=0}^{\infty} |u_j|^2 p^j < \infty$ $(p \in [1, \infty))$ when being expanded in the basis of Hermite polynomials. The internal description of this space is also possible. Denote by $\mathfrak{E}^2_{\min}(\mathbb{C}^1)$ the set of entire functions of minimal type on $\mathbb{C}^1$, with the order of growth at most 2. In other words, $\mathfrak{E}^2_{\min}(\mathbb{C}^1)$ is the set of entire functions $\mathbb{C}^1 \ni z \mapsto u(z) \in \mathbb{C}^1$ admitting, for any $\varepsilon > 0$, the estimate

$$|u(z)| \le c_{u,\varepsilon} e^{\varepsilon |z|^2} \qquad (z \in \mathbb{C}^1;\ c_{u,\varepsilon} > 0). \tag{4.11}$$

Let $\mathfrak{E}^2_{\min}(\mathbb{R}^1)$ denote also the set of functions obtained as a result of the restriction of functions from $\mathfrak{E}^2_{\min}(\mathbb{C}^1)$ to the real axis.

Theorem 4.1. $\mathcal{A}(\mathbb{R}^1) = \mathfrak{E}^2_{\min}(\mathbb{R}^1)$.

Proof. Let us describe the main steps of the proof, omitting the technical details of the procedure of establishing the required estimates.

I. Let $u \in \mathcal{A}(\mathbb{R}^1)$; we have

$$u(t) = \sum_{k=0}^{\infty} u_k h_k(t), \qquad \sum_{k=0}^{\infty} |u_k|^2 p^k < \infty \qquad (t \in \mathbb{R}^1;\ p \in [1, \infty)) \tag{4.12}$$

by definition. Employing the explicit form of the Hermite polynomials

$$h_k(t) = \sum_{j=0}^{[k/2]} (-1)^k \frac{\sqrt{k!}\, 2^{k/2}}{j!(k-2j)!2^{2j}} t^{k-2j}$$

we re-expand $u(t)$ in powers of t

$$u(t) = \sum_{j=0}^{\infty} \hat{u}_j t^j, \tag{4.13}$$

where

$$\hat{u}_j = \frac{2^{j/2}}{j!} \sum_{k=0}^{\infty} u_{j+2k} \frac{(-1)^k}{k!} \sqrt{(j+2k)!}\, 2^{-k}. \tag{4.14}$$

The next estimate follows from the explicit form of $\hat{u}_j$ and (4.12). We have

$$|\hat{u}_j| \le c_{u,p} j^{-j/2-1/4} \left(\frac{4e}{p}\right)^{j/2}. \tag{4.15}$$

This estimate enables us to extend $u(t)$ to the complex plane, as an entire function, according to

$$u(z) = \sum_{k=0}^{\infty} \hat{u}_k z^k \quad (z \in \mathbb{C}^1).$$

II. Let us estimate the order of growth of the entire function obtained as a result of extension. It is well known that the order ρ of the entire function is expressed in terms of the coefficients of the corresponding Taylor series by the relation

$$\rho = \limsup_{j\to\infty} \ln j / \ln (|\hat{u}_j|)^{-1/j}.$$

Taking (4.15) into account, we get that $\rho \le 2$.

III. If $\rho < 2$, then the estimate

$$|u(z)| \le c_{u,\delta}\, e^{|z|^{\rho+\delta}} \quad (z \in \mathbb{C}^1),$$

in which $\delta > 0$ is arbitrary, gives (4.1) and the inclusion $u \in \mathcal{E}^2_{\min}(\mathbb{C}^1)$. For $\rho = 2$, let us estimate the type τ of the function $u(z)$. The well-known formula

$$\rho\tau e = \limsup_{j\to\infty} j\, |\hat{u}_j|^{\rho/j}$$

together with (4.15), implies that $\tau \le 2/p$. Due to the arbitrariness of $p \in [1, \infty)$, this means that the entire function $u(z)$ is of the minimal type. Taking into account that $\rho = 2$, we obtain the inclusion $u \in \mathcal{E}^2_{\min}(\mathbb{C}^1)$ again.

IV. We have shown that $\mathcal{A}(\mathbb{R}^1) \subset \mathcal{E}^2_{\min}(\mathbb{R}^1)$. In order to prove the inverse inclusion, we define an integral operator acting in $L_2(\mathbb{R}^1, \gamma_1)$ by

$$(Gu)(t) = \int_{\mathbb{R}^1} G(t, s)\, u(s)\, d\gamma_1(s), \tag{4.16}$$

where

$$G(t, s) = \exp(its + t^2/4) \quad (t, s \in \mathbb{R}^1).$$

Clearly, $G(\cdot, \cdot) \in L_2(\mathbb{R}^1, \gamma_1 \times \gamma_1)$ and, hence, G is a Hilbert–Schmidt operator in $L_2(\mathbb{R}^1, \gamma_1)$. The well-known relations for the Hermite polynomials yield

$$(Gh_j)(t) = \frac{i^j t^j}{\sqrt{2^j j!}} \quad (t \in \mathbb{R}^1;\ j \in \mathbb{Z}_+).$$

Therefore,

$$G(\mathcal{A}(\mathbb{R}^1)) = \left\{ \sum_{j=0}^{\infty} u_j \frac{t^j}{\sqrt{2^j\ j!}} \;\middle|\; \sum_{j=0}^{\infty} |u_j|^2 p^j < \infty \quad (p \in [1, \infty)) \right\}. \tag{4.17}$$

V. Just as above, by use of the estimates for the coefficients of the Taylor series which define the functions from (4.17), we can establish the equality $G(\mathcal{A}(\mathbb{R}^1)) = \mathcal{E}^2_{\min}(\mathbb{R}^1)$ —here one should exploit the estimate for the behavior of the coefficients of the Taylor series, defining an entire function in terms of its order of growth and type.

VI. Let us show that the inclusion $\mathcal{A}(\mathbb{R}^1) \subset \mathcal{E}^2_{\min}(\mathbb{R}^1)$ is, in fact, the equality. Using the relation

$$G\left(\frac{t^k}{\sqrt{2^k\ k!}}\right) = \frac{i^k h_k(t)}{2^k} \quad (t \in \mathbb{R}^1;\ k \in \mathbb{Z}_+),$$

we obtain

$$G^2(\mathcal{A}(\mathbb{R}^1)) = G(G(\mathcal{A}(\mathbb{R}^1))) = \mathcal{A}_{2p}(\mathbb{R}^1).$$

Consequently,

$$G^2(\mathcal{A}(\mathbb{R}^1)) = G(\mathcal{E}^2_{\min}(\mathbb{R}^1)) = \mathcal{A}(\mathbb{R}^1).$$

It follows from the inclusion $\mathcal{A}(\mathbb{R}^1) \subset \mathcal{E}^2_{\min}(\mathbb{R}^1)$, already proved, that

$$G(\mathcal{A}(\mathbb{R}^1)) = \mathcal{E}^2_{\min}(\mathbb{R}^1) \subset G(\mathcal{E}^2_{\min}(\mathbb{R}^1)) = \mathcal{A}(\mathbb{R}^1).$$

But this means that

$$\mathcal{A}(\mathbb{R}^1) = \mathcal{E}^2_{\min}(\mathbb{R}^1).$$

■

4.3. The Spaces of Functions of Infinitely Many Variables

We turn to the construction of riggings of the space $L_2(\mathbb{R}^1, \gamma_1)$, $d\gamma_1(x) = \times_{k=1}^{\infty} d\gamma_1(x_k)$, $x = (x_k)_{k=1}^{\infty} \in \mathbb{R}^{\infty}$. For this purpose, we shall employ the isomorphism between $L_2(\mathbb{R}^{\infty}, \gamma_1)$ and

$$\bigotimes_{k=1;\, e}^{\infty} L_2(\mathbb{R}^1, \gamma_1(x_k)), \quad e = (e^{(k)})_{k=1}^{\infty}, \quad e^{(k)}(x_k) = h_0(x_k) = 1,$$

and realize the scheme of Subsection 4.1 with

$$H_{0,k} = L_2(\mathbb{R}^1, \gamma_1(x_k)), \qquad \Phi_k = \mathcal{A}(\mathbb{R}^1),$$

and $\mathrm{T}_k = [1, \infty)$ $(k \in \mathbb{N})$. Note that conditions (i)–(iii), imposed on the representation of Φ_k as a projective limit of Hilbert space, hold. In particular,we have $\| e^{(k)} \|_{A_{\tau_k}(\mathbb{R}^1)} = 1$ $(\tau_k \in [1, \infty);\ k \in \mathbb{N})$. Let us denote

$$\mathrm{T}_{\infty} = [1, \infty) \times [1, \infty) \times \ldots = [1, \infty)^{\infty}$$

and choose a subset $\mathrm{T} \subset \mathrm{T}_{\infty}$ possessing the following property of directedness

$$\forall\ \tau', \tau'' \in \mathrm{T}\ \exists\ \tau''' \in \mathrm{T} : \forall\ k \in \mathbb{N}\quad \tau'_k, \tau''_k \le \tau'''_k - 1. \tag{4.18}$$

For every $\tau \in \mathrm{T}$, we define the space of test functions by

$$A_{\tau}(\mathbb{R}^{\infty}) = \bigotimes_{k=1;\, e}^{\infty} A_{\tau_k}(\mathbb{R}^1) \subset \bigotimes_{k=1;\, e}^{\infty} L_2(\mathbb{R}^1, \gamma_1) = L_2(\mathbb{R}^{\infty}, \gamma_1),$$

where the imbedding is continuous, and its norm is equal to unity. This follows from the fact that for every $k \in \mathbb{N}$, the norm of the imbedding operator $A_{\tau_k}(\mathbb{R}^1) \to L_2(\mathbb{R}^1, \gamma_1(x_k))$ is equal to one. It follows from (4.18) that the family of Hilbert spaces $(A_{\tau}(\mathbb{R}^{\infty}))_{\tau \in \mathrm{T}}$ is directed by imbedding with respect to the given directedness of the set T. Therefore, the projective limit

$$\mathcal{A}(\mathbb{R}^1) = \operatorname*{pr\,lim}_{\tau \in \mathrm{T}} A_{\tau}(\mathbb{R}^{\infty})$$

is defined; it is a linear topologic space imbedded into $L_2(\mathbb{R}^{\infty}, \gamma_1)$ densely topologically.

It has been shown in Section 2 that the orthonormal basis in $L_2(\mathbb{R}^\infty, \gamma_1)$ is formed by the cylindrical polynomials

$$h_\alpha(x) = h_{\alpha_1}(x_1) \dots h_{\alpha_\nu}(x_\nu) \qquad (x \in \mathbb{R}^\infty),$$

$$\alpha = (\alpha_1, \dots, \alpha_\nu, 0, 0, \dots) \in \mathbb{Z}^\infty_{+,0}, \tag{4.19}$$

where $\mathbb{Z}^\infty_{+,0}$ is a set of finite multi-indices with components $\alpha_k \in \mathbb{Z}_+$ and $\nu = \nu(\alpha)$ is a length of a multi-index $\alpha \in \mathbb{Z}^\infty_{+,0}$. Each function $u \in L_2(\mathbb{R}^\infty, \gamma_1)$ is associated with the collection of Fourier coefficients $u_\alpha = (u, h_\alpha)_{L_2(\mathbb{R}^\infty, \gamma_1)}$ $(\alpha \in \mathbb{Z}^\infty_{+,0})$ which corresponds to the expansion of this function in the basis $(h_\alpha)_{\alpha\in\mathbb{Z}^\infty_{+,0}}$.

Theorem 4.2. *(i) The equality*

$$A_\tau(\mathbb{R}^\infty) = \Big\{u \in L_2(\mathbb{R}^\infty, \gamma_1) \,\Big|\, \sum_{\alpha\in\mathbb{Z}^\infty_{+,0}} |u_\alpha|^2 \tau_1^{\alpha_1} \dots \tau_\nu^{\alpha_\nu} < \infty\Big\} \tag{4.20}$$

holds, and the scalar product of the functions $u, v \in A_\tau(\mathbb{R}^\infty)$ in $A_\tau(\mathbb{R}^\infty)$ is given by

$$(u, v)_{A_\tau(\mathbb{R}^\infty)} = \sum_{\alpha\in\mathbb{Z}^\infty_{+,0}} u_\alpha v_\alpha \tau_1^{\alpha_1} \dots \tau_\nu^{\alpha_\nu}. \tag{4.21}$$

(ii) The space $\mathcal{A}_T(\mathbb{R}^\infty)$ is nuclear if and only if for any $\tau' \in T$ there exists $\tau'' \in T$ such that

$$\forall\, \tau' \in T\ \exists\, \tau'' \in T : \sum_{k=1}^\infty \tau'_k / \tau''_k < \infty. \tag{4.22}$$

Proof. (i) It follows from the definition of $A_{\tau_k}(\mathbb{R}^1)$ that the orthonormal basis in this Hilbert space is formed by the functions $\big(\tau_k^{-j/2} h_j(x_k)\big)_{j=0}^\infty$. Therefore,

$$A_\tau(\mathbb{R}^\infty) = \bigotimes_{k=1;e}^\infty A_{\tau_k}(\mathbb{R}^1)$$

is a Hilbert space spanned by the orthonormal basis

$$e_{\alpha,\tau} = \tau_1^{-\alpha_1/2} \dots \tau_\nu^{-\alpha_\nu/2} h_\alpha(x) \quad (x \in \mathbb{R}^\infty,\ \alpha \in \mathbb{Z}^\infty_{+,0}). \tag{4.23}$$

This proves that (4.20) and (4.21) hold.

(ii) Assume that (4.22) holds. Using (4.18), for arbitrary $\tau, \eta \in T$, let us choose $\tau' \in T$ such that $\tau_k < \tau'_k - 1$, $\eta_k < \tau'_k - 1$ for any $k \in \mathbb{N}$. For such $\tau' \in T$, we find $\tau'' \in T$ such that $\tau'_k < \tau''_k$ and $\sum_{k=1}^{\infty} \tau'_k / \tau''_k$ for any $k \in \mathbb{N}$. This choice gives the following inclusions $A_\tau(\mathbb{R}^\infty) \supset A_{\tau'}(\mathbb{R}^\infty)$, $A_\eta(\mathbb{R}^\infty) \supset A_{\tau'}(\mathbb{R}^\infty)$, and $A_{\tau'}(\mathbb{R}^\infty) \supset A_{\tau''}(\mathbb{R}^\infty)$. Let us show that the last inclusion is quasinuclear. Taking (4.21) and (4.23) into account, we obtain

$$|O_{\tau''\tau'}|^2 = \sum_{\alpha \in \mathbb{Z}_{+,0}^\infty} \| e_{\alpha,\tau''} \|^2_{A_{\tau'}(\mathbb{R}^\infty)} = \sum_{\alpha \in \mathbb{Z}_{+,0}^\infty} (\tau'_1 / \tau''_1)^{\alpha_1} \ldots (\tau'_\nu / \tau''_\nu)^{\alpha_\nu}$$

$$= \lim_{p \to \infty} \left(\sum_{\alpha_1, \ldots \alpha_p = 0}^{\infty} (\tau'_1 / \tau''_1)^{\alpha_1} \ldots (\tau'_p / \tau''_p)^{\alpha_p} \right)$$

$$= \lim_{p \to \infty} \prod_{k=1}^{p} (1 - \tau'_k / \tau''_k) = \prod_{k=1}^{\infty} (1 - \tau'_k / \tau''_k),$$

where the last infinite product converges by virtue of the convergence of the series $\sum_{k=1}^{\infty} \tau'_k / \tau''_k$. Thus, the imbeddings of $A_{\tau''}(\mathbb{R}^\infty)$ into $A_\eta(\mathbb{R}^\infty)$ and $A_\tau(\mathbb{R}^\infty)$ are quasinuclear, and the fact that $\tau, \eta \in T$ are arbitrary means that $\mathcal{A}_T(\mathbb{R}^\infty)$ is nuclear.

The necessity of (4.22) for the nuclearity of $\mathcal{A}_T(\mathbb{R}^\infty)$ follows immediately from the above formula for the Hilbert norm of the imbedding operator $|O_{\tau''\tau'}|$.

■

In order to emphasize the influence of the choice of the set of multi-indices $T \subset [1, \infty)^\infty$ upon the properties of the space $\mathcal{A}_T(\mathbb{R}^\infty)$, we consider the following example.

Example 4.2. Among the spaces $\mathcal{A}_T(\mathbb{R}^\infty)$, the space which corresponds to $T = [1, \infty)^\infty$ has the minimal collection of elements. Denote this space by $\mathcal{A}(\mathbb{R}^\infty)$. Clearly, condition (4.22) holds in this case and, therefore, $\mathcal{A}(\mathbb{R}^\infty)$ is nuclear. It is easy to see that $\mathcal{A}(\mathbb{R}^\infty)$ contains only cylindrical functions or, more exactly, that the following equality is true

$$\mathcal{A}(\mathbb{R}^\infty) = \bigcup_{n=1}^{\infty} \mathcal{A}(\mathbb{R}^n),$$

where

$$\mathcal{A}(\mathbb{R}^n) = \mathcal{A}(\mathbb{R}^1) \otimes \ldots \otimes \mathcal{A}(\mathbb{R}^1) \quad (n \text{ times}).$$

In fact, the inclusion $\bigcup_{n=1}^{\infty} \mathcal{A}(\mathbb{R}^n) \subset \mathcal{A}(\mathbb{R}^\infty)$ is obvious. In order to establish the inverse inclusion, we show that each function

$$u = \sum_{\alpha \in \mathbb{Z}^{\infty}_{+,0}} u_\alpha h_\alpha \in \mathcal{A}(\mathbb{R}^\infty)$$

is cylindrical. For this purpose, it suffices to prove that one can find ν_0 such that $u_\alpha = 0$ for $\nu_{(\alpha)} > \nu_0$. Assume the contrary. Then there exists an infinite number of nonempty sets among $U_n = \{u_\alpha, \nu(\alpha) = n\}$ $(n \in \mathbb{N})$ and, hence, one can select a sequence $u_\alpha^{(n_k)} \in U_{n_k}$, $u_\alpha^{(n_k)} \neq 0$, $n_k < n_{k+1}$ $(k \in \mathbb{N})$. Now fix $\tau \in [1, \infty)^\infty$ such that

$$\tau_{n_k} |u_\alpha^{(n_k)}|^2 \geq 1 \ (k \in \mathbb{N}).$$

Then

$$\| u \|^2_{A_{\tau'}(\mathbb{R}^\infty)} = \sum_{\alpha \in \mathbb{Z}^{\infty}_{+,0}} |u_\alpha| \, \tau_1^{\alpha_1} \dots \tau_\nu^{\alpha_\nu} \geq \sum_{k=1}^{\infty} |u_\alpha^{(n_k)}| \, \tau_{n_k} = \infty,$$

but this contradicts the inclusion $u \in \mathcal{A}(\mathbb{R}^\infty)$. Consequently, $u_\alpha = 0$ for $\nu(\alpha) > \nu_0$ but this means that $u \in \mathcal{A}(\mathbb{R}^{\nu_0})$, because $u \in \mathcal{A}(\mathbb{R}^\infty)$.

The cylindricity of functions from $\mathcal{A}(\mathbb{R}^\infty)$ is an essential defect for some applications. Even this remark shows that it is expedient to investigate the spaces $\mathcal{A}_T(\mathbb{R}^\infty)$, $T \neq [1, \infty)^\infty$, which may contain noncylindrical functions, too. This is, *a fortiori*, true for an arbitrary countable set T.

Let us introduce the Hilbert space $A_{-\tau}(\mathbb{R}^\infty)$ as a space dual to $A_\tau(\mathbb{R}^\infty)$ with respect to $L_2(\mathbb{R}^\infty, \gamma_1)$. The family of Hilbert spaces $(A_{-\tau}(\mathbb{R}^\infty))_{\tau \in T}$ is directed by increasing, and this enables us to consider the inductive limit $\operatorname{ind\,lim}_{\tau \in T} A_{-\tau}(\mathbb{R}^\infty)$. Using the equality

$$A_{-\tau}(\mathbb{R}^\infty) = \bigotimes_{k=1;\, p}^{\infty} A_{-\tau_k}(\mathbb{R}^1),$$

which has been proved for the general case of the tensor product of chains in Section 2 of Chapter 1, we get the following statement (in analogy with Lemma 4.3).

Lemma 4.4. *(i) The representation*

$$A_{-\tau}(\mathbb{R}^\infty) = \left\{ \xi = \sum_{\alpha \in \mathbb{Z}^{\infty}_{+,0}} \xi_\alpha h_\alpha \,\middle|\, \sum_{\alpha \in \mathbb{Z}^{\infty}_{+,0}} |\xi_\alpha|^2 \, \tau_1^{-\alpha_1} \dots \tau_\nu^{-\alpha_\nu} < \infty \right\} \tag{4.24}$$

is valid, and for $\xi, \eta \in A_{-\tau}(\mathbb{R}^\infty)$ *the scalar product in* $A_{-\tau}(\mathbb{R}^\infty)$ *is equal to*

$$(\xi, \eta)_{A_{-\tau}(\mathbb{R}^\infty)} = \sum_{\alpha \in \mathbb{Z}_{+,0}^\infty} \xi_\alpha \bar{\eta}_\alpha \tau_1^{-\alpha_1} \ldots \tau_\nu^{-\alpha_\nu}. \tag{4.25}$$

(ii) For $u \in A_\tau(\mathbb{R}^\infty)$, *the action of the generalized functions* $\xi \in A_{-\tau}(\mathbb{R}^\infty)$ *is given by*

$$(\xi, u)_{L_2(\mathbb{R}^\infty, \gamma_1)} = \sum_{\alpha \in \mathbb{Z}_{+,0}^\infty} \xi_\alpha \bar{u}_\alpha. \tag{4.26}$$

(iii) The space $\mathcal{A}'_T(\mathbb{R}^\infty)$ *dual to* $\mathcal{A}_T(\mathbb{R}^\infty)$ *in the Mackey topology* $\tau(\mathcal{A}'_T(\mathbb{R}^\infty), \mathcal{A}_T(\mathbb{R}^\infty))$ *coincides with* $\operatorname{ind\,lim}_{\tau \in T} A_{-\tau}(\mathbb{R}^\infty)$.

■

The representations (4.20) and (4.24) allow us to introduce a useful approximation of test and generalized functions. Denote by $\mathcal{P}_{\text{cyl}}(\mathbb{R}^\infty)$ the set of cylindrical polynomials on $\mathbb{R}^\infty$ constructed for the nuclear rigging

$$\mathbb{R}^\infty \supset l_2(\mathbb{R}^1) \supset \mathbb{R}_0^\infty, \tag{4.27}$$

according to the definition of $\mathcal{P}_{\text{cyl}}(\Phi')$ given in Section 2. In other words, $\mathcal{P}_{\text{cyl}}(\mathbb{R}^\infty)$ consists of all polynomials of finitely many variables $p(x) = p(x_1, \ldots, x_\nu)$ $(x \in \mathbb{R}^\infty,\ \nu \in \mathbb{N})$. By virtue of the fact that t^k can be expressed in terms of the Hermite polynomials $h_j(t)$ $(t \in \mathbb{R},\ j \le k)$, an arbitrary cylindrical polynomial can be represented in the form

$$p(x) = \sum_{\substack{\nu(\alpha) \le l \\ |\alpha| \le n}} p_\alpha h_\alpha(x) \quad (x \in \mathbb{R}^\infty;\ |\alpha| = \alpha_1 + \ldots + \alpha_\nu),$$

where l is the order of cylindricity, i.e., the largest number of coordinates $x_1, x_2, \ldots$ upon which $p(\cdot)$ depends, and n is the power of the polynomial $p(\cdot)$.

Let us topologize $\mathcal{P}_{\text{cyl}}(\mathbb{R}^\infty)$ by introducing the topology of the inductive limit (in l, $n \in \mathbb{N}$) of finite-dimensional spaces of polynomials in $x_1, \ldots, x_l$ whose power is at most n. There exist natural imbeddings of these spaces that can be used to order them by increasing. This enables us to introduce the inductive limit mentioned above. Note that the convergence of the sequence

$$(p^{(n)})_{n=1}^\infty \subset \mathcal{P}_{\text{cyl}}(\mathbb{R}^\infty)$$

in this topology means that the polynomials $p^{(n)}$ $(n \in \mathbb{N})$ are uniformly cylindrical and their powers are bounded, and that each number sequence of coefficients

$$(p_\alpha^{(n)} = (p^{(n)}, h_\alpha)_{L_2(\mathbb{R}^\infty, \gamma_1)})_{n=1}^\infty \quad (\alpha \in \mathbb{Z}_{+,0}^\infty)$$

converges. In this topology $\mathcal{P}_{\text{cyl}}(\mathbb{R}^\infty)$ is nuclear as an inductive limit of nuclear spaces.

We have constructed the nuclear space of test functions $\mathcal{P}_{\text{cyl}}(\mathbb{R}^\infty) \subset L_2(\mathbb{R}^\infty, \gamma_1)$ with the minimal collection of elements among all the spaces containing the basis of the Hermite polynomials in $L_2(\mathbb{R}^\infty, \gamma_1)$. It is easy to see that $\mathcal{P}_{\text{cyl}}(\mathbb{R}^\infty) \subset \mathcal{A}(\mathbb{R}^\infty)$ densely topologically and, moreover, the topological inclusion $\mathcal{P}_{\text{cyl}}(\mathbb{R}^\infty) \subset \mathcal{A}_T(\mathbb{R}^\infty)$ holds.

For all $l, n \in \mathbb{N}$, let us associate with each function $u \in A_\tau(\mathbb{R}^\infty)$, $\tau \in [1, \infty)^\infty$, a cylindrical polynomial

$$u_{l,n}(\cdot) = \sum_{\substack{\nu(\alpha) \le l \\ |\alpha| \le n}} u_\alpha h_\alpha(\cdot) \in \mathcal{P}_{\text{cyl}}(\mathbb{R}^\infty).$$

The polynomials $u_{l,n}$ $(l, n \in \mathbb{N})$ approximate $u \in A_\tau(\mathbb{R}^\infty)$ in the sense of the strong convergence in $A_\tau(\mathbb{R}^\infty)$. Indeed,

$$\| u - u_{l,n} \|^2_{A_\tau(\mathbb{R}^\infty)} = \sum_{(\nu(\alpha) > l) \vee (|\alpha| > n)} |u_\alpha| \, \tau_1^{\alpha_1} \ldots \tau_\nu^{\alpha_\nu} \to 0, \quad l, n \to \infty,$$

as the remainder of the convergent series. The sequence of cylindrical polynomials

$$(\xi_{l,n})_{l,n=1}^\infty \subset \mathcal{P}_{\text{cyl}}(\mathbb{R}^\infty),$$

which approximates a generalized function $\xi \in A_{-\tau}(\mathbb{R}^\infty)$ with respect to strong convergence, can be defined quite analogously.

■

As shown in Subsection 1.2, for every function $u \in L_2(\mathbb{R}^\infty, \gamma_1)$, one can define the sequence of cylindrical functions

$$u_l(x_1, \ldots, x_l) = \int_{\mathbb{R}^1} u(x) \, d\gamma_1(x_{l+1}) \, d\gamma_1(x_{l+2}) \ldots \quad (l \in \mathbb{N}).$$

Clearly,

$$u_l(x_1, \ldots, x_l) = \sum_{\nu(\alpha) \le l} u_\alpha h_\alpha(x) \quad (x \in \mathbb{R}^\infty) \tag{4.28}$$

and, consequently, $u_{l,n} \underset{n \to \infty}{\to} u_l$ for $u \in A_\tau(\mathbb{R}^\infty)$ in the sense of the strong convergence in $A_\tau(\mathbb{R}^\infty)$. It also follows from (4.28) that

$$u_l \in \bigotimes_{k=1}^{l} A_{\tau_k}(\mathbb{R}^1),$$

and $u_l \underset{l \to \infty}{\to} u$ strongly in $A_\tau(\mathbb{R}^\infty)$.

Similarly, for every generalized function $\xi \in A_{-\tau}(\mathbb{R}^\infty)$ and any $l \in \mathbb{N}$, one can define the cylindrical generalized function

$$\xi_l = \lim_{n \to \infty} \xi_{l,n} \in \bigotimes_{k=1}^{l} A_{-\tau_k}(\mathbb{R}^1).$$

The equality

$$\xi_l = \sum_{\nu(\alpha) \le l} \xi_\alpha h_\alpha \tag{4.29}$$

involves the convergence $\xi_l \underset{l \to \infty}{\to} \xi$ in $A_{-\tau}(\mathbb{R}^\infty)$.

Since

$$E_l \in \mathcal{L}(A_\tau(\mathbb{R}^\infty), A_\tau(\mathbb{R}^\infty))$$

is a selfadjoint projector in $L_2(\mathbb{R}^\infty, \gamma_1)$, the generalized function ξ_l may be equivalently determined as $E_l \xi$, where

$$E_l \in \mathcal{L}(A_{-\tau}(\mathbb{R}^\infty), A_{-\tau}(\mathbb{R}^\infty))$$

(this is the extension of E_l by continuity). Moreover, we have

$$(\xi_l, u)_{L_2(\mathbb{R}^\infty, \gamma_1)} = (\xi, u_l)_{L_2(\mathbb{R}^\infty, \gamma_1)} = (\xi_l, u_l)_{L_2(\mathbb{R}^\infty, \gamma_1)}$$

$$(\xi \in A_{-\tau}(\mathbb{R}^\infty), \quad u \in A_\tau(\mathbb{R}^\infty)).$$

This approximation of test and generalized functions will be called a canonical cylindrical approximation.

4.4. Some Properties of Test and Generalized Functions

Up till now, we have regarded test functions from $A_\tau(\mathbb{R}^\infty)$ as elements of the space $L_2(\mathbb{R}^\infty, \gamma_1)$. Below, we show that they are continuous functions on Hilbert subspaces of $\mathbb{R}^\infty$ depending on the multi-index $\tau \in [1, \infty)^\infty$.

Let us select some subspaces of $\mathbb{R}^\infty$ necessary for the study of the smoothness of test functions. For every $\tau \in [1, \infty)^\infty$, define the Hilbert space X_τ^+ as a set

$$X_\tau^+ = \left\{ \varphi = (\varphi_k)_{k=1}^\infty \in l_2(\mathbb{R}^1) \;\middle|\; \sum_{k=1}^\infty \varphi_k^2 (1 + \tau_k) < \infty \right\} \tag{4.30}$$

with the scalar product

$$(\varphi, \psi)_{X_\tau^+} = \sum_{k=1}^\infty \varphi_k \psi_k (1 + \tau_k) \quad (\varphi, \psi \in X_\tau^+). \tag{4.31}$$

The Hilbert space X_τ^- dual to X_τ^+ with respect to $l_2(\mathbb{R}^1)$ obviously admits the representation

$$X_\tau^- = \left\{ x \in \mathbb{R}^\infty \;\middle|\; \sum_{k=1}^\infty x_k^2 (1 + \tau_k)^{-1} < \infty \right\}. \tag{4.32}$$

The scalar product in this space is given by

$$(x, y)_{X_\tau^-} = \sum_{k=1}^\infty x_k y_k (1 + \tau_k)^{-1} \quad (x, y \in X_\tau^-).$$

Therefore, any $\tau \in [1, \infty)^\infty$ is associated with the rigging

$$X_\tau^- \supset l_2(\mathbb{R}^1) \supset X_\tau^+. \tag{4.33}$$

If the set of multi-indices $\mathrm{T} \subset [1, \infty)^\infty$ satisfies the condition of directedness (4.18), then one can define the space $\mathfrak{X}_\mathrm{T} = \operatorname{pr\,lim}_{\tau \in \mathrm{T}} X_\tau^+$ and the space $\mathfrak{X}'_\mathrm{T} = \operatorname{ind\,lim}_{\tau \in \mathrm{T}} X_\tau^-$ dual to $\mathfrak{X}_\mathrm{T}$ with respect to $l_2(\mathbb{R}^1)$. Moreover, condition (4.22), which guarantees the nuclearity of $A_\tau(\mathbb{R}^\infty)$, implies the nuclearity of $\mathfrak{X}_\mathrm{T}$. Hence, the rigging

$$\mathbb{R}^\infty \supset \mathfrak{X}'_\mathrm{T} \supset l_2(\mathbb{R}^1) \supset \mathfrak{X}_\mathrm{T} \supset \mathbb{R}_0^\infty \tag{4.34}$$

can be associated with each space $A_\tau(\mathbb{R}^\infty)$. In terms of this rigging, many properties of the spaces of test functions are formulated quite naturally. It is interesting to note that the rigging (4.34), in turn, generates the space $A_\tau(\mathbb{R}^\infty)$ (by a certain canonical procedure)[18].

Theorem 4.3. *For every* $\tau \in [1, \infty)^\infty$ *such that* $\sum_{k=1}^{\infty} \tau_k^{-1} < \infty$, $\tau_k > 1$ ($k \in \mathbb{N}$), *the functions from the space* $A_\tau(\mathbb{R}^\infty)$ *are continuous on the Hilbert space* X_τ^- *of full* γ_1*-measure.*

Proof. First, we show that for all $x \in X_\tau^-$ the functional

$$\delta_x(\cdot) = \sum_{\alpha \in \mathbb{Z}_{+,0}^\infty} h_\alpha(x)\, h_\alpha(\cdot)$$

belongs to $A_{-\tau}(\mathbb{R}^\infty)$. According to (4.25), we have

$$\|\delta_x\|^2_{A_{-\tau}(\mathbb{R}^\infty)} = \sum_{\alpha \in \mathbb{Z}_{+,0}^\infty} h_\alpha^2(x)\, \tau_1^{-\alpha_1} \ldots \tau_\nu^{-\alpha_\nu}$$

$$= \lim_{p \to \infty} \sum_{\nu(\alpha) \le l} h_\alpha^2(x)\, \tau_1^{-\alpha_1} \ldots \tau_\nu^{-\alpha_\nu} = \lim_{p \to \infty} \prod_{k=1}^{p} \sum_{j=0}^{\infty} h_j(x_k) \tau_k^{-1}$$

$$= \prod_{k=1}^{\infty} (1 - \tau_k^{-2})^{-1/2} \exp(2\|x\|^2_{X_\tau^-}) < \infty. \tag{4.35}$$

When computing this expression, we have used the bilinear generating function for the Hermite polynomial

$$\sum_{k=0}^{\infty} h_k^2(t)\, z^k\, (1 - z^2)^{-1/2} \exp(2t^2/1 + z^{-1}) \quad (t \in \mathbb{R}^1,\ z \in (0, 1)). \tag{4.36}$$

Taking into account (4.26) and the fact that $\delta_x \in A_{-\tau}(\mathbb{R}^\infty)$ for all $x \in X_\tau^-$, we obtain

$$(\delta_x, u)_{L_2(\mathbb{R}^\infty, \gamma_1)} = \sum_{\alpha \in \mathbb{Z}_{+,0}^\infty} \bar{u}_\alpha h_\alpha(x) = \bar{u}(x)$$

for $u \in A_\tau(\mathbb{R}^\infty)$. Thus, to prove the continuity of functions from $A_\tau(\mathbb{R}^\infty)$ at each point

18 The analysis of this correspondence in a much more general situation is presented in Section 5.

$x^{(0)} \in X_\tau^-$, it suffices to show that δ_x is weakly continuous at the point $x^{(0)}$. The convergence

$$X_\tau^- \ni x^{(n)} \underset{n\to\infty}{\to} x^{(0)} \in X_\tau^-$$

implies the coordinatewise convergence $x_k^{(n)} \underset{n\to\infty}{\to} x_k^{(0)}$ $(k \in \mathbb{N})$ and, therefore, $\delta_{x^{(n)}}$ converges weakly to $\delta_{x^{(0)}}$ as $n \to \infty$ on the set $\mathcal{P}_{\text{cyl}}(\mathbb{R}^\infty)$ dense in $A_\tau(\mathbb{R}^\infty)$. It follows from (4.35) that the norms

$$\|\delta_{x^{(n)}}\|_{A_{-\tau}(\mathbb{R}^\infty)} \le c \quad (n \in \mathbb{N})$$

are uniformly bounded. This fact, together with the weak convergence on the dense set, yields the weak convergence $\delta_{x^{(n)}} \underset{n\to\infty}{\to} \delta_{x^{(0)}}$ on the whole of $A_\tau(\mathbb{R}^\infty)$. Finally, the equality $\gamma_1(X_\tau^-) = 1$ is established by Theorem 1.3 in Section 1.

■

Corollary 1. *Suppose that* $\mathrm{T} \subset [1, \infty)^\infty$ *is such that its elements satisfy the conditions of Theorem* 4.3 *and the condition of directedness* (4.18). *Then the inclusion* $\mathcal{A}_\mathrm{T}(\mathbb{R}^\infty) \subset C(\mathfrak{X}'_\mathrm{T})$ *holds.*

This statement is obtained by applying Theorem 4.3 to each space $A_\tau(\mathbb{R}^\infty)$, $\tau \in \mathrm{T}$, and taking into account the definition of $C(\mathfrak{X}'_\mathrm{T})$ as a set of functions on $\mathfrak{X}'_\mathrm{T}$ whose restrictions to each X_τ^-, $\tau \in \mathrm{T}$, are continuous (see Section 3).

Corollary 2. *It follows from* (4.35) *that for* $\tau \in [1, \infty)^\infty$ *satisfying the conditions of Theorem* 4.3, *the estimate*

$$|u(x)| \le |(\delta_x, u)_{L_2(\mathbb{R}^\infty, \gamma_1)}| \le \|\delta_x\|_{A_{-\tau}(\mathbb{R}^\infty)} \|u\|_{A_\tau(\mathbb{R}^\infty)}$$

$$\le c_\tau \|u\|_{A_\tau(\mathbb{R}^\infty)} \exp(\|x\|^2_{X_\tau^-}) \tag{4.37}$$

holds for all $u \in A_\tau(\mathbb{R}^\infty)$ *and* $x \in X_\tau^-$; *here* $c_\tau = \prod_{k=1}^\infty (1 - \tau_k^{-2})^{-1/4}$.

■

The description of the spaces $A_\tau(\mathbb{R}^\infty)$, $\tau \in [1, \infty)^\infty$ in terms of the Fourier–Hermite coefficients (see Theorem 4.2) enables us to study the action of differential operators with polynomial coefficients in the variety of spaces of test functions.

Example 4.3. Let us investigate the action of the operator of partial derivative, $D_k = \partial / \partial x_k$ and the operator M_k of multiplication by x_k $(k \in \mathbb{N})$ in the spaces of test functions. These operators are defined on $\mathcal{P}_{\text{cyl}}(\mathbb{R}^\infty)$ and map it into itself. Let us take an arbitrary multi-index $\tau \in [1, \infty)^\infty$ and show that for every $\eta \in [1, \infty)^\infty$ such that $\eta_k \geq \tau_k + 1$ and $\eta_j \geq \tau_j$ $(j \in \mathbb{N},\ j \neq k)$, the operators D_k and M_k can be extended by continuity to the operators acting from $A_\eta(\mathbb{R}^\infty)$ to $A_\tau(\mathbb{R}^\infty)$.

The well-known relation for the Hermite polynomials

$$(h_j(t))' = \sqrt{2j}\, h_{j-1}(t) \quad (j \in \mathbb{N}), \quad (h_0(t))' = 0 \tag{4.38}$$

implies that

$$(D_k u)(x) = \sum_{\alpha \in \mathbb{Z}^\infty_{+,0}} u_\alpha \sqrt{2\alpha_k}\, h_{\alpha_1}(x_1) \ldots h_{\alpha_k}(x_k) \ldots h_{\alpha_\nu}(x_\nu)$$

for

$$u = \sum_{\alpha \in \mathbb{Z}^\infty_{+,0}} u_\alpha h_\alpha \in \mathcal{P}_{\text{cyl}}(\mathbb{R}^\infty).$$

This yields

$$(D_k u)_\alpha = \sqrt{2(\alpha_k + 1)}\, u_{\alpha + \delta_k},$$

where

$$\alpha + \delta_k = (\alpha_1, \ldots, \alpha_{k-1}, \alpha_k + 1, \alpha_{k+1}, \ldots).$$

Thus,

$$\|D_k u\|^2_{A_\tau(\mathbb{R}^\infty)} = \sum_{\alpha \in \mathbb{Z}^\infty_{+,0}} |(D_k u)_\alpha|^2\, \tau_1^{\alpha_1} \ldots \tau_\nu^{\alpha_\nu}$$

$$= 2 \sum_{\alpha \in \mathbb{Z}^\infty_{+,0}} (\alpha_k + 1) |u_{\alpha+\delta_k}|^2\, \tau_1^{\alpha_1} \ldots \tau_\nu^{\alpha_\nu}$$

$$= 2 \sum_{\alpha \in \mathbb{Z}^\infty_{+,0}}' |u_\alpha|^2 (\alpha_k \tau_k^{\alpha_k - 1})\, \tau_1^{\alpha_1} \ldots \tau_{k-1}^{\alpha_k - 1}\, \tau_{k+1}^{\alpha_k + 1} \ldots \tau_\nu^{\alpha_\nu}, \tag{4.39}$$

where Σ' does not contain terms which correspond to $\alpha \in \mathbb{Z}^\infty_{+,0}$ such that $\alpha_k = 0$. Since

$$2\, \alpha_k \tau_k^{\alpha_k - 1} \leq c(\tau_k)(1 + \tau_k)^{\alpha_k}\, c(\tau_k)\, \eta_k^{\alpha_k}$$

with some $c(\tau_k)$, the estimate (4.39) can be continued, and thus, replacing τ_j by $\eta_j \geq \tau_j$ $(j \neq k)$, we get

$$\| D_k u \|_{A_\tau(\mathbb{R}^\infty)} \leq c(\tau_k) \| u \|_{A_\eta(\mathbb{R}^\infty)} \quad (u \in \mathcal{P}_{\text{cyl}}(\mathbb{R}^\infty)).$$

This establishes the inclusion $D_k \in \mathcal{L}(A_\eta(\mathbb{R}^\infty), A_\tau(\mathbb{R}^\infty))$.

For $u \in \mathcal{P}_{\text{cyl}}(\mathbb{R}^\infty)$, we can also prove the inequality

$$\| M_k u \|_{A_\tau(\mathbb{R}^\infty)} \leq d(\tau_k) \| u \|_{A_\eta(\mathbb{R}^\infty)}$$

with some $d(\tau_k) > 0$ and the same $\tau, \eta \in [1, \infty)^\infty$ as above. The proof is quite analogous to that presented above and one should only employ the equality

$$t\, h_j(t) = \sqrt{j/2}\, h_{j-1}(t) + \sqrt{j+1/2}\, h_{j+1}(t) \quad (j \in \mathbb{Z}_+;\ h_{-1} = 0). \tag{4.40}$$

Let us consider the algebra $\mathfrak{A}_{\text{cyl}}$ spanned by the operators M_k and D_k $(k \in \mathbb{N})$. This means that we consider all possible finite linear combinations of finite products of these operators with coefficients from $\mathbb{C}^1$. Clearly,, each element of this algebra is a differential operator with cylindrical polynomial coefficients which contains only differentiation with respect to finitely many variables. Using the iterations of the assertion obtained above, for each $H \in \mathfrak{A}_{\text{cyl}}$ and $\tau \in [1, \infty)^\infty$, one can easily find the multi-index $\eta \in [1, \infty)^\infty$ such that $H \in \mathcal{L}(A_\eta(\mathbb{R}^\infty), A_\tau(\mathbb{R}^\infty))$, and η differs from τ only by finitely many components.

Remark 1. The norms of the operators D_k and M_k acting from $A_{\tau'}(\mathbb{R}^\infty)$ to $A_{\tau''}(\mathbb{R}^\infty)$ can be easily estimated in terms of $\tau', \tau'' \in [1, \infty)^\infty$. Using these estimates, we can investigate the action of the operators from the algebra $\mathfrak{A}$[19] in the variety of spaces of test functions. Under certain restrictions on the behavior of the coefficients of $H \in \mathfrak{A}$, one can guarantee the inclusion $H \in \mathcal{L}(A_{\tau(H)}(\mathbb{R}^\infty), A_\tau(\mathbb{R}^\infty))$ with certain

$$\tau(H) = (\tau_k(H))_{k=1}^\infty, \quad \tau_k(H) \geq \tau_k, \quad k \in \mathbb{N},$$

where $\tau_k(H) \geq \tau_k$, $k \in \mathbb{N}$. On the other hand, one can interpret the possibility of applying H to the functions from $A_{\tau(H)}(\mathbb{R}^\infty)$ as a certain smoothness of these functions. This gives us an analogue of the imbedding theorems for the spaces $A_\tau(\mathbb{R}^\infty)$: One can clarify what operators $H \in \mathfrak{A}$ can be applied to $u \in A_\tau(\mathbb{R}^\infty)$ and to what spaces these operators map. We shall not dwell upon these questions in more detail.

■

[19] This algebra consists of all formal infinite series spanned by the finite products of the operators D_k and M_k $(k \in \mathbb{N})$.

Let us establish a simple fact concerning the spaces $A_\tau(\mathbb{R}^\infty)$ that will be used below.

Lemma 4.5. *For all* $\varphi = (\varphi_k)_{k=1}^\infty \in X_\tau^+$, *the vector function*

$$X_\tau^+ \ni \varphi \mapsto \exp\,[i\,(\varphi, x)_{l_2})] \in A_\tau(\mathbb{R}^\infty), \tag{4.41}$$

given as a strong limit of cylindrical functions

$$u^{(n)}(x) = \exp\left(i \sum_{k=1}^{n} \varphi_k u_k \right) \quad (x \in \mathbb{R}^\infty,\ n \in \mathbb{N}) \tag{4.42}$$

in $A_\tau(\mathbb{R}^\infty)$, *is defined and continuous in the strong sense.*

Proof. The well-known formula

$$\int_{\mathbb{R}^1} \exp\,(i\,t\,s)\, h_k(s) \exp\,(-s^2)\, ds = \sqrt{\pi}\, i^k h^k (2^k k!)^{-1/2} e^{-1/4\,t^2} \quad (t \in \mathbb{R}^1,\ k \in \mathbb{Z}_+)$$

yields

$$u_\alpha^{(n)} = \int_{\mathbb{R}^\infty} \exp\left(i \sum_{k=1}^{n} \varphi_k u_k \right) h_\alpha(s)\, d\gamma_1(x)$$

$$= \frac{i^{|\alpha|} \varphi_1^{\alpha_1} \ldots \varphi_n^{\alpha_n}}{2^{|\alpha|/2} (\alpha_1! \ldots \alpha_n!)^{1/2}} \exp\left(-1/4 \sum_{k=1}^{n} \varphi_k^2 \right) \quad (\alpha \in \mathbb{Z}_+^n;\ \varphi_k \in \mathbb{R}^1).$$

Therefore, just as in (4.35), we get

$$\| u^{(n)} \|_{A_\tau(\mathbb{R}^\infty)}^2 = \sum_{\alpha \in \mathbb{Z}_{+,0}^\infty} \frac{\varphi_1^{2\alpha_1} \ldots \varphi_n^{2\alpha_n}}{\alpha_1! \ldots \alpha_n!\ 2^{|\alpha|}} \tau_1^{\alpha_1} \ldots \tau_n^{\alpha_n} \exp\left(-1/2 \sum_{k=1}^{n} \varphi_k^2 \right)$$

$$= \exp\left(1/2\, \| \varphi^{(n)} \|_{X_\tau^+}^2 - \| \varphi^{(n)} \|_{l_2}^2 \right),$$

where $\varphi^{(n)} = (\varphi_1, \ldots, \varphi_n, 0, 0, \ldots)$. Quite similarly, for $\varphi, \psi \in X_\tau^+$, we have

$$\left(\exp\left(i \sum_{k=1}^{n} \varphi_k x_k \right), \exp\left(i \sum_{k=1}^{n} \psi_k x_k \right) \right)_{A_\tau(\mathbb{R}^\infty)}$$

$$= \exp\left(1/2\, \| \varphi^{(n)} \|_{X_\tau^+} - \| \varphi^{(n)} \|_{l_2} \right). \tag{4.43}$$

Using (4.43), one can easily show that the sequence (4.42) is fundamental in $A_\tau(\mathbb{R}^\infty)$ and, thus, the vector function (4.41) is defined. Taking the limit $n \to \infty$ in (4.43), we obtain

$$\left(\exp[i(\varphi,\cdot)_{l_2}], \exp[i(\psi,\cdot)_{l_2}]\right)_{A_\tau(\mathbb{R}^\infty)} = \exp\left(1/2\,(\varphi,\psi)_{X_\tau^+} - (\varphi,\psi)_{l_2}\right) \tag{4.44}$$

$$(\varphi, \psi \in X_\tau^+),$$

whence

$$\|\exp[i(\varphi,\cdot)_{l_2}] - \exp[i(\psi,\cdot)_{l_2}]\|^2_{A_\tau(\mathbb{R}^\infty)}$$

$$= \left\{\exp[1/4(\|\varphi\|^2_{X_\tau^+} - 2\|\varphi\|^2_{l_2})] - \exp[1/4(\|\psi\|^2_{X_\tau^+} - 2\|\psi\|^2_{l_2})]\right\}$$

$$+ \left\{1 - \exp[-1/4(\|\varphi-\psi\|^2_{X_\tau^+} - 2\|\varphi-\psi\|^2_{l_2})]\right\}$$

$$\times \exp[1/4(\|\varphi\|^2_{X_\tau^+} + \|\psi\|^2_{X_\tau^+} - 2\|\varphi\|^2_{l_2} - 2\|\psi\|^2_{l_2})],$$

and this implies the strong continuity of the function (4.41).

■

Remark 2. *For any* $\varphi \in X_\tau^+$ *and* $p \in \mathcal{P}_{\mathrm{cyl}}(\mathbb{R}^\infty)$, *the function* $p(\cdot)\, e^{i(\varphi,\cdot)_{l_2}}$ *belongs to* $A_\tau(\mathbb{R}^\infty)$ *and depends on* $\varphi \in X_\tau^+$ *continuously in the norm of this space.*

Indeed, let M_p denote the operator of multiplication by the polynomial p. According to Example 4.3, the operator M_p acts continuously from $A_{\tau(p)}(\mathbb{R}^\infty)$ into $A_\tau(\mathbb{R}^\infty)$. Moreover, $\tau_k(p) \geq \tau_k$ $(k \in \mathbb{N})$ and $\tau_k(p) = \tau_k$ for all $k \in \mathbb{N}$, except finitely many numbers. Therefore, the space $X^+_{\tau(p)}$ is equivalent to X_τ^+ and, thus, according to Lemma 4.5, we have that

$$\exp[i(\varphi,\cdot)_{l_2}] \in A_{\tau(p)}(\mathbb{R}^\infty)$$

and continuously depends on $\varphi \in X^+_{\tau(p)}$ (i.e, on $\varphi \in X_\tau^+$) in the norm of this space. But then the function

$$p(\cdot)\exp[i(\varphi,\cdot)_{l_2}] = M_p \exp[i(\varphi,\cdot)_{l_2}]$$

possesses the desired properties by virtue of the continuity of the operator M_p.

■

Remark 3. It is sometimes necessary to consider the space of smooth functions analogous to $A_\tau(\mathbb{R}^\infty)$ in the case when the argument $z = x + iy \in \mathbb{C}^\infty$ $(x, y \in \mathbb{R}^\infty)$. For this purpose, we set

$$A_\tau(\mathbb{C}^\infty) = A_\tau(\mathbb{R}^\infty) \otimes A_\tau(\mathbb{R}^\infty)$$

for $\tau \in [1, \infty)^\infty$. The function

$$u(x, y) \in A_\tau(\mathbb{R}^\infty) \otimes A_\tau(\mathbb{R}^\infty)$$

is regarded as a function of a point $z = x + iy \in \mathbb{C}^\infty$. Here, the obvious analogue of Theorem 4.3 is valid. Under the same assumptions for $\tau \in [1, \infty)^\infty$, it states that the function $u \in A_\tau(\mathbb{C}^\infty)$ is continuous on $X^-_{\tau,c}$, and that the estimate

$$|u(z)| \leq c_\tau \|u\|_{A_\tau(\mathbb{R}^\infty)} \exp\left(\|\psi\|^2_{X^-_\tau}\right) \quad (z \in X^-_{\tau,c})$$

holds, where

$$X^-_{\tau,c} = X^-_\tau \oplus i X^-_\tau \subset \mathbb{C}^\infty$$

is the complexification of the Hilbert space $X^-_\tau \subset \mathbb{R}^\infty$. Just as in the proof of Lemma 4.5, we can establish the strong continuity of the vector function

$$X^+_\tau \ni \varphi \mapsto \exp\left[(\varphi, z)_{l_2}\right] = \exp\left[(\varphi, x)_{l_2}\right] \exp\left[(\varphi, x)_{l_2}\right] \in A_\tau(\mathbb{C}^\infty)$$

whereas Remark 1 ensures that, for every cylindrical polynomial

$$p \in \mathcal{P}_{\text{cyl}}(\mathbb{R}^\infty) \simeq \mathcal{P}_{\text{cyl}}(\mathbb{R}^\infty \times \mathbb{R}^\infty),$$

the function $p(z) \exp\left[(\varphi, z)_{l_2}\right]$ depends on $\varphi \in X^+_\tau$ continuously in the norm of the space $A_\tau(\mathbb{C}^\infty)$.

■

We now present two examples of generalized functions from $\mathcal{A}'(\mathbb{R}^\infty)$ whose singularities appear merely due to the fact that the space of the arguments is infinite-dimensional.

Example 4.4. The function

$$\|x\|^2_{l_2} = \sum_{k=1}^{\infty} x_k^2$$

is defined only on the set $l_2(\mathbb{R}^1) \subset \mathbb{R}^\infty$ such that $\gamma_1(l_2(\mathbb{R}^1)) = 0$ (see Subsection 1.3). Let us regularize this function by replacing each term in $\sum_{k=1}^\infty x_k^2$ by the Wick power

$$x_k^2 := x_k^2 - 1/2 = h_2(x_k).$$

The series $\sum_{k=1}^\infty h_2(x_k)$ obtained as a result of this procedure converges in the sense of the space of generalized functions $A_{-\tau}(\mathbb{R}^\infty)$ provided that $\sum_{k=1}^\infty \tau_k^{-2} < \infty$. Indeed, according to (4.24), we have

$$\left\| \sum_{k=1}^\infty h_2(x_k) \right\|^2_{A_{-\tau}(\mathbb{R}^\infty)} = \sum_{k=1}^\infty \tau_k^{-2} < \infty.$$

Denote

$$:\|x\|^2_{l_2}: = \sum_{k=1}^\infty h_2(x_k) \in A_{-\tau}(\mathbb{R}^\infty).$$

The generalized function $:\|x\|^2_{l_2}:$ can be formally written as

$$:\|x\|^2_{l_2}: = \|x\|^2_{l_2} - \sum_{k=1}^\infty 1/2.$$

Hence, the regularization of $\|x\|^2_{l_2}$ (consisting in the transition to $:\|x\|^2_{l_2}:$) is, in fact, reduced to the subtraction of an infinite constant.

This example illustrates the procedure of transition to the Wick ordering in constructive field theory (in the simplest case). This procedure enables us to lower the singularity of the interaction and give, at least, the meaning of generalized function to it. We shall return to the investigation of this question as applied to some models of the field theory.

Example 4.5. It has been shown in Subsection 2.4 that the shifted Gaussian measure $\gamma_1(\cdot + \varphi)$ is either equivalent to γ_1 (if $\varphi \in l_2(\mathbb{R}^1)$) or orthogonal to it (if $\varphi \notin l_2(\mathbb{R}^1)$). In the first case

$$\frac{d\gamma_1(x+\varphi)}{d\gamma_1(x)} = \exp\left[-\sum_{k=1}^\infty (2x_k\varphi_k + \varphi_k^2) \right] = \prod_{k=1}^\infty \exp(-2x_k\varphi_k - \varphi_k^2). \quad (4.45)$$

Let us show that for every $\varphi \in \mathbb{R}^\infty$, one can find $\tau \in [1, \infty)^\infty$ for which the infinite product in (4.45) converges in the sense of the space of generalized functions. Indeed, the function $\rho(\varphi_k, x_k) = \exp(-2x_k\varphi_k - \varphi_k)$ admits the representation

$$\rho(\varphi_k, x_k) = \sum_{j=0}^\infty (-1)^j \frac{\varphi_k^j 2^{j/2}}{\sqrt{j!}} h_j(x_k). \quad (4.46)$$

To obtain this representation, we have used the generating formula for the Hermite polynomials

$$\sum_{j=0}^{\infty} \frac{s^j\, h_j(t)}{\sqrt{j!}\; 2^{j/2}} = \exp(st - s^2/4) \quad (t, s \in \mathbb{R}^1). \tag{4.47}$$

The decomposition (4.46) and (4.24) together yield

$$\| \rho(\varphi_k, \cdot) \|^2_{A_{-\tau}(\mathbb{R}^1)} = \sum_{j=0}^{\infty} \frac{\varphi_k^j\, 2^{j/2}}{\sqrt{j!}} \tau_k^{-j} = \exp(2\varphi_k^2 / \tau_k). \tag{4.48}$$

If $\tau \in [1, \infty)^\infty$ is such that $\varphi \in X_\tau^-$, then the inclusion

$$\prod_{k=1}^{\infty} \rho(\varphi_k, x_k) \in A_{-\tau}(\mathbb{R}^\infty) = \bigotimes_{k=1;\, e}^{\infty} A_{-\tau_k}(\mathbb{R}^1)$$

follows easily from (4.48) and, moreover,

$$\left\| \prod_{k=1}^{\infty} \rho(\varphi_k, \cdot) \right\|_{A_{-\tau}(\mathbb{R}^\infty)} = \exp\Big(\sum_{k=1}^{\infty} \varphi_k^2 / \tau_k \Big) = \exp\,(\|\varphi\|^2_{X_\tau}).$$

Thus, we have shown that the Radon–Nikodym derivative

$$\rho(\varphi, x) = d\gamma_1(x + \varphi)/d\gamma_1(x)$$

can be also defined when $\varphi \notin l_2(\mathbb{R}^1)$. However, for $\varphi \notin l_2(\mathbb{R}^1)$, this is a generalized function from a properly chosen space $A_{-\tau}(\mathbb{R}^\infty)$. For $\rho(\varphi, \cdot)$, it is easy to establish the equality

$$(\rho(\varphi, \cdot), u)_{L_2(\mathbb{R}^\infty, \gamma_1)} = \int_{\mathbb{R}^\infty} \bar{u}(x)\, d\gamma_1(x + \varphi) \quad (u \in A_\tau(\mathbb{R}^\infty))$$

the left-hand side of which is regarded as an action of the generalized function $\rho(\varphi, \cdot)$ on the test function $u \in A_\tau(\mathbb{R}^\infty)$. In the case when $\varphi \in l_2(\mathbb{R}^1)$, we have $\rho(\varphi, \cdot) \in L_2(\mathbb{R}^\infty, \gamma_1)$, and this equality has the usual meaning.

4.5. The Space $\mathcal{G}(\mathbb{R}^\infty)$

The general procedure presented in Subsection 1 enables us to construct a large variety of spaces of test and generalized functions of infinitely many variables that are necessary for applications. Above we have studied in detail the concrete realization of this procedure which lead us to the spaces $\mathcal{A}_T(\mathbb{R}^\infty)$. Consider one more example of the use of this procedure for the construction of nuclear spaces of functions of an infinite-dimensional argument.

Let us denote by $G^l(\mathbb{R}^1)$ the completion of $C_0^\infty(\mathbb{R}^1)$ with respect to the scalar product

$$(u, v)_{G^l(\mathbb{R}^1)} = \sum_{k=0}^{l} \int_{\mathbb{R}^1} (D^k u)(t)\, \overline{(D^k v)(t)}\, d\gamma_1(t) \tag{4.49}$$

$$(u, v \in C_0^\infty(\mathbb{R}^1);\ l \in \mathbb{Z}_+).$$

This space is an analogue of the Sobolev space $W_2^l(\mathbb{R}^1)$ in which the Lebesgue measure dt is replaced by the Gaussian measure $d\gamma_1(t)$. The family of Hilbert spaces $(G^l(\mathbb{R}^1))_{i=0}^\infty$ is clearly directed by the imbedding

$$G^0(\mathbb{R}^1) = L_2(\mathbb{R}^\infty, \gamma_1) \supset G^1(\mathbb{R}^1) \supset \ldots \supset G^l(\mathbb{R}^1) \supset \ldots,$$

$$\|\cdot\|_{G^l(\mathbb{R}^1)} \le \|\cdot\|_{G^{l+1}(\mathbb{R}^1)} \quad (l \in \mathbb{Z}_+),$$

and this allows us to define the space

$$\mathcal{G}(\mathbb{R}^1) = \operatorname{pr} \lim_{l \to \infty} G^l(\mathbb{R}^1)$$

as their projective limit.

Let us obtain another representation of $\mathcal{G}(\mathbb{R}^1)$ as a projective limit of Hilbert spaces which is necessary for what follows. For each $l \in \mathbb{Z}_+$ we define the Hilbert space

$$H^l(\mathbb{R}^1) = \left\{ u \in L_2(\mathbb{R}^1, \gamma_1) \,\middle|\, \sum_{j=0}^{\infty} |u_j|^2 (1+j)^l < \infty \right\}, \tag{4.50}$$

$$(u, v)_{H^l(\mathbb{R}^1)} = \sum_{j=0}^{\infty} u_j \bar{v}_j (1+j)^l \quad (u, v \in H^l(\mathbb{R}^1)). \tag{4.51}$$

Here, as before, $(u_j)_{j=0}^\infty$ are the Fourier coefficients of the function u which correspond to its expansion in the basis $(h_j)_{j=0}^\infty$ of the Hermite polynomials in $L_2(\mathbb{R}^1, \gamma_1)$. The next lemma shows that the norms in the Hilbert spaces $G^l(\mathbb{R}^1)$ and $H^l(\mathbb{R}^1)$ are equivalent.

Lemma 4.3. *For every $l \in \mathbb{Z}_+$, there exist constants $c_l', c_l'' > 0$ such that*

$$c_l' \|u\|^2_{H^l(\mathbb{R}^1)} \le \|u\|^2_{G^l(\mathbb{R}^1)} \le c_l'' \|u\|^2_{H^l(\mathbb{R}^1)} \quad (u \in C_0^\infty(\mathbb{R}^1)). \tag{4.52}$$

Proof. Let $u \in C_0^\infty(\mathbb{R}^1)$ and let $u(t) = \sum_{j=0}^\infty u_j h_j(t)$ be the expansion of this function in the Hermite polynomials. Then, by virtue of (4.38), we get

$$(D u)\, t = \sum_{j=1}^\infty u_j \sqrt{2j}\; h_{j-1}(t) = \sum_{j=0}^\infty u_{j+1} \sqrt{2(j+1)}\; h_j(t) \quad (t \in \mathbb{R}^1).$$

By iterating this formula, we obtain

$$(D^k u)(t) = \sum_{j=0}^\infty u_{j+k} \left[2^k (j+1) \dots (j+k)\right]^{1/2} h_j(t) \quad (t \in \mathbb{R}^1;\ k \in \mathbb{N}).$$

Therefore, (4.49) can be rewritten as

$$(u, v)_{G^l(\mathbb{R}^1)} = \sum_{k=0}^{l} \left\{ \sum_{j=0}^\infty u_{j+k} \bar{v}_{j+k} [2^k (j+1) \dots (j+k)] \right\} \tag{4.53}$$

$$(u, v \in C_0^\infty(\mathbb{R}^1);\ l \in \mathbb{Z}_+).$$

For every $k = 0, \dots, l,$ we have

$$\sum_{j=0}^\infty |u_{j+k}|^2 [2^k (j+1) \dots (j+k)] \le 2^l \sum_{j=0}^\infty |u_{j+k}|^2 \; (j+k+1)^l$$

$$\le 2^l \sum_{j=0}^\infty |u_j|^2 (1+j)^l. \tag{4.54}$$

For $k = l,$ the left-hand side of (4.54) is bounded from below by

$$2^l \sum_{j=l} |u_j|^2 (1+j)^l$$

and, furthermore,

$$\sum_{j=0}^{l-1} |u_j|^2 (1+j)^l \le l^l \sum_{j=0}^\infty |u_j|^2.$$

The last two estimates and (4.54) imply that one can find $c_l', c_l'' > 0$ such that

$$c_l' \sum_{j=0}^\infty |u_j|^2 (1+j)^l \le \|u\|^2_{G^l(\mathbb{R}^1)} \le c_l'' \sum_{j=0}^\infty |u_j|^2 (1+j)^l,$$

i.e., the inequalities (4.52) hold. ■

Lemma 4.6 shows that the representation

$$\mathcal{G}(\mathbb{R}^1) = \operatorname{pr}\lim_{l\to\infty} H^l(\mathbb{R}^1)$$

holds. This representation will be employed to prove the next theorem which internally describes the space $\mathcal{G}(\mathbb{R}^1)$.

Theorem 4.4. *The space $\mathcal{G}(\mathbb{R}^1)$ is nuclear or, more exactly, for every $l \in \mathbb{Z}_+$, the space $G^{l+2}(\mathbb{R}^1)$ is quasinuclearly imbedded into $G^l(\mathbb{R}^1)$. The function u belongs to $\mathcal{G}(\mathbb{R}^1)$ if and only if $u(t)\, e^{-t^2/2} \in \mathcal{S}(\mathbb{R}^1)$. The mapping*

$$\mathcal{G}(\mathbb{R}^1) \ni u(t) \mapsto u(t)\, e^{-t^2/2} \in \mathcal{S}(\mathbb{R}^1)$$

is a topological isomorphism between the spaces $\mathcal{G}(\mathbb{R}^1)$ and $\mathcal{S}(\mathbb{R}^1)$.

Proof. Let us fix an arbitrary $l \in \mathbb{Z}_+$. We show that the imbedding of $H^{l+2}(\mathbb{R}^1)$ into $H^l(\mathbb{R}^1)$ is quasinuclear. According to Lemma 4.6, this is equivalent to the quasinuclearity of the imbedding of $G^{l+2}(\mathbb{R}^1)$ into $G^l(\mathbb{R}^1)$. By virtue of (4.50), the functions

$$((1+j)^{-(l+2)/2}\, h_j(\cdot))_{j=0}^{\infty}$$

form an orthonormal basis in the Hilbert space $H^{l+2}(\mathbb{R}^1)$. Therefore, for the imbedding operator $O_{l+2,\,l}\colon H^{l+2}(\mathbb{R}^1) \to H^l(\mathbb{R}^1)$, we have

$$\| O_{l+2,\,l} \|^2 = \sum_{j=0}^{\infty} (1+j)^{-2} < \infty,$$

and this gives the desired result. Since

$$\mathcal{G}(\mathbb{R}^1) = \operatorname{pr}\lim_{l\to\infty} G^l(\mathbb{R}^1),$$

the proved statement leads us to the conclusion that this space is nuclear.

In order to prove the remaining statements of the theorem, we employ Example 4.1. The definition (4.3) of the Hilbert space $H^l(\mathbb{R}^1)$ $(l \in \mathbb{Z}_+)$ implies that the mapping

$$H^l(\mathbb{R}^1) \ni u(t) \mapsto \pi^{-1/4} u(t)\, e^{-t^2/2} \in H_l(\mathbb{R}^1)$$

is an isometric isomorphism of the Hilbert spaces $H^l(\mathbb{R}^1)$ and $H_l(\mathbb{R}^1)$. However,

$$\mathcal{S}(\mathbb{R}^1) = \text{pr}\lim_{l\to\infty} H^l(\mathbb{R}^1), \quad \mathcal{G}(\mathbb{R}^1) = \text{pr}\lim_{l\to\infty} H^l(\mathbb{R}^1),$$

and consequently, the mapping from the statement of the theorem is a topological isomorphism.

■

We now pass to the construction of the spaces of infinitely many variables using the scheme of Subsection 4.1 with $L_2(\mathbb{R}^\infty, \gamma_1(x_k))$ and $\Phi_k = \mathcal{G}(\mathbb{R}^1)$. In this case, a purely infinite-dimensional effect appears to be connected with the fact that it is not indifferent what representation of $\mathcal{G}(\mathbb{R}^1)$ as a projective limit of Hilbert spaces is exploited. For example, we may use the representation $\mathcal{G}(\mathbb{R}^1) = \text{pr}\lim_{l\to\infty} G^l(\mathbb{R}^1)$. In this case, conditions (i)– (iii) of Subsection 4.1 are valid and, as above, the role of a stabilizing sequence is played by $e = \left(e^{(k)}(x_k)\right)_{k=1}^{\infty}$, $e^{(k)}(x_k) = 1$ $(x_k 2\ \mathbb{R}^1; k \in \mathbb{N})$. However, one can easily see that no choice exists of the set of weights $\mathrm{T} \subset \mathrm{T}_\infty = \mathbb{Z}_+^\infty = \mathbb{Z}_+ \times \mathbb{Z}_+ \times \ldots$ for which the condition of nuclearity (4.2) is satisfied. Therefore, the infinite tensor product of the nuclear spaces $\mathcal{G}(\mathbb{R}^1)$ obtained as a result is not a nuclear space. In order that the scheme of Subsection 4.1 lead to the nuclear space, it is necessary to renormalize the spaces $G^l(\mathbb{R}^1)$ $(l \in \mathbb{Z}_+)$ appropriately (by using Lemma 2.5 of Section 2, Chapter 1). This renormalization can be realized by using different methods. One of these consists in the transition to the spaces $H^l(\mathbb{R}^1)$, $l \in \mathbb{Z}_+$, such that

$$\mathcal{G}(\mathbb{R}^1) = \text{pr}\lim_{l\to\infty} H^l(\mathbb{R}^1).$$

Let us choose a subset $\mathrm{T} \subset \mathbb{Z}_+^\infty$ possessing the property of directedness (4.18). For $\tau = (\tau_k)_{k=1}^\infty \in \mathrm{T}$, we introduce the space of test functions

$$H^\tau(\mathbb{R}^\infty) = \bigotimes_{k=1;e}^{\infty} H^{\tau_k}(\mathbb{R}^\infty) \subset \bigotimes_{k=1;e}^{\infty} L_2(\mathbb{R}^\infty, \gamma_1(x_k)) = L_2(\mathbb{R}^\infty, \gamma_1).$$

Condition (4.18) implies that the family of Hilbert spaces $(H^\tau(\mathbb{R}^\infty))_{\tau\in\mathrm{T}}$ is directed by inclusion. This enables us to define their projective limit

$$\mathcal{G}_\mathrm{T}(\mathbb{R}^\infty) = \text{pr}\lim_{\tau\in\mathrm{T}} H^\tau(\mathbb{R}^\infty).$$

The following theorem is an analogue of Theorem 4.2 for the case of the space $\mathcal{G}_\mathrm{T}(\mathbb{R}^\infty)$.

Theorem 4.5. *(i) The equality*

$$H^{\tau}(\mathbb{R}^{\infty}) = \left\{ u \in L_2(\mathbb{R}^1, \gamma_1) \,\middle|\, \sum_{\alpha \in \mathbb{Z}^{\infty}_{+,0}} |u_\alpha|^2 (1+\alpha_1)^{\tau_1} \dots (1+\alpha_\nu)^{\tau_\nu} < \infty \right\} \tag{4.55}$$

holds, and the scalar product of the functions $u, v \in H^{\tau}(\mathbb{R}^{\infty})$ *in* $H^{\tau}(\mathbb{R}^{\infty})$ *is equal to*

$$(u, v)_{H^{\tau}(\mathbb{R}^{\infty})} = \sum_{\alpha \in \mathbb{Z}^{\infty}_{+,0}} u_\alpha \bar{v}_\alpha (1+\alpha_1)^{\tau_1} \dots (1+\alpha_\nu)^{\tau_\nu}. \tag{4.56}$$

(ii) The space $\mathcal{G}_{\mathrm{T}}(\mathbb{R}^{\infty})$ *is nuclear if and only if for any* $\tau' \in \mathrm{T}$ *there exists* $\tau'' \in \mathrm{T}$ *such that*

$$\sum_{k=1}^{\infty} (\xi(\tau''_k - \tau'_k) - 1) < \infty, \tag{4.57}$$

where $\xi(\cdot)$ *is the Riemann zeta function.*

Proof. (i) The equality

$$H^{\tau}(\mathbb{R}^{\infty}) = \bigotimes_{k=1;\, e}^{\infty} H^{\tau_k}(\mathbb{R}^{\infty})$$

and the explicit form of the orthonormal basis in the Hilbert space $H^{\tau_k}(\mathbb{R}^1)$ (introduced in the proof of Theorem 4.4) together imply that the functions

$$(1+\alpha_1)^{-\tau_1/2} \dots (1+\alpha_\nu)^{-\tau_\nu/2}\, h_\alpha(x) \quad (x \in \mathbb{R}^{\infty};\ \alpha \in \mathbb{Z}^{\infty}_{+,0}) \tag{4.58}$$

form an orthonormal basis in $H^{\tau}(\mathbb{R}^{\infty})$. This immediately yields the validity of (4.55) and (4.56).

(ii) Assume that $\tau', \tau'' \in \mathrm{T}$ are such that $\tau_k' < \tau_k''$ $(k \in \mathbb{N})$. Let us find the expression for the Hilbert norm of the imbedding operator $O_{\tau'\tau''}: H^{\tau''}(\mathbb{R}^{\infty}) \to H^{\tau'}(\mathbb{R}^{\infty})$. Using (4.58) and (4.56), we obtain

$$|O_{\tau'\tau''}|^2 = \sum_{\alpha \in \mathbb{Z}^{\infty}_{+,0}} (1+\alpha_1)^{\tau'_1 - \tau''_1} \dots (1+\alpha_\nu)^{\tau'_\nu - \tau''_\nu}$$

$$= \prod_{k=1}^{\infty} \Big[\sum_{\alpha_k = 0}^{\infty} (1+\alpha_k)^{-(\tau''_k - \tau'_k)} \Big] = \prod_{k=1}^{\infty} \xi(\tau''_k - \tau'_k),$$

where the last infinite product converges or diverges simultaneously with the series $\sum_{k=1}^{\infty}(\xi(\tau_k'' - \tau_k') - 1)$. By virtue of the condition (4.18) of directedness of T and the definition of nuclearity, this establishes the validity of statement (ii) of Theorem 4.5.

■

Example 4.6. In the case $T = \mathbb{Z}_+^{\infty}$, we denote

$$\mathcal{G}(\mathbb{R}^{\infty})_{\mathbb{Z}_+^{\infty}} = \mathcal{G}(\mathbb{R}^{\infty}).$$

Clearly, condition (4.57) now holds and, hence, $\mathcal{G}(\mathbb{R}^{\infty})$ is a nuclear space. Just as in Example 4.2, we establish the equality

$$\mathcal{G}(\mathbb{R}^{\infty}) = \bigcup_{n=1}^{\infty} \mathcal{G}(\mathbb{R}^{n}),$$

where $\mathcal{G}(\mathbb{R}^n) = \mathcal{G}(\mathbb{R}^1) \otimes \ldots \otimes \mathcal{G}(\mathbb{R}^1)$ (n times). This means that for every function $u \in \mathcal{G}(\mathbb{R}^{\infty})$, one can find $n \in \mathbb{N}$ and the function $\varphi \in \mathfrak{S}(\mathbb{R}^n)$ such that the following representation is valid

$$u(x_1, \ldots, x_n) = \varphi(x_1, \ldots, x_n) \exp\left(\sum_{k=1}^{n} x_k^2 / 2 \right).$$

Let us introduce the Hilbert space of generalized functions $H^{-\tau}(\mathbb{R}^{\infty})$ as a space dual to $H^{\tau}(\mathbb{R}^{\infty})$ with respect to $L_2(\mathbb{R}^{\infty}, \gamma_1)$. The representation

$$H^{-\tau}(\mathbb{R}^{\infty}) = \left\{ \xi = \sum_{\alpha \in \mathbb{Z}_{+,0}^{\infty}} \xi_\alpha h_\alpha \,\middle|\, \sum_{\alpha \in \mathbb{Z}_{+,0}^{\infty}} |\xi_\alpha|^2 (1 + \alpha_1)^{-\tau_1} \ldots (1 + \alpha_\nu)^{-\tau_\nu} < \infty \right\},$$

and the expression

$$(\xi, u)_{L_2(\mathbb{R}^{\infty}, \gamma_1)} = \sum_{\alpha \in \mathbb{Z}_{+,0}^{\infty}} \xi_\alpha \bar{u}_\alpha \quad (u_\alpha = (u, h_\alpha)_{L_2(\mathbb{R}^{\infty}, \gamma_1)})$$

for the action of the generalized function $\xi \in H^{-\tau}(\mathbb{R}^{\infty})$ upon the test function $u \in H^{\tau}(\mathbb{R}^{\infty})$ immediately follow from (4.58). The space $\mathcal{G}'_T(\mathbb{R}^{\infty})$ dual to $\mathcal{G}_T(\mathbb{R}^{\infty})$ in the Mackey topology $\tau(\mathcal{G}'_T(\mathbb{R}^{\infty}), \mathcal{G}_T(\mathbb{R}^{\infty}))$ coincides with $\operatorname{ind\,lim}_{\tau \in T} H^{-\tau}(\mathbb{R}^{\infty})$ (see Section 1, Chapter 1). Just as in Subsection 4.3, by using the description of the spaces $H^{\pm\tau}(\mathbb{R}^{\infty})$ in terms of the coefficients of expansions of their elements in the basis

$(h_\alpha)_{\alpha\in\mathbb{Z}^\infty_{+,0}}$ of Hermite polynomials in $L_2(\mathbb{R}^\infty, \gamma_1)$, we can define the approximation of test and generalized functions by cylindrical polynomials and establish the statements concerning the action of differential operators in the spaces of the family $(H^\tau(\mathbb{R}^\infty))_{\tau\in\mathrm{T}}$ (analogous with those presented in Subsection 4.4), etc. The only essential distinction from the case of the space $\mathcal{A}_\mathrm{T}(\mathbb{R}^\infty)$ consists in the fact that here the analogue of the imbedding Theorem 4.3 does not hold, in principle. More exactly, one cannot choose the multi-index $\tau\in\mathbb{Z}^\infty_+$ so that the functions from $H^\tau(\mathbb{R}^\infty)$ are continuous on a set of positive γ_1-measure in $\mathbb{R}^\infty$. The last assertion restricts our possibilities to employ the spaces introduced above in applications. For fixed $t\in\mathbb{N}^\infty$, we have the inclusions

$$A_{-\tau}(\mathbb{R}^\infty)\supset H^{-\tau}(\mathbb{R}^\infty)\supset L_2(\mathbb{R}^\infty,\gamma_1)\supset H^\tau(\mathbb{R}^\infty)\supset A_\tau(\mathbb{R}^\infty),$$

and for given set of multi-indices $\mathrm{T}\subset\mathbb{N}^\infty$, we have

$$\mathcal{A}'_\mathrm{T}(\mathbb{R}^\infty)\supset \mathcal{G}'_\mathrm{T}(\mathbb{R}^\infty)\supset L_2(\mathbb{R}^\infty,\gamma_1)\supset \mathcal{G}_\mathrm{T}(\mathbb{R}^\infty)\supset \mathcal{A}_\mathrm{T}(\mathbb{R}^\infty).$$

§5. Spaces of Test and Generalized Functions of Infinitely Many Variables in Invariant Form

Assume that the nuclear rigging

$$\Phi'\supset H_0\supset\Phi \tag{5.1}$$

of the real separable Hilbert space H_0 is given, and that $\gamma_1=\gamma_{H_0}$ is the canonical Gaussian measure on Φ' associated with H_0, i.e.,

$$\tilde{\gamma}_1(\varphi) = \exp(-1/4\,\|\varphi\|^2_{H_0})\quad(\varphi\in\Phi)$$

(see Section 1). The method of construction of the spaces of test and generalized functions examined in Section 4 cannot be directly employed to construct the riggings of $L_2(\Phi',\gamma_1)$, since in order for it to work, one should assume that the measure γ_1 can be represented as a product measure. In fact, the last statement means that a fixed coordinate system should be chosen in H_0 and Φ'. However, in many cases, this choice is inconvenient or unnatural. Therefore, it is desirable to find a method of construction of the spaces of test and generalized functions in the invariant form without using the coordinate representation of arguments.

Our method will be based on the construction of riggings of the Fock space $\mathfrak{F}(H_0)$ with subsequent application of the Segal isomorphism which maps $\mathfrak{F}(H_0)$ into $L_2(\Phi', \gamma_1)$. Below, it will be shown that the spaces of test functions, emerging as a result of this procedure, are the natural generalization of the spaces $\mathcal{A}_T(\mathbb{R}^\infty)$ and contain these as a special case.

5.1. Riggings of the Fock Space

Using the nuclear rigging (5.1) of the space H_0, one can construct riggings of the corresponding Fock space $\mathfrak{F}(H_0)$.

Let $\Phi = \text{pr lim}_{\tau\in T} H_\tau$ be the projective limit of the family of Hilbert spaces $(H_\tau)_{\tau\in T}$ with scalar products $(\cdot, \cdot)_{H_\tau}$, $\tau \in T$, which is directed by inclusion. One can always assume that $H_\tau \subset H_0$ $(\tau \in T)$ topologically. It is also convenient to suppose that the inequalities $\|\cdot\|_{H_0} \le \|\cdot\|_{H_\tau}$ $(\tau \in T)$ hold. For every $n \in \mathbb{N}$ denote

$$\Phi_c^{\otimes n} = \underset{\tau\in T}{\text{pr lim}}\, H_{\tau,c}^{\otimes n},$$

where $H_{\tau,c}$ is the complexification of H_τ. The scalar product in $H_{\tau,c}$ is denoted by $(\cdot\,,\cdot)_{H_{\tau,c}}$ (for the notations, exploited when the Fock space is constructed, see Section 2). Let us define the n-th symmetric tensor power $\mathfrak{F}_n(\Phi) \subset \mathfrak{F}_n(H_0)$ of the complexification Φ_c of the nuclear space Φ as the projective limit

$$\mathfrak{F}_n(\Phi) = \underset{\tau\in T}{\text{pr lim}}\, \mathfrak{F}_n(H_\tau).$$

Clearly, $\mathfrak{F}_n(\Phi) = P_n(\Phi_c^{\otimes n})$, where P_n is the projector onto the space $\mathfrak{F}_n(H_0)$ acting in $H_{0,c}^{\otimes n}$. Let us introduce the spaces $H_{-\tau}$ as dual to H_τ $(\tau \in T)$ with respect to H_0. Just as described in Section 1 of Chapter 1, we can now obtain a representation of the space Φ' with the Mackey topology $\tau(\Phi', \Phi)$ in the following form: $\Phi' = \text{ind lim}_{\tau\in T} H_{-\tau}$. For $n \in \mathbb{N}$, we set

$$\mathfrak{F}_n(\Phi') = \underset{\tau\in T}{\text{ind lim}}\, \mathfrak{F}_n(H_{-\tau}).$$

Let us complement the collection of spaces introduced above by setting

$$\mathfrak{F}_0(H_0) = \mathfrak{F}_0(\Phi) = \mathfrak{F}_0(\Phi') = \mathbb{C}^1$$

for $n = 0$. Thus, for every $n \in \mathbb{Z}_+$ we have the rigging

$$\mathfrak{F}_n(\Phi') \supset \mathfrak{F}_n(H_0) \supset \mathfrak{F}_n(\Phi). \tag{5.2}$$

In (5.2), the space $\mathfrak{F}_n(\Phi)$ is nuclear as a closed subspace of the nuclear space $\Phi_c^{\otimes n}$, and $\mathfrak{F}_n(\Phi')$ is dual to $\mathfrak{F}_n(\Phi)$ with respect to $\mathfrak{F}_n(H_0)$.

Let us introduce the space $\mathfrak{F}_{\text{fin}}(\Phi) = \oplus_{n=0}^{\infty}\mathfrak{F}_n(\Phi)$ as a topological direct sum of the nuclear spaces $\mathfrak{F}_n(\Phi)$, $n \in \mathbb{Z}_+$. The last statement means that elements of $\mathfrak{F}_{\text{fin}}(\Phi)$ have the form $\varphi = (\varphi_0, \varphi_1, \ldots, \varphi_n, 0, 0, \ldots)$, where $\varphi_k \in \mathfrak{F}_k(\Phi)$, and that the topology of $\mathfrak{F}_{\text{fin}}(\Phi)$ is the strongest locally convex topology on $\mathfrak{F}_{\text{fin}}(\Phi)$ with respect to which all the natural imbeddings

$$\mathfrak{F}_n(\Phi) \ni \varphi_n \mapsto (0, \ldots, 0, \varphi_n, 0, \ldots) \in \mathfrak{F}_{\text{fin}}(\Phi)$$

are continuous. In this topology, the basis of the neighborhoods of zero is formed by the sets

$$u = \left\{ (\lambda_n \varphi_n)_{n=0}^{\infty} \in \bigoplus_{n=0}^{\infty} \mathfrak{F}_n(\Phi) \,\middle|\, \sum_{n=0}^{\infty} |\lambda_n| < 1;\ \varphi_n \in u_n,\ n \in \mathbb{Z}_+ \right\}, \tag{5.3}$$

where $(u_n)_{n=0}^{\infty}$ is an arbitrary sequence of the base neighborhoods of zero in the spaces $\mathfrak{F}_n(\Phi)$, $n \in \mathbb{Z}_+$. We say that a sequence of vectors converges in $\mathfrak{F}_{\text{fin}}(\Phi)$ if these vectors are uniformly finite, and the sequence converges coordinate-wise in each generating space $\mathfrak{F}_n(\Phi)$, $n \in \mathbb{Z}_+$.

The space $\mathfrak{F}_{\text{fin}}(\Phi)$ is nuclear as a topological direct sum of nuclear spaces (see, for example, Schaefer [1], Chapter 3, Section 7). Moreover, it is densely topologically imbedded into the Hilbert direct sum $\mathfrak{F}(H_0) = \oplus_{n=0}^{\infty}\mathfrak{F}_n(H_0)$. The last statement follows from the fact that the imbeddings in the chain (5.2) are dense and topological. Denote by $\mathfrak{F}'_{\text{fin}}(\Phi)$ the space dual to $\mathfrak{F}_{\text{fin}}(\Phi)$ with respect to $\mathfrak{F}(H_0)$. This is the standard fact of the duality theory that $\mathfrak{F}'_{\text{fin}}(\Phi)$ (with the Mackey topology $\tau(\mathfrak{F}'_{\text{fin}}(\Phi), \mathfrak{F}_{\text{fin}}(\Phi))$ coincides with the topological product $\times_{n=0}^{\infty}\mathfrak{F}_n(\Phi')$. Recall that $\times_{n=0}^{\infty}\mathfrak{F}_n(\Phi')$ is the Cartesian product of the sets $\mathfrak{F}_n(\Phi')$, $n \in \mathbb{Z}_+$, and that the topology on this space is defined as the weakest topology on $\times_{n=0}^{\infty}\mathfrak{F}_n(\Phi')$ with respect to which all the projections

$$\mathop{\times}_{n=0}^{\infty} \mathfrak{F}_n(\Phi') \ni \xi = (\xi_n)_{n=0}^{\infty} \mapsto \xi_k \in \mathfrak{F}_k(\Phi') \quad (k \in \mathbb{Z}_+)$$

are continuous. We say that a sequence converges in this topology if it converges coordinate-wise.

We have thus constructed the nuclear rigging

$$\mathfrak{F}'_{\text{fin}}(\Phi) = \mathop{\times}_{n=0}^{\infty} \mathfrak{F}_n(\Phi') \supset \mathfrak{F}(H_0) \supset \bigoplus_{n=0}^{\infty} \mathfrak{F}_n(\Phi) = \mathfrak{F}_{\text{fin}}(\Phi), \tag{5.4}$$

induced by the nuclear rigging (5.1) of the one-particle space H_0. In many cases, the positive space $\mathfrak{F}_{\text{fin}}(\Phi)$ proves to be too narrow (according to its construction, it contains only finite vectors). For example, the coherent states, defined in Section 2 as vectors of the form $\Gamma(\varphi) = (1, \varphi\sqrt{1!}, \ldots, \varphi^{\otimes n}\sqrt{n!},\ldots)$, $\varphi \in \Phi$, do not belong to this space. This difficulty can be overcome as follows. Consider a family of Hilbert spaces $(\mathfrak{F}(H_\tau))_{\tau\in\mathrm{T}}$. We equip $\mathfrak{F}_{\text{fin}}(\Phi)$ with a projective topology with respect to the family of Hilbert spaces $(\mathfrak{F}(H_\tau))_{\tau\in\mathrm{T}}$ and natural imbeddings $\mathfrak{F}_{\text{fin}}(\Phi) \to \mathfrak{F}(H_\tau)$ generated by the imbeddings $\Phi \to H_\tau$, $\tau \in \mathrm{T}$. In other words, we introduce the weakest topology on $\mathfrak{F}_{\text{fin}}(\Phi)$ with respect to which all these imbeddings are continuous. The completion of $\mathfrak{F}_{\text{fin}}(\Phi)$ in this topology is denoted by $\mathfrak{F}(\Phi)$.

Note that in the general case $\mathfrak{F}(\Phi)$ cannot be defined as the projective limit of a family of Hilbert spaces $(\mathfrak{F}(H_\tau))_{\tau\in\mathrm{T}}$. This is so because the fact that the inclusion $H_{\tau_1} \subset H_{\tau_2}$ is topological, generally speaking, does not imply that $\mathfrak{F}(H_{\tau_1}) \subset \mathfrak{F}(H_{\tau_2})$ topologically and, therefore, the norms of the family $(\mathfrak{F}(H_\tau))_{\tau\in\mathrm{T}}$ may be not semi-ordered. However, in the case when this semi-ordering takes place, we have $\mathfrak{F}(\Phi) = \operatorname{pr\,lim}_{\tau\in\mathrm{T}} \mathfrak{F}(H_\tau)$. According to the definition of $\mathfrak{F}(H_\tau)$, for $\varphi \in \mathfrak{F}_{\text{fin}}(\Phi)$ we have

$$\|\varphi\|^2_{\mathfrak{F}(H_\tau)} = \sum_{n=0}^{\infty} \|\varphi_n\|^2_{\mathfrak{F}_n(H_\tau)} = \sum_{n=0}^{\infty} \|\varphi\|^2_{H^{\otimes n}_{\tau,c}} \geq \|\varphi\|^2_{\mathfrak{F}(H_0)}$$

since $\|\cdot\|_{H_0} \leq \|\cdot\|_{H_\tau}$ $(\tau \in \mathrm{T})$. This yields the continuity of the imbedding $\mathfrak{F}(\Phi) \subset \mathfrak{F}(H_0)$.

Our definition of $\mathfrak{F}(\Phi)$ depends on the choice of representation of Φ as a projective limit of a family of Hilbert spaces. As can be easily shown, in order that two families $(H_\tau)_{\tau\in\mathrm{T}}$ and $(\mathcal{H}_\sigma)_{\sigma\in S}$, such that

$$\Phi = \operatorname*{pr\,lim}_{\tau\in\mathrm{T}} H_\tau = \operatorname*{pr\,lim}_{\sigma\in S} \mathcal{H}_\sigma,$$

define the same space $\mathfrak{F}(\Phi)$, it is sufficient that the following conditions be satisfied

$$\forall\ \tau \in \mathrm{T}\ \exists\ \sigma_1, \sigma_2 \in S\colon \|\varphi\|_{\mathcal{H}_{\sigma_1}} \leq \|\varphi\|_{H_\tau} \leq \|\varphi\|_{\mathcal{H}_{\sigma_2}},$$

$$\forall\ \sigma \in S\ \exists\ \tau_1, \tau_2 \in \mathrm{T}\colon \|\varphi\|_{H_{\tau_1}} \leq \|\varphi\|_{\mathcal{H}_\sigma} \leq \|\varphi\|_{H_{\tau_2}}. \tag{5.5}$$

Such families of Hilbert norms are called strictly equivalent. It follows from the definition of the Fock space that (5.5) involves the validity of the inequalities

$$\|\varphi\|_{\mathcal{F}(H_{\tau_1})} \leq \|\varphi\|_{\mathcal{F}(\mathcal{H}_\sigma)} \leq \|\varphi\|_{\mathcal{F}(H_{\tau_2})},$$

$$\|\varphi\|_{\mathcal{F}(\mathcal{H}_{\sigma_1})} \leq \|\varphi\|_{\mathcal{F}(H_\tau)} \leq \|\varphi\|_{\mathcal{F}(\mathcal{H}_{\sigma_2})} \quad (\varphi \in \mathcal{F}_{\mathrm{fin}}(\Phi)),$$

i.e., the families $(\|\cdot\|_{\mathcal{F}(H_\tau)})_{\tau\in\mathrm{T}}$ and $(\|\cdot\|_{\mathcal{F}(\mathcal{H}_\sigma)})_{\sigma\in S}$ of Hilbert norms on $\mathcal{F}(\Phi)$ are strictly equivalent.

The next example shows that in the case when conditions (5.5) are violated, different representations of Φ may lead to different spaces $\mathcal{F}(\Phi)$ even in the case $\dim \Phi = 1$.

Example 5.1. Let $H_0 = \mathbb{R}^1$ with the natural scalar product

$$(f, g)_{H_0} = f \cdot g \quad (f, g \in \mathbb{R}^1).$$

We choose two families of Hilbert spaces $(H_k)_{k=1}^\infty$ and $(\mathcal{H}_k)_{k=1}^\infty$ as follows: For the first, we set $H_k = H_0$, $k \in \mathbb{N}$, and for the second one, we set $\mathcal{H}_k = \mathbb{R}^1$ and introduce another scalar product

$$(f, g)_{\mathcal{H}_k} = kfg \quad (f, g \in \mathbb{R}^1,\ k \in \mathbb{N}).$$

Clearly, we now have $\Phi = \mathbb{R}^1$ with the usual topology of real line. Since for every $n \in \mathbb{N}$, $\mathcal{F}_n(H_0) = \mathcal{F}_n(H_k) = \mathbb{C}^1$ with ordinary scalar product, we have $\mathcal{F}(H_0) = \mathcal{F}(H_k) = l_2(\mathbb{C}^1)$ $(k \in \mathbb{N})$. The space $\mathcal{F}_{\mathrm{fin}}(\Phi)$ is now a collection of finite complex sequences $\mathbb{C}_0^\infty$ and, hence, $\mathcal{F}(\Phi)$ constructed from the family $(H_k)_{k=1}^\infty$ coincides with $l_2(\mathbb{C}^1)$. In the case of the family $(\mathcal{H}_k)_{k=1}^\infty$, we have $\mathcal{F}_n(H_k) = \mathbb{C}^1$ with the scalar product

$$(f_n, g_n)_{\mathcal{F}_n(\mathcal{H}_k)} = k^n f_n \bar{g}_n \quad (f_n, g_n \in \mathbb{C}^1).$$

Therefore,

$$\mathcal{F}(\mathcal{H}_k) = \left\{ f = (f_n)_{n=0}^\infty \in l_2(\mathbb{C}^1) \,\middle|\, \sum_{n=0}^\infty k^n |f_n|^2 = \|f\|^2_{\mathcal{F}(\mathcal{H}_k)} < \infty \right\}.$$

Consequently, in this case the space $\mathfrak{F}_{\text{fin}}(\Phi) \subset \mathfrak{F}(\Phi) \subset l_2(\mathbb{C}^1)$ can be described as follows

$$\mathfrak{F}(\Phi) = \left\{ f = (f_n)_{n=0}^{\infty} \in l_2(\mathbb{C}^1) \,\middle|\, \sum_{n=0}^{\infty} k^n |f_n|^2 < \infty, \ k \in \mathbb{N} \right\}.$$

This means that the strict inclusion $\mathfrak{F}(\Phi)$ takes place.

Note, in addition, that in the first case the space $\mathfrak{F}(\Phi)$ is not nuclear, whereas in the second case each space $\mathfrak{F}(H_{k+1})$ is imbedded quasinuclearly into $\mathfrak{F}(H_k)$ for all $k \in \mathbb{N}$, and this involves the nuclearity of $\mathfrak{F}(\Phi)$.

The properties of the space $\mathfrak{F}(\Phi)$ depend essentially on the choice of representation of Φ in the form of a projective limit. In particular, the nuclearity of Φ, generally speaking, does not imply the nuclearity of $\mathfrak{F}(\Phi)$. However, $\mathfrak{F}(\Phi)$ can be made nuclear by imposing an additional condition on the family $(H_\tau)_{\tau \in \mathrm{T}}$.

Theorem 5.1. *Suppose that the family of Hilbert spaces $(H_\tau)_{\tau\in\mathrm{T}}$, realizing the representation of the nuclear space Φ as a projective limit, satisfies the following condition:*

For any $\tau', \tau'' \in \mathrm{T}$, there exists $\tau''' \in \mathrm{T}$ such that

$$H_{\tau'''} \subset H_{\tau'}, H_{\tau''}, \quad |O_{\tau'''\tau'}| < 1, \quad |O_{\tau'''\tau''}| < 1 \tag{5.6}$$

($O_{\tau'''\tau'} : H_{\tau'''} \to H_{\tau'}$ and $O_{\tau'''\tau'} : H_{\tau'''} \to H_{\tau''}$ are the imbedding operators). Then $\mathfrak{F}(\Phi) = \operatorname{pr} \lim_{\tau\in\mathrm{T}} \mathfrak{F}(H_\tau)$, and the space $\mathfrak{F}(\Phi)$ is nuclear.

Proof. Fix arbitrary $\tau', \tau'' \in \mathrm{T}$, and choose $\tau''' \in \mathrm{T}$ such that (5.6) is satisfied. Since

$$\| O_{\tau'''\tau'} \| \le |O_{\tau'''\tau'}| < 1, \quad \| O_{\tau'''\tau''} \| \le |O_{\tau'''\tau''}| < 1,$$

we obtain the following inequalities for the norms

$$\| \cdot \|_{H_{\tau'}} \le \| \cdot \|_{H_{\tau'''}} \quad \text{and} \quad \| \cdot \|_{H_{\tau''}} \le \| \cdot \|_{H_{\tau'''}}.$$

These inequalities imply that

$$\| \cdot \|_{\mathfrak{F}(H_{\tau'})} \le \| \cdot \|_{\mathfrak{F}(H_{\tau'''})} \quad \text{and} \quad \| \cdot \|_{\mathfrak{F}(H_{\tau''})} \le \| \cdot \|_{\mathfrak{F}(H_{\tau'''})}. \tag{5.7}$$

This means that the family of Hilbert spaces possesses a semi-ordering of norms.

According to the remark made after the definition of $\mathfrak{F}(\Phi)$, in this case we have $\mathfrak{F}(\Phi) = \mathrm{pr}\ \lim_{\tau\in\mathrm{T}} \mathfrak{F}(H_\tau)$.

Denote by $\Gamma(O_{\tau'''\tau'})$ the imbedding operator $\mathfrak{F}(H_{\tau'''}) \to \mathfrak{F}(H_{\tau'})$. By virtue of the definition of the Fock space, we have

$$\Gamma(O_{\tau'''\tau'}) \upharpoonright \mathfrak{F}_n(H_{\tau'''}) = O_{\tau'''\tau'} \otimes \ldots \otimes O_{\tau'''\tau'}: \mathfrak{F}_n(H_{\tau'''}) \to \mathfrak{F}_n(H_{\tau'}) .$$

Therefore,

$$\| \Gamma(O_{\tau'''\tau'}) \| \le 1 + \| O_{\tau'''\tau'} \| + \| O_{\tau'''\tau'} \|^2 + \ldots = (1 - \| O_{\tau'''\tau'} \|)^{-1} < \infty.$$

The quasinuclearity of the imbedding operator

$$\Gamma(O_{\tau'''\tau''}): \mathfrak{F}(H_{\tau'''}) \to \mathfrak{F}(H_{\tau''})$$

can be established similarly. Since $\tau', \tau'' \in \mathrm{T}$ are arbitrary, this proves the nuclearity of $\mathfrak{F}(\Phi)$.

■

Example 5.2. (the rigging by Köthe spaces).

Let $\mathrm{T} \subset [1, \infty)^\infty$ be a set of multi-indices directed in the sense (4.18) and satisfying condition (4.22), i.e.,

$$\forall\ \tau' \in \mathrm{T}\ \exists\ \tau'' \in \mathrm{T}: \sum_{k=1}^{\infty} \tau'_k / \tau''_k < \infty$$

(see Section 4). Just as in Section 4, for $\tau \in \mathrm{T}$, we define the Hilbert spaces $X_\tau^{\pm}$, the space $\mathfrak{X}_\mathrm{T} = \mathrm{pr}\ \lim_{\tau\in\mathrm{T}} X_\tau^{+}$, nuclear by virtue of (4.22), and the space $\mathfrak{X}'_\mathrm{T} = \mathrm{ind}\ \lim_{\tau\in\mathrm{T}} X_\tau^{-}$ dual to $\mathfrak{X}_\mathrm{T}$ with respect to $l_2(\mathbb{R}^1)$. It is now convenient to exploit the scalar product in X_τ^{+} which has the form

$$(x, y)_{X_\tau^+} = \sum_{k=1}^{\infty} x_k\, y_k\, \tau_k .$$

This scalar product differs from the one used before by the factors τ_k substituted in place of $1 + \tau_k$. The corresponding norm in X_τ^{+} is equivalent to the previous one and, moreover, the previous system of norms and the new one are strictly equivalent.

Let us fix the orthonormal basis $(e_k)_{k=1}^{\infty}$ in H_0, and realize H_0 as $l_2(\mathbb{R}^1)$ by use of this basis. Then the chain $\mathfrak{X}'_\mathrm{T} \supset l_2(\mathbb{R}^1) \supset \mathfrak{X}_\mathrm{T}$ can be regarded as a nuclear rigging of H_0

in this realization. As before, $X_\tau^\pm$ denotes the Hilbert spaces emerging when the inverse transition from $l_2(\mathbb{R}^1)$ to H_0 is considered. It is easy to see that the family of Hilbert spaces $(X_\tau^+)_{\tau\in T}$ satisfies the conditions of Theorem 5.1. Indeed, fix arbitrary $\tau', \tau'' \in T$. By using (4.18) and (4.22), we can find $\tau''' \in T$ such that

$$\begin{aligned}&\text{(i)}\quad \tau'_k + 1,\ \tau''_k + 1 < \tau'''_k \quad (k \in \mathbb{N});\\ &\text{(ii)}\quad \sum_{k=1}^{\infty} \tau'_k / \tau''_k < \infty,\quad \sum_{k=1}^{\infty} \tau''_k / \tau'''_k < \infty.\end{aligned} \tag{5.8}$$

Since the sums in (5.8) are finite, one can find $p \in \mathbb{N}$ such that

$$\sum_{k=1}^{\infty} \tau'_k(\tau'''_k + p) < 1,\quad \sum_{k=1}^{\infty} \tau''_k(\tau'''_k + p) < 1.$$

Inductively employing the condition of directness (4.18), we choose $\eta \in T$ such that $\eta_k > \tau'''_k + p$ $(k \in \mathbb{N})$. For the Hilbert norm of the imbedding operator $O_{\eta\tau'} : X_\eta^+ \to X_{\tau'}^+$, we have

$$\| O_{\eta\tau'} \| = \left(\sum_{k=1}^{\infty} \tau'_k / \eta_k \right)^{1/2} \le \left(\sum_{k=1}^{\infty} \tau'_k(\tau'''_k + p) \right)^{1/2} < 1.$$

Similarly, we obtain that $\| O_{\eta\tau''} \| < 1$.

Given the orthonormal basis $(e_k)_{k=1}^{\infty}$ in H_0, there exists the basis of occupation numbers $(e_\alpha)_{\alpha\in\mathbb{Z}_{+,0}^{\infty}}$ in $\mathfrak{F}(H_0)$ which corresponds to it (see Section 2)

$$e_\alpha = \left(\frac{|\alpha|!}{\alpha_1!\ldots\alpha_\nu!} \right)^{1/2} P_{|\alpha|}(e_1^{\otimes\alpha_1} \otimes \ldots \otimes e_\nu^{\otimes\alpha_\nu}),\quad \alpha \in \mathbb{Z}_{+,0}^{\infty}.$$

Since the orthonormal basis in X_τ^+ is formed by the vectors $(\tau_k^{-1/2} e_k)_{k=1}^{\infty}$, the basis in $\mathfrak{F}(X_\tau^+)$ is formed by the vectors

$$\begin{aligned} e_\alpha^{(\tau)} &= \left(\frac{|\alpha|!}{\alpha_1!\ldots\alpha_\nu!} \right)^{1/2} P_{|\alpha|}((\tau_1^{-1/2} e_1)^{\otimes\alpha_1} \otimes \ldots \otimes (\tau_\nu^{-1/2} e_\nu)^{\otimes\alpha_\nu}) \\ &= \tau_1^{-\alpha_1/2} \ldots \tau_\nu^{-\alpha_\nu/2} e_\alpha,\quad \alpha \in \mathbb{Z}_{+,0}^{\infty}. \end{aligned} \tag{5.9}$$

This allows us to describe the Hilbert spaces $\mathfrak{F}(X_\tau^+)$ in terms of the basis of occupation numbers in $\mathfrak{F}(H_0)$, namely

$$\mathcal{F}(X_\tau^+) = \left\{ f = \sum_{\alpha \in \mathbb{Z}_{+,0}^\infty} f_\alpha e_\alpha \| f \|^2_{\mathcal{F}(X_\tau^+)} = \sum_{\alpha \in \mathbb{Z}_{+,0}^\infty} |f_\alpha|^2 \tau_1^{\alpha_1} \dots \tau_\nu^{\alpha_\nu} < \infty \right\}. \quad (5.10)$$

Quite analogously, the basis in $\mathcal{F}(X_\tau^-)$ is formed by the vectors

$$e_\alpha^{(\tau)} = \tau_1^{\alpha_1/2} \dots \tau_\nu^{\alpha_\nu/2} e_\alpha, \quad \alpha \in \mathbb{Z}_{+,0}^\infty,$$

and, therefore,

$$\mathcal{F}(X_\tau^-) = \left\{ \xi = \sum_{\alpha \in \mathbb{Z}_{+,0}^\infty} \xi_\alpha e_\alpha \middle| \| \xi \|^2_{\mathcal{F}(X_\tau^-)} = \sum_{\alpha \in \mathbb{Z}_{+,0}^\infty} |\xi_\alpha|^2 \tau_1^{-\alpha_1} \dots \tau_\nu^{-\alpha_\nu} < \infty \right\}. \quad (5.11)$$

Finally,

$$\mathcal{F}(\mathcal{X}_T) = \operatorname{pr} \lim_{\tau \in T} \mathcal{F}(X_\tau^+)$$

$$= \left\{ f = \sum_{\alpha \in \mathbb{Z}_{+,0}^\infty} f_\alpha e_\alpha \middle| \forall\, \tau \in T \sum_{\alpha \in \mathbb{Z}_{+,0}^\infty} |f_\alpha|^2 \tau_1^{\alpha_1} \dots \tau_\nu^{\alpha_\nu} < \infty \right\},$$

and the space dual to $\mathcal{F}(\mathcal{X}_T)$ with respect to $\mathcal{F}(H_0)$ admits the following representation

$$\mathcal{F}'(\mathcal{X}_T) = \operatorname{ind} \lim_{\tau \in T} \mathcal{F}(X_\tau^-)$$

$$= \left\{ \xi = \sum_{\alpha \in \mathbb{Z}_{+,0}^\infty} \xi_\alpha e_\alpha \middle| \exists\, \tau \in T: \sum_{\alpha \in \mathbb{Z}_{+,0}^\infty} |\xi_\alpha|^2 \tau_1^{-\alpha_1} \dots \tau_\nu^{-\alpha_\nu} < \infty \right\}.$$

Thus, we have constructed the nuclear rigging

$$\mathcal{F}'(\mathcal{X}_T) \supset \mathcal{F}(H_0) \supset \mathcal{F}(\mathcal{X}_T), \quad (5.12)$$

which corresponds to the given nuclear rigging $\mathcal{X}'_T \supset H_0 \supset \mathcal{X}_T$ of the space H_0 by the Köthe spaces $\mathcal{X}_T$ and $\mathcal{X}'_T$.

Our consideration can be summarized as follows:

For the given nuclear rigging $\Phi' \supset H_0 \supset \Phi$ of the space H_0, by use of the representation of Φ as a projective limit of the Hilbert spaces $(H_\tau)_{\tau \in T}$ satisfying condition (5.6), we have constructed the chain

$$\mathcal{F}'_{\text{fin}}(\Phi) \supset \mathcal{F}'(\Phi) \supset \mathcal{F}(H_0) \supset \mathcal{F}(\Phi) \supset \mathcal{F}_{\text{fin}}(\Phi), \quad (5.13)$$

where

$$\mathfrak{F}_{\text{fin}}(\Phi) = \bigoplus_{n=0}^{\infty} \mathfrak{F}_n(\Phi) \text{ and } \mathfrak{F}(\Phi) = \operatorname{pr}\lim_{\tau\in T} \mathfrak{F}(H_\tau)$$

are nuclear spaces, and $\mathfrak{F}'_{\text{fin}}(\Phi)$ *and* $\mathfrak{F}'(\Phi)$ *are the spaces dual to these with respect to* $\mathfrak{F}(H_0)$. *They are given by*

$$\mathfrak{F}'_{\text{fin}}(\Phi) = \mathop{\times}_{n=0}^{\infty} \mathfrak{F}_n(\Phi') \text{ and } \mathfrak{F}'(\Phi) = \operatorname{ind}\lim_{\tau\in T} \mathfrak{F}(H_{-\tau}).$$

Furthermore, the space $\mathfrak{F}_{\text{fin}}(\Phi)$ *is uniquely determined for a given* Φ, *whereas* $\mathfrak{F}(\Phi)$ *depends on the concrete form of the representation of* Φ *as a projective limit.*

5.2. Functional Realizations of Riggings of the Fock Space

On Φ', we introduce a measure γ_1 canonically connected with the Hilbert space H_0. Recall that γ_1 is determined by its Fourier transform

$$\tilde{\gamma}_1(\varphi) = \exp(-1/4 \|\varphi\|^2_{H_0}) \quad (\varphi \in \Phi)$$

(see Subsection 1.8). The Segal isomorphism $I_1 \colon \mathfrak{F}(H_0) \to L_2(\Phi', \gamma_1)$ realizes the unitary mapping of $\mathfrak{F}(H_0)$ onto $L_2(\Phi', \gamma_1)$ which associates a Fock vector $f = (f_0, f_1, \ldots) \in \mathfrak{F}(H_0)$ with the function

$$(I_1 f)(x) = \sum_{n=0}^{\infty} \frac{2^{n/2}}{\sqrt{n!}} :(f_n, x^{\otimes n})_{H_0^{\otimes n}}: \quad (x \in \Phi').$$

This isomorphism has been studied in Subsection 2.3. In the notations used there, we now have $S = 1$ and, thus, $H_S = H_0$ and $\gamma_S = \gamma_1$. Our aim is to investigate the image of the chain (5.13) under the Segal isomorphism. On the one hand, this enables us to obtain some useful riggings of the space $L_2(\Phi', \gamma_1)$, and on the other hand, it will be necessary in what follows when studying the operators in the Fock space and their images in $L_2(\Phi', \gamma_1)$.

First, let us consider the functional realization of the chain (5.4). In $\mathfrak{F}_{\text{fin}}(\Phi)$, we select the dense subset

$$\mathring{\mathfrak{F}}_{\text{fin}}(\Phi) = \bigoplus_{n=0}^{\infty} \mathring{\mathfrak{F}}_n(\Phi),$$

where $\overset{\circ}{\mathfrak{F}}$ is the n-th algebraic symmetric tensor power of Φ_c. It has been shown in Subsection 2.3 that $I_1(\overset{\circ}{\mathfrak{F}}_{\text{fin}}(\Phi)) = \mathcal{P}_{\text{cyl}}(\Phi')$. Applying the operator I_1 to the spaces of the chain (5.4), we obtain the rigging

$$I_1(\mathfrak{F}'_{\text{fin}}(\Phi)) \supset L_2(\Phi', \gamma_1) \supset I_1(\mathfrak{F}_{\text{fin}}(\Phi)) \supset \mathcal{P}_{\text{cyl}}(\Phi'). \tag{5.14}$$

Since

$$\mathfrak{F}_{\text{fin}}(\Phi) = \bigoplus_{n=0}^{\infty} \mathfrak{F}_n(\Phi) \ni \varphi = (\varphi_0, \varphi_1, \dots, \varphi_m, 0, 0, \dots),\ \varphi_n \in \mathfrak{F}_n(\Phi)$$

$$(n = 0, 1, \dots, m;\ m \in \mathbb{N}),$$

the set $I_1(\mathfrak{F}_{\text{fin}}(\Phi))$ consists of the functions

$$u(x) = \sum_{k=0}^{m} \frac{2^{k/2}}{\sqrt{k!}} :(\varphi_k, x^{\otimes k})_{H_0^{\otimes k}}: \quad (x \in \Phi',\ \varphi_k \in \mathfrak{F}_k(\Phi)).$$

Hence, in order to describe $I_1(\mathfrak{F}_{\text{fin}}(\Phi))$, we must study the properties of the Wick monomials with smooth kernels.

Lemma 5.1. *(i) For $\varphi_n \in \mathfrak{F}_n(\Phi)$, the Wick monomial $:(\varphi_n, x^{\otimes n})_{H_0^{\otimes n}}:$ $(x \in \Phi')$ is continuous on Φ', i.e., it is continuous on every Hilbert space $H_{-\tau}$, $\tau \in \mathrm{T}$.*

(ii) For every $\tau \in \mathrm{T}$, the following estimate holds

$$|:(\varphi_n, x^{\otimes n})_{H_0^{\otimes n}}:| \le c^n \sqrt{n!}\, \| O_{\tau'\tau} \| \, \|\varphi_n\|_{H_{\tau',c}^{\otimes n}} (1 + \|x\|_{H_{-\tau}})^n, \tag{5.15}$$

where $c > 0$ is an absolute constant, and $\tau' \in \mathrm{T}$ is chosen according to the condition of the quasinuclearity of the imbedding operator $O_{\tau'\tau} : H_{\tau'} \to H_\tau$.

Proof. Let $\varphi_n = \varphi^{\otimes n}$, where $\varphi \in \Phi$. Equality (2.24) implies that

$$:(\varphi^{\otimes n}, x^{\otimes n})_{H_0^{\otimes n}}: \; = \; :(\varphi, x)_{H_0}^n: \; = 2^{-n} \|\varphi\|_{H_0}^n H_n(\|\varphi\|_{H_0}^{-1} (\varphi, x)_{H_0}), \tag{5.16}$$

and this yields the continuity of the Wick monomial on Φ'. Taking the explicit form of the Hermite polynomials into account

$$H_n(t) = n! \sum_{m=0}^{[n/2]} (-1)^m \frac{(2t)^{n-2m}}{m!(n-2m!)} \quad (t \in \mathbb{R}^1;\ n \in \mathbb{Z}_+),$$

we obtain

$$|:(\varphi, x)_{H_0}^n :| \le \frac{n!}{2^n}\sum_{m=0}^{[n/2]} \frac{2^{n-2m}}{m!(n-2m!)} |(\varphi, x)_{H_0}|^{n-2m} \|\varphi\|_0^{2m}$$

$$\le \|\varphi\|_{H_\tau}^n \sum_{m=0}^{[n/2]} \frac{n!}{2^{2m}\, m!(n-2m!)} \|x\|_{H_{-\tau}}^{n-2m} \quad (x \in H_{-\tau}). \tag{5.17}$$

Here we have used the inequality $\|\cdot\|_{H_0} \le \|\cdot\|_{H_\tau}$ $(\tau \in \mathrm{T})$, by virtue of which

$$\|\varphi\|_{H_\tau}^{n-2m} \|\varphi\|_{H_0}^{2m} \le \|\varphi\|_{H_\tau}^n .$$

For $\varphi_k = P_n(\psi_1 \otimes \ldots \otimes \psi_n)$, $\psi_k \in \Phi$, and $\|\psi_k\|_{H_\tau} = 1$ $(k = 1, \ldots, n)$, the polarization identity and (5.17) imply

$$|:(\varphi_n, x^{\otimes n})_{H_0^{\otimes n}} :| = \left| :\prod_{k=1}^{n} (\psi_k, x)_{H_0} :\right|$$

$$\le \frac{1}{n!}\sum_{k=0}^{n} \sum_{i_1<\ldots<i_k} |:(\psi_{i_1} + \ldots + \psi_{i_k}, x)_{H_0}^n :|$$

$$\le \frac{1}{n!}\sum_{k=0}^{n} \sum_{i_1<\ldots<i_k} \|\psi_{i_1} + \ldots + \psi_{i_k}\|_{H_\tau}^n \left(\sum_{m=0}^{[n/2]} \frac{\|x\|_{H_{-\tau}}^{n-2m}}{2^{2m}\, m!(n-2m!)}\right)$$

$$\le \frac{n^n}{n!}\sum_{k=0}^{n} C_n^k \left(\sum_{m=0}^{[n/2]} \frac{n!}{2^{2m}\, m!(n-2m!)} \|x\|_{H_{-\tau}}^{n-2m}\right)$$

$$\le (2e)^n \left(\sum_{m=0}^{[n/2]} \frac{n!}{2^{2m}\, m!(n-2m!)} \|x\|_{H_{-\tau}}^{n-2m}\right) \quad (x \in H_{-\tau}). \tag{5.18}$$

Using the equality

$$:\prod_{k=1}^{n} (\psi_k, x)_{H_0}: = \prod_{k=1}^{n} \|\varphi_k\|_{H_\tau} :\prod_{k=1}^{n} (\psi_k / \|\psi_k\|_{H_\tau}, x)_{H_0}:,$$

and (5.18), we get

$$|:(\psi_1 \otimes \ldots \otimes \psi_n, x^{\otimes n})_{H_0^{\otimes n}} :|$$

$$\le (2e)^n \| \psi_1 \otimes \ldots \otimes \psi_n \|_{H_\tau^{\otimes n}} \left(\sum_{m=0}^{[n/2]} \frac{n!}{2^{2m}\, m!(n-2m!)} \| x \|_{H_{-\tau}}^{n-2m} \right)$$

for all $\psi_1, \ldots, \psi_n \in \Phi$ and $x \in H_{-\tau}$.

Let us fix an arbitrary $\varphi_n \in \mathfrak{F}_n(\Phi)$. Given $\tau \in T$, we choose $\tau' \in T$, for which $H_{\tau'}$ is quasinuclearly imbedded into H_τ. In the Hilbert space $H_{\tau'}$ we choose an orthonormal basis $(e_j)_{j=1}^\infty \subset \Phi$. Then the following expansion in the basis $(e_{j_1} \otimes \cdots \otimes e_{j_n})_{j_1, \ldots, j_n = 1}^\infty$ in $H_{\tau', c}^{\otimes n}$ holds

$$\varphi_n = \sum_{j_1, \ldots, j_n = 1}^\infty c_{j_1 \ldots j_n} e_{j_1} \otimes \ldots \otimes e_{j_n}.$$

For the functions $:(\varphi_n, x^{\otimes n})_{H_0^{\otimes n}}:$, this expansion involves the following representation (as a series convergent in $L_2(\Phi', \gamma_1)$):

$$:(\varphi_n, x^{\otimes n})_{H_0^{\otimes n}}: = \sum_{j_1, \ldots, j_n = 1}^\infty c_{j_1 \ldots j_n} :(e_{j_1} \otimes \ldots \otimes e_{j_n}, x^{\otimes n})_{H_0^{\otimes n}}:. \tag{5.19}$$

Let us show that the series (5.19) converges uniformly on each ball

$$\tilde{B}_{R,-\tau}(0) = \{ x \in H_{-\tau} \mid \| x \|_{H_{-\tau}} \le R \}$$

(this fact yields the continuity of $:(\varphi_n, x^{\otimes n})_{H_0^{\otimes n}}:$ on $H_{-\tau}$). To prove this assertion, it suffices to employ the following estimate

$$\left| \sum_{j_1, \ldots, j_n = 1}^\infty c_{j_1 \ldots j_n} :(e_{j_1} \otimes \ldots \otimes e_{j_n}, x^{\otimes n})_{H_0^{\otimes n}}: \right| \le \left(\sum_{j_1, \ldots, j_n = 1}^\infty | c_{j_1 \ldots j_n} |^2 \right)^{1/2}$$

$$\times \left(\sum_{j_1, \ldots, j_n = 1}^\infty | :(e_{j_1} \otimes \ldots \otimes e_{j_n}, x^{\otimes n})_{H_0^{\otimes n}}: |^2 \right)^{1/2} \le (2e)^n \| \varphi_n \|_{H_{\tau', c}^{\otimes n}}$$

$$\times \left\{ \sum_{j_1, \ldots, j_n = 1}^\infty \| e_{j_1} \otimes \ldots \otimes e_{j_n} \|_{H_\tau^{\otimes n}}^2 \left(\sum_{m=0}^{[n/2]} \frac{n!}{2^{2m}\, m!(n-2m!)} \| x \|_{H_{-\tau}}^{n-2m} \right)^2 \right\}^{1/2}$$

$$\le (2e)^n \left| O_{\tau' \tau} \right|^n \| \varphi_n \|_{H_{\tau', c}^{\otimes n}} \sum_{m=0}^{[n/2]} \frac{n!}{2^{2m}\, m!(n-2m!)} \| x \|_{H_{-\tau}}^{n-2m}. \tag{5.20}$$

Since $\tau \in \mathrm{T}$ is arbitrary, $:(\varphi_n, x^{\otimes n})_{H_0^{\otimes n}}:$ is continuous on Φ'. The inequality

$$\sum_{m=0}^{[n/2]} \frac{n!}{2^{2m}\, m!(n-2m!)} |t|^{n-2m} \le e\sqrt{n!}\,(1+|t|)^n \quad (t \in \mathbb{R}^1)$$

allows us to continue (5.20) and, as a result, we obtain

$$|:(\varphi_n, x^{\otimes n})_{H_0^{\otimes n}}:| \le e^n \sqrt{n!}\, \| O_{\tau'\tau} \|^n \|\varphi_n\|_{H_{\tau',c}^{\otimes n}} (1+\|x\|_{H_{-\tau}})^n \quad (x \in H_{-\tau})$$

with the absolute constant $c > 0$.

■

Corollary 1. *The inclusion*

$$I_1(\mathfrak{F}_{\mathrm{fin}}(\Phi)) \subset \mathcal{P}(\Phi')$$

takes place. In particular, for $\varphi_k \in \Phi_c^{\otimes k}$, *we have*

$$:(\varphi_k, x^{\otimes k})_{H_0^{\otimes k}}: \;=\; \sum_{p=0}^{k} (\psi_p, x^{\otimes p})_{H_0^{\otimes p}}, \quad \psi_p \in \Phi_c^{\otimes p},\ x \in \Phi'.$$

Proof. Every element of $I_1(\mathfrak{F}_{\mathrm{fin}}(\Phi))$ has the form

$$\sum_{k=0}^{n} :(\varphi_k, x^{\otimes k})_{H_0^{\otimes k}}:, \quad \varphi_k \in \Phi_c^{\otimes k}, \quad k = 0, \dots, n;\ x \in \Phi'.$$

Moreover, the Wick monomial $:(\varphi_k, x^{\otimes k})_{H_0^{\otimes k}}:$ admits the expansion (5.19) with all the terms belonging to $\mathcal{P}_{\mathrm{cyl}}(\Phi')$ and having uniformly bounded powers. Therefore, $:(\varphi_k, x^{\otimes k})_{H_0^{\otimes k}}:$ is the polynomial on Φ' whose continuity is established by Lemma 5.1. Further, being a sum of homogeneous forms, each continuous polynomial on Φ' can be represented in the form (5.21). This can be shown by use of the kernel theorem and the nuclearity of Φ (see Section 2, Chapter 1).

■

Corollary 2. *For* $\varphi_n \in \Phi_c^{\otimes n}$ *the polynomial* $:(\varphi_n, x^{\otimes n})_{H_0^{\otimes n}}:$ $(x \in \Phi')$ *admits a natural extension to* Φ'_c *for which the estimate*

$$|:(\varphi_n, x^{\otimes n})_{H_0^{\otimes n}}:| \le c^n \sqrt{n!}\, \| O_{\tau'\tau} \|^n \|\varphi_n\|_{H_{\tau',c}^{\otimes n}} (1+\|z\|_{H_{-\tau,c}})^n$$

holds under conditions (ii) of Lemma 5.1.

Proof. The polynomial

$$:(\varphi_n, x^{\otimes n})_{H_0^{\otimes n}}: \;=\; \sum_{k=0}^{n} (\psi_k, x^{\otimes k})_{H_0^{\otimes k}} \quad (x \in \Phi';\ \psi_k \in \Phi_c^{\otimes k},\ \ k = 0, \ldots, n)$$

can be extended to Φ_c' by the formula

$$:(\varphi_n, z^{\otimes n})_{H_0^{\otimes n}}: \;=\; \sum_{k=0}^{n} (\psi_k, z^{\otimes k})_{H_0^{\otimes k}} \quad (z \in \Phi_c').$$

Here, as before, $(\psi_k, z^{\otimes k})_{H_0^{\otimes k}}$ is the two-linear extension of the scalar product in $H_0^{\otimes k}$. Clearly, the estimates (5.17) and (5.18) can be also established in the case $z \in \Phi_c'$. This enables us to estimate the sum of the series

$$:(\varphi_n, z^{\otimes n})_{H_0^{\otimes n}}: \;=\; \sum_{j_1,\ldots,j_n=1}^{\infty} c_{j_1 \ldots j_n} :(e_{j_1} \otimes \ldots \otimes e_{j_n}, z^{\otimes n})_{H_0^{\otimes n}}:$$

just like has been done above. As a result, we obtain the required extension of the estimate (5.15).

■

We shall need one more functional realization of the Fock space connected with the infinite-dimensional analogue of the Bargman space. First, let us consider the case when the space is one-dimensional.

Example 5.3. (the Bargman space).
Let $H_0 = \mathbb{R}^1$, then $\mathfrak{F}(H_0) = l_2(\mathbb{C}^1)$. On $\mathbb{C}^1$ we introduce the measure

$$d\nu_1(z) \;=\; \pi^{-1} \exp(-|z|^2)\, dx\, dy \quad (z = x + iy \in \mathbb{C}^1).$$

As can be easily shown by direct calculation, the functions $(z^k / \sqrt{k!})_{k=0}^{\infty}$ $(0! = 1)$ form an orthonormal (but not complete!) system of functions in the space $L_2(\mathbb{C}^1, \nu_1)$. Denote by $F(\mathbb{C}^1)$ the subspace of $L_2(\mathbb{C}^1, \nu_1)$ spanned by the vectors of this system. The correspondence

$$l_2(\mathbb{C}^1) \ni (f_k)_{k=0}^{\infty} \mapsto f(z) \;=\; \sum_{k=0}^{\infty} f_k \frac{z^k}{\sqrt{k!}} \in F(\mathbb{C}^1) \tag{5.21}$$

is a unitary isomorphism of $\mathfrak{F}(H_0)$ and the Hilbert space $F(\mathbb{C}^1)$ which is called the Bargman space. It is easy to see that the functions $f \in F(\mathbb{C}^1)$ are entire analytic functions. Below, we extend the correspondence (5.21) to the infinite-dimensional case.

On $\Phi'_c = \Phi' \oplus i\Phi' \ni z = x + iy;\ x, y \in \Phi'$, we introduce a Gaussian measure ν_1 canonically connected with the Hilbert space $H_{0,c}$, namely

$$\tilde{\nu}_1(\psi) = \exp(-1/4 \|\psi\|^2_{H_{0,c}} \quad (\psi \in \Phi_c)$$

(see Subsection 2.1). In other words, under the canonical isomorphism $\Phi'_c \ni z = x + iy \mapsto (x, y) \in \Phi' \times \Phi'$, the measure ν_1 turns into $\nu_1 \times \nu_1$. For an arbitrary vector $\varphi_n \in \mathfrak{F}_n(\Phi)$, consider the continuous n-linear homogeneous form

$$\varphi_n(z) = \left(\varphi_n, \frac{z^{\otimes n}}{\sqrt{n!}}\right)_{H_0^{\otimes n}} \quad (z \in \Phi'_c)$$

defined as a result of the two-linear extension of the scalar product in $H_0^{\otimes n}$.

Lemma 5.2. *For $\varphi_n \in \mathfrak{F}_n(\Phi)$, we have $\varphi_n(\cdot) \in L_2(\Phi'_c, \nu_1)$. Moreover,*

$$(\varphi_n(\cdot), \psi_m(\cdot))_{L_2(\Phi'_c, \nu_1)} = \delta_{nm} \|\varphi_n\|^2_{H_{0,c}^{\otimes n}}. \tag{5.22}$$

Proof. Let $(e_j)_{j=1}^{\infty}$ be the orthonormal basis in H_0 consisting of the elements of Φ. In the space $\mathfrak{F}(H_0)$, there exists the basis of occupation numbers $(e_\alpha)_{\alpha \in \mathbb{Z}_{+,0}^{\infty}}$ (defined in Subsection 2.3) which corresponds to $(e_j)_{j=1}^{\infty}$. For $\varphi_n \in \mathfrak{F}_n(\Phi)$, we have

$$\varphi_n = \sum_{|\alpha|=n} \varphi_n^{(\alpha)} e_\alpha .$$

It follows from the choice of the basis in H_0 that $\tilde{\nu}_1(\psi)$ $(\psi \in \Phi_c)$ can be expanded in the infinite product and, hence, that ν_1 can be realized as a product measure (see Subsection 1.2):

$$d\nu_1(z) \mapsto \mathop{\times}_{k=1}^{\infty} d\nu_1(z_k), \quad d\nu_1(z_k) = \pi^{-1} \exp(-|z_k|^2) dx_k\, dy_k = d\gamma_1(x_k)\, d\gamma_1(y_k),$$

where $z_k = (z, e_k)_{H_0} = x_k + i y_k$ $(k \in \mathbb{N})$. Taking into account the explicit form of the basis vectors $(e_\alpha)_{\alpha \in \mathbb{Z}_{+,0}^{\infty}}$, we get

$$\varphi_n(z) = \sum_{|\alpha|=n} \frac{\varphi_n^{(\alpha)}}{\sqrt{\alpha_1!\ldots\alpha_\nu!}} z_1^{\alpha_1}\ldots z_\nu^{\alpha_\nu} \quad (z \in \Phi_c'). \tag{5.23}$$

According to Example 5.3, the functions $\left(z_k^{\alpha_k}/\sqrt{\alpha_k!}\right)_{\alpha_k=0}^{\infty}$ form orthonormal systems of functions in each space $L_2(\mathbb{C}^1, \nu_1(z_k))$. Consequently, functions of the form

$$z_1^{\alpha_1}\ldots z_\nu^{\alpha_\nu}/\sqrt{\alpha_1!\ldots\alpha_\nu!} \quad (\alpha \in \mathbb{Z}_{+,0}^{\infty})$$

are orthonormal in $L_2(\Phi_c', \nu_1)$. This involves the orthogonality of $\varphi_n(\cdot)$ and $\varphi_m(\cdot)$ in $L_2(\Phi_c', \nu_1)$ for $n \neq m$ and the equality

$$\|\varphi_n(\cdot)\|^2_{L_2(\Phi_c', \nu_1)} = \sum_{|\alpha|=n} |\varphi_n^{(\alpha)}|^2 = \|\varphi_n\|^2_{H_{0,c}^{\otimes n}}.$$

■

Employing the approximation of vectors $f_n \in \mathcal{F}(H_0)$ by elements of $\mathcal{F}_n(\Phi)$ and Lemma 5.2, we obtain the following statement.

Corollary 1. *For all $f_n \in \mathcal{F}(H_0)$, the functions*

$$f_n(z) = \left(f_n, \frac{z^{\otimes n}}{\sqrt{n!}} \right)_{H_0^{\otimes n}}$$

are defined as measurable polynomials from $L_2(\Phi_c', \nu_1)$, and equality (5.22) remains valid.

■

With each Fock vector $f = (f_0, f_1, \ldots) \in \mathcal{F}(H_0)$, we associate the function

$$(J_1 f)(z) = \sum_{n=0}^{\infty} f_n(z) \quad (z \in \Phi_c') \tag{5.24}$$

which belongs to $L_2(\Phi_c', \nu_1)$ by virtue of Lemma 5.2 (the mapping J_1 is isometric). Denote by $F(H_{0,c})$ the closed subset of $L_2(\Phi_c', \nu_1)$ onto which the operator J_1 maps. It is easy to see that *the series (5.24) converges uniformly on each ball in $H_{0,c} \subset \Phi_c'$ and defines a locally bounded function.*

Therefore, $(J_1 f)(z)$ is an entire function of the bounded type. In fact, this follows from the estimate

$$|(J_1 f)(z)| \le \sum_{n=0}^{\infty} \left| \left(f_n, \frac{z^{\otimes n}}{\sqrt{n!}} \right)_{H_0^{\otimes n}} \right| \le \sum_{n=0}^{\infty} \|f_n\|_{H_{0,c}^{\otimes n}} \|z\|_{H_{0,c}}^n / \sqrt{n!}$$

$$\le \|z\|_{\mathfrak{F}(H_0)} \exp(1/2 \, \|z\|_{H_{0,c}^{\otimes n}}^2) \quad (z \in H_{0,c}).$$

■

Taking what has been said above into account, we shall call the space $F(H_{0,c})$ a holomorphic realization of the Fock space $\mathfrak{F}(H_0)$.

Lemma 5.3. *The equality*

$$J_1(\mathfrak{F}_{\text{fin}}(\Phi)) = \mathcal{P}(\Phi_c')$$

holds.

Proof. For $\varphi \in \mathfrak{F}_{\text{fin}}(\Phi)$ we have $\varphi = (\varphi_0, \dots, \varphi_n, 0, 0 \dots)$, where $\varphi_k \in \mathfrak{F}_k(\Phi)$. Hence, the function

$$(J_1 \varphi)(z) = \sum_{k=0}^{n} \left(\varphi_k, \frac{z^{\otimes k}}{\sqrt{k!}} \right)_{H_0^{\otimes k}} \quad (z \in \Phi_c')$$

is a continuous polynomial on Φ_c'. It suffices to show that any polynomial $p \in \mathcal{P}(\Phi_c')$ can be represented in such a way. For this purpose, we consider the following expansion

$$p(z) = \sum_{k=0}^{n} p_k(z),$$

where $p_k(z)$ are continuous k-linear homogeneous forms on Φ_c'. By virtue of the kernel theorem (Section 2 of Chapter 1), we have

$$p_k(z) = (q_k, z^{\otimes k})_{H_0^{\otimes k}}, \qquad q_k \in \Phi_c^{\otimes k} \ (k = 1, \dots, n).$$

The symmetrization operator $P_k : H_{0,c}^{\otimes k} \mapsto \mathfrak{F}_k(H_0)$ obviously has the continuous restriction $P_k: \Phi_c^{\otimes k} \mapsto \mathfrak{F}_k(\Phi)$. Therefore,

$$p_k(z) = (q_k, z^{\otimes k})_{H_0^{\otimes k}} = (q_k, P_k z^{\otimes k})_{H_0^{\otimes k}} = (P_k q_k, z^{\otimes k})_{H_0^{\otimes k}}$$

$$= (\tilde{q}_k, z^{\otimes k})_{H_0^{\otimes k}} \quad (\tilde{q}_k \in \mathfrak{F}_k(\Phi)).$$

Consequently, the polynomial $p \in \mathcal{P}(\Phi'_c)$ can be associated with the vector

$$q = (\tilde{q}_0, \tilde{q}_1, \dots, \tilde{q}_n, 0, 0 \dots) \in \mathcal{F}_{\text{fin}}(\Phi)$$

for which $(J_1 \varphi)(z) = p(z)$ $(z \in \Phi'_c)$.

■

The mapping J_1 admits a natural extension to $\mathcal{F}'_{\text{fin}}(\Phi)$. Indeed, with a vector $\xi = (\xi_0, \xi_1, \dots) \in \mathcal{F}'_{\text{fin}}(\Phi)$, we associate the formal power series

$$(J_1 \xi)(z) = \sum_{k=0}^{\infty} \left(\xi_k, \frac{z^{\otimes k}}{\sqrt{k!}} \right)_{H_0^{\otimes k}} \quad (z \in \Phi_c), \tag{5.25}$$

where each term is a k-linear homogeneous form on Φ_c. Denote the collection of formal series of the form (5.25) by $\mathcal{P}'(\Phi_c)$; this notation is reasonable, because the space dual to $\mathcal{P}'(\Phi_c)$ (with the natural nuclear topology) coincides with $\mathcal{P}'(\Phi_c)$. Under the holomorphic realization of the Fock space, we obtain the following image of the rigging (5.4):

$$\mathcal{P}'(\Phi_c) \supset F(H_{0,c}) \supset \mathcal{P}'(\Phi'_c), \tag{5.26}$$

where the positive space and the zero space consist of certain entire functions on Φ'_c and $H_{0,c}$, respectively, and the negative space consists of formal power series defined on Φ_c. The infinite-dimensional effect can be formulated as follows: When the "smoothness" of the elements of this chain decreases, the set, on which the corresponding functions are considered, becomes narrower.

5.3. Relation Between Functional Realizations of the Fock Space

We have two realizations of $\mathcal{F}(H_0)$ as a Hilbert space of functions. Denote by G the unitary operator which is generated by the operators I_1 and J_1 and maps $L_2(\Phi', \gamma_1)$ onto $F(H_{0,c})$ isomorphically. The inverse operator G^{-1} sets the isometric isomorphism of $F(H_{0,c})$ and $L_2(\Phi', \gamma_1)$. It follows from the definition of the Segal isomorphism that each Fock vector $f = (f_0, f_1, \dots) \in \mathcal{F}(H_0)$ is associated with the function

$$(I_1 f)(x) = \sum_{n=0}^{\infty} \frac{2^{n/2}}{\sqrt{n!}} :(f_n, x^{\otimes n})_{H_0^{\otimes n}}: \in L_2(\Phi', \gamma_1)$$

(see (2.28)). Under the holomorphic realization this vector turns into the entire function

$$(I_1 f)(x) = \sum_{n=0}^{\infty} \frac{1}{\sqrt{n!}} (f_n, z^{\otimes n})_{H_0^{\otimes n}} \in F(H_{0,c}).$$

Therefore, with the help of the correspondence

$$:(f_n, x^{\otimes n})_{H_0^{\otimes n}}: \mapsto 2^{-n/2} (f_n, z^{\otimes n})_{H_0^{\otimes n}}, \tag{5.27}$$

the operator G can be determined on the set of smooth Wick monomials total in $L_2(\Phi', \gamma_1)$, and being defined on this set, it can be extended to the whole of $L_2(\Phi', \gamma_1)$ by unitarity.

Let us obtain the useful representation for the operators G and G^{-1}. To do this, we introduce the function

$$G(x, z) = \exp\left(\sqrt{2}\,(x, z)_{H_0} - (z, z)_{H_0}\right) \tag{5.28}$$

for $x \in H_0$ and $z \in H_{0,c}$. Equality (2.32) implies that for every $z \in H_{0,c}$, $G(\cdot, z)$ can be extended to Φ' as a measurable function and, moreover,

$$G(x, z) = I_1\,\Gamma(z) = \sum_{n=0}^{\infty} \frac{2^{n/2}}{\sqrt{n!}} :(z, x)_{H_0}^n: \in L_2(\Phi', \gamma_1).$$

In the case when $z \in \Phi'_c$, the vector

$$\Gamma(z) = \left(1, \frac{z}{\sqrt{1!}}, \ldots, \frac{z^{\otimes n}}{\sqrt{n!}}, \ldots\right)$$

belongs to $\mathfrak{F}'_{\text{fin}}(\Phi)$. Therefore, $I_1\,\Gamma(z) = G(\cdot, z)$ is a generalized function from $I_1(\mathfrak{F}'_{\text{fin}}(\Phi)) \supset L_2(\Phi', \gamma_1)$. On the other hand, for fixed $x \in \Phi'$, the function $G(x, z)$ is analytic on Φ_c and, hence, belongs to the space of generalized functions $\mathcal{P}'(\Phi_c) \supset F(H_{0,c})$.

The next theorem gives the representation of G and G^{-1} as "integral" operators with the generalized kernel $G(\cdot, \cdot)$.

Theorem 5.2. *The following equalities hold*

$$(Gu)(z) = (G(\cdot, z), \bar{u}(\cdot))_{L_2(\Phi', \gamma_1)} \quad (u \in I_1(\mathfrak{F}'_{\text{fin}}(\Phi))), \tag{5.29}$$

$$(G^{-1}v)(z) = (G(x, \cdot), \bar{v}(\cdot))_{L_2(\Phi'_c, \nu_1)} \quad (v \in J_1(\mathfrak{F}'_{\text{fin}}(\Phi)) = \mathcal{P}(\Phi'_c)). \tag{5.30}$$

Furthermore, for $u \in \mathcal{P}(\Phi')$ *we have*

$$(Gu)(z) = \int_{\Phi'} u(x + z/\sqrt{2})\, d\gamma_1(x) \quad (z \in \Phi'_c). \tag{5.31}$$

Proof. Since equalities (5.29) and (5.30) can be established analogously, we shall prove only one of these, namely (5.29). Let

$$\varphi = (\varphi_0, \varphi_1, \dots, \varphi_n, 0, 0, \dots) \in \mathfrak{F}_{\text{fin}}(\Phi),$$

and let

$$u(x) = (I_1 \varphi)(x) = \sum_{k=0}^{n} \frac{2^{k/2}}{\sqrt{k!}} :(\varphi_k, x^{\otimes k})_{H_0^{\otimes k}}: \in I_1(\mathfrak{F}'_{\text{fin}}(\Phi)) \quad (x \in \Phi').$$

Then

$$(G(\cdot, z), \overline{u}(\cdot))_{L_2(\Phi', \gamma_1)} = (\Gamma(z), \overline{\varphi})_{\mathfrak{F}(H_0)}$$

$$= \sum_{k=0}^{n} \left(\varphi_k, \frac{z^{\otimes k}}{\sqrt{k!}} \right)_{H_0^{\otimes k}} = (J_1 \varphi)(z) = (Gu)(z) \in \mathcal{P}(\Phi'_c) = J_1(\mathfrak{F}_{\text{fin}}(\Phi)).$$

It suffices to establish (5.31) for

$$u(x) = :(\varphi_n, x^{\otimes n})_{H_0^{\otimes n}}: \ (x \in \Phi';\ \varphi_n \in \mathfrak{F}_n(\Phi);\ n \in \mathbb{Z}_+).$$

For $z \in \Phi'_c$ we have

$$u(x + z/\sqrt{2}) = \sum_{k+j=n} C_n^k\, 2^{-j/2} :(\varphi_n, x^{\otimes k} \otimes z^{\otimes j})_{H_0^{\otimes n}}: \ (x \in \Phi').$$

In this sum

$$:(\varphi_n, x^{\otimes k} \otimes z^{\otimes j})_{H_0^{\otimes n}}: \in \Gamma_k(\Phi', \gamma_1) \quad (k = 0, \dots, n).$$

Therefore,

$$\int_{\Phi'} :(\varphi_n, x^{\otimes k} \otimes z^{\otimes j})_{H_0^{\otimes n}}: d\gamma_1(x) = 0 \quad (k = 1, \dots, n).$$

For $k = 0$ we have

$$:\left(\varphi_n, \frac{z^{\otimes n}}{2^{n/2}}\right)_{H_0^{\otimes n}}: \;=\; 2^{-n/2} :(\varphi_n, z^{\otimes n})_{H_0^{\otimes n}}: .$$

Thus,

$$\int_{\Phi'} u(x + z/\sqrt{2})\, d\gamma_1(x) \;=\; 2^{-n/2} :(\varphi_n, z^{\otimes n})_{H_0^{\otimes n}} \;=\; (Gu)(z).$$

Here, we have taken (5.27) into account. It is now obvious that equality (5.31) can be extended from the dense set $I_1(\mathfrak{F}_{\text{fin}}(\Phi))$ to $\mathcal{P}(\Phi')$.

■

Corollary 1. *For the function*

$$u(x) \;=\; :(\varphi_n, x^{\otimes n})_{H_0^{\otimes n}}: \;\; (x \in \Phi';\;\; \varphi_n \in \mathfrak{F}_n(\Phi);\;\; n \in \mathbb{Z}_+),$$

we have

$$(Gu)(iy) \;=\; 2^{-n/2}\, i^n :(\varphi_n, y^{\otimes n})_{H_0^{\otimes n}}: \;\; (y \in \Phi'). \tag{5.32}$$

Proof. In fact, let $\varphi_n = \varphi^{\otimes n}$ and $\varphi \in \Phi$. Then

$$u(x + iy/\sqrt{2}) \;=\; (\varphi, x + iy/\sqrt{2})^n_{H_0}$$

$$= \sum_{k=0}^{n} C_n^k\, (\varphi, x)^k_{H_0}\, 2^{-(k-n)/2}\, i^{n-k}\, (\varphi, y)^{n-k}_{H_0}.$$

According to the formulas for moments of the Gaussian measure γ_1 (see Subsection 2.1), we have

$$\int_{\Phi'} (\varphi, x)^k_{H_0}\, d\gamma_1(x) \;=\; \begin{cases} 0, & k = 2p-1,\;\; p \in \mathbb{N}, \\ 2^{-p}\|\varphi\|^{2p}_{H_0}, & k = 2p,\;\; p \in \mathbb{Z}_+. \end{cases}$$

Taking this into account, we obtain

$$\int_{\Phi'} u(x + iy/\sqrt{2})\, d\gamma_1(x) \;=\; \sum_{\substack{k=0 \\ k=2p}}^{n} C_n^k 2^{-p}\, i^{n-2p}\, \|\varphi\|_0^{2p}\, 2^{(k-n)/2}\, (\varphi, y)^{n-2p}_{H_0}$$

$$= \sum_{\substack{k=0\\k=2p}}^{n} C_n^k 2^{-n/2} i^{-n} (-1)^{n-p} \| \varphi \|_{H_0}^{2p} (\varphi, y)_{H_0}^{n-2p}$$

$$= i^{-n} 2^{-n/2} \| \varphi \|_{H_0}^{n} \sum_{\substack{k=0\\k=2p}}^{n} C_n^k (-1)^{n-p} [2(\varphi / \| \varphi \|_{H_0}, y/2)_{H_0}]^{n-2p}$$

$$= i^{n} 2^{-n/2} \| \varphi \|_{H_0}^{n} \sum_{\substack{k=0\\k=2p}}^{n} C_n^k (-1)^{p} [2(\varphi / \| \varphi \|_{H_0}, y/2)_{H_0}]^{n-2p}$$

$$= i^{n} 2^{-n/2} \| \varphi \|_{H_0}^{n} H_n((\varphi / \| \varphi \|_{H_0}, y/2)_{H_0}).$$

By virtue of (2.24), the last expression equals

$$2^{-n/2} i^n 2^n : (\varphi, y/2)_{H_0}^n : \; = i^n 2^{-n/2} : (\varphi, y)_{H_0}^n : .$$

■

Corollary 2. $I_1(\mathfrak{F}_{\text{fin}}(\Phi)) = \mathcal{P}(\Phi')$.

Proof. According to Corollary 1 to Lemma 5.1, we have $I_1(\mathfrak{F}_{\text{fin}}(\Phi)) \subset \mathcal{P}(\Phi')$. Lemma 5.3 establishes that $J_1(\mathfrak{F}_{\text{fin}}(\Phi)) = \mathcal{P}(\Phi'_c)$. This yields

$$\{(Gu)(i\,\cdot) \mid u \in I_1(\mathfrak{F}_{\text{fin}}(\Phi))\} = \mathcal{P}(\Phi'). \tag{5.33}$$

It follows from (5.32) that

$$\{(Gu)(i\,\cdot) \mid u \in \mathcal{P}(\Phi')\} = I_1(\mathfrak{F}_{\text{fin}}(\Phi));$$

therefore, if we apply the operator G to both sets of the inclusion $I_1(\mathfrak{F}_{\text{fin}}(\Phi)) \subset \mathcal{P}(\Phi')$ and pass to the imaginary argument, then we get $\mathcal{P}(\Phi') \subset I_1(\mathfrak{F}_{\text{fin}}(\Phi))$. Thus, we have obtained the bilateral inclusion which proves that the sets $\mathcal{P}(\Phi')$ and $I_1(\mathfrak{F}_{\text{fin}}(\Phi))$ coincide.

■

Remark 1. The representation (5.31) resembles formula (2.50) which defines the Fourier–Wiener transform. However, the essential difference consists in the fact that in (5.31) we have the measure γ_1 but not γ_2 given by

$$\tilde{\gamma}_2(\varphi) = \exp(-1/2 \| \varphi \|_{H_0}^2) \qquad (\varphi \in \Phi).$$

This leads to the conclusion that the Fourier–Wiener transform W_1 is unitary in $L_2(\Phi', \gamma_1)$, and the operator G transforms $L_2(\Phi', \gamma_1)$ into the space $F(H_{0,c})$ of entire functions on $H_{0,c}$.

In the one-dimensional case, the family of Gaussian transformations is known

$$(G_\alpha f)(x) = \frac{1}{\sqrt{2\pi\alpha}} \int_{\mathbb{R}^1} f(y) \exp[-(x-y)^2/(2\alpha)]\, dy \quad (x \in \mathbb{R}^1;\ \alpha \in (0, 1/2]).$$

For $\alpha = 1/2$ we have

$$(G_{1/2} f)(x) = (Gf)(\sqrt{2}\, x).$$

This is why the operator G will be called the Gaussian transformation which corresponds to the measure γ_1.

■

5.4. Internal Description for Functional Realizations of the Riggings of the Fock Space

As has been established in Theorem 5.2 and its corollary, under the Segal isomorphism, the rigging

$$\mathfrak{F}'_{\mathrm{fin}}(\Phi) \supset \mathfrak{F}(H_0) \supset \mathfrak{F}_{\mathrm{fin}}(\Phi)$$

transforms into the chain

$$\mathcal{P}'(\Phi') \supset L_2(\Phi', \gamma_1) \supset \mathcal{P}(\Phi').$$

Now consider the analogous problem for the rigging

$$\mathfrak{F}'(\Phi) \supset \mathfrak{F}(H_0) \supset \mathfrak{F}(\Phi).$$

Assume that $\mathfrak{F}(\Phi)$ is constructed by means of the representation of the nuclear space $\Phi = \mathrm{pr}\lim_{j\to\infty} H_j$ as a countable projective limit of a family of Hilbert spaces $(H_j)_{j=1}^{\infty}$ with scalar products $(\cdot, \cdot)_{H_j}$ $(j \in \mathbb{N})$ such that the following conditions are satisfied:

(i) For any $i \in \mathbb{N}$, there exists $j \in \mathbb{N}$ such that H_j is quasinuclearly imbedded into H_i, and the Hilbert–Schmidt norm of the imbedding operator is less than 1;

(ii) For any $i \in \mathbb{N}$ and $\varepsilon > 0$, there exists $j \in \mathbb{N}$ such that

$$\|\varphi\|_{H_i} \le \varepsilon \|\varphi\|_{H_j} \quad (\varphi \in \Phi). \tag{5.34}$$

Note that given any nuclear countably normed space, conditions (5.34) can always be satisfied if we replace the Hilbert norms in H_j $(j \in \mathbb{N})$ by equivalent ones. Denote $\mathcal{A}(\Phi') = I_1(\mathfrak{F}(\Phi))$ and $\mathcal{A}'(\Phi') = I_1(\mathfrak{F}'(\Phi))$. Therefore, under the Segal isomorphism, the rigging

$$\mathfrak{F}'(\Phi) \supset \mathfrak{F}(H_0) \supset \mathfrak{F}(\Phi) \tag{5.35}$$

transforms into the chain

$$\mathcal{A}'(\Phi') \supset L_2(\Phi', \gamma_1) \supset \mathcal{A}(\Phi'). \tag{5.36}$$

Introducing the Hilbert spaces $A_{\pm j}(\Phi') = I_1(\mathfrak{F}(H_{\pm j}))$, we obtain the following representations (according to the construction of $\mathfrak{F}(\Phi)$ and $\mathfrak{F}'(\Phi)$):

$$\mathcal{A}(\Phi') = \operatorname{pr}\lim_{j\to\infty} A_{+j}(\Phi'), \quad \mathcal{A}'(\Phi') = \operatorname{ind}\lim_{j\to\infty} A_{-j}(\Phi').$$

We want to obtain the internal description of the rigging (5.36). First, let us consider the example which clarifies the notations exploited when we consider the spaces of the chain (5.35) and shows the connection between (5.35) and the spaces of test and generalized functions of infinitely many variables considered in Section 4.

Example 5.4. Let us return to the rigging (5.12) constructed in Example 5.2

$$\mathfrak{F}'(\mathcal{X}_T) \supset \mathfrak{F}(H_0) \supset \mathfrak{F}(\mathcal{X}_T).$$

According to Subsection 2.3, under the Segal isomorphism between $\mathfrak{F}(H_0)$ and $L_2(\mathbb{R}^\infty, \gamma_1)$, the basis of occupation numbers $(e_\alpha)_{\alpha\in\mathbb{Z}^\infty_{+,0}}$ in $\mathfrak{F}(H_0)$ turns into the basis of Hermite polynomials $(h_\alpha)_{\alpha\in\mathbb{Z}^\infty_{+,0}}$. But then functions such as $\tau_1^{-\alpha_1/2}\dots\tau_\nu^{-\alpha_\nu/2}\, h_\alpha(\cdot)$, $\alpha\in\mathbb{Z}^\infty_{+,\infty}$, form the basis in the space $I_1(\mathfrak{F}(X_\tau^+))$, $\tau \in T$. According to (4.23), this is an orthonormal basis in the Hilbert space $A_\tau(\mathbb{R}^\infty)$. Therefore, $I_1(\mathfrak{F}(X_\tau^+)) = A_\tau(\mathbb{R}^\infty)$ and

$$I_1(\mathfrak{F}(\mathcal{X}_T)) = \operatorname{pr}\lim_{\tau\in T} I_1(\mathfrak{F}(X_\tau^+)) = \operatorname{pr}\lim_{\tau\in T} A_\tau(\mathbb{R}^\infty) = \mathcal{A}_T(\mathbb{R}^\infty),$$

i.e. in the case when $\Phi = \mathcal{X}_T$, we have $\mathcal{A}(\Phi') = \mathcal{A}_T(\mathbb{R}^\infty)$. The last statement implies that $\mathcal{A}(\Phi')$ is a generalization of the space $\mathcal{A}_T(\mathbb{R}^\infty)$ for the invariant case.

To investigate the image of the rigging (5.35) under the Segal isomorphism, we need its holomorphic realization. To describe this realization, we introduce the space $\mathcal{E}^2_{\min}(\Phi'_c)$ of entire functions of a minimal type whose order of growth is at most 2. This means that, as a set, $\mathcal{E}^2_{\min}(\Phi'_c)$ consists of the functions entire on each H_{-j} $(j \in \mathbb{N})$ and such that

$$\forall\, u \in \mathcal{E}^2_{\min}(\Phi'_c),\ \ j \in \mathbb{N},\ \ \varepsilon > 0\ \exists\ c > 0:$$

$$|u(z)| \le c \exp(\varepsilon \|z\|^2_{H_{-j,c}}) \quad (z \in H_{-j,c}), \tag{5.37}$$

i.e., these are the functions of a minimal type on each $H_{-j,c}$ with the order of growth at most 2. We endow $\mathcal{E}^2_{\min}(\Phi'_c)$ with the projective topology with respect to the countable system of norms

$$|u|_{-j,k} = \sup_{z \in H_{-j,c}} \{|u(z)| \exp(-1/k \|z\|^2_{H_{-j,c}})\} \quad (k, j \in \mathbb{N}) \tag{5.38}$$

possessing the natural directness (see Section 1, Chapter 1). Since the first axiom of countability holds for $\mathcal{E}^2_{\min}(\Phi'_c)$ with this topology, it suffices to describe convergent sequences of functions from this space in order to define the topology under consideration. It follows from (5.38) that

$$\mathcal{E}^2_{\min}(\Phi'_c) \ni u_n \underset{n \to \infty}{\longrightarrow} 0$$

in $\mathcal{E}^2_{\min}(\Phi'_c)$ if and only if

(i) for any $j \in \mathbb{N}$ and $\varepsilon > 0$ there exists $c_{j,\varepsilon} > 0$ such that

$$|u_n(z)| \le c_{j,\varepsilon} \exp(\varepsilon \|z\|^2_{H_{-j,c}}) \quad (z \in H_{-j,c};\ n \in \mathbb{N}); \tag{5.39}$$

(ii) as $n \to \infty$, u_n converges to 0 uniformly on every ball

$$\{z \in H_{-j,c} \mid \|z\|_{H_{-j,c}} \le R\} \quad (R > 0)$$

in each $H_{-j,c}$ $(j \in \mathbb{N})$.

Theorem 5.3. *The space $\mathcal{E}^2_{\min}(\Phi'_c)$ is the image of the nuclear space $\mathcal{F}(\Phi)$ under the holomorphic realization of the Fock space $\mathcal{F}(H_0)$.*

Proof. According to (5.24), the function

$$u(z) = \sum_{n=0}^{\infty} \left(z_n, \frac{z^{\otimes n}}{\sqrt{n!}} \right)_{H_0^{\otimes n}} \quad (z \in \Phi'_c) \tag{5.40}$$

is the image of the vector

$$u = (u_n)_{n=0}^{\infty} \in \mathfrak{F}(\Phi) \quad (u_n \in \mathfrak{F}_n(\Phi))$$

under the isomorphism J_1. Since $(u_n)_{n=0}^{\infty} \in \mathfrak{F}(\Phi)$, the series

$$\sum_{n=0}^{\infty} \| u_n \|^2_{H_{j,c}^{\otimes n}} < \infty$$

for every $j \in \mathbb{N}$. Hence, the expansion (5.40) converges uniformly on each ball on $H_{-j,c}$, and the following estimate is valid

$$| u(z) | \le \| u \|_{\mathfrak{F}(H_j)} \exp(1/2 \, \| z \|^2_{H_{-j,c}}) \quad (z \in H_{-j,c}). \tag{5.41}$$

Due to (5.34), for any $\varepsilon > 0$ and $i \in \mathbb{N}$, one can find $j \in \mathbb{N}$ such that

$$\| z \|^2_{H_{-j,c}} \le 2\varepsilon \| z \|^2_{H_{-i,c}} \quad (z \in H_{-i,c}).$$

This, together with (5.41), yields the inclusion $u(\cdot) \in \mathcal{E}^2_{\min}(\Phi'_c)$.

If the sequence

$$u^{(k)} = (u_n^{(k)})_{n=0}^{\infty} \in \mathfrak{F}(\Phi) \quad (k \in \mathbb{N})$$

converges to 0 in $\mathfrak{F}(\Phi)$, then $\| u^{(k)} \|_{\mathfrak{F}(H_j)} \to 0$ as $k \to \infty$ for any $j \in \mathbb{N}$. Consequently,

$$| u^{(k)}(z) | \le \| u^{(k)} \|_{\mathfrak{F}(H_j)} \exp(1/2 \, \| z \|^2_{H_{-j,c}}) \le \sup_{k \in \mathbb{N}} \| u^{(k)} \|_{\mathfrak{F}(H_j)} \exp(\varepsilon \, \| z \|^2_{H_{-j,c}})$$

(here, $j \in \mathbb{N}$ is chosen for given $\varepsilon > 0$ and $i \in \mathbb{N}$, as mentioned above). Moreover, the same estimate implies that

$$\sup_{\{\|z\|_{H_{-j,c}} \le R\}} | u^{(k)}(z) | \le \| u^{(k)} \|_{\mathfrak{F}(H_j)} \exp(1/2 \, R^2) \to 0, \quad k \to \infty.$$

By virtue of (5.39), this proves that $(u^{(k)}(\cdot))_{k=1}^{\infty}$ converges in $\mathcal{E}^2_{\min}(\Phi'_c)$.

Let us prove the converse statement, i.e., that each function from $\mathcal{E}^2_{\min}(\Phi'_c)$ is the image of a uniquely defined vector from $\mathfrak{F}(\Phi)$, and that convergence in $\mathcal{E}^2_{\min}(\Phi'_c)$ implies convergence of the corresponding vectors in $\mathfrak{F}(\Phi)$. Let $u(\cdot) \in \mathcal{E}^2_{\min}(\Phi'_c)$. Then $u(\cdot)$ can be expanded in the Taylor series

$$u(z) = \sum_{n=0}^{\infty} \frac{1}{\sqrt{n!}} d^n u(0)(z) \quad (z \in \Phi'_c).$$

Since Φ is nuclear, we have the following representation of differentials

$$d^n u(0)(z) = (D^n u(0), z^{\otimes n})_{H_0^{\otimes n}},$$

where the kernel $D^n u(0) \in \Phi_c^{\otimes n}$ $(n \in \mathbb{Z}_+)$. Therefore,

$$u(z) = \sum_{n=0}^{\infty} \frac{1}{\sqrt{n!}} \left(D^n u(0), \frac{z^{\otimes n}}{\sqrt{n!}} \right)_{H_0^{\otimes n}} \quad (z \in \Phi'_c).$$

According to the definition of holomorphic realization and construction of $\mathfrak{F}(\Phi)$, it is now necessary to establish the inclusion

$$u = (1/\sqrt{n!}\, D^n u(0))_{n=0}^{\infty} \in \mathfrak{F}(H_j)$$

for all $j \in \mathbb{N}$.

For given $i \in \mathbb{N}$, we employ (5.34) to choose $j \in \mathbb{N}$ such that $H_{j,c} \subset H_{i,c}$, and for the Hilbert norm of the imbedding operator, we have $| O_{ji} | < 1$. Then for every kernel $a \in H_{j,c}^{\otimes n}$ $(n \in \mathbb{N})$, the inequality

$$\| a \|_{H_{i,c}^{\otimes n}} \leq | O_{ji} |^n \| a \|_{\mathfrak{F}(H_{-j,c}^n;\, \mathbb{C}^1)} \leq \| a \|_{\mathfrak{F}(H_{-j,c}^n;\, \mathbb{C}^1)}$$

holds, where $\| a \|_{\mathfrak{F}(H_{-j,c}^n;\, \mathbb{C}^1)}$ is the norm of the n-linear form

$$H_{-j,c}^n = \mathop{\times}_1^n H_{-j,c} \ni (\xi_1, \ldots, \xi_n) \mapsto (a, \xi_1 \otimes \ldots \otimes \xi_n)_{H_0^{\otimes n}} \in \mathbb{C}^1.$$

Indeed, for the orthonormal basis $(e_k)_{k=1}^{\infty}$ in $H_{-i,c}$, we have

$$\| a \|_{H_{i,c}^{\otimes n}} = \sum_{k_1, \ldots, k_n = 1}^{\infty} | (a, e_{k_1} \otimes \ldots \otimes e_{k_n})_{H_0^{\otimes n}} |^2$$

$$\leq \| a \|^2_{\mathfrak{F}(H_{-j,c}^n;\, \mathbb{C}^1)} \sum_{k_1, \ldots, k_n = 1}^{\infty} \| e_{k_1} \|^2_{H_{-j,c}} \cdots \| e_{k_n} \|^2_{H_{-j,c}} = \| a \|^2_{\mathfrak{F}(H_{-j,c}^n;\, \mathbb{C}^1)} | O_{ji} |^{2n}.$$

The Cauchy inequality for an entire function $u(\cdot)$ on the Hilbert space gives (Nachbin

[1, Chapter 2, Section 1])

$$\left\| \frac{1}{n!} D^n u(0) \right\|_{H_{i,c}^{\otimes n}} \le \left\| \frac{1}{n!} D^n u(0) \right\|_{\mathfrak{F}(H_{-j,c}^n;\, \mathbb{C}^1)}$$

$$\le \rho^{-n} \sup_{\|z\|_{H_{-j,c}} = \rho} |u(z)| \le c_{\varepsilon,j} \exp(\varepsilon \rho^2) \rho^{-n} \quad (\rho \ge 0).$$

Minimizing the last expression in $\rho > 0$ and applying the Stirling formula, we get

$$\left\| \frac{1}{n!} D^n u(0) \right\|_{H_{i,c}^{\otimes n}} \le c_{\varepsilon,j} (2\varepsilon e/n)^{n/2} \le c_{\varepsilon,j} (4\varepsilon)^{n/2} / \sqrt{n!}. \tag{5.42}$$

Choosing now $\varepsilon > 0$ such that $4\varepsilon > 1$, we find from (5.42) that

$$\sum_{n=0}^{\infty} \left\| \frac{1}{\sqrt{n!}} D^n u(0) \right\|_{\mathfrak{F}_n(H_i)}^2 \le c_{\varepsilon,j}^2 \sum_{n=0}^{\infty} (4\varepsilon)^n < \infty,$$

and this gives the desired inclusion

$$\left(\frac{1}{\sqrt{n!}} D^n u(0) \right)_{n=0}^{\infty} \in \mathfrak{F}(H_i).$$

If

$$\mathcal{E}^2_{\min}(\Phi_c') \ni u^{(n)}(\cdot) \underset{n\to\infty}{\to} 0$$

in $\mathcal{E}^2_{\min}(\Phi_c')$, then for given $\varepsilon, \delta > 0$ the estimate

$$|u^{(n)}(z)| \le \varepsilon \exp(\delta \|z\|^2_{H_{-j,c}}) \quad (z \in H_{-j,c}) \tag{5.43}$$

holds if $n > n_0(\varepsilon, \delta, j)$. In fact, according to the definition of convergence in $\mathcal{E}^2_{\min}(\Phi_c')$, for any $0 < \delta_1 < \delta$ and $j \in \mathbb{N}$, one can find $c_{\delta_1, j}$ such that

$$|u^{(n)}(z)| \le c_{\delta_1,j} \exp(\delta_1 \|z\|^2_{H_{-j,c}}) = c_{\delta_1,j} \exp(\delta \|z\|^2_{H_{-j,c}}) \exp(-(\delta - \delta_1) \|z\|^2_{H_{-j,c}})$$

$$\le \varepsilon \exp(\delta \|z\|^2_{H_{-j,c}}) \quad (\|z\|_{H_{-j,c}} \ge \rho = \rho(\varepsilon, \delta),\ n \in \mathbb{N}).$$

Since $(u^{(n)}(\cdot))_{n=1}^{\infty}$ converges uniformly on the ball

$$\{z \in H_{-j,c} \mid \|z\|_{H_{-j,c}} \leq \rho\} = \tilde{B}_{-j,\rho}(0),$$

we have

$$|u^{(n)}(z)| \leq \varepsilon \leq \varepsilon \exp(\delta \|z\|^2_{H_{-j,c}}) \quad (z \in \tilde{B}_{-j,\rho}(0), \ n > n_0(\varepsilon, \delta, j)).$$

These estimates together imply (5.43). As above, for given $i \in \mathbb{N}$, let us choose $j \in \mathbb{N}$ such that $H_j \subset H_i$ and $\| O_{ji} \| < 1$. Then we obtain

$$\left\| \frac{1}{m!} D^m u^{(n)}(0) \right\|_{H^{\otimes m}_{i,c}} \leq \rho^{-m} \sup_{\{\|z\|_{H_{-j,c}} \leq R\}} |u^{(n)}(z)|$$

$$\leq \varepsilon \rho^{-m} \exp(\delta \rho^2) \leq \varepsilon (4\delta)^{m/2} / \sqrt{m!} \quad (n > n_0(\varepsilon, \delta, i)).$$

Assuming that $\delta > 0$ is such that $4\delta > 1$, we get the following estimate

$$\|u^{(n)}\|_{\mathcal{F}(H_i)} = \sum_{m=0}^{\infty} \left\| \frac{1}{\sqrt{m!}} D^m u^{(n)}(0) \right\|^2_{\mathcal{F}_m(H_i)}$$

$$\leq \varepsilon \sum_{m=0}^{\infty} (4\delta)^m = \varepsilon / (1 - 4\delta) \quad (n > n_0(\varepsilon, i)); \qquad (5.44)$$

$$u^{(n)} = \left(\frac{1}{\sqrt{m!}} D^m u^{(n)}(0) \right)_{m=0}^{\infty}$$

($\delta > 0$ is fixed). Since $\varepsilon > 0$ and $i \in \mathbb{N}$ are arbitrary, inequality (5.44) shows that $u^{(n)} \to 0$ (as $n \to \infty$) in $\mathcal{F}(\Phi)$.

■

Let us introduce one more space of entire functions. Denote by $\mathcal{E}^2_{\max}(\Phi_c)$ the set of functions given on Φ_c and possessing the following properties:

(i) Each function is an entire function on some space $H_{j,c}$;
(ii) Regarded as an entire function on $H_{j,c}$, it has the finite type for the order of growth equal to 2.

In other words, for $\xi \in \mathcal{E}^2_{\max}(\Phi_c)$, the following estimate should hold

$$|\xi(z)| \leq c_{\xi,j} \exp(A \|z\|^2_{H_{j,c}}) \quad (z \in \Phi_c)$$

with some $c_{\xi,j}, A > 0$. Let us topologize $\mathcal{E}^2_{\max}(\Phi_c)$ by introducing the topology of inductive limit with respect to the family of norms

$$|\xi(z)|_{j,k} = \sup_{z\in\Phi_c} \{|\xi(z)| \exp(-k\|z\|^2_{H_{j,c}})\} \quad (j, k \in \mathbb{N}) \tag{5.45}$$

(see Section 1, Chapter 1). It is easy to see that in this topology the fact that $\xi^{(n)} \to 0$ (as $n \to \infty$) is equivalent to the facts that $(\xi^{(n)})_{n=1}^{\infty} \subset \mathcal{E}^2_{\max}(\Phi_c)$ are entire functions on the common Hilbert space $H_{j,c}$, that there exists the uniform majorization

$$|\xi^{(n)}(z)| \le c \exp(A\|z\|^2_{H_{j,c}}) \quad (z \in H_{j,c};\ n \in \mathbb{N}),$$

and that $\xi^{(n)}$ converges uniformly to 0 as $n \to \infty$ on every ball in $H_{j,c}$.

Theorem 5.4. *Under the holomorphic realization of $\mathfrak{F}(H_0)$, the image of $\mathfrak{F}'(\Phi)$ coincides with $\mathcal{E}^2_{\max}(\Phi_c)$ and, hence, the space $\mathcal{E}^2_{\max}(\Phi_c)$ is dual to $\mathcal{E}^2_{\min}(\Phi'_c)$ with respect to $F(H_{0,c})$.*

Proof. Just as in the proof of Theorem 5.3, the facts that each element of $J_1(\mathfrak{F}'(\Phi))$ belongs to $\mathcal{E}^2_{\max}(\Phi_c)$, and that the convergence in $\mathfrak{F}'(\Phi)$ involves the convergence of images in $\mathcal{E}^2_{\max}(\Phi_c)$, follow directly from the definition

$$\mathfrak{F}'(\Phi) = \operatorname{ind}\lim_{j\to\infty} \mathfrak{F}(H_{-j}).$$

Conversely, if $\xi \in \mathcal{E}^2_{\max}(\Phi_c)$, then

$$|\xi(z)| \le c \exp(A\|z\|^2_{H_{j,c}}).$$

We choose $\varepsilon > 0$ such that $4A\varepsilon < 1$. Using (5.34), we take $j \in \mathbb{N}$ for which $\|\cdot\|_{H_{i,c}} < \varepsilon \|\cdot\|_{H_{j,c}}$. Then

$$|\xi(z)| \le c \exp(A\varepsilon\|z\|^2_{H_{j,c}}),$$

and, as before, employing the Cauchy inequality for the entire function ξ on $H_{j,c}$, we obtain

$$\left\| \frac{1}{n!} D^n \xi(0) \right\|_{H^{\otimes n}_{-j,c}} \le c\,(4A\varepsilon)^{n/2}/\sqrt{n!},$$

whence for

$$J_1^{-1}\xi = \left(\frac{1}{\sqrt{n!}} D^n \xi(0) \right)_{n=0}^{\infty},$$

we get

$$\| J_1^{-1}\xi \|^2_{\mathfrak{F}(H_{-j})} = \sum_{n=0}^{\infty} \left\| \frac{1}{\sqrt{n!}} D^n \xi(0) \right\|^2_{H_{-j,c}^{\otimes n}} \le c^2 \sum_{n=0}^{\infty} (4A\varepsilon)^n < \infty.$$

The fact that the convergence in $\mathcal{E}^2_{\max}(\Phi_c)$ implies the convergence of the corresponding vectors in $\mathfrak{F}'(\Phi)$ can be established by analogy with the proof of the corresponding assertion for $\mathcal{E}^2_{\min}(\Phi'_c)$ in Theorem 5.3, namely, by use of the definition of the convergence in $\mathcal{E}^2_{\max}(\Phi_c)$ and the Cauchy inequality.

■

Thus, we have constructed the chain

$$\mathcal{E}^2_{\max}(\Phi_c) \supset F(H_{0,c}) \supset \mathcal{E}^2_{\min}(\Phi'_c), \tag{5.46}$$

in which all the spaces consist of entire functions. The entirety of generalized functions is a specific peculiarity of the holomorphic realization of the Fock space and its rigging (5.35). In this case, the singularity of generalized functions can be characterized by the following two properties:

(i) Let $H_{j,c}$ be the Hilbert space from the family $(H_{j,c})_{j=1}^{\infty}$ on which these functions are entire. Then the larger j is (i.e., the narrower is $H_{j,c}$) the more singular are the generalized functions.

(ii) The higher the type of a given entire function of the second order on $H_{j,c}$ is, the more the distant space $\mathfrak{F}(H_{-j})$ contains the vector from $\mathfrak{F}'(\Phi)$ that corresponds to it, i.e., the more singular is this function as a generalized function in the sense of the rigging (5.46).

We now have all the necessary tools for the internal description of the rigging (5.30). Denote by $\mathcal{E}^2_{\min}(\Phi'_c)$ the set of functions obtained by means of the restriction of the functions from $\mathcal{E}^2_{\min}(\Phi'_c)$ to Φ'. The next theorem is the infinite-dimensional generalization of Theorem 4.1.

Theorem 5.5. $\mathcal{A}(\Phi') = \mathcal{E}^2_{\min}(\Phi'_c).$

Proof. Let us show that every function $u \in \mathcal{A}(\Phi')$ admits the analytic extension onto Φ'_c, and that the function, obtained as a result of this extension, belongs to $\mathcal{E}^2_{\min}(\Phi'_c)$. According to the definition of $\mathcal{A}(\Phi')$, we have

$$u(x) = \sum_{n=0}^{\infty} \frac{2^{n/2}}{\sqrt{n!}} :(u_n, x^{\otimes n})_{H_0^{\otimes n}}: \quad (x \in \Phi'),$$

where

$$I_1^{-1} u = (u_n)_{n=0}^{\infty} \in \mathfrak{F}(\Phi).$$

We introduce the function

$$v(z) = \sum_{n=0}^{\infty} \left(v_n, \frac{z^{\otimes n}}{\sqrt{n!}} \right)_{H_0^{\otimes n}} \quad (z \in \Phi_c'),$$

where $v_n = (-i)^n 2^n u_n$, $n \in \mathbb{Z}_+$; hence, $(v_n)_{n=0}^{\infty} \in \mathfrak{F}(\Phi)$ due to the inclusion $I_1^{-1} u \in \mathfrak{F}(\Phi)$ and condition (5.34). Consequently, $v(\cdot) \in \mathcal{E}_{\min}^2(\Phi_c')$. It follows from (5.38) that the inclusion

$$v_{\mathrm{Re}} = v \upharpoonright \Phi' \in L_p(\Phi', \gamma_1) \quad (p \geq 1)$$

holds. Consider the function

$$w(z) = \int_{\Phi'} v(x + i z/\sqrt{2}) \, d\gamma_1(x) \quad (z \in \Phi_c'),$$

which is clearly analytic on Φ_c'. Furthermore, $w(\cdot) \in \mathcal{E}_{\min}^2(\Phi_c')$. In fact, this follows from the estimate

$$|v(x + i z/\sqrt{2})| \leq c_{j,\varepsilon} \exp(\varepsilon \, \| x + i z/\sqrt{2} \, \|_{H_{-j,c}}^2)$$

$$\leq c_{j,\varepsilon} \exp(2\,\varepsilon \, \| x \|_{H_{-j}}^2) \exp(\varepsilon \, \| x \|_{H_{-j,c}}^2) \quad (x \in H_{-j};\ z \in H_{-j,c}),$$

which yields

$$|w(z)| \leq \hat{c}_{j,\varepsilon} \exp(\varepsilon \, \| z \|_{H_{-j,c}}^2) \quad (\varepsilon > 0,\ j > j(\varepsilon),\ z \in H_{-j,c}).$$

On the other hand, $w \upharpoonright \Phi' = u$. This is a consequence of (5.32) which, in this case, gives

$$w(y) = (G v_{\mathrm{Re}})(i y) = \sum_{n=0}^{\infty} \frac{2^{n/2}}{\sqrt{n!}} :(u_n, y^{\otimes n})_{H_0^{\otimes n}}: = u(y) \quad (y \in \Phi').$$

The inclusion $\mathcal{A}(\Phi') \subset \mathcal{E}_{\min}^2(\Phi')$ is thus established. Quite analogously, in the case of an arbitrary function

$$v(x) = \sum_{n=0}^{\infty} \left(v_n, \frac{x^{\otimes n}}{\sqrt{n!}} \right)_{H_0^{\otimes n}} \in \mathcal{E}^2_{\min}(\Phi'),$$

we get

$$(Gv)(i\,y) = \sum_{n=0}^{\infty} \frac{2^{-n/2}\, i^n}{\sqrt{n!}} : (v_n, y^{\otimes n})_{H_0^{\otimes n}} : \in \mathcal{A}(\Phi'),$$

i.e.,

$$\{(Gv) \upharpoonright \Phi' \mid v \in \mathcal{E}^2_{\min}(\Phi')\} \subset \mathcal{A}(\Phi').$$

At the same time, $J_1(\mathfrak{F}(\Phi)) = \mathcal{E}^2_{\min}(\Phi'_c)$, and $I_1(\mathfrak{F}(\Phi)) = \mathcal{A}(\Phi')$; hence, $G(\mathcal{A}(\Phi'))$ $= \mathcal{E}^2_{\min}(\Phi'_c)$ and, therefore,

$$\mathcal{E}^2_{\min}(\Phi') = \{(Gu) \upharpoonright \Phi' \mid u \in \mathcal{A}(\Phi')\}.$$

If we apply the Gaussian transformation to each set of the inclusion $\mathcal{A}(\Phi') \subset \mathcal{E}^2_{\min}(\Phi')$ and consider the restrictions of the functional spaces thus obtained to Φ', we obtain

$$G(\mathcal{A}(\Phi')) \upharpoonright \Phi' = \mathcal{E}^2_{\min}(\Phi') \subset G(\mathcal{E}^2_{\min}(\Phi')) \upharpoonright \Phi' \subset \mathcal{A}(\Phi').$$

But this, together with the inverse inclusion proved earlier, involves the equality $\mathcal{A}(\Phi')$ $= \mathcal{E}^2_{\min}(\Phi')$.

■

Remark 1. A more thorough analysis, employing the relation between the convergence in $\mathcal{E}^2_{\min}(\Phi'_c)$ and the behavior of Taylor coefficients (similar to that carried out in the proof of Theorem 5.3), shows that the topology of $\mathcal{A}(\Phi')$ coincides with the topology induced on $\mathcal{E}^2_{\min}(\Phi')$ from the space $\mathcal{E}^2_{\min}(\Phi'_c)$. We shall not dwell upon this fact in detail.

■

Lemma 5.4. *Suppose that the function $G(\cdot,\cdot)$ is defined by* (5.28). *Then $G(x,\cdot)$ $\in \mathcal{E}^2_{\max}(\Phi_c)$ for every $x \in \Phi'$, and $G(\cdot, z) \in \mathcal{A}(\Phi')$ for every $z \in \Phi'_c$.*

Proof. For $x \in \Phi'$, one can find $i \in \mathbb{N}$ such that $x \in H_{-i}$. Then

$$G(x,z) = \exp(\sqrt{2}\,(z,x)_{H_0} - (z,z)_{H_0}) \quad (z \in H_{i,c})$$

is obviously an entire function of the finite type on $H_{i,c}$ with (at most) the second order of growth. Therefore, this function belongs to $\mathcal{E}^2_{\max}(\Phi_c)$. If $\xi \in \Phi_c$, we have

$$G(x, \xi) = I_1\Gamma(\xi) = \sum_{n=0}^{\infty} \frac{2^{n/2}}{\sqrt{n!}} :(\xi, x)^n_{H_0}: \in L_2(\Phi', \gamma_1).$$

For $z \in \Phi'_c$, let us find $j \in \mathbb{N}$ for which $z \in H_{-j,c}$ and fix $(\xi_k)_{k=1}^{\infty} \subset \Phi_c$ such that $\xi_k \to z$ as $k \to \infty$ strongly in $H_{-j,c}$. It is then easy to see that $\Gamma(\xi_k) \underset{k\to\infty}{\to} \Gamma(z)$ in $\mathfrak{F}(H_{-j})$, and this involves the convergence of $(G(\cdot\,, \xi_k))_{k=1}^{\infty}$ in $\mathcal{A}'(\Phi')$, since

$$\mathcal{A}'(\Phi') = \operatorname{ind}\lim_{j\to\infty} I_1(\mathfrak{F}(H_{-j})).$$

■

Corollary 1. *The Gaussian transformation admits the following representations*

$$(Gu)(z) = (G(\cdot, z), \bar{u}(\cdot))_{L_2(\Phi', \gamma_1)} \in \mathcal{E}^2_{\min}(\Phi'_c) \quad (u \in \mathcal{A}(\Phi')),$$

$$(G^{-1}v)(z) = (G(x, \cdot), \bar{v}(\cdot))_{L_2(\Phi'_c, \nu_1)} \in \mathcal{A}(\Phi') \;\; (v \in \mathcal{E}^2_{\min}(\Phi'_c)). \tag{5.47}$$

Proof. Equalities (5.47) have been established by Theorem 5.2 and corollaries to it on the dense sets $\mathcal{P}(\Phi') \subset \mathcal{A}(\Phi')$ and $\mathcal{P}(\Phi'_c) \subset \mathcal{E}^2_{\min}(\Phi'_c)$. Taking into account that Lemma 5.4 establishes the inclusion of the kernel $G(\cdot\,,\cdot)$ into the spaces of generalized functions $\mathcal{A}'(\Phi')$ or $\mathcal{E}^2_{\max}(\Phi_c)$, respectively, we obtain the desired statement by the limit transition in (5.30).

■

As shown above, the Gaussian transformation realizes a bijection between the corresponding spaces of the chains

$$\mathcal{P}'(\Phi') \supset \mathcal{A}'(\Phi') \supset L_2(\Phi', \gamma_1) \supset \mathcal{A}(\Phi') \supset \mathcal{P}(\Phi')$$

and

$$\mathcal{P}'(\Phi'_c) \supset \mathcal{E}^2_{\max}(\Phi_c) \supset F(H_{0,c}) \supset \mathcal{E}^2_{\min}(\Phi'_c) \supset \mathcal{P}(\Phi'_c).$$

Moreover, equalities (5.47) represent this transformation as "integral" with the generalized kernel $G(\cdot\,,\cdot)$. In many cases, the analogous representation of G in the space of generalized functions proves to be helpful.

Lemma 5.5. *The Gaussian transformation on the space of generalized functions*

$\mathcal{A}'(\Phi')$ is given by

$$\mathcal{A}'(\Phi') \ni \xi \mapsto (G\xi)(z) = (G(\cdot, z), \overline{\xi}(\cdot))_{L_2(\Phi', \gamma_1)} \in \mathcal{E}^2_{\max}(\Phi_c). \tag{5.48}$$

Proof. For every $z \in \Phi_c$, the function $G(x, z)$ $(x \in \Phi')$ belongs to $\mathcal{A}(\Phi')$ since

$$G(\cdot, z) = (I_1\Gamma(z))(\cdot) \in I_1(\mathcal{F}(H_j)) \quad (j \in \mathbb{N}).$$

Therefore, the right-hand side of (5.48) is meaningful for all $\xi \in \mathcal{A}'(\Phi')$. Formula (5.48) sets the Gaussian transformation on $\mathcal{A}(\Phi')$ dense in $\mathcal{A}'(\Phi')$ and, hence, it remains valid on the whole of $\mathcal{A}'(\Phi')$.

■

Remark 1. Under the Gaussian transformation, the space of generalized functions $\mathcal{A}'(\Phi')$ turns into the space of entire functions $\mathcal{E}^2_{\max}(\Phi_c)$ on Φ_c. Clearly, it is much simpler to consider the functions from $\mathcal{E}^2_{\max}(\Phi_c)$ than the generalized functions from $\mathcal{A}'(\Phi')$. This clarifies the important technical advantage of the Gaussian transformation — it is often simpler to investigate whether the function $G\xi$ is entire than to examine directly whether the inclusion $\xi \in \mathcal{A}'(\Phi')$ holds.

■

5.5. Examples of Generalized Functions of Infinitely Many Variables

In the infinite-dimensional case, the class formed by generalized functions given by measures proves to be important. The complete description of the measures on Φ', which generate generalized functions from $\mathcal{A}'(\Phi')$ by the formula

$$\mathcal{A}(\Phi') \ni u \mapsto \xi_\mu(u) = \int_{\Phi'} u(x)\, d\mu(x) \in \mathbb{C}^1 \tag{5.49}$$

is given by the following theorem.

Theorem 5.6. *In order that the measure μ on Φ' set a generalized function from $\mathcal{A}'(\Phi')$, it is necessary and sufficient that it have the moments*

$$s_n^\mu(\varphi_1, \ldots, \varphi_n) = \int_{\Phi'} (\varphi_1, x)_{H_0} \cdots (\varphi_n, x)_{H_0}\, d\mu(x) \quad (n \in \mathbb{Z}_+) \tag{5.50}$$

along the directions $\varphi_1, \ldots, \varphi_n \in \Phi_c$ *such that the estimates*

$$|s_n^{\mu}(\varphi_1, \ldots, \varphi_n)| \leq c^n \sqrt{n!}\, \|\varphi_1\|_{H_{j,c}} \cdots \|\varphi_n\|_{H_{j,c}} \tag{5.51}$$

hold for some $j \in \mathbb{N}$ *and* $c > 0$.

Remark 1. The generalized function ξ_μ generated by (5.49) will be called the generalized derivative of the measure μ with respect to γ_1 and denoted by $\xi_\mu = d\mu / d\gamma_1$. This notation can be motivated as follows: In the case when μ is absolutely continuous with respect to γ_1, the equality (5.49) can be rewritten in the form

$$(\xi_\mu, \bar{u})_{L_2(\Phi', \gamma_1)} = \int_{\Phi'} u(x)\,(d\mu / d\gamma_1)(x)\, d\gamma_1(x) \quad (u \in \mathcal{A}(\Phi')).$$

Thus, Theorem 5.6 states that the measure μ has a generalized derivative $d\mu / d\gamma_1 \in \mathcal{A}'(\Phi')$ if and only if its moments satisfy the estimates (5.51).

■

Remark 2. The original Gaussian measure γ_1 has moments satisfying (5.51) with $\|\cdot\|_{H_{j,c}} = \|\cdot\|_{H_{0,c}}$ (see Subsection 2.1). The measures on Φ' which satisfy (5.51) are called sub-Gaussian. By virtue of what has been said in Subsection 1.7, each of these measures is concentrated on every Hilbert space H_{-i} for which $H_{-j} \subset H_{-i}$ quasi-nuclearly. Moreover, the growth of its moments is not greater than the growth of the moments of some Gaussian measure on H_{-i} (one can take, for example, the Gaussian measure γ_{H_j} with the Fourier transform

$$\tilde{\gamma}_{H_j}(\varphi) = \exp(-1/4\, \|\varphi\|_{H_j}^2) \quad (\varphi \in \Phi)).$$

■

Proof of Theorem 5.6. *Necessity.* The existence of the moments (5.50) of the measure μ along the directions from Φ_c follows from the inclusion $\mathcal{P}(\Phi') \subset \mathcal{A}(\Phi')$. Taking into account that μ generates the generalized function $\xi_\mu \in \mathcal{A}'(\Phi')$ by the formula (5.49), one can find $i \in \mathbb{N}$ such that

$$\xi_\mu \in A_{-i}(\Phi') = I_1(\mathcal{F}(H_{-i})).$$

Therefore,

$$|s_n^{\mu}(\varphi_1, \ldots, \varphi_n)| = \left| \left(\xi_\mu, \prod_{k=1}^{n} (\varphi_k, \cdot)_{H_0} \right)_{L_2(\Phi', \gamma_1)} \right|$$

$$\leq \| \xi_\mu \|_{A_{-i}(\Phi')} \left\| \prod_{k=1}^{n} (\varphi_k, \cdot)_{H_0} \right\|_{A_i(\Phi')} . \tag{5.52}$$

The necessity of the conditions of Theorem 5.6 follows now from (5.52) and the next lemma.

Lemma 5.6. *For $\varphi_1, \ldots, \varphi_n \in \Phi_c$, the following estimate holds*

$$\left\| \prod_{k=1}^{n} (\varphi_k, \cdot)_{H_0} \right\|_{A_i(\Phi')} \leq \sqrt{n!} \prod_{k=1}^{n} \| \varphi \|_{H_{i,c}} \quad (n \in \mathbb{N}).$$

Proof. First, let us assume that $\varphi_1 = \ldots = \varphi_n = \varphi \in \Phi$. Then, according to what has been shown in Section 2 (see (2.24)), we have

$$:(\varphi, x)_{H_0}^k: \; = \; 2^{-k} \| \varphi \|_{H_0}^k H_k((\varphi / \| \varphi \|_{H_0}, x)_{H_0}) \quad (k \in \mathbb{Z}_+ ; x \in \Phi'). \tag{5.53}$$

Expanding $\| \varphi \|_{H_0}^{-n} (\varphi, x)_{H_0}^n$ into the series in $H_k((\varphi / \| \varphi \|_{H_0}, x)_{H_0})$ (according to the formula (2.18) for Hermite polynomials) and using relation (5.53), we obtain

$$(\varphi, x)_{H_0}^n = \sum_{k=0}^{[n/2]} \frac{n!}{k!(n-2k)!} (\| \varphi \|_{H_0} / 2) : (\varphi, x)_{H_0}^{n-2k} : .$$

Taking into account that

$$:(\varphi, x)_{H_0}^{n-2k}: \; = \; I_1 \left(2^{k-n/2} \sqrt{(n-2k)!} \, \varphi^{\otimes (n-2k)}\right),$$

we get

$$\| :(\varphi, \cdot)_{H_0}^{n-2k}: \|_{A_i(\Phi')}^2 = \| 2^{k-n/2} \sqrt{(n-2k)!} \, \varphi^{\otimes(n-2k)} \|_{\mathfrak{F}(H_i)}^2$$

$$= 2^{2k-n} (n-2k)! \, \| \varphi \|_{H_i}^{2(n-2k)} .$$

Due to the inequality $\| \varphi \|_{H_0} \leq \| \varphi \|_{H_i}$, this yields

$$\| :(\varphi, \cdot)_{H_0}^n : \|_{A_i(\Phi')}^2 \leq \| \varphi \|_{H_i}^{2n} \sum_{k=0}^{[n/2]} \left(\frac{n!}{k!} \right)^2 \frac{1}{(n-2k)!} 2^{-(n+2k)} \leq n! \, \| \varphi \|_{H_i}^{2n} .$$

Here, we have used the combinatorial identity

$$\sum_{k=0}^{[n/2]} \left(\frac{n!}{k! 2^k} \right)^2 \frac{1}{(n-2k)!} = 2^{-n} \frac{(2n)!}{n!} ,$$

and the inequality $(2n)! \le 2^{2n}(n!)^2$. In the case of different $\varphi_1, \dots, \varphi_n \in \Phi$, the statement of the lemma is obtained from the assertion already proved by applying the polarization identity (2.7). The results obtained can be easily extended to the case of complex $\varphi_k \in \Phi_c$ $(k = 1, \dots, n)$.

■

Sufficiency. If the moments of the measure μ on Φ' satisfy the estimates (5.51), then its Fourier transform is an entire function on $H_{j,c}$. In fact, by virtue of the inequalities for the moments, the expansion

$$\int_{\Phi'} e^{(z,x)_{H_0}} d\mu(x) = \sum_{n=0}^{\infty} \frac{1}{n!} \int_{\Phi'} (z,x)_{H_0} \, d\mu(x) = f_\mu(z) \quad (z \in \Phi_c)$$

can be extended to $H_{j,c}$, and the extended series converges uniformly on every ball in $H_{j,c}$. Furthermore, the estimate

$$|f_\mu(z)| \le \sum_{n=0}^{\infty} \frac{c^n}{\sqrt{n!}} \|z\|^n_{H_{j,c}} \le \left(\sum_{n=0}^{\infty} \frac{(c+1)^{2n}}{n!} \|z\|^{2n}_{H_{j,c}} \right)^{1/2} \left(\sum_{n=0}^{\infty} \left(\frac{c}{c+1} \right)^{2n} \right)^{1/2}$$

$$= c_1 \exp\left(\frac{(c+1)^2}{2} \|z\|^2_{H_{j,c}} \right) \quad (z \in H_{j,c})$$

shows that $f_\mu(z)$ is a function of finite type with the order of growth at most 2. Then

$$\int_{\Phi'} G(x,z) \, d\mu(x) = f_\mu(\sqrt{2}z) \exp((z,z)_{H_0}/2) \quad (z \in H_{j,c})$$

is an entire function on $H_{j,c}$ of finite type with the order of growth at most 2; it belongs to $\mathfrak{E}^2_{\max}(\Phi_c)$, i.e., it is the Gaussian transformation of the uniquely defined generalized function $\xi \in \mathcal{A}'(\Phi')$. The equality

$$(G\xi)(z) = (G(\cdot, z), \xi(\cdot))_{L_2(\Phi', \gamma_1)} = \int_{\Phi'} G(x,z) \, d\mu(x) \quad (z \in \Phi_c)$$

shows that $\xi = \xi_\mu = d\mu / d\gamma_1$.

■

Corollary 1. *If the measure* μ *satisfies the conditions of Theorem* 5.6, *then the imbedding*

$$\mathcal{A}(\Phi') \subset L_2(\Phi', \mu)$$

is dense topologically and, thus, the nuclear space $\mathcal{A}(\Phi')$ is suitable for the riggings of all the spaces $L_2(\Phi', \mu)$, where μ is a sub-Gaussian measure on Φ'.

■

Corollary 2. *Suppose that μ is a measure on Φ' whose moments satisfy the condition* (5.51). *Then $d\mu/d\gamma_1$ is a nonnegative generalized function from $\mathcal{A}'(\Phi')$ such that for any $u \in \mathcal{A}(\Phi')$, $u \geq 0$, we have*

$$(d\mu/d\gamma_1, u)_{L_2(\Phi', \gamma_1)} \geq 0. \tag{5.54}$$

Conversely, each nonnegative generalized function $\xi \in \mathcal{A}'(\Phi')$ defines a unique measure μ_ξ on Φ' such that $\xi = d\mu_\xi/d\gamma_1$.

Proof. For the measures from this class, the existence of $d\mu / d\gamma_1$ is established by Theorem 5.6. Its positiveness follows from the equality

$$(d\mu / d\gamma_1, u)_{L_2(\Phi', \gamma_1)} = \int_{\Phi'} u(x)\, d\mu(x) \geq 0,$$

since $u \geq 0$. Conversely, if $\xi \in \mathcal{A}'(\Phi')$, then

$$s_n(\varphi_1, \ldots, \varphi_n) = \left(\xi, \prod_{k=1}^{n} (\varphi_k, \cdot)_{H_0}\right)_{L_2(\Phi', \gamma_1)}$$

$$(\varphi_k \in \Phi,\ k = 1, \ldots, n;\ n \in \mathbb{Z}_+;\ s_0 = (\xi, 1)_{L_2(\Phi', \gamma_1)})$$

is the moment sequence of multilinear functionals (see Section 2, Chapter 5), and Lemma 5.6 guarantees the validity of the theorem on solvability of the moment problem presented in the same chapter. Thus, the reconstruction of the measure μ_ξ on Φ' such that

$$s_n(\cdot) = s_n^{\mu_\xi}(\cdot) \quad (n \in \mathbb{Z}_+)$$

is possible.

The measure μ_ξ satisfies (5.51) due to its construction. Therefore, Theorem 5.6 implies that $d\mu_\xi/d\gamma_1 \in \mathcal{A}'(\Phi')$. But the values of $d\mu_\xi/d\gamma_1$ and ξ coincide, by virtue of the definition of s_n $(n \in \mathbb{Z}_+)$ on the set $\mathcal{P}_{\text{cyl}}(\Phi')$ dense in $\mathcal{A}(\Phi')$. Consequently, $d\mu_\xi/d\gamma_1 = \xi$.

■

Corollary 2 to Theorem 5.6 enables us to construct the measures on Φ' from the given Gaussian measure γ_1 and the density (maybe generalized) from $\mathcal{A}'(\Phi')$. In order that this procedure define the measure on Φ', the corresponding density should be a positive generalized function. The example given below illustrates this approach.

Example 5.5. Fix arbitrary $\alpha \in \Phi'$. Then for $n \in \mathbb{N}$, one can define the generalized function $:(\alpha\,,\cdot)^n_{H_0}: \in \mathcal{A}'(\Phi')$. For this purpose, we choose a sequence $(\varphi_k)_{k=1}^{\infty} \subset \Phi$ ($\varphi_k \to \alpha$ as $k \to \infty$ in Φ') and consider the sequence of functions

$$(:(\varphi_k\,,\cdot)^n_{H_0}:)_{k=1}^{\infty} \subset \mathcal{A}(\Phi').$$

Since

$$:(\varphi_k\,,\cdot)^n_{H_0}: \;=\; I_1\,(2^{-n/2}\sqrt{n!}\;\varphi_k^{\oplus n}) \quad \text{and} \quad \varphi_j^{\oplus n} \underset{j\to\infty}{\to} \alpha^{\otimes n}$$

in $(\Phi')^{\otimes n}$, this sequence converges in $\mathcal{A}'(\Phi')$; denote its limit by $:(\alpha\,,\cdot)^n_{H_0}:$. This definition and (2.26) immediately yield the equality

$$\left(:(\alpha,\cdot)^n_{H_0}:,:\prod_{k=1}^{n}(\varphi_k,\cdot)_{H_0}:\right)_{L_2(\Phi',\gamma_1)} = 2^{-n}\,n!\prod_{k=1}^{n}(\alpha,\varphi_k)_{H_0} \tag{5.55}$$

$$(\varphi_k \in \Phi,\; k = 1, \dots, n).$$

Furthermore, for $\alpha \in H_{-j}$, the relation

$$:(\alpha\,,\cdot)^n_{H_0}: \;=\; I_1\,(2^{-n/2}\sqrt{n!}\;\alpha^{\otimes n})$$

implies that

$$\|:(\alpha\,,\cdot)^n_{H_0}:\|_{A_{-j}(\Phi')} = 2^{-n/2}\sqrt{n!}\;\|\alpha\|^n_{H_{-j}}. \tag{5.56}$$

Therefore,

$$:(\varphi_k\,,\cdot)^n_{H_0}: \underset{k\to\infty}{\to} :(\alpha\,,\cdot)^n_{H_0}:$$

strongly in $A_{-j}(\Phi')$.

Let F be the entire characteristic function, i.e., the entire function $F(z)$ $(z \in \mathbb{C}^1)$ such that there exists a finite measure $\mathfrak{B}(\mathbb{R}^1) \ni \beta \mapsto \sigma(\beta) \in [0, \infty)$ for which

$$F(z) = \int_{\mathbb{R}^1} e^{zs}\, d\sigma(s) \quad (z \in \mathbb{C}^1). \tag{5.57}$$

Denote by $f(t) = F(it)$ $(t \in \mathbb{R}^1)$ the restriction of F to the imaginary axis. Assume that F has the order of growth $\rho < 2$ and finite type. If

$$F(z) = \sum_{n=0}^{\infty} f_n z^n,$$

then

$$f(t) = \sum_{n=0}^{\infty} i^n f_n t^n \quad (t \in \mathbb{R}^1).$$

Lemma 5.7. *The function*

$$\xi_{F,\alpha} = :f((\alpha, \cdot)_{H_0}):$$

defined as a sum of the series

$$\sum_{n=0}^{\infty} i^n f_n :(\alpha\ ,\ \cdot)_{H_0}^n :$$

is a positive generalized function from $\mathcal{A}'(\Phi')$.

Proof. Since $:(\alpha, \cdot)_{H_0}^k :$ and $:(\alpha, \cdot)_{H_0}^l :$ are orthogonal in $A_{-j}(\Phi')$ for all $k \neq l$, it suffices to find $j \in \mathbb{N}$ such that

$$\sum_{n=0}^{\infty} |\, f_n\,|^2 \,\|:(\alpha\ ,\ \cdot)_{H_0}^n :\|_{A_{-j}(\Phi')}^2 = \sum_{n=0}^{\infty} |\, f_n\,|^2 \frac{2^{-n}}{n!} \|\alpha\|_{H_{-j}}^{2n} < \infty$$

in order to prove that the series from the condition of this lemma converges. The Taylor coefficients of the function F can be estimated in terms of the order of growth ρ and type τ, by using the well-known formula (see Markushevich [1, Section 1, Chapter 7]):

$$|\, f_n\,| \leq cn^{-n/\rho} (\tau\rho e)^{n/\rho} \quad (n \in \mathbb{Z}_+). \tag{5.58}$$

If we choose $j \in \mathbb{N}$ from condition (5.34) so that $\|\,\alpha\,\|_{H_{-j}} < (\rho\tau)^{-1/\rho}$ and insert (5.58) into the series under consideration, then we get

$$\|\,\xi_{F,\alpha}\|_{A_{-j}(\Phi')}^2 \leq c_1 \sum_{n=0}^{\infty} [(\tau\rho)^{1/\rho} \,\|\,\alpha\,\|_{H_{-j}}]^{2n} < \infty$$

(here, we have used the Stirling formula). Thus,

$$\xi_{F,\alpha} \in A_{-j}(\Phi') \subset \mathcal{A}'(\Phi').$$

To prove the positiveness of $\xi_{F,\alpha}$, let us consider another generalized function $\xi_{F,\varphi}$ (with $\varphi \in \Phi$) which is constructed similarly. By virtue of the formula

$$:(\varphi,\, x)^n_{H_0}: \;=\; 2^{-n}\,\|\varphi\|^n_{H_0}\, H_n\big(\|\varphi\|^{-1}_{H_0}(\varphi, x)_{H_0}\big)$$

and the integral representation for the Hermite polynomials

$$H_n(s) \;=\; \frac{e^{s^2}}{\sqrt{\pi}}\,(-2i)^n \int_{\mathbb{R}^1} t^n e^{2ist-t^2}\,dt \quad (s \in \mathbb{R}^1;\; n \in \mathbb{Z}_+),$$

the definition of $\xi_{F,\varphi}$ implies that

$$\xi_{F,\varphi} \;=\; \frac{\exp\Big((\varphi,\, x)^2_{H_0}\,\|\varphi\|^{-2}_{H_0}\Big)}{\sqrt{\pi}} \int_{\mathbb{R}^1} \exp\Big(2it(\varphi,\, x)_{H_0}\,\|\varphi\|^{-1}_{H_0}\Big)\, F(t\|\varphi\|_{H_0})\, e^{-t^2}\,dt.$$

Hence, $\xi_{F,\varphi}$ is a measurable smooth function (the convergence of the integral is guaranteed by the condition imposed on the order of growth of F). This function is positive as a product of a positive function and the Fourier transform of a positive-definite function. Therefore, for $u \in \mathcal{A}(\Phi')$ $(u \geq 0)$, we have

$$(\xi_{F,\varphi},\, u)_{L_2(\Phi',\gamma_1)} \;\geq\; 0.$$

But $\xi_{F,\varphi_k} \to \xi_{F,\alpha}$ as $k \to \infty$ weakly in $\mathcal{A}'(\Phi')$ whenever $\varphi_k \to \alpha$ as $k \to \infty$ in Φ'. This follows from the convergence of each term in the series which define these generalized functions. Hence, we have

$$(\xi_{F,\alpha},\, u)_{L_2(\Phi',\gamma_1)} \;\geq\; 0.$$

If we now employ Corollary 2 of Theorem 5.6, then for every entire characteristic function of the generalized direction $\alpha \in \Phi'$ which is of finite type and has the order of growth less than 2, we obtain the measure $\mu_{F,\alpha}$ on Φ' such that

$$d\mu_{F,\alpha}/d\gamma_1 \;=\; \xi_{F,\alpha} \in \mathcal{A}'(\Phi').$$

3. SPECTRAL THEOREM

The main part of this chapter will be devoted to the proof of a theorem on the expansion of an arbitrary family of commuting unbounded (in general) normal operators in generalized joint eigenvectors (the "spectral projection theorem"). This theorem singles out "projectors" on the generalized eigensubspaces. The spectral integrals themselves are continual. At the end of this chapter and in Chapter 4, we present applications of this theorem which illustrate the convenience of this form of spectral theorem for certain problems.

We clarify the foregoing. Let A be a selfadjoint operator defined in a Hilbert space H_0. It can be associated with a resolution of the identity (RI) E, i.e., with a projection-valued measure on the σ-algebra $\mathfrak{B}(\mathbb{R}^1)$ of Borel subsets of $\mathbb{R}^1$ for which we have the following spectral integral representations

$$1 = \int_{\mathbb{R}^1} dE(\lambda), \quad A = \int_{\mathbb{R}^1} \lambda dE(\lambda). \tag{0.1}$$

For an operator A with a discrete spectrum $(\lambda_j)_{j=1}^{\infty}$, we have

$$E(\alpha) = \sum_{\lambda_j \in \alpha} P(\lambda_j) \quad (\alpha \in \mathfrak{B}(\mathbb{R}^1)),$$

where $P(\lambda_j)$ is the projector onto the eigensubspace corresponding to λ_j, and the relations (0.1) turn into the equalities

$$1 = \sum_{j=1}^{\infty} P(\lambda_j) \quad \text{and} \quad A = \sum_{j=1}^{\infty} \lambda_j P(\lambda_j).$$

These equalities cannot be written in the case of a continuous spectrum. However, on the basis of the works by Gelfand and Kostyuchenko [1], Berezansky [1], and Kats [1] concerning generalized eigenfunction expansions, it is possible to generalize these equalities for separable H_0. This can be done as follows (see Berezansky [3], and [5], Chapter 5):

Let

$$H_- \supset H_0 \supset H_+ \supset D \tag{0.2}$$

be a rigging of the space H_0 by Hilbert spaces H_+ and H_- of "test" and "generalized" vectors with a quasinuclear (i.e., Hilbert–Schmidt) imbedding $H_+ \subset H_0$, and let D be a linear topological space topologically (i.e., densely and continuously) imbedded in H_+. Assume that the operator A is connected with the chain (0.2) in the standard way: its domain of definition $\mathfrak{D}(A)$ contains D, and the restriction $A \restriction D$ acts continuously from D to H_+. Then the RI E, regarded as an operator-valued measure whose values are operators acting from H_+ to H_-, can be differentiated with respect to a certain scalar measure $\mathfrak{B}(\mathbb{R}^1) \ni \alpha \mapsto \rho(\alpha) \in [0, \infty)$ (the so-called spectral measure): there exists

$$(dE/d\rho)(\lambda) = P(\lambda): H_+ \to H_-,$$

and the equalities (0.1) can be rewritten in the form

$$1 = \int_{\mathbb{R}^1} P(\lambda)\, d\rho(\lambda), \quad A = \int_{\mathbb{R}^1} \lambda P(\lambda)\, d\rho(\lambda). \tag{0.3}$$

It can be shown that for ρ-almost all $\lambda \in \mathbb{R}^1$, the range $\mathfrak{R}(P(\lambda))$ consists of generalized eigenvectors $\xi \in H_-$ corresponding to λ (i.e., $A\xi = \lambda\xi$ in the natural generalized sense). Thus, $P(\lambda)$ "projects" onto the corresponding generalized eigensubspace, and the equalities (0.3) generalize the equalities given above in the case of a discrete spectrum and constitute the substance of the spectral projection theorem for a single operator.

It is often convenient to consider a somewhat more special situation when the "quasinuclear" chain (0.2) is replaced by a "nuclear" chain

$$\Phi' \supset H_0 \supset \Phi, \tag{0.4}$$

where Φ is a nuclear space topologically imbedded in H_0 and Φ' is the dual space of continuous antilinear functionals on Φ. A picture analogous to (0.2) holds for a selfadjoint operator A defined in H_0 satisfying the more stringent restriction that $\Phi \subset \mathfrak{D}(A)$, and $A \restriction \Phi$ acts continuously in Φ. More precisely, the required chain (0.2) can be chosen for the given chain (0.4), and then the results (0.3) described above are preserved for A.

This is the situation in the case of a single selfadjoint operator. Now consider a family $A = (A_x)_{x \in X}$ of selfadjoint operators A_x acting in H_0, commuting in the sense that their RI's E_x commute, and connected with the chain (0.2) in the standard way. If $X = \{1, \dots, n\}$ is finite, then we can construct a joint RI for the family A: a projection-valued measure defined on the σ-algebra $\mathfrak{B}(\mathbb{R}^n)$ in the space $\mathbb{R}^n$ of points $\lambda = (\lambda_1, \dots, \lambda_n)$. The formulas (0.1) and (0.3) are preserved, one should only replace A by A_x and λ by λ_x $(x \in X)$ (Berezansky [3, 5]).

In this chapter, we obtain analogues of the representations (0.1) and (0.3) in the case when the index set X has arbitrary cardinality (for example, X may coincide with the space $\mathcal{D}(\mathbb{R}^d)$ of test functions. This is a typical quantum field situation). Difficulties now arise that are especially noticeable when we are dealing with a family of unbounded operators which is more than countable. Thus, the joint RI of the family A should evidently be a projection-valued measure on the space $\mathbb{R}^X$ of all possible functions $X \ni x \mapsto \lambda(x) \in \mathbb{R}^1$. This RI E can be constructed by use of the Kolmogorov theorem on extending a finite-dimensional distribution to a measure. However, it is defined on a σ-algebra $\mathfrak{R}$ poor enough in comparison with the σ-algebra $\mathfrak{B}(\mathbb{R}^X)$ of Borel sets ($\mathbb{R}^X$ is endowed with the Tykhonov topology). For instance, if X is more than countable, then the single-point sets do not belong to $\mathfrak{R}$.

This creates a number of difficulties, in particular, when working with the concept of a support $\operatorname{supp} E$ of a measure of this sort. The concept of a support is essential for the problems under discussion. For example, in the integrals (0.1) and (0.3), the domain of integration $\mathbb{R}^1$ can be replaced by the spectrum of the operator A, i.e., by $\operatorname{supp} E$. A similar transition to the spectrum would be natural and useful for general X, too (see also below). However, if X is more than countable and the operators A_x are unbounded, then the situation when $\operatorname{supp} E = \varnothing$ is possible (even though $E \neq 0$), and this transition cannot be realized.

Further, the role of a generalized eigenvector must now be played by a generalized joint eigenvector, i.e., a vector $\xi \in H_-$ such that $A_x \xi = \lambda(x)\xi$ in the generalized sense for every $x \in X$. The function $\lambda(\cdot) \in \mathbb{R}^X$ is the "eigenvalue" which corresponds to ξ. In the case of a single operator, the set $\mathfrak{R}(P(\lambda))$ consists of generalized eigenvectors for ρ-almost all λ. Similarly, for fixed x, there now exists a set $\beta_x \subset \mathbb{R}^X$ of full measure ρ (or E) such that for $\lambda(\cdot) \in \beta_x$ the set $\mathfrak{R}(P(\lambda))$ consists of the vectors ξ for which $A_x\xi = \lambda(x)\xi$ To obtain a similar equality for all $x \in X$, we would like to take the set $\pi = \bigcap_{x \in X} \beta_x$ and consider $P(\lambda(\cdot))$ for $\lambda(\cdot) \in \pi$. If X is at most countable, then π has full measure, and this procedure gives the required result. However, the "continuum difficulty" arises when X is more than countable: π can fail to be a set of full measure. Overcoming this difficulty involves constructing π in a proper way.

We have already mentioned that for general families it is not completely reasonable to regard the support $\operatorname{supp} E$ of the joint RI as the spectrum of a family A. Indeed, it is, generally speaking, impossible for the spectral integrals to be over $\operatorname{supp} E$. Thus, the following question arises: what set is a natural candidate for the role of a spectrum of a family from this point of view; or, more precisely, what is a minimal (in a certain sense) set $g(A) \subset \mathbb{R}^X$ for which, e.g., the following modification of the second formula in (0.1) holds

$$A_x = \int_{g(A)} \lambda(x)\, dE(\lambda(\cdot)) \quad (x \in X) \tag{0.5}$$

(here, the function $\mathbb{R}^X \supset g(A) \ni \lambda(\cdot) \mapsto \lambda(x) \in \mathbb{R}^1$ is being integrated)? It turns out that $g(A)$ can be regarded as the generalized spectrum of A, i.e., as the collection of all eigenvalues corresponding to generalized joint eigenvectors (for a fixed chain (0.2) or (0.4)).

Let us explain what consequences can be derived, in particular, from spectral representations such as (0.5); similar consequences are presented in Chapter 4.[20]

Suppose that the operators A_x are bounded and connected by certain relations for different $x \in X$. For example, let X be an Abelian group, and $A_{x+y} = A_x A_y$ $(x, y \in X)$, or let X be a linear space, and $A_{\alpha x+\beta y} = \alpha A_x + \beta A_y$ $(x, y \in X;\ \alpha, \beta \in \mathbb{C}^1)$, etc.; i.e., a function $X \ni x \mapsto A_x$ realizes a representation of one or another algebraic object, namely of a group, a linear space, an algebra, etc. Moreover, the relations may be of a less traditional type, for instance, $X = \mathbb{R}^1$, and $A_x A_y = \frac{1}{2}(A_{x+y} + A_{x-y})$ $(x, y \in \mathbb{R}^1)$, or A_x may satisfy some differential equation with respect to x, and so on.

Let us clarify how the representation (0.5) is now transformed. To do this, we note that applying the relations given above to the joint generalized eigenvector and taking into account the equality $A_x \xi = \lambda(x)\xi$ $(x \in X)$, we obtain equations of the form

$$\lambda(x+y) = \lambda(x)\lambda(y), \quad \lambda(\alpha x + \beta y) = \alpha\lambda(x) + \beta\lambda(y), \quad \lambda(x)\lambda(y) = \tfrac{1}{2}(\lambda(x+y) + \lambda(x-y)),$$

and so on, for the corresponding eigenvalues $\lambda(\cdot)$. Thus, the set $g(A)$ in (0.5) can be replaced by the space of all the solutions of the corresponding scalar equation (functional equation, differential equation, etc.). This, in general, increases the region of integration, and we set E to be zero on the corresponding sets. As a result, the "multiplicity" of the integral (0.5) essentially diminishes.

In a number of cases, these scalar equations can be solved explicitly. Thus, for classical Stone's theorem, $\lambda(x) = e^{i\lambda x}$ $(x \in X = \mathbb{R}^1)$, where $\lambda \in \mathbb{R}^1$, and we obtain the mapping $g(A) \ni \lambda(\cdot) = e^{i\lambda x} \mapsto \lambda \in \mathbb{R}^1$. Transporting the measure $E(\alpha)$ $(\alpha \subset g(A))$ to the RI F according to this mapping, we can rewrite (0.5) in the classical form

$$A_x = \int_{\mathbb{R}^1} e^{i\lambda x}\, dE(\lambda(\cdot)) \quad (x \in \mathbb{R}^1).$$

It is clear from the foregoing that the method developed here enables us to obtain broad generalizations of theorems of the Stone type and of the Sz.-Nagy – Hille type (on groups of selfadjoint operators), as well as generalizations of the results of other authors on spectral representations of families of operators. The next chapter contains these generalizations. Connections of these with spectral representations of positive definite functions and with the moment problem, and applications to noncommuting operators are studied in Chapters 4 and 5.

[20] Note first that everything said is also valid for normal commuting operators, one should only replace $\mathbb{R}^X$ by $\mathbb{C}^X$.

In Section 1, we construct and investigate a joint RI E by use of the RI's E_x of the operators A_x. The constructions presented here recall the constructions of measure theory and exploit these. However, they conceal some unexpected features: for example, two abstract commuting RI's cannot always be multiplied.

In Section 2, we present a procedure for differentiating a joint RI regarded as a measure whose values are operators from H_+ to H_-. The projection properties of the derivative $P(\lambda(\cdot)): H_+ \to H_-$ obtained as a result are then studied, and the main result of the chapter is proved, namely the spectral projection theorem.

When using our method of finding spectral representations for the families of operators A_x, we assume that these operators are connected with the riggings (0.2) or (0.4) in the standard way. In Section 3, we show that the converse statement is also valid: the existence of sufficiently good representations automatically implies the existence of riggings. In this section, we also discuss the following problem: There exists another approach to the theory of expansions in generalized joint eigenvectors for a family $A = (A_x)_{x \in X}$; it is based on the investigation of the C^*-algebra $\mathcal{A}$ associated with A (see Maurin [1]–[4], the "nuclear spectral theorem"). This approach is effective in the case of bounded operators A_x when $\mathcal{A}$ is constructed as the closure of the algebraic span of A in the operator norm[21]. It is now possible to identify $\operatorname{supp} E$ with the maximal ideal space of the algebra $\mathcal{A}$. The corresponding facts are presented in Subsection 3.5. But this approach becomes inconvenient when there exists a massive set of unbounded operators among A_x[22].

In papers [2] and [3], Richter outlined another method for the investigation of expansions for a family A which is based on the Choquet theorem. In both of these approaches, the spectral integrals for general families A are taken over a certain space which is mapped into the spectrum, and it proves to be impossible to rewrite these integrals as integrals of the type (0.5) over the generalized spectrum itself in the case of unbounded operators. This makes it difficult to apply such representations, for example, to theorems of the Stone type or Sz.-Nagy–Hille type.

Section 3 also contains a number of other facts connected with the construction of expansions (direct integrals of Hilbert spaces, expansions for Carleman operators, some important examples).

---

21 However, when X is more than countable, the results are somewhat less precise in this case, too. This means that, in the framework of this approach, the function $\lambda(\cdot) \mapsto P(\lambda(\cdot))$ is measurable with respect to a more extensive σ-algebra than in reality.

22 We note that if A_x is connected with the chain (0.2) or (0.4) in the standard way, then, as a rule, this is not true for a function $F(A_x)$ of it. This is why, for example, it is hard to construct the expansion in generalized eigenfunctions of a selfadjoint differential operator A by using bounded operators e^{itA} $(t \in \mathbb{R}^1)$.

§1. Joint Resolution of the Identity

A joint RI can be constructed as an operator-valued measure equal to an infinite product of RI's of commuting normal operators in a way similar to that for infinite products of probability measures. We shall need the product of commuting abstractly defined RI's.

1.1. General Resolution of the Identity

Let R be an abstract space, $\mathfrak{R}$ be a σ-algebra of subsets of it, and let $\mathfrak{R} \ni \alpha \mapsto E(\alpha)$ be an operator-valued function on $\mathfrak{R}$ whose values are (orthogonal) projectors in a fixed Hilbert space H. This function E is called an RI on $\mathfrak{R}$ provided that:

(a) $E(\varnothing) = 0$ and $E(R) = 1$, where 1 is the identity operator on H;

(b) the absolute additivity property holds, i.e., $E(\bigcup_{j=1}^{\infty} \alpha_j) = \sum_{j=1}^{\infty} E(\alpha_j)$ for any sequence $(\alpha_j)_{j=1}^{\infty}$ with $\alpha_j \in \mathfrak{R}$ and $\alpha_j \cap \alpha_k = \varnothing$ for $j \neq k$, where the series converges in the sense of strong operator convergence.

Remark 1. The orthogonality relation always holds for an RI: $E(\alpha \cap \beta) = E(\alpha)E(\beta)$ $(\alpha, \beta \in \mathfrak{R})$ (and, thus, the operators $E(\alpha)$ $(\alpha \in \mathfrak{R})$ are commuting). This is a simple consequence of additivity and the fact that a sum of two projectors is a projector only if they are orthogonal.

■

Remark 2. One can demand weak convergence of the series in condition (b) instead of strong convergence: Due to the orthogonality of terms, it will automatically converge strongly, too.

■

Remark 3. An RI can also be defined on some algebra $\mathfrak{R}$ (but not on a σ-algebra). The definition stays as before , but in (b) the requirement that $\bigcup_{j=1}^{\infty} \alpha_j \in \mathfrak{R}$ should be added. Remarks 1 and 2 are preserved for an RI on an algebra.

■

In the case of an RI corresponding to some selfadjoint (normal) operator acting in H, we have $R = \mathbb{R}^1$, $\mathfrak{R} = \mathfrak{B}(\mathbb{R}^1)$ $(R = \mathbb{C}^1$, $\mathfrak{R} = \mathfrak{B}(\mathbb{C}^1))$.

Theorem 1.1. *Let E be an RI on an algebra $\mathfrak{R}$, then there exists an exten-*

sion of it to an RI E_σ *on its* σ*-span* $\mathfrak{R}_\sigma$, *i.e., the restriction* $E_\sigma \restriction \mathfrak{R}$ *is E.* E_σ *is determined uniquely for given E.*

Proof. Fix $f \in H$ and consider the finite measure $\mathfrak{R} \ni \alpha \mapsto \rho_{f,f}(\alpha) = (E(\alpha) f, g)_H \in [0, \infty)$. According to the usual extension theory, there exists a measure $\mathfrak{R}_\sigma \ni \alpha \mapsto \hat{\rho}_{f,f}(\alpha) \in [0, \infty)$ such that $\hat{\rho}_{f,f}(\alpha) \restriction \mathfrak{R} = \rho_{f,f}$. By virtue of the polarization formula, the complex-valued measure (charge) $\mathfrak{R} \ni \alpha \mapsto \rho_{f,g}(\alpha) = (E(\alpha) f, g)_H \in \mathbb{C}^1$ can be expressed as a linear combination of four measures of the form $\rho_{h,h}$ $(h \in H)$. By extending each of these measures from $\mathfrak{R}$ to $\mathfrak{R}_\sigma$ and taking the corresponding linear combination, we obtain the extension of the charge $\rho_{f,g}$ to a charge $\mathfrak{R}_\sigma \ni \alpha \mapsto \hat{\rho}_{f,g}(\alpha) \in \mathbb{C}^1$.

For fixed $\alpha \in \mathfrak{R}_\sigma$, the mapping

$$H \times H \ni (f, g) \mapsto \hat{\rho}_{f,g}(\alpha) \in \mathbb{C}^1 \tag{1.1}$$

is bilinear. Indeed, it is well known that $\mathfrak{R}_\sigma$ coincides with the monotone span of $\mathfrak{R}$ and, thus, $\hat{\rho}_{f,f}$(and, consequently, $\hat{\rho}_{f,g}$) is constructed by successive monotone extensions, starting with the sets $\alpha \in \mathfrak{R}$. Since bilinearity is preserved for monotone extensions, the required property follows from the bilinearity of (1.1) for $\alpha \in \mathfrak{R}$.

The mapping (1.1) is continuous. Indeed, $\hat{\rho}_{f,f}(\alpha) \geq 0$ for $\alpha \in \mathfrak{R}_\sigma$. Therefore,

$$|\hat{\rho}_{f,g}(\alpha)| \leq \hat{\rho}_{f,f}(\alpha)\, \hat{\rho}_{g,g}(\alpha) \leq \hat{\rho}_{f,f}(R)\, \hat{\rho}_{g,g}(R) = \|f\|_H^2\, \|g\|_H^2$$

by virtue of the Cauchy–Schwarz–Bunyakovsky inequality. Denote by $E_\sigma(\alpha)$ the bounded operator in H corresponding to the continuous bilinear form (1.1). Hence,

$$\hat{\rho}_{f,g}(\alpha) = (E_\sigma(\alpha) f, g)_H \quad (f, g \in H;\ \alpha \in \mathfrak{R}_\sigma). \tag{1.2}$$

For $\alpha \in \mathfrak{R}$, $E_\sigma(\alpha) = E(\alpha)$ is a projector. The successive monotone extensions of $\hat{\rho}_{f,g}$ mean that it is extended from $\alpha \in \mathfrak{R}$ to $\alpha \in \mathfrak{R}_\sigma$ by weak limits of monotone sequences of projectors. Therefore, $E_\sigma(\alpha)$ is also a projector for any $\alpha \in \mathfrak{R}_\sigma$. Clearly, $E_\sigma(\varnothing) = 0$ and $E_\sigma(R) = 1$. The set function $\mathfrak{R} \ni \alpha \mapsto E_\sigma(\alpha)$ is absolutely additive in the sense of weak convergence of the corresponding series – according to (1.2), this is a rephrasing of the absolute additivity of the charge $\hat{\rho}_{f,g}$ $(f, g \in H)$. Thus, E_σ is an RI defined on $\mathfrak{R}_\sigma$. On $\mathfrak{R}$ it coincides with E by construction. The last statement of the theorem follows from the uniqueness of the extension of a charge from $\mathfrak{R}$ to $\mathfrak{R}_\sigma$.

■

1.2. Products of Finitely Many Resolutions of the Identity

Let R_j be a space, $\mathfrak{R}_j$ be a σ-algebra of its subsets, and let $\mathfrak{R}_j \ni \alpha_j \mapsto E_j(\alpha_j)$ be an RI ($j = 1, \ldots, n$). Assume that the RI's $E_1, \ldots, E_n$ commute, i.e.,

$$E_j(\alpha_j)\, E_k(\alpha_k) = E_k(\alpha_k)\, E_j(\alpha_j) \quad (\alpha_j \in \mathfrak{R}_j, \alpha_k \in \mathfrak{R}_k, \; j, k = 1, \ldots, n).$$

The question is whether it is possible (as in the case of ordinary measures) to define an RI E (given on the σ-algebra $\mathfrak{R} = \mathfrak{R}_1 \times \ldots \times \mathfrak{R}_n$ spanned by all the rectangles $\alpha_1 \times \ldots \times \alpha_n$) in the direct product $R = R_1 \times \ldots \times R_n$ so that

$$E(\alpha_1 \times \ldots \times \alpha_n) = E(\alpha_1) \ldots E(\alpha_n) \quad (\alpha_1 \in \mathfrak{R}_1, \ldots, \alpha_n \in \mathfrak{R}_n).$$

This RI E is called the product of the RI's $E_1, \ldots, E_n$, or the joint RI generated by these RI's; we denote it as follows: $E = E_1 \times \ldots \times E_n = \times_{k=1}^{n} E_k$. As has been already mentioned, it does not always exist (the corresponding counterexample is presented in Subsection 1.7). However, if the RI's to be multiplied are defined on "nice" σ-algebras, then E exists. Let us present the corresponding result. Surely, it suffices to investigate the case $n = 2$.

Let R be a complete separable metric space, and $\mathfrak{R} = \mathfrak{B}(R)$ be the σ-algebra of Borel subsets of it. An RI defined on $\mathfrak{B}(R)$ is called a Borel RI. The joint RI automatically exists if the RI's to be multiplied are Borel RI's. Before proving this, we mention the following important fact:

Every finite scalar measure $\mathfrak{B}(R) \ni \alpha \mapsto \mu(\alpha) \in [0, \infty)$ is automatically regular, i.e.,

$$\mu(\alpha) = \inf_{o \supset \alpha} \mu(o) \quad (\alpha \in \mathfrak{B}(R), \; o \text{ is open}). \tag{1.3}$$

By passing to complements, it is easy to show that (1.3) is equivalent to the relation

$$\mu(\alpha) = \sup_{\varphi \supset \alpha} \mu(\varphi) \quad (\alpha \in \mathfrak{B}(R), \; \varphi \text{ is closed}), \tag{1.4}$$

furthermore, the supremum in (1.4) can be taken only over the compact sets $\varphi \subset \alpha$ (Ulam's theorem, see Billingsley [1, Chapter 1, Section 1]).

Theorem 1.2 *Suppose that E_1 and E_2 are two commuting Borel RI's in the spaces R_1 and R_2, respectively. Then there exists a joint RI $E = E_1 \times E_2$ on $\mathfrak{R} = \mathfrak{B}(R_1, R_2)$ uniquely determined for given E_1 and E_2.*

Proof. Denote by $\mathfrak{R}'$ the σ-algebra generated by the collection of all the rectangles $\alpha_1 \times \alpha_2$ $(\alpha_1 \in \mathfrak{B}(R_1), \alpha_2 \in \mathfrak{B}(R_2))$. By additivity, the function $E(\alpha_1 \times \alpha_2) = E(\alpha_1) E(\alpha_2)$ can be uniquely extended to a finitely additive operator-valued function $\mathfrak{R}' \ni \alpha \mapsto E(\alpha)$. This can be proved by the usual argument as in the case of a scalar measure. Each value $E(\alpha)$ $(\alpha \in \mathfrak{R}')$ is a projector, because α can be represented as a union of finitely many disjoint rectangles $\alpha_1 \times \alpha_2$ and, therefore, $E(\alpha)$ is equal to a finite sum of projectors $E(\alpha_1 \times \alpha_2) = E(\alpha_1) E(\alpha_2)$ which are mutually orthogonal due to the orthogonality relations for E_1 and E_2. Hence, $\mathfrak{R}' \ni \alpha \mapsto E(\alpha)$ is a projector-valued finitely additive set function on the algebra $\mathfrak{R}'$; we also have $E(\varnothing) = 0$ and $E(R_1 \times R_2) = 1$. Let us prove that it is absolutely additive.

Note that for each rectangle $\alpha_1 \times \alpha_2$ $(\alpha_1 \in \mathfrak{B}(R_1), \alpha_2 \in \mathfrak{B}(R_2))$, $f \in H$, and $\varepsilon > 0$ one can find two rectangles of the form $o_1 \times o_2$ and $\varphi_1 \times \varphi_2$ ($o_1 \supset \alpha_1$ and $o_2 \supset \alpha_2$ are open, and $\varphi_1 \subset \alpha_1$ and $\varphi_2 \subset \alpha_2$ are compact), such that

$$(E(o_1 \times o_2) f, f)_H - (E(\alpha_1 \times \alpha_2) f, f)_H < \varepsilon,$$

$$(E(\alpha_1 \times \alpha_2) f, f)_H - (E(\varphi_1 \times \varphi_2) f, f)_H < \varepsilon. \tag{1.5}$$

Let us establish, for example, the first relation. By using what has been said about the regularity of scalar measures on $\mathfrak{B}(R_1)$ and $\mathfrak{B}(R_2)$, for given $\delta > 0$ we choose two open sets $o_1 \supset \alpha_1$ and $o_2 \supset \alpha_2$ so that $(E_1(o_1 \setminus \alpha_1) f, f)_H < \delta$ and $(E_2(o_2 \setminus \alpha_2) f, f)_H < \delta$. Since

$$(o_1 \times o_2) \setminus (\alpha_1 \times \alpha_2) = ((o_1 \setminus \alpha_1) \times o_2) \cup (\alpha_2 \times (o_2 \setminus \alpha_2)),$$

we have

$$\begin{aligned}(E(o_1 \times o_2) f, f)_H - (E(\alpha_1 \times \alpha_2) f, f)_H &= (E((o_1 \setminus \alpha_1) \times o_2) f, f)_H \\ + (E(\alpha_1 \times (o_2 \setminus \alpha_2)) f, f)_H &= (E_1(o_1 \setminus \alpha_1) E_2(o_2) f, f)_H \\ + (E_1(\alpha_1) E_2(o_2 \setminus \alpha_2) f, f)_H &\le \|f\|_H (\|E_1(o_1 \setminus \alpha_1) f\|_H \\ + \|E_2(o_2 \setminus \alpha_2) f\|_H) &= \|f\|_H ((E_1(o_1 \setminus \alpha_1) f, f)_H^{1/2} \\ + (E_2(o_2 \setminus \alpha_2) f, f)_H^{1/2}) &< 2 \|f\|_H \delta^{1/2}.\end{aligned}$$

By taking $\delta > 0$ sufficiently small, we obtain the required result.

Suppose now that $\alpha_1, \alpha_2, \ldots \in \mathfrak{R}'$ are disjoint and such that $\alpha = \bigcup_{k=1}^{\infty} \alpha_k \in \mathfrak{R}'$.

To prove that E is absolutely additive, it suffices to show that

$$(E(\alpha)f,f)_H \le \sum_{k=1}^{\infty} (E(\alpha_k)f,f)_H \tag{1.6}$$

for any $f \in H$.[23]

Taking into account that α is equal to the union of finitely many disjoint rectangles, we apply the second inequality in (1.5) to each of these rectangles. As a result, for given $\varepsilon > 0$, we find a compact set $\mathfrak{R}' \ni \varphi \subset \alpha$ such that

$$(E(\alpha)f,f)_H - (E(\varphi)f,f)_H < \varepsilon.$$

Similarly, by using the first inequality in (1.5), for every $k \in \mathbb{N}$ one can find an open set $\mathfrak{R}' \ni o_k \supset \alpha$ such that

$$(E(o_k)f,f)_H - (E(\alpha_k)f,f)_H < \varepsilon/2^k.$$

The family $(o_k)_{k=1}^{\infty}$ covers φ. Since φ is compact, there exists $n \in \mathbb{N}$ such that $\bigcup_{k=1}^{n} o_k \supset \varphi$. By using this fact, the monotonicity and finite semi-additivity of a scalar measure, and the last two inequalities, we obtain

$$(E(\alpha)f,f)_H < (E(\varphi)f,f)_H + \varepsilon \le \left(E\left(\bigcup_{k=1}^{n} o_k\right)f,\ f\right)_H + \varepsilon$$

$$\le \sum_{k=1}^{n} (E(o_k)f,f)_H + \varepsilon < \sum_{k=1}^{\infty} (E(\alpha_k)f,f)_H + 2\varepsilon.$$

Passing here to the limit as $\varepsilon \to 0$, we arrive at (1.6). The absolute additivity of E is thus proved.

Hence, we have constructed an RI E on the σ-algebra $\mathfrak{R}'$. According to Theorem 1.1, we extend it to an RI E_σ on $(\mathfrak{R}')_\sigma = \mathfrak{R}$; obviously, this RI is the required one. The equality $\mathfrak{R} = \mathfrak{B}(R_1 \times R_2)$ follows from the definition of the topology in $R_1 \times R_2$. ■

---

[23] The opposite inequality follows from the finite additivity and the monotonicity of the function $\mathfrak{R}' \ni \beta \mapsto (E(\beta)f,\ f)_H \in [0,\infty)$; therefore, (1.6) means that these expressions are equal.

1.3. Construction of a Joint Resolution of the Identity in the General Case

Let X be a set of indices x of arbitrary cardinality, and let $(R_x)_{x \in X}$ be a family of complete separable metric spaces. Assume that a Borel RI $E_x : \mathfrak{B}(R_x) \ni \alpha \mapsto E_x(\alpha)$ is given on each R_x, and that values of this RI are projectors $E_x(\alpha)$ in a fixed Hilbert space H. The family $(E_x)_{x \in X}$ is assumed to be a commuting family, i.e.,

$$E_x(\alpha)\, E_y(\beta) = E_y(\beta)\, E_x(\alpha) \quad (\alpha \in \mathfrak{B}(R_x),\ \beta \in \mathfrak{B}(R_y),\ x, y \in X).$$

We want to construct a corresponding joint RI $E = E_X = \times_{x \in X} E_x$; E may be called the product of the RI's E_x. This construction is quite similar to the method of construction of product measures presented in Subsection 1.1 in Chapter 2.

The joint RI E is an RI in the space $R_X = \times_{x \in X} R_x$ consisting of all the mappings $\lambda(\cdot)$ of the form

$$X \ni x \mapsto \lambda(x) \in R_x$$

(the topology in R_X is not yet introduced). Let us construct a σ-algebra of sets from R_X on which E is defined. Recall that a set $\mathfrak{C} \subset R_X$ is called cylindrical if it is determined by different points $x_1, \dots, x_n \in X$ (its coordinates) and a basis $\delta \in \mathfrak{B}(R_{x_1, \dots, x_n})$ $(R_{x_1, \dots, x_n} = R_{x_1} \times \dots \times R_{x_n})$ by means of the relation

$$\mathfrak{C} = \mathfrak{C}(x_1, \dots, x_n; \delta) = \{\lambda(\cdot) \in R_X \mid (\lambda(x_1), \dots, \lambda(x_n)) \in \delta\}. \tag{1.7}$$

As mentioned above (Chapter 2, Subsection 1.1), coordinates and a base of a given cylindrical set $\mathfrak{C}$ cannot be uniquely determined if this set is written in the form (1.7). A collection of all cylindrical sets forms an algebra $\mathfrak{C}(R_X)$, and $\mathfrak{C}_\sigma(R_X)$ is its σ-span.

Theorem 1.3. *Given a family $(E_x)_{x \in X}$ of commuting Borel RI's E_x, one can construct a unique RI $\mathfrak{C}(R_X) \ni \alpha \mapsto E(\alpha)$ such that*

$$E(\mathfrak{C}(x_1, \dots, x_n; \delta_1 \times \dots \times \delta_n)) = E_{x_1}(\delta_1) \dots E_{x_n}(\delta_n)$$

$$(x_1, \dots, x_n \in X;\ \delta_k \in \mathfrak{B}(R_{x_k}),\ k = 1, \dots, n;\ n \in \mathbb{N}), \tag{1.8}$$

E is called the joint RI constructed from $(E_x)_{x \in X}$.

Proof. According to Subsection 2, let us construct the joint RI

$$E_{x_1, \dots, x_n} = E_{x_1} \times \dots \times E_{x_n}$$

defined on $\mathfrak{B}(R_{x_1, \dots, x_n})$, and set

$$E(\mathfrak{C}(x_1, \dots, x_n ; \delta)) = E_{x_1, \dots, x_n}(\delta) \quad (\delta \in \mathfrak{B}(R_{x_1, \dots, x_n});\ n \in \mathbb{N}). \tag{1.9}$$

This definition is correct: Each cylindrical set $\mathfrak{C}$ is uniquely associated with the operator $E(\mathfrak{C})$. Indeed, each change of the form in which $\mathfrak{C}$ is written is connected either with the permutation of coordinates or with the direct multiplication of δ by R_y, or with the removal of this factor from δ. Since $E_y(R_y) = 1$ $(y \in X)$, it follows from the construction of $E_{x_1, \dots, x_n}(\delta)$ that $E(\mathfrak{C})$ is independent of the form in which $\mathfrak{C}$ is written. The set function $\mathfrak{C}(R_X) \ni \mathfrak{C} \mapsto E(\mathfrak{C})$ is such that $E(\varnothing) = 0$, $E(R_X) = 0$, and satisfies the property of finite additivity.[24] Furthermore, it satisfies the condition of absolute additivity, too: Assume that $f \in H$ is fixed. Then the nonnegative set function $\mathfrak{C}(R_X) \ni \mathfrak{C} \mapsto (E(\mathfrak{C})f, f)_H$ is a consistent system of finite-dimensional distributions which can be extended to a measure defined on $\mathfrak{C}_\sigma(R_X)$ by Kolmogorov's theorem (see Gikhman, Skorokhod [1, Chapter 1, Section 4]); but this means that, anyway, it is absolutely additive on $\mathfrak{C}(R_X)$ which is equivalent to condition (b). Thus, $\mathfrak{C}(R_X) \ni \mathfrak{C} \mapsto E(\mathfrak{C})$ is an RI defined on the algebra $\mathfrak{C}(R_X)$. According to Theorem 1.1, E can be extended to an RI E_σ on $\mathfrak{C}_\sigma(R_X)$, and this is obviously the required RI.

■

Remark 1. Every RI E defined on $\mathfrak{C}_\sigma(R_X)$ can be constructed with the help of Theorem 1.3 from some family $(E_x)_{x \in X}$ of commuting Borel RI's. Indeed, for each $x \in X$, it suffices to set

$$\mathfrak{B}(R_x) \ni \alpha \mapsto E_x(\alpha) = E(\mathfrak{C}(x ; \alpha))$$

and multiply these RI's.

■

If $R_x = C$ for any $x \in X$, then we use the notation $R_X = C^X$. Thus, the joint RI generated by RI's of selfadjoint (normal) operators is defined on $\mathbb{R}^X(\mathbb{C}^X)$. In the case when X is countable, we also write $C^X = C^\infty$.

---

24 The latter can be obtained from definition (1.9) if we write a finite collection of sets from the $\mathfrak{C}(R_X)$ under consideration by using a common set of coordinates.

1.4. Topologization

It is convenient to topologize R_X with the Tykhonov topology. The collection Σ of all cylindrical sets of the form

$$\mathfrak{C}_\delta = \mathfrak{C}(x_1, \ldots, x_n; u_1 \times \ldots \times u_n),$$

where u_k is an arbitrary open subset of the space R_{x_k} $(x_1, \ldots, x_n \in X;\ n \in \mathbb{N})$, is regarded as a base of neighborhoods in R_X. It is well known that this topologization transforms R_X into a regular topological space having a countable basis if and only if X is, at most, countable. It is compact if and only if R_x is compact for each $x \in X$.

We fix distinct points $x_1, \ldots, x_n \in X$ $(n \in \mathbb{N})$ and consider the "coordinate" mapping

$$R_X \ni \lambda(\cdot) \mapsto \pi_{x_1, \ldots, x_n}(\lambda(\cdot)) = (\lambda(x_1), \ldots, \lambda(x_n)) \in R_{x_1, \ldots, x_n}. \tag{1.10}$$

The continuity of $\pi_{x_1, \ldots, x_n}$ follows immediately from the way in which the Tykhonov topology is introduced. This mapping is open, i.e., it transforms open sets from R_X into open sets from $R_{x_1, \ldots, x_n}$. This follows from the fact that under the mapping $\pi_{x_1, \ldots, x_n}$ each base neighborhood

$$\mathfrak{C}_\delta = \mathfrak{C}(y_1, \ldots, y_m; u_1 \times \ldots \times u_m)$$

turns into some neighborhood in $R_{x_1, \ldots, x_n}$ (by adding (if necessary) coordinates y_j, one should write $\mathfrak{C}_\delta$ so that $\{y_1, \ldots, y_m\} \supset \{x_1, \ldots, x_n\}$).

Each cylindrical set $\mathfrak{C} = \mathfrak{C}(x_1, \ldots, x_n; \delta)$ can be written in the form $\mathfrak{C} = \pi^{-1}_{x_1, \ldots, x_n}(\delta)$, i.e., $\delta = \pi_{x_1, \ldots, x_n}(\mathfrak{C})$. Since $\pi_{x_1, \ldots, x_n}$ is continuous and open, the set $\mathfrak{C}(x_1, \ldots, x_n; \delta)$ is open in R_X if and only if its base δ is open in $R_{x_1, \ldots, x_n}$. Taking into account that

$$R_X \setminus \mathfrak{C}(x_1, \ldots, x_n; \delta) = \mathfrak{C}(x_1, \ldots, x_n; R_{x_1, \ldots, x_n} \setminus \delta),$$

we can make the same assertion concerning the closeness of $\mathfrak{C}$ and δ.

For fixed coordinates $x_1, \ldots, x_n$, the operations of union and difference of cylindrical sets $\mathfrak{C}(x_1, \ldots, x_n; \delta)$ reduce to the same operations with their bases δ. Therefore, $\mathfrak{C}(x_1, \ldots, x_n; \delta)$, where $\delta \in \mathfrak{B}(R_{x_1, \ldots, x_n})$, belongs to the σ-span of base neighborhoods of the form $\mathfrak{C}_\delta = \mathfrak{C}(x_1, \ldots, x_n; u_1 \times \ldots \times u_n)$, where u_k are varying. This implies that the σ-algebra $\mathfrak{C}_\sigma(R_X)$ coincides with the σ-span of the collection of all base neighborhoods $\mathfrak{C}_\delta$. Therefore, we always have $\mathfrak{C}_\sigma(R_X) \subset \mathfrak{B}(R_X)$.

If X is at most countable, then R_X has a countable base of neighborhoods: The neighborhoods

$$\hat{\mathfrak{C}}_\delta = \mathfrak{C}(x_1, \dots, x_n; v_1 \times \dots \times v_n) \quad (x_1, \dots, x_n \in X;\ n \in \mathbb{N}),$$

where v_k vary over a countable basis in R_{x_k}, form a countable system $\hat{\Sigma} \subset \Sigma$ giving the previous topology in R_X. In this case, each open set in R_X is a union of neighborhoods of the form $\hat{\mathfrak{C}}_\delta$ and, thus, coincides with an at most countable union of different sets $\hat{\mathfrak{C}}_\delta$. Hence, this set belongs to $\mathfrak{C}_\sigma(R_X)$. But then $\mathfrak{B}(R_X) \subset \mathfrak{C}_\sigma(R_X)$. Therefore, *if X is at most countable, then $\mathfrak{C}_\sigma(R_X) = \mathfrak{B}(R_X)$.*

If X is more than countable, then the σ-algebra $\mathfrak{C}_\sigma(R_X)$ is poor enough in comparison with $\mathfrak{B}(R_X)$. We now clarify the structure of the sets in it. Given a sequence $(x_k)_{k=1}^\infty$ of distinct points $x_k \in X$, let us construct the space $R_{x_1, x_2, \dots} = \times_{k=1}^\infty R_{x_k}$ which is a special case of R_X. It can be understood as a collection of all sequences $(\lambda(x_1), \lambda(x_2), \dots)$, where $\lambda(x_k) \in R_{x_k}$, and can always be topologized by the Tykhonov topology. A set from R_X having the form

$$\mathfrak{C}(x_1, \dots, x_n; \delta) = \{\lambda(\cdot) \in R_X \mid (\lambda(x_1), \lambda(x_2), \dots) \in \delta\}, \tag{1.11}$$

where the base δ lies in $\mathfrak{B}(R_{x_1, x_2, \dots}) = \mathfrak{C}_\sigma(R_{x_1, x_2, \dots})$, is called a generalized cylindrical set with coordinates $x_1, x_2, \dots$. As in the case of cylindrical sets, it is possible to change the order of coordinates in (1.11) and to add (or remove) new coordinates $y \in X$ by multiplying the base by R_y.

The σ-algebra $\mathfrak{C}_\sigma(R_X)$ coincides with the collection of all possible generalized cylindrical sets.

In fact, the collection of all sets of the form (1.11) forms a σ-algebra $\mathfrak{R} \supset \mathfrak{C}(R_X)$. This is proved in a way similar to the proof of the fact that $\mathfrak{C}(R_X)$ is an algebra, but when passing to common coordinates, it is necessary to "insert in δ" countable sequences of factors R_y, too. On the other hand, $\mathfrak{R} \subset \mathfrak{C}_\sigma(R_X)$. This follows from the fact that operations over bases involve the same operations over $\mathfrak{C}(x_1, x_2, \dots; \delta)$ (provided that the coordinates $x_1, x_2, \dots$ of these sets coincide). Therefore, the procedure of construction of $\delta \in \mathfrak{C}_\sigma(R_{x_1, x_2, \dots})$ from the sets belonging to $\mathfrak{C}(R_{x_1, x_2, \dots})$ implies the analogous procedure of construction of $\mathfrak{C}(x_1, x_2, \dots; \delta)$ from sets $\mathfrak{C}(x_1, x_2, \dots; \gamma)$ with $\gamma \in \mathfrak{C}(R_{x_1, x_2, \dots})$, and these sets lie in $\mathfrak{C}(R_X)$. Consequently, $\mathfrak{R} = \mathfrak{C}_\sigma(R_X)$.

■

It follows from what has been proved that the sets from R_X "distinguished by uncountably many conditions" may be found out of $\mathfrak{C}_\sigma(R_X)$. For example, if X is more than countable, then a closed subset of R_X of the form

$$\{\lambda(\cdot) \in R_X \mid \lambda(x) \in \alpha_x, \ \ x \in X\} = \underset{x \in X}{\times} \alpha_x,$$

where α_x is closed in R_x, $\alpha_x \neq R_x$, and $\alpha_x \neq \varnothing$, does not belong to $\mathfrak{C}_\sigma(R_X)$ (in particular, $\mathfrak{C}_\sigma(R_X)$ does not contain the set consisting of a single point $\lambda(\cdot)$).

Similar to (1.10), we can introduce a coordinate mapping $\pi_{x_1, x_2, \dots}$ for fixed distinct points $x_1, x_2, \dots \in X$ by the formula

$$R_X \ni \lambda(\cdot) \mapsto \pi_{x_1, x_2, \dots}(\lambda(\cdot)) = (\lambda(x_1), \lambda(x_2), \dots) \in R_{x_1, x_2, \dots}. \tag{1.12}$$

Clearly, this mapping is continuous and open (cf. (1.10)). Each generalized cylindrical set $\alpha = \mathfrak{C}(x_1, x_2, \dots; \delta)$ can be written in the form $\alpha = \pi^{-1}_{x_1, x_2, \dots}(\delta)$, i.e., $\delta = \pi_{x_1, x_2, \dots}(\alpha)$. Just as before, the fact that (1.12) is continuous and open implies that $\mathfrak{C}(x_1, x_2, \dots; \delta)$ is open (closed) in R_X if and only if its base δ is open (closed) in $R_{x_1, x_2, \dots}$.

Let $X = X_1 \cup X_2$, and $X_1 \cap X_2 = \varnothing$. Then $R_X = R_{X_1} \times R_{X_2}$ if $\lambda(\cdot) \in R_X$ is identified with the pair $(\lambda(\cdot) \restriction X_1, \lambda(\cdot) \restriction X_2)$[25]. In particular, if $x_1, x_2, \dots \in X$ are distinct, then

$$R_X = R_{x_1, x_2, \dots} \times R_{X_2} \quad (X_2 = X \setminus \{x_1, x_2, \dots\}),$$

and $\mathfrak{C}(x_1, x_2, \dots; \delta) = \delta \times R_{X_2}$. The σ-algebra $\mathfrak{C}_\sigma(R_X)$ can be constructed from $\mathfrak{C}_\sigma(R_{X_1})$ and $\mathfrak{C}_\sigma(R_{X_2})$ just as $\mathfrak{R}$ is constructed from $\mathfrak{R}_1$ and $\mathfrak{R}_2$ in Subsection 1.2. It is easy to see that $E_X = E_{X_1} \times E_{X_2}$.

1.5. Regularity of a Joint Resolution of the Identity

According to definition (1.3), a finite scalar measure

$$\mathfrak{C}_\sigma(R_X) \ni \alpha \mapsto \mu(\alpha) \in [0, \infty)$$

is called regular if one of the following two equivalent relations holds:

$$\mu(\alpha) = \inf_{o \supset \alpha} \mu(o) \quad (\alpha, o \in \mathfrak{C}_\sigma(R_X), \ o \text{ is open}),$$

25 The topology in R_X coincides with the direct product topology.

$$\mu(\alpha) = \sup_{\varphi \supset \alpha} \mu(\varphi) \quad (\alpha, \varphi \in \mathfrak{C}_\sigma(R_X),\ \varphi \text{ is closed}). \tag{1.13}$$

(Generally speaking, it is incorrect to consider only compact sets in the latter relation.)

Theorem 1.4. *A joint RI E is always a regular measure, i.e., for any $f \in H$, the finite scalar measure*

$$\mathfrak{C}_\sigma(R_X) \ni \alpha \mapsto (E(\alpha)f, f)_H \in [0, \infty)$$

is regular.

Proof. Fix $\alpha \in \mathfrak{C}_\sigma(R_X)$ and $f \in H$. Since a measure coincides with the corresponding outer measure on measurable sets, we can write

$$(E(\alpha)f, f)_H = \inf \sum_{k=1}^{\infty} (E(\mathbf{C}_k)f, f)_H,$$

where the infimum is taken over all sets $\mathbf{C}_k \in \mathfrak{C}(R_X)$ for which $\bigcup_{k=1}^{\infty} \mathbf{C}_k \supset \alpha$. For a given $\varepsilon > 0$, we choose sets $\mathbf{C}_k$ such that

$$\sum_{k=1}^{\infty} (E(\mathbf{C}_k)f, f)_H < (E(\alpha)f, f)_H + \varepsilon.$$

Assume that $\mathbf{C}_k = \mathbf{C}(x_1, \dots, x_{n_k}; \delta_k)$. Then

$$(E(\mathbf{C}_k)f, f)_H = (E_{x_1, \dots, x_{n_k}}(\delta_k)f, f)_H$$

according to (1.9). As mentioned in Subsection 1.2, the finite scalar measure

$$\mathfrak{B}(R_{x_1, \dots, x_{n_k}}) \ni \delta \mapsto (E_{x_1, \dots, x_{n_k}}(\delta)f, f)_H$$

is automatically regular. Therefore, $R_{x_1, \dots, x_{n_k}}$ contains an open set $o_k \supset \delta_k$ such that

$$(E_{x_1, \dots, x_{n_k}}(o_k)f, f)_H - (E_{x_1, \dots, x_{n_k}}(\delta_k)f, f)_H < \varepsilon/2^k.$$

Let us construct the open set $\mathbf{C}'_k = \mathbf{C}(x_1, \dots, x_{n_k}; o_k)$; then the last inequality can be rewritten in the form

$$(E(\mathbf{C}'_k)f, f)_H - (E(\mathbf{C}_k)f, f)_H < \varepsilon/2^k \quad (k \in \mathbb{N}).$$

Summing these inequalities, we obtain

$$\sum_{k=1}^{\infty} (E(\mathfrak{C}'_k)f,f)_H < \sum_{k=1}^{\infty} (E(\mathfrak{C}_k)f,f)_H + \varepsilon < (E(\alpha)f,f)_H + 2\varepsilon.$$

By virtue of the monotonicity of the measure, the absolute semiadditivity, and the above inequality, it is possible to write a relation for the open set

$$\mathfrak{C}_\sigma(R_X) \ni o = \bigcup_{k=1}^{\infty} \mathfrak{C}'_k \supset \bigcup_{k=1}^{\infty} \mathfrak{C}_k \supset \alpha,$$

which implies (1.13), namely,

$$(E(\alpha)f,f)_H \le (E(o)f,f)_H \le \sum_{k=1}^{\infty} (E(\mathfrak{C}'_k)f,f)_H < (E(\alpha)f,f)_H + 2\varepsilon.$$

■

1.6. The Concept of the Support of a Measure and Its Properties

We now give some simple general constructions from the measure theory. Let R be a Hausdorff topological space in which a neighborhood basis Σ containing R is fixed, and let $\mathfrak{R}$ be the σ-algebra generated by Σ. Clearly, $\mathfrak{R} \subset \mathfrak{B}(R)$, but not every open set must belong to $\mathfrak{R}$. Hence, the last inclusion can be strict. Such a situation occurs, for example, when $R = R_X$, Σ consists of all the base neighborhoods $\mathfrak{C}_\delta$ introduced in Subsection 1.4, and $\mathfrak{R} = \mathfrak{C}_\sigma(R_X)$.

Consider an operator-valued measure θ on $\mathfrak{R}$ (in particular, a finite scalar measure), i.e., an operator-valued function $\mathfrak{R} \ni \alpha \mapsto \theta(\alpha)$ (here $\theta(\alpha)$ is a nonnegative bounded operator on H such that $\theta(\varnothing) = 0$ and $\theta(R) \neq 0$) which is absolutely additive, that is

$$\theta\left(\bigcup_{k=1}^{\infty} \alpha_k\right) = \sum_{k=1}^{\infty} \theta(\alpha_k),$$

if $\alpha_k \in \mathfrak{R}$ and are mutually disjoint, the series converges weakly. Clearly, this measure is monotone (if $\alpha', \alpha'' \in \mathfrak{R}$ and $\alpha' \subset \alpha''$, then $\theta(\alpha') \le \theta(\alpha'')$) and absolutely semiadditive

$$\theta\left(\bigcup_{k=1}^{\infty} \alpha_k\right) \le \sum_{k=1}^{\infty} \theta(\alpha_k)$$

for any $\alpha_k \in \mathfrak{R}$ such that the series converges weakly. We often assume that the measure θ is regular, i.e., that for each finite measure

$$\mathfrak{R} \ni \alpha \mapsto \mu(\alpha) = (\theta(\alpha)f, f)_H \in [0, \infty) \qquad (f \in H)$$

one of the two equivalent relations (1.13) holds, with $\mathfrak{B}_\sigma(R_X)$ replaced by $\mathfrak{R}$. A joint RI is an example of a measure of this sort.

A set $\alpha \in \mathfrak{R}$ is called a set of full θ-measure if $\theta(\alpha) = \theta(R)$. The support $\operatorname{supp}\theta$ of a measure θ is defined as the intersection of all closed sets φ_ξ from $\mathfrak{R}$ of full measure: $\operatorname{supp}\theta = \bigcap_{\xi\in\Xi}\varphi_\xi$ and $\theta(\varphi_\xi) = \theta(R)$. The support always exists and is a closed set (however, it must not necessarily belong to $\mathfrak{R}$)$\bigcap_{\xi\in\Xi}\varphi_\xi$; there exist examples when $\operatorname{supp}\theta = \varnothing$ (see Subsection 1.7).

A measure θ is called proper if any open set $\mathfrak{R} \ni o \supset \operatorname{supp}\theta$ has full measure; if, in addition, the measure is regular, then any set $\mathfrak{R} \ni \alpha \supset \operatorname{supp}\theta$ has full measure. We give some simple general conditions on R which guarantee the properness of an arbitrary measure .

First, we note that if $u \in \mathfrak{R}$ is open, and $u \cap \operatorname{supp}\theta \neq \varnothing$, then $\theta(u) \neq 0$. In fact, assuming the contrary, we get that $\mathfrak{R} \setminus u$ is closed and of full measure. Therefore, it must be one of the sets φ_ξ in the definition of $\operatorname{supp}\theta$. But then $u \cap \operatorname{supp}\theta = \varnothing$.

Theorem 1.5. *Suppose that one of the two following conditions holds:*

(i) $R = \bigcup_{k=1}^{\infty} R_k$ *, where each* R_k *is a compact set;*

(ii) R *has a countable neighborhood basis in the system* Σ *.*

Then any measure θ *is proper. Moreover, if condition (ii) holds, then* $\operatorname{supp}\theta$ *belongs to* $\mathfrak{R}$ *and is of full measure.*

Proof. Consider case (i). Assume that $\mathfrak{R} \ni o \supset \operatorname{supp}\theta$ is open and that $\varphi_\xi \in \mathfrak{R}$ are closed sets from the definition of $\operatorname{supp}\theta$ $(\xi \in \Xi)$. The collection of open sets o and $(R \setminus \varphi_\xi)_{\xi\in\Xi}$ covers R and, in particular, R_k $(k = 1, 2, \ldots)$. We fix k and choose a finite subcovering from this covering. Assume that this is the sets o and $(R \setminus \varphi_{\xi_{k,j}})_{j=1}^{n_k}$. Now

$$R_k \setminus o \subset \bigcup_{j=1}^{n_k} (R \setminus \varphi_{\xi_{k,j}})$$

and, consequently,

$$R \setminus o = \bigcup_{k=1}^{\infty} (R_k \setminus o) \subset \bigcup_{k=1}^{\infty} \bigcup_{j=1}^{n_k} (R \setminus \varphi_{\xi_{k,j}}).$$

Since $\theta(\mathfrak{R} \setminus \varphi_{\xi_{k,j}}) = 0$, this implies that $\theta(\mathfrak{R} \setminus o) = 0$ by virtue of monotonicity and absolute semi-additivity. Hence, o has full measure, and this means that θ is proper.

Consider case (ii). Assume that $\varphi_\xi \in \mathfrak{R}$ $(\xi \in \Xi)$ are as before, then

$$R \setminus \operatorname{supp} \theta = \bigcup_{\xi \in \Sigma} (R \setminus \varphi_\xi).$$

Denote by $\hat{\Sigma}$ a countable collection of neighborhoods from the system Σ which determines the topology in R. The open set $\mathfrak{R} \setminus \varphi_\xi$ can be represented in the form $\mathfrak{R} \setminus \varphi_\xi = \bigcup_{\eta_\xi \in \Xi_\xi} u_{\eta_\xi}$ $(u_{\eta_\xi} \in \hat{\Sigma})$, whence $\mathfrak{R} \setminus \operatorname{supp} \theta = \bigcup_{\xi \in \Xi} \bigcup_{\eta_\xi \in \Xi_\xi} u_{\eta_\xi}$. Since there are at most countably many distinct neighborhoods u_{η_ξ}, we can retain only distinct neighborhoods in the last union and, as a result, get that $\mathfrak{R} \setminus \operatorname{supp} \theta = \bigcup_{n=1}^{\infty} u_n$, where u_n is one of the neighborhoods u_{η_ξ} (the union can also be finite). This implies that

$$\operatorname{supp} \theta = \bigcap_{n=1}^{\infty} (R \setminus u_n) = \bigcap_{n=1}^{\infty} \psi_n, \quad \text{where } \psi_n = \bigcap_{k=1}^{n} (R \setminus u_k) \in \mathfrak{R}.$$

For every $k \in \mathbf{N}$, one can find η_ξ such that $u_k = u_{\eta_\xi} \subset \mathfrak{R} \setminus \varphi_\xi$, therefore, $\theta(u_k) = 0$ and, consequently, ψ_n is a set of full measure. Taking into account that $\psi_1 \supset \psi_2 \supset \ldots$, we conclude that $\operatorname{supp} \theta \in \mathfrak{R}$ and

$$\theta(\operatorname{supp} \theta) = \lim_{n \to \infty} \theta(\psi_n) = \theta(R).$$

All the more so, $\theta(\alpha) = \theta(R)$ for every $\mathfrak{R} \ni \alpha \supset \operatorname{supp} \theta$.

■

Clearly, this theorem implies that every Borel RI is proper and defined on its support. Being applied to the joint RI E constructed from a family of commuting RI's on the spaces R_x, this theorem implies that E is proper in two cases:

(i) when each R_x is compact;

(ii) when X is at most countable.

Furthermore, it is possible to establish that E is proper when all the spaces R_x are compact except, at most, countably many spaces (see Subsection 1.10).

In the general case we have the inclusion

$$\operatorname{supp} E \subset \underset{x\in X}{\times} \operatorname{supp} E_x = \{\lambda(\cdot) \in R_X \mid \lambda(x) \in \operatorname{supp} E_x,\ x \in X\}. \tag{1.14}$$

Indeed, the closed subset $\mathfrak{C}(x; \operatorname{supp} E_x)$ $(x \in X)$ of R_X has full E-measure: According to (1.9), we have

$$E(\mathfrak{C}(x; \operatorname{supp} E_x)) = E_x(\operatorname{supp} E_x) = 1.$$

Therefore, it is one of the sets φ_ξ and, thus,

$$\operatorname{supp} E \subset \bigcap_{x\in X} \mathfrak{C}(x; \operatorname{supp} E_x).$$

The latter set coincides with the set on the right-hand side of (1.14).

■

As a rule, the inclusion in (1.14) is strict (see Subsection 1.7). At the same time, for the scalar measures, we have the equality. More precisely, let $(R_x)_{x\in X}$ be the same family of spaces as above, and let

$$\mathfrak{B}(R_n) \ni \alpha \mapsto \mu_x(\alpha) \in [0, 1]$$

be a scalar probability measure for every $x \in X$, i.e., $\mu_x(R_x) = 1$. In the usual way, we construct the product $\mu = \underset{x\in X}{\times} \mu_x$ of these measures in R_x. According to Theorem 1.4, μ is a regular measure. It is easy to show that μ is a *proper measure* and

$$\operatorname{supp} \mu = \underset{x\in X}{\times} \operatorname{supp} \mu_x. \tag{1.15}$$

In fact, let us first prove (1.15). The inclusion

$$\operatorname{supp} \mu \subset \underset{x\in X}{\times} \operatorname{supp} \mu_x = \varphi$$

is established just as (1.14). Let us prove the opposite inclusion. Assume that φ_ξ are closed sets of full measure from the definition of supp μ, then $\varphi \subset \varphi_\xi$ $(\xi \in \Xi)$. Indeed, if $\varphi \not\subset \varphi_\xi$ for some ξ then one can find a point $\lambda_0(\cdot)$ such that $\varphi \ni \lambda_0(\cdot) \notin \varphi_\xi$. Hence, some base neighborhood

$$\mathfrak{C}_\delta = \mathfrak{C}(x_1, \dots, x_n; u_1 \times \dots \times u_n)$$

of this point does not intersect φ_ξ and, thus, $\mu(\mathfrak{C}_\delta) = 0$. On the other hand, according to (1.8), we have

$$\mu(\mathfrak{C}_\delta) = \mu_{x_1}(u_1) \dots \mu_{x_n}(u_n),$$

because each $\mu_{x_k}(u_k)$ is positive by virtue of the fact that $u_k \cap \operatorname{supp} \mu_{x_k} \ni \lambda_0(x_k)$ and, hence, is nonempty. Consequently, $\varphi \subset \operatorname{supp} \mu$.

Let us prove that μ is proper. Consider an open set o such that $\mathfrak{C}_\sigma(R_X) \ni o \supset \operatorname{supp} \mu$. According to (1.11), $o = \mathfrak{C}(x_1, x_2, \dots; \delta)$, where δ is open in $R_{x_1, x_2, \dots}$. This generalized cylindrical set may contain $\operatorname{supp} \mu = \times_{x \in X} \operatorname{supp} \mu_x$ only in the case when $\delta \supset \times_{k=1}^{\infty} \operatorname{supp} \mu_{x_k}$. Therefore, by taking the formula[26]

$$\mu(\mathfrak{C}(x_1, x_2, \dots; \delta)) = \left(\mathop{\times}_{k=1}^{\infty} \mu_{x_k} \right)(\delta)$$

into account, we obtain

$$\mu(o) = \mu(\mathfrak{C}(x_1, x_2, \dots; \delta)) = \left(\mathop{\times}_{k=1}^{\infty} \mu_{x_k} \right)(\delta)$$

$$\geq \left(\mathop{\times}_{k=1}^{\infty} \mu_{x_k} \right)\left(\mathop{\times}_{k=1}^{\infty} \operatorname{supp} \mu_{x_k} \right) = \prod_{k=1}^{\infty} (\mu_{x_k}(\operatorname{supp} \mu_{x_k})) = 1 = \mu(R_X).$$

■

This argument does not work for E, and only the inclusion (1.14) holds in this case (here the fact that each $E_{x_k}(u_k) > 0$ does not, in general, imply that $E_{x_1}(u_1) \dots E_{x_n}(u_n) > 0$, since the projectors can be orthogonal). One can easily construct an example in which the inclusion in (1.14) is strict.

1.7. Families of Multiplication Operators

Consider an elementary example of a commuting family of RI's of an arbitrary cardinality which nevertheless shows many specific features of the theory constructed. Let M be an abstract space of points ω, and let $\mathfrak{M}$ be a σ-algebra of subsets on it. Assume also that $\mathfrak{M} \ni \alpha \mapsto \mu(\alpha) \in [0, \infty)$ is a σ-finite measure. Given an $\mathfrak{M}$-measurable almost everywhere finite complex-valued function a, we define the operator A of multiplication by a in the space $H = L_2(M, \mu) = L_2(M, \mathfrak{M}, \mu)$ by the formula

$$H \supset \mathfrak{D}(A) \ni f(\omega) \mapsto (Af)(\omega) = a(\omega) f(\omega) \in H,$$

---

26 See Remark 2 in Subsection 1.4; clearly, it is valid for μ, too.

$$\mathfrak{D}(A) = \{f \in H \mid a(\omega) f(\omega) \in H\}.$$

The domain of definition $\mathfrak{D}(A)$ is dense in H. Indeed, any function from H vanishing on $\alpha_n = \{\omega \in M \mid |a(\omega)| > n\}$ for some $n \in \mathbb{N}$, lies in $\mathfrak{D}(A)$. On the other hand, taking into account that $\mu(\alpha_n) \to 0$ as $n \to \infty$, for every $f \in H$ we find that

$$\mathfrak{D}(A) \ni f(\omega)\, \kappa_{M \setminus \alpha_n}(\omega) \to f(\omega)$$

as $n \to \infty$ in H (here κ_α is the characteristic function of the set α). The operator A is normal, and A^* is constructed similarly from the function $\overline{a(\omega)}$; it is bounded if and only if $a(\omega)$ is essentially bounded. The resolvent R_z is the operator of multiplication by the function $(a(\omega) - z)^{-1}$, where $z \in \mathbb{C}^1$ is such that this function is essentially bounded. This implies that the resolution of the identity $\mathfrak{B}(\mathbb{C}^1) \ni \alpha \mapsto E(\alpha)$ that corresponds to A is given by

$$H \ni f(\omega) \mapsto (E(\alpha) f)(\omega) = \kappa_\alpha(a(\omega)) f(\omega) = \kappa_{a^{-1}(\alpha)}(\omega) f(\omega).$$

Let $(a_x)_{x \in X}$ ($X = \{1, \dots, n\}$) be n functions of the indicated form, $(A_x)_{x \in X}$ be a corresponding family of normal commuting operators (i.e., their RI's E_x commute), and let $\mathfrak{B}(\mathbb{C}^n) \ni \alpha \mapsto E(\alpha)$ be the joint RI of the family $(E_x)_{x \in X}$. On rectangles, it is equal to

$$H \ni f(\omega) \mapsto (E(\alpha_1 \times \dots \times \alpha_n) f)(\omega) = (E_1(\alpha_1) \dots E_n(\alpha_n) f)(\omega)$$

$$= \kappa_{a_1^{-1}(\alpha_1)}(\omega) \dots \kappa_{a_n^{-1}(\alpha_n)}(\omega) f(\omega)$$

$$= \kappa_{(a_1^{-1}(\alpha_1) \cap \dots \cap a_n^{-1}(\alpha_n))}(\omega) f(\omega) \quad (\alpha_k \in \mathfrak{B}(\mathbb{C}^1);\ k = 1, \dots, n).$$

It is easy to show that the "*diagonal*"

$$\delta = \{(a_1(\omega) \dots a_n(\omega)) \in \mathbb{C}^n \mid \omega \in M\}^{\sim}$$

has full E-*measure.* In fact, it suffices to establish that any point of the open set $\mathbb{C}^n \setminus \psi$ has a neighborhood of zero E-measure. For this purpose, it suffices to verify that $E(o_1 \times \dots \times o_n) = 0$ if o_k are open in $\mathbb{C}^1$ and $(o_1 \times \dots \times o_n) \cap \delta = \varnothing$. Assume the contrary. Then

$$(a_1^{-1}(o_1)) \cap \dots \cap (a_n^{-1}(o_n)) \neq \varnothing.$$

Let

$$\omega_0 \in (a_1^{-1}(o_1)) \cap \ldots \cap (a_n^{-1}(o_n)).$$

For every $k \in \mathbb{N}$ we have $a_k(\omega_0) \in o_k$, i.e,

$$\delta \ni (a_1(\omega_0), \ldots, a_n(\omega_0)) \in o_1 \times \ldots \times o_n.$$

■

Therefore, supp $E \subset \delta$. One can easily formulate sufficient conditions under which this relation turns into the equality. For example, if M is a Hausdorff topological space, $\mathfrak{M} = \mathfrak{B}(M)$, the measure μ is positive on open sets, and the functions a_x, $x \in X = \{1, \ldots, n\}$, are continuous, then supp $E = \delta$. Since we now have

$$\mathop{\times}_{x=1}^{n} \operatorname{supp} E_x = \{(a_1(\omega_1) \ldots a_n(\omega_n)) \mid \omega_1, \ldots, \omega_n \in M\}^{\sim},$$

this set may coincide with δ only in exceptional cases. This implies that, as a rule, the inclusion in (1.14) is strict.

In the case when X is infinite, the equality supp $E = \delta$ does not hold. Moreover, in the case when X is uncountable, it may happen that supp $E = \varnothing$. We present an example, having observed first that all the foregoing can be reformulated in the natural way for the case of real-valued functions a_x.

Example 1.1. Let $H = L_2(\mathbb{R}^1, d\omega)$ with respect to the Lebesgue measure $d\omega$, and let $X = \{\xi\} \cup \mathbb{R}^1$, where ξ is some fixed abstract point. Assume that A_ξ and A_x are the operators of multiplication by the functions ω and $(\omega - x)^{-1}$ $(x \in \mathbb{R}^1)$, respectively.

It is stated that the joint RI E of this family $(A_x)_{x\in X}$ has empty support.

In fact, consider the joint two-dimensional RI $E_{\xi,x}$ $(x \in \mathbb{R}^1)$. Here $a_\xi(\omega) = \omega$, $a_x(\omega) = (\omega - x)^{-1}$ $(\omega \in \mathbb{R}^1 = M)$. The set

$$\psi_x = \{(a_\xi(\omega), a_x(\omega)) \in \mathbb{R}^2 \mid \omega \in \mathbb{R}^2\} = \{(\omega, (\omega - x)^{-1}) \in \mathbb{R}^2 \mid \omega \in \mathbb{R}^2\}$$

is the graph of a hyperbola and is closed in $\mathbb{R}^2$. According to the foregoing, supp $E_{\xi,x} \subset \psi_x$ (one can easily see that, in fact, here we have the equality). The closed cylindrical set $\varphi_x = \mathbb{C}(\xi, x; \psi_x)$ $(x \in \mathbb{R}^1)$ in the space $\mathbb{R}^X$ is of full E-measure, i.e.,

$$E(\mathbb{C}(\xi, x; \psi_x)) = E_{\xi,x}(\psi_x) = 1.$$

Therefore, supp $E \subset \bigcap_{x\in\mathbb{R}^1} \varphi_x = \varphi$. However, the intersection φ is empty. Assume the contrary, i.e., that there exists $\mathbb{R}^X \ni \lambda_0(\cdot) \in \varphi$. Then

$$\mathbb{R}^1 \supset \pi_\xi(\varphi) \ni \pi_\xi(\lambda_0(\cdot)) = \lambda_0(\xi).$$

On the other hand,

$$\varphi \subset \varphi_{\lambda_0(\xi)} = \mathbf{C}(\xi, x; \psi_{\lambda_0(\xi)}),$$

and thus,

$$\pi_\xi(\varphi) \subset \pi_\xi(\varphi_{\lambda_0(\xi)}) = \mathbb{R}^1 \setminus \{\lambda_0(\xi)\}.$$

Hence, $\varnothing = \varphi \supset \operatorname{supp} E$.

Remark 1. The above argument shows that $\operatorname{supp} E = \varnothing$ also in the case when $a_x(\omega)$ is an arbitrary real-valued function that is continuous on $\mathbb{R}^1 \setminus \{x\}$ and such that $\lim_{\omega \to x} |a_x(\omega)| = \infty$ $(x \in \mathbb{R}^1;\ a_\xi(\omega) = \omega)$. In particular, one can set $a_x(\omega) = |\omega - x|^{-\varepsilon}$ $(\varepsilon \in (0, 1/2))$. Now

$$\bigcap_{x \in X} \mathfrak{D}(A_x) \supset H \cap L_\infty(\mathbb{R}^1, d\omega)$$

is dense in H, in contrast to the example presented above. Note that uncountably many unbounded operators appear in all these examples. This circumstance is essential (see Theorem 1.6 below).

■

Example 1.2. We now present an example showing that, for two commuting RI's E_1 and E_2, the RI $E_1 \times E_2$ (see Subsection 1.2) does not always exist. We set $H = L_2 = L_2([0, 1], \mathfrak{M}, m)$, where m is the Lebesgue measure on [0, 1], and $\mathfrak{M}$ is the σ-algebra of sets measurable with respect to this measure. Assume that $[0, 1] = R_1 \cup R_2$ is a decomposition of the segment $[0, 1]$ into two disjoint nonmeasurable sets such that $m^*(R_1) = m^*(R_2) = 1$, where m^* is the outer measure (this decomposition is always possible). For any $j = 1, 2$, we consider a σ-algebra $\mathfrak{R}_j = \{\alpha_j = \alpha \cap R_j \mid \alpha \in \mathfrak{M}\}$ of subsets of the space R_j and define an RI $\mathfrak{R}_j \ni \alpha_j \mapsto E_j(\alpha_j)$, where $E_j(\alpha_j)$ is the operator of multiplication by the characteristic function $\kappa_\alpha(\cdot)$ of a set α defined in L_2. If $\alpha_j = \alpha \cap R_j = \alpha' \cap R_j$ are two representations of $\alpha_j \in \mathfrak{R}_j$ in terms of $\alpha, \alpha' \in \mathfrak{M}$, then

$$m((\alpha \setminus \alpha') \cup (\alpha' \setminus \alpha)) = 0$$

by virtue of the equality $m^*(R_j) = 1$. Therefore, the operator $E_j(\alpha_j)$ is correctly defined in

L_2. Clearly, E_1 and E_2 are two commuting resolutions of the identity in L_2.

Let us show that the RI $E_1 \times E_2$ does not exist. It suffices to establish that a finitely additive operator-valued measure E, given on the algebra $\mathfrak{R}'$ generated by all rectangles $\alpha_1 \times \alpha_2$ and such that

$$E(\alpha_1 \times \alpha_2) = E_1(\alpha_1) E_2(\alpha_2) \quad (\alpha_1 \in \mathfrak{R}_1, \alpha_2 \in \mathfrak{R}_2),$$

is not absolutely additive. For any $n \in \mathbb{N}$, we define a decomposition of [0, 1] into half-intervals of the same length by $[0, 1] = \bigcup_{k=1}^{2^n} \delta_{n,k}$. We set

$$\beta_n = \bigcup_{k=1}^{2^n} ((\delta_{n,k} \cap R_1) \times (\delta_{n,k} \cap R_2)) \in \mathfrak{R}'.$$

Then $E_j(\delta_{n,k} \cap R_j)$ coincides with the operator of multiplication by $\kappa_{\delta_{n,k}}(\cdot)$ $(j = 1, 2)$ and, hence, $E((\delta_{n,k} \cap R_1) \times (\delta_{n,k} \cap R_2))$ is also equal to this operator. Consequently, $E(\beta_n) = 1$ $(n \in \mathbb{N})$. On the other hand, $\beta_1 \supset \beta_2 \supset \ldots$, and $\bigcap_{n=1}^{\infty} \beta_n = \varnothing$ (since $R_1 \cap R_2 = \varnothing$). This shows that E is not absolutely additive.

1.8. Construction of a Measure on a Larger Space from a Measure on a Smaller Space. Compactification.

As in Subsection 1.6, suppose that R is a Hausdorff topological space, Σ is a basis of its neighborhoods containing R, and $\mathfrak{R}$ is the σ-algebra spanned by Σ. Fix a set $R' \subset R$ (which, generally speaking, does not belong to $\mathfrak{R}$) and topologize it by the relative topology of the space R: A basis of neighborhoods in R' is formed by the collection Σ' of all sets of the form $u' = u \cap R'$, where $u \in \Sigma$. Let $\mathfrak{R}'$ be the σ-algebra spanned by Σ'. The mapping

$$\mathfrak{R} \ni \alpha \mapsto \alpha' = \alpha \cap R' \in \mathfrak{R}' \tag{1.16}$$

obviously maps the whole of $\mathfrak{R}$ onto the whole of $\mathfrak{R}'$ (of course, coalescence is possible here).

Assume that an operator-valued measure $\mathfrak{R}' \ni \alpha' \mapsto \theta'(\alpha')$ is defined on $\mathfrak{R}'$. According to (1.16), we define a measure θ on $\mathfrak{R}$ by setting

$$\mathfrak{R} \ni \alpha \mapsto \theta(\alpha) = \theta'(\alpha \cap R'). \tag{1.17}$$

The following relation is valid:

$$\operatorname{supp} \theta' = (\operatorname{supp} \theta) \cap R'. \tag{1.18}$$

Indeed, let $\operatorname{supp} \theta = \bigcap_{\xi \in \Xi} \varphi_\xi$, where φ_ξ runs through the collection of all closed sets from $\mathfrak{R}$ of full θ-measure. Then $\varphi_\xi \cap R' \in \mathfrak{R}'$ is closed in the topology of R' and is of full θ-measure. Therefore, it belongs to the intersection which defines $\operatorname{supp} \theta'$. Consequently,

$$\operatorname{supp} \theta' \subset \bigcap_{\xi \in \Xi} (\varphi_\xi \cap R') = (\operatorname{supp} \theta) \cap R'.$$

Conversely, suppose that $(\varphi'_\eta)_{\eta \in \Xi'}$ is the collection of all closed subsets of R' which belong to $\mathfrak{R}'$ and are of full θ'-measure, i.e., $\operatorname{supp} \theta' = \bigcap_{\eta \in \Xi'} \varphi'_\eta$. Assume that instead of the equality (1.18), we have strict inclusion of the left-hand set in the right-hand set. Then there exists $\lambda' \in (\operatorname{supp} \theta) \cap R'$ such that $\lambda' \notin \operatorname{supp} \theta'$. Thus, $\lambda' \notin \varphi'_\eta$ for some $\eta \in \Xi'$. Suppose that $u' = u \cap R'$ is a neighborhood of the point λ' (in Σ') such that $u' \cap \varphi'_\eta = \varnothing$ $(u \in \Sigma)$. Since $u' \subset R' \setminus \varphi'_\eta$, and $\theta'(R' \setminus \varphi'_\eta) = 0$, this and (1.17) imply that $0 = \theta'(u') = \theta(u)$. On the other hand, $\lambda' \in u \cap \operatorname{supp} \theta$, and therefore, the last set is nonempty. But then $\theta(u) \neq 0$, that is absurd.

■

Let us clarify the relation between integrals with respect to the measures θ and θ'. If $R \ni \lambda \mapsto F(\lambda) \in \mathbb{C}^1$ is measurable with respect to the σ-algebra $\mathfrak{R}$, then its restriction $F \upharpoonright R'$ is measurable with respect to $\mathfrak{R}'$.

Let $f \in H$ be fixed. The function F is summable with respect to the measure

$$\mathfrak{R} \ni \alpha \mapsto (\theta(\alpha) f, f)_H \in [0, \infty)$$

if and only if $F \upharpoonright R'$ is summable with respect to

$$\mathfrak{R}' \ni \alpha \mapsto (\theta'(\alpha) f, f)_H \in [0, \infty),$$

and

$$\int_R F(\lambda)\, d(\theta(\lambda) f, f)_H = \int_{R'} (F \upharpoonright R')(\lambda')\, d(\theta'(\lambda') f, f)_H. \tag{1.19}$$

Indeed, the assertion will be proved if we establish the relation (1.19) for step functions F. But this relation is valid for step functions if it holds for the characteristic function $F(\lambda) = \kappa_\alpha(\lambda)$ of a set $\alpha \in \mathfrak{R}$, i.e., if

$$(\theta(\alpha)f, f)_H = (\theta'(\alpha \cap R')f, f)_H$$

(clearly, $\kappa_\alpha \upharpoonright R' = \kappa_{\alpha \cap R'}$). This equality immediately follows from (1.17).

■

If $o \in \mathfrak{R}$ is open in R, then $o' = o \cap R' \in \mathfrak{R}'$ is open in R'. This, (1.17), and (1.18) imply that θ' *is regular if* θ *is regular and that* θ *is proper if* θ' *is proper.* Generally speaking, the converse implications are not true.

It has been shown in Subsection 1.7 that a joint RI is not always a proper measure. This leads to a number of difficulties. We present one way of overcoming these difficulties which involves the compactification of the spaces R_x on which the RI's to be multiplied are defined. For simplicity we assume that $R_x = \mathbb{C}^1$ $(x \in X)$ (i.e., we consider RI's of normal operators).

We compactify $\mathbb{C}^1$ by adding a formal point ∞ at infinity, and denote the space thus obtained by $\mathbf{C}^1 = \mathbb{C}^1 \cup \{\infty\}$; any set $\mathbf{C}^1 \setminus \varphi$ with φ compact in $\mathbb{C}^1$ is regarded as a neighborhood of ∞. The space $\mathbf{C}^1$ is a compact separable complete metric space. The Tykhonov product $\mathbf{C}^X = \times_{x \in X} \mathbf{C}^1$ is a compact set containing $\mathbb{C}^X$. Its points and subsets are denoted by boldface letters. In particular, $X \ni x \mapsto \boldsymbol{\lambda}(x) \in \mathbf{C}^1$ is a point of $\mathbf{C}^X$.

If $\boldsymbol{\mathfrak{C}}_\delta$ *is a base neighborhood in* $\mathbf{C}^X$, *then* $\boldsymbol{\mathfrak{C}}_\delta \cap \mathbb{C}^X$ *is a base neighborhood in* $\mathbb{C}^X$. *Moreover all the base neighborhoods* $\mathfrak{C}_\delta$ *in* $\mathbb{C}^X$ *can be represented in this form.*

In fact, if $\boldsymbol{\mathfrak{C}}_\delta = \boldsymbol{\mathfrak{C}}(x_1, \dots, x_n; \boldsymbol{u}_1 \times \dots \times \boldsymbol{u}_n)$ is a base neighborhood in $\mathbf{C}^X$, then its intersection with $\mathbb{C}^X$ consists of all the functions $\lambda(\cdot) \in \mathbb{C}^X$ such that $\lambda(x_k) \in \boldsymbol{u}_k \setminus \{\infty\} = u_k$ $(k = 1, \dots, n)$, i.e., it coincides with the cylindrical set $\mathfrak{C}(x_1, \dots, x_n; u_1 \times \dots \times u_n)$ from $\mathbb{C}^X$. Since each $\boldsymbol{u}_k$ is open in $\mathbf{C}^1$, the set u_k is open in $\mathbb{C}^1$ and, thus, it is a base neighborhood $\mathfrak{C}_\delta$ from $\mathbb{C}^X$. Clearly, the neighborhoods constructed in this way run through the whole of the collection Σ of base neighborhoods in $\mathbb{C}^X$ (as $\boldsymbol{u}_k$ one should take neighborhoods from $\mathbf{C}^1$ which do not contain ∞).

■

Hence, the topology in $\mathbb{C}^X = R'$ coincides with the relative topology induced by the topology of $\mathbf{C}^X = R$. Moreover, the base neighborhoods $\boldsymbol{\mathfrak{C}}_\delta$ (in $\mathbf{C}^X$) and $\mathfrak{C}_\delta$ (in $\mathbb{C}^X = R'$) are connected just as required at the beginning of this subsection. Note that $\mathbb{C}^n$ is an open subset of $\mathbf{C}^n$, and $\mathbf{C}^n \setminus \mathbb{C}^n$ is nowhere dense. The situation is quite different in the case of infinite X: $\mathbb{C}^X$ is not open in $\mathbf{C}^X$, and both it and $\mathbf{C}^X \setminus \mathbb{C}^X$ are dense in $\mathbf{C}^X$.

We now apply the construction presented at the beginning of this subsection in the case when $R = \mathbf{C}^X$, $R' = \mathbb{C}^X$, $\Sigma = \{\text{arbitrary } \mathfrak{C}_\delta\}$, and $\Sigma' = \{\text{arbitrary } \mathfrak{C}_\delta\}$. As already mentioned, $\mathfrak{C}_\sigma(R_X)$ coincides with the σ-span of the corresponding base neighborhoods. Therefore, we now have $\mathfrak{R} = \mathfrak{C}_\sigma(\mathbf{C}^X)$ and $\mathfrak{R}' = \mathfrak{C}_\sigma(\mathbb{C}^X)$, and (1.16) establishes the connection between these two σ-algebras. We take the joint RI E constructed from the RI's E_x on $\mathbb{C}^1$ to be θ'. Then (1.17) defines a measure $\boldsymbol{E} = \theta$ on $\mathbf{C}^X$ which is obviously an RI. This measure $\boldsymbol{E}$ is called the compactified joint RI. Thus,

$$\mathfrak{C}_\sigma(\mathbf{C}^X) \ni \boldsymbol{\alpha} \mapsto \boldsymbol{E}(\boldsymbol{\alpha}) = E(\boldsymbol{\alpha} \cap \mathbb{C}^X). \tag{1.20}$$

We can also proceed in another way. Namely, let $R = \mathbf{C}^1$, $R' = \mathbb{C}^1$, $\Sigma = \mathfrak{B}(\mathbf{C}^1)$, and $\Sigma' = \mathfrak{B}(\mathbb{C}^1)$. (The required relations formulated at the beginning of this and, thus are clearly satisfied.) According to (1.17), for $E_x = \theta'$ one can define the compactified one-dimensional RI

$$\boldsymbol{E}_x = \theta\colon \mathfrak{B}(\mathbf{C}^1) \ni \boldsymbol{\alpha} \mapsto \boldsymbol{E}_x(\boldsymbol{\alpha}) = E_x(\boldsymbol{\alpha} \cap \mathbb{C}^1)$$

employing the construction (1.20) for $X = \{1\}$. Obviously, the RI's $\boldsymbol{E}_x$ and $\boldsymbol{E}_y$ ($x, y \in X$) commute. Therefore, it is possible to construct their joint RI $\boldsymbol{E} = \times_{x \in X} \boldsymbol{E}_x$.

It is easy to see that *this definition of* $\boldsymbol{E}$ *coincides with the definition* (1.20).

Indeed, denote the RI $\boldsymbol{E}$ just constructed by $\boldsymbol{F}$. Then, by virtue of of (1.8) and (1.20), we have

$$\boldsymbol{F}(\mathfrak{C}(x_1, \dots, x_n; \boldsymbol{\delta}_1 \times \dots \times \boldsymbol{\delta}_n)) = \boldsymbol{E}_{x_1}(\boldsymbol{\delta}_1) \times \dots \times \boldsymbol{E}_{x_n}(\boldsymbol{\delta}_n)$$

$$= E_{x_1}(\boldsymbol{\delta}_1 \cap \mathbb{C}^1) \times \dots \times E_{x_n}(\boldsymbol{\delta}_n \cap \mathbb{C}^1) = E(\mathfrak{C}(x_1, \dots, x_n; (\boldsymbol{\delta}_1 \cap \mathbb{C}^1) \times \dots$$

$$\times (\boldsymbol{\delta}_n \cap \mathbb{C}^1))) = E((\mathfrak{C}(x_1, \dots, x_n; \boldsymbol{\delta}_1 \times \dots \times \boldsymbol{\delta}_n)) \cap \mathbb{C}^X)$$

$$= \boldsymbol{E}(\mathfrak{C}(x_1, \dots, x_n; \boldsymbol{\delta}_1 \times \dots \times \boldsymbol{\delta}_n)) \quad (x_1, \dots, x_n \in X;\ \boldsymbol{\delta}_1, \dots, \boldsymbol{\delta}_n \in \mathfrak{B}(\mathbf{C}^1)).$$

■

Clearly, the procedure of compactification can be repeated also in the case when $R_x = \mathbb{R}^1$ ($x \in X$); the role of $\mathbf{C}^1$ is now played by the compactified line $\mathbf{R}^1 = \mathbb{R}^1 \cup \{\infty\}$.

1.9. Construction of a Measure on a Smaller Space from a Measure on a Larger Space. Modification of a Measure

Let $R, R', \Sigma, \Sigma', \mathfrak{R}$, and $\mathfrak{R}'$ be the same as at the beginning of Subsection 1.8. As before, relation (1.16) holds. However, we now assume that the operator-valued measure θ is given, but not θ', namely $\mathfrak{R} \ni \alpha \mapsto \theta(\alpha)$. If $R' \in \mathfrak{R}$, then it is possible to consider the restriction of θ to R':

$$(\theta \upharpoonright R')(\alpha') = \theta(\alpha') \quad (\alpha' \in \mathfrak{R}' = \{\alpha' \in \mathfrak{R}' \mid \alpha' \subset \mathfrak{R}'\}).$$

However, it is often necessary to define this restriction also in the case when $R' \notin \mathfrak{R}$. This can be realized if R' possesses the following property: Any $\mathfrak{R} \ni \alpha \supset R'$ has full θ-measure (we say that R' has full outer θ-measure).

For R' of this sort, we define a measure θ' on the σ-algebra $\mathfrak{R}'$ by (1.17), read from right to left: If $\alpha' \in \mathfrak{R}'$, then, according to (1.16), there exists $\alpha \in \mathfrak{R}$ such that $\alpha' = \alpha \cap R'$, and we set

$$\mathfrak{R}' \ni \alpha' \mapsto \theta'(\alpha') = \theta'(\alpha \cap R') = \theta(\alpha). \tag{1.21}$$

It is easy to see that *definition (1.21) is correct.*

Indeed, suppose that along with α there exists $\beta \in \mathfrak{R}$ such that $\alpha' = \beta \cap R'$. One must prove that $\theta(\alpha) = \theta(\beta)$. Since $\alpha \cap R' = \beta \cap R'$, we have

$$(\alpha \setminus \beta) \cap R' = (\beta \setminus \alpha) \cap R' = \varnothing$$

and, thus, $\mathfrak{R} \ni R \setminus (\alpha \setminus \beta)$, and $R \setminus (\beta \setminus \alpha) \supset R'$. Consequently,

$$\theta(R) = \theta(R \setminus (\alpha \setminus \beta)) = \theta(R \setminus (\beta \setminus \alpha)),$$

i.e., $\theta(\alpha \setminus \beta) = \theta(\beta \setminus \alpha) = 0$, and $\theta(\alpha) = \theta(\beta)$. It is clear that $\theta'(\varnothing) = 0$, and $\theta'(R') = \theta(R) \neq 0$. Let us verify that θ' is absolutely additive. Consider a sequence $(\alpha_j')_{j=1}^{\infty}$ such that $\alpha_j' \in \mathfrak{R}'$, and $\alpha_j' \cap \alpha_k' = \varnothing$ $(j \neq k)$. Let us choose $\alpha_j \in \mathfrak{R}$ so that $\alpha_j' = \alpha_j \cap R'$ $(j \in \mathbb{N})$. Also let $\beta_1 = \alpha_1$, $\beta_2 = \alpha_2 \setminus \alpha_1$, $\beta_3 = \alpha_3 \setminus (\alpha_1 \cup \alpha_2)$, ... These sets belong to $\mathfrak{R}$, they are mutually disjoint and such that $\beta_j \cap R' = \alpha_j'$ $(j \in \mathbb{N})$, and

$$\left(\bigcup_{j=1}^{\infty} \beta_j\right) \cap R' = \bigcup_{j=1}^{\infty} \alpha_j'.$$

Absolute additivity follows from the equality

$$\theta'\left(\bigcup_{j=1}^{\infty}\alpha'_j\right) = \theta\left(\bigcup_{j=1}^{\infty}\beta_j\right) = \sum_{j=1}^{\infty}\theta(\beta_j) = \sum_{j=1}^{\infty}\theta'(\alpha'_j).$$

■

The measure θ' is called the modified (by means of R') measure θ. Since the measures θ and θ' are connected by relation (1.17), one can regard θ as generated by θ'. Therefore, the facts presented in Subsection 1.8 are preserved both for the measure and its modification. If $R' \in \mathfrak{R}$, then $\theta' = \theta \upharpoonright R'$. A modification of an RI is again an RI which will be called a modified RI.

Let $(E_x)_{x\in X}$ be a family of commuting Borel RI's, with E_x given on the space R_x. The joint RI E was defined earlier in the space $R_X = \times_{x\in X} R_x$; it was a measure on the σ-algebra $\mathfrak{C}_\sigma(R_X)$. However, each E_x, generally speaking, could be regarded as given also on a smaller space $R'_x \subset R_x$ (for example, on $R'_x = \operatorname{supp} E_x$). In this case, a joint RI E' is defined in the space $R'_X = \times_{x\in X} R'_x$ on the σ-algebra $\mathfrak{C}_\sigma(R'_x)$. If X is more than countable, then R'_X can fail to belong to $\mathfrak{C}_\sigma(R_X)$ and, thus, the assertion $E' = E \upharpoonright R'_X$ is incorrect. At the same time, it is not hard to see that E' can be understood as E, modified by means of R'_X. More precisely, suppose that $\operatorname{supp} E_x \subset R'_x \subset R_x$ for every $x \in X$, where R'_x is closed in R_x and is topologized by the relative topology of R_x. The restriction $E'_x = E \upharpoonright R'_x$ is a Borel RI on R'_x, and the family $(E'_x)_{x\in X}$ is commutative. We construct from it a joint RI E':

$$\mathfrak{C}_\sigma(R'_X) \ni \alpha' \mapsto E'(\alpha)$$

on the space $R'_X = \times_{x\in X} R'_x$.

Lemma 1.1. *The set R'_X has full outer E-measure, and the RI E' coincides with the modified (by R'_X) RI E.*

Proof. Let us use the notations of Subsection 1.4 and set $R = R_X$, $\Sigma = \{$arbitrary $\mathfrak{C}_\delta\}$, and $\mathfrak{R} = \mathfrak{C}_\sigma(R_X)$. The intersection of

$$\mathfrak{C}_\delta = \mathfrak{C}(x_1, \dots, x_n; u_1 \times \dots \times u_n)$$

(u_k are open in R_{x_k}) with R'_X coincides with the cylindrical subset

$$\mathfrak{C}'_\delta = \mathfrak{C}(x_1, \dots, x_n; (u_1 \cap R'_{x_1}) \times \dots \times (u_n \cap R'_{x_n}))$$

of R'_X. The sets $u'_k = u_k \cap R'_{x_k}$ are open in R'_{x_k} and, thus, $\mathfrak{C}'_\delta$ is a neighborhood in R'_X. Clearly, any base neighborhood in R'_X can be obtained in the form of such an intersection. Hence, $\mathfrak{R}' = \mathfrak{C}_\sigma(R'_X)$.

Let us show that R'_X has full outer E-measure: If $\mathfrak{C}_\sigma(R_X) \ni \alpha \supset R'_X$, then $E(\alpha) = 1$. The set α is a generalized cylindrical set, i.e.,

$$\alpha = \mathfrak{C}(x_1, \dots, x_n; \delta) = \delta \times R_{X_2},$$

where $\delta \in \mathfrak{B}(R_{x_1, x_2, \dots})$, and $X_2 = X \setminus \{x_1, x_2, \dots\}$. Since

$$R'_X = \left(\mathop{\times}_{n=1}^{\infty} R'_{x_n} \right) \times \left(\mathop{\times}_{x \in X_2} R'_x \right),$$

the inclusion $\alpha \supset R'_X$ involves the inclusion

$$\delta \supset \mathop{\times}_{n=1}^{\infty} R'_{x_n} \supset \mathop{\times}_{n=1}^{\infty} (\operatorname{supp} E_{x_n}) \supset \operatorname{supp} E_{x_1, x_2, \dots}$$

(see (1.14)). Taking into account what has been mentioned at the end of Subsection 1.4 and the fact that $E_{x_1, x_2, \dots}$ is a proper measure, we get

$$E(\alpha) = E(\mathfrak{C}(x_1, \dots, x_n; \delta)) = E_{x_1, x_2, \dots}(\delta) = 1.$$

Let F' be the modification of E by R'_X; one should prove that $F' = E'$. This follows from the relation (see (1.21))

$$F'(\mathfrak{C}'_\delta) = F'(\mathfrak{C}(x_1, \dots, x_n; (u_1 \cap R'_{x_1}) \times \dots \times (u_n \cap R'_{x_n})))$$

$$= E(\mathfrak{C}(x_1, \dots, x_n; u_1 \times \dots \times u_n)) = E_{x_1}(u_1) \dots E_{x_n}(u_n)$$

$$= E'_{x_1}(u_1 \cap R'_{x_1}) \dots E'_{x_n}(u_n \cap R'_{x_n}) = E'(\mathfrak{C}'_\delta).$$

■

Below, we also use the ordinary restriction $\theta \restriction R'$, where $R' \in \mathfrak{R}$ (it may be not of full θ-measure), which is topologized by the relative topology of R. If we repeat the beginning of the proof of (1.18), we get the inclusion

$$\operatorname{supp}(\theta \restriction R') \subset (\operatorname{supp} \theta) \cap R'. \tag{1.22}$$

Simple examples of strict inclusion in (1.22) can be given.

Regularity of θ *implies the regularity of* $\theta \upharpoonright R'$.

Indeed, let $\mathfrak{R} \ni \alpha' \subset R'$, $f \in H$, and the sets $\mathfrak{R} \ni o_n \supset \alpha'$ open in R are such that

$$(\theta(o_n)f,f)_H = (\theta(\alpha')f,f)_H$$

as $n \to \infty$. Then

$$(\theta(\alpha')f,f)_H \le (\theta(o_n \cap R')f,f)_H \le (\theta(o_n)f,f)_H,$$

and

$$(\theta(o_n \cap R')f,f)_H \to (\theta(\alpha')f,f)_H.$$

■

1.10. Properness of a Joint Resolution of the Identity

It has been established in Subsection 1.6 that a joint RI E is a proper measure provided that each R_x is compact or X is at most countable. Furthermore, the following general assertion is valid.[27]

Theorem 1.6. *Let E be the joint RI constructed from a family $(E_x)_{x \in X}$ of commuting Borel RI's E_x on the spaces R_x, and let β be a fixed set in $\mathfrak{C}_\sigma(R_x)$. If all but at most a countable number of the sets* supp E_x *are compact, then the restriction $E \upharpoonright \beta$ is a proper measure.*

Lemma 1.2. *Suppose that R_j is a Hausdorff topological space, $\mathfrak{R}_j$ is the σ-algebra generated by a certain basis Σ_j of neighborhoods in R_j which contains R_j, and $\mathfrak{R}_j \ni \alpha_j \mapsto E_j(\alpha_j)$ is an RI ($j = 1, 2$). Assume that E_1 and E_2 commute and that the joint RI $\mathfrak{R} \ni \alpha \mapsto E(\alpha)$ constructed from E_1 and E_2 exists. Fix $\beta_1 \in \mathfrak{R}_1$ and consider the restriction $E \upharpoonright (\beta_1 \times R_2)$ which is assumed to be a nontrivial measure (i.e., $E(\beta_1 \times R_2) \neq 0$). Let R_2 be compact. Then the fact that a set of the form $o_1 \times R_2$ (open in $\beta_1 \times R_2$) contains* supp $(E \upharpoonright (\beta_1 \times R_2))$ *implies that* $o_1 \supset$ supp $(E_1 \upharpoonright \beta_1)$.

[27] In a certain sense, the countability condition given below is also a necessary condition; see Example 1.1.

Proof. Let $\operatorname{supp}(E \restriction (\beta_1 \times R_2)) = \bigcap_{\xi \in \Xi} \varphi_\xi$, where $\varphi_\xi \subset \beta_1 \times R_2$ runs through the collection of all sets from $\mathfrak{R}$ of full $E \restriction (\beta_1 \times R_2)$-measure that are closed in $\beta_1 \times R_2$. Assume the contrary, i.e., that $o_1 \not\supset \operatorname{supp}(E_1 \restriction \beta_1)$. Then there exists a point $\lambda_1^0 \in \operatorname{supp}(E_1 \restriction \beta_1)$ which does not belong to o_1. For any $\lambda_2 \in R_2$ we have $(\lambda_1^0, \lambda_2) \notin o_1 \times R_2$ and, hence,

$$(\lambda_1^0, \lambda_2) \notin \operatorname{supp}(E \restriction (\beta_1 \times R_2)) = \bigcap_{\xi \in \Xi} \varphi_\xi.$$

Therefore, for each $\lambda_2 \in R_2$, one can find $\xi(\lambda_2)$ such that $(\lambda_1^0, \lambda_2) \notin \varphi_{\xi(\lambda_2)}$ and, thus, there exists a neighborhood of the point (λ_1^0, λ_2) in $\beta_1 \times R_2$ which is disjoint with $\varphi_{\xi(\lambda_2)}$. This neighborhood can be taken in the form $(u_1^{(\lambda_2)} \cap \beta_1) \times u_2(\lambda_2)$ ($u_1^{(\lambda_2)} \in \Sigma_1$ is a neighborhood of the point λ_1^0); its $E \restriction (\beta_1 \times R_2)$-measure is zero, because $\varphi_{\xi(\lambda_2)}$ is of full measure. The neighborhoods $u_2(\lambda_2)$ cover the compact space R_2. Let us choose a finite subcovering $\bigcup_{k=1}^n u_2(\lambda_{2,k}) = R_2$ of this covering and consider the intersection

$$v_1 = \bigcap_{k=1}^n (u_1^{(\lambda_{2,k})} \cap \beta_1) \subset \beta_1.$$

Obviously, $v_1 \in \mathfrak{R}_1$, it is open in β_1 and such that

$$v_1 \times R_2 \subset \bigcup_{k=1}^n ((u_1^{(\lambda_{2,k})} \cap \beta_1) \times u_2(\lambda_{2,k})),$$

and

$$0 = (E \restriction (\beta_1 \times R_2))(v_1 \times R_2) = E_1(v_1) E_2(R_2) = (E_1 \restriction \beta_1)(v_1).$$

On the other hand, $\lambda_1^0 \in v_1$; hence, $v_1 \cap \operatorname{supp}(E_1 \restriction \beta_1) \neq \varnothing$. But then $(E_1 \restriction \beta_1)(v_1) \neq 0$. We have arrived at a contradiction.

■

***Proof of Theorem* 1.6. I.** Denote by $(x_k)_{k=1}^\infty$ a sequence of points $x_k \in X$ such that $\operatorname{supp} E_x$ is a compact set for $x \neq x_k$. Let $\beta = \mathbf{C}(y_1, y_2, \dots; \delta)$, $\delta \in \mathfrak{B}(R_{y_1, y_2, \dots})$, and $\{z_1, z_2, \dots\} = \{x_1, x_2, \dots\} \cup \{y_1, y_2, \dots\}$. Then $\beta = \mathbf{C}(z_1, z_2, \dots; \varepsilon)$, where $\varepsilon \in \mathfrak{B}(R_{z_1, z_2, \dots})$ is a new base of β. We set $R_x' = R_{z_k}$ for $x = z_k$ ($k \in \mathbb{N}$) and $R_x' = \operatorname{supp} E_x$ for the remaining $x \in X$; the latter space being topologized by the relative topology of R_x. According to Lemma 1.1, $R'_X = \times_{x\in X} R'_x$ is of full outer E-measure. Let E' be the modification of RI E by R'_X; E' can be also regarded as the joint RI in the space R'_X constructed from the family $(E'_X)_{x\in X}$ (on the σ-algebra $\mathfrak{C}_\sigma(R'_X)$), here $E'_X = E_x \upharpoonright R'_X$. Cylindrical and generalized cylindrical sets from the space R'_X will be marked by primes, for example,

$$\mathfrak{C}'(t_1, t_2, \dots; \delta') = \{\lambda'(\cdot) \in R'_X \mid (\lambda'(t_1), \lambda'(t_2), \dots) \in \delta'\}, \quad \delta' \in \mathfrak{B}(R_{t_1, t_2, \dots}).$$

II. Since $R'_{z_k} = R_{z_k}$ $(k \in \mathbb{N})$, we have

$$\mathfrak{C}_\sigma(R'_X) \ni \beta' = \beta \cap R'_X = \mathfrak{C}'(z_1, z_2, \dots; \varepsilon).$$

It will be proved below that the restriction $E' \upharpoonright \beta'$ is a regular proper measure on the space β' (topologized by the relative topology of R'_X). Assuming that this fact is already proved, we now derive the statement of the theorem: $E \upharpoonright \beta$ is a proper measure on the space β.

To prove this, we first note that β' is of full outer $E \upharpoonright \beta$-measure. Indeed, let

$$\mathfrak{C}_\sigma(R_X) \ni \alpha \subset \beta = \mathfrak{C}(z_1, z_2, \dots; \varepsilon),$$

and

$$\alpha \supset \beta' = \mathfrak{C}'(z_1, z_2, \dots; \varepsilon);$$

then

$$\mathfrak{C}_\sigma(R_X) \ni \beta \setminus \alpha \in R_X \setminus R'_X,$$

and

$$0 = E(\beta \setminus \alpha) = E(\beta) - E(\alpha) = (E \upharpoonright \beta)(\beta) - (E \upharpoonright \beta)(\alpha).$$

Denote by $(E \upharpoonright \beta)'$ the modification of $E \upharpoonright \beta$ by β', then

$$(E \upharpoonright \beta)' = E' \upharpoonright \beta'. \tag{1.23}$$

In fact, the measure $(E \upharpoonright \beta)'$ is defined on sets $\alpha' = \alpha \cap \beta'$ with $\mathfrak{C}_\sigma(R_X) \ni \alpha \subset \beta$ by the equality

$$(E \upharpoonright \beta)'(\alpha') = (E \upharpoonright \beta)(\alpha) = E(\alpha).$$

On the other hand, the measure $E' \upharpoonright \beta'$ is defined on sets $\alpha' \subset \beta'$ of the form $\alpha' = \alpha_1 \cap R'_X$, where $\alpha_1 \in \mathfrak{C}_\sigma(R_X)$, by the equality

$$(E' \upharpoonright \beta')(\alpha') = E'(\alpha') = E(\alpha_1).$$

Since $\alpha \subset \beta$, we have

$$\alpha \cap R'_X = (\alpha \cap \beta) \cap R'_X = \alpha \cap \beta' = \alpha' = \alpha_1 \cap R'_X,$$

and, therefore,

$$\mathfrak{C}_\sigma(R_X) \ni \alpha_1 \setminus \alpha, \quad \alpha \setminus \alpha_1 \subset R_X \setminus R'_X$$

Consequently, $E(\alpha_1 \setminus \alpha) = E(\alpha \setminus \alpha_1) = 0$, and $E(\alpha_1) = E(\alpha)$. Hence, (1.23) is valid.

Suppose that $\mathfrak{C}_\sigma(R_X) \ni \gamma \subset \beta$ and $\gamma \supset \operatorname{supp}(E \upharpoonright \beta)$. The properness of the measure $E \upharpoonright \beta$ will be proved if we establish the equality

$$(E \upharpoonright \beta)(\gamma) = (E \upharpoonright \beta)(\beta) = E(\beta).$$

Taking into account the general relation (1.18) and (1.23), we get

$$\gamma \cap \beta' \supset (\operatorname{supp}(E \upharpoonright \beta)) \cap \beta' = \operatorname{supp}(E \upharpoonright \beta)' = \operatorname{supp}(E' \upharpoonright \beta').$$

Moreover,

$$\gamma \cap \beta' = \gamma \cap (\beta \cap R'_X) = \gamma \cap R'_X,$$

therefore, the set $\gamma \cap \beta'$ belongs to the σ-algebra on which the measure $E' \upharpoonright \beta'$ is defined. Since this measure is regular and proper by assumption, we have

$$(E' \upharpoonright \beta')(\gamma \cap \beta') = (E' \upharpoonright \beta')(\beta').$$

But

$$(E' \upharpoonright \beta')(\gamma \cap \beta') = E'(\gamma \cap \beta') = E'(\beta \cap R'_X) = E(\gamma) = (E \upharpoonright \beta)(\gamma),$$

$$(E' \upharpoonright \beta')(\beta') = E'(\beta') = E'(\beta \cap R'_X) = E(\beta) = (E \upharpoonright \beta)(\beta).$$

Therefore, $(E \upharpoonright \beta)(\gamma) = (E \upharpoonright \beta)(\beta)$, and the properness of the measure $E \upharpoonright \beta$ is thus established.

III. To prove the theorem, it remains to establish that $E' \upharpoonright \beta'$ is a regular proper measure on β'. Its regularity follows from the regularity of E' (Theorem 1.4) and from Subsection 1.8. We now proceed to prove that $E' \upharpoonright \beta'$ is proper.

Suppose that $\mathfrak{C}_\sigma(R'_X) \ni o' \subset \beta'$ is open in β' and such that

$$o' \supset \operatorname{supp}(E' \upharpoonright \beta'),$$

one must prove that $(E' \upharpoonright \beta')(o') = (E' \upharpoonright \beta')(\beta')$, i.e., $E'(o') = E'(\beta')$. Clearly, it suffices to assume that the measure $E' \upharpoonright \beta'$ is nontrivial: $E'(\beta') \neq 0$. Let $o' = \mathbf{C}'(t_1, t_2, \ldots;\kappa')$, where $\kappa' \in \mathfrak{B}(R'_{t_1,t_2,\ldots})$. Consider $\{s_1, s_2, \ldots\} = \{z_1, z_2, \ldots\} \cup \{t_1, t_2, \ldots\}$ and rewrite β' and o' by using the coordinates $(s_k)_{k=1}^\infty$, namely, $\beta' = \mathbf{C}'(s_1, s_2, \ldots; \beta_1)$ and $o' = \mathbf{C}'(s_1, s_2, \ldots; o_1)$, where $\beta_1, o_1 \in \mathfrak{B}(R'_{s_1,s_2,\ldots})$ are the bases ε and κ', properly modified.

According to what has been said in Subsection 1.4, we set $X = X_1 \cup X_2$, where $X_1 = \{s_1, s_2, \ldots\}$ and $X_2 = X \setminus X_1$, and represent R'_X as the direct product $R'_X = R_1 \times R_2$ $(R_1 = R'_{X_1}, R_2 = R'_{X_2})$. The joint RI $E' = E'_X$ is equal to that constructed from the RI's

$$E_1 = E'_{X_1} = \mathop{\times}_{x \in X_1} E'_x \quad \text{and} \quad E_2 = E'_{X_2} = \mathop{\times}_{x \in X_2} E_x$$

in the spaces R_1 and R_2, respectively. Since

$$\beta' = \mathbf{C}'(s_1, s_2, \ldots; \beta_1) = \beta_1 \times R_2,$$

we have

$$E'(\beta_1 \times R_2) = E'(\beta') \neq 0.$$

Therefore, the restriction $E' \upharpoonright (\beta_1 \times R_2)$ is the nontrivial measure. The space $R_2 = R'_{X_2} = \times_{x \in X_2} R'_x$ is compact $(X_2 \subset X \setminus \{x_1, x_2, \ldots\}$ and, hence, the spaces $R'_x = \operatorname{supp} E_x$ being multiplied are compact). Thus, we find that Lemma 1.2 can be applied. By assumption, the set

$$o' = \mathbf{C}'(s_1, s_2, \ldots; o_1) = o_1 \times R_2$$

is open in $\beta' = \beta_1 \times R_2$ and contains the support of the measure $E' \upharpoonright \beta' = E' \upharpoonright (\beta_1 \times R_2)$. By virtue of this lemma, $o_1 \supset \operatorname{supp}(E_1 \upharpoonright \beta_1)$.

But X_1 is countable. Therefore, the system of base neighborhoods $\mathbf{C}_\delta$ in the space

$R_1 = R'_{X_1}$ contains a countable subsystem of neighborhoods $\hat{\mathbb{C}}_\delta$ which determine the same topology in R_1 as above (see Subsection 1.4). The collection of all intersections $\hat{\mathbb{C}}_{\delta \cap \beta_1}$ serves as a basis system in β_1 ($E_1 \upharpoonright \beta_1$ is defined on its σ-span). It contains the countable subsystem of all $\hat{\mathbb{C}}_{\delta \cap \beta_1}$ which determines the same topology. Consequently, by virtue of condition (ii) in Theorem 1.5, the measure $E_1 \upharpoonright \beta_1$ is proper. The measures E_1 and $E_1 \upharpoonright \beta_1$ are defined on

$$\mathfrak{C}_\sigma(R'_{X_1}) = \mathfrak{B}(R'_{X_1}) = \mathfrak{B}(R_1)$$

(see Subsection 1.4) and $\mathfrak{B}(\beta_1)$, respectively. Since $o_1 \times R_2$ is open in $\beta_1 \times R_2$, we conclude that o_1 is open in β_1 and, therefore, belongs to $\mathfrak{B}(\beta_1)$. Since $E_1 \upharpoonright \beta_1$ is proper, we have

$$(E_1 \upharpoonright \beta_1)(o_1) = (E_1 \upharpoonright \beta_1)(\beta_1),$$

i.e., $E_1(o_1) = E_1(\beta_1)$, whence

$$E'(o') = E'(o' \times R_2) = E_1(o_1)\, E_2(R_2) = E_1(\beta_1)\, E_2(R_2) = E'(\beta_1 \times R_2) = E'(\beta').$$

■

1.11. Spectral Representation of a Family of Commuting Normal Operators

Let $A = (A_x)_{x \in X}$ be a family of normal operators A_x acting in the Hilbert space H; each A_x has its RI $\mathfrak{B}(\mathbb{C}^1) \ni \alpha \mapsto E_x(\alpha)$. The operators A_x are assumed to be commuting. This means that the family of RI's $(E_x)_{x \in X}$ is commuting. Let

$$\mathfrak{C}_\sigma(\mathbb{C}^X) \ni \alpha \mapsto E(\alpha) = (\underset{x \in X}{\times} E_x)(\alpha)$$

be the joint RI constructed according to Theorem 1.3; it is called the joint RI of the family A.

The spectral integral

$$F = \int_R F(\lambda)\, dE(\lambda), \quad \mathfrak{D}(F) = \left\{ f \in H \mid \int_R |F(\lambda)|^2\, d(E(\lambda)f, f)_H < \infty \right\} \tag{1.24}$$

is defined in the usual way for a general RI $\mathfrak{R} \ni \alpha \mapsto E(\alpha)$ given on the σ-algebra

$\mathfrak{R}$ and described in Subsection 1.1. Here $F(\lambda)$ is a complex-valued function given almost everywhere with respect to the measure E on R and measurable with respect to $\mathfrak{R}$. The first of the integrals in (1.24) converges in the sense of strong convergence in H. The construction of the integral (1.24) is exactly the same as in the classical case of an RI corresponding to a single selfadjoint operator.

In particular, the integral (1.24) can be written with respect to the joint RI $E = \times_{x \in X} E_x$ of the family A. As a function $\mathbb{C}^X \ni \lambda(\cdot) \mapsto F(\lambda(\cdot)) \in \mathbb{C}^1$, we take a cylindrical function, i.e., a function constructed for a given function

$$\mathbb{C}^X \ni \lambda = (\lambda_1, \dots, \lambda_n) \mapsto G(\lambda) = G(\lambda_1, \dots, \lambda_n) \in \mathbb{C}^1$$

and fixed distinct points $x_1, \dots, x_n \in X$ by the formula $F(\lambda(\cdot)) = G(\lambda(x_1), \dots, \lambda(x_n))$. Each function $\mathbb{C}^X \ni \lambda(\cdot) \mapsto \lambda(x) \in \mathbb{C}^1$ $(x \in X)$ is measurable with respect to the σ-algebra $\mathfrak{C}_\sigma(\mathbb{C}^X)$; therefore, if G is a Borel function, then F is also measurable with respect to $\mathfrak{C}_\sigma(\mathbb{C}^X)$.

It is easy to see that

$$F = \int_{\mathbb{C}^X} F(\lambda(\cdot))\, dE(\lambda(\cdot)) = \int_{\mathbb{C}^X} G(\lambda(x_1), \dots, \lambda(x_n))\, dE(\lambda(\cdot))$$

$$= \int_{\mathbb{C}^n} G(\lambda_1, \dots, \lambda_n)\, dE_{x_1, \dots, x_n}(\lambda) = G(A_{x_1}, \dots, A_{x_n}) \tag{1.25}$$

(it suffices to verify this for the characteristic function $G(\lambda) = \kappa_\delta(\lambda)$ of a set $\delta \in \mathfrak{B}(\mathbb{C}^n)$, but in this case (1.25) turns into (1.9)). Thus, the left-hand integral in (1.25) gives a spectral representation of a function G of the normal commuting operators $A_{x_1}, \dots, A_{x_n}$. Its domain of definition is given by the second formula in (1.24).

In the case when $\mathbb{C}^1 \ni \lambda \mapsto G(\lambda) = \lambda \in \mathbb{C}^1$, formulas (1.24) and (1.25) give a representation of an arbitrary operator A_x from the family under consideration in the form of the "functional" spectral integral of the function $\mathbb{C}^X \ni \lambda(\cdot) \mapsto \lambda(x) \in \mathbb{C}^1$:

$$A_x = \int_{\mathbb{C}^X} \lambda(x)\, dE(\lambda(\cdot)),$$

$$\mathfrak{D}(A_x) = \Big\{ f \in H \,\Big|\, \int_{\mathbb{C}^X} |\lambda(x)|^2\, d(E(\lambda(\cdot))f, f)_H < \infty \Big\}. \tag{1.26}$$

Clearly, the adjoint operator A_x^* may be represented in the form of the first integral, where $\lambda(x)$ is replaced by $\overline{\lambda(x)}$, and $\mathfrak{D}(A_x^*) = \mathfrak{D}(A_x)$.

Remark 1. The representation (1.26) can be also written with E replaced by the compactified joint RI $\mathfrak{C}_\sigma(\mathbf{C}^X) \ni \boldsymbol{\alpha} \mapsto E(\boldsymbol{\alpha})$, $\mathbb{C}^X$ replaced by $\mathbf{C}^X$, and $\lambda(x)$ replaced by $\boldsymbol{\lambda}(x)$. Indeed,

$$E(\{\boldsymbol{\lambda}(\cdot) \in \mathbf{C}^X \mid \boldsymbol{\lambda}(x) = \infty\}) = E(\mathfrak{C}(x; \{\infty\})) = E_x(\{\infty\}) = 0.$$

Therefore, the mapping $\mathbf{C}^X \ni \boldsymbol{\lambda}(\cdot) \mapsto \boldsymbol{\lambda}(x) \in \mathbf{C}^1$ takes values in $\mathbb{C}^1$ almost everywhere with respect to E, and the integrals in (1.26) make sense with the indicated changes being realized. The equalities (1.26) hold by virtue of (1.20) and (1.19).

■

All what has been said in this subsection is also valid for a family of commuting self-adjoint operators; one should only replace $\mathbb{C}^X$ and $\mathbf{C}^X$ by $\mathbb{R}^X$ and $\mathbf{R}^X$, respectively.

§2. Spectral Projection Theorem

In Subsections 2.1 – 2.4, we prove a theorem of the Radon–Nikodym type on the differentiation of an operator-valued measure with respect to its trace ρ. We also present a corollary concerning the differentiation of an RI E. By using this fact, we establish the main result of Chapter 3 which can be formulated as follows: We prove that the "generalized projector" $P(\lambda(\cdot)) = (dE/d\rho)(\lambda(\cdot))$ for the joint RI E of a family $(A_x)_{x\in X}$ of commuting normal operators can actually be regarded as a certain "projector onto the subspace consisting of the generalized joint eigenvectors corresponding to the eigenvalue $\lambda(\cdot)$". The proof of this statement is first presented for the simple situation when X is at most countable, and then the general case is investigated.

2.1. Differentiation of an Operator-Valued Measure with Respect to Its Trace

We fix a chain

$$H_- \supset H_0 \supset H_+, \tag{2.1}$$

whose spaces are all separable (clearly, it suffices that H_+ should be separable). Recall (see Chapter 1, Subsection 1.1) that an operator $A : H_+ \to H_-$ is called nonnegative if

$$(A\varphi, \varphi)_{H_0} \ge 0 \quad (\varphi \in H_+).$$

The trace of a nonnegative operator A is by definition equal to

$$\operatorname{Tr}(A) = \sum_{j=1}^{\infty} (A e_j, e_j)_{H_0},$$

where $(e_j)_{j=1}^{\infty}$ is an orthonormal basis in H_+. The quantity $\operatorname{Tr}(A)$ does not depend on the choice of the basis: If I is the isometry associated with (2.1), then it follows from the relation

$$(\xi, \varphi)_{H_0} = (I\xi, \varphi)_{H_0} \quad (\xi \in H_-, \quad \varphi \in H_+),$$

that the nonnegativity of A is equivalent to the ordinary nonnegativity of $IA : H_+ \to H_+$, and $\operatorname{Tr}(A) = \operatorname{Tr}(IA)$.

Let R be an abstract space (it is not necessary to introduce a topology in it), and let $\mathfrak{R}$ be a σ-algebra of subsets of R. A function $\mathfrak{R} \ni \alpha \mapsto \theta(\alpha)$ is called an operator-valued measure with finite trace if the following conditions hold:

(i) $\theta(\alpha)$ is a nonnegative operator from H_+ to H_-, with $\theta(\varnothing) = 0$ and $\operatorname{Tr}(\theta(R)) < \infty$;

(ii) the absolute additivity property holds, namely, if $\alpha_j \in \mathfrak{R}$ $(j \in \mathbb{N})$ are mutually disjoint, then

$$\theta\left(\bigcup_{j=1}^{\infty} \alpha_j\right) = \sum_{j=1}^{\infty} \theta(\alpha_j),$$

where the series converges in the weak sense.

The additivity and nonnegativity of θ imply its monotonicity, namely, if $\alpha' \subset \alpha''$, then $\theta(\alpha') \le \theta(\alpha'')$. Therefore, $\theta(\alpha) \le \theta(R)$ and $\operatorname{Tr}(\theta(\alpha)) \le \operatorname{Tr}(\theta(R))$ $(\alpha \in \mathfrak{R})$.

We introduce the nonnegative numerical set function $\mathfrak{R} \ni \alpha \mapsto \rho(\alpha) = \operatorname{Tr}(\theta(\alpha))$. If $\alpha_j \in \mathfrak{R}$ $(j \in \mathbb{N})$ are mutually disjoint, then, by virtue of condition (ii) and the nonnegativity of the terms, we have

$$\rho\left(\bigcup_{j=1}^{\infty} \alpha_j\right) = \operatorname{Tr}\left(\theta\left(\bigcup_{j=1}^{\infty} \alpha_j\right)\right) = \operatorname{Tr}\left(\sum_{j=1}^{\infty} \theta(\alpha_j)\right) = \sum_{k=1}^{\infty}\left(\left(\sum_{j=1}^{\infty} \theta(\alpha_j) e_k, e_k\right)_{H_0}\right)$$

$$= \sum_{k=1}^{\infty} \sum_{j=1}^{\infty} ((\theta(\alpha_j) e_k, e_k)_{H_0}) = \sum_{j=1}^{\infty} \sum_{k=1}^{\infty} ((\theta(\alpha_j) e_k, e_k)_{H_0})$$

$$= \sum_{j=1}^{\infty} \operatorname{Tr}(\theta(\alpha_j)) = \sum_{j=1}^{\infty} \rho(\alpha_j).$$

Thus, $\mathfrak{R} \ni \alpha \mapsto \rho(\alpha)$ is a finite nonnegative numerical measure. The measure ρ will be called the trace measure for the measure θ.

Theorem 2.1. *An operator-valued measure* θ *with a finite trace can be differentiated with respect to its trace measure* ρ. *This means that there exists a weakly* $\mathfrak{R}$*-measurable operator-valued function* $Q(\lambda): H_+ \to H_-$ *defined for* ρ*-almost all* $\lambda \in R$ *such that* $Q(\lambda) \geq 0$, $\|Q(\lambda)\| \leq \operatorname{Tr}(Q(\lambda)) = 1$, *and*

$$\theta(\alpha) = \int_{\alpha} Q(\alpha)\, d\rho(\lambda) \quad (\alpha \in \mathfrak{R}) \tag{2.2}$$

(the integral converges with respect to the Hilbert–Schmidt norm). The function $Q(\lambda)$ *is uniquely determined to within its values on a set of* ρ*-measure zero and is called the Radon–Nikodym derivative* $(d\theta/d\rho)(\lambda) = Q(\lambda)$.

We note that the convergence of the integral in (2.2) in the Hilbert–Schmidt norm means its convergence in the Bochner sense if $Q(\lambda)$ is understood as a vector function with values in the space of Hilbert–Schmidt operators from H_+ to H_-.

Proof. Let $(e_j)_{j=1}^{\infty}$ be a fixed orthonormal basis in the space H_+. The measure θ is absolutely continuous with respect to ρ, i.e., if $\rho(\alpha) = 0$, then $\theta(\alpha) = 0$ $(\alpha \in \mathfrak{R})$:

$$|(\theta(\alpha) e_j, e_k)_{H_0}|^2 \leq (\theta(\alpha) e_j, e_j)_{H_0} (\theta(\alpha) e_k, e_k)_{H_0} \leq \rho^2(\alpha) = 0 \quad (j, k \in \mathbb{N}).$$

This implies that for any fixed $\varphi, \psi \in H_+$ the complex-valued measure $\mathfrak{R} \ni \alpha \mapsto (\theta(\alpha)\varphi, \psi)_{H_0} \in \mathbb{C}^1$ is also absolutely continuous with respect to ρ and, by the ordinary Radon–Nikodym theorem, we have

$$(\theta(\alpha)\varphi, \psi)_{H_0} = \int_{\alpha} q(\lambda; \varphi, \psi)\, d\rho(\lambda) \quad (\alpha \in \mathfrak{R},\ \varphi, \psi \in H_+), \tag{2.3}$$

where the derivative $q(\lambda; \varphi, \psi)$ is defined on a set $\beta_{\varphi,\psi} \subset R$ of full ρ-measure, is $\mathfrak{R}$-measurable, and is integrable; for $\varphi = \psi$ it is nonnegative. Denote by L the linear span of the vectors $(e_j)_{j=1}^{\infty}$ with rational complex coefficients; $\tilde{L} = H_+$. Since L is countable, the set $\bigcap_{\varphi,\, \psi \in L} \beta_{\varphi,\, \psi}$ is also of full measure; all the functions $q(\lambda;\ \varphi,\ \psi)$ $(\varphi,\ \psi \in L)$ are defined for λ belonging to this set, and $q(\lambda;\ \varphi,\ \varphi) \geq 0$ $(\varphi \in L)$.

The fact that the derivative is uniquely determined (up to its values on a set of measure zero) implies us that the bilinearity of the left-hand side of (2.3) with respect to $\varphi,\ \psi$ involves the bilinearity of $q(\lambda;\ \varphi,\ \psi)$. More precisely, there exists a set $\beta \subset \bigcap_{\varphi,\, \psi \in L} \beta_{\varphi,\, \psi}$ of full measure such that

$$q(\lambda; p_1\varphi_1 + p_2\varphi_2, r_1\psi_1 + r_2\psi_2) = p_1 \bar{r}_1\, q(\lambda; \varphi_1, \psi_1) + p_1 \bar{r}_2\, q(\lambda; \varphi_1, \psi_2)$$

$$+ p_2 \bar{r}_1\, q(\lambda; \varphi_2, \psi_1) + p_2 \bar{r}_2\, q(\lambda; \varphi_2, \psi_2)$$

for $\lambda \in \beta$, any $\varphi_1, \varphi_2, \psi_1, \psi_2 \in L$, and any complex rational numbers p_1, p_2, r_1, r_2. To prove this, we first take into account that $(\theta(\alpha)\varphi,\ \psi)_{H_0}$ is bilinear and that $\alpha \in \mathfrak{R}$ in (2.3) is arbitrary. This enables us to conclude that the last equality is valid for λ belonging to a set

$$\beta_{p_1, p_2, r_1, r_2, \varphi_1, \varphi_2, \psi_1, \psi_2} \subset \bigcap_{\varphi, \psi \in L} \beta_{\varphi,\, \psi}$$

of full measure. Then we take β to be the (countable) intersection of all possible sets of this sort. Moreover, it has been noted that $q(\lambda;\ \varphi,\ \varphi) \geq 0$ $(\varphi \in L)$ for these λ. In the usual way, this bilinearity and nonnegativity lead to the Cauchy–Schwarz–Bunyakovsky inequality

$$|q(\lambda; \varphi, \psi)|^2 \leq q(\lambda; \varphi, \varphi)\, q(\lambda; \psi, \psi) \quad (\lambda \in \beta;\ \varphi, \psi \in L). \tag{2.4}$$

With the help of the Fubini theorem, we obtain

$$\rho(\alpha) = \int_\alpha \left(\sum_{j=1}^{\infty} q(\lambda; e_j, e_j) \right) d\rho(\lambda) \qquad (\alpha \in \mathfrak{R})$$

by setting $\varphi = \psi = e_j$ in (2.3) and summing over $j \in \mathbb{N}$. This implies that for almost all $\lambda \in \beta$

$$\sum_{j=1}^{\infty} q(\lambda;\ e_j, e_j) = 1. \tag{2.5}$$

Assume that (2.5) is valid for all $\lambda \in \beta$ (if necessary, we can make the set β somewhat smaller). It follows from (2.4) and (2.5) that

$$\sum_{j,k=1}^{\infty} |q(\lambda;\ e_j, e_k)|^2 \le \sum_{j,k=1}^{\infty} q(\lambda;\ e_j, e_j)\, q(\lambda;\ e_k, e_k) = 1 \quad (\lambda \in \beta). \tag{2.6}$$

Fix $\lambda \in \beta$ and denote by $A(\lambda)$ the operator defined in H_+ which corresponds to the matrix $(a_{jk}(\lambda))_{j,k=1}^{\infty}$, in the basis $(e_j)_{j=1}^{\infty}$, where $a_{jk}(\lambda) = q(\lambda;\ e_k, e_j)$. By virtue of (2.6), this operator is well defined and is a Hilbert–Schmidt operator. The fact that each function $\beta \ni \lambda \mapsto q(\lambda;\ e_k, e_j)$ $(j \in \mathbb{N})$ is measurable implies that the operator-valued function $\beta \ni \lambda \mapsto A(\lambda)$ is weakly measurable. Let us introduce the continuous operator

$$Q(\lambda) = I^{-1} A(\lambda) \colon H_+ \to H_- .$$

We now show that it is just the desired operator.

Since $A(\lambda)$ is measurable, $\beta \ni \lambda \mapsto Q(\lambda)$ is weakly measurable. Further, for

$$\varphi = \sum_{k=1}^{\infty} p_k e_k, \quad \psi = \sum_{j=1}^{\infty} r_j e_j \in L$$

we have

$$(Q(\lambda)\varphi, \psi)_{H_0} = (I^{-1}A(\lambda)\varphi, \psi)_{H_0} = (A(\lambda)\varphi, \psi)_{H_+}$$

$$= \sum_{j,k=1}^{\infty} q(\lambda;\ e_k, e_j) p_k r_j = q(\lambda;\ \varphi, \psi). \tag{2.7}$$

In particular,

$$(Q(\lambda)\varphi, \varphi)_{H_0} = q(\lambda;\ \varphi, \varphi) \ge 0.$$

By passing to the limit, we conclude that this inequality is preserved for arbitrary $\varphi \in H_+$, i.e., $Q(\lambda) \ge 0$. According to (2.5),

$$\mathrm{Tr}\,(Q(\lambda)) = \mathrm{Tr}\,(A(\lambda)) = 1 \quad (\lambda \in \beta).$$

Hence, $\|Q(\lambda)\| \le 1$, and $Q(\lambda)$ $(\lambda \in \beta)$ is weakly measurable. Therefore, the integral

$$\int_{\alpha} Q(\lambda)\, d\rho(\lambda) \quad (\alpha \in \mathfrak{R})$$

converges in the Hilbert–Schmidt norm and thus exists. According to (2.7) and (2.3), for

$\varphi, \psi \in L$ and $\alpha \in \mathfrak{R}$, we have

$$\left(\left(\int_\alpha Q(\lambda)\,d\rho(\lambda)\right)\varphi, \psi\right)_{H_0} = \int_\alpha (Q(\lambda)\,\varphi, \psi)_{H_0}\,d\rho(\lambda)$$

$$= \int_\alpha q(\lambda; \varphi, \psi)\,d\rho(\lambda) = (\theta(\lambda)\,\varphi, \psi)_{H_0},$$

i.e., (2.2) holds.

Finally, we establish that $Q(\lambda)$ is unique. Suppose that, along with $Q(\lambda)$, there exists an operator-valued function $Q_1(\lambda)$ of the same type such that

$$\int_\alpha Q(\lambda)\,d\rho(\lambda) = \int_\alpha Q_1(\lambda)\,d\rho(\lambda) \quad (\alpha \in \mathfrak{R}).$$

This implies that

$$(Q(\lambda)\,\varphi, \psi)_{H_0} = (Q_1(\lambda)\,\varphi, \psi)_{H_0}$$

for every $\varphi, \psi \in L$ and λ from the set $\beta_{1;\,\varphi,\,\psi} \subset R$ of full measure. But then

$$(Q(\lambda)\,\varphi, \psi)_{H_0} = (Q_1(\lambda)\,\varphi, \psi)_{H_0}$$

for any $\varphi, \psi \in L$ and λ from the set

$$\beta_1 = \bigcap_{\varphi,\psi \in L} \beta_{1;\,\varphi,\,\psi}$$

of full measure. This and the continuity of the operators $Q(\lambda)$ and $Q_1(\lambda)$ for every fixed $\lambda \in \beta_1$ yields $Q(\lambda) = Q_1(\lambda)$.

■

In the case when, for example, $R = \mathbb{R}^p$ and $\mathfrak{R} = \mathfrak{B}(\mathbb{R}^p)$ ($p \in \mathbb{N}$), this proof becomes somewhat simpler. Moreover, we can obtain a useful formula for $Q(\lambda)$. Indeed, recall that if ω and ρ are two numerical measures defined on $\mathfrak{B}(\mathbb{R}^p)$ such that ρ is nonnegative and ω is absolutely continuous with respect to ρ, then the Radon–Nikodym derivative $(d\omega/d\rho)(\lambda)$ can be obtained by the following formula, valid for ρ-almost all λ,

$$\left(\frac{d\omega}{d\rho}\right)(\lambda) = \lim_{\delta_n \to \lambda} \frac{\omega(\delta_n)}{\rho(\delta_n)}. \tag{2.8}$$

Here δ_n is a rectangle in the nth partition of the space $\mathbb{R}^p$ into rectangles of the form $\times_{j=1}^{p}[a_j, b_j)$ whose diameters tend to zero as $n \to \infty$; $\delta_n \to \lambda$ means that the limit is taken for a sequence of rectangles $\delta_n \ni \lambda$ contracting to λ. (We shall not dwell upon the application of a formula of the type (2.8) which holds in more general situations (Gikhman and Skorokhod [2, Chapter 2, Section 2]).)

For the indicated R and $\mathfrak{R}$, Theorem 2.1 can be proved in the following way. As before, we obtain the representation (2.3). Applying (2.8), we can conclude that

$$q(\lambda; \varphi, \psi) = \lim_{\delta_n \to \lambda} \rho^{-1}(\delta_n)(\theta(\delta_n)\varphi, \psi)_{H_0}$$

for λ belonging to the set $\beta_{2;\varphi,\psi} \subset \mathbb{R}^p$ of full measure. Let L be an arbitrary countable dense subset of H_+ containing the basis $(e_j)_{j=1}^{\infty}$. Then the last inequality holds for all λ from the set $\beta_2 = \bigcap_{\varphi,\psi \in L} \beta_{2;\varphi,\psi}$ of full measure and any $\varphi, \psi \in L$. It is now easy to see that for $\lambda \in \beta_2$ the limit

$$\lim_{\delta_n \to \lambda} \rho^{-1}(\delta_n)\theta(\delta_n) \tag{2.9}$$

exists in the sense of weak convergence of operators from H_+ to H_-.

In fact, we have established that the limit

$$\lim_{\delta_n \to \lambda} (\rho^{-1}(\delta_n)\theta(\delta_n)\varphi, \psi)_{H_0}$$

exists. Therefore, in order to prove that the limit (2.9) exists, it suffices to show that

$$\| \rho^{-1}(\delta_n)\theta(\delta_n) \| \le c \quad (n \in \mathbb{N}).$$

But

$$\| \rho^{-1}(\delta_n)\theta(\delta_n) \| \le | \rho^{-1}(\delta_n)\theta(\delta_n) | \le \operatorname{Tr}(\rho^{-1}(\delta_n)\theta(\delta_n)) = 1$$

and, thus, the limit (2.9) exists. We denote this limit by $Q(\lambda) : H_+ \to H_-$. It is nonnegative, since the operators $\rho^{-1}(\delta_n)\theta(\delta_n)$ are nonnegative. Clearly,

$$(Q(\lambda)\varphi, \psi)_{H_0} = q(\lambda; \varphi, \psi) \quad (\varphi, \psi \in L).$$

Further, using (2.3) and the Fubini theorem, we get

$$\rho(\alpha) = \operatorname{Tr}\theta(\alpha) = \sum_{j=1}^{\infty} (\theta(\alpha)e_j, e_j)_{H_0} = \sum_{j=1}^{\infty} \int_{\alpha} q(\lambda; e_j, e_j)\, d\rho(\lambda)$$

$$= \sum_{j=1}^{\infty} \int_{\alpha} (Q(\lambda)\, e_j, e_j)_{H_0}\, d\rho(\lambda) = \int_{\alpha} \sum_{j=1}^{\infty} (Q(\lambda)\, e_j, e_j)_{H_0}\, d\rho(\lambda)$$

for any $\alpha \in \mathfrak{R}$, and this implies that $\mathrm{Tr}\,(Q(\lambda)) = 1$ for almost all $\lambda \in \beta_2$. It is also clear that (2.2) holds for the constructed $Q(\lambda)$.

Hence, the main part of Theorem 2.1 is proved in the particular case being discussed. We have also established that for ρ-almost all λ

$$Q(\lambda) = \lim_{\delta_n \to \lambda} \rho^{-1}(\delta_n)\, \theta(\delta_n) \tag{2.10}$$

in the sense of weak convergence of operators.

Remark 1. It is possible to consider an operator-valued measure θ with σ-finite trace. This means that there exists a sequence $(R_k)_{k=1}^{\infty}$, $R_k \in \mathfrak{R}$, such that $R_1 \subset R_2 \subset \ldots$, $\bigcup_{k=1}^{\infty} R_k = R$, and $\mathrm{Tr}\,\theta(R_k) < \infty$ $(k \in \mathbb{N})$. The formulation and proof of Theorem 2.1 remain the same, except that the trace measure is not finite but σ-finite, and the representation (2.2) holds for each $\mathfrak{R} \ni \alpha \subset R_k$ for some $k \in \mathbb{N}$.

■

Remark 2. The formulation of Theorem 2.1 can be made similar to that of the usual Radon–Nikodym theorem. Namely, let θ be an operator-valued measure with a σ-finite trace, and let $\mathfrak{R} \ni \alpha \mapsto \rho(\alpha) \in [0, \infty)$ be a σ-finite nonnegative numerical measure with respect to which θ is absolutely continuous: if $\rho(\alpha) = 0$ for some $\alpha \in \mathfrak{R}$, then $\theta(\alpha) = 0$. Then the representation (2.2) is valid, where $\mathfrak{R} \ni \alpha \subset R_k$ $(k \in \mathbb{N})$, and $Q(\lambda)$ is a weakly measurable operator-valued function defined for ρ-almost all $\lambda \in R$; its values are nonnegative operators from H_+ to H_-, each has finite trace and is integrable over R_k $(k \in \mathbb{N})$ with respect to ρ.

(One should write the representation (2.2) with the trace measure and then differentiate this measure with respect to ρ in this representation.)

■

2.2. Differentiation of a Resolution of the Identity. The Spectral Measure

Let R be an abstract space, $\mathfrak{R}$ be a σ-algebra of subsets of R, and $\mathfrak{R} \ni \alpha \mapsto E(\alpha)$ be a general RI acting in the space H_0. As a rule, the measure E does not have a finite or σ-finite trace; therefore, the direct application of Theorem 2.1 is now impossible.

However, it is convenient to proceed as follows.

Assume that we have a rigging (2.1) of the space H_0. The function

$$\mathcal{R} \ni \alpha \mapsto \theta(\alpha) = O^{+}E(\alpha)\,O, \tag{2.11}$$

whose values are continuous operators from H_+ to H_-, is clearly an operator-valued measure ($O^+ E(\alpha)\,O \geq 0$, since

$$(O^{+}E(\alpha)\,O\varphi, \varphi)_{H_0} = (E(\alpha)\,O\varphi, \varphi)_{H_0} \geq 0$$

for $\varphi \in H_+$).

Lemma 2.1. *If the rigging* (2.1) *is quasinuclear, then the operator-valued measure* (2.11) *has a finite trace.*

Note first that *if* $A : H_0 \to H_0$ *is nonnegative, then* $O^+A\, O: H_+ \to H_-$ *is nonnegative, too, and*

$$\operatorname{Tr}(O^{+}A\;O) \leq \|A\|\,|O|^2. \tag{2.12}$$

Indeed, the inequality $O^+A\, O \geq 0$ was just explained by the example $A = E(\alpha)$. Further, let $(e_j)_{j=1}^{\infty}$ be an orthonormal basis in H_+, then

$$\operatorname{Tr}(O^{+}A\;O) = \sum_{j=1}^{\infty} (O^{+}A\;Oe_j, e_j)_{H_0} = \sum_{j=1}^{\infty} (A\;Oe_j, Oe_j)_{H_0}$$

$$\leq \|A\| \sum_{j=1}^{\infty} \|Oe_j\|_{H_0}^2 = \|A\|\,|O|^2.$$

■

Proof of Lemma 2.1. According to (2.12),

$$\operatorname{Tr}(\theta(R)) = \operatorname{Tr}(O^{+}E(R)\,O) \leq |O|^2 < \infty.$$

■

Let us fix a quasinuclear rigging (2.1). The nonnegative finite measure

$$\mathcal{R} \ni \alpha \mapsto \rho(\alpha) = \operatorname{Tr}(O^{+}E(\alpha)\,O) \in [0, \infty)$$

is called the spectral measure of the RI E. Clearly, E and ρ are absolutely continuous with respect to each other: The equalities $E(\alpha) = 0$ and $\rho(\alpha) = 0$ are equivalent for some $\alpha \in \mathfrak{R}$. Applying Theorem 2.1 to (2.11) and ρ, we get the following assertion.

Theorem 2.2. *Let* $\mathfrak{R} \ni \alpha \mapsto E(\alpha)$ *be an RI acting in the space* H_0, (2.1) *be a fixed quasinuclear rigging, and* $\mathfrak{R} \ni \alpha \mapsto \rho(\alpha) \in [0, \infty)$ *be the corresponding spectral measure. Then the representation in the form of the integral*

$$O^+E(\alpha)\,O = \int_\alpha P(\lambda)\,d\rho(\lambda) \quad (\alpha \in \mathfrak{R}) \tag{2.13}$$

holds, and this integral converges in the Hilbert–Schmidt norm. Here $P(\lambda) : H_+ \to H_-$ *is a weakly* $\mathfrak{R}$*-measurable operator-valued function defined for* ρ*-almost all* $\lambda \in R$ *and such that* $P(\lambda) \geq 0$ *and* $\| P(\lambda) \| \leq \mathrm{Tr}\,(P(\lambda)) = 1$;$P(\lambda)$ *is called a generalized projector.*

We clarify that in the case of the RI E of a single selfadjoint operator acting in H_0 with a discrete spectrum $(\lambda_j)_{j=1}^{\infty}$, the equality

$$E(\alpha) = \sum_{\lambda_j \in \alpha} P(\lambda_j) \quad (\alpha \in \mathfrak{B}(\mathbb{R}^1))$$

holds, where $P(\lambda_j)$ is the projector onto the eigensubspace of A corresponding to the eigenvalue λ_j. The term "generalized projector" appeared just as a result of a comparison of this formula with (2.13).

Remark 1. According to Remark 2 in Subsection 2.1, it is also possible to introduce the concept of a general spectral measure corresponding to an RI E. This measure is defined to be a σ-finite nonnegative measure $\mathfrak{R} \ni \alpha \mapsto \rho(\alpha) \in [0, \infty)$ possessing the property that E and ρ are absolutely continuous with respect to each other. The representation (2.13) is preserved for a general spectral measure, too. The only difference is that the operator $P(\lambda)$ acquires a scalar factor.

■

2.3. Differentiation of a Joint Resolution of the Identity

Let $(E_x)_{x \in X}$ be a family of commuting RI's defined in the spaces R_x and acting in H_0, and let $\mathfrak{C}_\sigma(R_X) \ni \alpha \mapsto E(\alpha)$ be its joint RI. The general Theorem 2.2 is obviously applicable to E. We dwell upon some peculiarities connected with the fact that we now consider $E = \times_{x \in X} E_x$ (the quasinuclear rigging (2.1) is fixed).

Theorem 2.3. *The spectral measure* ρ *of the joint RI* E *is always regular, and* $\operatorname{supp} \rho = \operatorname{supp} E$. *The measure* ρ *is proper if and only if* E *is proper.*

Proof. Let us establish that ρ is regular. It suffices to prove that for any $\alpha \in \mathfrak{C}_\sigma(R_X)$ and any $\varepsilon > 0$, one can find an open set $\mathfrak{C}_\sigma(R_X) \ni o \supset \alpha$ such that $\rho(o \setminus \alpha) < \varepsilon$. Let $(e_j)_{j=1}^\infty$ be an orthonormal basis in H_+. Since

$$\sum_{j=1}^{\infty} \| O e_j \|_{H_0}^2 = |O|^2 < \infty,$$

one can find $n \in \mathbb{N}$ sufficiently large so that

$$\sum_{j=n+1}^{\infty} \| O e_j \|_{H_0}^2 < \varepsilon / 2.$$

We fix this number. The measure E is regular (Theorem 1.4). Therefore, for every $j = 1, \ldots, n$ one can find an open set $\mathfrak{C}_\sigma(R_X) \ni o_j \supset \alpha$ such that

$$(E(o_j \setminus \alpha) O e_j, O e_j)_{H_0} < \varepsilon / 2n.$$

Then $o = \bigcap_{j=1}^n o_j \in \mathfrak{C}_\sigma(R_X)$ is open, it contains α, and

$$\rho(o \setminus \alpha) = \operatorname{Tr}(O^+ E(o \setminus \alpha) O) = \sum_{j=1}^{\infty} (O^+ E(o \setminus \alpha) O e_j, O e_j)_{H_0}$$

$$\leq \sum_{j=1}^{n} (E(o_j \setminus \alpha) O e_j, O e_j)_{H_0} + \sum_{j=n+1}^{\infty} \| O e_j \|_{H_0}^2 < \varepsilon .$$

The regularity of ρ is proved.

The remaining assertions of the theorem follow from the mutual absolute continuity of the measures E and ρ.

■

Let $(A_x)_{x \in X}$ be a family of commuting normal operators acting in the space H_0, let $\mathfrak{C}_\sigma(\mathbb{C}^X) \ni \alpha \mapsto E(\alpha)$ and $\rho(\alpha)$ be the corresponding joint RI and spectral measure, respectively, and let $P(\lambda(\cdot))$ be the generalized projector. We can also define the analogous "compactified" objects, namely, $\mathfrak{C}_\sigma(\mathbf{C}^X) \ni \boldsymbol{\alpha} \mapsto \boldsymbol{E}(\boldsymbol{\alpha})$, $\operatorname{Tr}(O^+ \boldsymbol{E}(\boldsymbol{\alpha}) O) = \boldsymbol{\rho}(\boldsymbol{\alpha})$, and $\boldsymbol{P}(\boldsymbol{\lambda}(\cdot))$. According to (1.20), $\boldsymbol{E}(\boldsymbol{\alpha}) = E(\boldsymbol{\alpha} \cap \mathbb{C}^X)$ and, thus , $\boldsymbol{\rho}(\boldsymbol{\alpha}) = \rho(\boldsymbol{\alpha} \cap \mathbb{C}^X)$.

The following relation holds

$$P(\lambda(\cdot)) = (\boldsymbol{P} \upharpoonright \mathbb{C}^X)(\lambda(\cdot)) \quad (\lambda(\cdot) \in \mathbb{C}^X). \tag{2.14}$$

Indeed, denote by $P_1(\lambda(\cdot))$ the operator-valued function on the right-hand side of (2.14). By virtue of (1.19) and (2.13), $\boldsymbol{E}$ and E satisfy the following relation

$$\int_{\boldsymbol{\alpha} \cap \mathbb{C}^X} P_1(\lambda(\cdot))\, \varphi, \psi)_{H_0}\, d\rho(\lambda(\cdot)) = \int_{\boldsymbol{\alpha}} \boldsymbol{P}(\boldsymbol{\lambda}(\cdot))\, \varphi, \psi)_{H_0}\, d\boldsymbol{\rho}(\boldsymbol{\lambda}(\cdot))$$

$$= (O^+\boldsymbol{E}(\boldsymbol{\alpha})\, O\varphi, \psi)_{H_0} = (O^+E(\boldsymbol{\alpha} \cap \mathbb{C}^X)\, O\varphi, \psi)_{H_0}$$

$$= \int_{\boldsymbol{\alpha} \cap \mathbb{C}^X} P(\lambda(\cdot))\, \varphi, \psi)_{H_0}\, d\rho(\lambda(\cdot))$$

for arbitrary $\varphi, \psi \in H_+$ and $\boldsymbol{\alpha} \in \mathfrak{C}_\sigma(\mathbf{C}^X)$. Since $\boldsymbol{\alpha} \cap \mathbb{C}^X$ runs through the whole of the σ-algebra $\mathfrak{C}_\sigma(\mathbb{C}^X)$, this yields (2.14).

■

In the preceding subsection we have introduced the concept of a general spectral measure ρ corresponding to an RI E. We mention an important case connected with the joint RI generated by the family $A = (A_x)_{x \in X}$ under consideration when the role of ρ can be played by another measure different from $\operatorname{Tr}(O^+E(\alpha)\, O)$ and constructed by using the standard procedure.

Consider the product $A_{x_1}^{m_1} \cdots A_{x_p}^{m_p}$, where $x_1, \dots, x_p$ are distinct points from X, and $m_1, \dots, m_p$ are nonnegative integers; $p \in \mathbb{N}$. A unit vector $\Omega \in H_0$ is called a cyclic vector of the family $(A_x)_{x \in X}$ (or a vacuum) if there exist products $A_{x_1}^{m_1} \cdots A_{x_p}^{m_p}$ such that $\Omega \in \mathfrak{D}(A_{x_1}^{m_1} \cdots A_{x_p}^{m_p})$ and the linear span of all the vectors $A_{x_1}^{m_1} \cdots A_{x_p}^{m_p}\Omega$ is dense in H_0 (in particular, $x_j \in X$ and $m_j \in \mathbb{Z}_+$ may be arbitrary).

Theorem 2.4. *Assume that a family $(A_x)_{x \in X}$ of commuting normal operators with a joint RI E possesses a cyclic vector Ω. Then the finite nonnegative measure $\mathfrak{C}_\sigma(\mathbb{C}^X) \ni \alpha \mapsto \rho(\alpha) = (E(\alpha)\, \Omega, \Omega)_{H_0} \in [0, \infty)$ is one of the spectral measures for this family.*

Proof. If $E(\alpha) = 0$ for some $\alpha \in \mathfrak{C}_\sigma(\mathbb{C}^X)$, then $\rho(\alpha) = 0$. We prove the opposite implication. Suppose that

$$0 = \rho(\alpha) = (E(\alpha)\, \Omega, \Omega)_{H_0} = \| E(\alpha)\, \Omega \|_{H_0}^2,$$

i.e., $E(\alpha)\,\Omega = 0$. Then for any product under consideration

$$0 = A_{x_1}^{m_1} \cdots A_{x_p}^{m_p} E(\alpha)\,\Omega = E(\alpha)\, A_{x_1}^{m_1} \cdots A_{x_p}^{m_p}\, \Omega$$

and, hence, $E(\alpha) f = 0$, too, where f belongs to the linear span of the vectors $A_{x_1}^{m_1} \ldots A_{x_p}^{m_p}\Omega$. But this span is dense in H_0 by assumption and, thus , $E(\alpha) = 0$.

■

Clearly, the last two assertions also hold for a family of commuting selfadjoint operators.

2.4. The Case of a Nuclear Rigging

We can obtain results analogous to those presented in Subsections 2.1–2.3 if, instead of the quasinuclear rigging (2.1) of the Hilbert space H_0, we use a nuclear rigging of it

$$\Phi' \supset H_0 \supset \Phi \tag{2.15}$$

(recall that the rigging (2.15) is called a nuclear rigging if $\Phi = \text{pr lim}_{\tau \in \mathrm{T}} H_\tau$ is a nuclear space; see Chapter 1, Subsection 1.2). More precisely, the case of the chain (2.15) can be reduced to the case of the chain (2.1). We show this.

Consider the rigging (2.15) and denote by O and O^+ the imbedding operators $\Phi \subset H_0$ and $H_0 \subset \Phi'$, respectively. We shall consider continuous operators $A : \Phi \to \Phi'$. Such operators are called nonnegative ($A \geq 0$) if $(A\varphi, \varphi)_{H_0} \geq 0$ ($\varphi \in \Phi$). In particular, if $\mathfrak{R} \ni \alpha \mapsto E(\alpha)$ is the RI which has been studied in Subsection 2.2, then $O^+E(\alpha)O$: $\Phi \to \Phi'$ is just the nonnegative operator. The function $\mathfrak{R} \ni \alpha \mapsto \theta(\alpha) = O^+E(\alpha)O$ is a measure similar to that considered in Subsections 2.1 and 2.2, but with values in $\mathfrak{L}(\Phi, \Phi')$.

Theorem 2.5. *Suppose that* $\mathfrak{R} \ni \alpha \mapsto E(\alpha)$ *and* $\rho(\alpha)$ *are an RI acting in the space* H_0 *and a spectral measure for it, respectively, and let* (2.15) *be a fixed nuclear rigging. Then the representation* (2.13) *holds in the form of a weakly convergent integral, where* $0 \leq P(\lambda)$: $\Phi \to \Phi'$ *(a generalized projector) is a weakly* $\mathfrak{R}$*-measurable operator-valued function defined for* ρ*-almost all* $\lambda \in R$.

Proof. As explained in Subsection 1.10 of Chapter 1, each H_τ ($\tau \in \mathrm{T}$) is topologically imbedded in H_0, and $0 \in \mathrm{T}$. Choose τ such that the imbedding $H_\tau \subset H_0$ is quasinuclear. This is possible because Φ is nuclear. As a result, we get the chain

$$\Phi' \supset H_{-\tau} \supset H_0 \supset H_\tau \supset \Phi \tag{2.16}$$

with topological imbeddings. Now, take the middle part of the chain (2.16) as (2.1), and introduce the imbedding operators $O_1: H_\tau \to H_0$ and $O_1^+: H_0 \to H_{-\tau}$. Then, by virtue of Theorem 2.2 and Remark 1 in Subsection 2.2, we have

$$O_1^+ E(\alpha)\, O_1 = \int_\alpha P_1(\lambda)\, d\rho(\lambda) \quad (\alpha \in \mathfrak{R}), \tag{2.17}$$

where $P_1(\lambda): H_\tau \to H_{-\tau}$ is the corresponding generalized projector.

Consider the imbeddings $O_2: \Phi \to H_\tau$ and $O_3: H_{-\tau} \to \Phi'$. Multiplying (2.17) from the right and from the left by O_2 and O_3, respectively, we obtain the required equality (2.13) with $P(\lambda) = O_3 P_1(\lambda)\, O_2$. The convergence of the integral (2.17) in the Hilbert–Schmidt norm (from H_τ to $H_{-\tau}$) obviously involves the weak convergence of the integral (2.13).

■

2.5. The Concepts of a Generalized Joint Eigenvector and of a Generalized Spectrum of a Family of Operators

Let $A = (A_x)_{x \in X}$ be a family of commuting normal operators acting in the Hilbert space H_0, and let (2.1) be a rigging of H_0. Assume that there exists a linear topological space D topologically imbedded in H_+ such that $D \subset \mathfrak{D}(A_x)$, and the restrictions $A_x \restriction D$ and $A_x^* \restriction D$ act continuously from D to H_+ $(x \in X)$. Thus, instead of (2.1), there now appears the rigging (chain)

$$H_- \supset H_0 \supset H_+ \supset D. \tag{2.18}$$

We say that the family A of operators having the described properties is connected in a standard way with the rigging (2.18) (or that A admits (2.18)). The chain (2.18) is called an extension of (2.1). As before, (2.18) is quasinuclear by definition if the imbedding of the (separable) space H_+ in H_0 is quasinuclear.

A nonzero vector $\xi \in H_-$ is called a generalized joint eigenvector of the family A with the eigenvalue[28] $\lambda(\cdot) \in \mathbb{C}^X$ if for every $x \in X$ we have

$$(\xi, A_x^* \varphi)_{H_0} = \lambda(x)\, (\xi, \varphi)_{H_0}, \quad (\xi, A_x \varphi)_{H_0} = \overline{\lambda(x)}\, (\xi, \varphi)_{H_0} \quad (\varphi \in D). \tag{2.19}$$

[28] We stress that an eigenvalue is regarded as a function $X \ni x \mapsto \lambda(\cdot) \in \mathbb{C}^1$.

If $\xi \in \mathfrak{D}(A_x)$, then the operators A_x^* and A_x in (2.19) can be transposed to ξ. As a result, we get that $A_x\xi = \lambda(x)\xi$ and $A_x^*\xi = \overline{\lambda(x)}\xi$, i.e., ξ is an ordinary joint eigenvector of the family A.

The collection of all eigenvalues $\lambda(\cdot)$ of the family A is called its generalized spectrum $g(A) \subset \mathbb{C}^X$, and the support of the joint RI E of this family is called the spectrum $s(A)$ of A, namely, $s(A) = \operatorname{supp} E \subset \mathbb{C}^X$. Clearly, the definition of $g(A)$ depends essentially on the choice of the rigging (2.18).

Example 2.1. Even in the case of a family consisting of a single operator $A_1 = A$, the sets $s(A)$ and $g(A)$ (belonging to $\mathbb{C}^1$) are, generally speaking, distinct. Thus, let

$$H_0 = L_2(\mathbb{R}^1, dt) = L_2(\mathbb{R}^1, \mathfrak{B}(\mathbb{R}^1), dt)$$

with respect to the Lebesgue measure dt, and let A be the minimal operator generated by the expression

$$(\mathcal{L}\varphi)(t) = -i\varphi'(t),$$

i.e., the closure of the operator

$$H_0 \supset C_0^\infty(\mathbb{R}^1) \ni \varphi \mapsto -i\varphi'(t) \in H_0,$$

where $C_0^\infty(\mathbb{R}^1)$ is the collection of infinitely differentiable finite functions on $\mathbb{R}^1$. A is a selfadjoint operator acting in H_0, and $s(A) = \mathbb{R}^1$. We set $H_+ = L_2(\mathbb{R}^1, e^{t^2}dt)$ and $D = \mathcal{D}(\mathbb{R}^1)$, where $\mathcal{D}(\mathbb{R}^1)$ is a classical space $C_0^\infty(\mathbb{R}^1)$ of test functions (endowed, for example, with the projective topology). The requirements connected with the rigging (2.18) are satisfied, and $H_- = L_2(\mathbb{R}^1, e^{-t^2}dt)$. The function $\xi(t) = e^{i\lambda t}$ $(t \in \mathbb{R}^1)$ belongs to H_- and satisfies (2.19) for every $\lambda \in \mathbb{C}^1$. Thus, we have $g(A) = \mathbb{C}^1 \neq \mathbb{R}^1 = s(A)$, and $g(A) \supset s(A)$.

Now consider the same operator A, but set $H_+ = H_0$. In this case, $H_- = H_0$, and no vector $H_- \ni \xi \neq 0$ exists which satisfies (2.19). Indeed, every $\xi \in \mathcal{D}'(\mathbb{R}^1) \supset H_0$ satisfying (2.19) is a generalized solution of the equation $i\varphi' = \lambda\varphi$; hence, $\xi(t) = Ce^{i\lambda t}$ $(t \in \mathbb{R}^1$, $\mathbb{C}^1 \ni C \neq 0)$, but this function does not belong to $H_0 = H_-$ for any $\lambda \in \mathbb{C}^1$. Consequently, here we have $g(A) = \varnothing \neq \mathbb{R}^1 = s(A)$ and $g(A) \subset s(A)$.

If $A = (A_x)_{x \in X}$ is a family of commuting selfadjoint operators, then (2.19) can be reduced to the relation

$$(\xi, A_x\varphi)_{H_0} = \lambda(x)(\xi, \varphi)_{H_0} \quad (x \in X; \varphi \in D). \tag{2.20}$$

It is natural to understand the generalized spectrum $g(A)$ as the collection of all $\lambda(\cdot)$

$\in \mathbb{R}^X$ which appear in (2.20). Thus, $s(A)$, $g(A) \subset \mathbb{R}^X$ now.

Let us rephrase these definitions in terms of the rigging of H_0 by linear topological spaces. We say that a family A of commuting normal operators is connected with (2.15) in the standard way (or admits (2.15)) if $\Phi \subset \mathfrak{D}(A_x)$, and the operators $A_x \upharpoonright \Phi$ and $A_x^* \upharpoonright \Phi$ act continuously in Φ ($x \in X$). Then $\xi \in \Phi'$ and $\varphi \in \Phi$ in (2.19). The definitions of the spectra remain unchanged. Analogous changes should be made in (2.20). Note that if A is connected with (2.15), where $\Phi = \text{pr lim}_{\tau \in \mathrm{T}} H_\tau$, in the standard way, then A is also connected in the same way with any chain $H_{-\tau} \supset H_0 \supset H_\tau \supset \Phi = D$ of the form (2.18). If Φ is nuclear, then the rigging (chain) (2.15) is called nuclear.

We give three examples to illustrate our presentation.

Example 2.2. Let us study a normal operator A of multiplication by a function $a(\cdot) \in C^\infty(\mathbb{R}^1)$ acting in the space $H_0 = L_2(\mathbb{R}^1, dt)$ from Example 2.1, i.e.,

$$H_0 \supset \mathfrak{D}(A) \ni f(t) \mapsto (Af)(t) = a(t) f(t) \in H_0,$$

where

$$\mathfrak{D}(A) = \{ f \in H_0 \mid a(t) f(t) \in H_0 \}$$

(see Subsection 1.7). Let us rig H_0 with a linear topological space by setting $\Phi = \mathcal{D}(\mathbb{R}^1)$ (see Example 1.9 in Chapter 1). Obviously, the operator A is connected with the rigging $\mathcal{D}'(\mathbb{R}^1) \supset H_0 \supset \mathcal{D}(\mathbb{R}^1)$ in the standard way. According to (2.19), a vector $0 \neq \xi \in \mathcal{D}'(\mathbb{R}^1)$ is a generalized eigenvector of the operator A with a certain eigenvalue $\mathcal{D}(\mathbb{R}^1)$ if

$$(\xi, \overline{a(\cdot)}\, \varphi(\cdot))_{H_0} = \lambda(\xi, \varphi)_{H_0},$$

$$(\xi, a(\cdot)\, \varphi(\cdot))_{H_0} = \bar{\lambda}(\xi, \varphi)_{H_0} \quad (\varphi \in \Phi = \mathcal{D}(\mathbb{R}^1)). \tag{2.21}$$

In the last equalities, we take ξ to be the δ-function concentrated at the point $t \in \mathbb{R}^1$, namely, $\xi = \delta_t \in \mathcal{D}'(\mathbb{R}^1)$. The equalities (2.21) are satisfied if $\lambda = a(\ t)$, i.e., if δ_t is a generalized eigenvector of the operator A with the eigenvalue $a(t)$. In the case of a general function $a \in C^\infty(\mathbb{R}^1)$, the operator A may possess some other generalized eigenvectors ξ (e.g., if $a = 0$, then any $\xi \in \mathcal{D}'(\mathbb{R}^1)$ fits and $\lambda = 0$). However, if a function $\mathbb{R}^1 \ni t \mapsto a(t) \in \mathbb{C}^1$ takes distinct values at distinct points, then necessarily $\xi = c\delta_t$ with some $t \in \mathbb{R}^1$ and $c \neq 0$. Indeed, first assume that $\lambda \in \mathbb{C}^1$ is such that $a(t) \neq \lambda$ for any $t \in \mathbb{R}^1$. Then $\overline{(a(\cdot) - \lambda)}\, \varphi(\cdot)$ runs through the whole of $\mathcal{D}(\mathbb{R}^1)$ for $\varphi \in \mathcal{D}(\mathbb{R}^1)$. At the same time, it follows from (2.21) that

$$(\xi, \overline{(a(\cdot) - \lambda)}\, \varphi(\cdot))_{H_0} = 0 \quad (\varphi \in \mathcal{D}(\mathbb{R}^1)),$$

i.e., $\xi = 0$, but this is impossible. Therefore, there exists $t \in \mathbb{R}^1$ such that $\lambda = a(t)$, and this t is unique. Hence, the condition

$$(\xi, \overline{(a(\cdot)-\lambda)\varphi(\cdot)})_{H_0} = 0 \quad (\varphi \in \mathcal{D}(\mathbb{R}^1))$$

means that $\operatorname{supp} \xi = \{t\}$ and, according to the well-known theorem (e.g., see Vladimirov [1, Chapter 1, Section 2, Subsection 6]), we have

$$\xi = \sum_{k=0}^{p} c_k D^k \delta_t$$

with some $p \in \mathbb{Z}_+$ and $c_k \in \mathbb{C}^1$. It is easy to see that the equalities (2.21) yield $c_1 = \ldots = c_p = 0$, and this implies the desired result.

Hence, for the function a under consideration (it takes distinct values at different t), we have $g(A) = \{a(t) \mid t \in \mathbb{R}^1\} \subset \mathbb{C}^1$, and $s(A) = \operatorname{supp} E$ is equal to the closure of $g(A)$, as follows from Subsection 1.7.

Example 2.3. Consider the space of Example 2.2 with the same rigging. We want to investigate a family $A = (A_x)_{x \in X}$ of commuting normal operators A_x of multiplication by the function $a_x(\cdot) \in C^\infty(\mathbb{R}^1)$, where $a_x(t') \neq a_x(t'')$ for $t' \neq t''$ ($x \in X$. The indexing set X is arbitrary). The relations (2.19) transform into the equalities (2.21) in which $a(\cdot)$ and λ are replaced by $a_x(\cdot)$ and $\lambda(x)$, respectively; $x \in X$. The arguments of Example 2.2 show that generalized joint eigenvectors have the form $\xi = \delta_t$, where $t \in \mathbb{R}^1$; the generalized spectrum $g(A) \subset \mathbb{C}^X$ coincides with the collection of all functions of the form $X \ni x \mapsto a_x(t) \in \mathbb{C}^1$, where $t \in \mathbb{R}^1$.

Example 2.4 (the Schrödinger operator).

Let G be a bounded (or unbounded) region in the space $\mathbb{R}^d$ ($d \in \mathbb{N}$) with a regular enough boundary ∂G (the case when $G = \mathbb{R}^d$ is also possible), and let

$$H_0 = L_2(G, dx) = L_2(G, \mathfrak{B}(G), dx).$$

Consider an operator

$$H_0 \supset \mathfrak{D}(A') \ni f \mapsto A'f = \mathcal{L}f = -\Delta f + qf \in H_0,$$

acting in H_0, where $q \in C^l(\tilde{G})$ ($l = 2, 3, \ldots$) is a real-valued function and $\mathfrak{D}(A')$ consists of functions belonging to $W_2^2(G_R)$ ($R > 0$) on each part $G_R = \{x \in G \mid |x| < R\}$ of G, equal to zero on ∂G, and finite at infinity. This operator is clearly Hermitian. The closure A of this operator is selfadjoint provided that the behavior of $q(x)$ as $|x| \to \infty$ is sufficiently good. Denote by D a linear topological space consisting of all infinitely differentiable functions from $\mathfrak{D}(A')$ and provided with a topology which involves uniform convergence with all the derivatives and uniform finiteness of convergent sequences of functions at infinity. Let us take the completion of D with respect to a certain Sobolev space with a weight $W_2^l(G, p(x)dx)$ $\to\infty$ to be H_+ (the information concerning the Sobolev spaces is presented in Section 1 of Chapter 1 (Examples 1.2 and 1.5)). It is clear that the chain (2.18) obtained above is connected with A in the standard way. A generalized eigenvector (function) $\xi \in H_-$ satisfies the relation

$$(\xi, \mathcal{L}\varphi)_{H_0} = \lambda(\xi, \varphi)_{H_0} \quad (\varphi \in D). \tag{2.22}$$

It follows from the theorems on the regularity of generalized solutions of elliptic equations up to the boundary of the region (see Berezansky [5, Chapter 3, Section 4]) that $\xi \in W_2^2(G_R)$ for any $R > 0$ and satisfies zero boundary conditions on ∂G (in fact, ξ is even smoother. This depends on the smoothness of q). Hence, the only "generalized" feature of the eigenfunction ξ under consideration is that it does not belong to the space $L_2(G, dx)$ having an insufficient rate of decreasing at infinity. Similar considerations for more general differential expressions $\mathcal{L}$ and boundary conditions can be found in Berezansky [5, Chapter 6].

2.6. Spectral Theorem. The Case of At Most Countably Many Operators

We consider a family $A = (A_x)_{x\in X}$ of commuting normal operators acting in the space H_0; $\mathfrak{C}_\sigma(\mathbb{C}^X) \ni \alpha \mapsto E(\alpha)$ is its joint RI. We fix a quasinuclear rigging (2.18) connected with A in the standard way, and we let

$$\mathfrak{C}_\sigma(\mathbb{C}^X) \ni \alpha \mapsto \rho(\alpha) = \mathrm{Tr}\,(O^+(\alpha)\,O) \in [0, \infty)$$

be the corresponding spectral measure. According to Theorem 2.2, the representation (2.13) holds, namely

$$O^+E(\alpha)\,O = \int_\alpha P(\lambda(\cdot))\,d\rho(\lambda(\cdot)) \quad (\alpha \in \mathfrak{C}_\sigma(\mathbb{C}^X)), \tag{2.23}$$

where $P(\lambda(\cdot)) : H_+ \to H_-$ is a generalized projector.

Suppose that X is at most countable. Here $\mathfrak{C}_\sigma(\mathbb{C}^X) = \mathfrak{B}(\mathbb{C}^X)$, and the measures E and ρ are defined on the Borel subsets of the space $\mathbb{C}^X$. The proof of the spectral theorem in this case is not essentially different from the case of a single operator, and we do not employ most of the constructions presented in Section 1.

Theorem 2.6. *Suppose that X is at most countable, and assume that the space D in the quasinuclear chain* (2.18) *is separable. There exists a set*

$$\mathfrak{B}(\mathbb{C}^X) \ni \pi \subset s(A) \cap g(A)$$

of full spectral measure ρ *such that for every* $\lambda(\cdot) \in \pi$ *the range* $\mathfrak{R}(P(\lambda(\cdot)))$ *consists of generalized joint eigenvectors of the family* A *which correspond to the eigenvalue* $\lambda(\cdot)$, *and* $\mathrm{Tr}\,(P(\lambda(\cdot)) = 1$.

In the spectral representation (1.26), *the domain of integration and* $dE(\lambda(\cdot))$ *can be replaced by* π *and* $P(\lambda(\cdot))\, d\rho(\lambda(\cdot))$, *respectively (if* A_x *and* A_x^* *are regarded as restricted to* $\mathfrak{D}(A_x) \cap H$*).*

Let us make some remarks concerning the case of a general X.

The first statement of the theorem consists in the proof of the relations

$$(P(\lambda(\cdot))\varphi, A_x^*\psi)_{H_0} = \lambda(x)\, (P(\lambda(\cdot))\varphi, \psi)_{H_0},$$

$$(P(\lambda(\cdot))\varphi, A_x\psi)_{H_0} = \overline{\lambda(x)}\, (P(\lambda(\cdot))\varphi, \psi)_{H_0} \quad (\varphi \in H_+,\ \psi \in D) \tag{2.24}$$

for $\lambda(\cdot)$ from the properly chosen set π and $x \in X$.

Remark 1. It suffices to establish (2.24) with indicated $\lambda(\cdot)$ and x for φ and ψ varying over dense subsets of H_+ and D, respectively. This follows from the fact that, by virtue of the continuity of the operators $P(\lambda(\cdot))\colon H_+ \to H_-$, and $A_x^* \upharpoonright D$, $A_x \upharpoonright D\colon D \to H_+$, we can then pass to the limit in (2.24), written for φ and ψ from the indicated sets.

■

Lemma 2.2. *For every* $x \in X$ *there exists a set* $\beta_x \in \mathfrak{C}_\sigma(\mathbb{C}^X)$ *of full* ρ*-measure such that, for every* $\lambda(\cdot) \in \beta_x$ *the relations* (2.24) *hold, and* $\mathrm{Tr}\,(P(\lambda(\cdot))) = 1$.

Proof. Fix $\varphi, \psi \in D$. By using (2.23), we obtain

$$\int_\alpha (P(\lambda(\cdot))\varphi, A_x^*\psi)_{H_0} d\rho(\lambda(\cdot)) = \Big(\Big(\int_\alpha P(\lambda(\cdot))\, d\rho(\lambda(\cdot))\Big)\varphi, A_x^*\psi\Big)_{H_0}$$

$$= (O^+E(\alpha)\,O\varphi, A_x^*\psi)_{H_0} = (A_x E(\alpha)\varphi, \psi)_{H_0}$$

$$= \Big(\Big(\int_\alpha \lambda(x)\, dE(\lambda(\cdot))\Big)\varphi, \psi\Big)_{H_0}$$

$$= \int_\alpha \lambda(x)\, d(O^+E(\lambda(\cdot))\,O\varphi, \psi)_{H_0} \quad (\alpha \in \mathfrak{C}_\sigma(\mathbb{C}^X)). \tag{2.25}$$

Here we have exploited the relation

$$A_x E(\alpha) = \int_\alpha \lambda(x)\, dE(\lambda(\cdot)) \quad (\alpha \in \mathfrak{C}_\sigma(\mathbb{C}^X)),$$

which follows from (1.25). Further, replacing the differential in the last integral of (2.25) according to (2.23), we find that

$$\int_\alpha (P(\lambda(\cdot))\varphi, A_x^*\psi)_{H_0} d\rho(\lambda(\cdot)) = \int_\alpha \lambda(x)\, (P(\lambda(\cdot))\varphi, \psi)_{H_0} d\rho(\lambda(\cdot)) \quad (\alpha \in \mathfrak{C}_\sigma(\mathbb{C}^X)).$$

Since α is arbitrary, this implies that there exists a set $\beta_{\varphi,\psi,x} \in C_\sigma(\mathbb{C}^X)$ of full ρ-measure such that the first of the relations (2.24) holds for $\lambda(\cdot) \in \beta_{\varphi,\psi,x}$.

Suppose now that L is a countable set dense in the space D and, hence, in the space H_+. Consider the countable intersection $\bigcap_{\varphi,\psi\in L} \beta_{\varphi,\psi,x} = \beta'_x$. This is a set of full ρ-measure, and for $\lambda(\cdot) \in \beta'_x$ the first of the relations (2.24) holds for all $\varphi, \psi \in L$. According to Remark 1, this yields the validity of this relation for $\lambda(\cdot) \in \beta'_x$, $\varphi \in H_+$, and $\psi \in D$ as well.

Quite analogously, by replacing A_x^* by A_x in (2.25) and using the corresponding expression for $A_x^* E(\alpha)$, we establish that the second relation (2.24) holds for $\lambda(\cdot) \in \beta''_x$, $\varphi \in H_+$, and $\psi \in D$, where $\beta''_x \in C_\sigma(\mathbb{C}^X)$ is a set of full ρ-measure. The lemma is proved if we set

$$\beta_x = \beta'_x \cap \beta''_x \cap \{\lambda(\cdot) \in \mathbb{C}^X \mid \operatorname{Tr}(P\lambda(\cdot))) = 1\}.$$

■

Proof of the Theorem. Suppose that X is at most countable, and let β_x be the set from Lemma 2.2. The measure ρ is now defined on the set $s(A) = \operatorname{supp} E$ which is of full measure. The set

$$\pi = \Big(\bigcap_{x\in X} \beta_x\Big) \cap s(A) \in \mathfrak{B}(\mathbb{C}^X)$$

is also of full ρ-measure. For $\lambda(\cdot) \in \pi$ the relations (2.24) hold for all $x \in X$, i.e., $\mathfrak{R}(P(\lambda(\cdot)))$ possesses the required properties. Further, it is necessary to replace $\mathbb{C}^X$ by π in (1.25) (π is also of full E-measure) and to employ the equality

$$O^+ d\,E(\lambda(\cdot))\, O = P(\lambda(\cdot))\, d\rho(\lambda(\cdot)),$$

which is a corollary of (2.23).

■

At the end of this subsection, we present a formulation of the spectral projection theorem in the "classical" case of a single selfadjoint operator. This formulation follows from Theorem 2.6.

So, let us consider the quasinuclear chain (2.18) with a separable space D. Let A be a selfadjoint operator acting in H_0 connected with (2.18) in the standard way, i.e., $D \subset \mathfrak{D}(A)$ and $A \upharpoonright D$ acts continuously from D to H_+.

Then *there exists an operator-valued function* $P(\lambda)$, *defined for almost every* λ *from the spectrum* $s(A)$ *(in the sense of the spectral measure* ρ*), such that the values of it are nonnegative operators from* H_+ *to* H_- *with the Hilbert–Schmidt norm bounded by* 1 *(more precisely,* $|P(\lambda)| \leq \operatorname{Tr}(P(\lambda)) = 1$), *and it gives the following representations of the RI* E *of the operator* A *and of this operator itself:*

$$O^{+}E(\alpha)O = \int_{\alpha} P(\lambda)\, d\rho(\lambda), \quad A\varphi = \left(\int_{s(A)} \lambda P(\lambda) d\rho(\lambda)\right)\varphi$$

$$(\alpha \in \mathfrak{B}(\mathbb{R}^1);\ \varphi \in \mathfrak{D}(A) \cap H_+). \tag{2.26}$$

The range of values $\mathfrak{R}(P(\lambda)) \subset H_-$ *consists of the vectors* ξ *which are generalized eigenvectors of* A *with the eigenvalue* λ, *i.e.,*

$$(\xi, A\varphi)_{H_0} = \lambda\, (\xi, \varphi)_{H_0} \quad (\varphi \in D). \tag{2.27}$$

2.7. Spectral Theorem. The General Case

For an arbitrary indexing set X, an attempt to prove the spectral projection theorem by analogy with the argument presented in Subsection 2.6 leads to the following "continuum difficulty": the intersection $\pi = \bigcap_{x \in X} \beta_x$ may fail to be of full E-measure. Therefore, the preceding simple considerations now prove to be useless. Here we overcome this difficulty, but the set π of full outer E-measure constructed in a different way does not belong generally to the σ-algebra $C_\sigma(\mathbb{C}^X)$, and it is thus impossible to integrate over it.

This makes it necessary to modify the RI E and the spectral measure ρ by π. Let $\tau \subset \mathbb{C}^X$ be a set of full outer ρ-measure or, what is the same, of full outer E-measure. Denote by E_τ the RI modified by τ. Recall that E_τ is defined on the σ-algebra

$$C_\sigma(\tau) = \{\alpha' = \alpha \cap \tau \mid \alpha \in C_\sigma(\mathbb{C}^X)\}$$

by the equality

$$C_\sigma(\tau) \ni \alpha \cap \tau \mapsto E_\tau(\alpha \cap \tau) = E(\alpha),$$

if $\tau \in C_\sigma(\mathbb{C}^X)$, then $E_\tau \restriction \tau$. The modification of the spectral measure ρ is defined similarly; $E_{\mathbb{C}^X} = E$ and $\rho_{\mathbb{C}^X} = \rho$.

Theorem 2.7. *Let X be arbitrary. Suppose that the space D in the quasinuclear chain* (2.18) *is a separable projective limit of Hilbert spaces. There exists a set π of full outer ρ-measure which belongs to the generalized spectrum $g(A)$ and is such that for every $\lambda(\cdot) \in \pi$, the range $\mathfrak{R}(P(\lambda(\cdot)))$ consists of generalized joint eigenvectors of the family A corresponding to the eigenvalue $\lambda(\cdot)$, and* $\mathrm{Tr}\,(P(\lambda(\cdot))) = 1$.

The following representation in the form of a functional spectral integral holds:

$$A_x = \int_\pi \lambda(x)\, dE_\pi(\lambda(\cdot)),$$

$$\mathfrak{D}(A_x) = \Big\{ f \in H_0 \,\Big|\, \int_\pi |\lambda(x)|^2\, d(E_\pi(\lambda(\cdot)) f, f)_{H_0} < \infty \Big\} \quad (x \in X) \qquad (2.28)$$

(A_x^* can be represented as the integral of $\overline{\lambda(x)}$). If A_x and A_x^* are understood as the restrictions to $\mathfrak{D}(A_x) \cap H_+$, then $dE_\pi(\lambda(\cdot))$ can be replaced by $P(\lambda(\cdot))\, d\rho_\pi(\lambda(\cdot))$ in (2.28).

Remark 1. It is often convenient to employ a somewhat more general formulation of this theorem. We present this formulation.

Fix an arbitrary set $\tau \subset \mathbb{C}^X$ containing π. The representation (2.28) and the last result of the theorem are valid with π replaced by τ. Indeed, τ is of full outer ρ-measure. Therefore, E_τ and ρ_τ make sense. It is easy to see that the measures E_τ and ρ_τ may be modified by $\pi \subset \tau$. As a result, we again obtain E_π and ρ_π – this follows from the formulas defining the modified measures for given E and ρ. Writing the integrals in (2.28) for E_τ and ρ_τ (including the integrals with $P(\lambda(\cdot))d\rho_\tau(\lambda(\cdot))$) and employing the rule (1.19), we find that they are equal to the same integrals with τ replaced by π.

■

Remark 2. The measure ρ_τ $(\pi \subset \tau \subset \mathbb{C}^X)$ appearing above can be understood as the spectral measure of the RI E, namely,

$$\mathrm{Tr}\,(O^+ E_\tau(\alpha \cap \tau)\, O) = \mathrm{Tr}\,(O^+ E(\alpha)\, O) = \rho(\alpha) = \rho_\tau(\alpha \cap \tau) \quad (\alpha \in \mathfrak{C}_\sigma(\mathbb{C}^X)).$$

The set π has full outer ρ_τ-measure.

■

Let us establish some lemmas. We shall use the compactified objects (see Subsections 1.8 and 2.3). According to Theorem 2.2, the representation

$$O^{+}E(\boldsymbol{\alpha})\,O = \int_{\alpha} P(\boldsymbol{\lambda})\, d\boldsymbol{\rho}(\boldsymbol{\lambda}(\cdot)) \quad (\boldsymbol{\alpha} \in \mathfrak{C}_{\sigma}(\mathbf{C}^{X})) \tag{2.29}$$

holds along with (2.23).

Lemma 2.3. *For every $x \in X$, there exists a set $\boldsymbol{\beta}_x \in \mathfrak{C}_{\sigma}(\mathbf{C}^X)$ of full $\boldsymbol{\rho}$-measure such that for each $\boldsymbol{\lambda}(\cdot) \in \boldsymbol{\beta}_x$ we have $\boldsymbol{\lambda}(x) \neq \infty$, and the following relations hold*

$$(P(\boldsymbol{\lambda}(\cdot))\varphi, A_x^{*}\psi)_{H_0} = \boldsymbol{\lambda}(x)\,(P(\boldsymbol{\lambda}(\cdot))\varphi, \psi)_{H_0},$$

$$(P(\boldsymbol{\lambda}(\cdot))\varphi, A_x\psi)_{H_0} = \overline{\lambda(x)}\,(P(\boldsymbol{\lambda}(\cdot))\varphi, \psi)_{H_0} \quad (\varphi \in H_{+},\ \psi \in D), \tag{2.30}$$

$$\operatorname{Tr}(P(\boldsymbol{\lambda}(\cdot))) = 1.$$

The proof repeats the arguments in the proof of Lemma 2.2, except that instead of (2.23) and (1.25), one should exploit (2.29) and the representation (1.25) with E replaced by $\boldsymbol{E}$ (see Remark 1 in Subsection 1.11). In equalities of the type (2.25) or (2.28), α must be replaced by

$$\mathfrak{C}_{\sigma}(\mathbf{C}^X) \ni \boldsymbol{\alpha} \subset \mathbf{C}^X \setminus \{\boldsymbol{\lambda}(\cdot) \in \mathbf{C}^X \mid \boldsymbol{\lambda}(x) = \infty\}$$

(recall that the $\boldsymbol{\rho}$-measure of the set which is subtracted is equal to zero).

We shall use the following general "lemma on separability" (for convenience, the proof will be presented somewhat later).

Lemma 2.4. *Assume that Φ is a separable linear topological space, and $\Phi_0 = (\varphi_j)_{j=1}^{\infty}$ is a fixed countable dense subset of it. Assume that the dual space Φ' is also separable with respect to the weak topology. Consider an arbitrary set $\mathbf{A} \subset \mathfrak{L}(\Phi, H)$ of continuous linear operators acting from Φ to some separable Hilbert space H. Then $\mathbf{A}$ is always separable in the following sense: there exists a sequence of operators $(A_n)_{n=1}^{\infty}$ $(A_n \in \mathbf{A})$ such that for any operator $A \in \mathbf{A}$, one can find a subsequence $(A_{n_k})_{k=1}^{\infty}$ such that $\|(A - A_{n_k})\varphi_j\|_H \underset{k\to\infty}{\to} 0$ for every $j \in \mathbb{N}$.*

The main point in the proof of the theorem is the following lemma.

Lemma 2.5. *Suppose that the conditions of Theorem 2.7 hold. There exist a set $\mathfrak{B}(\mathbf{C}^X) \ni \boldsymbol{\gamma} \subset \operatorname{supp}\boldsymbol{\rho}$ of full outer measure and a mapping*

$$\mathbf{C}^X \supset \boldsymbol{\gamma} \ni \boldsymbol{\lambda}(\cdot) \mapsto (\operatorname{reg}\boldsymbol{\lambda}(\cdot))(\cdot) \in \mathbb{C}^X$$

such that for every $\boldsymbol{\lambda}(\cdot) \in \boldsymbol{\Upsilon}$, *the range* $\mathfrak{R}(\boldsymbol{P}(\boldsymbol{\lambda}(\cdot)))$ *consists of the generalized joint eigenvectors of the family* $\boldsymbol{A}$ *corresponding to the eigenvalue* $(\text{reg}\,\boldsymbol{\lambda}(\cdot))(\cdot)$, *and* $\text{Tr}\,(\boldsymbol{P}(\boldsymbol{\lambda}(\cdot))) = 1$.

The mapping reg *has the following form: We construct a sequence* $(y_n)_{n=1}^{\infty}$ *of points* $y_n \in X$ $(\boldsymbol{\lambda}(y_n) \neq \infty,\ \boldsymbol{\lambda}(\cdot) \in \boldsymbol{\Upsilon})$ *depending on* $\boldsymbol{A}$. *For every* $x \in X$, *there exists a subsequence* $(y_{n_k}(x))_{k=1}^{\infty}$ *of it such that*

$$(\text{reg}\,\boldsymbol{\lambda}(\cdot))\,(x) = \lim_{k\to\infty} \boldsymbol{\lambda}(y_{n_k}(x));$$

moreover,

$$(\text{reg}\,\boldsymbol{\lambda}(\cdot))\,(x) = \boldsymbol{\lambda}(x)$$

for each $x \in X$ *such that* $\boldsymbol{\lambda}(x) \neq \infty$.

Proof. **I.** Consider the direct sum $\Phi = D \dotplus D$. Like D, Φ is a separable projective limit of Hilbert spaces ($\Phi = \text{pr lim}_{\tau \in \mathrm{T}} F_\tau$) topologically imbedded in $F_0 = H_0 \oplus H_0$ ($0 \in \mathrm{T}$). The space Φ' is separable in the weak topology. Indeed, let Ψ be a countable set dense in Φ and, consequently, in each F_τ. Also let $\xi \in \Phi'$. According to (1.25) in Chapter 1, ξ belongs to a certain $F_{-\tau_0}$, i.e., $\xi(\varphi) = (a_\xi, \varphi)_{F_{\tau_0}}$, where $a_\xi \in F_{\tau_0}$ ($\varphi \in \Phi$). Suppose that the sequence $(\psi_n)_{n=1}^{\infty} \in \Psi$ is such that $\psi_n \to a_\xi$ in F_{τ_0} as $n \to \infty$. Then $(\psi_n, \varphi)_{F_{\tau_0}} \to \xi(\varphi)$ for every $\varphi \in \Phi$. In other words, the vectors ψ_n, regarded as functionals in Φ', approximate ξ weakly.

II. Let us employ the orthogonalization process to construct an orthonormal basis $(e_j)_{j=1}^{\infty}$ in the space H_+ consisting of vectors from the space D. Then we supplement it in an arbitrary way to form a countable set L dense in D. Now let us apply Lemma 2.4 with $\Phi = D \dotplus D$, $\Phi_0 = L \dotplus L$, $H = H_+ \oplus H_+$, and $\boldsymbol{A} = (B_x)_{x \in X} \subset \mathcal{L}(\Phi, H)$, where

$$\Phi = D \dotplus D \ni (\varphi, \psi) \mapsto B_x(\varphi, \psi) = (A_x'\varphi, A_x''\psi) \in H_+ \oplus H_+ = H,$$

$$A_x' = A_x^* \upharpoonright D \in \mathcal{L}(D, H_+),$$

and

$$A_x'' = A_x \upharpoonright D \in \mathcal{L}(D, H_+) \quad (x \in X)$$

(by virtue of Step I, the application of this lemma is possible). According to Lemma 2.4, one can find a sequence $(y_n)_{n=1}^{\infty}$ $(y_n \in X)$ such that for every $x \in X$, there exists a subsequence $(y_{n_k})_{k=1}^{\infty}$ of it for which

$$(A'_{y_{n_k}}\varphi, A''_{y_{n_k}}\psi) \to (A'_x\varphi, A''_x\psi) \quad (\varphi, \psi \in D)$$

in $H_+ \oplus H_+$ as $k \to \infty$, i.e.,

$$\| A'_{y_{n_k}}\varphi - A'_x\varphi \|_{H_+} \quad \text{and} \quad \| A''_{y_{n_k}}\varphi - A''_x\varphi \|_{H_+} \to 0.$$

In what follows, we fix the sequence $(y_n)_{n=1}^{\infty}$ and some subsequence $(y_{n_k(x)})_{k=1}^{\infty}$ of it of this sort (for every $x \in X$).

III. Let us construct the set $\boldsymbol{\gamma}$. We set

$$\boldsymbol{\beta} = \bigcap_{n=1}^{\infty} \boldsymbol{\beta}_{y_n} \in \mathfrak{C}_\sigma(\mathbf{C}^X),$$

where y_n are the points constructed on Step II, and $\boldsymbol{\beta}_{y_n}$ are the sets from Lemma 2.3. It is clear that $\boldsymbol{\beta}$ is of full $\boldsymbol{\rho}$-measure, and for $\boldsymbol{\lambda}(\cdot) \in \boldsymbol{\beta}$ the relations (2.30) hold with $x = y_n$ $(n \in \mathbb{N})$. The equality

$$\sum_{j=1}^{\infty} (\boldsymbol{P}(\boldsymbol{\lambda}(\cdot))e_j, e_j)_{H_0} = \operatorname{Tr}(\boldsymbol{P}(\boldsymbol{\lambda}(\cdot))) = 1 \quad (\boldsymbol{\lambda}(\cdot) \in \boldsymbol{\beta})$$

yields the decomposition

$$\boldsymbol{\beta} = \bigcup_{j,m=1}^{\infty} \{\boldsymbol{\lambda}(\cdot) \in \boldsymbol{\beta} \mid (\boldsymbol{P}(\boldsymbol{\lambda}(\cdot))e_j, e_j)_{H_0} \geq 1/m\}. \tag{2.31}$$

Let $(\boldsymbol{\beta}'_i)_{i=1}^{\infty}$ be a countable family of sets which appear under the union sign in (2.31); we assume that it is enumerated somehow. Consider the disjoint sets $\boldsymbol{\beta}'_1, \boldsymbol{\beta}'_2 \setminus \boldsymbol{\beta}'_1$, $\boldsymbol{\beta}'_3 \setminus (\boldsymbol{\beta}'_1 \cup \boldsymbol{\beta}'_2) \ldots$ and retain only those of them which have a nonzero $\boldsymbol{\rho}$-measure; denote these sets by $\boldsymbol{\beta}_1, \boldsymbol{\beta}_2, \ldots$ (the sequence $(\boldsymbol{\beta}_l)_{l=1}^{\infty}$ may be finite). Let $\boldsymbol{\beta}_0$ be the union of all other sets from the above sequence and $\mathbf{C}^X \setminus \boldsymbol{\beta}$. Thus, $\boldsymbol{\beta}_l$ belong to $\mathfrak{C}_\sigma(\mathbf{C}^X)$ and are mutually disjoint ($l \in \mathbb{Z}_+$); we also have $\boldsymbol{\rho}(\boldsymbol{\beta}_0) = 0$, $\boldsymbol{\rho}(\boldsymbol{\beta}_l) > 0$ $(l \in \mathbb{N})$, and $\mathbf{C}^X = \bigcup_{l=0}^{\infty} \boldsymbol{\beta}_l$. Each $\boldsymbol{\beta}_l$ belongs to some $\boldsymbol{\beta}'_{i_l}$. Therefore, according to (2.31), one can find $j_l, m_l \in \mathbb{N}$ such that

$$(P(\boldsymbol{\lambda}(\cdot))e_{j_l}, e_{j_l})_{H_0} \geq 1/m_l \quad (\boldsymbol{\lambda}(\cdot) \in \boldsymbol{\beta}_l;\ l \in \mathbb{N}). \tag{2.32}$$

According to Theorem 2.3, the measure $\boldsymbol{\rho}$ is regular; hence, for every $\boldsymbol{\beta}_l$ there exists a sequence $(\boldsymbol{\varphi}_{l,p})_{p=1}^{\infty}$ of sets $\mathfrak{C}_\sigma(\mathbf{C}^X) \ni \boldsymbol{\varphi}_{l,p} \subset \boldsymbol{\beta}_l$ closed in $\mathbf{C}^X$ and such that $\boldsymbol{\rho}(\boldsymbol{\varphi}_{l,p}) > 0$, and

$$\lim_{p\to\infty} \boldsymbol{\rho}(\boldsymbol{\varphi}_{l,p}) = \boldsymbol{\rho}(\boldsymbol{\beta}_l) \quad (l \in \mathbb{N}). \tag{2.33}$$

Consider the restriction $\boldsymbol{\rho} \upharpoonright \boldsymbol{\varphi}_{l,p}$ and let $\boldsymbol{\psi}_{l,p} = \operatorname{supp}(\boldsymbol{\rho} \upharpoonright \boldsymbol{\varphi}_{l,p})$ $(l, p \in \mathbb{N})$. We set

$$\mathfrak{B}(\mathbf{C}^X) \ni \boldsymbol{\gamma} = \bigcup_{l,p=1}^{\infty} \boldsymbol{\psi}_{l,p} \subset \bigcup_{l=1}^{\infty} \boldsymbol{\beta}_l \subset \boldsymbol{\beta}. \tag{2.34}$$

IV. By virtue of Theorem 2.3 and the general remark at the end of Subsection 1.9, the measure $\boldsymbol{\rho} \upharpoonright \boldsymbol{\varphi}_{l,p}$ is regular. It is proper by assertion (i) in Theorem 1.5; indeed, $\boldsymbol{\varphi}_{l,p}$ is a closed subset of the compact space $\mathbf{C}^X$ and is thus compact itself. Taking into account (1.22), we find that $\boldsymbol{\psi}_{l,p} \subset \operatorname{supp} \boldsymbol{\rho}$ $(l, p \in \mathbb{N})$; consequently, $\boldsymbol{\gamma} \subset \operatorname{supp} \boldsymbol{\rho}$.

V. The set (2.34) is of full outer $\boldsymbol{\rho}$-measure. Indeed, suppose that $\mathfrak{C}_\sigma(\mathbf{C}^X) \ni \boldsymbol{\alpha} \supset \boldsymbol{\gamma}$. Obviously,

$$\boldsymbol{\rho}(\boldsymbol{\alpha}) = \bigcup_{l=0}^{\infty} \boldsymbol{\rho}(\boldsymbol{\alpha} \cap \boldsymbol{\beta}_l)$$

and, hence, it suffices to establish that $\boldsymbol{\rho}(\boldsymbol{\alpha} \cap \boldsymbol{\beta}_l) = \boldsymbol{\rho}(\boldsymbol{\beta}_l)$ $(l \in \mathbb{N})$. But

$$\boldsymbol{\rho}(\boldsymbol{\alpha} \cap \boldsymbol{\varphi}_{l,p}) = \boldsymbol{\rho}(\boldsymbol{\alpha} \upharpoonright \boldsymbol{\varphi}_{l,p})(\boldsymbol{\varphi}_{l,p}).$$

Indeed, this equality is equivalent to

$$(\boldsymbol{\rho} \upharpoonright \boldsymbol{\varphi}_{l,p})(\boldsymbol{\alpha} \cap \boldsymbol{\varphi}_{l,p}) = \boldsymbol{\rho}(\boldsymbol{\alpha} \upharpoonright \boldsymbol{\varphi}_{l,p})(\boldsymbol{\varphi}_{l,p}),$$

and since

$$\boldsymbol{\alpha} \cap \boldsymbol{\varphi}_{l,p} \supset \boldsymbol{\varphi}_{l,p} = \operatorname{supp}(\boldsymbol{\rho} \upharpoonright \boldsymbol{\varphi}_{l,p}),$$

the last relation holds because the measure $\boldsymbol{\rho} \upharpoonright \boldsymbol{\varphi}_{l,p}$ is regular and proper. Passing to the limit as $p \to \infty$ in the equality obtained and taking (2.33) into account, we get that $\boldsymbol{\rho}(\boldsymbol{\alpha} \cap \boldsymbol{\beta}_l) = \boldsymbol{\rho}(\boldsymbol{\beta}_l)$.

VI. We define the mapping

$$\mathbf{C}^X \supset \boldsymbol{\gamma} \ni \boldsymbol{\lambda}(\cdot) \mapsto (\operatorname{reg} \boldsymbol{\lambda}(\cdot))(\cdot) \in \mathbb{C}^X.$$

Let $(y_n)_{n=1}^{\infty}$ be the sequence constructed on Step II, and let $\boldsymbol{\lambda}(\cdot) \in \boldsymbol{\gamma}$ and $x \in X$. If $\boldsymbol{\lambda}(x) \neq \infty$, then we set

$$(\operatorname{reg} \boldsymbol{\lambda}(\cdot))(x) = \boldsymbol{\lambda}(x),$$

and if $\boldsymbol{\lambda}(x) = \infty$, then

$$(\operatorname{reg} \boldsymbol{\lambda}(\cdot))(x) = \lim_{k\to\infty} \boldsymbol{\lambda}(y_{n_k(x)})$$

by definition.

Let us prove that in the second case, $\boldsymbol{\lambda}(y_{n_k(x)}) \neq \infty$ $(k \in \mathbb{N})$ and the indicated limit exists and is finite. In fact, the inequality follows from the inclusion $\boldsymbol{\gamma} \subset \boldsymbol{\beta} \subset \boldsymbol{\beta}_{y_{n_k}(x)}$ $(k \in \mathbb{N})$. Further, by virtue of (2.34), $\boldsymbol{\lambda}(\cdot) \in \boldsymbol{\psi}_{l,p} \subset \boldsymbol{\beta}_l$ for some l and p; therefore, (2.32) holds.

For $\varphi = \psi = e_{j_l}$ and $x = y_{n_k(x)}$, from the first inequality in (2.30), we obtain

$$\boldsymbol{\lambda}(y_{n_k(x)}) = (\boldsymbol{P}(\boldsymbol{\lambda}(\cdot))e_{j_l}, A'_{y_{n_k(x)}} e_{j_l})_{H_0} (\boldsymbol{P}(\boldsymbol{\lambda}(\cdot))e_{j_l}, e_{j_l})_{H_0}^{-1}$$

$$\underset{k\to\infty}{\to} (\boldsymbol{P}(\boldsymbol{\lambda}(\cdot))e_{j_l}, A'_x e_{j_l})_{H_0} (\boldsymbol{P}(\boldsymbol{\lambda}(\cdot))e_{j_l}, e_{j_l})_{H_0}^{-1}. \tag{2.35}$$

VII. We must verify that $\mathfrak{R}(\boldsymbol{P}(\boldsymbol{\lambda}(\cdot)))$ $(\boldsymbol{\lambda}(\cdot) \in \boldsymbol{\gamma})$ consists of generalized joint eigenvectors of the family A corresponding to the eigenvalue $(\operatorname{reg} \boldsymbol{\lambda}(\cdot))(\cdot)$. On Steps VII and VIII, we carry out some preliminary constructions.

First, let us show that for every $x \in X$ and $l \in \mathbb{N}$ the restrictions $(\pi_{y_{n_k(x)}} \restriction \boldsymbol{\beta}_l)(\boldsymbol{\lambda}(\cdot))$ converge uniformly in $\boldsymbol{\lambda}(\cdot) \in \boldsymbol{\beta}_l$ as $k \to \infty$ to a certain function $f_{x,l}(\boldsymbol{\lambda}(\cdot))$ $(\boldsymbol{\lambda}(\cdot) \in \boldsymbol{\beta}_l)$ which is continuous on $\boldsymbol{\beta}_l$ (endowed with the topology induced from $\mathbf{C}^X$) and measurable with respect to the σ-algebra $\{\boldsymbol{\alpha} \in \mathfrak{C}_\sigma(\mathbf{C}^X) \mid \boldsymbol{\alpha} \subset \boldsymbol{\beta}_l\}$ (recall that

$$\mathbf{C}^X \ni \boldsymbol{\lambda}(\cdot) \mapsto \pi_y(\boldsymbol{\lambda}(\cdot)) = \boldsymbol{\lambda}(y) \in \mathbb{C}^1,$$

and

$$(\pi_{y_{n_k(x)}} \restriction \boldsymbol{\beta}_l)(\boldsymbol{\lambda}(\cdot)) = \lambda(y_{n_k(x)}) \neq \infty,$$

according to Step VI).

Indeed, taking into account the first relation in (2.30), which holds for $x = y_{n_k(x)}$ and

$\boldsymbol{\lambda}(\cdot) \in \boldsymbol{\beta}_l \subset \boldsymbol{\beta}$, and (2.32), we obtain

$$(\pi_{y_{n_k(x)}} \upharpoonright \boldsymbol{\beta}_l)(\boldsymbol{\lambda}(\cdot)) = \boldsymbol{\lambda}(y_{n_k(x)})$$

$$= (P(\boldsymbol{\lambda}(\cdot))e_{j_l}, A'_{y_{n_k(x)}} e_{j_l})_{H_0} (P(\boldsymbol{\lambda}(\cdot))e_{j_l}, e_{j_l})_{H_0}^{-1} \quad (k \in \mathbb{N}). \tag{2.36}$$

Using (2.36) and (2.32), we conclude that the following estimate holds uniformly in $\boldsymbol{\lambda}(\cdot) \in \boldsymbol{\beta}_l$:

$$|(\pi_{y_{n_k(x)}} \upharpoonright \boldsymbol{\beta}_l)(\boldsymbol{\lambda}(\cdot)) - (\pi_{y_{n_i(x)}} \upharpoonright \boldsymbol{\beta}_l)(\boldsymbol{\lambda}(\cdot))|$$

$$= |(P(\boldsymbol{\lambda}(\cdot))e_{j_l}, (A_{y_{n_k(x)}} - A'_{y_{n_i(x)}})\, e_{j_l})_{H_0}|\, |(P(\boldsymbol{\lambda}(\cdot))e_{j_l}, e_{j_l})_{H_0}^{-1}|$$

$$\leq m_l \| (P(\boldsymbol{\lambda}(\cdot))e_{j_l} \|_{H_-} \| (A'_{y_{n_k(x)}} - A'_{y_{n_i(x)}})\, e_{j_l} \|_{H_+}$$

$$\leq m_l \| (A'_{y_{n_k(x)}} - A_{y_{n_i(x)}})\, e_{j_l} \|_{H_+} \quad (k, i \in \mathbb{N}) \tag{2.37}$$

(here we have used the fact that

$$\| (P(\boldsymbol{\lambda}(\cdot))e_{j_l} \|_{H_-} \leq |P(\boldsymbol{\lambda}(\cdot))| \leq \operatorname{Tr}(P(\boldsymbol{\lambda}(\cdot))) = 1$$

for $\boldsymbol{\lambda}(\cdot) \in \boldsymbol{\beta}$). The required convergence of the restrictions $\pi_{y_{n_k(x)}} \upharpoonright \boldsymbol{\beta}_l$ follows from (2.37). The properties of the limit are consequences of the properties of π_y.

VIII. Let us define a function $f_x(\boldsymbol{\lambda}(\cdot))$, where

$$\boldsymbol{\lambda}(\cdot) \in \bigcup_{l=1}^{\infty} \boldsymbol{\beta}_l \supset \boldsymbol{\lambda},$$

by setting $f_x(\boldsymbol{\lambda}(\cdot)) = f_{x,l}(\boldsymbol{\lambda}(\cdot))$ for $\boldsymbol{\lambda}(\cdot) \in \boldsymbol{\beta}_l$ $(x \in X)$. It is stated that

$$(f_x \upharpoonright \boldsymbol{\gamma})(\boldsymbol{\lambda}(\cdot)) = (\operatorname{reg}(\boldsymbol{\lambda}(\cdot))(x) \quad (\boldsymbol{\lambda}(\cdot) \in \boldsymbol{\gamma}). \tag{2.38}$$

Indeed, fix $x \in X$ and $\boldsymbol{\lambda}_0(\cdot) \in \boldsymbol{\gamma}$. If $\boldsymbol{\lambda}_0(x) = \infty$, then, according to Step VI, $(\operatorname{reg} \boldsymbol{\lambda}_0(\cdot))(x)$ is defined by (2.35), and $(f_x \upharpoonright \boldsymbol{\gamma})(\boldsymbol{\lambda}(\cdot))$ is always constructed by passing to the limit in (2.32), i.e., in exactly the same way. This means that (2.38) is valid.

Assume that $\lambda_0(x) \neq \infty$; according to Step VI,

$$(\text{reg } \lambda_0(\cdot))\,(x) = \lambda_0(x).$$

By virtue of (2.34), one can find $l, p \in \mathbb{N}$ such that $\lambda_0(\cdot) \in \psi_{l,p}$. Let us consider the corresponding $\varphi_{l,p} \supset \psi_{l,p}$ and $\rho \restriction \varphi_{l,p}$. Taking into account (1.25) (for the operators A_y written with the RI $\boldsymbol{E}$), (2.29), and (2.32), we get

$$\|(A'_{y_{n_k(x)}} - A'_x)\, e_{j_l}\|^2_{H_+} \geq \|(A^*_{y_{n_k(x)}} - A^*_x)\, e_{j_l}\|^2_{H_0}$$

$$= \int_{\mathbf{C}^X \setminus \delta} |\lambda(y_{n_k(x)}) - \lambda(x)|^2\, d(\boldsymbol{E}(\lambda(\cdot))e_{j_l}, e_{j_l})_{H_0}$$

$$= \int_{\mathbf{C}^X \setminus \delta} |\pi_{y_{n_k(x)}}(\lambda(\cdot)) - \pi_x(\lambda(\cdot))|^2\, d(\boldsymbol{E}(\lambda(\cdot))e_{j_l}, e_{j_l})_{H_0}$$

$$\geq \frac{1}{m_l} \int_{\varphi_{l,p} \setminus \delta} |(\pi_{y_{n_k(x)}} \restriction \varphi_{l,p})(\lambda(\cdot)) - (\pi_x \restriction \varphi_{l,p})(\lambda(\cdot))|^2\, d(\rho \restriction \varphi_{l,p})(\lambda(\cdot)). \quad (2.39)$$

Here $\delta \in \mathfrak{C}_\sigma(\mathbf{C}^X)$ is the union of the sets

$$\{\lambda_0(\cdot) \in \mathbf{C}^X \mid \lambda(y_{n_k(x)}) = \infty\} \quad \text{and} \quad \{\lambda_0(\cdot) \in \mathbf{C}^X \mid \lambda(x) = \infty\};$$

and $\rho(\delta) = 0$.

Let us pass to the limit as $k \to \infty$ in (2.35). According to Steps II and VII, the left-hand side of this relation tends to zero, and $\pi_{y_{n_k(x)}} \restriction \beta_l$ tends uniformly to $f_{x,l} = f_x \restriction \beta_l$. Therefore, as a result of this limit transition we obtain

$$0 = \int_{\varphi_{l,p} \setminus \delta} |(f_x \restriction \varphi_{l,p})(\lambda(\cdot)) - (\pi_x \restriction \varphi_{l,p})(\lambda(\cdot))|^2\, d(\rho \restriction \varphi_{l,p})(\lambda(\cdot)). \quad (2.40)$$

Denote the integrand in (2.40) by $g(\lambda(\cdot))$. This function is defined and continuous on β_l; its restriction to $\varphi_{l,p} \restriction \delta$ is continuous in the topology induced by $\mathbf{C}^X$. The point $\lambda_0(\cdot)$ belongs to $\varphi_{l,p} \setminus \delta$, because $\lambda_0(x) \neq \infty$. The relation (2.38) now means that $g(\lambda_0(\cdot)) = 0$. Assume the contrary. Then there exist $\varepsilon > 0$ and a neighborhood of the point $\lambda_0(\cdot)$ in the space $\varphi_{l,p} \setminus \delta$ (i.e., a set of the form $\boldsymbol{u} \setminus \delta$, where $\boldsymbol{u}$ is a neighborhood in $\varphi_{l,p}$) such that $g(\lambda(\cdot)) \geq \varepsilon$ for $\lambda(\cdot) \in \boldsymbol{u} \setminus \delta$. Since $\boldsymbol{u} \cap \operatorname{supp}(\rho \restriction \varphi_{l,p}) \neq \varnothing$ (this intersection contains $\lambda_0(\cdot)$), we have $(\rho \restriction \varphi_{l,p})(\boldsymbol{u}) > 0$. Therefore,

$$\int_{u \setminus \delta} g(\lambda(\cdot))\, d(\rho \upharpoonright \varphi_{l,p})(\lambda(\cdot)) \geq \varepsilon (\rho \upharpoonright \varphi_{l,p})(u \setminus \delta) = \varepsilon(\rho \upharpoonright \varphi_{l,p})(u) > 0,$$

which is absurd, because the integral in (2.40) is bounded from below by the last integral.

The relation (2.38) is now proved. Simultaneously, we have established that $(\text{reg}\, \lambda(\cdot))(x)$ can be defined by employing the limit $\lim_{k\to\infty} \lambda(y_{n_k(x)})$ also in the case when $\lambda(x) \neq \infty$. This is just the way which has been used to define the function f_x.

This clarifies the definition of reg given in the formulation of the lemma.

IX. Our goal is to prove the relations

$$(P(\lambda(\cdot))\varphi, A_x^* \psi)_{H_0} = (\text{reg}\, \lambda(\cdot))(x)\, (P(\lambda(\cdot))\varphi, \psi)_{H_0},$$

$$(P(\lambda(\cdot))\varphi, A_x \psi)_{H_0} = \overline{(\text{reg}\, \lambda(\cdot))(x)}\, (P(\lambda(\cdot))\varphi, \psi)_{H_0}. \qquad (2.41)$$

$$(\lambda(\cdot) \in \boldsymbol{\gamma}, x \in X;\ \varphi \in H_+, \psi \in D).$$

It follows from Remark 1 in Subsection 2.6 (which is also valid for (2.41)) that it suffices to establish these relations for $\varphi, \psi \in L$.

The proof of the first one is easy. Indeed, let us write the first equality (2.30) for x of the form $y_{n_k(x)}$, $\lambda(\cdot) \in \boldsymbol{\gamma}$, and $\varphi, \psi \in L$ (this is possible by Step III):

$$(P(\lambda(\cdot))\varphi, A'_{y_{n_k(x)}} \psi)_{H_0} = (\pi_{y_{n_k(x)}} \upharpoonright \boldsymbol{\gamma})\, (\lambda(\cdot))(P(\lambda(\cdot))\varphi, \psi)_{H_0}. \qquad (2.42)$$

Here we pass to the limit as $k \to \infty$. The left-hand side tends to

$$(P(\lambda(\cdot))\varphi, A'_x \psi)_{H_0} = (P(\lambda(\cdot))\varphi, A_x^* \psi)_{H_0},$$

due to the choice of $y_{n_k(x)}$. By virtue of Step VII, the factor $(\pi_{y_{n_k(x)}} \upharpoonright \boldsymbol{\gamma})\, (\lambda(\cdot))$ tends to $(f_x \upharpoonright \boldsymbol{\gamma})\, (\lambda(\cdot))$, since the inclusion $\lambda(\cdot) \in \boldsymbol{\gamma}$ yields the inclusion $\lambda(\cdot) \in \boldsymbol{\psi}_{l,p} \subset \boldsymbol{\beta}_l$ for some $l, p \in \mathbb{N}$. But

$$(f_x \upharpoonright \boldsymbol{\gamma})\, (\lambda(\cdot)) = (\text{reg}\, \lambda(\cdot))(x),$$

according to (2.38). Hence, we have arrived at the first equality in (2.41).

To prove the second equality we write (by virtue of Step III) the second equality in (2.30) for $x = y_{n_k(x)}$, $\lambda(\cdot) \in \boldsymbol{\gamma}$, and $\varphi, \psi \in L$. As a result, we obtain a relation similar to (2.42) in which $A'_{y_{n_k(x)}}$ is replaced by $A''_{y_{n_k(x)}}$ and $(\pi_{y_{n_k(x)}} \upharpoonright \boldsymbol{\gamma})\, (\lambda(\cdot))$ appears with a complex overbar. Passing to the limit as $k \to \infty$ and employing the relation (see Step II)

$$\| A''_{y_{n_k(x)}} \psi - A''_x \psi \|_{H_+} \to 0 \quad (\psi \in L),$$

we obtain the required result.

Finally, the equality $\mathrm{Tr}\,(\boldsymbol{P}(\boldsymbol{\lambda}(\cdot))) = 1$ $(\boldsymbol{\lambda}(\cdot) \in \boldsymbol{\gamma})$ follows from the inclusion $\boldsymbol{\gamma} \subset \boldsymbol{\beta}$ and Step III.

■

Proof of the Theorem. Let $\gamma = \mathrm{reg}\, \boldsymbol{\gamma} \subset \mathbb{C}^X$, where reg and $\boldsymbol{\gamma}$ are taken from Lemma 2.5.

On Steps I–III, we intend to show that the representations (2.28) are valid with π replaced by γ (one must only establish the formulas for A_x and $\mathfrak{D}(A_x)$, whereas the formula for A_x^* is obtained automatically).

I. Let us write (1.25) with the RI $\boldsymbol{E}$ (see Remark 1 in Subsection 1.11) and then modify $\boldsymbol{E}$ by $\boldsymbol{\gamma}$. For the modified RI $\boldsymbol{E}_{\boldsymbol{\gamma}}$ defined on the σ-algebra

$$\mathfrak{C}_\sigma(\boldsymbol{\gamma}) = \{\boldsymbol{\alpha}' = \boldsymbol{\alpha} \cap \boldsymbol{\gamma} \mid \boldsymbol{\alpha} \in \mathfrak{C}_\sigma(\mathbf{C}^X)\},$$

we get

$$A_x = \int_{\boldsymbol{\gamma}} \boldsymbol{\lambda}(x)\, d\boldsymbol{E}_{\boldsymbol{\gamma}}\,(\boldsymbol{\lambda}(\cdot)),$$

$$\mathfrak{D}(A_x) = \left\{f \in H_0 \,\middle|\, \int_{\boldsymbol{\gamma}} |\boldsymbol{\lambda}(x)|^2\, d(\boldsymbol{E}_{\boldsymbol{\gamma}}\,(\boldsymbol{\lambda}(\cdot)) f, f)_{H_0} < \infty\right\} \quad (x \in X). \tag{2.43}$$

By means of the mapping

$$\mathbf{C}^X \supset \boldsymbol{\gamma} \ni \boldsymbol{\lambda}(\cdot) \mapsto (\mathrm{reg}\, \boldsymbol{\lambda}(\cdot))\,(\cdot) \in \boldsymbol{\gamma} \subset \mathbb{C}^X,$$

we transform the measure $\boldsymbol{E}_{\boldsymbol{\gamma}}$ into a measure E' on the space γ (E' is an RI). The measure E' is defined on the σ-algebra

$$\mathfrak{R}' = \{\alpha \subset \lambda \mid \mathrm{reg}^{-1} \alpha \in \mathfrak{C}_\sigma(\boldsymbol{\gamma})\}$$

by the equality $E'(\alpha) = \boldsymbol{E}_{\boldsymbol{\gamma}}\,(\mathrm{reg}^{-1}\,\alpha)$.

II. To express the representation (2.43) in terms of E', we note that for every fixed $x \in X$, the function $\gamma \ni \lambda(\cdot) \mapsto \lambda(x) \in \mathbb{C}^1$ is measurable with respect to the σ-algebra $\mathfrak{R}'$. In fact, $\mathfrak{R}'$ is constructed so that it suffices to show that the function

$$\boldsymbol{\gamma} \ni \boldsymbol{\lambda}(\cdot) \mapsto (\text{reg}\, \boldsymbol{\lambda}(\cdot))\,(\cdot) \in \boldsymbol{\gamma} \subset \mathbb{C}^1$$

is $\mathfrak{C}_\sigma(\boldsymbol{\gamma})$-measurable. According to the definition of reg, there exists a sequence $(y_{n_k(x)})_{k=1}^{\infty}$ such that $\boldsymbol{\lambda}\,(y_{n_k(x)}) \neq \infty$ for every $\boldsymbol{\lambda}(\cdot) \in \boldsymbol{\gamma}$, and $\boldsymbol{\lambda}\,(y_{n_k(x)}) \to (\text{reg}\,\boldsymbol{\lambda}(\cdot))(x)$ as $k \to \infty$, i.e.,

$$(\pi_{y_{n_k(x)}} \restriction \boldsymbol{\gamma})\,(\boldsymbol{\lambda}(\cdot)) \mapsto (\text{reg}\,\boldsymbol{\lambda}(\cdot))(x) \quad (\boldsymbol{\lambda}(\cdot) \in \boldsymbol{\gamma}).$$

The $\mathfrak{C}_\sigma(\boldsymbol{\gamma})$-measurability of the functions before taking the limit (and, hence, the $\mathfrak{C}_\sigma(\boldsymbol{\gamma})$-measurability of the limiting function) follows from the $\mathfrak{C}_\sigma(\mathbf{C}^X)$-measurability of the function of $\boldsymbol{\lambda}(\cdot) \in \mathbf{C}^X$ equal to $(\pi_{y_{n_k(x)}} \restriction \boldsymbol{\gamma})\,(\boldsymbol{\lambda}(\cdot)$ if $\boldsymbol{\lambda}\,(y_{n_k(x)}) \neq \infty$ and equal to 0 otherwise.

Thus, the integral

$$\int_{\gamma} \lambda(x)\, dE'(\lambda(\cdot))$$

is meaningful. According to the rules for transforming integrals under mappings of measures, it is equal to

$$\int_{\boldsymbol{\gamma}} (\text{reg}\,\boldsymbol{\lambda}(\cdot))(x)\, d\boldsymbol{E}_{\boldsymbol{\gamma}}\,(\boldsymbol{\lambda}(\cdot)).$$

By definition,

$$\text{reg}\{\boldsymbol{\lambda}(\cdot) \in \boldsymbol{\gamma} \mid (\text{reg}\,\boldsymbol{\lambda}(\cdot))(x) \neq \lambda(x)\} \subset \{\boldsymbol{\lambda}(\cdot) \in \mathbf{C}^X \mid \lambda(x) = \infty\},$$

and, as has been just proved, the first of these sets belongs to $\mathfrak{C}_\sigma(\boldsymbol{\gamma})$. The $\boldsymbol{E}$-measure of the second set is equal to zero, but then the $\boldsymbol{E}_{\boldsymbol{\gamma}}$-measure of the first one is also zero. Therefore, the last integral is equal to

$$\int_{\boldsymbol{\gamma}} \boldsymbol{\lambda}(x)\, d\boldsymbol{E}_{\boldsymbol{\gamma}}\,(\boldsymbol{\lambda}(\cdot)) = A_x$$

(see (2.43)). The second integral in (2.43) can be transformed similarly. Accordingly,

$$A_x = \int_{\gamma} \lambda(x)\, dE'(\lambda(\cdot)),$$

$$\mathfrak{D}(A_x) = \left\{ f \in H_0 \;\middle|\; \int_{\gamma} |\lambda(x)|^2\, d(E'(\lambda(\cdot)) f, f)_{H_0} < \infty \right\} \quad (x \in X). \tag{2.44}$$

III. Let us show that (2.44) can be rewritten in the form (2.28) with $\pi = \gamma$. To do this, note that $\mathfrak{C}_\sigma(\boldsymbol{\gamma}) \subset \mathfrak{R}'$. Indeed, it suffices to establish the inclusion

$$\alpha' = \mathbb{C}(x_1, \dots, x_n; \kappa) \cap \mathfrak{R}'(x_1, \dots, x_n \in X; \kappa \in \mathfrak{B}(\mathbb{C}^n); n \in \mathbb{N}).$$

But

$$\alpha' = \{\lambda(\cdot) \in \gamma | (\lambda(x_1), \dots, \lambda(x_n)) \in \kappa\},$$

i.e., it is the full inverse image of κ under the mapping

$$\gamma \ni \lambda(\cdot) \mapsto (\lambda(x_1), \dots, \lambda(x_n)) \in \mathbb{C}^n.$$

This mapping is $\mathfrak{R}'$-measurable, because each coordinate of it $\gamma \ni \lambda(\cdot) \mapsto \lambda(x_k) \in \mathbb{C}^1$ $(k = 1, \dots, n)$ is measurable in this sense, according to Step II. Consequently, $\alpha' \in \mathfrak{R}'$, as was required.

Consider the restriction E'' of the RI E' to the σ-algebra $\mathfrak{C}_\sigma(\boldsymbol{\gamma})$. Since the function

$$\mathbb{C}^X \supset \gamma \ni \lambda(\cdot) \mapsto \lambda(x) \in \mathbb{C}^1$$

is measurable with respect to this σ-algebra, E' can be replaced by E'' in the integrals in (2.44), and the formulas remain valid. It is easy to see that E'' is equal to the modification of the original RI E by γ.

Thus, fix $x \in X$. The mapping

$$\mathbb{C}^X \supset \gamma \ni \lambda(\cdot) \mapsto \pi_x(\lambda(\cdot)) = \lambda(x) \in \mathbb{C}^1$$

is $\mathfrak{C}_\sigma(\boldsymbol{\gamma})$-measurable and transforms E'' into the RI

$$\mathfrak{B}(\mathbb{C}^1) \ni \kappa \mapsto E''_x(\kappa) = E''(\pi_x^{-1}(\kappa)) = E''(\mathbb{C}(x; \kappa) \cap \gamma),$$

whereas the first integral in (2.44) (with E' replaced by E'') turns into the integral $\int_{\mathbb{C}^1} \lambda \, dE''_x(\lambda)$. This implies that E''_x is the usual RI E_x of the operator A_x, i.e.,

$$E''(\mathbb{C}(x; \kappa) \cap \gamma) = E(\mathbb{C}(x; \kappa)) \quad (x \in \mathfrak{B}(\mathbb{C}^1)).$$

This equality enables us to conclude in the standard way that

$$E''(\mathbb{C}(x_1, \dots, x_n; \kappa) \cap \gamma) = E(\mathbb{C}(x_1, \dots, x_n; \kappa)) \;\; (x_1, \dots, x_n \in X; \kappa \in \mathfrak{B}(\mathbb{C}^n), n \in \mathbb{N})$$

and, hence,

$$E''(\alpha \cap \gamma) = E(\alpha) \quad (\alpha \in \mathfrak{C}_\sigma(\mathbb{C}^X))$$

(we first take κ to be rectangles and exploit the orthogonality of the RI). Taking $\mathfrak{C}_\sigma(\mathbb{C}^X) \ni \alpha \supset \gamma$ in the last equality, we get $E(\alpha) = E''(\gamma) = 1$, i.e., γ is of full outer E-measure. The equality itself means that E'' is the modification of E by γ, namely, $E'' = E_\gamma$ (see (1.21)).

Thus, the representation (2.44) holds with E' replaced by E_γ, i.e.,

$$A_x = \int_\gamma \lambda(x)\, dE_\gamma(\lambda(\cdot)),$$

$$\mathfrak{D}(A_x) = \left\{ f \in H_0 \mid \int_\gamma |\lambda(x)|^2\, d(E_\gamma(\lambda(\cdot)) f, f)_{H_0} < \infty \right\} \quad (x \in X). \tag{2.45}$$

IV. Let $\pi \in \mathfrak{C}_\sigma(\gamma)$ be some set such that $E_\gamma(\pi) = 1$. Then it is clear that π is of full outer E-measure. The representation (2.45) yields the analogous representations with γ replaced by π, i.e., (2.28). The inclusion $\pi \subset g(A)$ is valid. Indeed, let $\lambda(\cdot) \in \pi \subset \gamma$, then

$$\lambda(\cdot) = (\operatorname{reg} \boldsymbol{\lambda}(\cdot))(\cdot)$$

for some $\boldsymbol{\lambda}(\cdot) \in \gamma$ and, thus, by virtue of Lemma 2.5, $\lambda(\cdot)$ is an eigenvalue for any generalized joint eigenvector $P(\boldsymbol{\lambda}(\cdot))\,\varphi$ $(\varphi \in H_+)$.

V. Below (on Steps VI–VIII) we prove that there exists $\pi \in \mathfrak{C}_\sigma(\gamma)$ with $E_\gamma(\pi) = 1$ such that for every $\lambda(\cdot) \in \pi$ the range $\mathfrak{R}(P(\lambda(\cdot)))$ consists of generalized joint eigenvectors of the family A corresponding to $\lambda(\cdot)$, and $\operatorname{Tr}(P(\lambda(\cdot))) = 1$. In view of Step IV, this proves the theorem (some additional arguments are necessary only for the last assertion of it).

VI. Using (1.21) and (1.19), we get

$$O^+ E_\gamma(\alpha')\, O = O^+ E(\alpha)\, O = \int_\alpha P(\lambda(\cdot))\, d\rho(\lambda(\cdot))$$

$$= \int_{\alpha'} (P \restriction \gamma)(\lambda(\cdot))\, d\rho_\gamma(\lambda(\cdot)) \quad (\mathfrak{C}_\sigma(\gamma) \ni \alpha' = \alpha \cap \gamma,\ \alpha \in \mathfrak{C}_\sigma(\mathbb{C}^X)). \tag{2.46}$$

Denote by $\hat{E}$ the preimage of the measure E_γ under the mapping

$$\mathbf{C}^X \supset \boldsymbol{\gamma} \ni \boldsymbol{\lambda}(\cdot) \mapsto \operatorname{reg}(\boldsymbol{\lambda}(\cdot))\,(\cdot) \in \gamma \subset \mathbb{C}^X.$$

Since $\mathfrak{C}_\sigma(\gamma) \subset \mathfrak{R}'$, and $E_\gamma = E' \restriction \mathfrak{C}_\sigma(\gamma)$, we have $\hat{\boldsymbol{E}} = \boldsymbol{E}_\gamma \restriction \hat{\mathfrak{R}}$, where $\hat{\mathfrak{R}}$ is the σ-algebra

$$\{\hat{\boldsymbol{\alpha}} = \operatorname{reg}^{-1} \alpha' \mid \alpha' \in \mathfrak{C}_\sigma(\gamma)\} \subset \mathfrak{C}_\sigma(\boldsymbol{\gamma}).$$

Proceeding to these preimages in (2.46), we find

$$O^+ \hat{E}\ (\hat{\boldsymbol{\alpha}})\, O = \int_{\hat{\boldsymbol{\alpha}}} (P \restriction \gamma)\, ((\operatorname{reg} \boldsymbol{\lambda}(\cdot))\,(\cdot)\, d\hat{\boldsymbol{\rho}}\,(\boldsymbol{\lambda}(\cdot)) \quad (\hat{\boldsymbol{\alpha}} \in \hat{\mathfrak{R}}). \tag{2.47}$$

Here $\hat{\boldsymbol{\rho}}$ is the preimage of ρ_γ. At the same time, it is easy to see that $\hat{\boldsymbol{\rho}} = \boldsymbol{\rho}_{\boldsymbol{\gamma}} \restriction \hat{\mathfrak{R}}$, where $\boldsymbol{\rho}_{\boldsymbol{\gamma}}$ is the modification of $\boldsymbol{\rho}$ by $\boldsymbol{\gamma}$ and, thus ,

$$\hat{\boldsymbol{\rho}}\,(\boldsymbol{\alpha}) = \operatorname{Tr}\,(O^+ \hat{\boldsymbol{E}}\ (\hat{\boldsymbol{\alpha}})\, O) \quad (\hat{\boldsymbol{\alpha}} \in \hat{\mathfrak{R}}).$$

It follows from Lemma 2.1, Theorem 2.1, and (2.47) that there exists a derivative

$$(d(O^+ \hat{E}\ O) / d\hat{\boldsymbol{\rho}}\,)\,(\boldsymbol{\lambda}(\cdot)) = \hat{\boldsymbol{P}}\,(\boldsymbol{\lambda}(\cdot)),$$

which is uniquely determined for $\hat{\boldsymbol{\rho}}$ -almost all $\boldsymbol{\lambda}(\cdot) \in \boldsymbol{\gamma}$ and coincides with $(P \restriction \gamma)\ ((\operatorname{reg} \boldsymbol{\lambda}(\cdot))(\cdot))$.

It will be shown in Steps VII and VIII that there exists $\hat{\boldsymbol{\pi}} \in \hat{\mathfrak{R}}$ with $\hat{\boldsymbol{E}}(\hat{\boldsymbol{\pi}}) = 1$ such that for every $\boldsymbol{\lambda}(\cdot) \in \hat{\boldsymbol{\pi}}$, the range $\mathfrak{R}\,(\hat{\boldsymbol{P}}\,(\boldsymbol{\lambda}(\cdot)))$ consists of generalized joint eigenvectors of the family $\boldsymbol{A}$ corresponding to $(\operatorname{reg} \boldsymbol{\lambda}(\cdot))(\cdot)$, and $\operatorname{Tr}\,(\,\hat{\boldsymbol{P}}\,(\boldsymbol{\lambda}(\cdot))) = 1$. We set $\pi = \operatorname{reg} \hat{\boldsymbol{\pi}} \in \mathfrak{C}_\sigma(\lambda)$ and $E_\gamma(\pi) = \hat{\boldsymbol{E}}(\hat{\boldsymbol{\pi}}) = 1$. Since $(P \restriction \gamma)\ ((\operatorname{reg} \boldsymbol{\lambda}(\cdot))(\cdot)$ can be chosen to coincide with $\hat{\boldsymbol{P}}\,(\boldsymbol{\lambda}(\cdot))$, this yields the assertion of Step V.

VII. The assertion we have formulated is similar to Lemma 2.5. The difference is that the RI which is differentiated is defined not on the σ-algebra $\mathfrak{C}_\sigma(\mathbf{C}^X)$ or $\mathfrak{C}_\sigma(\boldsymbol{\gamma})$, but on the σ-subalgebra $\hat{\mathfrak{R}} \subset \mathfrak{C}_\sigma(\boldsymbol{\gamma})$. Therefore, the corresponding derivative $\hat{\boldsymbol{P}}\,(\boldsymbol{\lambda}(\cdot))$ is equal in a certain sense to the "conditional mathematical expectation" of $\boldsymbol{P}(\boldsymbol{\lambda}(\cdot))$.

Proceeding to the preimage of the measure E_γ in the integrals (2.45), we get

$$A_x = \int_{\boldsymbol{\gamma}} (\operatorname{reg} \lambda(\cdot))\,(x)\, d\hat{\boldsymbol{E}}\,(\boldsymbol{\lambda}(\cdot)),$$

$$\mathfrak{D}(A_x) = \left\{ f \in H_0 \,\middle|\, \int_{\boldsymbol{\gamma}} |(\operatorname{reg} \boldsymbol{\lambda}(\cdot))(x)|^2\, d(\hat{\boldsymbol{E}}\,(\boldsymbol{\lambda}(\cdot))\, f, f)_{H_0} < \infty \right\} \quad (x \in X). \tag{2.48}$$

Using (2.48), one can show (as in Lemma 2.2) that for every $x \in X$ there exists a set $\hat{\boldsymbol{\beta}}_x \in \hat{\mathfrak{R}}$ of full $\hat{\boldsymbol{\rho}}$ -measure such that $\operatorname{Tr}(\hat{P}(\boldsymbol{\lambda}(\cdot))) = 1$ for each $\boldsymbol{\lambda}(\cdot) \in \hat{\boldsymbol{\beta}}_x$, and

$$(\hat{P}(\boldsymbol{\lambda}(\cdot))\varphi, A_x^*\psi)_{H_0} = (\operatorname{reg}\boldsymbol{\lambda}(\cdot))(x)\,(\hat{P}(\boldsymbol{\lambda}(\cdot))\varphi, \psi)_{H_0},$$

$$(\hat{P}(\boldsymbol{\lambda}(\cdot))\varphi, A_x\psi)_{H_0} = \overline{(\operatorname{reg}\lambda(\cdot)(x)}\,(\hat{P}(\boldsymbol{\lambda}(\cdot))\varphi, \psi)_{H_0} \quad (\varphi \in H_+, \psi \in D). \quad (2.49)$$

Indeed, taking into account the equalities

$$O^+\hat{E}(\hat{\alpha})\,O = \int_{\hat{\boldsymbol{\alpha}}} \hat{P}(\boldsymbol{\lambda}(\cdot))\,(\cdot)\,d\hat{\boldsymbol{\rho}}(\boldsymbol{\lambda}(\cdot))$$

and

$$A_x\hat{E}(\hat{\alpha}) = \int_{\hat{\boldsymbol{\alpha}}} (\operatorname{reg}\boldsymbol{\lambda}(\cdot))(x)\,d\hat{E}(\boldsymbol{\lambda}(\cdot)) \quad (\hat{\boldsymbol{\alpha}} \in \hat{\mathfrak{R}}, x \in X)$$

(the latter is a consequence of (2.48)), as in the case of (2.25), we conclude that

$$\int_{\hat{\boldsymbol{\alpha}}} (\hat{P}(\boldsymbol{\lambda}(\cdot))\varphi, A_x^*\psi)_{H_0}\,d\hat{\boldsymbol{\rho}}(\boldsymbol{\lambda}(\cdot)) = \int_{\hat{\boldsymbol{\alpha}}} (\operatorname{reg}\boldsymbol{\lambda}(\cdot))(x)(\hat{P}(\boldsymbol{\lambda}(\cdot))\varphi, \psi)_{H_0}\,d\hat{\boldsymbol{\rho}}(\boldsymbol{\lambda}(\cdot))$$

$$(\hat{\boldsymbol{\alpha}} \in \hat{\mathfrak{R}}, \ \varphi \in H_+, \psi \in D).$$

As in Lemma 2.2, this and the analogous formula for A_x^* involve (2.45).

VIII. Let $\hat{\boldsymbol{\pi}} = \bigcap_{n=1}^{\infty} \hat{\boldsymbol{\beta}}_{y_n} \in \hat{\mathfrak{R}}$, where y_n are the points from Lemma 2.5. It is clear that $\hat{E}(\hat{\boldsymbol{\pi}}) = 1$ and $\operatorname{Tr}(\hat{P}(\boldsymbol{\lambda}(\cdot))) = 1$ for $\boldsymbol{\lambda}(\cdot) \in \hat{\boldsymbol{\pi}}$. The equalities (2.49) are valid for $x = y_n$. Moreover,

$$(\operatorname{reg}\boldsymbol{\lambda}(\cdot))(y_n) = \boldsymbol{\lambda}(y_n) = (\pi_{y_n} \upharpoonright \boldsymbol{\Upsilon})(\boldsymbol{\lambda}(\cdot)) \quad (n \in \mathbb{N}).$$

We fix $x \in X$ and write (2.49) for $\boldsymbol{\lambda}(\cdot) \in \hat{\boldsymbol{\pi}}$ and $y_n = y_{n_k(x)}$, where $(y_{n_k(x)})_{k=1}^{\infty}$ was introduced on Step II of the proof of Lemma 2.5.

According to Steps VII–IX of this proof, for every $\boldsymbol{\lambda}(\cdot) \in \boldsymbol{\Upsilon}$, we have

$$(\pi_{y_{n_k(x)}} \upharpoonright \boldsymbol{\Upsilon})(\boldsymbol{\lambda}(\cdot)) \to (\operatorname{reg}\lambda(\cdot))(x),$$

$$(\hat{P}(\boldsymbol{\lambda}(\cdot))\varphi, A_{y_{n_k(x)}}^*\psi)_{H_0} \to (\hat{P}(\boldsymbol{\lambda}(\cdot))\varphi, A_x^*\psi)_{H_0},$$

$$(\hat{P}(\boldsymbol{\lambda}(\cdot))\varphi, A_{y_{n_k(x)}}\psi)_{H_0} \to (\hat{P}(\boldsymbol{\lambda}(\cdot))\varphi, A_x\psi)_{H_0} \quad (\varphi, \psi \in L)$$

as $k \to \infty$ (the last two relations follow from the fact that $A'_{y_{n_k(x)}} \psi \to A'_x \psi$ and $A''_{y_{n_k(x)}} \times \psi \to A''_x \psi$ ($\psi \in L$) in H_+ and from the estimate $\| \hat{P}(\boldsymbol{\lambda}(\cdot)) \| \le 1$). But $\hat{\boldsymbol{\pi}} \subset \boldsymbol{\gamma}$, therefore, the last three limiting relations enable us to pass to the limit as $k \to \infty$ in (2.49) for $\varphi, \psi \in L$. As a result, we get (2.49) with arbitrary $\boldsymbol{\lambda}(\cdot) \in \hat{\boldsymbol{\pi}}$, $x \in X$, and $\varphi, \psi \in L$ and, hence, with arbitrary $\varphi \in H_+$ and $\psi \in D$, too, according to Remark 1 in Subsection 1.6.

Thus, we have proved the fact formulated on Step VI. According to Step V, the theorem is proved except for its last statement.

IX. The last statement of the theorem means that the following equality holds

$$O^+ E_\pi(\alpha')\, O = \int_{\alpha'} (P \upharpoonright \pi)(\lambda(\cdot))\, d\rho_\pi(\lambda(\cdot))$$

$$(\mathfrak{C}_\sigma(\pi) \ni \alpha' = \alpha \cap \pi,\ \alpha \in \mathfrak{C}_\sigma(\mathbb{C}^X)). \tag{2.50}$$

To derive (2.50), we must write

$$(O^+ E_\pi(\alpha')\, O\varphi, \psi)_{H_0} = (O^+ E(\alpha)\, O\varphi, \psi)_{H_0} \quad (\varphi, \psi \in H_+)$$

as an integral with respect to $d\rho(\lambda(\cdot))$ (according to (2.23)) and apply (1.19), with $R = \mathbb{C}^X$, $R' = \pi$, and

$$F(\lambda) = \kappa_\alpha(\lambda(\cdot)) \times (P(\lambda(\cdot))\varphi, \psi)_{H_0},$$

to this integral.

■

Proof of Lemma 2.4. Denote by $(\beta_m)_{m=1}^\infty$ a fixed dense subset of Φ', and let $(e_j)_{j=1}^\infty$ be an orthonormal basis in H. Consider the collection $\boldsymbol{B}$ of operators from Φ to H of the form

$$\Phi \ni \varphi \mapsto B\varphi = \sum_{j=1}^n \overline{\beta_{m_j}(\varphi)} e_j \in H,$$

where m_j are distinct ($n \in \mathbb{N}$). There are countably many operators in $\boldsymbol{B}$, and each operator is continuous. Let $\boldsymbol{B} = (B_p)_{p=1}^\infty \subset \mathfrak{L}(\Phi, H)$.

We fix $p, q \in \mathbb{N}$ and consider the set

$$\{A \in \boldsymbol{A} \mid \|(A - B_p)\varphi_k\|_H < q^{-1},\ k = 1, \dots, q\}.$$

If it is nonempty, then we choose some operator in it and denote this operator by $A_{p,q}$.

Arbitrarily varying $p, q \in \mathbb{N}$, we obtain an at most countable set of operators from $\boldsymbol{A}$. Let us show that it is the required family of operators[29].

We first show that for any $A \in \boldsymbol{A}$ and $q \in \mathbb{N}$, one can find an operator $B_{n_q} \in \boldsymbol{B}$ such that

$$\|(A - B_{n_q})\varphi_k\|_H < q^{-1}, \ (k = 1, \ldots, q). \tag{2.51}$$

The operator A belongs to $\mathfrak{L}(\Phi, H)$ and can thus be represented in the form

$$A\varphi = \sum_{j=1}^{\infty} \overline{\alpha_j(\varphi)} e_j \ (\varphi \in \Phi),$$

where $(\alpha_j)_{j=1}^{\infty}$ is a sequence of functionals from Φ' determined by A. We have

$$\|A\varphi_k\|_H^2 = \sum_{j=1}^{\infty} |\alpha_j(\varphi_k)|^2 < \infty,$$

therefore, for every $q \in \mathbb{N}$ there exists a number $l(q) \in \mathbb{N}$ so large that

$$\sum_{j=l(q)+1}^{\infty} |\alpha_j(\varphi_k)|^2 < (2q^2)^{-1} \quad (k = 1, \ldots, q).$$

The sequence $(\beta_m)_{m=1}^{\infty}$ is dense in Φ', hence, for any $j = 1, \ldots, l(q)$, we can find functionals $\beta_{m(j,q)}$ in it such that

$$|\alpha_j - \beta_{m(j,q)}(\varphi_k)|^2 < (2q^2\, l(q))^{-1} \quad (k = 1, \ldots, q);$$

it is clear that the indices $m(1, q), \ldots, m(l(q), q)$ can be regarded as distinct. The operator

$$\Phi \ni \varphi \mapsto B\varphi = \sum_{j=1}^{l(q)} \overline{\beta_{m(j,q)}(\varphi)} e_j \in H$$

belongs to $\boldsymbol{B}$ and thus has some number n_q, i.e., $B = B_{n_q}$. The estimate (2.51) holds for it, namely,

$$\|(A - B_{n_q})\varphi_k\|_H^2 = \sum_{j=1}^{l(q)} |\alpha_j - \beta_{m(j,q)}(\varphi_k)|^2 + \sum_{j=l(q)+1}^{\infty} |\alpha_j(\varphi_k)|^2 \le q^{-2}$$

$$(k = 1, \ldots, q).$$

---

29 It will be clear from what follows that $\boldsymbol{A}$ is nonempty.

The definition of the operators $A_{p,q}$ and the estimate (2.51) just established imply that for every $q \in \mathbb{N}$, there exists an operator $A_{n_q,q}$ (these operators may coincide for different q). Let us show that the sequence $(A_{n_q,q})_{q=1}^{\infty}$ approximates A in the required sense. In fact, the definition of $A_{p,q}$ yields

$$\| (A_{n_q,q} - B_{n_q}) \varphi_k \|_H \leq q^{-1}.$$

This inequality and (2.51) enable us to conclude that

$$\| (A - A_{n_q,q}) \varphi_k \|_H \leq 2q^{-1} \quad (k = 1, \ldots, q; \; q \in \mathbb{N}).$$

Consequently, $\|(A - A_{n_q,q})\varphi_j\|_H \to 0$ as $q \to \infty$ for every $j \in \mathbb{N}$.

■

2.8. Spectral Theorem. The Case of at Most Countably Many Unbounded Operators

Here, Theorem 2.7 is made somewhat more accurate, and its proof becomes essentially simpler. Namely, the following theorem holds.

Theorem 2.8. *Suppose that the assumptions of Theorem 2.7 hold. If there are at most countably many unbounded operators among the operators A_x from the family $A = (A_x)_{x \in X}$, then the set π from Theorem 2.7 can be chosen so that $\mathfrak{B}(\mathbb{C}^X) \ni \pi \subset s(A) \cap g(A)$.*

Proof. We repeat the proof of Lemma 2.5. Steps I and II remain unchanged. In Step III, we employ the generalized projector $P(\lambda(\cdot))$ and the spectral measure ρ instead of the compactified objects, and Lemma 2.2 instead of Lemma 2.3. We set $\beta = \bigcap_{n=1}^{\infty} \beta_{y_n} \in \mathfrak{C}_\sigma(\mathbb{C}^X)$ and define

$$\mathfrak{B}(\mathbb{C}^X) \ni \gamma \subset \operatorname{supp} E = s(A)$$

according to (2.34), namely, $\gamma = \bigcup_{l,p=1}^{\infty} \psi_{l,p}$,, where $\psi_{l,p} = \operatorname{supp}(\rho \restriction \varphi_{l,p})$ and $\varphi_{l,p}$ are constructed by analogy with $\varphi_{l,p}$. (One more simplification could be made: After (2.32) we set $\psi_l = \operatorname{supp}(\rho \restriction \beta_l)$ $(l \in \mathbb{N})$ and take $\gamma = \bigcup_{l=1}^{\infty} \psi_l$; the role of $\rho \restriction \varphi_{l,p}$ is played by $\rho \restriction \beta_l$.) We take $\pi = \gamma$ and verify the validity of all the required statements of the theorem.

As in Step IV, we conclude that $\rho \upharpoonright \varphi_{l,p}$ is regular. It is proper, by virtue of Theorem 1.6 (the possibility of applying this theorem simplifies the proof as compared with Theorem 2.7). Similar to Step V, we verify that γ is of full outer ρ-measure. Further, as in Step VII, we find that for every $x \in X$ and $l \in \mathbb{N}$ the restrictions $(\pi_{y_{n_k(x)}} \upharpoonright \beta_l)(\lambda(\cdot))$ converge uniformly in $\lambda(\cdot) \in \beta_l$ as $k \to \infty$ to a function $f_{x,l}(\lambda(\cdot))$ $(\lambda(\cdot) \in \beta)$ continuous on β_l and measurable with respect to the σ-algebra $\{\alpha \in \mathfrak{C}_\sigma(\mathbb{C}^X) \mid \alpha \subset \beta_l\}$. By analogy with Step VIII, for given $f_{x,l}$ we construct a function $f_x(\lambda(\cdot))$ on $\bigcup_{l=1}^{\infty} \beta_l \supset \gamma$, and then prove that

$$(f_x \upharpoonright \gamma)(\lambda(\cdot)) = (\pi_x \upharpoonright \gamma)(\lambda(\cdot)) = \lambda(x) \quad (\lambda(\cdot) \in \gamma). \tag{2.52}$$

The last statement is established by repeating the arguments of Step VIII, and the fact that $\psi_{l,p} = \operatorname{supp}(\rho \upharpoonright \varphi_{l,p})$ is essentially exploited. As in Step IX, we now easily verify that the relations (2.24) hold for $x \in X$, $\lambda(\cdot) \in \gamma$, $\varphi \in H_+$, and $\psi \in D$: We first write these relations with x replaced by $y_{n_k(x)}$ for $\varphi, \psi \in L$ and then proceed to the limit as $k \to \infty$ and use (2.52) and Remark 1 in Subsection 2.6.

Thus, the first part of Theorem 2.8, which corresponds to the first part of Theorem 2.7, is established. To prove the second part, we modify the RI E by $\pi = \gamma$ and apply (1.19) to the formulas (1.25). The last part is proved just as in Theorem 2.7 (Step IX of the proof).

■

Note that in the case when all the operators A_x in the family $A = (A_x)_{x \in X}$ are bounded, the proof of Theorem 2.8 becomes even simpler. Indeed, as explained in Subsection 1.9, we can now regard the compact set $\times_{x \in X} E_x$ instead of the space $\mathbb{C}^X$. Therefore, the elementary Theorem 1.5 can be employed in this proof instead of Theorem 2.6. In this case, most of the constructions presented in Section 1 are not necessary.

2.9. Continuity and Smoothness of Eigenvalues

Under one simply formulated assumption, the functions $X \ni x \mapsto \lambda(x) \in \mathbb{C}^1$ from the generalized spectrum $g(A)$ are continuous or even smooth. In this case, the set $\tau \supset g(A)$ in Remark 1 to Theorem 2.7 can be chosen to be only continuous (smooth) functions, and the integrals in (2.28) actually become "functional" integrals.

Thus, suppose that the index set X is a topological space, and the operators A_x depend continuously on $x \in X$ in the following sense: For every $\varphi \in D$, the vector function

$$X \ni x \mapsto A_x^* \varphi \in H_+ \tag{2.53}$$

is weakly continuous.

In this case, each eigenvalue $\lambda(\cdot) \in g(A)$ *is a continuous function.*

In fact, let $\xi \in H_-$ be a generalized joint eigenvector of the family A with the eigenvalue $\lambda(\cdot)$. Choose $\varphi \in D$ so that $(\xi, \varphi)_{H_0} \neq 0$. Then

$$\lambda(x) = (\xi, A_x^* \varphi)_{H_0} (\xi, \varphi)_{H_0}^{-1}$$

by virtue of the first equality in (2.19), and our assumption (2.53) involves the continuity of $\lambda(\cdot)$.

■

In exactly the same way, one can prove that if X is a differentiable manifold, and (2.53) is k times weakly continuously differentiable, then $\lambda(\cdot) \in g(A)$ is k times continuously differentiable.

Instead of the weak continuity (differentiability) of the function in (2.53), we can demand that the same property holds with A_x^* replaced by A_x. The conclusions remain valid; one should use the second equality (2.19).

■

2.10. Supplementary Remarks. Three Classical Examples of Expansion

Remark 1. Theorems 2.7 and 2.8 are preserved if D is a separable linear topological space such that D' is also separable. This follows from what has been said on Steps I and II of the proof of Lemma 2.5.

■

Remark 2. Assume that the family of commuting normal operators is connected in the standard way, not with a quasinuclear rigging, but with a nuclear rigging (2.3). The statements of Theorems 2.6–2.8 and those given in Subsection 2.9 are preserved in this case, provided that ρ is regarded as some spectral measure of the RI E, and the corresponding definition of a generalized joint eigenvector is used (see Subsection 2.5). Indeed, the situation can be now reduced to the case of a rigging (2.18) just as explained in the proof of Theorem 2.5.

■

Remark 3. Suppose that the family $A = (A_x)_{x \in X}$ of commuting normal operators is connected in the standard way with the rigging (2.18) in the sense that only the inclusion $A_x \upharpoonright D \in \mathfrak{L}(D, H_+)$ $(x \in X)$ holds (or $A_x^* \upharpoonright D \in \mathfrak{L}(D, H_+)$ $(x \in X)$). Then the results remain valid, except that in the definition (2.19) only the second (first) equality holds. An analogous situation occurs also in the case of a nuclear rigging.

■

Remark 4. In the case of a family of commuting selfadjoint operators, the space $\mathbb{C}^X$ can be replaced by $\mathbb{R}^X$ in the results of this section.

Let us present several classical examples of expansions (for a single selfadjoint operator).

Example 2.5 (Jacobian matrices).

Consider a difference expression $\mathcal{L}$ of the second order which acts on the sequences $x = (x_j)_{j=0}^{\infty}$ $(x_j \in \mathbb{C}^1)$ according to the law

$$(\mathcal{L}_x)_j = a_{j-1}x_{j-1} + b_j x_j + a_j x_{j+1} \quad (j \in \mathbb{Z}_+). \tag{2.54}$$

Here $a_j > 0$ and $b_j \in \mathbb{R}^1$ $(j \in \mathbb{Z}_+)$ are given coefficients. We assume that in (2.54) $x_{-1} = 0$ when $j = 0$; this plays the role of the boundary condition. The action of (2.54) can be rewritten as an action of a semi-infinite Jacobian matrix

$$\mathcal{L} = \begin{bmatrix} b_0 & a_0 & 0 & 0 & \dots & \dots & \dots & \dots \\ a_0 & b_1 & a_1 & 0 & 0 & \dots & \dots & \dots \\ 0 & a_1 & b_2 & a_2 & 0 & 0 & \dots & \dots \\ 0 & 0 & a_2 & b_3 & a_3 & 0 & 0 & \dots \\ \dots & \dots & \dots & \dots & \dots & \dots & \dots & \dots \end{bmatrix}. \tag{2.55}$$

The expression $\mathcal{L}$ generates the Hermitian operator A' acting in the space $H = l_2 = l_2(\mathbb{C}^1)$ of sequences $x = (x_j)_{j=0}^{\infty}$, namely, on the set $\mathbb{C}_0^{\infty}$ of finite sequences (in the vicinity of ∞) we set

$$l_2 \supset \mathbb{C}_0^{\infty} = \mathfrak{D}(A') \ni x \mapsto \mathcal{L}x \in \mathbb{C}_0^{\infty} \subset l_2$$

and let $A = (A')^{\sim}$. Under certain restrictions imposed on a_j and b_j the operator A is selfadjoint (for example, if $\sum_{j=0}^{\infty} a_j^{-1} = \infty$). Let us construct the expansion in generalized joint eigenvectors (sequences) of this operator. We take the chain

$$H_- = l_2(p^{-1}) \supset l_2 \supset l_2(p) = H_+ \supset \mathbb{C}_0^{\infty} = D \tag{2.56}$$

to play the role of (2.18); here $l_2\,(p)$ is a space l_2 with the weight $p = (p_j)_{j=1}^{\infty}$ $(p_j \ge 1)$ satisfying the following condition $\sum_{j=0}^{\infty} p_j^{-1} < \infty$, $p^{-1} = (p_j^{-1})_{j=0}^{\infty}$. Providing $\mathbb{C}_0^{\infty}$ with a topology, which guarantees pointwise convergence and the uniform finiteness of a convergent sequence, we find that (2.56) is a quasinuclear chain connected with A in the standard way (see Chapter 1, Section 1, Example 1.6).

Therefore, for the operator A under consideration and for its RI E, the formulas (2.26) hold. Let us describe $P(\lambda)$ and ρ from these formulas in detail. Since $P(\lambda): l_2(p) \to l_2(p^{-1})$ is positive (and, moreover, $\| P(\lambda) \| \le \operatorname{Tr}(P(\lambda)) = 1$), this operator is given by the positive matrix $(P_{jk}(\lambda))_{j,k=0}^{\infty}$:

$$(P(\lambda)\varphi)_j = \sum_{k=0}^{\infty} P_{jk}(\lambda)\,\varphi_k \quad (\varphi \in \mathbb{C}_0^{\infty});$$

$$\sum_{j,k=0}^{\infty} |P_{jk}(\lambda)|^2 p_j^{-1} p_k^{-1} \le \sum_{j=0}^{\infty} P_{jj}(\lambda) p_j^{-1} = 1. \tag{2.57}$$

Taking into account (2.27), which is valid for $\xi \in \mathfrak{R}(P(\lambda))$, and the fact that the matrix $(P_{jk}(\lambda))_{j,k=0}^{\infty}$ is Hermitian, we conclude that $P_{jk}(\lambda)$, regarded as a function of variable j or of variable k, in both cases satisfies the equation $(\mathcal{L}u)_j = \lambda u_j$ $(j \in \mathbb{Z}_+)$ and the boundary condition $u_{-1} = 0$.

Denote by $u_n = P_n(\lambda)$ the solution of the equation

$$(\mathcal{L}u)_n = \lambda u_n \quad (n \in \mathbb{Z}_+;\ \lambda \in \mathbb{C}^1) \tag{2.58}$$

with the initial conditions $u_{-1} = 0$ and $u_0 = 1$. This solution can be found recursively. It is clear that $P_n(\lambda)$ is a polynomial of the n-th power $(n \in \mathbb{Z}_+)$ with respect to λ (the so-called polynomial of the first kind connected with (2.55)). Any solution $v = (v_n)_{n=0}^{\infty}$ of equation (2.58) satisfying the condition $v_{-1} = 0$ is a multiple of $P_n(\lambda)$, namely, $v_n = v_0 P_n(\lambda)$ $(n \in \mathbb{Z}_+)$. Therefore,

$$P_{jk}(\lambda) = P_j(\lambda) P_k(\lambda) P_{00}(\lambda) \quad (j, k \in \mathbb{Z}_+).$$

If we substitute this expression to (2.57) and denote $P_{00}(\lambda)\, d\rho(\lambda)$ again by $d\rho(\lambda)$ (note that $P_{00}(\lambda) > 0$), then, by virtue of (2.26), we obtain (for $\alpha = \mathbb{R}^1$)

$$(f, g)_{l_2} = \int_{-\infty}^{\infty} \tilde{f}(\lambda)\, \overline{\tilde{g}(\lambda)}\, d\rho(\lambda), \quad \tilde{f}(\lambda) = \sum_{j=0}^{\infty} f_j P_j(\lambda) \quad (f, g \in l_2(p)),$$

$$(\mathcal{L}f)_j = \int_{-\infty}^{\infty} \lambda\, \tilde{f}(\lambda)\, d\rho(\lambda) \quad (f \in \mathbb{C}_0^{\infty}). \tag{2.59}$$

These are just the classical formulas of expansion connected with the Jacobian matrix (2.55). It is also clear from (2.26) and (2.59) how to write $E(\alpha)$ in terms of $P_j(\lambda)$. According to the first equality in (2.59), the mapping

$$l_2(p) \ni f \mapsto \tilde{f} \in L_2(\mathbb{R}^1, d\rho(\lambda)) = L_2(\mathbb{R}^1, \mathfrak{B}(\mathbb{R}^1), \rho)$$

is an isometry. Moreover, it is easy to see that its image is dense in $L_2(\mathbb{R}^1, d\rho(\lambda))$. Hence, extending this isometry by continuity to the whole of l_2, we obtain the isometry $l_2 \ni x \mapsto \tilde{x} \in L_2(\mathbb{R}^1, d\rho(\lambda))$ which is the Fourier transform with respect to the generalized eigenvectors of the operator A generated by $\mathcal{L}$. It follows from (2.59) that $(P_j(\lambda))_{j=0}^{\infty}$ form an orthonormal basis in $L_2(\mathbb{R}^1, d\rho(\lambda))$, that $\Omega = (1, 0, 0, \ldots) = \delta_0$ is a cyclic vector for A (see Subsection 2.3), and that

$$\rho(\alpha) = (E(\alpha)\,\delta_0, \delta_0)_{l_2} \quad (\alpha \in \mathfrak{B}(\mathbb{R}^1)).$$

In Subsection 3.1, we shall return to the analogues of formulas of the type (2.59) for the general situation of the family $A = (A_x)_{x \in X}$. The other derivations of the relations (2.59) and the generalizations of (2.59) for other difference expressions $\mathcal{L}$ can be found in Berezansky [5, Chapter 7]. Note that these formulas can be obtained also in the case when the operator A' is not essentially selfadjoint, and a certain selfadjoint extension of the operator A is considered instead of the operator itself.

Example 2.6 (the Schrödinger operator).
Consider the selfadjoint Schrödinger operator A defined in Example 2.4 as acting in the space $H_0 = L_2(G, dx)$ according to the differential expression

$$(\mathcal{L}u)(x) = -(\Delta u)(x) + q(x)\,u(x)$$

and zero boundary conditions on ∂G (if $G \neq \mathbb{R}^d$). Let $l = 2, 3, \ldots, l > d/2$, and $q \in C^l(G)$, and let ∂G belong to the class $C^{2+l}(G)$. One can choose a weight $p \in C^\infty(\tilde{G})$, $p(x) \geq 1$ $(x \in G)$ increasing towards ∞ (as $|x| \to \infty$) so fast that the imbedding $W_l^2(G, p(x)\,dx) \subset L_2(G, dx)$ is quasinuclear (see Chapter 1, Section 1, Example 1.5; Berezansky [18, Chapter 1, Subsection 3.8]). Then the role of (2.18) can be played by the quasinuclear chain

$$H_- = W_2^{-l}(G, p(x)\,dx) \supset L_2(G, dx) \supset W_l^2(G, p(x)\,dx) = H_+ \supset D, \qquad (2.60)$$

where D has been introduced in Example 2.4. The operator A is connected with (2.60) in the standard way.

The operator

$$P(\lambda)\colon W_l^2(G, p(x)\,dx) \to W_2^{-l}(G, p(x)\,dx),$$

appearing in (2.26) is an integral operator (as (2.57)), namely, for $f \in L_2(G, dx)$ which are finite at infinity, we have

$$(P(\lambda)f)(x) = \int_G P(x, y; \lambda)\, f(y)\, dy \quad (g \in G). \qquad (2.61)$$

The positive definite kernel $P(x, y; \lambda)$ is called a spectral kernel of the operator A. Regarded as a function of the variable (x, y), it belongs to the Sobolev space W_2^{2+l} on each bounded domain $G \times G$ and satisfies the equations

$$(\mathcal{L}P(\cdot, y; \lambda))(x) = \lambda P(x, y; \lambda),$$

$$(\mathcal{L}P(x, \cdot; \lambda))(y) = \lambda P(x, y; \lambda) \quad (x, y \in G) \tag{2.62}$$

with the boundary condition $P(x, y; \lambda) = 0$ for $x \in \partial G$ and $y \in \tilde{G}$ or $x \in \tilde{G}$ and $y \in \partial G$. This result follows from the fact that $\xi \in \mathfrak{R}(P(\lambda))$ satisfies (2.22) and is thus regular up to the boundary as a generalized solution of an elliptic equation (here we also employ the selfadjointness of $P(\lambda): H_+ \to H_-$). This fact is proved in the book by Berezansky [5, Chapter 6, Subsection 3].

Note that the restrictions of smoothness imposed on q can be made weaker. Thus, it suffices to assume that $q \in C(\tilde{G})$. This is connected with another approach to the theory of expansions which we shall touch upon in Subsection 3.2 (the theory of Carleman operators). In the framework of this approach, a chain of the type (2.56) appears instead of the chain (2.60) when the role of a positive space is played by the space $L_2(G, dx)$ with a weight.

The generalization of the results presented in this example to more general differential expressions $\mathcal{L}$ and boundary conditions can be found in Berezansky [5, Chapter 6]. The representation (2.61) also holds in the case when the potential q is such that the operator A' is only Hermitian. Then one should take some selfadjoint extension of this operator to play the role of A.

Example 2.7 (the Sturm–Liouville operator).

This operator is a special case of the Schrödinger operator (Examples 2.4 and 2.6) for $d = 1$. Here

$$(\mathcal{L}u)(x) = -u''(x) + q(x)u(x) \quad (x \in G \subset \mathbb{R}^1)$$

is an ordinary differential expression and, therefore, the equation $\mathcal{L}u = \lambda u$ possesses the fundamental system of solutions in terms of which any other solution can be expressed. This enables us to obtain formulas of the type (2.59).

First, consider the case of the semi-axis $G = \mathbb{R}^1_+ = (0, \infty)$ and null boundary condition at the point $\{0\} = \partial G$. Denote by $\psi(x; \lambda)$ the solution of the equation

$$(\mathcal{L}u)(x) = \lambda u(x) \qquad (x \in \mathbb{R}^1_+; \lambda \in \mathbb{C}^1), \tag{2.63}$$

satisfying the initial conditions $u(0) = 0$ and $u'(0) = 1$ (the solution of the sine type). Any solution v of this equation satisfying the condition $v(0) = 0$ has the form

$$v(x) = v'(0)\,\psi(x; \lambda) \quad (x \in [0, \infty)).$$

By virtue of (2.62) and the equality

$$P(0, y; \lambda) = P(x, 0; \lambda) = 0 \quad (x, y \in [0, \infty))$$

we have

$$P(x, y; \lambda) = c(\lambda)\,\psi(x; \lambda)\,\psi(y; \lambda) \quad (x, y \in [0, \infty)),$$

where $c(\lambda) = (\partial^2/\partial x\,\partial y\,P)(0, 0; \lambda)$. Substituting this expression to (2.61), denoting $c(\lambda)d\rho(\lambda)$ again by $d\rho(\lambda)$ (it is easy to understand that $c(\lambda) > 0$), and taking into account (2.26), we obtain the classical formulas for the expansion in eigenfunctions of the Sturm–Liouville operator

$$(f, g)_{L_2(\mathbb{R}^1_+,\, dx)} = \int_{-\infty}^{\infty} \tilde{f}(\lambda)\,\overline{\tilde{g}(\lambda)}\,d\rho(\lambda),$$

$$\tilde{f}(\lambda) = \int_{-\infty}^{\infty} f(x)\,\psi(x; \lambda)\,dx \quad (f, g \in L_{2,0}(\mathbb{R}^1_+, dx)),$$

$$(\mathcal{L}f) = \int_{-\infty}^{\infty} \lambda\,\tilde{f}(\lambda)\,\psi(x; \lambda)\,d\rho(\lambda) \quad (f \in C_0^2(\mathbb{R}^1_+)). \tag{2.64}$$

Here $L_{2,0}(\mathbb{R}^1_+, dx)$ and $C_0^2(\mathbb{R}^1_+)$ denote the classes of functions which are finite at ∞ and belong to the spaces $L_2(\mathbb{R}^1_+, dx)$ and $C^2(\mathbb{R}^1_+)$, respectively. As in the case of Jacobian matrices, the mapping

$$L_{2,0}(\mathbb{R}^1_+, dx) \ni f \mapsto \tilde{f} \in L_2(\mathbb{R}^1_+, d\rho(\lambda)) = L_2(\mathbb{R}^1, \mathfrak{B}(\mathbb{R}^1), \rho)$$

can be extended by continuity to the isometry between $L_2(\mathbb{R}^1_+, dx)$ and $L_2(\mathbb{R}^1_+, d\rho(\lambda))$.

In analogy with Examples 2.4 and 2.6, we may investigate the Schrödinger operator, not with the null boundary condition, but with the condition $(\partial u/\partial n)(x) = 0$ $(x \in \partial G)$, where $\partial/\partial n$ is a derivative with respect to the outer normal to ∂G. In the case of the Sturm–Liouville operator on the semi-axis $\mathbb{R}^1_+$, this means that the operator A' is constructed by using the functions f such that $f'(0) = 0$. All the results presented above remain valid, only in (2.64) one should replace the solution $\psi(x; \lambda)$ by the solution $\varphi(x;\lambda)$ of equation (2.63) satisfying the initial conditions $u(0) = 1$ and $u'(0) = 0$ (the solution of the cosine type; here $c(\lambda) = P(0, 0; \lambda)$).

Consider the case of the whole axis, namely, $G = \mathbb{R}^1$. Denote by $\chi_0(x; \lambda)$ the solution of equation (2.63), where $x \in \mathbb{R}^1$, satisfying the initial conditions $u(0) = 0$ and $u'(0) = 1$; and let $\chi_1(x; \lambda)$ be the analogous solution satisfying the initial conditions $u(0) = 1$ and $u'(0) = 0$. Any solution u of equation (2.63) has the form

$$u(x) = u'(0)\,\chi_0(x; \lambda) + u(0)\,\chi_1(x; \lambda) \quad (x \in \mathbb{R}^1).$$

Therefore, if we expand the kernel $P(x, y; \lambda)$ $(x, y \in \mathbb{R}^1)$, satisfying equations (2.62), in the solutions χ_0 and χ_1, then there appear four terms with the corresponding coefficients (we stress that this kernel does not satisfy any boundary conditions at the point 0. Generally speaking, one could take any point $a \in \mathbb{R}^1$ instead of 0). A simple calculation leads us to the following formulas for expansion

$$(f, g)_{L_2(\mathbb{R}^1_+, dx)} = \sum_{j,k=0}^{1} \int_{-\infty}^{\infty} \tilde{f}_k(\lambda) \overline{\tilde{g}_j(\lambda)} \, d\rho_{jk}(\lambda),$$

$$\tilde{f}_j(\lambda) = \int_{-\infty}^{\infty} f(x) \chi_j(x; \lambda) \, dx \quad (f, g \in L_{2,0}(\mathbb{R}^1, dx), \ j = 1, 2),$$

$$(\mathcal{L}f_j) = \sum_{k=0}^{1} \int_{-\infty}^{\infty} \lambda f_k(\lambda) \chi_j(x; \lambda) \, d\rho_{jk}(\lambda) \quad (f \in C_0^2(\mathbb{R}^1)). \tag{2.65}$$

Here $(d\rho_{jk}(\lambda))_{j,k=0}^{1}$ is a so-called spectral matrix; the elements of it are given by

$$d\rho_{jk}(\lambda) = ((\partial/\partial x)^j (\partial/\partial y)^k P)(0, 0; \lambda) \, d\rho(\lambda).$$

The fact that the operator $P(\lambda)$ is positive implies the positive definiteness of the kernel $P(x, y; \lambda)$ and, thus, the positive definiteness of the spectral matrix. Therefore, the operator-valued measure

$$\mathfrak{B}(\mathbb{R}^1) \ni \alpha \mapsto (d\rho_{jk}(\lambda))_{j,k=0}^{1} = \theta(\alpha)$$

is defined; the values of it are nonnegative operators acting in $\mathbb{C}^2$ (written in the matrix form). By analogy with the ordinary space $L_2(\mathbb{R}^1, d\rho(\lambda)) = L_2(\mathbb{R}^1, \mathfrak{B}(\mathbb{R}^1), \rho)$, we can construct a space

$$L_2(\mathbb{R}^1, d\theta(\lambda); \mathbb{C}^2) = L_2(\mathbb{R}^1, \mathfrak{B}(\mathbb{R}^1), \theta; \mathbb{C}^2),$$

consisting of the vector functions $\mathbb{R}^1 \ni \lambda \mapsto F(\lambda) \in \mathbb{C}^2$ "square summable with respect to θ". Then the relations (2.65) may be interpreted as an isometry

$$L_{2,0}(\mathbb{R}^1, dx) \ni f \mapsto \tilde{f}(\lambda) = (\tilde{f}_0(\lambda), \tilde{f}_1(\lambda)) \in L_2(\mathbb{R}^1, d\theta(\lambda); \mathbb{C}^2),$$

which can be extended by continuity to an isometry between $L_2(\mathbb{R}^1, dx)$ and $L_2(\mathbb{R}^1, d\theta(\lambda); \mathbb{C}^2)$. A more detailed presentation of this method of derivation (2.65) and generalizations of these formulas can be found in Berezansky [5, Chapter 6, Section 5].

The detailed investigation of formulas (2.26) and (2.62), which gives us (2.64) and (2.65), is possible due to the presence of the fundamental system of solutions for the ordinary equation (2.63). For the case of partial derivatives, this procedure is impossible; a somewhat more detailed study of these formulas can be found in Subsections 3.1 and 3.2.

§3. Supplementary Facts about Expansions

Consider some facts which are complementary to the spectral projection theorem which has been proved in Section 2:

(i) we diagonalize the operator $P(\lambda)$; this yields the decomposition of the original Hilbert spaces into the direct integral of "eigensubspaces";

(ii) we investigate the possibilities of constructing expansions in the case when the imbedding $H_+ \subset H_0$ is not quasinuclear;

(iii) we prove that the necessary condition for the existence of "good enough" spectral theorems for the family A is the existence of the quasinuclear chain connected with A in the standard way.

As an illustration, we consider two examples of expansions (the Fourier–Wiener transformation and the Segal isomorphism).

3.1. The Fourier Transform. The Direct Integral of Hilbert Spaces

Assume that the conditions, ensuring the validity of spectral projection Theorem 2.7 for the family $A = (A_x)_{x \in X}$, hold. Let us choose $\mathbb{C}^X \supset \tau \supset \pi$. By virtue of (2.28) (see Remark 1 in Subsection 2.7), we have

$$(E_\tau(\alpha)\varphi, \psi)_{H_0} = \int_\alpha (P(\lambda(\cdot))\varphi, \psi)_{H_0} \, d\rho_\tau(\lambda(\cdot))$$

$$(\varphi, \psi \in H_+; \ \alpha \in \mathfrak{C}_\sigma(\tau)). \tag{3.1}$$

In particular, $E_\tau(\tau) = 1$ for $\alpha = \tau$, and (3.1) yields the decomposition of $(\varphi, \psi)_{H_0}$ into the integral of $(P(\lambda(\cdot))\varphi, \psi)_{H_0}$. The last expression defines a scalar product, and (3.1) becomes a decomposition of H_0 into a direct integral of the corresponding Hilbert spaces. Let us dwell upon this in more detail.

Fix $\lambda(\cdot) \in \tau$ such that $\operatorname{Tr}(P(\lambda(\cdot))) = 1$. Clearly, these $\lambda(\cdot)$ form a set of full ρ_τ-measure. By virtue of the relations $P(\lambda(\cdot)) \geq 0$ and

$$\| P(\lambda(\cdot)) \| \leq \operatorname{Tr}(P(\lambda(\cdot))) = 1,$$

the operator $JP(\lambda(\cdot))J\colon H_0 \to H_0$ is a nonnegative Hilbert–Schmidt operator (see (2.1) and (2.2)). Let $h_\gamma(\lambda(\cdot)) \in H_0$ ($\gamma = 1, 2, \ldots, N_{\lambda(\cdot)} \leq \infty$) be an orthonormal sequence of eigenvectors of the operator $JP(\lambda(\cdot))J$ corresponding to the eigenvalues $\nu_\gamma(\lambda(\cdot))$. The dependence of $JP(\lambda(\cdot))J$ on the parameter $\lambda(\cdot)$ is weakly $\mathfrak{C}_\sigma(\tau)$-measurable, hence, as is well known, we can assume that $h_\gamma(\lambda(\cdot))$ is (weakly) measurable and $\nu_\gamma(\lambda(\cdot))$ is measurable ($\gamma = 1, 2, \dots, N_{\lambda(\cdot)}$). Then

$$(P(\lambda(\cdot))Jf, Jg)_{H_0} = (JP(\lambda(\cdot))Jf, g)_{H_0}$$

$$= \sum_{\gamma=1}^{N_{\lambda(\cdot)}} \nu_\gamma(\lambda(\cdot))\,(f, h_\gamma(\lambda(\cdot)))_{H_0}\,\overline{(g, h_\gamma(\lambda(\cdot)))_{H_0}}$$

$$= \sum_{\gamma=1}^{N_{\lambda(\cdot)}} (Jf, \xi_\gamma(\lambda(\cdot)))_{H_0}\,\overline{(Jg, \xi_\gamma(\lambda(\cdot)))_{H_0}} \quad (f, g \in H_0), \tag{3.2}$$

where

$$\xi_\gamma(\lambda(\cdot)) = \sqrt{\nu_\gamma(\lambda(\cdot))}\, J^{-1} h_\gamma(\lambda(\cdot))$$

$$= P(\lambda(\cdot))\,((\nu_\gamma(\lambda(\cdot)))^{-1/2} J h_\gamma(\lambda(\cdot))) \in \mathfrak{R}(P(\lambda(\cdot))) \subset H_-.$$

The vectors $\xi_\gamma(\lambda(\cdot))$ ($\lambda(\cdot) \in \tau$; $\gamma = 1, 2, \dots, N_{\lambda(\cdot)} \le \infty$) are "individual" generalized joint eigenvectors of the family $A = (A_x)_{x \in X}$.

It follows from (2.51) that

$$(P(\lambda(\cdot))\varphi, \psi)_{H_0} = \sum_{\gamma=1}^{N_{\lambda(\cdot)}} (\varphi, \xi_\gamma(\lambda(\cdot)))_{H_0}\,\overline{(\psi, \xi_\gamma(\lambda(\cdot)))_{H_0}} \quad (\varphi, \psi \in H_+) \tag{3.3}$$

for ρ_τ-almost every $\lambda(\cdot) \in \tau$.

Denote $l_2(\infty) = l_2 = l_2(\mathbb{C}^1)$ and $l_2(N) = \mathbb{C}^N$ ($N < \infty$), where the latter space is regarded as imbedded in l_2 (all the coordinates of the vector beginning with the $(N+1)$-th are zero). The mapping

$$H_+ \ni \varphi \mapsto \tilde{\varphi}(\lambda(\cdot)) = (\tilde{\varphi}_1(\lambda(\cdot)), \tilde{\varphi}_2(\lambda(\cdot)), \dots) \in l_2(N_{\lambda(\cdot)}), \tag{3.4}$$

$$\tilde{\varphi}_\gamma(\lambda(\cdot)) = (\varphi, \xi_\gamma(\lambda(\cdot)))_{H_0} \quad (\gamma = 1, 2, \dots, N_{\lambda(\cdot)})$$

is called the Fourier transform corresponding to the family A (the inclusion in $l_2(N_{\lambda(\cdot)})$ follows from (3.3) with $\psi = \varphi$). The Fourier transform $\tilde{\varphi}(\lambda(\cdot))$ of a vector φ is defined for ρ_τ-almost all $\lambda(\cdot) \in \tau$, and each of its coordinates $\tilde{\varphi}_\gamma(\lambda(\cdot))$ is $\mathfrak{C}_\sigma(\tau)$-measurable.

Substituting (3.3) in (3.1) and employing the notation (3.4), we obtain the Parseval equality for the Fourier transform

$$(E_\tau(\alpha)\varphi, \psi)_{H_0} = \int_\alpha (\tilde{\varphi}(\lambda(\cdot)), \tilde{\psi}(\lambda(\cdot)))_{l_2(N_{\lambda(\cdot)})} d\rho_\tau(\lambda(\cdot))$$

$$(\varphi, \psi \in H_+;\ \alpha \in \mathfrak{C}_\sigma(\tau)). \tag{3.5}$$

Let us note the important fact:

For fixed $\lambda(\cdot)$, *the set* $\{\tilde{\varphi}(\lambda(\cdot)) | \varphi \in H_+\}$ *coincides with* $l_2(N_{\lambda(\cdot)})$ *for* $N_{\lambda(\cdot)} < \infty$ *and contains all the finite vectors from* l_2 *when* $N_{\lambda(\cdot)} = \infty$.

Indeed, it suffices to show that every vector $(0, \ldots, 0, 1, 0, \ldots)$ with the unit at the κ-th place belongs to this set. We set

$$\varphi = (\nu_\kappa(\lambda(\cdot)))^{-1/2} J h_\kappa(\lambda(\cdot)) \in H_+,$$

then

$$(\varphi, \xi_\gamma(\lambda(\cdot)))_{H_0} = (\nu_\kappa(\lambda(\cdot)))^{-1/2} (J h_\kappa(\lambda(\cdot)), \xi_\gamma(\lambda(\cdot)))_{H_0}$$

$$= (\nu_\kappa(\lambda(\cdot)))^{-1/2} (h_\kappa(\lambda(\cdot)), J\xi_\gamma(\lambda(\cdot)))_{H_0} = (h_\kappa(\lambda(\cdot)), h_\gamma(\lambda(\cdot)))_{H_0} = \delta_{\kappa\gamma}$$

$$(\gamma = 1, 2, \ldots, N_{\lambda(\cdot)}).$$

■

Consider the direct integral of the Hilbert spaces $l_2(N_{\lambda(\cdot)})$ over τ (with respect to the measure ρ_τ), namely,

$$L_2 = \int_\tau \oplus\, l_2(N_{\lambda(\cdot)})\, d\rho_\tau(\lambda(\cdot)).$$

It is defined as the collection of all vector-valued functions

$$\tau \ni \lambda(\cdot) \mapsto F(\lambda(\cdot)) \in l_2(N_{\lambda(\cdot)}),$$

defined for ρ_τ-almost all $\lambda(\cdot)$, $\mathfrak{C}_\sigma(\tau)$-measurable in the sense that each coordinate $F_\gamma(\lambda(\cdot))$ $(\gamma = 1, 2, \ldots, N_{\lambda(\cdot)})$ is measurable, and such that

$$\int_\tau \| F(\lambda(\cdot)) \|^2_{l_2(N_{\lambda(\cdot)})} d\rho_\tau(\lambda(\cdot)) < \infty.$$

The direct integral is a Hilbert space with the scalar product

$$(F(\cdot), G(\cdot))_{L_2} = \int_{\tau} (F(\lambda(\cdot)), G(\lambda(\cdot)))_{l_2(N_{\lambda(\cdot)})} d\rho_\tau(\lambda(\cdot)) \tag{3.6}$$

$$(F(\cdot), G(\cdot) \in L_2).$$

Comparing (3.5) and (3.6), we conclude that for $\alpha = \tau$ the expression on the right-hand side of (3.5) defines the scalar product in the direct integral; therefore, the Parseval equality can be written in the form

$$(\varphi, \psi)_{H_0} = (\tilde{\varphi}(\cdot), \tilde{\psi}(\cdot))_{L_2} \quad (\varphi, \psi \in H_+).$$

Extending this equality by continuity to the whole of H_0, we obtain it in the form

$$(f, g)_{H_0} = \int_{\tau} (\tilde{f}(\lambda(\cdot)), \tilde{g}(\lambda(\cdot)))_{l_2(N_{\lambda(\cdot)})} d\rho_\tau(\lambda(\cdot)) \quad (f, g \in H_0), \tag{3.7}$$

where the Fourier transform $\tilde{f}(\lambda(\cdot))$ of the vector f is understood as the limit of the Fourier transforms (3.4) in the norm of the direct integral (clearly, the last formula of (3.4) cannot be written for $\tilde{f}(\lambda(\cdot))$).

Theorem 3.1. *If D is a core for every operator A_x $(x \in X)$, i.e., if the closure $A_x \upharpoonright D$ coincides with A_x in H_0, then the Fourier transforms $\tilde{\varphi}(\lambda(\cdot))$ $(\varphi \in H_+)$ are dense in the direct integral and, therefore, the Fourier transform $H_0 \ni f \mapsto \tilde{f}(\lambda(\cdot)) \in L_2$ implements an isomorphism between the spaces H_0 and L_2.*

Thus, in this sense, H_0 can be regarded as decomposed into the direct integral

$$H_0 = \int_{\tau} \oplus\, l_2(N_{\lambda(\cdot)})\, d\rho_\tau(\lambda(\cdot)). \tag{3.8}$$

Lemma 3.1. *Under the mapping* (3.4), *each operator $A_x \upharpoonright D$ transforms into the operator of multiplication by $\lambda(x)$, namely, for ρ_τ-almost all $\lambda(\cdot) \in \tau$, we have*

$$(A_x \varphi)^{\sim}(\lambda(\cdot)) = \lambda(x)\, \tilde{\varphi}(\lambda(\cdot))$$

$$= (\lambda(x)\, \tilde{\varphi}_1(\lambda(\cdot)), \lambda(x)\, \tilde{\varphi}_2(\lambda(\cdot)), \ldots) \quad (x \in X, \varphi \in D). \tag{3.9}$$

Proof. It should be established that, for every $x \in X$, $\varphi \in D$, and $\gamma \in \mathbb{N}$, the equality

$$f(x, \varphi, \gamma; \lambda(\cdot)) = (A_x \varphi)^{\sim}_{\gamma}(\lambda(\cdot)) - \lambda(x)\, \tilde{\varphi}_\gamma(\lambda(\cdot)) = 0$$

holds for ρ_τ-almost all $\lambda(\cdot) \in \tau$. Assuming the contrary and employing the fact that this function is $\mathfrak{C}_\sigma(\tau)$-measurable, we find x_0, φ_0, and γ_0 such that $\rho_\tau(\alpha) > 0$, where

$$\alpha = \{\lambda(\cdot) \in \tau \,|\, f(x_0, \varphi_0, \gamma_0; \lambda(\cdot)) \neq 0\} \in \mathfrak{C}_\sigma(\tau).$$

Let $\lambda(\cdot) \in \pi$, where π is the set from Theorem 2.7. According to (3.4), the inclusion $\xi_\gamma(\lambda(\cdot)) \in \mathfrak{R}(P(\lambda(\cdot)))$, Theorem 2.7, and the second equality in (2.19), we have

$$(A_{x_0}\varphi_0)\tilde{}_{\gamma_0}(\lambda(\cdot)) = (A_{x_0}\varphi_0, \xi_{\gamma_0}(\lambda(\cdot)))_{H_0}$$

$$= \lambda(x_0)\,(\varphi_0, \xi_{\gamma_0}(\lambda(\cdot)))_{H_0} = \lambda(x_0)\,(\tilde{\varphi}_0)_{\gamma_0}(\lambda(\cdot)),$$

whence $\lambda(\cdot) \notin \alpha$. Thus, $\pi \subset \tau \setminus \alpha$; but this contradicts the assumption that π is the set of full outer ρ_τ-measure.

■

Proof of the theorem. Without loss of generality, we can assume that the operators A_x are selfadjoint and that $\tau \subset \mathbb{R}^X$ (instead of $(A_x)_{x \in X}$, one may consider the family

$$\left(\frac{1}{2}(A_x + A_x^*),\ \frac{1}{2}i\,(A_x - A_x^*)\right)_{x \in X}$$

of commuting selfadjoint operators for which the corresponding direct integral is isomorphic to (3.6)).

Fix $x \in X$ and nonreal $z \in \mathbb{C}^1$. It is stated that

$$\varphi \in H_+ \ (\lambda(x) - z)^{-1}\, \tilde{\varphi}(\lambda(\cdot)) \in \tilde{H}_0 \subset L_2$$

for every $\varphi \in H_+$. Indeed, the boundedness of $(\lambda(x) - z)^{-1}$, regarded as a function of a point $\lambda(\cdot) \in \tau \subset \mathbb{R}^X$, implies that the operator of multiplication by this function is continuous in L_2 and, hence, according to (3.9) and (3.7), we have

$$\|\,(\lambda(x) - z)^{-1}\, \tilde{\varphi}(\lambda(\cdot)) - \tilde{\psi}(\lambda(\cdot))\|_{L_2} \le c\,\|\,\tilde{\varphi}(\lambda(\cdot)) - (\lambda(x) - z)\,\tilde{\psi}(\lambda(\cdot))\|_{L_2}$$

$$= c\,\|\,\tilde{\varphi}(\lambda(\cdot)) - ((A_x - z\,1)\,\psi)\tilde{}(\lambda(\cdot))\|_{L_2} = c\,\|\,\varphi - (A_x - z\,1)\,\psi\,\|_{H_0}$$

for $\psi \in D$. Since D forms a core of the selfadjoint operator A_x, the right-hand side of this estimate can be made as small as desired, and this proves the required result.

Suppose that $F(\lambda(\cdot)) \in L_2$ is orthogonal to $\tilde{H}_0$ in L_2. Then, in particular, for nonreal $z \in \mathbb{C}^1$ and $\varphi \in H_+$ chosen so that $\tilde{\varphi}(\lambda(\cdot)) = (1, 0, 0, \ldots)$ (this is possible, see the above discussion), we get

$$0 = \int_\tau (F(\lambda(\cdot)), \frac{1}{\lambda(x)-\bar{z}} \tilde{\varphi}(\lambda(\cdot)))_{l_2(N_{\lambda(\cdot)})} d\rho_\tau(\lambda(\cdot))$$

$$= \int_\tau \frac{1}{\lambda(x)-\bar{z}} F_1(\lambda(\cdot))\, d\rho_\tau(\lambda(\cdot)) = \int_{\mathbb{R}^1} \frac{1}{\lambda-\bar{z}}\, d\omega_x(\lambda),$$

where the complex-valued measure is given by

$$\omega_x(\delta) = \int_{\mathbb{C}(x;\delta)\cap\tau} F_1(\lambda(\cdot))\, d\rho_\tau(\lambda(\cdot)) \quad (\delta \in \mathfrak{B}(\mathbb{R}^1)).$$

Since z is arbitrary, we conclude that $\omega_x = 0$, and the arbitrariness of x implies that $F_1(\lambda(\cdot)) = 0$ ρ_τ-almost everywhere. An analogous argument can be applied to $F_2(\lambda(\cdot))$, and so on. Finally, we get that $F(\lambda(\cdot)) = 0$ in L_2 and, consequently, $\tilde{H}_0 = L_2$.

■

It follows from the above discussion that $N_{\lambda(\cdot)} = \dim \mathfrak{R}(P(\lambda(\cdot)))$, i.e., it is the "multiplicity" of the eigenvalue $\lambda(\cdot)$. In the case when $N_{\lambda(\cdot)} = 1$ for ρ_τ-almost all $\lambda(\cdot)$, i.e., when "the spectrum is simple", we have

$$\tilde{\varphi}(\lambda(\cdot)) = \tilde{\varphi}_1(\lambda(\cdot)) = (\varphi, \xi_1(\lambda(\cdot)))_{H_0} \in \mathbb{C}^1 \qquad (\varphi \in H_+),$$

and (3.8) gives the decomposition of H_0 into the direct integral of complex planes $\mathbb{C}^1 = l_2(1)$.

The expansion in the generalized eigenvectors of a Jacobian matrix or of a Sturm–Liouville operator on the semi-axis (see Examples 2.5 and 2.7) are the special cases of a decomposition of this sort with $N_{\lambda(\cdot)} = 1$. At the same time, in the case of the Sturm–Liouville operator on the whole axis, expansion (2.65) does not coincide with (3.8) (this is, as a rule, true in the case of more general ordinary differential operators). In fact, the spectral matrix $(d\rho_{jk})_{j,k=0}^1$ is not diagonalized. On the other hand, the transformation $f \mapsto \tilde{f}(\lambda)$ from Example 2.7 is "better" than the transformation (3.4), because here (in the case of a single operator A) $\tilde{\varphi}(\lambda)$, in general, does not depend on λ analytically.

Theorem 3.2. *Assume that the family $A = (A_x)_{x\in X}$ possesses a cyclic vector $\Omega \in D$ such that $A_{x_1}^{m_1} \cdots A_{x_p}^{m_p} \Omega \in D$ for all the products of the operators $A_{x_1}^{m_1} \cdots A_{x_p}^{m_p}$ in the definition of Ω, and the linear span of these vectors is dense not only in H_0 but* *also in H_+. Then the spectrum of this family is simple.*

A cyclic vector defined in Theorem 3.2 is called a *strong cyclic vector.*

Note that in the case of bounded operators A_x $(x \in X)$, the spectrum is simple if the indicated linear span is dense in H_0 (see Subsection 3.5).

Proof. Assume that $N_{\lambda(\cdot)} > 1$ for a set of $\lambda(\cdot)$ of positive ρ_τ-measure. Then there exists $\lambda(\cdot) \in \tau$ such that $N_{\lambda(\cdot)} > 1$, $\mathrm{Tr}(P(\lambda(\cdot))) = 1$, and $\Re(P(\lambda(\cdot)))$ consists of generalized joint eigenvectors corresponding to $\lambda(\cdot)$. For $f \in H_0$ and $g = J^{-1}\Omega$, it follows from (3.2) that

$$(P(\lambda(\cdot))\, Jf, \Omega)_{H_0} = \sum_{\gamma=1}^{N_{\lambda(\cdot)}} \nu_\gamma(\lambda(\cdot))\, (f, h_\gamma(\lambda(\cdot)))_{H_0}\, \overline{(J^{-1}\Omega, h_\gamma(\lambda(\cdot)))_{H_0}}$$

$$= \sum_{\gamma=1}^{N_{\lambda(\cdot)}} (f, h_\gamma(\lambda(\cdot)))_{H_0}\, a_\gamma. \tag{3.10}$$

Since $(h_\gamma(\lambda(\cdot)))_{\gamma=1}^{N_{\lambda(\cdot)}}$ is an orthonormal sequence in H_0, the vectors

$$b(f) = ((f, h_1(\lambda(\cdot)))_{H_0}, (f, h_2(\lambda(\cdot)))_{H_0}, \ldots) \in l_2(N_{\lambda(\cdot)})$$

run through the whole of $l_2(N_{\lambda(\cdot)})$ as f runs through H_0. This implies that for $N_{\lambda(\cdot)} > 1$ there exists $f_0 \in H_0$ such that $b(f_0) \neq 0$ and is orthogonal (in $l_2(N_{\lambda(\cdot)})$) to the vector $a = (a_1, a_2, \ldots)$ introduced in (3.10) (since $J^{-1}\Omega \in H_0$ and the factors $\nu_\gamma(\lambda(\cdot))$ are bounded, we have $a \in l_2(N_{\lambda(\cdot)})$). By setting $f = f_0$ in (3.10), we get

$$(P(\lambda(\cdot))\, Jf_0, \Omega)_{H_0} = 0.$$

Taking into account that $P(\lambda(\cdot))\, Jf_0$ is a generalized joint eigenvector of the family A corresponding to $\lambda(\cdot)$, by using the second equality in (2.18) successively, we obtain

$$(P(\lambda(\cdot))Jf_0,\ A_{x_1}^{m_1} \cdots A_{x_p}^{m_p}\, \Omega)_{H_0} = \overline{(\lambda(x_1))}^{m_1} \ldots \overline{(\lambda(x_p))}^{m_p} (P(\lambda(\cdot))\, Jf_0, \Omega)_{H_0} = 0$$

for all $x_j \in X$ and $m_j \in \mathbb{N}$ from the definition of the cyclic vector Ω. By virtue of the fact that the linear span of the vectors $A_{x_1}^{m_1} \cdots A_{x_p}^{m_p}\, \Omega$ is dense in H_+, this yields $P(\lambda(\cdot))\, Jf_0 = 0$.

By setting $f = f_0$ in (3.2), we get

$$0 = \sum_{\gamma=1}^{N_{\lambda(\cdot)}} \nu_\gamma(\lambda(\cdot))\, b_\gamma(f_0)\, \overline{b_\gamma(g)} \qquad (g \in H_0).$$

The vectors $b(g) \in l_2(N_{\lambda(\cdot)})$ run through the whole of $l_2(N_{\lambda(\cdot)})$ as g varies over H_0; therefore, the last equality and the relations $0 < \nu_\gamma(\lambda(\cdot)) \le c < \infty$ imply that $b(f_0) = 0$. We have obtained a contradiction.

■

All what has been said in this Subsection can be reformulated naturally for families of commuting selfadjoint operators, with $\mathbb{C}^X$ replaced by $\mathbb{R}^X$ (we have already exploited this fact).

3.2. The Case of an Imbedding That Is Not Quasinuclear

When constructing expansions in generalized joint eigenvectors of the family $A = (A_x)_{x \in X}$, we have assumed that the imbedding $H_+ \subset H_0$ in the chain (2.18) (connected with A in the standard way) is quasinuclear. In a certain sense, this restriction is also a necessary condition for the construction of these expansions. Thus, even in the case of a single selfadjoint operator A, one can prove the following fact.

Suppose that the chain (2.1) $H_- \supset H_0 \supset H_+$ is such that the representation (2.13) holds for an RI E of an *arbitrary* selfadjoint operator A acting in H_0, namely,

$$O^+E(\alpha)\,O = \int_\alpha P(\lambda)\, d\rho(\lambda) \quad (\alpha \in \mathfrak{B}(\mathbb{R}^1)), \tag{3.11}$$

where $\mathfrak{B}(\mathbb{R}^1) \ni \alpha \mapsto \rho(\alpha) \in [0, \infty)$ is a finite measure, and $\mathbb{R}^1 \ni \lambda \mapsto P(\lambda) \in \mathcal{L}(H_+, H_-)$ is an operator-valued function defined ρ-almost everywhere such that $\| P(\lambda) \| \le c < \infty$ for ρ-almost all $\lambda \in \mathbb{R}^1$. Then the imbedding $H_+ \subset H_0$ is quasinuclear (the proof of this fact can be found in Berezansky [18,Chapter 2, Section 5, Subsection 2]).

However, for fixed A, the choice of the chain (2.18) may turn out to be freer. We present some facts.

Thus, let A be a selfadjoint operator acting in H_0, and let E be its RI. If we construct the chain (2.18) (connected with A in the standard way) so that the operator-valued measure

$$\mathfrak{B}(\mathbb{R}^1) \ni \alpha \mapsto \theta(\alpha) = O^+E(\alpha)\,O$$

has a σ-finite trace $\rho(\alpha)$, then according to Remark 1 in Subsection 2.1, it can be differentiated with respect to this trace, and it is possible to repeat all the arguments in

Subsections 2.1–2.3 (for the case of a single operator). The formulations of Theorems 2.2 and 2.6 remain unchanged, except that the spectral measure

$$\mathfrak{B}(\mathbb{R}^1) \ni \alpha \mapsto \rho(\alpha) = \operatorname{Tr}(O^+E(\alpha)\,O)$$

is not finite in general.

Theorem 3.3. *Assume that the operator A and the chain (2.18) (connected with it in the standard way) are such that there exists a bounded continuous nonzero complex-valued function $a(\lambda)$ defined on the spectrum $s(A)$ of A such that the operator $a(A)\,O : H_+ \to H_0$ is quasinuclear.*

Then the operator-valued measure $\mathfrak{B}(\mathbb{R}^1) \ni \alpha \mapsto O^+E(\alpha)\,O$ has a σ-finite trace and, consequently, the chain (2.18) is suitable for constructing expansions in generalized eigenvectors of the operator A.

We first prove the following lemma.

Lemma 3.2. *If there exists a bounded continuous positive function $b(\lambda)$ defined on $s(A)$ such that $\operatorname{Tr}(JO^+\,b(A)\,OJ) < \infty$, then the measure $\mathfrak{B}(\mathbb{R}^1) \ni \alpha \mapsto O^+E(\alpha)\,O$ has a σ-finite trace.*

Proof. We set

$$s_n(A) = s(A) \cap [-n, n] \quad (n \in \mathbb{N}).$$

Then for every n, one can find $\varepsilon_n > 0$ such that $b(\lambda) \geq \varepsilon_n$ $(\lambda \in s_n(A))$. Therefore,

$$0 \leq \varepsilon_n E([-n, n]) = \varepsilon_n E(s_n(A)) \leq \int_{s_n(A)} b(\lambda)\, dE(\lambda)$$

$$\leq \int_{s(A)} b(\lambda)\, dE(\lambda) = b(A),$$

whence

$$0 \leq \varepsilon_n C^* E([-n, n])\, C \leq C^* b(A)\, C$$

for any $C \in \mathcal{L}(H_0, H_0)$. Setting $C = OJ$ and taking into account that $C^* = JO^+$, we get

$$0 \leq JO^+E([-n, n])\, OJ \leq \varepsilon_n^{-1} JO^+\, b(A)\, OJ$$

and, hence,

$$\operatorname{Tr}(JO^{+}E([-n,n])\,OJ) \le \varepsilon_n^{-1}\operatorname{Tr}(JO^{+}\,b(A)\,OJ) < \infty \quad (n \in \mathbb{N}). \tag{3.12}$$

On the other hand, if $C \in \mathcal{L}(H_+, H_-)$ is nonnegative, then $J\,CJ \in C \in \mathcal{L}(H_0, H_0)$ is clearly nonnegative, too, and

$$\operatorname{Tr} C = \operatorname{Tr}(J\,CJ). \tag{3.13}$$

Indeed, if $(e_j)_{j=1}^{\infty}$ is an orthonormal basis in the space H_+, then $(J^{-1}e_j)_{j=1}^{\infty}$ is an orthonormal basis in H_0, and we have

$$\operatorname{Tr}(C) = \sum_{j=1}^{\infty}(Ce_j, e_j)_{H_0} = \sum_{j=1}^{\infty}(J\,CJ\,J^{-1}e_j, J^{-1}e_j)_{H_0} = \operatorname{Tr}(J\,CJ).$$

Taking into account (3.13), we now get from (3.11) that

$$\operatorname{Tr}(O^{+}E([-n,n])\,OJ) = \operatorname{Tr}(JO^{+}\,E([-n,n])\,OJ) < \infty \quad (n \in \mathbb{N}).$$

■

Proof of the theorem. We use the following obvious relation:

$$\operatorname{Tr}(C^{*}C) = |C|^2 \le \infty$$

for any $C \in \mathcal{L}(H_0, H_0)$. By applying this to $C = a(A)\,OJ \in \mathcal{L}(H_0, H_0)$, we obtain the general formula

$$\operatorname{Tr}(JO^{+}(|a|^2(A))\,O) = \operatorname{Tr}(JO^{+}(a(A))^{*}\,a(A)\,OJ)$$

$$= \|a(A)\,OJ\|^2 = \|a(A)\,O\|^2 < \infty. \tag{3.14}$$

It remains to employ Lemma 3.2 with $b(\lambda) = |a(\lambda)|^2$.

■

We proceed to verify the condition $\|a(B)O\|^2 < \infty$. for a perturbation B of the operator A which itself satisfies the condition of this sort. We establish two lemmas.

Lemma 3.3. *Suppose that the conditions of Theorem 5.2 are satisfied and, in addition, that B is a selfadjoint operator acting in the space H_0 and $s(B)$ is its spectrum. Assume that there exists a nonnegative bounded function $c \in C(\mathbb{R}^1)$ that is positive on $s(B)$, does not exceed $|a(\lambda)|^2$ on $s(A)$, and is such that $c(B) < c(A)$ or, more generally,*

$$\operatorname{Tr}(JO^+c(B)OJ) \le \operatorname{Tr}(JO^+c(A)OJ).$$

Then $\| c^{1/2}(B)O \| < \infty$.

Proof. According to (3.14),

$$\| c^{1/2}(B)O \|^2 = \operatorname{Tr}(JO^+c(B)OJ) \le \operatorname{Tr}(JO^+c(A)OJ)$$

$$\le \operatorname{Tr}(JO^+(|a|^2(A))OJ) = \| a(A)O \|^2 < \infty.$$

■

Recall that a linear set $F \subset \mathfrak{D}(A)$ is called a core of a closed operator A if $(A \upharpoonright F)^\sim = A$.

Lemma 3.4. *Let H be a Hilbert space, and let A and B be nonnegative selfadjoint operators acting in it, such that their inverse operators A^{-1} and B^{-1} exist and are bounded. If $A \le B$ on a certain linear set F which is a core for both operators A and B (i.e., $(Af, f)_H \le (Bf, f)_H$, $f \in F$), then $B^{-1} \le A^{-1}$.*

This lemma is, in fact, a generalization of the fact well known in the finite-dimensional case to unbounded operators. Its proof can be found, for example, in Berezansky [18, Chapter 2, Section 5, Subsection 2].

■

The combination of the last two lemmas leads to the following theorem.

Theorem 3.4. *Suppose that a selfadjoint operator A semi-bounded from below by a number r and the chain* (2.18) *connected with it in the standard way, are such that $\| a(A)O \| < \infty$ for the function $a(\lambda) = (\lambda - z)^{-1/2}$ ($\lambda \in [r, \infty)$, and $z < r$ is fixed). Then the estimate $\| a(A)O \| < \infty$ holds for any selfadjoint operator B having a core common with A and such that $B \ge A$ on this core. Thus, the chain* (2.18) *is also suitable for constructing expansions in generalized eigenvectors of the operator B.*

Proof. Let F be a common core of A and B. Then it is a common core of the operators $A - z1$ and $B - z1$. Moreover, $A - z1 \le B - z1$ on F. According to Lemma 3.4, this implies that $(B - z1)^{-1} \le (A - z1)^{-1}$, i.e., the condition $c(B) \le c(A)$ is satisfied with the function $c(\lambda)$ equal to $(\lambda -)z^1$ for $\lambda \ge r$ and to $(r - z)^{-1}$ for $\lambda < r$ (we clarify that $s(A) \subset [r, \infty)$ by the assumption, and $s(B) \subset [r, \infty)$ due to the inequality $B \ge A$). This function is positive on $s(B)$, and $c(\lambda) = |a(\lambda)|^2$ on $s(A)$. The theorem now follows from Lemma 3.3.

■

3.3. The Expansion in Eigenfunctions of Carleman Operators

We now apply the results obtained in Subsection 3.2 to more concrete situations.

Let

$$H_0 = L_2(R, \mu) = L_2(R, \mathfrak{R}, \mu),$$

where R is a space with a measure μ given on a σ-algebra $\mathfrak{R}$ of subsets of R; $\mu(R) \leq \infty$. A selfadjoint operator A acting in this space is called a Carleman operator if there exists a bounded continuous nonzero complex-valued function $a(\lambda)$, defined on its spectrum, such that $a(A)$ is an integral Carleman operator. The latter means that there exists a $(\mu \times \mu)$-measurable kernel $K(x, y)$, defined for $(\mu \times \mu)$-almost all $(x, y) \in R \times R$, such that the representation

$$(a(A)f)(x) = \int_R K(x, y) f(y)\, d\mu(y)$$

holds for a certain set of functions f dense in $L_2(R, \mu)$, and

$$\int_R |K(x, y)|^2\, d\mu(x) < \infty \tag{3.15}$$

for μ-almost every $y \in R$.

Examples of Carleman operators (elliptic operators with smooth coefficients and others) are given in Berezansky [5, Chapter 6] and below.

Theorem 3.5. *Let A be a selfadjoint Carleman operator acting in the space $L_2(R, \mu)$. Then there exists a μ-measurable weight $p(x) \geq 1$ $(x \in R)$ such that the chain (see Example 1.1 in Chapter 1)*

$$L_2(R, p^{-1}(x)\, d\mu(x)) \supset L_2(R, d\mu(x)) \supset L_2(R, p(x)\, d\mu(x)) \tag{3.16}$$

becomes suitable for the construction of expansions in generalized eigenfunctions of the operator A being properly extended to (2.18).

Proof. Consider an arbitrary chain of the form (3.16). For this chain, we have

$$(If)(x) = p^{-1}(x) f(x) \quad (f \in L_2(R, \mu)),$$

therefore,

$$(Jf)(x) = p^{-1/2}(x) f(x) \quad (f \in L_2(R, \mu)). \tag{3.17}$$

Consequently, the operator $a(A)\,OJ$, acting continuously in the space $L_2(R, \mu)$, is associated with the kernel $K_1(x, y) = K(x, y)p^{-1/2}(y)$, and

$$|a(A)\,O|^2 = |a(A)\,OJ|^2 = \int_R \int_R |K(x, y)|^2\, p^{-1}(y)\, d\mu(x)\, d\mu(y) < \infty. \qquad (3.18)$$

According to (3.15), the function

$$k(y) = \int_R |K(x, y)|^2\, d\mu(x)$$

is measurable and finite almost everywhere. Therefore, one can choose a measurable weight $p(y) \geq 1$ so that

$$\int_R k(y)\, p^{-1}(y)\, d\mu(y) < \infty.$$

As follows from (3.18), the last condition means that $|a(A)\,O| < \infty$. We now see that the condition of Theorem 3.3 holds with the chain (3.16) and the weight $p(x)$ chosen as mentioned above.

■

We emphasize that the weight $p(x)$ in the above theorem is an arbitrary measurable weight $p(x) \geq 1$ $(x \in R)$ for which the integral (3.18) converges.

Hence, as in the case of Jacobian matrices, for Carleman operators, the fact that an eigenfunction is generalized means only that it belongs, not to $L_2(R, d\mu(x))$, but to the space $L_2(R, p^{-1}(x)\, d\mu(x))$ with the indicated weight $p(x)$. Expansions in individual eigenfunctions of it are carried out according to the general scheme of Subsection 3.1. We present four simple statements concerning Carleman operators A.

I. *The operator*

$$P(\lambda) \colon L_2(R, p(x)\, d\mu(x)) \mapsto L_2(R, p^{-1}(x)\, d\mu(x))$$

is an integral operator, i.e.,

$$(P(\lambda)u)(x) = \int_R P(x, y; \lambda)\, u(y)\, d\mu(y) \qquad (u \in L_2(R, p(x)\, d\mu(x))), \qquad (3.19)$$

and the kernel $P(x, y; \lambda)$ (the spectral kernel of A) satisfies the estimate

$$\int_R \int_R |P(x, y; \lambda)|^2\, p^{-1}(x)\, p^{-1}(y)\, d\mu(x)\, d\mu(y) \leq 1 \qquad (3.20)$$

for ρ*-almost all* $\lambda \in \mathbb{R}^1$ *and is jointly (with respect to all its variables)* $(\mu \times \mu \times \rho)$*-measurable.*

Indeed,

$$\lambda \| P(\lambda) \| \leq \mathrm{Tr}\,(P(\lambda)) = 1$$

for ρ-almost all λ. Consider the operator

$$J P(\lambda) J : L_2(R, d\mu) \mapsto L_2(R, d\mu),$$

such that the estimate

$$\| J P(\lambda) J \| = \| P(\lambda) \| \leq 1$$

holds ρ-almost everywhere. Therefore, for a proper fixed λ, this operator is an integral operator acting in $L_2(R, \mu)$ whose kernel $K(x, y)$ satisfies the estimate

$$\int_R \int_R |K(x, y)|^2 \, d\mu(x) \, d\mu(y) \leq 1.$$

By virtue of (3.17),

$$P(x, y; \lambda) = K(x, y)\, p^{1/2}(x)\, p^{1/2}(y)$$

is the kernel of the operator $P(\lambda)$, and this yields (3.19) and (3.20). The measurability of P follows from the paper by Orochko [8].

■

II. *Let* $\mathbb{R}^1 \ni \lambda \mapsto \gamma(\lambda) \in \mathbb{C}^1$ *be a Borel-measurable function such that* $|c(\lambda)| \leq |a(\lambda)|^2$ $(\lambda \in s(A))$. *Then*

$$\int_{\mathbb{R}^1} |c(\lambda)| \, d\rho(\alpha) < \infty.$$

Indeed, for $\alpha \in \mathfrak{B}(\mathbb{R}^1)$, we have

$$\rho(\alpha) = \mathrm{Tr}\,(O^+ E(\alpha)\, O) = \mathrm{Tr}\,(J\, O^+ E(\alpha)\, O J)$$

$$= \sum_{j=1}^{\infty} (J\, O^+ E(\alpha)\, O J e_j, e_j)_{H_0} = \lim_{n \to \infty} \rho_n(\alpha),$$

$$\rho_n(\alpha) = \sum_{j=1}^{n} (J\,O^+E(\alpha)\,OJe_j, e_j)_{H_0},$$

where $(e_j)_{j=1}^{\infty}$ is an orthonormal basis in $H_0 = L_2(R, \mu)$. By using Helly's theorems, the following limit transition can be easily justified

$$\int_{\mathbb{R}^1} |c(\lambda)|\, d\rho(\alpha) = \lim_{n\to\infty} \int_{\mathbb{R}^1} |c(\lambda)|\, d\rho_n(\alpha)$$

$$= \lim_{n\to\infty} \int_{\mathbb{R}^1} \sum_{j=1}^{n} |c(\lambda)|\, d(J\,O^+E(\alpha)\,OJe_j, e_j)_{H_0}$$

$$= \lim_{n\to\infty} \sum_{j=1}^{n} (J\,O^+(|c|(A))\,OJe_j, e_j)_{H_0}$$

$$\leq \lim_{n\to\infty} \sum_{j=1}^{n} (J\,O^+|a|^2(A)\,OJe_j, e_j)_{H_0}$$

$$= \mathrm{Tr}\,(J\,O^+|a|^2(A)\,OJ) = \| a(A) \|^2 < \infty.$$

■

III. *For every function $c(\lambda)$ in statement* **II**, *the operator*

$$O^+c(A)\,O\colon L_2(R, p(x)\,d\mu(x)) \mapsto L_2(R, p^{-1}(x)\,d\mu(x))$$

is an integral operator, and its kernel $K(x, y)$ belongs to

$$L_2(R\times R, (p^{-1}(x)\,d\mu(x)) \times (p^{-1}(y)\,d\mu(y))).$$

Indeed, by virtue of (3.11),

$$O^+c(A)\,O = \int_{\mathbb{R}^1} c(\lambda)\, d(O^+E(\lambda)\,O) = \int_{\mathbb{R}^1} c(\lambda)\,P(\lambda)d\rho(\alpha).$$

But according to statement I, $P(\lambda)$ is an integral operator and, thus, formally, $O^+c(A)\,O$ is also an integral operator with the kernel

$$K(x, y) = \int_{\mathbb{R}^2} c(\lambda)\,P(x, y; \lambda)\,d\rho(\lambda) \qquad (x, y \in R). \tag{3.21}$$

To prove the statement, it suffices to establish that the function (3.21) belongs to

$$L_2(R \times R, (p^{-1}(x)\, d\mu(x)) \times (p^{-1}(y)\, d\mu(y))).$$

By using the Cauchy–Schwarz–Bunyakovsky inequality and (3.20), we obtain

$$\int_R \int_R \left| \int_{\mathbb{R}^1} c(\lambda)\, P(x, y; \lambda)\, d\rho(\lambda) \right|^2 p^{-1}(x)\, p^{-1}(y)\, d\mu(x)\, d\mu(y)$$

$$\leq \int_{\mathbb{R}^1} |c(\lambda)|\, d\rho(\lambda) \int_R \int_R \left(\int_{\mathbb{R}^1} |c(\lambda)|\, |P(x, y; \lambda)|^2\, d\rho(\lambda) \right) p^{-1}(x)\, p^{-1}(y)\, d\mu(x)\, d\mu(y)$$

$$\leq \left(\int_{\mathbb{R}^1} |c(\lambda)|\, d\rho(\lambda) \right)^2 < \infty.$$

■

IV. *Assume that, for a selfadjoint operator A acting in the space $L_2(R, \mu)$ and some function $a(\lambda)$ satisfying the conditions formulated in the definition of the Carleman property, it is possible to find the rigging* (3.16) *such that*

$$\| a(A)\, OJ \|^2 = \mathrm{Tr}\, (J\, O^{+}\, |a|^2\, (A)\, OJ) < \infty.$$

Then A is a Carleman operator.

In fact, it is obvious that $\| a(A)\, OJ \| < \infty$. Therefore, the operator $a(A)\, OJ$, which acts in $L_2(R, \mu)$, is an integral operator whose kernel $K_1(x, y)$ is square-summable with respect to both variables. But then $K(x, y) = K_1(x, y)\, p^{1/2}(y)$ is just the required kernel of $a(A)$.

■

We proceed to the investigation (based on Theorem 3.4 and Lemma 3.3) of the Carleman property for a perturbation of a Carleman operator.

Theorem 3.6. *Let A be a selfadjoint operator acting in the space $L_2 = L_2(\mathbb{R}^d, dx)$ (with respect to the Lebesgue measure dx ($d \in \mathbb{N}$)) and semi-bounded from below by a number r. Assume that the resolvent $R_z(A)$ is an integral operator for some $z < r$, and that its kernel is locally bounded. Consider a selfadjoint operator B which acts in the same space, has a core common with A, and is such that $B \geq A$ on this core. Then B is a Carleman operator.*

Proof. It suffices to show that one can choose a chain of the form (3.16) so that the inequality $\| a(A)\, O \| < \infty$ holds, where $a(\lambda) = (\lambda - z)^{-1/2}$ $(\lambda \in [r, \infty))$. Then, according to Theorem 3.4, the same inequality is also valid when A is replaced by B. Hence, $\| a(B)\, O \| < \infty$ and, therefore, according to IV, B is a Carleman operator.

It follows from (3.14) that the weight $p(x) \geq 1$ $(x \in \mathbb{R}^d)$ must be chosen so that $\mathrm{Tr}\,(JO^+ COJ) < \infty$, where $C = R_z(A) > 0$.

We introduce the averaging operator

$$(S_\varepsilon f)(x) = (\omega_\varepsilon * f)(x),$$

acting in L_2. Here $*$ denotes the convolution and

$$\omega_\varepsilon(x) = (m(B_\varepsilon(0)))^{-1}\, \kappa_{B_\varepsilon(0)}(x)$$

($B_\varepsilon(0)$ is the open ball with radius ε centered at 0, $m(B_\varepsilon(0))$ is the Lebesgue measure of it, and $\kappa_{B_\varepsilon(0)}$ is the characteristic function of $B_\varepsilon(0)$). It is not hard to see that S_ε is a bounded operator with nonnegative symmetric kernel $S_\varepsilon(x, y)$, that $S_\varepsilon \underset{\varepsilon \to 0}{\to} 1$ in the strong sense, and that

$$\sup_{x \in G_1;\, 0 < \varepsilon \leq \delta} |(S_\varepsilon f)(x)| \leq \sup_{x \in G_2} |f(x)|$$

for each pair of domains $G_1 \subset G_2 \subset \mathbb{R}^d$, where $\delta = \rho\,(G_1, \mathbb{R}^d \backslash G_2) > 0$. The operator $JO^+ S_\varepsilon C S_\varepsilon OJ$ acts in $L_2(\mathbb{R}^d, dx)$ continuously and has the kernel

$$K_\varepsilon(x, y) = \int_{\mathbb{R}^d} \int_{\mathbb{R}^d} S_\varepsilon(x, s)\, C(s, t)\, S_\varepsilon(t, y)\, ds\, dt\, p^{-1/2}(x)\, p^{-1/2}(y) \quad (x, y \in \mathbb{R}^d), \tag{3.22}$$

where $C(x, y)$ is the kernel of the operator C. Assume that $p \in C(\mathbb{R}^d)$; then the kernel (3.22) continuously depends on (x, y) and is positive definite. Therefore,

$$\mathrm{Tr}\,(JO^+ S_\varepsilon C S_\varepsilon OJ) = \int_{\mathbb{R}^d} K_\varepsilon(x, x)\, dx \leq \infty \quad (\varepsilon > 0). \tag{3.23}$$

We estimate the integral (3.23). Let $G_1 \subset G_2 \subset \ldots$ be a sequence of balls $G_n = B_n(0)$, $n \in \mathbb{N}$. Denote $c_n = \sup_{x, y = G_n} |C(x, y)|$ $(n \in \mathbb{N})$ and consider the kernel

$$K_{1, \varepsilon}(x, y) = K_\varepsilon(x, y)\, p^{1/2}(x)\, p^{1/2}(y) \quad (x, y \in \mathbb{R}^d).$$

According to (3.22), it can be regarded as a result of applying the tensor product $S_\varepsilon \otimes S_\varepsilon$

of the two averaging operators to $C(t, s)$. Consequently,

$$\sup_{x, y \in G_n; 0<\varepsilon\le\delta} |K_{1,\varepsilon}(x, y)| \le c_{n+1} \quad (n \in \mathbb{N})$$

due to the properties of these operators. This yields $(G_0 = \varnothing)$

$$\int_{\mathbb{R}^d} K_\varepsilon(x, x)\, dx = \int_{\mathbb{R}^d} K_{1,\varepsilon}(x, x)\, p^{-1}(x)\, dx$$

$$\le \sum_{n=0}^{\infty} \int_{G_{n+1}\setminus G_n} K_{1,\varepsilon}(x, x)\, p^{-1}(x)\, dx \le \sum_{n=0}^{\infty} c_{n+2} \int_{G_{n+1}\setminus G_n} p^{-1}(x)\, dx$$

$$(\varepsilon \in (0, 1]). \tag{3.24}$$

Choose a function $p(x)$ which increases as $|x| \to \infty$ so rapidly that the series on the right-hand side of (3.24) converges. Then

$$\mathrm{Tr}\,(JO^+ S_\varepsilon C S_\varepsilon OJ) \le c < \infty \quad (\varepsilon \in (0, 1]). \tag{3.25}$$

Clearly, we can pass to the limit as $\varepsilon \to 0$ in the inequality (3.25); and this results in the required estimate $\mathrm{Tr}\,(JO^+ COJ) < \infty$. Indeed, if $(e_j)_{j=1}^{\infty}$ is an orthonormal basis in the space L_2, then the left-hand side of (3.25) is equal to

$$\sum_{j=1}^{\infty} (CS_\varepsilon OJe_j, S_\varepsilon OJe_j)_{L_2}.$$

For each term, we have

$$(CS_\varepsilon OJe_j, S_\varepsilon OJe_j)_{L_2} \to (COJe_j, OJe_j)_{L_2}.$$

as $\varepsilon \to 0$; therefore, by virtue of (3.25) and the fact that all the terms are nonnegative, we get

$$\mathrm{Tr}\,(JO^+ COJ) = \sum_{j=1}^{\infty} (COJe_j, OJe_j)_{L_2} \le c < \infty.$$

■

We emphasize that the chain of the form (3.16) with $p(x) \ge 1$ $(x \in \mathbb{R}^d)$ is continuous and is chosen so that the series (3.24) converges; hence, it is suitable for any operator B satisfying the conditions of the theorem.

Example 3.1. Consider the differential expression

$$(\mathfrak{M}u)(x) = \sum_{|\alpha| \le r} a_\alpha(x)(D^\alpha u)(x) + q(x)u(x) = (\mathfrak{L}u)(x) + q(x)u(x) \quad (3.26)$$

in the space $L_2 = L_2(\mathbb{R}^d, dx)$. Here $\mathfrak{L}$ is an elliptic, formally selfadjoint expression of the order $r > d$ with coefficients $a_\alpha \in C^\infty(\mathbb{R}^d)$; $q \in L_{2,\mathrm{loc}}(\mathbb{R}^d, dx)$ is nonnegative almost everywhere, and $\mathfrak{L}$ is assumed to be semi-bounded from below on $C_0^\infty(\mathbb{R}^d)$ by a number s, i.e.,

$$(\mathfrak{L}u, u)_{L_2} \ge s \|u\|_{L_2}^2, \quad u \in C_0^\infty(\mathbb{R}^d).$$

In L_2, we introduce the Hermitian operators $C_0^\infty(\mathbb{R}^d) \ni u \mapsto \mathfrak{L}u$ and $C_0^\infty(\mathbb{R}^d) \ni u \mapsto \mathfrak{M}u$ and assume that they have selfadjoint extensions to L_2 which are semi-bounded from below and have a common base. Denote these extensions by A and B, respectively. It is stated that the operators A and B are Carleman operators with common $a(\lambda) = (\lambda - z)^{-1/2}$ ($z < s$ is fixed; $\lambda \in [s, \infty)$) and common weight $p(x)$. This follows from the foregoing, since it is known (Berezansky [5, Chapter 3, Theorem 5.1]) that for $r > d$ the resolvent $R_z(A)$ of the operator A introduced just above is an integral operator whose kernel is locally bounded.

The scheme we have presented does not work in the case of differential expressions (3.26) whose order is not high enough (does not exceed d). This is so, because to prove the Carleman property in this case, one should take a function $a(\lambda)$ of the type $(\lambda^l - z)^{-1}$, where the integer $l > 0$ is sufficiently large, and the corresponding analogue of Lemma 3.4 is not true. We present a somewhat different approach.

Let A be a selfadjoint Carleman operator acting in the space $L_2(R, \mu)$ and semi-bounded from below, and let $a(\lambda)$ be the corresponding function. Assume that for any $t \in (0, 1]$, there exists $c_t > 0$ such that

$$|a(\lambda)|^2 \ge c_t e^{-t\lambda} \quad (\lambda \in s(A)). \quad (3.27)$$

According to Statement III given above, each operator

$$O^+ e^{-tA} O \colon L_2(R, p(x)\, d\mu(x)) \mapsto L_2(R, p^{-1}(x)\, d\mu(x))$$

is an integral operator; let

$$U(x, y; t) \in L_2(R \times R, (p^{-1}(x)\, d\mu(x)) \times (p^{-1}(y)\, d\mu(y))) \quad (t \in (0, 1])$$

be the corresponding kernel. We say that the operator A satisfies the semigroup positivity condition if the inequality $U(x, y; t) \geq 0$ holds for every $t \in (0, 1]$ and for $(\mu \times \mu)$-almost all $(x, y) \in R \times R$.

Theorem 3.7. *Assume that* A *is a Carleman operator acting in the space* $L_2(R, \mu)$ *which is semi-bounded from below and satisfies the semigroup positivity condition. Let* $R \ni x \mapsto v(x) \geq 0$ *be a measurable function such that there exists a core* F *of* A *for which the closure* V *of the operator* $F \ni f(x) \mapsto v(x)f(x)$ *is selfadjoint. Denote by* B *the closure of the operator* $F \ni f \mapsto Af + Vf$ *and suppose that* B *is selfadjoint.*

Then B *is a Carleman operator, and the chain* (3.16), *constructed for given* A, *is suitable for constructing expansions in generalized eigenfunctions of* B.

Proof. We employ a well-known multiplicative formula. Consider selfadjoint operators A and V acting in the Hilbert space H which are semi-bounded from below. Denote by B the closure of the operator

$$\mathfrak{D}(A) \cap \mathfrak{D}(A) \ni f \mapsto Af + Vf$$

and assume that B is selfadjoint. Then, for every $t > 0$ and $f \in H$, we have

$$\exp(-tB)f = \lim_{n\to\infty} (\exp(-\tfrac{t}{n}A) \exp(-\tfrac{t}{n}V))^n f \tag{3.28}$$

in the sense of strong convergence. Taking into account the estimate (3.27) with $t = 1$ and applying Lemma 3.3, where

$$c(\lambda) = \begin{cases} c_1 e^{-\lambda} & \text{for } \lambda \geq r \\ c_1 e^{-r} & \text{for } \lambda < r \end{cases},$$

we find that it suffices to prove the inequality

$$\mathrm{Tr}\,(JO^+e^{-B}OJ) \leq \mathrm{Tr}\,(JO^+e^{-A}OJ)$$

or, what is the same,

$$\| e^{-1/2B} OJ \| \leq \| e^{-1/2A} OJ \|$$

(see (3.14)).

By virtue of (3.28), for every $f \in L_2(R, \mu)$, we have

$$(e^{-1/2B} OJf)(x) = \lim_{n\to\infty} ((\exp(-\tfrac{1}{2n}A) \exp(-\tfrac{1}{2n}V))^n OJf)(x). \tag{3.29}$$

Denote

$$e_n(x) = \exp\left(-\frac{1}{2n}v(x)\right) \in [0, 1] \quad (x \in R; n \in \mathbb{N}).$$

By virtue of the Fubini theorem and the fact that the functions $U(x, y; t)$, $e_n(x) \le 1$, and $p(x)$ are nonnegative, for the kernel $K_n(x, y)$ of the operator under the limit sign on the right-hand side of (3.29), we obtain

$$0 \le K_n(x, y) = \int_R U(x, z_1; (2n)^{-1})\, e_n(z_1) \left(\int_R U(z_1, z_2; (2n)^{-1})\, e_n(z_2) \right.$$

$$\times \left(\ldots \left(\int_R U(z_n, y; (2n)^{-1})\, dz_n \right) \ldots \right) dz_2 \Bigg)\, dz_1\, e_n(y)\, p^{-1/2}(y)$$

$$= \int_R \ldots \int_R U(x, z_1; (2n)^{-1})\, e_n(z_1)\, U(z_1, z_2; (2n)^{-1})\, e_n(z_2)$$

$$\ldots U(z_n, y; (2n)^{-1})\, dz_1 \ldots dz_n\, e_n(y)\, p^{-1/2}(y)$$

$$\le \int_R \ldots \int_R U(x, z_1; (2n)^{-1})\, U(z_1, z_2; (2n)^{-1}) \ldots U(z_n, y; (2n)^{-1})$$

$$\times\, dz_1 \ldots dz_n\, e_n(y)\, p^{-1/2}(y)$$

for $(\mu \times \mu)$-almost all $(x, y) \in R \times R$. The last integral is the kernel of the operator

$$\left(\exp\left(-\frac{1}{2n}A\right)\right)^n OJ = \exp\left(-\frac{1}{2}A\right) OJ.$$

Denote it by $L(x, y)$. Hence, $O \le K_n(x, y) \le L(x, y)$ for $(\mu \times \mu)$-almost all $(x, y) \in R \times R$. This yields

$$\left\| \left(\exp\left(-\frac{1}{2n}A\right) \exp\left(-\frac{1}{2n}V\right)\right)^n OJ \right\|^2 = \int_R \int_R K_n^2(x, y)\, d\mu(x)\, d\mu(y)$$

$$\int_R \int_R L^2(x, y)\, d\mu(x)\, d\mu(y) = \left\| \exp\left(-\frac{1}{2}A\right) OJ \right\|^2.$$

This inequality and (3.28) imply that

$$\left\| \exp\left(-\frac{1}{2}B\right) OJ \right\| \le \left\| \exp\left(-\frac{1}{2}A\right) OJ \right\|.$$

■

Example 3.2. In the space $L_2(\mathbb{R}^d, dx)$, we consider the Schrödinger operator B with a singular potential semi-bounded from below. This operator is defined as a closure of the operator

$$C_0^\infty(\mathbb{R}^d) \ni u(x) \mapsto (\Delta u)(x) + q(x)\,u(x),$$

where $q \in L_{2,\text{loc}}(\mathbb{R}^d, dx)$ and $q(x) \geq r$ $(x \in \mathbb{R}^d)$. It is well known that B is selfadjoint (Kato [2], see also Subsection 1.5 in Chapter 5). By using Theorem 3.7, one can easily show that B is a Carleman operator.

Indeed, define A in the same way as B, but with the help of the differential expression $-\Delta u + ru$. It is known that A is selfadjoint, semi-bounded from below by the number r, and is a Carleman operator with $a(\lambda) = (\lambda^l - z)^{-1}$, where the integer l is greater than $d/4$ and $z < r$ (Berezansky [5, Chapter 6, Lemma 2.1]). Furthermore, A satisfies the semigroup positivity condition; in fact, (3.27) is obvious, and the required properties of the kernel $U(x, y; t)$ (this is the fundamental solution of the heat conduction equation, shifted by r) follow from the well-known representation of this kernel. Recall that for $r = 0$

$$U(x, y; t) = (2\sqrt{\pi t})^{-d} \exp(-|x-y|^2/4t) \quad (x, y \in \mathbb{R}^d;\ t \in (0, 1]).$$

Here $F = C_0^\infty(\mathbb{R}^d)$ and $v(x) = q(x) - r \geq 0$ $(x \in \mathbb{R}^d)$.

In the case of Carleman operators acting in the space $L_2(R, \mu)$, where R is a locally compact separable space, and μ is a measure defined on Borel sets, positive on open sets, and finite on compact sets, we can derive some useful conclusions from the *a priori* assumed continuity of the spectral kernel $P(x, y; \lambda)$ with respect to $(x, y) \in R \times R$[30]. We shall not dwell upon these results (Berezansky [5, Chapter 5, Section 4, Subsection 3]). We also shall not consider the estimates of growth at infinity for generalized eigenfunctions of Carleman operators and, in particular, of differential operators (see Berezansky [5, Chapter 5, Section 5], Simon [6], and Bibliographical Notes).

3.4. Existence of a Rigging

It was shown in Section 2 that if a family $A = (A_x)_{x \in X}$ of commuting normal operators admits a quasinuclear rigging, then the representation (2.28) holds and, clearly, its generalization, with π replaced by $\tau \supset \pi$ $(\tau \in \mathbb{C}^X)$, is also valid. We establish the converse statement:

The existence of a representation of this type with a "sufficiently good" space τ involves the existence of a quasinuclear rigging connected with A in the standard way[31].

Thus, let $\tau \in \mathbb{C}^X$ be a certain set, let

[30] For example, we can prove that the individual eigenfunctions $\xi_\gamma(x; \lambda)$ are continuous.

[31] The space τ should admit the construction of the space of test functions of a certain type.

$$\mathfrak{C}_\sigma(\tau) = \{\alpha' = \alpha \cap \tau \mid \alpha \in \mathfrak{C}_\sigma(\mathbb{C}^X)\}$$

be the same σ-algebra of its subsets as above, and let $\mathfrak{C}_\sigma(\tau) \ni \alpha' \mapsto E(\alpha')$ be a resolution of the identity in the space H_0. The function $\tau \ni \lambda(\cdot) \mapsto \lambda(x) \in \mathbb{C}^1$ ($x \in X$ is fixed) is $\mathfrak{C}_\sigma(\tau)$-measurable. Hence, as in Subsection 1.11, we can define a family $A = (A_x)_{x \in X}$ of operators by the formula

$$A_x = \int_\tau \lambda(x)\, dE(\lambda(\cdot)),$$

$$\mathfrak{D}(A_x) = \left\{f \in H_0 \mid \int_\tau |\lambda(x)|^2\, d(E(\lambda(\cdot))f, f)_{H_0} < \infty\right\}. \tag{3.30}$$

It is easy to prove directly that the operators A_x are normal and commuting and that A_x^* can be represented by the same integral (3.30) with $\lambda(x)$ replaced by $\overline{\lambda(x)}$ (this also follows from the preceding. Indeed, extend E to an RI on $\mathfrak{C}_\sigma(\mathbb{C}^X)$ by the procedure presented in Subsection 1.8; then employ Remark 1 of Subsection 1.3 and what has been said in Subsection 1.11).

Fix $\tau \subset \mathbb{C}^X$. Consider a chain

$$\mathcal{H}(\tau) \supset D(\tau), \tag{3.31}$$

where $\mathcal{H}(\tau)$ is a Hilbert space consisting of $\mathfrak{C}_\sigma(\tau)$-measurable functions

$$\tau \ni \lambda(\cdot) \mapsto a(\lambda(\cdot)) \in \mathbb{C}^1,$$

$D(\tau)$ is a linear topological space, and the imbedding in (3.31) is topological. Assume that the following conditions hold:

(i) The estimate

$$|a(\lambda(\cdot))| \le c_{\lambda(\cdot)} \|a\|_{\mathcal{H}(\tau)} \quad (a \in \mathcal{H}(\tau))$$

holds for each $\lambda(\cdot) \in \tau$.

(ii) The multiplication operators

$$D(\tau) \ni a(\lambda(\cdot)) \mapsto \lambda(x)\, a(\lambda(\cdot)),$$

and

$$\overline{\lambda(x)}\, a(\lambda(\cdot)) \in \mathcal{H}(\tau)$$

are defined and continuous for every $x \in X$.

(iii) Let $\varepsilon(\lambda(\cdot))$ be equal to 1 divided by the norm of the functional

$$\mathcal{H}(\tau) \ni a \mapsto a(\lambda(\cdot)) \in \mathbb{C}^1,$$

it is required that each finite charge

$$\mathcal{C}_\sigma(\tau) \ni \alpha \mapsto \omega(\alpha) \in \mathbb{C}^1$$

be equal to zero provided that

$$\int_\tau a(\lambda(\cdot))\, \varepsilon(\lambda(\cdot))\, d\omega(\lambda(\cdot)) = 0$$

for any $a \in \mathcal{H}(\tau)$.

The chain (3.31) for which conditions (i)–(iii) hold is called admissible. We clarify that the functional in condition (iii) is the delta-function $\delta_{\lambda(\cdot)}$ concentrated at the point $\lambda(\cdot)$; the existence of it follows from condition (i). The integrand in (iii) is bounded and, clearly, $\mathcal{C}_\sigma(\tau)$-measurable. In the simplest case of finite-dimensional τ, we can take, for example, a Sobolev space as $\mathcal{H}(\tau)$, and one of the classical spaces of infinitely differentiable test functions can be taken to play the role of $D(\tau)$ (for more detail see Theorem 3.9).

Theorem 3.8. *Suppose that an admissible chain* (3.31) *exists for a set* $\tau \in \mathbb{C}^X$. *Then it is possible to construct a quasinuclear rigging* (2.18) *connected with the family of operators* (3.30) *in the standard way, and* D *is a separable projective limit of Hilbert spaces.*

Proof. **I.** We mention a formula for calculating $\varepsilon(\lambda(\cdot))$. Let $(e_j)_{j=1}^\infty$ be an orthonormal basis in $\mathcal{H}(\tau)$. Regarding $\delta_{\lambda(\cdot)}$ as a vector from $\mathcal{H}(\tau)$, we get

$$\delta_{\lambda(\cdot)} = \sum_{j=1}^\infty (\delta_{\lambda(\cdot)}, e_j)_{\mathcal{H}(\tau)}\, e_j = \sum_{j=1}^\infty \overline{e_j(\lambda(\cdot))}\, e_j,$$

whence

$$\varepsilon(\lambda(\cdot)) = \| \delta_{\lambda(\cdot)} \|_{\mathcal{H}(\tau)}^{-1} = \left(\sum_{j=1}^\infty \left| e_j(\lambda(\cdot)) \right|^2 \right)^{-1/2} \qquad (\lambda(\cdot) \in \tau)$$

(this formula implies, in particular, that $\varepsilon(\cdot)$ is measurable).

II. Fix a vector $l_1 \in H_0$ with $\| l_1 \|_{H_0} = 1$, and consider the linear mapping

$$\mathcal{H}(\tau) \ni a \mapsto Q_1 a = \int_\tau a(\lambda(\cdot))\, \varepsilon(\lambda(\cdot))\, dE(\lambda(\cdot)), \quad l_1 \in H_0. \tag{3.32}$$

This is a Hilbert–Schmidt operator; indeed, by using Step I, we obtain

$$|Q_1|^2 = \sum_{j=1}^{\infty} \| Q_1 e_j \|_{H_0}^2 = \sum_{j=1}^{\infty} \int_\tau | e_j(\lambda(\cdot))|^2 \varepsilon^2(\lambda(\cdot))\, d(E(\lambda(\cdot))\, l_1, l_1)_{H_0} = 1.$$

Consider the kernel of the operator Q_1, i.e., the subspace $\{a \in \mathcal{H}(\tau) \mid Q_1 a = 0\}$. Let F be its orthogonal complement, and let $R = Q_1 \upharpoonright F$. Choosing an orthonormal basis $(e_j)_{j=1}^{\infty}$ so that the vectors of it are contained in this kernel and in F, we find that $|R| = |Q_1| = 1$ and, thus, $\| R \| \leq 1$. There exists the closed inverse operator R^{-1} defined on the range

$$\mathfrak{R}(R) = \mathfrak{R}(Q_1) \subset H_0$$

and

$$\| R^{-1} f \|_{\mathcal{H}(\tau)} \geq \| f \|_{H_0} \quad (f \in \mathfrak{R}(R)).$$

We transform $\mathfrak{R}(R) \subset H_0$ into a Hilbert space $H_{+,1}$ by setting

$$(\varphi, \psi)_{H_{+,1}} = \kappa_1 (R^{-1}\varphi, R^{-1}\psi)_{\mathcal{H}(\tau)} \quad (\varphi, \psi \in \mathfrak{R}(R)),$$

where $\kappa_1 \geq 1$ is a fixed number. Obviously, $H_{+,1}$ is complete, and $\|\cdot\|_{H_{+,1}} \geq \|\cdot\|_{H_0}$. In other words, if P is the projector onto F defined in $\mathcal{H}(\tau)$, then we set

$$(Q_1 a, Q_1 b)_{H_{+,1}} = \kappa_1 (Pa, Pb)_{\mathcal{H}(\tau)} \quad (a, b \in \mathcal{H}(\tau)). \tag{3.33}$$

The imbedding $O_1 : H_{+,1} \to H_0$ is quasinuclear, and $|O_1| = \kappa_1^{-1/2}$. Indeed, let $(e_j)_{j=1}^{\infty}$ be an orthonormal basis in F, then $(\kappa_1^{-1/2} R e_j)_{j=1}^{\infty}$ is an orthonormal basis in $H_{+,1}$. For this basis, we have

$$|Q_1|^2 = \sum_{j=1}^{\infty} \| \kappa_1^{-1/2} R e_j \|_{H_0}^2 = \kappa_1^{-1} |R|^2 = \kappa_1^{-1}.$$

III. By using the orthogonalization procedure, we construct an orthonormal basis $(h_j)_{j=1}^{\infty}$ in the space $\mathcal{H}(\tau)$ which consists of vectors from $D(\tau)$. Fix this basis. Let

$D_0(\tau)$ be the collection of all vectors from the space $\mathcal{H}(\tau)$ which have a finite sequence of coordinates in this basis; $D_0(\tau) \subset D(\tau)$ and is dense in $\mathcal{H}(\tau)$.

We set

$$D_1 = Q_1 D_0(\tau) \subset \mathfrak{R}(Q_1) = H_{+,1}.$$

It follows from (3.33) and the denseness of $D_0(\tau)$ in $\mathcal{H}(\tau)$ that D_1 is dense in $H_{+,1}$. Each vector in D_1 has the form

$$\varphi = \sum_{j=1}^{\infty} \alpha_j (Q_1 h_j),$$

where $(\alpha_j)_{j=1}^{\infty}$, $\alpha_j \in \mathbb{C}^1$, is an arbitrary finite sequence; $a = \sum_{j=1}^{\infty} \alpha_j h_j$ is an arbitrary vector from $D_0(\tau)$.

Generally speaking, the vectors $Q_1 h_1, Q_1 h_2, \ldots$ are linearly dependent. By shifting the elements of this sequence (moving from left to right), we can construct a subsequence $(Q_1 f_{1,k})_{k=1}^{\infty}$, $f_{1,k} = h_{j_k}$ $(k \in \mathbb{N})$ of it consisting of linearly independent vectors. Then the vectors from D_1 have the form

$$\varphi = \sum_{k=1}^{\infty} \beta_k (Q_1 f_{1,k}),$$

where $(\beta_k)_{k=1}^{\infty}$, $\beta_k \in \mathbb{C}^1$, is an arbitrary finite sequence uniquely determined for given φ (the sequence of coordinates of φ).

Let us define the convergence in D_1: We say that $D_1 \ni \varphi^{(n)} \to \varphi \in D_1$ as $n \to \infty$ if the corresponding coordinates $\beta_k^{(n)}$ are uniformly finite and $\beta_k^{(n)} \underset{n\to\infty}{\to} \beta_k$ for every $k \in \mathbb{N}$. It is clear that this convergence corresponds to the following topology. Consider the Hilbert space

$$G_{1,\sigma} = \left\{ \sum_{k=1}^{\infty} \beta_k (Q_1 f_{1,k}) \mid \sum_{k=1}^{\infty} |\beta_k|^2 \sigma_k < \infty \right\}$$

$(\sigma = (\sigma_k)_{k=1}^{\infty}$, $\sigma_k \geq 1$ is a weight) with the corresponding scalar product, then $D_1 = \operatorname{pr\,lim} G_{1,\sigma}$, where the projective limit is taken over all weights σ.

IV. The imbedding $D_1 \subset H_{+,1}$ is topological. Thus, suppose that $D_1 \ni \varphi^{(n)} \to 0$ in D_1 as $n \to \infty$, where

$$\varphi^{(n)} = \sum_{k=1}^{\infty} \beta_k^{(n)} (Q_1 f_{1,k}).$$

This means that $\beta_k^{(n)}$ are uniformly finite and $\beta_k^{(n)} \underset{n\to\infty}{\to} 0$ for each $k \in \mathbb{N}$. Therefore,

$$D_0(\tau) \ni \sum_{k=1}^{\infty} \beta_k^{(n)} f_{1,k} = a^{(n)} \to 0$$

in $D(\tau)$ as $n \to \infty$. Since the imbedding $D(\tau) \subset \mathcal{H}(\tau)$ is topological, this implies that $a^{(n)} \to 0$ in $\mathcal{H}(\tau)$. Consequently, $Pa^{(n)} \to 0$ in $\mathcal{H}(\tau)$, and

$$\| Q_1 a^{(n)} \|_{H_{+,1}} = \kappa_1^{1/2} \| P a^{(n)} \|_{\mathcal{H}(\tau)} \to 0$$

by virtue of (3.33), i.e., $\varphi^{(n)} = Q_1 a^{(n)} \to 0$ in $H_{+,1}$.

V. Show that $D_1 \subset \mathfrak{D}(A_x)$ $(x \in X)$. It follows from (5.30) that the vector

$$f = \int_{\tau} F(\lambda(\cdot))\, dE(\lambda(\cdot))\, l,$$

where $l \in H_0$ and

$$\tau \ni \lambda(\cdot) \mapsto F(\lambda(\cdot)) \in \mathbb{C}^1$$

is $\mathfrak{C}_\sigma(\tau)$-measurable, belongs to $\mathfrak{D}(A_x)$ if and only if

$$\int_{\tau} |\lambda(x) F(\lambda(\cdot))|^2\, d(E(\lambda(\cdot))\, l, l)_{H_0} < \infty,$$

$$A_x f = \int_{\tau} \lambda(x) F(\lambda(\cdot))\, dE(\lambda(\cdot))\, l. \tag{3.34}$$

Let $a \in D_0(\tau) \subset D(\tau)$. Condition (ii) implies that

$$\lambda(x)\; a(\lambda(\cdot)) \in \mathcal{H}(\tau)$$

and, consequently, the function

$$\tau \ni \lambda(\cdot) \mapsto \lambda(x)\, a(\lambda(\cdot))\, \varepsilon(\lambda(\cdot)) \in \mathbb{C}^1$$

is bounded. Therefore, condition (3.34) is satisfied for the vector $\varphi = Q_1 a = f$ given by (3.32) (the vectors φ run through the whole of D_1).

VI. Show that the operator $A_x \upharpoonright D_1$ acts continuously from D_1 to $H_{+,1}$ $(x \in X)$.

Suppose that $D_1 \ni \varphi^{(n)} \to 0$ in D_1 as $n \to \infty$, where

$$\varphi^{(n)} = \sum_{k=1}^{n} \beta_k^{(n)} (Q_1 f_{1,k}) = Q_1 a^{(n)}$$

with

$$a^{(n)} = \sum_{k=1}^{\infty} \beta_k^{(n)} f_{1,k} \in D_0(\tau).$$

Applying (3.32) to $\varphi^{(n)}$ and then acting with the operator A_x, we obtain

$$A_x \varphi^{(n)} = Q_1(\lambda(x)\, a^{(n)}(\lambda(\cdot)))$$

according to (3.34). Thus, by virtue of (3.33), we have

$$\| A_x \varphi^{(n)} \|_{H_{+,1}} = \kappa_1^{1/2} \| P(\lambda(x)\, a^{(n)}(\lambda(\cdot))) \|_{\mathcal{H}(\tau)}.$$

But, just as in Step IV, $a^{(n)} \to 0$ in $D(\tau)$ as $n \to \infty$ and, hence, by virtue of (ii), $\lambda(x)\, a^{(n)}(\lambda(\cdot)) \to 0$ in $\mathcal{H}(\tau)$. Consequently,

$$P(\lambda(x)\, a^{(n)}(\lambda(\cdot))) \to 0$$

in $\mathcal{H}(\tau)$. Therefore, $A_x \varphi^{(n)} \to 0$ in $H_{+,1}$, as is required.

By analogy, we establish that the operator $A_x^* \upharpoonright D_1$ also acts continuously from D_1 to $H_{+,1}$ $(x \in X)$.

VII. Summarizing Steps II–VI, we can conclude that the chain

$$H_0 \supset H_{+,1} \supset D_1, \quad \| \varphi \|_{H_0} \le \| \varphi \|_{H_{+,1}}, \quad (\varphi \in H_{+,1}) \tag{3.35}$$

is constructed, where the imbedding $O_1 \colon H_{+,1} \to H_0$ is quasinuclear and $| O_1 | = \kappa_1^{-1/2}$; the imbedding $D_1 \subset H_{+,1}$ is topological, and D_1 is a separable projective limit of Hilbert spaces. For every $x \in X$, we have $D_1 \subset \mathfrak{D}(A_x)$, and the restrictions $A_x \upharpoonright D_1$ and $A_x^* \upharpoonright D_1$ act continuously from D_1 to $H_{+,1}$.

VIII. Assume that $H_{+,1}$ is not dense in H_0, and let $l_2 \in H_0$, $\| l_2 \|_{H_0} = 1$, be orthogonal to $H_{+,1}$ in H_0. Repeating the arguments on Steps II–VI with l_1 replaced by l_2 and fixing another $\kappa_2 \in [1, \infty)$ instead of κ_1, we get the chain

$$H_0 \supset H_{+,2} \supset D_2 \tag{3.36}$$

(instead of (3.35)) which has analogous properties.

By using condition (iii), it can be easily proved that the linear sets $H_{+,1}$ and $H_{+,2}$ are orthogonal in the space H_0. Indeed, taking into account that $l_2 \perp H_{+,1}$ and (3.32), we obtain

$$0 = (Q_1 a, l_2)_{H_0} = \int_\tau a(\lambda(\cdot))\, \varepsilon(\lambda(\cdot))\, d(E(\lambda(\cdot))\, l_1, l_2)_{H_0} \quad (a \in \mathcal{H}(\tau)).$$

By (iii), this implies that the finite charge

$$\mathfrak{C}_\sigma(\tau) \ni \alpha \mapsto (E(\alpha)\, l_1, l_2)_{H_0} \in \mathbb{C}^1$$

is equal to zero. Therefore, denoting by Q_2 an operator similar to Q_1 but constructed from l_2, we get from (3.32) that

$$(Q_1 a, Q_2 b)_{H_0} = \int_\tau a(\lambda(\cdot))\, \overline{b(\lambda(\cdot))}\, \varepsilon^2(\lambda(\cdot))\, d(E(\lambda(\cdot))\, l_1, l_2)_{H_0} \quad (a, b \in \mathcal{H}(\tau)),$$

i.e., $H_{+,1} \perp H_{+,2}$.

IX. It follows from (3.35) and (3.36) that

$$H_0 \supset H_{+,1} \oplus H_{+,2} \supset D_1 \oplus D_2,$$

where $\oplus$ denotes the orthogonal sum in H_0. If $H_{+,1} \oplus H_{+,2}$ is dense in H_0, then the construction is finished. If not, then we choose the vector $l_3 \in H_0$, $\|l_3\|_{H_0} = 1$, orthogonal to $H_{+,1} \oplus H_{+,2}$ and a number $\kappa_3 \in [1, \infty)$ and construct a chain analogous to (3.35) and (3.36), and so on. As a result, we obtain the chain

$$H_0 \supset \bigoplus_{m=1}^{\infty} H_{+,m} \supset \bigoplus_{m=1}^{\infty} D_m, \tag{3.37}$$

where $\oplus_{m=1}^{\infty} H_{+,m}$ is now dense in H_0 (this sum may contain finitely many terms, but for definiteness we consider the case of infinitely many terms). The numbers κ_m are chosen so that $\sum_{m=1}^{\infty} \kappa_m^{-1} < \infty$.

Let us transform

$$H_+ = \bigoplus_{m=1}^{\infty} H_{+,m}$$

into a Hilbert space by setting

$$(\varphi, \psi)_{H_+} = \sum_{m=1}^{\infty} (\varphi_m, \psi_m)_{H_{+,m}}$$

for

$$H_+ \ni \varphi = (\varphi_m)_{m=1}^{\infty} \qquad (\varphi_m \in H_{+,m})$$

and analogous $\psi \in H_+$. Further, denote by D the collection of all finite sequences $\varphi = (\varphi_m)_{m=1}^{\infty}$ $(\varphi_m \in D_m)$ with the following convergence: $D \ni \varphi^{(n)} \to \varphi \in D$ as $n \to \infty$ if the corresponding coordinates $\varphi_m^{(n)}$ are uniformly finite and $\varphi_m^{(n)} \to \varphi_m$ in D_m for every $m \in \mathbb{N}$. Since each D_m is a separable projective limit of Hilbert spaces (see Step III), the same is true for D, and the corresponding Hilbert space is

$$G_\sigma = \left\{ \sum_{m,k=1}^{\infty} \beta_{m,k}(Q_m f_{m,k}) \mid \sum_{m,k=1}^{\infty} |\beta_{m,k}|^2 \sigma_{m,k} < \infty \right\},$$

where Q_m and $f_{m,k}$ are the operator Q_1 and the vectors $f_{1,k}$ associated with the m-th chain of the form (3.35), respectively; $\sigma = (\sigma_{m,k})_{m,k=1}^{\infty}$, $\sigma_{m,k} \geq 1$, is a weight. Here $D = \text{pr lim } G_\sigma$, where the projective limit is taken over all the weights σ.

Clearly, $\|\cdot\|_{H_0} \leq \|\cdot\|_{H_+}$. Hence, by virtue of Step IV, D is topologically imbedded in H_+.

Thus, according to (3.37), we have constructed the chain

$$H_- \supset H_0 \supset H_+ \supset D. \tag{3.38}$$

The imbedding $O: H_+ \to H_0$ is quasinuclear since $O = \bigoplus_{m=1}^{\infty} O_m$, where $O_m : H_{+,m} \to H_0$ and, therefore,

$$|O|^2 = \sum_{m=1}^{\infty} |O_m|^2 = \sum_{m=1}^{\infty} \kappa_m^{-1} < \infty.$$

X. Show that the family of operators (3.30) is connected with (3.38) in the standard way. In fact, $\varphi \in D$ has the form of a finite sequence $\varphi = (\varphi_m)_{m=1}^{\infty}$, where $\varphi_m \in D_m$. By virtue of Step V, $D \subset \mathfrak{D}(A_x)$ with

$$A_x \varphi = (A_x \varphi_m)_{m=1}^{\infty} \quad (x \in X).$$

Further, if

$$(\varphi_m^{(n)})_{m=1}^{\infty} = \varphi^{(n)} \to 0$$

in D as $n \to \infty$, then in view of the nature of this convergence, the sequence

$$A_x \varphi^{(n)} = (A_x \varphi_m)_{m=1}^{\infty}$$

is finite uniformly in n, and $A_x \varphi_m^{(n)} \to 0$ in the space $H_{+,m}$ for each $m \in \mathbb{N}$ (see Step VI). This implies that $A_x \varphi^{(n)} \to 0$ in H_+.

■

Remark 1. Instead of the operators (3.30), we may consider functions of these. For example, the operators A_x given by (3.30) with $\tau \ni \lambda(\cdot) \mapsto \lambda(x) \in \mathbb{C}^1$ being replaced by some $\mathfrak{C}_\sigma(\tau)$-measurable function $\tau \ni \lambda(\cdot) \mapsto F_x(\lambda(\cdot)) \in \mathbb{C}^1$ $(x \in X)$. It is easy to see that Theorem 3.8 is preserved in this case provided that $\lambda(x)$ is replaced by $F_x(\lambda(\cdot))$ in condition (ii) of admissibility of the chain (3.31).

Remark 2. Let X be a topological space, and let $(A_x)_{x \in X}$ be a family of operators (3.30) given by the functions $F_x(\lambda(\cdot))$. The requirement (2.53) ensuring the continuity of eigenvalues is satisfied under a natural additional condition imposed on these functions. Namely, it is easy to see from Step VI that this requirement holds if for every $a \in D(\tau)$, the vector-valued function

$$X \ni x \mapsto \overline{F_x(\lambda(\cdot))}\, a(\lambda(\cdot)) \in \mathcal{H}(\tau)$$

is weakly continuous. Indeed, it suffices to consider the chain (3.35). For every $\varphi \in D_1$, one can find $a \in D_0(\tau) \subset \mathcal{H}(\tau)$ such that $\varphi = Q_1 a$; therefore,

$$A_x^* \varphi = Q_1\, \overline{F_x(\lambda(\cdot))}\, a(\lambda(\cdot)),$$

and by virtue of (3.33),

$$\begin{aligned}(A_x^* \varphi, Q_1 b)_{H_{+,1}} &= \kappa_1(P(F_x(\lambda(\cdot))\, a(\lambda(\cdot))), P b)_{\mathcal{H}(\tau)} \\ &= \kappa_1(\overline{F_x(\lambda(\cdot))}\, a(\lambda(\cdot)), P b)_{\mathcal{H}(\tau)}.\end{aligned}$$

This yields the required result because $Q_1 b$ varies over the whole of $H_{+,1}$ as b varies over $\mathcal{H}(\tau)$.

■

Remark 3. The results presented in this Subsection are also valid for operators of the type (3.30) constructed from an RI E on an abstract space τ.

■

We now give an example of the application of Theorem 5.7.

Theorem 3.9. *For an arbitrary at most countable family* $(A_x)_{x\in X}$ $(X = \{1, \dots, n\},\ n \le \infty)$ *of commuting normal operators, there always exists a quasinuclear rigging connected with it in the standard way.*

Proof. Suppose first that $n < \infty$. According to Theorem 3.8 and Subsection 1.4, it suffices to construct an admissible chain (3.31) in the case when $\tau = \mathbb{C}^n$, $\mathfrak{C}_\sigma(\tau) = \mathfrak{B}(\mathbb{C}^n)$, and $\lambda(\cdot) = \lambda \in \mathbb{C}^n$. We identify $\mathbb{C}^n$ with $\mathbb{R}^{2n}$ and set

$$\mathcal{H}(\tau) = W_2^l(\mathbb{R}^{2n}, (1 + |\lambda|^2)^l d\lambda), \quad l > n$$

(the Sobolev space with the weight $(1 + |\lambda|^2)^l$), and $D(\tau) = \mathcal{D}(\mathbb{R}^{2n})$ (for the information about these spaces, see Section 1 of Chapter 1, Examples 1.8 and 1.9). Conditions (i) and (ii) are obviously satisfied (for (i) this is a consequence of the imbedding theorems). Condition (iii) is also satisfied, because any function in $C_0^\infty(\mathbb{R}^{2n})$ can be taken to play the role of $a(\lambda)\varepsilon(\lambda)$. The assertion is thus proved for the case of finitely many operators.

Suppose that $n = \infty$. Without loss of generality, we can assume that $(A_x)_{x\in X}$, $X = \{1, 2, \dots\}$, are commuting selfadjoint operators. Here $\tau = \mathbb{R}^\infty$, $\mathfrak{C}_\sigma(\tau) = \mathfrak{B}(\mathbb{R}^\infty)$, and $\lambda(\cdot) = \lambda = (\lambda_j)_{j=1}^\infty$. We now exploit the following fact[32]:

If $\mathfrak{B}(\mathbb{R}^\infty) \ni \alpha \mapsto E(\alpha)$ *is a resolution of the identity, then there exists a weight* $\sigma = (\sigma_j)_{j=1}^\infty$ $(\sigma_j > 0)$ *such that*

$$l_2(\sigma) = \left\{ \lambda \in \mathbb{R}^\infty \,\middle|\, \sum_{j=1}^{\infty} \lambda_j^2 \sigma_j < \infty \right\} \in \mathfrak{B}(\mathbb{R}^\infty)$$

is of full E-measure.

Due to this fact, we can replace $\tau = \mathbb{R}^\infty$ by $l_2(\sigma)$ in the integrals in (3.30) and assume that $\sum_{j=1}^\infty \sigma_j < \infty$, $\sigma_j < 1$. Thus, according to Theorem 3.8, it suffices to construct an admissible chain (3.31) in the case when $\tau = l_2(\sigma)$ and $\mathfrak{C}_\sigma(\tau) = \mathfrak{B}(l_2(\sigma))$. This chain is constructed with the help of the spaces $A_t(\mathbb{R}^\infty)$ of test functions of infinitely many variables introduced in Subsection 4.3 of Chapter 2. Namely,

---

32 This fact is known in the case when E is replaced by a nonnegative finite measure ρ; see Yu.S. Samoilenko [1, Chapter 1, Section 1, Lemma 3]. Taking ρ to be a spectral measure of the RI E, we get the desired result.

$$\mathcal{H}(\tau) = (A_t(\mathbb{R}^\infty)) \upharpoonright l_2(\sigma) \quad (t = (t_j)_{j=1}^\infty : t_j = \sigma_j^{-1} - 1, \ j \in \mathbb{N}),$$

and $D(\tau) = \mathcal{P}_{\text{cyl}}(\mathbb{R}^\infty)$, where $\mathcal{P}_{\text{cyl}}(\mathbb{R}^\infty)$ is the linear space of cylindrical polynomials of a point $\lambda \in \mathbb{R}^\infty$. The required properties of the chain (3.31) follow from Subsection 4.3 in Chapter 2. The detailed derivation of these properties is presented in Section 1 of Chapter 4 in the proof of Theorem 1.2 (the generalization of Stone's theorem).

■

Remark 4. For the family of operators in Theorem 3.9, there always exists a nuclear rigging connected with it in the standard way. The proof follows from Theorem 3.9 if we take different Sobolev spaces (with $n < \infty$) or spaces $(A_\tau(\mathbb{R}^\infty)) \upharpoonright l_2(\sigma)$ to play the role of $\mathcal{H}(\tau)$ and then consider the projective limit of the obtained Hilbert spaces in H_0.

■

We also mention the following useful fact.

Remark 5. For an arbitrary at most countable family $A = (A_x)_{x \in X}$ of bounded and (in general) noncommuting operators in H_0, there exists a nuclear rigging connected with it in the standard way (the definition of this rigging is the same as in Subsection 2.5). To prove this, it suffices to consider the case when X is infinite. Let $(B_j)_{j=1}^\infty$ be a collection of operators A_x and A_x^* $(x \in X)$ enumerated somehow. Fix a vector $l_1 \in H_0$, $\| l_1 \|_{H_0} = 1$, and consider a countable set of vectors

$$L_1 = \left(\bigcup_{n=1}^\infty \bigcup_{j_1, \ldots, j_n = 1}^\infty B_{j_1} \ldots B_{j_n} l_1 \right) \cup \{ l_1 \}.$$

We enumerate this set and construct a system of vectors $(e_k)_{k=1}^\infty$ by successive orthogonalization (there may be finitely many vectors of this sort). Denote by D_1 the collection of all vectors of the form

$$\varphi = \sum_{k=1}^\infty \beta_k e_k,$$

where $(\beta_k)_{k=1}^\infty$ is a finite sequence of numbers from $\mathbb{C}^1$, and introduce a topology in D_1 which is the image of the standard projective topology in $\mathbb{R}_0^\infty$ under the mapping $\mathbb{R}_0^\infty \ni (\beta_k)_{k=1}^\infty \mapsto \varphi \in D_1$ (see Step III in the proof of Theorem 3.8). The space D_1 is nuclear; it is easy to understand that each operator A_x or A_x^* acts continuously in it. If D_1 is dense in H_0, then the construction is completed. If not, then we choose a vector

$l_2 \in H_0$, $\| l_2 \|_{H_0} = 1$ orthogonal to D_1 and, for given l_2, construct L_2 and the space D_2, by using the same procedure as above. Clearly, D_1 and D_2 are mutually orthogonal. This follows from the mutual orthogonality of L_1 and L_2 which is established by transferring all the operators B_j to l_1 in the scalar product of two vectors from L_1 and L_2. If $D_1 \oplus D_2$ is not dense in H_0, then we construct D_3, and so on. As in Step IX in the proof of Theorem 3.8, we introduce a topology in a (nonclosed) orthogonal sum $\oplus_{m=1}^{\infty} D_m = \Phi$. The space Φ is just the required space.

■

In Sections 1 and 3 of Chapter 4, we shall return to Theorems 3.8 and 3.9 and Remarks 1–5.

3.5. Connection with the Theory of Commutative Normalized Algebras and with the Nuclear Spectral Theorem

Let us clarify the interrelation between Spectral Projection Theorem 2.7 and the nuclear spectral theorem established by Maurin in [1]–[4].

Let $A = (A_x)_{x \in X}$ be a family of *bounded* commuting normal operators. Without loss of generality, we can assume that A contains 1. Take the algebraic span of A and close it with respect to the operator norm convergence. As a result, we obtain a commutative C^*-algebra $\mathcal{A}$ of operators with identity.

Let M be the compact set of its maximal ideals ω. By virtue of the Gelfand–Naimark theorem, $\mathcal{A}$ is isometrically isomorphic to the algebra $C(M)$ of all complex-valued continuous functions on M with the usual algebraic operations and the uniform norm

By using this theorem, we can prove that, on the Borel subsets of M, there exists an RI $\mathfrak{B}(M) \ni \alpha \mapsto F(\alpha)$ which gives a representation of each operator in the algebra $\mathcal{A}$ in the form of a spectral integral

$$a = \int_M a(\omega)\, dF(\omega) \qquad (a \in \mathcal{A}). \tag{3.39}$$

Here $a(\omega)$ is the value of a on a maximal ideal ω, i.e., $a(\cdot)$ is the image of a under the isomorphism $\mathcal{A} \leftrightarrow C(M)$ (see Naimark [1], Chapter 3, Section 16, Subsection 2; Chapter 4, Section 17, Subsection 4).

We interpret a maximal ideal ω as a continuous linear multiplicative functional ω on $\mathcal{A}$, namely, $\omega(a) = a(\omega)$ $(a \in \mathcal{A})$. Recall that the topology in M is the relative topology induced by the Tykhonov topology in the space $\mathbb{C}^{\mathcal{A}} \supset M$.

Let $\omega \in M$. We set $\lambda_\omega(x) = \omega(A_x)$ $(x \in X)$; $\lambda_\omega = \omega \upharpoonright A$. The mapping $\mathbb{C}^{\mathcal{A}} \supset M$

$\ni \omega \mapsto \lambda_\omega(\cdot) \in \mathbb{C}^X$ is one-to-one, namely, if $\omega_1(A_x) = \omega_2(A_x)$ $(x \in X)$, then by linearity, multiplicativity, and continuity, this equality can be extended onto the whole of $\mathcal{A}$, i.e., $\omega_1 = \omega_2$. Assume that the compact set $\mu \subset \mathbb{C}^X$ is the image of M under this mapping topologized by the image of the topology. This means that, in fact, μ is the same compact set of maximal ideals. We introduce the RI $\mathfrak{B}(\mu) \ni \alpha \mapsto G(\alpha)$ which is the image of the RI F. For $a = A_x$ the representation (3.39) takes the form

$$A_x = \int_\mu \lambda(x)\, dG(\lambda(\cdot)) \quad (x \in X). \tag{3.40}$$

The intersections of cylindrical sets

$$\mathfrak{C}(\mathbb{C}^{\mathcal{A}}) = \mathfrak{C}(a_1, \ldots, a_n;\ u_1 \times \ldots \times u_n) \quad (a_k \in \mathcal{A},\ \ u_k \text{ are open in } \mathbb{C}^1)$$

in $\mathbb{C}^{\mathcal{A}}$ with M serve as base neighborhoods in M. Their images form a neighborhood basis $\Sigma(\mu)$ in the space μ. If $a_k = A_{x_k}$ $(k = 1, \ldots, n)$, then the corresponding image obviously coincides with a base neighborhood in the space μ topologized by the relative topology from $\mathbb{C}^X$. Let $\Sigma'(\mu)$ be the neighborhood basis for μ topologized in this way. Thus,

$$\Sigma'(\mu) \subset \Sigma(\mu). \tag{3.41}$$

Consider a σ-algebra

$$\mathfrak{C}_\sigma(\mu) = \{\alpha' = \alpha \cap \mu \mid \alpha \in \mathfrak{C}_\sigma(\mathbb{C}^X)\}$$

which coincides with the σ-span of the basis $\Sigma'(\mu)$ and belongs to $\mathfrak{B}(\mu)$, by virtue of (3.41). The function $\mathbb{C}^X \supset \mu \ni \lambda(\cdot) \mapsto \lambda(x) \in \mathbb{C}^1$ is $\mathfrak{C}_\sigma(\mu)$-measurable and, thus, the measure G (in (3.40)) can be replaced by its restriction $E' = G \restriction \mathfrak{C}_\sigma(\mu)$ which is an RI again:

$$A_x = \int_\mu \lambda(x)\, dE'(\lambda(\cdot)) \quad (x \in X). \tag{3.42}$$

It is not hard to see that E' is a modification of the joint RI E of the family A with respect to μ.

Indeed, according to Subsection 1.8, we can construct the RI E'' on $\mathfrak{C}_\sigma(\mathbb{C}^X)$ for the given RI E' on $\mathfrak{C}_\sigma(\mu)$, namely,

$$\mathfrak{C}_\sigma(\mathbb{C}^X) \ni \alpha \mapsto E''(\alpha) = E'(\alpha \cap \mu).$$

The integral (3.42) can be now written in the form (1.25) with E replaced by E''. But $E'' = E$. In fact, according to Remark 1 in Subsection 1.3, $E'' = \times_{x \in X} E''_x$, where $\mathfrak{B}(\mathbb{C}^1) \ni \alpha \mapsto E''_x(\alpha)$ is a certain RI. For E'', the equality (1.25) now yields

$$A_x = \int_{\mathbb{C}^1} \lambda \, d E''_x(\lambda),$$

whence $E''_x = E_x$ $(x \in X)$, and $E'' = E$. Thus,

$$\mathfrak{C}_\sigma(\mathbb{C}^X) \ni \alpha \mapsto E(\alpha) = E(\alpha \cap \mu),$$

and this implies that μ is of full outer E-measure, and the modification of E by μ leads to E'.

■

Generally speaking, the inclusion (3.41) is strict. However, the topologies determined by $\Sigma'(\mu)$ and $\Sigma(\mu)$ coincide. Thus, the following lemma takes place.

Lemma 3.5. *The topologization of* $\mu \subset \mathbb{C}^X$ *as the space of maximal ideals coincides with the relative topology induced by the space* $\mathbb{C}^X$.

Proof. It suffices to show that for any point $\omega_0 \in M$ and any neighborhood $\mathfrak{C}_\delta(\mathbb{C}^{\mathcal{A}}) \cap M$ of it in M, there exists a neighborhood of the point $\lambda_{\omega_0}(\cdot) \in \mu$ in μ of the form $\mathfrak{C}_\delta \cap \mu$, where

$$\mathfrak{C}_\delta = \mathfrak{C}(x_1, \dots, x_m; v_1 \times \dots \times v_m) \quad (x_k \in X,\ v_k \text{ are open in } \mathbb{C}^1),$$

such that if $\lambda_\omega(\cdot) \in \mathfrak{C}_\delta \cap \mu$, then $\omega \in \mathfrak{C}_\delta(\mathbb{C}^{\mathcal{A}}) \cap M$.

Let

$$u_j = \{ z \in \mathbb{C}^1 \mid |z - \omega_0(a_j)| < \varepsilon \} \quad (j = 1, \dots, n;\ \varepsilon > 0),$$

and let $\mathcal{A}'$ be the algebraic span of the family $(A_x)_{x \in X}$. For every a_j, one can find $b_j \in \mathcal{A}'$ such that

$$\| a_j - b_j \| < \varepsilon/3 \quad (j = 1, \dots, n),$$

suppose that $A_{x_1}, \dots, A_{x_m}$ are the operators whose algebraic span contains $b_1, \dots, b_n$. Hence,

$$b_j = \sum_{|\alpha| \le p_j} c_{\alpha,j} A_{x_1}^{\alpha_1} \dots A_{x_m}^{\alpha_m}, \tag{3.43}$$

$$\omega(b_j) = \sum_{|\alpha| \le p_j} c_{\alpha,j} \lambda_\omega^{\alpha_1}(x_1) \dots \lambda_\omega^{\alpha_m}(x_m) = P_j(\lambda_\omega(x_1), \dots, \lambda_\omega(x_m)),$$

$$(\alpha = (\alpha_1, \dots, \alpha_m) \in \mathbb{Z}_+^m,\ |\alpha| = \alpha_1 + \dots + \alpha_m;\ j = 1, \dots, n;\ \omega \in M),$$

where $c_{\alpha,j} \in \mathbb{C}^1$ are coefficients. The polynomial $P_j(z_1, \dots, z_m)$ $(z_k \in \mathbb{C}^1)$ takes the value $\omega_0(b_j)$ at the point

$$(z_1^0, \dots, z_m^0) = (\lambda_{\omega_0}(x_1), \dots, \lambda_{\omega_0}(x_m)).$$

Due to the continuity of it, there exists $\delta > 0$ such that

$$|\omega_0(b_j) - P_j(z_1, \dots, z_m)| < \varepsilon/3$$

for all $j = 1, \dots, n$ provided that $|z_k - z_k^0| < \delta$ $(k = 1, \dots, m)$. The neighborhood $\mathbb{C}_\delta$ with x_k chosen above and with $v_k = \{z \in \mathbb{C}^1 \mid |z_k - z_k^0| < \delta\}$ is just the required one. Indeed, if $\lambda_\omega(\cdot) \in \mathbb{C}_\delta \cap \mu$, then

$$|\omega_0(b_j) - \omega(b_j)| < \varepsilon/3 \quad (j = 1, \dots, n),$$

according to the second equality in (3.43), and

$$|\omega(a_j) - \omega_0(a_j)| \le |\omega(a_j) - \omega(b_j)| + |\omega(b_j) - \omega_0(b_j)|$$

$$+ |\omega_0(b_j) - \omega_0(a_j)| \le 2\|a_j - b_j\| + |\omega(b_j) - \omega_0(b_j)| < \varepsilon \quad (j = 1, \dots, n),$$

since the norm of a multiplicative functional is equal to 1; but this means that $\omega \in \mathbb{C}_\delta(\mathbb{C}^{\mathcal{A}}) \cap M$.

■

Theorem 3.10. *The support of the joint RI E of the family of operators under consideration coincides with the space of maximal ideals of the algebra $\mathcal{A}$; more precisely* supp $E = \mu$.

Proof. Suppose that $\varphi \in \mathfrak{C}_\sigma(\mathbb{C}^X)$ is of full E-measure and is closed in $\mathbb{C}^X$. Then $\varphi \supset \mu$. Indeed, assume the contrary. Then one can find a point $\lambda_0(\cdot) \in (\mathbb{C}^X \setminus \varphi) \cap \mu$. The last set is open in μ in the relative topology. Let $\mathbb{C}_\delta \cap \mu$ be a base neighborhood of the point $\lambda_0(\cdot)$ contained in $(\mathbb{C}^X \setminus \varphi) \cap \mu$. Since

$$E'((\mathbb{C}^X \setminus \varphi) \cap \mu) = E(\mathbb{C}^X \setminus \varphi) = 0,$$

we have

$$G(\mathbb{C}_\delta \cap \mu) = E'(\mathbb{C}_\delta \cap \mu) = 0.$$

Let o be the inverse image of $\mathbb{C}_\delta \cap \mu$ under the mapping $M \to \mu$. The set o is open in the topology of M, and $F(o) = 0$, but this contradicts (3.39) and the isomorphism $\mathcal{A} \leftrightarrow C(M)$. Therefore, $\varphi \supset \mu$ and, thus, $\operatorname{supp} E = \mu$.

On the other hand, as proved above, μ is a compact set in the relative topology of the space $\mathbb{C}^X$; hence, μ is closed in $\mathbb{C}^X$. Assume that $\operatorname{supp} E \neq \mu$. Then there exists a base neighborhood $\mathbb{C}_\delta \subset \mathbb{C}^X \setminus \mu$ having a nonempty intersection with $\operatorname{supp} E$. Therefore, $E(\mathbb{C}_\delta) \neq 0$, and the set $\mathfrak{C}_\sigma(\mathbb{C}^X) \ni \mathbb{C}^X \setminus \mathbb{C}_\delta \supset \mu$ is not of full E-measure. Consequently, μ is not of full outer E-measure, but this is absurd.

■

Thus, in the "algebraic" approach, the spectral representation (3.39) is valid for a family of bounded operators and this yields (3.42), i.e., (1.25). To proceed further with constructing a theory of expansions, we can use two schemes:

(i) differentiate the RI in (1.25) or in (3.42). This idea has been exploited in Section 2 (it proves to be less rational to differentiate F in (3.39) because the measure is now defined on the special subset M of $\mathbb{C}^{\mathcal{A}}$);

(ii) extend (3.39) to the full spectral theorem in the von Neumann form and use Fubini's theorem.

We outline scheme (ii). For simplicity, we assume that the family of operators possesses a cyclic vector $\Omega \in H_0$ (with norm 1), i.e., the set $\{a\Omega \mid a \in \mathcal{A}\}$ is dense in H_0. It is well known that in this case the isomorphism $\mathcal{A} \leftrightarrow C(M)$ generates an isometry

$$H_0 \ni f \mapsto \hat{f} = Gf \in L_2(M, d\sigma(\omega)) = L_2(M, \mathfrak{B}(M), d\sigma),$$

where

$$\mathfrak{B}(M) \ni \alpha \mapsto \sigma(\alpha) \in [0, \infty)$$

is a finite measure and $(af)^\wedge(\omega) = \omega(a)\hat{f}(\omega)$ $(f \in H_0,\ a \in \mathcal{A})$ for almost all ω with respect to σ (see Naimark [1, Sections 40 and 41 in Chapter 8] and Berezansky [18, Chapter 2, Section 3, Subsection 3]). Thus, the "Parseval inequality" of the type (3.7) (with $N_{\lambda(\cdot)} = 1$) holds, namely,

$$(f, g)_{H_0} = \int_M \hat{f}(\omega)\,\overline{\hat{g}(\omega)}\,d\sigma(\omega) \qquad (f, g \in H_0). \tag{3.44}$$

In a large part, the theory of expansions in generalized eigenvectors is reduced to obtaining representations of type (3.4) for $\hat{f}(\omega)$. Namely, for σ-almost all $\omega \in M$ and for f from some dense linear subset of H_0, the representation $\hat{f}(\omega) = (f, \xi(\omega))_{H_0}$ should hold, where $\xi(\omega)$ is a generalized eigenvector (in a certain sense). This representation can be obtained in the following simple way.

Assume that a quasinuclear chain (2.1) is given. Let $(e_j)_{j=1}^{\infty}$ be an orthonormal basis in H_+. Due to the fact that O is quasinuclear, the operator $GO : H_+ \to L_2(M, d\sigma(\omega))$ is quasinuclear, too; therefore, by Fubini's theorem, we find that

$$\infty > \sum_{j=1}^{\infty} \| GO \|^2_{L_2(M,\sigma)} = \sum_{j=1}^{\infty} \int_M |(GOe_j)(\omega)|^2\, d\sigma(\omega)$$

$$= \int_M \left(\sum_{j=1}^{\infty} |(GOe_j)(\omega)|^2 \right) d\sigma(\omega).$$

This enables us to conclude that $\sum_{j=1}^{\infty} |(GOe_j)(\omega)|^2 < \infty$ for σ-almost all $\omega \in M$. Therefore, the following vector is defined for ω:

$$\chi(\omega) = \sum_{j=1}^{\infty} \overline{(GOe_j)(\omega)}\, e_j \in H_+,$$

$$(\varphi, \chi(\omega))_{H_+} = \sum_{j=1}^{\infty} (\varphi, e_j)_{H_+} (GOe_j)(\omega) \quad (\varphi \in H_+). \tag{3.45}$$

But $\varphi = \sum_{j=1}^{\infty} (\varphi, e_j)_{H_+} e_j$. Applying the operator GO to this equality and using the second relation in (3.45), we get

$$\hat{\varphi}(\omega) = (GO\varphi)(\omega) = (\varphi, \chi(\omega))_{H_+} = (\varphi, \boldsymbol{I}^{-1}\chi(\omega))_{H_0} \quad (\varphi \in H_+).$$

Hence, if we set $\xi(\omega) = \boldsymbol{I}^{-1}\chi(\omega) \in H_-$, then

$$\hat{\varphi}(\omega) = (\varphi, \xi(\omega))_{H_0}, \quad \xi(\omega) \in H_-. \tag{3.46}$$

for σ-almost all $\omega \in M$.

Suppose that the chain (2.1) is extended to (2.18) and that the operator $a \in \mathcal{A}$ is such that $a^* \restriction D$ acts from D to H_+ (it is possible that this action is not continuous. If the family A is connected with (2.18) in the standard way, then this property is automatically satisfied for $a = A_x \in A$). The vector $\xi(\omega)$ is a generalized eigenvector of a with the eigenvalue $\omega(a)$; this should be understood in the following sense:

$$(\xi(\omega), a^*\varphi)_{H_0} = \omega(a)\,(\xi(\omega), \varphi)_{H_0} \quad (\varphi \in D).$$

Indeed, according to (3.46), we have

$$(\xi(\omega), a^*\varphi)_{H_0} = \overline{(a^*\varphi)^\wedge(\omega)} = \overline{\omega(a^*)\,\hat{\varphi}(\omega)} = \omega(a)\,(\xi(\omega), \varphi)_{H_0}.$$

We introduce the operator $P(\omega)\colon H_+ \to H_-$ by setting

$$P(\omega)\,\varphi = (\varphi, \xi(\omega))_{H_0}\,\xi(\omega) \quad (\varphi \in H_+).$$

Then

$$(P(\omega)\varphi, \psi)_{H_0} = (\varphi, \xi(\omega))_{H_0}\,\overline{(\psi, \xi(\omega))_{H_0}},$$

and (3.44) can be rewritten in a form similar to (3.1) (with $\alpha = \tau$), namely,

$$(\varphi, \psi)_{H_0} = \int_M (P(\omega)\varphi, \psi)_{H_0}\,d\sigma(\omega) \qquad (\varphi, \psi \in H_+). \tag{3.47}$$

Above, we have assumed that the family of operators possesses a cyclic vector Ω. In the general case, one should proceed as follows (this is the derivation of the spectral theorem in the von Neumann form):

The separable space H_0 is representable in the form of an orthogonal sum of Hilbert spaces $H_0^{(\gamma)}$, i.e.,

$$H_0 = \bigoplus_{\gamma=1}^{\infty} H_0^{(\gamma)}$$

(this sum may be finite), where each $H_0^{(\gamma)}$ contains a cyclic vector Ω. To prove this, fix a unit vector Ω_1 and let $H_0^{(1)}$ be the closure of $\{a\,\Omega_1 \mid a \in \mathcal{A}\}$. Then take a unit vector $\Omega_2 \perp H_0^{(1)}$ and construct $H_0^{(2)}$ in a similar way; $H_0^{(2)} \perp H_0^{(1)}$, since

$$(b\,\Omega_2, a\,\Omega_1)_{H_0} = (\Omega_2, b^*a\,\Omega_1)_{H_0} = 0$$

($b^*a \in \mathcal{A}$: $\mathcal{A}$ is a C^*-algebra), and so on. Each $H_0^{(\gamma)}$ is isometric to $L_2(M, d\sigma_\gamma(\omega))$

with some measure on $\mathfrak{B}(M)$ and, as a result, we get the isometry

$$H_0 \ni f \mapsto \hat{f} = Gf \in \bigoplus_{\gamma=1}^{\infty} L_2(M, d\sigma_\gamma(\omega)); \quad \hat{f}(\omega) = (\hat{f}_1(\omega), \hat{f}_2(\omega), \ldots).$$

It is easy to see that the measures σ_γ can be chosen to be absolutely continuous with respect to some single measure σ on $\mathfrak{B}(M)$. By introducing in $\hat{f}$ the corresponding factors connected with the derivatives $(d\sigma_\gamma / d\sigma)(\omega)$, we can pass to the orthogonal sum of the spaces $L_2(M, d\sigma(\omega))$ and write the Parseval inequality of the type (3.44) for for vector-valued functions $\hat{f}(\omega)$. Then one should repeat (with minor additions) the argument leading to formulas (3.46) and (3.47).

Let us modernize the last constructions by making them closer to scheme (i). As before, suppose first that there exists a cyclic vector Ω. The mapping

$$\mathbb{C}^{\mathcal{A}} \supset M \mapsto \lambda_\omega(\cdot) \in \mu \subset \mathbb{C}^X \quad (\lambda_\omega(x) = \omega(A_x),\ x \in X)$$

transforms $\mathfrak{B}(M)$ into $\mathfrak{B}(\mu)$, the measure σ is transformed (under this mapping) into some measure $\mathfrak{B}(\mu) \ni \alpha \mapsto \rho(\alpha) \in [0, \infty)$, and $L_2(M, d\sigma(\omega))$ is isometrically transformed into $L_2(\mu, d\rho(\lambda(\cdot))) = L_2(\mu, \mathfrak{B}(\mu), \rho)$. Taking this isometry into account, we can assume that

$$H_0 \ni f \mapsto Gf = \tilde{f}(\lambda(\cdot)) \in L_2(\mu, d\rho(\lambda(\cdot))) \qquad (\tilde{f}(\lambda(\cdot)) = \hat{f}(\omega) \upharpoonright \mu, \lambda(\cdot) \leftrightarrow \omega),$$

and (3.44), (3.46), and (3.47) take the following form

$$(f, g)_{H_0} = \int_\mu \tilde{f}(\lambda(\cdot)) \overline{\tilde{g}(\lambda(\cdot))} d\rho(\lambda(\cdot)) \qquad (f, g \in H_0),$$

$$\tilde{\varphi}(\lambda(\cdot)) = (\varphi, \xi(\lambda(\cdot)))_{H_0} \qquad (\varphi \in H_+,\ \xi(\lambda(\cdot)) \in H_-),$$

$$(\varphi, \psi)_{H_0} = \int_\mu (P(\lambda(\cdot))\varphi, \psi)_{H_0} d\rho(\lambda(\cdot)) \qquad (\varphi, \psi \in H_+). \tag{3.48}$$

The vector $\xi(\lambda(\cdot)) = \xi(\omega)\ (\lambda(\cdot) \leftrightarrow \omega)$ is a generalized eigenvector of the operator A_x with the eigenvalue $\omega(A_x) = \lambda(x)\ (x \in X)$. Therefore, formulas (3.48) are similar to (3.5) and (3.1) (with $\alpha = \tau = \pi$). In particular, the integrals are taken over the set of eigenvalues, as before. In the general case, when there is no cyclic vector, the situation is analogous.

It may seem that formulas (3.48) (and the corresponding more general formulas for the case when there is no cyclic vector) are the same as in the main scheme (i), and thereby, that this scheme may be avoided in the case of bounded operators A_x $(x \in X)$ if we consider the C^*-algebra generated by these operators.

However, *if X is more than countable, then this is not true.*

Indeed, when we exploit scheme (ii), the RI G and the measure ρ are defined on the σ-algebra $\mathfrak{B}(\mu)$ of Borel sets, and the generalized projector $\mu \ni \lambda(\cdot) \mapsto P(\lambda(\cdot))$ is weakly measurable with respect to this σ-algebra. In the case of scheme (i), E, ρ, and $P(\lambda(\cdot))$ are connected with the σ-algebra

$$\mathfrak{C}_\sigma(\mu) = \{\alpha' = \alpha \cap \mu | \alpha \in \mathfrak{C}_\sigma(\mathbb{C}^X)\} \subset \mathfrak{B}(\mu),$$

which is smaller than $\mathfrak{B}(\mu)$ (thus, for instance, a one-point set $\{\lambda(\cdot)\} \in \mathfrak{B}(\mu)$ does not belong to $\mathfrak{C}_\sigma(\mu)$ for sufficiently massive μ, and it does not belong to $\mathfrak{C}_\sigma(\mathbb{C}^X)$ when X is more than countable; see Subsection 1.4). This means that the information about the function $\mu \ni \lambda(\cdot) \mapsto P(\lambda(\cdot))$ obtained in scheme (ii) is poorer than in scheme (i). In fact, the collection of Lebesgue sets of this function, i.e., the collection of sets

$$\{\lambda(\cdot) \in \mu | \, (P(\lambda(\cdot))\varphi, \psi)_{H_0} \in \delta \in \mathfrak{C}_\sigma(\mathbb{C}^1); \;\; \varphi, \psi \in H_+\}$$

is automatically smaller than $\mathfrak{B}(\mu)$. (Certainly, one could first restrict G and ρ to $\mathfrak{C}_\sigma(\mu)$ and then perform the differentiation, but this is a repetition of scheme (i).)

■

Let us dwell briefly upon scheme (ii) in the general case when some of the operators A_x can be unbounded.

Below, we employ the well known facts presented, for example, in the books by Naimark [1, Chapter 8, Section 41] and Berezansky [18, Chapter 2, Section 3, Subsection 3]. Consider the algebraic span of the family of projectors $(E_x(\alpha))_{x \in X, \, \alpha \in \mathfrak{B}(\mathbb{C}^1)}$ and close it with respect to the convergence in operator norm. Let M be the maximal ideal space of the C^*-algebra $\mathcal{A}$ with identity obtained in this way. Assume that $\mathcal{A}$ has a cyclic unit vector $\Omega \in H_0$: the set $\{a\Omega \,|\, a \in \mathcal{A}\}$ is dense in H_0. As before, the isomorphism $\mathcal{A} \leftrightarrow C(M)$ generates the isometry

$$H_0 \ni f \mapsto Gf = \hat{f} \in L_2(M, d\sigma(\omega)),$$

and relations (3.44)–(3.47) remain valid in this case.

However, if the operator A_x is unbounded, then it does not belong to $\mathcal{A}$, and it is no longer possible to interpret $\xi(\omega)$ directly as a generalized eigenvector of this operator. At the same time, it is known that the correspondence between $a \in \mathcal{A}$ and the operators of multiplication by the functions $a(\omega) = \omega(a)$ in the space $L_2(M, d\sigma(\omega))$ can be extended to the operators A_x $(x \in X)$, and only the functions $A_x(\omega)$ corresponding to these operators are not continuous in ω (they are measurable and σ-almost everywhere finite). As before, this implies that $\xi(\omega)$ is a generalized eigenvector of A_x corresponding to the eigenvalue $A_x(\omega)$. If A does not possess a cyclic vector Ω, then we repeat the above argument concerning the decomposition

$$H_0 = \bigoplus_{\gamma=1}^{\infty} H_0^{(\gamma)},$$

and easily obtain the Parseval equality in the form (3.47).

It is no longer possible to go down from (3.47) to (3.48) and pass to the integration over the spectrum. Indeed, the function $X \ni x \mapsto A_x(\omega) \in \mathbb{C}^1$ cannot be extended by linearity, multiplicativity, and continuity to the whole of the algebra $\mathcal{A}$. Furthermore, for fixed ω, it is generally not well-defined, since $A_x(\omega)$ is defined only for σ-almost all ω for every $x \in X$. This is the difference between the representation (3.47) and the representation in the Spectral Projection Theorem 2.7.

3.6. The Example of a Family of Commuting Selfadjoint Operators: Operators Acting upon Different Variables

Operators acting upon different variables x_n are described by the following natural general scheme. Consider a sequence of Hilbert spaces $(H_{0,n})_{n=1}^{\infty}$, fix a sequence of unit vectors $e = (e^{(n)})_{n=1}^{\infty}$, $e^{(n)} \in H_{0,n}$, and construct the infinite tensor product $\mathcal{H}_{0,e} = \otimes_{n=1;e}^{n} H_{0,n}$ with a stabilizing sequence e. Assume that a selfadjoint operator A_n with a domain of definition $\mathfrak{D}(A_n)$ acts in each $H_{0,n}$. Given A_n, we construct an operator $\mathcal{A}_n'$ acting in the space $\mathcal{H}_{0,e}$ by setting

$$\mathcal{A}_n' = \underbrace{1 \otimes \dots \otimes 1}_{n-1} \otimes A_n \otimes 1 \otimes 1 \otimes \dots \quad (n \in \mathbb{N}).$$

This tensor product is defined, according to the constructions presented in Subsections 2.1 and 2.2 of Chapter 1, in the following natural way which generalizes the transition from d/dx_n to $\partial/\partial x_n$ (see (2.7) in Chapter 1):

$$\mathfrak{D}(\mathcal{A}_n') = \text{a.}(H_{0,n} \otimes \dots \otimes H_{0,n-1} \otimes \mathfrak{D}(A_n) \otimes H_{0,n+1} \otimes H_{0,n+2} \otimes \dots),$$

$$\mathcal{A}_n'(f^{(1)} \otimes \dots \otimes f^{(n-1)} \otimes f^{(n)} \otimes f^{(n+1)} \otimes f^{(n+2)} \otimes \dots)$$

$$= f^{(1)} \otimes \dots \otimes f^{(n-1)} \otimes (A_n f^{(n)}) \otimes f^{(n+1)} \otimes f^{(n+2)} \otimes \dots \quad (3.49)$$

$$(f^{(j)} \in H_{0,j},\ j \neq n;\ f^{(n)} \in \mathfrak{D}(A_n);\ n \in \mathbb{N})$$

(this definition involves the inclusion $e^{(n)} \in \mathfrak{D}(A_n)$ for any $n \in \mathbb{N}$). It is clear that $\mathcal{A}_n'$ is a Hermitian operator. Taking into account that

$$\mathcal{A}_n' - z1 = \underbrace{1 \otimes \ldots \otimes 1}_{n-1} \otimes (A_n - z1) \otimes 1 \otimes 1 \otimes \ldots \quad (z \in \mathbb{C}^1),$$

we conclude that each operator $\mathcal{A}_n'$ is essentially selfadjoint. Denote by $\mathcal{A}_n$ the closure of $\mathcal{A}_n'$. It is easy to see that the resolutions of the identity

$$\mathfrak{B}(\mathbb{R}^1) \ni \alpha \mapsto \mathcal{E}_n(\alpha) = \underbrace{1 \otimes \ldots \otimes 1}_{n-1} \otimes E_n(\alpha) \otimes 1 \otimes 1 \otimes \ldots \quad (n \in \mathbb{N}) \tag{3.50}$$

correspond to the operators $\mathcal{A}_n$. By virtue of (2.8) in Chapter 1, these RI commute.

Thus, for a given family $(A_n)_{n=1}^{\infty}$, we have constructed a family $\mathcal{A} = (\mathcal{A}_n)_{n=1}^{\infty}$ of commuting selfadjoint operators acting in $\mathcal{H}_{0,e}$. We say that they act "upon different variables". Here $X = \mathbb{N}$, $\mathbb{R}^X = \mathbb{R}^{\infty}$, and it is convenient to denote the points $\lambda(\cdot)$ of the space $\mathbb{R}^{\infty}$ by $\lambda = (\lambda_n)_{n=1}^{\infty}$. Denote by $\mathfrak{C}_{\sigma}(\mathbb{R}^{\infty}) \ni \alpha \mapsto \mathcal{E}(\alpha)$ the joint RI of the family $\mathcal{A}$.

Suppose that we know the expansion in generalized eigenvectors for each operator A_n $(n \in \mathbb{N})$. This means that we have a chain

$$H_{-,n} \supset H_{0,n} \supset H_{+,n} \tag{3.51}$$

with the quasinuclear imbedding $O_n : H_{+,n} \to H_{0,n}$ and, for each A_n, we know its spectral measure

$$\mathfrak{B}(\mathbb{R}^1) \ni \alpha \mapsto \rho_n(\alpha) = \operatorname{Tr}(O_n^+ E_n(\alpha)\, O_n),$$

and the operator of generalized projection $P_n(\lambda_n): H_{+,n} \to H_{-,n}$ $(\lambda_n \in \mathbb{R}^1)$. By using these data, we now construct an expansion for the whole family $(\mathcal{A}_n)_{n=1}^{\infty}$.

Assume that $e^{(n)} \in H_{+,n}$ and

$$\| e^{(n)} \|_{H_{+,n}} = \| e^{(n)} \|_{H_{0,n}} = 1 \quad (n \in \mathbb{N}).$$

Without loss of generality, we can suppose that

$$\sum_{n=1}^{\infty} (| O_n |^2 - 1) < \infty \tag{3.52}$$

(in order to guarantee (3.52), one should renormalize each $H_{+,n}$ according to Lemma 2.5 in Chapter 1).

Taking the tensor product of the chains (3.51), we get the chain

$$\mathcal{H}_{-,e} = \bigotimes_{n=1;e}^{\infty} H_{-,n} \supset \mathcal{H}_{0,e} = \bigotimes_{n=1;e}^{\infty} H_{0,n} \supset \mathcal{H}_{+,e} = \bigotimes_{n=1;e}^{\infty} H_{+,n} \qquad (3.53)$$

(see Lemma 2.4 in Chapter 1) with the quasinuclear (due to (3.52)) imbedding $\mathcal{H}_{+,e} \to \mathcal{H}_{0,e}$. By virtue of Subsection 2.2 in Chapter 1, the equality $\mathbb{O} = \otimes_{n=1}^{\infty} O_n$ holds for the imbedding operator $\mathbb{O}: \mathcal{H}_{+,e} \to \mathcal{H}_{0,e}$, i.e.,

$$\mathbb{O}\varphi = (O_1 \varphi^{(1)}) \otimes (O_2 \varphi^{(2)}) \otimes \ldots \in \text{a.}\left(\bigotimes_{n=1;e}^{\infty} H_{0,n}\right) \subset \mathcal{H}_{0,e}$$

for

$$\varphi = \varphi^{(1)} \otimes \varphi^{(2)} \otimes \ldots \in \text{a.}\left(\bigotimes_{n=1;e}^{\infty} H_{+,n}\right) \subset \mathcal{H}_{+,e}.$$

The analogous relations take place when $\mathbb{O}$ and O_n are replaced by $\mathbb{O}^+$ and O_n^+ or $\mathbb{O}^+\mathbb{O}$ and $O_n^+ O_n$, respectively. The chain (3.53) plays the role of the chain (2.1).

Theorem 3.11. *The spectral measure $\mathfrak{C}_\sigma(\mathbb{R}^\infty) \ni \alpha \mapsto \rho(\alpha)$ and the operator $\mathcal{P}(\lambda)$ of generalized projection for the family $\mathcal{A}$ are given by the following formulas*

$$\rho = \bigotimes_{n=1}^{\infty} \rho_n, \quad \mathcal{P}(\lambda) = \bigotimes_{n=1}^{\infty} P_n(\lambda_n) \quad (\lambda = (\lambda_n)_{n=1}^{\infty}). \qquad (3.54)$$

We now present some necessary explanations. The first product in (3.54) exists, since $\rho_n(\mathbb{R}^1) = \| O_n \|^2$, and the product $\prod_{n=1}^{\infty} \| O_n \|^2$ converges due to (3.52). The second product in (3.54) should be regarded as a weak limit as $p \to \infty$ of the products

$$\left(\bigotimes_{n=1}^{p} P_n(\lambda_n)\right) \otimes (O_{p+1}^+ O_{p+1}) \otimes (O_{p+2}^+ O_{p+2}) \otimes \ldots : \mathcal{H}_{+,e} \to \mathcal{H}_{-,e}$$

(each product (before passing to the limit) is defined naturally, as has just been explained for $\mathbb{O}$). Below, we establish the existence (in this sense) of the second product in (3.54) (in fact, the products (before passing to the limit) converge, and this convergence is strong, the proof is given in the book of Berezansky [18, Chapter 3, Section 1, Subsection 1]).

Proof. Fix $\lambda = (\lambda_n)_{n=1}^{\infty}$ such that $P_n(\lambda_n) \geq 0$, and $\mathrm{Tr}\,(P_n(\lambda_n)) = 1$ for any $n \in \mathbb{N}$ (by virtue of Theorem 2.2, ρ_n-almost all $\lambda_n \in \mathbb{R}^1$ possess this property). First, we show that for this λ the product in the second formula in (3.54) is well-defined.

Denote $B_n = P_n(\lambda_n)$ and

$$C_p = \left(\bigotimes_{n=1}^{p} B_n\right) \otimes (O_{p+1}^{+} O_{p+1}) \otimes (O_{p+2}^{+} O_{p+2}) \otimes \ldots$$

for any $n, p \in \mathbb{N}$. It is necessary to prove that there exists a weak limit $\lim_{p\to\infty} C_p$. We now use the notations introduced in Subsection 2.2 of Chapter 1. For e_α and e_β $(\alpha, \beta \in \mathbb{Z}_{+,0}^{\infty})$, we have

$$(C_p e_\alpha, e_\beta)_{\mathcal{H}_{0,e}} = \left(\prod_{n=1}^{m} (B_n e_{\alpha_n}^{(n)}, e_{\beta_n}^{(n)})_{H_{0,n}}\right)\left(\prod_{n=m+1}^{p} (B_n e^{(n)}, e^{(n)})_{H_{0,n}}\right) \quad (3.55)$$

with $p > m = \max\{\nu(\alpha), \nu(\beta)\}$ (here we have used the equality

$$(O_n^{+} O_n e^{(n)}, e^{(n)})_{H_{0,n}} = \|e^{(n)}\| = 1).$$

Since

$$0 \leq (B_n e^{(n)}, e^{(n)})_{H_{0,n}} \leq 1,$$

the second product in (3.55) has the limit as $p \to \infty$ (which may be equal to zero). This proves the existence of $\lim_{p\to\infty}(C_p e_\alpha, e_\beta)_{\mathcal{H}_{0,e}}$, and this, in turn, implies that there exists $\lim_{p\to\infty}(C_p \varphi, \psi)_{\mathcal{H}_{0,e}}$ for φ and ψ from a set dense in $\mathcal{H}_{+,e}$. To prove the assertion, it suffices to show that $C_p \geq 0$ and $\mathrm{Tr}\,(C_p) \leq c < \infty$ $(p \in \mathbb{N})$, where $c = \prod_{n=1}^{\infty} |O_n|^2$ (recall that

$$\|C_p\| \leq |C_p| \leq \mathrm{Tr}\,(C_p)).$$

Since C_p is continuous, it suffices to verify that it is nonnegative on the vectors

$$\varphi = \varphi_q \otimes e^{(q+1)} \otimes e^{(q+2)} \otimes \ldots \in \mathcal{H}_{+,e},$$

where

$$\varphi_q \in \bigotimes_{n=1}^{q} H_{+,n} \quad (q \in \mathbb{N}).$$

For simplicity, we denote $O_{p+1}^{+} O_{p+1}, O_{p+2}^{+} O_{p+2}, \ldots$ in the expression for C_p by

$B_{p+1}, B_{p+2}, \ldots,$ respectively. Then

$$(C_p \varphi, \varphi)_{\mathcal{H}_{0,e}} = \left(\left(\bigotimes_{n=1}^{q} B_n \right) \varphi_q, \varphi_q \right)_{\bigotimes_{n=1}^{q} H_{0,n}} \left(\bigotimes_{n=q+1}^{\infty} (B_n e^{(n)}, e^{(n)})_{H_{0,n}} \right) \geq 0.$$

Here we have used the nonnegativity of the operators $O_n^+ O_n$ and the fact that if B_n: $H_{+,n} \to H_{-,n}$ are nonnegative $(n = 1, \ldots, q)$, then

$$\bigotimes_{n=1}^{q} B_n : \bigotimes_{n=1}^{q} H_{+,n} \to \bigotimes_{n=1}^{q} H_{-,n}$$

is also nonnegative.

Let us establish the estimate $\operatorname{Tr}(C_p) \leq c$ for any $p \in \mathbb{N}$. By using the notations introduced above, we obtain

$$\operatorname{Tr}(C_p) = \sum_{\alpha \in \mathbb{Z}_{+,0}^{\infty}} (C_p e_\alpha, e_\alpha)_{\mathcal{H}_{0,e}} = \lim_{m \to \infty} \sum_{\nu(\alpha) \leq m} (C_p e_\alpha, e_\alpha)_{\mathcal{H}_{0,e}}$$

$$= \lim_{m \to \infty} \sum_{\alpha_1, \ldots, \alpha_m = 0}^{\infty} (C_p e_{\alpha_1}^{(1)} \otimes \ldots \otimes e_{\alpha_m}^{(m)} \otimes e^{(m+1)} \otimes e^{(m+2)} \otimes \ldots,$$

$$e_{\alpha_1}^{(1)} \otimes \ldots \otimes e_{\alpha_m}^{(m)} \otimes e^{(m+1)} \otimes e^{(m+2)} \otimes \ldots)_{\mathcal{H}_{0,e}}.$$

By virtue of the relations

$$(O_n^+ O_n e^{(n)}, e^{(n)})_{H_{0,n}} = 1, \quad \operatorname{Tr}(O_n^+ O_n) = |O_n|^2 \quad (n \in \mathbb{N})$$

and $\operatorname{Tr}(B_n) = 1$ $(n = 1, \ldots, p)$, the last sum under the limit sign is equal to

$$\prod_{n=1}^{m} \operatorname{Tr}(B_n) = \left(\prod_{n=1}^{p} \operatorname{Tp}(B_n) \right) \left(\prod_{n=p+1}^{m} |O_n|^2 \right) = \prod_{n=p+1}^{m} |O_n|^2$$

for $m \geq p$.

By passing to the limit as $m \to \infty$, we get

$$\operatorname{Tr}(C_p) = \prod_{n=p+1}^{\infty} |O_n|^2 \leq \prod_{n=1}^{\infty} |O_n|^2 = c$$

(recall that $|O_n| \geq 1$). The required estimate is thus obtained. This means that we have defined the product $\otimes_{n=1}^{\infty} P_n(\lambda_n)$ for the indicated $\lambda = (\lambda_n)_{n=1}^{\infty}$.

We prove the first formula in (3.54). The spectral measure ρ is defined by the relation

$$\rho(\beta) = \mathrm{Tr}\,(\mathbf{O}^{+}\mathfrak{E}(\beta)\mathbf{O}) \quad (\beta \in \mathfrak{C}_{\sigma}(\mathbb{R}^{\infty})).$$

Let us compute its value in the case when

$$\beta = \mathbf{C}\left(1, \ldots, p; \mathop{\times}_{n=1}^{p} \delta_n\right) = \left(\mathop{\times}_{n=1}^{p} \delta_n\right) \times \mathbb{R}^{\infty} \quad (\delta_n \in \mathfrak{B}(\mathbb{R}^1)).$$

By analogy with the above calculation, we get

$$\rho(\beta) = \rho\left(\left(\mathop{\times}_{n=1}^{p} \delta_n\right) \times \mathbb{R}^{\infty}\right) = \mathrm{Tr}\left(\mathbf{O}^{+}\mathfrak{E}\left(\left(\mathop{\times}_{n=1}^{p} \delta_n\right) \times \mathbb{R}^{\infty}\right)\mathbf{O}\right)$$

$$= \mathrm{Tr}\left(\mathbf{O}^{+}\left(\prod_{n=1}^{p} \mathfrak{E}_n(\delta_n)\right)\mathbf{O}\right) = \mathrm{Tr}\left(\mathbf{O}^{+}\left(\left(\mathop{\otimes}_{n=1}^{p} E_n(\delta_n)\right) \otimes 1 \otimes 1 \otimes \ldots\right)\mathbf{O}\right)$$

$$= \mathrm{Tr}\left(\left(\mathop{\otimes}_{n=1}^{p} O_n^{+} E_n(\delta_n) O_n\right) \otimes (O_{p+1}^{+} O_{p+1}) \otimes (O_{p+2}^{+} O_{p+2}) \otimes \ldots\right)$$

$$= \left(\prod_{n=1}^{p} \rho_n(\delta_n)\right)\left(\prod_{n=p+1}^{\infty} \rho_n(\mathbb{R}^1)\right) = \left(\mathop{\times}_{k=1}^{\infty} \rho_k\right)\left(\left(\mathop{\times}_{k=1}^{p} \delta_k\right) \times \mathbb{R}^{\infty}\right)$$

$$= \left(\mathop{\times}_{k=1}^{\infty} \rho_k\right)(\beta).$$

This equality implies that the measures ρ and $\times_{k=1}^{\infty} \rho_k$ coincide on any $\beta \in \mathfrak{C}_{\sigma}(\mathbb{R}^{\infty})$, and this means that the first relation in (3.54) holds.

We prove the second relation in (3.54). Using (2.13), the calculations carried out above, and the first equality in (3.54), we obtain

$$\int_{\left(\mathop{\times}_{n=1}^{p}\right) \times \mathbb{R}^{\infty}} \mathcal{P}(\lambda)\, d\rho(\lambda) = \mathbf{O}^{+}\mathfrak{E}\left(\left(\mathop{\times}_{n=1}^{p} \delta_n\right) \times \mathbb{R}^{\infty}\right)\mathbf{O}$$

$$= \left(\mathop{\otimes}_{n=1}^{p} O_n^{+} E_n(\delta_n) O_n\right) \otimes (O_{p+1}^{+} O_{p+1}) \otimes (O_{p+2}^{+} O_{p+2}) \otimes \ldots$$

$$= \left(\bigotimes_{n=1}^{p} \int_{\delta_n} P_n(\lambda_n)\, d\rho_n(\lambda_n) \right) \otimes \left(\int_{\mathbb{R}^1} P_{p+1}(\lambda_{p+1})\, d\rho_{p+1}(\lambda_{p+1}) \right)$$

$$\otimes \left(\int_{\mathbb{R}^1} P_{p+2}(\lambda_{p+2})\, d\rho_{p+2}(\lambda_{p+2}) \right) \otimes \ldots$$

$$= \int_{\left(\times_{n=1}^{p}\right)\times\mathbb{R}^\infty} \left(\bigotimes_{n=1}^{\infty} P_n(\lambda_n) \right) d\left(\times_{n=1}^{\infty} \rho_n \right)(\lambda) = \int_{\left(\times_{n=1}^{p}\right)\times\mathbb{R}^\infty} \left(\bigotimes_{n=1}^{\infty} P_n(\lambda_n) \right) d\rho(\lambda) \quad (3.56)$$

for any $\delta_n \in \mathfrak{B}(\mathbb{R}^1)$. Here we have also employed the already proven fact that the product $\otimes_{n=1}^{\infty} P_n(\lambda_n)$ converges in the indicated sense. Here $(\lambda_n)_{n=1}^{\infty} = \lambda$ is such that each λ_n changes over a certain set in $\mathbb{R}^1$ of full ρ_n-measure (this enables us to establish the fourth equality in (3.56)).

Since the set $(\times_{n=1}^{p} \delta_n) \times \mathbb{R}^\infty$ is arbitrary (to a certain extent), it follows from (3.56) that $\mathcal{P}(\lambda)$ and $\otimes_{n=1}^{\infty} P_n(\lambda_n)$ coincide for ρ-almost all $\lambda \in \mathbb{R}^\infty$.

■

From the first equality in (3.54) and relation (1.15), we obtain the following formula for the spectrum

$$s(\mathcal{A}) = \operatorname{supp} \mathcal{E} = \operatorname{supp} \rho$$

of the family under consideration

$$s(\mathcal{A}) = \times_{n=1}^{\infty} s(A_n). \quad (3.57)$$

We shall not construct the expansion in individual generalized eigenvectors of the family $\mathcal{A}$ for the general case (see Subsection 3.1), though this is not hard to do. We only note that all these eigenvectors belong to $\mathcal{R}(P(\lambda)) \subset \mathcal{H}_{-,e}$. Since the c.l.s. of the vectors $\varphi^{(1)} \otimes \varphi^{(2)} \otimes \ldots$ ($\varphi^{(n)} \in H_{+,n}$, and $\varphi^{(n)} = e^{(n)}$ beginning with a certain number) coincides with $\mathcal{H}_{+,e}$, these vectors belong to the c.l.s.of the vectors

$$\mathcal{P}(\lambda)(\varphi^{(1)} \otimes \varphi^{(2)} \otimes \ldots) = (P_1(\lambda_1))\, \varphi^{(1)} \otimes (P_2(\lambda_2))\, \varphi^{(2)} \otimes \ldots . \quad (3.58)$$

The last vectors are the infinite tensor products of the generalized eigenvectors of the operators A_n (note that, since $\otimes_{n=1}^{\infty} P_n(\lambda_n)$ is defined, the infinite tensor product of the vectors $P_n(\lambda_n)\varphi^{(n)}$ exists (in the evident sense) as a vector in $\mathcal{H}_{-,e}$).

We dwell briefly upon the construction of an extension of the rigging of the chain (3.53) from such extensions for (3.51). Suppose that for each $n \in \mathbb{N}$ we have an extension of the rigging

$$H_{-,n} \supset H_{0,n} \supset H_{+,n} \supset D_n, \tag{3.59}$$

connected with the operator A_n in the standard way. If D_n are Hilbert spaces containing $e^{(n)}$, and $\| e^{(n)} \|_{D_n} = 1$, then

$$\mathcal{H}_{-,e} \supset \mathcal{H}_{0,e} \supset \mathcal{H}_{+,e} \supset \mathcal{D}, \tag{3.60}$$

where $\mathcal{D} = \otimes_{n=1;\,e}^{\infty} D_n$, can be taken as an extension of the rigging of the chain (3.53). Clearly, it is suitable for the family $\mathcal{A}$. If D_n are nuclear spaces which are projective limits of Hilbert spaces (containing $e^{(n)}$ as a unit vector), then the role of $\mathcal{D}$ can be played by a weighted infinite tensor product of these (see Subsection 2.2 in Chapter 1).

Let us consider the more special situation when each operator A_n has a strong cyclic vector (in the sense of Subsection 3.1) coinciding with $e^{(n)}$. More precisely, assume that for each $n \in \mathbb{N}$ there exists an extension of the rigging (3.59) with nuclear D_n. Moreover, $A_n^m e^{(n)} \in D_n$ $(m \in \mathbb{Z}_+)$ and the l.s. of these vectors is dense in $H_{+,n}$ (such D_n can always be constructed; see Maurin [2, Chapter 17, Section 5] or the proof of Theorem 3.9).

Then *the family* $\mathcal{A}$ *of operators has the strong cyclic vector*

$$\Omega = e^{(1)} \otimes e^{(2)} \otimes \ldots : A_1^{m_1} \ldots A_p^{m_p} \Omega \in \mathcal{D} \quad (m_1, \ldots, m_p \in \mathbb{Z}_+;\; p \in \mathbb{N}),$$

and the l.s. of these vectors is dense in $\mathcal{H}_{+,e}$ (here $\mathcal{D}$ is the weighted tensor product of the spaces D_n). In fact,

$$\mathcal{A}_1^{m_1} \ldots \mathcal{A}_p^{m_p} \Omega = (A_1^{m_1} e^{(1)}) \otimes \ldots \otimes (A_p^{m_p} e^{(p)}) \otimes e^{(p+1)} \otimes e^{(p+2)} \otimes \ldots$$

$$\in \mathrm{a.}(D_1 \otimes \ldots \otimes D_p \otimes e^{(p+1)} \otimes e^{(p+2)} \otimes \ldots) \subset \mathcal{D}.$$

This immediately implies that the l.s. of these vectors is also dense in $H_{+,e}$.

■

According to Theorem 2.4,

$$\mathfrak{C}_\sigma(\mathbb{R}^\infty) \ni \alpha \mapsto \sigma(\alpha) = (\mathcal{E}(\alpha)\,\Omega, \Omega)_{\mathcal{H}_{0,e}}$$

can now serve as the spectral measure of the family $\mathcal{A}$. It is easy to see that this measure coincides with $\times_{n=1}^{\infty}\sigma_n$, where

$$\mathfrak{B}(\mathbb{R}^1) \ni \delta \mapsto \sigma_n(\delta) = (E_n(\delta_n)\, e^{(n)}, e^{(n)})_{H_{0,n}}$$

is the corresponding spectral measure of the operator A_n, i.e.,

$$\sigma = \mathop{\times}_{n=1}^{\infty} \sigma_n \tag{3.61}$$

(it suffices to establish the equality (3.61) on the sets

$$\left(\mathop{\times}_{n=1}^{p} \delta_n\right) \times \mathbb{R}^{\infty} \quad (\delta_n \in \mathfrak{B}(\mathbb{R}^1);\ p \in \mathbb{N}),$$

where it is obvious).

By virtue of Theorem 3.2, the multiplicity of the eigenvalues is now equal to 1, i.e., $N_\lambda = 1$ for ρ-almost all λ. Hence, the one-dimensional space $\mathfrak{R}(P(\lambda))$ lies in the c.l.s. of the vectors (3.58), and each $\mathfrak{R}(P_n(\lambda_n))$ is also one-dimensional. This yields the equality

$$\xi_1(\lambda) = \xi_1^{(1)}(\lambda_1) \otimes \xi_1^{(2)}(\lambda_2) \otimes \ldots \in \mathcal{H}_{-,e} \quad (\lambda = (\lambda_n)_{n=1}^{\infty}) \tag{3.62}$$

for the corresponding individual generalized eigenvectors ($\xi_1(\lambda)$ is a generalized joint eigenvector of the family $\mathcal{A}$, and $\xi_1^{(n)}(\lambda_n)$ is the same vector for the operator A_n).

The equality (3.62) implies a simple formula for the Fourier transform associated with $\mathcal{A}$. Thus, it follows from formulas (3.4), (3.57), and (3.62) that

$$\mathbb{R}^{\infty} \supset s(\mathcal{A}) = \mathop{\times}_{n=1}^{\infty} s(A_n) \ni \lambda = (\lambda_n)_{n=1}^{\infty} \mapsto \tilde{\varphi}(\lambda)$$

$$= \left(\varphi, \bigotimes_{n=1}^{\infty} \xi_1^{(n)}(\lambda_n)\right)_{\mathcal{H}_{0,e}} \quad (\varphi \in \mathcal{H}_{+,e}). \tag{3.63}$$

The Parseval equality (3.5) now has the form

$$(\mathcal{E}(\alpha)\,\varphi, \psi)_{\mathcal{H}_{0,e}} = \int_{\alpha} \tilde{\varphi}(\lambda)\, \overline{\tilde{\psi}(\lambda)}\, d\rho(\lambda)$$

$$= \int_{\alpha} \tilde{\varphi}(\lambda)\, \overline{\tilde{\psi}(\lambda)}\, (d\rho / d\sigma)\, d\sigma(\lambda) \quad (\alpha \in \mathfrak{C}_{\sigma}(\mathbb{R}^{\infty});\ \psi \in \mathcal{H}_{+,e}). \tag{3.64}$$

Clearly, it can be extended by to the vectors from $\mathcal{H}_{0,e}$ continuity.

All the constructions in this Subsection can be easily generalized to the case when each operator A_n is normal. As a result, we obtain a family $(\mathcal{A}_n)_{n=1}^{\infty}$ of commuting normal operators.

It is also important to study the functions of operators acting upon different variables. Some of them can be found in Chapter 6 (the operators admitting the separation of variables, i.e., operators of the form $\mathcal{A}_1 + \mathcal{A}_2 + \ldots$ and their perturbations). For further details, see Berezansky [18, Chapter 3, Sections 1 and 2] and Yu.S.Samoilenko [1, Chapter 1, Section 1, Subsection 7].

3.7. The Fourier–Wiener Transformation. Spectral Approach

Let us consider a concrete situation in Subsection 3.6, due to which the transformation (3.63) turns out to be the infinite-dimensional Fourier–Wiener transformation (introduced in Subsection 2.5 in Chapter 2). First, we establish the required facts in the case of a single variable.

Consider the space $H_0 = L_2(\mathbb{R}^1, \pi^{-1/2} e^{-t^2} dt)$. As above, the Gaussian measure is denoted by

$$\pi^{-1/2} e^{-t^2} dt = g(t)\, dt = d\gamma_1(t).$$

In this space, we define the operator A equal to the closure of the operator

$$C_0^\infty(\mathbb{R}^1) \ni u(t) \mapsto -i g^{-1/2}(t)\, (g^{1/2}(t)\, u(t)') = -i\,(u'(t) - t\, u(t)). \qquad (3.65)$$

The mapping

$$H_0 \ni u(t) \mapsto g^{1/2}(t)\, u(t) = \hat{u}(t) \in L_2(\mathbb{R}^1, dt) \qquad (3.66)$$

is an isometry between H_0 and $L_2(\mathbb{R}^1, dt)$ under which the expression $-i\hat{u}'(t)$ is the image of the differential expression in (3.65). Since $C_0^\infty(\mathbb{R}^1)$ is preserved under this mapping, the operator B, equal to the closure of the operator

$$C_1^\infty(\mathbb{R}^1) \ni \hat{u}(t) \mapsto -i\, \hat{u}'(t)$$

in $L_2(\mathbb{R}^1, dt)$, is the image of the operator A. It is well known that B is selfadjoint; therefore, the same is true for A.

We note that *the functions $u(t)$ from $C_0^\infty(\mathbb{R}^1)$ bounded together with their first derivatives belong to $\mathfrak{D}(A)$, and the action of A upon these functions is given by the same formula* (3.65).

To prove this fact, consider a sequence $(\chi_n(t))_{n=1}^{\infty}$ of functions $\chi_n \in C_0^{\infty}(\mathbb{R}^1)$ that are equal to 1 for $|t| \le n$ and to 0 for $|t| \ge n+1$. On the intervals $(-n-1, -n)$ and $(n, n+1)$, these functions take values from [0, 1] and are obtained by translating the appropriate pieces of the graph of the function $\chi_1(t)$. Then

$$u_n(t) = u(t)\chi_n(t) \in C_0^{\infty}(\mathbb{R}^1),$$

and it is easy to compute that $u_n \to u$ in H_0 as $n \to \infty$ and that $A u_n$ has a limit equal to the action of the differential expression (3.65) upon u.

■

It is known (see, for example, Berezansky [5, Chapter 6, Section 5, Subsection 3]) that B is a Carleman operator and that the following representation holds for its spectral measure ρ_B and spectral kernel $P_B(t, s; \lambda)$:

$$P_B(t, s; \lambda)\, d\rho_B(\lambda) = \frac{1}{2\pi} e^{i\lambda(t-s)} d\lambda \quad (t, s \in \mathbb{R}^1;\ \lambda \in s(B) = \mathbb{R}^1). \tag{3.67}$$

If we return to the operator A with the help of the mapping (3.66) introduced above, then we get from (3.67) the following formula

$$P(t, s; \lambda)\, d\rho(\lambda) = (g(t)\, g(s))^{-1/2} P_B(t, s; \lambda)\, d\rho_B(\lambda) = \frac{1}{2\sqrt{\pi}} e^{i\lambda(t-s) + 1/2(t^2+s^2)} d\lambda$$

$$(t, s \in \mathbb{R}^1;\ \lambda \in s(A) = s(B) = \mathbb{R}^1) \tag{3.68}$$

for its spectral measure ρ and spectral kernel $P(t, s; \lambda)$ (we clarify that the operator $P(\lambda)$ for A can be recovered from (3.68) according to (3.19), where the integration is taken with respect to the Gaussian measure $d\gamma_1(s)$).

The operator A possesses the cyclic vector $\Omega = 1 \in H_0$. Indeed, according to (3.68), we obtain

$$(E(\alpha)\, 1, 1)_{H_0} = \int_{\alpha} \left(\int_{\mathbb{R}^1} \int_{\mathbb{R}^1} P(t, s; \lambda)\, ds\, dt \right) d\rho(\lambda) = \frac{1}{\sqrt{\pi}} \int_{\alpha} e^{-\lambda^2} d\lambda \quad (\alpha \in \mathcal{B}(\mathbb{R}^1)) \tag{3.69}$$

for the RI E of the operator A (we have used the fact that

$$\int_{\mathbb{R}^1} e^{i\lambda x - 1/2 x^2} dx = \sqrt{2\pi} e^{-1/2\lambda^2} \quad (\lambda \in \mathbb{R}^1)).$$

Therefore,

$$\int_{\mathbb{R}^1} \lambda^{2m}\, d(E(\lambda)\, 1, 1)_{H_0} < \infty$$

for any $m = \mathbb{Z}_+$, i.e., $\Omega \in \bigcap_{m=0}^{\infty}\mathfrak{D}(A^m)$. Further, by induction on m, one can easily conclude from (3.65) that

$$\text{l.s.}(\Omega, A\,\Omega, \dots, A^m\,\Omega) = \text{l.s.}(1, t, \dots, t^m).$$

But it is well known that the collection of all polynomials is dense in H_0; therefore,

$$\text{c.l.s.}(\{A^m\Omega \mid m \in \mathbb{Z}_+\}) = H_0,$$

and this proves the required result. In our case, the spectral measure is given by the formula

$$\sigma(\alpha) = \frac{1}{\sqrt{\pi}} \int_{\alpha} e^{-\lambda^2}\, d\lambda, \tag{3.70}$$

according to Theorem 2.4 and (3.69).

Clearly, A is a Carleman operator, but here it proves to be useful to construct a rigging of the space H_0 not of the type (3.16), but having the form (2.18) with the quasinuclear imbedding $H_+ \to H_0$. The space $G^2(\mathbb{R}^1)$, introduced in Subsection 4.5 of Chapter 2, can be regarded as H_+. Then the quasinuclearity of $H_+ \to H_0$ is guaranteed by Theorem 4.4 in Chapter 2. Note that $\|\Omega\|_{H_0} = \|\Omega\|_{H_+} = 1$. By using Lemma 2.5 in Chapter 1, we renormalize H_+ ensuring the inequality $|O_n|^2 - 1 < n^{-2}$ ($n \in \mathbb{N}$ is fixed) for the imbedding operator O_n: $H_{+,n} \to H_0$ (it is convenient to note that this renormalization depends on n).

Hence, we have the rigging

$$H_{-,n} \supset L_2(\mathbb{R}^1, \gamma_1) \supset H_{+,n}, \tag{3.71}$$

where $H_{-,n}$ denotes the corresponding negative space. This rigging can be easily extended for A. For example, let

$$D = \text{pr}\lim_{\tau \in \mathbb{Z}_+} G^{\tau}(\mathbb{R}^1) = \mathcal{G}(\mathbb{R}^1).$$

Clearly, it follows from Theorem 4.4 in Chapter 2 that $A \upharpoonright D$ maps D onto D continuously. Indeed, one should only use the mapping (3.66) and the well-known fact that the operator

$$\mathcal{S}(\mathbb{R}^1) \ni \varphi \mapsto \varphi' \in \mathcal{S}(\mathbb{R}^1)$$

is continuous. Therefore, it is automatically true that $A \upharpoonright D$ maps D onto $G^2(\mathbb{R}^1)$ continuously, i.e., we have actually constructed an extension of the rigging (3.71) for A.

The cyclic vector Ω of the operator A is strong. It proves to be most convenient to verify this by using the mapping (3.66) and Theorem 4.4 in Chapter 2. Thus, all the vectors $A^m \Omega$ $(m \in \mathbb{Z}_+)$ belong to D, since this is equivalent to the statement that the function $\pi^{-1/4} e^{-t^2/2}$ and all its derivatives belong to $\mathfrak{S}(\mathbb{R}^1)$. Further, taking into account that

$$\text{l.s.}(\Omega, A\,\Omega, \ldots, A^m\Omega) = \text{l.s.}(1, t, \ldots, t^m),$$

applying (3.66), and observing that the collection of all polynomials in t, multiplied by $e^{-t^2/2}$, is dense in $\mathfrak{S}(\mathbb{R}^1)$, we conclude that $\text{l.s.}(\{A^m \Omega \,|\, m \in \mathbb{Z}_+\})$ is dense in D and, consequently, in $G^2(\mathbb{R}^1)$.

Taking into account (3.68) and (3.70), it is easy to write the transformation (3.4) and the Parseval equality (3.5) in this case. The corresponding transformation is called a one-dimensional Fourier–Wiener transformation and has the following form

$$\mathbb{R}^1 = s(A) \ni \lambda \mapsto \tilde{\varphi}(\lambda) = \frac{1}{\sqrt{2\pi}} \int_{\mathbb{R}^1} \varphi(t)\, e^{i\lambda t + 1/2(\lambda^2 - t^2)}\, dt$$

$$= \int_{\mathbb{R}^1} \varphi(t)\, e^{i\lambda t + 1/2\lambda^2} \frac{1}{\sqrt{2\pi}} e^{-1/2\, t^2}\, dt \quad (\varphi \in G^2(\mathbb{R}^1)), \tag{3.72}$$

while the Parseval equality (for $\alpha = \mathbb{R}^1$) has the form

$$\frac{1}{\sqrt{\pi}} \int_{\mathbb{R}^1} \varphi(t)\, \overline{\psi(t)}\, e^{-t^2}\, dt = \int_{\mathbb{R}^1} \tilde{\varphi}(\lambda)\, \overline{\tilde{\psi}(\lambda)}\, d\sigma(\lambda)$$

$$= \frac{1}{\sqrt{\pi}} \int_{\mathbb{R}^1} \tilde{\varphi}(\lambda)\, \overline{\tilde{\psi}(\lambda)}\, e^{-\lambda^2}\, d\lambda \quad (\varphi, \psi \in G^2(\mathbb{R}^1)). \tag{3.73}$$

The transformation (3.72) and the equality (3.73) can be extended by continuity to the functions $\varphi, \psi \in H_0$. A comparison between (3.72) and (2.50) in Chapter 2 shows that the definition of a one-dimensional Fourier–Wiener transformation given above coincides with the previous definition presented in Subsection 2.4 of Chapter 2.

We now pass to the infinite-dimensional case. Let $(H_{0,n})_{n=1}^{\infty}$ be a sequence of identical spaces $L_2(\mathbb{R}^1, d\gamma_1(x_n)) = H_{0,n}$ $(n \in \mathbb{N})$ and assume that the identical operators A_n, coinciding with the operator A constructed above, act in each of these spaces. Consider the stabilizing sequence $e = (e^{(n)}(x_n))_{n=1}^{\infty}$, where each $e^{(n)}(x_n) = 1$, and form the infinite tensor product

$$\mathcal{H}_{0,e} = \bigotimes_{n=1,e}^{\infty} H_{0,n} = L_2(\mathbb{R}^{\infty}, d\gamma_1).$$

Here, as in (1.10) of Chapter 2, γ_1 is a Gaussian measure on $\mathbb{R}^{\infty}$, i.e.,

$$d\gamma_1(x) = \mathop{\times}_{k=1}^{\infty} d\gamma_1(x_k) = \mathop{\times}_{k=1}^{\infty} \left(\frac{1}{\sqrt{\pi}} e^{-x_k^2}\, dx_k\right). \tag{3.74}$$

Given the operators A_n, we construct a family $\mathcal{A} = (\mathcal{A}_n)_{n=1}^{\infty}$ of commuting selfadjoint operators according to the general scheme presented in Subsection 3.6 (roughly speaking, each $\mathcal{A}_n$ is the operator (3.65) acting upon the n-th variable x_n); we also construct the corresponding joint resolution of the identity $\mathfrak{C}_\sigma(\mathbb{R}^{\infty}) \ni \alpha \mapsto \mathcal{E}(\alpha)$.

To construct a rigging of the space $L_2(\mathbb{R}^{\infty}, \gamma_1)$, one must multiply the chains (3.71) for $n \in \mathbb{N}$. We have

$$\mathcal{H}_{-,e} = \bigotimes_{n=1,e}^{\infty} H_{-,n} \supset L_2(\mathbb{R}^1, d\gamma_1) \supset \bigotimes_{n=1,e}^{\infty} H_{+,n} = \mathcal{H}_{+,e}. \tag{3.75}$$

In this chain, the imbedding of the positive space in the zero space is quasinuclear, because condition (3.52) holds due to the renormalization. To extend the rigging (3.75), we construct the infinite tensor product $\mathcal{G}(\mathbb{R}^{\infty})$ of the spaces $\mathcal{G}(\mathbb{R}^1)$; for a description of this product, see Subsection 4.5 in Chapter 2.

As noted above, each operator A_n has the strong cyclic vector $e^{(n)}(x_n) = 1$. But then, according to the general construction given in Subsection 3.6, the family $\mathcal{A}$ also has a strong cyclic vector $\Omega = 1 \otimes 1 \otimes \ldots = 1$. Therefore, in our case, not only formulas (3.54) and (3.57) hold, but also (3.61)–(3.63).

Let us write some relations obtained here. According to (3.57), we have $s(\mathcal{A}) = \mathbb{R}^{\infty}$. If we rewrite (3.68) in the form

$$P(t, s; \lambda) = e^{i\lambda(t-s) + 1/2(t^2+s^2) + 1/4\lambda^2} \left(\frac{1}{2\sqrt{\pi}} e^{-\lambda^2/4}\, d\lambda\right),$$

then (3.54) and this equality imply that the $\mathcal{P}(\lambda)$ is the "integral operator with the kernel"

$$\mathcal{P}(x, x; \lambda) = \exp\left[i\sum_{n=1}^{\infty} \lambda_n(x_n - y_n) + 1/2 \sum_{n=1}^{\infty} (x_n^2 + y_n^2)\right.$$

$$+ 1/4 \sum_{n=1}^{\infty} \lambda_n^2 \Bigg] \mathop{\times}_{n=1}^{\infty} \left(\frac{1}{2\sqrt{\pi}} e^{-\lambda_n^2/4} d\lambda_n \right) \tag{3.76}$$

$$(x = (x_n)_{n=1}^{\infty},\ y = (y_n)_{n=1}^{\infty},\ \lambda = (\lambda_n)_{n=1}^{\infty} \in \mathbb{R}^{\infty}).$$

The series in (3.76) converge in the sense that the "cutoff" kernels corresponding to summation from 1 to p (in (3.76)) generate a sequence of operators weakly convergent (as operators from $\mathcal{H}_{+,e}$ to $\mathcal{H}_{-,e}$) to $\mathcal{P}(\lambda)d\rho(\lambda)$ as $p \to \infty$.

It follows from relations (3.61) and (3.70) that the spectral measure σ is the Gaussian measure

$$d\sigma(\lambda) = \mathop{\times}_{n=1}^{\infty} \left(\frac{1}{\sqrt{\pi}} e^{-\lambda_n^2} d\lambda_n \right) = d\,\gamma_1(\lambda). \tag{3.77}$$

The Fourier transformation (3.63) means that each function $\varphi \in \mathcal{H}_{+,e}$ is associated with the function

$$\mathbb{R}^{\infty} \ni \lambda \mapsto \tilde{\varphi}(\lambda) = \int_{\mathbb{R}^{\infty}} \exp\left(i \sum_{n=1}^{\infty} \lambda_n x_n + 1/2 \sum_{n=1}^{\infty} \lambda_n^2 \right) d\gamma_2(x),$$

$$d\gamma_2(x) = \mathop{\times}_{n=1}^{\infty} \left(\frac{1}{\sqrt{2\pi}} e^{-x_n^2/2} dx_n \right) \tag{3.78}$$

(the exponential in the integrand converges in the same way as (3.62)). For φ, $\psi \in \mathcal{H}_{+,e}$, the Parseval equality (with $\alpha = \mathbb{R}^{\infty}$) has the form

$$\int_{\mathbb{R}^{\infty}} \varphi(x)\, \overline{\psi(x)}\, d\,\gamma_1(x) = \int_{\mathbb{R}^{\infty}} \tilde{\varphi}(\lambda)\, \overline{\tilde{\psi}(\lambda)}\, d\,\gamma_1(x). \tag{3.79}$$

The transformation (3.78) is called an infinite-dimensional Fourier–Wiener transformation (in the coordinate form). Comparing (3.78) with the construction in Subsection 2.5 of Chapter 2, we conclude that this is, in fact, another form of the transformation introduced above.

For the functions belonging to the class $\mathcal{H}_{+,e} = \otimes_{n=1,e}^{\infty} \mathcal{H}_{+,n}$, the integrals in (3.78) and (3.79) are absolutely convergent. We do not explain how to extend this class preserving absolute convergence. As already noted, if we pass to the mean square convergence of the integrals, then relations (3.78) and (3.79) can be extended to arbitrary φ, $\psi \in L_2(\mathbb{R}^{\infty}, \gamma_1)$. It is also clear that, with respect to the differentiation, the transformation (3.78) behaves just as the ordinary multi-dimensional Fourier transform; one should only replace the derivatives $-i(\partial/\partial x_n)$ by $-i((\partial/\partial x_n) - x_n)$.

3.8. The Creation and Annihilation Operators. The Segal Isomorphism (Spectral Approach)

We show that the Segal isomorphism, introduced in Subsection 2.2 of Chapter 2, admits a natural interpretation as a Fourier transform for the expansion in generalized joint eigenvectors of a certain family of commuting selfadjoint operators. These operators are constructed with the help of so-called creation and annihilation operators (this is a very important family of operators in the Fock space; in particular, the Hamiltonians of physical systems are expressed in terms of creation and annihilation operators).

As in Subsection 2.2 of Chapter 2, let H denote a real Hilbert space, let H_c be its complexification, and let $\mathfrak{F}(H) = \bigoplus_{n=0}^{\infty} \mathfrak{F}_n(H)$ be the corresponding Fock space. Here $\mathfrak{F}_n(H)$ is the n-particle subspace of $\mathfrak{F}(H)$ having the form $P_n(H_c^{\otimes n})$, where P_n is the symmetrization operator.

For each $h \in H$, we introduce a creation operator $a_+(h)$ by setting

$$a_+(h) f_n = \sqrt{n+1}\, P_{n+1}(h \otimes f_n) \quad (n \in \mathbb{Z}_+) \tag{3.80}$$

for $f_n \in \mathfrak{F}_n(H)$. The operator $a_+(h)$ given by (3.80) is a bounded mapping of $\mathfrak{F}_n(H)$ onto $\mathfrak{F}_{n+1}(H)$. One can easily compute that its norm is equal to $\sqrt{n+1}\, \| h \|_H$ and, hence, this operator determines the (unbounded) operator in $\mathfrak{F}(H)$ with the domain of definition $\mathfrak{D}(a_+(h)) = \mathfrak{F}_{\text{fin}}(H)$ which is mapped by the latter onto itself (recall that $\mathfrak{F}_{\text{fin}}(H)$ denotes the collection of all finite vectors from $\mathfrak{F}(H)$).

An annihilation operator $a_-(h)$ $(h \in H)$ is defined as follows:

$$a_-(h) = a_+^*(h) \upharpoonright \mathfrak{F}_{\text{fin}}(H),$$

$a_-(h)$ is a bounded operator from $\mathfrak{F}_n(H)$ onto $\mathfrak{F}_{n-1}(H)$ with the norm $\sqrt{n}\, \| h \|_H$, and $\mathfrak{R}(a_-(h)) \subset \mathfrak{F}_{\text{fin}}(H)$.

It easy to verify that the creation and annihilation operators satisfy the following canonical commutation relations:

$$[a_+(h_1), a_+(h_2)] = [a_-(h_1), a_-(h_2)] = 0,$$

$$[a_-(h_1), a_+(h_2)] = (h_1, h_2)_H 1 \quad (h_1, h_2 \in H). \tag{3.81}$$

Here $[A, B] = AB - BA$ denotes the commutator of the operators A and B; the equalities (3.81) should be regarded as an equality on vectors $f \in \mathfrak{F}_{\text{fin}}(H)$.

In $\mathfrak{F}(H)$, we also introduce an operator of the number of particles N with the domain of definition $\mathfrak{F}_{\text{fin}}(H)$ by setting

$$N \upharpoonright \mathfrak{F}_n(H) = n\,1 \quad (n \in \mathbb{N}).$$

It follows directly from this definition that the operator N is essentially selfadjoint.

We fix $h \in H$. For all $f \in \mathfrak{F}(H)$, we have

$$\| a_+(h)\,(1+N)^{-1/2} \|^2_{\mathfrak{F}_n(H)} = \sum_{n=0}^{\infty} \frac{1}{1+n} \| a_+(h) f_n \|^2_{\mathfrak{F}_{n+1}(H)}$$

$$\leq \| h \|^2_H \sum_{n=0}^{\infty} \| f_n \|^2_{\mathfrak{F}_n(H)} = \| h \|^2_H \, \| f \|^2_{\mathfrak{F}_n(H)},$$

but this means that the operator $a_+(h)\,(1+N)^{-1/2}$ is bounded in $\mathfrak{F}(H)$.

Similarly, one can establish that the operator $a_-(h)\,(1+N)^{-1/2}$ is bounded in $\mathfrak{F}(H)$. This immediately implies that

$$\sum_{n=0}^{\infty} \frac{|z|^n}{n!} \| a^n_{\pm}(h) f \|_{\mathfrak{F}(H)} < \infty \qquad (3.82)$$

for all $f \in \mathfrak{F}_{\text{fin}}(H)$ and $z \in \mathbb{C}^1$; but this means that $\mathfrak{F}_{\text{fin}}(H)$ is a dense set of analytical (and even entire) vectors for $a_{\pm}(h)$.

Recall that a vector $\varphi \in \mathcal{H}$ is called an analytical vector for an operator A acting in a Hilbert space $\mathcal{H}$ if $\varphi \in \bigcap_{n=1}^{\infty} \mathfrak{D}(A^n)$, and the power series

$$\sum_{n=0}^{\infty} \frac{\|A^n \varphi\|_{\mathcal{H}}}{n!} z^n \quad (z \in \mathbb{C}^1)$$

has a positive radius of convergence; it is called an entire vector if this radius of convergence is equal to infinity. In the case of an Hermitian operator A, the theorem, which states that it is essentially selfadjoint provided that there exists a set of analytical vectors total in $\mathcal{H}$, is valid (see Theorem 1.7 and the relation (1.39) in Chapter 5).

By using (3.81) and (3.82), we easily show that $\mathfrak{F}_{\text{fin}}(H)$ is a dense set of analytical vectors also for the so-called field operators

$$\varphi(h) = \frac{1}{\sqrt{2}}(a_+(h) + a_-(h)), \quad \mathfrak{D}(\varphi(h)) = F_{\text{fin}}(H) \quad (h \in H), \qquad (3.83)$$

which are already Hermitian. According to the theorem mentioned above, we can claim that each operator $\varphi(h)$ $(h \in H)$ is essentially selfadjoint.

Note that, along with $\varphi(h)$, one may also introduce the Hermitian field operators

$$\pi(h) = \frac{1}{i\sqrt{2}}(a_+(h) - a_-(h)), \quad \mathfrak{D}(\pi(h)) = \mathfrak{F}_{\text{fin}}(H).$$

These definitions and (3.81) imply that

$$[\varphi(h_1), \varphi(h_2)] = [\pi(h_1), \pi(h_2)] = 0,$$

$$[\pi(h_1), \varphi(h_2)] = -i\,(h_1, h_2)_H 1 \quad (h_1, h_2 \in H) \tag{3.84}$$

in the same sense as above.

The first equality in (3.84) shows that the operators $\varphi(h)$ $(h \in H)$ commute on $\mathfrak{F}_{\text{fin}}(H)$. Let us prove that

the closures $\varphi(h)^\sim$ $(h \in H)$ *of these operators form a family of commuting self-adjoint operators.*

In fact, assume that $h_1, h_2 \in H$ are given. As is known, to verify that $\varphi(h_1)^\sim$ and $\varphi(h_2)^\sim$ commute, it remains to establish that

$$\varphi(h_1) \restriction (\varphi(h_2) - z\,1)\,\mathfrak{F}_{\text{fin}}(H)$$

has a total set of analytical vectors for some nonreal z (see Theorem 1.15 in Chapter 5). It follows from the essential selfadjointness of $\varphi(h_2)$ that the set $(\varphi(h_2) - i\,1)\mathfrak{F}_{\text{fin}}(H)$ is dense in $\mathfrak{F}(H)$. At the same time, it belongs to $\mathfrak{F}_{\text{fin}}(H)$, because $\varphi(h_2)$ maps $\mathfrak{F}_{\text{fin}}(H)$ onto $\mathfrak{F}_{\text{fin}}(H)$ and, therefore, it consists of analytical vectors of the operator $\varphi(h_1)$. The fact that $\varphi(h_1)^\sim$ and $\varphi(h_2)^\sim$ commute is thus proved.

■

Let us choose a nuclear rigging $\Phi' \supset H \supset \Phi$. Fix an orthonormal basis $(e_j)_{j=1}^\infty \subset \Phi$ in H and set $\varphi_j = \varphi(e_j)$. According to what has been proved, the operators $(\varphi_j)_{j=1}^\infty$ form a countable set of commuting essentially selfadjoint operators. Consider the action of these operators on the vectors from the basis of occupation numbers constructed from $(e_j)_{j=1}^\infty$ in Subsection 2.2 of Chapter 2. By definition, we have

$$a_+(e_j)\,e_\alpha = \frac{(|\alpha|)^{1/2}}{(\alpha_1!\ldots\alpha_\nu!)^{1/2}}\, a_+(e_j)\, P_{|\alpha|}\,(e_1^{\otimes\alpha_1} \otimes \ldots \otimes e_\nu^{\otimes\alpha_\nu})$$

$$= \frac{(|\alpha|)^{1/2}}{(\alpha_1!\ldots\alpha_\nu!)^{1/2}}\,\frac{\sqrt{|\alpha|+1}}{|\alpha|!}\, P_{|\alpha|+1}\left(\sum_{s\in\sigma_n} e_j \otimes e_{k_{s(1)}} \otimes \ldots \otimes e_{k_{s(n)}}\right)$$

$$= \frac{(|\alpha|)^{1/2}}{(\alpha_1!\ldots\alpha_\nu!)^{1/2}}\,\sqrt{|\alpha|+1}\; P_{|\alpha|+1}\,(e_1^{\otimes\alpha_1} \otimes \ldots \otimes e_{j-1}^{\otimes\alpha_{j-1}} \otimes e_j^{\otimes\alpha_{j+1}}$$

$$\otimes e_{j+1}^{\alpha_{j+1}} \otimes \ldots \otimes e_{\nu}^{\alpha_\nu}) = \sqrt{\alpha_j + 1}\, e_{\alpha+1_j},$$

$$\alpha + 1_j = (\alpha_1, \ldots, \alpha_{j-1}, \alpha_j + 1, \alpha_{j+1}, \ldots, \alpha_\nu, 0, 0, \ldots) \in \mathbb{Z}_{+,0}^{\infty}$$

$$(\alpha \in \mathbb{Z}_{+,0}^{\infty};\; j \in \mathbb{N};\; n = |\alpha|).$$

Similarly,

$$a_-(e_j)\, e_\alpha = \sqrt{\alpha_j}\, e_{\alpha-1_j},$$

$$\alpha - 1_j = (\alpha_1, \ldots, \alpha_j - 1, \alpha_{j+1}, \ldots, \alpha_\nu, 0, 0, \ldots) \in \mathbb{Z}_{+,0}^{\infty}$$

$$(\alpha \in \mathbb{Z}_{+,0}^{\infty};\; j \in \mathbb{N}).$$

These relations should be supplemented by the following assertions: $a_+(e_j)\,\Omega = e_j$ and $a_-(e_j)\,\Omega = 0$ ($j \in \mathbb{N}$), where $\Omega = (1, 0, 0, \ldots) \in \mathfrak{F}(H)$ is a vacuum vector in the Fock space. Hence,

$$\varphi_j\, e_\alpha = \frac{1}{\sqrt{2}} \left(\sqrt{\alpha_j + 1}\, e_{\alpha+1_j} + \sqrt{\alpha_j}\, e_{\alpha-1_j}\right) \quad (\alpha \in \mathbb{Z}_{+,0}^{\infty}),$$

$$\varphi_j\,\Omega = \frac{1}{\sqrt{2}}\, e_j \quad (j \in \mathbb{N}). \tag{3.85}$$

It follows from (3.85) that *the vector* $\Omega = \Gamma(0)$ *is a cyclic vector for the family of operators* $(\varphi_j)_{j=1}^{\infty}$ *in the sense of Subsection 2.3.*

Indeed, it suffices to show that the linear span of the vectors of the form

$$\varphi_1^{\alpha_1} \ldots \varphi_\nu^{\alpha_\nu}\,\Omega \quad (\alpha \in \mathbb{Z}_{+,0}^{\infty}) \tag{3.86}$$

contains the basis vectors e_α ($\alpha \in \mathbb{Z}_{+,0}^{\infty}$). The last statement can be easily proved by induction. In fact, the linear span under consideration contains $\Omega = (1, 0, 0, \ldots)$, further, it also contains vectors of the form

$$\varphi_j\,\Omega = \frac{1}{\sqrt{2}}\, e_j \quad (j \in \mathbb{N}).$$

Now assume that the linear span of the vectors (3.86) (with $|\alpha| \le n$) contains all e_α with $|\alpha| \le n$ and show that all e_α with $|\alpha| \le n + 1$ belong to the linear span of the vectors (3.86) with $|\alpha| \le n + 1$. Indeed, if $|\alpha| \le n$, then this is true by the induction hypothesis, and if $|\alpha| \leq n+1$, then

$$\varphi_1^{\alpha_1} \dots \varphi_\nu^{\alpha_\nu}\Omega = a_+^{\alpha_1}(e_1) \dots a_+^{\alpha_\nu}(e_\nu)\,\Omega + f_n,$$

where

$$f_n \in \bigoplus_{k=0}^{n} \mathfrak{F}_k(H),$$

and this is the linear combination of the vectors e_β with $|\beta| \leq n$. But according to (3.85), we have

$$a_+^{\alpha_1}(e_1) \dots a_+^{\alpha_\nu}(e_\nu)\,\Omega = c e_\alpha \quad (c > 0).$$

By using the induction hypothesis, we conclude that e_α belongs to the linear span under consideration.

The properties of the vector $\Omega = \Gamma(0)$ described above allow us to clarify the meaning of the terms "vacuum" and "creation and annihilation operators". Namely, under the action of creation operators, the vacuum may be transformed into a state with an arbitrary number of particles, while the annihilation operators reduce the number of particles and, in particular, the result of the action of these operators upon vacuum is 0. The property of vacuum proved above shows that the field operators may also transform it into a state with an arbitrary number of particles.

We now proceed to study generalized eigenvectors of a family of selfadjoint operators $(\tilde{\varphi_j})_{j=1}^{\infty}$ and the corresponding Fourier transform. To show that the conditions of Theorem 2.6 and the Remark in Subsection 2.10 are satisfied, we now prove that the nuclear space $\mathfrak{F}_{\text{fin}}(\Phi)$ constructed in Subsection 5.1 of Chapter 2 is connected with the family $(\tilde{\varphi_j})_{j=1}^{\infty}$ in the standard way. To do this, we must establish the following facts:

(i) $\mathfrak{F}_{\text{fin}}(\Phi) \subset \mathfrak{D}(\tilde{\varphi_j})$ $(j \in \mathbb{N})$;

(ii) $\tilde{\varphi_j} : \mathfrak{F}_{\text{fin}}(\Phi) \to \mathfrak{F}_{\text{fin}}(\Phi)$ is continuous $(j \in \mathbb{N})$.

Since $e_j \in \Phi$, the mapping

$$\mathfrak{F}_n(\Phi) \ni f_n \mapsto e_j \otimes f_n \in \Phi_0^{\otimes(n+1)}$$

is continuous $(n \in \mathbb{Z}_+)$. This yields the continuity of the mapping

$$\mathfrak{F}_n(\Phi) \ni f_n \mapsto P_{n+1}(e_j \otimes f_n) \in \mathfrak{F}_{n+1}(\Phi) \quad (n \in \mathbb{Z}_+)$$

and, thus, the continuity of the creation operator

$$a_+(e_j)\colon \mathfrak{F}_n(\Phi) \to \mathfrak{F}_{n+1}(\Phi),$$

given by (3.80). By virtue of the definition of $\mathfrak{F}_{\text{fin}}(\Phi)$ as a topological direct sum $\oplus_{n=0}^{\infty}\mathfrak{F}_n(\Phi)$ (see Subsection 5.1 in Chapter 2), we obtain the inclusion

$$a_+(e_j) \in \mathfrak{L}(\mathfrak{F}_{\text{fin}}(\Phi), \mathfrak{F}_{\text{fin}}(\Phi)).$$

Similarly, we get that

$$a_-(e_j) \in \mathfrak{L}(\mathfrak{F}_{\text{fin}}(\Phi), \mathfrak{F}_{\text{fin}}(\Phi))$$

and, consequently,

$$\varphi_j = 2^{-1/2}(a_+(e_j) + a_-(e_j)) \in \mathfrak{L}(\mathfrak{F}_{\text{fin}}(\Phi), \mathfrak{F}_{\text{fin}}(\Phi)).$$

Hence, the conditions of Theorem 2.6 are satisfied. Moreover, our family of operators has a cyclic vector and, therefore, one can use the corresponding facts in Subsection 2.3 of Chapter 2 and Theorem 3.2. Let us find an explicit form of generalized joint eigenvectors for the family of operators $(\varphi_{\tilde{k}})_{k=1}^{\infty}$. For this purpose, we solve the set of equations

$$(\varphi_{\tilde{k}}u)(\lambda) = \lambda_k u(\lambda) \quad (\lambda = (\lambda_k)_{k=1}^{\infty} \in \mathbb{R}^{\infty};\ k \in \mathbb{N}). \tag{3.87}$$

Since $u(\lambda) \in \mathfrak{F}'_{\text{fin}}(\Phi)$, we conclude that $u(\lambda)$ has the form

$$u(\lambda) = (u_0(\lambda), u_1(\lambda), u_2(\lambda), \ldots),$$

where $u_n(\lambda) \in \mathfrak{F}_n(\Phi')$ $(n \in \mathbb{Z}_+)$. The set of equations (3.87) now gives a sequence of recurrence relations

$$\varphi_k u_n(\lambda) = \lambda_k u_n(\lambda) = \frac{1}{\sqrt{2}}(a_+(e_k)\, u_{n-1}(\lambda) + a_-(e_k)\, u_{n+1}(\lambda))$$

$$(n \in \mathbb{Z}_+;\ k \in \mathbb{N}). \tag{3.88}$$

Here we set $u_{-1}(\lambda) = 0$, $u_0(\lambda) = 1$. Expanding $u_n(\lambda)$ in the basis of occupation numbers and taking into account the values of the operators $a_\pm(e_k)$ on the vectors from this basis, we obtain a set of recurrence relations

$$\lambda_k u_n^{(\alpha)}(\lambda) = \frac{1}{\sqrt{2}}(\sqrt{\alpha_k}\, u_{n-1}^{(\alpha-1_k)}(\lambda) + \sqrt{\alpha_k+1}\, u_n^{(\alpha+1_k)}(\lambda))$$

$$(n \in \mathbb{Z}_+;\ k \in \mathbb{N};\ \alpha \in \mathbb{Z}_{+,0}^{\infty},\ |\alpha| = n)$$

for the coefficient functions

$$u_n^{(\alpha)}(\lambda) = (u_n(\lambda), e_\alpha)_{H_c^{\otimes n}} \quad (\alpha \in \mathbb{Z}_{+,0}^\infty,\ |\alpha| = n).$$

By induction, we now easily obtain

$$u_n^{(\alpha)}(\lambda) = h_{\alpha_1}(\lambda_1) \dots h_{\alpha_\nu}(\lambda_\nu) = h_\alpha(\lambda) \quad (\lambda = (\lambda_j)_{j=1}^\infty,\ |\alpha| = n),$$

where $h_{\alpha_j}(\lambda_j)$ are the Hermite polynomials normalized with respect to the measure $d\gamma_1(\lambda_j) = \pi^{-1/2} e^{-\lambda_j^2} d\lambda_j$ (see Subsection 2.2 in Chapter 2).

The Fourier transformation

$$\mathfrak{F}_{\text{fin}}(\Phi) \ni f \mapsto \tilde{f}(\lambda) = (f, u(\lambda))_{\mathfrak{F}(H)}$$

transforms vectors from $\mathfrak{F}_{\text{fin}}(\Phi)$ into the functions $\tilde{f}(\lambda)$ $(\lambda \in \mathbb{R}^\infty)$. Under this transformation, the basis vector $e_\alpha \in \mathfrak{F}_{\text{fin}}(\Phi)$ turns into the function $h_\alpha(\lambda)$ $(\lambda \in \mathbb{R}^\infty;\ \alpha \in \mathbb{Z}_{+,0}^\infty)$. The unique normalized measure defined on $\mathfrak{C}_\sigma(\mathbb{R}^\infty)$, for which the Parseval equality

$$(E(\delta)f, g)_{\mathfrak{F}(H)} = \int_\delta \tilde{f}(\lambda)\, \overline{\tilde{g}(\lambda)}\, d\rho(\lambda) \quad (f, g \in \mathfrak{F}_{\text{fin}}(\Phi);\ \delta \in \mathfrak{C}_\sigma(\mathbb{R}^\infty))$$

holds, has the form

$$d\rho(\lambda) = \mathop{\times}_{k=1}^{\infty} d\gamma_1(\lambda_k) = d\gamma_1(\lambda).$$

This follows from the facts that the functions $(h_\alpha(\lambda))_{\alpha \in \mathbb{Z}_{+,0}^\infty}$ form an orthonormal basis in $L_2(\mathbb{R}^\infty, d\gamma_1(\lambda))$ (see Subsection 2.2 in Chapter 2) and that the spectral measure is uniquely defined.

By virtue of the Parseval equality, the relation

$$\mathfrak{F}_{\text{fin}}(\Phi) \ni f \mapsto I_1 f = \tilde{f} \in L_2(\mathbb{R}^\infty, \gamma_1)$$

can be extended to the isometric isomorphism between the whole of the Fock space $\mathfrak{F}(H)$ and $L_2(\mathbb{R}^\infty, \gamma_1)$. The isomorphism just described coincides with the Segal isomorphism studied in Section 2 of Chapter 2. Under this isomorphism, the basis of occupation numbers in $\mathfrak{F}(H)$ turns into the basis of Hermite polynomials in $L_2(\mathbb{R}^\infty, \gamma_1)$ (see Subsection 2.3 of Chapter 2).

4. REPRESENTATIONS BY COMMUTING OPERATORS

As has been explained in the introduction to Chapter 3, by using the spectral projection theorem, one can obtain the spectral representations for the families $A = (A_x)_{x \in X}$ of commuting normal operators in H_0 satisfying certain relations. Recall that, according to this theorem, the following equality holds

$$A_x = \int_{g(A)} \lambda(x)\, dE(\lambda(\cdot)) \quad (x \in X), \tag{0.1}$$

where $g(A)$ is a generalized spectrum of the family. Assume that the operators $A(x)$ are not arbitrary but satisfy certain relations which can be written in the form $F(A\,.) = 0$. More exactly, this means that a certain functional or operator $F(f(\cdot))$ is defined on the space of complex-valued functions $f(x)$ of the variable $x \in X$, and this functional is such that the formal substitution of A_x for $f(x)$ in the corresponding expression is meaningful (for example,

$$X = \mathbb{R}^1 \text{ and } F(f(\cdot)) = f(x_1) + f(x_2) \ (x_1, x_2 \in \mathbb{R}^1); \ F(f(\cdot)) = -f''(\cdot) + q(\cdot) f(\cdot),$$

where $q \in C(\mathbb{R}^1)$ etc.). Then, by using the fact that ξ is a generalized joint eigenvector of the family A with the eigenvalue $\lambda(\cdot)$ (this means that $A_x \xi = \lambda(x)\, \xi$ for any $x \in X$), one can usually conclude that the equality $F(\lambda(\cdot)) = 0$ holds (for this purpose, we "convey" the vector ξ through the expression for F). In other words, every eigenvalue from $g(A)$ satisfies the equation $F(\lambda(\cdot)) = 0$ and, therefore, the region of integration in (0.1) may be replaced by a collection of all the solutions of this equation. If, in addition, this collection can be described somehow, then the functional integral from (0.1) may be transformed into a simpler and analyzable subject.

Instead of saying that the operators A_x satisfy the relation $F(A\,.) = 0$, one may say that there exists a representation of the formal relation $F(a.) = 0$ by commuting operators. Then the procedure described above leads to the spectral integrals for the operators A_x which realize this representation. The aim of this chapter is to prove spectral formulas describing the operators A_x when following this way. Here one can notice the following rule: If the space X is locally compact, these formulas are obtained without additional restrictions imposed on the operators A_x except the natural condition of their continuity with respect to x.

In the case when X is not locally compact (for example, when X is an infinite-dimensional Hilbert space), one should additionally demand that the space H_0 should possess a quasinuclear rigging connected with the operators A_x in a standard way. Moreover, the presence of this rigging is necessary in order that the spectral formulas be valid.[33]

In Section 1, we consider representations of this sort for the cases when X is a commutative group, semi-group, algebra, and linear space. In this chapter, we do not consider the important connections between the spectral representations obtained here and the theory of positive definite functions, with the moment problem, etc. These problems are investigated in Chapter 5.

In Section 2, we consider the representations of hypercomplex systems with a locally compact basis. These subjects generalize the concept of a usual commutative hypercomplex system (i.e., of a finite-dimensional commutative algebra with a chosen basis) in the same manner as the group algebra of a locally compact group generalizes the group algebra of a finite group. The main part of this section is devoted to the presentation of necessary facts from the theory of hypercomplex systems of this type including the construction of the theory of generalized functions on the basis of these systems. As examples, we obtain representations for the center of a group algebra of a compact group, for the algebra of orbital functions, for the algebra constructed by using the orthogonal polynomials or the Sturm–Liouville equation, etc.

Section 3 deals with applications of the spectral projection theorem which enables us to obtain representations of the families $B = (B_y)_{y \in Y}$ of operators connected by commutative relations. The scheme presented there can be formulated as follows. It may often happen that, along with B, one can construct a family $A = (A_x)_{x \in X}$ of commuting normal operators such that in terms of the Fourier transform connected with A, the actions of the original operators become more or less analyzable (it may often happen that A is even given). This enables us to describe the operators from the family B. This point of view is frequently used, for example, in the theory of unitary representations of groups (in this case, A consists of unitary commuting operators).

In Section 3, this procedure is extended to the families A of now unbounded (in general) normal commuting operators with arbitrary cardinality. We also consider several examples of representations (not necessarily unitary) of semi-direct products G of groups one of which is commutative, and of representations such as the canonical commutative relations. Note that in this case we have the same rule: If G is locally compact, then the required rigging exists automatically. In more general situations, this should be assumed (or, in some cases, proved).

---

[33] In the locally compact case, this rigging can be constructed automatically by using such procedure as the well known method for the construction of the Gårding domain in the theory of representations of groups.

§1. Representation of Some Algebraic Structures by Commuting Operators

If we consider an arbitrary family of commuting normal operators $A = (A_x)_{x \in X}$, then an arbitrary RI E on $\mathbb{C}^X$ can serve as its joint RI. However, in the case when operators A_x are connected by some algebraic relations, the RI E is not arbitrary; it is concentrated on certain sets from $\mathbb{C}^X$. Below, we consider special cases of these relations: X is a group, semi-group, algebra (all these are commutative), or a linear space, whereas A realizes a representation of these structures. In other words, the mapping $X \ni x \mapsto A_x$ should be such that the algebraic operations in X should transform into the corresponding operations over the operators A_x (for example, $A_{x+y} = A_x A_y$ in the case of a group). For simplicity of formulations, we should often employ a scheme with a nuclear rigging.

1.1. Representation of a Group

Let X be a commutative additive group. A complex-valued function $X \ni x \mapsto \chi(x) \in \mathbb{C}^1$ satisfying the condition

$$\chi(x+y) = \chi(x)\chi(y) \quad (x, y \in X)$$

is called a generalized character of this group. Furthermore, if $|\chi(x)| = 1$ $(x \in X)$, then χ is a character. A collection of all generalized characters (characters) is denoted by X_g (X); $X \subset X_g \subset \mathbb{C}^X$. In the case of a topological group X, it proves to be natural to consider characters which are, in addition, continuous functions. We set $X_{g,c} = X_g \cap C(X)$ and $X_c = X \cap C(X)$ ($C(X)$ is a collection of all the complex-valued continuous functions on X). Denote by $X_{g,\mathrm{Re}}$ and $X_{g,c,\mathrm{Re}}$ the corresponding collection of real-valued characters.

An operator-valued function $X \ni x \mapsto A x$ whose values are normal (generally speaking, unbounded) operators acting in a separable Hilbert space H_0 is called a representation of X if $A_{x+y} = A_x A_y$ $(x, y \in X)$. This representation is called nuclear if there exists a nuclear rigging $\Phi' \supset H_0 \supset \Phi$ standardly connected with $A = (A_x)_{x \in X}$ (this means that $\Phi \subset \mathfrak{D}(A_x)$ and $A_x \restriction \Phi \in \mathcal{L}(\Phi, \Phi)$ $(x \in X)$). Moreover, this representation is called nuclear continuous if, in addition, for every $\varphi \in \Phi$, the vector function $X \ni x \mapsto A_x\varphi \in \Phi$ is weakly continuous.

Theorem 1.1. *Let X be a commutative group, and let $X \ni x \mapsto A_x\varphi$ be its nuclear representation by normal operators. Then the following spectral integral representation holds*

$$A_x = \int_{X_g} \chi(x)\, dE(\chi),$$

$$\mathfrak{D}(A_x) = \Big\{f \in H_0 \,\Big|\, \int_{X_g} |\chi(x)|^2\, d(E(\chi) f, f)_{H_0} < \infty\Big\}, \tag{1.1}$$

where E is some RI on the σ-algebra

$$\mathfrak{C}_\sigma(X) = \{\alpha \cap X_g \mid \alpha \in \mathfrak{C}_\sigma(\mathbb{C}^X)\}.$$

If operators A_x are unitary (selfadjoint), then X_g should be replaced by X $(X_{g,\,\mathrm{Re}})$ in the above relation. In the case when the group X is topological and the representation is nuclear continuous, X_g should be replaced by $X_{g,\,c}$ and if, in addition, A_x are unitary (selfadjoint) operators, X_g is to be replaced by X_c $(X_{g,\,c,\,\mathrm{Re}})$.

Proof. We investigate the main case. Operators A_x commute, i.e., $A_x A_y = A_{x+y} = A_y A_x$ $(x, y \in X)$. But it is well known (see Devinatz, Nussbaum [1]) that if A, B and C are normal operators in the Hilbert space, and $AB = BA = C$, then the RI of operators A and B commute. We now take into account Remark 3 of Subsection 2.10 (Chapter 3) concerning the definition of the standard connection between a rigging and a family $A = (A_x)_{x\in X}$. Below, the generalized spectrum $g(A)$ is understood as a collection of all $\lambda(\cdot) \in \mathbb{C}^X$ such that only the second equality in (2.19) of Chapter 3 holds for some $\xi \in \Phi'$ and all $\varphi \in \Phi$. The following inclusion holds

$$g(A) \subset X_g. \tag{1.2}$$

In fact, let $\lambda(\cdot) \in g(A)$, and let $\xi \in \Phi'$ be the corresponding joint eigenvector. By using the second equality in (2.19) of Chapter 3 and the inclusion $\mathfrak{R}(A_x \restriction \Phi) \subset \Phi$, we obtain

$$\overline{\lambda(x+y)}\, (\xi, \varphi)_{H_0} = (\xi, A_{x+y}\varphi)_{H_0} = (\xi, A_x A_y \varphi)_{H_0}$$

$$= \overline{\lambda(x)}\, (\xi, A_y\varphi)_{H_0} = \overline{\lambda(x)\, \lambda(y)}\, (\xi, \varphi)_{H_0} \quad (x, y \in X) \tag{1.3}$$

for every $\varphi \in \Phi$. Choosing φ so that $(\xi, \varphi)_{H_0} \neq 0$, we find from (1.3) that

$$\lambda(x+y) = \lambda(x)\, \lambda(y) \quad (x, y \in X),$$

i.e., $\lambda(\cdot) \in X_g$, and (1.2) is valid.

Let us now apply Theorem 2.7 of Chapter 3 to A.[34] We set $\tau = X_g$. By virtue of (1.2), we have $\tau \supset g(A) \supset \pi$ and, therefore, one can write the representation (2.28) of Chapter 3, where π is replaced by $\tau = X_g$. Denoting E_τ by E, we get (1.1). In the main case, the theorem is proved.

The remaining representations are obtained similarly, one should only modify the inclusion (1.2) according to the additional information about the operators. Thus, if the operators A_x are selfadjoint, then $g(A)$ consists of real-valued functions (see Subsection 2.5 of Chapter 3). Consequently, (1.2) turns into

$$g(A) \subset X_{g,\,\mathrm{Re}}.$$

Then we operate as above with the only difference that $\mathbb{C}^X$ is replaced by $\mathbb{R}^X$ (see Remark 4 in Subsection 2.10 of Chapter 3).

If the group X is topological, and the representation is nuclear continuous, then for each eigenvalue $\lambda(\cdot)$, we have $\lambda(\cdot) \in C(X)$ by virtue of Subsection 2.9 (Chapter 3). Therefore,

$$g(A) \subset X_{g,c}.$$

This yields the required formula of type (1.1). In the case of a nuclear continuous representation by selfadjoint operators, we act analogously.

Now let a representation be unitary, i.e., A_x $(x \in X)$ are unitary. Then, according to Theorem 2.8, (1.14) of Chapter 3 and (1.2), the set π can be chosen so that

$$\pi \subset s(A) \cap g(A) \subset \{z \in \mathbb{C}^1 \mid |z| = 1\}^X \cap X_g = X.$$

We set $\tau = X$, and this yields the corresponding formula of type (1.1). The case of X_c can be investigated analogously.

■

1.2. Generalizations of Stone and Sz.-Nagy– Hille Theorems

Conditions of Theorem 1.1 are sufficient. This means that a formula of type (1.1) holds provided that these conditions are satisfied. It is a more delicate problem to obtain necessary and sufficient conditions of the validity of such representations. Thus, one should proceed from nuclear to quasinuclear chains. Let us present one result of this sort,

[34] More exactly, we apply Remark 1 to this theorem (taking into account what has been said in Remark 2 of Subsection 2.10 (Chapter 3)).

namely, the Stone theorem on the representation of a Hilbert space as an additive group. Note that the second part of the proof of this theorem (and of Theorems 1.3, 1.4, and 1.6) is closely connected with the constrictions presented in Subsection 3.4 of Chapter 3. More exactly, one can deduce this part of the proof from Theorem 3.8 of Chapter 3. However, this result is quite important and, therefore, we present a complex and independent version of the proof (see also the second part of the proof of Theorem 1.3). Hence, let us consider a chain

$$H_0 \supset H_+ \supset D, \tag{1.4}$$

where H_0 and H_+ are Hilbert spaces, H_+ is densely quasinuclearly imbedded into H_0, and D is a separable projective limit of Hilbert spaces topologically imbedded into H_+.

Theorem 1.2. *A unitary representation* $X \ni x \mapsto A_x \in \mathcal{L}(H_0, H_0)$ *of a real separable Hilbert space* X *has the form*

$$A_x = \int_X e^{i(\lambda, x)_X} dE(\lambda) \quad (x \in X), \tag{1.5}$$

where E *is some RI on the* σ*-algebra* $\mathfrak{B}(X)$ *of Borel sets from* X, *if and only if there exists a rigging* (1.4) *such that*

(i) $$A_x \upharpoonright D \in \mathcal{L}(D, H_+) \quad (x \in X);$$

(ii) for every $\varphi \in D$, *the vector function* $X \ni x \mapsto A_x\varphi \in H_+$ *is weakly continuous.*

The RI from (1.5) *is uniquely determined by* $(A_x)_{x \in X}$.

Proof. The family $(A_x)_{x \in X}$ of commuting unitary operators is standardly connected with the chain (1.4). According to Subsection 2.5 of Chapter 3, one should also show that $A_x^* \upharpoonright D \in \mathcal{L}(D, H_+)$, but this follows from the assumption that $A_x \upharpoonright D \in \mathcal{L}(D, H_+)$ and the equality $A_x^* = A_x^{-1} = A_{-x}$ $(x \in X)$.[35] However, the equality (1.3) is not valid in this case, since D is not invariant with respect to A_x. Let us act as follows.

Assume that $(\varphi_j)_{j=1}^{\infty}$ $((x_l)_{l=1}^{\infty} \ni 0)$ is a countable set of vectors dense in D (X). Then

$$H_+ \supset (A_{x_l}\varphi_j)_{j,l=1}^{\infty} \supset (\varphi_j)_{j=1}^{\infty} \quad (A_0 = 1)$$

is dense in H_+. In the set $(A_{x_l}\varphi_j)_{j,l=1}^{\infty}$, we select an arbitrary sequence of linearly independent vectors such that its linear span D_1 contains this set. Each vector from D_1

35 Clearly, this simplification of the definition of a standard connection for the case of unitary A_x can be realized in Theorem 1.1, as well.

has the form

$$\varphi = \sum_{k=1}^{\infty} \alpha_k A_{x_{l_k}} \varphi_{j_k},$$

where $(\alpha_k)_{k=1}^{\infty}$, $\alpha_k \in \mathbb{C}^1$ is an arbitrary finite sequence of coordinates φ. Let us transform D_1 into a linear topological space by introducing the following convergence: $D_1 \ni \varphi^{(n)} \to \varphi \in D_1$ as $n \to \infty$ if the corresponding coordinates $\alpha_k^{(n)}$ are uniformly finite, and $\alpha_k^{(n)} \to \alpha_k$ for every $k \in \mathbb{N}$. It is easy to see that D_1 will be a separable projective limit of Hilbert spaces with topological imbedding (cf. what has been said in steps III and IV of the proof of the existence of a rigging in Theorem 3.8 of Chapter 3). The quasinuclear chain

$$H_- \supset H_0 \supset H_+ \supset D_1$$

is standardly connected with A, i.e., let $(\alpha_k)_{k=1}^{\infty}$ be finite, then

$$A_x \varphi = A_x \left(\sum_{k=1}^{\infty} \alpha_k A_{l_k} \varphi_{j_k} \right) = \sum_{k=1}^{\infty} \alpha_k A_{x+x_{l_k}} \varphi_{j_k} \in H_+.$$

This also implies that $A_x \upharpoonright D_1 \in \mathcal{L}(D_1, H_+)$, and that the analogous relations hold for $A_x^* = A_{-x}$ $(x \in X)$. It is also clear that for every $\varphi \in D_1$, the function $X \ni x \mapsto A_x \varphi \in H_+$ is weakly continuous. According to Subsection 2.9 of Chapter 3, we have $g(A) \subset C(X)$, where $g(A)$ is understood as a generalized spectrum connected with the quasinuclear chain constructed.

Show that $g(A) \subset X_{g,c}$. Let $\lambda(\cdot) \in g(A)$, and $\xi \in H_-$ be the corresponding generalized joint eigenvector. The second equality in (2.19) of Chapter 3 gives

$$\overline{\lambda(x+x_l)}\, (\xi, \varphi_j)_{H_0} = (\xi, A_{x+x_l} \varphi_j)_{H_0} = (\xi, A_x A_{x_l} \varphi_j)_{H_0}$$

$$= \overline{\lambda(x)}\, (\xi, A_{x_l} \varphi_j)_{H_0} = \overline{\lambda(x)\,\lambda(x_l)}\, (\xi, \varphi_j)_{H_0} \quad (j, l \in \mathbb{N})$$

for arbitrary $x \in X$ (we have used the fact that φ_j, $A_{xl}\, \varphi_j \in D_1$). By virtue of the equality obtained, continuity of $\lambda(\cdot)$, and denseness of $(x_l)_{l=1}^{\infty}$ in X, we calculate that

$$\lambda(x+y) = \lambda(x)\, \lambda(y) \quad (x, y \in X),$$

i.e., $g(A) \subset X_{g,c}$.

Taking into account Theorem 2.8 and (1.14) of Chapter 3, and the statement just proved, we can choose the set π so that

$$\pi \subset s(A) \cap g(A) \subset X_c.$$

We put $\pi = X_c$ and write the corresponding formula of type (1.1). It is easy to see that each continuous character $\chi(X)$ on X is equal to $e^{i(\lambda, x)_X}$, where $\lambda \in X$ is a certain vector. This involves the existence of the bijection

$$X_c \ni \chi(\cdot) \mapsto e^{i(\lambda, \cdot)_X} \mapsto \lambda \in X.$$

Denoting the image of E under this mapping by E, again we proceed from the formula (1.1) written just above to (1.5).

One can easily find that the RI E obtained is given on $\mathfrak{C}_\sigma(X) = \{\alpha \cap X \mid \alpha \in \mathfrak{C}_\sigma(\mathbb{C}^X)\}$ and this σ-algebra coincides with $\mathfrak{B}(X)$. Indeed, since X is separable, it possesses a countable basis of neighborhoods in the weak topology. This means that $\mathfrak{C}_\sigma(X)$ coincides with the σ-algebra $\mathfrak{B}(X)$ of Borel sets in the topology X given by the norm.[36]

We proceed to the proof of the converse statement of the theorem. Here we shall employ the results of Subsection 4.4 (Chapter 2) and the notations of this Subsection.

Consider a sequence $\tau = (\tau_n)_{n=1}^\infty \in [1, \infty)^\infty$ such that $\tau = (\tau_n^{-1})_{n=1}^\infty \in l_1$ (hereinafter $l_p = l_p(\mathbb{C}^1)$; $p = 1, 2$). Let us also introduce the space X_τ^+ according to (4.30) of Chapter 2. Since X is isomorphic to X_τ^+ (as any two separable real Hilbert spaces), one can assume that there exists a representation (1.5) of this theorem, where $X = X_\tau^+$. For future discussions, it is convenient to identify X' not with X but with a negative space X_τ^- of the chain (4.33) of Chapter 2 (here one should employ the relation

$$(x, y)_{X_\tau^+} = (\boldsymbol{I}^{-1}x, y)_{l_2} \quad (x, y \in X_\tau^+),$$

where the operator $\boldsymbol{I}$ is connected with this chain). Let F be the image of the RI E under the mapping $X_\tau^+ \ni x \mapsto \boldsymbol{I}^{-1}x, \in X_\tau^-$, i.e., F is defined by the formula

$$\mathfrak{B}(X_\tau^-) \ni \alpha \mapsto F(\alpha) = E(\boldsymbol{I}(\alpha)).$$

Clearly, F is the RI in the space X_τ^-. The equality (1.5) can be now rewritten as

$$A_x = \int_{X_\tau^-} e^{i(\lambda, x)_{l_2}} \, dF(\lambda) \quad (x \in X_\tau^+). \tag{1.6}$$

---

[36] Here one should clarify that each ball $\{x \in X \mid \|x\|_X \le 1\}$ can be obtained as an intersection of countably many subspaces $\{x \in X \mid (x, y_n)_X \le 1\}$, where $(y_n)_{n=1}^\infty$ is a dense sequence of points on its surface.

Thus, one should prove that the existence of the representation (1.6) guarantees that the rigging (1.4) with the required properties can be constructed. We shall prove this in several steps.

I. Fix a vector $l_1 \in H_0$, $\| l_1 \|_{H_0} = 1$, and consider a linear mapping

$$A_\tau(\mathbb{R}^\infty) \ni a \mapsto Q_1 a = \int_{X_\tau^-} a(\lambda) \, \| \delta_\lambda \|^{-1}_{A_{-\tau}(\mathbb{R}^\infty)} \, dF(\lambda) \, l_1. \tag{1.7}$$

We clarify that $a(\lambda)$ is continuous for $\lambda \in X_\tau^-$ (see Theorem 4.3 of Chapter 2). By virtue of (4.37) of Chapter 2, the function under the integral sign in (1.7) is also continuous and does not exceed $\| a \|_{A_\tau(\mathbb{R}^\infty)}$ in modulus. Hence, the integral exists, and we can write the estimate

$$\| Q_1 a \|^2_{H_0} = \int_{X_\tau^-} | a(\lambda) |^2 \, \| \delta_\lambda \|^2_{A_{-\tau}(\mathbb{R}^\infty)} \, d(F(\lambda) \, l_1, l_1)_{H_0} \le \| a \|^2_{A_{-\tau}(\mathbb{R}^\infty)}. \tag{1.8}$$

Therefore, Q_1 is continuous and $\| Q_1 \| \le 1$. Furthermore, Q_1 is a Hilbert–Schmidt operator, and $|\!|\!| Q_1 |\!|\!| = 1$. In fact, let us consider an orthonormal basis

$$e_\alpha^{(\tau)}(\lambda) = \tau_1^{-\alpha_1/2} \ldots \tau_\nu^{-\alpha_\nu/2} h_\alpha(\lambda) \quad (\alpha \in \mathbb{Z}^\infty_{+,0})$$

in the space $A_\tau(\mathbb{R}^\infty)$. Taking into account (4.21) and (4.35) of Chapter 2, we get

$$|\!|\!| Q_1 |\!|\!|^2 = \sum_{\alpha \in \mathbb{Z}^\infty_{+,0}} \| Q_1 e_\alpha^{(\tau)} \|^2_{H_0}$$

$$= \sum_{\alpha \in \mathbb{Z}^\infty_{+,0}} \int_{X_\tau^-} h_\alpha^2(\lambda) \, \tau_1^{-\alpha_1} \ldots \tau_\nu^{-\alpha_\nu} \, \| \delta_\lambda \|^2_{A_{-\tau}(\mathbb{R}^\infty)} \, d(F(\lambda) \, l_1, l_1)_{H_0} = 1.$$

II. Consider the kernel of the operator Q_1, i.e., the subspace $\{ a \in A_\tau(\mathbb{R}^\infty) \mid Q_1 a = 0 \}$, and denote by $\hat{A}_\tau(\mathbb{R}^\infty)$ its orthogonal complement. Let $\hat{Q}_1$ be the restriction of Q_1 to $\hat{A}_\tau(\mathbb{R}^\infty)$. By choosing an orthonormal basis in $A_\tau(\mathbb{R}^\infty)$ so that its vectors belong to $\hat{A}_\tau(\mathbb{R}^\infty)$ and to this kernel, we find that $|\!|\!| \hat{Q}_1 |\!|\!| = |\!|\!| Q_1 |\!|\!| = 1$. On $\mathfrak{R}(Q_1) = \mathfrak{R}(\hat{Q}_1)$ there exists the inverse operator $\hat{Q}_1^{-1}$ which is closed (since $\hat{Q}_1$ is continuous) and such that

$$\| \hat{Q}_1^{-1} f \|_{A_\tau(\mathbb{R}^\infty)} \ge \| f \|_{H_0} \quad (f \in \mathfrak{R}(Q_1)). \tag{1.9}$$

We now transform $\mathfrak{R}(Q_1) \subset H_0$ into a Hilbert space $H_{+,1}$ by setting

$$(f, g)_{H_{+,1}} = (\hat{Q}_1^{-1} f, \hat{Q}_1^{-1} g)_{A_\tau(\mathbb{R}^\infty)} \quad (f, g \in \mathfrak{R}(Q_1)).$$

Since $\hat{Q}_1^{-1}$ is closed and the estimate (1.9) holds, the space $H_{+,1} = \mathfrak{R}(Q_1)$ is complete. The imbedding $O_1 \colon H_{+,1} \to H_0$ is quasinuclear and $|O_1| = 1$. Indeed, let $(a_n)_{n=1}^\infty$ be an orthonormal basis in $\hat{A}_\tau(\mathbb{R}^\infty)$, then $(\hat{Q}_1 a_n)_{n=1}^\infty$ is an orthonormal basis in $H_{+,1}$. For $(\hat{Q}_1 a_n)_{n=1}^\infty$, we have

$$|O_1|^2 = \| \hat{Q}_1 a_n \|_{H_0}^2 = |\hat{Q}_1| = 1.$$

We set

$$D_1 = Q_1 \mathcal{P}_{\mathrm{cyl}}(\mathbb{R}^\infty) \subset H_{+,1},$$

where $\mathcal{P}_{\mathrm{cyl}}(\mathbb{R}^\infty)$ is a collection of all the cylindrical polynomials. In other words, D_1 consists of all the vectors of the form

$$\varphi = \sum_{\alpha \in \mathbb{Z}_{+,0}^\infty} \varphi_\alpha Q_1 e_\alpha^{(\tau)},$$

where the function $\mathbb{Z}_{+,0}^\infty \ni \alpha \mapsto \varphi_\alpha \in \mathbb{C}^1$ is finite. Thus, we have obtained the chain

$$H_0 \supset H_{+,1} \supset D_1, \tag{1.10}$$

in which the imbedding $O_1 \colon H_{+,1} \to H_0$ is quasinuclear, and D_1 is dense in $H_{+,1}$.

III. If $H_{+,1}$ is dense in H_0, then the construction is already completed. If this is not true, then we choose a vector $l_2 \in H_0$, $\|l_2\|_{H_0} = 1$ orthogonal to $H_{+,1}$ and repeat the construction of I and II with l_1 replaced by l_2. As a result, we receive the chain

$$H_0 \supset H_{+,2} \supset D_2, \tag{1.11}$$

with the quasinuclear imbedding $O_2 \colon H_{+,2} \to H_0$; D_2 is dense in $H_{+,2}$. The operator (1.7), with l_1 replaced by l_2, is denoted by Q_2.

The space $H_{+,2}$ is orthogonal to $H_{+,1}$ with respect to the scalar product $(\cdot, \cdot)_{H_0}$. In fact, for $a, b \in A_\tau(\mathbb{R}^\infty)$ we have

$$(Q_1 a, Q_2 b)_{H_0} = \int_{X_\tau^-} a(\lambda) \overline{b(\lambda)} \, \| \delta_\lambda \|_{A_{-\tau}(\mathbb{R}^\infty)}^{-2} \, d(F(\lambda) l_1, l_2)_{H_0}.$$

This implies that it suffices to establish the equality

$$\int\limits_{\alpha} \| \delta_\lambda \|^{-1}_{A_{-\tau}(\mathbb{R}^\infty)} \, d(F(\lambda)\, l_1, l_2)_{H_0} = 0$$

for every

$$\alpha \in \mathfrak{B}(X_\tau^-) = \{\beta \cap X_\tau^- \mid \beta \in \mathfrak{C}_\sigma(\mathbb{R}^\infty)\}$$

and, consequently, for every $\alpha \in \mathfrak{B}(X_\tau^-) = \{\beta \cap X_\tau^- \mid \beta \in \mathfrak{C}(\mathbb{R}^\infty)\}$. The last set has the form

$$\alpha = \{\lambda \in X_\tau^- \mid (\lambda_1, \ldots, \lambda_p) \in \delta\},$$

where $\delta \in \mathfrak{B}(\mathbb{R}^p)$ with some $p \in \mathbb{N}$. Therefore, it suffices to show that a complex-valued finite measure

$$\mathfrak{B}(\mathbb{R}^p) \ni \gamma \mapsto \omega(\gamma) = \int\limits_{\{\lambda \in X_\tau^- \mid (\lambda_1, \ldots, \lambda_p) \in \gamma\}} d_\tau^{-1} \exp\left(- \sum_{k=p+1}^{\infty} \lambda_k^2 (1+\tau_k)^{-1} \right)$$

$$\times \, d(F(\lambda)\, l_1, l_2)_{H_0}$$

is zero. Due to (4.35) of Chapter 2, we have

$$\int\limits_{\{\lambda \in X_\tau^- \mid (\lambda_1, \ldots, \lambda_p) \in \gamma\}} \| \delta_\lambda \|^{-1}_{A_{-\tau}(\mathbb{R}^\infty)} \, d(F(\lambda)\, l_1, l_2)_{H_0}$$

$$= \int\limits_{\gamma} \exp\left(-\sum_{k=1}^{p} \lambda_k^2 (1+\tau_k)^{-1} \right) d\omega(\lambda_1, \ldots, \lambda_p).$$

In order to prove this fact, we note that the vector l_2 is orthogonal in H_0 to the space $H_{+,1}$ by its construction. Hence, it is orthogonal to all the vectors $Q_1 l_\alpha$ ($\alpha \in \mathbb{Z}^\infty_{+,0}$). In particular, it is orthogonal to all the vectors $Q_1(h_{\alpha_1}(\lambda_1) \ldots h_{\alpha_p}(\lambda_p))$ ($\alpha_1, \ldots, \alpha_p \in \mathbb{Z}^\infty_{+,0}$), i.e.

$$0 = (Q_1(h_{\alpha_1}(\lambda_1) \ldots h_{\alpha_p}(\lambda_p)), l_2)_{H_0}$$

$$= \int\limits_{X_\tau^-} h_{\alpha_1}(\lambda_1) \ldots h_{\alpha_p}(\lambda_p) \| \delta_\lambda \|^{-1}_{A_{-\tau}(\mathbb{R}^\infty)} \, d(F(\lambda)\, l_1, l_2)_{H_0}$$

$$= \int\limits_{\mathbb{R}^p} h_{\alpha_1}(\lambda_1) \ldots h_{\alpha_p}(\lambda_p) \exp\left(-\sum_{k=1}^{p} \lambda_k^2 (1+\tau_k)^{-1} \right) d\omega(\lambda_1, \ldots, \lambda_p). \quad (1.12)$$

Below, we shall prove Lemma 1.1 which implies that relation (1.12) involves the equality $w = 0$. Thus, $H_{+,2}$ is orthogonal to $H_{+,1}$ in H_0.

IV. It follows from (1.10) and (1.11) that

$$H_0 \supset H_{+,1} \oplus H_{+,2} \supset D_1 \otimes D_2, \tag{1.13}$$

where $\oplus$ denotes an orthogonal sum in H_0. If $H_{+,1} \oplus H_{+,2}$ is dense in H_0, then the construction is thus completed. But if this is not so, we choose a vector $l_3 \in H_0$, $\| l_3 \|_{H_0} = 1$ orthogonal to $H_{+,1} \oplus H_{+,2}$, and construct a chain similar to (1.10) and (1.11), etc. As a result, we obtain the chain

$$H_0 \supset \bigoplus_{n=1}^{\infty} H_{+,n} \supset \bigoplus_{n=1}^{\infty} D_n, \tag{1.14}$$

where the orthogonal summation is considered in H_0, and $\oplus_{n=1}^{\infty} H_{+,n}$ is now dense in H_0.[37]

We now replace $\oplus_{n=1}^{\infty} H_{+,n}$ in (1.14) by a "weighted" orthogonal sum of these spaces $\oplus_{n=1;\,\delta}^{\infty} H_{+,n}$, where $\delta = (\delta_n)_{n=1}^{\infty}$, $\delta_n > 0$ is a certain weight. This sum is defined as a Hilbert space of sequences $f = (f_n)_{n=1}^{\infty}$, $f_n \in H_{+,n}$ such that

$$\sum_{n=1}^{\infty} \| f_n \|^2_{H_{+,n}} \delta_n < \infty$$

with natural linear operations and a scalar product. As before, $\oplus_{n=1;\,\delta}^{\infty} H_{+,n}$ is dense in H_0, and for the corresponding imbedding operator $O = \oplus_{n=1}^{\infty} O_n$, we have

$$| O |^2 = \sum_{n=1}^{\infty} \delta_n^{-1} | O_n |^2 = \sum_{n=1}^{\infty} \delta_n^{-1}. \tag{1.15}$$

Let us choose δ_n rapidly decreasing so that the series (1.15) converge, and set $H_+ = \oplus_{n=1;\,\delta}^{\infty} H_{+,n}$. Hence, the imbedding $H_+ \to H_0$ is now quasinuclear.

Let us regard a collection of all finite sequences $\varphi = (\varphi_n)_{n=1}^{\infty}$, $\varphi_n \in D_n$ as D. We topologize D by assuming that $\varphi^{(k)} \in D$ converge to $\varphi \in D$ as $k \to \infty$ if all $\varphi^{(k)}$ are finite uniformly in k, and $\varphi_k^{(n)} \to \varphi_n$ for every $n \in \mathbb{N}$ in the sense of convergence of the corresponding coordinates $(\varphi_k^{(n)})(\alpha)$ in the expansion of the vector $\varphi_k^{(n)}$ in the

[37] Here, the number of terms may be finite, too. For definiteness, we investigate the more difficult case of infinitely many terms.

basis vectors $(Q_n e_\alpha^{(\tau)})_{\alpha\in\mathbb{Z}^\infty_{+,0}}$ (here Q_n denotes the operator Q_1 for the n-th step). Moreover, the functions $\mathbb{Z}^\infty_{+,0} \ni \alpha \mapsto (\varphi_k^{(n)})(\alpha) \in \mathbb{C}^1$ should be finite uniformly in k. It is easy to see that one may realize this topologization of D more accurately (indicating a system of neighborhoods). To do this, we consider a Hilbert space

$$G_{n,\sigma_n} = \left\{ \sum_{\alpha\in\mathbb{Z}^\infty_{+,0}} f_\alpha(Q_n e_\alpha^{(\tau)}) \mid \sum_{\alpha\in\mathbb{Z}^\infty_{+,0}} |f_\alpha|^2 \sigma_n(\alpha) < \infty \right\}$$

$$(\sigma_n = (\sigma_n(\alpha))_{\alpha\in\mathbb{Z}^\infty_{+,0}}, \ \sigma_n(\alpha) \geq 1)$$

with a natural scalar product and construct the weighted orthogonal sum $\oplus_{n=1;\,\rho}^\infty G_{n,\sigma_n}$, where $\rho = (\rho_n)_{n=1}^\infty$, $\rho_n \geq 1$ is a weight. Then $D = \operatorname{pr\,lim} \oplus_{n=1;\,\rho}^\infty G_{n,\sigma_n}$, where the projective limit is taken in all the weights σ_n and ρ. Clearly, the imbedding $D \to H_+$ is topological and, moreover, D is dense in H_+.

Thus, the chain (1.4) is constructed. Now show that conditions (i) and (ii) in the statement of the theorem hold.

V. In order to prove the inclusion

$$A_x \upharpoonright D \in \mathcal{L}(D, H_+) \quad (x \in X_\tau^+)$$

it suffices to show that

$$A_x \upharpoonright D \in \mathcal{L}(D, H_{+,n}) \quad (n \in \mathbb{N},\ x \in X_\tau^+).$$

For example, let $n = 1$ and $\varphi \in D_1$, then $\varphi = Q_1 p$, where $p \in \mathcal{P}_{\mathrm{cyl}}(\mathbb{R}^\infty)$. According to (1.6) and (1.7), for fixed $x \in X_\tau^+$, we have

$$A_x \varphi = \int_{X_\tau^-} e^{i(\lambda, x)_{l_2}} dF(\lambda) \left(\int_{X_\tau^-} p(\mu) \| \delta_\lambda \|^{-1}_{A_{-\tau}(\mathbb{R}^\infty)} dF(\mu)\, l_1 \right)$$

$$= \int_{X_\tau^-} e^{i(\lambda, x)_{l_2}} p(\lambda) \| \delta_\lambda \|^{-1}_{A_{-\tau}(\mathbb{R}^\infty)} dF(\mu)\, l_1. \tag{1.16}$$

According to Lemma 4.2 of Chapter 2 and the Remark to it, we have $e^{i(\cdot, x)_{l_2}} p(\cdot) \in A_\tau(\mathbb{R}^\infty)$ (λ and x now exchange their roles). Therefore,

$$A_x \varphi = Q_1(e^{i(\cdot, x)_{l_2}} p(\cdot)) \in \mathcal{R}(Q_1) = H_{+,1}.$$

The operator $D_1 \ni \varphi \mapsto A_x \varphi \in H_{+,1}$ is obviously continuous because the fact that φ converges means the coordinatewise convergence in the expansion of φ in the basis $(Q_n e_\alpha^{(\tau)})_{\alpha \in \mathbb{Z}_{+,0}^\infty}$ and, moreover, the coordinates should be uniformly finite. Condition (i) is thus established.

Condition (ii) states that a vector function $X_\tau^+ \ni x \mapsto A_x \varphi \in H_+$ is weakly continuous for every $\varphi \in D$. It suffices to consider $\varphi \in D_n$. Then $A_x \varphi \in H_{+,n}$. Let $n = 1$. According to (1.16), we have

$$A_x \varphi = Q_1(e^{i(\cdot, x)_{l_2}} p(\cdot)).$$

But due to Lemma 4.5 of Chapter 2, the vector function

$$X_\tau^+ \ni x \mapsto e^{i(\cdot, x)_{l_2}} p(\cdot) \in A_\tau(\mathbb{R}^\infty)$$

is strongly continuous. By virtue of the fact that

$$Q_1 \in \mathcal{L}(A_\tau(\mathbb{R}^\infty), H_0),$$

the function

$$X_\tau^+ \ni x \mapsto Q_1(e^{i(\cdot, x)_{l_2}} p(\cdot)) = A_x \varphi \in H_{+,1}$$

is even strongly continuous.

We now prove the last statement of the theorem, namely the fact that E is uniquely defined for a given family $(A_x)_{x \in X}$. Let us choose an orthonormal basis in X and realize this space as a real space l_2. Then (1.5) can be rewritten as

$$A_x \varphi = \int_{l_2} e^{i(\lambda, x)_{l_2}} \, dE(\lambda) \quad (x \in l_2). \tag{1.17}$$

Fix $p \in \mathbb{N}$ and consider $x^{(p)} \in \mathbb{R}^p$. Let $x = (x_1^{(p)}, \ldots, x_p^{(p)}, 0, 0, \ldots)$. Inserting this x into (1.17) and denoting $A_x = A_{x^{(p)}}^{(p)}$, we obtain the following formula

$$A_{x^{(p)}}^{(p)} = \int_{\mathbb{R}^p} \exp\left(i \sum_{n=1}^{p} \lambda_n x_n \right) dE^{(p)}(\lambda) \quad (x^{(p)} \in \mathbb{R}^p) \tag{1.18}$$

for the unitary representation $\mathbb{R}^p \ni x^{(p)} \mapsto A_{x^{(p)}}^{(p)}$, where

$$\mathcal{B}(\mathbb{R}^p) \ni \delta \mapsto dE^{(p)}(\delta) = E(\mathbb{C}(1, \ldots, p; \delta) \cap L_2).$$

By virtue of the usual p-dimensional Stone's theorem, the resolution of the identity $E^{(p)}$ from (1.18) is uniquely determined in terms of $(A^{(p)}_{x^{(p)}})_{x^{(p)} \in \mathbb{R}^p}$ (and, hence, in terms of $(A_x)_{x \in l_2}$). But it is obvious that $E^{(p)}$ $(p \in \mathbb{N})$ determine E uniquely.

■

It remains to prove the following general lemma.

Lemma 1.1. *Assume that* $\mathfrak{B}(\mathbb{R}^p) \ni \delta \mapsto \omega(\delta)$ *is a complex-valued finite charge such that for some* $\varepsilon_1, \ldots, \varepsilon_p > 0$, *we have*

$$\int\limits_{\mathbb{R}^p} h_{\alpha_1}(\lambda_1) \ldots h_{\alpha_p}(\lambda_p) \exp\left(-\sum_{k=1}^{p} x_k^2 \varepsilon_k\right) d\omega(x_1, \ldots, x_n) = 0$$

$$(\alpha_1, \ldots, \alpha_p \in \mathbb{Z}_+).$$

Then $\omega = 0$.

Proof. Suppose that the lemma is proved for $p = 1$ and $p = n - 1$. Show that then it is also valid for $p = n$ $(n = 2, 3, \ldots)$. Fix $\delta_0 \in \mathfrak{B}(\mathbb{R}^1)$ and consider a finite complex-valued measure

$$\mathfrak{B}(\mathbb{R}^{n-1}) \ni \delta \mapsto \int\limits_{\delta_0 \times \delta} d\omega(x_1, x_2, \ldots, x_p) = \omega_{\delta_0}(\delta).$$

Then we fix $\alpha_2, \ldots, \alpha_n \in \mathbb{Z}_+$ and consider a similar measure

$$\mathfrak{B}(\mathbb{R}^1) \ni \delta' \mapsto \int\limits_{\mathbb{R}^{n-1}} h_{\alpha_2}(x_2) \ldots h_{\alpha_n}(x_n) \exp\left(-\sum_{k=2}^{n} x_k^2 \varepsilon_k\right) d\omega_{\delta'}(x_2, \ldots, x_n)$$

$$= \int\limits_{\delta' \times \mathbb{R}^{n-1}} h_{\alpha_2}(x_2) \ldots h_{\alpha_n}(x_n) \exp\left(-\sum_{k=2}^{n} x_k^2 \varepsilon_k\right) d\omega(x_1, \ldots, x_n)$$

$$= \omega_{\alpha_2, \ldots, \alpha_n}(\delta'). \tag{1.19}$$

We have

$$0 = \int\limits_{\mathbb{R}^n} h_{\alpha_1}(x_1) \ldots h_{\alpha_n}(x_n) \exp\left(-\sum_{k=1}^{n} x_k^2 \varepsilon_k\right) d\omega(x_1, \ldots, x_n)$$

$$= \int\limits_{\mathbb{R}^1} h_{\alpha_1}(x_1) \exp(-x_1^2 \varepsilon_1)\, d\omega_{\alpha_2, \ldots, \alpha_n}(x_1) \quad (\alpha_1 \in \mathbb{Z}_+).$$

whence, according to the assumption, $\omega_{\alpha_2,\ldots,\alpha_n} = 0$, i.e.,

$$\omega_{\alpha_2,\ldots,\alpha_n}(\delta') = 0 \quad (\delta_1 \in \mathfrak{B}(\mathbb{R}^1)).$$

Fix δ'. Then taking into account the last equality for $\alpha_2, \ldots, \alpha_n \in \mathbb{Z}_+$, (1.19), and our assumption, we can conclude that $\omega_{\delta'} = 0$, i.e.,

$$\omega_{\delta'}(\delta) = 0 \quad (\delta \in \mathfrak{B}(\mathbb{R}^{n-1})).$$

Due to the arbitrariness of δ', this yields $\omega = 0$.

It is now clear that it suffices to prove Lemma 1.1 for $p = 1$. We have

$$0 = \int_{\mathbb{R}^1} h_{\alpha_1}(x_1) \exp(-x_1^2 \varepsilon_1)\, d\omega(x_1) = \int_{\mathbb{R}^1} h_{\alpha_1}(x_1) \exp(-x_1^2)\, d\omega_1(x_1),$$

$$\mathfrak{B}(\mathbb{R}^1) \ni \delta \mapsto d\omega_1(\delta) = \int_\delta \exp\,[x_1^2(1-\varepsilon_1)]\, d\omega(x_1) \quad (\alpha_1 \in \mathbb{Z}_+).$$

Let $0 < \varepsilon_1 < 1$. The function

$$\omega_1(x_1) = \int_0^{x_1} \exp\,[x^2(1-\varepsilon_1)]\, d\omega(t)$$

is, generally speaking, unbounded but satisfies the inequality

$$|\,\omega_1(x_1)\,| \le c \exp\,[x_1^2(1-\varepsilon_1)] \quad (x \in \mathbb{R}^1).$$

Therefore, by integrating by parts and using the equalities (4.38) and (4.40) of Chapter 2, we obtain that

$$0 = \int_{\mathbb{R}^1} h_{\alpha_1}(x_1) \exp(-x_1^2)\, d\omega_1(x_1) = -\int_{\mathbb{R}^1} \omega_1(x_1)\, d(h_{\alpha_1}(x_1) \exp(-x_1^2))$$

$$= \sqrt{2(\alpha_1+1)} \int_{\mathbb{R}^1} \omega_1(x_1)\, h_{\alpha_1+1}(x_1) \exp(-x_1^2)\, dx_1 \quad (\alpha_1 \in \mathbb{Z}_+).$$

This implies that $\omega_1(x_1) = c \in \mathbb{C}^1$ $(x_1 \in \mathbb{R}^1)$, i.e., $\omega_1(\delta) = 0$ $(\delta \in \mathfrak{B}(\mathbb{R}^1))$, and hence $\omega = 0$. The case $\varepsilon_1 \ge 1$ is investigated similarly.

■

Remark 1. Let $X = \mathbb{R}^n$ ($n \in \mathbb{N}$) be finite-dimensional. Then for the given unitary representation $\mathbb{R}^n \ni x \mapsto A_x \in \mathcal{L}(H_0, H_0)$, we construct the rigging (1.4) with the required properties (i) and (ii) by using constructions similar to those described in Theorem 3.9 of Chapter 3 (here the space $\mathcal{D}(\mathbb{R}^n)$ is exploited and, clearly, one should demand that, for any $\varphi \in H_0$, the function $\mathbb{R}^n \ni x \mapsto A_x\varphi \in H_0$ should be weakly continuous). Therefore, in this case, Theorem 1.2 transforms into the usual Stone's theorem: Conditions connected with the rigging are automatically satisfied. A similar situation takes place if X is a locally compact separable commutative group with a countable base of neighborhoods. However, the choice of the desired rigging is now much more complicated. This choice will be realized in Subsection 2.4 for the general case of a hypercomplex system with a locally compact basis.

■

By analogy with Theorem 4.2, we consider a representation of a real separable Hilbert space X by normal, in general, nonnegative operators acting in H_0 (this is the generalization of the Sz.-Nagy–Hille theorem similar to the previous one). Thus, assume that $A = (A_x)_{x\in X}$ is a family of normal operators in H_0 such that $A_{x+y} = A_x A_y$ ($x, y \in X$). According to the result of Devinatz and Nussbaum already mentioned in Subsection 1, the operators A_x are commuting. For the sake of brevity, we shall say that there exists a normal representation of X. Below, we shall consider the complexification X_c of the space X; $\lambda = \lambda_1 + i\lambda_2$ ($\lambda_1, \lambda_2 \in X$) are its points, and

$$(\lambda, x)_X = (\lambda_1, x)_X \; i\,(\lambda_2, x)_X .$$

Theorem 1.3. *Assume that $X \ni x \mapsto A_x$ is a normal representation of a real separable space X by operators in H_0. There exists a decomposition of the space H_0 into the orthonormal sum $H_0 = H_1 \oplus H_2$ of subspaces invariant with respect to A_x such that the operators A_x ($x \in X$) act as zero operators in H_2, and admit the following spectral representation in H_1*

$$A_x = \int_{X_c} e^{(\lambda, x)_X}\, dE(\lambda),$$

$$\mathfrak{D}(A_x) = \left\{ f \in H_1 \;\middle|\; \int_{X_c} |e^{(\lambda, x)_X}|^2\, d(E(\lambda)f, f)_{H_1} < \infty \right\} \quad (x \in X), \qquad (1.20)$$

where E is some RI on $\mathfrak{B}(X_c)$ if and only if there exists a rigging (1.4) *for which*

(i) $A_x \restriction D$, $A_x^* \restriction D \in \mathcal{L}(D, H_+)$;

(ii) for every $\varphi \in D$, the vector function $X \ni x \mapsto A_x^*\varphi \in H_+$ is weakly continuous. The RI from (1.20) *is uniquely defined for a given $(A_x)_{x\in X}$.*

Proof. This is, in fact, an advanced version of the proof of Theorem 1.2. Thus, let us show that representation (1.20) exists under conditions (i), (ii). The chain $H_- \supset H_0 \supset H_+ \supset D_1$ is constructed as earlier with the only distinction that D_1 is the linear span of linearly independent vectors from the set $(A^*_{x_l}\varphi_j)_{j,l=1}^{\infty} \cup (\varphi_j)_{j=1}^{\infty}$, where $(\varphi_j)_{j=1}^{\infty}$ and $(x_l)_{l=1}^{\infty}$ are dense in D and X, respectively. Since $A^*_{x+y} = A^*_x A^*_y$ $(x, y \in X)$, we have $A^*_x \upharpoonright D \in \mathcal{L}(D_1, H_+)$ for any $x \in X$, and the chain constructed is connected with A in the standard way (in the sense of Remark 3, Subsection 2.10, Chapter 3). In this case, the first equality (2.19) (Chapter 3) holds. This equality, condition (ii), and what has been said in Subsection 2.9 (Chapter 3) enable us to prove that $g(A) \subset X_{g,c}$ (just as above).

Every function $\chi \in X_{g,c}$ which is not identically zero has the form $\chi(x) = \exp((\lambda, x)_X)$ with some $\lambda \in X_c$. In other words, one should establish that each nonzero continuous complex-valued solution $\chi(\cdot)$ of the functional equation

$$\chi(x+y) = \chi(x)\chi(y) \quad (x, y \in X)$$

has this form. This can be easily shown (see also the proof of Theorem 1.4). Thus, we have a bijection $X_{g,c}\setminus\{0(\cdot)\} \leftrightarrow \lambda \in X_c$. We set $H_2 = \mathfrak{R}(E(\{0(\cdot)\}))$, then this bijection enables us to rewrite the integrals (1.1) (which are over $X_{g,c}$ but not over $X_{g,c}\setminus\{0(\cdot)\}$) in the form (1.20). Hence, we have established the representation (1.20).

The univalence of the definition of E in terms of $(A_x)_{x\in X}$ is proved in Theorem 1.2 by reducing to the corresponding finite-dimensional result.

Now let us show how representation (1.20) involves the existence of the rigging (1.4) with the properties (i) and (ii). We employ Theorem 3.8 of Chapter 3. First, it proves to be expedient to show *that the second part of Theorem* 1.2 *follows from this theorem.*

Thus, consider representations (1.5) or (1.6) (this is the same)[38]. By identifying $\lambda \in X_m^-$ with a function $X_m^+ \ni x \mapsto \exp[i(\lambda, x)_{l_2}] = \lambda(x)$, we can rewrite this representation in the form of the first integral from (3.30) in Chapter 3, where the role of $\tau \subset \mathbb{C}^{X_m^+}$ $(X = X_m^+)$ is played by the set of all functions of this sort $\forall \lambda \in X_m^-$. Hence, the RI F turns into a new RI E on the σ-algebra $\mathfrak{C}_\sigma(\tau)$. We take $A_m(\mathbb{R}^\infty)$ as a Hilbert space $\mathcal{H}(\tau)$ of the chain (3.31), Chapter 3. Here $a \in A_m(\mathbb{R}^\infty)$ is first restricted to $\lambda \in X_m^-$ and then considered as being given on the function $\lambda(\cdot) = \exp[i(\lambda, \cdot)_{l_2}] \in \tau$ associated with this λ. As $D(\tau)$ from the same chain, we take the space $\mathcal{P}_{\text{cyl}}(\mathbb{R}^\infty)$ whose functions are interpreted as functions on τ, as just mentioned above.

It is easy to see that conditions (a)–(c) of Subsection 3.4 in Chapter 3 are now satisfied. Thus, condition (a) means that the inequality (4.37) of Chapter 2 holds. Condi-

---

[38] Henceforth, in the notations of the spaces X_τ^+, X_τ^-, etc., τ is replaced by m in order not to confuse with the set τ from (3.30), Chapter 3.

tion (b) follows from Example 4.3 of Chapter 2. Finally, condition (c) is just the contents of the statement of Lemma 1.1. One should only note that $\varepsilon(\lambda(\cdot))$ has the form of the exponent from the formulation of Lemma 1.1 (here we employ formula (4.35) of Chapter 3 which yields the expression for $\|\delta_\lambda\|_{A_{-m}(\mathbb{R}^\infty)}$).

Therefore, if we have the representation (1.5), then we can construct the chain (3.31) with the properties (a)–(c), and this means that, by virtue of Theorem 3.8 of Chapter 3, we can construct the quasinuclear rigging (1.4) connected with the family $(A_x)_{x\in X}$ in the standard way. Thus, we have obtained the second version of the proof of the converse statement of Theorem 1.2 (in fact, this is clearly the same proof as above).

■

Let us return to Theorem 1.3. Just as in the case of Theorem 1.2, we fix a sequence $m = (m_n)_{n=1}^{\infty} \in [1, \infty)^\infty$ such that $(m_n^{-1})_{n=1}^{\infty} \in l_1$ and introduce spaces X_m^+ and X_m^- dual with respect to l_2. Let $X_{m,c}^-$ be a complexification of X_m^-, and let $\lambda = \lambda_1 + i\lambda_2$ $(\lambda_1, \lambda_2 \in X_m^-)$ be its points. Now

$$(\lambda, x)_{l_2} = (\lambda_1, x)_{l_2} + i(\lambda_2, x)_{l_2} \quad (X \in X_m^+).$$

By analogy with the transition from (1.5) to (1.6), let us pass from (1.20) to the representation

$$A_x = \int_{X_{m,c}^-} e^{(\lambda, x)_{l_2}}\, dF(\lambda),$$

$$\mathfrak{D}(A_x) = \left\{ f \in H_1 \;\middle|\; \int_{X_{m,c}^-} |e^{(\lambda, x)_{l_2}}|^2\, d(F(\lambda)f, f)_{H_1} < \infty \right\} \quad (x \in X_m^+). \tag{1.21}$$

Acting then as in the proof of Theorem 1.3 above, we pass from (1.21) to (3.30) in Chapter 3. When realizing this procedure, points $\lambda \in X_{m,c}^-$ are identified with the functions

$$X_m^+ \ni x \mapsto e^{(\lambda, x)_{l_2}} = \lambda(x),$$

$\tau \subset \mathbb{C}^{X_m^+}$ $(X = X_m^+)$, and the RI F turns into a new RI E on $\mathfrak{C}_\sigma(\tau)$. As $\mathcal{H}(\tau)$ from the chain (3.31), Chapter 3, we take $A_m(\mathbb{R}^\infty) \otimes A_m(\mathbb{R}^\infty)$ (see Chapter 2, Subsection 4.4). Here, as earlier,

$$a(\lambda_1, \lambda_2) \in A_m(\mathbb{R}^\infty) \otimes A_m(\mathbb{R}^\infty)$$

is first restricted to

$$(\lambda_1, \lambda_2) \in X_m^- \times X_m^- \cong X_{m,c}^-,$$

and then understood as given on the function

$$\lambda(\cdot) = e^{(\lambda, x)_{l_2}} \in \tau \quad (\lambda = \lambda_1 + i\lambda_2 \in X_{m,c}^-).$$

Let $D(\tau) = \mathcal{P}_{\text{cyl}}(\mathbb{R}^\infty \times \mathbb{R}^\infty)$ be a set of cylindrical polynomials of a point $(\lambda_1, \lambda_2) \in \mathbb{R}^\infty \times \mathbb{R}^\infty$. Its elements can be obviously regarded as functions on τ. Conditions (a)–(c) of Subsection 3.4 in Chapter 3 now hold, too. Just as in the case of Theorem 1.2, this follows from Remark 2 in Subsection 4.4, Chapter 2. Consequently, we can now apply Theorem 3.8 of Chapter 3, and this involves the existence of the desired rigging for the family $(A_x)_{x \in X}$ provided that for the operators from this family, the representation (1.20) takes place.

■

Remark 2. In Theorem 1.3, operators A_x may be unitary. Then $H_1 = H_0$ and (1.20) turns into (1.5). This follows from the fact that the set $X_{g,c}$ can be replaced by X due to the unitarity of these operators. Therefore, Theorem 1.3 covers Theorem 1.2. In the case of selfadjoint operators A_x $(x \in X)$, X_c should be replaced by X in (1.20).

■

Remark 3. As in Remark 1, for the case $X = \mathbb{R}^n$ $(n \in \mathbb{N})$, the rigging (1.4) standardly connected with the normal representation of the group $\mathbb{R}^n$ exists automatically. In the simplest situation of bounded operators, the result can be formulated as follows (Sz.-Nagy–Hille theorem) :

Let $\mathbb{R}^n \ni x \mapsto A_x \in \mathcal{L}(H_0, H_0)$ be a normal representation such that for any $\varphi \in H_0$, the function $\mathbb{R}^n \ni x \mapsto A_x\varphi \in H_0$ is weakly continuous. Then (1.20) is valid, and E is uniquely defined for the given $(A_x)_{x \in \mathbb{R}^n}$. This theorem follows from Theorem 1.3 if we employ the construction of a rigging outlined in Theorem 3.9 of Chapter 3.

An analogous theorem for a commutative, locally compact separable group X with a countable basis of neighborhoods is also valid; it is presented in Subsection 2.4 as a special case of more general constructions.

■

1.3. Representation of Semi-Groups

Let X be a commutative semi-group, i.e., a set in which an associative and commutative operation $X \times X \ni (x, y) \mapsto x + y \in X$ is introduced; zero element (i.e., the element $0 \in X$ such that $x + 0 = x$, $x \in X$) may belong to this semi-group or may not.

If, in addition, X is a topological space and the mapping $X \times X \to X$ just written above is continuous, then this semi-group is called topological. The definitions of a generalized character, character, and so on, presented in Subsection 1.1, remain valid for semi-groups without any changes. The same is true for the representations of X. Theorem 1.1 can be also reformulated for semi-groups without any changes in the statement and proof.

The situation is somewhat more complicated for Theorem 1.3 and formulas (1.20). We now present one of the possible results. Let X be a semi-group which coincides with an open positive octant of a certain real separable Hilbert space K. In other words, we consider some fixed orthonormal basis $(e_j)_{j=1}^{\infty}$ in K, and

$$X = K_+ = \left\{ x = \sum_{j=1}^{\infty} x_j e_j \,\Big|\, \sum_{j=1}^{\infty} x_j^2 < \infty,\ x_j > 0,\ j \in \mathbb{N} \right\}$$

with the imbedding and topology induced from $K = l_2 = l_2(\mathbb{R}^1)$. A family $A = (A_x)_{x \in K_+}$ of normal operators realizes a (normal) representation K_+ if $A_{x+y} = A_x A_y$ $(x, y \in K_+)$. Just as in Subsection 1.2, we conclude that the operators A_x are commuting.

Theorem 1.4. *Theorem* 1.3 *is preserved in the original form for a normal representation of a semi-group* $K_+ \ni x \mapsto A_x$ *by operators acting in* H_0. *In* (1.20), $e^{(\lambda, x)_X}$ *should be replaced by* $e^{(\lambda, x)_K}$ *and the region of integration* X_c *should be replaced by the complexification* K_c.

Proof. One should simply repeat the proof of Theorem 1.3. An additional explanation is necessary only when we solve a functional equation for generalized continuous characters of the semi-group K_+. This equation has the form

$$\chi(x + y) = \chi(x)\,\chi(y) \quad (x, y \in K_+), \tag{1.22}$$

where $K_+ \ni x \mapsto \chi(x) \in \mathbb{C}^1$ is continuous. We must find all its solutions different from the function identically equal to zero.

First note that (1.22) yields

$$\chi(\alpha x) = (\chi(x))^{\alpha} \quad (x \in K_+;\ \alpha \in (0, \infty)) \tag{1.23}$$

(for $\alpha \in \mathbb{N}$ this is obvious; this implies the validity of (1.23) for rational α; and finally, by limit transition, it can be proved for all $\alpha \in (0, \infty)$). Show that if there exists $x \in K_+$ such that $\chi(x) \neq 0$, then $\chi(x) \neq 0$ for some $y \in K_+$. Then, by virtue of (1.22) and (1.23), we set

$$\chi(x+\alpha y) = \chi(x)\,(\chi(y))^{\alpha} = 0.$$

Therefore, as a result of the limit transition as $\alpha \to 0$, we can conclude that $\chi(x) = 0$, too; but this is nonsense. Hence, the desired solution $\chi(\cdot)$ of Eqn.(1.22) is such that $\chi(x) \neq 0$ for any $x \in K_+$.

We regard K_+ as a cone in the space K, and let $z \in \tilde{K}_+ \setminus K_+$ be points of its boundary. It follows from (1.22) that there exists

$$\lim_{K_+ \ni y \to z} \chi(y) = \chi(x+z)\,(\chi(x))^{-1} \neq 0,$$

where $x \in K_+$ is arbitrary. Therefore, $\chi(\cdot)$ can be extended by continuity to a function $\chi \in C(\tilde{K}_+)$ such that $\chi(0) = 1$ which is nonzero everywhere and satisfies the equation

$$\varphi(x+y) = \varphi(x)\,\varphi(y) \quad (x, y \in \tilde{K}_+). \tag{1.24}$$

We set

$$\chi(x) = |\chi(x)|\, e^{i \arg \chi(x)} \quad (x \in \tilde{K}_+;\ \arg \chi(x) \in [0, 2\pi)). \tag{1.25}$$

It is clear that $\psi(\cdot) = |\chi(\cdot)| \in C(\tilde{K}_+)$ satisfies (1.24) and, therefore,

$$\psi(\alpha y) = (\psi(x))^{\alpha} \quad (x \in \tilde{K}_+;\ \alpha \in (0, \infty))$$

according to (1.23). In other words, the functional

$$K_+ \ni x \mapsto f(x) = \ln \psi(x) = \ln|\chi(x)| \in \mathbb{R}^1$$

is additive and homogeneous, i.e.,

$$f(x+y) = f(x) + f(y)\,,\ f(\alpha x) = \alpha f(x) \quad (x, y \in \tilde{K}_+;\ \alpha \in (0, \infty)).$$

It is bounded in the sense of the cone $\tilde{K}_+$, that is

$$\sup_{x \in \tilde{K}_+,\, \|x\|_K = 1} |f(x)| < \infty.$$

Indeed, one should verify that the following situations are impossible

(i) There exists $(x_n)_{n=1}^{\infty}$, $x_n \in \tilde{K}_+$, $\|x_n\|_K = 1$ such that $\psi(x_n) \to 0$ as $n \to \infty$;

(ii) there exists $(x_n)_{n=1}^{\infty}$, $x_n \in \tilde{K}_+$, $\|x_n\|_K = 1$ such that $\psi(x_n) \to +\infty$ as $n \to \infty$.

In the first case, let us find $\varepsilon_n > 0$, $\varepsilon_n \to 0$ such that $(\psi(x_n))^{\varepsilon_n} \underset{n\to\infty}{\to} 0$. But

$$(\psi(x_n))^{\varepsilon_n} = \psi(\varepsilon_n x_n) \to 1,$$

since $\varepsilon_n x_n \to 0$ and $\psi(0) = 1$, and we arrive at a contradiction. In the second case, we choose $\varepsilon_n > 0$, $\varepsilon_n \to 0$ so that $(\psi(x_n))^{\varepsilon_n} \underset{n\to\infty}{\to} +\infty$ and this again leads to a contradiction.

Hence, f is bounded in the sense of $\tilde{K}_+$. This cone is generating , i.e., each $x \in K$ can be represented in the form $x = x' - x''$, where $x', x'' \in \tilde{K}_+$.

But, the additive, homogeneous, and bounded functional given on a generating cone can be extended to a linear continuous functional on K (see, for example, S. G. Krein [2, Chapter 8, Section 2, Subsection 4]). Therefore, there exists $\lambda_1 \in K$ such that $f(x) = (\lambda_1, x)_K$ $(x \in K)$. Consequently,

$$|\chi(x)| = e^{(\lambda_1, x)_K} \quad (x \in \tilde{K}_+). \tag{1.26}$$

Let us find the expression for the second factor in (1.25). An equality

$$\arg\chi(x+y) = \arg\chi(x)\arg\chi(y) \pmod{2\pi} \quad (x, y \in \tilde{K}_+)$$

follows from the fact that the function (1.25) satisfies (1.24). Since $\chi(0) = 1$, we have $\arg\chi(0) = 0$ and, by virtue of the continuity of χ, there exists $r > 0$ so small that

$$\arg\chi(x+y) = \arg\chi(x)\arg\chi(y)$$

for any $x, y \in \{x_n \in \tilde{K}_+ \mid \|x\|_K \le r\} = K_r$. Moreover, by virtue of (1.23), we have

$$\arg\chi(\alpha x) = \alpha\arg\chi(x) \quad (x \in K_r,\ \alpha \in [0, 1).$$

We fix such r and extend the function $\arg\chi$ "by homogeneity" from K_r to $\tilde{K}_+$ by setting

$$f(x) = \frac{\|x\|_K}{r}\arg\chi\left(\frac{rx}{\|x\|_K}\right) \quad (x \in \tilde{K}_+ \setminus \{0\},\ f(0) = 0). \tag{1.27}$$

For $x \in K_r$ $rx\|x\|_K^{-1} \in K_r$, we have $\|x\|_K r^{-1} = \alpha \in [0, 1]$ and, therefore, $f(x) = \arg\chi$ by virtue of the property of the homogeneity of $\arg\chi$ presented just above. Hence, formula (1.27) really extends $\arg\chi$ to $\tilde{K}_+ \setminus K_r$. It is easy to see that the functional $\tilde{K}_+ \ni x \mapsto f(x) \in \mathbb{R}^1$ is additive and homogeneous. Thus, its homogeneity (for $\alpha \in [0, \infty)$) immediately follows from (1.27). Let us prove its additivity. Let $x, y \in \tilde{K}_+$. Then $\|x\|_K, \|y\|_K \le \|x+y\|_K$, and by using the above-mentioned fact that $\arg\chi$ is additive and homogeneous, we obtain

$$f(x+y) = \frac{\|x+y\|_K}{r} \arg\chi\left(\frac{r(x+y)}{\|x+y\|_K}\right)$$

$$= \frac{\|x+y\|_K}{r}\left[\arg\chi\left(\frac{rx}{\|x+y\|_K}\right) + \arg\chi\left(\frac{ry}{\|x+y\|_K}\right)\right] = f(x)+f(y).$$

It is also clear that the functional (1.27) is bounded in the sense of a cone $\tilde{K}_+$. Therefore, by virtue of the arguments already employed, it admits an extension to K and a representation

$$f(x) = (\lambda_2, x)_K \quad (x \in K)$$

with some $\lambda_2 \in K$. Taking (1.27) with $x \in K_r$ into account, we get that

$$\arg\chi(x) = f(x) = (\lambda_2, x)_K.$$

Now let $x \in K_+$ be arbitrary, and let $\alpha \in (0, \infty)$ be so small that $\alpha x \in K_r$. Then it follows from (1.23) (written for $x \in \tilde{K}_+$) and (1.25) that

$$(e^{i\arg\chi(x)})^\alpha = e^{i\arg\chi(\alpha x)} = e^{i(\lambda_2, \alpha x)_K},$$

which yields

$$\arg\chi(x) = (\lambda_2, x)_K \quad (x \in \tilde{K}_+).$$

This relation, (1.25), and (1.26) together imply that

$$\chi(x) = e^{i(\lambda, x)_K} \quad (x \in \tilde{K}_+), \tag{1.28}$$

where $\lambda = \lambda_1 + i\lambda_2 \in K_c$ is a vector that corresponds to $\chi(\cdot)$. Now employing (1.28) instead of the analogous formula for the case of a group and repeating the argument presented in the proof of Theorem 1.3, we prove Theorem 1.4.

■

1.4. Representations of Linear Spaces

Let X be a linear complex space (without topology), and let X_a^* be a space of all linear functionals $X \ni x \mapsto l(x) \in \mathbb{C}^1$ on X. If X is a linear topological space, then X^* is a dual space of its linear continuous functionals. Let $\Phi' \supset H_0 \supset \Phi$ be a nuclear rigging of a separable Hilbert space H_0. An operator-valued function $X \ni x \mapsto A_x$ whose values are commuting normal operators in H_0, connected in a standard way with this rigging in the sense that $\Phi \subset \mathfrak{D}(A_x)$, and $A_x \upharpoonright \Phi \in \mathcal{L}(\Phi, \Phi)$ $(x \in X)$, is called a nuclear representation of X, provided that

$$A_{\alpha x + \beta y}\, \varphi = \alpha A_x \varphi + \beta A_y \varphi \quad (x, y \in X;\ \alpha, \beta \in \mathbb{C}^1;\ \varphi \in \Phi).$$

By definition, this representation is nuclear and continuous if X is a linear topological space, and the vector function $X \ni x \mapsto A_x\varphi \in \Phi$ is weakly continuous for every $\varphi \in \Phi$.

Theorem 1.5. *Let X be a linear space, and $X \ni x \mapsto A_x$ be its nuclear representation by commuting normal operators. Then the following representation in the form of a spectral integral holds*

$$A_x = \int_{X_a^*} l(x)\, dE(l),$$

$$\mathfrak{D}(A_x) = \left\{ f \in H_0 \,\middle|\, \int_{X_a^*} |l(x)|^2\, d(E(l) f, f)_{H_0} < \infty \right\}, \tag{1.29}$$

where E is a RI on the σ-algebra $\mathcal{C}_\sigma(X_a^*) = \{\alpha \cap X_a^* \mid \alpha \in \mathcal{C}_\sigma(\mathbb{C}^X)\}$. If X is a linear topological space, and the representation is nuclear and continuous, then X_a^* should be replaced by X^* in the above formulas.

Proof. In fact, this proof repeats the scheme of the proof of Theorem 1.1. Thus, consider the case of a nontopological X. As above, let $g(A)$ be a generalized spectrum of a commuting family $A = (A_x)_{x \in X}$[39]. The inclusion (1.2) must be now replaced by the inclusion $g(A) \subset X_a^*$. Indeed, let $\lambda(\cdot) \in g(A)$, and let $\xi \in \Phi'$ be the corresponding generalized joint eigenvector. By using the second equality (2.19), Chapter 3, we obtain

$$\overline{\lambda(\alpha x + \beta y)}\, (\xi, \varphi)_{H_0} = (\xi, A_{\alpha x + \beta y}\, \varphi)_{H_0} = (\xi, (\alpha A_x + \beta A_y)\varphi)_{H_0}$$

$$= \overline{\lambda(\alpha\lambda(x) + \beta\lambda(y))}\, (\xi, \varphi)_{H_0} \quad (x, y \in X;\ \alpha, \beta \in \mathbb{C}^1;\ \varphi \in \Phi), \tag{1.30}$$

whence

$$\lambda(\alpha x + \beta y) = \alpha\lambda(x) + \beta\lambda(y) \quad (x, y \in X;\ \alpha, \beta \in \mathbb{C}^1).$$

This means that $\lambda(\cdot) \in X_a^*$, and $g(A) \subset X_a^*$. Now applying Theorem 2.7 of Chap-

[39] Here, we take into account Remark 3 of Subsection 2.10 in Chapter 3.

ter 3 [40], we find that (2.28), Chapter 3, with π replaced by $\tau = X_a^* \supset g(A) \supset \pi$, is now valid. Finally, we get (1.29).

In the case of a nuclear continuous representation $g(A) \subset X^*$, one should, in addition, employ the statement of Subsection 2.9 of Chapter 3 which involves the continuity of $\lambda(\cdot) \in g(A)$.

■

Remark 1. Acting by analogy with Theorem 1.5, one may consider a representation of a real linear X by selfadjoint operators or a representation of a complex X with an evolution $X \ni x \mapsto x^* \in X$ by commuting normal operators. This means that we demand, in addition, that $A_{x^*} = A_x^*$ $(x \in X)$. In this case, the integration in (1.29) is carried out over $l \in X_a^*$ such that $l(x^*) = \overline{l(x)}$ $(x \in X)$. Here the standard character of the connection between the rigging and A should be understood just as in Subsection 2.5 of Chapter 3.

■

Show that the analogue of Theorem 1.2 is also valid.

Theorem 1.6. *Consider the chain* (1.4). *Let X be a real Hilbert space, and let $(A_x)_{x\in X}$ be a family of commuting selfadjoint operators acting in H_0 with domains of definitions $\mathfrak{D}(A_x) \supset D$. Assume that these operators are such that*

$$A_{\alpha x + \beta y}\varphi = \alpha A_x\varphi + \beta A_y\varphi \quad (x, y \in X;\ \alpha, \beta \in \mathbb{R}^1;\ \varphi \in \Phi).$$

If conditions (i) and (ii) of Theorem 1.2 *hold for A_x, then this family admits a representation*

$$A_x = \int_X (\lambda, x)_X \, dE(\lambda),$$

$$\mathfrak{D}(A_x) = \left\{ f \in H_0 \mid \int_X (\lambda, x)_X^2 \, d(E(\lambda)f, f)_{H_0} < \infty \right\} \quad (x \in X), \tag{1.31}$$

where E is a RI on $\mathfrak{B}(X)$. Conversely, the fact that a family $(A_x)_{x\in X}$ possesses the representation (1.31) *involves the existence of the chain* (1.4) *and the other conditions of the theorem. The RI E from* (1.31) *is uniquely determined for a given $(A_x)_{x\in X}$.*

Proof. The validity of the representation (1.31) follows from the modification of Theorem 1.5 applied to a real Hilbert space X. The fact that here $\mathfrak{C}_\sigma(X) = \mathfrak{B}(X)$ has been clarified in the proof of Theorem 1.2. The proof of the converse statement of the

[40] More exactly, we apply Remark 1 to this theorem and Remark 2, Subsection 2.10, Chapter 3.

theorem generally repeats the corresponding part of the proof of Theorem 1.2. Namely, we rewrite representation (1.31) in the form

$$A_x = \int_{X_\tau^-} (\lambda, x)_{l_2}\, dF(\lambda) \quad (x \in X_\tau^+),$$

which is analogous with (1.6), and then repeat steps I – IV. Step V is to be changed in order to employ the strong continuity of the vector function $X_\tau^+ \ni x \mapsto (\cdot, x)_{l_2}\, p(\cdot) \in A_\tau(\mathbb{R}^\infty)$. This fact is proved just as in the Remark 1 after Lemma 4.5 of Chapter 2. Thus, we employ an analogue of Lemma 4.5, Chapter 2, in which $e^{i(\lambda, x)_{l_2}}$ is replaced by $(\lambda, x)_{l_2}$. The proof of this fact is similar to the proof of Lemma 4.5; we employ the formula

$$\int_{\mathbb{R}^1} t\, h_j(t)\, d\gamma_1(t) = \frac{1}{\sqrt{2}}\, \delta_{j,1} \quad (j \in \mathbb{Z}_+)$$

(see (4.40), Chapter 2). We do not present the corresponding calculation. The uniqueness of E is established by the standard procedure.

■

Note that we cannot obtain Theorem 1.2 from Theorem 1.6 and vice versa by taking the corresponding functions of operators. This is so because, for example, the fact that conditions (i) and (ii) hold for A_x does not, generally speaking, involve the validity of these conditions for e^{iA_x} $(x \in X)$. As has been stated above, the second part of Theorem 1.6 can be derived from the general Theorem 3.8, Chapter 3.

The results presented in this Subsection may be also interpreted as follows : By analogy with generalized random processes, we say that a generalized selfadjoint quantum (commutative) process is given if we have a family $(A_x)_{x \in X}$ of commuting selfadjoint operators, acting in the space H_0, indexed by elements of a real linear topological space X, and such that

$$A_{\alpha x + \beta y}\, \varphi = \alpha A_x \varphi + \beta A_y \varphi \quad (x, y \in X;\ \alpha, \beta \in \mathbb{R}^1),$$

and this family possesses a cyclic vector Ω. Here, φ is an arbitrary element of a linear set $D \subset \mathfrak{D}(A_x)$ dense in H_0.

Note that, as a rule, a usual selfadjoint quantum process $(A_t)_{t \in \mathbb{R}^n}$, i.e., a family of commuting selfadjoint operators A_t, can be imbedded into a generalized process in a standard way, namely one should consider a real linear topological space X of test functions $\mathbb{R}^n \ni t \mapsto x(t) \in \mathbb{R}^1$ and form the operators

$$A_x = \int_{\mathbb{R}^n} A_t\, x(t) \quad (x \in X).$$

It is also clear how to define a generalized process which corresponds to normal operators A_t.

Theorems 1.5, 1.6 and their already mentioned modifications for different X lead us to spectral representations of these processes which are close to the well known representation of random processes in the nongeneralized stationary case (Gikhman, Skorokhod [2, Chapter 4, Section 5]) and the generalized stationary case (Gelfand, Vilenkin [1, Chapter 3, Section 3, Subsection 4]). We now present a result with the simplest formulation which follows from the corresponding modification of Theorem 1.5 :

Let $(A_x)_{x \in X}$ be a generalized selfadjoint quantum process. Assume that H_0 is a separable space, D is a nuclear space and, moreover, the restriction $A_x \upharpoonright D$ belongs to $\mathfrak{L}(D, D)$ $(x \in X)$ and the vector function $X \ni x \mapsto A_x \varphi \in D$ is weakly continuous for every $\varphi \in D$. Then the following representation is valid

$$A_x = \int_{X^*} \lambda(x)\, dE(\lambda),$$

where E is some RI on $\mathfrak{C}_\sigma(X^*)$.

1.5. Representations of an Algebra

Let us also investigate a representation of a commutative algebra X over a field of complex numbers. A functional $\omega \in X_a^*$ is called multiplicative if

$$\omega(xy) = \omega(x)\,\omega(y) \quad (x, y \in X).$$

Let M_a be a collection of functionals of this sort. If X is a topological algebra, then by M we denote a collection of all continuous multiplicative functionals. Assume that H_0 is the same space as before, and that its nuclear rigging is given. An operator-valued function $X \ni x \mapsto A_x$, whose values are normal operators in H_0 standardly connected with the rigging in the sense pointed out at the beginning of Subsection 1.4, is called a nuclear representation of X if

$$A_{\alpha x + \beta y}\,\varphi = \alpha A_x \varphi + \beta A_y \varphi$$

and

$$A_{xy} = A_x A_y \quad (x, y \in X;\ \alpha, \beta \in \mathbb{C}^1;\ \varphi \in \Phi).$$

As above, in the case of a topological algebra X, we introduce a concept of a nuclear continuous representation, i.e., for any $\varphi \in \Phi$ the vector function $X \ni x \mapsto A_x \varphi \in \Phi$ is weakly continuous.

Similarly to Theorems 1.1 and 1.5, we can prove that *the following representation in the form of a spectral integral holds*

$$A_x = \int_{M_a} \omega(x)\, dE(l),$$

$$\mathfrak{D}(A_x) = \left\{ f \in H_0 \mid \int_{M_a} |\omega(x)|^2\, d(E(\omega)f, f)_{H_0} < \infty \right\} \quad (x \in X), \tag{1.32}$$

where E is a RI in the σ*-algebra* $\mathfrak{C}_\sigma(M_a) = \{\alpha \cap M_a \mid \alpha \in \mathfrak{C}_\sigma(\mathbb{C}^X)\}$. *In the case of a nuclear continuous representation,* M_a *must be replaced in the above formulas by* M.[41]

Clearly, in this case we can also consider nuclear $*$-representation, e.g., if X is a $*$-algebra, etc. Thus, let X be a $*$-algebra, i.e., there exists an involution bijection $X \ni x \mapsto x^* \in X$, $(x^*)^* = x$ $(x \in X)$ in X such that $(xy)^* = x^* y^*$ $(x, y \in X)$. A nuclear representation of X is called a $*$-representation if $A_{x^*} = (A_x)^*$ $(x \in X)$. In this case, the integration in (1.32) is carried out, not over M_a, but over the set $M_{a,H}$ of multiplicative Hermitian functionals. This means that ω is such that $\omega(x^*) = \overline{\omega(x)}$ $(x \in X)$.

In the case of a topological $*$-algebra X, we demand that the mapping $*$ should be a homeomorphism. If there exists a nuclear continuous $*$-representation of such an algebra, then the integration in (1.32) is carried out over the set $M_H = M \cap M_{a,H}$ of continuous multiplicative Hermitian functionals.

The last results generalize classical facts concerning the representations of commutative C^*-algebras, we do not need to demand the existence of the rigging connected with the representation in a standard way, since it automatically exists (this follows, for example, just from the representations of type (1.32) and general Theorem 3.8 of Chapter 3; we can also give a direct proof of the existence of this rigging but, for the sake of brevity, we shall not do this).

We make a remark concerning the case when Theorems 1.2–1.4 and 1.6 deal with a representation of a finite-dimensional real Hilbert space $X = \mathbb{R}^n$. In this situation, one can always construct a nuclear rigging $\Phi' \supset H_0 \supset \Phi$ of the space H_0 which is connected with the operators of a given representation in a standard way. This construction can be realized if we actually repeat the proof of necessity in Theorem 1.2 using the nuclear spaces of functions on X instead of Hilbert spaces. This is clarified by the following example which is of independent interest.

Example 1.1. Let A be a selfadjoint nonnegative operator in the Hilbert space H_0. *One can construct a nuclear rigging* $\Phi' \supset H_0 \supset \Phi$ *such that* $\Phi \subset \mathfrak{D}(A_x)$, $A_x \restriction \Phi$

---

[41] By calculations such as (1.3) and (1.30), we establish the inclusion $g(A) \subset M_a$, M, and then apply Theorem 2.7.

$\in \mathcal{L}(\Phi, \Phi)$, *and* $e^{-tA} \in \mathcal{L}(\Phi, \Phi)$ *for any* $t \geq 0$. Show how the scheme of the proof of necessity in Theorem 1.2 can be modified in order to obtain the required rigging.

Let E be a RI of the operator A. For fixed $l_1 \in H_0$, we introduce a mapping

$$\mathcal{A}(\mathbb{R}^1) \ni a \mapsto Q_1 a = \int_0^\infty a(\lambda)\, e^{-\lambda^2}\, dE(\lambda)\, l_1 \in H_0$$

(cf. (1.7)). Here $\mathcal{A}(\mathbb{R}^1)$ is a nuclear space of test functions on $\mathbb{R}^1$ studied in Subsection 4.2, Chapter 2; Q_1 is correctly defined by virtue of Theorem 4.1, Chapter 2. A topology on $\mathfrak{R}(Q_1)$ is induced by the topology of the nuclear space $\mathcal{A}(\mathbb{R}^1)$ (by analogy with Step II in Theorem 1.2). This topology transforms $\mathfrak{R}(Q_1)$ into a nuclear countably Hilbert space $\Phi_1 \subset H_0$ where the imbedding is topological. If Φ_1 is dense in H_0, then the construction is already finished. Otherwise, we choose $l_2 \in H_0$ orthogonal to Φ_1 and construct the corresponding nuclear space $\Phi_2 \subset H_0$, etc. As a result, we obtain a finite or infinite collection of nuclear spaces Φ_k, $k = 1, \ldots, N$, $N \leq \infty$ mutually orthogonal in H_0. The desired nuclear space is constructed as a topological direct sum of nuclear spaces Φ_k, i.e., $\Phi = \oplus_{k=1}^{N} \Phi_k \subset H_0$ (see Subsection 5.1, Chapter 2).

According to the construction, the inclusions $A \in \mathcal{L}(\Phi, \Phi)$ and $e^{-tA} \in \mathcal{L}(\Phi, \Phi)$ follow from the fact that the operators

$$\mathcal{A}(\mathbb{R}^1) \ni a(\lambda) \mapsto \lambda\, a(\lambda) \in \mathcal{A}(\mathbb{R}^1),$$

$$\mathcal{A}(\mathbb{R}^1) \ni a(\lambda) \mapsto e^{-\lambda t} a(\lambda) \in \mathcal{A}(\mathbb{R}^1) \quad (t \geq 0)$$

are continuous. Here we use the arguments similar to those of Step V in the proof of Theorem 1.2. Note that in the case of finite multiplicity of the spectrum of the operator A, we necessarily have $N < \infty$, and the space Φ just constructed is a nuclear countably Hilbert space.

§2. Hypercomplex Systems with Locally Compact Basis. Representations of These Systems by Commuting Operators

In the previous section, we have obtained spectral representations of the family $(A_x)_{x \in X}$ of commuting normal operators connected by traditional algebraic relations. However, these relations may be nontraditional. The simplest example is $X = \mathbb{R}^1$

$$A_x A_y = \frac{1}{2}(A_{x+y} - A_{x-y}) \quad (x, y \in \mathbb{R}^1).$$

We now investigate one class of relations of this type generated by hypercomplex systems (h.s.) with a locally compact basis.

A finite-dimensional algebra with a fixed basis of a linear space formed by its elements is called a finite-dimensional h.s. Clearly, in order to define the operation of multiplication in this algebra, it suffices to give the rule for how to multiply the basis vectors. We shall consider h.s. only over a field of complex members.

So, assume that a h.s. of this sort is given, with $d \in \mathbb{N}$ being dimensionality, and that its elements are vectors of the space $\mathbb{C}^d$. If we fix a basis Q in this space, then each vector x of our h.s. can be interpreted as a complex-valued function on Q which consists of d points. Thus, an ordinary d-dimensional h.s. may be regarded as a space of functions $Q \ni p \mapsto x(p) \in \mathbb{C}^1$ with operations of addition of functions and multiplication by a scalar and with a multiplication

$$(x * y)(r) = \sum_{p,q \in Q} x(p)\, y(q)\, c(p, q, r) \quad (r \in Q),$$

where $c(p, q, r)$ is a function ("cubic matrix of structural constants") which defines a multiplication and possesses certain properties that guarantee the associativity of multiplication (and its commutativity, if necessary).

The generalization of this h.s. presented below consists in the transition from a finite basis Q to a certain locally compact space Q. Taking into account the available examples, it is now appropriate to replace c by a "structure measure" $c(\alpha, \beta, r)$ $(\alpha, \beta \subset Q;$ $r \in Q)$, but not by a function on $Q \times Q \times Q$. The most complete results of the harmonic analysis are attained in the case of a commutative h.s. with nonnegative c whose properties resemble the properties of a group algebra of a locally compact commutative group G (these are so-called normal h.s. with a basis identity O).

In this section, we introduce a class of h.s. necessary for our considerations, establish some properties of these h.s., construct the theory of generalized functions on the basis Q of h.s. necessary for what follows, and then obtain spectral representations of a family $(A_p)_{p \in Q}$ of commuting normal operators which realizes a representation (in a certain sense) of an h.s. Note that the situation when Q is a locally compact commutative group is covered by this scheme.

2.1. Concept of a Hypercomplex System with Locally Compact Basis

Let Q be a complete separable locally compact metric space of points p, q, r; $\mathfrak{B}(Q)$ be a σ-algebra of its Borel sets; and $\mathfrak{B}_0(Q)$ be a sub-ring of $\mathfrak{B}(Q)$ consisting of sets with compact closures. We shall consider Borel measures, i.e., nonnegative measures on $\mathfrak{B}(Q)$ which are finite on compacts. Measures μ of this sort are regular in the following sense: for every $\alpha \in \mathfrak{B}(Q)$, we have $\mu(\alpha) = \sup \mu(\varphi)$, where sup is taken over all compacts $\varphi \subset \alpha$.

An h.s. with a basis Q is given by its structural (s.) measure $c(\alpha, \beta, r)$ ($\alpha, \beta \in \mathfrak{B}(Q)$; $r \in Q$). An s. measure $c(\alpha, \beta, r)$ is a Borel measure with respect to $\alpha(\beta)$ for fixed β and $r(\alpha, r)$. This measure satisfies the following conditions :

(N1) $c(\alpha, \beta, r)$ is a continuous finite function in r for every $\alpha, \beta \in \mathfrak{B}_0(Q)$;

(N2) for every $\alpha, \beta, \gamma \in \mathfrak{B}_0(Q)$ and $s \in Q$, the following associativity relation holds

$$\int c(\alpha, \beta, r)\, d_r\, c(\varepsilon_r, \gamma, s) = \int c(\alpha, \gamma, r)\, d_r\, c(\alpha, \varepsilon_r, s) \tag{2.1}$$

(here and below, the omitted designation of the region of integration means that the integration is taken over the whole of Q; it is convenient to select the variable with respect to which we integrate, not by the point r, but by the set ε_r);

(N3) for every $\alpha, \beta \in \mathfrak{B}_0(Q)$ and $r \in Q$, the commutativity relation $c(\alpha, \beta, r) = c(\beta, \alpha, r)$ is valid.

A Borel measure μ which is additive on open sets and such that

$$\int c(\alpha, \beta, r)\, dm(r) = m(\alpha)\, m(\beta) \quad (\alpha, \beta \in \mathfrak{B}_0(Q)) \tag{2.2}$$

is called a multiplicative (m.) measure.

(N4) Assume that there exists at least one m. measure.

One can prove (see Berezansky, S.Krein [1] and Kalyuzhny [2]) that condition (N4) is *a fortiori* satisfied provided that $c(Q, \beta, r)$ is a continuous bounded function in r for every $\beta \in \mathfrak{B}_0(Q)$ and, moreover, $c(Q, o, r) > 0$ for every open $o \in \mathfrak{B}_0(Q)$. We fix this measure and denote the integration with respect to it by dp (concepts such as "almost everywhere", etc., are related to this measure). Let us also denote

$$c(\alpha, \beta, \gamma) = \int_\gamma c(\alpha, \beta, r)\, dr \quad (\alpha, \beta, \gamma \in \mathfrak{B}_0(Q)).$$

Consider the space $L_1 = L_1(Q, \mathfrak{B}(Q), m)$. Let us introduce an operation of (generalized) convolution

$$(x * y)(r) = \int x(p)\, d_p \left(\int y(p)\, d_p\, c(\varepsilon_p, \varepsilon_q, r)\right) \quad (r \in Q). \tag{2.3}$$

for $x, y \in L_1$. It is easy to establish that the integral (2.3) exists for almost all r, belongs to L_1, and that the following estimate holds

$$\| x * y \|_1 \leq \| x \|_1 \| y \|_1 \quad (x, y \in L_1; \quad \| \cdot \|_1 = \| \cdot \|_{L_1}).$$

In order to prove this inequality, one should estimate $\| |x| * |y| \|_1$ by use of the natural version of the Fubini theorem for the integration with respect to measures depending on parameters. Conditions (N2) and (N3) guarantee the associativity and commutativity of the convolution (2.3). Therefore, the space L_1 now turns into a commutative normalized algebra with the multiplication (2.3). This algebra will be called a h.s. with a basis Q. Note that

$$c(\alpha, \beta, r) = (\kappa_\alpha * \kappa_\beta)(r) \quad (\alpha, \beta, \in \mathfrak{B}_0(Q); \; r \in Q),$$

where $\kappa_\gamma(p)$ is a characteristic function of the set γ.

Let us realize an auxiliary construction. Fix a point $r \in Q$ and set

$$m_r(\alpha \times \beta) = c(\alpha, \beta, r) \quad (\alpha, \beta, \in \mathfrak{B}(Q)). \tag{2.4}$$

By additivity, the set function $m_r(\alpha \times \beta)$ can be uniquely extended to the algebra $\mathfrak{R}'$ generated by the rectangles

$$\alpha \times \beta \;\; (\alpha, \beta, \in \mathfrak{B}(Q)) \colon \mathfrak{R}' \in \gamma \mapsto m_r(\gamma) \in [0, \infty].$$

One can easily show that m_r is *absolutely additive* on $\mathfrak{R}'$. Thus, m_r is regular, i.e., $m_r(\gamma) = \sup m_r(\gamma)$ for every $\gamma \in \mathfrak{R}'$. Here sup is taken over the compacts $\mathfrak{R}' \ni \varphi \subset \gamma$. Indeed, it suffices to verify this for the rectangle $\gamma = \alpha_0 \times \beta_0$. Due to the fact that the measure $\mathfrak{B}(Q) \ni \alpha \mapsto c\,(\alpha, \beta_0, r)$ is regular for fixed β_0, for given $\varepsilon > 0$ we can find a compact $\varphi_0 \subset \alpha_0$ such that

$$c(\alpha_0, \beta_0, r) - c(\varphi_0, \beta_0, r) < \varepsilon.$$

Similarly, one can find a compact $\psi_0 \subset \beta_0$ such that

$$c(\varphi_0, \beta_0, r) - c(\varphi_0, \psi_0, r) < \varepsilon.$$

These two inequalities yield

$$m_r(\alpha_0 \times \beta_0) - m_r(\varphi_0 \times \psi_0) < 2\varepsilon,$$

i.e., m_r is regular. It remains to exploit the fact that an additive regular set function on a certain algebra of sets from $Q \times Q$ is absolutely additive.

■

Hence, m_r can be extended to all measurable sets and, in particular, to the σ-algebra $\mathfrak{B}(Q \times Q)$ of Borel sets from the space $Q \times Q$. Below, we regard m_r as being extended to this σ-algebra; m_r is clearly a Borel measure. The convolution (2.3) can be written as follows

$$(x * y)(r) = \int\int x(p)\, y(q)\, dm_r(r, q) \quad (x, y \in L_1;\ r \in Q). \tag{2.5}$$

Normal h.s. with a basis identity (b.i.) are defined as follows:

(N5) a h.s. is called normal if there exists an involutive homeomorphism $Q \ni p \mapsto p^* \in Q$ such that

$$m(\delta^*) = m(\delta), \quad c(\alpha, \beta, \gamma) = c(\gamma, \beta^*, \alpha) \quad (\alpha, \beta, \gamma \in \mathfrak{B}_0(Q)).$$

(N6) a normal h.s. possesses a weak b.i. O provided that there exists in Q a point O such that

$$\mathrm{O}^* = \mathrm{O} \quad \text{and} \quad c(\alpha, \beta, \mathrm{O}) = m(\alpha^* \cap \beta).$$

In a normal h.s., the correspondence

$$L_1 \ni x = x(p) \mapsto \overline{x(p^*)} = x^* \in L_1$$

is an involution.

Indeed, all the properties of an involution are obvious with the exception of the equality

$$(x * y)^* = x^* * y^* \quad (x, y \in L_1).$$

It suffices to prove this equality on characteristic functions of sets from $\mathfrak{B}_0(Q)$. Here it is equivalent to the relation $c(\alpha, \beta, \gamma^*) = c(\alpha^*, \beta^*, \gamma)$ $(\alpha, \beta, \gamma \in \mathfrak{B}_0(Q))$ which is established by using (N3) and (N5) as follows

$$c(\alpha, \beta, \gamma^*) = c(\gamma^*, \beta^*, \alpha) = c(\beta^*, \gamma^*, \alpha) = c(\alpha, \gamma, \beta^*)$$

$$= c(\gamma, \alpha, \beta^*) = c(\beta^*, \alpha^*, \gamma) = c(\alpha^*, \beta^*, \gamma).$$

■

It follows from what has been said above that the normality of an h.s. is closely connected with the existence of an involutive homeomorphism $Q \ni p \mapsto p^* \in Q$ for which the mapping $L_1 \ni x \mapsto x^* \in L_1$ introduced above is an involution in an h.s.

The following inequality holds in a normal h.s.

$$m_r(\alpha \times \beta) = c(\alpha, \beta, r) \le m(\alpha) \quad (\alpha, \beta \in \mathfrak{B}(Q);\ r \in Q). \tag{2.6}$$

In fact, it suffices to establish (2.6) for $\alpha, \beta \in \mathfrak{B}_0(Q)$. By virtue of (N5) and (2.2), for any $\gamma \in \mathfrak{B}_0(Q)$, we have

$$c(\alpha, \beta, \gamma) = c(\gamma, \beta^*, \alpha) \le c(\gamma, \beta^*, Q) = m(\gamma)\, m(\beta^*) = m(\beta)\, m(\gamma).$$

The inequality (2.6) follows from the arbitrariness of γ, (N4), and (2.4).

■

The fact that a normal h.s. with a discrete basis possesses a weak b.i. O *means that* O *is an identity of the h.s. located in the basis* (more precisely, a function $\chi_{\{0\}}(p)$ is an identity of the h.s.).

Thus, by virtue of (2.6) and the normality of $\gamma \in \mathfrak{B}_0(Q)$ for fixed β, there exists a derivative $c(p, \beta, \gamma)$ of the measure $c(\alpha, \beta, \gamma)$ with respect to $m(\alpha)$ equal to $c(\alpha, \beta^*, p)$ $(p \in Q)$. By using this equality and (2.6), we conclude that for fixed $p \in Q$ and $\beta \in \mathfrak{B}_0(Q)$ there exists a derivative $c(p, \beta, r)$ of the measure $c(p, \beta, \gamma)$ with respect to $m(\gamma)$, and it is defined almost everywhere. For $p = O$, we get

$$c(O, \beta, \gamma) = c(\gamma, \beta^*, O) = c(\beta^*, \gamma, O) = m(\beta \cap \gamma) \quad (\beta, \gamma \in \mathfrak{B}(Q))$$

and, consequently, $c(O, \beta, r) = \kappa_\beta(r)$ ("the convolution of O with κ_β equals κ_β"). This implies that O is an identity of the h.s. if the basis Q of a normal h.s. with a weak b.i. is discrete.

■

We strengthen condition (N6) somewhat : A certain "continuity under multiplication by O" should hold. Namely,

(N7) a weak b.i. O of a normal h.s. is called a b.i. if for any $\beta \in \mathfrak{B}_0(Q)$ and any neighborhood $u \subset \beta$ of it, there exists a neighborhood o of the point O such that

$$c(o, \beta, r) \subset u.$$

Later, we shall investigate the h.s. satisfying all conditions (N1)–(N7). At the same time, many facts from Subsection 2.2 are valid under fewer conditions. This can be easily seen from the proofs of the corresponding facts.

Example 2.1. (a group algebra of a locally compact commutative group).

Let G be a group of this sort being, in addition, separable and having a countable base of neighborhoods (it is thus a metric space). Consider an invariant measure $\mathfrak{B}(G) \ni \alpha \mapsto m(\alpha) \in [0, \infty)$ on G and the corresponding space $L_1 = L_1(G, \mathfrak{B}(G), m)$. This space is a commutative normalized algebra with respect to a convolution

$$(x * y)(r) = \int_G x(rp^{-1})\, y(p)\, dm(p) \quad (x, y \in L_1;\ r \in G) \tag{2.7}$$

(it is now convenient to designate the group operation in the multiplicative form). It is easy to show that condition (2.7) can be rewritten in the form 2.3 with $Q = G$; it suffices to set

$$c(\alpha, \beta, r) = (\kappa_\alpha * \kappa_\beta)(r) = m(\alpha^{-1} r \cap \beta) \quad (\alpha, \beta \in \mathfrak{B}_0(Q)). \tag{2.8}$$

Conditions (N1)-(N7) are now satisfied, and the role of an m. measure is played by the invariant measure $p^* = p^{-1}$ $(p \in Q)$, $0 = e$, where e is the identity element of the group. Consequently, the group algebra under consideration can be regarded as an example of an h.s.

As follows from (2.8), in the case of a discrete group, $c(\{p\},\{q\},r)$ is equal to zero for all $r \neq pq$ and to unit for $r = pq$. Hence, "product $p * q$ of two points of the basis" is not "dispersed" but is concentrated just at the point $p\,q$. In an h.s. it becomes "dispersed" over the whole Q with nonnegative coefficients. In a certain sense, the product of the points p and q now consists of several points. This clarifies the notion of "hypergroup" which is used as another name for a h.s.

2.2. Some Properties of Hypercomplex Systems. Operators of a Generalized Shift

In the case of the h.s. constructed above, there exists the highly developed harmonic analysis which generalizes the harmonic analysis on a locally compact commutative group[42]. However, we shall not present the corresponding theory. We only consider several of the results necessary for the study of operators A_p $(p \in Q)$.

Denote $L_a = L_a(Q, \mathfrak{B}(Q), m)$, $\|\cdot\|_a = \|\cdot\|_{L_a}$, and $L_a' = L_{a'}$, where $a^{-1} + a'^{-1} = 1$ $(a \in [1, \infty])$.

Lemma 2.1. *For $f \in L_a$ and $g \in L_{a'}$ $(a \in [1, \infty])$, there exists a convolution $f * g$ which is a bounded function, moreover,*

$$\|f * g\|_\infty \leq \|f\|_a \times \|g\|_{a'}.$$

This function is continuous for $a \in (1, \infty)$. In order that the convolution $f * g$ is continuous for $a = 1$, it suffices that, in addition, $g \in L_b$ for some $b \in [1, \infty)$.

42 In fact, the axiomatics of h.s. was constructed with a view to preserving this analysis.

Proof. Let $a \in (1, \infty)$, and let k and l be step functions, i.e., linear combinations of characteristic functions of sets from $\mathfrak{B}_0(Q)$. By virtue of (2.5), the Hölder inequality, and (2.6), we obtain

$$|(k * l)(r)| = |\int \int k(p)\, l(q)\, dm_r(p, q)|$$

$$\leq \left(\int \int |k(p)|^a\, dm_r(p, q)\right)^{1/a} \times \left(\int \int |l(q)|^{a'}\, dm_r(p, q)\right)^{1/a'}$$

$$\leq \left(\int |k(p)|^a\, dp\right)^{1/a} \left(\int |l(q)|^{a'}\, dq\right)^{1/a'} = \|k\|_a \times \|l\|_{a'}, \tag{2.9}$$

and the function $k * l$ is continuous due to the continuity of $(\kappa_\alpha * \kappa_\beta)(r) = c(\alpha, \beta, r)$, where $\alpha, \beta \in \mathfrak{B}_0(Q)$. Approximating the functions f and g by the functions k and l (in the metrics L_a and $L_{a'}$, respectively) and employing estimate (2.9), we conclude that $(k * l)(r)$ converges uniformly to $(f * g)(r)$. Therefore, the latter function is continuous and the estimate (2.9) holds for it. Lemma 2.1 is thus proved for the case $a \in (1, \infty)$.

Let $a = 1$, $f \in L_1$, and $g \in L_\infty$. By analogy with (2.9), we get

$$|(f * g)(r)| = |\int \int f(p)\, g(q)\, dm_r(p, q)| \leq \|g\|_\infty \int \int |f(p)|\, dm_r(p, q) \leq \|g\|_\infty \|f\|_1.$$

This yields the existence and boundedness of $(f * g)(r)$. Assume, in addition, that $g \in L_b$ for some $b \in [1, \infty)$. It suffices to show that one can find sequences k_n and l_n of step finite functions such that

$$\|k_n * l_n - f * g\|_\infty \to 0 \quad \text{as} \quad n \to \infty.$$

To prove this, we choose sequences k_n and l_n so that

$$\|k_n - f\|_1 < n^{-1} \quad (n \in \mathbb{N})$$

and

$$\|l_n - g\|_b < n^{-1} \|k_n\|_{b'}^{-1} \quad (n \in \mathbb{N}).$$

Then, by using the Hölder inequality and (2.6), we obtain

$$|(f * g)(r) - (k_n * l_n)(r)| \leq |(f * g)(r) - (k_n * g)(r)| + |(k_n * g)(r) - (k_n * l_n)(r)|$$

$$= |\int \int (f - k_n)(p)\, g(q)\, dm_r(p, q)| + |\int \int k_n(p)\,(g - l_n)(q)\, dm_r(p, q)|$$

$$\leq n^{-1}(\|g\|_\infty + 1) \underset{n\to\infty}{\longrightarrow} 0.$$

■

Corollary 1. *If a normal h.s. contains an identity element, then the basis Q of it consists of at most a countable number of points with a discrete topology.*

Indeed, let $e \in L_1$ be the identity element. Then by virtue of Lemma 2.1, for $a = 1$ and any $g \in L_\infty \cap L_1$, the function $e * g$ is continuous and coincides with g almost everywhere. But if an arbitrary function $g \in L_\infty \cap L_1$ coincides with a continuous function almost everywhere, then the topology in Q must be discrete. It remains to exploit the fact that Q is separable.

■

Lemma 2.2 *For $x \in L_1$ and $f \in L_2$, the convolution $(f * g)(r)$ exists for almost all $r \in Q$ and belongs to L_2; moreover,*

$$\|x * f\|_2 \leq \|x\|_1 \, \|f\|_2.$$

Proof. Let $\alpha, \beta \in \mathfrak{B}_0(Q)$. Then, by virtue of (N5) and (N3),

$$(\kappa_\alpha * \kappa_\beta, \kappa_\gamma)_{L_2} = c(\alpha, \beta, \gamma) = c(\gamma, \beta^*, \alpha) = (\kappa_\alpha, \overset{*}{\kappa_\beta} * \kappa_\gamma)_{L_2}.$$

For step functions k, l, and m, this yields

$$(k * l, m)_{L_2} = (k, l^* * m)_{L_2}. \tag{2.10}$$

Using the estimate from Lemma 2.1 for $a = 2$, we find

$$|(k * l, m)_{L_2}| \leq \|k\|_1 \, \|l^* * m\|_\infty \leq \|k\|_1 \, \|l^*\|_2 \, \|m\|_2 = \|k\|_1 \, \|l\|_2 \, \|m\|_2.$$

Since m is arbitrary, this implies that

$$\|k * l\|_\infty \leq \|k\|_1 \, \|l\|_2.$$

Approximating x and f by the functions k and l (in the metrics L_1 and L_2, respectively), we obtain the statement of the lemma.

■

It follows from Lemma 2.2 that in the space L_2 one can define a convolution operator with fixed $x \in L_1$, namely

$$L_2 \ni f \mapsto x * f = T_x f \in L_2. \tag{2.11}$$

This operator is bounded and $\|T_x\| \le \|x\|_1$ $(x \in L_1)$. It is easy to see that

$$T_{\lambda x + \mu y} = \lambda T_x + \mu T_y,$$

and

$$T_{x*y} = T_x T_y \quad (x, y \in L_1;\ \lambda, \mu \in \mathbb{C}^1).$$

For the adjoint operator T_x^*, we have $T_x^* = T_{x^*}$ $(x \in L_1)$. This equality is equivalent to the relation

$$(x * f, g)_{L_2} = (f, x^* * g)_{L_2} \quad (f, g \in L_2),$$

which can be obtained by the limit transition. One should only direct l to x in the space L_1 and k and m to f and g, respectively, in L_2 and then employ Lemma 2.2.

Parallel to the convolution operators T_x $(x \in L_2)$, one can define another family of operators in the space L_2 which are closely connected with h.s. These are operators T_p $(p \in Q)$ of a generalized shift. Operators T_p are defined in L_2 by the bilinear form

$$(T_p f, g)_{L_2} = (f * g^*)(p^*) \quad (f, g \in L_2;\ p \in Q). \tag{2.12}$$

It follows from Lemma 2.1 (for $a = 2$) that this form is continuous (therefore, the definition of the operators is corrected and that $\|T_p\| \le 1$. According to (2.12), we have

$$(f, T_{p^*} g)_{L_2} = \overline{(g * f^*)(p)} = ((g * f^*)^*)(p^*) = (f * g^*)(p^*) = (T_p f, g)_{L_2}.$$

Consequently,

$$T_p^* = T_{p^*} \quad (p \in Q).$$

(In the case of Example 2.1, definition (2.12) obviously implies that

$$(T_{p^*} f)(q) = f(qp^{-1}) \quad (p, q \in Q)).$$

Clearly, *the following equality holds*

$$T_x = \int x(p)\, T_p\, dp \quad (x \in L_1), \tag{2.13}$$

i.e., the convolution operator is an operator of a generalized shift "averaged by the function x", and $T_p f$ is equal to a "convolution of f with the δ-function concentrated at the point p". Let us clarify that (2.12) and the continuity of $f * g^*$ together yield the weak continuity of the function $Q \ni p \mapsto T_p \in \mathfrak{L}(L_2, L_2)$. The integral in (2.13) should be regarded as a weak integral in L_2. However, this integral may be also regarded as a Bochner integral with respect to the σ-algebra $\mathfrak{B}(Q)$. This follows from the well known facts concerning these integrals (see, e.g., Yosida [1, Chapter 5, Subsections 4, 5]).

We prove (2.13). Let $f, g \in L_2$. Then, by virtue of (2.12), we have

$$\Big(\Big(\int x(p)\, T_p\, dp\Big) f, g\Big)_{L_2} = \int x(p)\, (T_p f, g)_{L_2}\, dp = \int x(p)\, (f * g^*)(p^*)\, dp$$

$$= (x, f^* * g)_{L_2}. \tag{2.14}$$

Passing in (2.10) to the limit when k tends to x in L_1 and l and m tend to f and g, respectively, in L_2, we get

$$(x * f, g)_{L_2} = (x, f^* * g)_{L_2}.$$

Hence, (2.14) can be extended as follows

$$\Big(\Big(\int x(p)\, T_p\, dp\Big) f, g\Big)_{L_2} = (x * f, g)_{L_2} = (T_x f, g)_{L_2} \quad (f, g \in L_2).$$

And this yields (2.13).

■

Since $T_x f = x * f$ for any $f \in L_2$, the equality (2.13) leads to the following natural formula where the integral is understood properly

$$(x * f)(r) = \int x(p)\, (T_p f)(r)\, dp \quad (x \in L_1,\ f \in L_2). \tag{2.15}$$

Let $(\mathrm{o}_n)_{n=1}^{\infty}$ be a sequence of bounded neighborhoods of O contracting to this point. A sequence $(e_n)_{n=1}^{\infty}$ of functions from L_1 such that $e_n(p) \geq 0$ $(p \in Q)$, $\operatorname{supp} e_n \subset \mathrm{o}_n$ and $\| e_n \|_1 = 1$, is called an approximative identity element of a h.s. (in particular, one may set

$$e_n(p)\ m^{-1}(\mathrm{o}_n)\, \kappa_{\mathrm{o}_n}(p) \quad (p \in Q;\ n \in \mathbb{N})).$$

Lemma 2.3. *Assume that $(e_n)_{n=1}^{\infty}$ is an approximative identity element of a h.s. Then $e_n * x \underset{n\to\infty}{\longrightarrow} x$ weakly in L_1, and $e_n * f \underset{n\to\infty}{\longrightarrow} f$ strongly in L_2 for every $x \in L_1$ and $f \in L_2$.*

Proof. Let $l \in L_1' = L_\infty$. To prove the first assertion, one must establish that

$l(e_n * x) \to l(x)$ as $n \to \infty$. It suffices to investigate the case $x = \kappa_\alpha$ $(\alpha \in \mathfrak{B}(Q))$. Then this relation also holds for step functions k which are dense in L_1. The required result follows from the inequality

$$| l(e_n * x) - l(x) | \le | l(e_n * k) - l(k) | + 2\| l \|_\infty \| x - k \|_1$$

(here we have used the inequality

$$\| e_n * (x - k) \|_1 \le \| e_n \|_1 \| x - k \|_1 = \| x - k \|_1).$$

Thus, let $x = \kappa_\alpha$. Denote by $\varphi \subset Q$ a compact so large that $\varphi \supset o_n$ $(n \in \mathbb{N})$. By virtue of (N1), one can find a compact $\psi \supset \alpha$ such that $c(\varphi, \alpha, r) = 0$ outside ψ. Let l_ψ be a function equal to $l(r)$ when $r \in \psi$ and equal to zero otherwise. Using (N5), we obtain

$$l(e_n * \kappa_\alpha) = \int (e_n * \kappa_\alpha)(r)\, l(r)\, dr = \int (e_n * \kappa_\alpha)(r)\, l_\psi(r)\, dr$$

$$= \int l_\psi(r)\, d_r\Big(\int (e_n(p)\, d_p c(\varepsilon_p, \alpha, \varepsilon_r)\Big) = \int l_\psi(r)\, d_r\Big(\int (e_n(p)\, d_p\, c(\varepsilon_r, \alpha^*, \varepsilon_p)\Big)$$

$$= \int e_n(p)\, d_p\Big(\int l_\psi(r)\, d_r c(\varepsilon_r, \alpha^*, \varepsilon_p)\Big) = \int e_n(p)\, (l_\psi * \kappa_\alpha)(p)\, dp.$$

The function $l_\psi * \kappa_\alpha$ is continuous by Lemma 2.1. Therefore, employing the standard argument and (N6), we can conclude from the last equality that

$$l(e_n * \kappa_\alpha) \underset{n\to\infty}{\to} (l_\psi * \kappa_{\alpha^*})(O) = \int l_\psi(p)\, d_p c(\varepsilon_p, \alpha^*, O)$$

$$= \int l_\psi(p)\, d_p m(\varepsilon_p \cap \alpha) = \int l_\psi(p)\, dp = l(\kappa_\alpha).$$

We prove the second assertion. First, let us show that $e_n * f \to f$ weakly in L_2. Let $g \in L_2$. Then using the old notations, we obtain

$$| (e_n * f, g)_{L_2} - (f, g)_{L_2} | \le | (e_n * k, g)_{L_2} - (k, g)_{L_2} | + 2\| g \|_{L_2} \| f - k \|_{L_2}.$$

Here we have used the inequality

$$\| e_n * (f - k) \|_{L_2} \le \| e_n \|_1 \| f - k \|_{L_2} = \| f - k \|_{L_2}.$$

(see Lemma 2.2). This and the relation $l(e_n * \kappa_\alpha) \underset{n\to\infty}{\to} l(\kappa_\alpha)$ which has been already proved, yield the required weak convergence. Strong convergence now follows from the relation

$$\| e_n * f - f \|_{L_2}^2 = \| e_n * f \|_{L_2}^2 - 2\,\mathrm{Re}\,(e_n * f, f)_{L_2} + \| f \|_{L_2}^2$$

$$\leq 2 \| f \|_{L_2}^2 - 2\,\mathrm{Re}\,(e_n * f, f)_{L_2} \underset{n\to\infty}{\to} 0$$

(we have used the estimate

$$\| e_n * f \|_2 \leq \| e_n \|_1 \|f\|_2 = \|f\|_2,$$

and the weak convergence established just above).

■

A measurable bounded function $Q \ni p \mapsto \chi(p) \in \mathbb{C}^1$ which differs from zero almost everywhere is called a character of an h.s. if

$$\int c(\alpha, \beta, r)(r)\, \chi(r)\, dr = \int_\alpha \chi(p)\, dp = \int_\beta \chi(q)\, dq \quad (\alpha, \beta \in \mathfrak{B}_0(Q)). \tag{2.16}$$

A set of characters of an h.s. is in correspondence with a set of maximal ideals of a normalized algebra L_1 [43]. A maximal ideal M and a character χ which correspond to each other, satisfy the following relation

$$x(M) = \int x(p)\chi(p)\, dp = x(\chi) \quad (x \in L_1) \tag{2.17}$$

(it follows from (2.17) that $\|\chi\|_\infty = \|M\| = 1$). The proof is quite simple: A maximal ideal is identified with the corresponding multiplicative functional. Then this functional is represented (as a linear continuous functional on L_1) in the form of an integral with a certain function (see (2.16)), and this function is just the desired character. The relation (2.16) means that

$$(\kappa_\alpha * \kappa_\beta)(M) = \kappa_\alpha(M)\, \kappa_\beta(M) \quad (\alpha, \beta \in \mathfrak{B}_0(Q)).$$

Denote by X a set of all characters of a h.s. topologized by the topology of the space of maximal ideals, namely, X is a complete separable, locally compact metric space in the case when the algebra L_1 does not contain an identity element. In the case when L_1 contains this element, X is a compact space. X is called a dual basis. Note that a function $\chi(p) = 1$ $(p \in Q)$ is always a character of a h.s. by virtue of (2.2).

Theorem 2.1. *Characters of a normal h.s. are continuous functions (more precisely, coincide with these functions almost everywhere). If Q is a compact, then*

43 Or of an algebra $\mathfrak{L}_1$ which is obtained by the formal introduction of an identity element into L_1 if L_1 does not possess, this element. In this last case, we consider ideals different from L_1.

the characters χ are Hermitian, i.e., $\chi(p^*) = \overline{\chi(p)}$ $(p \in Q)$. The number of these characters is at most countable, they are mutually orthogonal, and form an idempotent system of an algebra L_1 (to within certain factors)

$$(\kappa_j, \kappa_k)_{L_2} = \delta_{jk} \|\chi_j\|_2^2, \quad \kappa_j * \kappa_k = \delta_{jk} \chi_j \|\chi_j\|_2^2 \quad (j, k = 1, \ldots, N; \;\; N \le \infty).$$

Proof. Assume that $\alpha, \beta \in \mathfrak{B}_0(Q)$ and β is such that $\chi(\beta) \neq 0$, where we denote

$$\chi(\alpha) = \int_\alpha \chi(p)\, dp \quad (\alpha \in \mathfrak{B}_0(Q)).$$

Let us construct a compact φ so that $c(\alpha, \beta, r) = 0$ for $r \,\overline{\in}\, \varphi$ and set $\chi^\varphi(r) = \chi(r)$ for $r \in \varphi$, and $\chi^\varphi(r) = 0$ otherwise. According to (2.16), (N5), and Lemma 2.1 (the case $a = 1$), we have

$$\chi(\gamma)\chi(\beta) = \int c(\gamma, \beta, r)\chi(r)\, dr = \int c(\gamma, \beta, r)\chi^\varphi(r)\, dr$$

$$= \int \chi^\varphi(r)\, d_r c(\gamma, \beta, \varepsilon_r) = \int \chi^\varphi(r)\, d_r c(\varepsilon_r, \beta, \gamma) = \int_\gamma (\chi^\varphi * \kappa_{\beta^*})(p)\, dp$$

for $\gamma \subset \alpha$. Since γ is arbitrary, for almost all $r \in \alpha$ we have

$$\chi(r) = \frac{1}{\chi(\beta)} (\chi^\varphi * \kappa_{\beta^*})(r). \tag{2.19}$$

Since α is arbitrary, this equality and Lemma 2.1 (for $a = 1$) yield the continuity of χ.

Let Q be a compact set. The Hermitian character of χ follows from the relation

$$\chi(\alpha)\|\chi\|_2^2 = \int \overline{\chi(p)}\, d_p \Big(\int c(\varepsilon_p, \alpha, r)\chi(r)\, dr\Big)$$

$$= \int \overline{\chi(p)}\, d_p \Big(\int \chi(r)\, d_r c(\varepsilon_p, \alpha, \varepsilon_r)\Big) = \int \overline{\chi(p)}\, d_p \Big(\int \chi(r)\, d_r c(\varepsilon_r, \alpha^*, \varepsilon_p)\Big)$$

$$= \Big(\int \chi(r)\, d_r \Big(\int \overline{\chi(p)}\, d_p c(\varepsilon_r, \alpha^*, \varepsilon_p)\Big) = \overline{\chi(\alpha^*)}\|\chi\|_2^2 \quad (\alpha \in \mathfrak{B}_0(Q)).$$

Further, to prove the orthogonality of two characters χ' and χ'' $(\chi' \neq \chi'')$, we choose $\beta \in \mathfrak{B}_0(Q)$ so that

$$\chi'(\beta) \neq \chi''(\beta) = \overline{\chi''(\beta^*)}.$$

Using (2.19), we obtain

$$(\chi'(\beta)-\overline{\chi''(\beta^*)})\int \chi'(p)\,\overline{\chi''(p)}dp = \int (\chi' * \kappa_{\beta^*})(p)\,\chi''(p)\,dp$$

$$-\int \chi'(p)\,\overline{(\chi''*\kappa_\beta)}\,(p)\,dp = (\chi' * \kappa_{\beta^*}, \chi'')_{L_2} - (\chi', \chi'' * \kappa_{\beta^*})_{L_2}$$

$$= (T_{\kappa_{\beta^*}}\chi', \chi'')_{L_2} - (\chi', T_{\kappa_{\beta^*}}\chi'')_{L_2} = 0,$$

whence $(\chi', \chi'')_{L_2} = 0$.

The separability of L_2 and the orthogonality of characters imply that there exist at most countably many characters. The last assertion of the theorem follows from the relation

$$(\chi_j * \chi_k, \kappa_\alpha)_{L_2} = (T_{\chi_j}\chi_k, \kappa_\alpha)_{L_2} = (\chi_k, T_{\chi_j}\kappa_\alpha)_{L_2}$$

$$= (\chi_k, \chi_j * \kappa_\alpha)_{L_2} = (\chi_k, \chi_j)_{L_2}\,\chi_j(\alpha) \quad (\alpha \in \mathfrak{B}_0(Q))$$

(see (2.19)).

■

One can show that a radical R of a normal h.s., i.e., an intersection of all its maximal ideals, coincides with a collection of its annihilators, namely

$$R = \{a \in L_1 \mid a * x = 0,\ x \in L_1\}.$$

Thus, if a normal h.s. contains a weak b.i., then $R = 0$ by Lemma 2.3. This means that the h.s. is a semi-simple normalized algebra with an involution. Generally speaking, this is not a symmetric algebra, i.e., the equality $M = M^*$ does not hold for all its maximal ideals (or, in other words, there exist the same characters of it which are not Hermitian).

However, if Q is a compact, then each character is Hermitian by Theorem 2.1. This means that a normal h.s. with a compact basis is a symmetric algebra. If, in addition, it contains a weak b.i., then it is semi-simple and, hence, its characters form a complete orthogonal system in L_2 (if $f \in L_2 \subset L_1$ is such that

$$(f, \chi_j)_{L_2} = 0 \quad (j = 1, \ldots, N;\ N \le \infty),$$

then according to (2.17), we have $\tilde{f}(M) = 0$ for all maximal ideals M; therefore $\tilde{f} \in R$ and, consequently, $f = 0$).

This implies that if the original basis Q of a normal h.s. with a weak b.i. is compact (consisting of infinitely many points), then the dual basis X is a countable discrete space. If Q is a countable discrete space, then X is a compact. If Q is locally compact, then X is locally compact, too. On the dual basis X, one can construct the dual h.s. However, here we shall not present the corresponding constructions and the facts from the harmonic analysis necessary for realizing these constructions (one can find these facts in Berezansky, Kalyuzhny [4]).

We now present several simple results concerning the h.s. under consideration.

I. *Assume that Q is a compact, and a h.s. is normal. The function $\chi \in C(Q)$ is a character if and only if*

$$(T_p \chi)(q) = \chi(p^*)\chi(q) \quad (p, q \in Q). \tag{2.20}$$

Indeed, let χ be a character, then $\chi \in C(Q) \subset L_2$ and, according to (2.13), (2.11), (2.3), the definition of normality, and (2.16), we have

$$\int_\alpha \int_\beta (T_p \chi)(q)\, dp\, dq = \int_\beta (T_{\kappa_\alpha} \chi)(q)\, dq = \int_\beta (\kappa_{\alpha^*} \chi)(q)\, dq$$

$$= \int \chi(p)\, d_p\, c(\varepsilon_p, \alpha, \beta) = \int \chi(p)\, d_p\, c(\beta, \alpha^*, \varepsilon_p) = \chi(\beta)\chi(\alpha^*)$$

for any $\alpha, \beta \in \mathfrak{B}_0(Q)$. This yields (2.20). The converse statement is established analogously.

■

Remark 1. The operator T_p can be defined, for example, on $C(Q)$ for the case of noncompact Q, too; the relation (2.20) can be also established. To do this, we employ the fact that $(T_p f)(q) = 0$ for fixed $p, q \in Q$ and $f \in L_2$ provided that supp$f$ lies sufficiently far from O. Therefore, $(T_p f)(q)$ can be defined for $f \in C(Q)$ as $(T_p f')(q)$, where f' coincides with f in a ball with a sufficiently large radius (depending on p and q) centered at O and is zero outside this ball. This property of T_p follows from the fact that

$$\int_\gamma (T_{\kappa_\alpha} \kappa_\beta)(r)\, dr = c(\alpha, \beta, \gamma) = c(\beta, \alpha, \gamma) = c(\gamma, \alpha^*, \beta) = 0$$

for any $\alpha, \beta, \gamma \in \mathfrak{B}_0(Q)$ provided that β lies sufficiently far from α and γ.

■

II. A normal h.s. is called Hermitian if $x^* = x$ for every element of it. Given an arbitrary normal h.s. L_1, we can construct an Hermitian h.s. by a natural procedure. Indeed, let $x(p)$ be some function on Q, we set

$$x'(p) = x(p^*) = \overline{x^*(p)} \quad (p \in Q).$$

We call the function x even if $x' = x$ and odd if $x' = -x$. Clearly, every function on

Q can be represented in a unique manner as a sum of the odd and even components of it, namely

$$x = \tfrac{1}{2}(x + x') + \tfrac{1}{2}(x - x').$$

Consider a mapping $x \mapsto x'$ in the space L_1. It is linear involutive and, moreover, $(x * y)' = x' * y'$. Therefore, a collection of all even functions from the h.s. L_1 forms a subalgebra of it. This subalgebra is denoted by E_1 and called an even subsystem of L_1.

An even subsystem can be regarded as an Hermitian h.s. whose basis is a locally compact space $\mathbf{Q}$, *which is obtained from the original basis* Q *by identifying the points* p *and* p^* *as a class* $\mathbf{p}$.

Indeed, let us denote the full preimages of the sets $\boldsymbol{\alpha}, \boldsymbol{\beta}, \boldsymbol{\gamma}, \ldots \in \mathfrak{B}(\mathbf{Q})$ under the mapping $p \mapsto \mathbf{p}$ $(p \in \mathbf{p})$ by $\alpha, \beta, \gamma, \ldots \in \mathfrak{B}(Q)$; these preimages are invariant with respect to $*$. On $\mathbf{Q}$, we introduce a measure $\mathbf{m}$ by setting $\mathbf{m}(\boldsymbol{\alpha}) = m(\alpha)$ $(\boldsymbol{\alpha} \in \mathfrak{B}(\mathbf{Q}))$ and $d\mathbf{m}(\mathbf{p}) = d\mathbf{p}$. Every function $\mathbf{x} \in L_1(\mathbf{Q}, \mathfrak{B}(\mathbf{Q}), \mathbf{m})$ generates some even function, and vice versa. Moreover,

$$\int_{\boldsymbol{\alpha}} \mathbf{x}(\mathbf{p})\, d\mathbf{p} = \int_{\alpha} x(p)\, dp$$

for any $\boldsymbol{\alpha} \in \mathfrak{B}(\mathbf{Q})$. The results presented above immediately imply that the convolution of two functions from E_1 can be written in the form (2.3), where Q is replaced by $\mathbf{Q}$, and the measure $\mathbf{c}$ is defined by the equality $\mathbf{c}(\boldsymbol{\alpha}, \boldsymbol{\beta}, \mathbf{r}) = c(\alpha, \beta, r)$. Therefore, E_1 is a h.s. which is Hermitian. The latter follows from the relations $\mathbf{c}(\boldsymbol{\alpha}, \boldsymbol{\beta}, \boldsymbol{\gamma}) = c(\alpha, \beta, \gamma)$ and $\beta = \beta^*$.

■

III. *If* χ *is a character of a h.s.* L_1, *then the function*

$$\boldsymbol{\chi}(\mathbf{p}) = \tfrac{1}{2}(\chi(p) + \chi(p^*)) \quad (p \in \mathbf{p} \in \mathbf{Q}) \tag{2.21}$$

is a character of its even subsystem E_1. *Conversely, in the case of a symmetric algebra* E_1, *all the characters of it are given by formula* (2.21).

In fact, let M be a maximal ideal of the algebra $\mathfrak{L}_1$ which corresponds to the character χ[44]. The mapping $E_1 \ni x \mapsto x(M)$ is a homomorphism of the algebra E_1 over the field of complex numbers. This homomorphism does not transform the whole of E_1 into O. This is so because one can find an odd function $x \in L_1$ such that $x(M) \neq 0$. But then

$$(x * x)(M) = |x(M)|^2 \neq 0,$$

44 For definiteness, we assume that L_1 does not contain the identity element and that $\mathfrak{L}_1$ is the corresponding extension of L_1

which contradicts the assertion, since $x * x$ is even. Therefore, the homomorphism $E_1 \ni x \mapsto x(M)$ generates a certain character of an even subsystem. We denote it by $\boldsymbol{\chi}$. The equality (2.21) is obtained from the relation

$$\int_{\mathbf{Q}} x(\boldsymbol{p})\boldsymbol{\chi}(\boldsymbol{p})\,d\boldsymbol{p} = \int x(p)\chi(p)\,dp \quad (x \in E_1).$$

The converse statement follows from the well-known fact that each maximal ideal of a symmetric subalgebra can be extended to the maximal ideal of the whole algebra (Gelfand, Raikov, Shilov [1, Chapter 2, Section 12].

IV. *The following statement is valid: Let E_2 be a collection of even functions from L_2. Then*

$$T_{\boldsymbol{p}} = \tfrac{1}{2}(T_p + T_{p^*}) \restriction E_2 \quad (p \in \boldsymbol{p} \in \mathbf{Q}). \tag{2.22}$$

Indeed, let $x \in E_1$, then by virtue of (2.11), for any $f \in E_2$, expression (2.11) can be understood either as an action of the operator T_x (with respect to the h.s. E_1 with the representation (2.13) in terms of $T_{\boldsymbol{p}}$) or as a restriction of T_x (with respect to the h.s. E_1 with the representation (2.13) in terms of T_p (restricted to E_2)) to E_2. Writing both these representations for any $x \in E_1$, we obtain (2.22).

■

Example 2.2. Let L_1 be a group algebra of the group $G = Q$ from Example 2.1. Then

$$(T_p f)(q) = f(qp^{-1}) \quad (p, q \in Q).$$

Its characters are ordinary continuous characters of a group, and the statement presented above are well known in this case. According to (2.22), an operator of a generalized shift for the even subsystem E_1 of this h.s. L_1 has the form

$$T_{\boldsymbol{p}} = \tfrac{1}{2}(T_p + T_p^{-1}) \restriction E_2 \quad (p \in \mathbf{Q})$$

and is closely connected with the form of the relations between operators which have been mentioned in the introduction to this section.

2.3. Nuclear Space of Functions on a Basis of a Hypercomplex System

Let us construct a space of test functions on Q which resembles the classic space $\mathcal{D}(\mathbb{R}^d)$ and is invariant with respect to the operators of convolution and generalized shift. This space will give us the required riggings when we consider the representation of the h.s. In the case of locally compact groups, this space gives us those automatically existing riggings which have been investigated in Subsection 1.2 when studying Stone's theorem and the Sz.-Nagy–Hille theorem. Below, we consider only normal h.s. with b.i.

Let us introduce some notations. Let $\rho(p, q)$ $(p, q \in Q)$ be a distance in Q, and let B_a be an open ball in Q centered at O with radius $a > 0$. According to (N1), $c(\tilde{B}_{a_1}, \tilde{B}_{a_2}, r)$ is a continuous finite function of a point $r \in Q$. Denote by $d(a_1, a_2)$ the radius of the minimal open ball centered at O outside of which this function vanishes. Clearly, $d(a_1, a_2) = d(a_2, a_1)$; and $c(\alpha, \beta, r) = 0$ for $\rho(r, 0) > d(a_1, a_2)$ if $\alpha, \beta \in \mathfrak{B}_0(Q)$ and $\alpha \subset \tilde{B}_{a_1}$, $\beta \subset \tilde{B}_{a_2}$. It follows from (N7) that for any $a_2 > 0$ and $\varepsilon > 0$, there exists $\delta > 0$ such that

$$d(a_1, a_2) \le a_2 + \varepsilon$$

for $a_1 < \delta$.

Further, we shall regard the space

$$L_2(B_a, dp) = L_2(B_a, \mathfrak{B}(B_a), m)$$

as a subspace of L_2 extending every function $f \in L_2(B_a, dp)$ as equal to zero outside B_a. Denote by $L_{2,0}$ a linear set of finite functions from L_2; $L_{2,0}$ is equipped with a natural convergence, namely,

$$L_{2,0} \ni \xi_n \to \xi \in L_{2,0}$$

as $n \to \infty$ if $\|\xi_n - \xi\|_2 \to 0$ and ξ_n are uniformly finite.

Lemma 2.4. *Let* $\xi \in L_2$, $\operatorname{supp} \xi \subset B_\delta$ *and, consequently,* $\xi \in L_1$. *The operator* $T_\xi \restriction L_2(B_a, dp)$ *transforms the space* $L_2(B_a, dp)$ *into* $L_2(B_{d(a,\delta)}, dp)$; *it is a Hilbert– Schmidt operator* ($\delta > 0$ *and* $a > 0$ *are fixed*). *In particular,* $L_{2,0}$ *is invariant with respect to* T_x, *where* $\xi \in L_{2,0}$.

Proof. Let $\alpha, \beta \in \mathfrak{B}_0(Q)$, $\alpha \subset B_\delta$, and $\beta \subset B_\delta$. Then

$$(\kappa_\alpha * \kappa_\beta)(r) = c(\alpha, \beta, r)$$

vanishes outside $B_{d(a,\delta)}$ and, hence, $(k * l)(r)$ also vanishes outside these sphere (here k and l are step functions such that $\operatorname{supp} k \subset B_\delta$ and $\operatorname{supp} l \subset B_a$). Approximating ξ and f by the functions k and l in the metrics L_1 and L_2, respectively, and employing Lemma 2.2, we conclude that $\xi * f$ also vanishes almost everywhere outside $B_{d(a,\delta)}$.

Thus, we have established how the operator $T_\xi \restriction L_2(B_\rho, dp)$ acts.

Let us prove the second statement. Let $f \in L_2(B_a, dp)$ and $g \in L_2(B_{d(a,\delta)}, dp)$. With the help of (2.12) we get

$$(T_\xi f, g)_{L_2} = (\xi * f, g)_{L_2} = \int (\xi * f)(r)\, \overline{g(r)}\, dr = \int (T_{r^*}\xi, f^*)_{L_2}\, \overline{g(r)}\, dr$$

$$= \int \Big(\int (T_{r^*}\xi)(p) f(p^*)\, dp \Big) g(r)\, dr$$

$$= \int_{B_{d(a,\delta)}} \left(\int_{B_a} (T_{r^*}\xi)(p^*) f(p)\, dp \right) \overline{g(r)}\, dr;$$

$$\int_{B_{d(a,\delta)}} \int_{B_a} |(T_{r^*}\xi)(p^*)|^2\, dp\, dr = \int_{B_{d(a,\delta)}} \|(T_{r^*}\xi)\|_2^2\, dr \le \|\xi\|_2^2\, m(B_{d(a,\delta)}) < \infty$$

(we have used the inequality $\| T_{r^*} \| \le 1$).

■

Let us construct a function on Q "infinitely many times differentiable" in a certain sense. Fix a sequence $\Xi = (\xi_n)_{n=1}^\infty$ of functions $\xi_n \in L_2$ such that $\operatorname{supp} \xi_n \in B_{\delta_n}$ $(\delta_n \in (0, \infty))$. A function $f \in L_2$ is called infinitely many times differentiable (with respect to Ξ) if for every $n \in \mathbb{N}$ we have $f \in \mathfrak{R}\,(T_{\xi_1} \dots T_{\xi_n})$. By virtue of Lemma 2.1 (for $a = 2$), every function which is infinitely many times differentiable is necessarily continuous and bounded. The numbers δ_n $(n \in \mathbb{N})$ from these expressions are chosen in a certain way which depends upon the character of the convolution in the h.s.; this character is described by the function d. Namely, we take $\delta_1 = 2^{-1}$ and the numbers $\delta_n > 0$ with $n = 2, 3, \dots$ are chosen to be so small that

$$d\left(\delta_n, \sum_{k=1}^{n-1} 2^{-k} \right) \le \sum_{k=1}^{n} 2^{-k}.$$

We can assume that the sequence $(\xi_n)_{n=1}^\infty$ monotonically decreases to zero. Further, we assume that Ξ is such that the product $\prod_{n=1}^\infty \|\xi_n\|_1$ converges.

Lemma 2.5. *For every $f \in L_2(B_a, dp)$ $(a \in (0, \infty))$, the sequence $(T_{\xi_1} \dots T_{\xi_n} f)_{n=1}^\infty$ is precompact, and each limiting point of it is an infinitely differentiable finite function from the space $L_2(B_{d(a,1)}, dp)$. Furthermore, for every $n \in \mathbb{N}$, this limiting point belongs to $\mathfrak{R}((T_{\xi_1} \dots T_{\xi_n}) \restriction L_{2,0})$.*

Proof. Clearly,

$$T_{\xi_1} \dots T_{\xi_n} f = T_{\xi_1 * \dots * \xi_n} f.$$

The function $\xi_1 * \dots * \xi_n$ is finite, and

$$\operatorname{supp}(\xi_1 * \dots * \xi_n) \subset B_{1-2^{-n}}.$$

Indeed, $\operatorname{supp} \xi_1 \subset B_{2^{-1}}$ $(\delta_1 = 2^{-1})$. Further, according to Lemma 2.4, we have

$$\operatorname{supp}(\xi_1 * \xi_2) \subset B_{d(\delta_2, 2^{-1})} \subset B_{2^{-1}+2^{-2}},$$

since

$$d(\delta_2, 2^{-1}) \leq 2^{-1} + 2^{-2};$$

analogously,

$$\operatorname{supp}(\xi_1 * \xi_2 * \xi_3) \subset B_{d(\delta_3, 2^{-1}+2^{-2})} \subset B_{2^{-1}+2^{-2}+2^{-3}},$$

etc. By Lemma 2.2, we have $\xi_1 * \dots * \xi_n \in L_2$. Applying Lemma 2.4, we obtain the required inclusion

$$\operatorname{supp}(T_{\xi_1} \dots T_{\xi_n} f) = \operatorname{supp}(T_{\xi_1 * \dots * \xi_n} f)$$

$$\subset B_{d(a, 1-2^{-n})} \subset B_{d(a, 1)} \quad (n \in \mathbb{N}). \tag{2.23}$$

The sequence $(T_{\xi_2} \dots T_{\xi_n} f)_{n=2}^{\infty}$ is bounded since

$$\| T_{\xi_2} \dots T_{\xi_n} f \| \leq \prod_{k=2}^{n} \| \xi_k \|_1 \leq c \quad (n = 2, 3, \dots).$$

According to (2.23), we have

$$T_{\xi_2} \dots T_{\xi_n} f \in L_2(B_{d(a, 1)}, dp),$$

and thus, by virtue of Lemma 2.4 applied to the restriction of the operator T_{ξ_1} to this subspace, the sequence $(T_{\xi_1} \dots T_{\xi_n} f)_{n=1}^{\infty}$ is precompact. Let f' be one of its limiting points. Taking into account (2.23) and the following limiting transition, we conclude that $\operatorname{supp} f' \subset B_{d(a,1)}$, i.e., $f' \in (B_{d(a,1)}, dp)$.

Let us show that f' is infinitely differentiable, i.e., that for every $m \in \mathbb{N}$, there exists $g'_m \in L_2$ such that $f' = T_{\xi_1} \dots T_{\xi_m} g'_m$. Assume that $f' = \lim_{j\to\infty} T_{\xi_1} \dots T_{\xi_{n_j}} f$. Fix j_0 sufficiently large such that $n_{j_0} \geq m$ and consider the sequence $(T_{\xi_{n_{j_0+1}}} \dots T_{\xi_{n_j}} f)_{j=j_0+1}^{\infty}$. The argument which was applied above to the sequence $(T_{\xi_2} \dots T_{\xi_n} f)_{n=2}^{\infty}$ and the operator T_{ξ_1} can be also employed when considering this sequence and the operator $T_{\xi_{n_{j_0}}}$. As a result, we obtain that the sequence $(T_{\xi_{n_{j_0}}} \dots T_{\xi_{n_j}} f)_{j=j_0}^{\infty}$ is precompact. Let g''_m be its limiting point, i.e.

$$g''_m = \lim_{l\to\infty} T_{\xi_{n_{j_0}}} \dots T_{\xi_{n_{j_l}}} f.$$

Then

$$T_{\xi_1} \dots T_{\xi_{n_{j_0-1}}} g''_m = \lim_{l\to\infty} T_{\xi_1} \dots T_{\xi_{n_{j_l}}} f = f',$$

$$f' = T_{\xi_1} \dots T_{\xi_m} g'_m,$$

and

$$g'_m = T_{\xi_{m+1}} \dots T_{\xi_{n_{j_0-1}}} g''_m \in L_2.$$

We now prove the last assertion of the Lemma. One must establish that for every $m \in \mathbb{N}$, the function g'_m introduced above is finite. Indeed, g''_m is finite just as f' and, therefore, by virtue of Lemma 2.4, the function

$$T_{\xi_{m+1}} \dots T_{\xi_{n_{j_0-1}}} g''_m = g'_m.$$

is also finite.

■

We denote by $C_0^\infty(Q, \Xi)$ the space of all functions given on Q which are infinitely many times differentiable in the above-mentioned sense and belong to $\mathfrak{R}((T_{\xi_1} \dots T_{\xi_n}) \restriction L_{2,0})$ for every $n \in \mathbb{N}$ (this means that they are finite). Clearly, $C_0^\infty(Q, \Xi)$ is a linear set. The important property of it is invariance with respect to the operators T_ξ, namely

$$T_\xi(C_0^\infty(Q, \Xi)) \subset C_0^\infty(Q, \Xi),$$

where $\xi \in L_{2,0}$. In fact, if

$$f \in \Re((T_{\xi_1} \dots T_{\xi_n}) \upharpoonright L_{2,0}) \quad (n \in \mathbb{N}),$$

then $T_\xi f$ also has the same property due to the commutativity of T_ξ and T_{ξ_k}. Show that $C_0^\infty(Q, \Xi)$ can be transformed into a nontrivial nuclear space by a proper choice of Ξ and the topology.

Lemma 2.6. *For any neighborhood u of the point* O, *there exists a continuous nonnegative function* $\xi = \xi^*$, $\operatorname{supp} \xi \subset u$, $\|\xi\|_1 = 1$ *such that* $\operatorname{Ker} T_\xi = \{0\}$.

Proof. First, we note that (N7) implies that for any neighborhood v of the point O there exists a neighborhood $o \subset v$ of this point such that $\operatorname{supp} c(o, o^*, r)$. Indeed, the continuity of the mapping $Q \ni p \mapsto p^* \in Q$ and the equality $O^* = O$ yield the existence of the neighborhood w of the point O such that $(w^*)^\sim \subset v$. Choose a neighborhood o of the point O to be so small that $\operatorname{supp} c(o, (w^*)^\sim, r) \subset v$ and $o \subset w$. But then $\operatorname{supp} c(o, o^*, r) \subset v$, since the s. measure is nonnegative.

Employing this remark step by step, we construct a sequence of neighborhoods o_n of the point O such that

$$\operatorname{supp} c(o, \overset{*}{o}_n, r) \subset o_{n-1} \quad (n \in \mathbb{N};\ o_0 = u),$$

and $\bigcap_{n=1}^\infty o_n = O$. Let us set

$$e_n(p) = m^{-1}(o_n)\, \kappa_{o_n}(p)$$

and

$$(e_n * \overset{*}{e}_n)(p) = m^{-2}(o_n)\, c(o, o^*, p) \quad (p \in Q).$$

Define a function ξ by the equality

$$\xi(p) = \sum_{n=1}^{\infty} a_n (e_n * \overset{*}{e}_n)(p),$$

where $a_n > 0$ will be chosen just below. Clearly $\xi^* = \xi$.

■

Due to the properties of the s. measure, the functions $e_n * \overset{*}{e}_n$ are continuous and nonnegative and $\operatorname{supp}(e_n * \overset{*}{e}_n) \subset o_{n-1} \subset u$. By virtue of (2.6) and (2.2), we have $\|e_n * \overset{*}{e}_n\|_\infty \leq m^{-1}(o_n)$ and $\|e_n * \overset{*}{e}_n\|_1 = 1$ $(n \in \mathbb{N})$. Choose $a_n > 0$ so that $\sum_{n=1}^\infty a_n m^{-1}(o_n) < \infty$

and $\sum_{n=1}^{\infty} a_n = 1$. The series determining ξ now converges uniformly. The function ξ is continuous and nonnegative; supp $\xi \subset u$, and

$$\operatorname{supp} \xi \subset u \quad \text{and} \quad \|\xi\|_1 = \sum_{n=1}^{\infty} a_n \|e_n * e_n^*\|_1.$$

We show that Ker $T_\xi = \{0\}$.

Assume $f \in L_2$ is such that $T_\xi f = \xi * f = 0$. It follows from the form of ξ and the equality

$$(e_n * e_n^* * f, f)_{L_2} = (e_n^* * f, e_n^* * f)_{L_2}$$

that $e_n^* * f = 0$ and, thus, $e_n * f^* = 0$ $(n \in \mathbb{N})$. According to Lemma 2.3, we have $f^* = 0$ and, therefore, $f = 0$.

■

We now complete the choice of the sequence $\Xi = (\xi_n)_{n=1}^{\infty}$ which defines the concept of infinite differentiability. Namely, as ξ_n we take the function ξ from Lemma 2.6 associated with the neighborhood $u = B_{\delta_n}$. Hence, *the sequence* $\Xi = (\xi_n)_{n=1}^{\infty}$ *is fixed and each function* $\xi_n = \xi_n^*$ *in it is continuous and nonnegative;* supp $\xi_n \subset B_{\delta_n}$, $\|\xi\|_1 = 1$ *and* Ker $T_{\xi_n} = \{0\}$ (the numbers δ_n were chosen above). Henceforth, the sequence Ξ is always regarded as constructed in this manner.

Lemma 2.7. *The space* $C_0^{\infty}(Q, \Xi)$ *is dense in* L_2 *and thus nontrivial.*

Proof. It suffices to show that $C_0^{\infty}(Q, \Xi)$ contains at least one function φ such that Ker $T_\varphi = \{0\}$. In fact, if $\xi \in L_{2,0}$, then by virtue of the invariance of $C_0^{\infty}(Q, \Xi)$ with respect to T_ξ, we can conclude that $T_\xi \in C_0^{\infty}(Q, \Xi)$. But $T_\xi \varphi = \xi * \varphi = \varphi * \xi = T_\varphi \xi$ and, thus, $C_0^{\infty}(Q, \Xi) \supset T_\varphi(L_{2,0})$. To prove denseness, it suffices to establish that $T_\varphi(L_{2,0})$ is dense in L_2. Assume that

$$0 = (T_\varphi \xi, f)_{L_2} = (\xi, T_\varphi^* f)_{L_2} \quad (\xi \in L_{2,0})$$

for $f \in L_2$. This yields

$$0 = T_\varphi^* f = T_{\varphi^*} f = \varphi^* * f,$$

hence, $\varphi * f^* = 0$. But Ker $T_\varphi = \{0\}$. Consequently, $f^* = 0$ and, therefore, $f = 0$, as well.

To construct the function φ we act as follows. We modify the proof of Lemma 2.5 and consider

$$e_{n+1}(p) = m^{-1}(B_{\delta_{n+1}})\kappa_{B_{\delta_{n+1}}}(p)$$

instead of f. This means that we examine the limiting points of the sequence $(T_{\xi_1} \dots T_{\xi_n} e_{n+1})_{n=1}^{\infty}$ in L_2. The fact that this sequence is precompact is proved just as in Lemma 2.5: It suffices to show that the sequence $(T_{\xi_2} \dots T_{\xi_n} e_{n+1})_{n=2}^{\infty}$ is bounded in L_2 and that $\operatorname{supp}(T_{\xi_2} \dots T_{\xi_n} e_{n+1}) \subset B_a$ with some $a > 0$. Using Lemma 2.2, we establish boundedness ($\| T_{\xi_k} \| \le \| \xi_k \|_1 = 1$)

$$\| T_{\xi_2} \dots T_{\xi_n} e_{n+1} \|_2 = \| T_{\xi_3} \dots T_{\xi_n} (\xi_2 * e_{n+1}) \|_2$$

$$\le \| \xi_2 * e_{n+1} \|_2 \le \| \xi_2 \|_2 \| e_{n+1} \|_1 = \| \xi_2 \|_2 \quad (n = 3, 4, \dots).$$

Further, as in the proof of (2.23), we find that

$$\operatorname{supp}(T_{\xi_2} \dots T_{\xi_n} e_{n+1}) \subset B_{d(\delta_{n+1}, 1-2^{-n})} \subset B_{d(\delta_3, 2^{-1})} \quad (n = 2, 3, \dots). \tag{2.24}$$

Assume that φ is one of the limiting points of the sequence under consideration, i.e., $\varphi = \lim_{j\to\infty} T_{\xi_1} \dots T_{\xi_{n_j}} e_{n_j+1}$. Clearly, φ is finite. The fact that it is infinitely differentiable is verified just as in Lemma 2.5. Here we only write the expression for the function $\psi'_m \in L_2$ such that

$$\varphi = T_{\xi_1} \dots T_{\xi_m} \psi'_m \quad (m \in \mathbb{N}).$$

By analogy with the proof of Lemma 2.5, we obtain

$$\psi'_m = T_{\xi_{m+1}} \dots T_{\xi_{n_{j_0}-1}} \psi''_m ,$$

where $n_{j_0} \ge m$, and ψ''_m is one of the limiting points of the sequence $(T_{\xi_{n_{j_0}}} \dots T_{\xi_{n_j}} \times e_{n_j+1})_{j=j_0}^{\infty}$. Therefore,

$$\psi'_m = \lim_{l\to\infty} T_{\xi_{m+1}} \dots T_{\xi_{n_{j_l}}} e_{n_{j_l}+1}, \tag{2.25}$$

where the sequence depends on $m \in \mathbb{N}$. It is obvious that ψ'_m is finite. This follows from the finiteness of ψ''_m which can be established as in the case of φ. Hence, $\varphi \in C_0^\infty(Q, \Xi)$.

We prove that $\operatorname{Ker} T_\varphi = \{0\}$. Let $g \in L_2$ be such that

$$0 = T_\varphi g = \varphi' * g = T_{\xi_1} \dots T_{\xi_m} (\psi'_m * g).$$

It follows from the relation

$$\operatorname{Ker} (T_{\xi_1} \dots T_{\xi_m}) = \{0\}$$

that $\psi'_m * g = 0$ $(m \in \mathbb{N})$. It suffices to prove that $(\psi'_m)_{m=1}^\infty$ is an approximative identity (see Lemma 2.3), since in this case the last equality yields $g = 0$. It follows from (2.25) that ψ'_m is nonnegative. The relation (2.24) involves the inclusion

$$\operatorname{supp} \left(T_{\xi_{m+1}} \dots T_{\xi_{n_{j_l}}} e_{n_{j_l}+1}\right) \subset B_{d(\delta_{m+1}, 2^{-m+1})}$$

and, hence,

$$\operatorname{supp} \psi'_m \subset B_{d(\delta_{m+1}, 2^{-m+1})}.,$$

By virtue of the fact established at the beginning of the proof of Lemma 2.6, we have

$$\lim_{\rho(a_1, a_2) \to 0} d(a_1, a_2) = 0,$$

therefore, $\operatorname{supp} \psi'_m \subset B_{r_m}$, where $r_m \to 0$ as $m \to \infty$. Further, ψ'_m is nonnegative. Taking into account the definition of m. measure, we obtain

$$\int (T_{\xi_{m+1}} \dots T_{\xi_{n_{j_l}}} e_{n_{j_l}+1})(p)\, dp = \int (\xi_{m+1} * \dots * \xi_{n_{j_l}} * e_{n_{j_l}+1})(p)\, dp$$

$$= \prod_{k=m+1}^{n_{j_l}} \int \xi_m(p)\, dp \int e_{n_{j_l}+1}(p)\, dp .$$

Thus, by virtue of (2.25), we also have $\int \psi'_m(p)\, dp = 1$. Hence, $(\psi'_m)_{m=1}^\infty$ is an approximative identity.

■

We introduce an analogue of a Sobolev space with a weight. If

$$\operatorname{Ker} T_x = \operatorname{Ker} T_y = \{0\},$$

then

$$\operatorname{Ker}(T_x T_y) = \operatorname{Ker}(T_{x*y}) = \{0\}.$$

Therefore, the operator of a "derivative"

$$D_k (T_{\xi_1} \dots T_{\xi_k})^{-1} = T^{-1}_{\xi_1 * \dots * \xi_k} = T^{-1}_{\xi_1} \dots T^{-1}_{\xi_k},$$

defined on $\mathfrak{D}(D_k) = \mathfrak{R}(T_{\xi_1} \dots T_{\xi_k})$ is meaningful.

In particular, the operator D_k $(k \in \mathbb{Z}_+)$ is also defined on $C_0^\infty(Q, \Xi)$. We also set $D_0 = 1$. Since $f \in C_0^\infty(Q, \Xi)$ belongs to $\mathfrak{R}((T_{\xi_1} \dots T_{\xi_k}) \upharpoonright L_{2,0})$ for every $n \in \mathbb{N}$, the function $D_k f$ $(k \in \mathbb{Z}_+)$ is finite. Assume that $f \in C_0^\infty(Q, \Xi)$. Then for any $k \in \mathbb{Z}_+$, we have

$$f = T_{\xi_1} \dots T_{\xi_k} g = \xi_1 * \dots * \xi_k * g$$

with some $g \in L_{2,0}$ which is uniquely determined for given f. But the last equality implies that $g = D_k f$. Therefore, one can always write

$$f = (T_{\xi_1} \dots T_{\xi_k}) D_k f = \xi_1 * \dots * \xi_k * (D_k f).$$

Let $\xi \in L_{2,0}$, then

$$C_0^\infty(Q, \Xi) \ni f * \xi = (\xi_1 * \dots * \xi_k * (D_k f)) * \xi = (\xi_1 * \dots * \xi_k) * ((D_k f) * \xi)$$

for any $k \in \mathbb{Z}_+$. Since $(D_k f) * \xi \in L_{2,0}$, the last equality means that

$$(D_k f) * \xi = D_k (f * \xi).$$

Henceforth, we shall use the following simple relations:

$$f = D_k (T_{\xi_1} \dots T_{\xi_k} f) = (T_{\xi_1} \dots T_{\xi_k}) D_k f$$

$$= D_k (\xi_1 * \dots * \xi_k * f) = \xi_1 * \dots * \xi_k * (D_k f),$$

$$D_k (f * \xi) = (D_k f) * \xi,$$

$$(D_k f)^* = D_k f^* \quad (f \in C_0^\infty(Q, \Xi),\ \xi \in L_{2,0},\ k \in \mathbb{Z}_+). \tag{2.26}$$

We clarify that in order to obtain the last equality, one should employ the fact that ξ_n^*

$= \xi_n$ ($n \in \mathbb{N}$). Then, by the first relation in (2.26), we have

$$f^* = \xi_1 * \ldots * \xi_k * (D_k f)^*,$$

and this yields $(D_k f)^* = D_k f^*$. Clearly, the invariance of $C_0^\infty(Q, \Xi)$ with respect to the operator of conjugation $*$ also follows from the above argument.

Let $Q \ni p \mapsto \tau_2(p) \in [1, \infty)$ be a weight which is a continuous function and $\tau = (\tau_1, \tau_2(p))$ ($\tau_1 \in \mathbb{Z}_+$). We construct the Hilbert "Sobolev" space $\mathcal{H}_\tau = W_2^{\tau_1}(Q, \Xi, \tau_2(p)\, dp)$ as the completion of $C_0^\infty(Q, \Xi)$ with respect to the scalar product

$$(f, g)_{\mathcal{H}_\tau} = \sum_{k=0}^{\tau_1} \int (D_k f)(p)\, \overline{(D_k g)(p)}\tau_2(p)\, dp \quad (f,\ g \in C_0^\infty(Q, \Xi)). \qquad (2.27)$$

We now construct the projective limit of the spaces $\mathcal{H}_\tau$ (see Chapter 1, Subsection 1.2 (in particular, Example 1.7)).

Let T be a set of all indices τ introduced here; for $\tau', \tau'' \in$ T we write $\tau'' \geq \tau'$ if $\tau_1'' \geq \tau'$ and $\tau_2''(p) \geq \tau_1'(p)$ ($p \in Q$).

One can easily see that *for* $\tau'' \geq \tau'$, *we have* $\mathcal{H}_{\tau''} \subset \mathcal{H}_{\tau'}$, *and the imbedding is topological*.

In fact, one should only verify the compatibility of norms in $\mathcal{H}_{\tau''}$ and $\mathcal{H}_{\tau'}$ (see Chapter 1, Subsection 3.1). Assume that the sequence $(f_n)_{n=1}^\infty$, $f_n \in C_0^\infty(Q, \Xi)$ is fundamental in $\mathcal{H}_{\tau''}$ and $\|f_n\|_{\mathcal{H}_{\tau'}} \to 0$ as $n \to \infty$. One must show that $\|f_n\|_{\mathcal{H}_{\tau''}} \to 0$ as $n \to \infty$. Clearly, $\|f_n\|_2 \to 0$ as $n \to \infty$ and the sequence $(D_k f_n)_{n=1}^\infty$ is fundamental in L_2 for $k = 1, \ldots, \tau'$. Since the operator D_k is closed, we have $\|D_k f_n\|_{L_2} \to 0$ as $n \to \infty$. Moreover, the boundedness of $(f_n)_{n=1}^\infty$ in $\mathcal{H}_{\tau''}$ implies the boundedness of $(D_k f_n)_{n=1}^\infty$ in $L_2(Q, \tau_2'(p)\, dp)$. This enables us to proceed to the limit in expressions of the type (2.27) for $\|\cdot\|_{\mathcal{H}_{\tau''}}$ and to conclude that $\|f_n\|_{\mathcal{H}_{\tau''}} \to 0$ as $n \to \infty$.

■

Denote the pair (0, 1) as follows: (0, 1) = 0 ∈ T. By virtue of Lemma 2.7, $\mathcal{H}_0 = L_2$. Clearly, $\mathcal{H}_0 \supset \mathcal{H}_\tau$ ($\tau \in$ T) topologically. The family of spaces $(\mathcal{H}_\tau)_{\tau \in \mathrm{T}}$ satisfies the following condition of directedness : For $\tau', \tau'' \in$ T, there exists $\tau''' \in$ T such that $\mathcal{H}_{\tau'''} \subset \mathcal{H}_{\tau'}$ and $\mathcal{H}_{\tau'''} \subset \mathcal{H}_{\tau''}$ topologically (it suffices to set

$$\tau_1''' = \max(\tau_1', \tau_1''), \quad \tau_2'''(p) = \max(\tau_2'(p), \tau_2''(p)) \quad (p \in Q)).$$

Therefore, it is possible to define a projective limit

$$\mathcal{D}(Q, \Xi) = \operatorname{pr}\lim_{\tau \in \mathrm{T}} \mathcal{H}_\tau.$$

As a set, $\mathcal{D}(Q,\Xi)=\bigcap_{\tau\in T}\mathcal{H}_\tau$ and the basis of neighborhoods in $\mathcal{D}(Q,\Xi)$ consists of sets of the form

$$U(\varphi;\tau;\varepsilon) = \{\varphi\in\mathcal{D}(Q,\Xi)\mid \|\psi-\varphi\|_{\mathcal{H}_\tau}<\varepsilon\}\quad(\varphi\in\mathcal{D}(Q,\Xi),\ \tau\in T,\ \varepsilon>0).$$

Theorem 2.2 *Assume that* L_1 *is a normal h.s. with a basis* Q *and a basis identity, i.e., that the axioms* (N1) – (N7) *hold. Then the projective limit*

$$\mathcal{D}(Q,\Xi) = \operatorname{pr}\lim_{\tau\in T}\mathcal{H}_\tau$$

of the "Sobolev" spaces $\mathcal{H}_\tau$ *is a nuclear space which coincides (as a set) with the collection* $C_0^\infty(Q,\Xi)$ *of all functions on* Q *which are "infinitely differentiable" with respect to* Ξ *and finite together with all their "derivatives".*

The space $\mathcal{D}(Q,\Xi)$ *is invariant with respect to the action of the operators* T_ξ $(\xi\in L_{2,0})$, *and the vector function*

$$L_{2,0}\times\mathcal{D}(Q,\Xi)\ni(\xi,\varphi)\mapsto T_\xi\varphi = \xi*\varphi\in\mathcal{D}(Q,\Xi)$$

is separately continuous. This space is also invariant with respect to the action of the conjugation operator, namely

$$\mathcal{D}(Q,\Xi)\ni\varphi(p)\mapsto\varphi^*(p) = \overline{\varphi(p^*)}\ \mathcal{D}(Q,\Xi),$$

and this operator is continuous.

Proof. The proof is split into several steps.

I. A function $f\in\mathcal{H}_\tau$ belongs to $\mathcal{R}(T_{\xi_1}\dots T_{\xi_{\tau_1}})\subset\mathcal{D}(D_{\tau_1})$: Assume that $C_0^\infty(Q,\Xi)\ni f_n\underset{n\to\infty}{\to}f$ in $\mathcal{H}_\tau$. It follows from the form of $(\cdot,\cdot)_{\mathcal{H}_\tau}$ that $f_n\to f$ in L_2 and $D_{\tau_1}f_n\to h\in L_2$. It remains to employ the fact that D_{τ_1} is closed. Therefore, $\varphi\in\mathcal{D}(Q,\Xi)=\bigcap_{\tau\in T}\mathcal{H}_\tau$ belongs to $\mathcal{R}(T_{\xi_1}\dots T_{\xi_n})$ for each $n\in\mathbb{N}$, i.e., this function is infinitely differentiable.

II. We show that $\varphi\in\mathcal{D}(Q,\Xi)$ is finite. Assume the contrary. Then we can construct a sequence of points $(p_n)_{n=1}^\infty$, $p_n\in Q$ such that $\rho(p_n,O)\to\infty$ as $n\to\infty$ and $\varphi(p_n)\neq0$ $(n\in\mathbb{N})$. For every n, one can find a neighborhood u_n of the point p_n such that $|\varphi(p)|\geq\varepsilon_n>0$ $(p\in u_n)$, and the neighborhoods u_n are disjoint. Let us construct a continuous function $\tau_2(p)\geq1$ so that

$$\tau_2(p)\geq\varepsilon_n^{-1}m^{-1}(u_n)\quad(p\in u_n;\ n\in\mathbb{N}).$$

Then

$$\int |\varphi(p)|^2 \tau_2(p)\,dp = \infty,$$

but this contradicts the inclusion $\varphi \in \mathcal{H}_{(0,\tau_2)}$.

The derivative $D_1\varphi$ is also finite. In fact, since $\varphi \in \mathcal{R}(T_{\xi_1} T_{\xi_2})$, the following representation holds

$$\varphi = T_{\xi_1} T_{\xi_2} \psi = \xi_1 * \xi_2 * \psi,$$

where $\psi \in L_2$. Therefore, $D_1\varphi = \xi_2 * \psi$ is continuous by Lemma 2.1. Then, assuming, as above, that $D_1\varphi$ is not finite, we arrive at a contradiction with the inclusion $\varphi \in \mathcal{H}_{(1,\tau_2)}$. Analogously, the inclusions $\varphi \in \mathcal{R}(T_{\xi_1} T_{\xi_2} T_{\xi_3})$ and $\varphi \in \mathcal{H}_{(2\tau_2)}$ imply that $D_2\varphi$ is finite, etc. Hence, for $\varphi \in \mathcal{D}(Q, \Xi)$ all the functions φ, $D_1\varphi$, $D_2\varphi$ are finite or, in other words, $\varphi \in L_{2,0}$ and $\varphi \in \mathcal{R}((T_{\xi_1} \ldots T_{\xi_n}) \restriction L_{2,0})$. Thus, $\mathcal{D}(Q, \Xi) \subset C_0^\infty(Q, \Xi)$ and, consequently, $\mathcal{D}(Q, \Xi) = C_0^\infty(Q, \Xi)$.

Somewhat later, we shall prove that $\mathcal{D}(Q, \Xi)$ is nuclear. Let us first establish the last statements of the theorem. Note that the invariance of $\mathcal{D}(Q, \Xi) = C_0^\infty(Q, \Xi)$ with respect to T_ξ ($\xi \in L_{2,0}$) has been already mentioned when introducing the space $C_0^\infty(Q, \Xi)$.

III. Let us show that the vector function $L_{2,0} \ni \xi \mapsto T_\xi \varphi \in \mathcal{D}(Q, \Xi)$ is continuous for every fixed $\varphi \in \mathcal{D}(Q, \Xi)$. Since it is linear in ξ, it suffices to investigate the point $\xi = 0$ and establish the inequality

$$\| T_\xi \varphi \|_{\mathcal{H}_\tau} \le c_\delta \| \xi \|_2 \quad (c_\delta > 0;\ \xi \in L_2(B_\delta, dp))$$

for every $\tau \in \mathrm{T}$ and $\delta \in (0, \infty)$.

We set $\varphi_k = D_k\varphi$ and choose $a > 0$ large enough so that $\operatorname{supp} \subset B_a$ $(k = 0, \ldots, \tau_1)$. Employing the commutativity of T_ξ and D_k and acting just as in the proof of Lemma 2.4, we find from (2.12) that

$$(D_k T_\xi \varphi)(r) = (\xi * \varphi_k)(r) = \int_{B_a} (T_{r^*} \xi)(p^*)\, \varphi_k(p)\, dp$$

vanishes outside $B_{d(a,d)}$ and, therefore,

$$\| T_\xi \varphi \|^2_{\mathcal{H}_\tau} = \sum_{k=0}^{\tau_1} \int_{B_{d(a,\delta)}} \Big| \int_{B_a} (T_{r^*} \xi)(p^*)\, \varphi_k p)\, dp \Big|^2 \tau_2(r)\, dr$$

$$\leq c \int_{B_{d(a,\delta)}} \left(\int_{B_a} |(T_{r^*}\xi)(p^*)|^2 \, dp \right) \tau_2(r)\, dr \leq c \int_{B_{d(a,\delta)}} \| T_{r^*}\xi \|_2^2 \, \tau_2(r)\, dr$$

$$\leq \|\xi\|_2^2 \, c \int_{B_{d(a,\delta)}} \tau_2(r)\, dr = c_\delta^2 \|\xi\|_2^2 \quad (\xi \in L_2(B_\delta, dp)). \tag{2.28}$$

IV. We show that the vector function $\mathcal{D}(Q, \Xi) \ni \varphi \mapsto T_\xi \varphi \in \mathcal{D}(Q, \Xi)$ is continuous for each fixed $\xi \in L_{2,0}$. One should prove that for any $\tau = (\tau_1, \tau_2) \in \mathrm{T}$, one can find $\tau' = (\tau_1', \tau_2') \in \mathrm{T}$ such that

$$\| \xi * \varphi \|_{\mathcal{H}_\tau} \leq c \| \varphi \|_{\mathcal{H}_{\tau'}} \quad (c > 0;\ \varphi \in \mathcal{D}(Q, \Xi)).$$

Let us fix some $\tau \in \mathrm{T}$. Using (2.26) and (2.15), we get

$$|((D_k(\xi * \varphi))\, \tau^{1/2}, g)_{L_2}| = |\int (D_k\varphi * \xi)(p)\, \overline{g(p)} \tau_2^{1/2}(p)\, dp|$$

$$= |\int\int (D_k\varphi)(r)\,(T_r\xi)(p)\, dr\, \overline{g(p)} \tau_2^{1/2}(p)\, dp|$$

$$\leq \int |(D_k\varphi)(r)| \int |(T_r\xi)(p) \overline{g(p)}| \tau_2^{1/2}(p)\, dp\, dr$$

$$\leq \|g\|_{L_2} \int |(D_k\varphi)(r)| \left(\int |(T_r\xi)(p)|^2 \tau_2(p)\, dp \right)^{1/2} dr$$

for every $k = 0, \ldots, \tau_1$ and an arbitrary function $g \in L_2$.

The finiteness of the function $(T_x\xi)(\cdot) = (x * \xi)(\cdot)$ for $x \in L_{2,0}$ and (2.13) imply that $(T_r\xi)(\cdot)$ is finite for any $r \in Q$ (for more details, see step II of the proof of Theorem 2.3). This and (2.28) yield that

$$\eta(r) = \int |(T_r\xi)(p)|^2 \tau_2(p)\, dp$$

is locally summable. Therefore, one can choose a continuous weight τ_2' so that

$$\int \eta(r)\,(\tau_2'(r))^{-1}\, dr = c^2 < \infty.$$

This means that the above estimate can be continued as follows

$$|((D_k(\xi * \varphi))\, \tau^{1/2}, g)_{L_2}|$$

$$\leq \|g\|_2 \int |(D_k\varphi)(r)|\, (\tau_2'(r))^{1/2}\, \eta^{1/2}(r)\, (\tau_2'(r))^{-1/2}\, dr$$

$$\leq c \| g \|_2 \Big(\int |(D_k \varphi)(r)|^2 \, \tau_2'(r) \, dr \Big)^{1/2} = c \| g \|_2 \| D_k \varphi \|_{L_2(Q, \tau_2'(r)\,dr)}.$$

Since $g \in L_2$ is arbitrary, this implies that

$$\| D_k (\xi * \varphi) \| \phi \|_{L_2(Q, \tau_2(r)\,dr)} = \| D_k (\xi * \varphi) \, \tau_2^{1/2} \|_2 \leq c \| D_k \varphi \|_{L_2(Q, \tau_2'(r)\,dr)}. \quad (2.29)$$

To complete the proof, it remains to sum up the squared inequality (2.29)[45] over $k = 0, \ldots, \tau_1$; $\tau' = (\tau_1, \tau_2')$.

V. The fact that $\mathcal{D}(Q, \Xi) \subset C_0^\infty(Q, \Xi)$ is invariant with respect to the conjugation operator has already been noted when proving (2.26). It is also easy to understand that this operator is continuous in the topology of $\mathcal{D}(Q, \Xi)$. In fact, by virtue of the last equality in (2.26), the relation (2.27) implies that

$$\| f^* \|_{\mathcal{H}_\tau} \leq \| f \|_{\mathcal{H}_{\tau^*}}$$

for any $f \in C_0^\infty(Q, \Xi)$, where $\tau^* = (\tau_1, \tau_2(p))^* = (\tau_1, \tau_2(p^*))$. Since $Q \ni p \mapsto \tau_2(p^*)$ is again a weight, we have $\tau^* \in \mathrm{T}$, and this yields the required continuity.

We now proceed to the proof of nuclearity. It is based on the following lemma.

Lemma 2.8. *Assume that* $\beta(p) \geq \alpha(p) \geq 1$ $(p \in Q)$ *are two continuous weights and* $\xi \in L_{2,0}$ *is one of the functions from the sequence* $(\xi_n)_{n=1}^\infty = \Xi$ *introduced above. Denote by L and W the Hilbert spaces which are the completions of* $C_0^\infty(Q, \Xi)$ *with respect to the scalar products*

$$(f, g)_L = \int f(p) \overline{g(p)} \alpha(p) \, dp,$$

and

$$(f, g)_W = \int f(p) \overline{g(p)} \beta(p) \, dp$$

$$+ \int (T_\xi^{-1} f)(p) \overline{(T_\xi^{-1} g)(p)} \, \beta(p) \, dp \quad (f, g \in C_0^\infty(Q, \Xi)), \quad (2.30)$$

respectively. It is stated that β *can be chosen so that the imbedding* $W \subset L$ *is quasinuclear.*

Proof. The existence of the imbedding $W \subset L$ and the fact that this imbedding is topological are established just as in the case of the spaces $\mathcal{H}_\tau$. Clearly, $L \subset$

[45] Note that the estimate (2.29) is of independent interest.

$L_2(Q, \alpha(p)\,dp)$. Being closed, the mapping $C_0^\infty(Q, \Xi) \ni f \mapsto \beta^{1/2} f = \varphi$ transforms L into the completion H_0 of the collection Φ of these functions φ with respect to the scalar product in $L_2(Q, \delta(p)\,dp)$, where $\delta = \alpha\beta^{-1}$ $(H_0 \subset L_2(Q, \delta(p)\,dp))$, whereas W is transformed into the completion H_+ of the same collection of functions with respect to

$$(\varphi, \psi)_{H_+} = \int \varphi(p)\,\overline{\psi(p)}\alpha(p)\,dp + \int (\beta^{1/2} T_\xi^{-1/2}\, \beta^{-1/2}\varphi)(p)\,\overline{(\beta^{1/2}\, T_\xi\, \beta^{-1/2}\,\psi)}dp. \quad (2.31)$$

It suffices to establish the quasinuclearity of the imbedding $H_+ \subset H_0$ for the proper choice of β. We take H_0 and H_+ as a zero and positive space, respectively, and construct a chain $H_- \supset H_0 \supset H_+$. Let us calculate the operator $I: H_0 \to H_+$ associated with this chain. To do this, we denote by $G \subset L_2$ the completion of Φ with respect to the scalar product in L_2. Since T_ξ^{-1} transforms $C_0^\infty(Q, \Xi)$ into itself, the operator $\Phi \ni \varphi \mapsto \beta^{1/2} T_\xi^{-1} \beta^{-1/2}\varphi$ preserves Φ and, hence, can be considered as a densely defined operator acting in G whose inverse operator exists and is bounded. We denote the closure of it by S. It follows from (2.31) that $H_+ = \mathfrak{D}(S) \subset G$, and the scalar product in H_+ coincides with the scalar product of the graph of S. For I we have

$$(\varphi, \psi)_{H_0} = (\delta\varphi, \psi)_G = (I\varphi, \psi)_{H_+} = (I\varphi, \psi)_G + (SI\varphi, S\psi)_G \quad (\varphi \in H_0,\ \psi \in H_+).$$

Since $\delta\varphi \in G$, this equality yields

$$SI\varphi \in \mathfrak{D}(S^*), \quad \delta\varphi = (1 + S^*S)I\varphi \quad (\varphi \in H_0).$$

Denote by R the bounded inverse operator S^{-1}, we get

$$RR^*\,\delta\varphi = (RR^* + 1)\,I\varphi,$$

i.e.,

$$I\varphi = (1 + RR^*)^{-1} RR^*\,\delta\varphi. \quad (2.32)$$

Let $O: H_+ \to H_0$ be the imbedding operator; it is adjoint to the operator I. Therefore, the quasinuclearity of the imbedding $H_+ \subset H_0$ is equivalent to the quasinuclearity of the operator $I: H_0 \to H_+$. We set $1 + RR^* = A: G \to G$, $B = RR^*\delta: H_0 \to G$, and $\mathfrak{R}(B) = H_+$. Then, by virtue of (2.32), $I = AB$. We show that the quasinuclearity of I follows from the nuclearity of the operator $OB: H_0 \to H_0$. Indeed, if $\psi \in H_+$, then

$$A^{-1}\psi = (1 + RR^*)\,\psi \in H_+,$$

and

$$(A^{-1}\psi, \psi)_{H_+} = ((1+RR^*)\psi, \psi)_G + (S(1+RR^*)\psi, S\psi)_G$$

$$= 2\|\psi\|_G^2 + \|S\psi\|_G^2 + \|R^*\psi\|_G^2 \geq \|\psi\|_{H_+}^2.$$

Thus,

$$(OB\varphi, \varphi)_{H_0} = (A^{-1}I\varphi, I\varphi)_{H_+} \geq \|\varphi\|_{H_+}^2$$

for $\varphi \in H_0$, and this easily implies the required result.

It remains to show that the nonnegative operator OB can be made nuclear by the proper choice of β. The operator R is an extension by continuity in G of the operator inverse to $S \upharpoonright \Phi$, i.e., this is the mapping $\Phi \ni \varphi \mapsto \beta^{1/2} T_\xi \beta^{-1/2}\varphi \in \Phi$. The operator R^* is an extension by continuity in G of the mapping $\Phi \ni \varphi \mapsto \beta^{-1/2} T_{\xi^*}\beta^{1/2}\varphi \in \Phi$, and OB is an extension by continuity in H_0 of the mapping

$$\Phi \ni \varphi \mapsto \beta^{1/2} T_\xi \beta^{-1} T_{\xi^*}\beta^{1/2}\delta\varphi \in \Phi.$$

Let $(\varepsilon_j)_{j=1}^{\infty}$ be an orthonormal basis in H_0, where $\varepsilon_j = \beta^{1/2} e_j \in \Phi$. Then $e_j \in C_0^\infty(Q, \Xi)$ form an orthonormal basis in the space $L \subset L_2(Q, \alpha(p)\,dp)$. This and the fact that the operators T_ξ and $T_\xi^* = T_{\xi^*}$ are generated in the space L_2 by the kernels $K(p, q) = (T_{p^*}\xi)(q^*)$ and $K^*(p, q) = \overline{K(p, q)}$ $(p, q \in Q)$, together imply that

$$\operatorname{Tr}(OB) = \sum_{j=1}^{\infty} (OB\varepsilon_j, \varepsilon_j)_{H_0} = \sum_{j=1}^{\infty} \int \left| \int K(p, s)\,\overline{e_j(p)}\,\alpha(p)\,dp \right|^2 \beta^{-1}(s)\,ds$$

$$\leq \int \left(\int |K(p, s)|^2 \alpha(p)\,dp \right) \beta^{-1}(s)\,ds. \tag{2.33}$$

It follows from (2.12), (N1), and Lemma 2.1 that for $f, g \in L_{2,0}$ the function $Q \ni p \mapsto (T_p f, g)_{L_2}$ is finite and continuous. Therefore, the kernel $K(p, s)$, with s taking its values on a sphere, is uniformly finite in p; moreover,

$$\int |K(p, s)|^2\,ds = \|T_{p^*}\xi\|_2^2 = \|\xi\|_2^2.$$

But then the function $\int_{B_a} |K(p, s)|^2$ of a point $p \in Q$ is bounded and finite for every $a > 0$. Consequently, the function $\int |K(p, s)|^2 \alpha(p)\,dp$ of a point $s \in Q$ is locally summable. This implies that there exists a continuous function $\beta(s) \geq \alpha(s) \geq 1$ $(s \in Q)$ which grows to $+\infty$ as $\rho(s, O) \to \infty$ so quickly that the last integral in (2.33) is convergent.

■

VI. To complete the proof of the theorem, we must show *that for every* $\tau = (\tau_1, \tau_2) \in \mathrm{T}$, *there exists* $\tau' = (\tau_1', \tau_2') \in \mathrm{T}$, $\tau_1 \geq \tau$, *for which the imbedding* $\mathcal{H}_{\tau'} \subset \mathcal{H}_{\tau}$ *is quasinuclear.* This fact is established by the standard procedure with the help of the following two assertions which can be easily proved (see Berezansky [18, Chapter 1, Section 3, Subsection 4, Remark 2 to Lemma 3.1 and Lemma 3.2]):

(a) Assume that E is a linear set on which a linear operator $E \ni f \mapsto Tf \in E$, $\operatorname{Ker} T = \{0\}$ and two scalar products $(\cdot,\cdot)_{H_1}$ and $(\cdot,\cdot)_{G_1}$ are given. Denote by H_1 and G_1 the completions of E with respect to $(\cdot,\cdot)_{H_1}$ and $(\cdot,\cdot)_{G_1}$. Suppose that $H_1 \subset G_1$ topologically. It is stated that if the imbedding $O_1\colon H_1 \to G_1$ is quasinuclear, then the imbedding $O_2\colon H_2 \to G_2$, where H_2 and G_2 are completions of E with respect to the scalar products

$$(f, g)_{H_2} = (Tf, Tg)_{H_1}$$

and

$$(f, g)_{G_2} = (Tf, Tg)_{G_1} \quad (f, g \in E)$$

is also quasinuclear.

(b) Assume that E is a linear set on which the scalar products $(\cdot,\cdot)_{H_k}$ and $(\cdot,\cdot)_{G_k}$ $(k = 1, \ldots, n)$ are given. We set

$$(\cdot,\cdot)_H = \sum_{k=1}^{n} (\cdot,\cdot)_{H_k}, \quad (\cdot,\cdot)_G = \sum_{k=1}^{n} (\cdot,\cdot)_{G_k}.$$

Let H_k, G_k, H, and G be the corresponding completion of E. Suppose that $H_k \subset G_k$ and the imbedding $H_k \mapsto G_k$ are quasinuclear $(k = 1, \ldots, n)$. Then the imbedding $H \mapsto G$ is also quasinuclear.

We make some necessary explanations. Let us set $\tau_1' = \tau_1$ and take $\max(\beta_0(p), \ldots, \beta_{\tau_1}(p))$ $(p \in Q)$ as $\tau_2'(p)$. Here β_k is a weight β from Lemma 2.8 constructed for $\alpha = \tau_2$ and $\xi = \xi_{k+1}$. We fix k. According to this lemma, the imbedding $W \subset L$, where L and W are constructed for $\alpha = \tau_2$ and $\beta = \tau_2'$, $\xi = \xi_{k+1}$, respectively, is quasinuclear. Then we set $E = C_0^\infty(Q, \Xi)$, $T = D_k$, $H_1 = W$, and $G_1 = H$ and apply statement (a). As a result, we establish the quasinuclearity of the imbedding $H_2 \subset G_2$, where H_2 and G_2 are Hilbert spaces obtained by the completion of $C_0^\infty(Q, \Xi)$ with respect to the corresponding scalar products

$$(f, g)_{H_2} = \int (D_k f)(p)\, \overline{(D_k g)(p)}\, \tau_2'(p)\, dp$$

$$+ \int (D_{k+1} f)(p) \overline{(D_{k+1} g)(p)} \, \tau_2'(p) \, dp = (f, g)_{H_{2,k}},$$

and

$$(f, g)_{G_2} = \int (D_k f)(p) \overline{(D_k g)(p)} \, \tau_2(p) \, dp = (f, g)_{G_{2,k}} \quad (f, g \in C_0^\infty(Q, \Xi)), \quad (2.34)$$

generated by (2.30) (note that $D_k T_{\xi_{k+1}}^{-1} = D_{k+1}$). Now apply statement (b). Let us set $E = C_0^\infty(Q, \Xi)$; $(\cdot, \cdot)_{H_k}$ and $(\cdot, \cdot)_{G_k}$ are the scalar products (2.34), respectively $(k = 0, \ldots, \tau_1)$. Statement (b) yields the quasinuclearity of the imbedding $H \subset G$, where H and G are the completions of $C_0^\infty(Q, \Xi)$ with respect to the products

$$(f, g)_H = \sum_{k=0}^{\tau_1} \Big(\int (D_k f)(p) \overline{(D_k g)(p)} \, \tau_2'(p) \, dp$$

$$+ \int (D_{k+1} f)(p) \overline{(D_{k+1} g)(p)} \, \tau_2'(p) \, dp \Big),$$

$$(f, g)_G = \sum_{k=0}^{\tau_1} \int (D_k f)(p) \overline{(D_k g)(p)} \, \tau_2(p) \, dp = (f, g)_{\mathcal{H}_\tau} \quad (f, g \in C_0^\infty(Q, \Xi)).$$

The first of this products is obviously equivalent to $(f, g)_{\mathcal{H}_{\tau'}}$. Hence, the quasinuclearity of the imbedding $H \subset G$ is equivalent to the required quasinuclearity of the imbedding $\mathcal{H}_{\tau'} \subset \mathcal{H}_\tau$.

■

Since $\mathcal{D}(Q, \Xi) \subset C_0^\infty(Q, \Xi) \subset L_{2,0}$, the following statement is valid.

Corollary 1. *The space $\mathcal{D}(Q, \Xi)$ is a subalgebra (with involution $*$) of an h.s. L_1. The mapping*

$$\mathcal{D}(Q, \Xi) \times \mathcal{D}(Q, \Xi) \ni (\varphi, \psi) \mapsto \varphi * \psi \in \mathcal{D}(Q, \Xi)$$

is separately continuous, whereas the mapping $\mathcal{D}(Q, \Xi) \ni \varphi \mapsto \varphi^* \in \mathcal{D}(Q, \Xi)$ is continuous in the topology of $\mathcal{D}(Q, \Xi)$.

■

Let us establish an assertion concerning the action of the operators T_p of a generalized shift in the space $\mathcal{D}(Q, \Xi)$. This result is important for our purposes.

Theorem 2.3. *The space $\mathcal{D}(Q, \Xi)$ is invariant with respect to the action of the operators T_p $(p \in Q)$ of a generalized shift. The vector function*

$$Q \times \mathcal{D}(Q, \Xi) \ni (p, \varphi) \mapsto T_\varphi \varphi \in \mathcal{D}(Q, \Xi)$$

is separately continuous.

Proof.

I. First, we note that the operators T_p $(p \in Q)$ commute with the convolution operators T_ξ $(\xi \in L_{2,0})$. In other words,

$$T_p(\xi * f) = \xi * (T_p f) \quad (f \in L_2).$$

Indeed, according to (2.11) and (2.12), for any $f, g \in L_2$ we have

$$(T_\xi T_p f, g)_{L_2} = (T_p f, T_\xi^* g)_{L_2} = (T_p f, T_{\xi^*} g)_{L_2} = (T_p f, \xi^* * g)_{L_2}$$

$$= (f * (\xi^* * g)^*)(p^*) = (\xi * f * g^*)(p^*) = (T_p T_\xi f, g)_{L_2}.$$

II. Now let us show that $L_{2,0}$ is invariant with respect to T_p $(p \in Q)$. Assume that $\xi \in L_{2,0}$, $\operatorname{supp} \xi \subset B_{a_1}$, and the point p lies inside the ball B_{a_2} $(a_1, a_2 > 0)$. It suffices to establish that $\operatorname{supp}(T_p\xi)(\cdot) \subset B_{d(a_1, a_2)}$. The formula (2.13) implies that the operator T_p is a weak limit of the operators T_{e_n}, where $e_n(r) = m^{-1}(u_n)\,\kappa_{u_n}(r)$, and $(u_n)_{n=1}^\infty$ is a sequence of neighborhoods of the point p contradicting to this point. Beginning with a certain number $u_n \subset B_{a_2}$, by virtue of the Lemma 2.4, we have $\operatorname{supp} T_{e_n}\xi \subset B_{d(a_1, a_2)}$. Consequently, $\operatorname{supp}(T_p\xi)(\cdot) \subset B_{d(a_1, a_2)}$, too.

III. The invariance of $\mathcal{D}(Q, \Xi)$ with respect to T_p $(p \in Q)$ easily follows from step II. In fact, let $\varphi \in \mathcal{D}(Q, \Xi)$, i.e., $\varphi \in \mathfrak{R}((T_{\xi_1} \dots T_{\xi_n}) \restriction L_{2,0})$ $(n \in \mathbb{N})$. By virtue of the commutativity of the operators T_p and T_{ξ_k}, for any n we find

$$T_p\varphi = T_p(T_{\xi_1} \dots T_{\xi_n} D_n \varphi) = T_{\xi_1} \dots T_{\xi_n}(T_p D_n \varphi).$$

But $D_n\varphi \in L_{2,0}$ and $T_p D_n\varphi \in L_{2,0}$ according to what has been proved. Hence, $T_p\varphi \in \mathfrak{R}((T_{\xi_1} \dots T_{\xi_n}) \restriction L_{2,0})$ $(n \in \mathbb{N})$, i.e. $T_p\varphi \in \mathcal{D}(Q, \Xi)$.

IV. Let us show that the vector function $\mathcal{D}(Q, \Xi) \ni \varphi \mapsto T_p\varphi \in \mathcal{D}(Q, \Xi)$ is continuous for every fixed $p \in Q$. Let $\tau = (\tau_1, \tau_2) \in \mathrm{T}$. We choose a weight τ_2' so that the inequality (2.29) holds for $k = 0, \dots, \tau_1$. Since $\varphi \in \mathcal{D}(Q, \Xi)$, we have

$$\varphi = \xi_{k+1} * (T^{-1}_{\xi_{k+1}} \varphi) = (T^{-1}_{\xi_{k+1}} \varphi) * \xi_{k+1},$$

where $T^{-1}_{\xi_{k+1}}\varphi \in L_{2,0}$. According to Step I,

$$T_p\varphi = (T^{-1}_{\xi_{k+1}}\varphi) * T_p\xi_{k+1}.$$

Thus, by virtue of (2.29),

$$\| D_k(T_p\varphi)\|_{L_2(Q,\tau_2(r)dr)} = \| D_k((T^{-1}_{\xi_{k+1}}\varphi) * T_p\xi_{k+1})\|_{L_2(Q,\tau_2(r)dr)}$$

$$\leq c\,\| D_k T^{-1}_{\xi_{k+1}}\varphi\|_{L_2(Q,\tau_2'(r)dr)} = c\,\| D_{k+1}\varphi\|_{L_2(Q,\tau_2'(r)dr)}.$$

If we take the second power of this inequality and sum up over $k = 0, \ldots, \tau_1$, we obtain the desired estimate

$$\| T_p\varphi\|_{\mathcal{H}_\tau} \leq c\,\|\varphi\|_{\mathcal{H}_\tau}.$$

V. We show that the vector function $Q \ni p \mapsto T_p\varphi \in \mathcal{D}(Q,\Xi)$ is continuous for every fixed $\varphi \in \mathcal{D}(Q,\Xi)$. Let $(p_n)_{n=1}^{\infty}$ be a sequence of points from Q which converges to a fixed point $p_0 \in Q$. By virtue of the separability of Q, it suffices to show that

$$\| T_{p_n}\varphi\|_{\mathcal{H}_\tau} \underset{n\to\infty}{\to} \| T_{p_0}\varphi\|_{\mathcal{H}_\tau}$$

for any $\tau = (\tau_1, \tau_2) \in \mathrm{T}$.

We first show that for any $k = 0, \ldots, \tau_1$ the function

$$g_k(p) = \int |D_k(T_p\varphi)(r)|^b\,\tau_2(r)\,dr,$$

where $b > 0$ is fixed, is locally bounded. Indeed, as proved on Step II, if supp $\xi \subset B_{a_1}$ and $p \in B_{a_2}$, then supp $(T_p\xi)(\cdot) \subset B_{d(a_1,a_2)}$. By virtue of Step I and Lemma 2.1, we have

$$|g_k(p)| = \int |D_k(T_p\varphi)(r)|^b\,\tau_2(r)\,dr$$

$$= \int_{B_{d(a_1,a_2)}} |(D_{k+1}\varphi * T_p\xi_k)(r)|^b\,\tau_2(r)\,dr$$

$$\leq \| D_{k+1}\varphi\|_2^b\,\| T_p\xi_{k+1}\|_2^b \int_{B_{d(a_1,a_2)}} \tau_2(r)\,dr$$

$$\leq \| D_{k+1}\varphi \|_2^b \, \| \xi_{k+1} \|_2^b \int_{B_{d(a_1, a_2)}} \tau_2(r)\, dr \quad (p \in B_{a_2}) \tag{2.35}$$

for some $a_1 > 0$. Hence, $g_k(p)$ is locally bounded.

This enables us to pass to the limit under the integral sign

$$\int | D_k (T_{p_n} \varphi)(r) |^2 \tau_2(r)\, dr \underset{n\to\infty}{\rightarrow} \int | D_k (T_{p_0} \varphi)(r) |^2 \tau_2(r)\, dr, \tag{2.36}$$

because, by virtue of (2.35), for $\varepsilon > 0$ the integrals

$$\int (| (D_k (T_{p_n} \varphi)(r) |^2)^{1+\varepsilon} \tau_2(r)\, dr$$

are uniformly bounded in n and are taken, in fact, over the ball $B_{d(a_1, a_2)}$. Summing up (2.36) over $k = 0, \ldots, \tau_1$, we obtain the required relation

$$\| T_{p_n} \varphi \|_{\mathcal{H}_\tau} \underset{n\to\infty}{\rightarrow} \| T_{p_0} \varphi \|_{\mathcal{H}_\tau}.$$

The theorem is thus proved. ■

2.4. Representation of Hypercomplex Systems

Let L_1 be a normal h.s. with a basis Q and a b.i. O. We want to study the representation of this h.s. by bounded normal operators.

Consider the Hilbert space H_0 and a family $A = (A_p)_{p \in Q}$ of bounded normal operators acting in this space. We say that this family is a representation of a h.s. if the following conditions hold :

(i) $A_O = 1$;

(ii) $A_p^* = A_{p^*}$ $(p \in Q)$;

(iii) for every $f \in H_0$ the vector function $Q \ni p \mapsto A_p f \in H_0$ is weakly continuous;

(iv) the main algebraic requirement

$$\int c(\alpha, \beta, r) A_r\, dr = \int_\alpha A_p\, dp \int_\beta A_q\, dq \quad (\alpha, \beta \in \mathfrak{B}_0(Q)) \tag{2.37}$$

holds. We clarify that the integrals in (2.37) are understood in Bochner's sense with the σ-algebra $\mathfrak{B}(Q)$. These are the integrals of the vector function $Q \ni p \mapsto A_p \in \mathfrak{L}(H_0, H_0)$ or of this vector function multiplied by a continuous function. The fact that these integrals exist follow from condition (iii) and the general facts concerning Bochner integrals (here one should take into account that, by virtue of the Banach–Steinhaus theorem, condition (iii) implies that the function $Q \ni p \mapsto \|A_p\|$ is locally bounded (see, e.g., Yosida [1, Chapter 5, Subsection 4, 5]).

A measurable (generally speaking, unbounded) function $Q \ni p \mapsto \chi(p) \in \mathbb{C}^1$ which is nonzero almost everywhere is called a generalized character of an h.s. provided that it is locally integrable and the relation (2.16) holds. Therefore, a generalized character which is not an ordinary one does not correspond to the maximal ideal; on the part of L_1 it generates a multiplicative functional which is not continuous. Repeating the beginning of the proof of Theorem 2.1, we easily find that each generalized character is continuous. Hence, if Q is a compact set, then every generalized character is ordinary. Denote the collection of all generalized characters of our h.s. by X_g; $X \subset X_g \subset \mathbb{C}^Q$. A generalized character χ is called Hermitian if $\chi(p^*) = \overline{\chi(p)}$ $(p \in Q)$. The set of all Hermitian generalized characters is denoted by $X_{g,h}$; and X_h is the collection of all ordinary Hermitian characters.

Theorem 2.4. *Assume that L_1 is a normal h.s. with a basis Q and a b.i. O, and that $Q \ni p \mapsto A_p$ is a representation of L_1 by bounded normal operators. Then the following representation in the form of a spectral integral holds:*

$$A_p = \int_{X_{g,h}} \chi(p)\, dE(\chi) \quad (p \in Q), \tag{2.38}$$

where Q is an RI on a σ-algebra

$$\mathfrak{C}_\sigma(X_{g,h}) = \{\alpha \cap X_{g,h} \mid \alpha \in \mathfrak{C}_\sigma(\mathbb{C}^Q)\},$$

concentrated on a set

$$\{\chi \in X_{g,h} \mid |\chi(p)| \le \|A_p\|,\ p \in Q\}$$

If the function $Q \ni p \mapsto \|A_p\|$ is bounded, then conditions (i)–(iv) are necessary, in order that the representation (2.38) *exist. Moreover, E is concentrated on the set X_h and is given on the σ-algebra $\mathfrak{B}(X_h)$, where $X_h \subset X$ is topologized by the topology of the space of maximal ideals induced from X.*

Proof. We divide the proof into several steps. The existence of the rigging (Steps II–X) is established to a large extent by using arguments similar to those presented in the proof of Theorem 3.8 (Chapter 3). However, here our discussion is based on formula (2.39) given below, but not on the spectral representation (3.30, Chapter 3).

I. By virtue of the equality $c(\alpha, \beta, r) = c(\beta, \alpha, r)$, (2.37) yields

$$\int_\alpha A_p\, dp \int_\beta A_q\, dq = \int_\beta A_q\, dq \int_\alpha A_p\, dp .$$

If we apply this equality to $f \in H_0$ and divide it by $m(\alpha)\, m(\beta)$, then after a proper limit transition, we obtain $A_p A_q = A_q A_p$ $(p, q \in Q)$ (here one should employ condition (iii)). Thus, $(A_p)_{p \in Q}$ is a family of bounded normal commuting operators.

II. Let $\varphi \in \mathcal{D}(Q, \Xi)$. Consider the Bochner integral

$$A_\varphi = \int \varphi(p)\, A_p\, dp \tag{2.39}$$

and a fixed vector $l_1 \in H_0$, $\| l_1 \|_{H_0} = 1$. We construct a linear mapping

$$\mathcal{D}(Q, \Xi) \ni \varphi \mapsto Q_\varphi = A_\varphi l_1 \in H_0.$$

It is stated that there exists a weight τ_2 such that this mapping acts continuously from the space $\mathcal{H}_\tau = \mathcal{H}_{(0,\tau_2)} = L_2(Q, \tau_2(p)\, dp)$ to the space H_0. Indeed, we have already mentioned that the function $Q \ni p \mapsto \| A_p \|$ is locally bounded. We take the weight τ_2 which grows (as $\rho(p, O) \to \infty$) so fast that $\|A_p\| \tau_2^{-1/2}(p) \in L_2$. We now obtain the required inequality

$$\| Q_1' \varphi \|_{H_0} = \| A_\varphi l_1 \|_{H_0} \le \int | \varphi(p) | \, \| A_p \| \, dp$$

$$\left(\int \| A_p \|^2 \tau_2^{-1}(p)\, dp\right)^{1/2} \| \varphi \|_{\mathcal{H}_\tau} \quad (\varphi \in \mathcal{D}(Q, \Xi)). \tag{2.40}$$

III. According to Theorem 2.2, we can choose the index $\tau' = (\tau_1', \tau_2') \in \mathrm{T}$ so that the imbedding $\mathcal{H}_{\tau'} \subset \mathcal{H}_\tau$ is quasinuclear. Fix this τ'. Then Q_1', regarded as a mapping of a dense set from $\mathcal{H}_{\tau'}$ to H_0, is quasinuclear; it is the product of a quasinuclear imbedding $\mathcal{H}_{\tau'} \subset \mathcal{H}_\tau$ and the continuous mapping constructed on Step II. Closing it by continuity, we obtain the Hilbert–Schmidt operator $Q_1 \in \mathcal{S}(\mathcal{H}_{\tau'}, H_0)$.

IV. The operator Q_1 is nonzero. Indeed, assume the contrary. Then

$$0 = (Q_1\, \varphi, f)_{H_0} = \int \varphi(p)\, (A_p l_1, f)_{H_0}\, dp$$

for any $\varphi \in \mathcal{D}(Q, \Xi)$ and $f \in H_0$. Taking into account that $\mathcal{D}(Q, \Xi) = C_0^\infty(Q, \Xi)$ is dense in L_2 (Lemma 2.7), that the function $Q \ni p \mapsto (A_p l_1, f)_{H_0}$ is continuous, and that f is arbitrary, we find that $A_p l_1 = 0$ $(p \in Q)$. But this contradicts the requirement $A_O = 1$.

V. Consider the kernel

$$\{\varphi \in \mathcal{H}_{\tau'} \,|\, Q_1\varphi = 0\} = \operatorname{Ker} Q_1$$

of the operator Q_1. Let F be the orthogonal complement of it with respect to $\mathcal{H}_{\tau'}$, and let $R = Q_1 \restriction F$ (by virtue of Step IV, $\dim F > 0$). By choosing an orthonormal basis $(\varepsilon_j)_{j=1}^{\infty}$ in the space $\mathcal{H}_{\tau'}$ so that the vectors of it belong to this kernel and F_1, we obtain $\|R\| \le |R|\,|Q_1| = c < \infty$ (c does not depend on the choice of the unit vector l_1). On the range of values $\mathcal{R}(R) = \mathcal{R}(Q_1) \subset H_0$ there exists the closed inverse operator R^{-1}. Moreover,

$$\|R^{-1}f\|_{\mathcal{H}_{\tau'}} \ge c^{-1}\|f\|_{H_0} \quad (f \in \mathcal{R}(R)).$$

Let us transform $\mathcal{R}(R) \subset H_0$ into a Hilbert space $H_{+,1} \subset H_0$ by setting

$$(u, v)_{H_{+,1}} = \kappa_1^{-1}(R_1^{-1}u, R^{-1}v)_{\mathcal{H}_{\tau'}} \quad (u, v \in \mathcal{R}(R)),$$

where $\kappa_1 \ge c^2$ is a fixed number. Thus, $H_{+,1}$ is isometric to $\mathcal{H}_{\tau'}/\operatorname{Ker} Q_1$. Clearly, $\dim H_{+,1} > 0$, $H_{+,1}$ is complete and $\|\cdot\|_{H_{+,1}}$.

The imbedding $O_1 \colon H_{+,1} \to H_0$ is quasinuclear and $|O_1| = \kappa_1^{-1/2}c_1$. In fact, let $(\varepsilon_j)_{j=1}^{\infty}$ be an orthonormal basis in F. Then $(\kappa_1^{-1/2}R\varepsilon_j)_{j=1}^{\infty}$ is an orthonormal basis in $H_{+,1}$. For this basis, we have

$$|O_1|^2 = \sum_{j=1}^{\infty} \|\kappa_1^{-1/2}R\,\varepsilon_j\|_{H_0}^2 = \kappa_1^{-1}|R|^2 = \kappa_1^{-1}c^2.$$

VI. Taking into account that $\mathfrak{D}(Q, \Xi) \subset \mathcal{H}_{\tau'}$ topologically, that $\operatorname{Ker} Q_1' = (\operatorname{Ker} Q_1) \cap \mathfrak{D}(Q, \Xi)$, and that $Q_1' \in \mathcal{L}(\mathfrak{D}(Q, \Xi), H_0)$ by (2.40), we can say that

$$\mathfrak{D}(Q, \Xi)/\operatorname{Ker} Q_1' \subset \mathcal{H}_{\tau'}/\operatorname{Ker} Q_1.$$

topologically. Denote by D_1 the image of $\mathfrak{D}(Q, \Xi)/\operatorname{Ker} Q_1'$ in H_0 under the action of the operator Q' which is dropped on classes. This image is topologized by the topology of the preimage, i.e., $D_1 = \mathcal{R}(Q_1') \subset \mathcal{R}(Q_1) \subset H_0$, where some elements of the form $A_\varphi l_1$ are assembled into classes for different φ. Obviously, $D_1 \subset H_{+,1}$ topologically. As a result, we get the chain

$$H_0 \supset H_{+,1} \supset D_1, \tag{2.41}$$

where the imbedding $D_1 \subset H_{+,1}$ is topological and the imbedding $H_{+,1} \subset H_0$ is quasinuclear. Furthermore, it follows directly from the definition of D_1 that it is separable. Taking into account that

$$\mathcal{D}(Q, \Xi) / \operatorname{Ker} Q_1' = \{l \in \mathcal{D}'(Q, \Xi) \mid l(\varphi) = 0, \ \varphi \in \operatorname{Ker} Q_1'\}$$

and $\mathcal{D}'(Q, \Xi)$ is separable with respect to the weak topology, we conclude that D_1' is also separable with respect to the weak topology.

Note that this step of the proof, where we introduce the space D_1, is distinct from Step III in the proof of Theorem 3.8 in Chapter 3.

VII. Let us show that for any $p \in Q$, the operator $A_p \upharpoonright D_1$ satisfies the following inclusion

$$A_p \upharpoonright D_1 \in \mathcal{L}(D_1, D_1) \subset \mathcal{L}(D_1, H_{+,1})$$

(the last inclusion is a consequence of the fact that the imbedding $D_1 \subset H_{+,1}$ is topological). This fact follows from the equality which will be established somewhat later (see Step XI), namely

$$A_p A_\varphi = A_{T_p \varphi} \quad (\varphi \in \mathcal{D}(Q, \Xi); \ p \in Q). \tag{2.42}$$

In fact, let $p \in Q$ be fixed, let $u \in D_1$, and let $Q' \varphi = A_\varphi l_1$ ($\varphi \in \mathcal{D}(Q, \Xi)$) be an element from the class u. The operator $A_p \upharpoonright D_1$ transforms the class u into the class $(A_p \upharpoonright D_1)\, u \in D_1$ generated by the element $A_p A_\varphi l_1 = A_{T_p \varphi} l_1$. Since the topology on classes is induced by the topology $\mathcal{D}(Q, \Xi)$, the continuity of the mapping

$$D_1 \ni u \mapsto (A_p \upharpoonright D_1)\, u \in D_1$$

follows from the continuity of the mapping

$$\mathcal{D}(Q, \Xi) \ni \varphi \mapsto T_p \varphi \in \mathcal{D}(Q, \Xi)$$

(Theorem 2.3). Note also that the construction given here is correct, because $\operatorname{Ker} Q_1'$ is invariant under the action of the operator T_p.

The inclusion

$$A_p^* \upharpoonright D_1 \in \mathcal{L}(D_1, D_1) \subset \mathcal{L}(D_1, H_{+,1})$$

holds, since $A_p^* = A_{p^*}$ $(p \in Q)$.

VIII. For every $u \in D_1$ the vector function $Q \ni p \mapsto A_{p*} u = A_p^* u$ is continuous.

Taking into account (2.42) and the argument given on Step VII, it suffices to establish the continuity of the vector function $Q \ni p \mapsto T_{p*}\varphi \in \mathcal{D}(Q, \Xi)$ for any $\varphi \in \mathcal{D}(Q, \Xi)$. But this is again the statement of Theorem 2.3 (here we take into account the fact that the mapping $Q \ni p \mapsto p^* \in Q$ is continuous). Clearly, the vector function $Q \ni p \mapsto \overset{*}{A}_p u \in H_{+,1}$ is also continuous.

IX. Consider the chain (2.41). If $H_{+,1}$ is dense in H_0, then the construction terminates on this step. If this is not true, then we choose a vector $l_2 \in H_0$, $\| l_2 \|_{H_0} = 1$, orthogonal in H_0 to $H_{+,1}$. Repeating Steps II–VII for this vector, we obtain the chain

$$H_0 \supset H_{+,2} \supset D_2,$$

possessing the properties analogous to those discussed on Steps VI–VII. It is essential that $H_{+,2}$ is orthogonal to $H_{+,1}$ in H_0. This fact follows from the next relations which will be proved on Step XI, namely,

$$\overset{*}{A}_\varphi = A_{\varphi^*}, \quad A_\varphi A_\psi = A_{\varphi * \psi} \quad (\varphi, \psi \in \mathcal{D}(Q, \Xi)). \tag{2.43}$$

Indeed, it suffices to show that the vectors $A_\varphi l_1$ and $A_\varphi l_2$ are orthogonal in H_0. But, by virtue of (2.4), we have

$$(A_\varphi l_1, A_\psi l_2)_{H_0} = (\overset{*}{A}_\psi A_\varphi l_1, l_2)_{H_0} = (A_{\psi^* * \varphi} l_1, l_2)_{H_0} = 0,$$

since $\psi^* * \varphi \in \mathcal{D}(Q, \Xi)$.

X. By continuing this construction, we obtain the sequence of chains (infinite, for definiteness)

$$H_0 \supset H_{+,m} \supset D_m \quad (m \in \mathbb{N}),$$

possessing properties similar to those established on Steps VI–VIII. Moreover, the orthogonal sum $\oplus_{m=1}^{\infty} H_{+,m}$ in H_0 is dense in H_0. We regard this sum as H_+ and assume that $\sum_{m=1}^{\infty} \kappa_m^{-1} < \infty$. The last statement implies the quasinuclearity of the imbedding $H_+ \subset H_0$. The space D is constructed as a collection of all finite sequences $u = (u_m)_{m=1}^{\infty}$, where $u_m \in D_m$, with convergence, which means the uniform finiteness and convergence of the coordinate u_m (in D_m) for any $m \in \mathbb{N}$. It follows from arguments similar to those given in Steps IX–X of the proof of Theorem 3.8 in Chapter 3, that finally we shall get a quasinuclear chain

$$H_- \supset H_0 \supset H_+ \supset D \tag{2.44}$$

connected with the family A in the standard way, where D is a linear topological separable space for which D' is also separable in the weak topology.

By virtue of Step VII, the chain (2.44) has an important additional property, namely, D is invariant with respect to A_p $(p \in Q)$ and $A_p \upharpoonright D \in \mathcal{L}(D, D)$ (and, hence, as usually, $A_p \upharpoonright D \in \mathcal{L}(D, H_+)$). Clearly, the same is also true for $A_p^* = A_{p^*}$. Moreover, Step VIII implies that the vector function $Q \ni p \mapsto A_p u \in H_+$ is weakly continuous for any $u \in D$.

XI. Let us now prove the relations (2.43) and (2.42). They are valid even for $\varphi, \psi \in L_{2,0} \supset \mathcal{D}(Q, \Xi)$, and we shall prove them for functions of this sort. Taking into account (2.39), the equality $A_p^* = A_{p^*}$ $(p \in Q)$, the invariance of the m. measure with respect to $*$, and the properties of the Bochner integral, we obtain

$$A_\varphi^* = \int \overline{\varphi(p)}\, A_p^* dp = \int \overline{\varphi(p)} A_{p^*}\, dp = \int \varphi^*(p) A_p\, dp = A_{\varphi^*}.$$

We establish the second equality in (2.43). The relation (2.37) can be rewritten in the form $A_{\kappa_{\alpha^*} \kappa_\beta} = A_{\kappa_\alpha} A_{\kappa_\beta}$. This yields the required equality for step-functions φ, ψ. Approximating arbitrary functions $\varphi, \psi \in L_{2,0}$ by step functions (in L_2) and passing to the limit in this equality, we obtain the desired result. Here one must exploit Lemma 2.1 and the fact that it is possible to pass to the limit under the Bochner integral sign.

We prove (2.42). Let $\psi \in L_{2,0}$. Then, according to (2.39), (2.15), and the second equality in (2.43), we have

$$\int A_{T_p \varphi} \psi(p) = \int \Big(\int (T_p \varphi)(r) A_r\, dr \Big)\, \varphi(p)\, dp$$

$$= \int (\varphi * \psi)(r) A_r\, dr = \int \psi(p) A_p\, dp \int \varphi(q) A_p\, dq.$$

Since ψ is arbitrary, this yields (2.42).

XII. Henceforth, we shall need the following fact which is similar to that presented at the end of Step X: $A_\varphi \upharpoonright D \in \mathcal{L}(D, D)$ for any $\varphi \in \mathcal{D}(Q, \Xi)$. This inclusion follows from the fact that $A_\varphi \upharpoonright D_1 \in \mathcal{L}(D_1, D_1)$. The latter is proved in the same way as Step VII; one should only employ the second equality in (2.43) and Theorem 2.2 instead of (2.42).

XIII. We have established the existence of the quasinuclear chain (2.44) which has the properties described in Steps X and XII and is standardly connected with the family $A = (A_p)_{p \in Q}$ of bounded normal operators. We proceed to the proof of the representation (2.38).

For this purpose, we exploit spectral Theorem 2.7 of Chapter 3 in the same manner as in Subsection 1.1. The chain (2.44) fits for this, since the space D is such that the assertion of Remark 1 to Subsection 2.10 of Chapter 3 holds. Thus, to prove the representation (2.30), it suffices to show that the generalized spectrum $g(A)$ lies in $X_{g,h}$. It consists of the eigenvalues $\lambda(\cdot) \in \mathbb{C}^Q$ satisfying the equalities

$$(\xi, A_p^* u)_{H_0} = \lambda(p)\,(\xi, u)_{H_0}, \quad (\xi, A_p u)_{H_0} = \overline{\lambda(p)}(\xi, u)_{H_0} \quad (p \in Q;\ u \in D) \tag{2.45}$$

with a corresponding generalized joint eigenvector $0 \neq \xi \in H_-$.

By virtue of Step X and Subsection 2.9 (Chapter 3), each eigenvalue λ belongs to $C(Q)$. Further, since $A_p^* = A_{p^*}$, the relations (2.45) imply that this operator is Hermitian : $\lambda(p^*) = \overline{\lambda(p)}$ $(p \in Q)$. By setting $p = O$ in (2.45), we find that $\lambda(O) = 1$ and, thus, $\lambda(\cdot) \neq 0$.

Let us prove that $\lambda(\cdot)$ is a generalized character of a h.s. It is necessary to establish the validity of the equality (2.16). In Step XIV, we shall show that relations (2.45) (they are equivalent) can be "integrated over p" or, more exactly, that the following equality holds

$$(\xi, A_\varphi u)_{H_0} = \int \overline{\lambda(p)\varphi(p)}\, dp\, (\xi, u)_{H_0} \quad (\varphi \in \mathcal{D}(Q, \Xi);\ u \in D) \tag{2.46}$$

((2.46) is meaningful by virtue of Step XII). Taking into account that D is invariant with respect to A_φ (Step XII) and employing the second relation (2.43) and (2.46), we get

$$\begin{aligned}(\xi, A_{\varphi*\psi} u)_{H_0} &= (\xi, A_\varphi A_\psi u)_{H_0} = \int \overline{\lambda(p)\varphi(p)}\, dp\, (\xi, A_\psi u)_{H_0} \\ &= \int \overline{\lambda(p)\varphi(p)}\, dp \int \overline{\lambda(q)\psi(q)}\, dq\, (\xi, u)_{H_0}\end{aligned}$$

for any $\varphi, \psi \in \mathcal{D}(Q, \Xi)$. On the other hand, according to (2.46) and (2.3), we have

$$\begin{aligned}(\xi, A_{\varphi*\psi} u)_{H_0} &= \int \overline{\lambda(r)(\varphi * \psi)(p)}\, dr\, (\xi, u)_{H_0} \\ &= \int \overline{\lambda(r)} \left(\int \overline{\varphi(p)} d_p \left(\int \psi(q)\, d_q c(\varepsilon_p, \varepsilon_q, r) \right) \right) dr\, (\xi, u)_{H_0},\end{aligned}$$

whence

$$\int \lambda(r) \left(\int \varphi(p) d_p \left(\int \psi(q) d_q c(\varepsilon_p, \varepsilon_q, r) \right) \right) = \int \lambda(p)\varphi(p)\, dp \int \lambda(q)\varphi(q)\, dq$$

$$(\varphi, \psi \in \mathcal{D}(Q, \Xi)). \tag{2.47}$$

Approximating the functions κ_α and κ_β $(\alpha, \beta \in \mathfrak{B}_0(Q))$ by the functions φ and ψ in

L_2 and passing to the limit in (2.47), we obtain the equality (2.16) for $\lambda(\cdot)$ (one should use Lemmas 2.7 and 2.1).

XIV. We establish (2.46). Fix $u \in D$. The weak continuity of the vector function $Q \ni p \mapsto A_p u \in H_+$ yields the local boundedness of the function $Q \ni p \mapsto \|A_p u\|_{H_+}$ and the existence of the Bochner integral $\int \varphi(p)\, A_p u\, dp \in H_+$ for any $\varphi \in \mathcal{D}(Q, \Xi) \subset L_{2,0}$ in the sense of norm $\|\cdot\|_{H_+}$. At the same time, this integral clearly equals $A_\varphi u$. Multiplying the second equality (2.45) by $\varphi(p)$ and integrating over p, we obtain (2.46). This is so because, due to the character of the convergence of the Bochner integral, it can be permuted with the action of $\xi \in H_- = (H_+)'$.

XV. To complete the proof of the theorem, we must make several simple remarks. First, since the operators A_p are bounded, we can take the product of circles $\times_{p \in Q} \{|z| \le \|A_p\|\}$ instead of $\mathbb{C}^Q$ (see Subsection 2.8 of Chapter 3), and this implies that E is concentrated on the set $\{\chi \in X_{g,h} \mid |\chi(p)| \le \|A_p\|, \ p \in Q\}$.

Further, if the function $Q \ni p \mapsto \|A_p\|$ is bounded by a number c, then, by virtue of the above remark, E is concentrated on the generalized characters for which $\|\chi(p)\| \le c$ $(p \in Q)$, i.e., on the ordinary characters. In this case, the equality $\mathfrak{C}_\sigma(X_h) = \mathfrak{B}(X_h)$ is established quite analogously with the corresponding equality in Theorem 1.2. The necessity of conditions (i)–(iv) follows directly from the representation (2.38).

■

Remark 1. It may seem that the representation of h.s. by the bounded normal operators studied above can be reduced to the theory of representations of a commutative normalized algebra L_1 if, for a function $Q \ni p \mapsto A_p \in \mathcal{L}(H_0, H_0)$, we introduce the "integrated" function

$$L_1 \ni x \mapsto A_x = \int x(p)\, A_p\, dp \in \mathcal{L}(H_0, H_0).$$

However, the latter should be defined for all $x \in L_1$. Therefore, in order that the integral exists, the behavior of the norm $\|A_p\|$ cannot be arbitrary. In the case when this norm behaves arbitrarily, we proceed to representation (2.38) with integration over the set of generalized characters. If the function $Q \ni p \mapsto \|A_p\|$ is bounded, then A_x exists for any $x \in L_1$, and we are in the situation of the representation of a normalized algebra described in Subsection 1.5; this is justified by the second part of Theorem 2.4, too.

■

2.5. Examples

We give several examples of the h.s. for which Theorem 2.4 on representations holds. We shall not repeat the formulation of the theorem in each case, since this can be easily done by the reader. The representation of the examples is concise (the last of these is not connected with an h.s.).

Example 2.3. If we apply Theorem 2.4 to an h.s. which is a group algebra of a locally compact commutative group $G = Q$ (Example 2.1), then we obtain the Sz.-Nagy-Hille theorem and Stone's theorem for G. These theorems were considered in Remark 2.3 to Section 1. Recall that here all the characters are Hermitian, i.e., $\chi(p^{-1}) = \overline{\chi(p)}$ $(p \in Q)$.

Example 2.4. Consider an even subsystem E_1 of an h.s. from Example 2.3, i.e., the h.s. from Example 2.2. One can easily understand that the representation of it is a family of bounded selfadjoint operators $(A_p)_{p \in Q}$ acting in H_0 such that the function $Q \ni p \mapsto A_p \in \mathfrak{L}(H_0, H_0)$ is even, i.e., $A_{p^{-1}} = A_p$ $(p \in Q)$, $A_e = 1$, for any $f \in H_0$ the vector function $Q \ni p \mapsto A_p f \in H_0$ is weakly continuous, and the equality

$$A_p A_q = \tfrac{1}{2}(A_{pq^{-1}} + A_{pq}) \quad (p, q \in Q = G) \tag{2.48}$$

holds. According to property III in Subsection 2.2, the characters of an h.s. E_1 are the functions

$$\operatorname{Re}\chi(p) = \tfrac{1}{2}(\chi(p) + \chi(p^{-1})) \quad (p \in Q),$$

when we write the representation for A_p, the integral in (2.38) is taken over the collection of these functions.

Example 2.5 (center of the group algebra of a compact group). Assume that G is a group of this sort which is, in addition, separable and possesses a countable base of neighborhoods. For every $g \in G$, we construct a class of adjoint elements $\{h^{-1} g h \in G \mid h \in G\} = p(g) = p$ which contains it. The whole of G is split into these classes, and their collection Q becomes a compact if we induce on Q the topology from G by the mapping

$$G \ni g \mapsto p(g) \in Q. \tag{2.49}$$

Consider the group algebra $L_1(G) = L_1(G, \mathfrak{B}(G), \mu)$ of this group (μ is an invariant measure). The operation of multiplication is given by

$$(\xi * \eta)(g) = \int_G \xi(gh^{-1})\, \eta(h)\, d\mu(h) \quad (\xi, \eta \in L_1(G);\ g \in G). \tag{2.50}$$

It can be shown that the center Z_1 of $L_1(G)$ (i.e., the collection of all $\zeta \in L_1(G)$ such that $\zeta * \xi = \xi * \zeta$ for any $\xi \in L_1(G)$), consists of the functions constant on the classes of adjoint elements and, hence, may be regarded as $L_1 = L_1(Q, \mathfrak{B}(Q), m)$ where a measure m is an image of μ under the mapping (2.49). The space $L_1 = Z_1$ with the operation (2.50) is an h.s. with a compact basis Q whose s. measure has the form

$$c(\alpha, \beta, r) = (\kappa_{p^{-1}(\alpha)} * \kappa_{p^{-1}(\beta)})(g) \quad (\alpha, \beta \in \mathfrak{B}(Q);\ g \in r \in Q).$$

Using this s. measure, convolution (2.50) can be written in the form (2.3). This h.s. is normal and has a b.i., namely, if $p = p(g)$, then $p^* = p(g^{-1})$ $(g \in G)$ by definition; $\mathrm{O} = e = p(e)$. The generalized characters of it are ordinary. According to Theorem 2.1, there exists a countable number of these characters (or a finite number, if G is finite). They are Hermitian and have the form $\chi(p) = n^{-1}(\theta)\,\theta(g)$ $(g \in r \in Q)$, where Q is a character of an irreducible $n(\theta)$-dimensional representation of the group G.

For the representation A of this h.s. by normal operators, (2.38) holds with $X_{g,h}$ being a countable set of characters of irreducible representations of G.

We can also investigate the even subsystem of this h.s.

Example 2.6 (an algebra of orbital functions).

Let G be the same compact group as in the previous example, and let H be a closed subgroup of it. The group G is split into the set $\Gamma = G/H$ of left co-sets

$$\{fg \in G \mid h \in H\} = fH = \gamma_f = \gamma \quad (f \in G).$$

This set can be regarded as a uniform space with a group of motion, namely, for any $g \in G$, we set $\Gamma \ni \gamma = \gamma_f \mapsto g\gamma = \gamma_{gf} \in \Gamma$. Inducing the topology from G to Γ by the mapping $G \ni f \mapsto \gamma_f \in \Gamma$, we transform Γ into a compact on which G acts continuously. The action of this group of motion is always transitive, namely, for any φ, $\psi \in \Gamma$, one can find $g \in G$ such that $\psi = g\varphi$. A homogeneous space Γ is called symmetric if, for any φ and $\psi \in \Gamma$, there exists a motion $g \in G$ which exchanges the position of these elements: $g\varphi = \psi$, $g\psi = \varphi$.

We now consider the motions $h \in H \subset G$. The point (class) $\gamma_e = H = \mathrm{O} \in \Gamma$ is fixed under the motions $h \in H$. Other points $\varphi \in \Gamma$ move along the orbits

$$\{h\varphi \in \Gamma \mid h \in H\} = p(\varphi) = p$$

under the action of $h \in H$, and Γ is split into the collection Q of these orbits. By analogy with the above argument, we can induce in Q the topology from Γ by means of the mapping $\Gamma \ni \varphi \mapsto p(\varphi) \in Q$, and this topology transforms Q into a compact. Note that the subgroup H and the point O are rigidly connected, i.e., $H = \{g \in G \mid g\mathrm{O} = \mathrm{O}\}$. H is called a stationary subgroup which corresponds to O.

One can proceed to the compact Q of orbits by considering the decomposition of the group G into dual co-sets with respect to H, namely, $Q = G /\!/ H$, where classes are understood as sets

$$\{h_1 g h_2 \in G \mid h_1, h_2 \in H\} = HgH = p(g) = p \quad (g \in G).$$

Clearly, this is just the construction described above. The (same as above) topology in Q is induced from G by the mapping

$$G \ni g \mapsto p(g) \in Q. \tag{2.51}$$

A typical example of the scheme under consideration can be described as follows: Γ is a unit sphere in $\mathbb{R}^3$, G are all possible rotations of this sphere (they allow to move each point φ on the sphere to any other point on it in many different ways). O is the North pole, H are all the movements (rotations) which preserve O fixed. Orbits are all parallels on this sphere, and Q is one-half of a certain meridian. The uniform space Γ is symmetric, i.e., for any two points on the sphere, there exists a motion (from G) which permute their places.

The same operation as in Example 2.5 can be applied to the group algebra $L_1(G) = L_1(G, \mathfrak{B}(G), \mu)$ of the group G. One must only take the mapping (2.51) just constructed instead of (2.49): Let the measure m be an image of μ under the mapping (2.51). The space $L_1 = L_1(Q, \mathfrak{B}(Q), m)$ may be regarded as a subspace K_1 of the space $L_1(G)$ consisting of functions constant on dual co-sets. We must verify that $L_1 = K_1$ is commutative with respect to the operation (2.50). If we assume, as above, that

$$c(\alpha, \beta, r) = (\kappa_{p^{-1}(\alpha)} * \kappa_{p^{-1}(\beta)})(g) \quad (\alpha, \beta \in \mathfrak{B}(Q);\ g \in r \in Q),$$

where p is the mapping (2.51), then, by using this s. measure, the convolution (2.50) of functions from $L_1 = K_1$ can be written in the form (2.3), whereas L_1 transforms into an h.s. with a compact basis Q. Functions on Q, i.e., functions on Q taking constant values on dual co-sets, are called orbital, and the h.s. constructed is called an algebra of orbital functions.

This h.s. is Hermitian — the mapping

$$Q \ni p = p(g) \mapsto p^* = p(g^{-1}) \in Q$$

should be taken as involution (for any $g \in G$, the classes HgH and $Hg^{-1}H$ coincide). It has the b.i. $O = p(e) = H$. Generalized characters of it are ordinary. According to Theorem 2.1, there are countably many (or finitely many), characters of this sort. They are real-valued and coincide with so-called spherical functions, i.e., functions $0 \neq \chi \in C(D)$ for which the following equality holds

$$\int_H \chi(g_1 h g_2)\, dh = \chi(g_1)\chi(g_2) \quad (g_1, g_2 \in G), \tag{2.52}$$

where the integral is taken over the invariant measure of the group H. The equality (2.52) immediately implies that χ takes constant values on each dual co-set. Therefore, we can assume that $\chi \in C(Q)$.

For the representation A of the h.s. under consideration by normal operators, the formula (2.38) holds in which $X_{g,h}$ is a countable set of spherical functions which correspond to the pair G, H.

Constructions presented in this example allow various generalizations. Thus, G cannot be a compact group but a locally compact one. In this case, H must be a compact subgroup and μ must be a left invariant measure on G. Naturally, the h.s. L_1 obtained now possesses a locally compact basis Q. Spherical functions are defined by the same relation (2.52), but may be unbounded — in this case, they are the generalized characters of the h.s. L_1. As before, bounded spherical functions (and only these functions) are the characters of the h.s.

Another generalization: Instead of the definition of symmetry Γ presented above, one may also exploit another (distinct) definition which, as earlier, leads to the commutativity of K_1 and guarantees the transformation of $K_1 = L_1$ into an h.s.

Note also that Example 2.5 may be regarded as a specific case of this example. One should investigate $G \times G$ instead of G and consider a subgroup $H = G$ of it (G is compact).

Example 2.7 (an h.s. constructed in terms of orthogonal polynomials).

Assume that $\delta \subset \mathbb{R}^1$ is a bounded closed set, $\mathfrak{B}(\delta) \ni \alpha \mapsto \sigma(\alpha) \in [0, 1]$ is a probability measure ($\sigma(\delta) = 1$) having infinitely many growth points, and $(P_n(t))_{n=0}^{\infty}$ is a sequence of orthonormal polynomials in σ ($P_0(\cdot) = 1$). Denote $a = \sup_{t \in \delta} t$. By virtue of the well-known fact, we have $P_n(a) > 0$ ($n \in \mathbb{Z}_+$). Let us set

$$c(p, q, r) = P_p(a) P_q(a) P_r^{-1}(a) \int_\delta P_p(t) P_q(t) P_r(t)\, d\sigma(t),$$

$$m(p) = (P_p(a))^2 \quad (p, q, r \in \mathbb{Z}_+). \tag{2.53}$$

One can easily show that if the measure σ is such that the integrals in (2.53) are nonnegative, then the numbers $c(p, q, r) = c(\{p\}, \{q\}, r)$ define a s. measure on the discrete basis $Q = \mathbb{Z}_+$, and $m(p) = m(\{p\})$ define the corresponding m. measure (we clarify that $c(p, q, r) = 0$ for all $r > p + q$ and $r < |p - q|$). Thus, the space l_1 of sequences $x = (x(p))_{p=0}^{\infty}$, for which $\sum_{p=0}^{\infty} |x(p)|\, m(p) < \infty$, transforms into an h.s. with the basis Q and convolution

$$(x * y)(r) = \sum_{p,q=0}^{\infty} x(p)\, y(q)\, c(p, q, r) \quad (r \in Q). \tag{2.54}$$

This h.s. is Hermitian ($p^* = p$; $p \in Q$) and possesses the b.i. $O = 0$ which is, in fact, the unit. Each character of this h.s. (and only these characters) has the form $\chi(p) = P_p^{-1}(a) P_p(z)$ ($p \in Q$), where $z \in \mathbb{C}^1$ is such that the sequence $(P_p^{-1}(a) P_p(z))_{p=0}^{\infty}$ is bounded. Therefore, X can be now identified with a certain set on the complex plain; $X_h = X \cap \mathbb{R}^1$. Arbitrary sequences $(P_p^{-1}(a) P_p(z))_{p=0}^{\infty}$, $X_g = \mathbb{C}^1$, $X_{g,h} = \mathbb{R}^1$ are the generalized characters of this h.s.

The above argument clarifies the form of representation (2.38) for the sequence of normal operators $A = (A_p)_{p=0}^{\infty}$ which is a representation of the h.s. under consideration. The integration in it is carried out over a certain set on the complex plain.

We do not give examples of measures σ for which the integrals (2.53) are nonnegative. We only note that, in the case when the signs of these integrals are arbitrary, an h.s. can be constructed analogously, but with a s. measure with alternating signs. Here Theorem 2.4 and a number of other facts from Subsections 2.1–2.3 remain valid.

Example 2.8 (an h.s. constructed by using the Sturm–Liouville equation on a semi-axis).

This example is a continual analogue of Example 2.7.

Consider the Sturm–Liouville equation on a semi-axis

$$(Ly)(t) = -y'' + v(t)\,y = zy \quad (t \in \mathbb{R}^1_+ = [0, \infty);\ z \in \mathbb{C}^1), \tag{2.55}$$

where $v \in C(\mathbb{R}^1_+)$ is a nonnegative nonincreasing function. We intend to construct an h.s. such that the solutions $\varphi(t; z)$ of eqn.(2.55), satisfying the initial conditions $\varphi(0; z) = 1$ and $\varphi'(0; z) = 0$, will be closely connected with its characters.

Here it proves to be convenient to construct a convolution (an analogue of the convolution (2.54)) by using the operators of the generalized shift connected with (2.55). In order to construct these, we consider the hyperbolic equation on the plane

$$-\frac{\partial^2 u}{\partial t^2} + v(t)\,y = -\frac{\partial^2 u}{\partial s^2} + v(s)\,u, \quad u = u(t, s) \quad (t, s \in \mathbb{R}^1), \tag{2.56}$$

where v is extended as an even function to the whole of the axis $\mathbb{R}^1$. Let $R(t, s; \tau, \sigma)$ be the Riemann function of equation (2.56) and let

$$W(t, s, \tau) = -\,\partial R(t, s; \tau, 0)/\partial\sigma \quad (t, s; \tau, \sigma \in \mathbb{R}^1).$$

The role of the basis of the h.s. is played by the semi-axis $\mathbb{R}^1_+$ with the ordinary topology. An operator of generalized shift T_p $(p \in Q)$ is defined first on twice continuously differentiable functions $x(p)$ $(p \in Q)$ which are extended as even twice continuously differentiable functions to the axis $\mathbb{R}^1$ by the formula

$$(T_p x)(q) = \frac{1}{\mu(p)\,\mu(q)}\, u(p, q) \quad (p, q \in Q), \tag{2.57}$$

where $u(t, s)$ is a solution of the Cauchy problem for (2.56) with the initial conditions

$$u(t, 0) = x(t)\,\mu(t), \quad \partial u(t, 0)/\partial s = 0 \quad (t \in \mathbb{R}^1),$$

and $\mu(t) = \varphi(t; 0)$ extended as an even function to $\mathbb{R}^1$. If we write u from (2.57) by using the Riemann function, then we obtain the following formula for T_p

$$(T_p x)(q) = \frac{1}{2\mu(p)\,\mu(q)}\,(x(q-p)\,\mu(q-p) + x(q+p)\,\mu(q+p))$$

$$+ \frac{1}{2\mu(p)\,\mu(q)} \int_{q-p}^{q+p} W(q, p, r)\, x(r)\,\mu(r)\, dr \quad (p, q \in Q) \tag{2.58}$$

This formula extends the operator T_p to the collection $L_{1,0}$ of arbitrary functions locally integrable with respect to the Lebesgue measure.

A m. measure is now defined as $\mu^2(p)dp$, and L_1 is the corresponding space of summable functions. One can prove that there exists a convolution which can be expressed in terms of operators T_p by the usual formula (2.15), namely

$$(x * y)(r) = \int_0^\infty x(p)(T_p y)(r)\,dp \quad (x, y \in L_{1,0};\ r \in Q). \tag{2.59}$$

A s. measure is defined by the equality

$$c(\alpha, \beta, r) = (\kappa_\alpha * \kappa_\beta)(r) \quad (\alpha, \beta \in \mathfrak{B}_0(\mathbb{R}^1_+);\ r \in Q).$$

It can be proved that this s. measure is nonnegative and satisfies all necessary requirements. Note that the nonnegativity of it is a consequence of the fact that the potential v in (2.55) is nondecreasing. As a result, we obtain the h.s. for which the convolution (2.3) is another form of the convolution (2.59).

The h.s. L_1 is Hermitian $(p^* = p;\ p \in Q)$ and contains the b.i. $O = 0$. Each character of it (and only these characters) has the form $\chi(p) = \mu^{-1}(p)\,\varphi(p; z)$ $(p \in Q)$, where $z \in \mathbb{C}^1$ is such that the function $\mu^{-1}(p)\,\varphi(p; z)$ is bounded. The same functions with arbitrary $z \in \mathbb{C}^1$ are generalized characters of this h.s. Thus, we now have $X \subset \mathbb{C}^1$, $X_h = X \cap \mathbb{R}^1$, $X_g \in \mathbb{C}^1$, and $X_{g,h} = \mathbb{R}^1$.

As in Example 2.7, the integration in formula (2.38) for the representation of the h.s. constructed is carried out over a certain set on the complex plane.

A construction similar to that presented above can be realized also in the case of a nonmonotonic potential v in (2.55). One should demand that the variation of this potential is locally bounded. The role of μ is now played by the solution of equation (2.55) (with $z = 0$ and the same initial conditions as above) in which v is replaced by a nonnegative nonincreasing function $\tilde{v} \in C(\mathbb{R}^1_+)$ such that

$$|v(t') - v(t'')| \le \tilde{v}(t') - \tilde{v}(t'') \quad (0 \le t' < t'' < \infty).$$

In this case, the s. measure is not nonnegative, but a number of facts presented in Subsections 2.1-2.4 (including Theorem 2.4) remain valid for it.

Example 2.9. Consider the situation of Example 2.3 in the case $G = \mathbb{R}^1$, but here the operators A_p are suppose to be unbounded. This means that Theorem 2.4 cannot be applied immediately. Thus, let $X = \mathbb{R}^1$ and let $A = (A_x)_{x \in \mathbb{R}^1}$ be a family of commuting normal operators in the space H_0 connected with the nuclear rigging $\Phi' \supset H_0 \supset \Phi$ in the standard way (in the sense of Subsection 1.4) and satisfying the relation

$$A_x A_y = \tfrac{1}{2}(A_{x+y} + A_{x-y}) \quad (x, y \in \mathbb{R}^1), \tag{2.60}$$

i.e., (2.48). Acting as in (1.3) and (1.30), we obtain the following functional equation for the eigenvalue $\lambda(\cdot)$:

$$\lambda(x)\,\lambda(y) = \tfrac{1}{2}\,(\lambda(x+y)+\lambda(x-y)) \quad (x, y \in \mathbb{R}^1);$$

the spectral integral for A_x has the form (1.1), where integration is carried out over all solutions $\mathbb{R}^1 \ni x \mapsto \chi(x) \in \mathbb{C}^1$ of this functional equation.

If A_x depends on x continuously (in the case of Subsection 1.4), then it suffices to consider only continuous solutions of the functional equation. Assume, in addition, that the operators A_x are selfadjoint and $A_0 = 1$. Substituting $x = 0$ to the relation (2.60), we find $A_{-x} = A_x$ $(x \in \mathbb{R}^1)$. The spectral representation includes only continuous real-valued (even) solutions of the above equation such that $\chi(0) = 1$. They have the form

$$\chi(x) = \chi_\lambda(x) = \cos\sqrt{\lambda}\,x \quad (x \in \mathbb{R}^1),$$

where $\lambda \in \mathbb{R}^1$. Transporting the RI from the set of these solutions to $\mathbb{R}^1$, we obtain the representation in the form of a spectral integral

$$A_x = \int_{\mathbb{R}^1} \cos\sqrt{\lambda}\,x \; dE(\lambda),$$

$$\mathfrak{D}(A_x) = \left\{ f \in H_0 \mid \int_{\mathbb{R}^1} (\cos\sqrt{\lambda}\,x)^2\, d(E(\lambda)f, f)_{H_0} < \infty \right\} \quad (x \in \mathbb{R}^1). \tag{2.61}$$

We can now assume that X is a separable real Hilbert space and obtain a theorem similar to Theorem 1.2 which generalizes representation (2.61). On the other hand, this example is a special case of the h.s. in Example 2.8 (for the potential $v = 0$). At the same time, it can be easily modified so that it does not fit the requirement of the theory of h.s., for example, the sign "+" between the operators in (2.60) can be replaced by "–". By using the method just described, this relation can be easily investigated.

Example 2.10. Consider the relation between the operators A_x which is of a completely different nature. Let $X = G$, where $G \subset \mathbb{R}^d$ $(d \in \mathbb{N})$ is a certain domain, and let $A = (A_x)_{x \in G}$ be a family of bounded (for simplicity) normal commuting operators acting in H_0 connected with the nuclear rigging $\Phi' \supset H_0 \supset \Phi$ in the standard way (in the sense of Subsection 1.4). The function $G \ni x \mapsto A_x \in \mathcal{L}(H_0, H_0)$ is assumed to be r times strongly continuously differentiable and the relation between the operators has the form of a differential equation of the r-th order

$$\sum_{|\alpha| \le r} a_\alpha(x)\, D^\alpha A_x = f(x)\,1 \quad (x \in G), \tag{2.62}$$

where α is a multi-index, D^α is a mixed derivative, and $G \ni x \mapsto a_\alpha(x)$ and $f(x) \in \mathbb{C}^1$ are complex-valued coefficients. The spectral representation for A can be described as follows. Denote by $\Lambda \subset \mathbb{C}^X$ a collection of all complex-valued r times continuously differentiable solutions $G \ni x \mapsto \lambda(x) \in \mathbb{C}^1$ of the equation

$$\sum_{|\alpha| \le r} a_\alpha(x)\,(D^\alpha \lambda)(x) = f(x).$$

Then there exists an RI E on the σ-algebra $\mathfrak{C}_\sigma(\Lambda) = \{\alpha \cap \Lambda \mid \alpha \in \mathfrak{C}_\sigma(\mathbb{C}^G)\}$ such that

$$A_x = \int_\Lambda \lambda(x)\, dE(\lambda(\cdot)) \quad (x \in G). \tag{2.63}$$

The proof of (2.63) is carried out by use of the same scheme as above, namely, we apply Theorem 2.7 (Chapter 3), Remark 1 to this theorem, and Remarks 2 and 3 in Subsection 2.10 of Chapter 3. We now have $\Lambda \supset g(A)$, because each eigenvalue $\lambda(\cdot) \in g(A)$ is a sufficiently smooth function (see Subsection 2.9 of Chapter 3) satisfying the desired scalar differential equation. This follows from the second equality (2.19) in Chapter 3 by virtue of (2.62) (cf. (1.3), (1.30)). Therefore, we can set $\tau = \Lambda$, and this leads us to (2.63).

It is clear that if $G \ni x \mapsto A_x \in \mathfrak{L}(H_0, H_0)$ satisfy conditions of a certain sort (for example, boundary conditions), then Λ includes scalar solutions satisfying these boundary conditions. It is also evident that one can also consider functional equations differential-functional equations, etc.

§3. Applications to Representations of Commutation Relations

In this section, we show how the spectral projection theorem can be applied to obtain representations of operators connected by certain commutation relations. The scheme can be described as follows. Assume that a family B of operators satisfying commutation relations of a certain type is given. The minimal program is to find the representation of this family, i.e., to find concrete operators satisfying these relations. It is often possible to act as follows. The family B is associated with another family A of commuting self-adjoint (or normal) operators, the rules of commutation of which with the operators from B appear to be simple enough in terms of the Fourier transform connected with A. Then this space is represented by the operators acting in the space of these Fourier transforms. Below, we realize this scheme and present several examples.

3.1. Change of Variables in Spectral Integrals

First, let us clarify how the results of Section 2 in Chapter 3 are modified in the case when a certain invertible transformation F acts in the space $\mathbb{C}^X$. In other words, instead of the original family A of commuting operators, we consider a "function $F(A)$" of it.

Here, it is convenient for us to consider the case of a nuclear rigging and to restrict ourselves to the case of selfadjoint operators which is most important for our purposes.

Consider a family $A = (A_x)_{x\in X}$ of commuting selfadjoint operators acting in the Hilbert space H_0 and connected with the nuclear rigging

$$\Phi' \supset H_0 \supset \Phi, \quad \Phi = \operatorname{pr}\lim_{t\in T} H_t \tag{3.1}$$

in the standard way. We fix a certain positive space $H_+ = H_t$ quasinuclearly imbedded into H_0 and consider a quasinuclear chain

$$\Phi' \supset H_- \supset H_0 \supset H_+ \supset \Phi \tag{3.2}$$

(clearly, (3.2) is also connected with A in the standard way; now $D = \Phi$).

Let $\mathfrak{C}_\sigma(\mathbb{R}^X) \ni \alpha \mapsto E(\alpha)$ and $\rho(\alpha) = \operatorname{Tr}(O^+E(\alpha)O)$ be the corresponding joint RI and the spectral measure, respectively ($O\colon H_+ \mapsto H_0$ and $O_+ : H_0 \mapsto H_-$ are the imbeddings). We employ the spectral projection Theorem 2.7 (Chapter 3) in the following form. This theorem states that there exists a set $\pi \subset \mathbb{R}^X$ (belonging to the generalized spectrum $g(A)$ of the family A) of full outer ρ-measure such that

$$\mathbb{R}^X \supset \pi \ni \lambda(\cdot) \mapsto P(\lambda(\cdot)) \in \mathfrak{L}(H_+, H_-)$$

is a generalized projector corresponding to the eigenvalue $\lambda(\cdot)$, and $\operatorname{Tr}(P(\lambda(\cdot))) = 1$.

We fix a certain set $\tau \subset \mathbb{R}^X$ containing π. It has a full outer ρ-measure. Let E_τ and ρ_τ be the modifications of E and ρ by $\tau(\rho_{\mathbb{R}^X} = \rho)$. Then the following equalities hold (see Remark 1 to Theorem 2.7 in Chapter 3)

$$O^+E_\tau(\alpha)\,O = \int_\alpha P(\lambda(\cdot))\,d\rho_\tau(\lambda(\cdot)), \quad A_x\varphi = \int_\tau \lambda(x)\,P(\lambda(\cdot))\,\varphi\,d\rho_\tau(\lambda(\cdot)),$$

$$(\alpha \in \mathfrak{C}_\sigma(\tau) = \{\alpha = \alpha' \cap \tau \mid \alpha' \in \mathfrak{C}_\sigma(\mathbb{R}^X)\};\ \varphi \in \Phi \subset \mathfrak{D}(A_x);\ x \in X). \tag{3.3}$$

For $\lambda(\cdot) \in \pi \subset \tau$, the range $\mathfrak{R}(P(\lambda(\cdot))$ consists of generalized joint eigenvectors of the family A corresponding to the eigenvalue $\lambda(\cdot)$, and $\operatorname{Tr}(P(\lambda(\cdot))) = 1$.

Let us recall how to rewrite formulas (3.3) in terms of a direct integral of Hilbert spaces and the Fourier transform with respect to generalized joint eigenvectors of the family A (see Chapter 3, Subsection 3.1) (this is important for what follows).

Thus, H_0 is isometric to the direct integral of Hilbert spaces $l_2(N_{\lambda(\cdot)})$ $(N_{\lambda(\cdot)} \le \infty)$, i.e.

$$H_0 = \int_\tau \otimes\, l_2(N_{\lambda(\cdot)})\,d\rho_\tau(\lambda(\cdot)),$$

$$(\varphi, \psi)_{H_0} = \int_\tau (\tilde{\varphi}(\lambda(\cdot)), \tilde{\psi}(\lambda(\cdot)))_{l_2(N_{\lambda(\cdot)})}\,d\rho_\tau(\lambda(\cdot)) \quad (\varphi, \psi \in \Phi). \tag{3.4}$$

Here

$$\Phi \ni \varphi \mapsto \tilde{\varphi}(\lambda(\cdot)) = (\tilde{\varphi}_1(\lambda(\cdot)), \tilde{\varphi}_2(\lambda(\cdot)), \ldots)$$

is a Fourier transform of the vector φ defined for $\lambda(\cdot)$ belonging to a certain set of full ρ_τ-measure (which does not depend on φ and is equal to

$$\{\lambda(\cdot) \in \mathbb{R}^X \mid \operatorname{Tr} P(\lambda(\cdot)) = 1\} \cap \tau \supset \pi).$$

For $\lambda(\cdot)$ belonging to this set and any $\gamma = 1, \ldots, N_{\lambda(\cdot)}$, the coordinates of $\tilde{\varphi}(\lambda(.))$ are given by the formula

$$\tilde{\varphi}_\gamma(\lambda(\cdot)) = (\varphi, \xi_\gamma(\lambda(\cdot)))_{H_0},$$

$$\xi_\gamma(\lambda(\cdot)) = \sqrt{\nu_\gamma(\lambda(\cdot))}\, J^{-1} h_\gamma(\lambda(\cdot))$$

$$= P(\lambda(\cdot))\,(\nu_\gamma(\lambda(\cdot)))^{-1/2} J h_\gamma(\lambda(\cdot)) \in \mathfrak{R}(P(\lambda(\cdot))) \subset H_-, \qquad (3.5)$$

where $h_\gamma(\lambda(\cdot))$ are the orthonormal eigenvectors in H_0 corresponding to the eigenvalues $\nu_\gamma(\lambda(\cdot))$ of a quasinuclear operator $J\,P(\lambda(\cdot))\,J\colon H_0 \to H_0$ whose trace is equal to unit. For fixed $\lambda(\cdot)$ from the same set, the vectors $\tilde{\varphi}(\lambda(\cdot))$ $(\varphi \in \Phi)$ are dense in $l_2(N_{\lambda(\cdot)})$. For any $x \in X$, the Fourier image of the operator A_x is equal to the operator of multiplication by the function $\tau \ni \lambda(\cdot) \mapsto \lambda(x)$ in the direct integral.

We proceed to a change of variables. Fix a one-to-one mapping $\mathbb{R}^X \ni \lambda(\cdot) \mapsto F(\lambda(\cdot))(\cdot) \in \mathbb{R}^X$ preserving the σ-algebra $\mathfrak{C}_\sigma(\mathbb{R}^X)$. Then, for any $x \in X$, the functions $\mathbb{R}^X \ni \lambda(\cdot) \mapsto F(\lambda(\cdot))(x)$ and $F^{-1}(\lambda(\cdot))(x) \in \mathbb{R}^1$ are measurable with respect to $\mathfrak{C}_\sigma(\mathbb{R}^X)$. According to Subsection 1.11 of Chapter 3, the selfadjoint operator

$$F_x(A) = \int_{\mathbb{R}^X} F(\lambda(\cdot))(x)\, dE(\lambda(\cdot)),$$

$$\mathfrak{D}(F_x(A)) = \left\{ f \in H_0 \mid \int_{\mathbb{R}^X} |F(\lambda(\cdot))(x)|^2 d(E(\lambda) f, f)_{H_0} < \infty \right\} \qquad (3.6)$$

is defined for any $x \in X$. Moreover, any two operators $F_x(A)$ and $F_y(A)$ $(x, y \in X)$ commute.

This means that for the given mapping F, we have constructed the second family $F(A) = (F_x(A))_{x \in X}$ of commuting selfadjoint operators $F_x(A)$ acting in H_0 (if F is the identity mapping, then $F(A) = A$). Let us calculate the joint RI of the family $F(A)$.

To do this, we change the variables in the first integral of (3.6) as follows

$$\mathbb{R}^X \ni \lambda(\cdot) \mapsto F(\lambda(\cdot)) = \mu(\cdot) \in \mathbb{R}^X.$$

Then this integral takes the form

$$F_x(A) = \int_{\mathbb{R}^X} \mu(x)\, dE_F(\mu(\cdot)) \quad (x \in X), \tag{3.7}$$

where the operator-valued measure E_F is the image of the RI E, namely,

$$\mathfrak{C}_\sigma(\mathbb{R}^X) \ni \alpha \mapsto E_F(\alpha) = E(F^{-1}\alpha). \tag{3.8}$$

Clearly, E_F is also an RI, and (3.7) shows that E_F is the joint RI of the family $F(A)$.

Assume that the chain (3.1) and the family $F(A)$ are connected in the standard way (this is not automatically true for arbitrary F). Since a spectral measure is defined as a trace of an RI surrounded by the operators O_+ and O, relation (3.8) involves the following relation between the spectral measures ρ_F and ρ of the families $F(A)$ and A, respectively,

$$\rho_F(\alpha) = \rho(F^{-1}\alpha) \quad (\alpha \in \mathfrak{C}_\sigma(\mathbb{R}^X)). \tag{3.9}$$

For the operators of a generalized projection P_F and P, this implies that for ρ_F -almost all $\lambda(\cdot)\,(\cdot) \in \mathbb{R}^X$, we have

$$P_F(\lambda(\cdot)) = P(F^{-1}(\lambda(\cdot))). \tag{3.10}$$

Indeed, one must write the following equality (see (2.13) of Chapter 3, (3.8), and (3.9)) and employ the fact that $\alpha \in \mathfrak{C}_\sigma(\mathbb{R}^X)$ is arbitrary. We obtain

$$\int_\alpha P_F(\lambda(\cdot))\, d\rho_F(\lambda(\cdot)) = O^+E_F(\alpha)\, O = O^+E(F^{-1}\alpha)\, O$$

$$= \int_{F^{-1}\alpha} P(\lambda(\cdot))\, d\rho(\lambda(\cdot)) = \int_\alpha P(F^{-1}(\mu(\cdot))\,(\cdot))\, d\rho_F(\mu(\cdot)).$$

Remark 1. Assume that the set $\tau \subset \mathbb{R}^X$ is of full outer ρ-measure. Then F_τ is of full outer ρ_F -measure. The equalities (3.8) and (3.9) remain valid if we replace E, ρ, E_F, and ρ_F by the modified objects E_τ, ρ_τ, $E_{F;\,F_\tau}$, and $\rho_{F;\,F_\tau}$, respectively. The equality (3.10) holds for $\rho_{F;\,F_\tau}$-almost all $\lambda(\cdot) \in \tau$. These assertions immediately follow from the definition of a modification of the measure (see (1.21), Chapter 3).

■

Remark 2. All of the above discussion is also clearly valid for the family A of normal commuting operators.

■

3.2. A Commutation Relation and Its Representation

Let us investigate a commutation relation which can be often encountered. It can be described as follows. Let $A = (A_x)_{x \in X}$ be a family of commuting selfadjoint operators, let $F(A) = (F_x(A))_{x \in X}$ be the second family of the operators of this sort constructed above for given A and F, and let A and $F(A)$ be connected with the nuclear chain (3.1) in the standard way. Suppose that a closed operator B acting in H_0 is given, such that Φ belongs to its domain of definition $\mathfrak{D}(B)$ and is a base of B. Furthermore, Φ is invariant with respect to B, and $B \upharpoonright \Phi$ acts continuously in Φ. Assume that the same conditions are imposed on B^*. Let B be connected with the operators A_x by the commutation relation

$$A_x B \varphi = B F_x(A) \varphi \quad (\varphi \in \Phi;\ x \in X). \tag{3.11}$$

We clarify the form of this relation in terms of the Fourier transform (3.4) constructed for the given family A (later, we shall study families of operators B connected with A_x by the equalities of the type (3.11)).

Let us now employ a concept which will be considered in detail in Subsection 1.3 of Chapter 5. Let C be an operator acting in H_0. A vector $\varphi \in H_0$ is called an entire vector (with respect to C) if it belongs to $\bigcap_{n=1}^{\infty} \mathfrak{D}(C^n)$, and the power series

$$\sum_{n=0}^{\infty} \frac{\|C^n \varphi\|_{H_0}}{n!} z^n \tag{3.12}$$

converges for all $z \in \mathbb{C}^1$. If C is selfadjoint, then the unitary operators e^{itC} $(t \in \mathbb{R}^1)$ are defined for it in terms of the spectral integral. On entire vectors φ, these operators may be calculated with the help of the series

$$e^{itC} \varphi = \sum_{n=0}^{\infty} \frac{(it)^n}{n!} C^n \varphi \quad (t \in \mathbb{R}^1). \tag{3.13}$$

In fact, the series on the right-hand side of (3.13) converges in the norm of the space

H_0. Therefore, if we write C^n in terms of the spectral integral and carry it out of the sum, then we obtain the required result. However, if C is bounded, then every vector $\varphi \in H_0$ will be an entire vector for it.

Let us establish the following three lemmas which can be proved without any assumption concerning the action of the operator B^* on Φ.

Lemma 3.1. *We fix $x \in X$. Let $\Psi_x \subset H_0$ be a set of entire vectors of the operators A_x and $F_x(A)$ belong to $\mathfrak{D}(B)$. Assume that under the action of the operator B, this set transforms into itself. If* (3.11) *holds for $\varphi \in \Psi_x$, then the following equality is valid*

$$E_x(\delta)\, B\, \varphi = B\, F_{F;\,x}(\delta)\, \varphi \quad (\varphi \in \Psi_x;\ \delta \in \mathfrak{B}(\mathbb{R}^1)), \tag{3.14}$$

where E_x and $E_{F;x}$ are the RI of the operators A_x and $F_x(A)$, respectively (in (3.14) *the inclusion $E_{F;\,x}(\delta)\,\varphi \in \mathfrak{D}(B)$ holds automatically).*

Proof. Applying relation (3.11) successively, we obtain

$$A_x^n\, B\, \varphi = A_x^{n-1}(A_x B\, \varphi) = A_x^{n-1}\, B\, F_x(A)\, \varphi$$

$$= A_x^{n-2}(A_x B\, F_x(A)\, \varphi) = \ldots = B\,(F_x(A))^n\, \varphi \quad (\varphi \in \Psi_x) \tag{3.15}$$

for any $n \in \mathbb{N}$. This yields

$$e^{itA_x}\, B\, \varphi = B\, e^{itF_x(A)}\, \varphi \quad (\varphi \in \Psi_x,\ t \in \mathbb{R}^1). \tag{3.16}$$

Indeed, by using (3.13) and (3.15), we get that

$$e^{itA_x}\, B\, \varphi = \lim_{m \to \infty} B f_m$$

in H_0, where

$$f_m = \sum_{n=0}^{m} (it)^n\, (n!)^{-1}\, (F_x(A))^n\, \varphi.$$

Since $f_m \to \exp(itF_x(A))\,\varphi$, and the operator B is closed, we find that $\exp(itF_x(A))\,\varphi \in \mathfrak{D}(B)$, and (3.16) is valid.

Rewriting (3.16) with the help of the corresponding spectral integral, we obtain

$$\int_{\mathbb{R}^1} e^{it\lambda}\, d(F_x(\lambda)\, B\varphi, g)_{H_0} = (e^{itA_x} B\varphi, g)_{H_0} = (Be^{itF_x(A)}\varphi, g)_{H_0}$$

$$= (Be^{itF_x(A)}\varphi, B^*g)_{H_0} = \int_{\mathbb{R}^1} e^{it\lambda}\, d(E_{F;x}(\lambda)\varphi, B^*g)_{H_0} \quad (t \in \mathbb{R}^1)$$

for any $g \in \mathfrak{D}(B^*)$. By virtue of the theorem on the uniqueness of a Fourier transform, this yields

$$(E_x(\gamma)B\varphi, g)_{H_0} = (E_{F;x}(\delta)\varphi, B^*g)_{H_0} \quad (g \in \mathfrak{D}(B^*))$$

for any $\delta \in \mathfrak{B}(\mathbb{R}^1)$. Consequently, $E_{F;x}(\delta)\varphi \in \mathfrak{D}(B)$, and (3.14) holds.

∎

Lemma 3.2. *Assume that the space Φ of the chain* (3.1) *consists of entire vectors for each operator A_x or $F_x(A)$ $(x \in X)$. Then* (3.11) *implies the equality*

$$E(\alpha)\, B\,\varphi = B\, E_F(\alpha) \quad (\varphi \in \Phi;\ \alpha \in \mathfrak{C}_\sigma(\mathbb{R}^X)) \tag{3.17}$$

(here, $E_F(\alpha)\varphi \in \mathfrak{D}(B)$, automatically).

Proof. Fix $x_1 \in X$ and apply Lemma 3.1 for $x = x_1$ and $\Psi_x = \Phi$. As a result, we obtain

$$E_{x_1}(\delta_1)\, B\,\varphi = B\, E_{F;x}(\delta_1)\,\varphi \quad (\varphi \in \Phi;\ \delta_1 \in \mathfrak{B}(\mathbb{R}^1)). \tag{3.18}$$

Fix $x_2 \in X$ $(x_2 \neq x_1)$ and $\delta_1 \in \mathfrak{B}(\mathbb{R}^1)$ and apply Lemma 3.1 for $x = x_2$ and $\Psi_x = E_{F;x_1}(\delta_1)\,\Phi$. It is easy to see that this choice of Ψ_x is correct. Indeed, $\Psi_x = E_{F;x_1}(\delta_1)\,\Phi \subset \mathfrak{D}(B)$ by Lemma 3.1. According to (3.18), we have

$$B\,\Psi_x \subset E_{x_1}(\delta_1)\, B\,\Phi \subset E_{x_1}(\delta_1)\,\Phi = \Psi_x.$$

Each vector from Ψ_x is an entire vector with respect to the operator A_{x_2}. This is so, because it has the form $\psi = E_{F;x_1}(\delta)\,\varphi$ and, thus, due to the fact that A_{x_2} and $E_{F;x_1}(\delta)$ commute, we have

$$\| A_{x_2}^n \psi \|_{H_0} \le \| A_{x_2}^n \varphi \|_{H_0}$$

and $\varphi \in \Phi$ is an entire vector for A_{x_2}. By analogy, we conclude that ψ is an entire vector with respect to $F_{x_2}(A)$.

Applying the Lemma, we obtain the equality

$$E_{x_2}(\delta_2) \times B\, E_{F;x_1}(\delta_1)\,\varphi = B\, E_{F;x_2}(\delta_2)\; E_{F;x_1}(\delta_1)\,\varphi \quad (\varphi \in \Phi;\ \delta_2 \in \mathfrak{B}(\mathbb{R}^1))$$

the left-hand side of which can be rewritten in the form $E_{x_2}(\delta_2)\, E_{x_1}(\delta_1)\, B\varphi$, by virtue of (3.18). As usual, we denote by E_{x_1, x_2} and $E_{F;x_1, x_2}$ the joint RI's of the operators A_{x_1}, A_{x_2} and $F_{x_1}(A)$, $F_{x_2}(A)$, respectively. Then the just-obtained equality can be rewritten as

$$E_{x_1, x_2}(\delta_1 \times \delta_2)\, B\,\varphi = B\, E_{F;x_1, x_2}(\delta_1 \times \delta_2)\,\varphi$$

$$(\varphi \in \Phi;\ \delta_1, \delta_2 \in \mathfrak{B}(\mathbb{R}^1);\ x_1, x_2 \in X) \tag{3.19}$$

We fix $x_3 \in X\ (x_3 \neq x_1, x_2)$ and $\delta_1, \delta_2 \in \mathfrak{B}(\mathbb{R}^1)$ and apply Lemma 3.1 for $x = x_3$ and $\Psi_x = E_{F;x_1, x_2}(\delta_1 \times \delta_2)\Phi$. Repeating the above arguments with (3.18) replaced by (3.19), we get the equality (3.19) for E_{x_1, x_2, x_3}. If we continue this procedure, then we obtain the following relation

$$E_{x_1, \dots, x_n}(\delta_1 \times \dots \times \delta_n) B\,\varphi = B\, E_{F;x_1, \dots, x_n}(\delta_1 \times \dots \times \delta_n)\,\varphi$$

$$(\varphi \in \Phi;\ \delta_1, \dots, \delta_n \in \mathfrak{B}(\mathbb{R}^1);\ n \in \mathbb{N}) \tag{3.20}$$

for distinct points $x_1, \dots, x_n \in X$. Here, $E_{F;x_1, \dots, x_n}(\delta_1 \times \dots \times \delta_n) \in \mathfrak{D}(B)$, automatically.

Suppose that $\varphi \in \Phi$ and $g \in \mathfrak{D}(B^*)$ are fixed. Consider two charges

$$\mathfrak{C}_\sigma(\mathbb{R}^X) \ni \alpha \mapsto \omega_1(\alpha) = (E(\alpha)\, B\,\varphi, g)_{H_0}$$

and

$$\omega_2(\alpha) = (E_F(\alpha)\,\varphi, B^* g)_{H_0} \in \mathbb{C}^1.$$

It follows from (3.20) that $\omega_1(\beta) = \omega_2(\beta)$, where β is an arbitrary cylindrical set of the form $\beta = \mathbb{C}(x_1, \dots, x_n; \delta_1 \times \dots \times \delta_n)$. But then these two charges coincide, i.e.,

$$(E(\alpha)\, B\,\varphi, g)_{H_0} = (E_F(\alpha)\,\varphi, B^* g)_{H_0}$$

for any $\alpha \in \mathfrak{C}_\sigma(\mathbb{R}^X)$. Since $g \in \mathfrak{D}(B^*)$ is arbitrary, we have $E_F(\alpha)\varphi \in \mathfrak{D}(B)$ (in this equality), and (3.17) holds.

■

The statement converse to the statement of this lemma is also valid.

Lemma 3.3. *Assume that* $E_F(\alpha)\varphi \in \mathfrak{D}(B)$ *for any* $\alpha \in \mathfrak{C}_\sigma(\mathbb{R}^X)$ *and* $\varphi \in \Phi$ *and that the equality* (3.17) *is valid. Then* (3.11) *holds.*

Proof. Let $\varphi \in \Phi$ and $g \in \mathfrak{D}(B^*)$. Then, by virtue of (3.17), we have

$$(E(\alpha)\, B\, \varphi, g)_{H_0} = (E_F(\alpha)\, \varphi, B^* g)_{H_0}$$

and

$$(A_x B\, \varphi, g)_{H_0} = \int_{\mathbb{R}^X} \lambda(x)\, d(E(\lambda(\cdot))\, B\, \varphi, g)_{H_0}$$

$$= \int_{\mathbb{R}^X} \lambda(x)\, d(F_F(\lambda(\cdot))\, \varphi, B^* g)_{H_0} = (F_x(A)\, \varphi, B^* g)_{H_0} = (BF_x(A)\, \varphi, g)_{H_0}.$$

Since $\mathfrak{D}(B)$ is dense in H_0, this equality yields (3.11).

■

Remark 1. *The relations* (3.11) *and* (3.17) *are equivalent to the relations*

$$B^* A_x \varphi = F_x(A) B^* \varphi \quad (\varphi \in \Phi;\ x \in X) \tag{3.21a}$$

and

$$B^* E(\alpha)\, \varphi = E_F(\alpha) B^* \varphi \ (E(\alpha)\, \varphi \in \mathfrak{D}(B^*);\ (\varphi \in \Phi;\ \alpha \in \mathfrak{C}_\sigma(\mathbb{R}^X)), \quad (3.21.b)$$

respectively.

Indeed, multiply the equality (3.11) by $\psi \in \Phi$ (in the sense of a scalar product in H_0) and transfer the operators to act upon ψ. This transfer is possible, according to the agreement concerning the action of the operators A_x, $F_x(A)$, B and B^* in the space Φ. Since $\varphi \in \Phi$ is arbitrary, the relation obtained above yields (3.21a), where φ is replaced by ψ. Further, for $\varphi \in \Phi$ the relation (3.17) implies that

$$(\varphi, E_F(\alpha)\, B^* \psi)_{H_0} = (B\, E_F(\alpha)\, \varphi, \psi)_{H_0} = (E(\alpha)\, B\, \varphi, \psi)_{H_0} = (B\, \varphi, E(\alpha)\, \psi)_{H_0}$$

for any $\varphi \in \Phi$. Since Φ is a base of the operator B, this yields $E(\alpha)\psi \in \mathfrak{D}(B^*)$ and

$$B^* E(\alpha)\, \psi = E_F(\alpha) B^* \psi \quad (\alpha \in \mathfrak{C}_\sigma(\mathbb{R}^X)).$$

The two inverse implications can be established analogously.

■

Remark 2. In equalities (3.17) and (3.21), the RI's E and E_F can be replaced by the RI's E_τ and $E_{F;F_\tau}$ ($\mathbb{R}^X \supset \tau \supset \pi$) modified by τ (see Remark 1 in Subsection 3.1).

■

Remark 3. The above results hold for a family $A = (A_x)_{x\in X}$ of bounded commuting normal operators. In this case, A_x and $F_x(A)$ in (3.21.a) should be replaced by A_x^* and $(F_x(A))^*$, respectively.

Indeed, in the case of a bounded A_x, the operator $\exp(itA_x)$ is correctly defined in terms of the spectral integral and, therefore, all the above arguments can be repeated here.

■

Let us now write the action of the operator B in terms of the Fourier transform (3.4). The main presentation is carried out in the case $\tau = \mathbb{R}^X$.

Lemma 3.4. *Assume that* 0 *is not an eigenvalue of the operator B. Then the measure ρ_F is absolutely continuous with respect to ρ and, consequently, a non-negative derivative*

$$r(\lambda(\cdot)) = \left(\frac{d\rho_F}{d\rho}\right)(\lambda(\cdot)) = \frac{d\rho(F^{-1}(\lambda(\cdot)))}{d\rho(\lambda(\cdot))}, \tag{3.22}$$

summable with respect to ρ, exists for ρ-almost all $\lambda(\cdot) \in \mathbb{R}^X$.

Proof. The condition imposed on B means that $\mathfrak{R}(B^*)$ is dense in H_0. Since Φ is a base of B^*, this implies that $\mathfrak{R}(B^* \upharpoonright \Phi)$ is dense in H_0, too. Now let $\alpha \in \mathfrak{C}_\sigma(\mathbb{R}^X)$ be such that $\rho(\alpha) = 0$ and, consequently, $E(\alpha) = 0$. Then it follows from (3.21b) that $E_F(\alpha)\, B^*\varphi = 0$, and the above-mentioned denseness yields $E_F(\alpha) = 0$, i.e., $\rho_F(\alpha) = 0$. The equality (3.22) is written on the basis of the relation (3.9).

■

The next theorem gives a "commutative model" of the operator B.

Theorem 3.1. *Assume that the chain* (3.1) *is connected with the family of operators $A = (A_x)_{x\in X}$, $F(A) = (F_x(A))_{x\in X}$ in the standard way, that for any $x \in X$ the space Φ consists of the entire vectors of the operators A_x, $F_x(A)$, and that* 0 *is not an eigenvalue of the operator B. Then, for given B, one can choose a positive space $H_+ = H_t$ of the chain* (3.2) *such that the commutation relation* (3.11) *is equivalent to the following action of the operator B in terms of the Fourier transform:*

$$(B\varphi)\tilde{\ }(\lambda(\cdot)) = \frac{d\rho(F^{-1}(\lambda(\cdot)))}{d\rho(\lambda(\cdot))} B(\lambda(\cdot), F^{-1}(\lambda(\cdot)))\, \tilde{\varphi}(F^{-1}(\lambda(\cdot))) \tag{3.23}$$

for $\varphi \in \Phi$.

Here $\lambda(\cdot)$ *changes inside a certain subset of* $\mathbb{R}^X$ *of full* ρ*-measure (independent of* φ*), and* $B(\lambda(\cdot), F^{-1}(\lambda(\cdot)))$ *is a linear operator with dense domain of definition acting from the space* $l_2(N_{F^{-1}(\lambda(\cdot))})$ *to* $l_2(N_{\lambda(\cdot)})$ $(N_{F^{-1}(\lambda(\cdot))}, N_{\lambda(\cdot)} \leq \infty)$*; it depends on* $\lambda(\cdot)$ *in the sense of weak measurability with respect to the* σ*-algebra* $\mathfrak{C}_\sigma(\mathbb{R}^X)$.

First, we realize one general construction. Let us extend the action of the operator B onto the vector from Φ' in the chain (3.1). This can be done by use of the standard procedure described in Subsection 2.5 of Chapter 3. Thus, let the operator C belong to $\mathfrak{L}(\Phi, \Phi)$. Then its adjoint operator C_+ (in the common sense) belongs to $\mathfrak{L}(\Phi', \Phi')$. We now set $B' = (B^* \upharpoonright \Phi)^+ \in \mathfrak{L}(\Phi', \Phi')$. This is just the required extension of the operator B (the fact that B' is an extension of B can be immediately verified. Indeed, assume that $f \in \mathfrak{D}(B) \subset H_0$, then

$$(B'f, \varphi)_{H_0} = ((B^* \upharpoonright \Phi)^+ f, \varphi)_{H_0} = (f, B^* \varphi)_{H_0} = (Bf, \varphi)_{H_0}$$

for any $\varphi \in \Phi$, whence $B'f = Bf$). Replacing the operator B in the above relation by B^*, we get the extension of B^* to the operator $B^{*\prime} \in \mathfrak{L}(\Phi', \Phi')$ (the conditions imposed earlier upon B and B^* are symmetric).

Lemma 3.5. *Assume that the assumptions of Theorem* 3.1 *hold. The commutation relation* (3.11) *is equivalent to the equality*

$$B^{*\prime} P(\lambda(\cdot))\, \varphi = \frac{d\rho(F^{-1}(\lambda(\cdot)))}{d\rho(\lambda(\cdot))} P(F^{-1}(\lambda(\cdot)))\, B^* \varphi \quad (\varphi \in \Phi), \tag{3.24}$$

which is valid for ρ*-almost all* $\lambda(\cdot) \in \mathbb{R}^X$ *(independently of* φ*).*

Proof. The relation (3.11) is equivalent to (3.21b). This follows from Lemma 3.2, Remark 1, and Lemma 3.3. Using formulas of type (2.13) in Chapter 3, we can rewrite this equality as follows

$$\int_\alpha (B^{*\prime} P(\lambda(\cdot))\, \varphi, \psi)_{H_0}\, d\rho(\lambda(\cdot)) = \int_\alpha (P(\lambda(\cdot))\, \varphi, B\, \psi)_{H_0}\, d\rho(\lambda(\cdot))$$

$$= \left(\left(\int_\alpha P(\lambda(\cdot))\, d\rho(\lambda(\cdot))\right)\varphi, B\, \psi\right)_{H_0} = (O^+ E(\alpha)\, O\varphi, B\, \psi)_{H_0}$$

$$= (B^* E(\alpha)\, \varphi, B\, \psi)_{H_0} = (E_F(\alpha)\, B^* \varphi, B\, \psi)_{H_0} = (O^+ E_F(\alpha)\, O B^* \varphi, \psi)_{H_0}$$

$$= \left(\left(\int\limits_{\alpha} P_F(\lambda(\cdot))\, d\rho_F(\lambda(\cdot))\right) B^*\varphi, \psi\right)_{H_0} = \int\limits_{\alpha} (P_F(\lambda(\cdot))\, B^*\varphi, \psi)_{H_0}\, d\rho_F(\lambda(\cdot)) \quad (3.25)$$

for any $\alpha \in \mathfrak{C}_\sigma(\mathbb{R}^X)$. Here, α is arbitrary and, hence, taking into account Lemma 3.4, we find

$$(B^{*\prime} P(\lambda(\cdot))\, \varphi, \psi)_{H_0} = r(\lambda(\cdot))\, (P_F(\lambda(\cdot))\, B^*\varphi, \psi)_{H_0} \quad (3.26)$$

for $\lambda(\cdot)$ from a certain set $\beta_{\varphi,\psi} \in \mathfrak{C}_\sigma(\mathbb{R}^X)$ of full ρ-measure. Now assume that in (3.26), φ and ψ run through a countable set L which is dense in Φ. Due to the facts that L is dense in Φ and that the operators $B^{*\prime} P(\lambda(\cdot)), P_F P(\lambda(\cdot))\, B^*: \Phi \to \Phi'$ are continuous, we can conclude for $\lambda(\cdot) \in \bigcap_{\varphi,\psi \in L} \beta_{\varphi,\psi} = \beta$ that (3.24) is valid.

The equality (3.21b) (and, hence, (3.11)) follows from (3.24) if we integrate the latter and perform the calculation (3.25) in the inverse direction.

■

Remark 4. If we repeat the proof of Lemma 3.5 for the equality (3.17) (instead of (3.21b), then we obtain a relation similar to (3.24) (but with the operators B and B') which is equivalent to (3.11). We shall not use this relation.

■

Proof of Theorem 3.1. By the condition of the theorem, we have $B^* \restriction \Phi \in \mathfrak{L}(\Phi, \Phi)$. Therefore, according to the definition of the topology in Φ and the linearity of this operator, for any $t' \in \mathrm{T}$ there exists $t'' \in \mathrm{T}$ such that $B^* \restriction \Phi$ acts continuously from Φ, equipped with the topology of $H_{t''}$, to Φ, equipped with the topology $H_{t'}$. Choose t' so that the imbedding $H_{t'} \subset H_0$ is quasinuclear. Let $H_{t'''}$ be such that $H_{t'}, H_{t''} \supset H_{t'''}$ topologically. The imbedding $H_{t'''} \subset H_0$ is also quasinuclear. Denote $H_1 = H_{t'}$ and $H_2 = H_{t'''}$. The operator $B^* \restriction \Phi$ acts continuously from H_2 to H_1. Let C be an extension of this operator by continuity to the whole of H_2; $C \in \mathfrak{L}(H_2, H_1)$.

Consider the chain

$$\Phi' \supset H_{-2} \supset H_{-1} \supset H_0 \supset H_1 \supset H_2 \supset \Phi, \quad (3.27)$$

and denote by $O_{j,k}$ the operator of the imbedding of the space H_j in H_k ($j, k = -2, \ldots, 2$). Let us construct the operator $P(\lambda(\cdot))$ taking $\Phi' \supset H_{-1} \supset H_0 \supset H_1 \supset \Phi$ to play the role of the chain (3.2). Denote this operator by $P_1(\lambda(\cdot))$. By $P_2(\lambda(\cdot))$ we denote an operator of the same sort constructed for $\Phi' \supset H_{-2} \supset H_0 \supset H_2 \supset \Phi$. Since, for any $\alpha \in \mathfrak{C}_\sigma(\mathbb{R}^X)$, we have

$$O_{0,-2}\, E(\alpha)\, O_{0,2} = O_{-1,-2}\,(O_{0,-1}\, E(\alpha)\, O_{1,0})\, O_{2,1}$$

it easily follows from Theorem 2.2 in Chapter 3 that for E-almost all $\lambda(\cdot) \in \mathbb{R}^X$

$$P_2(\lambda(\cdot)) = \left(\frac{d\rho_1}{d\rho_2}\right)(\lambda(\cdot))\, O_{-1,-2} P_1(\lambda(\cdot))\, O_{2,1} \tag{3.28}$$

(ρ_1 and ρ_2 are the corresponding spectral measures absolutely continuous with respect to each other).

For the j-th chain (one of the two chains written above), formulas (3.5) can be rewritten as follows

$$\tilde{\varphi}_{\gamma,j}(\lambda(\cdot)) = (\varphi, \xi_{\gamma,j}(\lambda(\cdot)))_{H_0}\, \xi_{\gamma,j}(\lambda(\cdot))$$

$$= P_j(\lambda(\cdot))\, ((\nu_{\gamma,j}(\lambda(\cdot)))^{-1/2} J_j\, h_{\gamma,j}(\lambda(\cdot)) \in \mathfrak{R}(P_j(\lambda(\cdot))) \subset H_{-j} \tag{3.29}$$

($j = 1, 2$). The equality (3.28) implies that $N_{\lambda(\cdot)}$ is independent of the choice of the chain.

We start deriving (3.23). Let us write the equality (3.24) for the case $P = P_1$ and employ (3.28). As a result, for $\lambda(\cdot)$ from a certain set $\beta \subset \mathbb{R}^X$ of full ρ_1-measure, we obtain

$$\left(\frac{d\rho_2}{d\rho_1}\right)(\lambda(\cdot))\, B^{*\prime} P_2(\lambda(\cdot))\, \varphi = \frac{d\rho_1(F^{-1}(\lambda(\cdot)))}{d\rho_1(\lambda(\cdot))} P_1 F^{-1}(\lambda(\cdot)))\, C\varphi \quad (\varphi \in \Phi). \tag{3.30}$$

Using the limiting transition, we establish this equality for all $\varphi \in H_2$ (one should exploit the inclusions $C \in \mathfrak{L}(H_2, H_1)$, $P_1 F^{-1}(\lambda(\cdot))) \in \mathfrak{L}(H_1, H_{-1})$ and $P_2(\lambda(\cdot)) \in \mathfrak{L}(H_2, H_{-2})$ and the fact that Φ is dense in H_2).

According to (3.29), for any $\varphi \in \Phi$, $\lambda(\cdot) \in \beta$, and $\gamma = 1, \ldots, N_{\lambda(\cdot)}$, we have

$$(B\varphi)\tilde{}_{\gamma,2}(\lambda(\cdot)) = (B\varphi, \xi_{\gamma,2}(\lambda(\cdot)))_{H_0} = (\varphi, B^{*\prime}\xi_{\gamma,2}(\lambda(\cdot)))_{H_0}.$$

By virtue of (3.29) and (3.30) (with $\varphi \in H_2$), we have

$$B^{*\prime}\xi_{\gamma,2}(\lambda(\cdot)) = B^{*\prime} P_2(\lambda(\cdot))\, (\nu_{\gamma,2}(\lambda(\cdot)))^{-1/2} J_2\, h_{\gamma,2}(\lambda(\cdot))$$

$$= (\nu_{\gamma,2}(\lambda(\cdot)))^{-1/2} \frac{d\rho_1(F^{-1}(\lambda(\cdot)))}{d\rho_2(\lambda(\cdot))} P_1 F^{-1}(\lambda(\cdot)))\, C J_2\, h_{\gamma,2}(\lambda(\cdot)). \tag{3.31}$$

Let us expand $CJ_2 h_{\gamma,2}(\lambda(\cdot)) \in H_1$ in the system of vectors $(J_1 h_{\kappa,2}(F^{-1}(\lambda(\cdot))))_{\kappa=1}^{N_{F^{-1}(\lambda(\cdot))}}$ which is orthonormal in H_1. As a result, we get

$$CJ_2 h_{\gamma,2}(\lambda(\cdot)) = \sum_{\kappa=1}^{N_{F^{-1}(\lambda(\cdot))}} c_{\gamma\kappa} J_1 h_{\kappa,1}(F^{-1}(\lambda(\cdot))) + \psi,$$

$$c_{\gamma\kappa} = (CJ_2 h_{\gamma,2}(\lambda(\cdot)),\ J_1 h_{\kappa,1}(F^{-1}(\lambda(\cdot))))_{H_1}. \tag{3.32}$$

Here the vector $\psi \in H_1$ is such that $P_1(F^{-1}(\lambda(\cdot)))\,\psi = 0$. Substituting (3.32) in (3.31) and employing (3.29), we find

$$B^{*\prime}\xi_{\gamma,2}(\lambda(\cdot)) = (\nu_{\gamma,2}(\lambda(\cdot)))^{-1/2}\frac{d\rho_1(F^{-1}(\lambda(\cdot)))}{d\rho_2(\lambda(\cdot))}$$

$$\times \sum_{\kappa=1}^{N_{F^{-1}(\lambda(\cdot))}} c_{\gamma\kappa} P_1(F^{-1}(\lambda(\cdot)))\, J_1 h_{\kappa,1}(F^{-1}(\lambda(\cdot)))$$

$$= \frac{d\rho_1(F^{-1}(\lambda(\cdot)))}{d\rho_2(\lambda(\cdot))}\sum_{\kappa=1}^{N_{F^{-1}(\lambda(\cdot))}}\left[\frac{\nu_{\kappa,1}(F^{-1}(\lambda(\cdot)))}{\nu_{\gamma,2}(\lambda(\cdot))}\right]^{1/2} c_{\gamma\kappa}\xi_{\kappa,1}(F^{-1}(\lambda(\cdot))). \tag{3.33}$$

Using (3.33), for any $\varphi \in \Phi$, $\lambda(\cdot) \in \beta$, and $\gamma = 1, \ldots, N_{\lambda(\cdot)}$, we obtain

$$(B^*\varphi)\tilde{}_{\gamma,2}(\lambda(\cdot)) = (\varphi, B^{*\prime}\xi_{\gamma,2}(\lambda(\cdot)))_{H_0}$$

$$= \frac{d\rho_1(F^{-1}(\lambda(\cdot)))}{d\rho_2(\lambda(\cdot))}\sum_{\kappa=1}^{N_{F^{-1}(\lambda(\cdot))}}\left[\frac{\nu_{\kappa,1}(F^{-1}(\lambda(\cdot)))}{\nu_{\gamma,2}(\lambda(\cdot))}\right]^{1/2} c_{\gamma\kappa}\tilde{\varphi}_{\kappa,1}(F^{-1}(\lambda(\cdot)))$$

$$= \frac{d\rho_1(F^{-1}(\lambda(\cdot)))}{d\rho_1(\lambda(\cdot))}\sum_{\kappa=1}^{N_{F^{-1}(\lambda(\cdot))}}\left(\frac{d\rho_1}{d\rho_2}\right)(\lambda(\cdot))\left[\frac{\nu_{\kappa,1}(F^{-1}(\lambda(\cdot)))}{\nu_{\gamma,2}(\lambda(\cdot))}\right]^{1/2} c_{\gamma\kappa}$$

$$\times\, (G(\tilde{\varphi}_{1,2}(F^{-1}(\lambda(\cdot))),\ \tilde{\varphi}_{2,2}(F^{-1}(\lambda(\cdot))),\ \ldots))_{\kappa}. \tag{3.34}$$

In (3.34), G is defined as a linear operator in the space $l_2(N_{F^{-1}(\lambda(\cdot))})$ acting on a dense set of vectors of the form $(\tilde{\varphi}_{\sigma,2}(F^{-1}(\lambda(\cdot))))_{\sigma=1}^{N_{F^{-1}(\lambda(\cdot))}})$, where $\varphi \in \Phi$, and transforming each vector of this sort into the vector $(\tilde{\varphi}_{\sigma,1}(F^{-1}(\lambda(\cdot))))_{\sigma=1}^{N_{F^{-1}(\lambda(\cdot))}})$. Generally speaking, this operator is unbounded. Note that the matrix with the elements $c_{\gamma\kappa}$ is the matrix of a bounded operator $C \in \mathfrak{L}(H_2, H_1)$.

As the final result, (3.34) can be written in the form (3.23), in general, with an unbounded operator $B(\lambda(\cdot), F^{-1}(\lambda(\cdot)))$ acting from the space $l_2(N_{F^{-1}(\lambda(\cdot))})$ into $l_2(N_{\lambda(\cdot)})$.

We outline the proof of the converse statement. Assume that there exists an operator B of the sort described at the beginning of this Subsection, and that (3.23) holds for it instead of (3.11). We show that (3.11) is also valid in this case. We use the fact that for any $x \in X$, the Fourier image of the operator A_x is equal to the operator of multiplication by the function $\mathbb{R}^X \ni \lambda(\cdot) \mapsto \lambda(x) \in \mathbb{R}^1$ on the vectors $\tilde{\varphi}(\lambda(\cdot))$ ($\varphi \in \Phi$). In analogy with Lemma 3.1 in Chapter 3 (using Theorem 2.7 in Chapter 3 for the family $F(A)$), we prove that, on the same vectors, the Fourier image of the operator $F_x(A)$ is equal to the operator of multiplication by the function $\mathbb{R}^X \ni \lambda(\cdot) \mapsto F(\lambda(\cdot)) \in \mathbb{R}^1$.

Since $B\varphi \in \Phi$ for any $\varphi \in \Phi$, we have

$$(A_x B\varphi)^{\sim}(\lambda(\cdot)) = \lambda(x)\,(B\varphi)^{\sim}(\lambda(\cdot)),$$

where $(B\varphi)^{\sim}(\lambda(\cdot))$ is calculated according to (3.23). On the other hand,

$$(F_x(A)\varphi)^{\sim}(\lambda(\cdot)) = F(\lambda(\cdot))(x)\tilde{\varphi}(\lambda(\cdot))$$

and, therefore, according to formula (3.23),

$$(BF_x(A)\varphi)^{\sim}(\lambda(\cdot)) = \frac{d\rho(F^{-1}(\lambda(\cdot)))}{d\rho(\lambda(\cdot))}\,B(\lambda(\cdot), F^{-1}(\lambda(\cdot)))\,(\lambda(x)\,\tilde{\varphi}(F^{-1}(\lambda(\cdot)))$$

$$= \lambda(x)\,(B\varphi)^{\sim}(\lambda(\cdot)) = (A_x B\varphi)^{\sim}(\lambda(\cdot))$$

(we have transferred the scalar $\lambda(x)$ through the operator $B(\lambda(\cdot), F^{-1}(\lambda(\cdot)))$. This means that (3.11) is valid.

Note that instead of (3.11), one could verify the validity of relation (3.17), using the equality (3.5) in Chapter 3; when this verification is carried out, it is not necessary to clarify the action of the operator $F_x(A)$ upon the Fourier images beforehand.

■

Remark 5. Theorem 3.1 remains valid in the case when the Fourier transform (3.4) with $\tau \supset \pi$, $\tau \subset \mathbb{R}^X$ is used. All the arguments remain unchanged. One should only repeat Lemmas 3.1–3.5 for modified objects (see Remark 2). This theorem holds also in the case of a family $A = (A_x)_{x \in X}$ of bounded commuting normal operators (see Remark 3).

■

Remark 6. One can consider a family $B = (B_y)_{y \in Y}$ of operators in H_0 each of them satisfying the conditions imposed earlier on the operator B. Assume that the families A and B are connected by the commutation relations

$$A_x B_x \varphi = B_y F_x^y(A) \quad (\varphi \in \Phi;\ x \in X;\ y \in Y), \tag{3.35}$$

where $(F^y)_{y\in Y}$ is a family of one-to-one mappings of $\mathbb{R}^X$ onto itself which preserve $\mathfrak{C}_\sigma(\mathbb{R}^X)$. For every B_y, one can write the representation (3.23) — its commutative model.

■

3.3. Examples of Commutative Models

We present, in a concise form, several examples of applications of Theorem 3.1 and its generalizations described in Remarks 5 and 6.

Example 3.1 (representations of semi-direct products of groups).

Let X be a group and let K be a certain group of its automorphisms k $(X \ni x \mapsto k(x) \in X)$. Recall that a group of its ordered pairs (x, k) with the law of composition $(x, k)\,(x', k') = (x k(x'), k k')$ $(x, x' \in X; k, k' \in K)$ is called a semi-direct product $G = X \otimes K$. (The original example of a semi-direct product is as follows: $X = \mathbb{R}^2$ with the group operation of composition of vectors, and $K = SO(2)$ is a group of rotations of $\mathbb{R}^2$ about the origin). Then $\mathbb{R}^2 \otimes SO(2)$ consists of the movements of the plane

$$\mathbb{R}^2 \ni a \mapsto (x, k)\,a = k\,a + x \in \mathbb{R}^2,$$

and the product of these movements gives the above-mentioned law of composition). One can easily understand that X is a normal subgroup in G and each element $g \in G$ can be expressed in a unique manner in the form $g = x\,k$, where $x \in X$; and $k \in K$, i.e., $G = XK$ (we clarify that we identify $X = \{\,(x, 1) \mid x \in X\,\}$ and $K = \{\,(e, k) \mid k \in K\,\}$, where 1 is the identity automorphism, and e is the identity of the group X). If X and K are topological groups, then a topology is introduced in G by means of the decomposition $G = XK$, and this group becomes topological.

Let $A = (A_g)_{g\in G}$ be a family of normal operators in a separable Hilbert space H_0 (generally speaking, they are unbounded and noncommuting). A mapping $G \ni g \mapsto A_g$ is called a representation if $A_{g_1 g_2} = A_{g_1} A_{g_2}$. By analogy with Subsection 1.1, we introduce the notion of a nuclear representation, namely, for any $g \in G$, the operators A_g, A_g^* are connected with the chain (3.1) in the standard way. The representation is nuclear and continuous if, in addition, for any $\varphi \in \Phi$, the vector function $G \ni g \mapsto A_g\varphi \in \Phi$ is weakly continuous. Since X is a normal subgroup in G, the following relation holds

$$A_x A_g = A_{xg} = A_{gg^{-1}xg} = A_g A_{x[g]} \quad (x \in X,\ g \in G), \tag{3.36}$$

where we denote $x[g] = g^{-1}\,x\,g$; we have $x[g] \in X$, because X is normal. Henceforth, we assume that the subgroup X is commutative. Suppose that there exists a nuclear representation $G \ni g \mapsto A_g$ of the group g by normal operators. Moreover, we suppose that the space Φ from the chain (3.1) consists of entire vectors for each operator A_x $(x \in X)$.

According to Theorem 1.1, we expand H_0 into a direct integral (3.4) over the family $A = (A_x)_{x\in X}$ of commuting normal operators (one should set $\tau = X_g$; for convenience the set X_g of generalized characters of a commutative group X is denoted below by X_{gen}). The role of $\lambda(\cdot)$ is now played by generalized characters χ , and the expansion itself has the form

$$H_0 = \int\limits_{X_{\text{gen}}} \oplus\, l_2(N_\chi)\, d\rho_{X_{\text{gen}}}(\chi); \quad (\varphi, \psi)_{H_0} = \int\limits_{X_{\text{gen}}} (\tilde{\varphi}(\chi), \tilde{\psi}(\chi))_{l_2(N_\chi)}\, d\rho_{X_{\text{gen}}}(\chi),$$

$$\tilde{\varphi}(\chi) = ((\varphi, \xi_1(\chi))_{H_0}, (\varphi, \xi_2(\chi))_{H_0}, \ldots) \quad (\varphi, \psi \in \Phi), \tag{3.37}$$

where $\xi_\gamma(\chi) \in H_-$ is a generalized joint eigenvector of the family A corresponding to the eigenvalue χ.

Below, we employ the expansion (3.37) in the cases when the operators A_g $(g \in G)$ are selfadjoint or unitary. In these cases, X_{gen} in (3.37) is replaced by $X_{\text{gen, Re}}$ and X, respectively. If, in addition, the representation is nuclear and continuous, then the last sets are replaced by $X_{\text{gen, c, Re}}$ and X_c (by virtue of Remark 3, one may also consider bounded normal operators A_g). For definiteness, we now assume that the operators A_x $(x \in X)$ are selfadjoint.

Fix $g \in G$ and consider a mapping $\mathbb{R}^X \ni \lambda(\cdot) \mapsto (F^g(\lambda(\cdot)))(\cdot)$, where

$$(F^g(\lambda(\cdot)))(x) = \lambda(g^{-1} x g) = \lambda(x[g]).$$

More precisely, we shall need its restriction to $X_{\text{gen, Re}} \subset \mathbb{R}^X$. It follows from the formula (3.6) that $F^g_x(A) = A_{x[g]}$ $(x \in X)$, therefore, equality (3.36) means that commutation relation (3.11) holds for $B = A_g$. We assume that 0 is not an eigenvector of the operator A_g. Then all the conditions of Theorem 3.1 are satisfied, and we can write the representation (3.23) (one should take into account that $(F^g)^{-1} = F^{g^{-1}}$).

Thus, if we have a nuclear representation $G \ni g \mapsto A_g$ of the group G by normal operators such that A_x are selfadjoint for $x \in X$, and 0 is not an eigenvalue of A_k for any $k \in K$, then the expansion of H_0 into the direct integral (3.37) holds, where X_{gen} is replaced by a set of generalized real characters of a commutative group X, namely, by $X_{\text{gen, Re}} = Z$. The operators A_g act upon the Fourier images as follows

$$(A_x \varphi)^{\sim}(\chi) = \chi(x)\tilde{\varphi}(\chi),$$

$$(A_k \varphi)^{\sim}(\chi) = \frac{d\rho_Z(\chi[k^{-1}])}{d\rho_Z(\chi)} B(\chi, \chi[k^{-1}])\, \tilde{\varphi}(\chi[k^{-1}]), \tag{3.38}$$

$$(x \in X,\ k \in K;\ \varphi \in \Phi;\ \chi \in Z).$$

Here the character $\chi[k] \in Z$ is given by the formula $(\chi[k])(x) = \chi(x[k])$. The operator

$B(\chi, \chi[k^{-1}])$ is densely defined in $l_2(N_{\chi[k^{-1}]})$ and acts from this space into $l_2(N_\chi)$. Taking into account that A_g realizes the representation G, we find, in particular, that $A_{k_1 k_2} = A_{k_1} A_{k_2}$, for any $k_1, k_2 \in K$. This relation can be easily rewritten in terms of the operators $B(\cdot, \cdot)$. For ρ-almost every $\chi \in Z$, it has the form

$$B(\chi, \chi[(k_1 k_2)^{-1}]) = B(\chi, \chi[k_1^{-1}])\, B(\chi[k_1^{-1}], \chi[(k_1 k_2)^{-1}]) \quad (k_1, k_2 \in K) \quad (3.39)$$

(one must use the equality $\chi[k_1 k_2] = (\chi[k_2])[k_1]$ which follows from $x[k_1 k_2] = (x[k_2])[k_1]$). Formulas (3.38) and (3.39) give the general for of the representations G under consideration if we add the condition of normality of the operators A_k $(k \in K)$ written in the proper form.

If, in addition, the representation is continuous, then we must replace $Z = X_{\text{gen, Re}}$ in the above formulas by $X_{\text{gen, c, Re}}$. In the case of representation by unitary operators, we write the analogous formulas in which Z is replaced by X or X_c. It is now easy to write the condition for the operators $B(\cdot, \cdot)$ which guarantees the unitarity of the representation. These operators, equipped with a factor equal to the square root of the derivative of the measure in (3.38), transform the spaces into each other isometrically (see also (3.41)).

Finally, note that if G is locally compact, then, for any representation of it by bounded normal operators A_g, there exists a nuclear rigging (3.1) connected with A_g $(g \in G)$ in the standard way. This is proved by a proper modification of the classical Gording construction with the help of the Kats theorem [4] on the existence of the required spaces of test functions on G (cf. Steps II–X in the proof of Theorem 2.4; in the commutative case, this fact is clearly proved in Steps II–X; see also Example 2.3). For unitary representations of nonlocally compact groups, the Gorging region may not exist (see Kosyak [1], Ostrovsky and Yu. Samoilenko [1]).

Let us illustrate the constructions presented in this example by three more concrete situations.

Example 3.2 (the representation of the infinite-dimensional Heisenberg group).

The elements of the group G are triples (t, s, α) $(t, s \in \mathbb{R}_0^\infty,\ \alpha \in \mathbb{R}^1)$ with a group operation

$$(t, s, \alpha)\,(t', s', \alpha') = (t + t', s + s', \alpha + \alpha' + (t, s')_{l_2})$$

and a natural topology (as a set, $G = \mathbb{R}_0^\infty \times \mathbb{R}_0^\infty \times \mathbb{R}^1$). One can easily calculate that $G = X \otimes K$, where X and K are commutative subgroups of G consisting of the elements $x = (0, s, \alpha)$ $(s \in \mathbb{R}_0^\infty,\ \alpha \in \mathbb{R}^1)$ and $k = (t, 0, 0)$ $(t \in \mathbb{R}_0^\infty)$, respectively. Moreover, k acts upon X as an internal automorphism $X \ni x \mapsto k x k^{-1}$ (the action of it is reduced to the transformation

$$(0, s, \alpha) \mapsto k((0, s, \alpha)) = (0, s, \alpha + (t, s)_{l_2}),$$

if $k = (t, 0, 0)$; X is clearly normal.

For an arbitrary unitary strongly continuous representation of the group G, the existence of a nuclear rigging (3.1) connected with the operators of the representation in the standard way has been proved by Kosyak [2] (the method of calculation presented in this paper can be also applied to some other types of representations presented below). See also the earlier paper by Reed [1]. Therefore, the formulas from Example 3.1 are valid for the group under consideration.

Let us write these formulas for a special case of the representation. Assume that $G \ni g \mapsto A_g$ is a strongly continuous unitary representation such that $A_{(0,0,\alpha)} = e^{i\alpha} 1$ $(\alpha \in \mathbb{R}^1)$. It is not hard to verify that the groups of unitary operators $U(t) = A_{(t,0,0)}$ and $V(s) = A_{(0,s,0)}$ $(t, s \in \mathbb{R}_0^\infty)$ satisfy the relation

$$U(t)\, V(s) = \exp\left[i\,(t, s)_{l_2}\right] V(s)\, U(t) \quad (t, s \in \mathbb{R}_0^\infty). \tag{3.40}$$

This relation is called canonical commutation relations (CCR) with countably many degrees of freedom written in Weyl's form; for more detail, see Example 3.8 (note that the above-mentioned Reed's constructions (Reed, [1]) are based on the formulas (3.61) in this example). Writing the corresponding modification of formulas (3.37) - (3.39), we obtain

$$H_0 = \int_{\mathbb{R}^\infty} \oplus\, l_2(N_x)\, d\rho(x),$$

$$(U(t)\,\varphi)^\sim(x) = \left[\frac{d\rho(x+t)}{d\rho(x)}\right]^{1/2} C(x, x+t)\, \tilde{\varphi}(x+t),$$

$$(V(s)\,\varphi)^\sim(x) = \exp\left[i\,(s, x)_{l_2}\right] \tilde{\varphi}(x) \quad (t, s \in \mathbb{R}_0^\infty;\ \varphi \in \Phi), \tag{3.41}$$

where $C(x, x+t)$ are the operator functions defined for ρ -almost all $x \in \mathbb{R}^\infty$ which map the whole of $l_2(N_{x+t})$ onto the whole of $l_2(N_x)$ isometrically and satisfy the relation

$$C(x, x+t_1+t_2) = C(x, x+t_1)\, C(x+t_1, x+t_1+t_2) \quad (t_1, t_2 \in \mathbb{R}_0^\infty).$$

Let us clarify (3.41). We have written (3.37) - (3.39) with X_{gen} and Z being replaced by the set X_c of continuous characters (the representation is unitary). Since $A_{(0,0,\alpha)} = e^{i\alpha} 1$, we can restrict ourselves to the characters of a subgroup of X consisting of triples $\{(0, s, 0) \mid s \in \mathbb{R}_0^\infty\} = \mathbb{R}_0^\infty$. They have the form $\chi(s) = e^{i\,(s,x)_{l_2}}$ $(x \in \mathbb{R}_0)$. In (3.41), the measure is transported from the set of these characters to $\mathbb{R}^\infty$. Moreover, it is convenient to denote by $C(\cdot,\cdot)$ the operator $B(\cdot,\cdot)$ equipped with a factor equal to the square root of the derivative of the measure in (3.38). The existence of the derivative in (3.41), i.e., the quasi-invariance of the measure ρ under the shifts by $t \in \mathbb{R}_0^\infty$ immediately follows from Theorem 3.1. For any $t \in \mathbb{R}_0^\infty$ we have $N_x = N_{x+t}$ for ρ -almost all x.

Note that similar constructions can be also realized for the Heisenberg group of flows whose elements are the triples (t, s, α), where $\alpha \in \mathbb{R}^1$, and t and s are functions from $C^l_{0,\text{Re}}(\mathbb{R}^d)$ $(d \in \mathbb{N};\ l = 0, \ldots, \infty$ are fixed) or $\mathcal{S}(\mathbb{R}^d)$. The operation of multiplication is the same as above, but $(\cdot,\cdot)_{l_2}$ is replaced by $(\cdot,\cdot)_{L_2(\mathbb{R}^d)}$. The problem of the existence of the required nuclear rigging has been investigated in Hegerfeldt [1].

Example 3.3. Let $B(\infty, \mathbb{R}^1)$ be a group of real upper-triangle infinite matrices with finitely nonzero elements in each column and row, with units in the principal diagonal and with ordinary multiplication. Assume that $B(\infty, \mathbb{R}^1)$ is topologized as $\mathbb{R}_0^\infty$. Denote by X a normal commutative subgroup of $B(\infty, \mathbb{R}^1)$ consisting of the matrices with nonzero elements only in the first row (and in the principal diagonal), and by $B_1(\infty, \mathbb{R}^1)$ we denote a subgroup of $B(\infty, \mathbb{R}^1)$ consisting of the matrices with all the elements of the first row being zero (except the first one). It is easy to see that

$$B(\infty, \mathbb{R}^1) = X \otimes B_1(\infty, \mathbb{R}^1).$$

Consider a strongly continuous unitary representation $B(\infty, \mathbb{R}^1) \ni g \mapsto A_g$. The method developed in the paper by Kosyak [2] enables us to prove the existence of the nuclear rigging connected with A_g $(g \in B(\infty, \mathbb{R}^1))$ in the standard way. Hence, we can write formulas of the type (3.37) - (3.41) in this case, too. The group X is clearly isomorphic to $\mathbb{R}_0^\infty$ and, therefore, the group of continuous characters of it is isomorphic to $\mathbb{R}^\infty$, as in Example 3.2. This results in the fact that the integration in (3.37) is carried out over $\mathbb{R}^\infty$ as in (3.41). The detailed presentation of this example and relative facts, as well as the corresponding generalizations, can be found in Ostrovsky [1,2].

Example 3.4. Consider a real space $\mathcal{S}_{\mathrm{Re}}(\mathbb{R}^d) = X$ $(d \in \mathbb{N}$ is fixed) as a commutative topological group with respect to the addition of functions $x(t)$ $(t \in \mathbb{R}^d)$. Let $\mathbb{R}^d \ni t \mapsto k(t) \in \mathbb{R}^d$ be a diffeomorphism of the space $\mathbb{R}^d$ such that, for any $x \in \mathcal{S}_{\mathrm{Re}}(\mathbb{R}^d)$, the function $(x \circ k)(t) = x(k(t))$ belongs to $\mathcal{S}_{\mathrm{Re}}(\mathbb{R}^d)$. The collection of all diffeomorphism of this sort forms a group $\mathrm{Diff}\,(\mathbb{R}^d) = K$ with respect to the operation $k \circ k'$ of successive applications of these diffeomorphisms. This group may be regarded as a certain group of automorphisms of $\mathcal{S}_{\mathrm{Re}}(\mathbb{R}^d)$, namely, for $k \in \mathrm{Diff}\,(\mathbb{R}^d)$, we must take $k(x) = x \circ k$ $(x \in \mathcal{S}_{\mathrm{Re}}(\mathbb{R}^d))$. Consider a semi-direct product

$$G = \mathcal{S}_{\mathrm{Re}}(\mathbb{R}^d) \otimes \mathrm{Diff}\,(\mathbb{R}^d)$$

consisting of pairs $g = (x, k)$ with the law of action

$$(x, k)(x', k') = (x + x' \circ k, k \circ k'),$$

the topology in $\mathrm{Diff}\,(\mathbb{R}^d)$, and G is not introduced. Recall that $\mathcal{S}_{\mathrm{Re}}(\mathbb{R}^d)$ and $\mathrm{Diff}\,(\mathbb{R}^d)$ can be imbedded in G and $G = \mathcal{S}_{\mathrm{Re}}(\mathbb{R}^d)\, \mathrm{Diff}\,(\mathbb{R}^d)$, respectively.

Assume that $G \ni g \mapsto A_g$ is a nuclear representation of G by unitary operators in H_0 such that, for any $\varphi \in \Phi$, the vector function $\mathcal{S}_{\mathrm{Re}}(\mathbb{R}^d) \ni x \mapsto A_x\varphi \in \Phi$ is weakly continuous. Here, when constructing the direct integral (3.41) over the family $(A_x)_{x \in X}$ $(X = \mathcal{S}_{\mathrm{Re}}(\mathbb{R}^d))$, one can take a collection of all the continuous characters of the group $\mathcal{S}_{\mathrm{Re}}(\mathbb{R}^d)$ to play the role of a set $\tau \subset \mathbb{C}^X$. Each of these characters has the form $\chi(x) = \exp i(\xi, x)_{L_2(\mathbb{R}^d)}$ $(x \in \mathcal{S}_{\mathrm{Re}}(\mathbb{R}^d))$, where $\xi \in \mathcal{S}'_{\mathrm{Re}}(\mathbb{R}^d)$. Therefore, by the transportation of the measure, the corresponding integrals can be extended to be over $\mathcal{S}'_{\mathrm{Re}}(\mathbb{R}^d)$. Then the following formulas become clear (they hold by virtue of Theorem 3.1):

$$H_0 = \int_{\mathcal{S}'_{\mathrm{Re}}(\mathbb{R}^d)} \oplus\, l_2(N_\xi)\, d\rho(\xi), \quad (A_x \varphi)^\sim(\xi) = \exp\left[i\,(\xi, x)_{L_2(\mathbb{R}^d)}\right] \tilde{\varphi}(\xi),$$

$$(A_k \varphi)^\sim(\xi) = \left[\frac{d\rho(\xi \circ k^{-1})}{d\rho(\xi)}\right]^{1/2} C(\xi, \xi \circ k^{-1})\, \tilde{\varphi}(\xi \circ k^{-1}),$$

$$(x \in \mathcal{S}'_{\mathrm{Re}}(\mathbb{R}^d),\ k \in \mathrm{Diff}\,(\mathbb{R}^d);\ \varphi \in \Phi). \tag{3.42}$$

Here, for any $\xi \in \mathcal{S}'_{\mathrm{Re}}(\mathbb{R}^d)$ and $k \in \mathrm{Diff}\,(\mathbb{R}^d)$, a generalized function $\xi \circ k \in \mathcal{S}'_{\mathrm{Re}}(\mathbb{R}^d)$ is defined by the equality

$$(\xi \circ k, x)_{L_2(\mathbb{R}^d)} = (\xi, x \circ k)_{L_2(\mathbb{R}^d)} \quad (x \in \mathcal{S}_{\mathrm{Re}}(\mathbb{R}^d)).$$

The operators $C(\xi, \xi \circ k^{-1})$ are defined for ρ-almost all $\xi \in \mathcal{S}'_{\mathrm{Re}}(\mathbb{R}^d)$. They act isometrically from the whole of $l_2\ (N_{\xi \circ k^{-1}})$ into the whole $l_2(N_\xi)$ and possess the property

$$C(\xi, \xi \circ (k_1 \circ k_2)^{-1}) = C(\xi, \xi \circ k_1^{-1})\, C(\xi \circ k_1^{-1}, \xi \circ (k_1 \circ k_2)^{-1}).$$

As in (3.410, in (3.42) we have passed from $B(\cdot, \cdot)$ to the operators $C(\cdot, \cdot)$.

Clearly, in this example it is also possible to consider other types of representations of the group G.

Example 3.5 (a one-dimensional quantum spin system with countably many degrees of freedom).

In a $*$-algebra of the local observables of this system, one can choose generators which, being represented by the operators in H_0, give the following relations: There exists two families $A = (A_j)_{j=1}^\infty$ and $B = (B_j)_{j=1}^\infty$, each consisting of commuting selfadjoint operators satisfying the equalities

$$[A_j, B_k] = 0 \quad (j, k \in \mathbb{N};\ j \neq k),$$

$$\{A_k, B_k\} = 0, \quad A_k^2 = B_k^2 = 1 \quad (k \in \mathbb{N}),$$

$$[A, B] = AB - BA, \quad \{A, B\} = AB + BA. \tag{3.43}$$

We choose A to be a family in terms of which we construct the expansion of the space H_0 in the direct integral. Since the operators A_j and B_j are bounded, the required nuclear rigging (3.1) exists. This follows from Remark 5 in Subsection 3.4 of Chapter 3 applied to the family $(A_j, B_j)_{j=1}^\infty$. We fix $k \in \mathbb{N}$ and consider the mapping

$$\mathbb{R}^\infty \ni \lambda = (\lambda_j)_{j=1}^\infty = \lambda(\cdot) \mapsto (F^k(\lambda(\cdot)))(\cdot) \in \mathbb{R}^\infty,$$

where $(F^k(\lambda(\cdot)))(j)$ is equal to λ_j for $j \neq k$ and to $-\lambda_k$ for $j = k$. According to (3.6), we have $F_j^k(A) = A_j$ for $j \in \mathbb{N}, j \neq k$ and $F_k^k(A) = -A_k$. Therefore, the first two relations in (3.43) can be written in the form (3.11) for the operator $B = B_k$ and the indicated mapping $F = F^k$. Henceforth, for $\lambda = (\lambda_j)_{j=1}^\infty \in \mathbb{R}^\infty$, we denote

$$\lambda[k] = (\lambda_1, \ldots, \lambda_{k-1}, -\lambda_k, \lambda_{k+1}, \lambda_{k+2}, \ldots) \in \mathbb{R}^\infty.$$

Consequently, $(F^k(\lambda(\cdot)))(\cdot) = \lambda[k]$. Note that $(\lambda[k])[k] = \lambda$ and $(F^k)^{-1} = F^k$. Thus, by virtue of the equality $B_k^2 = 1$, zero is not an eigenvalue of B_k $(k \in \mathbb{N})$.

Since there are countably many operators A_j, there appear to be no difficulties connected with nonmeasurability. Hence, one can apply Theorem 2.6 in Chapter 3 instead of Theorem 2.7 in Chapter 3. This gives us the possibility to exploit the extension (3.4), where τ coincides with $\{-1, 1\} \times \{-1, 1\} \times \ldots = \{-1, 1\}^\infty$. We clarify that for $k \in \mathbb{N}$ we have $A_k^2 = 1$ and, therefore, the spectrum of the operator A_j belongs to $\{-1, 1\}$. Thus, by virtue of the inclusion (1.14) in Chapter 3, we have $s(A) \subset \{-1, 1\}^\infty$ and, hence, $\pi \subset \{-1, 1\}^\infty$, too.

Applying Theorem 3.1 for every $k \in \mathbb{N}$, we obtain the following representation for the operators under consideration

$$H_0 = \int_{\{-1,1\}^\infty} \oplus\, l_2(N_\lambda)\, d\rho(\lambda), \quad (A_j \varphi)\tilde{}(\lambda) = \lambda_j \tilde{\varphi}(\lambda),$$

$$(B_k \varphi)\tilde{}(\lambda) = \left[\frac{d\rho(\lambda[k])}{d\rho(\lambda)}\right]^{1/2} C(\lambda, \lambda[k])\, \tilde{\varphi}(\lambda[k]) \quad (j, k \in \mathbb{N}). \tag{3.44}$$

Here the operators $C(\lambda, \lambda[k])$ defined for ρ-almost all $\lambda \subset \{-1, 1\}^\infty$ act isometrically from the whole of $l_2(N_{\lambda[k]})$ to the whole of $l_2(N_\lambda) = l_2(N_{\lambda[k]})$ and possess the properties

$$C(\lambda, \lambda[k])\, C(\lambda[k], \lambda) = 1,$$

and

$$C(\lambda, \lambda[k])\, C(\lambda[k], \lambda[j]) = C(\lambda, \lambda[j])\, C(\lambda[j], \lambda[k]).$$

We explain that the equality $l_2(N_\lambda) = l_2(N_{\lambda[k]})$, and further relations for $C(\cdot, \cdot)$, reflect (3.43). As above, we now introduce $C(\cdot, \cdot)$ equal to the operators $B(\cdot, \cdot)$ multiplied by the square root of the derivative measure ρ.

The set $\{-1, 1\}$ can be understood as a group with respect to the ordinary multiplication, then $\{-1, 1\}^\infty$ transforms into the direct product of these groups. It follows from the fact that the derivative in (3.44) exists, that N_λ is invariant, and the measure π is quasi-invariant under "finite" group shifts of the point $\lambda \subset \{-1, 1\}^\infty$, i.e., under the shifts by $\mu =$

$(\mu_j)_{j=1}^{\infty} \subset \{-1, 1\}^{\infty}$ with a finite number $\mu_j = -1$.

Example 3.6 (canonical anti-commutation relations (CAR).

We say that a family $(\varphi_j)_{j=1}^{\infty}$ of bounded operators in the space H_0 satisfies CAR, if for anti-commutators, we have

$$\{\psi_j, \psi_k\} = 0, \quad \{\psi_j, \psi_k^*\} = \delta_{jk} 1 \quad (j, k \in \mathbb{N}). \tag{3.45}$$

Consider the selfadjoint operators

$$U_j = \psi_j + \psi_j^*, \quad V_j = i^{-1}(\psi_j - \psi_j^*) \quad (j \in \mathbb{N}). \tag{3.46}$$

In terms of these operators, the equalities (3.45) can be rewritten in the following equivalent form

$$\{U_j, U_k\} = \{V_j, V_k\} = 0 \quad (j, k \in \mathbb{N};\ j \neq k),$$

$$U_j^2 = V_j^2 = 1 \ (j \in \mathbb{N}), \quad \{U_j, V_k\} = 0 \quad (j, k \in \mathbb{N}).$$

One can easily verify that for any $j \in \mathbb{N}$, the operator $A_j = \psi_j^* \psi_j$ ("the operator of a number of particles in a state with number j") is a projector and, moreover, that all these projectors commute. The expansion of H_0 into the direct integral is constructed for the family $A = (A_j)_{j=1}^{\infty}$. As a nuclear space Φ, we take the space, constructed in Remark 5, Subsection 3.4, Chapter 3 in terms of a countable family of bounded operators $(A_j, U_j, V_j)_{j=1}^{\infty}$. It can be easily shown that the family A and the operators U_k are connected as follows

$$[A_j, U_k] = 0 \quad (j, k \in \mathbb{N};\ j \neq k), \quad \{A_k, U_k\} = 1 \quad (k \in \mathbb{N}). \tag{3.48}$$

We fix $k \in \mathbb{N}$ and consider the mapping

$$\mathbb{R}^{\infty} \ni \lambda = (\lambda_j)_{j=1}^{\infty} = \lambda(\cdot) \mapsto (F^k(\lambda(\cdot)))(\cdot) \in \mathbb{R}^{\infty},$$

where $(F^k(\lambda(\cdot)))(j)$ is equal to λ_j for $j \neq k$ and to $1 - \lambda_k$ for $j = k$. Thus, according to (3.6), we have $F_j^k(A) = A_j$ for $j \in \mathbb{N}, j \neq k$ and $F_k^k(A) = 1 - A_k$. The relations (3.48) can be written in the form (3.11) for the operator $B = U_k$ and the mapping $F = F^k$ just introduced. As in the previous example, we have $(F^k)^{-1} = F^k$, and zero is not an eigenvalue of U_k $(k \in \mathbb{N})$.

Let us construct the expansion of the space H_0 into a direct integral for the given family $A = (A_j)_{j=1}^{\infty}$, here we employ Theorem 2.6 in Chapter 3. Each operator A_j is a projector. Therefore, the spectrum of it belongs to $\{0, 1\}$. According to (1.14) in Chapter 3, we have

$$s(A) \subset \{0, 1\} \times \{0, 1\} \times \ldots = \{0, 1\}^{\infty},$$

and the expansion (3.4) with $\tau = \{0, 1\}^\infty$ may be employed. The set $\{0, 1\}$ is an additive group modulo 2. Let us regard $\{0, 1\}^\infty$ as a direct product of this groups and denote the group operation by $\lambda + \mu$. Therefore, this sequence has the form

$$((\lambda_j + \mu_j) \,(\mathrm{mod}\, 2))_{j=1}^\infty \quad (\lambda = (\lambda_j)_{j=1}^\infty,\ \mu = (\mu_j)_{j=1}^\infty \in \{0, 1\}^\infty).$$

We set $\delta_k = (\delta_{jk})_{j=1}^\infty$ $(k \in \mathbb{N})$.

Applying Theorem 3.1 for every $k \in \mathbb{N}$, we obtain the following representation for the operators A_j and U_k:

$$H_0 = \int_{\{0,1\}^\infty} \oplus\, l_2(N_\lambda)\, d\rho(\lambda), \quad (A_j \varphi)^\sim(\lambda) = \lambda_j \tilde{\varphi}(\lambda),$$

$$(U_k \varphi)^\sim(\lambda) = j_k(\lambda) \left[\frac{d\rho(\lambda + \delta_k)}{d\rho(\lambda)}\right]^{1/2} C(\lambda, \lambda + \delta_k)\, \tilde{\varphi}(\lambda + \delta_k),$$

$$j_k(\lambda) = (-1)^{\lambda_1 + \ldots + \lambda_{k-1}}, \quad j, k \in \mathbb{N}. \tag{3.49}$$

Here for ρ -almost all $\lambda = \{0, 1\}^\infty$, the operators $C(\lambda, \lambda + \delta_k)$ just defined act isometrically from the whole of $l_2(N_{\lambda+\delta_k})$ onto the whole of $l_2(N_\lambda) = l_2(N_{\lambda+\delta_k})$ and possess the properties

$$C(\lambda, \lambda + \delta_k)\, C(\lambda + \delta_k, \lambda) = 1 \quad (k \in \mathbb{N}),$$

and

$$C(\lambda, \lambda + \delta_k)\, C(\lambda + \delta_k, \lambda + \delta_j) = C(\lambda, \lambda + \delta_j)\, C(\lambda + \delta_j, \lambda + \delta_k) \quad (j, k \in \mathbb{N}, j \neq k).$$

The measure ρ is quasi-invariant under the shift of $\lambda = \{0, 1\}^\infty$ by a finite (in the usual sense) sequence $\mu \in \{0, 1\}^\infty$, and N_λ is invariant under this shift. It can be easily seen that these properties of $C(\cdot, \cdot)$, of the measure ρ, and of N_λ are the consequence of the relations (3.48) and the equalities

$$\{U_j, U_k\} = 0 \quad (j, k \in \mathbb{N};\ j \neq k), \quad U_j^2 = 1 \ (j \in \mathbb{N}).$$

Now let us consider a representation of the operators V_k . It follows directly from (3.46) that $V_k U_k = i^{-1} (2A_k - 1)$ and, therefore,

$$V_k = i^{-1} (2A_k - 1)\, U_k \quad (k \in \mathbb{N}). \tag{3.50}$$

This implies that V_k are connected with A by the relations (3.48) in which U_k is replaced by V_k. Hence, a representation similar to (3.49) holds for V_k. More precisely, this representation has the form

$$(V_k \varphi)^{\sim}(\lambda) = i^{-1} j_{k+1}(\lambda) \left[\frac{d\rho(\lambda+\delta_k)}{d\rho(\lambda)}\right]^{1/2} C(\lambda, \lambda+\delta_k)\, \tilde{\varphi}(\lambda+\delta_k) \quad (k \in \mathbb{N}). \quad (3.51)$$

The other form of a scalar factor in (3.50) is a consequence of the formula (3.50) (cf. (3.49)).

The formulas (3.49) and (3.50), together with (3.46), give us the representation of the original family $(\psi_j)_{j=1}^{\infty}$.

Example 3.7 (unitary representation of a group of finite $SU(2)$ flows).

Consider a compact group $SU(2)$ of unitary complex 2×2 matrices with the determinant equal to unity. Let us form a product

$$SU(2)^{\infty} = (SU(2)) \times (SU(2)) \times \ldots,$$

consisting of all possible sequences $g = (g_j)_{j=1}^{\infty}$, where $g_j \in SU(2)$. This product is called a group of (countable) flows if we introduce a coordinate-wise multiplication

$$f \cdot g = (f_j)_{j=1}^{\infty} (g_j)_{j=1}^{\infty} = (f_j g_j)_{j=1}^{\infty} \quad (f_j,\ g_j \in SU(2)).$$

Let us consider a subgroup $SU(2)_0^{\infty}$ of it consisting of finite flows, i.e., of the sequences such that all their coordinates except at most finitely many, coincide with the identity of the group $SU(2)$. Let $SU(2)_0^{\infty} \ni g \mapsto U(g)$ be a unitary irreducible representation of this subgroup in the space H_0. We construct a commutative model for the generators of this representation.

Let $su(2)$ be a real Lie algebra of the group $SU(2)$. We can choose a basis a_1, a_2, a_3 in it so that

$$[a_1, a_2] = a_3, \quad [a_2, a_3] = a_1, \quad [a_3, a_1] = a_2. \quad (3.52)$$

Consider $su(2)^{\infty} = (su(2)) \times (su(2)) \times \ldots$ and denote by $su(2)_0^{\infty}$ a collection of finite (in the usual sense) sequences from $su(2)_0^{\infty}$. By introducing the operations on $su(2)_0^{\infty}$ coordinate-wise, we transform $su(2)_0^{\infty}$ into a Lie algebra which can be naturally regarded as a Lie algebra of the group $SU(2)_0^{\infty}$. The basis of this algebra consists of sequences from $su(2)_0^{\infty}$ such that all their elements are equal to zero except the element on the j- th place, where we have a_s $(s = 1, 2, 3)$. This element of the basis is denoted by $a_s^{(j)}$ $(j \in \mathbb{N})$. The commutation law for $a_s^{(j)}$ reflects (3.52) and the fact that the operation of multiplication of two elements $f, g \in SU(2)_0^{\infty}$ is commutative, provided that their elements, which differ from the identity element, occupy distinct places. As a result, we get the following relations

$$[a_1^{(j)}, a_2^{(k)}] = \delta_{jk} a_3^{(k)}, \quad [a_2^{(j)}, a_3^{(k)}] = \delta_{jk} a_1^{(k)},$$

$$[a_3^{(j)}, a_1^{(k)}] = \delta_{jk} a_2^{(k)} \quad (j, k \in \mathbb{N}). \quad (3.53)$$

When constructing the representation $SU(2)_0^{\infty} \ni g \mapsto U(g)$, the base elements $a_s^{(j)}$ of the Lie algebra of $su(2)_0^\infty$ transform into a bounded skew-selfadjoint operators $A_s^{(j)}$ in the space H_0, and the operation $[\cdot,\cdot]$ transforms into an ordinary commutator. Taking (3.53) into account, we get

$$[A_1^{(j)}, A_2^{(k)}] = \delta_{jk} A_3^{(k)}, \quad [A_2^{(j)}, A_3^{(k)}] = \delta_{jk} A_1^{(k)},$$

$$[A_3^{(j)}, A_1^{(k)}] = \delta_{jk} A_2^{(k)} \quad (j, k \in \mathbb{N}). \tag{3.54}$$

Let us introduce the operators $E_0^{(j)} = i A_3^{(j)}$, $E_+^{(j)} = i\,(A_1^{(j)} + i\,A_2^{(j)})$, and $E_-^{(j)} = i\,(A_1^{(j)} - i\,A_2^{(j)})$ and rewrite the commutation conditions (3.54) in the form

$$[E_0^{(j)}, E_+^{(k)}] = \delta_{jk} E_+^{(k)}, \quad [E_0^{(j)}, E_-^{(k)}] = -\delta_{jk} E_-^{(k)},$$

$$[E_+^{(j)}, E_-^{(k)}] = 2\,\delta_{jk} E_0^{(k)},$$

$$(E_0^{(j)})^* = E_0^{(j)}, \quad (E_+^{(j)})^* = E_-^{(j)} \quad (j, k \in \mathbb{N}). \tag{3.55}$$

It follows from the irreducibility of the representation that, for the Casimir operators, we have

$$\Delta_j = \tfrac{1}{2}\,(E_-^{(j)} E_+^{(j)} + E_-^{(j)} E_+^{(j)}) + (E_0^{(j)})^2 = l_j\,(l_j + 1)\,1,$$

where l_j is a certain integer or half-integer number. Moreover, the spectrum s_{l_j} of the operator $E_0^{(j)}$ has the form $s_{l_j} = \{-l_j, -l_j + 1, \ldots, l_j - 1, l_j\}$ $(j \in \mathbb{N})$.

It is clear from (3.55) that the family $A = (A_j)_{j=1}^\infty$, where $A_j = E_0^{(j)}$, consists of commuting selfadjoint operators. Let us construct a decomposition of H_0 into the direct integral (3.4) by taking the set $s_{j_1} \times s_{j_2} \times \ldots$ to play the role of τ (for a fixed irreducible representation). Since the operators which appear in (3.55) are bounded, there are no difficulties connected with the construction of the rigging. The first two equalities in (3.55) have the form (3.11) with a proper choice of the mapping $F(\cdot)$ which maps the space $\mathbb{R}^\infty$ of points $\lambda = (\lambda_j)_{j=1}^\infty$ onto itself. The third equality in (3.55) should be taken into account when we describe the properties of the operators $C(\cdot,\cdot)$ in the commutative model. As a result, we obtain the following representation: Suppose that $\delta_k = (\delta_{jk})_{j=1}^\infty$ $(k \in \mathbb{N})$, then

$$H_0 = \int\limits_{sl_1 \times sl_2 \times \ldots} \oplus\, l_2(N_\lambda)\, d\rho(\lambda), \quad (E_0^{(k)}\varphi)\tilde{}(\lambda) = \lambda_k\, \tilde{\varphi}(\lambda), \tag{3.56}$$

$$(E_+^{(k)}\varphi)\tilde{}(\lambda) = \sqrt{l_k\,(l_k - 1) - \lambda_k\,(\lambda_k - 1)}\, \left[\frac{d\rho(\lambda + \delta_k)}{d\rho(\lambda)}\right]^{1/2} C(\lambda, \lambda + \delta_k)\, \tilde{\varphi}(\lambda + \delta_k),$$

$$(E_{-}^{(k)}\varphi)^{\sim}(\lambda) = \sqrt{l_k(l_k+1) - \lambda_k(\lambda_k+1)}\left[\frac{d\rho(\lambda-\delta_k)}{d\rho(\lambda)}\right]^{1/2} C^*(\lambda-\delta_k,\lambda)\,\tilde{\varphi}(\lambda-\delta_k)$$

$$(\varphi \in \Phi;\ k \in \mathbb{N}).$$

Here and below, we assume, for extreme values, that $\lambda_k + 1 = -l_k$ for $\lambda_k = l_k$ and $\lambda_k - 1 = l_k$ for $\lambda_k = -l_k$. The measure ρ in (3.56) is quasi-invariant with respect to the shift $\lambda \in s_{j_1} \times s_{j_2} \times \ldots$ by δ_k, and the dimensionalities N_λ of the spaces $l_2(N_\lambda)$ are invariant with respect to this shift ($k \in \mathbb{N}$). The operators $C(\cdot,\cdot)$ map the whole of $l_2(N_{\lambda+\delta_k})$ to the whole of $l_2(N_\lambda)$ isometrically. They are connected by the equality

$$C(\lambda,\lambda+\delta_k)\,C(\lambda+\delta_k,\lambda+\delta_j) = C(\lambda,\lambda+\delta_j)\,C(\lambda+\delta_j,\lambda+\delta_k) \quad (j,k \in \mathbb{N};\ j \neq k).$$

In more detail, this example is presented in the book by Yu. S. Samoilenko [2, Chapter 3, Section 7]. Note that one can also investigate a group of flows whose elements g are, for example, continuous mapping $\mathbb{R}^1 \ni t \mapsto g(t) \in SU(2)_0^\infty$. In this case, it is necessary to assume that the rigging exists (see Berezansky, Ostrovsky, and Yu. S. Samoilenko [1]).

Example 3.8 (the representation of canonical commutation relations (CCR)).

Consider two families of Hermitian operators $(\varphi_j)_{j=1}^\infty$ and $(\pi_j)_{j=1}^\infty$ in the Hilbert space H_0. We say that they satisfy the CCR if, for the commutators, we have

$$[\varphi_j,\varphi_k]f = 0,\quad [\pi_j,\pi_k]f = 0,\quad [\varphi_j,\pi_k]f = -i\,\delta_{k}f \quad (j,k \in \mathbb{N}), \tag{3.57}$$

where f runs through a certain linear set $\mathfrak{D}$ (dense in H) such that all the operator equalities (3.57) hold on it.

To obtain the representations of the operators φ_j and π_j, one should demand that $\mathfrak{D}$ satisfies some additional conditions. For example, assume that for any $j \in \mathbb{N}$, the set $\mathfrak{D}$ is invariant under the action of the operator φ_j and π_j and consists of the entire vectors with respect to each operator of this sort (recall that the definition of an entire vector is given in Subsection 1.3 of Chapter 5). The closures of the operators φ_j and π_j given on $\mathfrak{D}$ will be denoted by the same letters. Theorem 1.15 in Chapter 5 easily implies that each family ($(\varphi_j)_{j=1}^\infty$ or $(\pi_j)_{j=1}^\infty$) consists of commuting selfadjoint operators and, moreover, φ_j and π_j $(j \neq k)$ also commute (for example, consider the operators φ_1 and φ_2; we must replace $\mathfrak{D}$ from this theorem by $\mathfrak{D}$ defined above. The conditions of this theorem are satisfied, since for any $z \in \mathbb{C}^1 \setminus \mathbb{R}^1$ $(\varphi_2 - z\,1)\,\mathfrak{D} \subset \mathfrak{D}$ and is dense in H_0 due to the fact that φ_2 is selfadjoint).

Let $\mathfrak{C}_\sigma(\mathbb{R}^\infty) \ni \alpha \mapsto E(\alpha)$ be the joint RI of the family $(\varphi_j)_{j=1}^\infty$. Consider a group of unitary operators

$$\mathbb{R}_0^\infty \ni t \mapsto U(t) = \int_{\mathbb{R}^\infty} \exp\left[i\,(t,\lambda)_{l_2}\right] dE(\lambda). \tag{3.58}$$

The analogous group $\mathbb{R}_0^\infty \ni s \mapsto V(s)$ can be constructed for the family $(\pi_j)_{j=1}^\infty$.

These two groups just constructed satisfy the CCR in Weyl's form (see (3.40))

$$U(t)\,V(s) = \exp\left[i\,(t,s)_{l_2}\right] V(s)\,U(t) \quad (t, s \in \mathbb{R}_0^\infty). \tag{3.59}$$

It can be easily verified that, conversely, the relations (3.59) yield CCR in the form (3.57) for the generators of this group.

The equality (3.59) can be proved as follows. First, we note that the rules of commutation of φ_j and π_k established above imply that each family

$$(U((\underbrace{0,\ldots,0}_{j-1}, t_j, 0, 0, \ldots)))_{j=1}^\infty, \quad (V((\underbrace{0,\ldots,0}_{k-1}, s_k, 0, 0, \ldots)))_{k=1}^\infty$$

consists of unitary commuting operators, and the operators of different families commute for $j \neq k$. Due to this fact, it suffices to verify (3.59) for $t = (t_1, 0, 0, \ldots)$, $s = (s_1, 0, 0, \ldots)$. The corresponding operators and generators of these families are denoted by $U_1(t_1)$, $V_1(s_1)$ $(t_1, s_1 \in \mathbb{R}^1)$ and φ_1, π_1, respectively.

Therefore, we must prove that

$$U_1(t_1)\,V_1(s_1) = \exp(i\,t_1\,s_1)\,V_1(s_1)\,U_1(t_1)$$

or (this is the same) that

$$(U_1(t_1)\,V_1(s_1)f, g)_{H_0} = \exp(i\,t_1\,s_1)\,(V_1(s_1)\,U_1(t_1)f, g)_{H_0} \tag{3.60}$$

$$(t_1, s_1 \in \mathbb{R}^1;\ f, g \in \mathcal{D}).$$

Since f is an entire vector, the expression $V_1(s_1)f$ can be represented in the form of the corresponding power series. The same is true for $U_1^*(t_1)\,g = U_1(-t_1)\,g$. Multiplying this series in the sense of the scalar product, we obtain that (3.60) is equivalent to the equality

$$\sum_{n,m=0}^{\infty} \frac{1}{n!m!}\,((i\,t_1\,\varphi_1)^n\,(i\,s_1\,\pi_1)^m f, g)_{H_0}$$

$$= \exp(i\,t_1\,s_1) \int_{n,m=0}^{\infty} \frac{1}{n!m!}\,((i\,s_1\,\pi_1)^m\,(i\,t_1\,\varphi_1)^n f, g)_{H_0} \quad (t_1, s_1 \in \mathbb{R}^1;\ f, g \in \mathcal{D}).$$

which can be immediately verified due to the commutation relation $[\,i\,t_1\varphi_1, i\,s_1\pi_1] = i\,t_1 s_1 1$. (See also the Baker-Hausdorff formula

$$\exp A \exp B = \exp\left(A + B + \tfrac{1}{2}[A, B]\right) = \exp([A, B])\exp B \exp A\,,$$

which is valid for arbitrary two operators A and B whose commutator is a multiple of the identity, and the series which determine the exponents converge; this formula can be found in Barut, Rączka [1, Chapter 20, Section 2]). Thus, (3.60) and, consequently, the CCR in Weyl's form (3.59) are proved.

■

As shown in the paper by Kosyak [2] (the detailed presentation can be found in the book by Yu. S. Samoilenko [2, Chapter 3, Section 6, Subsection 5]), the CCR in Weyl's form yield the existence of the nuclear rigging (3.1) connected with φ_j and π_j in the standard way, and the space Φ consists of entire vectors for each of these operators ($j \in \mathbb{N}$). Let us introduce the operators of creation $a_j^* = (\varphi_j + i\pi_j)/\sqrt{2}$, of annihilation $a_j = (\varphi_j - i\pi_j)/\sqrt{2}$, and of the number of particles $A_j = a_j^* a_j = \frac{1}{2}(\varphi_j^2 + \pi_j^2 - 1)$ (the latter is often denoted by N_j, $j \in \mathbb{N}$). One can show that, for any $j \in \mathbb{N}$, the operator a_j^* is really adjoint to a_j and the operator A_j is selfadjoint and possesses a discrete spectrum which coincides with $\mathbb{Z}_+$. Furthermore, the above-mentioned rigging (3.1) is connected with a_j^*, a_j, and A_j in the standard way and consists of the entire vectors for A_j.

Using Theorem 1.15 of Chapter 5, we conclude (as earlier) that the operators A_j of the family $A = (A_j)_{j=1}^\infty$ are commuting and selfadjoint. It can be easily calculated that the operators of creation and annihilation are connected with the family A by the relations

$$[A_j, a_k^*]\,\varphi = \delta_{jk}\, a_k^*\, \varphi,$$

and

$$[A_j, a_k]\,\varphi = -\delta_{jk}\, a_k\, \varphi \quad (j, k \in \mathbb{N};\ \varphi \in \Phi)$$

which have the form (3.11) with a mapping $F(\cdot)$ of the space $\mathbb{R}^\infty$ onto itself which is properly defined.

As in Examples 3.5 and (3.6), we can take the space $\mathbb{R}_+^\infty = \mathbb{Z}_+ \times \mathbb{Z}_+ \times \ldots$ to play the role of the set τ in the expansion in the direct integral (3.4). The points of this space are denoted by $\lambda = (\lambda_j)_{j=1}^\infty$. If we apply Theorem 3.1, then, after calculations similar to those carried out above, we arrive at the following representation, we assume that $\delta_k = (\delta_{jk})_{j=1}^\infty$ $(k \in \mathbb{N})$ and then

$$H_0 = \int_{\mathbb{Z}_+^\infty} \oplus\, l_2(N_\lambda)\, d\rho(\lambda), \tag{3.61}$$

$$(a_k^* \varphi)^\sim(\lambda) = \sqrt{\frac{\lambda_k + 1}{2}} \left[\frac{d\rho(\lambda + \delta_k)}{d\rho(\lambda)}\right]^{1/2} C(\lambda, \lambda + \delta_k)\, \tilde{\varphi}(\lambda + \delta_k),$$

$$(a_k \varphi)^\sim(\lambda) = \sqrt{\frac{\lambda_k}{2}} \left[\frac{d\rho(\lambda - \delta_k)}{d\rho(\lambda)}\right]^{1/2} C^*(\lambda - \delta_k, \lambda)\, \tilde{\varphi}(\lambda - \delta_k) \quad (\lambda_k > 0),$$

$$(a_k \varphi)^\sim(\lambda) = 0, \quad \lambda_k = 0 \quad (\varphi \in \Phi;\ k \in \mathbb{N}).$$

Here the measure ρ is quasi-invariant under the shift of λ by $\mu \in \mathbb{Z}_+^\infty \cap \mathbb{R}_+^\infty$, and the dimensionalities N_λ of the spaces $l_2(N_\lambda)$ are invariant under this shift. The operators $C(\lambda, \lambda + \delta_k)$ map the entire $l_2(N_\lambda + \delta_k)$ isometrically onto the entire $l_2(N_\lambda)$. They are connected by the equality

$$C(\lambda, \lambda + \delta_k)\, C(\lambda + \delta_k, \lambda + \delta_j) = C(\lambda, \lambda + \delta_j)\, C(\lambda + \delta_j, \lambda + \delta_k) \quad (j, k \in \mathbb{N};\ j \neq k).$$

By passing from a^*_k and a_k to the operators $\varphi_k = 2^{-1/2}(a^*_k + a_k)$ and $\pi_k = i^{-1}\, 2^{-1/2}(a^*_k + a_k)$ in (3.61), we set the representation of the original CCR (3.57). Another representation of actually the same CCR is given by formulas (3.41) in Example 3.2.

Detailed information about CCR (3.57) and their connection with the CCR in Weyl's form can be found in the following books: Emch [1], Reed and Simon [1,2], Birman and Solomyak [1], Barut and Rączka [1], and Yu. S. Samoilenko [2].

Bibliographical Notes

Volume 1

Here we present references which, being added to those given in the main text, make our reference list more detailed and comprehensive.

First, let us note some books that contain information necessary (to a certain extent) for the reader to understand our book. Thus, the facts concerning topological spaces are given in the book of Pontryagin [1], and it seems useful to be also acquainted with the corresponding sections in the book of Arkhangelsky and Ponomaryov [1]. The presentation of the measure and integration theory can be found in the books of Halmos [1] and Gikhman and Skorokhod [1]; for the theory of linear topological spaces, see Schaefer [1]. We employ the facts from the theory of operators in Hilbert spaces which are mainly contained in the book of Akhiezer and Glazman [1]; many points of view and approaches useful for us can be found in the monographs of Plesner [1], Maurin [2], Birman and Solomyak [1], Dunford and Schwartz [1], and Reed and Simon [1–4]. The required aspect of the theory of generalized functions is presented in the books of Sobolev [1], Schwartz [2], Gelfand and Shilov [1, 2], and Vladimirov [1].

CHAPTER 1

§ 1. Spaces with negative norm first appeared in the papers of Leray [1] and Lax [1] for the particular cases of special Sobolev spaces; later they became a tool frequently used in different areas of analysis. The history of the development of the corresponding concepts and references related to this subject can be found in the book by Berezansky [5]. A more detailed presentation of the constructions in Section 1 and many additional facts (especially those connected with examples of spaces with negative norm) are contained in the books by Berezansky [5, 18, 26]. Note also the book of V. I. Gorbachuk and M. L. Gorbachuk [1], where, in particular, wide classes of inductive and projective limits of spaces with negative norm are constructed; they are important in the investigation of boundary properties of solutions of differential equations. The book of Simon [2] deals with the application of these spaces to quantum field theory.

§ 2. The concept of an infinite tensor product goes back to the work of von Neumann [1]. The "coordinate" construction of a separable subspace of the complete von Neumann tensor product given here is the same as in Berezansky [18, 26]; these books also contain the presentation of a complete tensor product from this point of view. The infinite tensor products of chains and the weighted infinite tensor product were introduced and investigated for the purposes of spectral theory in Berezansky, Gali, and Zhuk [1], Berezansky and Gali [1], Berezansky and Us [1, 2]. For the tensor product of infinitely many operators connected with this construction, see the papers of Berezansky and Us [1, 2] and the books of Berezansky [18, 26]. Another results concerning tensor products of infinitely many operators, including unbounded operators, can be found in Nakagami [1, 2], Araki and Nakagami [1]. In Subsection 2.3, we present a version of the Schwartz kernel theorem with a rigging by Hilbert spaces (L. Schwartz [2]). The presentation is basically the same as in the books of Berezansky [5, 18, 26]; these books also contain many supplementary facts concerning infinite tensor products and the kernel theorem. A. V. Marchenko [1–3] gave a generalization of the concept of the infinite tensor product of Hilbert spaces presented above. Example 2.1 and Lemma 2.3 are taken from the paper of Segal [3].

§ 3. The general theory of bilinear forms in Hilbert spaces was created in the fifties by many mathematicians. For complete presentation of it, see the books of T. Kato [1], Schechter [1], Reed and Simon [1], and Faris [2], where, in particular, one can find the corresponding references. On the other hand, the theory of Hilbert riggings was developed almost at the same time. Certainly, many researchers understood that there existed connections and analogies between these two theories, but these connections were not fully exploited. In Section 3, we follow the work of Berezansky [21] and present the theory of bilinear forms on the basis of Hilbert riggings. Subsection 3.5 (nonclosable form) is written under the influence of the papers of Simon [4] and Koshmanenko [1]. Some of their results are given here. Note also the paper of Koshmanenko [2] where the author classifies singular forms perturbing a given self-adjoint nonnegative operator.

CHAPTER 2

§ 1. The construction of product measures and their properties are discussed in Halmos [1], Shilov and Fan Dyk Tin' [1], Skorokhod [1], and in a great number of books on probability theory, where these measures appear in the course of studying families of independent random variables (see, for example, Shiryaev [1]). The Kakutani theorem (in more general formulation than in Example 1.1) can be found in the books of Skorokhod [1] and Hida [2]. For the first proof of the Kolmogorov–Khinchin criterion, see Kolmogorov–Khinchin [1].

Information about the measures in Hilbert spaces (and, in particular, about Gaussian measures) can be found in the books of Gikhman and Skorokhod [2], Skorokhod [1], Kuo [1], and Yu. L. Daletsky and Fomin [1]. Our presentation is adapted for use in other parts of the book and contains many technical improvements and refinements necessary for this purpose; it generally repeats the presentation in Berezansky and V. G. Samoilenko [3]. The Minlos–Sazonov theorem appeared independently in the works of Minlos [1] and Sazonov [1]. The measures in nuclear riggings were considered in the book of Gelfand and Vilenkin [1]; see also the review of Smolyanov and Fomin [1] and the book of Yu. L. Daletsky and Fomin [1].

§ 2. The measurable linear and multilinear functionals on spaces with Gaussian measures and measurable polynomials are studied in detail in the books of Gikhman and Skorokhod [2], Skorokhod [1], and in the review of Dobrushin and Minlos [1]. Our approach is much simpler, because we consider only the Gaussian case (we do not try to obtain it as a result of more general but complicated investigations). Note that in this approach, an important role is played by the set of smooth polynomials which may not even be introduced when using the probabilistic point of view.

The Fock space was introduced into quantum field theory by V. A. Fock in 1932. The classical reference concerning the description of its mathematical structure is the work of Cook [1]. The connection between Gaussian processes and the Fock space was first mentioned by Segal [1]. In the special cases of the Wiener measure, the polynomials, orthogonal with respect to this measure, were used by Wiener [1] and Itô [1]; the latter constructed the polynomials by using the so-called multiple stochastic integrals. In the book of Simon [1], similar problems are discussed using the methods of constructive field theory.

The conditions of mutual absolute continuity of Gaussian measures are discussed in the books of Skorokhod [1] and Yu. L. Daletsky and Fomin [1]. The Fourier–Wiener transformation was introduced in Cameron and Martin [1] for the case of the Wiener measure. For a canonical Gaussian measure on $\mathbb{R}^\infty$ and on a Hilbert space, this transformation was studied by Guseinov [1, 2] and in the books of Yu. L. Daletsky and Fomin [1] and Hida [2], respectively. Our presentation includes the case of the Gaussian measure with an arbitrary correlation operator.

§ 3. Subsections 3.1 and 3.2 contain a complementary investigation of some chapters of the theory of differential functions on Hilbert spaces, based on the paper of Berezansky and V. G. Samoilenko [3]. For general facts from differential calculus in normalized spaces, see Dieudonné [1] and Shilov [1]. In the presentation of the properties of smooth functions in nuclear riggings and constructing perfect spaces consisting of functions of this sort, we follow the work of Kondratyev and Tsykalenko [1]. Note that this construction is a modification of the scheme used for constructing the spaces of smooth functions which was described in the book of Gelfand and Shilov [2].

§ 4. The construction of the infinite tensor product of nuclear spaces given here is a version of the investigations presented in Berezansky and Yu. S. Samoilenko [1], where the concept of a weighted infinite tensor product of nuclear spaces was introduced, and the theory of generalized functions of infinitely many variables was constructed on the basis of this concept (earlier constructions of this type can be found in Berezansky [7, 9] and Berezansky and Shifrin [1]). The important classes of such test and generalized functions of infinitely many variables, namely, the spaces $\mathcal{A}(\mathbb{R}^\infty)$, $\mathcal{G}(\mathbb{R}^\infty)$ and dual to these, were introduced by Kondratyev and Yu. S. Samoilenko [1]. Note that the Sobolev spaces of functions of infinitely many variables with fixed smoothness were introduced and studied by Frolov [1, 2]; however, some results (such as the quasinuclearity of imbedding) are not valid in these spaces. A less modern presentation of many of the results from this section is contained in the books of Berezansky [18, 26].

§ 5. The contents of this section are based on the works of Kondratyev [3, 4]. The riggings of the Fock space were also considered by Kristensen, Mejlbo, and Poulsen [1]; and by Napiórkowski [1], but in these works the riggings were not nuclear. The Bargman space was introduced in his work [1]. A holomorphic realization of the Fock space was used by Krée and Rączka [1] when studying the operators of quantum field theory. Theorem 5.6 generalizes the result of Kondratyev and Yu. S. Samoilenko [3]. The facts from the theory of holomorphic functions on infinite-dimensional spaces which have been exploited in our presentation can be found in the book of Nachbin [1].

Another method of introducing generalized functions of infinitely many variables, based on the analogy with the finite-dimensional situation, is developed by Krée [1, 2]. He employs the concept of a prodistribution; simplifying somehow, one can say that this concept is connected with the definition of a generalized function given in this book just as the collection of finite-dimensional distributions is connected with the measure on infinite-dimensional space which generates these distributions.

We dwell upon another branch in infinite-dimensional analysis which has now been intensively developed. This approach is called White Noise Analysis; it deals with the construction of the theory of generalized functions of infinitely many variables connected with the rigging of the space $L_2(\mathcal{S}'_{\mathrm{Re}}(\mathbb{R}^1),\gamma)$, where γ is a Gaussian measure on $\mathcal{S}'_{\mathrm{Re}}(\mathbb{R}^1)$ associated canonically with the real Hilbert space $L_{2,\mathrm{Re}}(\mathbb{R}^1)$. In another terminology, the measure γ corresponds to a white noise process. In the situation under consideration, the concept of a generalized function of an infinite-dimensional argument appeared in lectures by Hida [1] when he considered the so-called generalized Brownian functionals. This approach was developed in a series of papers by Kubo and Takenaka [14]; they employed the construction of spaces of test and generalized functions, based on the functional realization of the Fock space rigging, close to that introduced in the works of Kondratyev and Yu. S. Samoilenko [1, 2] and [3, 4]. At the moment, a great number of works exist, which are devoted to White Noise Analysis and its applications. Here we mention only a few reviews: Albeverio, Hida, Potthoff, Röckner and Streit [1, 2], Kuo [2], Hida [3], Hida, Potthoff, and Streit [1]. Many related papers (both reviews and papers devoted to the investigation of special problems) can be found in the proceedings of the conference "White Noise Analysis. Mathematics and Application". (9–15 July, 1989, Bielefeld), Editors T. Hida, H. H. Kuo, J. Potthoff, and L. Streit, World Scientific, Singapore, 1990. See also the book of Hida [2].

Note that for a long period of time, White Noise Analysis and a version of the theory of generalized functions of infinitely many variables presented in Sections 4 and 5 have been developed independently, and this is the reason why a series of results have been obtained independently in both of these approaches. Thus, Theorem 5.3 on the description of the space of test functions $\mathcal{A}(\Phi')$ in the case $\Phi = \mathcal{S}_{\mathrm{Re}}(\mathbb{R}^1)$ is contained in the work of Lee [1], a version of Theorem 5.6 on the description of measures generating generalized functions is also proved in this work. Corollary 2 of this theorem, which is important for applications, is the main result of the papers by Yokoi [1, 2]. The analogue of Theorem 5.4 has been proved in the White Noise Analysis by Potthoff and Streit [1].

CHAPTER 3

§ 1. The approach (developed here) to the investigation of commuting normal operators by means of the construction of a joint resolution of the identity for a countable family of operators can be found in the earlier works of Plesner and Rokhlin [1] and of Plesner [1]. The joint RI for a more than countable family was introduced and exploited in Berezansky [7, 9]. In the proof of Theorem 1.2, we follow the book of Birman and Solomyak [1, Chapter 5, Section 2]. Example 1.2 is taken from Birman, Vershik, and Solomyak [1]. The concept of the support of a joint RI is essential for Section 2. At the same time, the support of a measure is investigated, as a rule, for the case of measures on locally compact spaces; for measures on Tykhonov products, unusual situations may occur (see Subsection 1.7). Example 1.1 is a modification of the example from the book of Berezansky [18, Chapter 2, Section 2, Subsection 6] which can be found in Kosyak and Yu. S. Samoilenko [1]; basically the same example was considered in the earlier paper of Vershik and Sudakov [1]. The main results of this section are contained in the papers [8, 17, 24] and books [18, 26] of Berezansky. They are obtained by using the technique which is, in fact, standard for the theory of random processes (see Gikhman and Skorokhod [2], Ventzel [1]); however, the Tikhonov topologization, which is very important for us, is rarely considered in this theory, and the corresponding consideration has some specific features.

The first presentation of the Kolmogorov theorem can be found in his book [1] and the proof of the Tykhonov theorem (Tykhonov, [1]) is given in the book of Pontryagin [1]. For the investigation of measures on the Tykhonov products, see the book of Bourbaki [2] and the works of Prokhorov [1], Vershik and Sudakov [1], and Smolyanov and Fomin [1]. The noncommutative constructions of Belavkin [1] are connected with the constructions in this section. To read Section 1, it is useful to be acquainted with the corresponding sections of the books of Birman and Solomyak [1], Gikhman and Skorokhod [2], and Ventzel [1].

Also note the following results connected with the constructions in this section. In the work of Kolomytsev and Yu. S. Samoilenko [1], it is established that, for countable families $A = (A_n)_{n=1}^{\infty}$ of commuting selfadjoint operators A_n acting in H, the intersection $\bigcap_{n=1}^{\infty} \mathfrak{D}(A_n)$ is always dense in H. Furthermore, if the joint spectrum of the family A is E-simple (i.e., there exists a vector $f \in H$ such that c.l.s. $(E(\alpha) f) = H$, where E is the joint RI of the family A, and $\alpha \in \mathfrak{C}_\sigma(\mathbb{R}^\infty)$ is arbitrary), then A always possesses a cyclic vector Ω. If the number of unbounded operators is more that countable, then the situation may change. Thus, in the work of Kosyak and Yu. S. Samoilenko [1] one can find examples of selfadjoint representations of the Schwartz space $\mathcal{S}(\mathbb{R}^1) \ni x \mapsto A_x$ such that the joint spectrum of the representation operators $(A_x)_{x \in \mathcal{S}(\mathbb{R}^1)}$ is E-simple, but there is no cyclic vector for this family. The authors also give examples of families of commuting selfadjoint operators $(A_x)_{x \in X}$ such that $\bigcap_{x \in X} \mathcal{D}(A_x) = \{0\}$. These results can be also found in the book of Yu. S. Samoilenko [2, Chapter 1, Section 1, Subsections 3 and 6].

§ 2. The results presented in this section were obtained (in a less perfect form) and applied to the representations of the Wightman-type functionals in Berezansky [7–9]; in Berezansky [17, 23], these results were announced in a form close to that in Section 2. We stress that an improvement of the results is attained only for the situation when there are more than countably many operators. In this case, there appear to be some essential difficulties. An important stimulating role was played by the paper of Maurin [4]. Lemma 2.4, which generalizes the so-called separability lemma (see Naimark [1, Chapter 8, Section 40, Subsection 2]), belongs to Kondratyev. The whole Section 2 is an extended version of the paper by Berezansky [24] and is close to the corresponding section of his book [26]. Unlike the book of Berezansky [18], here the spectral projection theorem (Theorem 2.7) is proved for an arbitrary family of operators (a similar result has been only established before for a family with countably many unbounded operators; in the general case, the problem was to prove that $\mathcal{R}(P(\lambda(\cdot)))$ consists of generalized joint eigenvectors of the family A, see Berezansky [18, p.197]).

The first paper, where the expansions in generalized eigenvectors of a selfadjoint operator were considered, was the work of Gelfand and Kostyuchenko [1]. Later, the paper of Berezansky [1] appeared where expansions with a rigging by Hilbert spaces were studied. In the books of Berezansky [5, 18, 26], one can find a detailed bibliography of the papers in this field. We now only mention that the constructions in Chapter 3 were inspired by the works of Povzner [1], Gårding [2], Maurin [1], and Kats [1–3]; many interesting problems were discussed in the book of Berezin and Shubin [1] and in the review by Simon [6]. For the expansions in eigenfunctions of the Sturm–Liouville operator, see the summarizing books of V. A. Marchenko [1] and Levitan [3].

§ 3. In Subsections 3.4 and 3.5, we present the results obtained in Berezansky [25] (see also his book [26]). The idea of construction of a rigging (Subsection 3.4) goes back to the classical construction in Gårding [1] and generalizes the method of proving the necessity in Theorems 1.2–1.4, and 1.6 in Section 1 of Chapter 4. It is also connected with the construction (Maurin [3]) of a rigging of H_0 by linear topological spaces invariant under the action of a given operator in H_0 and with the results of Yu. S. Samoilenko [2, Chapter 1, Section 1, Theorem 11]. In their paper [2], Kosyak and Yu. S. Samoilenko give the construction of a rigging for representations of inductive limits of commuting locally compact groups; they follow the construction of Gårding, but instead of an invariant measure absent on these groups, they employ a quasi-invariant measure dependent on the representation. The work of Yakovlev [1] contains a general fact which clarifies, in particular, the restriction of the σ-algebra with respect to which the generalized projector $P(\lambda(\cdot))$ is measurable in the case of bounded operators A_x $(x \in X)$ within an "algebraic" approach to the theory of expansions (see Subsection 3.5). For the kernel spectral theorem discussed in Subsection 3.5, see Maurin [1–4], and for the construction of expansions by means of the Chauquet theorem, see Richter [2, 3]. The construction of the Segal isomorphism by using the expansion in generalized eigenvectors (Subsection 3.8) follows the paper of Koshmanenko and Yu. S. Samoilenko [1].

The other subsections of Section 3 are, in fact, modernized and up-to-date versions of the corresponding sections of the books by Berezansky [18, 26]. The direct integrals of Hilbert spaces are considered in the paper of Naimark and Fomin [1], and in the book of Naimark [1]. The result mentioned at the beginning of Subsection 3.2 belongs to Kats [1, 3]. The other results in this subsection belong to Berezansky and Geidarov (Geidarov [1]). These results are improvement of the corresponding facts from the book of Berezansky [5] (concerning the expansion in eigenfunctions of Carleman operators) which enables us to employ these facts for broader classes of operators and, in particular, for the Schrödinger-type operators with a singular potential which were mentioned in Subsection 3.3. Somewhat later, the expansion in eigenfunctions of the Schrödinger operator with a singular potential was obtained in Semenov [2] (see also the earlier work of Gestrin [1]). Then, the local properties of the eigenfunctions of the Schrödinger operator with a singular potential and their behavior at infinity were studied by Orochko [6–9], [13], Bely, Kovalenko, and Semenov [1], Kovalenko and Semenov [1], Kovalenko, Perelmuter, and Semenov [1], and Simon [6]; the work of Perelmuter [1] deals with problems close to these. Note that the main idea of Subsection 3.2 belongs to Kats and Berezansky (see Berezansky [5, Chapter 5, Section 4]) . When verifying the Carleman property for an operator, it may be useful to become acquainted with the works of Dunford and Pettis [1], and Korotkov [1], see also the review of Bukhvalov [1].

Estimates of the growth mentioned at the end of Subsection 3.3 have the following form. For the Schrödinger operator A in $\mathbb{R}^d$ with a continuous potential q semi-bounded from below, almost every (with respect to a spectral measure) generalized eigenfunction $\varphi_\gamma(x;\lambda)$ increases as $|x| \to \infty$ no faster than $c_\varepsilon |x|^{d/2+\varepsilon}$ $(\varepsilon > 0$ is arbitrary). For $d = 1$, this result was obtained by Shnol [1] who employed the classical methods of the theory of the Sturm–Liouville equation. For $d = 2$, the same result was obtained on the basis of the theory of expansions in generalized eigenfunctions by Berezansky, Kats, and Kostyuchenko (see Berezansky [3, 5]). These estimates determine the spectrum of the operator A. Indeed, if for some λ there exists a solution of the equation

$$-\Delta u + q(x)u = \lambda u \quad (x \in \mathbb{R}^d),$$

satisfying this estimate (or even less restrictive), then λ belongs to the spectrum of A (Shnol [1], see also the book of Glazman [1]). From the physical point of view, it is desirable to replace $c_\varepsilon |x|^{d/2+\varepsilon}$ by a constant. The question of whether this is possible is still open and has a rather dramatic history which has been partially described in the review by Simon [6] (Faris and Simon made their own contribution to this drama).

The results presented here are connected with earlier estimates of the integrals of eigenfunctions of a Carleman operator (and, in particular, of an elliptic operator) which were obtained by Berezansky and with the subsequent estimates of averaged eigenfunctions by Berezansky and Orochko, and by Orochko (Berezansky [5, Chapter 5, Section 5], Berezansky and Orochko[1], Orochko [3]).

Recently, V. P. Maslov, S. A. Molchanov, and A. Ya. Gordon have established the incorrectness of the assertion that almost all generalized eigenfunctions of this operator are bounded. Further, there exists a cycle of papers where the above estimates of Shnol–Berezansky–Kats–Kostyuchenko are proved for singular potentials (the form of the estimates is preserved) (Kovalenko and Semenov [1], Kovalenko, Perelmuter, and Semenov [1], and Simon [6]). Finally, for potentials which are not semibounded below, results such as in Shnol [1] (that the presence of the estimate involves the fact that the corresponding eigenvalue belongs to the spectrum) are contained in the paper of Simon [5]. In the book of Berezin and Shubin [1], devoted to the detailed study of the Schrödinger operator, one can also find the presentation of many problems connected with Subsections 3.1–3.3 and with the above-mentioned estimates of growth. The results in Subsections 3.6 and 3.7 belong to Berezansky and Us [1, 2] and to Us [2, 3].

CHAPTER 4

§ 1. The approach to the construction of spectral representations for families of commuting normal operators connected by certain relations which is studied here, was announced in the papers of Berezansky [17, 19]; Theorems 1.2 and 1.6 were announced in Berezansky [7] and Berezansky, Gali, and Kondratyev [1]. For the concepts which appear in this section, see the books of Schaefer [1], Pontryagin [1], Gelfand, Raikov, and Shilov [1], and Naimark [1]. Many classical results concerning groups and semigroups of operators (the Stone theorem and Sz.-Nagy–Hille theorem) are contained in the books of Riesz and Sz.-Nagy [1] and Hille and Phillips [1]. For the original presentation of the Sz.-Nagy–Hille theorem, see Sz.-Nagy [1, 2] and Hille [1]. The argument, which makes it possible to reduce the set where the joint RI is concentrated in the case of the family of operators $(A_x)_{x \in X}$ with algebraic relations, was employed by Korsynsky [1, 2] (in these papers, the author used the facts about expansions which are weaker than those given in Section 2 of Chapter 3). Theorems 1.1 and 1.2 and the results of Subsection 1.6 (formula (1.32)) are well known in the cases of a locally compact commutative group and of a normalized commutative algebra (see Naimark [1, Chapter 6, Section 3, Subsection 7; Chapter 4, Section 17, Subsection 4]). Theorem 1.3 and a part of Theorem 1.4 can be found in Tishchenko [1], [2]. In addition to Stone's theorem and the Sz.-Nagy–Hille theorem, in Section 1, we discuss and generalize spectral representations studied in a number of papers devoted to the development of the indicated theorems, namely, in Phillips [1], Devinatz [1], Nussbaum [1, 2], Getoor [1], Ionescu Tulcea [1], and Devinatz and Nussbaum [2]. In a less complete form, the results of Section 1 were given in the paper of Berezansky [24] and in his books [18], [26]. The papers of Yakovlev [2, 3], where the representations of a commutative nuclear $*$-algebra by unbounded operators have been studied in detail, are connected with the results of Subsection 1.5. General topological algebras, their representations, and algebras of unbounded operators are investigated in Powers [1], Lassner [1], and Borchers and Yngvanson [1].

When speaking about the contents of Section 1, we must mention the book of Yu. S. Samoilenko devoted to the spectral problems of countable families of operators connected by commutation relations or by more general algebraic relations, and the paper of Nussbaum devoted to semigroups of Hermitian operators.

The paper of Vershik [1] adjoins the contents of Section 1. It contains investigation of problems with difficulties close to those appearing in Section 2 of Chapter 3.

§ 2. In 1950, when investigating the theory of the Delsarte–Levitan operators of generalized shift, Berezansky and S. G. Krein introduced a concept of a (commutative) hypercomplex system with a continuous basis, and constructed the harmonic analysis on it (see the review of Berezansky and S. G. Krein [1] and the book of Levitan [2]). These works dealt mainly with hypercomplex systems having compact or discrete bases; the theory for a locally compact case has been recently developed by Berezansky and Kalyuzhny [1] and Kalyuzhny [2]. On the other hand, in the seventies, many mathematicians independently introduced the concept of a hypergroup and constructed the theory of hypergroups (Dunkl [1, 2], Jewett [1], Spector [1], and others, see also the review by Ross [1]). Later, it has become clear that the commutative hypergroups and the so-called convolution algebras are covered by hypercomplex systems. For the results obtained in these theories and for the correlation between them, see the reviews by Litvinov [1], Vainerman [1], and the paper of Berezansky and Kalyuzhny [4] (the small part of the last one is presented in Subsections 2.1 and 2.2). The construction of the theory of generalized functions on the basis of a hypercomplex system (Subsection 2.3) belongs to Berezansky and Kalyuzhny [1]; it generalizes and makes more precise the construction of Pytlik [1] for groups. Another method of construction of the spaces of test functions on locally compact groups was proposed by Kats [4].

The results in Subsection 2.4 belong to Berezansky and Kalyuzhny [2, 3]; they generalize the results in Ionescu Tulcea and Simon [1], Kurepa [1], and Maltese [1, 2]. Note that Theorem 2.4, being formulated as in Section 2, remains valid also for unbounded operators A_p. However, if it is formulated as in Berezansky and Kalyuzhny [2, 3], then one must demand, in addition, that the representation $Q \ni p \mapsto A_p$ should be expandable into a sum of cyclic representations, i.e., that there should exist an expansion $H_0 = \bigoplus_{k=1}^{\infty} H_k$ such that for the linear set $\mathfrak{D}$ defined in these papers, the set $\mathfrak{D} \cap H_k$ is invariant with respect to A_p $(p \in Q)$, and in each H_k there exists a vector $\varphi_k \in \mathfrak{D} \cap H_k$ such that c.l.s. $\{A_p\varphi_k | p \in Q\} = H_k$. If there exists at most a countable subset $Q_0 \subset Q$ such that $\bigcap_{p\in Q_0} \mathfrak{D}(A_p) = \bigcap_{p\in Q} \mathfrak{D}(A_p)$, then the representation A_p is expandable into a sum of cyclic representations. This simple sufficient condition is due to Ionescu Tulcea [1]. The references concerning Examples 2.3–2.9 can be found in the review by Berezansky and S. G. Krein [1]. We only note that the construction of hypercomplex systems in Examples 2.5 and 2.6 has been stimulated by the paper of M.G.Krein [4]; Example 2.7 is also considered in the review by Berezansky and Kalyuzhny [5]. Example 2.10 is taken from the paper of Berezansky [19].

§ 3. The results in Subsections 3.1 and 3.2 belong to Berezansky, Ostrovsky, and Yu. S. Samoilenko [1]. With the help of the spectral projection theorem, these results extend the methods of constructing different commutative models to families with arbitrary cardinality consisting, in general, of unbounded operators. These models have appeared in different contexts in the papers of Gårding and Wightman [1], Araki [1], Gelfand and Vilenkin [1], Vershik, Gelfand, and Graev [1–3], Hegerfeldt and Melsheimer [1], Kirillov [1, 2], Golodets [1], Ismagilov [2, 3], Albeverio and Høegh-Krohn [3], Strătilă and Voiculescu [1], and others.

As far as the general Example 3.1 is concerned, see the works of Mackey [1, 2] in which locally compact groups are considered. The representation (3.41) in Example 3.2 belongs to Gelfand and Vilenkin [1]. Example 3.4 and its physical interpretation are studied, for example, in Menikoff and Sharp [1], and Bogolyubov (jr) and Prikarpatsky [1]. Example 3.5 belongs to Kolomytsev and Yu. S. Samoilenko [1], and Yu. S. Samoilenko [2]. Examples 3.6 and 3.8 are the classical results of Gårding and Wightman [1, 2] presented on the basis of the technique developed here, another exposition of Example 3.6 can be found in Golodets [1]. Example 3.7 is studied by Yu. S. Samoilenko [2]. This section is closely connected with other results from the book of Yu. S. Samoilenko and the review by Ostrovsky and Yu. S. Samoilenko [1]. For the concepts of the theory of representations of groups which appear in this book and, in particular, in this chapter, see the book of Barut and Rączka [1].

References

de Acosta, A. D.

[1] *Existence and convergence of probability measures in Banach spaces,* Trans. Amer. Math. Soc. **152** (1970), 273–298.

Akhiezer, N.I.

[1] The Classical Moment Problem and Some Related Problems in Analysis, Fizmatgiz, Moscow, 1961; ***English transl.*** Hafner, New York, 1965.

Akhiezer, N.I. and Glazman, I.M.

[1] The Theory of Linear Operators in Hilbert Space, Nauka, Moscow, 1966; ***English transl. of 1st edn.*** Ungar, New York, 1961.

Albeverio, S., Hida,T., Potthoff, J., Röckner, M., and Streit, L.

[1] *Dirichlet forms in terms of white noise analysis* **I**: *Construction and QFT examples,* Reviews in Math. Phys. **1** (1990), 291–312.

[2] *Dirichlet forms in terms of white noise analysis* **II**: *Closability and diffusion processes,* Reviews in Math. Phys. **1** (1990), 313–323.

Albeverio, S. and Høegh-Krohn, R.

[1] *Dirichlet forms and diffusion processes on rigged Hilbert spaces,* Z. Wahrsch. und verw. Geb. **40** (1977), 1–57.

[2] *Hunt processes and analytic potential theory on rigged Hilbert spaces,* Ann. Inst. H. Poincaré **A 13** (1977), 269–291.

[3] *The energy representation of Sobolev–Lie groups,* Compositio Math. **36** (1978), 37–52.

[4] *Diffusion fields, quantum fields, and fields with values in Lie groups,* Stochastic Analysis and Applications. Dekker, New York, 1985, pp. 1–98.

Albeverio, S., Høegh-Krohn, R., and Streit, L.

[1] *Energy forms, Hamiltonians, and Brownian paths,* J. Math. Phys. **18** (1977), 907–917.

[2] *Regularization of Hamiltonians and processes,* J. Math. Phys. **21** (1980), 1636–1642.

Araki, H.

[1] *Hamiltonian formalism and canonical relations in quantum field theory,* J. Math. Phys. **1** (1960), 492–504.

Araki, H. and Nakagami, Y.

[1] *A remark on an infinite tensor product of von Neumann algebras,* Publ. Res. Inst. Math. Sci. Kyoto Univ. **8** (1972), 363–374.

Arkhangelsky, A.V. and Ponomaryov, V.I.

[1] Foundations of General Topology in Problems and Exercises, Nauka, Moscow, 1974. (*Russian*)

Bargman, V.

[1] *Hilbert spaces of analytic functions,* Commun. Pure and Appl. Math. **14** (1961), 187–214.

Barut, A. and Rączka, R.

[1] Theory of Group Representations and Applications, PWN, Warsaw, 1977.

Bateman, H. and Erdeyi, A.

[1] Higher Transcendental Functions, vol. **2**, Nauka, Moscow, 1966. (*Russian translation*)

Belavkin, V.P.,

[1] *A reconstruction theorem for a random quantum field,* Uspekhi Mat. Nauk **39** (1984), No. 2 (236), 137–138; ***English transl.*** in Russian Math. Surveys **39** (1984).

Belichkova, N.V.

[1] *Selfadjointness of degenerating elliptic operators,* Dokl. Akad. Nauk Ukrain. SSR Ser. A, No. 1 (1984), 3–6. *(Russian)*

Bely, A.G., Kovalenko, V.F., and Semenov, Yu.A.

[1] *On the continuity of generalized eigenfunctions of Schrödinger operators,* Rep. Math. Phys. **12** (1977), 307–310.

Berezansky, Yu.M.

[1] *On expansion in eigenfunctions of general selfadjoint differential operators,* Dokl. Akad. Nauk SSSR **108** (1956), 379–382. *(Russian)*

[2] *Generalization of the Bochner theorem to expansions in eigenfunctions of partial differential equations,* Dokl. Akad. Nauk SSSR **110** (1956), 893–896. *(Russian)*

[3] *On expansion in eigenfunctions of selfadjoint operators,* Ukrain. Mat. Zh. **11** (1959), 16–24; ***English transl.*** in Amer. Math. Soc. Transl. (2) **93** (1970).

[4] *A generalization of the multi-dimensional Bochner theorem,* Dokl. Akad. Nauk SSSR **136** (1961), 1011–1014; ***English transl.:*** Soviet Math. Dokl. **2** (1961), 143–147.

[5] Expansions in eigenfunctions of selfadjoint operators, Naukova Dumka, Kiev, 1965; ***English transl.:*** Amer. Math. Soc. Transl. vol. **17**, Providence, R.I., 1968.

[6] *On selfadjointness of field operators and on integral representations of Wightman type functionals,* Ukrain. Mat. Zh. **18** (1966), No. 3, 3–12. *(Russian)*

[7] *Integral representations of positive definite functionals of Wightman type,* Ukrain. Mat. Zh. **19** (1967), 89–95; ***English transl.*** in Ukrain. Math. J. **19** (1967).

[8] *Expansion in generalized eigenvectors and integral representation of positive definite kernels in the form of a functional integral,* Sibirsk. Mat. Zh. **9** (1968), 998–1013; ***English transl.*** in Siberian Math. J. **9** (1968).

[9] *Representation of functionals of Wightman type by functional integrals,* Funk. Anal. i Prilozh. **3** (1969), No. 2, 1–13; ***English transl.*** in Funct. Anal. Appl. **3** (1969).

[10] *The generalized power moment problem,* Trudy Mosk. Mat. Obshch. **21** (1970), 47–102; ***English transl.*** in Trans. Mosc. Math. Soc. **21** (1970).

[11] *On the generalized power moment problem,* Ukrain. Mat. Zh. **22** (1970), 435–460; ***English transl.*** in Ukrain. Math. J. **22** (1970).

[12] *A general theorem on moment matrices,* Dokl. Akad. Nauk Ukrain. SSR Ser. A (**1971**), No. 4, 291–294. *(Russian)*

[13] *Generalized power moment problems,* Hilbert Space Operators and Operator Algebras (Tihany, Hungary, 1970), North-Holland, Amsterdam, 1972, pp. 11–20.

[14] *Selfadjointness of elliptic operators with a singular potential,* Ukrain. Mat. Zh. **26** (1974), 579–590; ***English transl.*** in Ukrain. Math. J. **26** (1974).

[15] *A remark about essential selfadjointness of powers of an operator,* Ukrain. Mat. Zh. **26** (1974), 790–793; ***English transl.*** in Ukrain. Math. J. **26** (1974).

[16] *Selfadjointness of elliptic operators with infinitely many variables,* Ukrain. Mat. Zh. **27** (1975), 729–742; ***English transl.*** in Ukrain. Math. J. **27** (1975).

[17] *On expansion in joint generalized eigenvectors of an arbitrary family of commuting normal operators,* Dokl. Akad. Nauk SSSR **229** (1976), 531–533; ***English transl.*** in Soviet Math. Dokl. **17** (1976).

[18] Selfadjoint Operators in Spaces of Functions of Infinitely Many Variables, Naukova Dumka, Kiev, 1978; ***English transl.*:** see Berezansky [26].

[19] *Spectral representations of solutions of certain classes of functional and differential equations,* Dokl. Akad. Nauk Ukrain. SSR Ser. A (**1978**), No. 7, 579–583. *(Russian)*

[20] *Representations of positive definite kernels by functional integrals,* Operators in Mathematical Physics and Infinite-Dimensional Analysis, Inst. Mat. Akad. Nauk Ukrain. SSR, Kiev, 1979, pp. 5–35. *(Russian)*

[21] *Bilinear forms and Hilbert riggings,* Spectral Analysis of Differential Operators, Inst. Mat. Akad. Nauk Ukrain. SSR, Kiev, 1980, pp. 83–106; ***English transl.*** **:** Selecta Math. Sovietica **9** (1990), 235–253.

[22] *On selfadjointness of infinite-dimensional elliptic differential operators with singular potentials,* Funk. Anal. i Prilozh. **16** (1982), No. 4, 55–56; ***English transl.*** in Funct. Anal. Appl. **16** (1982).

[23] *The spectral theory of commuting normal operators and applications of it,* Uspekhi Mat. Nauk **38** (1983), No. 5 (233), 138. *(Russian)*

[24] *The spectral projection theorem,* Uspekhi Mat. Nauk **39** (1984), No. 4 (238), 3–52; ***English transl.*** in Russian Math. Surveys **39** (1984).

[25] *On the spectral projection theorem,* Ukrain. Mat. Zh. **37** (1985), 146–154. *(Russian)*

[26] Selfadjoint Operators in Spaces of Functions of Infinitely Many Variables, Amer. Math. Soc. Transl. vol. **63**, Providence, R.I., 1986.

Berezansky, Yu.M. and Gali, I.M.

[1] *Positive definite functions of infinitely many variables on a layer,* Ukrain. Mat. Zh. **24** (1972), 435–464; ***English transl.*** in Ukrain. Math. J. **24** (1972).

Berezansky, Yu.M., Gali, I.M., and Kondratyev, Yu.G.

[1] *Stone's theorem for the additive group of a Hilbert space,* Funk. Anal. i Prilozh. **11** (1977), No. 4, 68–69; ***English transl.*** in Funct. Anal. Appl. **11** (1977).

Berezansky, Yu.M., Gali, I.M., and Zhuk, V.A.

[1] *On positive definite functions of infinitely many variables,* Dokl. Akad. Nauk SSSR **203** (1972), 13–15; ***English transl.*** in Soviet Math. Dokl. **13** (1972).

Berezansky, Yu.M. and Gorbachuk, M.L.

[1] *On extension of positive definite functions of two variables,* Ukrain. Mat. Zh. **17** (1965), No. 5, 96–102. *(Russian)*

Berezansky, Yu.M. and Kalyuzhny, A.A.

[1] *Nuclear spaces of functions on the basis of a hypercomplex system,* Ukrain. Mat. Zh. **35** (1983), 9–17; ***English transl.*** in Ukrain. Math. J. **35** (1983).

[2] *Representations of hypercomplex systems with locally compact basis,* Ukrain. Mat. Zh. **36** (1984), 417–421; ***English transl.*** in Ukrain. Math. J. **36** (1984).

[3] *Spectral decompositions of representations of hypercomplex systems,* Spectral Theory of Operators and Infinite-Dimensional Analysis, Inst. Mat. Akad. Nauk Ukrain. SSR, Kiev, 1984, pp. 4–19. *(Russian)*

[4] *Hypercomplex systems with locally compact bases,* Selecta Math. Sovietica **4** (1985), 151–200.

[5] *Hypercomplex systems constructed from orthogonal polynomials,* Ukrain. Mat. Zh. **38** (1986), 275–284. *(Russian)*

Berezansky, Yu.M., Kondratyev, Yu.G., and Samoilenko, Yu.S.

[1] *Generalized functions of infinitely many variables and their applications in spectral theory,* Generalized Functions and Their Applications in Mathematical Physics (Proc. Internat. Conf., Moscow, 1980), Vychisl. Tsentr Akad. Nauk SSSR, Moscow, 1981, pp. 50–70. *(Russian)*

Berezansky, Yu.M. and Krein, S.G.

[1] *Hypercomplex systems with a continuous basis,* Uspekhi Mat. Nauk **12** (1957), No. 1 (73) 147–152; ***English transl.*** in Amer. Math. Soc. Transl. (2) **16** (1960).

Berezansky, Yu.M., Lassner, G., and Yakovlev, V.S.

[1] *On decomposition of positive functionals on commutative nuclear $*$-algebras,* Ukrain. Mat. Zh. **39** (1987), 638–641. *(Russian)*

Berezansky, Yu.M. and Orochko, Yu.B.

[1] *A remark concerning the growth of eigenfunctions of selfadjoint operators,* Ukrain. Mat. Zh. **14** (1962), No. 2, 180–184. *(Russian)*

Berezansky, Yu.M., Ostrovsky, V.L., and Samoilenko, Yu.S.

[1] *Expansions in eigenfunctions of the families of commuting operators and the representations of commutation relations,* Ukrain. Mat. Zh. **40** (1988), 106–109. *(Russian)*

Berezansky, Yu.M. and Samoilenko, V.G.

[1] *On selfadjointness of differential operators with finitely or infinitely many variables,* Dokl. Akad. Nauk Ukrain. SSR Ser. A (**1979**), No. 9, 691–695. *(Russian)*

[2] *Selfadjointness of differential operators with finitely or infinitely many variables and evolution equations,* Uspekhi Mat. Nauk **36** (1981), No. 5 (221), 3–56; ***English transl.*** in Russian Math. Surveys **36** (1981).

[3] *Selfadjointness of infinite-dimensional elliptic differential operators,* Trudy Mosk. Mat. Obshch. **50** (1987), 3–54. *(Russian)*

Berezansky, Yu.M. and Samoilenko, Yu.S.

[1] *Nuclear spaces of functions of infinitely many variables,* Ukrain. Mat. Zh. **25** (1973), 723–737; ***English transl.*** in Ukrain. Math. J. **25** (1973).

Berezansky, Yu.M. and Shifrin, S.N.

[1] *The generalized symmetric power moment problem,* Ukrain. Mat. Zh. **23** (1971), 291–306; ***English transl.*** in Ukrain. Math. J. **23** (1971).

Berezansky, Yu.M. and Us, G.F.

[1] *On expansions in eigenfunctions of selfadjoint operators admitting separation of infinitely many variables,* Dokl. Akad. Nauk SSSR **213** (1973), 1005–1008; ***English transl.*** in Soviet Math. Dokl. **14** (1973).

[2] *Eigenfunction expansions of operators admitting separation of an infinite number of variables,* Rep. Math. Phys. **7** (1975), 103–126.

Berezansky, Yu.M., Zhernakov, N.V., and Us, G.F.

[1] *Operator stochastic integrals,* Ukrain. Mat. Zh. **39** (1987), 144–149. *(Russian)*

Berezin, F.A.

[1] The Second Quantization Method, Nauka, Moscow, 1986. *(Russian)*

Berezin, F.A. and Shubin, M.A.

[1] The Schrödinger Equation, Izdat. Moskov. Univ., Moscow, 1983. *(Russian)*

Billingsley, P.

[1] Convergence of Probability Measures, Wiley & Sons, New York, 1968.

Birman, M.Sh. and Solomyak, M.Z.

[1] The Spectral Theory of Selfadjoint Operators in Hilbert Spaces, Izdat. Leningrad. Univ., Leningrad, 1980. *(Russian)*

Birman, M.Sh., Vershik, A.M., and Solomyak, M.Z.

[1] *A product of commutative spectral measures can fail to be countably additive,* Funk. Anal. i Prilozh. **13** (1979), No. 1, 61–62; ***English transl.*** in Funct. Anal. Appl. **13** (1979).

Bogolyubov, N.N. and Bogolyubov, N.N. (Jr.)

[1] Introduction to Quantum Statistical Mechanics, Nauka, Moscow, 1984. *(Russian)*

Bogolyubov, N.N., Logunov, A.A., Oksak, A.I., and Todorov, I.T.

[1] General Principles of Quantum Field Theory, Nauka, Moscow, 1987. *(Russian)*

Bogolyubov, N.N. (Jr.) and Prikarpatsky, A.K.

[1] *Quantum method of N.N. Bogolyubov's generating functionals in statistical physics: the Lie algebra of currents, its representations and functional equations,* Ukrain. Mat. Zh. **38** (1986), 284–289. *(Russian)*

Borchers, H.J. and Yngvason, J.

[1] *On the algebra of field operators. The weak commutant and integral decomposition of states,* Comm. Math. Phys **42** (1975), 231–252.

[2] *Integral representations for Schwinger functionals and the moment problem over nuclear spaces,* Comm. Math. Phys **43** (1975), 255–271.

[3] *Necessary and sufficient conditions for integral representations of Wightman functionals at Schwinger points,* Comm. Math. Phys **47** (1976), 197– 213.

Borchers, H.J. and Zimmerman, W.

[1] *On the self-adjointness of field operators,* Nuovo Cim. **31** (1964), 1047–1059.

Botwich, D.D. and Malyshev, V.A.

[1] *Unitary equivalence of temperature dynamics for ideal and locally perturbed Fermi-gas,* Comm. Math. Phys **91** (1983), 301–312.

Bourbaki, N.

[1] General Topology. Real Numbers in General Topology. Functional Spaces, Nauka, Moscow, 1975. *(Russian translation)*

[2] Integration. Measures on Locally-Compact Spaces. Extension of a Measure. Integration of Measures. Measures on Separable Spaces, Nauka, Moscow, 1977 *(Russian translation)*

Brezis, H. and Kato, T.

[1] *Remarks on the Schrödinger operators with singular complex potentials,* J. Math. Pures Appl. **58** (1979), 137–151.

Brüning, E.

[1] *On the characterization of relativistic quantum field theories in terms of finitely many vacuum expectation values* **I**, Comm. Math. Phys **58** (1978), 139–166; **II**, Comm. Math. Phys **58** (1978), 167–194.

Bukhvalov, V.B.

[1] *Applications of the methods in the theory of order-bounded operators to the theory of operators in the spaces L^p*, Uspekhi Mat. Nauk **38** (1983), No. 6 (234), 37–83; ***English transl.*** in Russian Math. Surveys **38** (1983).

Cameron, R. H. and Martin, W.T.

[1] *Fourier–Wiener transforms of analytic functionals,* Duke Math. J. **12** (1945), 489–507.

Chaus, N. N.

[1] *On classes of uniqueness for solutions of the Cauchy problem and representations of positive definite kernels,* Dokl. Akad. Nauk SSSR **163** (1965), 36–39; ***English transl.*** in Soviet Math. Dokl. **6** (1965).

[2] *Classes of uniqueness for solutions of the Cauchy problem and representations of positive definite kernels,* Proc. Sem. on Functional Analysis, vol. 1, Inst. Mat. Akad. Nauk Ukrain. SSR, Kiev, 1968, pp. 176–270. *(Russian)*

Chernoff, P. D.

[1] *Some remarks on quasi-analytic vectors,* Trans. Amer. Math. Soc. **167** (1972), 105–113.

[2] *Essential self-adjointness of powers of generators of hyperbolic equations,* J. Funct. Anal. **12** (1973), 401–414.

[3] *Quasi-analytic vectors and quasi-analytic functions,* Bull. Amer. Math. Soc. **81** (1975), 637–646.

[4] *Schrödinger and Dirac operators with singular potentials and hyperbolic equations,* Pacific J. Math. **72** (1977), 361–382.

Chumak, A.A.

[1] *Selfadjointness of the Beltrami–Laplace operator on a complete paracompact Riemannian manifold without boundary,* Ukrain. Mat. Zh. **25** (1973), 784–791; ***English transl.*** in Ukrain. Math. J. **25** (1973).

Cook, J. M.

[1] *The mathematics of second quantization,* Trans. Amer. Math. Soc. **81** (1956), 106–134.

Daletsky, A. Yu.

[1] *Representations of translation groups of a nuclear space,* Dokl. Akad. Nauk Ukrain SSR Ser. A (1982), No. 12, 9–11. *(Russian)*

[2] *The moment problem on finite-dimensional Lie algebras,* Spectral Theory of Operators in the Problems of Mathematical Physics, Inst. Mat. Akad. Nauk Ukrain. SSR, Kiev, 1983, pp. 93–97. *(Russian)*

[3] *Integration of representations of nuclear Lie algebras of smooth currents,* Spectral Theory of Operators and Infinite-Dimensional Analysis, Inst. Mat. Akad. Nauk Ukrain. SSR, Kiev, 1984, pp. 77–92. *(Russian)*

Daletsky, A.Yu. and Samoilenko, Yu.S.

[1] *The noncommutative moment problem,* Funk. Anal. i Prilozh. **21** (1987), No. 2, 72–73. *(Russian)*

Daletsky, Yu.L.

[1] *Infinite-dimensional elliptic operators and the parabolic equations connected with them,* Uspekhi Mat. Nauk **22** (1967), No. 4 (136), 3–54; ***English transl.*** in Russian Math. Surveys **22** (1967).

[2] *On selfadjointness and maximal dissipativeness of differential operators for functions of an infinite-dimensional argument,* Dokl. Akad. Nauk SSSR **227** (1976), 784–787; ***English transl.*** in Soviet Math. Dokl. **17** (1976).

[3] *Probability methods in some problems of infinite-dimensional analysis,* Random Operators and Distributions in Banach Spaces, Inst. Mat. Akad. Nauk Ukrain. SSR, Kiev, 1977, pp. 108–124. *(Russian)*

Daletsky, Yu.L. and Fomin, S.V.

[1] Measures and Differential Equations in Infinite-Dimensional Spaces, Nauka, Moscow, 1983; ***English transl.:*** Kluwer AP, Dordrecht, 1991.

Devinatz, A.

[1] *A note on semi-groups of unbounded self-adjoint operators,* Proc. Amer. Math. Soc. **5** (1954), 101–102.

[2] *Essential selfadjointness of certain partial differential operators on R^n,* Proc. Amer. Math. Soc. **60** (1976), 235–242.

[3] *Selfadjointness of second-order elliptic and degenerate elliptic differential operators,* Differential Equations (Proc. Internat. Conf., Uppsala, 1977), Almqvist & Wiksell, Stockholm, 1977, pp. 37–51.

[4] *Essential self-adjointness of Schrödinger-type operators,* J. Funct. Anal. **25** (1977), 58–69.

[5] *Selfadjointness of second order degenerate-elliptic operators,* Indiana Univ. Math. J. **27** (1978), 255–266.

[6] *On an inequality of Tosio Kato for degenerate-elliptic operators,* J. Funct. Anal. **32** (1979), 312–335.

Devinatz, A. and Nussbaum, A.E.

[1] *On the permutability of normal operators,* Ann. of Math. **65** (1957), 144–152.

[2] *Real characters of certain semi-groups with applications,* Duke Math. J. **28** (1961), 221–238.

Dieudonné, J.

[1] Fundamentals of Modern Analysis, Mir, Moscow, 1964. *(Russian translation)*

Dobrushin, R.L. and Minlos, R.A.

[1] *Polynomials of linear random functions,* Uspekhi Mat. Nauk **32** (1977), No. 2 (194), 67–122. *(Russian)*

Dunford, N. and Pettis, B.J.

[1] *Linear operations on summable functions,* Trans. Amer. Math. Soc. **47** (1940), 323–392.

Dunford, N. and Schwartz, J.T.

[1] Linear Operators, vol. **I**: General Theory, Interscience, New York and London, 1958.

[2] Linear Operators, vol. **II**: Spectral Theory. Selfadjoint Operators in Hilbert Spaces, Interscience, New York and London, 1963.

Dunkl, C.F.

[1] *The measure algebra of a locally compact hypergroup,* Trans. Amer. Math. Soc. **179** (1973), 331–348.

[2] *Structure hypergroups for measure algebras,* Pacific J. Math. **47** (1973), 413–425.

Emch, G.G.

[1] Algebraic Methods in Statistical Mechanics and Quantum Field Theory, Wiley–Interscience, New York, London, 1972.

Eskin, G.I.

[1] *A sufficient condition of solvability of the multi-dimensional moment problem,* Dokl. Akad. Nauk SSSR **133** (1960), 540–543. *(Russian)*

Faddeev, L.D.

[1] *On the separation of self-action and scattering effects, according to the perturbation theory,* Dokl. Akad. Nauk SSSR **152** (1963), 573–580. *(Russian)*

Faris, W.G.

[1] *Quadratic forms and essential self-adjointness,* Helv. Phys. Acta **45** (1972), 1074–1088.

[2] *Self-adjoint operators,* Lecture Notes in Math., vol. **437**, Springer-Verlag, Berlin, 1975.

Feissner, G.

[1] *Hypercontractive semigroups and Sobolev's inequality,* Trans, Amer. Math. Soc. **210** (1975), 51–62.

Feller, W.

[1] An Introduction to Probability Theory and Its Applications, 3rd edn., vol. **2**, Wiley & Sons, New York, London, Sydney, Toronto, 1971.

Friedrich, J.

[1] *A note on the two-dimensional moment problem,* Math. Nachr. **121** (1985), 285–286.

Friedrich, J. and Klotz, L.

[1] *On extensions of positive definite operator-valued functions,* Preprint, Karl-Marx-Univ., Leipzig, 1986.

Fröhlich, J.

[1] *Schwinger functions and their generating functionals.* **I**, Helv. Phys. Acta **47** (1974), 265–306.

[2] *Verification of axioms for Euclidean and relativistic fields and Haag's theorem in a class of $P(\varphi)_2$ models,* Ann. Inst. H. Poincaré **A 21** (1974), 271–317

Frolov, N.N.

[1] *Imbedding theorems for functions of countably many variables and their application to the Dirichlet problem,* Dokl. Akad. Nauk SSSR **203** (1972), 39–42; ***English transl.*** in Soviet Math. Dokl. **13** (1972).

[2] *On a coercivity inequality for an elliptic operator with infinitely many independent variables,* Mat. Sbornik **90** (**132**) (1973), 403–414; ***English transl.*** in Math. USSR Sb. **19** (1973).

[3] *On essential selfadjointness of an infinite-dimensional differential operator,* Mat. Zametki **24** (1978), 241–248; ***English transl.*** in Math. Notes **24** (1978).

[4] *Selfadjointness of an elliptic operator with infinitely many variables,* Funk. Anal. i Prilozh. **14** (1980), No. 1, 85–86; ***English transl.*** in Funct. Anal. Appl. **14** (1980).

[5] *On selfadjointness of a Schrödinger operator with infinitely many variables,* Sibirsk. Mat. Zh. **22** (1981), No. 1, 198–204; ***English transl.*** in Siberian Math. J. **22** (1981).

Fukushima, M.

[1] *Energy forms and diffusion processes,* Mathematics + Physics. Lectures on Recent Results, vol. **1**, World Scientific, Singapore, 1985, pp. 65–98.

Gachok, V. P.

[1] *Selfadjointness of field operators and the moment problem,* Ukrain. Mat. Zh. **17** (1965), No. 5, 3–13. *(Russian)*

[2] *Quasi-analytical functionals in the quantum field theory,* Dokl. Akad. Nauk SSSR **168** (1966), 1030–1032. *(Russian)*

Gårding, L.

[1] *Notes on continuous representations of Lie groups,* Proc. Nat. Acad. Sci. U.S.A. **33** (1947), 331–332.

[2] *Eigenfunction expansions,* Supplement **I** to **L. Bers, F. John, and M. Schechter,** Partial Differential Equations (Summer Sem., Boulder, Colo., 1957), Interscience, 1964; reprint, Amer. Math. Soc., Providence, R.I., 1974, pp. 301–325.

Gårding, L. and Wightman, A.S.

[1] *Representations of the anticommutation relations,* Proc. Nat. Acad. Sci. U.S.A. **40** (1954), 617–621.

[2] *Representations of the commutation relations,* Proc. Nat. Acad. Sci. U.S.A. **40** (1954), 622–625.

Geidarov, A.G.

[1] *On expansion in eigenfunctions of selfadjoint elliptic operators with singular potential,* Proc. All-Union Conf. Partial Differential Equations, Moscow, 1976; Izdat. Moskov. Univ., Moscow, 1978, pp. 287–288. *(Russian)*

Gelfand, I.M. and Kostyuchenko, A.G.

[1] *On expansion in eigenfunctions of differential and other operators,* Dokl. Akad. Nauk SSSR **103** (1955), 349–352. *(Russian)*

Gelfand, I.M., Raikov, D.A., and Shilov, G.E.

[1] Commutative Normed Rings, Fizmatgiz, Moscow, 1960; ***English transl.*:** Chelsea, New York, 1964.

Gelfand, I.M. and Shilov, G.E.

[1] Generalized Functions and Operations on Them, 2nd edn., Fizmatgiz, Moscow, 1959; ***English transl. of 1st edn.*:** Academic Press, New York, 1964.

[2] Spaces of Test and Generalized Functions, Fizmatgiz, Moscow, 1958; ***English transls.*:** Academic Press; Gordon and Breach, New York, 1968.

Gelfand, I.M. and Vilenkin, N.Ya.

[1] Some Applications of Harmonic Analysis. Rigged Hilbert Spaces, Fizmatgiz, Moscow, 1961; ***English transl.*:** Academic Press, New York, 1964.

Gestrin, G.P.

[1] *On expansion in eigenfunctions of the Schrödinger operator with singular potential,* Mat. Zametki **15** (1974), 458–465; ***English transl.*** in Math. Notes **15** (1974).

Getoor, R.K.

[1] *On semi-groups of unbounded normal operators,* Proc. Amer. Math. Soc. **7** (1956), 387–391.

Gikhman, I.I. and Skorokhod, A.V.

[1] Introduction to the Theory of Random Processes, Nauka, Moscow, 1965; ***English transl.*:** Saunders, Philadelphia, 1969.

[2] The Theory of Stochastic Processes, vol. **1**, Nauka, Moscow, 1971; ***English transl.*** **:** Springer-Verlag, Berlin, 1974.

Glazman, I.M.

[1] Direct Methods of Qualitative Spectral Analysis of Singular Differential Operators, Fizmatgiz, Moscow, 1963; ***English transls.*** **:** Israel Program Sci. Transls., Jerusalem, 1965, and Davey, New York, 1966.

Glimm, J. and Jaffe, A.

[1] Quantum Physics. A Functional Integral Point of View, Springer-Verlag, Berlin, 1981.

Golodets, V.Ya.

[1] *Description of representations of anticommutation relations,* Uspekhi Mat. Nauk **24** (1969), No. 4 (148), 3–64. *(Russian)*

Gorbachuk, V.I. and Gorbachuk, M.L.

[1] Boundary Value Problems for Operator Differential Equations, Naukova Dumka, Kiev, 1984; ***English transl.*** **:** Kluwer AP, Dordrecht, 1991.

Gross, L.

[1] *Harmonic analysis on Hilbert space,* Mem. Amer. Math. Soc. **46** (1963), 1–62.

[2] *Logarithmic Sobolev inequalities,* Amer. J. Math. **97** (1976), 1061–1083.

Guichardet, A.

[1] *Produits tensoriels continus d'espaces et d'algébres de Banach,* Comm. Math. Phys. **5** (1967), 262–287.

Guseinov, R.V.

[1] *On the theory of the Fourier–Wiener transformation,* Vestnik Moskov. Univ. Ser. I: Mat., Mekh. (1969), No. 4, 17–25; ***English transl.*** in Moscow Univ. Math. Bull. **24** (1969).

[2] *Some properties of the Fourier–Wiener transformation,* Vestnik Moskov. Univ. Ser. I: Mat., Mekh. (1970), No. 4, 39–49; ***English transl.*** in Moscow Univ. Math. Bull. **25** (1970).

Halmos, P.R.

[1] Measure Theory, Van Nostrand, Princeton, N.J., 1950.

Hasegawa, M.

[1] *On quasi-analytic vectors for dissipative operators,* Proc. Amer. Math. Soc. **29** (1971), 81–84.

Hegerfeldt, G.C.

[1] *Gårding domain and analytic vectors for quantum fields,* J. Math. Phys. **13** (1972), 821–827.

Hegerfeldt, G.C. and Melsheimer, O.

[1] *The form of representations of CCR for Bose fields and connection with finitely many degrees of freedom,* Comm. Math. Phys. **12** (1969), 304–323.

Hepp, K.

[1] Théorie de la renormalisation, Springer-Verlag, New York, 1970.

Hida, T.

[1] *Analysis of Brownian functionals,* Carleton Math. Lecture Notes, vol. **13**, 1975.

[2] Brownian Motion, Springer-Verlag, Berlin, 1980.

[3] *White noise and random fields — old and new,* Proc. of Preseminar for Internat. Conf. on Gaussian Random Fields, Nagoya, 1990, Pt. **3**, Nagoya Univ., Nagoya, 1991, pp. 1–10.

Hida, T., Potthoff, J., and Streit, L.

[1] *White noise analysis and applications,* Mathematics + Physics. Lectures on Recent Results, vol. **3**, World Scientific, Singapore, 1990, pp. 143–178.

Hille, E.

[1] *On semi-groups of transformations in Hilbert space,* Proc. Nat. Acad. Sci. U.S.A. **24** (1938), 159–161.

Hille, E. and Phillips, R.S.

[1] Functional Analysis and Semi-Groups, rev. edn., Amer. Math. Soc., Providence, R.I., 1957.

Høegh-Krohn, R.

[1] *On the spectrum of the space cutoff* $P(\phi)_2$ *Hamiltonian,* Comm. Math. Phys. **21** (1971), 256–260.

Ionescu Tulcea, C.

[1] *Spectral representations of semigroups of normal operators,* Proc. Nat. Acad. Sci. U.S.A. **44** (1958), 44–46.

[2] *Spectral representations of certain semi-groups of operators,* J. Math. and Mech. **8** (1958), 95–109.

Ionescu Tulcea, C. and Simon, A.B.

[1] *Spectral representations and unbounded convolution operators,* Proc. Nat. Acad. Sci. U.S.A. **45** (1959), 1765–1767.

Ismagilov, R.S.

[1] *Selfadjoint extensions of a system of commuting symmetric operators,* Dokl. Akad. Nauk SSSR **133** (1960), 511–514; ***English transl.*** in Soviet Math. Dokl. **1** (1960).

[2] *On unitary representation of groups of diffeomorphisms of compact manifold,* Izvestiya Akad. Nauk SSSR. Ser. Mat. **86** (1972), 180–208 (Russian).

[3] *On unitary representations of group* $C_0^\infty(X, G)$, $G = SU$, Mat. Sbornik **100**

Itô, K.

[1] *Complex multiple Wiener integral,* Japan J. Math. **22** (1952), 63–86.

Jewett, R.I.

[1] *Spaces with an abstract convolution of measures,* Adv. Math. **18** (1975), 1–101.

Kalyuzhny, A.A.

[1] *On integral representation of exponentially convex functions,* Ukrain. Mat. Zh. **34** (1982), 370–373; ***English transl.*** in Ukrain. Math. J. **34** (1982).

[2] *A theorem on the existence of a multiplicative measure,* Ukrain. Mat. Zh. **35** (1983), 369–371; ***English transl.*** in Ukrain. Math. J. **35** (1983).

Kato, T.

[1] Perturbation Theory for Linear Operators, Springer-Verlag, Berlin, 1966.

[2] *Schrödinger operators with singular potentials,* Israel J. Math. **13** (1972), 135–148.

[3] *A remark to the preceding paper by Chernoff,* J. Funct. Anal. **12** (1973), 415–417.

Kato, Yu. and Mugibayashi, M.

[1] *Regular perturbations in quantum field theory,* Progr. Theoret. Phys. **30** (1963), 103–133.

Kats, G.I.

[1] *On expansions in eigenfunctions of selfadjoint operators,* Dokl. Akad. Nauk SSSR **119** (1958), 19–22. *(Russian)*

[2] *Generalized elements of a Hilbert space,* Ukrain. Mat. Zh. **12** (1960), No. 1, 13–24. *(Russian)*

[3] *Spectral decompositions of selfadjoint operators in generalized elements of a Hilbert space,* Ukrain. Mat. Zh. **13** (1961), No. 4, 13–33. *(Russian)*

[4] *Generalized functions on a locally compact group and decompositions of unitary representations,* Trudy Mosk. Mat. Obshch. **10** (1961), 3–41. *(Russian)*

Kirillov, A.A.

[1] *Dynamical systems, factors, and representations of groups,* Uspekhi Mat. Nauk **22** (1967), No. 5 (137), 67–80. *(Russian)*

[2] *Representations of certain infinite-dimensional Lie algebras,* Vestnik Moskov. Univ. Ser. I: Mat., Mekh. (1974), No. 1, 75–83; ***English transl.*** in Moscow Univ. Math. Bull. **29** (1974).

Klein, A. and Landau, L.J.

[1] *Singular perturbations of positivity preserving semigroups via path space techniques,* J. Funct. Anal. **20** (1975), 44–82.

Kolmogorov, A.N.

[1] Fundamental Concepts of Probability Theory, ONTI, Moscow–Leningrad, 1936; ***in German*: Kolmogoroff, A.,** Grundbegriffe der Wahrscheinlichkeitsrechnung, Springer-Verlag, Berlin, 1933; ***English transl.*:** Chelsea, New York, 1950.

Kolmogorov, A.N. and Khinchin, A.Ya. [Kolmogoroff, A. and Khintchine, A.]

[1] *Über Konvergenz von Reihen, deren Glieder durch den Zufalle bestimmt werden,* Mat. Sbornik **32** (1925), 668–677.

Kolomytsev, V.I. and Samoilenko, Yu. S.

[1] *On a countable set of commuting selfadjoint operators and the algebra of local observables,* Ukrain. Mat. Zh. **31** (1979), 365–371; ***English transl.*** in Ukrain. Math. J. **31** (1979).

Kondratyev, Yu.G.

[1] *Positive definite functions on certain spaces of sequences,* Ukrain. Mat. Zh. **28** (1976), 27–35; ***English transl.*** in Ukrain. Math. J. **28** (1976).

[2] *The Wick powers of generalized Gaussian random processes,* Methods of Functional Analysis in the Problems of Mathematical Physics, Inst. Mat. Akad. Nauk Ukrain. SSR, Kiev, 1978, pp. 129–158. *(Russian)*

[3] *Nuclear spaces of entire functions in the problems of infinite-dimensional analysis,* Dokl. Akad. Nauk SSSR **254** (1980), 1325–1329; ***English transl.*** in Soviet Math. Dokl. **22** (1981).

[4] *Spaces of entire functions of infinitely many variables associated with the riggings of the Fock spaces,* Spectral Analysis of Differential Operators, Inst. Mat. Akad. Nauk Ukrain. SSR, Kiev, 1980, pp. 18–37. *(Russian)*

[5] *The second quantization operators and perturbations of them,* Preprint No. 82.28, Inst. Mat. Akad. Nauk Ukrain SSR, Kiev, 1982. *(Russian)*

[6] *Perturbations of the second quantization operators,* Funk. Anal. i Prilozh. **16** (1982), No. 4, 76–77; ***English transl.*** in Funct. Anal. Appl. **16** (1982).

[7] *The Kato inequality for the second quantization operators,* Ukrain. Mat. Zh. **35** (1983), 753–756; ***English transl.*** in Ukrain. Math. J. **35** (1983).

[8] *Dirichlet operators and smoothness of solutions of infinite-dimensional elliptic equations,* Dokl. Akad. Nauk SSSR **282** (1985), 269–273. *(Russian)*

[9] *Construction of dynamics for quantum lattice systems,* Dokl. Akad. Nauk SSSR **283** (1985), 82–85. *(Russian)*

Kondratyev, Yu.G. and Koshmanenko, V.D.

[1] *The scattering problem for operators associated with Dirichlet forms,* Dokl. Akad. Nauk SSSR **267** (1982), 285–288; ***English transl.*** in Soviet Math. Dokl. **26** (1983).

Kondratyev, Yu.G. and Samoilenko, Yu.S.

[1] *The integral representation of generalized positive definite kernels of infinitely many variables,* Dokl. Akad. Nauk SSSR **227** (1976), 800–803; ***English transl.*** in Soviet Math. Dokl. **17** (1976).

[2] *The spaces of trial and generalized functions of infinite number of variables,* Rep. Math. Phys. **14** (1978), 325–350.

[3] *Generalized derivatives of probability measures on* $\mathbb{R}^\infty$, Methods of Functional Analysis in the Problems of Mathematical Physics, Inst. Mat. Akad. Nauk Ukrain. SSR, Kiev, 1978, pp. 159–176. *(Russian)*

Kondratyev, Yu.G. and Tsykalenko, T.V.

[1] *The spaces of test and generalized functions of infinitely many variables,* Problems in the Theory of Probability Distributions, Inst. Mat. Akad. Nauk Ukrain. SSR, Kiev, 1983, pp. 123–133. *(Russian)*

[2] *Infinite-dimensional hyperbolic equations associated with Dirichlet operators,* Dokl. Akad. Nauk SSSR **288** (1986), 814–817. *(Russian)*

[3] *Dirichlet operators and differential equations associated with them,* Preprint No. 86.37, Inst. Mat. Akad. Nauk Ukrain. SSR, Kiev, 1986. *(Russian)*

Korotkov, V.B.

[1] *On integral operators with Carleman kernels,* Dokl. Akad. Nauk SSSR **165** (1965), 748–751; ***English transl.*** in Soviet Math. Dokl. **6** (1965).

Korsunsky, L.M.

[1] *Expansions in generalized eigenvectors and integral representations of invariant positive-definite kernels,* Ukrain. Mat. Zh. **24** (1972), 341–351; ***English transl.*** in Ukrain. Math. J. **24** (1972).

[2] *Integral representations of invariant positive-definite kernels,* Spectral Theory of Operators and Infinite-Dimensional Analysis, Inst. Mat. Akad. Nauk Ukrain. SSR, Kiev, 1976, pp. 26–87. *(Russian)*

Koshmanenko, V.D.

[1] *Operator representations for nonclosable quadratic forms and the scattering theory,* Dokl. Akad. Nauk SSSR **245** (1979), 295–298. *(Russian)*

[2] *Classification of singular perturbations of selfadjoint operators,* Preprint No. 82.34, Inst. Mat. Akad. Nauk Ukrain. SSR, Kiev, 1982. *(Russian)*

Koshmanenko, V.D. and Samoilenko, Yu.S.

[1] *On an isomorphism between the Fock space and a space of functions of infinitely many variables,* Ukrain. Mat. Zh. **27** (1975), 669–674; ***English transl.*** in Ukrain. Math. J. **27** (1975).

Kostyuchenko, A.G. and Mityagin, B.S.

[1] *Positive definite functionals on nuclear spaces,* Trudy Mosk. Mat. Obshch. **9** (1960), 283–316. *(Russian)*

Kosyak, A.V.

[1] *Analytic and entire vectors for families of operators,* Spectral Analysis of Differential Operators, Inst. Mat. Akad. Nauk Ukrain. SSR, Kiev, 1980, pp. 3–18. *(Russian)*

[2] *The Gårding domain for representations of canonical commutation relations,* Ukrain. Mat. Zh. **36** (1984), 709–715; ***English transl.*** in Ukrain. Math. J. **36** (1984).

Kosyak, A.V. and Samoilenko, Yu.S.

[1] *On families of commuting selfadjoint operators,* Ukrain. Mat. Zh. **31** (1979), 555–558; ***English transl.*** in Ukrain. Math. J. **31** (1979).

[2] *The Gårding domain and entire vectors for inductive limits of commuting locally compact groups,* Ukrain. Mat. Zh. **35** (1983), 427–434; ***English transl.*** in Ukrain. Math. J. **35** (1983).

Kovalenko, V.F., Perelmuter, M.A., and Semenov, Yu.A.

[1] *Schrödinger operators with* $L_{\omega}^{1/2}(\mathbb{R}^1)$*-potentials,* J. Math. Phys. **22** (1981), 1033–1044.

Kovalenko, V.F. and Semenov, Yu.A.

[1] *Some questions about expansions in generalized eigenfunctions of the Schrödinger operator with strongly singular potentials,* Uspekhi Mat. Nauk **33** (1978), No. 4 (202), 107–140; ***English transl.*** in Russian Math. Surveys **33** (1978).

Krée, P.

[1] *Prodistributions,* Séminaire Paul Krée 1974/1975, Inst. H. Poincaré, Paris, 1975, pp. 1–14.
[2] *Calcul d'integrales et de dérivées en dimension infinie,* J. Funct. Anal. **31** (1979), 150–186.

Krée, P. and Rączka, R.

[1] *Kernels and symbols of operators in QFT,* Ann. Inst. H. Poincaré **A 28** (1978), 41–73.

Krein, M.G.

[1] *On the problem of extension of Hermitian-positive continuous functions,* Dokl. Akad. Nauk SSSR **26** (1940), 17–21. *(Russian)*
[2] *On a general method of decomposing positive definite kernels into elementary products,* Dokl. Akad. Nauk SSSR **53** (1946), 3–6. *(Russian)*
[3] *On Hermitian operators with directing functionals,* Zbirnyk Prats Inst. Mat. Akad. Nauk Ukrain. RSR **10** (1948), pp. 83–106. *(Ukrainian)*
[4] *Hermitian positive kernels on homogeneous spaces,* **I:** Ukrain. Mat. Zh. **1** (1949), No. 4, 64–98; **II:** Ukrain. Mat. Zh. **2** (1950), No. 1, 10–59; ***English transl. of Part*** **I:** Amer. Math. Soc. Transl. (2) **34** (1963), 69–108.

Krein, M.G. and Krasnoselsky, M.A.

[1] *Fundamental theorems on the extension of Hermitian operators and some applications of these to the theory of orthogonal polynomials and to the moment problem,* Uspekhi Mat. Nauk, **2** (1947), No. 3 (19), 60–106. *(Russian)*

Krein, S.G.

[1] Linear Differential Equations in Banach Space, Nauka, Moscow, 1967; ***English transl.*:** Amer. Math. Soc., Providence, R.I., 1971.
[2] (editor) Functional Analysis, 2nd rev. edn., Nauka, Moscow, 1972; ***English transl. of 1st edn.*** **:** Noordhoff, 1972.

Kristensen, P., Mejlbo, L., and Poulsen, E.T.

[1] *Tempered distributions in infinitely many dimensions.* 1. *Canonical field operators,* Comm. Math. Phys. **1** (1965), 175–214.

Kubo, I. and Takenaka, S.

[1] *Calculus on Gaussian white noise.* **I**, Proc. Japan Acad. **56(A)** (1980), 376–380.

[2] *Calculus on Gaussian white noise.* **II**, Proc. Japan Acad. **56(A)** (1980), 411–416.

[3] *Calculus on Gaussian white noise.* **III**, Proc. Japan Acad. **57(A)** (1981), 433–437.

[4] *Calculus on Gaussian white noise.* **IV**, Proc. Japan Acad. **58(A)** (1982), 186–189.

Kuo, Hui-Hsiung

[1] *Gaussian measures in Banach spaces,* Lecture Notes in Math., vol. 463, Springer-Verlag, Berlin, 1975.

[2] *Lectures on white noise analysis,* Proc. of Preseminar for Internat. Conf. on Gaussian Random Fields, Nagoya, 1990, Pt. **1**, Nagoya Univ., Nagoya, 1991, pp. 1–65.

Kurepa, S.

[1] *A cosine functional equation in Hilbert space,* Canad. J. Math. **12** (1960), 45–50.

Kusuoka, S.

[1] *Dirichlet forms and diffusion processes on Banach spaces,* J. Fac. Sci. Univ. Tokyo Sec. 1A **29** (1982), 79–95.

Lanford, O. and Lebowitz, J.

[1] *Time evolution of harmonic systems,* Lecture Notes in Phys., vol. 38, Springer-Verlag, Berlin, 1975, pp. 144–177.

Lassner, G.

[1] *Topological algebras of operators,* Rep. Math. Phys. **3** (1972), 279–293.

Lax, P.D.

[1] *On Cauchy's problem for hyperbolic equations and the differentiability of solutions of elliptic equations,* Comm. Pure Appl. Math. **8** (1955), 615–633.

Lee, Y.-J.

[1] *Analytic version of test functionals, Fourier transform, and a characterization of measures in white noise calculus,* Preprint, National Univ. Tainan, Taiwan, 1989.

Leray, J.

[1] Lectures on Hyperbolic Equations with Variable Coefficients, Inst. Advanced Study, Princeton, N.J., 1952.

Levin, B.Ya. and Ovcharenko, I.E.

[1] *Extension of Hermitian-positive functions given on a band,* Teor. Funktsij Funk. Anal. i Prilozh., Issue 5, (1967), 68–83. *(Russian)*

Levitan, B.M.

[1] *On a theorem of Titchmarsh and Sears,* Uspekhi Mat. Nauk **16** (1961), No. 4 (100), 175–178. *(Russian)*
[2] Theory of Operators of Generalized Shift, Nauka, Moscow, 1973. *(Russian)*
[3] The Inverse Sturm–Liouville Problems, Nauka, Moscow, 1984. *(Russian)*

Lipkin, H.

[1] Quantum Mechanics, Mir, Moscow, 1977. *(Russian translation)*

Litvinov, G.L.

[1] *Hypergroups and hypergroup algebras,* VINITI Series on Contemporary Problems in Mathematics. The Latest Achievements, vol. **26**, VINITI, Moscow, 1985, pp. 57–106. *(Russian)*

Mackey, G.W.

[1] *Imprimitivity for representations of locally compact groups.* **I**, Proc. Nat. Acad. Sci. U.S.A. **35** (1949), 537–545.
[2] *Induced representations of locally compact groups.* **I**, Ann. Math. **55** (1952), 101–139.

Maltese, G.

[1] *Spectral representations for solutions of certain abstract functional equations,* Compositio Math. **15** (1962), 1–22.

[2] *Spectral representations for some unbounded normal operators,* Trans. Amer. Math. Soc. **110** (1964), 79–87.

Malyshev, V.A. and Minlos, R.A.

[1] Gibbsian Random Fields, Nauka, Moscow, 1985. *(Russian)*

Mandelbrojt, S.

[1] Séries Adhérentes, Régularisation des Suites, Applications, Gauthier-Villars, Paris, 1952.

Maradudin, A., Montroll, E., and Weiss, G.

[1] Dynamical Theory of Crystal Lattice in Harmonic Approximation, Mir, Moscow, 1965. *(Russian translation)*

Marchenko, A.V.

[1] *On inductive limits of linear spaces and operators, and their applications,* Vestnik Moskov. Univ. Ser. I: Mat., Mekh. (1974), No. 2, 26–33; ***English transl.*** in Moscow Univ. Math. Bull. **29** (1974).

[2] *Selfadjoint differential operators with infinitely many independent variables,* Mat. Sbornik **96** (**138**) (1975), 276–293; ***English transl.*** in Math. USSR Sb. **25** (1975).

[3] *Spaces of functions of infinitely many variables as inductive limits of locally convex spaces of functions,* Teor. Funktsij Funk. Anal. i Prilozh., Issue 24, (1975), 86–98. *(Russian)*

Marchenko, V.A.

[1] The Sturm–Liouville Operators and Applications of Them, Nauk. Dumka, Kiev, 1977. *(Russian)*

Markushevich, A.I.

[1] Theory of Analytical Functions, vol. **2**, Nauka, Moscow, 1968. *(Russian)*

Maurin, K.

[1] *Eine Bemerkung zur allgemeinen Eigenfunktionsentwicklungen für vertauschbare Operatorensysteme beliebiger Mächtigkeit,* Bull, Acad. Polon. Sci. Sér.: Sci. Math. Astr. Phys. **8** (1960), 381–384.

[2] Methods of Hilbert Spaces, PWN, Warsaw, 1959; ***English transl.,*** 1967; ***Russian transl.*** Mir, Moscow, 1965.

[3] General Eigenfunction Expansions and Unitary Representations of Topological Groups, PWN, Warsaw, 1968.

[4] *A remark on Berezanski version of spectral theorem,* Studia Math. **34** (1970), 165–167.

Menicoff, R. and Sharp, D.H.

[1] *Representations of a local current algebras: their dynamical determination,* J. Math. Phys. **16** (1975), 2341–2360.

Messon, D. and McClary, W.

[1] *Classes of C^∞ vectors and essential self-adjointness,* J. Funct. Anal. **10** (1972), 19–32.

Minlos, R.A.

[1] *Generalized random processes and their extensions to measures,* Trudy Moskov. Mat. Obshch. **8** (1959), 497–518; ***English transl.*** in Selected Transl. Math. Statist. and Probab., vol. **3**, Amer. Math. Soc., Providence, R.I., 1963.

Nachbin, L.

[1] Topology on Spaces of Holomorphic Mappings, Springer-Verlag, Berlin, 1969.

Naimark, M.A.

[1] Normed Rings, Nauka, Moscow, 1968; ***English transl.*:** Normed Algebras, Noordhoff, 1972.

Naimark, M.A. and Fomin, S.V.

[1] *Continuous direct sums of Hilbert spaces and some applications of them,* Uspekhi Mat. Nauk **10** (1955), No. 2 (64), 111–142; ***English transl.*** in Amer. Math. Soc. Transl. (2) **5** (1957).

Nakagami, Yo.

[1] *Infinite tensor product of von Neumann algebras.* **I**, Kodai Math. Semin. Rep. **22** (1970), 341–354; **II**, Publ. Res. Inst. Math. Sci. Kyoto Univ. **6** (1970), 257–292.

[2] *Infinite tensor product of operators,* Publ. Res. Inst. Math. Sci. Kyoto Univ. **10** (1974), 111–145.

Napiórkowski, K.

[1] *Continuous tensor products of Hilbert spaces and product operators,* Studia Math. **39** (1971), 307–327.

Nelson, E.

[1] *Analytic vectors,* Ann. of Math. (2) **70** (1959), 572–615.
[2] *The free Markoff fields,* J. Funct. Anal. **12** (1973), 97–112.

von Neumann, J.

[1] *On infinite direct products,* Compositio Math. **6** (1939), 1–77.

Nussbaum, A.E.

[1] *The Hausdorff–Bernstein–Widder theorem for semi-groups in locally compact Abelian groups,* Duke Math. J. **22** (1955), 573–582.
[2] *Integral representation of semi-groups of unbounded selfadjoint operators,* Ann. of Math. (2) **69** (1959), 133–141.
[3] *Quasi-analytic vectors,* Ark. Mat. **6** (1965), 179–191.
[4] *A note on quasi-analytic vectors,* Studia Math. **33** (1969), 305–309.
[5] *Spectral representation of certain one-parametric families of symmetric operators in Hilbert space,* Trans. Amer. Math. Soc. **152** (1970), 419–429.
[6] *Radial exponentially convex functions,* J. d'Analyse Math. **25** (1972), 277–288.
[7] *Integral representation of functions and distributions positive definite relative to the orthogonal group,* Trans. Amer. Math. Soc. **175** (1973), 355–387.
[8] *On functions positive definite relative to the orthogonal group and the representation of functions as Hankel–Stieltjes transforms,* Trans. Amer. Math. Soc. **175** (1973), 389–408.
[9] *Extension of positive definite functions and representation of functions in terms of spherical functions in symmetric spaces of noncompact type of rank 1,* Math. Ann. **215** (1975), 97–116.

Orochko, Yu.B.

[1] *A uniqueness theorem for a selfadjoint extension of a Schrödinger operator,* Dopovidi. Akad. Nauk. Ukrain. RSR (**1966**), No. 11, 1391–1394. *(Ukrainian)*

[2] *Some estimates at infinity for eigenfunctions of the Schrödinger operator,* Ukrain. Mat. Zh. **19** (1967), No. 3, 39–52; ***English transl.*** in Ukrain. Math. J. **19** (1967).

[3] *An estimate of the growth of averaged eigenfunctions of selfadjoint operators,* Ukrain. Mat. Zh. **20** (1968), 400–405; ***English transl.*** in Ukrain. Math. J. **20** (1968).

[4] *A remark on the essential selfadjointness of the Schrödinger operator with a singular potential,* Mat. Zametki **20** (1976), 571–580; ***English transl.*** in Math. Notes **20** (1976).

[5] *A sufficient condition for the essential selfadjointness of polynomials in the Schrödinger operator,* Mat. Sbornik **99(141)** (1976), 192–210; ***English transl.*** in Math. USSR Sb. **28** (1976).

[6] *The Carleman estimates for the Schrödinger operator with a locally semibounded strongly singular potential,* Mat. Sbornik **104(146)** (1977), 162–174; ***English transl.*** in Math. USSR Sb. **33** (1977).

[7] *The hyperbolic equation method in the spectral theory of a Schrödinger operator with a real singular potential,* Proc. All-Union Conf. Partial Differential Equations, Moscow, 1976; Izdat. Moskov. Univ., Moscow, 1978, pp. 403–404. *(Russian)*

[8] *On the theory of expansions in generalized eigenfunctions of selfadjoint operators,* Ukrain. Mat. Zh. **30** (1978), 63–69; ***English transl.*** in Ukrain. Math. J. **30** (1978).

[9] *Some properties of expansions in generalized eigenfunctions of selfadjoint Cárleman operators,* Sibirsk. Mat. Zh. **21** (1980), No. 6, 96–107. *(Russian)*

[10] *The operator cosine method in the problem of essential selfadjointness of a symmetric operator which is not semibounded,* Ukrain. Mat. Zh. **33** (1981), 348–355; ***English transl.*** in Ukrain. Math. J. **33** (1981).

[11] *On the property of a finite global rate of propagation of a second-order elliptic differential expression,* Diff. Uravneniya **18** (1982), 1764–1772; ***English transl.*** in Diff. Equations **18** (1982).

[12] *Finite local rate of propagation of a hyperbolic equation in the problem of selfadjointness of powers of a second-order elliptic differential operator,* Izvestiya Akad. Nauk SSSR Ser. Mat. **47** (1983), 298–314. *(Russian)*

[13] *On the Carleman resolvent estimate of Schrödinger type operator with arbitrary potential,* J. Oper. Theory **9** (1983), 27–52.

Osterwalder, K. and Schrader, R.

[1] *Axioms for Euclidean Green's functions.* **II**, Comm. Math. Phys. **42** (1975), 281–305.

Ostrovsky, V.L.

[1] *An analogue of the Nelson theorem for nuclear nilpotent Lie algebras of currents,* Spectral Theory of Operators and Infinite-Dimensional Analysis, Inst. Mat. Akad. Nauk Ukrain. SSR, Kiev, 1984, pp. 120–131. *(Russian)*

[2] *Irreducible representations of the group of infinite upper triangular matrices,* Ukrain. Mat. Zh. **38** (1986), 255–259. *(Russian)*

Ostrovsky, V.L. and Samoilenko, Yu.S.

[1] *An application of the spectral projection theorem to noncommutative families of operators,* Ukrain. Mat. Zh. **40** (1988), 478–493. *(Russian)*

Ovcharenko, I.E.

[1] *Integral representations of semi-definite many-dimensional power moment sequences,* Dokl. Akad. Nauk Ukrain. SSR Ser. A (**1983**), No. 3, 17–20. *(Russian)*

[2] *Moment sequences with representing measures free of masses in certain domains,* Dokl. Akad. Nauk SSSR **273** (1983), 41–45. *(Russian)*

[3] *On two-dimensional power moment sequences,* Ukrain. Mat. Zh. **36** (1984), 51–56. *(Russian)*

Paclet, Ph.

[1] *Espaces de Dirichlet en dimension infinie,* Compt. Rend. Acad. Sci. A **288** (1979), 981–983.

Perelmuter, M.A.

[1] *Perturbations of operators with integral resolvent,* Funk. Anal. i Prilozh. **14** (1980), No. 2, 75–76; ***English transl.*** in Funct. Anal. Appl. **14** (1980).

[2] *Positivity preserving operators and one criterion of essential self-adjointness,* J. Math. Anal. Appl. **82** (1981), 406–419.

[3] *The hyperbolic equation method in L^p-spaces,* Ukrain. Mat. Zh. **40** (1988), 494–500. *(Russian)*

Perelmuter, M.A. and Semenov, Yu.A.

[1] *Selfadjointness of elliptic operators with finitely and infinitely many variables,* Funk. Anal. i Prilozh. **14** (1980), No. 1, 81–82; ***English transl.*** in Funct. Anal. Appl. **14** (1980).

[2] *On finiteness of the rate of propagation of a perturbation for hyperbolic equations,* Ukrain. Mat. Zh. **36** (1984), 56–63; ***English transl.*** in Ukrain. Math. J. **36** (1984).

Phillips, R.S.

[1] *Spectral theory for semi-groups of linear operators,* Trans. Amer. Math. Soc. **71** (1951), 393–415.

Plesner, A.I.

[1] Spectral Theory of Linear Operators, Nauka, Moscow, 1965; ***English transl.*** **:** vols. **I**, **II**, Ungar, New York, 1969.

Plesner, A.I. and Rokhlin, V.A.

[1] *Spectral theory of linear operators,* Uspekhi Mat. Nauk **1** (1946), No. 1 (11), 71–191; ***English transl.*** in Amer. Math. Soc. Transl. (2) **62** (1967).

Pontryagin, L.S.

[1] Continuous Groups, Gostekhizdat., Moscow, 1954. *(Russian)*

Potthoff, J. and Streit, L.

[1] *A characterization of Hida distributions,* Preprint BiBoS, 1989.

Povzner, A.Ya.

[1] *On expansions of arbitrary functions in eigenfunctions of the operator* $-\Delta u + cu$, Mat. Sbornik **32 (74)** (1953), 109–156; ***English transl.*** in Amer. Math. Soc. Transl. (2) **60** (1967).

Powers, R.T.

[1] *Self-adjoint algebras of unbounded operators.* **I**, Comm. Math. Phys. **21** (1971), 85–124; **II**, Trans. Amer. Math. Soc. **187** (1974), 261–293.

Prokhorov, Yu.V.

[1] *Convergence of random processes and limit theorems in probability theory,* Teor. Veroyatn. i Primen. **1** (1956), 177–238; ***English transl.*** in Theor. Probab. Appl. **1** (1956).

Pytlik, T.

[1] *Nuclear spaces on a locally compact group,* Studia Math. **50** (1975), 225–243.

Rauch, J. and Taylor, M.

[1] *Essential selfadjointness of powers of generators of hyperbolic mixed problem,* J. Funct. Anal. **12** (1973), 491–493.

Reed, M.

[1] *A Gårding domain for quantum fields,* Comm. Math. Phys. **14** (1969), 336–346.

[2] *On self-adjointness in infinite tensor product spaces,* J. Funct. Anal. **5** (1970), 94–124.

Reed, M. and Simon, B.

[1] Methods of Modern Mathematical Physics, vol. **1**, Academic Press, New York, San Francisco, London, 1972.

[2] Methods of Modern Mathematical Physics, vol. **2**, Academic Press, New York, San Francisco, London, 1975.

[3] Methods of Modern Mathematical Physics, vol. **3**, Academic Press, New York, San Francisco, London, 1979.

[4] Methods of Modern Mathematical Physics, vol. **4**, Academic Press, New York, San Francisco, London, 1978.

Richter, P.

[1] *Zur Extremalzerlegung von Zustanden,* Wiss. Z. Karl-Marx-Univ. Leipzig. Math.-Naturwiss. Reihe **27** (1978), 293–297.

[2] *Zerlegung einer Familie symmetrischer Operatoren nach gemeinsamen verallgemeinerten Eigenvektoren,* Preprint, Karl-Marx-Univ., Leipzig, 1979.

[3] *Zerlegung positiv definiter Kerne und Entwicklung nach gemeinsamen verallgemeinerten Eigenfunktionen für Familien streng kommutierender symmetrischer Operatoren,* Wiss. Z. Karl-Marx-Univ. Leipzig. Math. Naturwiss. Reihe **31** (1982), 63–68.

Riesz, F. and Szökefalvi-Nagy, B.

[1] Leçons d'Analyse Fonctionelle, 2nd edn., Akad. Kiadó, Budapest, 1952; ***English. transl.*:** Ungar, New York, 1955.

Robinson, D.

[1] *Return to equilibrium,* Comm. Math. Phys. **31** (1973), 171–189.

Rofe-Beketov, F.S.

[1] *On positive differential operators (deficiency indices, factorization, and perturbations),* Proc. Roy. Soc. Edinburgh Sect. **A 97** (1984), 247–257.
[2] *Necessary and sufficient conditions of finite rate of propagation for elliptic operators,* Ukrain. Mat. Zh. **37** (1985), 668–670. *(Russian)*

Ross, K.A.

[1] *Hypergroups and centers of measure algebras,* Ist. Naz. Alta Mat. (Symposia Math.) **22** (1977), 189–203.

Rudin, W.

[1] *An extension theorem for positive-definite functions,* Duke Math. J. **37** (1970), 49–53.

Rudinsky, I.I.

[1] *On integral representations of even positive definite functions on nuclear spaces,* Ukrain. Mat. Zh. **36** (1984), 522–524. *(Russian)*

Sakhnovich, L.A.

[1] *Effective construction of inextensible Hermitian-positive functions of several variables,* Funk. Anal. i Prilozh. **14** (1980), No. 1, 55–60; ***English transl.*** in Funct. Anal. Appl. **14** (1980).

Samoilenko, V.G.

[1] *On wave operators for a Schrödinger-type operator in the space of functions of infinitely many variables,* Operators in Mathematical Physics and Infinite-Dimensional Analysis, Inst. Mat. Akad. Nauk Ukrain. SSR, Kiev, 1979, pp. 103–109. *(Russian)*
[2] *On second-order elliptic operators with infinitely many variables,* Ukrain. Mat. Zh. **32** (1980), 405–409. *(Russian)*
[3] *On selfadjointness of a second-order elliptic operator with infinitely many variables,* Ukrain. Mat. Zh. **34** (1982), 647–650. *(Russian)*
[4] *On the spectrum of a differential operator with infinitely many variables,* Sibirsk. Mat. Zh. **23** (1982), 147–159. *(Russian)*

[5] *On the boundedness of a multiplication operator with respect to a second quantization operator,* Dokl. Akad. Nauk SSSR **281** (1985), 1316–1319. *(Russian)*

Samoilenko, Yu.S.

[1] *Matrix-valued kernels of the Wightman functional type,* Methods of Functional Analysis in Problems of Mathematical Physics, Inst. Mat. Akad. Nauk Ukrain. SSR, Kiev, 1971, pp. 201–254. *(Russian)*

[2] Spectral Theory of Families of Self-Adjoint Operators, Naukova Dumka, Kiev, 1984; ***English transl.*:** Kluwer AP, Dordrecht, 1991.

Samoilenko, Yu.S. and Us, G.F.

[1] *On differential operators with constant coefficients in spaces of functions of countably many variables,* Spectral Analysis of Differential Operators, Inst. Mat. Akad. Nauk Ukrain. SSR, Kiev, 1980, pp. 49–67. *(Russian)*

[2] *Differential operators with constant coefficients on functions of countably many variables,* Uspekhi Mat. Nauk **39** (1984), No. 1 (235), 155–156; ***English transl.*** in Russian Math. Surveys **39** (1984).

Sazonov, V.V.

[1] *A remark concerning characteristic functionals,* Teor. Veroyatn. i Primen. **3** (1958), 201–205; ***English transl.*** in Theor. Probab. Appl. **3** (1958).

Schaefer, H.

[1] Topological Vector Spaces, Macmillan, London, 1966.

Schechter, M.

[1] Principles of Functional Analysis, Academic Press, New York, London, 1971.

Schmüdgen, K.

[1] *Positive cones in enveloping algebras,* Rep. Math. Phys. **14** (1978), 385–404.

Schoenberg, I.J.

[1] *Metric spaces and completely monotone functions,* Ann. Math. **39** (1938), 811–841.

Schwartz, L.

[1] Théorie des Distributions, vols. **1**, **2**, Hermann, Paris, 1950–51.

[2] *Théorie des noyaux,* Proc. Internat. Congr. Math. (Cambridge, Mass., 1950), vol. **1**, Amer. Math. Soc., Providence, R.I., 1952, pp. 220–230.

Sears, D.B.

[1] *Note on the uniqueness of the Green's functions associated with certain differential equations,* Canad. J. Math. **2** (1950), 314–325.

Segal, I.

[1] *Tensor algebras over Hilbert spaces.* **I**, Trans. Amer. Math. Soc. **81** (1956), 106–134.

[2] Mathematical Problems of Relativistic Physics, American Mathematical Society, Providence, R.I., 1963.

[3] *Construction of non-linear local quantum process,* Ann. Math. **92** (1970), 462–481.

Semenov, Yu.A.

[1] *On the Lie–Trotter theorems in L_p-spaces,* Lett. Math. Phys. **1** (1977), 379–385.

[2] *On the theory of eigenfunctions expansion of the Schrödinger operator,* Lett. Math. Phys. **1** (1977), 463–465.

Shakhbagyan, R.L.

[1] *The general boundary value problem for high-order elliptic operators on infinite-dimensional manifolds with boundary,* Izvestiya Akad. Nauk Armyan. SSR Ser. Mat. **19** (1984), No. 1, 3–18. *(Russian)*

Shifrin, S.N.

[1] *On infinite-dimensional symmetric analogues of the Stieltjes moment problem,* Ukrain. Mat. Zh. **26** (1974), 696–700. *(Russian)*

[2] *The infinite-dimensional smooth symmetric power moment problem on nuclear spaces,* Ukrain. Mat. Zh. **28** (1976), 793–802. *(Russian)*

Shilov, G.E.

[1] Mathematical Analysis. Functions of Several Real Variables, Nauka, Moscow, 1972. *(Russian)*

Shilov, G.E. and Tin', Fan Dyk

[1] Integral, Measure, and Derivative on Linear Spaces, Nauka, Moscow, 1967. *(Russian)*

Shiryaev, A.N.

[1] Probability, Nauka, Moscow, 1980. *(Russian)*

Shnol, E.E.

[1] *On the behavior of eigenfunctions of the Schrödinger equation,* Mat. Sbornik **42** (**84**) (1957), 273–286. *(Russian)*

Shohat, J.A. and Tamarkin, J.D.

[1] The Problem of Moments, Amer. Math. Soc., vol. 14, Providence, R.I., 1943.

Shwartz, A.S.

[1] Mathematical Foundations of Quantum Field Theory, Atomizdat., Moscow, 1975. *(Russian)*

Simon, B.

[1] *The theory of semi-analytic vectors: A new proof of a theorem of Masson and McClary,* Indiana Univ. Math. J. **20** (1970/71), 1145–1151.

[2] The $P(\phi)_2$ Euclidean (Quantum) Field Theory, Princeton University Press, Princeton, N.J., 1974.

[3] *An abstract Kato's inequality for generators of positivity preserving semigroups,* Indiana Univ. Math. J. **26** (1977), 1067–1073.

[4] *A canonical decomposition for quadratic forms with applications to monotone convergence theorem,* J. Funct. Anal. **28** (1978), 377–385.

[5] *Spectrum and continuum eigenfunctions of Schrödinger operators,* J. Funct. Anal. **42** (1981), 347–355.

[6] *Schrödinger semigroups,* Bull. Amer. Math. Soc. (N.S.) **7** (1982), 447–526.

Simon, B. and Høegh-Krohn, R.

[1] *Hypercontractive semigroups and two-dimensional self-coupled Bose fields,* J. Funct. Anal. **9** (1972), 121–180.

Skorokhod, A.V.

[1] Integration in Hilbert Space, Nauka, Moscow, 1975; ***English transl.***: Springer-Verlag, Berlin, 1974.

Smolyanov, O.G. and Fomin, S.V.

[1] *Measures on linear topological spaces,* Uspekhi Mat. Nauk **31** (1976), No. 4 (190), 3–56; ***English transl.*** in Russian Math. Surveys **31** (1976).

Sobolev, S.L.

[1] Introduction to the Theory of Cubic Formulas, Nauka, Moscow, 1974. *(Russian)*

Spector, R.

[1] *Apercu de la théorie des hypergroups,* Lecture Notes in Math., vol. **497**, Springer-Verlag, Berlin, 1975, pp. 643–673.

Strătilă, S. and Voiculescu, D.

[1] *Representations of AF-algebras and of the group* $U(\infty)$, Lecture Notes in Math., vol. **486**, Springer-Verlag, Berlin, 1975.

Streater, R.F. and Wulfsohn, A.

[1] *Continuous tensor products of Hilbert spaces and generalized random fields,* Nuovo Cim. **57** (1968), 330–339.

Streit, L.

[1] *Test function spaces for direct product representations of the canonical commutation relations,* Comm. Math. Phys. **4** (1967), 22–31.

Sukhov, Yu.M.

[1] *Linear boson models of time evolution in quantum statistical mechanics,* Izvestiya Akad. Nauk SSSR Ser. Mat. **48** (1981), 155–191. *(Russian)*

Szökefalvi-Nagy, B.

[1] *Über messbare Darstellungen Liescher Gruppen,* Math. Ann. **112** (1936), 286–296.

[2] *On semi-groups of selfadjoint transformations in Hilbert space,* Proc. Nat. Acad. Sci. U.S.A. **24** (1938), 559–560.

Takeda, M.

[1] *On the uniqueness of Markovian self-adjoint extension of diffusion operators on infinite-dimensional spaces,* Osaka J. Math. **22** (1984), 733–742.

Tishchenko, S.V.

[1] *Spectral representation of an infinite-dimensional contractive semigroup of selfadjoint operators,* Operators in Mathematical Physics and Infinite-Dimensional Analysis, Inst. Mat. Akad. Nauk Ukrain. SSR, Kiev, 1979, pp. 115–120. *(Russian)*

[2] *Spectral representations of countably parametric semi-groups,* Ukrain. Mat. Zh. **34** (1982), 532–535; ***English transl.*** in Ukrain. Math. J. **34** (1982).

[3] *On the connection of the Minlos–Sazonov theorem with Stone's theorem for an additive group in Hilbert space,* Abstracts of the School on Theory of Operators in Functional Spaces, Minsk, 1982, p. 188. *(Russian)*

Titchmarsh, E.C.

[1] *On the uniqueness of the Green's functions associated with a second-order differential equation,* Can. J. Math. **1** (1949), 191–198.

Tsykalenko, T.V.

[1] *The infinite-dimensional Schrödinger operator and potential perturbations of it,* Ukrain. Mat. Zh. **38** (1986), 534–538. *(Russian)*

Tykhonov, A.N.

[1] *Sur les espaces abstraits,* Compt. Rend. Acad. Sci. **182** (1926), 1519–1520.

Us, G.F.

[1] *The truncated generalized symmetric power moment problem,* Ukrain. Mat. Zh. **26** (1974), 348–358; ***English transl.*** in Ukrain. Math. J. **26** (1974).

[2] *Spectral decomposition of selfadjoint operators which are functions of operators acting upon different variables,* Dokl. Akad. Nauk SSSR **229** (1976), 812–815; ***English transl.*** in Soviet Math. Dokl. **17** (1976).

[3] *On functions of operators acting upon different variables,* Operators in Mathematical Physics and Infinite-Dimensional Analysis, Inst. Mat. Akad. Nauk Ukrain. SSR, Kiev, 1979, pp. 121–138. *(Russian)*

Vainerman, L.I.

[1] *The duality of algebras with involution and the operators of generalized shift,* VINITI Series on Mathematical Analysis, vol. **24**, VINITI, Moscow, 1986, pp. 165–205. *(Russian)*

Vakhaniya, N.N., Tarieladze, V.I., and Chobanyan, S. A.

[1] Probability Distributions on Banach Spaces, Nauka, Moscow, 1985.

Ventzel, A.D.

[1] A Course of the Theory of Random Processes, Nauka, Moscow, 1975; ***German transl.*****: A.D.Wentzell,** Theorie zufällige Prozesse, Akademie-Verlag, Berlin and Birkhäuser, Basel and Boston, Mass., 1979.

Vershik, A.M.

[1] *Measurable realizations of groups of automorphisms and integral representations of positive operators,* Sibirsk. Mat. Zh. **28** (1987), 52–60 *(Russian)*

Vershik, A.M., Gelfand, I.M., and Graev, M.I.

[1] *Representations of the group $SL(2, R)$, where R is a ring of functions,* Uspekhi Mat. Nauk **28** (1973), No. 5 (173), 83–128. *(Russian)*

[2] *Remarks on the representation of the groups of functions with values in a compact Lie group* **I**, Compositio Math. **35** (1977), 299–334; **II**, Compositio Math. **42** (1981), 217–243.

[3] *A commutative model of the representation of the group of currents $SL(2,\mathbb{R})^X$ connected with a unipotent subgroup,* Funk. Anal. i Prilozh. **17** (1983), No. 2, 70–72. *(Russian)*

Vershik, A.M. and Sudakov, V.N.

[1] *Probability measures on infinite-dimensional spaces,* Zapiski Nauch. Seminarov Lenigrad. Otdel. Mat. Inst. Akad. Nauk SSSR **12** (1969), 7–67. *(Russian)*

Vishik, M.I.

[1] *Parametrix of elliptic operators with an infinite number of independent variables,* Uspekhi Mat. Nauk **26** (1971), No. 2 (158), 155–174. ***(Russian)***

[2] *Fundamental solutions of infinite-dimensional elliptic operators of arbitrary order with constant coefficients,* Dokl. Akad. Nauk SSSR **208** (1973), 764–767. ***(Russian)***

Vishik, M.I. and Marchenko, A.V.

[1] *Boundary value problems for elliptic and parabolic operators of the second order on infinite-dimensional manifolds with an edge,* Mat. Sbornik **90** (132) (1973), 331–371. ***English transl.*** in Math. USSR Sb. **19** (1973).

Vladimirov, V.S.

[1] Generalized Functions in Mathematical Physics, Nauka, Moscow, 1976; ***English transl. of 2nd edn.*** **:** Mir, Moscow, 1979.

Vul, E.B.

[1] *On uniqueness theorems for a class of integral representations,* Dokl. Akad. Nauk SSSR **129** (1959), 722–725. ***(Russian)***

Wiener, N.

[1] Nonlinear Problems in the Theory of Random Processes, Izdat. Inostr. Lit., Moscow, 1961. ***(Russian translation)***

Woronowicz, S.L.

[1] *The quantum problem of moments.* **I**, Rep. Math. Phys. **1** (1970), 135–145; **II**, Rep. Math. Phys. **1** (1970), 175–183.

Yadrenko, M.I.

[1] Spectral Theory of Random Fields, Vyshcha Shkola, Kiev, 1980. ***(Russian)***

Yakovlev, V.S.

[1] *On measurability of functions with respect to spaces with resolutions of the identity,* Spectral Theory of Operators and Infinite-Dimensional Analysis, Inst. Mat. Akad. Nauk Ukrain. SSR, Kiev, 1984, pp. 138–143. ***(Russian)***

[2] *On representations of some commutative nuclear ∗-algebras,* Ukrain. Mat. Zh. **39** (1987), 235–243. *(Russian)*

[3] *On representations of some ∗-algebras of smooth functions,* Ukrain. Mat. Zh. **39** (1987), 400–403. *(Russian)*

Yokoi, Y.

[1] *Positive generalized white noise functionals,* Hiroshima Math. J. **20** (1990), 137–157.

[2] *Properties of Gelfand triplet in white noise analysis and a characterization of positive Hida distributions,* Proc. of Preseminar for International Conf. on Gaussian Random Fields (Nagoya, 1990), Pt. **2**, Nagoya Univ., Nagoya, 1991, pp. 34–48.

Yosida, K.,

[1] Functional Analysis, Mir, Moscow, 1967. *(Russian translation)*

Contents

Volume 2

Mathematical Physics and Applied Mathematics

Publications

1. M. Flato, Z. Maric, A. Milojevic, D. Sternheimer and J.P. Vigier (eds.): *Quantum Mechanics, Determinism, Causality, and Particles.* An International Collection of Contributions in Honor of Louis de Broglie on the Occasion of the Jubilee of His Celebrated Thesis. 1976
ISBN 90-277-0623-9
2. K. Maurin and R. Rączka (eds.): *Mathematical Physics and Physical Mathematics.* Proceedings of an International Symposium held in Warsaw (1974). 1976 ISBN 90-277-0537-2
3. M. Cahen and M. Flato (eds.): *Differential Geometry and Relativity.* A Volume in Honour of André Lichnerowicz on His 60th Birthday. 1976
ISBN 90-277-0745-6
4. M. Flato, C. Fronsdal and K.A. Milton (eds.): *Selected Papers (1937-1976) of Julian Schwinger.* 1979 ISBN Hb: 90-277-0974-2; Pb 90-277-0975-0
5. J.A. Wolf, M. Cahen and M. de Wilde (eds.): *Harmonic Analysis and Representations of Semisimple Lie Groups.* 1980 ISBN 90-277-1042-2
6. E. Tirapegui (ed.): *Field Theory, Quantization and Statistical Physics.* In Memory of Bernard Jouvet. 1981 ISBN 90-277-1128-3
7. V.P. Maslov and M.V. Fedoriuk: *Semi-classical Approximation in Quantum Mechanics.* 1981 ISBN 90-277-1219-0
8. V.N. Popov: *Functional Integrals in Quantum Field Theory and Statistical Physics.* 1983 ISBN 90-247-1471-1
9. F.A. Berezin: *Introduction to Superanalysis.* 1987 ISBN 90-277-1668-4
10. N.N. Bogolubov, A.A. Logunov, A.I. Oksak and I.T. Todorov: *General Principles of Quantum Field Theory.* 1990 ISBN 0-7923-0540-X
11. L.D. Faddeev and S.P. Merkuriev: *Quantum Scattering Theory for Several Particle Systems.* 1993 ISBN 0-7923-2414-5
12. Y.M. Berezansky and Y.G. Kondratiev: *Spectral Methods in Infinite-Dimensional Analysis.* 1995 ISBN 0-7923-2847-7 (Vol. 1);
ISBN 0-7923-2848-5 (Vol. 2);
ISBN 0-7923-2849-3 (Set)

Kluwer Academic Publishers – Dordrecht / Boston / London

GPSR Compliance

The European Union's (EU) General Product Safety Regulation (GPSR) is a set of rules that requires consumer products to be safe and our obligations to ensure this.

If you have any concerns about our products, you can contact us on ProductSafety@springernature.com

In case Publisher is established outside the EU, the EU authorized representative is:

Springer Nature Customer Service Center GmbH
Europaplatz 3
69115 Heidelberg, Germany

Batch number: 10823775

Printed by Printforce, the Netherlands

SPECTRAL METHODS IN INFINITE-DIMENSIONAL ANALYSIS

Mathematical Physics and Applied Mathematics

Editor:

M. Flato, *Université de Bourgogne, Dijon, France*

The titles published in this series are listed at the end of this volume.

Volume 12/2

SPECTRAL METHODS IN INFINITE-DIMENSIONAL ANALYSIS

Volume 2

by

Y. M. Berezansky

and

Y. G. Kondratiev

Institute of Mathematics,
Ukrainian Academy of Science,
Kiev, Ukraine

SPRINGER-SCIENCE+BUSINESS MEDIA, B.V.

Library of Congress Cataloging-in-Publication Data

Berezanskiĭ, IU. M. (IUriĭ Makarovich)
[Spektral'nye metody v beskonechnomernom analize. English]
Spectral methods in infinite-dimensional analysis / by Yurij M. Berezansky and Yurij G. Kondratiev.
p. cm. -- (Mathematical physics and applied mathematics ; v. 12)
Includes bibliographical references and index.
ISBN 978-94-010-4227-7 ISBN 978-94-011-0509-5 (eBook)
DOI 10.1007/978-94-011-0509-5
1. Quantum field theory--Mathematics. 2. Degrees of freedom. 3. Statistical physics. 4. Spectral theory (Mathematics) 5. Infinite groups. 6. Mathematical physics. I. Kondrat'ev, IU. G. (IUriĭ Grigor'evich) II. Title. III. Series.
QC174.52.D43B47 1994
530.1'55782--dc20 94-14885

ISBN 978-94-010-4227-7

This is the updated and revised translation of the original Russian work,
Spectral Methods in Infinite-Dimensional Analysis.
Translated by P.V. Malyshev and D. V. Malyshev.
Published by Naukova Dumka, Kiev, © 1988.

Printed on acid-free paper

Contents

Volume 2

5. APPLICATION OF THE THEORY OF EXPANSIONS TO HARMONIC ANALYSIS

In this chapter, we present a method of construction of spectral representations of positive definite kernels by the eigenfunctions of operators (theorems of Bochner's type on positive definite functions, representations of moment sequences, and so on). The main idea of this construction can be explained most simply if we take a positive definite function of a single variable as an example.

Recall that a continuous function $(-2l, 2l) \ni x \mapsto k(x) \in \mathbb{C}^1$ $(0 < l \leq \infty$ is fixed) is called positive definite if the kernel $K(x, y) = k(y - x)$ $(x, y \in (-l, l))$ is positive definite. In order to obtain the representation of this function in terms of elementary positive definite functions $k_\lambda(x) = e^{i\lambda x}$ $(\lambda \in \mathbb{R}^1$ is a parameter enumerating these functions), we can act as follows.

For given k, we introduce a quasiscalar product

$$\langle \varphi, \psi \rangle = \int_{-l}^{l} \int_{-l}^{l} k(y - x) \varphi(y) \overline{\psi(x)} \, dx \, dy, \tag{0.1}$$

where $\varphi, \psi \in C_0^\infty((-l, l))$. After identification and completion, we obtain a Hilbert space H_k. Changing the variables and integrating by parts, we easily find that $\langle -i\varphi', \psi \rangle = \langle \varphi, -i\psi' \rangle$ $(\varphi, \psi \in C_0^\infty((-l, l)))$. Therefore, the mapping

$$C_0^\infty((-l, l)) \ni \varphi \mapsto -i\varphi' \in C_0^\infty((-l, l)) \tag{0.2}$$

induces an Hermitian operator A' in H_k which has a selfadjoint extension A in H_k (if $l = \infty$, then $A = \tilde{A}'$; for $l < \infty$, the essential selfadjointness of A' is also possible).

Applying the spectral projection theorem to A, we can write

$$1 = \int_{\mathbb{R}^1} P(\lambda)\, d\rho(\lambda), \tag{0.3}$$

where the generalized projector $P(\lambda)$ "projects" onto the generalized eigenvectors $e^{i\lambda x}$ of the operator A. Applying (0.3) to φ and taking the scalar product of φ and ψ (in the sense of (0.1)), we get

$$\int_{-l}^{l}\int_{-l}^{l} k(y-x)\varphi(y)\overline{\psi(x)}\, dx\, dy = \int_{\mathbb{R}^1} \langle P(\lambda)\varphi, \psi\rangle\, d\rho(\lambda)$$

$$(\varphi, \psi \in C_0^\infty((\mathbb{R}^1))). \tag{0.4}$$

Computing $\langle P(\lambda)\varphi, \psi\rangle$, one can easily find that this expression is equal to

$$c(\lambda)\int_{-l}^{l}\int_{-l}^{l} e^{i\lambda(y-x)}\varphi(y)\overline{\psi(x)}\, dx\, dy \quad (c(\lambda) \geq 0).$$

Substituting the last integral in (0.4) and taking into account that φ and ψ are arbitrary, we obtain the equality

$$k(t) = \int_{\mathbb{R}^1} e^{i\lambda t}\, d\sigma(\lambda) \quad (t \in (-2l, 2l)) \tag{0.5}$$

(we have denoted $y-x$ by t and $c(\lambda)d\rho(\lambda)$ by $d\sigma(\lambda)$).

Formula (0.5) contains two classical theorems of harmonic analysis, namely, the Bochner theorem (for the case $l = \infty$), and M. Krein's theorem (see M. G.Krein [1]) on the possibility of the extension of a positive definite function defined on the interval to a positive definite function on the whole axis (for the case $l < \infty$). The integral in (0.4) is defined for all $t \in \mathbb{R}^1$ and is a positive definite function. Note that there exist as many extensions of the function as selfadjoint extensions A of the operator A'.

For the classical moment problem, the construction is similar: For a given moment sequence $s = (s_n)_{n=0}^\infty$ $(s_n \in \mathbb{R}^1)$, by analogy with (0.1), we construct a quasinuclear product

$$\langle\varphi, \psi\rangle = \sum_{j,k=0}^{\infty} s_{j+k}\varphi_k\overline{\psi_j}, \tag{0.6}$$

where $\varphi = (\varphi_j)_{j=0}^\infty$ and $\psi = (\psi_j)_{j=0}^\infty$ are finite sequences of complex numbers (condition $\langle\varphi, \psi\rangle \geq 0$ is just the definition of the moment character of s). Further, for the product (0.6), we create a Hilbert space H_s in which the operation of shift

$$\mathcal{A}\varphi = \mathcal{A}(\varphi_0, \varphi_1, \varphi_2, \ldots) = (0, \varphi_0, \varphi_1, \varphi_2, \ldots), \tag{0.7}$$

defined on finite sequences, is considered instead of (0.2). Repeating the above argument, we get the solution of the moment problem, i.e., we obtain the representation

$$s_n = \int_{\mathbb{R}^1} \lambda^n d\sigma(\lambda) \quad (n = 0, 1, \ldots). \tag{0.8}$$

These two schemes are applicable in many other cases when both the objects (k and s), which are represented and operations (0.2) and (0.7), are changed.

The approach presented is due to M. G. Krein [2-4] with the only difference that instead of the spectral projection theorem, Krein used its equivalent, namely, the method of directing functionals suitable in the case of finite multiplicity of the spectrum of the operator A. The representation used here belongs to Berezansky [2–4; 5, Chapter 8]; in these papers the operator A is an arbitrary selfadjoint operator and, moreover, the finite-dimensional spectral projection theorem (0.3) corresponding to multi-dimensional harmonic analysis is employed.

Our aim is to obtain results analogous with those mentioned above and connected not with a single operator but with an infinite family of commuting selfadjoint operators $(A_x)_{x \in X}$ and with the corresponding spectral projection theorem. When realizing this program, we establish the theorems on representation (and extension) of positive definite functions of infinitely many variables on a layer, study the infinite-dimensional moment problem, etc. The role of the operators is played by a commuting family of differentiations (0.2) with respect to different variables, infinite families of shifts of type (0.7), and so on.

The general scheme itself is presented in Section 5. Certainly, it is possible to begin with it and then use the corresponding results in our constructions. However, this method leads to the accentuation of details which does not seem to be appropriate. Therefore, we begin with general important examples, realizing all the constructions for them independently.

Thus, in Section 2, we study a general version of the infinite-dimensional moment problem when s_n is an element of the nth tensor power of a certain fixed space. We give different special examples of this problem and some generalizations of it.

In Section 3, we prove the theorem on representation and, hence, on the extension of a function positive definite in the integral sense and given on a layer in the space $\mathbb{R}^\infty$ equipped with a product measure.

In Section 4, with the help of the results from Section 3, we establish the Minlos – Sazonov theorem for positive definite functions on a layer in a Hilbert space and, in particular, prove that these functions can be extended to the whole space.

Many criteria of selfadjointness of operators in the Hilbert space are presented in Section 1. These results are employed both in Chapter 5 and in subsequent chapters. They are concentrated around evolutionary criteria and quasianalytical criteria which follow from the evolutionary ones (the criteria, which guarantee the selfadjointness of operators on the basis of a certain knowledge about the dynamics associated with these operators, are called evolutionary criteria).

§1. Some Criteria of the Selfadjointness of Operators

In this section, we give the methods for proving the selfadjointness of operators based mainly on the study of corresponding evolutionary equations. We also present some criteria of selfadjointness which can be derived when realizing this program (in particular, the "quasianalytical" criteria) and the facts concerning the verification of the commutation properties of selfadjoint operators. Many results given below are employed somewhere in this book; the others are included to make our survey of this problem more complete.

1.1. Schrödinger and Hyperbolic Criteria of Selfadjointness

Let H be a Hilbert space. Assume that an operator B with an everywhere dense domain of definition $\mathfrak{D}(B)$ acts in this space. Let $I \subset \mathbb{R}^1$ be a finite or infinite interval closed, open, or half-open, and let $r \in \mathbb{N}$. We say that a vector function $I \ni t \mapsto u(t) \in H$ is a strong solution of the equation

$$\left(\frac{d^r u}{dt^r}\right)(t) + Bu(t) = 0 \quad (t \in I) \tag{1.1}$$

on I, if it is r times strongly continuously differentiable (i.e., has r strong derivatives on I the last of which is continuous) and such that $u(t) \in \mathfrak{D}(B)$ for every $t \in I$ and the equality (1.1) is satisfied.

A vector function $I \ni t \mapsto u(t) \in H$, r times strongly continuously differentiable, is a strong solution of the equation

$$\left(\frac{d^r u}{dt^r}\right)(t) + B^* u(t) = 0 \quad (t \in I) \tag{1.2}$$

if and only if the following "weak" equality holds

$$\left(\left(\frac{d^r u}{dt^r}\right)(t), f\right)_H + (u(t), Bf)_H = 0 \quad (f \in \mathcal{D}(B);\ t \in I). \tag{1.3}$$

This statement immediately follows from the definition of a selfadjoint operator because (1.3) implies that $u(t) \in \mathcal{D}(B^*)$ for every $t \in I$ by virtue of the inclusion $u^{(r)}(t) \in H$.

We say that the uniqueness of strong solutions of the Cauchy problem for Equation (1.1) on $I = [0, b)$ $(0 < b \leq \infty)$ takes place of each strong solution $u(t)$ of this equation on $[0, b)$, such that

$$u(0) = \ldots = u^{(r-1)}(0) = 0$$

vanishes also for all $t \in (0, b)$. Clearly, *if the uniqueness takes place on* $[0, b)$ *for some* $b > 0$, *then it holds also on* $[0, \infty)$. Indeed, assume that $[0, \infty) \ni t \mapsto u(t)$ is a strong solution of Equation (1.1) on $[0, \infty)$ such that

$$u(0) = \ldots = u^{(r-1)}(0) = 0.$$

Due to the uniqueness which is assumed, we can conclude that $u(t) = 0$ for $t \in (0, b)$ and, in particular, that $u(t) = 0$ in the vicinity of the point $c = b/2$; therefore,

$$u(c) = \ldots = u^{(r-1)}(c) = 0.$$

The function

$$[0, \infty) \ni t \mapsto u_1(t) = u(t + c)$$

is a strong solution of (1.1) on $[0, \infty)$ such that

$$u_1(0) = u(t + c) = 0, \ldots, \quad u_1^{(r-1)}(0) = u^{(r-1)}(c) = 0$$

and, therefore, $u_1(t) = 0$ for $t \in (0, b)$. Repeating the previous arguments, we conclude that the function

$$[0, \infty) \ni t \mapsto u_2(t) = u_1(t + c) = u(t + 2c)$$

vanishes for $t \in (0, b)$. Then we construct the function $u_3(t)$ and so on. As a result, we get that $u(t) = 0$ $(t \in (0, \infty))$.

■

In what follows, we assume that r is equal to 1 or 2. The following "Schrödinger" criterion of selfadjointness holds.

Theorem. 1.1. *Assume that A is an Hermitian operator acting in H. In order that this operator be essentially selfadjoint, it is necessary that for both equations*

$$\left(\frac{du}{dt}\right)(t) \pm (iA)^* u(t) = 0 \qquad (t \in [0, b)), \tag{1.4}$$

the uniqueness of strong solutions of the Cauchy problem on $[0, b)$ takes place for any $b \in (0, \infty]$ and it is sufficient that it take place for some b from this interval.

Proof. We split the proof into several steps.

I. We establish the sufficiency under the assumption that A has equal deficiency numbers. Assume the contrary. Let the closure $\tilde{A}$ be not selfadjoint. Then A has two different selfadjoint extensions A_1 and A_2 in H. Let E_1 and E_2 be the corresponding resolutions of the identity. For any $g \in \mathfrak{D}(A) \subset \mathfrak{D}(A_1)$, the integral

$$\int_{\mathbb{R}^1} \lambda^2 d(E_1(\lambda)g, g)_H$$

converges; therefore, the vector function

$$[0, \infty) \ni t \mapsto u_1(t) = \int_{\mathbb{R}^1} e^{i\lambda t} dE_1(\lambda)\, g \tag{1.5}$$

is once strongly continuously differentiable, and

$$u_t'(t) = i \int_{\mathbb{R}^1} \lambda e^{i\lambda t} dE_1(\lambda)\, g.$$

One can easily see that this function is a strong solution of Equation (1.4) with the sign "+" on $[0, \infty)$. Indeed, one should verify the corresponding weak equality (1.3) which now has the form

$$\left(\left(\frac{du_1}{dt}\right)(t),\, f\right)_H + (u_1(t), iAf)_H = 0 \qquad (f \in \mathfrak{D}(A); \quad t \in [0, \infty)).$$

Since

$$d(E_1(\lambda)\, g, Af)_H = d(E_1(\lambda)\, g, A_1 f)_H$$

$$= d\left(\int_{-\infty}^{\lambda} \mu d(E_1(\mu) g, f)_H\right) = \lambda d(E_1(\lambda) g, f)_H ,$$

we have

$$\left(\left(\frac{du_1}{dt}\right)(t), f\right)_H + (u_1(t), iAf)_H$$

$$= i \int_{\mathbb{R}^1} \lambda e^{i\lambda t} d(E_1(\lambda) g, f)_H - i \int_{\mathbb{R}^1} e^{i\lambda t} d(E_1(\lambda) g, Af)_H = 0$$

$$(f \in \mathfrak{D}(A);\ t \in [0, \infty)),$$

i.e., the required relation is satisfied. Analogously, the function $u_2(t)$, constructed according to (1.5) from E_2, is a strong solution of the same equation; $u_1(0) = u_2(0) = g$. Hence, $u(t) = u_1(t) - u_2(t)$ is a strong solution of Equation (1.4) with the sign "+" on $[0, \infty)$ and, moreover, $u(0) = 0$. Due to the assumption made above, uniqueness takes place on $[0, \infty)$, therefore $u(t) = 0$ for $t \in [0, \infty)$. Whence

$$\int_{\mathbb{R}^1} e^{i\lambda t} d((E_1(\lambda) - E_2(\lambda)) g, h)_H = 0 \quad (g \in \mathfrak{D}(A),\ h \in H,\ t \in [0, \infty)). \tag{1.6}$$

Now consider Equation (1.4) with the sign "–". We can repeat the above argument replacing $e^{i\lambda t}$ by $e^{-i\lambda t}$ in (1.5). As a result, we obtain the relation (1.6) in which the some substitution is made. Hence, if we introduce a function of bounded variation

$$\omega(\lambda) = ((E_1(\lambda) - E_2(\lambda)) g, h)_H \quad (\lambda \in \mathbb{R}^1),$$

then, according to (1.6) and the last modification of it, we have

$$\int_{\mathbb{R}^1} e^{i\lambda t} d\omega(\lambda) = 0$$

for all $t \in \mathbb{R}^1$. By virtue of the uniqueness theorem for the Fourier–Stieltjes transformation, this implies that $d\omega(\lambda) = 0$, i.e.,

$$((E_1(\lambda) - E_2(\lambda)) g, h)_H = 0 \quad (\lambda \in \mathbb{R}^1).$$

Taking into account that here $g \in \mathfrak{D}(A)$, $\mathfrak{D}(A)$ is dense in H, and $h \in H$, we obtain $E_1(\lambda) = E_2(\lambda)$ $(\lambda \in \mathbb{R}^1)$. Thus, we get a contradiction.

II. In the case of distinct deficiency numbers of the operator A, we use the following simple lemma.

Lemma 1.1. *Let* $\mathcal{H} = H \oplus H$ *be a space of vectors* $f = (f_1, f_2)$ $(f_1, f_2 \in H)$. *We construct an operator* $\mathcal{A}$ *acting in this space and having the everywhere dense domain of definition* $\mathcal{D}(\mathcal{A}) = \mathcal{D}(A) \oplus \mathcal{D}(A)$ *by setting* $\mathcal{A}f = (Af_1, -Af_2)$ $(f \in \mathcal{D}(\mathcal{A}))$. *Consider the equation*

$$\left(\frac{du}{dt}\right)(t) \pm (i\mathcal{A})^* u(t) = 0 \qquad (t \in [0, b)) \tag{1.7}$$

on vector functions with values in $\mathcal{H}$. *It is stated that if the uniqueness of strong solutions of the Cauchy problem takes place on* $[0, b)$ *for both Equations* (1.4), *then strong solutions of Equation* (1.7) *are unique in the same sense, and vice versa.*

Proof. Let

$$[0, b) \ni t \mapsto u(t) = (u_1(t), u_2(t)) \in \mathcal{H}$$

be a strong solution of the Cauchy problem for Equation (1.7). Since

$$\mathcal{A}^* f = (A^* f_1, -A^* f_2) \quad (f \in \mathcal{D}(\mathcal{A}^*) = \mathcal{D}(A^*) \oplus \mathcal{D}(A^*)),$$

we conclude that $[0, b) \ni t \mapsto u_1(t) \in H$ and $[0, b) \ni t \mapsto u_2(t) \in H$ are strong solutions of Equation (1.4) with the signs "+" and "–", respectively. This and the uniqueness of strong solutions of the Cauchy problem for (1.4) on $[0, b)$ which has been assumed, yield the uniqueness for (1.7). The converse statement is derived in the same simple manner.

III. We now prove sufficiency in the theorem for the operator A with deficiency index (m, n). As in Lemma 1.1, let us construct the operator $\mathcal{A}$. It is easy to verify that the deficiency index of this operator is equal to $(m + n, m + n)$ (see, e.g., Akhiezer and Glazman [1, Chapter 9, Subsection III]). By virtue of Lemma 1.1, the uniqueness for (1.7) takes place on $[0, b)$. Using this lemma in the case when A is replaced by $-A$, we conclude that the existence in the same sense, takes place also for Equation (1.7) in which the sign "+" is replaced by "–". Since the deficiency numbers of the operator $\mathcal{A}$ are identical and equal to $m + n$, we can apply the argument of Step I to this operator and conclude that $\mathcal{A}$ is essentially selfadjoint. But then $m + n = 0$, whence $m = n = 0$, i.e., A is also essentially selfadjoint.

IV. To prove necessity, we establish a general lemma reflecting the Holmgren principle for our situation. Below, this lemma will be also employed for some other purposes.

Lemma 1.2. *Consider Equation* (1.2) *on* $[0, b)$ $(b \in (0, \infty])$. *Assume that there exists a set* Φ *dense in* H *such that for any* $T \in (0, b)$ *and* $\varphi_0, \dots, \varphi_{r-1} \in \Phi$ *there exists a strong solution of the Cauchy problem*

$$\left(\frac{d^r\varphi}{dt^r}\right)(t) = (-1)^r B\varphi(t) = 0 \quad (t \in [0, T]),$$

$$\varphi(T) = \varphi_0, \dots, \quad \varphi^{(r-1)}(T) = \varphi_{r-1}. \tag{1.8}$$

Then the strong solutions of the Cauchy problem for (1.2) *are unique on* $[0, b)$.

Proof. Let us first prove the lemma for the case of $r = 2$. We simply verify the following formula for integration by parts: Assume that $[0, T] \ni t \mapsto \alpha(t)$ and $\beta(t) \in H$ are twice strongly continuously differentiable vector functions. Then

$$\int_0^T (\alpha''(t), \beta(t))_H \, dt$$

$$= \int_0^T (\alpha(t), \beta''(t))_H \, dt + [(\alpha'(t), \beta(t))_H - (\alpha(t), \beta'(t))_H]_0^T. \tag{1.9}$$

Let $u(t)$ be a strong solution of the Cauchy problem for Equation (1.2) with $r = 2$ on $[0, b)$ such that $u(0) = u'(0) = 0$, and let $\varphi(t)$ be a strong solution mentioned in the formulation of the lemma. Using (1.9), we get

$$\int_0^T ((u''(t), \varphi(t))_H - (u(t), \varphi''(t))_H) \, dt = (u'(T), \varphi_0)_H - (u(T), \varphi_1)_H. \tag{1.10}$$

Taking into account that $\varphi(s) \in \mathfrak{D}(B)$ for every $s \in [0, T]$, we can write

$$(u''(t), \varphi(s))_H + (u(t), B\varphi(s))_H = 0 \quad (t \in [0, b]),$$

according to the equality (1.3) with $f = \varphi(s)$. Setting here $t = s$, and then changing s by t, we obtain

$$(u''(t), \varphi(t))_H = -(u(t), B\varphi(t))_H \quad (t \in [0, T]).$$

By virtue of (1.8) with $r = 2$, we have

$$(u(t), \varphi''(t))_H = -(u(t), B\varphi(t))_H \quad (t \in [0, T]).$$

These two equalities imply that the expression on the left-hand side of (1.10) vanishes and, therefore,

$$(u'(T), \varphi_0)_H - (u(T), \varphi_1)_H = 0 \quad (\varphi_0, \varphi_1 \in \Phi).$$

It now follows from the denseness of Φ in H that $u(T) = u'(T) = 0$. Since $T \in (0, b)$ is arbitrary, this yields the required statement.

For the case of $r = 1$, the proof is similar; one should only employ the formula for integration by parts

$$\int_0^T (\alpha'(t), \beta(t))_H \, dt = -\int_0^T (\alpha(t), \beta'(t))_H \, dt + [(\alpha(t), \beta(t))_H]_0^T, \qquad (1.11)$$

valid for vector functions $[0, T] \ni t \mapsto \alpha(t)$ and $\beta(t) \in H$ continuously differentiable once. In the case of general r, one must iterate the formula (1.11) r times (the formula (1.10) is, in fact, the equality (1.11) iterated two times).

■

V. We now prove the necessity condition. Let $\tilde{A}$ be a selfadjoint operator, and let E be its resolution of the identity. We apply Lemma 1.2, setting $r = 1$, $B = (iA)^* = -i\tilde{A}$, and

$$\Phi = \bigcup_{n=1}^{\infty} E((-n, n))H.$$

The strong solution of the Cauchy problem (1.8), which now has the form

$$\varphi'(t) + i\tilde{A}\varphi(t) = 0 \quad (t \in [0, T]), \quad \varphi(T) = \varphi_0,$$

exists and is equal to

$$\varphi(t) = \int_{\mathbb{R}^1} e^{-i\lambda(t-T)} dE(\lambda)\varphi_0 \quad (t \in [0, T]) \qquad (1.12)$$

(since $\varphi_0 \in \Phi$, the integral in (1.12) is, in fact, taken over a finite interval and, therefore, the function $[0, T] \ni t \mapsto \varphi(t)$ is once continuously differentiable; clearly, it is a solution of the required problem). Thus, by virtue of this lemma, the uniqueness takes place for Equation (1.4) with the sign "+" on $[0, b)$. Equation (1.4) with sign "–" is investigated in a similar way. Now $B = -(iA)^* = i\tilde{A}$.

■

The following two theorems represent the contents of the "hyperbolic" criterion of selfadjointness.

Theorem 1.2. *Let A be an Hermitian operator acting in H. In order that this operator be essentially selfadjoint, it is necessary that the Cauchy problem for the equation*

$$\left(\frac{d^2u}{dt^2}\right)(t) + A^* u(t) = 0 \quad (t \in [0, b)) \tag{1.13}$$

have the unique strong solution on $[0, b)$ for any $b \in (0, \infty]$, and it is sufficient that A be semi-bounded below and that this uniqueness of strong solutions take place for some $l > 0$.

Proof. *Sufficiency.* Assume that $\tilde{A}$ is not selfadjoint. Then there exist two different selfadjoint extensions (A_1 and A_2) of A in H bounded below by a number $c > -\infty$. Let E_1 and E_2 be the corresponding resolutions of the identity. For any $g \in \mathfrak{D}(A) \subset \mathfrak{D}(A_1)$, the integral $\int_{\mathbb{R}^1} \lambda^2 \, d(E_1(\lambda) g, g)_H$ converges and, hence, the vector function

$$[0, \infty) \ni t \mapsto u_1(t) = \int_c^\infty \cos(\sqrt{\lambda}\, t) dE_1(\lambda)\, g \tag{1.14}$$

is twice strongly continuously differentiable. As in the proof of Theorem 1.1, we can easily establish that it is a strong solution of Equation (1.13) on $[0, \infty)$. To do this, one must verify the validity of the corresponding weak equality of the type (1.3). In addition, $u_1(0) = g$ and $u_1'(0) = 0$. Similarly, replacing E_1 by E_2 in (1.14), we construct the function $u_2(t)$. The difference $u(t) = u_1(t) - u_2(t)$ is also a strong solution of Equation (1.13) on $[0, \infty)$ such that $u(0) = u'(0) = 0$. By virtue of the assumed uniqueness of the strong solutions of the Cauchy problem, we have $u(t) = 0$ for $t \geq 0$. Scalarly multiplying this equality scalarly by $h \in H$, we obtain

$$\int_c^\infty \cos(\sqrt{\lambda}\, t) d((E_1(\lambda) - E_2(\lambda)), h)_H = 0 \quad (t \in [0, \infty)).$$

By virtue of the fact that the measure $d\omega(\lambda)$ is uniquely defined by a given Fourier–Stieltjes transform of it, we conclude that

$$\omega(\lambda) = ((E_1(\lambda) - E_2(\lambda))\, g, h)_H = 0 \quad (\lambda \in \mathbb{R}^1).$$

Due to the arbitrariness of $g \in \mathfrak{D}(A)$ and $h \in H$, we get that $E_1(\lambda) = E_2(\lambda)$ $(\lambda \in \mathbb{R}^1)$, which is not possible.

Necessity. Assume that $\tilde{A}$ is selfadjoint and that E is its resolution of the identity. We apply Lemma 1.2, setting $r = 2$, $B = A^* = \tilde{A}$, and

$$\Phi = \bigcup_{n=1}^{\infty} E((-n, n))H.$$

The Cauchy problem (1.8) now has the form

$$\varphi''(t) + \tilde{A}\varphi(t) = 0 \quad (t \in [0, T]), \quad \varphi(T) = \varphi_0, \quad \varphi'(T) = \varphi_1.$$

A strong solution of it exists and is equal to

$$\varphi(t) = \int_{\mathbb{R}^1} \cos(\sqrt{\lambda}(t-T))\, dE(\lambda)\varphi_0 + \int_{\mathbb{R}^1} \frac{1}{\sqrt{\lambda}} \sin(\sqrt{\lambda}(t-T))\, dE(\lambda)\varphi_1$$

(as in (1.12), here the integration is, in fact, carried out over a finite interval). Thus, according to this lemma, the Cauchy problem for (1.13) has the unique strong solution on $[0, b)$ for every $b \in (0, \infty]$.

■

Remark 1. It follows from the proof of necessity in Theorems 1.1 and 1.2 that the essential selfadjointness of A yields the uniqueness of strong solutions of the Cauchy problem for the equation

$$\left(\frac{d^r u}{dt^r}\right)(t) + z\tilde{A}u(t) = 0 \quad (r \in \mathbb{N};\ z \in \mathbb{C}^1)$$

on $[0, b)$ for any $b \in (0, \infty]$.

■

Usually, it is convenient to employ Theorem 1.2 in a simple combination with Lemma 1.2. We formulate the corresponding result in the form of a theorem.

Theorem 1.3. *Let A be an Hermitian operator semi-bounded below and acting in H. Assume that there exists a linear set $\Phi \subset H$ dense in H and such that the Cauchy problem*

$$\left(\frac{d^2\varphi}{dt^2}\right)(t) + A\varphi(t) = 0 \quad (t \in [0, T]), \quad \varphi(T) = \varphi_0, \quad \varphi'(T) = \varphi_1 \qquad (1.15)$$

has a strong solution for some $b > 0$ and for arbitrary $T \in (0, b)$ and $\varphi_0, \varphi_1 \in \Phi$. Then the operator A is essentially selfadjoint.

Proof. By virtue of Lemma 1.2, it follows from the condition of the theorem that the Cauchy problem for Equation (1.13) has a unique strong solution on $[0, b)$. But then, according to Theorem 1.2, $\tilde{A}$ is selfadjoint.

■

Clearly, a similar theorem is valid also in the case of Theorem 1.1. We also note that the set Φ in Lemma 1.2 and Theorem 1.3 can be dependent on T.

Example 1.1. This is an illustration of the application of Theorem 1.3 to the simplest situation. In the space $L_2(\mathbb{R}^d, dx)$ $(d \in \mathbb{N})$ we consider the Schrödinger operator A, i.e., the closure of the operator

$$C_0^\infty(\mathbb{R}^d) \ni u(x) \mapsto (\mathfrak{L}u)(x) = -(\Delta u)(x) + q(x)u(x),$$

where q is a real-valued potential from $C(\mathbb{R}^d)$. We prove that if A is semi-bounded, then it is selfadjoint. Indeed, consider the Cauchy problem for the hyperbolic equation

$$\left(\frac{d^2 u}{dt^2}\right)(x, t) + (\mathfrak{L}u)(x, t) = 0,$$

$$u(x, T) = \varphi_0, \quad \left(\frac{du}{dt}\right)(x, T) = \varphi_1 \quad (\varphi_0, \varphi_1 \in C_0^\infty(\mathbb{R}^d)) \tag{1.16}$$

on $[0, T]$. By virtue of the classical theorems of the theory of partial differential equations, the problem (1.16) has a solution $u \in C^2(\mathbb{R}^2 \times [0, T])$ and, moreover, this solution $u(x, t)$ is finite with respect to x for every $t \in [0, T]$ (because the rate of propagation of perturbation is finite for (1.16)). Therefore, $u(\cdot, t)$ may be interpreted as a vector function $\varphi(t)$ with values in $L_2(\mathbb{R}^d, dx)$ which is a strong solution of the Cauchy problem for (1.15) with $\Phi = C_0^\infty(\mathbb{R}^d)$. By Theorem 1.3, the operator A is selfadjoint.

1.2. Some Generalizations of the Hyperbolic Criterion

We present three results which extend the applicability of Theorem 1.3 to operators A which are not semi-bounded, to operators for which the Cauchy problem (1.15) is hard to solve, and to the powers of the operator A.

Theorem 1.4. *Let A be an Hermitian operator acting in H with equal defici-*

ency numbers. Suppose that there exists a set Φ *dense in* H *such that for any* $\varphi_0, \varphi_1 \in \Phi$ *the Cauchy problem*

$$\left(\frac{d^2\varphi}{dt^2}\right)(t) + A\varphi(t) = 0 \quad (t \in [0, \infty]), \ \varphi(0) = \varphi_0, \quad \varphi'(0) = \varphi_1 \qquad (1.17)$$

has a strong solution. In addition, assume that for every fixed $\varphi_0 \in \Phi$, *the solution* $\varphi(t)$ *of the Cauchy problem* (1.17) *with* $\varphi_1 = 0$ *is such that*

$$(\varphi(t), \varphi_0)_H = O(e^{ct^2}), \quad t \to \infty, \qquad (1.18)$$

with some $c = c_{\varphi_0} \geq 0$. *Then* A *is essentially selfadjoint.*

Proof. First, we note that the condition of the theorem implies the existence of a strong solution $\varphi(t)$ of the Cauchy problem (1.15) for any $T \in [0, \infty)$. Indeed, if we denote a strong solution of the problem (1.17) by $\psi(t)$, then $\varphi(t) = \psi(T - t)$ ($t \in [0, T]$) is the required solution of the problem (1.15).

Let $[0, \infty) \ni t \mapsto \varphi(t)$ be a strong solution of the problem (1.17) such that $\varphi_1 = 0$. The following representation holds

$$\varphi(t) = \int_{\mathbb{R}^1} \cos(\sqrt{\lambda}\, t)\, dE_1(\lambda)\varphi_0 \quad (t \in [0, \infty)), \qquad (1.19)$$

where E_1 is a resolution of the identity corresponding to an arbitrary selfadjoint extension A_1 of the operator A in H. In fact, fix $\alpha \in \mathfrak{B}(\mathbb{R}^1)$ with a compact closure. The vector function

$$[0, \infty) \ni t \mapsto \chi_1(t) = E_1(\lambda)\varphi(t)$$

is twice strongly continuously differentiable, because this is true for the function $\varphi(t)$, and the operator $E_1(\alpha)$ is continuous. Moreover,

$$\chi_1''(t) = E_1(\alpha)\varphi''(t) \quad (t \in [0, \infty)).$$

In addition, for every $t \in [0, \infty)$, we have

$$\chi_1(t) = E_1(\alpha)\varphi(t) \in \mathfrak{D}(A_1) \subset \mathfrak{D}(A^*)$$

and

$$A^*\chi_1(t) = A_1\chi_1(t) = A_1E_1(\alpha)\varphi(t) = E_1(\alpha)A_1\varphi(t) = E_1(\alpha)A\,\varphi(t).$$

Taking into account these relations and (1.17), we can write

$$\chi_1''(t) + A^*\chi_1(t) = E_1(\alpha)\,\varphi''(t) + E_1(\alpha)\,A\,\varphi(t)$$

$$= E_1(\alpha)(\varphi''(t) + A\,\varphi(t)) = 0 \quad (t \in [0,\infty));$$

$$\chi_1(0) = E_1(\alpha)\varphi(0) = E_1(\alpha)\varphi_0,$$

and

$$\chi'(0) = E_1(\alpha)\varphi'(0) = E_1(\alpha)\varphi_1.$$

Therefore, $\chi_1(t)$ is a strong solution of the Cauchy problem

$$\chi''(t) + A^*\chi(t) = 0 \quad (t \in [0,\infty))$$

and

$$\chi(0) = E_1(\alpha)\varphi_0, \quad \chi'(0) = 0.$$

Clearly, the vector function

$$[0,\infty) \ni t \mapsto \chi_2(t) = \int_\alpha \cos(\sqrt{\lambda}\,t)\,dE_1(\lambda)\varphi_0$$

is another strong solution (see the proof of sufficiency in Theorems 1.2 and 1.1). By virtue of Lemma 1.2, these solutions must coincide, because the Cauchy problem (1.16) has a strong solution, as has been already clarified. Hence,

$$E_1(\alpha)\,\varphi(t) = \int_\alpha \cos(\sqrt{\lambda}\,t)\,dE_1(\lambda)\varphi_0 \quad (t \in [0,\infty)). \tag{1.20}$$

In (1.20), we set $\alpha = (-n, n)$ and pass to the limit as $n \to \infty$. For every t in H, we have

$$E(-n,n))\varphi(t) \underset{n\to\infty}{\to} \varphi(t),$$

therefore there exists a strong limit

$$\lim_{n\to\infty} \int_{-n}^{n} \cos(\sqrt{\lambda}\,t)\,dE_1(\lambda)\varphi_0,$$

equal to the integral on the right-hand side of (1.19) by definition. Thus, (1.19) is now established.

Assume that $\tilde{A}$ is not selfadjoint. Then A has two different selfadjoint extensions A_1 and A_2 in H. Let E_1 and E_2 be the corresponding RI's. According to what has been proved, formula (1.19) is preserved both for E_1 and E_2. Consequently, due to (1.18), one can write

$$\int_{\mathbb{R}^1} \cos(\sqrt{\lambda}\, t)\, d(E_1(\lambda)\varphi_0, \varphi_0)_H = \int_{\mathbb{R}^1} \cos(\sqrt{\lambda}\, t)\, d(E_2(\lambda)\varphi_0, \varphi_0)_H$$

$$= (\varphi(t), \varphi_0)_H = O(e^{ct^2}), \quad t \to \infty$$

for every $t \in [0, \infty)$. It is well known (see below) that this yields the coincidence of measures, namely,

$$(E_1(\alpha)\varphi_0, \varphi_0)_H = (E_2(\alpha)\varphi_0, \varphi_0)_H \quad (\alpha \in \mathcal{B}(\mathbb{R}^1);\ \varphi_0 \in \Phi).$$

Since Φ is dense in H, the last equality implies that $E_1 = E_2$, which is nonsense.

■

Above, we have used the following result which strengthens the theorem on the uniqueness of reconstruction of a measure from a given cosine Fourier–Stieltjes transform of it: Namely, if $\mathbb{R}^1 \ni \lambda \mapsto \omega(\lambda) \in \mathbb{C}^1$ is a function of bounded variation such that

$$\int_{\mathbb{R}^1} \cos(\sqrt{\lambda}\, t)\, d\omega(\lambda) = 0 \ (t \in \mathbb{R}^1), \quad \int_{\mathbb{R}^1} \cos(\sqrt{\lambda}\, t)\, |d\omega(\lambda)| \underset{t\to\infty}{=} O(e^{ct^2}) \qquad (1.21)$$

for some $c \geq 0$, then $d\omega(\lambda) = 0$ (see, for example, Berezansky [5, Chapter 8. Theorem 3.18 and its corollary]); one should set, in the formulas presented above,

$$\omega(\lambda) = (E_1(\lambda)\varphi_0, \varphi_0)_H - (E_2(\lambda)\varphi_0, \varphi_0)_H \,.$$

This result can be strengthened (Vul [1] and Chaus [1, 2, Theorem 7]) as follows: The constant c in (1.21) may by replaced by a slowly increasing function $c(t)$, i.e., by a twice continuously differentiable function $(0, \infty) \ni t \mapsto c(t) > 0$ such that

(i) $c'(t) \geq 0 \quad (t \in (0, \infty))$ and $\lim_{t\to\infty} c(t) = \infty$;

(ii) for every $\varepsilon > 0$, one can find t_ε, $d_\varepsilon > 0$ such that $c(t) \leq d_\varepsilon t^\varepsilon$ for $t \geq t_\varepsilon$;

(iii) there exists finite or infinite limit $\lim_{t\to\infty} tc''(t) \times (c'(t))^{-1}$;

(iv) the function $tc'(t)\,(c(t))^{-1}$ is nonincreasing for sufficiently large t;

(v) the following condition

$$\int_1^{\infty} t^{-1}(c(t))^{-1}dt = \infty$$

holds.

One can easily see from the proof of Theorem 1.4 that the theorem remains valid, if condition (1.18) is replaced by

$$(\varphi(t), \varphi_0)_H = O(e^{c(t)t^2}), \quad t \to \infty, \tag{1.22}$$

where $c(t)$ is a strongly increasing function.

This theorem, in fact, generalized Theorem 1.3 for operators A which are not semi-bounded. The next statement generalizes this theorem to the case when it is not exactly known whether the Cauchy problem (1.5) has a strong solution.

Theorem 1.5. *Assume that* A *is an Hermitian operator acting in* H *and semi-bounded from below. Suppose that there exist a sequence of operators* $(A_n)_{n=1}^{\infty}$ *acting in* H *with domains of definitions* $\mathfrak{D}(A_n)$, *a set* Φ *dense in* H, *and a number* $b > 0$ *having the following properties:*

(i) for arbitrary $T \in (0, b)$ *and* $\varphi_0, \varphi_1 \in \Phi$, *there exist sequences of vectors from* H $(\varphi_{0,n})_{n=1}^{\infty}$ *and* $(\varphi_{1,n})_{n=1}^{\infty}$ *such that* $\varphi_{0,n} \mapsto \varphi_0$ *and* $\varphi_{1,n} \mapsto \varphi_1$ *in* H *as* $n \to \infty$, *and the strong solutions of the Cauchy problems*

$$\left(\frac{d^2\varphi_n}{dt^2}\right)(t) + A_n\varphi_n(t) = 0 \quad (t \in [0, T]),$$

$$\varphi_n(T) = \varphi_{0,n}, \quad \varphi_n'(T) = \varphi_{1,n} \quad (n \in \mathbb{N}) \tag{1.23}$$

and, moreover, $\varphi_n(t) \in \mathfrak{D}(A)$ $(t \in [0, T])$;

(ii) for arbitrary $T \in (0, b)$ *and* $\varphi_0, \varphi_1 \in \Phi$, *we have*

$$\int_0^T (u(t), (A_n - A)\varphi_n(t))_H\, dt \to 0, \quad n \to \infty, \tag{1.24}$$

where $u(t)$ *is a strong solution of Equation* (1.13) *such that* $u(0) = u'(0) = 0$. *Then* A *is essentially selfadjoint.*

Proof. According to Theorem 1.2, it suffices to show that $u(t) = 0$ for $t \in (0, b)$. Assume that $T \in (0, b)$ and $\varphi_0, \varphi_1 \in \Phi$. Let us construct the corresponding approximating sequences $(\varphi_{0,n})_{n=1}^{\infty}$ and $(\varphi_{1,n})_{n=1}^{\infty}$ and the solutions $\varphi_n(t)$. It follows from (1.9) that

$$\int_0^T ((u''(t), \varphi_n(t))_H - (u(t), \varphi_n''(t))_H)\, dt$$

$$= (u'(T), \varphi_{0,\, n})_H - (u(T), \varphi_{1,\, n})_H \quad (n \in \mathbb{N}). \tag{1.25}$$

Taking into account (1.3) and the fact that $\varphi_n(t) \in \mathfrak{D}(A)$ and acting as in the proof of Lemma 1.2, we obtain

$$(u''(t), \varphi_n(t))_H = -(u(t), A\, \varphi_n(t))_H \quad (t \in [0, T]).$$

By virtue of (1.23), we also have

$$(u(t), \varphi_n''(t))_H = -(u(t), A_n\, \varphi_n(t))_H \quad (t \in [0, T]).$$

These two relations imply that the left-hand side of the equality (1.25) is equal to the left-hand side of (1.24) and, hence,

$$(u'(T), \varphi_0)_H - (u(T), \varphi_1)_H = \lim_{n\to\infty} [(u'(T), \varphi_{0,\, n})_H - (u(T), \varphi_{1,\, n})_H] = 0.$$

Due to the arbitrariness of $\varphi_0, \varphi_1 \in \Phi$ and the denseness of Φ in H, this yields $u(T) = 0$.

■

Roughly speaking, the next theorem shows that if we prove the essential selfadjointness of A by using Theorem 1.3, then we automatically prove the essential selfadjointness of an arbitrary power of this operator.

Theorem 1.6. *Assume that $\mathfrak{D} \subset H$ is a dense set in a Hilbert space H and that an Hermitian operator A semi-bounded from below is defined on this set so that $A\,\mathfrak{D} \subset \mathfrak{D}$. Suppose that for some $b > 0$ and every $T \in (0, b)$, $\varphi_0, \varphi_1 \in \mathfrak{D}$, there exists a vector function $[0, T] \ni t \mapsto \varphi(t) \in \mathfrak{D}$ such that*

(i) for every $n \in \mathbb{Z}_+$, the vector function $[0, T] \ni t \mapsto A^n \varphi(t) \in H$ is twice strongly continuously differentiable;

(ii) the following equation with initial conditions is satisfied:

$$\left(\frac{d^2\varphi}{dt^2}\right)(t) + A\varphi(t) = 0 \quad (t \in [0, T]),\ \varphi(T) = \varphi_0,\quad \varphi'(T) = \varphi_1. \tag{1.26}$$

Under these assumptions, each operator $\mathcal{D} \ni f \mapsto A^n f$ $(n \in \mathbb{N})$ *in* H *is essentially selfadjoint.*

Proof. Let $A \geq c\,1$ $(c \in \mathbb{R}^1)$. We set $B = A + a\,1$, where $a = -c + 1$. Clearly, $B \geq 1$. We fix $m \in \mathbb{Z}_+$ and introduce a scalar product on $\mathcal{D}$ by

$$(f, g)_{\mathcal{H}_m} = (B^m f, B^m g)_H \quad (f, g \in \mathcal{D}). \tag{1.27}$$

Since $\|Bf\|_H \geq \|f\|_H$, we have $\|B^m f\|_H \geq \|f\|_H$ $(f \in \mathcal{D})$. Therefore, (1.27) indeed determines the scalar product in $\mathcal{D}$. Let $\mathcal{H}_m$ be the corresponding completion of $\mathcal{D}$ $(\mathcal{H}_0 = H)$. Given A, we construct an operator $\mathcal{A}$ in $\mathcal{H}_m$ taking $\mathcal{D}$ to be its domain of definition and setting $\mathcal{A}f = Af$ $(f \in \mathcal{D}(\mathcal{A}) = \mathcal{D})$. This operator is Hermitian in $\mathcal{H}_m$ and semi-bounded:

$$(\mathcal{A}f, g)_{\mathcal{H}_m} = (B^m A f, B^m g)_H = (B^m f, B^m A g)_H = (f, \mathcal{A}g)_{\mathcal{H}_m} \quad (f, g \in \mathcal{D}),$$

$$(\mathcal{A}f, f)_{\mathcal{H}_m} = (B^m A f, B^m f)_H = (A B^m f, B^m f)_H \geq c\,\|B^m f\|_H^2 = c\,\|f\|_{\mathcal{H}_m}^2 \quad (f, g \in \mathcal{D}).$$

It is also essentially selfadjoint. Indeed, according to Theorem 1.3, in order that $\mathcal{A}$ be essentially selfadjoint, it suffices that there exist a set Φ dense in $\mathcal{H}_m$ such that for any $T \in (0, b)$ and $\varphi_0, \varphi_1 \in \Phi$ the Cauchy problem

$$\left(\frac{d^2\varphi}{dt^2}\right)(t) + \mathcal{A}\varphi(t) = 0 \quad (t \in [0, T]),\ \varphi(T) = \varphi_0, \quad \varphi'(T) = \varphi_1 \tag{1.28}$$

has the strong solution $[0, T] \ni t \mapsto \varphi(t) \in \mathcal{H}_m$. It is easy to understand that the function $\varphi(t)$ in the condition of the theorem is just the solution required and, moreover, $\Phi = \mathcal{D}$. In fact, one can regard this function as a vector function $\varphi(t)$: $[0, T] \mapsto \mathcal{H}_m$; furthermore, it is twice strongly continuously differentiable (the last assertion follows from the fact that

$$B^m \varphi(t) = A^m \varphi(t) + m a A^{m-1} \varphi(t) + \ldots + a^m \varphi(t):\ [0, T] \to H$$

is twice strongly continuously differentiable). For every $t \in [0, T]$, we have $\varphi(t) \in \mathcal{D} = \mathcal{D}(\mathcal{A})$, and (1.28) is satisfied by virtue of (1.26).

Hence, $\mathcal{A}$ is essentially selfadjoint and, therefore, for every $z \notin [c, \infty)$ the range of values of $\mathcal{A} - z\,1$, i.e., $(\mathcal{A} - z\,1)\,\mathcal{D}$ is dense in $\mathcal{H}_m$ $(m \in \mathbb{Z}_+)$.

We fix $n \in \mathbb{N}$ and a nonreal number z such that any of the n roots $z_1, \ldots,$ and z_n of this number does not belong to $[c, \infty)$. To prove the theorem, it suffices to establish that $(\mathcal{A} - z\,1)\,\mathcal{D}$ is dense in H. Clearly, for $f \in \mathcal{D}$, the following decomposition holds:

$$(A^n - z\,1)f = \left(\prod_{j=1}^{n} (A - z_j\,1)\right) f.$$

Consequently, by using the triangle inequality, we can write the estimate

$$\|f_0 - (A^n - z\,1)f_n\|_H \le \|f_0 - (A^n - z_1 1)f_1\|_H$$
$$+ \sum_{m=1}^{n-1} \left\| \left(\prod_{j=1}^{m} (A - z_j 1)\right) (f_m - (A - z_{m+1} 1)\, f_{m+1}) \right\|_H \qquad (1.29)$$

for any $f_0 \in H$ and $f_1, \ldots, f_n \in \mathcal{D}$.

For $\xi \in \mathbb{C}^1$ and $g \in \mathcal{D}$, we have

$$\|(A - \xi 1)g\|_H = \|(B - (\xi + a)\,1)g\|_H \le (1 + |\xi + a|)\ \|Bg\|_H.$$

Iterating this inequality and using (1.27) and the fact that A and B commute, we can extend (1.29) in the following way:

$$\|f_0 - (A^n - z\,1)f_n\|_H$$
$$\le \|f_0 - (A^n - z_1 1)f_1\|_H \sum_{m=1}^{n-1} \left(\prod_{j=1}^{m} (1 + |z_j + a|)\right) \|B^m (f_m - (A - z_{m+1} 1)f_{m+1}\|_H$$
$$\le c \sum_{m=0}^{n-1} \|f_m - (A - z_{m+1} 1)f_{m+1}\|_{\mathcal{H}_m}, \qquad (1.30)$$

where c is a constant dependent on a and z_j.

Assume that $f_0 \in H$ and $\varepsilon > 0$ are given. Due to the denseness of $(\mathcal{A} - z\,1)\,\mathcal{D}$ in $\mathcal{H}_0$, we can choose $f_1 \in \mathcal{D}$ such that

$$\|f_0 - (A - z_1\,1)f_1\|_{\mathcal{H}_0} < \varepsilon.$$

Then, due to the denseness of $(A - z_2\,1)\,\mathcal{D}$ in $\mathcal{H}_1$, we choose $f_2 \in \mathcal{D}$ such that

$$\|f_1 - (A - z_2\,1)f_2\|_{\mathcal{H}_1} < \varepsilon,$$

etc., $f_n \in \mathcal{D}$ is chosen so that

$$\| f_{n-1} - (A - z_n 1) f_n \|_{\mathcal{H}_{n-1}} < \varepsilon.$$

As a result, we obtain the following estimate from (1.30):

$$\| f_0 - (A^n - z 1) f_n \|_H < cn\varepsilon,$$

ani this proves that $(\mathcal{A} - z\,1)\mathcal{D}$ is dense in H.

∎

In the applications of this theorem to differential operators, the region $\mathcal{D}$ consists of infinitely differentiable functions; therefore, the condition $A\mathcal{D} \subset \mathcal{D}$ involves the requirement of infinite differentiability of coefficients. However, we are interested in essential selfadjointness of a certain nth power of A, then this requirement may be avoided. To this end, one should somewhat weaken the assumptions of the theorem somewhat (the proof remains the same). Let A be an Hermitian operator in H with dense domain of definition $\mathcal{D}(A)$. We set

$$\mathcal{D}(A^m) = \{ f \in \mathcal{D}(A) \,|\, Af \in \mathcal{D}(A), \ldots, A^{m-1} f \in \mathcal{D}(A) \}.$$

The operator A^m is defined on $\mathcal{D}(A^m)$ $(m \in \mathbb{N})$ in a natural way. Clearly, $\mathcal{D}(A) \supset \mathcal{D}(A^2) \supset \ldots$. Assume that $n \in \mathbb{N}$ is fixed and A is an Hermitian operator in H semi-bounded from below whose domain of definition is $\mathcal{D}(A)$, and $\mathcal{D}(A^n)$ is dense in H. Assume that for some $b > 0$ and for every $T \in (0, b)$ and $\varphi_0, \varphi_1 \in \mathcal{D}(A^{n-1})$, there exists a vector function $[0, T] \ni t \mapsto \varphi(t) \in \mathcal{D}(A^n)$ such that

(i) for every $m = 0, \ldots, n$, the vector function $[0, T] \ni t \mapsto A^m \varphi(t) \in H$ is twice strongly continuously differentiable;

(ii) condition (ii) in Theorem 1.6 holds.

Then the assertion of this theorem preserves for A^n.

We note that a theorem similar to Theorem 1.6 also holds for evolutionary equations of the first order, i.e., under conditions of Theorem 1.1.

1.3. Quasianalytical Criteria of Selfadjointness

Let us first recall some facts from the theory of quasianalytical functions (see, e.g., Mandelbrojt [1, Chapter 4]).

Let $[a, b] \subset \mathbb{R}^1$ be a finite segment and let $(m_n)_{n=1}^{\infty}$ be a fixed sequence of positive numbers. A linear collection of all functions $f \in C^\infty([a, b])$ is called a class $C\{m_n\}$ if the following estimates are valid for each function from this collection:

$$|(D^n f)(t)| \le K_f^n m_n \quad (t \in [a, b];\ n \in \mathbb{N}), \tag{1.31}$$

where K_f is a constant depending on f.

As is well known, the class of functions analytical on $[a, b]$ can be characterized by the estimates (1.31) in which we set $m_n = n!$. The following fact obviously holds for the class $C\{n!\}$. If $f \in C\{n!\}$ is such that $(D^n f)(t_0) = 0$ at the fixed point $t_0 \in [a, b]$ for all $n \in \mathbb{N}$ and $f(t_0) = 0$, then $f(t) = 0$ for $t \in [a, b]$. To generalize this situation, we introduce the following definition. A class $C\{m_n\}$ is called quasianalytical if the fact that the equalities

$$(D^n f)(t_0) = 0 \quad (n \in \mathbb{N}) \quad \text{and} \quad f(t_0) = 0$$

hold at a certain fixed point $t_0 \in [a, b]$ for $f \in C\{m_n\}$ implies that $f(t) = 0$ $(t \in [a, b])$. The following Denjoy–Càrleman theorem is valid:

A class $C\{m_n\}$ is quasianalytical if and only if

$$\sum_{n=1}^{\infty} \left(\inf_{k=n,\, n+1,\, \dots} m_k^{1/k} \right)^{-1} = \infty. \tag{1.32}$$

Let H be a Hilbert space and let A be an Hermitian operator acting in it. The vector $\varphi \in H$ is called quasianalytical (with respect to A) if $\varphi \in \bigcap_{n=1}^{\infty} \mathfrak{D}(A^n)$ and the class $C\{\|A^n \varphi\|_H\}$ is quasianalytical.

Lemma 1.3. *A vector $\varphi \in \bigcap_{n=1}^{\infty} \mathfrak{D}(A^n)$ is quasianalytical if and only if*

$$\sum_{n=1}^{\infty} \|A^n \varphi\|_H^{-\frac{1}{n}} = \infty. \tag{1.33}$$

Proof. Clearly,

$$C\{\|A^n \varphi\|_H\} = C\{\|A^n(\lambda\varphi)\|_H\},$$

where $\lambda > 0$ is fixed. This implies that it suffices to verify the assertion of the lemma for a vector φ whose norm $\|\varphi\|_H$ is equal to 1. For this vector, the sequence

$$(\|A^n \varphi\|_H^{1/n})_{n=1}^{\infty} \tag{1.34}$$

is nondecreasing. Indeed,

$$\|A\varphi\|_H^2 = (A\varphi, A\varphi)_H = (A^2\varphi, \varphi)_H \le \|A^2\varphi\|_H \|\varphi\|_H$$

i.e.,

$$\|A\varphi\|_H \le \|A^2\varphi\|_H^{\frac{1}{2}}.$$

Assume that the inequality

$$\|A^n\varphi\|_H^{\frac{1}{n}} \le \|A^{n+1}\varphi\|_H^{\frac{1}{n+1}}$$

is already established. We now prove that

$$\|A^{n+1}\varphi\|_H^{\frac{1}{n+1}} \le \|A^{n+2}\varphi\|_H^{\frac{1}{n+2}} \quad (n \in \mathbb{N}).$$

Employing the inequality whose validity has been assumed above, we obtain

$$\|A^{n+1}\varphi\|_H^2 = (A^{n+1}\varphi, A^{n+1}\varphi)_H = (A^{n+2}\varphi, A^n\varphi)_H$$

$$\le \|A^{n+2}\varphi\|_H \|A^n\varphi\|_H \le \|A^{n+2}\varphi\|_H \|A^{n+1}\varphi\|_H^{\frac{n}{n+1}},$$

whence

$$\|A^{n+1}\varphi\|_H^{\frac{n+2}{n+1}} \le \|A^{n+2}\varphi\|_H,$$

as required. Hence, the sequence (1.34) in nondecreasing.

Let us apply the Denjoy–Càrleman criterion to the class $C\{\|A^n\varphi\|_H\}$ ($\|\varphi\|_H = 1$). Since the sequence (1.34) is nonnegative, we have

$$\inf_{k=n,n+1,\dots} \|A^k\varphi\|_H^{\frac{1}{k}} = \|A^n\varphi\|_H^{\frac{1}{n}}.$$

Therefore, the condition (1.32) of quasianalyticity of this class (i.e., of quasianalyticity of a vector φ) can be rewritten in the form (1.33).

■

Theorem 1.7. *Suppose that A is a closed Hermitian operator acting in H. It is selfadjoint if and only if the space H contains a total set consisting of quasianalytical vectors.*

Proof. In one direction, this statement is trivial. Indeed, let A be selfadjoint; then it suffices to establish the quasianalyticity of each vector φ of the form $\varphi = E((a, b))f$, where E is the RI corresponding to A, $a, b \in \mathbb{R}^1$ $(a < b)$, $f \in H$. Clearly, $\varphi \in \bigcap_{n=1}^{\infty} \mathfrak{D}(A^n)$. Further, we have

$$\| A^n \varphi \|_H^2 = \int_a^b \lambda^{2n} d(E(\lambda) f, f)_H \leq c^{2n} \| f \|_H^2$$

$$(c = \max(|a|, |b|); \ \ n \in \mathbb{N}),$$

therefore, the series (1.33) diverges, and according to Lemma 1.3, the vector φ is quasianalytical.

Now assume that A possesses a total set M of quasianalytical vectors φ. Since A is closed, it suffices to establish its essential selfadjointness or, according to Theorem 1.1, the uniqueness of strong solutions of the Cauchy problem for Equations (1.4) with $b = \infty$. Suppose that $u(t)$ is a strong solution of the problem

$$\left(\frac{du}{dt}\right)(t) - (\zeta A)^* u(t) = 0 \quad (t \in [0, \infty), \ u(0) = 0), \tag{1.35}$$

where $\zeta = \pm i$. It suffices to establish that $u(t) = 0$ for any $T > 0$ and $t \in [0, T]$.

For (1.35) with quasianalytical vector $f = \varphi \in \bigcap_{n=1}^{\infty} \mathfrak{D}(A^n)$, the "weak" equality (1.3) gives

$$\frac{d}{dt}(u(t), \varphi)_H = \left(\left(\frac{du}{dt}\right)(t), \ \varphi\right)_H = (u(t), (\zeta A)\varphi)_H \quad (t \in [0, T]).$$

But $(\zeta A)\varphi \in \bigcap_{n=1}^{\infty} \mathfrak{D}(A^n)$, therefore

$$\frac{d}{dt}(u(t), (\zeta A)\varphi)_H = (u(t), (\zeta A)^2 \varphi)_H \quad (t \in [0, T]),$$

etc. This yields $(u(t), \varphi)_H \in C^{\infty}([0, T])$ and

$$D^n (u(t), \varphi)_H = D^{n-1}(u(t), (\zeta A)\varphi)_H = \ldots = (u(t), (\zeta A)^n \varphi)_H$$

$$(t \in [0, T]; \ n \in \mathbb{Z}_+). \tag{1.36}$$

Since the values of $u(t)$ on $[0, T]$ are bounded, it follows from (1.36) that

$$| D^n (u(t), \varphi)_H | \leq c \| (\zeta A)^n \varphi \|_H = c \| A^n \varphi \|_H \quad (t \in [0, T]; \ n \in \mathbb{Z}_+),$$

i.e., the scalar function

$$[0, T] \ni t \mapsto f(t) = (u(t), \varphi)_H$$

belongs to the class $C\{\| A^n \varphi \|_H\}$. It follows from (1.36) and the equality $u(0) = 0$ that $(D^n f)(0) = 0$ $(n \in \mathbb{Z}_+)$. Consequently, by virtue of the quasianalyticity of $C\{\| A^n \varphi \|_H\}$, the equality

$$f(t) = (u(t), \varphi)_H = 0 \quad (t \in [0, T])$$

holds. Since the set M of vectors φ is total, we have $u(t) = 0$ $(t \in [0, T])$.

■

The following definition turns out to be useful. As before, let A be an Hermitian operator acting in the Hilbert space H. A vector $\varphi \in H$ is called analytic (with respect to A) if $\varphi \in \bigcap_{n=1}^{\infty} \mathfrak{D}(A^n)$ and the power series

$$\sum_{n=0}^{\infty} \frac{\|A^n \varphi\|_H}{n!} z^n \tag{1.37}$$

has a nonzero radius of convergence; it is called an entire vector if this radius is equal to infinity. Clearly, every analytical vector is, at the same time, quasianalytical; however, the converse statement is not true. Note that when proving the first part of Theorem 1.7, we have established the following stronger fact:

If A is selfadjoint, then it possesses a total set of entire vectors (each vector $\varphi = E((a, b))f$ considered in this proof satisfies the estimate

$$\| A^n \varphi \|_H \leq c^n \|f\|_H \quad (n \in \mathbb{N}),$$

thus, it is an entire vector).

Let us establish a theorem which specifies Theorem 1.7 for operators semi-bounded from below.

Let A be an Hermitian operator acting in H. A vector $\varphi \in H$ is called a Stieltjes vector (with respect to A) if $\varphi \in \bigcap_{n=1}^{\infty} \mathfrak{D}(A^n)$ and the class $C\{\| A^n \varphi \|_H^{1/2}\}$ is quasi-analytical or, in other words,

$$\sum_{n=1}^{\infty} \| A^n \varphi \|_H^{-\frac{1}{2n}} = \infty \tag{1.38}$$

(the equivalence of the Stieltjes property and the condition (1.38) follows from the fact that the sequence $(\| A^n \varphi \|_H^{\frac{1}{2}})_{n=1}^{\infty}$ is nondecreasing for $\| \varphi \|_H = 1$ together with (1.34); then one should employ the Denjoy–Càrleman criterion).

Clearly, every quasianalytical vector is, at the same time, a Stieltjes vector, but the converse assertion is not true. Hence, if we denote the sets of all entire, analytical, quasianalytical, and Stieltjes vectors with respect to the operator A by $\mathfrak{E}(A)$, $\mathfrak{A}(A)$, $\mathfrak{Q}(A)$, and $\mathfrak{S}(A)$, respectively, then we obtain the inclusions

$$\mathfrak{E}(A) \subset \mathfrak{A}(A) \subset \mathfrak{Q}(A) \subset \mathfrak{S}(A). \tag{1.39}$$

Theorem 1.8. *Assume that A is a closed Hermitian operator semi-bounded from below. If H contains a total set consisting of Stieltjes vectors, then A is self-adjoint.*

The converse statement is obvious by virtue of the results proved before and (1.39).

Proof. According to Theorem 1.2, it suffices to prove the uniqueness of strong solutions of the Cauchy problem for Equation (1.13) with $b = \infty$. Let $u(t)$ be a strong solution of this problem such that $u(0) = u'(0) = 0$. We show that $u(t) = 0$ $(t \in [0, T])$ for any $T > 0$. Let M be a total set of Stieltjes vectors φ appearing in the condition of the theorem. We set

$$[0, T] \ni t \mapsto f(t) = (u(t)), \varphi)_H,$$

where $\varphi \in M$. The relation (1.3), written for (1.13), implies that $f \in C^2([0, T])$ and

$$\left(\frac{d^2 f}{dt^2}\right)(t) = -(u(t), A\varphi)_H \quad (t \in [0, T]).$$

Since $A\varphi \in \bigcap_{n=1}^{\infty} \mathfrak{D}(A^n)$, by analogy we conclude that

$$(u(t), A\varphi)_H \in C^2([0, T])$$

and

$$\frac{d^2}{dt^2}(u(t), A\varphi)_H = -(u(t), A^2\varphi)_H \quad (t \in [0, T]),$$

etc. As a result, we find that $f \in C^{\infty}([0, T])$ and

$$\begin{aligned}(D^{2k} f)(t) &= D^{2k}(u(t), A\varphi)_H = -D^{2(k-1)}(u(t), A\varphi)_H \\ &= \ldots = (-1)^k(u(t), A^k\varphi)_H \quad (t \in [0, T];\ k \in \mathbb{Z}_+).\end{aligned} \tag{1.40}$$

A similar equality can be written also for odd derivatives. Indeed, differentiating (1.40), we obtain

$$(D^{2k+1}f)(t) = (-1)^k(u'(t), A^k\varphi)_H \quad (t \in [0,T];\ k \in \mathbb{Z}_+). \tag{1.41}$$

The values of $u(t)$ and $u'(t)$ on $[0,T]$ are bounded. Therefore, it follows from (1.40), (1.41), the Cauchy–Bunyakovsky inequality, and from the fact that A is Hermitian, that

$$|(D^{2k}f)(t)|,\ |(D^{2k+1}f)(t)| \le c\,\|A^k\varphi\|_H \le c\,\|\varphi\|_H^{\frac{1}{2}}\|A^{2k}\varphi\|_H^{\frac{1}{2}}$$

$$(t \in [0,T];\ k \in \mathbb{Z}_+),$$

i.e., $f \in C\{m_n\}$, where $(m_n)_{n=1}^{\infty}$ is a sequence of numbers

$$(\|\varphi\|_H^{\frac{1}{2}}, \|\varphi\|_H^{\frac{1}{2}}, \|A\varphi\|_H^{\frac{1}{2}}, \|A\varphi\|_H^{\frac{1}{2}}, \|A^2\varphi\|_H^{\frac{1}{2}}, \|A^2\varphi\|_H^{\frac{1}{2}}, \ldots).$$

The class $C\{m_n\}$ does not change if we norm φ. But then, as has been already mentioned, the sequence $(\|A^n\varphi\|_H^{\frac{1}{2}})_{n=1}^{\infty}$ is nondecreasing. This implies that the Denjoy–Càrleman condition (1.32) for the sequence under consideration can be written in the form (1.38). Therefore, the class $C\{m_n\}$ is quasianalytical.

On the other hand, the relations (1.40) and (1.41) and the condition $u(0) = u'(0) = 0$ yield $(D^n f)(0) = 0$ $(n \in \mathbb{Z}_+)$. Therefore,

$$(u(t), \varphi)_H = f(t) = 0 \quad (t \in [0,T]).$$

It follows from the totality of M that $u(t) = 0$ $(t \in [0,T])$.

■

Example 1.2. Let $s = (s_n)_{n=1}^{\infty}$ be a fixed sequence of real numbers such that for any finite sequence $\varphi = (\varphi_j)_{j=1}^{\infty}$ $(\varphi_j \in \mathbb{C}^1)$ (the set of these sequences is denote by C_0), we have

$$\sum_{j,k=0}^{\infty} s_{j+k}\overline{\psi}_j\,\varphi_k \ge 0. \tag{1.42}$$

According to (1.42), for given s, we introduce a quasiscalar product on C_0 by setting

$$\langle\varphi, \psi\rangle = \sum_{j,k=0}^{\infty} s_{j+k}\overline{\psi}_j\,\varphi_k \quad (\varphi, \psi \in C_0).$$

After factorization and completion, we obtain the Hilbert space H_s. It is easy to see that the mapping

$$C_0 \ni \varphi = (\varphi_0, \varphi_1, \dots) \mapsto (0, \varphi_0, \varphi_1, \dots) \in C_0$$

generates an Hermitian operator; we denote its closure by A. It is stated that if a class $C\{s_{2n}^{1/2}\}$ is quasianalytical, then A is selfadjoint. This result follows from Theorem 1.7. Indeed, a simple calculation implies that each vector from H_s, which is associated with $\varphi \in C_0$ under factorization, is quasianalytical for A.

This example is connected with the classical (one-dimensional) moment problem. The infinite-dimensional generalization of it is considered in Section 2; concerning this example, see the proof of Theorem 2.1 (Steps I and II).

1.4. Parabolic Criteria of Selfadjointness

In Subsection 1.1, we have considered the "hyperbolic" and "Schrödinger" criteria of selfadjointness. The "parabolic" criteria (when the evolutionary equation is parabolic) also prove to be quite useful.

We now present the corresponding results which are mainly similar to those in Subsection 1.1. For simplicity, we formulate these results in the case of the interval $[0, b) = [0, \infty)$.

Let H be a Hilbert space, and let A be an Hermitian operator acting in it with the domain of definition $\mathcal{D}(A)$. Consider the differential equation

$$\left(\frac{du}{dt}\right)(t) - A^* u(t) = 0 \qquad (t \in [0, \infty)). \tag{1.43}$$

Theorem 1.9. *Let A be an Hermitian operator acting in H. In order that this operator be essentially selfadjoint, it is necessary that Equation* (1.43) *have the unique strong solution of the Cauchy problem. If A is semi-bounded from below, this condition is also sufficient.*

Proof. Let us assume that the closure $\tilde{A}$ is selfadjoint and that E is its RI. We now apply Lemma 1.2 setting $r = 1$, $B = \tilde{A}$, and $\Phi = \bigcup_{n=1}^{\infty} E((-n, n))\, H$. The strong solution of the corresponding Cauchy problem, which now has the form

$$\varphi'(t) - \tilde{A}\,\varphi(t) = 0 \quad (t \in [0, T]),\ \varphi(T) = \varphi_0 \in \Phi,$$

exists and is equal to

$$\varphi(t) = \int_{\mathbb{R}^1} e^{\lambda(t-T)} dE(\lambda)\, \varphi_0 \ \ (t \in [0, T]) \tag{1.44}$$

(the integral in (1.44) is, in fact, taken over a finite interval and, therefore, the function $[0, T] \ni t \mapsto \varphi(t)$ is strongly continuously differentiable and is a solution of the required problem). The above-mentioned lemma yields the uniqueness of the strong solutions of the Cauchy problem for (1.43).

Now we prove sufficiency. Assume that $\tilde{A}$ is not selfadjoint. Then there exist two distinct selfadjoint extensions of A in H bounded from below by a number $c > -\infty$. Denote these extensions by A_1 and A_2. Let E_1 and E_2 be the corresponding RI. For any $g \in \mathcal{D}(A) \subset \mathcal{D}(A_1)$, the integral

$$\int_c^\infty \lambda^2 \, d(E_1(\lambda) g, g)_H$$

converges. Hence, the vector function

$$[0, \infty) \ni t \mapsto u_1(t) = \int_c^\infty e^{-\lambda t} dE_1(\lambda) g \tag{1.45}$$

is once strongly continuously differentiable, and $u_1(t) \in \mathcal{D}(A_1) \subset \mathcal{D}(A^*)$. The derivative $u_1'(t)$ is expressed as the integral (1.45) with the factor $-\lambda$ before $e^{-\lambda t}$. The expression $A^* u_1(t) = A_1 u_1(t)$ also has the same form. Therefore, (1.45) is a strong solution of Equation (1.43), and $u_1(0) = g$. Similarly, by substituting E_2 for E_1 in (1.45), we construct the function $u_2(t)$. The difference

$$u(t) = u_1(t) - u_2(t)$$

is also a strong solution of Equation (1.43) for which $u(0) = 0$. By virtue of the assumed uniqueness, we have $u(t) = 0$ $(t \in [0, \infty))$. Multiplying this equality scalarly by $h \in H$, we obtain

$$\int_c^\infty e^{-\lambda t} d((E_1(\lambda) - E_2(\lambda)) g, h)_H = 0 \quad (t \in [0, \infty)).$$

Since the measure is uniquely determined by its Laplace–Stieltjes transform, this implies that

$$((E_1(\lambda) - E_2(\lambda)) g, h)_H = 0 \quad (\lambda \in \mathbb{R}^1; \; g \in \mathcal{D}(A), \; h \in H).$$

Therefore, $E_1 = E_2$, which is absurd.

■

We stress that if A is a selfadjoint operator semi-bounded from below, then the strong solution $u(t)$ of Equation (1.43) satisfying the initial condition $u(0) = u_0 \in \mathcal{D}(A)$, exists and can be expressed with the help of the semi-group generated by A:

$$u(t) = e^{-tA}u_0 = \int_c^\infty e^{-\lambda t}\, dE(\lambda)u_0 \qquad (t \in [0, \infty))$$

(here E is the RI corresponding to A, and $c \in \mathbb{R}^1$ is such that $A \geq c\,1$).

The uniqueness of strong solutions of the Cauchy problem for (1.43) and, thereby, the essential selfadjointness of A, can be often established with the help of the following assertion analogous to Theorem 1.5.

Theorem 1.10. *Let A be an Hermitian operator acting in H and semi-bounded from below. Assume that there exist a sequence $(A_n)_{n=1}^\infty$ of operators acting in H with the domains of definition $\mathcal{D}(A_n)$, a set Φ dense in H, and a number $b \in (0, \infty]$ possessing the following properties:*

(i) for arbitrary $T \in (0, b)$ and $\varphi_0 \in \Phi$, there exist a sequence $(\varphi_{0,n})_{n=1}^\infty$ of vectors from H such that $\varphi_{0,n} \to \varphi_0$ as $n \to \infty$ in H and strong solutions of the Cauchy problems

$$\left(\frac{d\varphi_n}{dt}\right)(t) - A_n \varphi_n(t) = 0 \qquad (t \in [0, T]); \tag{1.46}$$

$$\varphi_n(T) = \varphi_{0,n}, \qquad \varphi_n(t) \in \mathcal{D}(A);$$

(ii) for arbitrary $T \in (0, b)$ and $\varphi_0 \in \Phi$,

$$\int_0^T (u(t), (A_n - A)\varphi_n(t))_H\, dt \to 0 \qquad \text{as } n \to \infty \tag{1.47}$$

where $u(t)$ is and arbitrary strong solution of Equation (1.43) *such that* $u(0) = 0$.

Then A is essentially selfadjoint.

Proof. First we show that $u(t) = 0$ $(t \in (0, b))$. Let $T \in (0, b)$ and $\varphi_0 \in \Phi$. We construct an approximating sequence $(\varphi_{0,n})_{n=1}^\infty$ and the corresponding solution $\varphi_n(t)$. Integrating by parts, we get

$$\int_0^T ((u'(t), \varphi_n(t))_H + (u(t), \varphi_n'(t))_H)\, dt = (u(T), \varphi_{0,n})_H. \tag{1.48}$$

Taking into account (1.43) and the inclusion $\varphi_n(t) \in \mathfrak{D}(A)$, we find

$$(u'(t), \varphi_n(t))_H = -(A^* u(t), \varphi_n(t))_H = -(u(t), A\varphi_n(t))_H .$$

This and (1.46) imply that the left-hand side of (1.48) is equal to the integral (1.47). Passing in (1.48) to the limit as $n \to \infty$, we conclude that

$$(u(T), \varphi_0)_H = 0 \quad (\varphi_0 \in \Phi).$$

Hence, $u(T) = 0$.

Denote $c = b/2$. The function

$$[0, \infty) \ni t \mapsto u_1(t) = u(t+c)$$

is a strong solution of Equation (1.43) such that $u_1(0) = 0$. Repeating the above-mentioned arguments, we obtain $u_1(t) = 0$ $(t \in (0, b))$. Further, we consider the function

$$[0, \infty) \ni t \mapsto u_2(t) = u_1(t+c) = u(t+2c),$$

and repeat this argument again. Finally, we conclude that $u(t) = 0$ $(t \in [0, \infty))$. But then, according to Theorem 1.9, A is essentially selfadjoint.

■

1.5. Selfadjointness of Perturbations of an Operator by a Potential

Let us consider the constructions of Subsection 1.4 in a more specific case when $H = L_2(R, \mathfrak{R}, \mu)$, where R is a certain space of points x with a σ-finite measure given on the σ-algebra $\mathfrak{R}$ of subsets of R. Below in this subsection, we denote the space $L_p(R, \mathfrak{R}, \mu)$ $(p \in [1, \infty])$ by L_p.

Theorem 1.11. *Assume that A_0 is a nonnegative selfadjoint operator acting in the space L_2 for which $D = \mathfrak{D}(A_0) \cap L_p$ is a base for some fixed $p \in (2, \infty]$, and the semigroup e^{-tA_0} is contracting in L_p, i.e.,*

$$\| e^{-tA_0}\varphi \|_{L_p} \le \| \varphi \|_{L_p} \quad (\varphi \in L_2 \cap L_p;\ t \in [0, \infty)).$$

Suppose that the real-valued function $V \in L_r$, where $r = 2p(p-2)^{-1}$, is semi-bounded from below, i.e., $\inf_{x \in R} V(x) > -\infty$. Then the operator

$$D \ni f \mapsto (Af)(x) = (A_0 f)(x) + V(x)f(x) \tag{1.49}$$

acting in L_2 is essentially selfadjoint.

Proof. We split the proof into four steps.

I. Assume that $p = \infty$ and the potential V is nonnegative. Here $r = 2$ and, therefore, the operator (1.49) with $V \in L_2$ is densely defined on $D = \mathfrak{D}(A_0) \cap L_\infty$ and, obviously, is nonnegative. Let us verify the conditions of Theorem 1.10 for this operator. As A_n, we take the closures of operators determined by the relation (1.49) in which $V(x)$ is replaced by $V_n(x) = \min(V(x), n)$. Here $b = \infty$, $\Phi = D$, and a sequence $(\varphi_{0,n})_{n=1}^{\infty}$ which approximates the vector $\varphi_0 \in D$ coincides with it, i.e., $\varphi_{0,n} = \varphi_0$. Clearly, each operator A_n is selfadjoint and nonnegative and, therefore, the solution of the Cauchy problem (1.46) exists and has the form

$$\varphi_n(t) = e^{(t-T)A_n}\varphi_0 \quad (t \in [0, \infty);\ n \in \mathbb{N}). \tag{1.50}$$

Since the operator of multiplication by a potential V_n is bounded, we have

$$\mathfrak{D}(A_n) = \mathfrak{D}(A_0) \quad (n \in \mathbb{N}). \tag{1.51}$$

It follows from (1.50) and (1.51) that the following inclusion holds

$$\varphi_n(t) \in \mathfrak{D}(A_0) \quad (t \in [0, \infty);\ n \in \mathbb{N}). \tag{1.52}$$

We now employ the estimate

$$\| \varphi_n(t) \|_{L_\infty} \le \| \varphi_0 \|_{L_\infty} \quad (t \in [0, \infty);\ n \in \mathbb{N}), \tag{1.53}$$

which will be established at Step II. This estimate and (1.52) yield the inclusion

$$\varphi_n(t) \in \mathfrak{D}(A_0) \cap L_\infty = D \quad (t \in [0, \infty);\ n \in \mathbb{N}),$$

necessary to make it possible to apply Theorem 1.10.

It remains to verify condition (ii) of Theorem 1.10. Due to the continuity of the vector function $[0, T] \ni t \mapsto u(t) \in L_2$, we have

$$\| u(t) \|_{L_2} \le c \quad (t \in [0, T]).$$

This and (1.53) imply that

$$\left| \int_0^T (u(t), (A_n - A)\varphi_n(t))_H \, dt \right| \le \int_0^T \| (u(t) \|_{L_2} \| V_n - V \|_{L_2} \| \varphi_n(t) \|_{L_\infty} dt$$

$$\leq Tc\|\varphi_0\|_{L_\infty}\|V_n - V\|_{L_2} \to 0, \quad n \to \infty. \tag{1.54}$$

Thus, all conditions of Theorem 1.10 are verified, and this means that we have established (provided that (1.53) holds) the essential selfadjointness of the operator A in the considered case.

II. Let us prove the estimate (1.53). By virtue of (1.50), it is a consequence of a more general inequality

$$\|e^{-sA_n}f\|_{L_\infty} \leq \|f\|_{L_\infty} \quad (f \in L_2 \cap L_\infty;\ s \in [0,\infty);\ n \in \mathbb{N}), \tag{1.55}$$

which is established below. We fix n. Taking into account that $A_n = A_0 + C_n$, where the operators A_0, C_n, and A_n acting in L_2 are nonnegative and selfadjoint, and C_n is bounded, we can apply the multiplicative formula (see Reed and Simon [2, Theorem VIII.31]). We have

$$e^{-sA_n}f = \lim_{m\to\infty}(e^{-\frac{s}{m}A_0}e^{-\frac{s}{m}C_n})^m f \quad (f \in L_2 \cap L_\infty;\ s \in [0,\infty)) \tag{1.56}$$

in the sense of convergence in the norm of the space L_2.

The operator $e^{-\frac{s}{m}C_n}$ regarded as an operator of multiplication by the function $e^{-\frac{s}{m}V_n(\cdot)}$ has the norms (in the spaces L_2 and L_∞) which do not exceed 1. By virtue of the conditions of the theorem, the same is true for the operator $e^{-\frac{s}{m}A_0}$. Therefore,

$$e^{-\frac{s}{m}A_0}e^{-\frac{s}{m}C_n}f \in L_2 \cap L_\infty$$

and

$$\|e^{-\frac{s}{m}A_0}e^{-\frac{s}{m}C_n}f\|_{L_\infty} \leq \|f\|_{L_\infty}.$$

Iterating this estimate, we obtain the inequality

$$\|g_m\|_{L_\infty} \leq \|f\|_{L_\infty} \quad (m \in \mathbb{N}), \tag{1.57}$$

where g_m denotes the vector under the sign lim in (1.56). In L_2, we have $g_m \to e^{-sA_n}f$ as $m \to \infty$; therefore, we can choose a sequence such that $g_{m_k}(x) \to (e^{-sA_n}f)(x)$ as $k \to \infty$ for μ-almost all $x \in R$. This and the estimate (1.57) yield (1.55).

III. Assume that $p \in (2,\infty)$ and $V(x) \geq 0$ $(x \in R)$. The operator (1.49) is densely defined on $D = \mathfrak{D}(A_0) \cap L_p$, because

$$\| Vf\|_{L_2} \le \| V\|_{L_r} \|f\|_{L_p} \quad (f \in D)$$

by the Hölder inequality. Clearly, it is nonnegative. Then we repeat the argument of Step I employing the relation

$$\| \varphi_n(t)\|_{L_p} \le \| \varphi_0\|_{L_p} \qquad (t \in [0, \infty);\ n \in \mathbb{N}) \tag{1.58}$$

(it will be proved below) instead of (1.53). When verifying condition (ii) of Theorem (1.10), instead of (1.54), we employ the estimate which is obtained with the help of the Hölder inequality and (1.58), namely

$$\begin{aligned} &\left| \int_0^T (u(t), (A_n - A)\varphi_n(t))_{L_2} \, dt \right| \\ &\qquad \le Tc \Big(\int_R |(V_n(x) - V(x))\, \varphi_n(x,t)|^2 \, d\mu(x) \Big)^{1/2} \\ &\qquad \le Tc \, \| V - V_n\|_{L_r} \| \varphi_n(t)\|_{L_p} \\ &\qquad \le Tc \, \| \varphi_0\|_{L_p} \| V - V_n\|_{L_r} \to 0, \quad n \to \infty. \end{aligned} \tag{1.59}$$

We prove (1.58). As above, this inequality follows from the more general estimate

$$\| e^{-sA_n} f\|_{L_p} \le \|f\|_{L_p} \quad (f \in L_2 \cap L_p;\ s \in [0, \infty);\ n \in \mathbb{N}). \tag{1.60}$$

The inequality (1.60) is proved, as in Step II; one should only replace L_∞ by L_p. Instead of (1.57), we obtain

$$\| g_m\|_{L_p} \le \|f\|_{L_p} \quad (m \in \mathbb{N}).$$

In analogy with Step II, we have $g_{m_k}(x) \to (e^{-sA_n} f)(x)$ as $k \to \infty$ for μ-almost all $x \in R$. Consequently,

$$| g_{m_k}(x)|^p \underset{k \to \infty}{\to} |(e^{-sA_n} f)(x)|^p$$

for the same x. Applying the Fatou Lemma to the last sequence, we obtain

$$\| e^{-sA_n} f\|_{L_p}^p \le \varliminf_{k \to \infty} \| g_{m_k}\|_{L_p}^p \le \|f\|_{L_p}^p .$$

The estimate (1.60) is thus established.

IV. Let us show how the case of potential V semi-bounded from below can be investigated. We set

$$V_1(x) = \min(V(x), 0) \quad \text{and} \quad V_2(x) = \max(V(x), 0).$$

Then

$$V(x) = V_1(x) + V_2(x) \quad (x \in R),$$

where the function $V_1(x)$ is bounded, and the function $V_2(x)$ is nonnegative. The fact that V belongs to a certain L_q $(q \in [1, \infty])$ involves the inclusion $V_1, V_2 \in L_q$. For example, let $p = \infty$. Then $V_2 \in L_2$ and the operator

$$D \ni f \mapsto A_2 f = A_0 f + V_2 f$$

is essentially selfadjoint. But then the operator (1.49) is also essentially selfadjoint because it differs from A_2 by a bounded operator.

■

Remark 1. When verifying the contractibility of a semi-group, it is useful to keep in mind the following fact:

Let A_0 be a nonnegative selfadjoint operator acting in L_2. If the semi-group e^{-tA_0} f is contracting in a certain space L_p $(p \in [1, \infty])$, i.e.,

$$\| e^{-tA_0} f \|_{L_p} \le \| f \|_{L_p} \quad (f \in L_2 \cap L_p;\ t \in [0, \infty)),$$

then it is also contracting in each L_q, where $q \in [p', p]$, $p^{-1} + (p')^{-1} = 1$ (if $p \ge 2$) or $q \in [p, p']$ (if $p < 2$).

Indeed, let us first prove the contractibility in $L_{p'}$. For definiteness, let $p > 2$ and $g \in L_2 \cap L_{p'}$. Then

$$|(e^{-tA_0} g, f)_{L_2}| = |(g, e^{-tA_0} f)_{L_2}| \le \| g \|_{L_{p'}} \| e^{-tA_0} f \|_{L_p} \le \| g \|_{L_{p'}} \| f \|_{L_p}.$$

This and the arbitrariness of $f \in L_2 \cap L_p$ imply that

$$\| e^{-tA_0} g \|_{L_{p'}} \le \| g \|_{L_{p'}}.$$

Therefore,

$$\| e^{-tA_0} f \|_{L_0} \le \| f \|_{L_q} \quad (f \in L_2 \cap L_q;\ t \in [0, \infty))$$

for $q = p$ and $q = p' < p$. By the Riesz–Thorin theorem (see Reed and Simon [2, Theorem IX.17]), the same estimate holds also for every $q \in [p', p]$, i.e., the semigroup e^{-tA_0} is contracting in each L_q where $q \in [p', p]$.

■

Consider the situation when μ is a probability measure, i.e., $\mu(R) = 1$. Here $L_p \supset L_q$ and $\|\cdot\|_{L_p} \le \|\cdot\|_{L_q}$ for $1 \le p \le q \le \infty$. In this case, Theorem 1.11 can be generalized for potentials V for which we demand (instead of the semi-boundedness of the potentials V themselves) that the whole operator A should be semi-bounded from below (and that V should be summable in a certain sense). Thus, the following theorem holds.

Theorem 1.12. *Assume that a nonnegative selfadjoint operator A_0 (for which $D = \mathfrak{D}(A_0) \cap L_p$ is a base for certain fixed $p \in (2, \infty)$ and $\varepsilon > 0$) and a semi-group e^{-tA_0} contracting in $L_{p+\varepsilon}$ act in the space L_2 constructed by a probability measure. Suppose that a real-valued function V belongs to L_r, where $r = 2p(p-2)^{-1}$, and that $e^{-\delta V} \in L_1$ for some $\delta > 0$. If V is such that the operator (1.49) is semi-bounded from below, then it is essentially selfadjoint.*

Proof. The proof is similar to the proof of Theorem 1.11 with the following changes: here $b > 0$ is finite and v_n is determined by the relation $V_n(x) = V(x)$ for $x \in R$ such that $|V(x)| \le n$ and by $V_n(x) = n$ otherwise. Let us determine $h > 1$ from the equality $h^{-1} + p_1^{-1} = p^{-1}$, where $p_1 = p + \varepsilon$. Below, we establish the following estimate for the solution (1.50) of Equation (1.46):

$$\|\varphi_n(t)\|_{L_p} \le c\,\|\varphi_0\|_{L_{p_1}} \tag{1.61}$$

$$(\varphi_0 \in D;\ t \in [0, T];\ T \in (0, \delta h^{-1});\ n \in \mathbb{N}).$$

By using this estimate, we can verify the conditions of Theorem 1.9 just as in the proof of Theorem 1.11. Here, $b = \delta h^{-1}$.

We proceed to the derivation of (1.61). First, we note that by Remark 1, the semigroup e^{-tA_0} is contracting in each L_q with $q \in [p_1', p_1]$. The estimate (1.61) has the form (see (1.50))

$$\|e^{-sA_n}\varphi_0\|_{L_p} \le c\,\|\varphi_0\|_{L_{p_1}} \quad (\varphi_0 \in D;\ s \in [0, \delta h^{-1}];\ n \in \mathbb{N}). \tag{1.62}$$

We set

$$A_{n,m} = (e^{-\frac{s}{m}V_n} e^{-\frac{s}{m}A_0})^m,$$

where $s \in [0, \delta h^{-1}]$ and $m \in \mathbb{N}$; n is fixed. We regard this operator as acting from L_{p_1} to L_p. The operator $A_{n,m}$ can be understood as a superposition of the following operators

$$\begin{array}{lllll}
L_{p_1} \xrightarrow{e^{-\frac{s}{m}A_0}} & L_{p_1} & & & \\
 & \big\downarrow e^{-\frac{s}{m}V_n} & & & \\
 & L_{(p_1^{-1}+m^{-1}h^{-1})^{-1}} & \xrightarrow{e^{-\frac{s}{m}A_0}} & L_{(p_1^{-1}+m^{-1}h^{-1})^{-1}} & \\
 & & & \big\downarrow e^{-\frac{s}{m}V_n} & \\
 & & & L_{(p_1^{-1}+2m^{-1}h^{-1})^{-1}} \rightarrow & \\
 & & & & \vdots \\
 & & & & \big\downarrow e^{-\frac{s}{m}V_n} \\
 & & & & L_p
\end{array}$$

Since $p \in [p_1', p_1]$, the semigroup e^{-sA_0} is contracting in each space L_q with $q \in [p, p_1]$. In particular, it is contracting for

$$q = (p_1^{-1} + jm^{-1}h^{-1})^{-1} \quad (j = 0, 1, \dots, m).$$

Further, it follows from the Hölder inequality that each mapping

$$e^{-\frac{s}{m}V_n} : L_{(p_1^{-1}+jm^{-1}h^{-1})^{-1}} \to L_{(p_1^{-1}+(j+1)m^{-1}h^{-1})^{-1}} \qquad (j = 0, 1, \dots, m-1)$$

does not exceed $\| e^{-\frac{s}{m}V_n} \|_{L_{mh}}$ in norm. Hence,

$$\| A_{n,m}\varphi_0 \|_{L_p} \le \| e^{-\frac{s}{m}V_n} \|_{L_{mh}}^m \| \varphi_0 \|_{L_{p_1}}$$

$$= \Big(\int_R e^{-\frac{s}{m}V_n(x)hm} \, d\mu(x) \Big)^{1/h} \| \varphi_0 \|_{L_{p_1}} = \| e^{-sV_n} \|_{L_h} \| \varphi_0 \|_{L_{p_1}}. \tag{1.63}$$

Let us estimate $\| e^{-sV_n} \|_{L_h}$ for $s \in [0, \delta h^{-1}]$. Denote

$$V_{n,+}(x) = \max(V_n(x), 0), \quad V_{n,-}(x) = -\min(V_n(x), 0),$$

$$V_-(x) = -\min(V(x), 0) \quad (x \in R); \qquad sh = \sigma \in [0, \delta].$$

Then

$$\| e^{-sV_n} \|_{L_h}^h = \int_R e^{-\sigma V_n(x)} d\mu(x)$$

$$= \int_R e^{-\sigma V_{n,+}(x)} e^{-\sigma V_{n,-}(x)} d\mu(x) \le \int_R e^{\delta V_{n,-}(x)} d\mu(x)$$

$$\le \int_R e^{\delta V_-(x)} d\mu(x) \le \int_R (1 + e^{-\delta V(x)} d\mu(x) = 1 + \| e^{-\delta V} \|_{L_1}.$$

This estimate and (1.63) yield

$$\| A_{n,m} \varphi_0 \|_{L_p} \le c \| \varphi_0 \|_{L_p} \tag{1.64}$$

$$(c = (1 + \| e^{-\delta V} \|_{L_1})^{1/h}; \quad m \in \mathbb{N}).$$

According to the multiplicative formula which has already been used, we have

$$A_{n,m} \varphi_0 \to e^{-sA_n} \varphi_0$$

as $m \to \infty$ in the sense of convergence in the norm of L_2. Therefore, one can extract a subsequence such that

$$(A_{n,m_k} \varphi_0)(x) \to (e^{-sA_n} \varphi_0)(x)$$

as $k \to \infty$ for almost all $x \in R$. Consequently,

$$|(A_{n,m_k} \varphi_0)(x)|^p \underset{k \to \infty}{\to} |(e^{-sA_n} \varphi_0)(x)|^p$$

for the same x. Applying the Fatou Lemma to the last sequence and employing (1.64), we conclude that the estimate (1.61) holds even for $s \in [0, \delta h^{-1}]$.

■

In connection with the verification of the conditions of the theorems in this subsection, we mention the following general fact (see Reed and Simon [2, Theorem X.55]) which, in many cases, allows us to establish the required contractibility of semigroups easily.

Let A_0 be a nonnegative selfadjoint operator in the space L_2 constructed by a probability measure. Assume that the semigroup e^{-tA_0} preserves positivity (i.e., the fact that $f \in L_2$ and $f \ge 0$ almost everywhere implies that $e^{-tA_0} \ge 0$ almost every-

where for $t \in [0, \infty))$, *and* $e^{-tA_0} 1 = 1$ $(t \in [0, \infty))$. *Then this semi-group is contracting in each* L_p $(p \in [1, \infty])$.

1.6. Another Evolutionary Criterion of Selfadjointness

Let us prove the following theorem which is similar to Theorems 1.1 and 1.6.

Theorem 1.13. *Let* $\mathcal{D} \subset H$ *be a dense set in a Hilbert space* H *on which an Hermitian operator* A *is defined such that* $A\mathcal{D} \subset \mathcal{D}$. *Assume that there exists a group of unitary operators* $\mathbb{R}^1 \ni t \mapsto U_t$ *in* H *such that*

(i) $U_t \mathcal{D} \subset \mathcal{D}$ $(t \in \mathbb{R}^1)$;

(ii) $U_t A f = A U_t f$ $(f \in \mathcal{D};\ t \in \mathbb{R}^1)$;

(iii) $\dfrac{d}{dt} U_t f = iAU_t f$ $(f \in \mathcal{D};\ t \in \mathbb{R}^1)$.

Then each operator $\mathcal{D} \ni f \mapsto A^n f$ $(n \in \mathbb{N})$ *in* H *is essentially selfadjoint.*

Proof. We fix $n \in \mathbb{N}$ and denote $B = A^n$. To prove the theorem, it suffices to show that the equations $B^* g = \pm i g$ have only trivial solutions.

Thus, assume that $B^* g = ig$. We fix $f \in \mathcal{D}$ and set

$$u(t) = (U_t f, g)_H \quad (t \in \mathbb{R}^1).$$

This function is bounded and satisfies the equation

$$u^{(n)}(t) = (i^n A^n U_t f, g)_H = (i^n B U_t f, g)_H = (i^n U_t f, B^* g)_H = -i^{n+1} u(t) \quad (t \in \mathbb{R}^1).$$

Each solution of this equation is equal to the linear combination of $e^{\lambda_j t}$, where λ_j are the roots of the equation $\lambda^n = -i^{n+1}$ $(j = 1, \dots, n)$. The last equation has no imaginary roots, and therefore, each solution under consideration can be bounded on the whole of the axis $\mathbb{R}^1$ only when it is identically zero. But this means that $g = 0$.

■

1.7. Approximation Theorem

We continue our investigation of potential perturbations of operators initiated in Subsection 1.5. Here, we use the notations introduced in this subsection.

Suppose that a nonnegative selfadjoint operator A_0 acts in the space L_2. With the help of the operator A_0, we define the bilinear form

$$a_{A_0}(f, g) = (A_0 f, g)_{L_2} \qquad (f, g \in \mathcal{D}(A_0)).$$

The closure of it is a form $\tilde{a}_{A_0}$ with the domain of definition $\mathcal{D}(\tilde{a}_{A_0}) = \mathcal{D}(A_0^{1/2})$ (see Chapter 1, Subsection 3.3). If $V \geq 0$ is a measurable function on R, then we associate with it the operator of multiplication by V (denoted by the same symbol) with the natural domain of definition $\mathcal{D}(V) = \{f \in L_2 \mid Vf \in L_2\}$. The corresponding bilinear form is

$$a_V(f, g) = (Vf, g)_{L_2} \qquad (f, g \in \mathcal{D}(V)).$$

The closure of it is the form $\tilde{a}_V$ with $\mathcal{D}(\tilde{a}_V) = \mathcal{D}(V^{1/2})$. Assume that the set $\mathcal{D}(\tilde{a}_{A_0}) \cap \mathcal{D}(\tilde{a}_V)$ is dense in l_2. Then the form

$$a(f, g) = \tilde{a}_{A_0}(f, g) + \tilde{a}_V(f, g) \quad (f, g \in \mathcal{D}(a) = \mathcal{D}(\tilde{a}_{A_0}) \cap \mathcal{D}(\tilde{a}_V))$$

is positive and closed. According to Theorem 3.1 of Chapter 1, it is associated with the selfadjoint operator $A_0 + V$ (the form sum of the operators A_0 and V) for which $\mathcal{D}(a) = \mathcal{D}((A_0 + V)^{1/2})$ and

$$a(f, g) = ((A_0 + V)f, g)_{L_2} \quad (f, g \in \mathcal{D}(A_0 + V), \; g \in \mathcal{D}(a)).$$

We stress that, under given assumptions, the operator sum $A_0 + V$ may be meaningful only on 0.

To take into account negative potentials, we use the KLMN-theory studied in Subsection 3.4 of Chapter 1. Namely, for a measurable function $W \leq 0$, we assume that the inclusion

$$\mathcal{D}(|W|^{1/2}) \supset \mathcal{D}(A_0^{1/2})$$

and the estimate

$$(|W|f, f)_{L_2} \leq p(A_0^{1/2}f, A_0^{1/2}f)_{L_2} + q(f, f)_{L_2}$$

$$(f \in \mathcal{D}(A_0^{1/2}); \; p \in [0, 1), \; q \in \mathbb{R}^1) \tag{1.65}$$

hold. This means that (1.65) with A_0 replaced by $A_0 + V$ is all the more valid, and thus, according to Theorem 3.4 of Chapter 1, the form sum

$$(A_0 \dotplus V) \dotplus W = A$$

is defined. Moreover, $A \geq -q\,1$, and for the closed bilinear form $\tilde{a}_A$ which corresponds to A, we have

$$\mathfrak{D}(\tilde{a}_A) = \mathfrak{D}(\tilde{a}_{A_0}) \cap \mathfrak{D}(\tilde{a}_V) = \mathfrak{D}(A_0^{1/2}) \cap \mathfrak{D}(V^{1/2})$$

For applications, it is important to study the question of convergence (in some sense) to $(A_0 \dot{+} V) \dot{+} W$ of the operators obtained when the potentials V and W are approximated by more regular potentials. We analyze this problem below in the case of approximation by standard cutoffs

$$V_n(x) = \begin{cases} V(x), & V(x) \leq n, \\ n, & V(x) > n; \end{cases} \qquad W_n(x) = \begin{cases} W(x), & W(x) \geq -n, \\ -n, & W(x) < -n \end{cases}$$

$$(x \in R;\ n \in \mathbb{N}).$$

For any $k, l \in \mathbb{N}$, we denote

$$A_{kl} = A_0 + V_k + W_l, \qquad A_k = (A_0 + V_k) \dot{+} W. \tag{1.66}$$

Theorem 1.14. *Let A_0 be a selfadjoint nonnegative operator in L_2, and let $V \geq 0$ and $W \leq 0$ be measurable functions for which*

(i) the set $\mathfrak{D}(A_0^{1/2}) \cap \mathfrak{D}(V^{1/2})$ is dense in L_2;

(ii) $\mathfrak{D}(A_0^{1/2}) \subset \mathfrak{D}(|W|^{1/2})$;

(iii) there exist $p \in [0, 1)$ and $q \in \mathbb{R}^1$ such that

$$\|\,|W|^{1/2} f\|_{L_2}^2 \leq p\,\|A_0^{1/2} f\|_{L_2}^2 + q\,\|f\|_{L_2}^2.$$

for any $f \in \mathfrak{D}(A_0^{1/2})$. Assume that the operators A_{kl} and A_k are defined by (1.66) *and*

$$A = (A_0 \dot{+} V) \dot{+} W.$$

Then, for any $\lambda > q$, we have

$$(A_{kl} + \lambda 1)^{-1} \underset{l \to \infty}{\rightarrow} (A_k + \lambda 1)^{-1},$$

and

$$(A_k + \lambda 1)^{-1} \underset{k \to \infty}{\rightarrow} (A + \lambda 1)^{-1}$$

strongly in L_2. Furthermore, for any $t \geq 0$,

$$e^{-tA_{kl}} \underset{l\to\infty}{\to} e^{-tA_k},$$

and

$$e^{-tA_k} \underset{k\to\infty}{\to} e^{-tA}$$

strongly in L_2.

Proof. As has been already mentioned, condition (iii) implies $A_0 \dot{+} W \geq -q\,1$, hence, for $\lambda > q$, we have

$$A_0 \dot{+} W + \lambda\,1 \geq c\,1,$$

$c = \lambda - q > 0$. We include the additional term $\lambda\,1$ in the operator A_0; then

$$c\,1 \leq A_0 \dot{+} W \leq A_k = (A_0 + V_k) \dot{+} W \leq A_{kl} = A_0 + V_k + W_l \tag{1.67}$$

for any $k, l \in \mathbb{N}$. According to the construction, we have

$$\mathcal{D}(\tilde{a}_{A_k}) = \mathcal{D}(\tilde{a}_{A_{kl}}) = \mathcal{D}(A_0^{1/2}) \quad (l \in \mathbb{N})$$

for fixed $k \in \mathbb{N}$. Since

$$\mathcal{D}(|W|^{1/2}) \supset \mathcal{D}(A_0^{1/2}),$$

we have

$$\tilde{a}_{A_{kl}}(f,f) = \|A^{1/2}f\|^2_{L_2} + \|V_k^{1/2}f\|^2_{L_2} - \||W_l|^{1/2}f\|^2_{L_2}$$

$$\underset{l\to\infty}{\to} \tilde{a}_{A_k}(f,f) = \|A^{1/2}f\|^2_{L_2} + \|V_k^{1/2}f\|^2_{L_2} - \||W|^{1/2}f\|^2_{L_2} \tag{1.68}$$

for any $f \in \mathcal{D}(A_0^{1/2})$. Let us show that this convergence implies the strong convergence $A_{kl}^{-1} \underset{l\to\infty}{\to} A_k^{-1}$ in L_2. Taking into account the inequalities

$$A_{kl}^{-1} \leq A_k^{-1} \leq c^{-1}\,1,$$

which follow from (1.67), we get

$$|\tilde{a}_{A_k}((A_k^{-1} - A_{kl}^{-1})f, g)| = |((A_k^{-1} - A_{kl}^{-1})f, A_k g)_{L_2}| \leq |((A_{kl} - A_k)A_{kl}^{-1}f,\ g)_{L_2}|$$

$$\leq ((A_{kl} - A_k)A_{kl}^{-1}f,\ A_{kl}^{-1}f)^{1/2}_{L_2}\,((A_{kl} - A_k)\,A_{kl}^{-1}g,\ A_{kl}^{-1}g)^{1/2}_{L_2}$$

$$\leq (f,\ A_{kl}^{-1} f)_{L_2}^{1/2} \, ((A_{kl} - A_k)g,\ g)_{L_2}^{1/2}$$

$$\leq c^{-1} \|f\|_{L_2} \left[\tilde{a}_{A_{kl}}(g, g) - \tilde{a}_{A_k}(g, g)\right]^{1/2} \to 0, \quad l \to \infty \qquad (1.69)$$

for $f \in L_2$ and

$$g \in \mathfrak{D}(A_k) \supset \mathfrak{D}(A_0^{1/2}).$$

Therefore, for a set of vectors of the form $h = A_k g,\ g \in \mathfrak{D}(A_k)$ ($A_k \geq c\,1$ is selfadjoint) which is dense in L_2, we have

$$(A_{kl}^{-1} f, h)_{L_2} \to (A_k^{-1} f,\ g)_{L_2}, \quad l \to \infty.$$

It remains to note that

$$\|A_{kl}^{-1} f\|_{L_2} \leq \|A_k^{-1} f\|_{L_2} \leq c^{-1} \|f\|_{L_2} \qquad (l \in \mathbb{N})$$

and, hence, $A_{kl}^{-1} \underset{l \to \infty}{\to} A_k^{-1}$ weakly. Since $A_{kl}^{-1} < A_k^{-1}$, this yields the strong convergence

$$\|(A_k^{-1} - A_{kl}^{-1}) f\|_{L_2}^2 \leq \|(A_k^{-1} - A_{kl}^{-1})^{1/2}\| \, \|(A_k^{-1} - A_{kl}^{-1})^{1/2} f\|_{L_2}$$

$$\leq c^{-1} (f, (A_k^{-1} - A_{kl}^{-1}) f)_{L_2} \to 0, \quad l \to \infty.$$

To prove the strong convergence $A_k^{-1} \to A^{-1}$ as $k \to \infty$, we apply the theorem on monotone operator convergence (see Kato [1, Chapter 8, Theorem 3.13]). Recall that this theorem establishes the required convergence of our monotonically increasing sequence $(A_k)_{k=1}^{\infty}$ of selfadjoint operators under the condition that the inclusion $f \in \mathfrak{D}(\tilde{a}_{A_k})$ and the boundedness of a number sequence $\tilde{a}_{A_k}(f, f)$, $k \in \mathbb{N}$, imply that $f \in \mathfrak{D}(\tilde{a}_A)$ and

$$\tilde{a}_{A_k}(f, f) \to \tilde{a}_A(f, f), \quad k \to \infty.$$

To check this condition, we choose an arbitrary vector such that

$$f \in \mathfrak{D}(\tilde{a}_{A_k}) = \mathfrak{D}(A_0^{1/2}).$$

Then the sequence $(V_k f, f)_{L_2}$, $k \in \mathbb{N}$ is bounded and, therefore, we have

$$(V_k f, f)_{L_2} \underset{k \to \infty}{\to} (Vf, f)_{L_2},$$

i.e., $f \in \mathfrak{D}(V^{1/2})$. Hence,

$$f \in \mathcal{D}(A_0^{1/2}) \cap \mathcal{D}(V^{1/2}) = \mathcal{D}(\tilde{a}_A)$$

and

$$\tilde{a}_{A_k}(f,f) \underset{k\to\infty}{\to} \tilde{a}_A(f,f).$$

The last statement of the theorem follows from the standard fact concerning the relation between strong convergences of resolvents and semigroups.

■

1.8. Verification of the Commutativity of Selfadjoint Operators

Assume that essentially selfadjoint operators A_1 and A_2 which are, generally speaking, unbounded, act in a Hilbert space H. Recall that they are called commuting if

$$E_1(\alpha_1)E_2(\alpha_2) = E_2(\alpha_2)\,E_1(\alpha_1)$$

for any $\alpha_1, \alpha_2 \in \mathfrak{B}(\mathbb{R}^1)$, where E_1 and E_2 are the RIs of the operators A_1 and A_2, respectively. The verification of this commutativity is often connected with serious difficulties. We now give some facts which make this verification easier. First, we establish the following well-known lemma.

Lemma 1.4. *Suppose that A_1 and A_2 are two selfadjoint operators. These operators are commuting if and only if their resolvents $R_{z_1}(A_1)$ and $R_{z_2}(A_2)$ are commuting for some fixed z_1 and z_2.*

Proof. If A_1 and A_2 commute, then the fact that their resolvents are commuting follows from the representation

$$R_{z_j}(A_j) = \int\limits_{\mathbb{R}^1} (\lambda - z_j)^{-1}\, dE_j(\lambda) \quad (j = 1, 2).$$

Conversely, assume that $R_{z_1}(A_1)$ and $R_{z_2}(A_2)$ commute. First, we prove that $R_{\zeta_1}(A_1)$ and $R_{\zeta_2}(A_2)$ are commuting for arbitrary nonreal ζ_1 and ζ_2. Indeed, let B be an operator acting in H, and let $z \in \mathbb{C}^1$ and $\zeta \in \mathbb{C}^1$ be two regular points of it. Then the operator $(1-(z-\zeta)R_\zeta(B))^{-1}$ exists and is equal to $1+(z-\zeta)R_z$. This can be easily verified with the help of the Hilbert identity. Using this identity and the fact established above, we obtain

$$R_{\zeta_1}(A_1) - R_{z_1}(A_1) = (\zeta_1 - z_1)R_{\zeta_1}(A_1)R_{z_1}(A_1),$$

$$R_{\zeta_1}(A_1)(1-(\zeta_1-z_1)R_{z_1}(A_1)) = R_{z_1}(A_1),$$

and

$$R_{\zeta_1}(A_1) = (1-(\zeta_1-z_1)R_{z_1}(A_1))^{-1}R_{z_1}(A_1).$$

The last equality and the commutativity of $R_{z_1}(A_1)$ and $R_{z_2}(A_2)$ imply that $R_{\zeta_1}(A_1)$ and $R_{z_2}(A_2)$ are commuting. Similarly, this yields that $R_{\zeta_1}(A_1)$ and $R_{\zeta_2}(A_2)$ are commuting, too.

We now employ the well-known formula

$$\tfrac{1}{2}(E_j(\bar{\alpha}_j)+E_j(\alpha_j)) = \frac{1}{2\pi i}\lim_{\varepsilon_j\to+0}\int_{\alpha_j+i\varepsilon_j}(R_{\zeta_j}(A_j)-R_{\bar{\zeta}_j}(A_j))\,d\zeta_j \quad (j=1,2) \tag{1.70}$$

in the sense of strong convergence for finite open intervals α_1 and α_2. By virtue of what has been proved, the integrals in (1.70) for $j=1$ and $j=2$ commute. But then their strong limits also commute and, hence, one can easily find that $E_1(\alpha_1)$ and $E_2(\alpha_2)$ (i.e., A_1 and A_2) are commuting as well.

■

Theorem 1.15. *Let A_1 and A_2 be two Hermitian operators acting in H with the domains of definition $\mathcal{D}(A_1)$ and $\mathcal{D}(A_2)$; and let $\mathcal{D}$ be a linear set such that $\mathcal{D} \subset \mathcal{D}(A_1) \cap \mathcal{D}(A_2)$. Assume that these operators commute on $\mathcal{D}$, i.e., $A_1\mathcal{D} \subset \mathcal{D}(A_2)$, $A_2\mathcal{D} \subset \mathcal{D}(A_1)$, and $A_1A_2f = A_2A_1f$ ($f \in \mathcal{D}$).*

In addition, assume that A_1, A_2, and the restriction $A_1 \upharpoonright ((A_2 - z\,1)\mathcal{D})$ possess total sets of quasianalytical vectors for some nonreal z. Then the closures $\tilde{A}_1$ and $\tilde{A}_2$ are selfadjoint and commuting.

Proof. According to Theorem 1.7, the operators $\tilde{A}_1$ and $\tilde{A}_2$ are selfadjoint. Let $R_\zeta(\tilde{A}_1)$ and $R_\zeta(\tilde{A}_2)$ be their resolvents. By virtue of Lemma 1.4, to prove that $\tilde{A}_1$ and $\tilde{A}_2$ are commuting, it suffices to verify the commutativity of $R_z(\tilde{A}_1)$ and $R_z(\tilde{A}_2)$. By virtue of the fact that A_1 and A_2 commute, for $f \in \mathcal{D}$ we have

$$R_z(\tilde{A}_1)R_z(\tilde{A}_2)(A_1-z\,1)(A_2-z\,1)f = R_z(\tilde{A}_1)R_z(\tilde{A}_2)(A_2-z\,1)(A_1-z\,1)f$$

$$= f = R_z(\tilde{A}_2)R_z(\tilde{A}_1)(A_1-z\,1)(A_2-z\,1)f.$$

Therefore, to prove that $R_z(\tilde{A}_1)$ and $R_z(\tilde{A}_2)$ are commuting, it suffices to show that $(A_1-z\,1)(A_2-z\,1)\mathcal{D}$ is dense in H. But this set coincides with the range of values

$$\mathfrak{R}(A_1 \restriction ((A_2 - z\,1)\mathcal{D}) - z\,1),$$

and the operator $A_1 \restriction ((A_2 - z\,1)\mathcal{D})$ is essentially selfadjoint by the same Theorem 1.7. Hence, the indicated range of values is dense in H.

■

We now investigate a situation connected with the spectral representation of positive definite kernels when it is necessary to verify the commutativity of selfadjoint operators. Let H_1 and H_2 be separable Hilbert spaces. We construct their tensor product $H = H_1 \otimes H_2$ and consider a quasiscalar product $\langle \cdot, \cdot \rangle$ in H such that

$$\langle\langle f \rangle\rangle = \langle f, f \rangle^{1/2} \le c \| f \|_H \quad (f \in H) \tag{1.71}$$

for some $c > 0$. For given H, with the help of $\langle \cdot, \cdot \rangle$ we define a Hilbert space $\mathcal{H}$ in the standard way by means of identifying and completing.

Assume that the operator A_j with the dense domain of definition $\mathcal{D}(A_j)$ is given in H_j. Having given A_1 and A_2, we construct the operators acting in H with dense domains of definition $\mathcal{D}(A_j)$ by setting

$$A_{(1)} = A_1 \otimes 1, \quad \mathcal{D}(A_{(1)}) = \text{a.}\,(\mathcal{D}(A_1) \otimes H_2),$$

and

$$A_{(2)} = 1 \otimes A_2, \quad \mathcal{D}(A_{(2)}) = \text{a.}\,(H_1 \otimes \mathcal{D}(A_2)).$$

It is clear that these operator commute on

$$\mathcal{D} = \text{a.}\,(\mathcal{D}(A_1) \otimes \mathcal{D}(A_2)) \subset \mathcal{D}(A_{(1)}) \cap \mathcal{D}(A_{(2)}),$$

i.e., we have $A_{(1)}\mathcal{D} \subset \mathcal{D}(A_{(2)})$, $A_{(2)}\mathcal{D} \subset \mathcal{D}(A_{(1)})$, and

$$A_{(1)}A_{(2)}f = A_{(2)}A_{(1)}f \quad (f \in \mathcal{D}).$$

Assume that, for $j = 1, 2$, the operator $A_{(j)}$ is Hermitian with respect to $\langle \cdot, \cdot \rangle$, i.e.,

$$\langle A_{(j)}f, g \rangle = \langle f, A_{(j)}g \rangle \quad (f, g \in \mathcal{D}(A_{(j)})).$$

Then the mapping

$$H \supset \mathcal{D}(A_{(j)}) \ni f \mapsto A_{(j)}f \in H$$

generates the operator $\mathcal{A}_j$ in $\mathcal{H}$ in a natural way.

Indeed, the fact that $\langle f, f \rangle = 0$ implies that

$$\langle A_{(j)}f, A_{(j)}f\rangle = 0 \quad (f, g \in \mathcal{D}(A_{(j)})).$$

We prove this. For arbitrary $g \in \mathcal{D}(A_{(j)})$, we have

$$|\langle A_{(j)}f, g\rangle|^2 = |\langle f, A_{(j)}g\rangle|^2 \le \langle f,f\rangle \langle A_{(j)}\, g, A_{(j)}g\rangle = 0,$$

whence $\langle A_{(j)}f, g\rangle = 0$. Due to the denseness of $\mathcal{D}(A_{(j)})$ in H, we approximate the vector $A_{(j)}$ in H by the vectors $g_n \in \mathcal{D}(A_{(j)})$. Then, by virtue of (1.71), we have

$$\langle A_{(j)}\, f, A_{(j)}g\rangle = \lim_{n\to\infty} \langle A_{(j)}f,\, g_n\rangle = 0.$$

Thus, if we denote by $\hat{f}, \hat{g}, \ldots$ the classes which contain elements $f, g, \ldots$ from H under factorization with respect to $\{h \in H \mid \langle h, h\rangle = 0\}$, then we can set $\mathcal{A}_j \hat{f} = (A_{(j)}f)^\wedge$ and $\mathcal{D}(\mathcal{A}_j) = (\mathcal{D}(A_{(j)}f))^\wedge$. It is clear that the operators $\mathcal{A}_1$ and $\mathcal{A}_2$ are Hermitian and commute on $\hat{\mathcal{D}} \subset \mathcal{D}(\mathcal{A}_1) \cap \mathcal{D}(\mathcal{A}_2)$, i.e., $\mathcal{A}_1\hat{\mathcal{D}} \subset \mathcal{D}(\mathcal{A}_2)$, $\mathcal{A}_2\hat{\mathcal{D}} \subset \mathcal{D}(\mathcal{A}_1)$, and $\mathcal{A}_1\mathcal{A}_2 f = \mathcal{A}_2\mathcal{A}_1 f$ $(f \in \hat{\mathcal{D}})$. The problem is to clarify the conditions under which we can state that the operators $\mathcal{A}_1$ and $\mathcal{A}_2$ are essentially selfadjoint and commuting. Surely, in many cases we can apply Theorem 1.5; nevertheless, we establish two more results.

To formulate the first result, we consider the following additional construction. We fix $f_2 \in H_2$ and consider a quasiscalar product

$$\langle f_1, g_1\rangle_{f_2} = \langle f_1 \otimes f_2, g_1 \otimes f_2\rangle \quad (f_1, g_1 \in H_1)$$

in H_1. After identifying and completing with respect to $\langle\cdot,\cdot\rangle_{f_2}$, we construct a Hilbert space $\mathcal{H}_{1,f_2}$. The operator A_1 is Hermitian with respect to $\langle\cdot,\cdot\rangle_{f_2}$. Indeed,

$$\langle A_1 f_1, g_1\rangle_{f_2} = \langle (A_1 f_1) \otimes f_2, g_1 \otimes f_2\rangle = \langle A_{(1)}(f_1 \otimes f_2), g_1 \otimes f_2\rangle$$

$$= \langle f_1 \otimes f_2, A_{(1)}(g_1 \otimes f_2)\rangle = \langle f_1, A_1 g_1\rangle_{f_2} \quad (f_1, g_1 \in \mathfrak{D}(A_1)).$$

Therefore, by analogy with the previous case, it generates an Hermitian operator $\mathcal{A}_{1,f_2}$ in the space $\mathcal{H}_{1,f_2}$. Similarly, we can fix $f_1 \in H_1$ and construct a space $\mathcal{H}_{2,f_1}$ and an Hermitian operator $\mathcal{A}_{2,f_1}$ acting in it.

Theorem 1.16. *Assume that the operators $\mathcal{A}_{1,f_2}$ and $\mathcal{A}_{2,f_1}$ are essentially selfadjoint for f_1 and f_2 running through dense sets M_1 and M_2 in the spaces H_1 and H_2, respectively. Then the operators $\mathcal{A}_1$ and $\mathcal{A}_2$ are also essentially self-*

adjoint. They are commuting provided that, for some nonreal z, each operator $\mathcal{A}_{1,h_2}$ *is essentially selfadjoint for* $h_2 \in (A_2 - z\,1)\mathcal{D}$ *and*

$$\operatorname{Ker}(A_j - z\,1) = \{g_j \in \mathcal{D}(A_j) \,|\, (A_j - z\,1)\, g_j = 0\} = \{0\}.$$

Proof. We establish the essential selfadjointness of $\mathcal{A}_1$. Let $f_1 \in H_1$ and $f_2 \in M_2$. It suffices to show that for fixed $\varepsilon > 0$ and nonreal z, one can choose $g_j \in \mathcal{D}(A_1)$ such that

$$\| (f_1 \otimes f_2)^\wedge - (\mathcal{A}_1 - z\,1)(g_1 \otimes f_2)^\wedge \|_{\mathcal{H}} < \varepsilon.$$

But this norm is equal to

$$\langle\langle f_1 \otimes f_2 - (A_{(1)} - z\,1)\,(g_1 \otimes f_2)\rangle\rangle$$

$$= \langle\langle\,(f_1 - (A_1 - z\,1)\,g_1) \otimes f_2)\rangle\rangle = \| \hat{f}_1 - (\mathcal{A}_{1,f_2} - z\,1)\,\hat{g}_1 \|_{\mathcal{H}_{1,f_2}}, \quad (1.72)$$

where $\hat{f}_1$ and $\hat{g}_1$ are classes which appear in the construction of the space $\mathcal{H}_{1,f_2}$ and contain f_1 and g_1, respectively. By the proper choice of $g_1 \in \mathcal{D}(A_1)$, the right-hand side of (1.72) can be made as small as desired because $\mathcal{A}_{1,f_2}$ is essentially selfadjoint. The statement is proved. The essential selfadjointness of $\mathcal{A}_2$ can be proved similarly.

We prove that $\mathcal{A}_1$ and $\mathcal{A}_2$ are commuting. First, we establish the fact that the vectors

$$(((A_1 - z\,1)g_1) \otimes ((A_2 - z\,1)g_2))^\wedge \quad (g_1 \in \mathcal{D}(A_1),\ g_2 \in \mathcal{D}(A_2))$$

are dense in $\mathcal{H}$. Let $f_1 \in M_1$ and $f_2 \in H_2$. In natural notations, we have

$$\| (f_1 \otimes f_2)^\wedge - (((A_1 - z\,1)g_1) \otimes ((A_2 - z\,1)g_2))^\wedge \|_{\mathcal{H}}$$

$$= \langle\langle f_1 \otimes f_2 - ((A_1 - z\,1)g_1) \otimes ((A_2 - z\,1)g_2\rangle\rangle$$

$$\leq \langle\langle f_1 \otimes (f_2 - (A_2 - z\,1)g_2\rangle\rangle$$

$$+ \langle\langle (f_1 - (A_1 - z\,1)g_1 \otimes ((A_2 - z\,1)g_2\rangle\rangle$$

$$= \| \hat{f}_2 - (\mathcal{A}_{2,f_1} - z\,1)\hat{g}_2 \|_{\mathcal{H}_{2,f_1}}$$

$$+ \| \hat{f}_1 - (\mathcal{A}_{1,(A_2 - z1)g_2} - z\,1)\hat{g}_1 \|_{\mathcal{H}_{1,(A_2 - z1)g_2}}. \quad (1.73)$$

By virtue of the essential selfadjointness of $\mathcal{A}_{2,f_1}$, the first term can be made as small as desired by the proper choice of $g_2 \in \mathcal{D}(A_2)$. The same is true for the second term because $\mathcal{A}_{1,(A_2-z1)g_2}$ is essentially selfadjoint. The required denseness is thus established.

Since $\operatorname{Ker}(A_1 - z\,1) = \{0\}$, there exists an algebraically inverse operator $(A_1 - z\,1)^{-1}$ in H_1. But then the algebraically inverse operator

$$(A_{(1)} - z\,1)^{-1} = (A_1 - z\,1)^{-1} \otimes 1$$

also exists in H. Clearly, the action of the resolvent $R_z(\tilde{\mathcal{A}}_1)$ upon a vector $(f_1 \otimes f_2)^\wedge$ $(f_1 \in H_1, f_2 \in H_2)$ is equal to $((A_{(1)} - z\,1)^{-1}(f_1 \otimes f_2))^\wedge$. In the case of the operator A_2, the situation is similar. Therefore, we have

$$\begin{aligned}
&R_z(\tilde{\mathcal{A}}_2)R_z(\tilde{\mathcal{A}}_1)(((A_1 - z\,1)g_1) \otimes ((A_2 - z\,1)g_2))^\wedge \\
&\quad = R_z(\tilde{\mathcal{A}}_2)(((A_1 - z\,1)^{-1} \otimes 1)((A_1 - z\,1)g_1) \otimes ((A_2 - z\,1)g_2)))^\wedge \\
&\quad = R_z(\tilde{\mathcal{A}}_2)(g_1 \otimes ((A_2 - z\,1)g_2)^\wedge = ((1 \otimes (A_2 - z\,1)^{-1}(g_1 \otimes ((A_2 - z\,1)g_2))^\wedge \\
&\quad = (g_1 \otimes g_2)^\wedge = R_z(\tilde{\mathcal{A}}_1)R_z(\tilde{\mathcal{A}}_2)(((A_1 - z\,1)g_1) \otimes ((A_2 - z\,1)g_2))^\wedge
\end{aligned} \tag{1.74}$$

for $g_1 \in \mathcal{D}(A_1)$ and $g_2 \in \mathcal{D}(A_2)$. By virtue of the denseness of the vectors

$$(((A_1 - z\,1)g_1) \otimes ((A_2 - z\,1)g_2))^\wedge$$

in $\mathcal{H}$, the relation (1.74) means that $R_z(\tilde{\mathcal{A}}_1)$ and $R_z(\tilde{\mathcal{A}}_2)$ are commuting. It remains to apply Lemma 1.4.

■

The next theorem is closely connected with Theorem 1.1.

Theorem 1.17. *Assume that the uniqueness of strong solutions of the equations*

$$\left(\frac{du}{dt}\right)(t) \pm (iA_j^*)u(t) = 0 \quad (t \in [0, b_j)) \tag{1.75}$$

in H_j takes place for some $b_j \in (0, \infty]$. Then the operator $\mathcal{A}_j$ is essentially selfadjoint $(j = 1, 2)$. In addition, if $\operatorname{Ker}(A_j - z_j\,1) = \{0\}$ $(j = 1, 2)$ *for some nonreal z_j, then $\tilde{\mathcal{A}}_1$ and $\tilde{\mathcal{A}}_2$ commute.*

Proof. Let us establish the selfadjointness of $\tilde{\mathcal{A}}_1$; the selfadjointness of $\tilde{\mathcal{A}}_2$ can be established analogously. As a preliminary, we note that for every fixed $f_2 \in H_2$ there exists a continuous operator $\mathcal{H} \ni w \mapsto T_{f_2}w \in H_1$ such that

$$\langle w, (f_1 \otimes f_2)^\wedge \rangle = (T_{f_2} w, f_1)_{H_1}. \tag{1.76}$$

Indeed, the expression to the left in (1.76) is a bilinear form with respect to $w \in \mathcal{H}$ and $f_1 \in H_1$ which is continuous by virtue of the estimate (1.71) and the Cauchy–Bunyakovsky inequality for $\langle \cdot, \cdot \rangle$. Hence, it can be written in the form (1.76).

Consider a strong solution $[0, b_1) \ni t \mapsto v(t) \in \mathcal{H}$ of the equation

$$\left(\frac{dv}{dt}\right)(t) + (i\mathcal{A}_1)^* v(t) = 0, \quad v(0) = 0$$

in $\mathcal{H}$. If we write this equation in a weak form of the type (1.3), where the role of f is played by $(f_1 \otimes f_2)^\wedge$ $(f_1 \in \mathcal{D}(A_1),\ f \in H_2)$, then we get

$$\left\langle \left(\frac{dv}{dt}\right)(t), (f_1 \otimes f_2)^\wedge \right\rangle + \langle v(t), i\mathcal{A}_1 (f_1 \otimes f_2)^\wedge \rangle = 0 \quad (t \in [0, b_1)). \tag{1.77}$$

The function

$$[0, b_1) \ni t \mapsto u(t) = T_{f_2} v(t) \in H_1$$

is once strongly continuously differentiable. By using (1.76) and the relation

$$\mathcal{A}_1 (f_1 \otimes f_2)^\wedge = ((A_1 f_1) \otimes f_2)^\wedge$$

we can write the equality (1.77) as follows

$$\left(\left(\frac{du}{dt}\right)(t), f_1 \right)_{H_1} + (u(t), iA_1 f_1)_{H_1} = 0, \quad u(0) = T_{f_2} v(0) = 0$$

$$(f_1 \in \mathcal{D}(A_1);\ (t \in [0, b_1)).$$

In other words, we obtain Equation (1.75) written in the weak form for $j = 1$ with the sign "+". It follows from the assumed uniqueness that $u(t) = 0$ $(t \in [0, b_1))$. If we return to $v(t)$ with the help of (1.76), then we get

$$\langle v(t), (f_1 \otimes f_2)^\wedge \rangle = 0 \quad (t \in [0, b_1))$$

for all $f_1 \in \mathcal{D}(A_1)$ and $f_2 \in H_2$. This yields $v(t) = 0$ $(t \in [0, b_1))$. In the case of the sign "–", the uniqueness is established analogously. To prove the selfadjointness of $\tilde{\mathcal{A}}_1$, it remains to apply Theorem 1.1.

For this proof of the selfadjointness of $\tilde{\mathcal{A}}_1$, it is important that $f_1 \in \mathcal{D}(A_1)$; the vector f_2 may run, not through the whole of the space H_2, but only through a subset of this space such that the vectors $(f_1 \otimes f_2)^\wedge$ $(f_1 \in \mathcal{D}(A_1))$ form a total set in $\mathcal{H}$. In particular, one can take $f_2 \in (A_2 - z_2 1)\mathcal{D}(A_2)$ because, in this case, the vectors $(f_1 \otimes f_2)^\wedge$ run through a collection of the vectors of the form $(\mathcal{A}_2 - z_2 1)(f_1 \otimes g_2)^\wedge$, where $g_1 \in \mathcal{D}(A_2)$, and by virtue of the essential selfadjointness of $\mathcal{A}_2$, the desired totality takes place. This remark leads us to the conclusion that the restriction

$$\mathcal{A}_1 \restriction (\text{a. } (\mathcal{D}(A_1) \otimes \mathfrak{R}(A_2 - z_2 1)))^\wedge$$

is essentially selfadjoint. This implies that the vectors

$$(((A_1 - z_1 1)g_1) \otimes ((A_2 - z_2 1)g_2))^\wedge \quad (g_1 \in \mathcal{D}(A_1),\ g_2 \in \mathcal{D}(A_2)) \tag{1.78}$$

form a dense set in $\mathcal{H}$.

Calculating $R_{z_2}(\tilde{\mathcal{A}}_2)R_{z_1}(\tilde{\mathcal{A}}_1)$ and $R_{z_1}(\tilde{\mathcal{A}}_1)R_{z_2}(\tilde{\mathcal{A}}_2)$ on the vectors (1.78) by analogy with (1.74), we obtain the same result $(g_1 \otimes g_2)^\wedge$. Therefore, the operators $R_{z_1}(\tilde{\mathcal{A}}_1)$ and $R_{z_2}(\tilde{\mathcal{A}}_2)$ commute on a set dense in $\mathcal{H}$. By virtue of Lemma 1.4, this yields the commutativity of $\tilde{\mathcal{A}}_1$ and $\tilde{\mathcal{A}}_2$.

■

Note that, in Theorem 1.17, the uniqueness can often be proved by use of Lemma 1.2.

The case when the operators $\tilde{\mathcal{A}}_1$ and $\tilde{\mathcal{A}}_2$ are not selfadjoint after being closed, but admit selfadjoint commuting extensions, was studied (in a somewhat different formulation) in the book of Berezansky [5, Chapter 8, Section 2, Subsection 3].

§2. Generalized Power Moment Problem

The classical power moment problem, i.e., the problem of representation of a sequence of real numbers $s = (s_n)_{n=0}^{\infty}$ in the form

$$s_n = \int_{\mathbb{R}^1} \lambda^n \, d\sigma(\lambda) \quad (n \in \mathbb{Z}_+),$$

where σ is a certain measure, is generalized in this section for the case of "infinitely many variables": The role of s_n $(n \in \mathbb{Z}_+)$ is played by elements of a tensor product of n copies of a certain space (in particular, s_n may be a function of n variables $s_n = s_n(x_1, \dots, x_n)$). We begin with the investigation of the simplest and the most remarkable symmetric case; for the functions, this means that each s_n is symmetric with respect to all variables $x_1, \dots, x_n$.

2.1. Generalized Symmetric Power Moment Problem

All the spaces considered below are assumed to be separable. The nth ($n \in \mathbb{N}$) tensor power of a Hilbert space H is denoted by $H^{\otimes n}$; for $n = 0$, $H^{\otimes n} = \mathbb{C}^1$. Let Φ be a nuclear space which is a projective limit of a family $(H_\tau)_{\tau \in T}$ of Hilbert spaces, i.e., $\Phi =$ pr $\lim_{\tau \in T} H_\tau$. Without loss of generality, we can assume that the zero space H_0 ($0 \in T$) is distinguished; $(H_\tau)' = H_{-\tau}$ and $\xi(\varphi) = (\xi, \varphi)_{H_0}$ ($\xi \in \Phi'$, $\varphi \in \Phi$). Consider the nth tensor power $\Phi^{\otimes n}$ ($z \in \mathbb{Z}_+$) of the space Φ which is defined as pr $\lim_{\tau \in T} H_\tau^{\otimes n}$ (for these notations, see Sections 1 and 2 of Chapter 1). Then the dual space is

$$(\Phi^{\otimes n})' = \bigcup_{\tau \in T} (H_\tau^{\otimes n})' = \bigcup_{\tau \in T} H_{-\tau}^{\otimes n};$$

it is equipped with a weak topology, and

$$\xi(\varphi) = (\xi, \varphi)_{H_0^{\otimes n}} \quad (\xi \in (\Phi^{\otimes n})', \ \varphi \in \Phi^{\otimes n}).$$

Assume that an involution $\Phi \ni \varphi \mapsto \bar{\varphi} \in \Phi$ is defined in Φ and that it can be extended by continuity to an involution in each H_τ ($\tau \in T$). Constructing tensor powers and passing to adjoint operators, we can extend this involution in a natural way to involutions in tensor powers and dual spaces; for these convolutions, we preserve the notation "–". The notation "Re" indicates that we consider real elements of this space with respect to the involution. For example, Φ'_{Re} is a space of all $\xi \in \Phi'$ such that $\bar{\xi} = \xi$. Clearly, the operations "–" and "⊗" are commuting, and the same is true for "′" and "Re".

Consider a sequence $s = (s_n)_{n=0}^\infty$, where $s_n \in (\Phi^{\otimes n})'$ and is completely symmetric, i.e., $s_n(\varphi^{(1)} \otimes \ldots \otimes \varphi^{(n)})$ is a symmetric function of the variables $\varphi^{(1)}, \ldots, \varphi^{(n)} \in \Phi$. We call s a moment sequence if the condition of positive definiteness is satisfied, namely, for any finite sequence $\varphi = (\varphi_j)_{j=0}^\infty$, $\varphi_j \in \Phi^{\otimes j}$ (their collection is denoted by C_0), we have

$$\sum_{j,k=0}^{\infty} s_{j+k}(\varphi_j \otimes \bar{\varphi}_k) \geq 0. \tag{2.1}$$

Example 2.1 (the finite-dimensional power moment problem). Let $T = \{0\}$ and $H_0 = \Phi = \mathbb{C}^d$ ($d \in \mathbb{N}$ is fixed); the involution is an ordinary transition to complex conjugation. Here $s_n \in \left(\Phi^{\otimes n}\right)' = \mathbb{C}^{nd}$; by virtue of its symmetry, this vector has the following set of coordinates $s_{n_1,\dots,n_d} = s_{\boldsymbol{n}}$, where $n_1,\dots,n_d \in \mathbb{Z}_+$, $\boldsymbol{n} = (n_1,\dots,n_d) \in \mathbb{Z}_+^d$, and $|\boldsymbol{n}| = n_1 + \dots + n_d = n$ (if $e_1,\dots,e_d$ is a standard basis in $\mathbb{C}^d$, then $s_{n_1,\dots,n_d}$ are coordinates of the vector s_n corresponding to the unit vector obtained by the symmetrization of unit vectors $e_\alpha = e_{\alpha_1} \otimes \dots \otimes e_{\alpha_n} \in \mathbb{C}^{nd}$, provided that the set $\alpha = (\alpha_1,\dots,\alpha_n)$ contains n_1 indices 1, n_2 indices 2, etc.). Condition (2.1) transforms into an ordinary condition for a moment sequence, namely, for any finite d-dimensional sequence $x = (x_j)_{j\in\mathbb{Z}_+^d}$ of complex numbers x_j, the following condition should hold

$$\sum_{j,k\in\mathbb{Z}_+^d} s_{j+k}\bar{x}_j x_k \geq 0. \tag{2.2}$$

A sequence of the form

$$s_{\boldsymbol{n}} = \int_{\mathbb{R}^d} \lambda^{\boldsymbol{n}}\, d\sigma(\lambda) \quad (\lambda^{\boldsymbol{n}} = \lambda_1^{n_1}\dots\lambda_d^{n_d};\quad \boldsymbol{n} = n_1 + \dots + n_d \in \mathbb{Z}_+^d), \tag{2.3}$$

where $\mathcal{B}(\mathbb{R}^d) \ni \alpha \mapsto \sigma(\alpha) \geq 0$ is a certain measure satisfying the condition of integrability of all functions $\mathbb{R}^d \ni \lambda \mapsto \lambda^{\boldsymbol{n}} \in \mathbb{R}^1$ $(n \in \mathbb{Z}_+^d)$, is an example of a d-dimensional sequence $s = (s_{\boldsymbol{n}})_{\boldsymbol{n}\in\mathbb{Z}_+^d}$ for which condition (2.2) is satisfied. In the one-dimensional case, i.e. for $d=1$, the condition of momentness (2.2) is also sufficient for the existence of the representation (2.3). In many-dimensional case, to guarantee the validity of (2.3) one should impose on $s_{\boldsymbol{n}}$ (in addition, to the condition of momentness) certain restrictions on growth as $n \to \infty$ ensuring the selfadjointness and commutativity of the corresponding shift operators (see (2.8) below). For a more detailed presentation of the finite-dimensional moment problem, see the work of M. G. Krein and Krasnoselsky [1], the book of Shohat and Tamarkin [1], Akhiezer [1], and Berezansky [5, Chapter 8, Section 5]. Here we only note that, in the case of a one-dimensional moment sequence $s = (s_n)_{n=0}^\infty$, the shift operator is selfadjoint, and the measure σ in the representation (2.3) (for $d=1$) is uniquely determined for given s if the class $C\{s_{2n}^{1/2}\}$ is quasianalytic. See also Examples 1.2 and 2.2.

Our purpose is to obtain a representation of the type (2.3) for a general moment sequence s by imposing certain additional restrictions upon the growth of s_n as $n \to \infty$ which generalize the corresponding conditions for a finite-dimensional moment problem. It seems to be useful to note as a preliminary that condition (2.1) involves the fact that s_n is real, i.e., the inclusion

$$s_n \in ((\Phi_{\mathrm{Re}})^{\otimes n})' = ((\Phi^{\otimes n})_{\mathrm{Re}})' \quad (z \in \mathbb{Z}_+).$$

To prove this, we fix $\varphi = (\varphi_j)_{j=0}^\infty \in C_0$, denote $a_{jk} = s_{j+k}(\varphi_j \otimes \bar{\varphi}_k)$, and then replace φ_l by $x_l\varphi_l$ with $x_l \in \mathbb{C}^1$ in (2.1). As a result, we get the inequality

$$\sum_{j,k=0}^{\infty} a_{jk}\bar{x}_j x_k \geq 0,$$

which yields the relation $a_{jk} = \bar{a}_{kj}$ $(j, k \in \mathbb{Z}_+)$. This and the arbitrariness of φ_l imply the required result. Therefore, just in the definition of a moment sequence $s = (s_n)_{n=0}^{\infty}$, one may demand that s_n $(n \in \mathbb{Z}_+)$ should be real.

We now give the required restrictions on the growth of s_n. Since

$$s_n \in (\Phi^{\otimes n})' = \bigcup_{\tau \in T} (H_\tau^{\otimes n})' = \bigcup_{\tau \in T} H_{-\tau}^{\otimes n},$$

one can find $\tau = \tau(n)$ such that $s_n \in H_{-\tau(n)}^{\otimes n}$ $(n \in \mathbb{Z}_+)$. The sequence $(\tau(n))_{n=0}^{\infty}$ often appears in what follows. Fix a set E of real vectors from Φ whose linear span is dense in Φ and denote

$$d(\tau, \mathrm{E}) = \sup_{\varphi \in \mathrm{E}} \| \varphi \|_{H_\tau} < \infty \tag{2.4}$$

for any $\tau \in T$. We say that s is definite if the class

$$C\{m_n\}, \quad m_n = (d(\tau(2n), \mathrm{E}))^n \| s_{2n} \|^{1/2}_{H_{-\tau(2n)}^{\otimes 2n}} \quad (n \in \mathbb{N}) \tag{2.5}$$

is quasianalytical. We shall also study the more general case of quasidefinite s when all the above-mentioned conditions are satisfied with a modification consisting in the facts that the codimension of the closure of the linear span E in Φ is equal to 1 and that $\| s_{2n} \|^{1/2}_{H_{-\tau(2n)}^{\otimes 2n}}$ is replaced by $\| \hat{s}_{2n} \|^{1/2}_{H_{-\tau(2n)}^{\otimes 2n}}$ in the definition (2.5) of m_n; here the "projection on $H_0^{\otimes 2n}$" of $\hat{s}_{2n} \in H_{-\tau(2n)}^{\otimes 2n}$ is defined by the relation

$$(\hat{s}_{2n}, \varphi)_{H_0^{\otimes 2n}} = (s_{2n}, \varphi)_{H_0^{\otimes 2n}} \qquad \Big(\varphi = \underbrace{e \otimes \ldots \otimes e}_{2n}, \quad e \in \mathrm{E}\Big).$$

Below, we present another version of the concept of definiteness (Remark 3). Here we only note that in the case of definiteness, all the operators associated with the problem are selfadjoint and commuting, while in the case of quasidefiniteness, this is true for all operators but one.

Theorem 2.1. *Let $s = (s_n)_{n=0}^{\infty}$ be a moment sequence satisfying the conditions of definiteness or quasidefiniteness. Then there exists a finite measure $\mathfrak{B}(\Phi'_{\mathrm{Re}}) \ni \alpha \mapsto \sigma(\alpha) \geq 0$ such that*

$$\int_{\Phi'_{\mathrm{Re}}} \underbrace{\lambda \otimes \ldots \otimes \lambda}_{n}\, d\sigma(\lambda) \quad (n \in \mathbb{Z}_+). \tag{2.6}$$

Here, we integrate the vector function

$$\Phi'_{\mathrm{Re}} \ni \lambda \mapsto \underbrace{\lambda \otimes \ldots \otimes \lambda} \in (\Phi^{\otimes n})',$$

and the integral converges weakly. If s is definite, then the measure σ can be uniquely determined (generally speaking, there is no uniqueness in the quasidefinite case). Conversely, every sequence of the form (2.6) *is a moment sequence.*

Note that it is often convenient to write the equality (2.6) in the form

$$s_n(\varphi^{(1)}, \ldots, \varphi^{(n)}) = \int_{\Phi'_{\mathrm{Re}}} \lambda(\varphi^{(1)}) \ldots \lambda(\varphi^{(n)})\, d\sigma(\lambda)$$

$$(\varphi^{(1)}, \ldots, \varphi^{(n)} \in \Phi;\ n \in \mathbb{Z}_+).$$

Proof. The proof is split into steps.

I. Recall that C_0 denotes a linear set of all finite sequences of the form $\varphi = (\varphi_j)_{j=0}^{\infty}$, $\varphi_j \in \Phi^{\otimes j}$. On C_0, we introduce a quasiscalar product by setting

$$\langle \varphi, \psi \rangle = \sum_{j,k=0}^{\infty} (s_{j+k}, \psi_j \otimes \overline{\varphi}_k)_{H_0^{\otimes(j+k)}}$$

$$(\varphi, \psi \in C_0;\ (s_{j+k}, \psi_j \otimes \overline{\varphi}_k)_{H_0^{\otimes(j+k)}} = s_{j+k}(\psi_j \otimes \overline{\varphi}_k)). \tag{2.7}$$

After the factorization $\varphi \mapsto \hat{\varphi}$ with respect to $\{\varphi \in C_0 \mid \langle \varphi, \psi \rangle = 0\}$ and subsequent completion, we obtain a Hilbert space H_s. Let $e = \bar{e} \in \Phi$. We consider an operation of shift

$$\mathcal{A}_e \varphi = \mathcal{A}_e(\varphi_0, \varphi_1, \varphi_2, \ldots) = (0, e \otimes \varphi_0, e \otimes \varphi_1, e \otimes \varphi_2, \ldots) \tag{2.8}$$

on the sequences $\varphi = (\varphi_j)_{j=0}^{\infty}$ $(\varphi_j \in \Phi^{\otimes j})$. By virtue of the symmetricity of s_n, it is easy to verify that

$$\langle \mathcal{A}_e \varphi, \psi \rangle = \langle \varphi, \mathcal{A}_e \psi \rangle, \quad \langle \mathcal{A}_e \mathcal{A}_l \varphi, \psi \rangle = \langle \mathcal{A}_l \mathcal{A}_e \varphi, \psi \rangle,$$

$$(\varphi, \psi \in C_0;\ e = \bar{e},\ l = \bar{l} \in \Phi). \tag{2.9}$$

To do this, it suffices to consider a vector φ all components of which are zero, except the nth one which is equal to

$$\varphi^{(1)} \otimes \ldots \otimes \varphi^{(n)} \quad (\varphi^{(1)}, \ldots, \varphi^{(n)} \in \Phi)$$

for $n \in \mathbb{N}$ and $\varphi_0 \in \mathbb{C}^1$ for $n = 0$; ψ is an analogous vector. The first equality (2.9) shows that the correspondence $C_0 \ni \varphi \mapsto \mathcal{A}_e \varphi \in C_0$ generates an Hermitian operator in H_s (see Subsection 1.8); its closure is denoted by A_e. Clearly, $A_e \hat{C}_0 \subset \hat{C}_0$ and the second inequality in (2.9) implies that the operators A_e and A_l commute on $\hat{C}_0$.

II. Consider the case of a definite moment sequences. Let us show that in this case any two operators A_e and A_l $(e, l \in \mathrm{E})$ are selfadjoint and commute.

We employ Theorem 1.15 setting in it $H = H_s, A_1 = A_e, A_2 = A_l$, and

$$\mathfrak{D} = \hat{C}_0 \subset \mathfrak{D}(A_1) \cap \mathfrak{D}(A_2).$$

Let

$$\varphi = \Big(\underbrace{0, \ldots, 0}_{p}, \varphi_p, 0, 0, \ldots\Big) \in C_0,$$

where $\varphi_p = \varphi^{(1)} \otimes \ldots \otimes \varphi^{(p)}$ and $\varphi^{(1)}, \ldots, \varphi^{(p)} \in \mathrm{E}$. The linear span of vectors of the form $\hat{\varphi}$, where $p \in \mathbb{Z}_+$ (for $p = 0$, we have $\varphi = (\varphi_0, 0, 0, \ldots)$), is dense in H_s, and $\varphi \in \bigcap_{n=1} \mathfrak{D}(A_e^n)$. Each vector of this sort is quasianalytical with respect to A_e. Indeed, we have (see (2.4) and (2.5))

$$\| A_e^n \hat{\varphi} \|_{H_s} = \langle \mathcal{A}_e^n \varphi, \mathcal{A}_e^n \varphi \rangle^{1/2} = \Big(s_{2(p+n)}, \underbrace{e \otimes \ldots \otimes e}_{n} \otimes \varphi_p$$

$$\otimes \overline{\underbrace{e \otimes \ldots \otimes e}_{n} \otimes \varphi_p} \Big)^{1/2}_{H_0^{\otimes 2(p+n)}} \le \| s_{2(p+n)} \|^{1/2}_{H^{\otimes 2(p+n)}_{-\tau(2(p+n))}}$$

$$\times \| e \|^n_{H_{\tau(2(p+n))}} \prod_{j=1}^{p} \| \varphi^{(j)} \|_{H_{\tau(2(p+n))}}$$

$$\le \| s_{2(p+n)} \|^{1/2}_{H^{\otimes 2(p+n)}_{-\tau(2(p+n))}} (d(\tau(2(p+n)), \mathrm{E}))^{p+n} = m_{p+n} \quad (n \in \mathbb{N}). \tag{2.10}$$

However, the quasianalyticity of the class $C\{m_n\}$ implies the quasianalyticity of $C\{m_{p+n}\}$ ($p \in \mathbb{Z}_+$ is fixed). This follows from the criterion (1.32) of quasianalyticity because one can always assume that $m_n \ge 1$ $(n \in \mathbb{N})$ and, thus,

$$m_{p+k}^{1/k} \leq m_{p+k}^{1/p+k} \qquad (k \in \mathbb{N}).$$

But then the estimate (2.10) means that the vector $\hat{\varphi}$ is quasianalytical with respect to A_e (and also with respect to A_l). Thus, A_e and A_l possess a total set of quasianalytical vectors.

We now consider the restriction of A_e to $(A_l - 1\, z)\hat{C}_0$ $(z \in \mathbb{C}^1 \setminus \mathbb{R}^1)$, and denote it by B. Assume that a fixed φ has the previous form

$$\psi = (\mathcal{A}_l - z\,1)\varphi = \Big(\underbrace{0, \ldots, 0}_{p}, -z\,\varphi_p, \, l \otimes \varphi_p, 0, 0, \ldots\Big).$$

Then

$$\hat{\psi} \in \bigcap_{n=1}^{\infty} \mathcal{D}(B^n).$$

Indeed, by virtue of (2.9), for arbitrary $\varphi \in C_0$ we have

$$\mathcal{A}_e (\mathcal{A}_l - z\,1)\varphi = (\mathcal{A}_l - z\,1)\,\mathcal{A}_e\,\varphi + \theta,$$

where $\langle \theta, \theta \rangle = 0$, $\mathcal{A}_e\,\varphi \in C_0$. Hence, $B\,\mathcal{D}(B) \subset \mathcal{D}(B)$ and

$$\bigcap_{n=1}^{\infty} \mathcal{D}(B^n) \supset \mathcal{D}(B) \ni \hat{\psi}.$$

Further, by virtue of (2.9) and the Cauchy–Bunyakovsky inequality for $\langle \cdot, \cdot \rangle$, we get

$$\begin{aligned} \| B^n \hat{\psi} \|_{H_s}^2 &= \langle \mathcal{A}_e^n \psi, \mathcal{A}_e^n \psi \rangle = \langle \mathcal{A}_e^n \mathcal{A}_l \varphi, \mathcal{A}_e^n \mathcal{A}_l \varphi \rangle \\ &- 2 \operatorname{Re} z \langle \mathcal{A}_e^n \mathcal{A}_l \varphi, \mathcal{A}_e^n \varphi \rangle + |z|^2 \langle \mathcal{A}_e^n \varphi, \mathcal{A}_e^n \varphi \rangle \leq \langle \mathcal{A}_e^n \mathcal{A}_l \varphi, \mathcal{A}_e^n \mathcal{A}_l \varphi \rangle \\ &+ 2\,|z|\,\langle \mathcal{A}_e^n \mathcal{A}_l \varphi, \mathcal{A}_e^n \mathcal{A}_l \varphi \rangle^{1/2} \langle \mathcal{A}_e^n \varphi, \mathcal{A}_e^n \varphi \rangle^{1/2} + |z|^2 \langle \mathcal{A}_e^n \varphi, \mathcal{A}_e^n \varphi \rangle \\ &\leq c_1 (\langle \mathcal{A}_e^n \mathcal{A}_l \varphi, \mathcal{A}_e^n \mathcal{A}_l \varphi \rangle + \langle \mathcal{A}_e^n \varphi, \mathcal{A}_e^n \varphi \rangle) \qquad (n \in \mathbb{N}). \end{aligned} \tag{2.11}$$

But

$$\langle \mathcal{A}_e^n \varphi, \mathcal{A}_e^n \varphi \rangle^{1/2n} \leq \langle \mathcal{A}_e^{n+1} \varphi, \mathcal{A}_e^{n+1} \varphi \rangle^{1/2(n+1)} \langle \varphi, \varphi \rangle^{1/2(n+1)} \qquad (n \in \mathbb{N}),$$

this inequality is proved just as the fact that the sequence (1.34) is nondecreasing (see the proof of Lemma 1.3; this fact is a consequence of the first relation (2.9) and the

Cauchy–Bunyakovsky inequality for $\langle \cdot, \cdot \rangle$; φ should be normalized). Therefore, in the case of

$$\langle \mathcal{A}_e^{n+1} \varphi, \mathcal{A}_e^{n+1} \varphi \rangle \geq 1,$$

the relation (2.11) can be continued as follows

$$\| B^n \hat{\psi} \|_{H_s}^2 \leq c_1 (\langle \mathcal{A}_e^n \mathcal{A}_l \varphi, \mathcal{A}_e^n \mathcal{A}_l \varphi \rangle + \langle \mathcal{A}_e^{n+1} \varphi, \mathcal{A}_e^{n+1} \varphi \rangle \langle \varphi, \varphi \rangle^{1/n+1}),$$

and in the case of

$$\langle \mathcal{A}_e^{n+1} \varphi, \mathcal{A}_e^{n+1} \varphi \rangle < 1,$$

we have

$$\| B^n \hat{\psi} \|_{H_s}^2 \leq c_1 (\langle \mathcal{A}_e^n \mathcal{A}_l \varphi, \mathcal{A}_e^n \mathcal{A}_l \varphi \rangle + \langle \varphi, \varphi \rangle^{1/n+1}).$$

In the general case, by using the estimate of the type (2.10), we obtain

$$\| B^n \hat{\psi} \|_{H_s}^2 \leq c_2 \| s_{2(p+n+1)} \|_{H_{-\tau(2(p+n+1))}^{\otimes 2(p+n+1)}} (d(\tau(2(p+n+1)), \mathrm{E}))^{2(p+n+1)} + c_3$$

$$= c_2 m_{p+1+n}^2 + c_3 \qquad (n \in \mathbb{N})$$

for some $c_2, c_3 > 0$. This estimate and the quasianalyticity of the class $C\{m_n\}$ imply that the class $C\{\| B^n \hat{\psi} \|_{H_s}\}$ is quasianalytical, i.e., the vector $\hat{\psi}$ is quasianalytical with respect to B. Due to the selfadjointness of the operator A_l, the linear span of the vectors $\hat{\psi}$ with variable vectors φ of the indicated structure is dense in H_s. Therefore, conditions of Theorem 1.15 are satisfied, and this proves the selfadjointness of A_e and A_l and their commutativity.

Below, we construct the decomposition of a certain countable family of commuting selfadjoint operators in generalized joint eigenvectors. In the case of a definite moment sequence s, this family has the form $A = (A_{e_\alpha})_{\alpha=1}^\infty$, where $(e_\alpha)_{\alpha=1}^\infty$ is some sequence of vectors $e_\alpha \in \mathrm{E}$ such that the linear span of $((e_\alpha)_{\alpha=1}^\infty)$ is dense in Φ. By virtue of what has been proved above, the operators A_{e_α} are selfadjoint and commuting.

We also note the following fact. In the space H_s, the involution $H_s \ni f \mapsto f^\circ \in H_s$ is defined; it is constructed by means of the transition to a factor space and the extension from the mapping

$$C_0 \ni (\varphi_j)_{j=0}^\infty \mapsto (\overline{\varphi}_j)_{j=0}^\infty \in C_0$$

by continuity (here we should take into account that s_n are real, i.e., $\overline{s}_n = s_n, n \in \mathbb{Z}_+$). Due to the fact that $\overline{e}_\alpha = e_\alpha$, the operators A_{e_α} $(\alpha \in \mathbb{N})$ are real with respect to this involution.

III. Consider the case of a quasidefinite moment sequence s. Assume that a sequence $(e_\alpha)_{\alpha=2}^{\infty}$ from E and a vector $e_1 = \bar{e}_1$ are such that the linear span of $((e_\alpha)_{\alpha=1}^{\infty})$ is dense in Φ. Let us construct a family of Hermitian operators $(A_{e_\alpha})_{\alpha=1}^{\infty}$. It follows from Step II that the subfamily $(A_{e_\alpha})_{\alpha=2}^{\infty}$ consists of selfadjoint commuting operators real with respect to the involution "$\circ$". Further, the following lemma is valid.

Lemma 2.1. *The operator A_{e_1} can be extended to a selfadjoint operator A_{e_1} which commutes with each operator A_{e_α} $(\alpha = 2, 3, \ldots)$ and is real with respect to the involution* "$\circ$".

It is convenient to present the proof of this lemma later.

Here we take $A = (A_{e_\alpha})_{\alpha=1}^{\infty}$ to be a family of commuting selfadjoint operators, and A_{e_1} is understood as the operator B_{e_1} appearing in Lemma 2.1.

Below, we give the proof for the case of definite s with the corresponding remarks concerning the quasidefinite case.

IV. Let us construct a quasinuclear chain

$$H_{-,s} \supset H_s \supset H_{+,s} \supset D_s, \tag{2.12}$$

connected with A in the standard way. To do this, we denote by $(\tau'(n))_{n=0}^{\infty}$ a sequence of indices $\tau'(n) \in \mathrm{T}$ such that the imbedding $H_{\tau'(n)} \subset H_{\tau(n)}$ is quasinuclear for any n. Let d_n be a Hilbert–Schmidt norm of this imbedding operator. Consider a sequence of numbers $(p_j)_{j=0}^{\infty}$ and $(q_j)_{j=0}^{\infty}$ for which

$$p_j,\ q_j \geq 1,\ \sum_{j=0}^{\infty} \frac{1}{p_j} \leq 1,\quad \sum_{j=0}^{\infty} \| s_{2j} \|_{H_{-\tau(2j)}^{\otimes 2j}} d_{2j}^{2j} \frac{p_j}{q_j}, \tag{2.13}$$

and construct (by the completion of C_0) a Hilbert space H_+ of sequences $\varphi = (\varphi_j)_{j=0}^{\infty}$ $(\varphi_j \in H_{\tau'(2j)}^{\otimes j})$ with the scalar product

$$(\varphi, \psi)_{H_+} = \sum_{j=0}^{\infty} (\varphi_j, \psi_j)_{H_{\tau'(2j)}^{\otimes j}} q_j.$$

We now show that the factorization $\varphi \mapsto \hat{\varphi}$ of the space H_+ with respect to $\{\varphi \in H_+ \mid \langle \varphi, \psi \rangle = 0\}$ is possible and gives the required space $H_{+,s}$.

Let us construct an auxiliary space G_+ of sequences

$$\varphi = (\varphi_j)_{j=0}^{\infty}\ (\varphi_j \in H_{\tau(2j)}^{\otimes j})$$

with the scalar product

$$(\varphi, \psi)_{G_+} = \sum_{j=0}^{\infty} (\varphi_j, \psi_j)_{H^{\otimes j}_{\tau(2j)}} \| s_{2j} \|_{H^{\otimes 2j}_{-\tau(2j)}} p_j .$$

By virtue of (2.13) and the estimate

$$|(s_{j+k}, \varphi_j \otimes \overline{\varphi}_k)_{H_0^{\otimes(j+k)}}|^2 \le (s_{2k}, \varphi_k \otimes \overline{\varphi}_k)_{H_0^{\otimes 2k}} (s_{2j}, \varphi_j \otimes \overline{\varphi}_j)_{H_0^{\otimes 2j}}$$

$$(\varphi_k \in \Phi^{\otimes k}, \ \varphi_j \in \Phi^{\otimes j}; \ j, \ k \in \mathbb{Z}_+),$$

which follows from the Cauchy–Bunyakovsky inequality, we obtain

$$\langle \varphi, \varphi \rangle = \sum_{j,k=0}^{\infty} (s_{j+k}, \varphi_j \otimes \overline{\varphi}_k)_{H_0^{\otimes(j+k)}} \le \sum_{j,k=0}^{\infty} |(s_{j+k}, \varphi_j \otimes \overline{\varphi}_k)_{H_0^{\otimes(j+k)}}|$$

$$\le \left(\sum_{j=0}^{\infty} (s_{2j}, \varphi_j \otimes \overline{\varphi}_j)^{1/2}_{H_0^{\otimes 2j}} \right)^2 \le \left(\sum_{j=0}^{\infty} \| s_{2j} \|^{1/2}_{H^{\otimes 2j}_{-\tau(2j)}} \| \varphi_j \|_{H^{\otimes j}_{\tau(2j)}} \right)^2$$

$$\le \left(\sum_{j=0}^{\infty} \frac{1}{p_j} \right) \left(\sum_{j=0}^{\infty} \| \varphi_j \|^2_{H^{\otimes j}_{\tau(2j)}} \| s_{2j} \|_{H^{\otimes 2j}_{-\tau(2j)}} p_j \right) \le \| \varphi \|^2_{G_+}$$

for $\varphi \in C_0$. Therefore, the mapping $C_0 \ni \varphi \mapsto \hat{\varphi} \in \hat{C}_0$ acts from G_+ into H_s continuously and has the norm $n \le 1$. Extending it to the whole of G_+ by continuity, we get the mapping $G_+ \to H_s$ with the same properties.

Further, according to (2.13),

$$\| \varphi \|^2_{G_+} \le \sum_{j=0}^{\infty} \| \varphi_j \|^2_{H^{\otimes j}_{\tau'(2j)}} d_{2j}^{2j} \| s_{2j} \|_{H^{\otimes 2j}_{-\tau(2j)}} p_j$$

$$\le \sum_{j=0}^{\infty} \| \varphi_j \|^2_{H^{\otimes j}_{\tau'(2j)}} q_j = \| \varphi \|^2_{H_+} \qquad (\varphi \in H_+),$$

and, therefore, the imbedding $H_+ \to G_+$ is continuous and has the norm ≤ 1. Let us show that it is also quasinuclear with the Hilbert–Schmidt norm ≤ 1. Denote by $(e_j^k)_{k=1}^{\infty}$ an orthonormal basis in the space $H^{\otimes j}_{\tau'(2j)}$ $(j \in \mathbb{N})$; then the vectors

$$(1, 0, 0, \ldots), \ e_{(j,k)} = \Big(\underbrace{0, \ \ldots, \ 0}_{j}, \ q_j^{-1/2} e_j^k, 0, 0, \ldots \Big) \quad (j, k \in \mathbb{N})$$

form an orthonormal basis in H_+. The required statement follows from the inequality (see (2.13))

$$\sum_{j,k=1}^{\infty} \| e_{(j,k)} \|_{G_+}^2 = \sum_{j,k=1}^{\infty} \| e_j^k \|_{H_{\tau(2j)}^{\otimes j}}^2 \| s_{2j} \|_{H_{-\tau(2j)}^{\otimes 2j}} p_j / q_j$$

$$= \sum_{j=1}^{\infty} \left(\sum_{k=1}^{\infty} \| e_j^k \|_{H_{\tau(2j)}^{\otimes j}}^2 \right) \| s_{2j} \|_{H_{-\tau(2j)}^{\otimes 2j}} p_j / q_j$$

$$= \sum_{j=1}^{\infty} d_{2j}^{2j} \| s_{2j} \|_{H_{-\tau(2j)}^{\otimes 2j}} p_j / q_j \leq 1.$$

The above-mentioned facts imply that the mapping $H_+ \to G_+ \to H_s$ is quasinuclear (with norm ≤ 1) as a superposition of the quasinuclear mapping $H_+ \to G_+$ and the continuous mapping $G_+ \to H_s$. On the other hand, this yields the continuity of a bilinear form

$$H_+ \times H_+ \ni (\varphi, \psi) \mapsto \langle \varphi, \psi \rangle \in \mathbb{C}^1$$

and, hence, the fact that the subspace

$$\{ \varphi \in H_+ \mid \langle \varphi, \varphi \rangle = 0 \}$$

is closed in H_+. Let us denote by $H_{+,s}$ the factor space of H_+ with respect to this subspace. We can assume that it is imbedded in H_s densely and quasinuclearly; then

$$\| \hat{\varphi} \|_{H_s} \leq \| \hat{\varphi} \|_{H_{+,s}} \quad (\hat{\varphi} \in H_{+,s}).$$

This means that the quasinuclear chain $H_{-,s} \supset H_s \supset H_{+,s}$ is constructed.

To construct D_s, we take C_0 with the natural topology of uniformly finite coordinatewise convergence (each jth coordinate converges in $\Phi^{\otimes j}$) and then factorize C_0 with respect to $\{ \varphi \in C_0 \mid \langle \varphi, \varphi \rangle = 0 \}$, introducing a topology of factor space in $\hat{C}_0$. We set $D_s = \hat{C}_0$; obviously, D_s is separable. The chain (2.12) obtained above is connected with the family A in the standard way (in connection with what has been said on Step III, we note that $B_{e_1} \upharpoonright D_s$ coincides with the initial Hermitian operator).

We have given a somewhat schematic exposition of the final steps in the construction of the chain (2.12). This construction is presented in more detail for the general case in Subsections 5.1 and 5.2.

V. Let us apply Theorem 2.6 of Chapter 3 to the family of operators $(A_{e_\alpha})_{\alpha=1}^{\infty}$. If we write the representation (2.21) of Chapter 3, then we get

$$O^+ O = \int_{\mathbb{R}^\infty} P(\lambda(\cdot))\, d\rho(\lambda(\cdot)), \tag{2.14}$$

where $P(\lambda(\cdot)) : H_{+,s} \to H_{-,s}$ is a generalized projector, and $\mathfrak{B}(\mathbb{R}^\infty) \ni \alpha \mapsto \rho(\alpha)$ is a spectral measure. There exists a set $\pi \in \mathfrak{B}(\mathbb{R}^\infty)$ of full spectral measure such that, for $\lambda(\cdot) \in \pi$ (see (2.20), Chapter 3), we have

$$(P(\lambda(\cdot))\, \hat{\varphi}, A_{e_\alpha} \hat{\psi})_{H_s} = \lambda(\alpha)(P(\lambda(\cdot))\, \hat{\varphi}, \hat{\psi})_{H_s} \tag{2.15}$$

$$(\alpha \in \mathbb{N};\ \hat{\varphi} \in H_{+,s},\ \hat{\psi} \in D_s).$$

VI. In Step VII, we shall write the relation (2.15) in the required form. For this purpose, parallel with the chain (2.12), we consider a chain

$$H_- \supset G_0 \supset H_+, \tag{2.16}$$

where

$$G_0 = \bigoplus_{j=0}^{\infty} H_0^{\otimes j},$$

the space H_- is constructed like H_+, but $H_{\tau'(2j)}^{\otimes j}$ and q_j should be replaced by $H_{-\tau'(2j)}^{\otimes j}$ and q_j^{-1}, respectively. The mapping

$$(\varphi_j)_{j=0}^{\infty} = \varphi \mapsto \overline{\varphi}\ (\overline{\varphi}_j)_{j=0}^{\infty}$$

transforms the chain (2.16) into a chain with the involution — in the natural way. We construct its tensor square

$$H_- \otimes H_- \supset G_0 \otimes G_0 \supset H_+ \otimes H_+;$$

the elements of $H_- \otimes H_-$ are the matrices $L = (L_{jk})_{j,k=0}^{\infty}$, where

$$L_{jk} \in H_{-\tau'(2j)}^{\otimes j} \otimes H_{-\tau'(2k)}^{\otimes k},$$

and

$$\| L \|_{H_- \otimes H_-}^2 = \sum_{j,k=0}^{\infty} \| L_{jk} \|_{H_{-\tau'(2j)}^{\otimes j} \otimes H_{-\tau'(2k)}^{\otimes k}} \frac{1}{q_j q_k}. \tag{2.17}$$

Let us transform the expression $(P(\lambda(\cdot))\, \hat{\varphi}, \hat{\psi})_{H_s}$ $(\lambda \in \pi;\ \hat{\varphi}, \hat{\psi} \in H_{+,s})$. Assume that $I_1: H_{-,s} \to H_{+,s}$ is an isometric operator connected with the chain (2.12) and Q is an operator of orthogonal projection $\varphi \mapsto \hat{\varphi}$ in H_+ onto $H_{+,s}$ (we note that $H_{+,s}$ can be regarded as an orthogonal completion to the subspace $\{\varphi \in H_+ \mid \langle\varphi, \varphi\rangle = 0\}$ in H_+). Then

$$(P(\lambda(\cdot))\,\hat{\varphi}, \hat{\psi})_{H_s} = (I_1 P(\lambda(\cdot))\,\hat{\varphi}, \hat{\psi})_{H_{+,s}} = ((QI_1 P(\lambda(\cdot))\,Q\hat{\varphi}, \hat{\psi})_{H_+}$$

$$(\varphi, \psi \in H_+).$$

The operator

$$QI_1 P(\lambda(\cdot))\,Q: H_+ \to H_+$$

is quasinuclear. Consequently, there exists $F(\lambda(\cdot)) \in H_+ \otimes H_+$ such that

$$(QI_1 P(\lambda(\cdot))\,Q\varphi, \psi)_{H_+} = (F(\lambda(\cdot)), \psi \otimes \overline{\varphi})_{H_+ \otimes H_+} \quad (\varphi, \psi \in H_+).$$

Denoting an isometric operator, associated with the corresponding chain, by

$$I_2: H_- \otimes H_- \to H_+ \otimes H_+,$$

we obtain

$$(F(\lambda(\cdot)), \psi \otimes \overline{\varphi})_{H_+ \otimes H_+} = (I_2^{-1} F(\lambda(\cdot)), \psi \otimes \overline{\varphi})_{G_0 \otimes G_0}$$

$$= (\Omega(\lambda(\cdot)), \psi \otimes \overline{\varphi})_{G_0 \otimes G_0},$$

where

$$\Omega(\lambda(\cdot)) = I_2^{-1} F(\lambda(\cdot)) \in H_- \otimes H_-$$

are so-called elementary kernels. Finally,[1]

$$(P(\lambda(\cdot))\,\hat{\varphi}, \hat{\psi})_{H_s} = (\Omega(\lambda(\cdot)), \psi \otimes \overline{\varphi})_{G_0 \otimes G_0} \quad (\varphi, \psi \in H_+). \tag{2.18}$$

VII. We rewrite (2.15) in terms of elementary kernels. Let $\lambda(\cdot) \in \pi$. Since $\Omega(\lambda(\cdot)) \in H_- \otimes H_-$, according to (2.17), we obtain

$$\Omega(\lambda(\cdot)) = (\Omega_{jk}(\lambda(\cdot)))_{j,k=0}^{\infty}, \quad \Omega_{jk}(\lambda(\cdot)) \in H_{-\tau'(2j)}^{\otimes j} \otimes H_{-\tau'(2k)}^{\otimes k}; \tag{2.19}$$

these are positive definite in the sense (2.1) due to the positive definiteness of $P(\lambda(\cdot))$.

[1] In Subsection 5. 2, a similar transition from $P(\lambda(\cdot))$ to $\Omega(\lambda(\cdot))$ is presented for the more general situation.

Further, rewriting the relation (2.15) by using (2.18) and taking into account the form (2.8) of the operation A_{e_α}, we get

$$(\Omega_{j+1\,k}(\lambda(\cdot)), e_\alpha \otimes \psi_j \otimes \varphi_k)_{H_0^{\otimes(j+k+1)}} = \lambda(\alpha)(\Omega_{jk}(\lambda(\cdot)), \psi_j \otimes \varphi_k)_{H_0^{\otimes(j+k)}}, \quad (2.20)$$

$$(\Omega_{j\,k+1}(\lambda(\cdot)), \psi_j \otimes e_\alpha \otimes \varphi_k)_{H_0^{\otimes(j+k+1)}} = \lambda(\alpha)(\Omega_{jk}(\lambda(\cdot)), \psi_j \otimes \varphi_k)_{H_0^{\otimes(j+k)}} \quad (2.21)$$

$$(\varphi_k \in H^{\otimes k}_{\tau'(2k)},\ \psi_j \in H^{\otimes j}_{\tau'(2j)};\ j, k \in \mathbb{Z}_+;\ \alpha \in \mathbb{N};\ \lambda(\cdot) \in \pi)$$

(the second of these inequalities follows from the first one because $\Omega(\lambda(\cdot))$ is Hermitian, namely,

$$\Omega_{jk}(\lambda(\cdot)) = \overline{\Omega_{kj}(\lambda(\cdot))} \qquad (j, k \in \mathbb{Z}_+).$$

As mentioned in Step II, the operators A_{e_α} $(\alpha \in \mathbb{N})$ are real with respect to the involution "∘", and therefore, $P(\lambda(\cdot))$ is also real, i.e.,

$$\overline{\Omega_{jk}(\lambda(\cdot))} = \Omega_{jk}(\lambda(\cdot)) \quad (j, k \in \mathbb{Z}_+;\ \lambda(\cdot) \in \pi).$$

Hence, one may use the indices Re in the notations of spaces in (2.19).

VIII. We proceed to solve the set of equations (2.20) and (2.21). It is convenient to formulate the result in the form of the following lemma.

Lemma 2.2. *Assume that the matrix* (2.19) *with real elements satisfies* (2.20) *and* (2.21) *for fixed* $\lambda(\cdot) \in \mathbb{R}^\infty$. *Then* $\Omega_{jk}(\lambda(\cdot))$ *can be extended to elements in* $H^{\otimes(j+k)}_{-\tau'(2),\,\mathrm{Re}}$ *and have the following form*

$$\Omega_{jk}(\lambda(\cdot)) = \underbrace{\lambda \otimes \ldots \otimes \lambda}_{j+k}\, \Omega_{00}(\lambda(\cdot)) \ (j, k \in \mathbb{Z}_+), \quad (2.22)$$

where

$$\lambda = \Omega_{10}(\lambda(\cdot)) = \Omega_{01}(\lambda(\cdot)) \in H_{-\tau'(2),\,\mathrm{Re}};$$

and

$$\lambda(\alpha) = (\lambda, e_\alpha)_{H_0} \quad (\alpha \in \mathbb{N}).$$

Proof. The positive definiteness of $\Omega(\lambda(\cdot))$ implies that $\Omega_{00}(\lambda(\cdot)) \geq 0$. If $\Omega_{00}(\lambda(\cdot)) = 0$, then, as will be clear from what follows, we have $\Omega_{jk}(\lambda(\cdot)) = 0$ $(j, k \in \mathbb{Z}_+)$ and, hence, (2.22) is valid. In the case of $\Omega_{00}(\lambda(\cdot)) > 0$ which is studied below, without loss of generality, we can assume that $\Omega_{00}(\lambda(\cdot)) = 1$.

Setting $j = k = 0$ in (2.20) and (2.21), we obtain

$$(\Omega_{10}(\lambda(\cdot)), e_\alpha)_{H_0} = \lambda(\alpha) = (\Omega_{01}(\lambda(\cdot)), e_\alpha)_{H_0} \quad (\alpha \in \mathbb{N}). \tag{2.23}$$

Taking into account the denseness of the linear span of $((e_\alpha)_{\alpha=1}^\infty)$ in Φ_{Re} (and, hence, in $H_{\tau'(2),\mathrm{Re}}$) and the inclusions $\Omega_{10}(\lambda(\cdot)), \Omega_{01}(\lambda(\cdot)) \in H_{-\tau'(2),\mathrm{Re}}$, we conclude from (2.23) that

$$\Omega_{10}(\lambda(\cdot)) = \Omega_{01}(\lambda(\cdot)) = \lambda \in H_{-\tau'(2),\mathrm{Re}}$$

and

$$\lambda(\alpha) = (\lambda, e_\alpha)_{H_0} \quad (\alpha \in \mathbb{N}).$$

Thus, (2.22) is established for $j + k = 1$.

Consider now the cases $j + k = 2, 3, \ldots$.

The case of $j + k = 2$. We set $j = 1$ and $k = 0$ in (2.20) and replace ψ_1 by e_β. Taking (2.23) into account, we get

$$(\Omega_{20}(\lambda(\cdot)), e_\alpha \otimes e_\beta)_{H_0^{\otimes 2}} = \lambda(\alpha)(\Omega_{01}(\lambda(\cdot)), e_\beta)_{H_0} = \lambda(\alpha)\lambda(\beta)$$

$$= (\lambda, e_\alpha)_{H_0} (\lambda, e_\beta)_{H_0} = (\lambda \otimes \lambda, e_\alpha \otimes e_\beta)_{H_0^{\otimes 2}}$$

$$(\alpha, \beta \in \mathbb{N}). \tag{2.24}$$

Since the real linear span of $((e_\alpha \otimes e_\beta)_{\alpha,\beta=1}^\infty)$ is dense in $H_{\tau'(2),\mathrm{Re}}^{\otimes 2}$ and $H_{\tau'(4),\mathrm{Re}}^{\otimes 2}$, and $\Omega_{20}(\lambda(\cdot)) \in H_{-\tau'(4),\mathrm{Re}}^{\otimes 2}$ and $\lambda \otimes \lambda \in H_{-\tau'(2),\mathrm{Re}}^{\otimes 2}$, it follows from (2.24) that $\Omega_{20}(\lambda(\cdot)) = \lambda \otimes \lambda$.

We set $j = 0$ and $k = 1$ in (2.20) and replace φ_1 by e_β. Taking (2.23) into account, we obtain

$$(\Omega_{11}(\lambda(\cdot)), e_\alpha \otimes e_\beta)_{H_0^{\otimes 2}} = \lambda(\alpha)(\Omega_{01}(\lambda(\cdot)), e_\beta)_{H_0} = \lambda(\alpha)\lambda(\beta)$$

$$= (\lambda \otimes \lambda, e_\alpha \otimes e_\beta)_{H_0^{\otimes 2}} \quad (\alpha, \beta \in \mathbb{N}). \tag{2.25}$$

Since the real linear span of $((e_\alpha \otimes e_\beta)_{\alpha,\beta=1}^\infty)$ is dense in $H_{\tau'(2),\mathrm{Re}}^{\otimes 2}$ and $\Omega_{11}(\lambda(\cdot))$, $\lambda \otimes \lambda \in H_{-\tau'(2),\mathrm{Re}}^{\otimes 2}$, it follows from (2.25) that $\Omega_{11}(\lambda(\cdot)) = \lambda \otimes \lambda$.

We set $j = 0$ and $k = 1$ in (2.21) and replace φ_1 by e_β. Taking into account

(2.23), we get

$$(\Omega_{02}(\lambda(\cdot)), e_\alpha \otimes e_\beta)_{H_0^{\otimes 2}} = \lambda(\alpha)(\Omega_{01}(\lambda(\cdot)), e_\beta)_{H_0} = \lambda(\alpha)\lambda(\beta)$$

$$= (\lambda \otimes \lambda, e_\alpha \otimes e_\beta)_{H_0^{\otimes 2}} \quad (\alpha, \beta \in \mathbb{N}).$$

As in the case when $j = 1$ and $k = 0$, we conclude that $\Omega_{02}(\lambda(\cdot)) = \lambda \otimes \lambda$.

The case of $j + k = 3$. We set $j = 2$ and $k = 0$ in (2.20) and replace ψ_2 by $e_\beta \otimes e_\gamma$. Taking into account (2.24) and (2.23), we obtain

$$(\Omega_{30}(\lambda(\cdot)), e_\alpha \otimes e_\beta \otimes e_\gamma)_{H_0^{\otimes 3}}$$

$$= \lambda(\alpha)(\Omega_{20}(\lambda(\cdot)), e_\beta \otimes e_\gamma)_{H_0^{\otimes 2}} = \lambda(\alpha)\lambda(\beta)\lambda(\gamma)$$

$$= (\lambda \otimes \lambda \otimes \lambda, e_\alpha \otimes e_\beta \otimes e_\gamma)_{H_0^{\otimes 3}} \quad (\alpha, \beta, \gamma \in \mathbb{N}). \tag{2.26}$$

Since the real linear span of $((e_\alpha \otimes e_\beta \otimes e_\gamma)_{\alpha, \beta, \gamma=1}^{\infty})$ is dense in $H_{\tau'(6), \mathrm{Re}}^{\otimes 3}$ and $H_{\tau'(2), \mathrm{Re}}^{\otimes 3}$, and

$$\Omega_{30}(\lambda(\cdot)) \in H_{-\tau'(6), \mathrm{Re}}^{\otimes 3} \quad \text{and} \quad \lambda \otimes \lambda \otimes \lambda \in H_{-\tau'(2), \mathrm{Re}}^{\otimes 3},$$

it follows from (2.26) that $\Omega_{30}(\lambda(\cdot)) = \lambda \otimes \lambda \otimes \lambda$.

It is now clear from the above argument how to complete the proof of (2.22) for $j + k = 3$ and how to investigate the general case $j + k = 4, 5, \ldots$.

■

IX. Using (2.7), (2.18), and (2.22), from formula (2.14) written for vectors, we find that [2]

$$\sum_{j,k=0}^{\infty} (s_{j+k}, \psi_j \otimes \overline{\varphi}_k)_{H_0^{\otimes(j+k)}} = \langle \hat{\varphi}, \hat{\psi} \rangle_{H_s}$$

$$= \int_{\mathbb{R}^\infty} (P(\lambda(\cdot))\hat{\varphi}, \hat{\psi})_{H_s} d\rho(\lambda(\cdot)) = \int_{\pi} (P(\lambda(\cdot))\hat{\varphi}, \hat{\psi})_{H_s} d\rho(\lambda(\cdot))$$

$$= \int_{\pi} (\Omega(\lambda(\cdot)), \psi \otimes \overline{\varphi})_{G_0 \otimes G_0} d\rho(\lambda(\cdot))$$

[2] In this calculations, all permutations of integrals, sums, and actions of functionals are valid due to the estimate $\| P(\lambda(\cdot)) \| \le 1$ and the estimates for $\Omega(\lambda(\cdot))$ which follow from it.

$$= \sum_{j,k=0}^{\infty} \int_{\pi} (\Omega_{jk}(\lambda(\cdot)), \psi_j \otimes \overline{\varphi}_k)_{H_0^{\otimes(j+k)}} d\rho(\lambda(\cdot))$$

$$= \sum_{j,k=0}^{\infty} \left(\int_{\pi} \underbrace{\lambda \otimes \ldots \otimes \lambda}_{j+k} \Omega_{00}(\lambda(\cdot))\, d\rho(\lambda(\cdot)), \psi_j \otimes \overline{\varphi}_k \right)_{H_0^{\otimes(j+k)}}$$

for any $\varphi, \psi \in C_0$, i.e.,

$$s_n = \int_{\pi} \underbrace{\lambda \otimes \ldots \otimes \lambda}_{n} \Omega_{00}(\lambda(\cdot))\, d\rho(\lambda(\cdot))$$

$$= \int_{\pi} \underbrace{\lambda \otimes \ldots \otimes \lambda}_{n}\, d\sigma_1(\lambda(\cdot)) \qquad (n \in \mathbb{Z}_+), \tag{2.27}$$

where the finite measures σ_1 is given by the equality

$$\sigma_1(\alpha) = \int_{\alpha} \Omega_{00}(\lambda(\cdot))\, d\rho(\lambda(\cdot)) \geq 0 \qquad (\alpha \in \mathfrak{B}(\mathbb{R}^{\infty}),\ \alpha \subset \pi).$$

The representation (2.27) can be now easily rewritten in the form (2.6). Consider a mapping

$$\mathbb{R}^{\infty} \supset \pi \ni \lambda(\cdot) \mapsto \varphi(\lambda(\cdot)) = \Omega_{10}(\lambda(\cdot)) = \lambda \in H_{-\tau'(2),\mathrm{Re}};$$

$$\lambda(\alpha) = (\lambda, e_{\alpha})_{H_0} \quad (\alpha \in \mathbb{N}),$$

which exists according to Lemma 2.2. Using the mapping φ, we transport the measure σ_1 into the measure σ_2 given on the corresponding σ-algebra in the space $\varphi(\pi) \subset H_{-\tau'(2),\mathrm{Re}}$. This σ-algebra consists of the intersections of cylindrical or Borel (this is just the same) subsets of the space $H_{-\tau'(2),\mathrm{Re}}$ with $\varphi(\pi)$ (one should employ the fact that $\mathfrak{B}(\mathbb{R}^{\infty}) = \mathfrak{C}_{\sigma}(\mathbb{R}^{\infty})$ and the general facts given in Subsection 1.4 of Chapter 2). It remains to set

$$\sigma(\alpha) = \sigma_2(\alpha \cap \lambda(\alpha) = \varphi(\pi)) \qquad (\alpha \in \mathfrak{B}(H_{-\tau'(2),\mathrm{Re}})),$$

and to extend this measure to $\Phi'_{\mathrm{Re}} \setminus H_{-\tau'(2),\mathrm{Re}}$ as zero.

X. Let us prove that the measure σ in the representation (2.6) is unique in the case of the definite moment problem. Assume that $\varphi \in \Phi_{\mathrm{Re}}$ is fixed. By virtue of (2.6), a number moment sequence

$$s_{\varphi, n} = (s_n, \underbrace{\varphi \otimes \ldots \otimes \varphi}_{n})_{H_0^{\otimes n}}$$

admits the representation

$$s_{\varphi, n} = \Bigg(\int_{\Phi'_{\mathrm{Re}}} \underbrace{\lambda \otimes \ldots \otimes \lambda}_{n} \, d\sigma(\lambda), \underbrace{\varphi \otimes \ldots \otimes \varphi}_{n} \Bigg)_{H_0^{\otimes n}}$$

$$= \int_{\Phi'_{\mathrm{Re}}} (\lambda, \varphi)_{H_0}^n \, d\sigma(\lambda) = \int_{\mathbb{R}^1} t^n \, d\sigma_\varphi(t) \qquad (n \in \mathbb{Z}_+) \tag{2.28}$$

(we have changed the variables

$$\Phi'_{\mathrm{Re}} \ni \lambda \mapsto (\lambda, \varphi)_{H_0} = t \in \mathbb{R}^1).$$

In the case of φ from the real linear span of E, the moment problem (2.28) is definite. Thus, here $\|\varphi\|_{H_\tau} \le c d(\tau, \mathrm{E})$ for any $\tau \in \mathrm{T}$, whence,

$$s_{\varphi,2n}^2 \le s_{\varphi, 0}\, s_{\varphi, 4n} = s_{\varphi,0} (s_{4n}, \underbrace{\varphi \otimes \ldots \otimes \varphi}_{4n})_{H_0^{\otimes 4n}}$$

$$\le s_{\varphi, 0} \| s_{4n} \|_{H_{-\tau(4n)}^{\otimes 4n}} \| \varphi \|_{H_{\tau(4n)}}^{4n}$$

$$\le s_{\varphi, 0} \| s_{4n} \|_{H_{-\tau(4n)}^{\otimes 4n}} c^{4n} d^{4n} (\tau(4n), \mathrm{E})$$

and, consequently, the class $C\{(s_{\varphi,2n})^{1/2}\}$ is quasianalytical (see (2.5)). By virtue of the argument in Example 2.1, the measure σ_φ is uniquely determined for given $s = (s_n)_{n=0}^\infty$ and φ.

Assume that there exist two representations (2.6) of the same s with measures σ' and σ''. It suffices to prove that these measures are equal on an arbitrary cylindrical set

$$\mathbb{C}(\varphi_1, \ldots, \varphi_n; \delta) = \{\lambda \in \Phi'_{\mathrm{Re}} \mid ((\lambda, \varphi_1)_{H_0}, \ldots, (\lambda, \varphi_n)_{H_0}) \in \delta\}$$

$$(\varphi_1, \ldots, \varphi_n \in \Phi_{\mathrm{Re}};\ \delta \in \mathfrak{B}(\mathbb{R}^n);\ n \in \mathbb{N}).$$

Furthermore, since the real linear span of E is dense in Φ_{Re}, it suffices to consider φ_j belonging to this linear span. We fix these $\varphi_1, \ldots, \varphi_n$. Then the measures

$$\mathfrak{B}(\mathbb{R}^n) \ni \delta \mapsto \sigma'(\mathbb{C}(\varphi_1, \ldots, \varphi_n; \delta)),\ \sigma''(\mathbb{C}(\varphi_1, \ldots, \varphi_n; \delta))$$

take the same values on an arbitrary half-space in $\mathbb{R}^n$; this follows from the equality $\sigma'_\varphi = \sigma''_\varphi$ ($\varphi \in \Phi_{\mathrm{Re}}$) established above and the regularity of these measures. But then, due to the well-known theorem (for example, see Gelfand and Vilenkin [1, Chapter 4, Section 1, Subsection 3]), these measures coincide for any $\delta \in \mathfrak{B}(\mathbb{R}^n)$.

XI. The fact that each sequence (2.6) is a moment sequence is verified by direct testing.

■

Proof of Lemma 2.1. Consider the Cayley transformations

$$U_{z_\alpha}^{(\alpha)} = (A_{e_\alpha} - \bar{z}_\alpha 1)(A_{e_\alpha} - z_\alpha 1)^{-1}$$

($z_\alpha \in \mathbb{C}^1 \setminus \mathbb{R}^1$ are fixed; $\alpha \in \mathbb{N}$). For $\alpha = 2, 3, \dots$, these operators are unitary, and the operator $U_{z_1}^{(1)}$ transforms $\mathfrak{R}(A_{e_1} - z_1 1)$ into $\mathfrak{R}(A_{\alpha_1} - \bar{z}_1 1)$ isometrically. We first show that each of these subspaces is invariant with respect to $U_{z_\alpha}^{(\alpha)}$ ($\alpha = 2, 3, \dots$), and that the above-mentioned operators are unitary on these subsets. It suffices to consider $\mathfrak{R}(A_{e_1} - z_1 1)$ and $U_{z_2}^{(2)}$. As in Step II, we conclude from (2.3) that

$$A_{e_1}(A_{e_1} - z_1 1)\hat{C}_0 \subset (A_{e_1} - z_1 1)\hat{C}_0 \subset \mathfrak{R}(A_{e_1} - z_1 1).$$

This enables us to regard the restriction of A_{e_2} to $(A_{e_1} - z_1 1)\hat{C}_0$ as an operator in the space $\mathfrak{R}(A_{e_1} - z_1 1)$. According to the argument in the proof of the result of Step II, the closure of this operator is selfadjoint in $\mathfrak{R}(A_{e_1} - z_1 1)$. Therefore, the set of vectors $(A_{e_2} - z_2 1)(A_{e_1} - z_1 1)\hat{C}_0$ is dense in $\mathfrak{R}(A_{e_1} - z_1 1)$. The operator $U_{z_2}^{(2)}$ transforms this set isometrically into the set $(A_{e_2} - \bar{z}_2 1)(A_{e_1} - z_1 1)\hat{C}_0$ which is also dense in $\mathfrak{R}(A_{e_1} - z_1 1)$ by virtue of the same argument. This yields the required result.

Therefore, both the products $U_{z_\alpha}^{(\alpha)} U_{z_1}^{(1)}$ and $U_{z_1}^{(1)} U_{z_\alpha}^{(\alpha)}$ make sense on $\mathfrak{R}(A_{e_1} - z_1 1)$. Let us show that they are equal. Assume that $\alpha = 2$; it suffices to verify this equality on $(A_{e_2} - \bar{z}_2 1)(A_{e_1} - z_1 1)\hat{C}_0$. For $\varphi \in C_0$, due to the commutativity of A_{e_2} and A_{e_1} on $\hat{C}_0$, we obtain

$$\begin{aligned} &U_{z_2}^{(2)} U_{z_1}^{(1)} (A_{e_2} - z_2 1)(A_{e_1} - z_1 1)\hat{\varphi} \\ &\qquad = U_{z_2}^{(2)} U_{z_1}^{(1)} (A_{e_1} - z_1 1)(A_{e_2} - z_2 1)\hat{\varphi} \\ &\qquad = U_{z_2}^{(2)} (A_{e_1} - \bar{z}_1 1)(A_{e_2} - z_2 1)\hat{\varphi} \end{aligned}$$

$$= U_{z_2}^{(2)}(A_{e_2} - z_2 1)(A_{e_1} - \bar{z}_1 1)\hat{\varphi}$$

$$= (A_{e_2} - \bar{z}_2 1)(A_{e_1} - \bar{z}_1 1)\hat{\varphi}.$$

Similarly, we prove that

$$U_{z_1}^{(1)} U_{z_2}^{(2)}(A_{e_2} - z_2 1)(A_{e_1} - z_1 1)\hat{\varphi} = (A_{e_2} - \bar{z}_2 1)(A_{e_1} - \bar{z}_1 1)\hat{\varphi},$$

and this yields the desired commutativity.

Denote by $N_{z_1} = H_s \ominus \mathfrak{R}(A_{e_1} - z_1 1)$ and $N_{\bar{z}_1} = H_s \ominus \mathfrak{R}(A_{e_1} - \bar{z}_1 1)$ the defect subspaces of the operator A_{e_1}. Let W be an operator which maps the whole of N_{z_1} onto the whole of $N_{\bar{z}_1}$ isometrically. Then $U_{z_1}^{(1)} \oplus W$ is the Cayley transform $V_{z_1}^{(1)}$ of some selfadjoint extension B_{e_1} of the operator A_{e_1} determined by the operator W. In order that B_{e_1} and A_{e_α} $(\alpha = 2, 3, \ldots)$ be commuting, it suffices to choose W such that $V_{z_1}^{(1)} = U_{z_1}^{(1)} \oplus W$ and $U_{z_\alpha}^{(\alpha)}$ $(\alpha = 2, 3, \ldots)$ are commuting. But, the restrictions of $U_{z_\alpha}^{(\alpha)}$ $(\alpha = 2, 3, \ldots)$ to $\mathfrak{R}(A_{e_1} - z_1 1)$ and $\mathfrak{R}(A_{\alpha_1} - \bar{z}_1 1)$ act in these spaces, and $U_{z_\alpha}^{(\alpha)} U_{z_1}^{(1)} = U_{z_1}^{(1)} U_{z_\alpha}^{(\alpha)}$. Therefore, for this sort of commutativity, it suffices that the equalities

$$U_{z_\alpha}^{(\alpha)} Wf = WU_{z_\alpha}^{(\alpha)} f \quad (f \in N_{z_1};\ \alpha = 2, 3, \ldots)$$

should hold (clearly, the restrictions of $U_{z_\alpha}^{(\alpha)}$ to N_{z_1} and $N_{\bar{z}_1}$ act in these subspaces unitarily). In other words, it is sufficient that the equalities hold

$$E_\alpha(\delta) Wf = WE_\alpha(\delta) f \quad (f \in N_{z_1};\ \delta \in \mathfrak{B}(\mathbb{R}^1);\ \alpha = 2, 3, \ldots),$$

or, which is the same, that the following equality holds

$$E(\delta) Wf = WE(\delta) f \qquad (f \in N_{z_1};\ \delta \in \mathfrak{B}(\mathbb{R}^\infty)), \tag{2.29}$$

where E_α is the RI of the operator A_{e_α} and E is the joint RI of the family $(A_{e_\alpha})_{\alpha=2}^{\infty}$. Since the operators A_{e_α} $(\alpha = 2, 3, \ldots)$ are real with respect to the involution "$\circ$", the same is true for their RI's E_α and, hence, for the joint RI E.

Further, the operator A_{e_1} is real with respect to the involution "$\circ$", and thus, $N_{z_1}^{\circ} = N_{\bar{z}_1}$. Hence, the operator W should act from the whole of N_{z_1} to the whole of $N_{z_1}^{\circ}$ isometrically and satisfy the relation (2.29). Moreover, $U_{z_1}^{(1)} \oplus W$ should be real with respect to "$\circ$" (this guarantees the validity of the same property for B_{e_1}).

Let us construct the operator W. Fix $f_1 \in N_{z_1}$ and consider

$$N_{z_1}^{(1)} = \text{c.l.s.}(\{E(\delta)f_1 \mid \delta \in \mathfrak{B}(\mathbb{R}^\infty)\}) \subset N_{z_1}.$$

If $N_{z_1}^{(1)} \neq N_{z_1}$, then we consider a vector $0 \neq f_2 \in N_{z_1}$, $f_2 \perp N_{z_1}^{(1)}$ and construct

$$N_{z_1}^{(2)} = \text{c.l.s.}(\{E(\delta)f_2 \mid \delta \in \mathfrak{B}(\mathbb{R}^\infty)\}) \subset N_{z_1}.$$

Clearly, $N_{z_1}^{(2)} \perp N_{z_1}^{(1)}$. If $N_{z_1}^{(1)} \oplus N_{z_1}^{(2)} \neq N_{z_1}$, then we consider a vector $0 \neq f_3 \in N_{z_1}$, $f_3 \perp N_{z_1}^{(1)} \oplus N_{z_1}^{(2)}$ and construct

$$N_{z_1}^{(3)} = \text{c.l.s.}(\{E(\delta)f_3 \mid \delta \in \mathfrak{B}(\mathbb{R}^\infty)\}),$$

etc. Clearly

$$N_{z_1}^{(1)} \oplus N_{z_1}^{(2)} \oplus \ldots = N_{z_1}.$$

Taking the involution "$\circ$" over this equality, we obtain

$$(N_{z_1}^{(1)})^\circ \oplus (N_{z_1}^{(2)})^\circ \oplus \ldots = N_{z_1}^\circ.$$

The invariance of $N_{z_1}^{(j)}$ and $(N_{z_1}^{(j)})^\circ$ with respect to any $E(\delta)$, implies that in order to construct the operator W, it is sufficient to construct the operator which maps the whole of $N_{z_1}^{(j)}$ onto the whole of $(N_{z_1}^{(j)})^\circ$ and satisfies (2.29) for $f \in N_{z_1}^{(j)}$ $(j \in \mathbb{N})$. It suffices to define this operator on vectors of the form

$$f = \sum_{k=1}^{n} c_k E(\delta_k) f_j \qquad (c_k \in \mathbb{C}^1;\ \delta_k \in \mathfrak{B}(\mathbb{R}^\infty);\ j \in \mathbb{N}).$$

We set

$$W\left(\sum_{k=1}^{n} c_k E(\delta_k) f_j\right) = \sum_{k=1}^{n} c_k E(\delta_k) f_j^\circ = \left(\sum_{k=1}^{n} \bar{c}_k E(\delta_k) f_j\right)^\circ \in (N_{z_1}^{(j)})^\circ.$$

The following equality is valid

$$\left(W\left(\sum_{k=1}^{n} c_k E(\delta_k) f_j\right), W\left(\sum_{l=1}^{n} d_l E(\gamma_l) f_j\right)\right)_{H_s} = \left(\left(\sum_{k=1}^{n} c_k E(\delta_k) f_j\right)^\circ,\right.$$

$$\left(\sum_{l=1}^{n} \bar{d}_l \, E(\gamma_l) f_j \right)^{\circ} \Bigg)_{H_s} = \overline{\left(\sum_{k=1}^{n} \bar{c}_k \, E(\delta_k) \, f_j \, , \, \sum_{l=1}^{n} \bar{d}_l \, E(\gamma_l) \, f_j \right)_{H_s}}$$

$$= \sum_{k,l=1}^{n} c_k \, \bar{d}_l \, \overline{(E(\delta_k) \, f_j \, , E(\gamma_l) \, f_j)_{H_s}} = \sum_{k,l=1}^{n} c_k \, \bar{d}_l \, \overline{(E(\delta_k \cap \gamma_l) \, f_j \, , f_j)_{H_s}}$$

$$= \sum_{k,l=1}^{n} c_k \, \bar{d}_l \, (E(\delta_k \cap \gamma_l) f_j , \, f_j)_{H_s} = \left(\sum_{k=1}^{n} c_k \, E(\delta_k) f_j \, , \, \sum_{l=1}^{n} d_l \, E(\gamma_l) f_j \right)_{H_s}$$

On the other hand, this equality shows that W is defined correctly: If

$$\sum_{k=1}^{n} c_k \, E(\delta_k) f_j = 0,$$

then

$$\left(\sum_{k=1}^{n} c_k \, E(\delta_k) f_j \right)^{\circ} = 0.$$

On the other hand, it yields the isometricity of W. Further, it is clear that W maps the whole of $N_{z_1}^{(j)}$ onto the whole of $(N_{z_1}^{(j)})^{\circ}$. Finally,

$$E(\delta) \, W \left(\sum_{k=1}^{n} c_k \, E(\delta_k) f_j \right) = E(\delta) \sum_{k=1}^{n} c_k \, E(\delta_k) f_j^{\circ} = \sum_{k=1}^{n} c_k \, E(\delta \cap \delta_k) f_j^{\circ},$$

$$W \left(E(\delta) \sum_{k=1}^{n} c_k \, E(\delta_k) f_j \right) = W \left(\sum_{k=1}^{n} c_k \, E(\delta \cap \delta_k) f_j \right) = \sum_{k=1}^{n} c_k \, E(\delta \cap \delta_k) f_j^{\circ},$$

i.e., (2.29) holds on

$$f = \sum_{k=1}^{n} c_k \, E(\delta_k) f_j$$

and, therefore, everywhere on $N_{z_1}^{(j)}$. The required operator W and, hence, the extension of $U_{z_1}^{(1)}$ to the unitary operator $V_{z_1}^{(1)}$ commuting with each $U_{z_\alpha}^{(\alpha)}$ $(\alpha = 2, 3, \ldots)$ are thus constructed.

■

Remark 1. It follows from the proof presented above that the measure σ in the representation (2.6) is, in fact, concentrated not on Φ'_{Re} but on some negative space

$H_{-\tau(s),\,\mathrm{Re}}$ where the index $\tau(s) \in \mathrm{T}$ depends on the moment space s (more precisely, the measure σ is given on $\mathfrak{B}(H_{-\tau(s),\,\mathrm{Re}})$). We clarify that this remark does not contradict the fact that every sequence s of the form (2.6) is a moment sequence. Indeed, in order that this remark hold, it is necessary that s be, in addition, definite or quasi-definite. The imbedding

$$H_{0,\,\mathrm{Re}} \subset H_{-\tau(s),\,\mathrm{Re}} = H_{-\tau'(2),\,\mathrm{Re}}$$

is quasinuclear.

■

Remark 2. The proof of sufficiency in Theorem 2.1 consists of two independent parts:

(i) establishing the possibility to extend the shift operators (2.8) to commuting selfadjoint operators (Steps I–III);

(ii) proving the representation (2.6); it should be mentioned that the form of conditions imposed on s to guarantee the existence of extensions is not essential for this second part of the proof.

■

Remark 3. We now give one more convenient condition of definiteness. Assume that the estimate

$$|(s_{2n}, \varphi^{(1)} \otimes \ldots \otimes \varphi^{(2n)})_{H_0^{\otimes 2n}}| \le m_n^2 \prod_{j=1}^{2n} \|\varphi^{(j)}\|_{H_{\tau(s)}} \tag{2.30}$$

$$(\varphi^{(1)}, \ldots, \varphi^{(2n)} \in \Phi;\ n \in \mathbb{N})$$

holds for a given moment sequence $s = (s_n)_{n=0}^{\infty}$ and some $\tau(s) \in \mathrm{T}$; here the class $C\{m_n\}$ is quasianalytical. The result of Step II in the proof of Theorem 2.1 takes place in this case, too. One can easily verify this by repeating the calculation of type (2.11). This means that Theorem 2.1 remains valid for the concept of definiteness introduced in this remark.

■

2.2. Examples

Example 2.2. (continuation of Example 2.1.).

Let us clarify the form of the results of Subsection 2.1 for the d-dimensional moment problem. In the basis $e_{\alpha_1} \otimes \ldots \otimes e_{\alpha_n}$ $(\alpha_j = 1, \ldots, d)$, the representation (2.6) has the form

$$s_{n;\,\alpha_1,\,\dots,\,\alpha_n} = (s_n, e_{\alpha_1} \otimes \dots \otimes e_{\alpha_n})_{\mathbb{C}^{nd}} = \int_{\mathbb{R}^d} \lambda_{\alpha_1} \dots \lambda_{\alpha_n} \, d\sigma(\lambda) \qquad (2.31)$$

$$(\lambda = (\lambda_1, \dots, \lambda_d);\ \alpha_1, \dots, \alpha_n = 1, \dots, d;\ n \in \mathbb{N};\ s_0 = \sigma(\mathbb{R}^d)).$$

Employing the symmetry of s_n and proceeding to the symmetrized basis l_n in (2.31), we obviously get the ordinary representation (2.3).

Let us write the condition of definiteness. We set $\mathrm{E} = \{e_1, \dots, e_d\}$; then, according to (2.4) and (2.5), we have $d(E, 0) = 1$, and

$$m_n = \| s_{2n} \|^{1/2}_{\mathbb{C}^{2nd}} \le d^{n/2} \max_{n_1 + \dots + n_d = 2n} | s_{n_1, \dots, n_d} |^{1/2} \quad (n \in \mathbb{N}).$$

Since the factor d^n is inessential, this condition is reduced to the requirement that the class $C\left\{\max_{|\boldsymbol{k}| = 2n} | s_{\boldsymbol{k}} |^{1/2}\right\}$ should be quasianalytical. The condition of quasi-definiteness has the analogous form: The class

$$C\left\{ \max_{|(0,\, \boldsymbol{k}')|} | s(0, \boldsymbol{k}) |^{1/2} \right\}$$

should be quasianalytical. Clearly, these conditions can be made more precise (see references given in Example 2.1). Note that if we take into account the condition of definiteness of the finite-dimensional moment problem obtained above, then, in fact, we make our presentation in Subsection 2.1 closed. One should only establish (see Step X of the proof of Theorem 2.1) the uniqueness of the definition of measure σ in the case of a one-dimensional definite moment problem. This can be shown as in the proof of Theorem 3.1 in Chapter 3 (see also Subsection 2.5).

Example 2.3 (the infinite-dimensional scalar moment problem). Denote by T a collection of all weights $\tau = (\tau_j)_{j=1}^{\infty}$ $(\tau_j \ge 1)$;

$$H_\tau = l_2(\tau) = \left\{ (\varphi_j)_{j=1}^{\infty},\ \varphi_j = \mathbb{C}^1 \ \middle|\ \sum_{j=1}^{\infty} |\varphi_j|^2 \tau_j < \infty \right\};$$

$$H_{-\tau} = l_2(\tau^{-1}) \ (\tau^{-1} = (\tau_j^{-1})_{j=1}^{\infty}).$$

Then the nuclear space

$$\Phi = \mathrm{pr} \lim_{\tau \in \mathrm{T}} H_\tau = \mathbb{C}_0^\infty,\ \Phi' = \mathbb{C}^\infty.$$

Further, for any $n \in \mathbb{N}$, $\Phi^{\otimes n}$ consists of all finite n-dimensional sequences $\varphi_n = (\varphi_{\alpha_1, \dots, \alpha_n})_{\alpha_1, \dots, \alpha_n = 1}^{\infty}$ of complex numbers ($\varphi_{\alpha_1, \dots, \alpha_n}$ vanishes for sufficiently large

$\alpha_1 + \alpha_n$), and $(\Phi^{\otimes n})'$ consists of the same sequences φ_n but without the condition of finiteness (in this connection, see Example 1.6 and Subsection 2.1 in Chapter 1).

Writing the elements of the spaces $\Phi^{\otimes n}$ and $(\Phi^{\otimes n})'$ in the form of sequences, we, in fact, write them in the basis $e_{\alpha_1} \otimes \ldots \otimes e_{\alpha_n}$, where $(e_j)_{j=1}^{\infty}$ is a standard basis in $l_2 = l_2(1)$. If $\varphi_n \in (\Phi^{\otimes n})'$ and is symmetric with respect to all variables, then it is reasonable to introduce a symmetrized basis (as in Examples 2.1 and 2.2). Thus, let $\boldsymbol{n} = (n_j)_{j=1}^{\infty}$ $(n_j \in \mathbb{Z}_+)$, i.e., $\boldsymbol{n} \in \mathbb{Z}_+^{\infty}$; $|\boldsymbol{n}| = n_1 + n_2 + \ldots = n$. As above, denote by $l_{\boldsymbol{n}}$ a vector of a new basis obtained by the symmetrization of all unit vectors $e_{\alpha_1} \otimes \ldots \otimes e_{\alpha_n}$ which contain n_1 indices 1, n_2 indices 2, etc., in the collection $(\alpha_1, \ldots, \alpha_n) = \alpha$.

Consider a moment sequence $s = (s_n)_{n=0}^{\infty}$, $\overline{s_n} \in \left(\Phi^{\otimes n}\right)'$, where conjugation means an ordinary transition to a complex conjugate. Rewriting each s_n in the basis $(l_{\boldsymbol{n}})_{\boldsymbol{n} \in \mathbb{Z}_+^{\infty}}$ $(|\boldsymbol{n}| = n)$, we get this moment sequence in the form $s = (s_{\boldsymbol{n}})_{\boldsymbol{n} \in \mathbb{Z}_+^{\infty}}$ $\left(s_{\boldsymbol{n}} \in \mathbb{R}^1\right)$ (or $s = ((s_{\boldsymbol{n}})_{\boldsymbol{n} \in \mathbb{Z}_+^{\infty}})_{|\boldsymbol{n}|=0}^{\infty}$). Due to the symmetry of s_{j+k} in (2.1), one can take symmetric φ_j and φ_k in this condition. If we also write them in the basis $(l_{\boldsymbol{n}})_{\boldsymbol{n} \in \mathbb{Z}_+^{\infty}}$, then we get condition (2.1) in the form

$$\sum_{\boldsymbol{j}, \boldsymbol{k} \in \mathbb{Z}_+^{\infty}} s_{\boldsymbol{j}+\boldsymbol{k}} \, \overline{\varphi}_{\boldsymbol{j}} \varphi_{\boldsymbol{k}} \geq 0, \tag{2.32}$$

where $(\varphi_{\boldsymbol{j}})_{\boldsymbol{j} \in \mathbb{Z}_+^{\infty}}$ is an arbitrary finite sequence of complex numbers (finiteness is to be understood in the following sense: $\varphi_{\boldsymbol{j}}$ vanishes provided that $\boldsymbol{j} = (j_l)_{l=1}^{\infty}$, $j_l \in \mathbb{Z}_+$ is such that $j_l + l$ is sufficiently large).

It is easy to understand that the representation (2.6) now has the following form (as in Examples 2.1 and 2.2)

$$s_{\boldsymbol{n}} = \int_{\mathbb{R}^{\infty}} \lambda^{\boldsymbol{n}} \, d\sigma(\lambda)$$

$$(\lambda^{\boldsymbol{n}} = \lambda_1^{n_1} \lambda_2^{n_2} \ldots; \ \boldsymbol{n} = n_1 + n_2 + \ldots \in \mathbb{Z}_+^{\infty}; \ |\boldsymbol{n}| = n_1 + n_2 + \ldots < \infty). \tag{2.33}$$

In order that (2.33) hold, it is necessary to demand that condition (2.32) hold and, in addition, that s be definite or quasi-definite in one of the forms (2.5) or (2.30). Unlike the finite-dimensional moment problem, an important role is played here by a sequence of weights $(\tau(n))_{n=1}^{\infty}$ such that

$$s_n \in H_{-\tau(n)}^{\otimes n} = (l_2(\tau^{-1}(n)))^{\otimes n};$$

the measure σ is concentrated on a certain $l_2(\tau^{-1})$.

We also note that, since

$$\mathbb{R}^\infty = \bigcup_{\tau\in T} H_{-\tau,\mathrm{Re}} = \bigcup_{\tau\in T} l_{2,\mathrm{Re}}(\tau^{-1}),$$

one can construct a discrete measure σ concentrated at the points from different $l_{2,\mathrm{Re}}(\tau^{-1})$. This gives an example of a sequence of the form (2.33), i.e., of a moment sequence for which the condition of definiteness or quasi-definiteness does not hold.

Example 2.4 (the moment problem on a class of tempered generalized functions). Assume that $T=\{0,1,\dots\}=\mathbb{Z}_+$, H_τ is a Sobolev space, and $W_2^\tau\big(\mathbb{R}^d, (1+|x|^2)^\tau dx\big)$ is the completion of the class $C_0^\infty\big(\mathbb{R}^d\big)$ $(d\in\mathbb{N})$ of finite infinitely differentiable functions with respect to the scalar product

$$(\varphi,\psi)_{W_2^\tau(\mathbb{R}^d,(1+|x|^2)^\tau dx)} = \sum_{|\alpha|\le\tau}\int_{\mathbb{R}^d}(D^\alpha\varphi)(x)\overline{(D^\alpha\psi)(x)}\,(1+|x|^2)^\tau dx. \qquad (2.34)$$

The projective limit $\Phi=\mathrm{pr}\lim_{\tau\in T}H_\tau$ is a nuclear space coinciding with the Schwartz space $\mathcal{S}\big(\mathbb{R}^d\big)$; $\Phi^{\otimes n}=\mathcal{S}\big(\mathbb{R}^{dn}\big)$ $(n\in\mathbb{N})$ (as far as these facts are concerned, see Section 1 of Chapter 1, Example 1.8). Below, the conjugation means ordinary transition to a complex bar.

The sequence $s=(s_n)_{n=0}^\infty$ now consists of tempered generalized functions, i.e., $s_n\in\mathcal{S}'_{\mathrm{Re}}\big(\mathbb{R}^{dn}\big)$. Let us consider the conditions of definiteness. First, we note that

for any $\varepsilon>0$, *it is possible to choose* $E\subset\mathcal{S}_{\mathrm{Re}}\big(\mathbb{R}^{dn}\big)$ *such that its linear span is dense in* $\mathcal{S}\big(\mathbb{R}^{dn}\big)$, *and*

$$d(\tau,E)\le\tau^{(d+\varepsilon)\tau}\quad(\tau\in T). \qquad (2.35)$$

We outline the proof of this fact. In the case of the general space $\Phi=\mathrm{pr}\lim_{\tau\in T}H_\tau$, where $T=\mathbb{Z}_+$, E can be chosen as follows. Let $\delta=(\delta_\tau)_{\tau=0}^\infty$ $(\delta_\tau>0)$ be a weight. We introduce the scalar product

$$(\varphi,\psi)_{S(\delta)} = \sum_{\tau=0}^\infty(\varphi,\psi)_{H_\tau}\delta_\tau,$$

defined on $\varphi\in\Phi$ for which $(\varphi,\varphi)_{S(\delta)}<\infty$; these φ form a Hilbert space $S(\delta)$. One may take a ball $\|\cdot\|_{S(\delta)_{\mathrm{Re}}}<1$ to play the role of E, then

$$d(\tau,E) = \sup_{\varphi\in E}\|\varphi\|_{H_\tau}\le\delta_\tau^{-1/2}\quad(\tau\in T).$$

Moreover, if $S(\delta)$ is, in this case, dense in Φ, then the same is true for the linear span of E.

Hence, to prove estimate (2.35), it suffices to show that the space $S(\delta)$ constructed for given

$$H_\tau = W_2^\tau(\mathbb{R}^d, (1+|x|^2)^\tau dx)$$

and $\delta_\tau = \tau^{-2(d+\varepsilon)\tau}$ is dense in $\mathcal{S}(\mathbb{R}^d)$. Consider the space S_α^β of functions of a point $x \in \mathbb{R}^d$ with d-dimensional vector indices α and β (see Gelfand and Shilov [2, Chapter 4, Section 9]). When estimating the norm $\|\varphi\|_{S(\delta)}$, one can verify that $S_\alpha^\beta \subset S(\delta)$ for $|\alpha|+|\beta|< d+\varepsilon$. However, it follows from Sections 8 and 9 in Chapter 4 of the above-mentioned book that S_α^β is dense in $\mathcal{S}_{\mathrm{Re}}(\mathbb{R}^d)$ for

$$\alpha = \beta = (1/2+\varepsilon_1, \dots, 1/2+\varepsilon_1) \quad (\varepsilon_1 > 0),$$

and this yields the desired statement.

Estimate (2.35) enables us to write the condition of definiteness (2.5) or condition (2.30) for any moment sequence considered here. The obtained representation has the form (2.6), where Φ'_{Re} is replaced by $\mathcal{S}'_{\mathrm{Re}}(\mathbb{R}^d)$ or, more precisely, by certain negative $W_2^{-\tau}\left(\mathbb{R}^d, (1+|x|^2)^\tau dx\right)$.

Example 2.5. As in the previous example, we can consider the case where $\Phi = \mathcal{D}(\mathbb{R}^d)$ $(d \in \mathbb{N})$ (concerning this space, see Example 1.9 in Section 1 of Chapter 1). Here T consists of pairs $\tau = (\tau_1, \tau_2(x))$, where $\tau_1 \in \mathbb{Z}_+$, and $C^\infty(\mathbb{R}^d) \ni \tau_2(x) \ge 1$ $(x \in \mathbb{R}^d)$ is a weight. The Sobolev space $W_2^{\tau_1}(\mathbb{R}^d, \tau_2(x)dx)$ plays the role of H_τ (in expression (2.34), one should replace τ and $(1+|x|^2)^\tau$ by τ_1 and $\tau_1(x)$, respectively).

Example 2.6. Let us study the case of moment sequences $s = (s_n)_{n=0}^\infty$ constructed from the elements of negative Hilbert spaces. Let $H_- \supset H_0 \supset H_+$ be a chain with involution. It can always be regarded as a part of the following chain with involution

$$\bigcup_{\tau\in\mathbb{N}} H_{-\tau} = \Phi' \supset \dots \supset H_{-2} \supset H_{-1} \supset H_0 \supset H_1 \supset H_2 \supset \dots \supset \Phi = \operatorname{pr}\lim_{\tau\in\mathbb{N}} H$$

$$(H_1 = H_+, \ H_{-1} = H_-),$$

where Φ is nuclear. Therefore, if $s_n \in H_-^{\otimes n}$, then $s_n \in (\Phi^{\otimes n})'$ and one can write representation (2.6) with the measure σ concentrated on $H_{-\tau(s),\mathrm{Re}}$. The arbitrariness in the choice of Φ enables us to take any space $H_{-2,\mathrm{Re}}$ for which the corresponding positive space H_2 is quasinuclearly imbedded in H_1 to play the role of $H_{-\tau(s),\mathrm{Re}}$ (this follows from the proof of Theorem 2.1). Since, in this case, $\tau(n) = 1$ $(n \in \mathbb{Z}_+)$, the condition of definiteness (2.5) can be formulated as follows. The class

$$C\left\{\|s_{2n}\|_{H_-^{\otimes 2n}}^{1/2}\right\} \tag{2.36}$$

is quasianalytical. Clearly, this result can be reformulated in the following form.

Theorem 2.2. *Let $H_- \supset H_0 \supset H_+$ be a chain with involution. A sequence $s = (s_n)_{n=0}^\infty$, where $s_n \in H_-^{\otimes 2n}$, is called a moment sequence if s_n is symmetric for any $n \in \mathbb{N}$, and the condition of positive definiteness holds, namely,*

$$\sum_{j,k=0}^{\infty} (s_{j+k}, \varphi_j \otimes \overline{\varphi}_k)_{H_0^{\otimes(j+k)}} \geq 0 \tag{2.37}$$

for any finite sequence $\varphi = (\varphi_j)_{j=0}^\infty$ $(\varphi_j \in H_+^{\otimes j})$.

Assume that s is definite, i.e., that the class (2.36) is quasianalytical. Then the representation

$$s_n = \int_{G_{\mathrm{Re}}} \underbrace{\lambda \otimes \dots \otimes \lambda}_{n} \, d\sigma(\lambda) \quad (n \in \mathbb{Z}_+) \tag{2.38}$$

holds, where G_{Re} is an arbitrary real Hilbert space such that $H_{-,\mathrm{Re}}$ is dense in G_{Re} and the imbedding $H_{-,\mathrm{Re}} \subset G_{\mathrm{Re}}$ is quasinuclear, and

$$\mathfrak{B}(G_{\mathrm{Re}}) \ni \alpha \mapsto \sigma(\alpha) \geq 0$$

is some measure. The integral in (2.38) converges in the norm of the space $G_{\mathrm{Re}}^{\otimes n}$; the measure σ is uniquely defined for given s.

It is clear that one can now formulate the result for the case of quasi-definite moment sequence, as well.

For example, one can take negative Sobolev spaces to play the role of H_-. However, one can also take positive Sobolev spaces; this will be explained in detail in Subsection 5.4. We now only study into what form representation (2.38) transforms in this case. Thus, if the moment sequences $s = (s_n)_{n=0}^\infty$ consists of smooth enough functions $s_n = s_n(x_1, \dots, x_n)$ of the points $x_1, \dots, x_n \in \mathbb{R}^d$, then under a certain condition similar to the condition of definiteness, it can be represented in the form

$$s_n(x_1, \dots, x_n) = \int_{C_{\mathrm{Re}}(\mathbb{R}^d)} \lambda(x_1) \dots \lambda(x_n) \, d\sigma(\lambda) \tag{2.39}$$

$$(x_1, \dots, x_n \in \mathbb{R}^d;\ n \in \mathbb{Z}_+),$$

where the integration is carried out over the space of continuous real functions on $\mathbb{R}^d$, and the integral converges absolutely and uniformly when $(x_1, \dots, x_n)$ changes on a compact set in $\mathbb{R}^{nd}$.

Example 2.7 (moment sequence of the Osterwalder–Schrader type).
As mentioned above, the representation of a moment sequence in Example 2.4 has the form

$$s_n = \int_{\mathcal{S}'_{\mathrm{Re}}(\mathbb{R}^d)} \underbrace{\lambda \otimes \ldots \otimes \lambda}_{n} \, d\sigma(\lambda) \quad (n \in \mathbb{Z}_+) \tag{2.40}$$

(provided that all necessary conditions are satisfied).

Consider an example the same as 2.4, but with another involution which will be denoted by θ for convenience. Namely, for a point $x \in \mathbb{R}^d$, we set $x = (\boldsymbol{x}, t)$, where $\boldsymbol{x} \in \mathbb{R}^{d-1}$, $t \in \mathbb{R}^1$ (the spatial and temporal parts of the point x; $d \in \mathbb{N}$). The involution θ on $\mathcal{S}(\mathbb{R}^d)$ is introduced by

$$\varphi^\theta(x) = \varphi^\theta(\boldsymbol{x}, t) = \overline{\varphi(\boldsymbol{x}, -t)},$$

where the overbar denotes the ordinary complex conjugation $(\varphi \in \mathcal{S}(\mathbb{R}^d);\ \boldsymbol{x} \in \mathbb{R}^{d-1},\ t \in \mathbb{R}^1)$. It follows from (2.34) that θ can be extended to an involution by continuity in each

$$W_2^\tau(\mathbb{R}^d, (1 + |x|^2)^\tau dx) \qquad (\tau \in \mathbb{Z}_+),$$

and therefore, it is possible to formulate the condition of positive definiteness (2.1) with the involution θ. It is clear that everything what has been said in Example 2.4 remains true for this involution. The representation for the moment sequence $s = (s_n)_{n=0}^\infty$ now has the form (2.40), too, but the integration is carried out not over $\mathcal{S}'_{\mathrm{Re}}(\mathbb{R}^d)$ but over a space $\mathcal{S}'_{\theta,\mathrm{Re}}(\mathbb{R}^d)$ which is a part of the space $\mathcal{S}'(\mathbb{R}^d)$ consisting of generalized functions $\xi \in \mathcal{S}'(\mathbb{R}^d)$ which are real with respect to the involution θ, i.e., such that

$$\xi(\varphi^\theta) = \overline{\xi(\varphi)} \qquad (\varphi \in \mathcal{S}(\mathbb{R}^d)).$$

The involution θ and the corresponding definition of a moment sequence play an important role in the Euclidean quantum field theory. The main objects of this theory, namely, the so-called Schwinger functions $s_n \in \mathcal{S}'(\mathbb{R}^{nd})$, form a sequence $s = (s_n)_{n=0}^\infty$ which possesses the property of positive definiteness (2.1) with $\overline{}$ replaced by θ. Unfortunately, this inequality does not hold for all

$$\varphi_n \in \mathcal{S}(\mathbb{R}^{nd}) = (\mathcal{S}(\mathbb{R}^d))^{\otimes n};$$

it holds only on $\varphi_n \in \left(\mathcal{S}_+(\mathbb{R}^d)\right)^{\otimes n}$, where $\mathcal{S}_+(\mathbb{R}^d)$ consists of functions from the space $\mathcal{S}(\mathbb{R}^d)$ with supports in

$$\mathbb{R}^{d-1} \times (0, \infty) \quad (n \in \mathbb{N};\ \varphi_0 \in \mathbb{C}^1).$$

This is the positive definiteness of the Osterwalder–Schrader (OS) type. We can assume that s possesses all the other required properties. It is easy to verify that it is impossible to prove a theorem similar to Theorem 2.1 directly for this concept of a moment sequence.

Example 2.8 (Schwinger functions of a free field). For given $m_0 > 0$, the operator $\left(-\Delta + m_0^2\right)^{-1} > 0$ is bounded in $L_2(\mathbb{R}^d)$; therefore, a Gaussian measure ν_{m_0} with mean value zero and the correlation operator $S = 2\left(-\Delta + m_0^2\right)^{-1}$ is defined on $\mathfrak{B}\left(\mathcal{S}'_{\mathrm{Re}}(\mathbb{R}^d)\right)$ (see Subsection 1.9 in Chapter 2). In other words, ν_{m_0} is a canonical Gaussian measure associated with a Hilbert space $\mathcal{H}_{m_0}$ which is the completion of $\mathcal{S}_{\mathrm{Re}}(\mathbb{R}^d)$ with respect to the scalar product

$$(\varphi_1, \varphi_2)_{\mathcal{H}_{m_0}} = 2\left(\varphi_1, (-\Delta + m_0^2)^{-1} \varphi_2\right)_{L_2(\mathbb{R}^d)}$$

$$= 2 \int_{\mathbb{R}^d} \frac{\tilde{\varphi}_1(k)\, \overline{\tilde{\varphi}_2(k)}}{|k|^2 + m_0^2}\, dk \quad (\varphi_1, \varphi_2 \in \mathcal{S}_{\mathrm{Re}}(\mathbb{R}^d)) \tag{2.41}$$

The measure ν_{m_0} is called an Euclidean measure of the free field with mass m_0 in the space-time with dimensionality d; its moments $\left(s_n^{\nu_{m_0}}\right)_{n=0}^{\infty}$ are Schwinger functions of a free field. Formula (2.5) in Chapter 2 for moments of a Gaussian measure guarantees the validity of the equality

$$s_2^{\nu_{m_0}}(\varphi_1, \varphi_2) = \int_{\mathcal{S}'_{\mathrm{Re}}(\mathbb{R}^d)} (\varphi_1, \lambda)_{L_2(\mathbb{R}^d)} (\varphi_2, \lambda)_{L_2(\mathbb{R}^d)}\, d\nu_{m_0}(\lambda) = \frac{1}{2} (\varphi_1, \varphi_2)_{\mathcal{H}_{m_0}},$$

$$\left(s_2^{\nu_{m_0}}(\varphi_1, \varphi_2) = \left(s_2^{\nu_{m_0}}, \overline{\varphi}_1 \otimes \overline{\varphi}_2\right)_{L_2(\mathbb{R}^{2d})}\right),$$

in which $\varphi_1, \varphi_2 \in \mathcal{S}(\mathbb{R}^d)$, and on the right-hand side one can find a bilinear extension of a scalar product in $\mathcal{H}_{m_0}$. For the other moments, the same formula yields

$$s_{2n+1}^{\nu_{m_0}}(\varphi_1, \ldots, \varphi_{2n+1}) = 0 \quad (\varphi_k \in \mathcal{S}0\mathbb{R}^d);\ k = 1, \ldots, 2n+1;\ n \in \mathbb{Z}_+),$$

$$s_{2n}^{\nu_{m_0}}(\varphi_1, \ldots, \varphi_{2n}) = \sum s_2^{\nu_{m_0}}(\varphi_{i_1}, \varphi_{j_1}) \ldots s_2^{\nu_{m_0}}(\varphi_{i_n}, \varphi_{j_n}), \tag{2.42}$$

where summation is carried out over all pairings $(i_1, j_1), \ldots, (i_n, j_n)$ of the numbers $1, \ldots, 2n$. The representation (2.42) enables us to reduce the investigation of properties of Schwinger functions of free field to the study of a two-point function $s_2^{\nu_{m_0}}$, and this is the characteristic feature of the Gaussian nature of the measure ν_{m_0}, namely, a Gaussian measure (with mean value zero) is uniquely determined by its correlation operator, and the latter can be reconstructed from the second moment.

In particular, the above-mentioned property relates to the analysis of the property of positive definiteness with respect to the involution θ. First, we show that for

$$\varphi_1, \varphi_2 \in \mathcal{S}(\mathbb{R}^d) \quad \text{and} \quad \varphi_k(x, t) = \psi_k(x)\, \psi_k^0(t) \quad (k = 1, 2)$$

we have the following formula

$$s_2^{\nu_{m0}}(\varphi_1, \overline{\varphi}_2) = (s_2^{\nu_{m0}}, \overline{\varphi}_1 \otimes \varphi_2)_{L_2(\mathbb{R}^{2d})}$$

$$= \int_{\mathbb{R}^2} (\psi_1, A_{m_0}^{-1} e^{-|t_1 - t_2| A_{m_0}} \psi_2)_{L_2(\mathbb{R}^{d-1})}\, \psi_1^0(t_1)\, \overline{\psi_2^0}(t_2)\, dt_1\, dt_2, \qquad (2.43)$$

where A_{m_0} is the operator $(-\Delta + m_0^2)^{-1}$ in $L_2(\mathbb{R}^{d-1})$. Indeed, it follows from (2.41) that

$$s_2^{\nu_{m0}}(\varphi_1, \overline{\varphi}_2) = 2 \int_{\mathbb{R}^{d-1} \times \mathbb{R}^1} \frac{\tilde{\psi}_1(\boldsymbol{k})\, \overline{\tilde{\psi}_2(\boldsymbol{k})}}{|\boldsymbol{k}|^2 + k_0^2 + m_0^2}\, \tilde{\psi}_1^0(k_0)\, \overline{\tilde{\psi}_2^0}(k_0)\, d\boldsymbol{k}\, dk_0$$

$$(k = (\boldsymbol{k}, k_0) \in \mathbb{R}^d = \mathbb{R}^{d-1} \times \mathbb{R}^1).$$

Let us make a substitution:

$$\tilde{\psi}_1^0(k_0)\, \overline{\tilde{\psi}_2^0}(k_0) = \frac{1}{2\pi} \int_{\mathbb{R}^2} e^{ik_0(t_1 - t_2)}\, \psi_1^0(t_1)\, \overline{\psi_2^0}(t_2)\, dt_1\, dt_2$$

in the last equality; integrating it over $k_0 \in \mathbb{R}^1$ by use of the relation

$$\frac{1}{\pi} \int_{\mathbb{R}^1} \frac{e^{ik_0 s}}{k_0^2 + a^2}\, dk_0 = \frac{e^{-a|s|}}{a} \qquad (a > 0;\ s \in \mathbb{R}^1),$$

we get (2.43).

For

$$\varphi_1 = \varphi = \psi(x)\, \psi^0(t) \in \mathcal{S}_+(\mathbb{R}^d) \quad \text{and} \quad \overline{\varphi}_2 = \varphi^\theta,$$

it follows from (2.43) that

$$s_2^{\nu_{m0}}(\varphi_1, \varphi^\theta) = \int_{\mathbb{R}^2} (\psi_1, A_{m_0}^{-1} e^{-|t_1 - t_2| B_{m_0}} \psi)_{L_2(\mathbb{R}^{d-1})}\, \psi^0(t_1)\, \overline{\psi^0(-t_2)}\, dt_1\, dt_2$$

$$= \int_{\mathbb{R}^2} (\psi_1, A_{m_0}^{-1} e^{-(t_1 + t_2) A_{m_0}} \psi)_{L_2(\mathbb{R}^{d-1})}\, \psi^0(t_1)\, \overline{\psi^0}(t_2)\, dt_1\, dt_2$$

$$= \int_{m_0}^{\infty} \lambda^{-1} \left| \int_0^{\infty} e^{-t\lambda} \psi^0(t)\, dt \right|^2 d(E(\lambda)\psi, \psi)_{L_2(\mathbb{R}^{d-1})} \geq 0,$$

where E is the RI of the operator A_{m_0}. In general, the inequality $\overset{\vee}{s}{}_2^{m_0}(\varphi, \varphi^{\theta}) \geq 0$ is not valid without the condition $\operatorname{supp} \varphi^0 \subset [0, +\infty)$. If we assume that it holds, then, taking into account that ψ is arbitrary and applying, by analogy, the spectral theorem to the operator A_{m_0} in (2.43) with $\varphi_1 = \varphi$ and $\overline{\varphi}_2 = \varphi^{\theta}$, we get the inequality

$$\int_{\mathbb{R}^2} \psi^0(t_1)\, \overline{\psi^0}(t_2)\, e^{-\lambda|t_1 + t_2|}\, dt_1 dt_2 \geq 0$$

for all $\varphi^0 \in \mathcal{S}(\mathbb{R}^1)$, $\lambda \geq m_0$. But this is evidently incorrect because the kernels $e^{-\lambda|t_1 + t_2|}$ are not positive definite. Therefore,, the positive definiteness of the Schwinger functions of the free field with respect to the involution θ holds only on the class of test functions described above. Therefore,, one cannot assume that this positive definiteness may take place for all test functions in the general case of the Schwinger functions of Euclidean quantum field theory.

2.3. Two More Proofs of the Theorem on Representation

We now outline two more proofs of Theorem 2.1 (however, not of its complete version) which reflect other approaches to obtaining integral representations of a moment sequence.

Let us show how one can prove Theorem 2.1 in a certain case (without Remark 1) by a "limit transition as $d \to \infty$" from the d-dimensional moment problem. Without loss of generality, we can assume that $s_0 = 1$. Let F be a finite-dimensional subspace of Φ, invariant with respect to the involution $^{-}$. Consider the restriction of a functional s_n to

$$F^{\otimes n} \subset \Phi^{\otimes n}: \; s_{n,F} = s_n \upharpoonright F^{\otimes n} \qquad (n \in \mathbb{Z}_+).$$

Clearly, $s_F = (s_{n,F})_{n=0}^{\infty}$ is a definite moment sequence, and the representation (2.6) for it is equivalent to the representation (2.3) for a definite finite-dimensional moment sequence. Therefore, we can assume that we have a formula of the form (2.6) for s_F, and, obviously, it can be rewritten as

$$s_{n,F} = \int_{\Phi'_{\text{Re}} / \hat{F}_{\text{Re}}} \underbrace{\lambda \otimes \ldots \otimes \lambda}_{n}\, d\sigma_F(\lambda) \qquad (n \in \mathbb{Z}_+). \tag{2.44}$$

Here, $\hat{F}_{\text{Re}}$ is a subspace of functionals from Φ'_{Re} orthogonal to F_{Re}, i.e.,

$$\hat{F}_{\text{Re}} = \{\xi \in \Phi'_{\text{Re}} \mid \xi(\varphi) = 0,\ \varphi \in F_{\text{Re}}\}.$$

The measure σ_F induces the measure γ on cylindrical sets from Φ'_{Re} with the bases

$$\delta \subset \Phi'_{\text{Re}} / \hat{F}_{\text{Re}} = \mathbb{R}^d \quad (d = \dim F),$$

i.e., on sets of the form

$$\mathbb{C}(\varphi_1, \dots, \varphi_d; \delta) = \{\xi \in \Phi'_{\text{Re}} \mid (\xi(\varphi_1, \dots, \xi(\varphi_d)) \in \delta\},$$

where $\varphi_1, \dots, \varphi_d \in F_{\text{Re}}$; namely, we set

$$\gamma(\mathbb{C}(\varphi_1, \dots, \varphi_d; \delta)) = \sigma_F(\delta).$$

By arbitrary changes of F, we obtain a system of measures of cylindrical sets. The consistency of this system easily follows from the uniqueness of the definition of the measure σ_F for given $s_{n,F}$ in (2.44). Therefore, we can assume that we have a cylindrical measure γ on Φ'_{Re}. The measure γ determined by the corresponding measure σ_F is continuous in the following sense. Let

$$F^{(m)} = \text{l.s.}((\varphi_j^{(m)})_{j=1}^d) \qquad (\varphi_j^{(n)} \in \Phi;\ m \in \mathbb{N};\ d \in \mathbb{N} \text{ is fixed}),$$

and $\varphi_j^{(m)} \underset{m\to\infty}{\to} \varphi_j$ in Φ for every $j = 1, \dots, d$. Then $\sigma_{F^{(m)}} \to \sigma_F$, where $F = \text{l.s.}((\varphi_j)_{j=1}^d)$, in the sense of weak convergence of measures. Indeed, let us write the representation (2.44) for $S_{n,F^{(m)}}$ and $S_{n,F}$ in the coordinate form taking systems of vectors $(\varphi_j^{(m)})_{j=1}^d$ and $(\varphi_j)_{j=1}^d$ as bases (one can regard each this system as linearly independent). According to (2.44) and (2.3), we get

$$s_n^{(m)} = \int_{\mathbb{R}^d} t^n \, d\sigma_m(t), \quad s_n = \int_{\mathbb{R}^d} t^n \, d\sigma(t) \quad (n \in \mathbb{Z}_+^d;\ m \in \mathbb{N}), \tag{2.45}$$

where σ_m and σ are measures on $\mathbb{R}^d$ induced by the measures $\sigma_{F^{(m)}}$ and σ_F under the transition to the coordinate form.

Taking into account that for any $n \in \mathbb{Z}_+^d$, we obviously have $s_n^{(m)} \underset{m\to\infty}{\to} s_n$ and $\sigma_m(\mathbb{R}^d) = \sigma(\mathbb{R}^d) = 1$ $(s_0 = 1)$, we can conclude from (2.45), the Helly theorem, and the uniqueness of the definition of the measure σ for given s_n that $\sigma_m \to \sigma$ weakly as $m \to \infty$, i.e., $\sigma_{F^{(m)}} \to \sigma_F$. Hence, continuity is proved.

Thus, the cylindrical measure γ in the space dual to the nuclear space Φ is continuous in the above-mentioned sense. Then, according to the Minlos theorem[3], it admits the restriction to a measure σ on a σ-span of cylindrical sets under consideration. Rewriting (2.44) in terms of this measure σ and taking into account that F is arbitrary, we obtain (2.6).

■

Remark 1. In fact, we have just proved the version of Theorem 2.1 under consideration for the more general concept of the definiteness of $s = (s_n)_{n=0}^{\infty}$; namely, for each finite-dimensional $F \subset \Phi$, the finite-dimensional moment sequence $s_F = (s_{n,F})_{n=0}^{\infty}$ should be definite. The proof given in Subsection 2.1 can be also used in this case because Step II of the proof of Theorem 2.15 remains valid for this case, too (this can be established by the use of Theorem 1.15).

■

We now give the third method of proving Theorem 2.1 with the help of the results of Subsection 1.4 in Chapter 4. We also study the related problems in Subsection 2.5.

In the notations of Theorem 2.2, the general representation (1.31) in Chapter 4 can be rewritten in the form

$$A_u = \int_{G_{\mathrm{Re}}} (\lambda, u)_{H_0} dE(\lambda), \quad \mathfrak{D}(A_u) = \left\{ f \in H_s \,\middle|\, \int_{G_{\mathrm{Re}}} (\lambda, u)_{H_0}^2 d(E(\lambda)f, f)_{H_s} < \infty \right\}. \tag{2.46}$$

Here, for given negative space G_{Re}, we have introduced the chain

$$G_{\mathrm{Re}} \supset H_{0,\mathrm{Re}} \supset G_{+,\mathrm{Re}}.$$

In (2.46), we have that $u \in G_{+,\mathrm{Re}}$ (the scalar product $(\lambda, x)_{G_{\mathrm{Re}}}$ in (1.31) is written in the form $(\lambda, u)_{H_0}$, where $u = I x \in G_{+,\mathrm{Re}}$). The role of the space H_0 is played by H_s.

Let us now apply (2.46) in the situation when the role of A_u is played by the operators A_e introduced on Step I of the proof of Theorem 2.1 applied to the case of Theorem 2.2. Now we can take a vector from $H_{+,\mathrm{Re}}$ to be e. However, it is more convenient to assume that $e \in G_{+,\mathrm{Re}}$ (we clarify that

$$G_{\mathrm{Re}} = H_{-2,\mathrm{Re}} \supset H_{-1,\mathrm{Re}} \supset H_{0,\mathrm{Re}} \supset H_{+,\mathrm{Re}} \supset H_{2,\mathrm{Re}} \supset G_{+,\mathrm{Re}}).$$

The role of the chain (1.4) can be played by the chain (2.12), where $\tau'(n) = 2$; $\tau(n) = 1$, and $H_1 = H_+$ $(n \in \mathbb{Z}_+)$. It is connected in the standard way with the family of commu-

[3] See, for example, the book by Gelfand and Vilenkin [1, Chapter 4, Section 2, Theorem 3]; compare also with Theorem 1.7 in Chapter 2 and Subsection 4.3.

ting selfadjoint operators $A = (A_e)_{e \in G_{+,\mathrm{Re}}}$ which satisfy the equality

$$A_{\alpha e+\beta l}\hat{\varphi} = \alpha A_e \hat{\varphi} + \beta A_l \hat{\varphi} \quad (e, l \in G_{+,\mathrm{Re}};\ a, b \in \mathbb{R}^1;\ \varphi \in C_0)$$

according to definition (2.8). It can be easily seen that all other requirements of Theorem 1.6 in Chapter 4 are also satisfied. Therefore, according to this theorem, the representation (1.31) in Chapter 4 is valid for these operators, i.e., (2.46) holds.

According to (2.8), the definition of A_e, and (2.46), the vector

$$\Omega = (1, 0, 0, \ldots)^\wedge \tag{2.47}$$

belongs to $\mathfrak{D}(A_{e_1} \ldots A_{e_n})$, where $e_1, \ldots, e_n \in G_{+,\mathrm{Re}}$ $(n \in \mathbb{N})$. It follows from (2.46) and the orthogonality of E that for the operators A_{e_j} we have

$$A_{e_1} \ldots A_{e_n}\Omega = \int_{G_{\mathrm{Re}}} (\lambda, e_1)_{H_0} \ldots (\lambda, e_n)_{H_0}\, dE(\lambda)\,\Omega.$$

Multiplying this equality by Ω scalarly in H_s and using (2.7) and the definition of A_{e_j}, we get

$$\int_{G_{\mathrm{Re}}} (\lambda, e_1)_{H_0} \ldots (\lambda, e_n)_{H_0}\, d(E(\lambda)\,\Omega, \Omega)_{H_s} = (A_{e_1} \ldots A_{e_n}\Omega, \Omega)_{H_s}$$

$$= (s_n, e_1 \otimes \ldots \otimes e_n)_{H_0^{\otimes n}} \quad (e_1, \ldots, e_n \in G_{+,\mathrm{Re}};\ n \in \mathbb{N}).$$

This yields representation (2.38) with

$$\mathfrak{B}(G_{\mathrm{Re}}) \ni \alpha \mapsto \sigma(\alpha) = (E(\lambda)\,\Omega, \Omega)_{H_s}.$$

This means that we have proved Theorem 2.2 with the help of Theorem 1.6 in Chapter 4 and the results of Steps I, II, and IV in the proof of Theorem 2.1.

■

Clearly, (2.47) is a strong cyclic vector of the family $(A_e)_{e \in G_{+,\mathrm{Re}}}$ in the sense of the definition in Theorem 3.2 of Chapter 3.

2.4. Generalized Power Moment Problem

We now give a generalization of the construction presented in Subsection 2.1 connected with the fact that s_n are no longer completely symmetric. This implies the fact that not all operations of the type (2.8) are Hermitian and commute, and representation (2.6) itself becomes much more complicated.

Suppose that there exist spaces H_τ, $H_\tau^{\otimes n}$, Φ, and $\Phi^{\otimes n}$ such as in Subsection 2.1. At the same time, an involution in these spaces is introduced in a somewhat different manner, namely, for any $n \in \mathbb{N}$, there exists an involution $\leftarrow$ in $\Phi^{\otimes n}$ which depends on n (we omit the index n in the notation of involution) and can be extended by continuity to the involution $\leftarrow$ in $H_\tau^{\otimes n}$ $(\tau \in \mathrm{T})$ and then in $H_{-\tau}^{\otimes n}$ and $(\Phi^{\otimes n})'$. Assume that this countable family of involutions possesses the property

$$\overleftarrow{\varphi_j \otimes \varphi_k} = \overleftarrow{\psi}_k \otimes \overleftarrow{\varphi}_j \quad (\varphi_j \in \Phi^{\otimes j},\ \psi_k \in \Phi^{\otimes k};\ j, k \in \mathbb{N}). \tag{2.48}$$

An example of a family of involutions of this sort on the spaces of functions is given by

$$\mathcal{S}(\mathbb{R}^{nd}) = (\mathcal{S}(\mathbb{R}^d))^{\otimes n} \ni \varphi_n(x_1, \dots, x_n) \mapsto \overleftarrow{\varphi}_n(x_1, \dots, x_n)$$
$$= \overline{\varphi_n(x_n,\dots,x_1)},$$

where the overbar denotes the complex conjugation $(x_1, \dots, x_n \in \mathbb{R}^d;\ n \in \mathbb{N})$ (recall that here

$$(\varphi_j \otimes \psi_k)(x_1, \dots, x_{j+k}) = \varphi_j(x_1, \dots, x_j)\psi_k(x_{j+1}, \dots, x_{j+k})).$$

For $\varphi_0 \in \mathbb{C}^1$, we assume that $\overleftarrow{\varphi}_0 = \overline{\varphi}_0$.

Consider a matrix

$$K = (K_{jk})_{j,\,k=0}^{\infty},$$

where

$$K_{jk} \in (\Phi^{\otimes(j+k)})'.$$

It is called positive definite if

$$\sum_{j,k=0}^{\infty} K_{jk}(\varphi_j \otimes \overleftarrow{\varphi}_k) \geq 0 \quad (\varphi \in C_0), \tag{2.49}$$

as in (2.1) (recall that C_0 denotes a linear set of finite sequences of the form $\varphi = (\varphi_j)_{j=0}^{\infty}$, $\varphi_j \in \Phi^{\otimes j}$). We fix a subspace M of the space Φ which is invariant with respect to the involution $\leftarrow$. A positive definite matrix K is called a generalized moment matrix if

$$K_{j+1\,k}(\varphi_j \otimes \chi \otimes \psi_k) = K_{j\,k+1}(\varphi_j \otimes \chi \otimes \psi_k),$$

$$K_{j+1\,k+1}(\varphi_j \otimes \chi \otimes \omega \otimes \psi_k) = K_{j+1\,k+1}(\varphi_j \otimes \omega \otimes \chi \otimes \psi_k)$$

$$(\chi, \omega \in \mathrm{M};\ \varphi_j \in \Phi^{\otimes j},\ \psi_k \in \Phi^{\otimes k};\ j, k \in \mathbb{Z}_+). \tag{2.50}$$

In the case of $\mathrm{M} = \Phi$, (2.50) immediately implies that K_{jk} depends only on the sum of indices, i.e., $K_{jk} = s_{j+k}$ $(j, k \in \mathbb{Z}_+)$, and the sequence $s = (s_n)_{n=0}^{\infty}$ is a moment sequence.

By analogy with Subsection 2.1, we can introduce the notion of definiteness for K in the following form. Since

$$K_{jk} \in (\Phi^{\otimes(j+k)})' = \bigcup_{\tau \in \mathrm{T}} H_{-\tau}^{\otimes(j+k)},$$

for any $j, k \in \mathbb{Z}_+$ there exists $\tau = \tau(j, k) \in \mathrm{T}$ such that $K_{jk} \in H_{-\tau(j,k)}^{\otimes(j+k)}$ and we fix a set $\mathrm{E} \subset \mathrm{M}$ real with respect to the involution $\leftarrow$ whose linear span is dense in M. We say that K is definite if the class

$$C\{m_n\}, \quad m_n = (d(\tau(n,n), \mathrm{E}))^n \, \|K_{nn}\|_{H_{-\tau(n,n)}^{\otimes 2n}}^{1/2} \qquad (n \in \mathbb{N}) \tag{2.51}$$

is quasianalytical (d is determined by formula (2.4) as before). By analogy with Subsection 2.1, we could also introduce a notion of quasidefiniteness for this case and prove the theorem formulated below.

To obtain the representation of the type (2.6), it is now convenient to proceed from the matrix $K = (K_{jk})_{j,k=0}^{\infty}$ to the matrix "with changed order of variables" $K = (\tilde{K}_{jk})_{j,k=0}^{\infty}$, where $\tilde{K}_{jk} \in (\Phi^{\otimes(j+k)})'$ is determined for given K by

$$\tilde{K}_{jk}(\varphi_j \otimes \psi_k) = \overline{K_{jk}(\bar{\varphi}_j \otimes \bar{\psi}_k)} \quad (\varphi_j \in \Phi^{\otimes j}, \psi_k \in \Phi^{\otimes k};\ j, k \in \mathbb{Z}_+). \tag{2.52}$$

Clearly, the matrices K and $\tilde{K}$ determine each other. The following theorem which generalizes Theorem 2.1 (for a certain case) is valid.

Theorem 2.3. *Let $K = (K_{jk})_{j,k=0}^{\infty}$ be a generalized moment matrix satisfying the condition of definiteness. Then the following representation is valid:*

$$\tilde{K}_{jk} = \sum_{p=0}^{j} \sum_{q=0}^{k} \int_{\Phi'_{\mathrm{Re}} / \hat{\mathrm{M}}_{\mathrm{Re}}} \underbrace{\xi(\lambda) \otimes \ldots \otimes \xi(\lambda)}_{j-p} \otimes d\sigma_{pq}(\lambda)$$

$$\otimes \underbrace{\xi(\lambda) \otimes \ldots \otimes \xi(\lambda)}_{k-q} \qquad (j, k \in \mathbb{Z}_+). \tag{2.53}$$

Here $\hat{M}_{Re}$ *denotes a subspace of functionals from* Φ'_{Re} *which are orthogonal to* M_{Re}, *and*

$$\Phi'_{Re} / \hat{M}_{Re} \ni \lambda \mapsto \xi(\lambda) \in \Phi'_{Re}$$

is a fixed mapping which is measurable with respect to $\mathfrak{B}(\Phi'_{Re} / \hat{M}_{Re})$ *and is a left inverse mapping of the natural homomorphism* $\Phi'_{Re} \to \Phi'_{Re} / \hat{M}_{Re}$.

The integration is carried out with respect to a matrix vector valued measure

$$\mathfrak{B}(\Phi'_{Re} / \hat{M}_{Re}) \ni \alpha \mapsto (\sigma_{pq}(\alpha))_{p,q=0}^{\infty},$$

where $\sigma_{pq}(\alpha) \in (\Phi^{\otimes(p+q)})'$; *for any* α, *the matrix* $(\sigma_{pq}(\alpha))_{p,q=0}^{\infty}$ *is positive definite in the sense of* (2.49). *Furthermore, the relation*

$$(\sigma_{pq}(\alpha))(\chi \otimes \varphi_{p+q-1}) = (\sigma_{pq}(\alpha))(\varphi_{p+q-1} \otimes \chi) = 0$$

$$(\chi \in M;\ \varphi_{p+q-1} \in \Phi^{\otimes(p+q-1)};\ p, q \in \mathbb{Z}_+,\ p+q \geq 1) \tag{2.54}$$

holds.

The integral (2.53) *is understood in the weak sense. The measure* $(\sigma_{pq}(\alpha))_{p,q=0}^{\infty}$ *is uniquely determined by* (2.53). *Conversely, after the transition from* $\tilde{K}$ *to* K, *an arbitrary matrix of the form* (2.53) *is a generalized moment matrix.*

Let us clarify this formulation. The mapping $\xi(\cdot)$ is a fixed mapping of a given form. The changes of it involve changes to the matrix measure in (2.53), and

$$\hat{M}_{Re} = \{\eta \in \Phi'_{Re} \mid \eta(\varphi) = 0,\ \varphi \in M\}.$$

The integral in (2.53) is understood as a weak limit (in the space $(\Phi^{\otimes(j+k)})'$) of finite sums of terms of the form

$$\underbrace{\xi(\lambda_\nu) \otimes \ldots \otimes \xi(\lambda_\nu)}_{j-p} \otimes\ \sigma_{pq}(\lambda) \otimes \underbrace{\xi(\lambda_\nu) \otimes \ldots \otimes \xi(\lambda_\nu)}_{k-q},$$

where $\lambda_\nu \in \alpha_\nu$, $\alpha_1, \ldots, \alpha_n$ is a decomposition of $\Phi'_{Re} / \hat{M}_{Re}$ into Borel sets. However, this integral can be also understood in the sense which becomes clear from the proof of the theorem; this statement is clarified in Remark 1.

If $M = \Phi$, then it follows from (2.54) that $\sigma_{pq} = 0$ for $p + q \geq 1$, and (2.53) transforms into (2.6) for s_{j+k} after a proper change of the order of variables.

Proof. The main steps are analogous with the proof of Theorem 2.1, and we only note all the distinctions.

I. The quasiscalar product is introduced by the formula

$$\langle\varphi, \psi\rangle = \sum_{j,k=0}^{\infty} (\tilde{K}_{jk}, \psi_j \otimes \bar{\varphi}_k)_{H_0^{\otimes(j+k)}} = \sum_{j,k=0}^{\infty} \overline{(K_{jk}, \bar{\psi}_j \otimes \varphi_k)}_{H_0^{\otimes(j+k)}}, \quad (2.55)$$

and the operation of shift $\mathcal{A}_e$ is introduced by the same formula (2.8) as above with $e = \bar{e} \in \mathrm{M}$. Then equalities (2.50) guarantee that relations (2.9) hold. As above, the mapping $C_0 \ni \varphi \mapsto \mathcal{A}_e \varphi \in C_0$ generates an Hermitian operator in the space H_K, where H_K is constructed from (2.55), just as H_s is constructed from (2.7).

II. With the help of the condition of definiteness, we prove the selfadjointness of A_e and A_l and their commutativity for $e, l \in \mathrm{M}_{\mathrm{Re}}$. Acting by analogy with Subsection 2.1. We then construct a family $A = (A_{e_\alpha})_{\alpha=1}^{\infty}$ of commuting selfadjoint operators A_{e_α}, where $e_\alpha \in \mathrm{E} \subset \mathrm{M}$ are such that the linear span of $((e_\alpha)_{\alpha=1}^{\infty})$ is dense in M.

IV. A quasinuclear chain

$$H_{-,K} \supset H_K \supset H_{+,K} \supset D_K$$

is constructed just as (2.12); one should only replace s_{2j} by K_{jj} in (2.13) and employ the estimate

$$|(\tilde{K}_{jk}, \varphi_j \otimes \bar{\varphi}_k)_{H_0^{\otimes(j+k)}}|^2 \le (\tilde{K}_{kk}, \varphi_j \otimes \bar{\varphi}_k)_{H_0^{\otimes 2k}} (\tilde{K}_{jj}, \varphi_j \otimes \bar{\varphi}_j)_{H_0^{\otimes 2j}} \quad (2.56)$$

$$(\varphi_k \in \Phi^{\otimes k},\ \varphi_j \in \Phi^{\otimes j};\ j, k \in \mathbb{Z}_+)$$

(the inequality (2.56) follows from the positive definiteness of the matrix $(a_{j\,k})_{j,k=1}^{\infty}$, where

$$a_{jk} = (\tilde{K}_{jk}, \varphi_j \otimes \varphi_k)_{H_0^{\otimes(j+k)}},$$

and this positive definiteness follows from (2.49) if we replace φ_l by $x_l \varphi_l$, $x_l \in \mathbb{C}^1$ in this inequality).

V–VII. These steps of the proof, in fact, repeat the corresponding steps in Subsection 2.1. The chain (2.16) transforms into the chain $H_- \supset G_0 \supset H_+$ with the involution $\leftarrow$ introduced with the help of the mapping

$$(\varphi_j)_{j=0}^{\infty} = \varphi \mapsto \bar{\varphi} = (\bar{\varphi}_j)_{j=0}^{\infty}.$$

Equality (2.18) has the form

$$(P(\lambda(\cdot))\hat{\varphi}, \hat{\psi})_{H_K} = (\Omega(\lambda(\cdot)), \psi \otimes \bar{\varphi})_{G_0 \otimes G_0} \quad (\varphi, \psi \in H_+), \tag{2.57}$$

where $\Omega(\lambda(\cdot)) \in H_- \otimes H_-$ is associated with matrix (2.19) (according to (2.17)). The fact that matrix (2.19) is Hermitian which follows from (2.57) means that

$$\Omega_{jk}(\lambda(\cdot)) = \bar{\Omega}_{kj}(\lambda(\cdot)) \quad (j, k \in \mathbb{Z}_+).$$

Equations (2.20) and (2.21) now have the form

$$(\Omega_{j+1\,k}(\lambda(\cdot)), e_\alpha \otimes \psi_j \otimes \varphi_k)_{H_0^{\otimes(j+k+1)}} = \lambda(\alpha)(\Omega_{jk}(\lambda(\cdot)), \psi_j \otimes \varphi_k)_{H_0^{\otimes(j+k)}}, \tag{2.58}$$

$$(\Omega_{j\,k+1}(\lambda(\cdot)), \psi_j \otimes e_\alpha \otimes \varphi_k)_{H_0^{\otimes(j+k+1)}} = \lambda(\alpha)(\Omega_{jk}(\lambda(\cdot)), \psi_j \otimes \varphi_k)_{H_0^{\otimes(j+k)}} \tag{2.59}$$

$$(\varphi_k \in H_{\tau'(2k)}^{\otimes k}, \psi_j \in H_{\tau'(2j)}^{\otimes j};\ j, k \in \mathbb{Z}_+;\ \alpha \in \mathbb{N};\ \lambda(\cdot) \in \pi).$$

Here the second equation (2.59) also follows from (2.58) due to the above-mentioned fact that matrix (2.19) is Hermitian. Unlike (2.20) and (2.21), the vectors e_α which appear in (2.58) and (2.59) form a total set only in $\mathbf{M}$ but not in the whole of Φ. Moreover, the elements K_{jk} of the matrix K generally speaking are not real with respect to the involution $\leftarrow$ in the space $(\Phi^{\otimes(j+k)})'$. Therefore, the quasiscalar product (2.55) gives no possibility of introducing an involution of the type $\circ$ in H_K, having given the involution $\leftarrow$ in the chain (2.16). It is also impossible to say that the operations $\mathcal{A}_e$ are real with respect to $\circ$ and that the elements $\Omega_{jk}(\lambda(\cdot))$ are real with respect to $\leftarrow$, etc.

We proceed to solve the system of equations (2.58), (2.59), i.e., we want to derive a result similar to Lemma 2.2.

It follows from the positive definiteness of $\Omega(\lambda(\cdot))$ that $\Omega_{00}(\lambda(\cdot)) \geq 0$.

In (2.58), we set $j = 0$ and $k = 0$. Then we get the equality

$$(\Omega_{10}(\lambda(\cdot)), e_\alpha)_{H_0} = \lambda(\alpha)\Omega_{00}(\lambda(\cdot)) \quad (\alpha \in \mathbb{N})$$

for the vector $\Omega_{10}(\lambda(\cdot)) \in H_{-\tau'(2)}$. This enables us to conclude that there exist vectors $\xi(\lambda(\cdot)), \omega_{10}(\lambda(\cdot)) \in H_{-\tau'(2)}$ such that

$$\Omega_{10}(\lambda(\cdot)) = \xi(\lambda(\cdot))\Omega_{00}(\lambda(\cdot)) + \omega_{10}(\lambda(\cdot)), \quad (\xi(\lambda(\cdot)), e_\alpha)_{H_0}$$

$$= \lambda(\alpha) \quad (\alpha \in \mathbb{N}), \quad (\omega_{10}(\lambda(\cdot)), \chi)_{H_0} = 0 \quad (\chi \in \mathrm{M}) \tag{2.60}$$

(more precisely, $\xi(\lambda(\cdot)) \in H_{-\tau'(2), \mathrm{Re}}$, because e_α is real with respect to $\leftarrow$ for any $\alpha \in \mathbb{N}$, and $\lambda(\alpha) \in \mathbb{R}^1$). Expression (2.60) is a general solution of Equation (2.58) for $j = 0$ and $k = 0$; in (2.60), a vector $\omega_{10}(\lambda(\cdot))$ of the indicated form is arbitrary ("an arbitrary constant").

In (2.58), we now set $j = 1$ and $k = 0$. Then, taking into account (2.60), we obtain the equality

$$(\Omega_{20}(\lambda(\cdot)), e_\alpha \otimes \psi_1)_{H_0^{\otimes 2}} = \lambda(\alpha)(\Omega_{10}(\lambda(\cdot)), \psi_1)_{H_0}$$

$$= \lambda(\alpha)(\xi(\lambda(\cdot)), \psi_1)_{H_0} \Omega_{00}(\lambda(\cdot)) + \lambda(\alpha)(\omega_{10}(\lambda(\cdot)), \psi_1)_{H_0^{\otimes 2}}$$

$$= (\xi(\lambda(\cdot)) \otimes \xi(\lambda(\cdot))\Omega_{00}(\lambda(\cdot)) + \xi(\lambda(\cdot)) \otimes \omega_{10}(\lambda(\cdot)), e_\alpha \otimes \psi_1)_{H_0^{\otimes 2}}$$

$$(\alpha \in \mathbb{N}, \quad \psi_1 \in H_{\tau'(2)})$$

for the vector $\Omega_{20}(\lambda(\cdot)) \in H_{-\tau'(4)}$. This gives the general solution of Equation (2.58) for $j = 1$ and $k = 0$, namely, there exists a vector $\omega_{20}(\lambda(\cdot)) \in H_{-\tau'(4)}$ ("an arbitrary constant") such that

$$\Omega_{20}(\lambda(\cdot)) = \xi(\lambda(\cdot)) \otimes \xi(\lambda(\cdot))\Omega_{00}(\lambda(\cdot)) + \xi(\lambda(\cdot)) \otimes \omega_{10}(\lambda(\cdot)) + \omega_{20}(\lambda(\cdot)),$$

$$(\omega_{20}(\lambda(\cdot)), \chi \otimes \psi_1)_{H_0^{\otimes 2}} = 0 \ (\chi \in \mathrm{M}, \ \psi_1 \in H_{\tau'(2)}). \tag{2.61}$$

By analogy, we obtain the expression for $\Omega_{01}(\lambda(\cdot))$ and $\Omega_{02}(\lambda(\cdot))$ (the factor ξ appear to the right), for $\Omega_{11}(\lambda(\cdot))$ (the factors appear on both sides), and so on. It is easy to check that the general solution of the system (2.58), (2.59) has the form

$$\Omega_{jk}(\lambda(\cdot)) = \sum_{p=0}^{j} \sum_{q=0}^{k} \underbrace{\xi(\lambda(\cdot)) \otimes \ldots \otimes \xi(\lambda(\cdot))}_{j-p}$$

$$\otimes\, \omega_{pq}(\lambda(\cdot)) \otimes \underbrace{\xi(\lambda(\cdot)) \otimes \ldots \otimes \xi(\lambda(\cdot))}_{k-q} \quad (j, k \in \mathbb{Z}_+; \ \lambda(\cdot) \in \pi), \tag{2.62}$$

where $\omega_{pq}(\lambda(\cdot)) \in H_{-\tau'(2p)}^{\otimes p} \otimes H_{-\tau'(2q)}^{\otimes q}$ satisfy the condition

$$(\omega_{pq}(\lambda(\cdot)))(\chi \otimes \varphi_{p+q-1}) = (\omega_{pq}(\lambda(\cdot)))(\varphi_{p+q-1} \otimes \chi) = 0 \tag{2.63}$$

$$(\chi \in \mathrm{M};\ \ \varphi_{p+q-1} \in \Phi^{\otimes(p+q-1)};\ \ p, q \in \mathbb{Z}_+,\ p+q \geq 1).$$

We note that, in (2.60) and (2.61), conditions (2.63) have been written more precisely, namely, φ_{p+q-1} has been taken from the appropriate positive Hilbert spaces. Clearly, this fact is not very important because one can proceed to these φ_{p+q-1} by a limit transition (for example, in the first expression of (2.63), we have $\varphi_{p+q-1} \in H^{\otimes(p-1)}_{\tau'(2(p-1))} \otimes H^{\otimes q}_{\tau'(2q)}$).

The weak measurability of $\Omega_{jk}(\lambda(\cdot))$ $(j, k \in \mathbb{Z}_+)$ follows from the measurability of $P(\lambda(\cdot))$ with respect to $\lambda(\cdot)$ and from the relations (2.57) and (2.19). Then employing the representations (2.60)–(2.62) and moving recursively, we establish the weak measurability of the vector function

$$\pi \ni \lambda(\cdot) \mapsto \omega_{pq}(\lambda(\cdot)) \in \Phi^{\otimes(p+q)}\ \ (p, q \in \mathbb{Z}_+)$$

with respect to the σ-algebra $\mathfrak{B}(\mathbb{R}^\infty)$ (restricted to π).

IX. It follows from (2.14), (2.57), (2.19), and (2.55) that for any $\varphi, \psi \in C_0$, we have

$$\sum_{j,k=0}^{\infty} (\tilde{K}_{jk}, \psi_j \otimes \bar{\varphi}_k)_{H_0^{\otimes(j+k)}} = \langle \varphi, \psi \rangle = (\hat{\varphi}, \hat{\psi})_{H_K}$$

$$= \int_\pi (P(\lambda(\cdot))\, \hat{\varphi}, \hat{\psi})_{H_K}\, d\rho(\lambda(\cdot))$$

$$= \int_\pi (\Omega(\lambda(\cdot)), \psi \otimes \bar{\varphi})_{G_0 \otimes G_0}\, d\rho(\lambda(\cdot))$$

$$= \sum_{j,k=0}^{\infty} \Big(\int_\pi \Omega_{jk}(\lambda(\cdot))\, d\rho(\lambda(\cdot)), \psi_j \otimes \bar{\varphi}_k \Big)_{H_0^{\otimes(j+k)}}. \tag{2.64}$$

Inserting expression (2.62) for $\Omega_{jk}(\lambda(\cdot))$ in (2.64), denoting

$$\mu_{pq}(\delta) = \int_\delta \omega_{pq}(\lambda(\cdot))\, d\rho(\lambda(\cdot)) \quad (\pi \supset \delta \in \mathfrak{B}(\mathbb{R}^\infty)),$$

and taking into account that φ and ψ are arbitrary, we obtain the following representation

$$\tilde{K}_{jk} = \sum_{p=0}^{j} \sum_{q=0}^{k} \int_\pi \underbrace{\xi(\lambda(\cdot)) \otimes \ldots \otimes \xi(\lambda(\cdot))}_{j-p}$$

$$\otimes\, d\mu_{pq}(\lambda(\cdot))\, \underbrace{\xi(\lambda(\cdot)) \otimes \ldots \otimes \xi(\lambda(\cdot))}_{k-q}, \tag{2.65}$$

which is, in fact, the representation (2.53). Due to the indicated measurability of $\omega_{pq}(\lambda(\cdot))$, the procedure of introducing a vector valued measure μ_{pq} is correct.

To pass to (2.53), it is necessary to consider a mapping

$$\mathbb{R}^{\infty} \supset \pi \ni \lambda(\cdot) \mapsto \xi(\lambda(\cdot)) = \lambda \in H_{-\tau'(2)},$$

where $\xi(\lambda(\cdot))$ is determined in (2.60) as above. The vectors λ do not run through the whole of $H_{-\tau'(2),\mathrm{Re}}$ because $(e_\alpha)_{\alpha=1}^{\infty}$ is total only in M. The domain to which these vectors belong can be obviously regarded as $\Phi'_{\mathrm{Re}} / \hat{M}_{\mathrm{Re}}$, and we can denote by $\xi(\lambda)$ the same vector interpreted in a natural way as a vector from Φ'_{Re} (we take a representative of a class). This mapping in which λ is interpreted as an element of $\Phi'_{\mathrm{Re}} / \hat{M}_{\mathrm{Re}}$ transports the measure μ_{pq} given on π into a certain measure σ_{pq} on $\Phi'_{\mathrm{Re}} / \hat{M}_{\mathrm{Re}}$. At the same time, (2.65) turns into (2.53) (for another choice of a representative, we have another transportation and other measures σ_{pq}).

The properties (2.54) of the matrix measure $(\sigma_{pq})_{p,\,q=0}^{\infty}$ clearly follow from the equalities (2.63). The positive definiteness of it (in the sense that inequality (2.49) should be satisfied) can be proved in the following way. Equality (2.57) and the fact that the operator $P(\lambda(\cdot))$ is nonnegative imply that the matrix $(\Omega_{jk}(\lambda(\cdot)))_{j,\,k=0}^{\infty}$ is positive definite in the indicated sense. The matrix $(\omega_{pq}(\lambda(\cdot)))_{p,\,q=0}^{\infty}$ is the matrix of the system (2.58), (2.59); therefore, the positive definiteness of the solution involves the positive definiteness of this matrix. This yields the positive definiteness of the matrix $(\sigma_{pq}(\lambda(\delta)))_{p,\,q=0}^{\infty}$, because it is obtained by the integration of $(\omega_{pq}(\lambda(\cdot)))_{p,\,q=0}^{\infty}$ with respect to the measure $d\rho(\lambda(\cdot))$ and subsequent mapping.

X. The uniqueness of the definition of a measure by (2.53) can be established in a way similar to the proof of Theorem 3.1 in Chapter 3. However, we leave this proof to the reader.

XI. The generalized moment property of the matrix (2.53) after the transition from $\tilde{K}$ to K can be established by direct test.

■

Remark 1. It follows from the proof of Theorem 2.3 that the integral in (2.53) can be regarded as a weak integral of the vector function

$$\Phi'_{\mathrm{Re}} / \hat{M}_{\mathrm{Re}} \ni \lambda \mapsto \underbrace{\xi(\lambda) \otimes \ldots \otimes \xi(\lambda)}_{j-p} \otimes\, \omega_{pq}(\lambda) \otimes \underbrace{\xi(\lambda) \otimes \ldots \otimes \xi(\lambda)}_{k-q} \in (\Phi^{\otimes (j+k)})'$$

with respect to a scalar measure $d\rho(\lambda(\cdot))$ on $\Phi'_{\mathrm{Re}} / \hat{M}_{\mathrm{Re}}$; this vector function is obtained from the term in (2.62) by the mapping $\lambda(\cdot) \mapsto \lambda$.

∎

Remark 2. Unlike Theorem 2.1, we do not only have the space $H_{-\tau(s),\,\mathrm{Re}}$ on the right-hand side of the representation (2.53) (see Remark 1 in Subsection 2.1): The values of the vector valued measures σ_{pq} lie in

$$H_{-\tau'(2p)}^{\otimes p} \otimes H_{-\tau'(2q)}^{\otimes q} \subset (\Phi^{\otimes(p+q)})'.$$

∎

2.5. A Family of Shift Operators and the Corresponding Fourier Transform

Consider the results obtained in Subsection 2.1 from a somewhat different point of view. Let $s = (s_n)_{n=0}^{\infty}$ $(s_n \in (\Phi^{\otimes n})')$ be a *definite* moment sequence. The operators from the family $A = (A_e)_{e \in \Phi_{\mathrm{Re}}}$ act in the space H_s constructed for the given quasiscalar product (2.7). Each of these operators is generated by the operation (2.8), i.e., it is a "shift operator"; it is selfadjoint and connected with the chain (2.12) in the standard way. Any two operators of this sort commute. Moreover, due to (2.8), we have the relation

$$A_{\alpha e + \beta l}\hat{\varphi} = \alpha A_e \hat{\varphi} + \beta A_l \hat{\varphi} \quad (e, l \in \Phi_{\mathrm{Re}};\ \alpha, \beta \in \mathbb{R}^1;\ \hat{\varphi} \in D_s), \tag{2.66}$$

and for any $\hat{\varphi} \in D_s$, the mapping $\Phi \ni e \mapsto A_e \hat{\varphi} \in H_s$ is weakly continuous. In other words, the family A realizes a nuclear continuous representation Φ_{Re} by selfadjoint operators in H_s. Thus, by virtue of Theorem 1.6 of Chapter 4 and Remark 1 to this theorem, the following spectral representation holds

$$A_e = \int_{\Phi'_{\mathrm{Re}}} \lambda(e)\, dE(\lambda), \quad \mathfrak{D}(A_e) = \left\{ f \in H_s \,\middle|\, \int_{\Phi'_{\mathrm{Re}}} |\lambda(e)|^2 d(E(\lambda)f, f)_{H_s} < \infty \right\}$$

$$(e \in \Phi_{\mathrm{Re}}). \tag{2.67}$$

One can derive the representation (2.6) for s from (2.67) just as (2.38) is derived from (2.46) in Subsection 2.3. However, in this case we cannot immediately pass in (2.6) to the integration over a negative Hilbert space (Remark 1 to Theorem 2.1).

Let us consider the Fourier transform with respect to generalized joint eigenvectors connected with decomposition (2.67). It is now convenient to use an approach which is formally distinct from that discussed in Subsection 3.1 of Chapter 3: roughly speaking, the family of these eigenvectors admits an analytic parametrization with respect to $\lambda(\cdot)$ (similar to the case of Jacobian matrices or Sturm–Liouville operators; see Examples 2.5 and 2.7 in Chapter 3).

Hence, we now employ representation (2.6). According to the above-mentioned remark, in this representation one can replace Φ'_{Re} by a certain Hilbert space $H_{-\tau(s),\,\mathrm{Re}}$ (the imbedding $H_{0,\,\mathrm{Re}} \subset H_{-\tau(s),\,\mathrm{Re}}$ is quasinuclear). Let $\varphi = (\varphi_j)_{j=0}^{\infty}$. A vector function

$$C_0 \ni \varphi = (\varphi_0, \varphi_1, \varphi_2, \ldots) \mapsto \tilde{\varphi}(\lambda) = \varphi_0 + (\varphi_1, \lambda)_{H_0}$$

$$+ (\varphi_2, \lambda \otimes \lambda)_{H_0^{\otimes 2}} + \ldots; \quad \tilde{\hat{\varphi}}(\lambda) = \tilde{\varphi}(\lambda) \qquad (2.68)$$

of a point $\lambda \in \Phi'_{\mathrm{Re}} \supset H_{-\tau(s),\,\mathrm{Re}}$ is called its Fourier transform. We note that φ is finite and, therefore, the series to the right in (2.68) terminates, and $\tilde{\varphi}(\lambda)$ is a "polynomial in $\lambda \in \Phi'_{\mathrm{Re}}$ with complex coefficients". This type of polynomial was investigated in Subsection 1.1 of Chapter 2. A collection of polynomials of this sort was denoted by $\mathcal{P}(\Phi'_{\mathrm{Re}})$.

To obtain the Parseval equality corresponding to the Fourier transform (2.68), it is necessary to exploit (2.7) and (2.6). For $\varphi, \psi \in C_0$, we obtain

$$(\hat{\varphi}, \hat{\psi})_{H_s} = \langle \varphi, \psi \rangle = \sum_{j,k=0}^{\infty} (s_{j+k}, \psi_j \otimes \overline{\varphi}_k)_{H_0^{\otimes(j+k)}}$$

$$= \sum_{j,k=0}^{\infty} \int_{\Phi'_{\mathrm{Re}}} (\underbrace{\lambda \otimes \ldots \otimes \lambda}_{j+k}, \psi_j \otimes \overline{\varphi}_k)_{H_0^{\otimes(j+k)}} d\sigma(\lambda)$$

$$= \int_{\Phi'_{\mathrm{Re}}} \Big(\sum_{j=0}^{\infty} (\underbrace{\lambda \otimes \ldots \otimes \lambda}_{j}, \psi_j)_{H_0^{\otimes j}} \Big) \Big(\sum_{k=0}^{\infty} (\underbrace{\lambda \otimes \ldots \otimes \lambda}_{k}, \overline{\varphi}_k)_{H_0^{\otimes k}} \Big) d\sigma(\lambda)$$

$$= \int_{\Phi'_{\mathrm{Re}}} \tilde{\varphi}(\lambda)\, \overline{\tilde{\psi}}(\lambda) d\sigma(\lambda).$$

Hence, the Parseval equality has the form

$$\langle \varphi, \psi \rangle = (\hat{\varphi}, \hat{\psi})_{H_s} = \int_{\Phi'_{\mathrm{Re}}} \tilde{\hat{\varphi}}(\lambda)\, \overline{\tilde{\hat{\psi}}}(\lambda) d\sigma(\lambda)$$

$$= \int_{H_{-\tau(s),\,\mathrm{Re}}} \tilde{\hat{\varphi}}(\lambda)\, \overline{\tilde{\hat{\psi}}}(\lambda) d\sigma(\lambda) \quad (\varphi, \psi \in C_0). \qquad (2.69)$$

Below, we show that Fourier transforms $\tilde{C}_0 = \mathcal{P}(\Phi'_{\text{Re}})$ form a dense set in the space

$$L_2(\Phi'_{\text{Re}}, \mathcal{B}(\Phi'_{\text{Re}}), \sigma) = L_2(\Phi'_{\text{Re}}, \sigma).$$

Therefore, extending the mapping (2.67) to the whole of H_s by continuity, we obtain an isometry $H_s \ni f \mapsto \tilde{f}(\cdot) \in L_2(\Phi'_{\text{Re}}, \sigma)$ between H_s and $L_2(\Phi'_{\text{Re}}, \sigma)$.

Let us compare these constructions with the construction presented in Subsection 3.1 of Chapter 3. It is easy to understand that the family of operators $A = (A_e)_{e \in \Phi_{\text{Re}}}$ possesses a strong cyclic vector in the sense of the definition given in Theorem 3.2 in Chapter 3. Clearly, the same vector (2.47) plays a role of this strong cyclic vector. This fact follows from the equality

$$A_{e_1} \dots A_{e_n} \Omega = (0, \dots, 0, e_1 \otimes \dots \otimes e_n, 0, 0, \dots)^{\wedge},$$

from the character of the set $\{e_1 \otimes \dots \otimes e_n | e_1, \dots, e_n \in \Phi_{\text{Re}}\}$ in the space $H^{\otimes n}_{\tau'(2n)}$ for any $n \in \mathbb{N}$, and from the method of construction of the chain (2.12). Therefore, according to Theorem 3.2 in Chapter 3, the spectrum of the family A is simple, and in the expression (3.4) of Chapter 3 we have only $\tilde{\varphi}_1(\lambda(\cdot))$.

The Parseval equality (3.7) in Chapter 3 takes the form

$$(f, g)_{H_s} = \int_{\tau} \tilde{f}_1(\lambda(\cdot)) \overline{\tilde{g}_1(\lambda(\cdot))} \, d\rho_{\tau}(\lambda(\cdot)) \quad (f, g \in H_s). \tag{2.70}$$

Here, the set τ is a subset of $\mathbb{R}^{\Phi_{\text{Re}}}$ and contains the set π from the spectral projection Theorem 2.7 in Chapter 3. However, if we take into account that there exists a linear dependence (2.66) between the operators A_e, then, according to what has been said in Subsection 1.4 of Chapter 4, one can choose Φ'_{Re} to play the role of τ. Then formula (2.70) with $f = \hat{\varphi}$ and $g = \hat{\psi}$ turns into (2.69). However, to prove that the measures $\rho_{\Phi'_{\text{Re}}}$ and σ coincide, one should be sure that $\tilde{C}_0 = \tilde{\hat{C}}_1$ is dense in $L_2(\Phi'_{\text{Re}}, \sigma)$.

The denseness of $\tilde{C}_0$ in $L_2(\Phi'_{\text{Re}}, \sigma)$ is proved just as Theorem 3.1 in Chapter 3. Here the fact that the space D_s from (2.12) is a base of the operator A_e $(e \in \Phi_{\text{Re}})$ follows from the definition of A_e, and the analogue of Lemma 3.1 in Chapter 3 is a consequence of (2.68) and (2.8). Indeed, the validity of this analogue means that the following equality holds:

$$(A_e \hat{\varphi})^{\sim}(\lambda) = (\lambda, e)_{H_0} \tilde{\hat{\varphi}}(\lambda) \quad (\varphi \in C_0, \ e \in \Phi_{\text{Re}}, \ \lambda \in \Phi'_{\text{Re}}), \tag{2.71}$$

i.e., that the equality

$$(\mathcal{A}_e\varphi)^{\sim}(\lambda) = (\lambda, e)_{H_0}\tilde{\varphi}(\lambda)$$

is valid, but the latter follows immediately from (2.68) and (2.8).

Hence, one can assume that *for a family of shift operators* $A = (A_e)_{e\in\Phi_{\mathrm{Re}}}$ *under consideration (which possesses the strong cyclic vector* $\Omega = (1, 0, 0, \ldots)^{\wedge}$), *the space* Φ'_{Re} *(rather,* $H_{-\tau(s),\,\mathrm{Re}}$) *is its generalized spectrum, the measure* σ *from the representation* (2.6) *is its spectral measure, and the mapping* (2.68) *is its Fourier transform on* $\hat{C}_0$. *The Parseval equality* (2.69) *holds; being extended by continuity, this equality establishes an isometry between* H_s *and the whole of* $L_2(\Phi'_{\mathrm{Re}}, \sigma)$ (note that our formulation is not completely precise, i.e., we can prove that the indicated set contains a generalized spectrum).

We do not give the expression for the generalized projector $P(\lambda)$ which can be obtained by use of the statement proved above.

Analogous constructions can be also realized for a generalized moment sequence studied in Subsection 2.4. However, in this case, a family $A = (A_e)_{e\in \mathrm{M}_{\mathrm{Re}}}$ does not possess a strong cyclic vector, the spectrum is not simple, and formulas (2.68) and (2.69) are much more complicated; for details, see the papers of Berezansky [7, 9, 10].

Example 2.9. The above-mentioned isometry between H_s and $L_2(\Phi'_{\mathrm{Re}})$ can be explained, for example, as follows. Consider a definite moment sequence $s = (s_n)_{n=0}^{\infty}$ ($s_n \in \mathcal{S}'(\mathbb{R}^{dn})$) ($d \in \mathbb{N}$) from Example 2.4. Assume that the Euclidean group

$$E_d^+ :\ \mathbb{R}^d \ni x \mapsto gx = \Lambda x + a \in \mathbb{R}^d$$

acts in $\mathbb{R}^d$, where Λ is a rotation by 0, and $a \in \mathbb{R}^d$ is a shift; $g = (a, \Lambda) \in E_d^+$. The space $\mathcal{S}(\mathbb{R}^d)$ is invariant under the action of E_d^+, i.e.,

$$E_d^+ :\ \mathcal{S}(\mathbb{R}^d) \ni \varphi \mapsto (T_g\varphi)(x) = \varphi(g^{-1}x) = \varphi(\Lambda^{-1}(x-a)) \in \mathcal{S}(\mathbb{R}^d).$$

For $\varphi_n \in \mathcal{S}(\mathbb{R}^{dn})$, we set

$$(T_g\varphi_n)(x_1, \ldots, x_n) = \varphi_n(g^{-1}x_1, \ldots, g^{-1}x_n) \quad ((x_1, \ldots, x_n \in \mathcal{S}(\mathbb{R}^d);\ g \in E_d^+).$$

Let $\varphi = (\varphi_j)_{j=0}^{\infty}$. Denote

$$T_g\varphi = (T_g\varphi_j)_{j=0}^{\infty} \quad (g \in E_d^+,\ T_g\varphi_0 = \varphi_0).$$

Let us consider a sequence s invariant with respect to E_d^+, i.e., for any $n \in \mathbb{N}$, we have

$$(T_g\varphi_n) = s_n(\varphi_n) \quad (\varphi_n \in \mathcal{S}(\mathbb{R}^{dn});\ g \in E_d^+).$$

The equality

$$\langle T_g\varphi, T_g\psi\rangle = \langle\varphi, \psi\rangle \quad (\varphi, \psi \in C_0;\ \ g \in E_d^+)$$

follows from (2.7). Clearly, $T_g\varphi,\ T_g\psi \in C_0$. This implies that T_g transforms a class from C_0 into a class (under factorization with respect to $\{\varphi \in C_0 | \langle\varphi, \varphi\rangle = 0\}$); therefore, the operator $T_g\hat{\varphi} = (T_g\varphi)^\wedge$ $\left(\varphi \in C_0;\ g \in E_d^+\right)$ is defined and its extension by continuity generates a unitary operator U_g $\left(g \in E_d^+\right)$ in H_s. Clearly, the mapping $E_d^+ \ni g \mapsto U_g$ is a unitary representation of E_d^+ in H_s.

Consider the representation (2.6) for s

$$s_n = \int_{\mathfrak{S}'_{\mathrm{Re}}(\mathbb{R}^d)} \underbrace{\lambda \otimes \ldots \otimes \lambda}_{n}\, d\sigma(\lambda) \quad (n \in \mathbb{Z}_+). \tag{2.71}$$

It is obvious that in order to guarantee the invariance of s, the measure in (2.71) should be invariant in a certain sense, as well. Namely, let us set

$$(T_g\lambda)(\varphi) = \lambda(T_g\varphi) \quad (\varphi \in \mathfrak{S}_{\mathrm{Re}}(\mathbb{R}^d);\ \ g \in E_x^+)$$

for $\varphi \in \mathfrak{S}'_{\mathrm{Re}}(\mathbb{R}^d)$; the mapping

$$\mathfrak{S}'_{\mathrm{Re}}(\mathbb{R}^d) \ni \lambda \mapsto T_g\lambda \in \mathfrak{S}'_{\mathrm{Re}}(\mathbb{R}^d)$$

transforms $\mathfrak{S}'_{\mathrm{Re}}(\mathbb{R}^d)$ into itself. It is easy to understand that s *is invariant if and only if the spectral measure* σ *is invariant with respect to the transformation* T_g *in* $\mathfrak{S}'_{\mathrm{Re}}(\mathbb{R}^d)$ $(g \in E_d^+)$. Indeed, to prove this fact one should note that

$$(T_g\hat{\varphi})^\sim = ((T_g\varphi)^\wedge)^\sim = (T_g\varphi)^\sim = \left(\left(T_g\varphi_j, \underbrace{\lambda \otimes \ldots \otimes \lambda}_{j}\right)_{H_0^{\otimes j}}\right)_{j=0}^{\infty}$$

$$= \left(\left(\varphi_j, \underbrace{(T_g\lambda) \otimes \ldots \otimes (T_g\lambda)}_{j}\right)_{H_0^{\otimes j}}\right)_{j=0}^{\infty} = \tilde{\varphi}(T_g\lambda) = \tilde{\hat{\varphi}}(T_g\lambda), \tag{2.72}$$

according to (2.68) (we clarify that it suffices to check the equality

$$\left(T_g\varphi_j, \underbrace{\lambda \otimes \ldots \otimes \lambda}_{j}\right)_{H_0^{\otimes j}} = \left(\varphi_j, \underbrace{(T_g\lambda) \otimes \ldots \otimes (T_g\lambda)}_{j}\right)_{H_0^{\otimes j}}$$

for $\varphi_j = \varphi^{(1)} \otimes \ldots \otimes \varphi^{(j)}$, where $\varphi^{(1)}, \ldots, \varphi^{(j)} \in \mathfrak{S}_{\mathrm{Re}}(\mathbb{R}^d)$, and, on these vectors, it follows from the definition of $T_g\lambda$). Substituting the vectors $T_g\hat{\varphi}$ and $T_g\hat{\psi}$ for $\hat{\varphi}$ and $\hat{\psi}$ in (2.69) and using the unitary of U_g, the formula (2.72), and the fact that $\hat{\hat{C}}_0$ is dense in $L_2(\Phi'_{\mathrm{Re}}, \sigma)$, we obtain the required result.

Example 2.10. Let us consider the definite moment sequence $s=(s_n)_{n=0}^{\infty}$ from Example 2.4 with $d=2, 3, \ldots$ and assume that it is invariant only with respect to the inversion of time. More precisely, we choose $\eta=(0, \Lambda_0)\in E_d^+$, where Λ_0 is a fixed rotation of points (x, t) in the space $\mathbb{R}^d=\mathbb{R}^{d-1}\times\mathbb{R}^1$ which maps the point $(0, 1)$ onto $(0, -1)$ and, therefore,

$$\mathbb{R}^d \ni (x, t) \mapsto \eta(x, t) = (x, -t) \in \mathbb{R}^d.$$

Assume that s is invariant in the sense of Example 2.9, but with respect to the only one transformation η of the group E_d^+. Taking into account that s is real with respect to the involution generated by the complex overbar (as a moment sequence from Example 2.4) and invariant with respect to η, we find that it is real with respect to the involution θ in Example 2.7. Assume, in addition, that s is an OS positive definite sequence. By virtue of (2.69), this property influences the character of its spectral measure σ in the following way: This measure is invariant with respect to the transformation T_η in $\mathcal{S}'_{\mathrm{Re}}(\mathbb{R}^d)$ and

$$\int_{\mathcal{S}'_{\mathrm{Re}}(\mathbb{R}^d)} \tilde{\varphi}(\lambda)\,\overline{\tilde{\varphi}(T_\eta\lambda)}\,d\sigma(\lambda) = \langle\varphi, T_\eta\varphi\rangle \ge 0$$

for every sequence $\varphi=(\varphi_j)_{j=0}^{\infty}\in C_0$ such that $\varphi_j\in\left(\mathcal{S}_+(\mathbb{R}^d)\right)^{\otimes j}$.

Example 2.11 (the Osterwalder–Schrader positiveness for a free Euclidean measure). Since the Schwinger functions $s_n^{\nu_{m_0}}$, $n\in\mathbb{Z}_+$, of a free Euclidean field from Example 2.8 satisfy all the assumptions of Example 2.10, the measure ν_{m_0} on $\mathfrak{B}\left(\mathcal{S}'_{\mathrm{Re}}(\mathbb{R}^d)\right)$ generating these functions possesses the important property of OS positiveness (or physical positiveness if we use other terminology). Namely, denote by $\mathcal{P}_+\left(\mathcal{S}'_{\mathrm{Re}}(\mathbb{R}^d)\right)$ a set of continuous polynomials on $\mathcal{S}'_{\mathrm{Re}}(\mathbb{R}^d)$ with generating kernels from $\left(\mathcal{S}_+(\mathbb{R}^d)\right)^{\otimes j}$, i.e., for

$$f\in\mathcal{P}_+\left(\mathcal{S}'_{\mathrm{Re}}(\mathbb{R}^d)\right)\quad f(\lambda) = \varphi_0 + \sum_{j=1}^{\infty}(\varphi_j, \lambda^{\otimes j})_{L_2(\mathbb{R}^{dj})}\quad (\lambda\in\mathcal{S}'_{\mathrm{Re}}(\mathbb{R}^d)),$$

we have $\varphi_j\in\left(\mathcal{S}_+(\mathbb{R}^d)\right)^{\otimes j}$ $(j=1,\ldots, n;\ n\in\mathbb{N})$, $\varphi_0\in\mathbb{C}^1$. The assertion of the previous example can be now written as follows

$$\int_{\mathcal{S}'_{\mathrm{Re}}(\mathbb{R}^d)} f(\lambda)\,\overline{f(T_\eta\lambda)}\,d\nu_{m_0} \ge 0\quad (f\in\mathcal{P}_+(\mathcal{S}'_{\mathrm{Re}}(\mathbb{R}^d))). \tag{2.73}$$

Sometimes it is convenient to write the property of OS positiveness by using other sets of functions from $\mathcal{S}'_{\mathrm{Re}}(\mathbb{R}^d)$. For example, let us introduce a collection $\mathfrak{T}(\mathcal{S}'_{\mathrm{Re}}(\mathbb{R}^d))$ of trigonometric polynomials on $\mathcal{S}'_{\mathrm{Re}}(\mathbb{R}^d)$ as a linear span of

$$\{\exp i(\varphi, \lambda)_{L_2(\mathbb{R}^d)} \mid \varphi \in \mathcal{S}_{\mathrm{Re}}(\mathbb{R}^d),\ \lambda \in \mathcal{S}'_{\mathrm{Re}}(\mathbb{R}^d)\}$$

and choose a subset $\mathfrak{T}_+(\mathcal{S}'_{\mathrm{Re}}(\mathbb{R}^d))$ in it which is determined similarly to $\mathfrak{T}(\mathcal{S}'_{\mathrm{Re}}(\mathbb{R}^d))$ but for $\varphi \in \mathcal{S}'_{+,\mathrm{Re}}(\mathbb{R}^d)$. Then (2.73) is valid for all $f \in \mathfrak{T}_+(\mathcal{S}'_{\mathrm{Re}}(\mathbb{R}^d))$; in this form, one can give a definition of the OS positiveness of a measure on $\mathfrak{B}(\mathcal{S}'_{\mathrm{Re}}(\mathbb{R}^d))$ without assuming that the moments of it exist.

2.6. Some Additional Remarks

Let us now consider a generalization for moment sequences of two problems in the classical power moment problem studied in Subsection 2.1, i.e., the truncated moment problem and the Stieltjes moment problem.

1. We fix $m \in \mathbb{Z}_+$ and consider a finite sequence $s = (s_n)_{n=0}^{2m}$, where $s_n \in (\Phi^{\otimes n})'$, and Φ is a nuclear space with involution — the same as in Subsection 2.1. It is called a moment sequence if the condition of positive definiteness holds, i.e., for any finite sequence $\varphi = (\varphi_j)_{j=0}^m$ ($\varphi_j \in \Phi^{\otimes j}$, $C_0^{(m)}$ denotes a collection of these), we have

$$\sum_{j,k=0}^{m} s_{k+j}(\varphi_j \otimes \overline{\varphi}_k) \geq 0. \tag{2.74}$$

The truncated moment problem consists in clarifying the possibility of establishing the formula (2.6) with $n = 0, \dots, 2m$ for s (i.e., the possibility of extending s to a moment space $(s_n)_{n=0}^{\infty}$) and in describing all spectral measures σ involved in this representation.

One can try to solve this problem according to the scheme of Subsection 2.1: Introduce in $C_0^{(m)}$ a quasiscalar product $\langle\varphi, \psi\rangle$ of the type (2.7) for the given quadratic form (2.74), and then construct the space H_s by identifying and completion. The operation of shift (2.8) should be introduced by the formula

$$\mathcal{A}_e\varphi = \mathcal{A}_e(\varphi_0, \dots, \varphi_{m-1}, 0) = (0, e \otimes \varphi_0, \dots, e \otimes \varphi_{m-1}) \quad (e \in \Phi_{\mathrm{Re}}). \tag{2.75}$$

Then the equalities

$$\langle\mathcal{A}_e\varphi, \psi\rangle = \langle\varphi, \mathcal{A}_e\psi\rangle, \quad \langle\mathcal{A}_e\varphi, \mathcal{A}_l\psi\rangle = \langle\mathcal{A}_l\varphi, \mathcal{A}_e\psi\rangle$$

$$(\varphi, \psi \in C_0^{(m)}\,; \;\; e,\, l \in \Phi_{\mathrm{Re}})$$

are analogues with (2.9). As above, $\mathcal{A}_e$ generates an Hermitian operator A_e in H_s but with a nondense domain of definition (see (2.75)); A_e and A_l commute in a certain sense (see the last of the equalities given above). We choose a sequence $(e_\alpha)_{\alpha=1}^{\infty}$ $(e_\alpha \in \Phi_{\mathrm{Re}})$ total in Φ and construct the corresponding sequence of operators $(A_{e_\alpha})_{\alpha=1}^{\infty}$. If it is possible to extend the operators A_{e_α} to commuting operators B_{e_α} which are selfadjoint in H_s, then we can employ the scheme of Subsection 2.1 and obtain the required representation (2.6).

Unfortunately, this extension can be realized without additional assumptions only in the rare cases:

(i) when $\Phi = \mathbb{C}^1$, and we study the classical truncated moment problem (here we have a single operator A_1);

(ii) when $m = 1$, i.e., our moment sequence has the form $s = (s_0, s_1, s_2)$ and is nondegenerate (this means that if $\varphi \neq 0$ in (2.74), then the inequality is strict).

In the second case, it suffices to consider $s_0 = 1$, when the inequality (2.74) means that a bilinear real valued form

$$\Phi_{\mathrm{Re}} \times \Phi_{\mathrm{Re}} \ni (\varphi, \psi) \mapsto a(\varphi, \psi) = s_2(\varphi, \psi) - s_1(\varphi)\, s_1(\psi)$$

is such that $a(\varphi, \varphi) > 0$ $(\varphi \neq 0)$. This enables us to find a Gaussian measure γ in Φ'_{Re} whose first and second moments are equal to s_1 and s_2, respectively (see Gelfand and Vilenkin [1, Chapter 3, Section 2, Theorem 2]; this can be also derived from what has been said in Subsection 1.5 and 1.9 of Chapter 2). This means that it suffices to set $\sigma = \gamma$. Having this solution, we can also describe all other spectral measures which give the representation (2.6) with $n = 0, 1, 2$. These and related results of G. Us are presented in more detail in his paper [1]. Another approach to similar problems can be found in the papers of Brüning [1, 2].

2. Let us dwell briefly upon the generalization of the Stieltjes moment problem. In the classical situation, the problem is reduced to establishing (for given moment sequence $s = (s_n)_{n=0}^{\infty}$ $(s_n \in \mathbb{R}^1)$) the condition under which the support of its spectral measure σ is concentrated on the semi-axis $[0, \infty)$, i.e., when the corresponding shift operator A_1 is nonnegative. This condition can be reduced to an additional condition of positive definiteness, namely, for any finite sequence $\varphi = (\varphi_j)_{j=0}^{\infty}$ $(\varphi_j \in \mathbb{C}^1)$, we have

$$\sum_{j,\,k=0}^{\infty} s_{j+k+1}\, \bar{\varphi}_j \varphi_k \geq 0. \tag{2.76}$$

It is easy to see that an inequality similar to (2.76) also ensures the nonnegativity of the operator A_e ($e \in \Phi_{\mathrm{Re}}$) in the proof of Theorem 2.1: This operator is nonnegative provided that

$$\sum_{j,k=0}^{\infty} s_{j+k+1}(e \otimes \varphi_j \otimes \bar{\varphi}_k) \geq 0 \quad (\varphi \in C_0). \tag{2.77}$$

If we assume that, for a definite moment sequence, conditions (2.77) hold with one or another set of vectors $e \in \mathrm{E} \subset \Phi_{\mathrm{Re}}$, then we find that the spectral measure σ corresponding to s has its support on a certain part of Φ'_{Re} (for example, on a cone). A more detailed presentation can be found in the paper by Shifrin [1].

§3. Positive Definite Functions of Infinitely Many Variables

Below, we obtain a spectral representation for positive definite (p.d.) functions $k(x)$ of points $x \in \mathbb{R}^\infty$ which generalizes the classical Bochner theorem. The positive definiteness of a function will be understood in the integral sense. For this purpose, we first introduce a certain product measure on $\mathbb{R}^\infty$.

Another point of view based on the pointwise definition of the positive definiteness function k and of its continuity in a certain topology is discussed in Section 4.

3.1. Some Auxiliary Facts

We fix a sequence of continuous positive probability weights on $\mathbb{R}^1$: $p = (p_\alpha)_{\alpha=1}^\infty$ ($p_\alpha(t) > 0$ ($t \in \mathbb{R}^1$), $p_\alpha \in C(\mathbb{R}^1)$ and

$$\int_{\mathbb{R}^1} p_\alpha(t)\,dt = 1).$$

Given this sequence, we construct a product measure $\mu = \times_{\alpha=1}^\infty \mu_\alpha$, where $d\mu_\alpha(x_\alpha) = p_\alpha(x_\alpha)\,dx_\alpha$, on the space $\mathbb{R}^\infty$ (see Subsections 1.2 and 1.3 in Chapter 2).

Together with p, we consider a sequence of weights $p^* = (p^*_\alpha)_{\alpha=1}^\infty$, where p^*_α is equal to the convolution of $p_\alpha(t)$ and $p_\alpha(-t)$, namely

$$p^*_\alpha(t) = \int_{\mathbb{R}^1} p_\alpha(s) p_\alpha(s-t) ds. \tag{3.1}$$

Just as p_α, the weights (3.1) are positive continuous probability weights. Denote $\mu^* = \times_{\alpha=1}^\infty \mu^*_\alpha$, where $d\mu^*_\alpha(x_\alpha) = p^*_\alpha(x_\alpha) dx_\alpha$. It is easy to prove that *if* $\mathbb{R}^\infty \ni x \mapsto f(x) \in \mathbb{C}^1$ *is measurable and summable with respect to the measure* μ^*, *then* $g(x, y) = f(y-x)$ *is a function of point* $(x, y) \in \mathbb{R}^\infty \times \mathbb{R}^\infty$ *which is measurable and summable with respect to* $\mu \times \mu$; *moreover,*

$$\int_{\mathbb{R}^\infty} \int_{\mathbb{R}^\infty} f(y-x) d\mu(x) d\mu(y) = \int_{\mathbb{R}^\infty} f(x) d\mu^*(x). \tag{3.2}$$

Indeed, let $f_n(x) = u_{f_n}(x_1, \ldots, x_{m_n})$ $(n \in \mathbb{N})$ be a sequence of cylindrical functions which converges in

$$L_1(\mathbb{R}^\infty, \mu^*) = L_1(\mathbb{R}^\infty, \mathfrak{C}_\sigma(\mathbb{R}^\infty), \mu^*)$$

to f. The equality (3.2) is immediately verified for $f = f_n$. In particular, it implies that the sequence $g_n(x, y) = f_n(y-x)$ is fundamental in

$$L_1(\mathbb{R}^\infty \times \mathbb{R}^\infty, \mu \times \mu) = L_1(\mathbb{R}^\infty \times \mathbb{R}^\infty, \mathfrak{C}_\sigma(\mathbb{R}^\infty \times \mathbb{R}^\infty), \mu \times \mu).$$

Therefore, g_n converges in this space to a certain limiting function, and it can be easily shown that the latter coincides with $g(x, y)$. Hence, g is measurable and summable. The equality (3.2) can be now obtained by the limit transition as $n \to \infty$ from the analogous equalities with $f = f_n$.

■

This, in particular, implies that if $\mu^*(\alpha) = 0$ $(\alpha \in \mathfrak{C}_\sigma(\mathbb{R}^\infty))$, then we also have

$$(\mu \times \mu)(\{(x, y) \in \mathbb{R}^\infty \times \mathbb{R}^\infty \mid y - x \in \alpha\}) = \mu^*(\alpha) = 0.$$

Therefore, if, for example, f is bounded almost everywhere with respect to μ^*, then $f(y-x)$ has the same property with respect to $\mu \times \mu$.

Denote by $\mathbb{R}^\infty(p)$ the set of all sequences $\lambda = (\lambda_\alpha)_{\alpha=1}^\infty$ for which the infinite product of Fourier transforms

$$\prod_{\alpha=1}^{\infty} |\tilde{p}_\alpha(\lambda_\alpha)| \neq 0, \infty \qquad \left(\tilde{p}_\alpha(\lambda_\alpha) = \int_{\mathbb{R}^1} e^{i\lambda_\alpha t} p_\alpha(t)\,dt\right) \tag{3.3}$$

converges (this set is nonempty, because always $(0, 0, \ldots) \in \mathbb{R}^\infty(p)$).

For every $\lambda \in \mathbb{R}^\infty(\lambda)$, *there exists a strong limit*

$$\lim_{n\to\infty} \exp\left(i \sum_{\alpha=1}^{n} \lambda_\alpha x_\alpha\right) \tag{3.4}$$

(in the sense of convergence in the space $L_2(\mathbb{R}^\infty, \mu^*)$*) which can be naturally denoted by* $e^{i(\lambda, x)_{l_2}}$*; this limit determines the vector function*

$$\mathbb{R}^\infty(p) \ni \lambda \mapsto e^{i(\lambda, x)_{l_2}} \in L_2(\mathbb{R}^\infty, \mu^*);$$

$|e^{i(\lambda, x)_{l_2}}| = 1$ $(\lambda \in \mathbb{R}^\infty(p))$ *for almost all* x *with respect to the measure* μ^*.

Indeed, let $n > m$, then

$$\left\| \exp\left(i \sum_{\alpha=1}^{n} \lambda_\alpha x_\alpha\right) - \exp\left(i \sum_{\alpha=1}^{m} \lambda_\alpha x_\alpha\right) \right\|^2_{L_2(\mathbb{R}^\infty, \mu^*)}$$

$$= 2 - 2\,\mathrm{Re} \int_{\mathbb{R}^\infty} \exp\left(i \sum_{\alpha=m+1}^{n} \lambda_\alpha x_\alpha\right) d\mu^*(x)$$

$$= 2\left(1 - \left(\prod_{\alpha=m+1}^{\infty} |\tilde{p}_\alpha(\lambda_\alpha)|\right)^2\right) \tag{3.5}$$

(we have used the equality $\tilde{p}^*_\alpha(\lambda) = |\tilde{p}_\alpha(\lambda)|^2$). Since the product (3.3) converges, we have

$$\prod_{\alpha=m+1}^{\infty} |\tilde{p}_\alpha(\lambda_\alpha)| \underset{n,m\to\infty}{\rightarrow} 1$$

and, therefore, the expression (3.5) tends to zero as $n, m \to \infty$. This proves the existence of the strong limit (3.4). We fix $\lambda \in \mathbb{R}^\infty(p)$ and extract from the sequence

$$f_n(x) = \exp\left(i \sum_{\alpha=1}^{n} \lambda_\alpha x_\alpha\right),$$

which converges to $e^{i(\lambda,x)_{l_2}}$ in the norm of the space $L_2(\mathbb{R}^\infty, \mu^*)$, a subsequence $(f_{n_\beta}(x))_{\beta=1}^\infty$ convergent μ^*-almost everywhere. Since $|f_{n_\beta}(x)| = 1$ $(x \in \mathbb{R}^\infty)$, we also have that $|e^{i(\lambda,x)_{l_2}}| = 1$ μ^*-almost everywhere.

■

Quite similarly, we prove that the strong limit

$$\lim_{n\to\infty} \exp\left(i \sum_{\alpha=1}^{n} \lambda_\alpha (y_\alpha - x_\alpha)\right) = e^{i(\lambda,y-x)_{l_2}} \tag{3.6}$$

(in the sense of convergence in $L_2(\mathbb{R}^\infty \times \mathbb{R}^\infty, \mu \times \mu)$) exists for all $\lambda \in \mathbb{R}^\infty(p)$; this limit determines the vector function

$$\mathbb{R}^\infty(p) \ni \lambda \mapsto e^{i(\lambda,y-x)_{l_2}} \in L_2(\mathbb{R}^\infty \times \mathbb{R}^\infty, \mu \times \mu).$$

The functions $e^{i(\lambda,y-x)_{l_2}}$ and $e^{i(\lambda,x)_{l_2}}$ (determined according to (3.4)) are connected by the relation (3.2).

Note that the existence of the limits (3.4) and (3.6) is a fact analogous to the existence of measurable linear functions on a space with a Gaussian measure (see Subsection 2.2 in Chapter 2).

We now give examples of computing the set $\mathbb{R}^\infty(p)$ in classical cases.

Example 3.1. Assume that μ is equal to a canonical Gaussian product measure γ_1, i.e., for any $\alpha \in \mathbb{N}$, we have $p_\alpha(t) = \pi^{-1/2} e^{t^2}$ (see Subsection 1.2 in Chapter 2). Then

$$\tilde{p}_\alpha(\lambda_\alpha) = \exp(-1/4\, \lambda_\alpha^2),$$

and condition (3.3) is satisfied if and only if $\sum_{\alpha=1}^\infty \lambda_\alpha^2 < \infty$. Hence, we now have $\mathbb{R}^\infty(p) = l_2$. Further,

$$p_\alpha^*(t) = (2\pi)^{-1/2} e^{-t^2/2} \qquad (t \in \mathbb{R}^1,\ \lambda \in \mathbb{N}),$$

and $\mu^* = \gamma_2$, where $2 = (2, 2, \dots)$.

Example 3.2. Assume that

$$p_\alpha(t) = \pi^{-1}(1 + t^2)^{-1} \qquad (t \in \mathbb{R}^1)$$

is a density of the Cauchy distribution for any $\alpha \in \mathbb{N}$. Here $\tilde{p}_\alpha(\lambda_\alpha) = e^{-|\lambda_\alpha|}$, and (3.3) leads to the equality $\mathbb{R}^\infty(p) = l_1$ and

$$p_\alpha^*(t) = 2/\pi(4+t^2)^{-1} \quad (t \in \mathbb{R}^1,\ \lambda \in \mathbb{N}).$$

Example 3.3. It is known (see, for example, Feller [1, Chapter 15, Section 3]) that for each fixed $\alpha \in (1, 2)$, the function $e^{-|s|^a}$ $\left(s \in \mathbb{R}^1\right)$ is a Fourier transform of a certain continuous positive probability weight. If we take a sequence of these weights to play a role of p, then $\mathbb{R}^\infty(p)$ is equal to l_α. Therefore, the set $\mathbb{R}^\infty(p) \subset \mathbb{R}^\infty$ may coincide with any space l_α, where $a \in [1, 2]$, provided that p is chosen properly.

3.2. Positive Definite Functions and Their Spectral Representation

Consider the space $\mathbb{R}^\infty$ with the measures μ and μ^* introduced on it in Subsection 3.1. A complex-valued function $k(x)$ $(x \in \mathbb{R}^\infty)$ which is measurable and almost everywhere bounded with respect to the measure μ^* is called positive definite if for any cylindrical function $\varphi(x) = u_\varphi(x_1, \dots, x_m)$ $(u_\varphi \in C_0^\infty(\mathbb{R}^\infty);\ m \in \mathbb{N})$, the inequality

$$\int_{\mathbb{R}^\infty}\int_{\mathbb{R}^\infty} k(y-x)\varphi(y)\overline{\varphi(x)}\,d\mu(x)d\mu(y) \geq 0 \tag{3.7}$$

holds (the fact that this definition is correct follows from what has been said in Subsection 3.1).

Below, we prove a Bochner type theorem for these positive definite functions. Henceforth, we always assume, in addition, that the weights $p_\alpha(t)$ are functions from $C^1(\mathbb{R}^1)$ bounded on the axis $\mathbb{R}^1$ and satisfying the following requirement

$$\int_{\mathbb{R}^1} (p'_\alpha(t))^2 p_\alpha^{-1}(t)dt < \infty \quad (\alpha \in \mathbb{N}). \tag{3.8}$$

Theorem 3.1. *Any positive definite function* $k(x)$ $(x \in \mathbb{R}^\infty)$ *admits a representation*

$$k(x) = \int_{\mathbb{R}^\infty(p)} e^{i(\lambda, x)_{l_2}} d\sigma(\lambda), \tag{3.9}$$

where $\sigma(\lambda)$ is a nonnegative finite measure on a σ-algebra

$$\mathfrak{C}_\sigma(\mathbb{R}^\infty(p)) = \{\alpha \cap \mathbb{R}^\infty(p) \mid \alpha \in \mathfrak{C}_\sigma(\mathbb{R}^\infty)\},$$

the integral of a vector function

$$\mathbb{R}^\infty(p) \ni \lambda \mapsto e^{i(\lambda, x)_{l_2}} \in L_2(\mathbb{R}^\infty, \mu^*)$$

converges strongly. The equality in (3.9) *is regarded as an equality for almost all $x \in \mathbb{R}^\infty$ with respect to the measure μ^*. Conversely, each integral of the form* (3.9) *is a positive definite function of the point $x \in \mathbb{R}^\infty$. The measure σ is uniquely determined for given k.*

Proof. Theorem is proved according to the scheme which has been already exploited in Section 2 to obtain the representations of moment sequences. First, we construct the corresponding spaces and operators.

Given $k(x)$, let us construct a Hilbert space H_k. We introduce in $L_2(\mathbb{R}^\infty, \mu^*)$ a quasiscalar product

$$\langle\varphi, \psi\rangle = \int_{\mathbb{R}^\infty}\int_{\mathbb{R}^\infty} k(y-x)\varphi(y)\overline{\psi(x)}\,d\mu(x)d\mu(y) \quad (\varphi, \psi \in L_2(\mathbb{R}^\infty, \mu^*)), \tag{3.10}$$

and then factorize and complete it; recall that $k(y-x)\mu \times \mu$ is almost everywhere bounded, and $\langle\varphi, \varphi\rangle \geq 0$, by virtue of (3.7). Since

$$\langle\varphi, \varphi\rangle \leq c\,\|\varphi\|^2_{L_2(\mathbb{R}^\infty, \mu)} \qquad (\varphi \in L_2(\mathbb{R}^\infty, \mu^*)),$$

we have that

$$\{\varphi \in L_2(\mathbb{R}^\infty, \mu) \mid \langle\varphi, \varphi\rangle = 0\}$$

forms a subspace in $L_2(\mathbb{R}^\infty, \mu)$. Denoting by $\hat{L}_2(\mathbb{R}^\infty, \mu)$ its orthogonal complement, we obtain $H_k \supset \hat{L}_2(\mathbb{R}^\infty, \mu)$ and, moreover, the imbedding $\hat{L}_2(\mathbb{R}^\infty, \mu) \to H_k$ is continuous.

On the collection $C^\infty_{\text{cyl}}(\mathbb{R}^\infty)$ of cylindrical functions $\varphi(x) = u_\varphi(x_1, \ldots, x_m)$ such that $u_\varphi \in C^\infty(\mathbb{R}^m)$, we introduce a differential operation by setting

$$C^\infty_{\text{cyl}}(\mathbb{R}^\infty) \ni \varphi(x) \mapsto \frac{i}{p(x_\alpha)}\left(\frac{\partial}{\partial x_\alpha} p(x_\alpha)\varphi\right)(x) \quad (\alpha \in \mathbb{N}). \tag{3.11}$$

Denote

$$D_\alpha = \{\varphi \in C^\infty(\mathbb{R}^\infty) \,|\, \varphi, (p_\alpha \varphi)' p_\alpha^{-1} \in L_2(\mathbb{R}^1, p_\alpha(t)\,dt)\} \quad (\alpha \in \mathbb{N}),\ e = (1, 1, \dots).$$

The set

$$\text{a.} \bigotimes_{\alpha=1;\, e}^{\infty} D_\alpha \subset \bigotimes_{\alpha=1;\, e}^{\infty} L_2(\mathbb{R}^1, p_\alpha(t)\,dt) = L_2(\mathbb{R}^\infty, \mu)$$

is correctly defined and dense in $L_2(\mathbb{R}^\infty, \mu)$.

Lemma 3.1. *The restriction of the operation* (3.11) *to* a. $\otimes_{\alpha=1}^{\infty} D_\alpha$ *is an Hermitian operator* A_α' *on* $L_2(\mathbb{R}^\infty, \mu)$.

Proof. It follows from (3.8) that $1 \in D_\alpha$ for any $\alpha \in \mathbb{N}$. Therefore, operation (3.11) transforms a. $\otimes_{\alpha=1;\,e}^{\infty} D_\alpha$ into $L_2(\mathbb{R}^\infty, \mu)$ and, hence, it can be regarded as an operator in $L_2(\mathbb{R}^\infty, \mu)$. Since a. $\otimes_{\alpha=1;\,e}^{\infty} D_\alpha$ is dense in $L_2(\mathbb{R}^\infty, \mu)$ (and thus in H_k), the lemma will be proved if we show that A_α' is Hermitian with respect to $\langle\cdot, \cdot\rangle$. Clearly, it suffices to establish this fact on the functions of the form $\varphi^{(1)}(x_1) \dots \varphi^{(m)}(x_m)$ $(\varphi^{(k)} \in D_k)$. We can assume that $m \geq \alpha$.

Without loss of generality, we can set $\alpha = 1$. Let $\varphi^{(k)}, \psi^{(k)} \in D_k$. Consider the expression

$$\langle A_1'(\varphi^{(1)}(x_1) \dots \varphi^{(m)}(x_m)), \psi^{(1)}(x_1) \dots \psi^{(m)}(x_m)\rangle = I.$$

We introduce a bounded function

$$f(t) = \int_{\mathbb{R}^\infty} \int_{\mathbb{R}^\infty} k(t, y_2 - x_2, y_3 - x_3, \dots)\, \varphi^{(2)}(y_2) \dots \varphi^{(m)}(y_m)$$

$$\times\, \overline{\psi^{(2)}(x_2)} \dots \overline{\psi^{(m)}(x_m)}\, d\mu((x_2, x_3, \dots))\, d\mu((y_2, y_3, \dots)),$$

and transform I in the following way

$$I = \int_{\mathbb{R}^1} \int_{\mathbb{R}^1} f(y_1 - x_1) \frac{i}{p_1(y_1)} (p_1(y_1)\varphi^{(1)}(y_1))' \overline{\psi^{(1)}(x_1)} p_1(x_1) p_1(y_1)\, dx_1\, dy_1$$

$$= i \int_{\mathbb{R}^1} f(y_1) \left(\int_{\mathbb{R}^1} (p_1(y_1 + x_1)\varphi^{(1)}(y_1 + x_1))' \overline{p_1(x_1)\psi^{(1)}(x_1)}\, dx_1 \right) dy_1$$

$$= i \int_{\mathbb{R}^1} f(y_1) \left(\int_{\mathbb{R}^1} p_1(y_1 + x_1) \varphi^{(1)}(y_1 + x_1)) \overline{(p_1(x_1)\, \psi^{(1)}(x_1))'} dx_1 \right) dy_1$$

$$= \langle \varphi^{(1)}(x_1) \ldots \varphi^{(m)}(x_m), A_1'(\psi^{(1)}(x_1) \ldots \psi^{(m)}(x_m)) \rangle .$$

To justify integration by parts in the above calculation, it suffices to note that the form of D_1 and the boundedness of the function $p_1(t)$ from above yield the validity of the inclusions

$$p_1(y_1 + \cdot\,)\varphi^{(1)}(y + \cdot\,), \ \ (p_1(y_1 + \cdot\,)\varphi^{(1)}(y_1 + \cdot\,))' \in L_2(\mathbb{R}^1, dt),$$

$$p_1(\cdot\,)\psi^{(1)}(\cdot\,), \ \ (p_1(\cdot\,)\psi^{(1)}(\cdot\,))' \in L_2(\mathbb{R}^1, dt),$$

and to use the selfadjointness in $L_2(\mathbb{R}^1, dt)$ of the maximal operator generated by the expression $i\, du/dt$.

Denote by A_α the closure of the operator A_α' *descended to the space* H_k ($\alpha \in \mathbb{N}$). The next lemma actually follows from the general Theorem 1.16, but it is convenient to present a sketch of its direct proof.

Lemma 3.2. *The operators* A_α ($\alpha \in \mathbb{N}$) *are selfadjoint and commuting.*

Proof. It suffices to establish the selfadjointness of the operators A_1 and A_2 and the fact that their resolvents $R_{z_1}(A_1)$ and $R_{z_2}(A_2)$ are commuting for some nonreal z_1 and z_2 (see Lemma 1.4). Let F be a linear set from $L_2(\mathbb{R}^1, p(x_2)dx_2)$. Denote by A_1'' the restriction of the operation (3.11) (for $\alpha = 1$) to a. $(D_1 \otimes F \otimes D_3 \otimes D_4 \otimes \ldots)$ (clearly, the product is taken with the stabilization e). By repeating the proof of Lemma 3.1, we find that A_1'' is Hermitian in H_k

We now prove that the closure A_1'' is selfadjoint in H_k provided that

$$\text{a.}\,(D_1 \otimes F \otimes D_3 \otimes D_4 \otimes \ldots)$$

is dense in H_k. Indeed, fix $\chi^{(2)}(x_2) \in F$, $\chi^{(3)}(x_2) \in D_3$, and $\chi^{(4)}(x_4) \in D_4, \ldots$, and form a product

$$\text{a.}\,(L_2(\mathbb{R}^1, p(x_1)dx_1) \otimes \chi^{(2)} \otimes \ldots \otimes \chi^{(m)} \otimes 1 \otimes 1 \otimes \ldots)$$

for fixed $m = 2, 3, \ldots$. Then we construct the quasiscalar product (3.10) on this linear set, factorize it and complete. As a result, we obtain a subspace $H_{k_{\chi^{(2)}, \ldots, \chi^{(m)}}}$ which is constructed, just as H_k, from $L_2(\mathbb{R}^1, p(x_1)\,dx_1)$ with the help of a measurable bounded positive definite function

$$k_{\chi^{(2)},\ldots,\chi^{(m)}}(x_1)$$

$$= \int_{\mathbb{R}^\infty}\int_{\mathbb{R}^\infty} k(x_1, y_2 - x_2, y_3 - x_3, \ldots)\,\chi^{(2)}(y_2)\ldots\chi^{(m)}(y_m)$$

$$\times\,\overline{\chi^{(2)}(x_2)\ldots\chi^{(m)}(x_m)}\,d\mu((x_2, x_3, \ldots))\,d\mu((y_2, y_3, \ldots))\quad (x_1 \in \mathbb{R}^1),$$

given on the axis (the quasiscalar product is defined on $\varphi(y_1), \psi(x_1) \in L_2(\mathbb{R}^1, p(t)\,dt)$ by formula (3.10) in which $\mathbb{R}^\infty$ and $k(y-x)\,d\mu(x)\,d\mu(y)$ are replaced by $\mathbb{R}^1$ and

$$k_{\chi^{(2)},\ldots,\chi^{(m)}}(y_1 - x_1)\,p_1(x_1)\,p_1(y_1)\,dx_1\,dy_1,$$

respectively.

Let us restrict the operator A_1'' to the linear set

$$\text{a.}\ (D_1 \otimes \chi^{(2)} \otimes \ldots \otimes \chi^{(m)} \otimes 1 \otimes 1 \otimes \ldots).$$

Being regarded as an operator in $H_{k_{\chi^{(2)},\ldots,\chi^{(m)}}}$, the operator obtained is generated by the mapping

$$D_1 \ni \varphi(x_1) \mapsto \frac{i}{p_1(x_1)}\left(\frac{\partial}{\partial x_1}p_1(x_1)\,\varphi\right)(x_1), \tag{3.12}$$

and it is easy to understand that it becomes selfadjoint after being closed. Indeed, if we introduce the functions $\varphi(x_1) \in L_2(\mathbb{R}^1, p(x_1)\,dx_1)$ instead of the function $f(x_1) = p(x_1)\,\varphi(x_1)$, then we find that this operator turns into an operator

$$f(x_1) \mapsto i\left(\frac{\partial}{\partial x_1} f\right)(x_1), \tag{3.13}$$

which should be regarded in a space constructed for given $k_{\chi^{(2)},\ldots,\chi^{(m)}}$ just as $H_{k_{\chi^{(2)},\ldots,\chi^{(m)}}}$ except that, in the scalar product, the measure $p(x_1)\,p(y_1)\,dx_1\,dy_1$ should be replaced by $dx_1\,dy_1$. The mapping (3.13) is *a priori* given on $C_0^\infty(\mathbb{R}^1)$. The closure of this operator is selfadjoint. This follows from the fact that, according to the ordinary Bochner theorem, the spectral measure is defined uniquely for given positive definite function (in this connection, see the book of Berezansky [5, Chapter 8, Section 1, Theorem 1.1; Section 3, Subsection 6]). But then the closure of an operator generated by

(3.12) is also selfadjoint.

Therefore, for $z \in \mathbb{C}^1 \setminus \mathbb{R}^1$, a linear set

$$\mathrm{a}.\left(\left(\left(\frac{i}{p_1(x_1)}\frac{\partial}{\partial x_1}p_1(x_1) - z\,1\right)D\right) \otimes \chi^{(2)} \otimes \ldots \otimes \chi^{(m)}\right)$$

is dense in the closure of a linear set

$$\mathrm{a}.\,(L_2(\mathbb{R}^1, p(x_1)\,dx_1) \otimes \chi^{(2)} \otimes \ldots \otimes \chi^{(m)})$$

in H_k. Sorting out all possible $\chi^{(k)}(x_k)$ and $m = 2, 3, \ldots$, we obtain that

$$\mathrm{a}.\left(\left(\left(\frac{i}{p(x_1)}\frac{\partial}{\partial x_1}p_1(x_1) - z\,1\right)D_1\right) \otimes F \otimes D_3 \otimes D_4 \otimes \ldots\right)$$

is dense in the closure of the product

$$(L_2(\mathbb{R}^1, p(x_1)\,dx_1) \otimes F \otimes D_3 \otimes D_4 \otimes \ldots)$$

in H_k, and this closure by assumption coincides with H_k. The last statement means that the closure of A''_1 in H_k is selfadjoint.

Setting above $F = D_2$, we find that A_1 is selfadjoint in H_k. Similarly, we can prove that A_2 is selfadjoint.

The commutativity of $R_{z_1}(A_1)$ and $R_{z_2}(A_2)$ on

$$\mathrm{a}.\,(((A'_1 - z_1\,1)D_1) \otimes (A'_2 - z_2\,1)D_2) \otimes D_3 \otimes D_4 \otimes \ldots)$$

is obvious; consequently, to prove the lemma it remains to show that this set is dense in H_k. But

$$\mathrm{a}.\,(D_1 \otimes ((A'_2 - z_2\,1)D_2) \otimes D_3 \otimes D_4 \otimes \ldots)$$

is dense in H_k. This follows from the selfadjointness of the operator A_2. Let us now construct the operator A''_1 by setting $F = (A'_2 - z_2 1)D_2$. According to what has been proved, the closure of A''_1 is selfadjoint in H_k, and this means that the required denseness takes place.

The proof of the theorem is based on the expansion in generalized joint eigenvectors of the family of selfadjoint operators $A = (A_\alpha)_{\alpha=1}^{\infty}$ in H_k by use of the spectral projection theorem in Chapter 3. For this purpose, we must construct the chain connected with A in the standard way.

First, we consider the space $L_2(\mathbb{R}^1, p_\alpha(x_\alpha)\,dx_\alpha)$ for fixed $\alpha \in \mathbb{N}$, and choose a positive weight $q_\alpha \in C(\mathbb{R}^1)$ such that the imbedding

$$W_2^1(\mathbb{R}^1, q_\alpha(x_\alpha)\,dx_\alpha) \subset L_2(\mathbb{R}^1, p_\alpha(x_\alpha)\,dx_\alpha)$$

is quasinuclear. This is possible by virtue of what has been said in Example 1.5 in Section 1 of Chapter 1. Then we renormalize the spaces $W_2^1(\mathbb{R}^1, q_\alpha(x_\alpha)\,dx_\alpha)$ with the help of Lemma 2.7 in Section 2 of Chapter 1 to guarantee that, for any $\alpha \in \mathbb{N}$, the function $\varphi(x_\alpha) = 1$ has, in this space, the norm equal to 1. After this, we consider the tensor product

$$\bigotimes_{\alpha=1;\,e}^{\infty} W_2^1(\mathbb{R}^1, q_\alpha(x_\alpha)\,dx_\alpha) \quad (e = (1, 1, \ldots))$$

of the spaces $W_2^1(\mathbb{R}^1, q_\alpha(x_\alpha)\,dx_\alpha)$ renormalizing them once again with the help of Lemma 2.5 in Section 2 of Chapter 1 in order that the imbedding

$$H_+ = \bigotimes_{\alpha=1;\,e}^{\infty} W_2^1(\mathbb{R}^1, q_\alpha(x_\alpha)\,dx_\alpha) \subset \bigotimes_{\alpha=1;\,e}^{\infty} L_2(\mathbb{R}^1, p_\alpha(x_\alpha)\,dx_\alpha) = L_2(\mathbb{R}^\infty, \mu) \quad (3.14)$$

be quasinuclear (another possibility is to take a proper weighted tensor product of these spaces renormalized by using only the above-mentioned Lemma 2.7).

As mentioned above, $\{\varphi \in L_2(\mathbb{R}^\infty, \mu) \mid \langle\varphi, \varphi\rangle = 0\}$ is a subspace in $L_2(\mathbb{R}^\infty, \mu)$, the intersection of it with H_+ obviously forms a subspace in H_+. Let $H_{+,k}$ be an orthogonal complement in H_+ to this subspace and let $H_+ \ni \varphi \to \hat{\varphi} \in H_{+,k}$ be the corresponding projector. Clearly, $H_{+,k} \subset \hat{L}_2(\mathbb{R}^\infty, \mu)$, and the imbedding $H_{+,k} \to \hat{L}_2(\mathbb{R}^\infty, \mu)$ is quasinuclear by virtue of (3.14). But $\hat{L}_2(\mathbb{R}^\infty, \mu) \to H_k$, and this imbedding is continuous. therefore, the imbedding $H_{+,k} \to H_k$ is quasinuclear. Taking the spaces $H_{+,k}$ and H_k as positive and zero space, respectively, we construct the chain

$$H_{-,k} \supset H_k \supset H_{+,k}. \quad (3.15)$$

After an appropriate extension, this chain can be taken as the desired chain.

Let us extend the rigging (3.15) by finding a linear topological space $D_k \subset H_{+,k}$ (the inclusion is topological) which lies in the domain of definition of each operator A_α $(\alpha \in \mathbb{N})$ and is transformed by this operator in $H_{+,k}$ continuously. Thus, we set

$$C_0 = \mathrm{a.}(C_0^\infty(\mathbb{R}^1) \otimes C_0^\infty(\mathbb{R}^1) \otimes \ldots) \subset C_{\mathrm{cyl}}^\infty(\mathbb{R}^\infty) \subset L_2(\mathbb{R}^\infty, \mu),$$

and assume that $C_0 \ni \varphi_n(x) \to 0$ as $n \to \infty$, provided that

(i) all functions $\varphi_n(x)$ are uniformly cylindrical, i.e.,

$$\varphi_n(x) = u_{\varphi_n}(x_1, \ldots, x_m) \quad (n \in \mathbb{N});$$

(ii) $u_{\varphi_n} \to 0$ in the sense of convergence in the Schwartz space $\mathcal{D}(\mathbb{R}^m)$.

It is clear that $C_0 \subset H_+$ and $C_0 \subset \mathcal{D}(A_\alpha)$ topologically (because $C_0^\infty(\mathbb{R}^1) \subset D_\alpha$ for any $\alpha \in \mathbb{N}$); it is also easy to understand that A_α transforms C_0 in H_+ continuously. If we factorize $\varphi \to \hat{\varphi}$ the space C_0 with respect to the linear set $\{\varphi \in C_0 | \langle\varphi, \varphi\rangle = 0\}$, then we get the required $D_k = \hat{C}_0$ (as far the similar factorization is concerned, see also Lemma 5.2 below). Hence, the role of the chain connected with A in the standard way is played here by the chain

$$H_{-,k} \supset H_k \supset H_+ \supset D_k. \tag{3.16}$$

Proof of Theorem. I. Let us apply Theorem 2.6 in Section 2 of Chapter 3 (Remark 4 in Subsection 2.10 of Chapter 3) to the family of operators $A = (A_\alpha)_{\alpha=1}^\infty$. The Parseval equality now has the form

$$\langle \hat{\varphi}, \hat{\psi} \rangle_{H_k} = \int_{\mathbb{R}^\infty} (P(\lambda)\hat{\varphi}, \hat{\psi})_{H_k} d\rho(\lambda) \quad (\hat{\varphi}, \hat{\psi} \in H_{+,k}), \tag{3.17}$$

where $\mathcal{C}_\sigma(\mathbb{R}^\infty) \ni \delta\rho \mapsto \rho(\delta)$ is a spectral measure, and $P(\lambda)$ is a generalized projector. There exists a set $\pi \in \mathcal{C}_\sigma(\mathbb{R}^\infty)$ of full ρ-measure such that, for $\lambda \in \pi$, $\mathfrak{R}(P(\lambda))$ consists of generalized joint eigenvectors of the operators A_α with the eigenvalues λ_α, i.e.,

$$(P(\lambda)\hat{\varphi}, A_\alpha\hat{\psi})_{H_k} = \lambda_\alpha(P(\lambda)\hat{\varphi}, \hat{\psi})_{H_k} \quad (\hat{\varphi} \in H_{+,k}, \ \hat{\psi} \in D_k; \ \alpha \in \mathbb{N}). \tag{3.18}$$

Parallel with (3.15), we consider the tensor square of the chain

$$H_- \supset G_0 = L_2(\mathbb{R}^\infty, \mu) \supset H_+,$$

generated by (3.14), i.e., the chain

$$H_- \otimes H_- \supset G_0 \otimes G_0 = L_2(\mathbb{R}^\infty \times \mathbb{R}^\infty, \mu \times \mu) \supset H_+ \otimes H_+. \tag{3.19}$$

The elements of the zero space of this chain are kernels $K(x, y)$ $(x, y \in \mathbb{R}^\infty)$ square summable with respect to the measure $\mu \times \mu$. Under certain conditions of "smoothness", these kernels form $H_+ \otimes H_+$; the elements of the negative space $H_- \otimes H_-$ may be interpreted as generalized kernels (in a certain sense).

By using the procedure already exploited when proving Theorem 2.1 (see also general Lemma 5.3 presented below) for given family of operators $P(\lambda)$, we can construct a family of elementary kernels $\Omega(\lambda) \in H_- \otimes H_-$, moreover,

$$\|\Omega(\lambda)\|_{H_- \otimes H_-} \leq c < \infty$$

uniformly in λ. The connection between $P(\lambda)$ and $\Omega(\lambda)$ is given by the equality

$$(P(\lambda)\hat{\varphi}, \hat{\psi})_{H_k} = (\Omega(\lambda), \psi \otimes \overline{\varphi})_{L_2(\mathbb{R}^\infty \times \mathbb{R}^\infty, \mu \times \mu)} \quad (\varphi, \psi \in H_+). \tag{3.20}$$

By virtue of (3.20) and the inclusion $C_0 \in H_+$, the positive definiteness of $P(\lambda)$ yields the positive definiteness of the kernel $\Omega(\lambda)$, i.e.,

$$(\Omega(\lambda), \psi \otimes \overline{\varphi})_{L_2(\mathbb{R}^\infty \times \mathbb{R}^\infty, \mu \times \mu)} \geq 0 \qquad (\varphi \in C_0). \tag{3.21}$$

Similarly, it follows from (3.18) and the form of A_α that the following relations hold

$$\left(\Omega(\lambda), \frac{i}{p_\alpha(x_\alpha)}\left(\frac{\partial}{\partial x_\alpha} p_\alpha(x_\alpha))\,\psi\right)(x)\,\overline{\varphi(y)} - \lambda_\alpha \psi(x)\overline{\varphi(y)}\right)_{L_2(\mathbb{R}^\infty \times \mathbb{R}^\infty, \mu\times\mu)} = 0,$$

$$\left(\Omega(\lambda), \psi(x)\overline{\frac{i}{p_\alpha(y_\alpha)}\left(\frac{\partial}{\partial y_\alpha}(p_\alpha(y_\alpha)\,\varphi)\,(y)\right)} - \lambda_\alpha \psi(x)\overline{\varphi(y)}\right)_{L_2(\mathbb{R}^\infty \times \mathbb{R}^\infty, \mu\times\mu)} = 0 \tag{3.22}$$

$$(\varphi, \psi \in C_0; \;\; \alpha \in \mathbb{N})$$

(the second equality in (3.22) follows from the first one and from the fact that $\Omega(\lambda)$ is Hermitian). Taking into account (3.17), (3.4), and (3.20), we obtain the following representation of $k(y-x)$ in the form of an integral convergent in norm of the space

$$k(y-x) = \int_{\mathbb{R}^\infty} \Omega(\lambda)\, d\rho(\lambda). \tag{3.23}$$

II. We now regard (3.22) as a system of equations (in generalized functions) with respect to unknown $\Omega(\lambda)$. Let us find its general solution.

Lemma 3.3. *Assume that* $\Omega(\lambda) \in H_- \otimes H_-$ *satisfy the system* (3.22) *for some* $\lambda \in \mathbb{R}^\infty$. *Then*

$$\Omega(\lambda) = c(\lambda) \lim_{n\to\infty} \frac{\exp\left(i\sum_{\alpha=1}^{n} \lambda_\alpha (y_\alpha - x_\alpha) \right)}{|\tilde{p}(\lambda_1) \dots \tilde{p}(\lambda_n)|^2} \quad (c(\lambda) \geq 0). \tag{3.24}$$

in the sense of convergence in norm of the space $H_- \otimes H_-$.

Proof. Fix $n \in \mathbb{N}$. The expression

$$(\Omega(\lambda), \psi(x_1, \dots, x_n)\overline{\varphi(y_1, \dots, y_n)})_{L_2(\mathbb{R}^\infty \times \mathbb{R}^\infty, \mu\times\mu)}$$

depends continuously on

$$\psi \otimes \overline{\varphi} \in \left(\bigotimes_{\alpha=1}^{n} W_2^1(\mathbb{R}^1, q_\alpha(x_\alpha)\, dx_\alpha) \right) \otimes \left(\bigotimes_{\alpha=1}^{n} W_2^1(\mathbb{R}^1, q_\alpha(x_\alpha)\, dx_\alpha) \right),$$

and, therefore, coincides with $(\Omega_n(\lambda), \psi \otimes \overline{\varphi})_{L_2 \otimes L_2}$, where

$$L_2 = L_2(\mathbb{R}^n, p_1(x_1) \dots p_n(x_n) dx_1 \dots dx_n),$$

$\Omega_n(\lambda) \in H_- \otimes H_-$, and H_- is a negative space obtained as a product of n negative spaces of the chains

$$H_{-,\alpha} \supset L_2(\mathbb{R}^1, p_\alpha(x_\alpha)\, dx_\alpha) \supset W_2^1(\mathbb{R}^1, q_\alpha(x_\alpha)\, dx_\alpha) \quad (\alpha = 1, \dots, n).$$

In particular, this coincidence occurs on the functions $\varphi, \psi \in (C_0^\infty(\mathbb{R}^1))^{\otimes n} \subset C_0$. This and (3.22) for $\alpha = 1, \dots, n$, yield the following system of equations of unknown $\Omega_n(\lambda)$:

$$\left(\Omega_n(\lambda), \frac{i}{p_\alpha(x_\alpha)} \left(\frac{\partial}{\partial x_\alpha} p_\alpha(x_\alpha)) \psi \right) \otimes \overline{\varphi} - \lambda_\alpha \psi \otimes \overline{\varphi} \right)_{L_2 \otimes L_2} = 0,$$

$$\left(\Omega_n(\lambda), \psi \otimes \overline{\frac{i}{p_\alpha(y_\alpha)} \left(\frac{\partial}{\partial y_\alpha} (p_\alpha(y_\alpha)\varphi) \right)} - \lambda_\alpha \psi \otimes \overline{\varphi} \right)_{L_2 \otimes L_2} = 0$$

$$(\varphi, \psi \in (C_0^\infty(\mathbb{R}^1))^{\otimes n};\ \alpha = 1, \dots, n).$$

If we pass in this system from the scalar products $(\cdot, \cdot)_{L_2 \otimes L_2}$ to the scalar products $(\cdot, \cdot)_{L_2(\mathbb{R}^{2n}, dx)}$ and employ the well-known facts concerning the smoothness of generalized solutions of ordinary differential equations (see Berezansky [5, Chapter 6, Section 5, Subsection 1; Chapter 8, Section 3, Subsection 3; Chapter 8, Section 4, Subsection 2]), then we conclude that $\Omega_n(\lambda)$ is an ordinary function of $2n$ variables which has the form

$$\Omega_n(\lambda; x_1, \dots, x_n, y_1, \dots, y_n) = \Omega_n(\lambda; 0, \dots, 0) \exp\left(i \sum_{\alpha=1}^{n} \lambda_\alpha (y_\alpha - x_\alpha) \right). \quad (3.25)$$

Multiplying (3.25) by $p_1(x_1) \dots p_n(y_n)$ and integrating over $x_1, \dots, y_n \in \mathbb{R}^1$, we get

$$\Omega_n(\lambda; 0, \dots, 0) | \tilde{p}_1(\lambda_1) \dots \tilde{p}_n(\lambda_n)|^2 = (\Omega_n(\lambda), 1 \otimes 1)_{L_2 \otimes L_2}$$

$$= (\Omega(\lambda), 1 \otimes 1))_{L_2(\mathbb{R}^\infty \times \mathbb{R}^\infty, \mu \times \mu)} = c(\lambda) \ge 0.$$

Determining $\Omega_n(\lambda; 0, \dots, 0)$ from this relation and inserting it in (3.25), we obtain the expression for $\Omega_n(\lambda; x_1, \dots, x_n, y_1, \dots, y_n)$ identical to the right-hand side of (3.24).

On the other hand, $\Omega_n(\lambda) \to \Omega(\lambda)$ as $n \to \infty$ in the sense of $H_- \otimes H_-$. This follows from the fact that $\Omega_n(\lambda)$ may be regarded as an orthogonal projection in this space onto the isometric image of the space

$$\left(\bigotimes_{\alpha=1}^{n} W_2^1(\mathbb{R}^1, q_\alpha(x_\alpha)\, dx_\alpha) \right) \otimes \left(\bigotimes_{\alpha=1}^{n} W_2^1(\mathbb{R}^1, q_\alpha(x_\alpha)\, dx_\alpha) \right),$$

regarded as a subspace of $H_+ \otimes H_+$.

Note that it is not essential for formula (3.24) that $\tilde{p}(\lambda_k)$ may be 0, because this formula in fact contains

$$c(\lambda) | \tilde{p}_1(\lambda_1) \dots \tilde{p}_n(\lambda_n)|^{-2} = \Omega_n(\lambda; 0, 0) \ge 0.$$

Substituting (3.24) in (3.23), we get

$$k(y-x) = \lim_{n \to \infty} \int_{\mathbb{R}^\infty} c(\lambda) \frac{\exp\left(i \sum_{\alpha=1}^{n} \lambda_\alpha (y_\alpha - x_\alpha) \right)}{|\tilde{p}(\lambda_1) \dots \tilde{p}(\lambda_n)|^2} \, d\rho(\lambda). \quad (3.26)$$

Here, the integral and the limit are understood in the sense of the norm of the space $H_- \otimes H_-$.

■

III. **Lemma 3.4.** *The infinite product* (3.3) *converges for almost all* $\lambda \in \mathbb{R}^\infty$ *with respect to the measure* ρ, *i.e.,* $\rho(\mathbb{R}^\infty(p)) = 1$.

Proof. Since $| \tilde{p}_\alpha(\lambda)| \le 1$ for any $\alpha \in \mathbb{N}$, there exists a finite positive (or equal to $+\infty$) limit

$$f(\lambda) = \lim_{n\to\infty} |\tilde{p}_1(\lambda_1) \dots \tilde{p}_n(\lambda_n)|^{-2} = \left(\prod_{\alpha=1}^{\infty} |\tilde{p}_\alpha(\lambda_\alpha)| \right)^{-2} \tag{3.27}$$

for any $\lambda \in \mathbb{R}^\infty$. We must show that

$$\rho(\{\lambda \in \mathbb{R}^\infty \,|\, f(\lambda) = +\infty\}) = 0.$$

Let $\varphi(x_1, \dots, x_m) \in C_0^\infty(\mathbb{R}^m)$ be nonnegative and $\tilde{\varphi}(\lambda_1, \dots, \lambda_m)$ be an m-dimensional Fourier transform of φ. Denote

$$\psi(x_1, \dots, x_m) = \varphi(x_1, \dots, x_m) p_1^{-1}(x_1) \dots p_m^{-1}(x_m);$$

we regard this function as a cylindrical function $\psi(x)$ from $H_+ \otimes H_+$. By virtue of (3.26), we obtain

$$\int_{\mathbb{R}^\infty} \int_{\mathbb{R}^\infty} k(y-x)\psi(y)\overline{\psi(x)}\, d\mu(x)\, d\mu(y)$$

$$= \lim_{n\to\infty} \int_{\mathbb{R}^\infty} \frac{c(\lambda)}{|\tilde{p}_1(\lambda_1) \dots \tilde{p}_n(\lambda_n)|^2} \left(\exp\left(i\sum_{\alpha=1}^{n} \lambda_\alpha (y_\alpha - x_\alpha) \right), \right. \tag{3.28}$$

$$\left. \psi(y)\overline{\psi(x)} \right)_{L_2(\mathbb{R}^\infty \times \mathbb{R}^\infty,\, \mu\times\mu)} d\rho(\lambda)$$

$$= \int_{\mathbb{R}^\infty} \frac{c(\lambda)}{|\tilde{p}_1(\lambda_1) \dots \tilde{p}_m(\lambda_m)|^2} |\tilde{\varphi}(\lambda_1, \dots, \lambda_m)|^2 \, d\rho(\lambda)$$

for $n \ge m$. We now take a δ-shaped sequence $\varphi_\beta(x_1, \dots, x_m)$, contracting to the δ-function at 0 with respect to the Lebesgue measure and ordinary topology in $\mathbb{R}^m$, to play the role of $\varphi(x_1, \dots, x_m)$. Then the left-hand side of (3.28) can be estimated above uniformly in β by

$$c_1 = \operatorname{sup\,ess} |k(y-x)|$$

(since

$$\int_{\mathbb{R}^\infty} \varphi_\beta(x)\, d\mu(x) = 1),$$

and the integrand on the right-hand side converges to $c(\lambda)\,|\,\tilde{p}_1(\lambda_1)\,\ldots\,\tilde{p}_m(\lambda_m)|^{-2}$ as $\beta\to\infty$. By virtue of the Fatou lemma, we conclude that

$$\int\limits_{\mathbb{R}^\infty} \frac{c(\lambda)}{|\tilde{p}_1(\lambda_1)\,\ldots\,\tilde{p}_m(\lambda_m)|^2}\,d\rho(\lambda) \le c_1 < \infty \qquad (m\in\mathbb{N}). \tag{3.29}$$

Taking into account that $0 < f(\lambda) \le +\infty$ for every $\lambda\in\mathbb{R}^\infty$, passing to the limit in (3.29) as $m\to\infty$, and employing the Fatou lemma once again, we conclude that $c(\lambda)f(\lambda)$ is summable and, therefore, $f(\lambda) < +\infty$ ρ-almost everywhere.

■

Note that we have also established the summability of $c(\lambda)\,f(\lambda)$ over $\mathbb{R}^\infty$ with respect to ρ.

IV. For $\lambda\in\mathbb{R}^\infty(p)$, there exists a positive finite limit

$$f(\lambda) = \lim_{n\to\infty} |\,\tilde{p}_1(\lambda_1)\,\ldots\,\tilde{p}_n(\lambda_n)|^2.$$

But then Lemma 3.3 implies that there exists a limit

$$\lim_{n\to\infty}\exp\left(i\sum_{\alpha=1}^{n}\lambda_\alpha(y_\alpha - x_\alpha)\right),$$

in the sense of the norm of the space $H_-\otimes H_-$, and $\Omega(\lambda)$ is equal to this limit multiplied by $c(\lambda)f(\lambda\varpi)$. Replacing $\mathbb{R}^\infty$ by $\mathbb{R}^\infty(p)$ in (3.23) (this is possible due to Lemma 3.4), using this equality for $\Omega(\lambda)$, and denoting

$$c(\lambda)f(\lambda)\,d\rho(\lambda) = d\sigma(\lambda) \quad (\sigma(\mathbb{R}^\infty(p)) < \infty),$$

we obtain

$$k(y-x) = \lim_{n\to\infty}\int\limits_{\mathbb{R}^\infty(p)}\exp\left(i\sum_{\alpha=1}^{n}\lambda_\alpha(y_\alpha - x_\alpha)\right)d\sigma(\lambda). \tag{3.30}$$

The limit and the integral in (3.30) exist in the norm of $H_-\otimes H_-$.

V. It remains to pass through several simple steps to derive the representation (3.9) from (3.30). Consider a vector function

$$\mathbb{R}^\infty(p) \ni \lambda \mapsto e^{i(\lambda,x)_{l_2}} \in L_2(\mathbb{R}^\infty, \mu^*),$$

defined by the relation (2.4). Clearly, it is measurable with respect to $\mathfrak{C}_\sigma(\mathbb{R}^\infty)$ and

$$\| e^{i(\lambda,\cdot)_{l_2}} \|_{L_2(\mathbb{R}^\infty, \mu^*)} = 1.$$

Therefore, there exists a strong integral $\int_{\mathbb{R}^\infty(p)} e^{i(\lambda,\cdot)_{l_2}} \, d\sigma(\lambda)$, where σ has been introduced, and we can write

$$\int_{\mathbb{R}^\infty(p)} e^{i(\lambda,x)_{l_2}} \, d\sigma(\lambda) = \lim_{n\to\infty} \int_{\mathbb{R}^\infty(p)} \exp\left(i \sum_{\alpha=1}^{n} \lambda_\alpha (y_\alpha - x_\alpha) \right) d\sigma(\lambda) \tag{3.31}$$

in the sense of strong convergence in $L_2(\mathbb{R}^\infty, \mu^*)$.

Let $\varphi(x) = u_\varphi(x_1, \ldots, x_m)$ ($u_\varphi \in C_0^\infty(\mathbb{R}^m)$). Using (3.2), the inclusion $\varphi(y-x) \in H_+ \otimes H_+$, (3.30), and again (3.2), we get from (3.31)

$$\int_{\mathbb{R}^\infty} \left(\int_{\mathbb{R}^\infty(p)} e^{i(\lambda,x)_{l_2}} \, d\sigma(\lambda) \right) \overline{\varphi(x)} \, d\mu^*(x)$$

$$= \lim_{n\to\infty} \int_{\mathbb{R}^\infty(p)} \left(\int_{\mathbb{R}^\infty} \exp\left(i \sum_{\alpha=1}^{n} \lambda_\alpha x_\alpha \right) \overline{\varphi(x)} \, d\mu^*(x) \right) d\sigma(\lambda)$$

$$= \lim_{n\to\infty} \int_{\mathbb{R}^\infty(p)} \left(\int_{\mathbb{R}^\infty} \int_{\mathbb{R}^\infty} \exp\left(i \sum_{\alpha=1}^{n} \lambda_\alpha (y_\alpha - x_\alpha) \right) \overline{\varphi(y-x)} \, d\mu(x) \, d\mu(y) \right) d\sigma(\lambda)$$

$$= \int_{\mathbb{R}^\infty} \int_{\mathbb{R}^\infty} k(y-x) \, \overline{\varphi(y-x)} \, d\mu(x) \, d\mu(y) = \int_{\mathbb{R}^\infty} k(x) \, \overline{\varphi(x)} \, d\mu^*(x).$$

The validity of the equality (3.9) for μ^*-almost all $x \in \mathbb{R}^\infty$ follows from the arbitrariness of $\varphi(x)$.

VI. Let us establish the converse statement of the theorem. Above, we have shown that the integral in (3.9) exists in the space $L_2(\mathbb{R}^\infty, \mu^*)$. It is easy to see that the function $k(x)$, determined by this integral, is bounded μ^*-almost everywhere. Indeed,

$$k(x) = \lim_{n\to\infty} \int_{\mathbb{R}^\infty(p)} \exp\left(i \sum_{\alpha=1}^{n} \lambda_\alpha x_\alpha \right) d\sigma(\lambda)$$

in the sense of the strong convergence in $L_2(\mathbb{R}^\infty, \mu^*)$. From this sequence, one can extract a subsequence convergent to $k(x)$ μ^*-almost everywhere. At the same time, the terms of this sequence can be estimated (in modulus) from above by the number $\sigma(\mathbb{R}^\infty(p)) < \infty$, and this proves the boundedness.

It is easy to understand that for $k(x)$ under consideration, we can write the equality

$$k(y-x) = \int\limits_{\mathbb{R}^\infty(p)} e^{i(\lambda,\, y-x)_{l_2}}\, d\sigma(\lambda), \tag{3.32}$$

where $e^{i(\lambda,\, y-x)_{l_2}}$ is a vector function of the point $\lambda \in \mathbb{R}^\infty(p)$ defined by the relation (3.6) and taking values in the space $L_2(\mathbb{R}^\infty \times \mathbb{R}^\infty, \mu \times \mu)$; the integral converges in the norm of this space. Inserting the expression (3.6) for $e^{i(\lambda,\, y-x)_{l_2}}$ in (3.32), we obtain an equality of the form (3.30), but with the convergence in norm of $L_2(\mathbb{R}^\infty \times \mathbb{R}^\infty, \mu \times \mu)$. With the help of this equality, inequality (3.7) is verified trivially. Thus, any function of the form (3.9) is positive definite.

VII. Let us show that (3.9) uniquely determines a measure. Assume that

$$\int\limits_{\mathbb{R}^\infty(p)} e^{i(\lambda,x)_{l_2}}\, d\sigma'(\lambda) = \int\limits_{\mathbb{R}^\infty(p)} e^{i(\lambda,x)_{l_2}}\, d\sigma''(\lambda) \tag{3.33}$$

for μ^*-almost all $x \in \mathbb{R}^\infty$. We now establish that $\sigma' = \sigma''$ or (which is the same) that

$$d\tau'(\lambda) = f^{-1}(\lambda)\, d\sigma'(\lambda) = f^{-1}(\lambda)\, d\sigma''(\lambda) = d\tau''(\lambda)$$

($f(\lambda)$ is determined by the relation (3.27)). If we multiply (3.33) by

$$\varphi(x) = u(x_1, \ldots, x_m) p_1^{-1}(x_1) \ldots p_m^{-1}(x_m),$$

where $u \in C_0^\infty(\mathbb{R}^m)$, and integrate over μ^*, then we get

$$\int\limits_{\mathbb{R}^\infty(p)} \tilde{u}(\lambda_1, \ldots, \lambda_m) |\, \tilde{p}_1(\lambda_1) \ldots \tilde{p}_m(\lambda_m)|^2\, d(\tau' - \tau'')(\lambda) = 0.$$

Due to the arbitrariness of $u \in C_0^\infty(\mathbb{R}^m)$, this implies that $\tau'(\alpha \times \mathbb{R}^\infty) = \tau''(\alpha \times \mathbb{R}^\infty)$ for arbitrary $\alpha \in \mathfrak{B}(\mathbb{R}^m)$. The last equality implies the coincidence of measures $d\tau'(\lambda)$ and $d\tau''(\lambda)$.

■

3.3. Positive Definite Functions on a Layer

It is well known that a positive definite function of a single variable given, not on the whole axis $\mathbb{R}^1$, but on a certain interval $(-2l, 2l)$ admits an extension (generally speaking, nonunique) to a positive definite function on $\mathbb{R}^1$.[4] Let us show that a similar fact holds also for positive definite functions of infinitely many variables.

Consider the space $\mathbb{R}^\infty$ with the measures μ and μ^* introduced in Subsection 3.1. As before, we assume that condition (3.8) on weights holds. A set

$$\{x = (x_\alpha)_{\alpha=1}^\infty = (x_1, x') \in \mathbb{R}^\infty \,|\, x_1 \in (-l, l)\} = (-l, l) \times \mathbb{R}^\infty$$

is called a layer $\mathbb{R}_l^\infty$ ($l \in (0, \infty)$) in the space $\mathbb{R}^\infty$. Clearly, $\mathbb{R}_l^\infty \in \mathfrak{C}(\mathbb{R}^\infty)$, and $\mu(\mathbb{R}_l^\infty)$, $\mu^*(\mathbb{R}_l^\infty) > 0$.

A complex-valued function $k(x)$ $(x \in \mathbb{R}_{2l}^\infty)$ measurable and almost everywhere bounded with respect to the measure μ^*, is called a positive definite function on a layer $\mathbb{R}_{2l}^\infty$ provided that, for any cylindrical function

$$\varphi(x) = u_\varphi(x_1, \dots, x_m) \quad (u_\varphi \in C_0^\infty(\mathbb{R}^m);\; m \in \mathbb{N}),$$

the inequality

$$\int_{\mathbb{R}_l^\infty} \int_{\mathbb{R}_l^\infty} k(y - x)\, \overline{\varphi(x)}\, d\mu(x) d\mu(y) \geq 0 \tag{3.34}$$

holds (cf. (3.7)).

Theorem 3.2. *Every function* $k(x)$ $(x \in \mathbb{R}_{2l}^\infty)$ *positive definite on a layer admits the representation* (3.9). *This means that it can be extended (generally speaking, ambiguously) to a positive definite function on the whole of* $\mathbb{R}^\infty$.

Proof. The proof is similar to that of Theorem 2.1 in quasidefinite case; namely, we employ the scheme developed in Subsection 3.2 and the fact concerning the existence of the required extension of the operator A analogous to Lemma 2.1 (here this operator is only Hermitian).

I. Given the positive definite function $k(x)$ $(x \in \mathbb{R}_{2l}^\infty)$, we construct a Hilbert space H_k by introducing in $L_2(\mathbb{R}_l^\infty, m \times \mu')$, where $\mu' = \times_{\alpha=2}^\infty \mu_\alpha$, m is a Lebesgue measure, a quasiscalar product with the help of the formula (3.10) in which $\mathbb{R}^\infty$ and

---

[4] This means that the spectral Bochner representation can be written for this function.

$d\mu(z)$ are replaced by $\mathbb{R}_l^\infty$ and $dz_1 \times \mu'(z')$, respectively. The chain (3.15) is constructed as before by multiplying the chains (3.14) for $\alpha = 2, 3, \ldots$, and the chain

$$W_2^1((-l, l), dx_1) \subset L_2((-l, l), dx_1)$$

for $\alpha = 1$. It now has the form

$$H_{-,k} \supset H_k \supset (W_2^1((-l, l), dx_1) \otimes H'_+)^\wedge. \tag{3.35}$$

Here $\varphi \mapsto \hat{\varphi}$ denotes the factorization with respect to

$$\{\varphi \in L_2(\mathbb{R}_l^\infty, m \times \mu') \mid \langle \varphi, \varphi \rangle = 0\}.$$

Clearly, in (3.35), the imbedding of a positive space in a zero space is quasinuclear. To continue the riggings (3.35), we take

$$C_0 = \mathrm{a.}(\, C_0^\infty((-l, l)) \otimes C_0^\infty(\mathbb{R}^1) \otimes C_0^\infty(\mathbb{R}^1) \otimes \ldots)$$

with natural topology and then realize the factorization $D = \hat{C}_0$.

II. The operators A'_α are defined on

$$\mathrm{a.}(\, C_0^\infty((-l, l)) \otimes D_2 \otimes D_3 \otimes \ldots)$$

for $\alpha = 2, 3, \ldots$ by the operation (3.11); for $\alpha = 1$ they are defined by

$$C_{\mathrm{cyl}}^\infty(\mathbb{R}^\infty) \ni \varphi(x) \mapsto i\left(\frac{\partial \varphi}{\partial x_1}\right)(x) \in C_{\mathrm{cyl}}^\infty(\mathbb{R}^\infty).$$

The fact that they are Hermitian for $\alpha = 2, 3, \ldots$ is verified just as in Lemma 3.1, while the fact that A'_1 is Hermitian is established in an even simpler way by integrating by parts. Denoting by A_α the closure of A'_α, we obtain a family of Hermitian operators $(A_\alpha)_{\alpha=1}^\infty$ in the space H_k.

For $\alpha = 2, 3, \ldots$, the operators A_α are selfadjoint and commuting; this is proved in exactly the same way as Lemma 3.2. The operator A_1 is only Hermitian, but we can show that it admits a selfadjoint extension B_1 to H_k which commutes with all A_α ($\alpha = 2, 3, \ldots$) (see Step IV below).

III. We now construct an expansion in generalized joint eigenvectors of this family of operators $(B_1, A_2, A_3, \ldots)$ by making use of the chain (3.35) and the extended rigging indicated above. As in the case $l = \infty$, we establish formula (3.23) with elementary kernels $\Omega(\lambda) \in H_- \otimes H_-$; the negative space H_- is now introduced by the chain

$$H_{-} \supset L_2(\mathbb{R}_l^{\infty}, m \times \mu') \supset W_2^1((-l, l), dx_1) \otimes H_{+}^{'}.$$

These kernels satisfy the system of equations (3.22) with the following changes: For $\alpha = 1$, we have $p_1(x_1) = 1$, the scalar product is taken in $L_2(\mathbb{R}_l^{\infty} \times \mathbb{R}_l^{\infty}, m \times \mu' \times m \times \mu')$, and

$$C_0 = \text{a.}(\, C_0^{\infty}((-l, l)) \otimes C_0^{\infty}(\mathbb{R}^1) \otimes C_0^{\infty}(\mathbb{R}^1) \otimes \dots).$$

This modified system (3.23) is solved exactly as the original one. The answer is given by formula (3.24) in which $\tilde{p}_1(\lambda_1)$ is replaced by

$$\int_{-l}^{l} e^{it\lambda_1}\, dt = 2l \sin(l\lambda_1)$$

and the limit is taken in the norm of the space $G_{-} \otimes G_{-}$ (a possible situation where $\sin(l\lambda_1)$ turns to zero is not essential; see the remark after the proof of Lemma 3.3). Multiplying and dividing the answer obtained by $|\tilde{p}_1(\lambda_1)|^2$, we can write it in the form (3.24) with modified $c(\lambda)$.

As in the case $l = \infty$, we now prove Lemma 3.4, derive formula (3.30), and then establish the representation (3.9) with the help of this formula.

IV. Let us show that the required extension of the operator A_1 is possible. Let $\varphi \in C_{\text{cyl}}^{\infty}(\mathbb{R}^{\infty})$, i.e., $\varphi(x) = u_{\varphi}(x_1, \dots, x_m)$, where $u_{\varphi} \in C_0^{\infty}(\mathbb{R}^m)$ with some $m \in \mathbb{N}$. We set

$$\varphi^{\circ}(x) = \overline{u_{\varphi}(-x_1, -x_2, \dots, -x_m)}\, p_2(-x_2) p_2^{-1}(x_2) \dots p_m(-x_m) p_m^{-1}(x_m).$$

After being extended by continuity, the mapping $\varphi \mapsto \varphi^{\circ}$ becomes an involution "∘" in the space H_k (one should employ the fact that $k(-x) = \overline{k(x)}$; this equality follows from the positive definiteness). The operators A_{α} $(\alpha \in \mathbb{N})$ are obviously real with respect to this involution.

Lemma 3.5. *The operator A_1 can be extended to a selfadjoint operator B_1, and this extension is such that B_1 commutes with each operator A_{α} $(\alpha = 2, 3, \dots)$ and is real with respect to the involution "∘".*

Proof. The proof is similar to that of Lemma 2.1. Consider the Kelly transformations

$$U_{z_{\alpha}}^{(\alpha)} = (A_{\alpha} - \bar{z}_{\alpha} 1)(A_{\alpha} - z_{\alpha} 1)^{-1} \quad (z_{\alpha} \in \mathbb{C}^1 \setminus \mathbb{R}^1 \text{ are fixed, } \alpha \in \mathbb{N}).$$

For $\alpha = 2, 3, \dots$, these operators are unitary, and $U_{z_1}^{(1)}$ transforms isometrically $\mathfrak{R}(A_1$

$-z_1 1)$ into $\mathfrak{R}(A_1 - \bar{z}_1 1)$. It is stated that each of these subspaces is invariant with respect to $U_{z_\alpha}^{(\alpha)}$ $(\alpha = 2, 3, \ldots)$. It suffices to consider $\mathfrak{R}(A_1 - z_1 1)$ and $U_{z_2}^{(2)}$. A set of vectors $(A_2 - z_2 1) \times (A_1 - z_1 1)\hat{D} \in H_k$, where $D = \text{a.}\ (D_1 \otimes D_2 \otimes \ldots)$, is dense in $\mathfrak{R}(A_1 - z_1 1)$ by virtue of the argument which was used in the proof of Lemma 3.2. The same is true for a set $(A_2 - \bar{z}_2 1) \times (A_1 - z_1 1)\hat{D}$. This obviously implies that $U_{z_2}^{(2)}$ transforms unitary $\mathfrak{R}(As_1 - z_1 1)$ into itself.

Further, as in the proof of Lemma 2.1, we conclude that both products $U_{z_\alpha}^{(\alpha)} U_{z_1}^{(1)}$ and $U_{z_1}^{(1)} U_{z_\alpha}^{(\alpha)}$ $(\alpha = 2, 3, \ldots)$ are meaningful and equal on $\mathfrak{R}(A_1 - z_1 1)$. As in the indicated lemma, the remaining part of the proof is reduced to choosing the operator W which transforms isometrically the defect space N_{z_1} for the operator A_1 into $N_{\bar{z}_1}$ and satisfies relation (2.29). This choice is realized in exactly the same way as in the case of Lemma 2.1 by use of the involution "∘" introduced above.

■

The proof of the theorem is now completed.

■

Remark 1. If the operator A_1, which appears in the proof of Theorem 3.2, is selfadjoint, then the measure σ in the representation (3.9) is uniquely defined for given k (this means that k can be extended to $\mathbb{R}^\infty$ in a unique manner). This follows from the fact that the presence of representation (3.9) implies the existence of a selfadjoint extension B_1 of the operator A_1 which commutes with A_α $(\alpha = 2, 3, \ldots)$ (see the book of Berezansky [5, Chapter 8, Section 1, Subsections 2 and 3] and Subsection 5.3).

■

Remark 2. We can also consider positive definite functions on a set

$$\mathbb{R}^\infty_{2l_1, \ldots, 2l_m} = \{x = (x_\alpha)_{\alpha=1}^\infty \in \mathbb{R}^\infty \mid x_1 \in (-2l_1, 2l_1), \ldots, x_m \in (-2l_m, 2l_m)\},$$

where $l_1, \ldots, l_m \in (0, \infty)$, $m \in \mathbb{N}$. The definition of positive definiteness is the same as above (see (3.34)); one should only replace $\mathbb{R}^\infty_l$ by $\mathbb{R}^\infty_{l_1, \ldots, l_m}$. Generally speaking, for $m > 1$, this positive definite function cannot be extended to a positive definite function on $\mathbb{R}^\infty$. In fact, the operators $A_1, \ldots, A_m$ can be extended to selfadjoint operators commuting with each other and with A_α $(\alpha = m + 1, m + 2, \ldots)$ not always.

■

Example 3.4. As in Example 3.1, we assume that μ is a canonical Gaussian measure γ_1. Then, $\mu^* = \gamma_2$ and $\mathbb{R}^\infty(p) = l_2$; condition (3.8) is satisfied. Here representation

(3.9), for a positive definite function $k(x)$ $\left(x \in \mathbb{R}_{2l}^{\infty},\ 0 < l \leq \infty\right)$, has the form

$$k(x) = \int_{l_2} e^{i(\lambda, x)_{l_2}} d\sigma(\lambda) \tag{3.36}$$

for γ_2-almost all $x \in \mathbb{R}_{2l}^{\infty}$. This means that a positive definite function $k(x)$ can be associated with its "regularized values" on l_2 (the values of the integral on the right-hand side of (3.35) for $x \in l_2$) which completely define this function (note that $\gamma_2(l_2) = 0$). This fact is studied in more details in Section 4.

Representations similar to (3.35) can be also written for positive definite functions connected with Examples 3.2 and 3.3; in these cases, the integration in (3.35) is carried out over l_1 and l_a $(a \in (1, 2))$, respectively.

§4. Positive Definite Functions on a Hilbert Space

Let us show how the theory of positive definite functions $k(x)$ given on a real Hilbert space H can be reduced to the results presented in Section 3. Together with classical theorems, we shall obtain a theorem on the extension of a positive definite function from a layer in the Hilbert space to the whole of H. This is why it is convenient to begin our exposition with the investigation of the functions given on a layer. We shall also give the corresponding facts for positive definite functions on a nuclear space.

4.1. Positive Definite Functions on a Layer in the Hilbert Space

Let H be a real separable Hilbert space, e_1 be a certain unit vector in H, and let H' be an orthogonal complement to $\{\lambda e_1 \mid \lambda \in \mathbb{R}^1\}$ in H. Let $x = (x_1, x')$ $(x_1 \in \mathbb{R}^1, x' \in H')$ denote the vector $x_1 e_1 + x'$. A set $\{(x_1, x') \mid x_1 \in (-l, l), x' \in H'\}$ is called a layer $H_l \subset H$ $(0 < l < \infty)$ in the space H; $H_\infty = H$. A complex-valued function $k(x)$ $(x \in H_{2l})$ is called positive definite if the inequality

$$\sum_{j,k=1}^{n} k(x^{(j)} - x^{(k)}) \xi_j \bar{\xi}_k \geq 0 \tag{4.1}$$

holds for any $x^{(1)}, \ldots, x^{(n)} \in H_l$ and $\xi_1, \ldots, \xi_n \in \mathbb{C}^1$ $(n \in \mathbb{N})$.

A topology given by neighborhoods of zero of the form $\{x \in H \mid (Ax, x)_H < \varepsilon\}$ $(\varepsilon > 0)$, where A are nonnegative nuclear operators, is called the J-topology in H. We say that a function $k(x)\,(x \in H_{2l})$ is continuous at the origin in the J-topology if it is continuous at 0 in the topology induced by the J-topology on H_{2l}.

We now derive from Theorems 3.1 and 3.2 the following theorem which coincides with the Minlos–Sazonov theorem in the case $l = \infty$.

Theorem 4.1. *Any positive definite function $k(x)$ $(x \in H_{2l},\ l \le \infty)$ continuous at 0 in the J-topology, admits the representation*

$$k(x) = \int_H e^{i(x,\lambda)_H}\, d\sigma(\lambda) \quad (x \in H_{2l}), \tag{4.2}$$

where σ is a nonnegative finite measure defined on the σ-algebra $\mathfrak{B}(H)$ of Borel subsets of H. Conversely, each integral of the form (4.2) is a positive definite function in H continuous at the origin in the J-topology. In the case $l = \infty$, the measure is uniquely defined for a given k; in the case $l < \infty$, generally speaking, this is not true.

Note that it is impossible to prove this theorem by using a traditional argument employing the restriction of $k(x)$ $(x \in H_{2l})$ to an arbitrary finite-dimensional layer $(-2l, 2l) \times G'$ $(G' \subset H',\ \dim G' < \infty)$. This is so because the measure on a finite-dimensional space corresponding to this restriction, cannot be, generally, determined uniquely and, therefore, it is impossible even to speak about the consistency of an obtained system of measures of cylindrical sets (cf. Subsection 2.3).

Proof. We split the proof into several steps. In the case $l = \infty$, Steps II, V, and IX are unnecessary.

I. There exists a positive nuclear operator A in H such that $k(x)$ $(x \in H_{2l})$ is uniformly continuous in the following sense: For any $\varepsilon > 0$, one can find $\delta > 0$ such that

$$|k(x^{(1)}) - k(x^{(2)})| < \varepsilon$$

for $x^{(1)}, x^{(2)} \in H_{2l}$ provided that

$$(A(x^{(1)} - x^{(2)}), x^{(1)} - x^{(2)})_H < \delta,$$

and $x^{(1)} - x^{(2)} \in H_{2l}$.

Indeed, by virtue of the continuity of k at 0 in the J-topology, one can find a sequence of nonnegative nuclear operators $(A_\alpha)_{\alpha=1}^{\infty}$ such that $|k(0) - k(x)| < 1/\alpha$ for $x \in H_{2l}$ provided that $(A_\alpha x, x)_H < 1$. We set

$$A = \sum_{\alpha=1}^{\infty} c_\alpha A_\alpha + C,$$

where $c_\alpha > 0$ are rapidly decreasing as $\alpha \to \infty$, so that

$$\sum_{\alpha=1}^{\infty} c_\alpha \operatorname{Tr}(A_\alpha) < \infty,$$

and C is a positive nuclear operator. Since $\|A_\alpha\| \le \operatorname{Tr}(A_\alpha)$, the series for A converges in the operator norm and defines an operator of the required type. Assume that $\varepsilon > 0$ is given. We find α_0 such that $1/\alpha_0 < \varepsilon$ and set $\delta = c_{\alpha_0}$. If $x^{(1)}, x^{(2)} \in H_{2l}$ are such that $x^{(1)} - x^{(2)} \in H_{2l}$ and the inequality

$$\delta = c_{\alpha_0} > (A(x^{(1)} - x^{(2)}), x^{(1)} - x^{(2)})_H \ge c_{\alpha_0} (A_{\alpha_0}(x^{(1)} - x^{(2)}), x^{(1)} - x^{(2)})_H,$$

holds, then

$$(A_{\alpha_0}(x^{(1)} - x^{(2)}), x^{(1)} - x^{(2)})_H < 1,$$

and, hence,

$$|k(0) - k(x^{(1)} - x^{(2)})| < 1/\alpha_0 < \varepsilon.$$

However, the function k (and, in general, any positive definite function on a certain commutative group (or on a part of it which contains 0)), satisfies the important inequality

$$|k(x^{(1)}) - k(x^{(2)})|^2 \le 2k(0)(k(0) - \operatorname{Re} k(x^{(1)} - x^{(2)})) \tag{4.3}$$

$$(x^{(1)}, x^{(2)},\ x^{(1)} - x^{(2)} \in H_{2l}),$$

which shows that *the continuity of a positive definite function at* 0 *implies its uniform continuity.* This yields

$$|k(x^{(1)}) - k(x^{(2)})| < \sqrt{2k(0)\varepsilon}$$

for the same variables $x^{(1)}$ and $x^{(2)}$ as above.

Let us clarify the inequality (4.3). Let Q be an abstract space, and let $Q \times Q \ni (p, q) \mapsto K(p, q) \in \mathbb{C}^1$ be a kernel positive definite in the sense that for any $p_1, \dots, p_n \in Q$ and $\xi_1, \dots, \xi_n \in \mathbb{C}^1$ $(n \in \mathbb{N})$ the inequality

$$\sum_{j,k=1}^{n} K(p_j, p_k)\xi_j \bar{\xi}_k \ge 0$$

holds. With each point $x \in Q$, we associate a formal vector e_x and consider a linear span L of these vectors. For $\xi = \sum_j \xi_j e_{p_j}$ and $\eta = \sum_k \eta_k e_{q_k}$ $(\xi_j, \eta_k \in \mathbb{C}^1$; the sums are finite), we introduce a quasiscalar product by setting

$$\langle \xi, \eta \rangle = \sum_{j,k} K(p_j, q_k) \xi_j \bar{\eta}_k.$$

After identifying $L \ni \xi \mapsto \hat{\xi}$ and completing this set, we obtain the Hilbert space L_K. For $x, y, z \in Q$, we have

$$|K(x,z) - K(y,z))|^2 = |(\hat{e}_x, \hat{e}_z)_{L_K} - (\hat{e}_y, \hat{e}_z)_{L_K})|^2 = |(\hat{e}_x - \hat{e}_y, \hat{e}_z)_{L_K})|^2$$

$$\leq \| \hat{e}_x - \hat{e}_y \|^2_{L_K} \| \hat{e}_z \|^2_{L_K} = (\hat{e}_z, \hat{e}_z)_{L_K} ((\hat{e}_x, \hat{e}_x)_{L_K} - 2 \operatorname{Re} (\hat{e}_x - \hat{e}_y)_{L_K}$$

$$+ (\hat{e}_y, \hat{e}_y)_{L_K}) = K(z,z)(K(x,x) - 2 \operatorname{Re} K(x,y) + K(y,y)).$$

This means that we have proved the following useful lemma.

Lemma 4.1. *For an arbitrary positive definite kernel* $Q \times Q \ni (x, y) \mapsto K(x, y) \in \mathbb{C}^1$, *we have*

$$|K(x,z) - K(y,z))|^2 \leq K(z,z)(K(x,x)$$

$$- 2 \operatorname{Re} K(x,y) + K(y,y)) \quad (x, y, z \in Q).$$

Setting in the lemma $K(x, y) = k(x - y)$ and $z = 0$, we get (4.3).

■

Note also that

$$|K(x,y)|^2 \leq (K(x,x)K(y,y) \quad (x, y \in Q),$$

whence $|k(x)| \leq k(0)$ $(x \in H_{2l})$.

II. The operator A on Step I can be chosen so that the vector e_1 is its eigenvector. Indeed, let $(e_\alpha)_{\alpha=2}^\infty$ be an orthonormal basis in H'. If we denote by $(x_\alpha)_{\alpha=1}^\infty$ and $(a_{\alpha,\beta})_{\alpha,\beta=1}^\infty$ the coordinates of the vector $x \in H$ and the matrix of the operator A in the basis $(e_\alpha)_{\alpha=1}^\infty$, respectively, then, by virtue of the positive definiteness of A, we get

$$(Ax, x)_H = \sum_{\alpha,\beta=2}^\infty a_{\alpha\beta} x_\alpha x_\beta + a_{11} x_1^2 + \sum_{\beta=2}^\infty a_{1\beta} x_1 x_\beta$$

$$\leq \sum_{\alpha,\beta=2}^{\infty} a_{\alpha\beta}x_\alpha x_\beta + a_{11}x_1^2 + 2\sqrt{a_{11}x_1^2}\sqrt{\sum_{\alpha,\beta=2}^{\infty} a_{\alpha\beta}x_\alpha x_\beta}$$

$$\leq \sum_{\alpha,\beta=2}^{\infty} 2a_{\alpha\beta}x_\alpha x_\beta + 2a_{11}x_1^2. \tag{4.4}$$

We introduce an operator B which acts in H and has a matrix in the basis $(e_\alpha)_{\alpha=1}^{\infty}$ coinciding with the matrix $(2a_{\alpha,\beta})_{\alpha,\beta=1}^{\infty}$ in which $2a_{\alpha\beta}$ for $(\alpha, \beta) = (1, 2), (1, 3), \ldots ; (2, 1), (3,1), \ldots$ are replaced by zeros. The operator B is positive and nuclear, and e_1 is its eigenvector. At the same time, inequality (4.4) shows that $(Ax, x)_H \leq (Bx, x)_H$ $(x \in H)$ and, therefore, the statement is proved if we replace A by B on Step I.

II. We introduce a scalar product in H by setting

$$(x, y)_{H_A} = (Ax, y)_H .$$

Let H_A be the completion of H with respect to $(\cdot, \cdot)_{H_A}$. Let us describe H_A by coordinates. For this purpose, we choose an orthonormal basis in H consisting of the eigenvectors of the operator A and realize H as a space $l_2 = l_2(\mathbb{R}^1)$ with respect to decomposition in this basis. Then the space H_A is realized as a Hilbert space

$$l_2(A) = \left\{ x \in \mathbb{R}^\infty \,\middle|\, \sum_{\alpha=1}^{\infty} a_\alpha x_\alpha^2 < \infty \right\}$$

with the scalar product

$$(x, y)_{l_2(A)} = \sum_{\alpha=1}^{\infty} a_\alpha x_\alpha y_\alpha \qquad (x, y \in l_2(A))$$

(here $(a_\alpha \delta_{\alpha\beta})_{\alpha,\beta=1}^{\infty}$ is the matrix of the operator A in the basis under consideration; $a_\alpha > 0$). It is essential that, for any Gaussian measure, we have $\gamma_\kappa(l_2(A))$ $(\kappa = (\kappa, \kappa, \ldots))$; this follows from the Kolmogorov–Khinchin criterion (Chapter 2, Theorem 1.3), since

$$\sum_{\alpha=1}^{\infty} a_\alpha = \operatorname{Tr} A < \infty.$$

IV. Let $l = \infty$. Then it follows from Step I that $k(x)$ $(x \in H)$ is uniformly continuous with respect to the norm $\|\cdot\|_{H_A} = (\cdot, \cdot)_{H_A}^{1/2}$ and, hence, can be extended by continuity to a uniformly continuous function $k_1(x)$ on the whole of H_A. Passing to the coordinate form of H_A, we find that $k_1(x)$ is defined on a set $l_2(A) \subset \mathbb{R}^\infty$ of a full Gaussian measure γ_2 and is obviously bounded (since (4.1) yields the inequality $|k(x)| \le k(0)$ $(x \in H)$). Due to the continuity of $k_1(x)$ in the topology of $l_2(A)$, the sequence of cylindrical functions $f_n(x) = k_1(x_1, \dots, x_n, 0, 0, \dots)$ converges to $k_1(x)$ at any point $x \in l_2(A)$. This implies the measurability of $k_1(x)$ $(x \in l_2(A))$ with respect to the σ-algebra

$$\mathfrak{C}_\sigma(l_2(A)) = \{\alpha \in \mathfrak{C}_\sigma(\mathbb{R}^\infty) \mid \alpha \subset l_2(A)\}.$$

We define $k_1(x)$ to be zero on $\mathbb{R}^\infty \setminus l_2(A)$. As a result, we obtain the measurable function $k_2(x)$ $(x \in \mathbb{R}^\infty)$ almost everywhere bounded with respect to the measure γ_2. On Step VI we show that it is positive definite in the sense of Subsection 3.1 with $\mu = \gamma_1$ and $\mu^* = \gamma_2$ (clearly, the results remain valid if we replace γ_1 by any γ_κ, $\kappa = (\kappa, \kappa, \dots)$).

V. Here we present a construction similar to that of Step V for the case $l < \infty$. The operator A is taken in the form indicated on Step II. Then the layer H_{2l} coincides with the set

$$l_{2,2l} = \left\{ x \in \mathbb{R}^\infty \mid x_1 \in (-2l, 2l), \sum_{\alpha=2}^{\infty} x_\alpha^2 < \infty \right\},$$

and its closure $H_{2l;A}$ in the topology of H_A coincides with

$$l_{2,2l}(A) = \left\{ x \in \mathbb{R}^\infty \mid x_1 \in [-2l, 2l], \sum_{\alpha=2}^{\infty} a_\alpha x_\alpha^2 < \infty \right\}.$$

It follows from Step I that $k(x)$ $(x \in H_{2l})$ is uniformly continuous with respect to the norm $\|\cdot\|_{H_A}$: If $x^{(1)}, x^{(2)} \in H_{2l}$ are such that

$$\|x^{(1)} - x^{(2)}\|_{H_A} < \min(\delta, 2l\sqrt{a_1}),$$

where δ is taken from Step I, then $|k(x^{(1)}) - k(x^{(2)})| < \varepsilon$. Extending $k(x)$ by continuity, we obtain a uniformly continuous function $k_1(x)$ $(x \in H_{2l;A} = l_{2;2l}(A))$. As on Step IV, $k_1(x)$ is bounded and measurable with respect to $\mathfrak{C}_\sigma(l_{2;2l}(A))$. It is clear that $\gamma_2(\mathbb{R}^\infty_{2l} \setminus l_{2;2l}(A)) = 0$; therefore, extending $k_1(x)$ to $\mathbb{R}^\infty_{2l} \setminus l_{2;2l}(A)$ by setting it to be zero, we obtain a measurable function $k_2(x)$ $(x \in \mathbb{R}^\infty_{2l})$ almost everywhere bounded with respect to the measure γ_2.

VI. The functions $k_2(x)$ constructed on Steps IV and V are positive definite in the sense of Subsection 3.1 with $\mu = \gamma_1$ and $\mu^* = \gamma_2$.

Indeed, assume, for example, that $l < \infty$. Passing to the limit in the inequality (4.1), we find that it is valid also for $k_1(x)$; here $x^{(j)} \in l_{2;\,l}(A)$. Consider a cylindrical function $\varphi(x) = u_\varphi(x_1, \ldots, x_m)$ $(u_\varphi \in C_0^\infty((-l, l) \times \mathbb{R}^{m-1}))$ and set $\xi_j = \varphi(x^{(j)})$ in this inequality. We get

$$\sum_{j,k=1}^{n} k_1(x^{(j)} - x^{(k)})\,\varphi(x^{(j)})\,\overline{\varphi(x^{(k)})} \geq 0 \quad (x^{(j)} \in l_{2;\,l}(A)). \tag{4.5}$$

Let us integrate (4.5) with respect to each point $x^{(1)}, \ldots, x^{(n)} \in l_{2;\,l}(A)$ over the measure γ_1. Taking into account that $\gamma_1(\mathbb{R}_l^\infty \setminus l_{2;l}(A)) = 0$, substituting k_2 for k_1, gathering similar terms, and cancelling $\gamma_1(\mathbb{R}_l^\infty) > 0$, we obtain

$$n\,k_2(0) \int_{\mathbb{R}_l^\infty} |\varphi(x)|^2\, d\gamma_1(x)$$

$$+\; n(n-1) \int_{\mathbb{R}_l^\infty} \int_{\mathbb{R}_l^\infty} k_2(y-x)\varphi(y)\,\overline{\varphi(x)}\, d\gamma_1(x)\,d\gamma_1(y) \geq 0.$$

We divide this inequality by n^2 and pass to the limit as $n \to \infty$. As a result, we get (2.1) with $k = k_2$ and $\mu = \gamma_1$.

VII. Taking into account what has been said in Example 3.4, let us apply Theorem 3.2 to the constructed function $k_2(x)$ $(x \in \mathbb{R}_{2l}^\infty)$. We find that for γ_2-almost all $x \in \mathbb{R}_{2l}^\infty$, the following representation is valid

$$k_2(x) = \int_{l_2} e^{i(\lambda, x)_{l_2}}\, d\sigma(\lambda), \tag{4.6}$$

where σ is a finite measure on the σ-algebra $\mathfrak{C}_\sigma(l_2) = \mathfrak{B}(l_2)$. To prove the theorem we must show that (4.6) implies the following relation

$$k(z) = \int_{l_2} e^{i(z,\lambda)_{l_2}}\, d\sigma(\lambda) \quad (z \in l_2) \tag{4.7}$$

((4.7) does not follow immediately from (4.6), since $\gamma_2(l_2) = 0$).

First, we establish the following general lemma.

Lemma 4.2. *The sequence*

$$\delta_n(x) = \prod_{\alpha=1}^{\infty} (1 + 2na_\alpha)^{1/2} e^{-n\|x\|^2_{H_A}} \left(x \in \mathbb{R}^\infty;\ \int_{\mathbb{R}^\infty} \delta_n(x) d\gamma_2(x) = 1;\ n \in \mathbb{N} \right) \quad (4.8)$$

is δ-shaped and contracting to zero with respect to the measure γ_2 and the topology $l_2(A)$. In other words, let $f(x)$ $(x \in \mathbb{R}^\infty)$ be a γ_2-almost everywhere bounded complex-valued function continuous at $x = 0$ in the sense of $l_2(A)$; then for any $\varepsilon \geq 0$ one can find $n(\varepsilon)$ such that

$$\left| f(0) - \int_{\mathbb{R}^\infty} f(x) \delta_n(x) d\gamma_2(x) \right| < \varepsilon \quad (4.9)$$

for $n > n(\varepsilon)$. Moreover, relation (4.9) also holds for any function $f(x) = e^{i(\lambda, x)_{l_2}}$, where $\lambda \in l_2$.

Proof. We have

$$\int_{\mathbb{R}^\infty} e^{-n\|x\|^2_{H_A}} d\gamma_2(x) = \lim_{m\to\infty} \int_{\mathbb{R}^m} \exp\left(-\sum_{\alpha=1}^{m} \left(na_\alpha + \frac{1}{2} \right) x_\alpha^2 \right) \frac{dx_1}{\sqrt{2\pi}} \cdots \frac{dx_m}{\sqrt{2\pi}}$$

$$= \lim_{m\to\infty} \frac{1}{(\sqrt{2\pi})^m} \prod_{\alpha=1}^{m} \int_{\mathbb{R}^1} \exp\left(-\left(\frac{1}{2} + na_\alpha \right) t^2 \right) dt$$

$$= \prod_{\alpha=1}^{\infty} (1 + 2na_\alpha)^{1/2}.$$

Hence, the last relation in (4.8) is valid.

Let us show that for any $r, \varepsilon > 0$ one can find $n(r, \varepsilon)$ such that for $n > n(r, \varepsilon)$ and $\|x\|_{H_A} \geq r$, the estimate $\delta_n(x) < \varepsilon$ holds. Indeed, let us fix some $p \in \mathbb{N}$. For $n > p$, we get

$$\delta_n(x) = \prod_{\alpha=1}^{p} (1 + 2na_\alpha)^{1/2} \exp\left(\frac{1}{2} \sum_{\alpha=p+1}^{\infty} \ln(1 + 2na_\alpha) - n\|x\|^2_{H_A} \right)$$

$$< \prod_{\alpha=1}^{p} (1 + 2na_\alpha)^{1/2} \exp\left(n \sum_{\alpha=p+1}^{\infty} a_\alpha - n\|x\|^2_{H_A} \right).$$

We now choose sufficiently large p such that $\sum_{\alpha=p+1}^{\infty} a_\alpha < r^2/2$. Then we have

$$\delta_n(x) < \prod_{\alpha=1}^{p} (1+2na_\alpha)^{1/2} \exp(-nr^2/2)$$

for $\|x\|_{H_A} \geq r$, and this yields the required result.

Let f be a function indicated in the formulation of the lemma. Denote $c = \text{ess sup}|f|$. Given $\varepsilon > 0$, we choose $r > 0$ so small that $|f(0) - f(x)| < \varepsilon$ for $\|x\|_{H_A} < r$. Then

$$\left| f(0) - \int_{\mathbb{R}^\infty} f(x)\,\delta_n(x)\,d\gamma_2(x) \right| = \left| \int_{\mathbb{R}^\infty} (f(0) - f(x))\,\delta_n(x)\,d\gamma_2(x) \right|$$

$$\leq \int_{\|x\|_{H_A} < r} |f(0) - f(x)|\,\delta_n(x)\,d\gamma_2(x)$$

$$+ 2c \int_{\|x\|_{H_A} \geq r} \delta_n(x)\,d\gamma_2(x) < (1+2c)\varepsilon$$

for $n > n(r, \varepsilon)$. Thus, $\delta_n(x)$ is a δ-shaped sequence.

Let us verify the last statement of the lemma (it does not follow from the argument presented above, since $f(x) = \exp\{i(\lambda, x)_{l_2}\}$ is not continuous at 0 with respect to $l_2(A)$ for $\sum_{\alpha=1}^{\infty} \lambda_\alpha^2 / a_\alpha = \infty$). With the help of the formula

$$(e^{-\kappa t^2})^\sim = (\pi/\kappa)^{1/2} e^{-\lambda^2/4\kappa} \quad (\kappa > 0),$$

by direct calculation, we obtain

$$\int_{\mathbb{R}^\infty} e^{i(\lambda, x)_{l_2}} \delta_n(x)\,d\gamma_2(x) = \prod_{\alpha=1}^{\infty} (1+2na_\alpha)^{1/2} \lim_{m\to\infty} \frac{1}{(\sqrt{2\pi})^m}$$

$$\times \int_{\mathbb{R}^m} \exp\left(\sum_{\alpha=1}^{m} \left(i\lambda_\alpha x_\alpha - \left(\frac{1}{2} + na_\alpha\right) x_\alpha^2\right)\right) dx_1 \ldots dx_m$$

$$= \prod_{\alpha=1}^{\infty} (1+2na_\alpha)^{1/2} \lim_{m\to\infty} \frac{1}{(\sqrt{2\pi})^m} \prod_{\alpha=1}^{m} \int_{\mathbb{R}^1} \exp\left(i\lambda_\alpha t - \left(\frac{1}{2} + na_\alpha\right) t^2\right) dt$$

$$= \prod_{\alpha=1}^{\infty} (1+2na_\alpha)^{1/2} \lim_{m\to\infty} \prod_{\alpha=1}^{m} (1+2na_\alpha)^{-1/2} \exp\left(\frac{-\lambda_\alpha^2}{2(1+2na_\alpha)}\right)$$

$$= \exp\left(-\frac{1}{2}\sum_{\alpha=1}^{\infty}\frac{\lambda_\alpha^2}{1+2na_\alpha}\right). \tag{4.10}$$

Given $\varepsilon > 0$, we find sufficiently large p such that $\sum_{\alpha=p+1}^{\infty}\lambda_\alpha^2 < \varepsilon$. Denote $a = \min_{\alpha=1,\dots,p} a_\alpha > 0$. Then we get

$$\sum_{\alpha=1}^{\infty}\frac{\lambda_\alpha^2}{1+2na_\alpha} < \sum_{\alpha=1}^{p}\frac{\lambda_\alpha^2}{1+2na_\alpha} + \varepsilon < \frac{1}{2na}\sum_{\alpha=1}^{p}\lambda_\alpha^2 + \varepsilon < 2\varepsilon$$

for sufficiently large n. This implies that the integral (4.10) tends to $1 = e^{i(\lambda,0)_{l_2}}$ as $n \to \infty$.

VIII. We prove (4.7) in the case $l = \infty$. Let $z \in l_2$ be fixed. It follows from (4.6) that for γ_2-almost all $x \in \mathbb{R}^\infty$ we have

$$k_2(x+z) = \int_{l_2} e^{i(\lambda, x+z)_{l_2}}\, d\sigma(\lambda) = \int_{l_2} e^{i(\lambda, z)_{l_2}} e^{i(\lambda, x)_{l_2}}\, d\sigma(\lambda). \tag{4.11}$$

Here we have used the fact that the measures $\gamma_2(\alpha)$ and $\gamma_2(\alpha + z)$ $(\alpha \in \mathfrak{C}_\sigma(\mathbb{R}^\infty))$ are absolutely continuous one with respect to the other for $z \in l_2$ (see Theorem 2.3 in Chapter 2) and the equality

$$e^{i(\lambda, x+z)_{l_2}} = e^{i(\lambda, z)_{l_2}} e^{i(\lambda, x)_{l_2}},$$

valid for γ_2-almost all $x \in \mathbb{R}^\infty$ (this equality is a consequence of definition (3.4) for the case $\mathbb{R}^\infty(p) = l_2$). It follows from (4.11) that

$$\int_{\mathbb{R}^\infty} k_2(x+z)\delta_n(x)\,d\gamma_2(x) = \int_{l_2}\left\{\int_{\mathbb{R}^\infty} e^{i(\lambda,x)_{l_2}}\delta_n(x)\,d\gamma_2(x)\right\} e^{i(\lambda, z)_{l_2}}\, d\sigma(\lambda).$$

Passing here to the limit as $n \to \infty$ and applying Lemma 4.2, we obtain (4.7) (note that $k_2(z) = k_1(z) = k(z)$ for $z \in l_2$). The limit transition under the integral sign is correct, since $|e^{i(\lambda,x)_{l_2}}| = 1$ for for γ_2-almost all $x \in \mathbb{R}^\infty$ and, consequently,

$$\left|\int_{\mathbb{R}^\infty} e^{i(\lambda,x)_{l_2}}\delta_n(x)\,d\gamma_2(x)\right| \le 1 \quad (\lambda \in l_2;\ n \in \mathbb{N}).$$

IX. Let us prove (4.7) in the case $l < \infty$. We fix $z_1 \in (-2l, 2l)$ and construct a δ-shaped sequence $\delta_n(x)$ $(x \in \mathbb{R}^\infty_{2l})$ contracting to the point $(z_1, 0) \in \mathbb{R}^\infty_{2l}$ with respect to the measure $dx_1 \times d\gamma_2(x')$ and the topology induced on $\mathbb{R}^\infty_{2l}$ by the topology of H_A, i.e., by the topology of the space $(-2l, 2l) \times l_2(A')$, where

$$l_{2,A'} = \left\{ (x_\alpha)_{\alpha=2}^\infty \in \mathbb{R}^\infty \mid \sum_{\alpha=2}^\infty a_\alpha x_\alpha^2 < \infty \right\}.$$

Let $\chi_n(x_1)$ be a characteristic function of the interval $(z_1 - 1/n, z_1 + 1/n)$ divided by $2/n$. The desired sequence has the form

$$\delta_n(x) = \delta_n((x_1, x_n')) = \chi_n(x_1)\, \delta_n(x') \quad (x \in \mathbb{R}^\infty_{2l}),$$

where $\delta_n(x')$ is a δ-shaped sequence (4.8) in which x and A are replaced by $x' = (x_\alpha)_{\alpha=2}^\infty$ and $A' = (a_\alpha \delta_{\alpha\beta})_{\alpha,\beta=2}^\infty$, respectively.

Let us derive (4.7). Let $z = (z_1, z')$, where $z_1 \in (-2l, 2l)$ and $z' \in l_2$. It follows from (4.6) that the equality

$$k_1(x + (0, z')) = \int_{l_2} e^{i(\lambda, x + (0, z'))_{l_2}} d\sigma(\lambda) = \int_{l_2} e^{i(\lambda, (0, z'))_{l_2}} e^{i(\lambda, x)_{l_2}} d\sigma(\lambda) \qquad (4.12)$$

holds for γ_2-almost all $x \in \mathbb{R}^\infty$. As in Step VIII, we have employed the fact that under the shift by the vector $z' \in l_2$, the measure γ_2 is transformed into a measure absolutely continuous with respect to the original one. Multiplying (4.12) by the sequence $\delta_n(x)$ constructed above, integrating with respect to $dx_1 \times d\gamma_2(x')$ over $\mathbb{R}^\infty_{2l}$, and passing to the limit as $n \to \infty$, we get (4.7), just as in Step VIII.

X. Consider the last two assertions of the theorem. By direct substitution of (4.2) into (4.1), we find that the integral (4.2) is a positive definite function on H. To prove that the measure σ is uniquely determined for given k (for $l = \infty$), one must pass from H to l_2 in (4.2) and set $x = (x_1, \ldots, x_m, 0, 0, \ldots)$. We obtain the Bochner representation for an ordinary many-dimensional positive definite function

$$k_m(x_1, \ldots, x_m) = k(x_1, \ldots, x_m, 0, 0, \ldots)$$

with the measure

$$\mathfrak{B}(\mathbb{R}^m) \ni \alpha \mapsto \sigma_m(\alpha) = \sigma(\alpha \times \mathbb{R}^\infty).$$

It is well known that the measure σ_m is uniquely defined for the given k_m and, hence, for the given k. On the other hand, the measures σ_m determine the measure σ uniquely for arbitrary m.

Finally, let us prove the continuity of the function (4.2) at 0 in the J-topology. Taking into account the inequality (4.3) for fixed $r > 0$, we obtain

$$\begin{aligned}|k(x)-k(0)|^2 &\le 2k(0)(k(0)-\operatorname{Re} k(x)) \\ &= 2k(0)\int_H (1-\cos(\lambda, x)_H)\,d\sigma(\lambda) \\ &\le 2k(0)\int_{\|\lambda\|\le r} 2\sin^2\frac{1}{2}(\lambda, x)_H\,d\sigma(\lambda) + 4k(0)\int_{\|\lambda\|>r} d\sigma(\lambda) \\ &\le k(0)\int_{\|\lambda\|\le r} (\lambda, x)_H^2\,d\sigma(\lambda) + 4k(0)\int_{\|\lambda\|>r} d\sigma(\lambda).\end{aligned} \tag{4.13}$$

A bilinear form

$$H \times H \ni (x, y) \mapsto a_r(x, y) = \int_{\|\lambda\|\le r} (\lambda, x)_H (\lambda, y)_H\,d\sigma(\lambda) \in \mathbb{R}^1$$

is continuous and, hence, has the form $(A_r x, y)_H$, where $A_r \in \mathcal{L}(H, H)$. Clearly, this operator is nonnegative. It is nuclear: For an orthonormal basis $(e_j)_{j=1}^{\infty}$ in H, we have

$$\begin{aligned}\sum_{j=1}^{\infty} (A_r e_j, e_j)_H &= \sum_{j=1}^{\infty} \int_{\|\lambda\|\le r} (\lambda, e_j)_H^2\,d\sigma(\lambda) \\ &= \int_{\|\lambda\|\le r} \sum_{j=1}^{\infty} (\lambda, e_j)_H^2\,d\sigma(\lambda) = \int_{\|\lambda\|\le r} \|\lambda\|_H^2\,d\sigma(\lambda) < \infty.\end{aligned}$$

Therefore, the estimate (4.13) can be rewritten in the form

$$|k(x)-k(0)|^2 \le k(0)(A_r x, x)_H + 4k(0)\,\sigma(\{\lambda \in H \mid \|\lambda\|_H > r\}).$$

Given $\varepsilon > 0$, we choose sufficiently large r such that the second term in the last sum becomes less than $\varepsilon/2$; then we choose a sufficiently small neighborhood of 0 determined by the operator A_r. As a result, we obtain the required estimate $|k(x)-k(0)| < \varepsilon$ for x from the constructed neighborhood. Continuity is proved.

■

Example 4.1. Assume that a positive nuclear operator S is given on H. Consider a Gaussian measure γ_S with a correlation operator S and mean value zero (see Subsection 1.6 in Chapter 2). Taking this measure to be σ in formula (4.2) and employing the equality (1.39) in Chapter 2, we get

$$\exp\left(-\frac{1}{4}(Sx, x)\right) = \int e^{i(\lambda, x)_H}\, d\gamma_S(\lambda) \quad (x \in H). \tag{4.14}$$

In particular, this formula shows that the function

$$k(x) = \exp\left(-\frac{1}{4}(Sx, x)_H\right) \quad (x \in H)$$

is positive definite. It is continuous in the J-topology. As a neighborhood of 0, one may take $\{x \in H \mid (Sx, x)_H < \varepsilon\}$. The equality (4.14) is a representation (4.2) of this function according to Theorem 4.1.

4.2. Extension of a Positive Definite Function from a Layer in the Hilbert Space to the Whole of the Space. Connection with the Moment Problem

The next essential result follows immediately from Theorem 4.1.

Theorem 4.2. *Any positive definite function $k(x)$ ($x \in H_{2l}$, $l < \infty$) on a layer in the Hilbert space H which is continuous at 0 in the J-topology admits an extension to a positive definite function on the whole of H with the same property of continuity.*

Generally speaking, this extension is not unique, even in the simplest case $H = \mathbb{R}^2$, i.e., for a positive definite function of two variables. We now give a condition which guarantees uniqueness; it is related to the investigation of the corresponding infinite-dimensional moment problem.

Let us clarify this relation to the moment problem.

We say that a complex-valued function $f(x)$ defined in the neighborhood U of 0 of the real Hilbert space H, has $n \in \mathbb{N}$ point derivatives at zero provided that there exist k-linear symmetric continuous forms

$$H^k = \mathop{\times}_{j=1}^{k} H \ni (x_1, \ldots, x_k) \mapsto a_k(x_1, \ldots, x_k) \in \mathbb{C}^1$$

$$(k = 1, \ldots, n;\ a_0 \in \mathbb{C}^1)$$

such that

$$f(x) = \sum_{k=0}^{n} \frac{1}{k!} a_k(x_1, \ldots, x_k) + \varepsilon_n(0; x),$$

$$\varepsilon_n(0; x) = o(\|x\|_H^n), \quad \|x\|_H \to 0. \tag{4.15}$$

The forms a_k in (4.15) are defined uniquely; they are called the kth point derivatives of f at zero: $(D^k f)(0) = a_k$. The function f is called n times point differentiable at zero.

Shifting (4.15) by $y \in H$, we obtain the decomposition for $f(y + x)$ which defines the derivatives $(D^k f)(y)$ of f at the point y. The equalities which are established below can be derived also for $(D^k f)(y)$. However, we need only the relations for $(D^k f)(0)$ and, thus, we write them only in the indicated case.

If a function $f(x)\,(x \in U)$ is n times continuously differentiable in U in the ordinary sense (see Subsection 3.1 in Chapter 2), then the Taylor formula holds for it in the form (3.4) of Chapter 2 with $x = 0$. Therefore, this function possesses n point derivatives at 0. At the same time, the existence of these derivatives does not, in general, guarantee its differentiability (simple examples for the case of a finite-dimensional H can be found, for example, in the book of Shilov [1, Chapter 2, Section 2.4, Subsection 2.42]). However, the concept of point derivatives is now more convenient for our purposes.

Consider a chain of real Hilbert spaces

$$H_{-1} \supset H_0 \supset H_{+1} \tag{4.16}$$

with a quasinuclear imbedding $H_{+1} \to H_0$. Denote by

$$H_{-1,c} \supset H_{0,c} \supset H_{+1,c}$$

its complexification. Let $f(x)$ be a complex-valued function defined in a neighborhood of "0" of the space H_0 and n times point differentiable at 0. It follows from the kernel theorem (Theorem 2.3 in Chapter 1) that, for each form a_k $(k = 1, \ldots, n)$, there exists an element ("kernel") $\theta_k \in H_{-1,c}^{\otimes k}$ such that

$$a_k(\varphi_1, \ldots, \varphi_k) = (\theta_k, \varphi_1 \otimes \cdots \otimes \varphi_k)_{H_{0,c}^{\otimes k}}.$$

Clearly, the kernel θ_k is symmetric, i.e., $(\theta_k, \varphi_1 \otimes \cdots \otimes \varphi_k)_{H_{0,c}^{\otimes k}}$ is symmetric with respect to $\varphi_1, \ldots, \varphi_k$. Thus, the derivatives $(D^k f)(0)$ can be now interpreted as symmetric kernels from $H_{-1,c}^{\otimes k}$; for these kernels, we preserve the same notations $(D^k f)(0)$. Formula (4.15) can be rewritten as follows

$$f(x) = \sum_{k=0}^{n} \frac{1}{k!} ((D^k f)(0), \varphi \otimes \cdots \otimes \varphi)_{H_{0,c}^{\otimes k}} + \varepsilon_n(0; \varphi),$$

$$\varepsilon_n(0;\varphi) \underset{\|\varphi\|_{H_0}\to 0}{=} o(\|\varphi\|^n_{H_0}) \quad (\varphi \in H_{+1}). \tag{4.17}$$

Let us express the derivatives $(D^k f)(0)$ in terms of f by using symmetric differences. Let $r > 0$. We set

$$(\Delta_{\varphi,r} f)(\psi) = \frac{1}{2r}[f(\psi + r\varphi) - f(\psi - r\varphi)] \quad (\varphi, \psi \in H_{+1}) \tag{4.18}$$

and denote the kth iteration of this operation by $\Delta^k_{\varphi,r}$. It is easy to show that *each derivative* $(D^k f)(0)$ $(k \in \mathbb{N})$ *can be calculated with the help of the formula*

$$((D^k f)(0), \underbrace{\varphi \otimes \ldots \otimes \varphi}_{k})_{H^{\otimes k}_{0,c}} = \lim_{r\to 0} (\Delta^k_{\varphi,r} f)(0) \quad (\varphi \in H_{+1}). \tag{4.19}$$

Indeed, first we note that $(D^k f)(0)$ is completely defined by formula (4.19). This follows from the fact that the k-linear symmetric form $a_k(\varphi_1, \ldots, \varphi_k)$ is defined by its diagonal values $a_k(\varphi, \ldots, \varphi)$. We have

$$(\Delta_{\varphi,r} f)(0) = \frac{1}{2r}[f(r\varphi) - f(-r\varphi)].$$

By using (4.17) (with $n = 1$), we obtain

$$f(r\varphi) - f(-r\varphi) = 2r((D^1 f)(0), \varphi)_{H_{0,c}},$$

and this yields (4.19) with $k = 1$. For $k = 2$, we have

$$(\Delta^2_{\varphi,r} f)(0) = \frac{1}{4r^2}[f(2r\varphi) - 2f(0) + f(-2r\varphi)].$$

Using (4.17) once again (for $n = 2$), we find

$$f(2r\varphi) - 2f(0) + f(-2r\varphi) = 4r^2(D^2 f)(0), \varphi \otimes \varphi)_{H^{\otimes 2}_{0,c}},$$

and this implies (4.19) for $k = 2$. For $k \geq 3$, we act analogously.

■

Let $k(x)$ $(x \in H)$ be a positive definite function continuous in the J-topology, and let A be a positive nuclear operator in the space $H_0 = H$ introduced for given $k(x)$ in Step I of the proof of Theorem 4.1. We set $H_{+1} = \mathfrak{D}(A^{-1/2})$ and

$$(\varphi, \psi)_{H_{+1}} = (A^{-1/2}\varphi, A^{-1/2}\psi)_{H_0} \quad (\varphi, \psi \in H_{+1}),$$

and construct the chain (4.16). Clearly, $H_{-1} = H_A$, where the space H_A is introduced on Step III of this proof; the imbedding $H_{+1} \to H_0$ is quasinuclear. Below we regard (4.16) as the chain just constructed above. For this chain, the operator OJ coincides with $A^{1/2}$ (see (1.18) and (1.19) in Chapter 1).

Let us now change the variables in (4.2) (for $l = \infty$), namely, $x = A^{-1/2}\varphi$, where $\varphi \in H_{+1}$. We get

$$k(A^{-1/2}\varphi) = \int_{H_0} e^{i(\lambda, A^{-1/2}\varphi)_{H_0}} d\sigma(\lambda)$$

$$= \int_{H_A} e^{i(A^{1/2}\xi, A^{-1/2}\varphi)_{H_0}} d\sigma(A^{1/2}\xi)$$

$$= \int_{H_{-1}} e^{i(\xi, \varphi)_{H_0}} d\sigma(A^{1/2}\xi) \quad (\varphi \in H_{+1}). \tag{4.20}$$

Denote

$$p(\varphi) = k(A^{-1/2}\varphi) \quad (\varphi \in H_{+1}), \quad d\tau(\xi) = d\sigma(A^{1/2}\xi) \quad (\xi \in H_{-1}), \tag{4.21}$$

then (4.20) transforms into the equality

$$p(\varphi) = \int_{H_{-1}} e^{i(\xi, \varphi)_{H_0}} d\tau(\xi) \quad (\varphi \in H_{+1}). \tag{4.22}$$

More precisely, we consider a mapping $H_0 \ni x \mapsto J^{-1}x \in H_-$, where an operator J is connected with (4.16). Then the measure τ is an image of σ under this mapping. Since σ is given on $\mathfrak{B}(H_0)$, the measure τ is given on $\mathfrak{B}(H_{-1})$. The property of uniform continuity of k in the topology of the space H_A is equivalent to the uniform continuity of p in H_0. This gives us the possibility of extending p from H_{+1} to H_0 by continuity (clearly, this is the same extension as in Subsection 4.1, Step IV). The extended function is also denoted by $p(x)$ $(x \in H_0)$.

Lemma 4.3. *Let us extend the chain* (4.16) *to the chain*

$$H_{-2} \supset H_{-1} \supset H_0 \supset H_{+1} \supset H_{+2}, \tag{4.23}$$

where H_{+2} is a real Hilbert space dense in H_{+1} such that the imbedding $H_{+2} \to H_{+1}$ is quasinuclear. Assume that for some $m \in \mathbb{Z}_+$, the extended function $p(x)$ $(x \in H_0)$ has a point derivative at the origin $(D^{2m}p)(0)$ which is regarded as an element of

$H_{-1,c}^{\otimes 2m}$. *Then the measure* τ *in* (4.22) *is such that the following integrals exist*

$$i^k \int_{H_{-1}} \underbrace{\xi \otimes \ldots \otimes \xi}_{k} \, d\tau(\xi) \quad (k = 0, \ldots, 2m), \tag{4.24}$$

which are understood as weak integrals of the vector function

$$H_{-1} \ni \xi \mapsto \underbrace{\xi \otimes \ldots \otimes \xi}_{k} \in H_{-2}^{\otimes k}.$$

The value of each integral (4.24) *belongs to* $H_{-1,c}^{\otimes k}$ *and coincides with the derivative* $(D^k p)(0)$ *the existence of which is guaranteed.*

Proof. It follows from (4.18) that

$$(\Delta_{\varphi,r}^{j} e^{i(\xi,\cdot)_{H_0}})(0) = \left(\frac{i}{r}\sin(r(\xi,\varphi)_{H_0}\right)^j \quad (j \in H_{-1}, \varphi \in H_{+1};\ r>0;\ j \in \mathbb{N}).$$

Using this formula, and applying the operation $\Delta_{\varphi,r}^{2m}$ to the equality (4.22), we obtain

$$(\Delta_{\varphi,r}^{2m} p)(0) = (-1)^m \int_{H_{-1}} \left(\frac{1}{r}\sin(r(\xi,\varphi)_{H_0}\right)^{2m} d\tau(\xi) \quad (\varphi \in H_{+1};\ r>0). \tag{4.25}$$

Passing here to the limit as $r \to 0$ and employing (4.19), we find

$$((D_p^{2m})(0), \underbrace{\varphi \otimes \ldots \otimes \varphi}_{2m})_{H_{0,c}^{\otimes 2m}} = (-1)^m \int_{H_{-1}} (\xi,\varphi)_{H_0}^{2m} \, d\tau(\xi) \quad (\varphi \in H_{+1}). \tag{4.26}$$

The limit transition under the integral sign realized above can be easily justified: First, by using the Fatou Lemma, we prove that the function $(\xi,\varphi)_{H_0}^{2m}$ is summable and then note that the integrand in (4.25) can be majorized by this function.

By using the general inequality

$$\left|\int f_1(x)\ldots f_k(x)\,d\mu(x)\right| \le \prod_{j=1}^{k}\left(\int |f_j(x)|^k d\mu(x)\right)^{1/k},$$

the Hölder inequality, and the relation (4.26), we obtain

$$\int_{H_{-1}} |(\xi,\varphi_1)_{H_0}\cdots(\xi,\varphi_k)_{H_0}|\,d\tau(\xi) \le \prod_{j=1}^{k}\left(\int_{H_{-1}} |(\xi,\varphi)_{H_0}|^k d\tau(\xi)\right)^{1/k}$$

$$\leq \prod_{j=1}^{k} \left(\int_{H_{-1}} |(\xi, \varphi_j)_{H_0}^{2m} \, d\tau(\xi) \right)^{1/2m}$$

$$\leq \prod_{j=1}^{k} (|((D_p^{2m})(0), \underbrace{\varphi_j \otimes \ldots \otimes \varphi_j}_{2m})_{H_{0,c}^{\otimes 2m}}|)^{1/2m}$$

$$\leq \|(D_p^{2m})\|_{H_{-1,c}^{\otimes 2m}}^{k/2m} \|\varphi_1\|_{H_{+1}} \cdots \|\varphi_k\|_{H_{+1}}$$

$$\leq c_m \|\varphi_1\|_{H_{+1}} \cdots \|\varphi_k\|_{H_{+1}} \tag{4.27}$$

$$(c_m < \infty;\ \varphi_1, \ldots, \varphi_k \in H_{+1})$$

for $k = 1, \ldots, 2m$.

We now establish the existence of the weak integral (4.24). Consider an orthonormal basis $(e_j)_{j=1}^{\infty}$ in the space $H_{+2} \subset H_{+1}$. Then any vector $\varphi \in H_{+2}^{\otimes k}$ $(k = 1, \ldots, 2m)$ can be written in the form

$$\varphi = \sum_{j_1, \ldots, j_k = 1}^{\infty} \varphi_{j_1, \ldots, j_k} e_{j_1} \otimes \ldots \otimes e_{j_k}.$$

By using (4.27), we get

$$\int_{H_{-1}} |(\underbrace{\xi \otimes \ldots \otimes \xi}_{k}, \varphi)|_{H_0^{\otimes k}} \, d\tau(\xi)$$

$$= \int_{H_{-1}} \left| \sum_{j_1, \ldots, j_k = 1}^{\infty} \varphi_{j_1, \ldots, j_k} (\xi, e_{j_1})_{H_0} \cdots (\xi, e_{j_k})_{H_0} \right| d\tau(\xi)$$

$$\leq \sum_{j_1, \ldots, j_k = 1}^{\infty} |\varphi_{j_1, \ldots, j_k}| \int_{H_{-1}} |(\xi, e_{j_1})_{H_0} \cdots (\xi, e_{j_k})_{H_0}| \, d\tau(\xi)$$

$$\leq c_m \sum_{j_1, \ldots, j_k = 1}^{\infty} |\varphi_{j_1, \ldots, j_k}| \, \|e_{j_1}\|_{H_{+1}} \cdots \|e_{j_k}\|_{H_{+1}}$$

$$\leq c_m \left(\sum_{j_1, \ldots, j_k = 1}^{\infty} \varphi_{j_1, \ldots, j_k}^2 \right)^{1/2} \left(\sum_{j=1}^{\infty} \|e_j\|_{H_{+1}}^2 \right)^{k/2}$$

$$= c_m \|\varphi\|_{H_{+2}^{\otimes k}} |O|^k < \infty,$$

where O is an operator of the imbedding $H_{+2} \to H_{+1}$. Thus, the integral (4.24) exists.

Consider the equality (4.25) in which $2m$ is replaced by $k = 1, \ldots, 2m$ and pass it to the limit as $r \to \infty$. Here the limit transition under the integral sign can be easily realized, because the function $|(\xi, \varphi)_{H_0}|^k$ is summable (this follows from (4.27)). Therefore, there exists $\lim_{r\to\infty}(\Delta^k_{\varphi,r}p)(0)$ $(\varphi \in H_{+1})$, and this is equivalent to the existence of the derivative $(D^k p)(0)$. Finally, we obtain

$$((D^k p)(0), \underbrace{\varphi \otimes \ldots \otimes \varphi}_{k})_{H_{0,c}^{\otimes k}} = i^k \int_{H_{-1}} (\xi, \varphi)^k_{H_0}\, d\tau(\xi) \quad (\varphi \in H_{+1}). \tag{4.28}$$

Since $(D^k p)(0) \in H_{-1,c}^{\otimes k}$, it remains to prove that the integral (4.24) coincides with $(D^k p)(0)$. This coincidence is equivalent to the equality

$$\left(i^k \int_{H_{-1}} \underbrace{\xi \otimes \ldots \otimes \xi}_{k}\, d\tau(\xi), \varphi_1 \otimes \ldots \otimes \varphi_k \right)_{H_{0,c}^{\otimes k}}$$

$$= i^k \int_{H_{-1}} (\xi, \varphi_1)_{H_0} \cdots (\xi, \varphi_k)_{H_0}\, d\tau(\xi)$$

$$= ((D^k p)(0), \varphi_1 \otimes \ldots \otimes \varphi_k)_{H_{0,c}^{\otimes k}} \quad (\varphi_1, \ldots, \varphi_k \in H_{+2}). \tag{4.29}$$

Due to the symmetry of the second integral in (4.29) with respect to $\varphi_1, \ldots, \varphi_k$ and the symmetry of $(D^k p)(0)$, it suffices to establish the equality (4.29) on the diagonal $\varphi_1 = \ldots = \varphi_k = \varphi$, but it has been already proved for this case.

■

Recall in a somewhat modified form the necessary facts from the theory of the symmetric power moment problem in the case of Hilbert spaces (see Subsection 2.2). Consider a chain (4.23) of real Hilbert spaces; the imbedding $H_{+2} \to H_{+1}$ is assumed to be quasinuclear and the quasinuclearity of the imbedding $H_{+1} \to H_{+0}$ is not necessary. A sequence $s = (s_n)_{n=0}^{\infty}$, where $s_n \in H_{-1}^{\otimes n}$, is called a moment sequence provided that s_n is symmetric and the condition of positive definiteness holds, namely,

$$\sum_{j,k=0}^{\infty} (s_{j+k}, \varphi_j \otimes \varphi_k)_{H_0^{\otimes(j+k)}} \geq 0$$

for any finite sequence $(\varphi_j)_{j=0}^{\infty}$ $(\varphi_j \in H_{+1}^{\otimes j})$. We consider definite moment sequences, i.e., such that the class $C\{\|s_{2n}\|^{1/2}_{H_{-1}^{\otimes 2n}}\}$ is quasianalytical. These sequences admit a representation

$$s_n = \int_{H_{-2}} \underbrace{\xi \otimes \ldots \otimes \xi}_{n} \, d\tau(\xi) \quad (n \in \mathbb{Z}_+), \tag{4.30}$$

where τ is a nonnegative finite measure given on the σ-algebra $\mathfrak{B}(H_{-2})$ of Borel subsets of H_{-2}. Conversely, each sequence of the form (4.30) is a moment sequence. Due to the requirement of definiteness, the measure τ is uniquely constructed for the given s (we clarify that the set E in (2.4) now coincides with the ball

$$\{x \in H_{+2} \,|\, \|x\|_{H_{+2}} < 1\}).$$

Let us return to the representation (4.22) for a positive definite function. Since $H_{-2} \supset H_{-1}$, we can assume that the measure τ is given on the σ-algebra $\mathfrak{B}(H_{-2})$ of Borel subsets of H_{-2} and define it to be zero on $H_{-2} \setminus H_{-1}$. Introducing the previous notation of this measure, we can rewrite (4.22) in the form

$$p(\varphi) = k(A^{-1/2}\varphi) = \int_{H_{-2}} e^{i(\xi,\varphi)_{H_0}} \, d\tau(\xi) \quad (\varphi \in H_{+1}). \tag{4.31}$$

Theorem 4.3. *Assume that a positive definite function* $p(x)$ $(x \in H_0)$ *extended to* H_0 *by continuity, is point differentiable at* 0 *infinitely many times. Then a sequence* $s = (s_n)_{n=0}^{\infty}$, *where*

$$s_n = (-i)^n (D^n p)(0) \in H_{-1}^{\otimes n}$$

is a moment sequence, and the following representation holds for it

$$s_n = (-i)^n (D^n p)(0) = \int_{H_{-2}} \underbrace{\xi \otimes \ldots \otimes \xi}_{n} \, d\tau(\xi) \quad (n \in \mathbb{Z}_+). \tag{4.32}$$

Here the integral of the vector function

$$H_{-2} \ni \xi \mapsto \underbrace{\xi \otimes \ldots \otimes \xi}_{n} \in H_{-2}^{\otimes n}$$

converges weakly, and the measure τ *is a nonnegative finite measure on* $\mathfrak{B}(H_{-2})$ *from the representation* (4.31).

Proof. The integral (4.32) coincides (to within a factor) with the integral (4.24) in which τ is extended as zero to H_{-2}. Therefore, according to Lemma 4.3, it converges in the indicated manner. The equality (4.32) follows from the second part of this lemma. The fact that the sequence (4.32) is a moment sequence follows from the fact that sequences of the form (4.30) are moment.

■

One can now easily establish the sufficient condition of the uniqueness of the extension of a positive definite function.

Theorem 4.4. *Let k be a positive definite function defined on a layer H_{2l} $(0 < l < \infty)$ in the Hilbert space H. Assume that k is continuous at 0 in the J-topology, i.e., in the equivalent formulation, we assume that there exists a positive nuclear operator A in the space H such that k is continuous at 0 with respect to the convergence in the space H_A constructed by use of the scalar product*

$$(x, y)_{H_A} = (Ax, y)_H \quad (x, y \in H).$$

We set

$$p(\varphi) = k(A^{-1/2}\varphi) \quad (\varphi \in \mathfrak{D}(A^{-1/2})).$$

If $p(\varphi)$ is infinitely point differentiable at 0 in the topology of the space H and the class $C\{\|(D^{2n}p)(0)\|^{1/2}_{H_A^{\otimes 2n}}\}$ is quasianalytical, then k can be uniquely extended from H_{2l} to a positive definite function in H continuous in the J-topology.

Proof. Suppose that k possesses two distinct extensions k_1 and k_2 and construct the functions

$$p_s(\varphi) = k_s(A^{-1/2}\varphi) \quad (s = 1, 2),$$

where $\varphi \in H_{+1} = \mathfrak{D}(A^{-1/2})$. Let τ_s be the corresponding measures in the representation (4.31). In a certain neighborhood of 0 in the space H_{+1}, we have $p_1(\varphi) = p_2(\varphi) = p(\varphi)$. But then, by virtue of the extension by continuity, this equality also holds for the extended functions p_1, p_2, and p in a certain neighborhood of 0 in the space H_0. Due to the assumed infinite differentiability of $p(x)$ $(x \in H_0)$ at 0, we can write

$$s_n = (-i)^n (D^n p)(0) = (-i)^n (D^n p_s)(0) \quad (s = 1, 2),$$

whence, according to (4.31),

$$s_n = \int_{H_{-2}} \underbrace{\xi \otimes \ldots \otimes \xi}_{n} \, d\tau_1(\xi) = \int_{H_{-2}} \xi \otimes \ldots \otimes \xi \, d\tau_2(\xi) \quad (n \in \mathbb{Z}_+). \tag{4.33}$$

By assumption, the class $C\{\|s_{2n}\|^{1/2}_{H_{-1}^{\otimes 2n}}\}$ is quasianalytical; therefore, $s = (s_n)_{n=0}^{\infty}$ is a definite moment sequence. Relation (4.33) and Theorem 2.1 now imply that the equality $\tau_1 = \tau_2$ holds. But then it follows from (4.31) that $p_1(\varphi) = p_2(\varphi)$ $(\varphi \in H_{+1})$

and, hence, $k_1(x) = k_2(x)$ $(x \in H)$.

■

Remark 1. It is easy to understand that the other conditions which guarantee the uniqueness of the extension of $k(x)$ $(x \in H_{2l})$ can be obtained as follows. We extend $k(x)$ from H_{2l} to $\mathbb{R}^{\infty}_{2l}$ as a function $k_2(x)$ by using the method described in the proof of Theorem 4.1 (Step V). Then, for the given k_2, we construct the space H_{k_2} and an operator A_1 in it (see Subsection 3.2). If this operator A_1 is selfadjoint, then the extension of $k(x)$ onto H is unique. Indeed, according to what has been said in Subsection 3.3, the measure σ in representation (4.6) is uniquely determined. This yields the uniqueness of the measure σ in (4.2), because each integral of the form (4.2), according to Step IV of the proof of Theorem 4.1, gives the extension of $k_2(x)$ from $\mathbb{R}^{\infty}_{2l}$ to $\mathbb{R}^{\infty}$.

We do not give concrete sufficient conditions of the uniqueness of extension of $k(x)$ $(x \in H_{2l})$ which can be obtained by using the described method. They are similar to the corresponding statements in the one-dimensional case (see the book of Berezansky [5, Chapter 8, Section 3, Subsection 8]).

■

Remark 2. In the case when the extension $k(x)$ from H_{2l} to H is not unique, we can describe all possible extensions, just as is done in the finite-dimensional case (see Berezansky [5, Chapter 8, Section 4, Subsection 4]).

■

4.3. Other Formulations of the Results Concerning Positive Definite Functions on Hilbert Spaces

We now present some other formulations of the results obtained in Section 4 which are sometimes convenient. First, we consider the case $l = \infty$, i.e., of positive definite functions given on the whole of H. Theorem 4.1 can be formulated as follows.

Theorem 4.5. *Every positive definite function $k(x)$ $(x \in H)$ admits the representation*

$$k(x) = \int_H e^{i(x,\lambda)_H} \, d\sigma(\lambda) \quad (x \in H), \tag{4.34}$$

where σ is a finite nonnegative measure on $\mathfrak{B}(H)$, if and only if one can find a Hilbert space $F \supset H$ with a dense quasinuclear imbedding such that k is continuous in *the topology of* F. *The measure* σ *is uniquely defined for the given* k.

Proof. Assume that the required F exists. Then one can suppose that $H = H_0$ and $F = H_-$, and construct the chain $H_- \supset H_0 \supset H_+$ (see Lemma 1.1 in Chapter 1). According to (1.11) in Chapter 1, we have

$$(\xi, \eta)_{H_-} = (\boldsymbol{I}\xi, \eta)_{H_0}$$

for any $\xi, \eta \in H_-$ and, thus,

$$(x, y)_F = (x, y)_{H_-} = (\boldsymbol{I}x, y)_{H_0} = ((OJ)^2 x, y)_{H_0}$$

for any $x, y \in H_0$. The quasinuclearity of $H = H_0 \to H_- = F$ is equivalent to the quasinuclearity of the imbedding $H_+ \to H_0$ or, which is the same, of the operator OJ: $H_0 \to H_0$. But then $A = (OJ)^2$ is nuclear and the relation

$$(x, y)_F = (Ax, y)_H \quad (x, y \in H)$$

shows that the continuity of k in the topology of F implies the continuity in the J-topology and, therefore, by virtue of Theorem 4.1, representation (4.34) holds.

Conversely, if we have representation (4.34), then k is continuous with respect to the J-topology by Theorem 4.1. But, for this case, a positive nuclear operator A was constructed in the proof of Theorem 4.1 (Step I) such that k is continuous with respect to the topology of the corresponding space H_A. But this space can be taken as H_- (see Subsection 4.2), and the imbedding $H = H_0 \to H_-$ is quasinuclear. It remains to set $F = H_-$ and employ the fact that the continuity of a positive definite function at 0 involves its continuity at any other point (see the inequality (4.3)).

■

The equality (4.34) means that k is a Fourier transform of the measure σ: $k = \tilde{\sigma}$ (see Section 1 in Chapter 2). Let us show how we can use Theorem 4.5 to prove the fact concerning the extendability of a cylindrical measure to a measure in terms of the Fourier transform presented at the end of Subsection 1.6 in Chapter 2.

Proof of Theorem 1.7 in Chapter 2. Assume that the cylindrical measure γ can be extended to the measure σ on $\mathfrak{B}(H)$. Then formula (1.40) in Chapter 1 can be rewritten in the form

$$\tilde{\gamma}(y) = \tilde{\sigma}(y) \quad \Big(y \in \bigcup_{K \in \mathcal{K}} K \subset H \Big),$$

and the same is true for $\tilde{\gamma}$, since $\tilde{\sigma}$ is continuous in the topology of F according to Theorem 4.5.

Conversely, suppose that the function $\tilde{\gamma}(y)$, defined by formula (1.40 (Chapter 2)) on a linear set $L = \bigcup_{K \in \mathcal{K}}$ dense in H, is continuous in the topology of the space F. Clearly, $\tilde{\gamma}$ satisfies the condition of positive definiteness (4.1) on L and, therefore, according to (4.3), it is uniformly continuous in the topology of F. Since L is dense in F, it can be extended by continuity to a positive definite function k continuous on F. According to Theorem 4.5, the function k admits the representation (4.34); in particular, this yields the equality

$$\int_H e^{i(y,\lambda)_H} d\gamma(\lambda) = \tilde{\gamma}(\lambda) = \int_H e^{i(\lambda, y)_H} d\gamma(\lambda) \quad (y \in L). \tag{4.35}$$

It is not hard to conclude from (4.35) that, for any cylindrical set $\mathfrak{C} = \mathfrak{C}(K; \delta)$ $(K \in \mathcal{K}; \delta \in \mathfrak{B}(K))$, we have $\gamma(\mathfrak{C}) = \sigma(\mathfrak{C})$, i.e., γ can be extended to the measure σ. Indeed, let us choose an orthonormal basis $(e_j)_{j=1}^{\infty}$ in H such that $(e_j)_{j=1}^{\infty}$ form a basis in K. Let $(y_j)_{j=1}^{\infty}$ be the coordinates of a vector $y \in H$ in this basis. Then (4.35) implies that

$$\int_{\mathbb{R}^m} e^{i(y_1\lambda_1 + \ldots + y_m\lambda_m)} d\gamma_K(\lambda_1, \ldots, \lambda_m) = \tilde{\gamma}(y_1, \ldots, y_m, 0, 0, \ldots)$$

$$= \int_{\mathbb{R}^m} e^{i(y_1\lambda_1 + \ldots + y_m\lambda_m)} d\sigma_K(\lambda_1, \ldots, \lambda_m)$$

$$(y_1, \ldots, y_m \in \mathbb{R}^1). \tag{4.36}$$

Here, $\gamma_K(\delta) = \gamma(\mathfrak{C}(K; \delta))$ is a measure of the base of a cylindrical set induced by the cylindrical measure, and $\sigma_K(\delta) = \sigma(\mathfrak{C}(K; \delta))$ $(\delta \in \mathfrak{B}(K))$. The required equality $\gamma(\mathfrak{C}) = \sigma(\mathfrak{C})$ follows from (4.36) and the fact that the measure is uniquely defined in the finite-dimensional Bochner theorem.

■

Theorems 4.5, 1.7 in Chapter 2, and 4.1 (for $l = \infty$) are different formulations of the Minlos–Sazonov theorem concerning the Fourier transform of measures on Hilbert spaces.

It is easy to understand that the argument with the help of which Theorem 4.5 was derived from Theorem 4.1 (for $l = \infty$) enables us to reformulate Theorem 4.2 in the following way, which is close to Theorem 4.5.

Theorem 4.6. *Any positive definite function* $k(x)$ $(x \in H_{2l},\ 0 < l < \infty)$ *on a* *layer in a Hilbert space H continuous in the topology of a certain Hilbert space $F \supset H$ with dense and quasinuclear imbedding, admits an extension to a positive definite function on the whole of H which has the same property of continuity.*

It is also easy to formulate Theorem 4.4 in terms of the space F.

Example 4.2. In Example 4.1, we have obtained the representation (4.14) of a positive definite function

$$k(x) = \exp\left(-\frac{1}{4}(Sx, x)\right) \quad (x \in H),$$

where S is a positive nuclear operator in H. The problem appears of how to write These representations for the case of S which is not nuclear or, more generally, for a positive definite function of the form

$$k(x) = \exp(-(x, x)_F) \quad (x \in F),$$

where F is a Hilbert space (the condition of positive definiteness (4.1) can be immediately verified for this function). According to Theorem 4.5, we can find a Hilbert space $H \subset F$ such that this imbedding is dense and quasinuclear. Then, for the restriction $k \Subset H$, one can write representation (4.34) with a certain measure σ on $\mathfrak{B}(H)$ depending on H, i.e., $\sigma = \sigma_H$. Clearly, this measure is Gaussian, and its correlation operator S can be found from the equality

$$\frac{1}{4}(Sx, y)_H = (x, y)_F \quad (x, y \in H)$$

(see also Subsection 1.8 in Chapter 2).

4.4. Connection between Unitary Representations of a Hilbert Space and Positive Definite Functions on this Space

It is well known that Stone's classical theorems concerning a group of unitary operators A_x $(x \in \mathbb{R}^1)$ and Bochner's theorems concerning positive definite functions $k(x)$ $(x \in \mathbb{R}^1)$ can be derived each from other. We now outline the connection of this sort for the case when the real axis $\mathbb{R}^1$ is replaced by a real Hilbert space. First, we prove the following result.

Theorem 4.7. *Theorem* 4.1 *(for $l = \infty$) follows from Theorem* 1.2 *in Chapter* 4.

The proof of this fact is analogous to the derivation of Theorem 2.1 (on integral representation for the moment problem (in a certain case)) from Theorem 1.6 in Chapter 4 (concerning the representation of Hilbert space by selfadjoint operators) presented in Subsection 2.3.

Proof. Consider a positive definite function $H \ni x \mapsto k(x) \in \mathbb{C}^1$ continuous in the J-topology. Let us obtain the representation (4.2) for this function from the representation (1.5) in Chapter 4 for the properly chosen operators A_x ($x \in H$). Below, it is convenient to denote the space H by X.

I. In Steps I, III, and IV of the proof of Theorem 4.1, we have shown that after a proper choice of an orthonormal basis in X, we can extend k by continuity to a positive definite function $k_1(x)$,

$$x \in X_A = l_2(A) = \left\{ x \in \mathbb{R}^\infty \mid \sum_{\alpha=1}^{\infty} a_\alpha x_\alpha^2 < \infty \right\},$$

where A is a positive nuclear operator constructed for the given k, and $(a_\alpha \delta_{\alpha\beta})_{\alpha,\beta=1}^{\infty}$ is its matrix; $X = l_2$. Consider a standard Gaussian measure γ_1 given on $l_2(A) \subset \mathbb{R}^\infty$; $\gamma_1(l_2(A)) = 1$. In step VI of this proof, we have shown that the expression

$$\langle \varphi, \psi \rangle = \int_{l_2(A)} \int_{l_2(A)} k(y-x)\varphi(y)\overline{\psi(x)}\, d\gamma_1(x) d\gamma_1(y) \quad (\varphi, \psi \in C_{b,\text{cyl}}^{\infty}(\mathbb{R}^\infty)) \tag{4.37}$$

is a quasiscalar product on $C_{b,\text{cyl}}^{\infty}(\mathbb{R}^\infty)$. For simplicity, we now assume that the product (4.37) is nondegenerate, i.e., it is a scalar product. Let H_k be a Hilbert space obtained as a result of a completion; $(\cdot, \cdot)_{H_k} = \langle \cdot, \cdot \rangle$.

II. In Step III, we shall construct a family of unitary operators acting in H_k. The role of the space H_0 in Theorem 1.2 of Chapter 4 will be played just by H_k, namely, $H_0 = H_k$. Let us now construct the required rigging (1.4) of it (Chapter 4). To do this, we set

$$H_+ = L_2(l_2(A), \mathfrak{B}(l_2(A)), \gamma_1) = L_2,$$

and assume that D consists of the restrictions of functions from the space $\mathcal{A}(\mathbb{R}^\infty)$ to $l_2(A) \subset \mathbb{R}^\infty$ (see Example 4.2 in Chapter 2) and has the topology of $\mathcal{A}(\mathbb{R}^\infty)$ (this way of introducing a topology is correct because functions $\varphi \in \mathcal{A}(\mathbb{R}^\infty)$ and $\varphi \restriction l_2(A)$ are indistinguishable as elements of L_2).

The properties of the space $\mathcal{A}(\mathbb{R}^\infty)$ imply the properties of the space D and the imbedding $D \subset L_2 = H_+$ necessary for Theorem 1.2 in Chapter 4. Further, it follows from (4.37) that $H_+ \subset H_k = H_0$ topologically. This imbedding is quasinuclear. Indeed, the scalar product (4.37) has the form

$$(\varphi, \psi)_{H_k} = (K\varphi, \psi)_{L_2},$$

where K is an integral operator in L_2 generated by the continuous kernel $K(x, y) = k(y-x)$. Since

$$\int_{l_2(A)} K(x, x)\, d\gamma_1(x) = k(0) < \infty,$$

this operator is nuclear. On the other hand, for the chain $H_- \supset H_0 \supset H_+$, we have

$$(\varphi, \psi)_{H_0} = (I\varphi, \psi)_{H_+} \quad (\varphi, \psi \in H_+);$$

therefore, $K = O(I \upharpoonright H_+)$, and the nuclearity of K yields the quasinuclearity of the imbedding $H_+ \to H_0$. Hence, we have constructed the chain

$$H_k = H_0 \supset H_+ = L_2 \supset D = \mathcal{A}(\mathbb{R}^\infty). \tag{4.38}$$

III. Let us construct a family of unitary operators A_x $(x \in l_2 = X)$ acting in the space H_k. These are the operators of the shift of functions by t with a certain factor which appears because the Gaussian measure γ_1 is not invariant under such shifts. We employ some facts from Subsection 2.4 in Chapter 2. For γ_1, the correlation operator S is equal to 1 and, therefore, this measure and the measure $\gamma_{1,t}$ shifted by $t \in l_2$ $(d\gamma_{1,t}(\cdot) = d\gamma_1(\cdot + t))$ are absolutely continuous with respect to each other, and we have

$$\left(\frac{d\gamma_{1,t}}{d\gamma_1}\right)(\cdot) = \rho_{\gamma_1}(t, \cdot) = \exp\left(-\|t\|_{l_2}^2 - 2(t, \cdot)_{l_2}\right) \tag{4.39}$$

for the derivative (see Theorem 2.3 and formula (2.43) in Chapter 2).

Let us first define the operator A_t on the functions

$$\varphi \in C_{b,\mathrm{cyl}}^\infty(\mathbb{R}^\infty) = \bigcup_{n=1}^\infty C_b^\infty(\mathbb{R}^n)$$

by the equality

$$(A_t\varphi)(x) = \varphi(x-t)\left(\frac{d\gamma_{1,-t}}{d\gamma_1}\right)(x) \quad (t \in l_2;\ x \in l_2(A)). \tag{4.40}$$

For any $t \in l_2$, by the simple change of variables, we get

$$(A_t\varphi, A_t\psi)_{H_k} = (\varphi, \psi)_{H_k} \quad (\varphi, \psi \in C^\infty_{b,\mathrm{cyl}}(\mathbb{R}^\infty)).$$

Thus, each operator A_t is isometric and, therefore, it can be extended to an isometric operator on the whole of H_k by continuity ; we preserve the notation A_t for the latter. Similarly, we can verify the equalities

$$(A_t\varphi, \psi)_{H_k} = (\varphi, A_{-t}\psi)_{H_k}$$

and

$$(A_t\varphi, A_s\psi)_{H_k} = (A_{t-s}\varphi, \psi)_{H_k} \quad (\varphi, \psi \in C^\infty_{b,\mathrm{cyl}}(\mathbb{R}^\infty)),$$

which imply that $A_t^* = A_{-t}$ and $A_tA_s = A_{t+s}$ $(t, s \in l_2)$. Hence, we have constructed the group of unitary operators $(A_x)_{x\in l_2}$.

IV. Let us show that the family $(A_t)_{t\in l_2}$ just constructed and the chain (4.38) are connected by conditions (i) and (ii) of Theorem 1.2 in Chapter 4. To establish the validity of condition (i), one should verify that, for fixed $t \in l_2$, there exists a multi-index $\tau \in [1, \infty)^\infty$ such that

$$\| A_t\varphi \|^2_{l_2} \le c_1 \| \varphi \|^2_{A_t(\mathbb{R}^\infty)} \quad (\varphi \in \mathcal{A}(\mathbb{R}^\infty)).$$

For this purpose, we choose $\tau = (\tau_k)_{k=1}^\infty$ such that $\sum_{k=1}^\infty \tau_k^{-1} < \infty$ and $\tau_k \ge 2$. Then, by using Corollary 2 of Theorem 4.3 in Chapter 2 and formula (4.36) in Chapter 2, we obtain the estimate

$$|\varphi(x)|^2 = |(\varphi, \delta_x)_{l_2}|^2 \le \| \varphi \|^2_{A_\tau(\mathbb{R}^\infty)} \| \delta_x \|^2_{A_{-t}(\mathbb{R}^\infty)}$$

$$= c_2 \| \varphi \|^2_{A_\tau(\mathbb{R}^\infty)} \exp\left(2 \sum_{k=1}^\infty (1+\tau_k)^{-1} x_k^2 \right)$$

$$\left(x \in \mathbb{R}^\infty,\ c_2 = \prod_{k=1}^\infty (1-\tau_k^{-2})^{1/2} \right). \tag{4.41}$$

Taking into account (4.40), (4.41), and (4.39), we find

$$\| A_t\varphi \|^2_{l_2} = \int_{\mathbb{R}^\infty} \left| \varphi(x-t)^2 \left(\frac{d\gamma_{1,-t}}{d\gamma_1} \right)(x) \right|^2 d\gamma_1(x)$$

$$= \int_{\mathbb{R}^\infty} |\varphi(x-t)|^2 \left(\frac{d\gamma_{1,-t}}{d\gamma_1} \right)(x)\, d\gamma_{1,-t}(x)$$

$$= \int_{\mathbb{R}^\infty} |\varphi(x)|^2 \left(\frac{d\gamma_1}{d\gamma_{1,t}}\right)(x)\, d\gamma_1(x) \le \|\varphi\|^2_{A_\tau(\mathbb{R}^\infty)} \int_{\mathbb{R}^\infty} \|\delta_x\|^2_{A_{-\tau}(\mathbb{R}^\infty)}$$

$$\times \exp\left(\|t\|^2_{l_2} - 2(t,x)_{l_2}\right) d\gamma_1(x)$$

$$= \|\varphi\|^2_{A_\tau(\mathbb{R}^\infty)} c_2 \prod_{k=1}^\infty \left(\frac{1}{\sqrt{\pi}} \int_{\mathbb{R}^1} \exp\left(2(1+\tau_k)^{-1} x_k^2 + t_k^2 + 2t_k x_k - x_k^2\right) dx_k \right)$$

$$= \|\varphi\|^2_{A_\tau(\mathbb{R}^\infty)} c_2 \exp\left(\|t\|^2_{l_2}\right)$$

$$\times \prod_{k=1}^\infty \left(\frac{1}{\sqrt{\pi}} \int_{\mathbb{R}^1} \exp\left(-(1-2(1+\tau_k)^{-1}) x_k^2 + 2t_k x_k\right) dx_k \right)$$

$$= \|\varphi\|^2_{A_\tau(\mathbb{R}^\infty)} c_2 \exp\left(\|t\|^2_{l_2}\right)$$

$$\times \prod_{k=1}^\infty (1-2(1+\tau_k)^{-1})^{-1/2} \exp\left((1-2(1+\tau_k)^{-1})^{-1}\, t_k^2\right)$$

$$= \|\varphi\|^2_{A_\tau(\mathbb{R}^\infty)} c_2 \exp\left(\|t\|^2_{l_2}\right) \exp\left(\sum_{k=1}^\infty (1-2(1+\tau_k)^{-1})^{-1}\, t_k^2 \right)$$

$$\times \prod_{k=1}^\infty (1-2(1+\tau_k)^{-1})^{-1/2} = c_3 \|\varphi\|^2_{A_\tau(\mathbb{R}^\infty)} \quad (\varphi \in \mathcal{A}(\mathbb{R}^\infty);\ c_3 < \infty). \tag{4.42}$$

Condition (i) is thus established.

The verification of condition (ii) is reduced to the proof of the continuity of the function

$$l_2 \ni t \mapsto (A_t\varphi, \psi)_{L_2} \in \mathbb{C}^1,$$

where $\varphi \in \mathcal{A}(\mathbb{R}^\infty)$ and $\psi \in L_2$ are arbitrary and fixed. First, we note that the constant c_3 in (4.42) can be estimated from above by $c_4 \exp(c_5\|t\|^2_{l_2})$, where c_4 and c_5 do not depend on t. Therefore, it follows from (4.42) that, for fixed $\varphi \in \mathcal{A}(\mathbb{R}^\infty)$, the norm $\|A_t\varphi\|_{L_2}$ is bounded in t on each ball of the space l_2.

By using the elementary estimates, we can conclude from this fact that it suffices to establish the continuity of

$$l_2 \ni t \mapsto (A_t\varphi, \psi)_{L_2} \in \mathbb{C}^1$$

for a certain set of functions ψ dense in L_2. Let us take $C^{\infty}_{b,\text{cyl}}(\mathbb{R}^{\infty})$ to play the role of this set. According to (4.40), for $\varphi \in \mathcal{A}(\mathbb{R}^{\infty})$ and $\psi \in C^{\infty}_{b,\text{cyl}}(\mathbb{R}^{\infty})$, we have

$$(A_t\varphi, \psi)_{L_2} = \int_{\mathbb{R}^{\infty}} \varphi(x-t)\left(\frac{d\gamma_{1,-t}}{d\gamma_1}\right)(x)\,\psi(x)\,d\gamma_1(x)$$

$$= \int_{\mathbb{R}^{\infty}} \varphi(x-t)\,\psi(x)\,d\gamma_{1,-t}(x) = \int_{\mathbb{R}^{\infty}} \varphi(x)\,\psi(x+t)\,d\gamma_1(x) \quad (t \in l_2). \quad (4.43)$$

But ψ is cylindrical and smooth and, thus, the vector-valued function $l_2 \ni t \mapsto \psi(\cdot + t) \in L_2$ is continuous. By virtue of (4.43), this yields the continuity of $(A_t\varphi, \psi)_{L_2}$.

Therefore, the conditions of Theorem 1.2 in Chapter 4 are satisfied, and we can write the representation

$$A_x = \int_X e^{i(\lambda, x)_X}\, dE(\lambda) \quad (x \in X), \tag{4.44}$$

for the introduced operators A_x in the space H_k. Here, E is an RI on $\mathfrak{B}(X)$.

V. For any $x \in X_A$, the space H_k includes the δ-function δ_x concentrated at the point x, and $(\delta_x, \delta_y)_{H_k} = k_1(x-y)$ $(x, y \in X_A)$. More precisely, by using Lemma 1.1 in Chapter 1, for given negative space H_k and zero space L_2, we can construct a chain $H_k \supset L_2 \supset H_+$. As a result, we obtain the space H_+ consisting of continuous bounded functions on X_A. The space H_k includes vectors δ_x such that $(\delta_x, \varphi)_{L_2} = \overline{\varphi(x)}$ $(x \in X_A;\, \varphi \in H_+)$, and the indicated equality holds. This assertion is a consequence of the facts that the kernel

$$X_A \times X_A \ni (x, y) \mapsto K(x, y) = k_1(x-y) \in \mathbb{C}^1$$

is continuous and bounded and that the measure γ_1 is positive on any ball in the space X_A. It can be proved just as in the case of a positive definite continuous kernel $K(x, y)$ on a precompact domain $G \subset \mathbb{R}^d$ and Lebesgue measure (see the book of Berezansky [18, Chapter 1, Section 3, Subsection 12]).

It is easy to understand that $A_t\delta_0 = \delta_t$ for any $t \in X$. Indeed, let $(\chi_n(\cdot))_{n=1}^{\infty}$ be a sequence of functions from L_2 which converges to δ_0 in H_k, and let each of these functions be equal to the characteristic function of the ball centered at 0 in X_A with the radius n^{-1} divided by the measure γ_1 of this ball. Clearly, the action of the operator A_t on χ_n is given by formula (4.40). For $\varphi \in H_+$, we have

$$(A_t \delta_0 \varphi)_{L_2} = \lim_{n\to\infty} (A_t \chi_n \varphi)_{L_2}$$

$$= \lim_{n\to\infty} \int_{X_A} \chi_n(x-t)\left(\frac{d\gamma_{1,-t}}{d\gamma_1}\right)(x)\,\overline{\varphi(x)}\,d\gamma_1(x)$$

$$= \lim_{n\to\infty} \int_{X_A} \chi_n(x-t)\,\overline{\varphi(x)}\,d\gamma_{1,-t}(x)$$

$$= \lim_{n\to\infty} \int_{X_A} \chi_n(x)\,\overline{\varphi(x+t)}\,d\gamma_1(x) = \overline{\varphi(x)} = (\delta_t, \varphi)_{L_2},$$

and this yields the equality $A_t\delta_0 = \delta_t$.

After these remarks, representation (4.2) immediately follows from (4.44): One should apply the last equality to δ_0 and multiply by δ_0 scalarly in H_k. Here

$$\sigma(\alpha) = (E(\alpha)\delta_0, \delta_0)_{H_k} \quad (\alpha \in \mathfrak{B}(X)).$$

VI. We now briefly dwell upon the general case when the product (4.37) is degenerate. Clearly, the space H_k is now constructed by the preliminary factorization of the space $C^\infty_{b,\text{cyl}}(\mathbb{R}^\infty)$ with respect to the set $\{\varphi \in C^\infty_{b,\text{cyl}}(\mathbb{R}^\infty) \mid \langle\varphi, \varphi\rangle = 0\}$ and by the subsequent completion. Due to their isometry, the operators (4.40) generate operators on classes which are then extended by continuity to operators on H_k. As a result, we obtain a group of unitary operators on H_k indexed by points $x \in X$.

To construct the corresponding rigging of the type (4.38), one should factorize (4.38) with respect to the subspace $N = \{\varphi \in L_2 \mid \langle\varphi, \varphi\rangle = 0\}$ of the space L_2 (the bilinear form (4.37) continuously depends on $\varphi, \psi \in L_2$ and, therefore, N is closed in L_2). We shall not study this factorization in detail, because it is similar to the analogous constructions which appear in the proofs of Theorems 2.1 and 3.1. In Subsections 5.1 and 5.2, we shall also encounter similar constructions in the general situation. The proof is completed by analogy with Step V.

■

It proves impossible to establish a theorem converse to Theorem 4.7, i.e., to derive Theorem 1.2 in Chapter 4 directly from Theorem 4.1 ($l = \infty$). Nevertheless, we make some remarks.

Let $X \ni x \mapsto A_x \in \mathcal{L}(H_0, H_0)$ be a unitary representation of a real Hilbert space X by unitary operators. We say that it is continuous at 0 in the J-topology if, for any $f \in H_0$, the same is true for the function $X \ni x \mapsto (A_x f, f)_{H_0} \in \mathbb{C}^1$. Stone's Theorem 1.2 in Chapter 4 for the representation of a Hilbert space can be also formulated in the following form.

Theorem 4.8. *The unitary representation* X *has the form* (1.5) *of Chapter* 4 *if and only if it is continuous at* 0 *in the J-topology.*

Proof. The representation (1.5) in Chapter 4 can be derived in the way similar to the classical derivation of Stone's theorem from the Bochner theorem. Namely, for any $f \in H_0$, the function $k_f(x) = (A_x f, f)_{H_0}$ $(x \in X)$ is positive definite and continuous at 0 in the J-topology; therefore, according to Theorem 4.1 $(l = \infty)$, it admits the representation

$$(A_x f, f)_{H_0} = \int_X e^{i(\lambda, x)_X} d\sigma_{f,f}(\alpha) \quad (x \in X), \tag{4.45}$$

Taking into account the facts that the measure $\sigma_{f,f}$ is uniquely defined for the given k_f and that on the left-hand side of (4.45) we have a quadratic form of the bilinear form $(A_x f, g)_{H_0}$ and using the polarization identity, we easily conclude that $\sigma_{f,f}(\alpha)$ is a quadratic form of some Hermitian bilinear form $\sigma_{f,g}(\alpha)$ $(f, g \in H_0)$ for any $\alpha \in \mathfrak{B}(X)$. This bilinear form is a charge with respect to α, and the corresponding integral (4.45) is equal to $(A_x f, g)_{H_0}$. It is easy to conclude that it is continuous for any $\alpha \in \mathfrak{B}(X)$ (note that

$$\sigma_{f,f}(\alpha) \le k_f(0) = \|f\|^2_{H_0})$$

and, hence, has the form

$$\sigma_{f,g}(\alpha) = (E(\alpha)f, g)_{H_0},$$

where $E(\alpha) \in \mathfrak{L}(H_0, H_0)$. The fact that the family $(A_x)_{x \in X}$ forms a group of unitary operators enables us to conclude that $\mathfrak{B}(X) \ni \alpha \mapsto E(\alpha)$ is an RI. As a result, we come to formula (1.5) in Chapter 4.

Conversely, the continuity of a representation of the form (1.5) in Chapter 4 in the J-topology is an immediate consequence of the second part of Theorem 4.1.

■

***Corollary* 1.** *For a unitary representation of a real Hilbert space, the presence of a rigging connected in the standard way and satisfying conditions of Theorem* 1.2 *in Chapter* 4, *is equivalent to the continuity at* 0 *in the J-topology. This statement follows from a comparison of Theorem* 1.2 *in Chapter* 4 *and Theorem* 4.8. *However, we cannot establish this equivalence directly.*

■

4.5. Some Additional Remarks

Let us consider several problems closely connected with the results of this section.

1. As in Subsection 4.1, we can define a positive definite function $k(x)$, not on a Hilbert space, but on a real Banach space. In the case when this space coincides with $l_q = l_q(\mathbb{R}^1)$, where $q \in [1, 2]$, the results generalizing what has been said in Subsections 4.1 and 4.2 are valid.

Thus, a set

$$l_{q;a} = \{x = (x_\alpha)_{\alpha=1}^\infty \in l_q \mid x_1 \in (-a, a)\}$$

is called a layer in the space l_q; positive definiteness is defined by the inequality (4.1). A collection of sets of the form

$$\left\{ x \in l_q \mid \sum_{\alpha=1}^\infty a_\alpha x_\alpha < \varepsilon \right\} \quad (\varepsilon > 0),$$

where $a_\alpha \geq 0$ are such that $\sum_{\alpha=1}^\infty a_\alpha^{q'/2} < \infty$ (q' is the index conjugated with q, i.e., $q^{-1} + q'^{-1} = 1$), is understood as a set of base neighborhoods of the origin of the space l_q in the J-topology. Representation (4.2) takes the form

$$k(x) = \int_{l_{q'}} e^{i\langle\lambda, x\rangle} \, d\sigma(\alpha) \quad (x \in l_{q;2a};\ \mathfrak{B}(l_{q'}) \ni \alpha \mapsto \sigma(\alpha) \geq 0). \tag{4.46}$$

The formulation of the theorem on the representation of a positive definite function $k(x)$ $(x \in l_{q;2a})$ on a layer in the space l_q $(q \in [1, 2])$ coincides with the formulation of Theorem 4.1 with the only exception being that, generally speaking, the converse statement does not hold, i.e., function (4.46) is clearly positive definite but it may be discontinuous at 0 in the J-topology.

This theorem is proved with the help of Theorems 3.1 and 3.2 using the method presented in Subsection 4.1 (the proof belongs to Kondratyev [1]). The role of $p(t) = p_1(t) = p_2(t) = \ldots$ is now played by the Fourier transform of the function $\exp(-|\lambda|)$; for $q \in (1, 2)$, it cannot be explicitly computed, but the well-known expression which gives it in the form of a series (see, for example, Feller [1, Chapter 17, Section 6]) enables us to realize all necessary investigations. Note that here we also have the following generalization of the Kolmogorov–Khinchin criterion (Theorem 1.3 in Chapter 2) concerning the measure $\times_{\alpha=1}^\infty dp(x_\alpha)$ on $\mathbb{R}^\infty$:

The set

$$\left\{ x \in \mathbb{R}^{\infty} \mid \sum_{\alpha=1}^{\infty} a_{\alpha} x_{\alpha}^{2} < \infty \right\}$$

is of full measure if and only if $\sum_{\alpha=1}^{\infty} a_{\alpha}^{q'/2} < \infty$.

If a layer coincides with the whole of the space l_q, then the Kondratyev theorem formulated above transforms into the de Acosta theorem [1]. The analogue of Theorem 4.4 is also valid in this case. The measures on the spaces l_q are studied in the book of Vakhania, Tarieladze, and Chobanyan [1].

2. We can consider the positive definite functions $k(x)$ on a real nuclear space

$$\Phi = \operatorname{pr} \lim_{\tau \in \mathrm{T}} H_{\tau} : \Phi \ni x \mapsto k(x) \in \mathbb{C}^{1},$$

the condition of positive definiteness (4.1) remains the same. It is convenient to introduce the chain $\Phi' \supset H_0 \supset \Phi \; (0 \in \mathrm{T})$ The following Minlos theorem is valid:

Any continuous positive definite function $k(x)$ *on a nuclear space* Φ *admits the representation*

$$k(x) = \int_{\Phi'} e^{i(\lambda, x)_{H_0}} \, d\sigma(\alpha) \quad (x \in \Phi). \tag{4.47}$$

Here, $\mathfrak{C}_{\sigma}(\Phi') \ni \alpha \mapsto \sigma(\alpha) \geq 0$ *is a finite measure which is uniquely determined for a given* k.

The proof of this theorem can be found in the paper by Minlos [1] and in the books of Gelfand and Vilenkin [1, Chapter 4, Section 4, Theorem 2] and Daletsky and Fomin [1, Chapter 3, Theorem 1.3].

Conversely, under certain restrictions imposed on Φ, the positive definite function (4.47) is continuous, too. The Minlos theorem presented in Chapter 2 (Theorems 1.11 and 1.12) and the theorem given above, are the different formulations of the same fact.

3. It is well known that a function $\mathbb{R}^1 \ni x \mapsto k(x) \in \mathbb{R}^1$ is called exponentially convex if the kernel

$$\mathbb{R}^1 \times \mathbb{R}^1 \ni (x, y) \mapsto K(x, y) = k(x + y) \in \mathbb{R}^1$$

is positive definite. For continuous exponentially convex functions, the classical Bernstein theorem holds (for example, see the book of Berezansky [5, Chapter 8, Theorem 3.17]):

Any continuous exponentially convex function admits a representation

$$k(x) = \int_{\mathbb{R}^1} e^{\lambda x} d\sigma(\alpha) \quad (x \in \mathbb{R}^1), \tag{4.48}$$

where $\mathfrak{B}(\mathbb{R}^1) \ni \alpha \mapsto \sigma(\alpha) \geq 0$ *is a finite measure uniquely defined by* k. *Conversely, each function of the form* (4.48) *is continuous and exponentially convex.*

The Bernstein theorem can be generalized for a nuclear space, just as the Minlos theorem generalizes the Bochner theorem. Namely, using the same notations as above, we consider a function $\Phi \ni x \mapsto k(x) \in \mathbb{C}^1$ which satisfies the condition of exponential convexity

$$\sum_{j,k=1}^{n} k(x^{(j)} + x^{(k)}) \xi_j \bar{\xi}_k \geq 0$$

for any $x^{(1)}, \dots, x^{(n)} \in \Phi$ and $\xi_1, \dots, \xi_n \in \mathbb{C}^1$ $(n \in \mathbb{N})$. If Φ is nuclear and k is continuous, then the latter admits the representation

$$k(x) = \int_{\Phi'} e^{i(\lambda, x)_{H_0}} d\sigma(\lambda) \quad (x \in \Phi), \tag{4.49}$$

where $\mathfrak{C}_\sigma(\Phi') \ni \alpha \mapsto \sigma(\alpha) \geq 0$ is a finite measure uniquely determined for a given k. As in the case of the Minlos theorem, here we can also formulate the converse statement. This theorem belongs to Kalyuzhny [1] and can be proved in a way similar to the proof of the Minlos theorem. The close result holds also for exponentially convex functions given on a real Hilbert space and continuous in the J-topology.

4. As shown in Subsection 4.2 (in particular, in Theorem 4.3), the derivatives of a positive definite function on a Hilbert space at 0 form a moment sequence (provided that this function is differentiable).

In quantum field theory, an essential role is played by a positive definite function k (for example, on the space $\mathfrak{S}(\mathbb{R}^d)(d = 2, 3, \dots))$ which guarantees that the derivatives at the origin form the Osterwalder–Schrader moment sequence (see Examples 2.7 and 2.11). The condition of positive definiteness of this sort (the Fröhlich positive definiteness) can be formulated as follows. Let $\mathfrak{S}(\mathbb{R}^d) \ni \varphi \mapsto k(\varphi) \in \mathbb{C}^1$, and let θ be an involution considered in these examples. Then it is necessary that the inequality

$$\sum_{j,k=1}^{n} k(\varphi^{(j)} - (\varphi^{(k)})^\theta) \xi_j \bar{\xi}_k \geq 0 \tag{4.50}$$

hold for any $\xi_1, \dots, \xi_n \in \mathbb{C}^1$ and any $\varphi^{(1)}, \dots, \varphi^{(n)} \in \mathfrak{S}(\mathbb{R}^d)$ whose supports lie in $\mathbb{R}^{d-1} \times [0, \infty)$. As in the case of the Osterwalder–Schrader moment sequence, here it is essential that the inequality (4.50) does not hold on all $\varphi^{(j)} \in \mathfrak{S}(\mathbb{R}^d)$ (the functions k

for which (4.50) holds for all $\varphi^{(j)} \in \mathcal{S}(\mathbb{R}^d)$ admit a representation which can be obtained by the methods presented in Section 3 and Subsection 4.1). In connection with the definition of positive definiteness given above, see the papers by Fröhlich [1–3].

§5. General Scheme of the Representation of Positive Definite Kernels in Terms of Elementary Ones

Comparing the constructions in Section 2 with those in Section 3, we can easily notice that they are quite similar, namely, for any given object of investigation, we construct a Hilbert space H_s or H_k, study in it a family of commuting selfadjoint operators corresponding to the problem under consideration, and then construct a decomposition in the generalized joint eigenvectors of this family. The Parseval equality leads to the required representation. In this section, we describe the principal steps of this construction for the general case.

5.1. Hilbert Space Constructed for a Given Positive Definite Kernel and Its Rigging

Recall some of the general concepts introduced in Subsections 2.3 and 2.4 of Chapter 1. Consider a chain of Hilbert spaces with involution "*"

$$G_- \supset G_0 \supset G_+ . \tag{5.1}$$

In other words, the involution $G_- \ni \xi \mapsto \xi^* \in G_-$ is defined in the negative space G_-; the restriction of this involution to the zero space G_0 or to the positive space G_+ is an involution in the corresponding space. We form a tensor square of the chain (5.1)

$$G_- \otimes G_- \supset G_0 \otimes G_0 \supset G_+ \otimes G_+ \tag{5.2}$$

and fix a certain generalized kernel $\mathrm{K} \in G_- \otimes G_-$. It is called positive definite if

$$(\mathrm{K}, \varphi \otimes \varphi^*)_{G_0 \otimes G_0} \geq 0 \quad (\varphi \in G_+), \tag{5.3}$$

i.e., a continuous bilinear form

$$G_+ \otimes G_+ \ni (\varphi, \psi) \mapsto a(\varphi, \psi) = (\mathrm{K}, \psi \otimes \varphi^*)_{G_0 \otimes G_0} \in \mathbb{C}^1$$

is nonnegative. The kernel $\mathrm{K} \in G_- \otimes G_-$ is called Hermitian if this form is Hermitian, i.e., if

$$(\mathrm{K}, \psi \otimes \varphi^*)_{G_0 \otimes G_0} = \overline{(\mathrm{K}, \varphi \otimes \psi^*)}_{G_0 \otimes G_0}.$$

Every positive definite kernel is Hermitian.

Our aim is to obtain spectral representations of positive definite kernels with the help of the spectral analysis of operators acting in the Hilbert space generated by a kernel of this sort. This space is constructed in the standard way: Given a positive definite kernel K, we introduce a quasiscalar product in G_+ by

$$\langle \varphi, \psi \rangle = (\mathrm{K}, \psi \otimes \varphi^*)_{G_0 \otimes G_0} \quad (\varphi, \psi \in G_+). \tag{5.4}$$

Identifying all $\varphi \in G_+$ for which $\langle \varphi, \varphi \rangle = 0$ with zero and then completing a collection of classes

$$\hat{\varphi} = \{\chi \in G_+ \mid \langle \varphi - \chi, \varphi - \chi \rangle = 0\},$$

we obtain a space which (for convenience) will be denoted by H_K (but not by G_K).

In more detail, this procedure can be described as follows. We set

$$N = \{\varphi \in G_+ \mid \langle \varphi, \varphi \rangle = 0\}.$$

By virtue of the inequality

$$\langle \varphi, \varphi \rangle \leq \|\mathrm{K}\|_{G_- \otimes G_-} \|\varphi\|^2_{G_+} \quad (\varphi \in G_+), \tag{5.5}$$

N is a subspace of G_+. A class $\hat{\varphi}$ corresponding to $\varphi \in G_+$ coincides with the hyperplane $\varphi + N$. The form (5.4) does not change if φ and ψ run through the corresponding classes $\hat{\varphi}$ and $\hat{\psi}$ and defines their scalar product $(\hat{\varphi}, \hat{\psi})_{H_\mathrm{K}}$. The space H_K is a completion of this pre-Hilbert space of classes. It is clear that the picture described is isomorphic to the following one: On the orthogonal (in G_+) completion of $G_+ \ominus N$, formula (5.4) generates a scalar product, and the completion with respect to it gives H_K. The isomorphism is defined by the correspondence (before the completion) between a class $\hat{\varphi}$ and the projection (in G_+) of a vector φ onto $G_+ \ominus N$ (all projections of vectors $\varphi \in \hat{\varphi}$ coincide).

We shall employ (and, in fact, we have already done this in Sections 2 and 3) the following procedure of constructing the rigging of the space H_K from the rigging of G_0. First, we consider the case of Hilbert riggings. Assume that there exists a chain with involution

$$H_- \supset G_- \supset G_0 \supset G_+ \supset H_+, \tag{5.6}$$

where the Hilbert space H_+ is topologically imbedded in G_+. Our aim is to construct the rigging of the space H_K from (5.6).

To do this, we note that $N \cap H_+$ is a subspace of H_+ (due to (5.5) and the continuity of the imbedding $H_+ \to G_+$) and consider $H_+ \ominus (N \cap H_+)$ (the orthogonal subtraction in H_+).

Vectors of the space $H_+ \ominus (N \cap H_+)$ can be identified with a collection of all classes $\hat{\varphi}$, where $\varphi \in H_+$; each $\hat{\varphi}$ has one and only one common vector φ_N with $H_+ \ominus (N \cap H_+)$, namely, the projection of φ onto $H_+ \ominus (N \cap H_+)$ in H_+.

Indeed, let φ_1 be the desired projection. Then

$$\varphi - \varphi_1 \in N \cap H_+ \subset N,$$

i.e., $\varphi_1 \in \hat{\varphi}$. Now let

$$\varphi_1, \varphi_2 \in \hat{\varphi} \cap (H_+ \ominus (N \cap H_+)),$$

then $\varphi_1, \varphi_2 \in H_+$ and $\varphi_1 - \varphi_2 \in N$, whence $\varphi_1 - \varphi_2 \in N \cap H_+$. We also have $\varphi_1 - \varphi_2 \in H_+ \ominus (N \cap H_+)$ and, thus, $\varphi_1 - \varphi_2 = 0$.

■

Denote by $H_{+,K}$ a linear collection of all classes $\hat{\varphi}$, where $\varphi \in H_+$, and transform it into a Hilbert space by setting

$$(\hat{\varphi}, \hat{\psi})_{H_{+,K}} = (\varphi_N, \psi_N)_{H_+} \quad (\varphi, \psi \in H_+). \tag{5.7}$$

This definition is obviously correct and the obtained space $H_{+,K}$ is isometric to $H_+ \ominus (N \cap H_+)$.

Clearly, the space $H_{+,K}$ is imbedded in H_K, and it can be easily understood that the imbedding operator O can be represented in the form of a product $O = S\,O_+$, where $O_+ : H_+ \to G_+$ is an operator which imbeds H_+ into G_+, and $S: G_+ \to H_K$ is an operator which transforms a vector $\varphi \in G_+$ into the corresponding class $\hat{\varphi} \in H_K$. Therefore,

$$\|\hat{\varphi}\|_{H_K} \le \|S\,O_+\| \, \|\hat{\varphi}\|_{H_{+,K}} \le \|K\|^{1/2}_{G_-\otimes G_-} \|O_+\| \, \|\hat{\varphi}\|_{H_{+,K}} \quad (\hat{\varphi} \in H_{+,K})$$

(see (5.5)). Obviously, $H_{+,K}$ is dense in H_K. This follows from the denseness of H_+ in G_+ and inequality (5.5). This means that we can take H_K and $H_{+,K}$ as zero and positive space, respectively, and construct the required chain

$$H_{-,\mathrm{K}} \supset H_{\mathrm{K}} \supset H_{+,\mathrm{K}}. \tag{5.8}$$

It follows from the representation $O = S O_+$ that *the chain* (5.8) *is quasinuclear if the imbedding* $H_+ \to G_+$ *is quasinuclear.*

Clearly, in the case of a nondegenerate kernel K, i.e., when $N = \{0\}$, the described procedure for the construction of the chain (5.8) becomes quite trivial. Here $H_{+,\mathrm{K}} = H_+$.

We now proceed from the riggings (5.6) to riggings by linear topological spaces. Consider the chain

$$\Phi' \supset G_- \supset G_0 \supset G_+ \supset \Phi, \tag{5.9}$$

with involution $*$, where $\Phi = \mathrm{pr}\ \lim_{\tau \in \mathrm{T}} H_\tau$ is a projective limit of Hilbert spaces topologically imbedded in G_+. We can assume that each H_τ is topologically imbedded in G_+, $\|\varphi\|_{G_+} \le \|\varphi\|_{H_\tau}$ for any $\tau \in \mathrm{T}$ and $\varphi \in H_\tau$, and that there exists $\tau = 0$ such that $H_0 = G_+$ (see Subsection 1.2 in Chapter 1); $*$ is an involution in each H_τ $(\tau \in \mathrm{T})$. Therefore, according to (5.6) and (5.8), for any $\tau \in \mathrm{T}$, we can consider the chains

$$H_{-\tau} \supset G_- \supset G_0 \supset G_+ \supset H_\tau \quad \text{and} \quad H_{-\tau,\mathrm{K}} \supset H_{\mathrm{K}} \supset H_{\tau,\mathrm{K}}. \tag{5.10}$$

Lemma 5.1. *A family of Hilbert spaces $(H_{\tau,\mathrm{K}})_{\tau \in \mathrm{T}}$ is suitable for the construction of the projective limit $\Phi_{\mathrm{K}} = \mathrm{pr}\ \lim_{\tau \in \mathrm{T}} H_{\tau,\mathrm{K}}$. Therefore, we can construct the chain*

$$\Phi'_{\mathrm{K}} \supset H_{\mathrm{K}} \supset \Phi_{\mathrm{K}}, \tag{5.11}$$

where Φ_{K} can be regarded as a space of classes $\hat{\varphi}$ with $\varphi \in \Phi$. If the space Φ is nuclear, then Φ_{K} is also nuclear.

Proof. Let us establish the first statement of the lemma. The intersection $\bigcap_{\tau \in \mathrm{T}} H_{\tau,\mathrm{K}}$ is dense in each $H_{\tau_0,\mathrm{K}}$ $(\tau_0 \in \mathrm{T})$. Indeed, this intersection contains the classes $\hat{\varphi}$, where $\varphi \in \bigcap_{\tau \in \mathrm{T}} H_\tau = \Phi$ and, therefore, it suffices to show that the collection of these classes is dense in $H_{\tau_0,\mathrm{K}}$. Let $\varphi \in H_{\tau_0}$ and $\Phi \ni \varphi_n \underset{n \to \infty}{\to} \varphi$ in H_{τ_0}. Then the corresponding projections onto $H_{\tau_0} \ominus (N \cap H_{\tau_0})$ in H_{τ_0} (we denote these projections by $(\varphi_n)_{N,\tau_0}$) also converge to the projection φ_{N,τ_0} in H_{τ_0}. In other words, the classes $\hat{\varphi}_n$ converge to the class $\hat{\varphi}$ in the metric of the space $H_{\tau_0,\mathrm{K}}$ (see (5.7)), and this means that the required denseness takes place.

Further, the family $(H_{\tau,\mathrm{K}})_{\tau \in \mathrm{T}}$ is directed by imbedding. Indeed, for any $\tau', \tau'' \in \mathrm{T}$, there exists $\tau''' \in \mathrm{T}$ such that $H_{\tau'''}$ is topologically imbedded in $H_{\tau'}$ and in $H_{\tau''}$. Then $H_{\tau''',\mathrm{K}}$ is dense in $H_{\tau',\mathrm{K}}$; this follows from an argument similar to that presented above. Let $\varphi \in H_{\tau'''}$, then

$$\| \hat{\varphi} \|_{H_{\tau''',K}} = \| \varphi_{N,\tau'''} \|_{H_{\tau'''}} \geq c_{\tau',\tau'''} \| \varphi_{N,\tau'''} \|_{H_{\tau'}}$$

$$\geq c_{\tau',\tau'''} \| (\varphi_{N,\tau'''})_{N,\tau'} \|_{H_{\tau'}} = c_{\tau',\tau'''} \| \hat{\varphi} \|_{H_{\tau',K}}$$

(here we have used the fact that $\varphi_{N,\tau'''} \in \hat{\varphi}$ and defines the same class). Hence, the imbedding $H_{\tau''',K} \to H_{\tau',K}$ is continuous.

Analogous conclusions can be also made for the index τ''. This means that Φ_K can be defined as a projective limit of the spaces $H_{\tau,K}$ $(\tau \in T)$. Due to the second chain in (5.10), we get (5.11). It is also clear that Φ_K may be regarded as a collection of classes.

Let us show that the nuclearity of Φ yields the nuclearity of Φ_K. Assume that $\tau \in T$ and $\tau' \in T$ are such that the imbedding $H_{\tau'} \subset H_\tau$ is quasinuclear. Obviously, $H_{\tau',K} \subset H_{\tau,K}$. Let us choose an orthonormal basis $(e_j)_{j=1}^\infty$ in the space $H_{\tau'} \ominus (N \cap H_{\tau'})$. Then the classes $(\hat{e}_j)_{j=1}^\infty$ form an orthonormal basis in $H_{\tau',K}$, and we must prove that

$$\sum_{j=1}^{\infty} \| \hat{e}_j \|_{H_{\tau,K}}^2 = \sum_{j=1}^{\infty} \| (e_j)_{N,\tau} \|_{H_\tau}^2 < \infty$$

(see (5.7)). But

$$\| (e_j)_{N,\tau} \|_{H_\tau} \leq \| e_j \|_{H_\tau},$$

and the series $\sum_{j=1}^{\infty} \| e_j \|_{H_\tau}^2$ converges due to the quasinuclearity of the imbedding $H_{\tau'} \subset H_\tau$. This implies the convergence of the series under consideration.

■

Remark 1. The original positive definite kernel K was regarded as an element of the tensor square $G_- \otimes G_-$ of a negative Hilbert space. However, one may consider the chain $\Psi' \supset H_0 \supset \Psi$, where $\Psi = \mathrm{pr\,lim}_{\tau \in T} G_\tau$, instead of (5.1) and assume that $K \in \Psi' \otimes \Psi'$. Recall[5] that

$$\Psi \otimes \Psi = \underset{(\tau_1,\tau_2) \in T \times T}{\mathrm{pr\,lim}} G_{\tau_1} \otimes G_{\tau_2}$$

by definition; then

[5] See Subsection 2.1 in Chapter 1.

$$\Psi' \otimes \Psi' = \bigcup_{(\tau_1,\tau_2)\in \mathrm{T}\times\mathrm{T}} G_{-\tau_1} \otimes G_{-\tau_2} .$$

The inclusion $\mathrm{K} \in \Psi' \otimes \Psi'$, the last equality, and the directedness of T together imply that there exists $\tau \in \mathrm{T}$ such that $\mathrm{K} \in G_{-\tau} \otimes G_{-\tau}$. In fact, we arrived at the situation of the chain (5.1), and all that has been said above concerning the rigging H_{K} can be now reformulated correspondingly.

■

Example 5.1. Let us clarify how the moment problem investigated in Subsection 2.1 can be considered within the framework of a general scheme. Let $s = (s_n)_{n=0}^{\infty}$ $\left(s_n \in (\Phi^{\otimes n})'\right)$ be a moment sequence; it has been noted that $s_n \in H_{-\tau(n)}^{\otimes n}$. Here $G_0 = \oplus_{j=0}^{\infty} H_0^{\otimes j}$ and G_+ is a space of sequences $\varphi = (\varphi_j)_{j=0}^{\infty}$ $\left(\varphi_j \in H_{\tau(2j)}^{\otimes j}\right)$ with the scalar product $(\varphi, \psi)_{G_+}$ introduced on Step IV of the proof of Theorem 2.1. Then s can be regarded as an element K from $G_- \otimes G_-$. In this case, the involution $*$ is a natural extension of the already existing involution to the sequences $\left(\varphi^* = (\overline{\varphi}_j)_{j=0}^{\infty}\right)$. Clearly, the space H_s which appears in Section 2 coincides with the space H_{K} introduced above.

The generalized power moment problem presented in Subsection 2.4 can be investigated in the framework of the general scheme analogously.

Example 5.2. Clearly, the problem of the spectral representation of the positive definite (in the integral sense) function $k(x)$ $(x \in \mathbb{R}^{\infty})$ considered in Subsection 3.1 can be also studied with the help of the described scheme. Here

$$G_- = G_0 = G_+ = L_2(\mathbb{R}^{\infty}, d\mu(x)),$$

the kernel $K(x, y) = k(y - x)$ $(x, y \in \mathbb{R}^{\infty})$, and the involution $*$ is a transition to the complex overbar, $H_k = H_{\mathrm{K}}$.

5.2. Operators Acting in the Space Generated by a Positive Definite Kernel

Assume that the chain (5.1) and a positive definite kernel $\mathrm{K} \in G_- \otimes G_-$ are given. Consider the operator A acting in G_+ with the dense domain of definition $\mathfrak{D}(A)$. The question then arises as to when it can be reduced to the operator $\hat{A}$ acting in the space H_{K}, i.e., when A transforms a class into another class. Then we take $\hat{A}\hat{\varphi} = (A\varphi)^{\wedge}$ $(\varphi \in \hat{\varphi}$, $\varphi \in \mathfrak{D}(A)$; $\mathfrak{D}(\hat{A}) = (\mathfrak{D}(A))^{\wedge})$. This problem can be easily solved in the following two cases, the first of which has been already considered several times (see Subsections 1.8, 2.1, 2.4, and 3.1).

I. *The operator A is Hermitian with respect to* (5.4), *i.e.*,

$$\langle A\varphi, \psi\rangle = \langle \varphi, A\psi\rangle \quad (\varphi, \psi \in \mathfrak{D}(A)).$$

Indeed, one should prove that for any $\varphi \in \mathfrak{D}(A)$, the equality $\langle \varphi, \varphi\rangle = 0$ yields $\langle A\varphi, A\varphi\rangle = 0$. Let $\psi \in \mathfrak{D}(A)$, then

$$|\langle A\varphi, \psi\rangle|^2 = |\langle \varphi, A\psi\rangle|^2 \le \langle \varphi, \varphi\rangle \langle A\psi, A\psi\rangle,$$

whence $\langle A\varphi, A\psi\rangle = 0$. Since $\mathfrak{D}(A)$ is dense in G_+, we can approximate the vector $A\varphi$ in G_+ by the vectors $\psi_n \in \mathfrak{D}(A)$. By virtue of the inequality (5.5), we have

$$\langle A\varphi, A\varphi\rangle = \lim_{n\to\infty} \langle A\varphi, \psi_n\rangle = 0.$$

■

II. *The operator A is isometric with respect to* (5.4), *i.e.*,

$$\langle A\varphi, A\varphi\rangle = \langle \varphi, \varphi\rangle \quad (\varphi \in \mathfrak{D}(A)).$$

Indeed, the equality $\langle \varphi, \varphi\rangle = 0$ now yields $\langle A\varphi, A\varphi\rangle = 0\ (\varphi \in \mathfrak{D}(A))$.

■

Consider a family $(A_x)_{x\in X}$ of Hermitian operators in G_+ (with respect to $\langle \cdot, \cdot\rangle$) admitting a rigging in the following sense: There exists a chain

$$G_0 \supset G_+ \supset H_+ \supset D \tag{5.12}$$

(H_+ is a Hilbert space topologically imbedded in G_+ and D is a projective limit of Hilbert spaces which is also imbedded in H_+) such that $D \in \mathfrak{D}(A_x)$ for any $x \in X$, and $A_x \restriction D$ acts continuously from D to H_+.

According to the rules described in Subsection 5.1, we now construct the spaces H_{K}, $H_{+,\mathrm{K}}$, and $H_{-,\mathrm{K}}$ (starting from the chain

$$H_- \supset G_- \supset G_0 \supset G_+ \supset H_+)$$

and the projective limit D_{K} (starting from the chain

$$D' \supset G_- \supset G_0 \supset G_+ \supset D).$$

Lemma 5.2. *The spaces just constructed form a chain*

$$H_{-,\mathrm{K}} \supset H_{\mathrm{K}} \supset H_{+,\mathrm{K}} \supset D_{\mathrm{K}} \tag{5.13}$$

(where each subspace to the right is topologically imbedded into the subspace to the left) connected in the standard way with a family of Hermitian operators $(\hat{A}_x)_{x \in X}$ *in* H_K, *i.e., for any* $x \in X$, $D_K \in \mathfrak{D}(\hat{A}_x)$, *and* $\hat{A}_x \restriction D_K \in \mathfrak{L}(D_K, H_{+,K})$. *Each element in* D_K *may be regarded, not only as an element of the projective limit, but also as being constructed with the help of the procedure of constructing* $H_{+,K}$ *from* $H_+ \supset D$. *If the imbedding* $H_+ \subset G_+$ *is quasinuclear, then chain* (5.13) *is also quasinuclear.*

Proof. The assertions concerning the three spaces to the left in the chain (5.13) obviously follow from what has been said in Subsection 5.1.

Let us show that the imbedding $D_K \subset H_{+,K}$ is topological. The space D_K is constructed from the chain (5.9), where we set $\Phi = D$, by use of Lemma 5.1; the space $H_{+,K}$ is constructed from the chain (5.6) by the method described in Subsection 5.1. However, by virtue of (5.12), the space H_+ may be regarded as one of the spaces H_τ whose projective limit $\operatorname{pr\,lim}_{\tau \in T} H_\tau$ is D, and we can assume that $H_{\tau_0} = H_+$ for some $\tau_0 \in T$. But then the fact that the imbedding $D_K \subset H_{+,K}$ is topological follows from the proof of the first part of Lemma 5.1. By the same argument, we show the validity of the second assertion of the lemma.

The inclusion $D_K \subset \mathfrak{D}(\hat{A}_x)$ is obvious, and it remains only to establish the continuity of $\hat{A}_x \restriction D_K : D_K \to H_{+,K}$ $(x \in X)$. Suppose that $D_K \ni \hat{\varphi}_n \to \hat{\varphi} \in D_K$ as $n \to \infty$ in D_K. Then the last convergence also takes place in $H_{\tau,K}$ for some $\tau \in T$. In other words, by virtue of (5.7), we have $(\varphi_n)_{N,\tau} \to \varphi_{N,\tau}$ in H_τ. Thus, by virtue of the continuity of $A_x \restriction D : D \to H_+$, we obtain

$$A_x(\varphi_n)_{N,\tau} \to A_x \varphi_{N,\tau}$$

in H_{τ_0}. Hence, we have the convergence in H_{τ_0} and the convergence of the projections onto

$$H_{\tau_0} \ominus (N \cap H_{\tau_0}) : (A_x(\varphi_n)_{N,\tau})_{N,\tau_0} \to (A_x \varphi_{N,\tau})_{N,\tau_0},$$

i.e.,

$$(A_x(\varphi_n)_{N,\tau})^\wedge \to (A_x \varphi_{N,\tau})^\wedge$$

in $H_{+,K}$. Since $(\varphi_n)_{N,\tau} \in \hat{\varphi}_n$ and $\varphi_{N,\tau} \in \hat{\varphi}$, the last statement means that $\hat{A}_x \hat{\varphi}_n \to \hat{A}_x \hat{\varphi}$ in $H_{+,K}$.

■

It is easy to understand that the analogous situation takes place if we exploit the chain (5.9) instead of (5.6). More precisely, assume that a family $(A_x)_{x \in X}$ of Hermitian operators in G_+ (with respect to $\langle \cdot, \cdot \rangle$) admits a rigging in the following sense: There exists a chain

$$G_0 \supset G_+ \supset \Phi$$

(Φ is a projective limit of Hilbert spaces topologically imbedded in G_+) such that $\Phi \subset \mathfrak{D}(A_x)$ for any $x \in X$ and $A_x \upharpoonright \Phi \in \mathfrak{L}(\Phi, \Phi)$. Then the family $(\hat{A}_x)_{x \in X}$ is connected with the chain (5.11) in the standard way.

Example 5.3. Let us clarify the form of the families of operators and the chains connected with these families in the standard way in the corresponding constructions of Sections 2 and 3.

In the case of the moment problem, the closures of the operators $\hat{A}_x$ in $H_s = H_K$ were constructed with the help of the operations of shift (2.8) on Steps I–III of the proof of Theorem 2.1 (here $x \in X = \mathbb{N}$). At the same time, it was not necessary to introduce the operators A_x acting in G_+ and the operators $\hat{A}_x$ lowered to classes, because we have directly constructed the closures $(\hat{A}_x)^\sim$ of these (in a certain case) which were denoted by A_{e_α}. The chain (2.12) played the role of the chain (5.13) and was constructed similarly from the space of sequences H_+ introduced on Step IV; the role of D was played by the space C_0. In the quasi-definite case, the operator $(\hat{A}_1)^\sim$ is then extended in the required manner to a selfadjoint operator.

An analogous scheme has been also used in Subsection 2.4.

In the case of positive definite functions considered in Section 3, we first introduce (with the help of the operation (3.11)) the operators A'_α acting in the space $L_2(\mathbb{R}^\infty, \mu) = G_0 = G_+$ (see Lemma 3.1); then we construct the closures A_α in H_k of the operators A'_α $(\alpha = x \in X = \mathbb{N})$ lowered to this space. Therefore, in Subsection 3.1, A_α also denotes the operator $(\hat{A}_\alpha)^\sim$ considered here. The space H_+ has the form (3.14), and the chain (3.16) plays the role of the chain (5.13), $D = C_0$.

We now establish the lemma which will be necessary when proving Theorem 5.1 (similar facts have been already employed in Subsections 2.1, 2.4, 3.2, and 3.3).

Lemma 5.3. *Assume that the chain* (5.6) *is given. Each quasinuclear operator* $C\colon H_{+,K} \to H_{-,K}$ *is associated with the kernel* $\Theta \in H_- \otimes H_-$ *such that*

$$(C\hat{\varphi}, \hat{\psi})_{H_K} = (\Theta, \psi \otimes \varphi^*)_{G_0 \otimes G_0} \quad (\varphi, \psi \in H_+);$$

$$\|\Theta\|_{H_- \otimes H_-} \le k\, |C| \quad (k < \infty). \tag{5.14}$$

Proof. The proof is based on the following general statement which can be established, just as Lemma 2.8 in Chapter 1. Suppose that H is a Hilbert space with involution $*$ and $A: H \to H$ is a bounded operator. It is stated that the form $(Af, g)_H$ can be written as $(F, g \otimes f^*)_{H \otimes H}$ $(f, g \in H)$, where $F \in H \otimes H$, if and only if A is a quasinuclear operator and $\|F\|_{H \otimes H} = |\!|\!| A |\!|\!|$.

Thus, let S be an operator which transforms a vector $\varphi \in G_+$ into the corresponding class $\hat{\varphi} \in H_{-,K}$; its restriction $S \upharpoonright H_+$ acts continuously from H_+ into $H_{+,K}$, since

$$\|\hat{\varphi}\|_{H_{+,K}} = \|\varphi_N\|_{H_+} \le \|\varphi\|_{H_+}.$$

Denote by I_1 the operator I constructed from the chain (5.8). For any $\varphi, \psi \in H_+$, we have

$$(C\hat{\varphi}, \hat{\psi})_{H_K} = (I_1 C\varphi, S\psi)_{H_K} = ((S \upharpoonright H_+)^* I_1 C\varphi, S\psi)_{H_K} = (A\varphi, \psi)_{H_K},$$

where the operator

$$A = (S \upharpoonright H_+)^* I_1 CS : H_+ \to H_+$$

is quasinuclear due to the quasinuclearity of C. By virtue of the statement formulated above, there exists $F \in H_+ \otimes H_+$ such that

$$(C\hat{\varphi}, \hat{\psi})_{H_K} = (F, \psi \otimes \varphi^*)_{H_+ \otimes H_+} \quad (\varphi, \psi \in H_+). \tag{5.15}$$

Denote by I_2 the operator I constructed from the chain

$$H_- \otimes H_- \supset G_0 \otimes G_0 \supset H_+ \otimes H_+.$$

Then

$$(F, \psi \otimes \varphi^*)_{H_+ \otimes H_+} = (I_2^{-1} F, \psi \otimes \varphi^*)_{G_0 \otimes G_0} \quad (\varphi, \psi \in H_+).$$

Setting $\Theta = I_2^{-1} F \in H_- \otimes H_-$ and employing the last equality, we obtain from (5.15) the required representation in (5.14).

It is also clear that

$$\|\Theta\|_{H_- \otimes H_-} = \|F\|_{H_+ \otimes H_+} = |\!|\!| A |\!|\!| \le k |\!|\!| C |\!|\!|.$$

■

5.3. Spectral Representation of Positive Definite Kernels in Terms of Elementary Ones

Consider the chain (5.1) with involution $*$. Let $K \in G_- \otimes G_-$ be some positive definite kernel. Assume that a family $A = (A_x)_{x \in X}$ of operators $A_x\colon G_+ \to G_+$ with dense domains of definition $\mathfrak{D}(A_x)$ is given; assume also that these operators are Hermitian with respect to $\langle \cdot, \cdot \rangle$, i.e.,

$$(\mathrm{K}, \psi \otimes (A_x\varphi)^*)_{G_0 \otimes G_0} = (\mathrm{K}, (A_x)\psi \otimes \varphi^*)_{G_0 \otimes G_0} \quad (\varphi, \psi \in \mathfrak{D}(A_x);\, x \in X). \quad (5.16)$$

Suppose that the family A admits a rigging in the sense of Subsection 5.2, i.e., there exists a chain (5.12) such that $D \subset \mathfrak{D}(A_x)$ for any $x \in X$ and $A_x \upharpoonright D \in \mathfrak{L}(D, H_+)$. We also suppose that the space H_+ (and, hence, H_-) is invariant with respect to $*$.

Consider the kernel $\Omega \in H_+ \otimes H_-$. For Θ, we introduce the notion of positive definiteness as usual:

$$(\Theta, \psi \otimes \varphi^*)_{G_0 \otimes G_0} \geq 0 \quad (\varphi \in H_+).$$

This kernel is called an elementary kernel of the family A corresponding to an eigenvalue $\lambda(\cdot)$ (where $\lambda(\cdot)$ is a function of the form $x \ni X \mapsto \lambda(\cdot) \in \mathbb{R}^1$) provided that the equality

$$(\Theta, \psi \otimes (A_x\varphi)^*)_{G_0 \otimes G_0} = (\Theta, (A_x)\psi \otimes \varphi^*)_{G_0 \otimes G_0} = \lambda(x)\, (\Theta, \psi \otimes \varphi^*)_{G_0 \otimes G_0}$$

$$(\varphi, \psi \in \mathfrak{D}(A_x);\, x \in X). \quad (5.17)$$

holds. Below, we prove Theorem 5.1 which shows that, under certain conditions, a positive definite kernel K can be "assembled" from the elementary positive definite kernels Ω.

Example 5.4. Let us show that the elementary kernels $\Omega(\lambda(\cdot))$ which appear in the proofs of Theorems 2.1, 2.3, and 3.1 are elementary, just in the sense (5.17). Inded, in the case of Theorem 2.1, $\Omega(\lambda(\cdot)) \in H_- \otimes H_-$ and, therefore, it has the form of a matrix $(\Omega_{jk}(\lambda(\cdot)))_{j,k=0}^{\infty}$ (see (2.19)). The equalities (2.20) and (2.21) coincide exactly with (5.17) $(x \in \mathbb{N} = X)$, because the operators A_x are now equal to the operators of shift in the space G_+ generated by (2.8) (with $e = e_1, e_2, \ldots$). In the case of Theorem 2.3, equalities (2.58) and (2.59) coincide with (5.17). In the case of Theorem 3.1, (5.17) is equivalent to the equations the solution of which is sought in Lemma 3.3.

Theorem 5.1. *Assume that the imbedding* $H_+ \subset G_+$ *in the chain* (5.12) *is quasinuclear and that* $A = (A_x)_{x \in X}$ *is an introduced family of operators. We lower each operator* A_x *to an Hermitian operator* $\hat{A}_x$ *in* H_K *and denote by* B_x *its closure in* H_K. *Assume that the family* $B = (B_x)_{x \in X}$ *obtained consists of selfadjoint commuting operators.*

Then the following decomposition of the kernel $K \in G_- \otimes G_-$ *into the elementary kernels* $\Omega(\lambda(\cdot))$ (*corresponding to the eigenvalues* $\lambda(\cdot)$ *and defined for* ρ-*almost all* $\lambda(\cdot)$) *is valid:*

$$K = \int_{\mathbb{R}^X} \Omega(\lambda(\cdot))\, d\rho(\lambda(\cdot)). \tag{5.18}$$

Here ρ *is a nonnegative finite measure defined on* $\mathfrak{C}_\sigma(\mathbb{R}^X)$; $\|\Omega(\lambda(\cdot))\|_{H_- \otimes H_-}$ *are bounded uniformly in* $\lambda(\cdot)$, *and the integral in* (5.18) *converges in norm of the space* $H_- \otimes H_-$.

Clearly, the corresponding representations which appear in the proofs of Theorems 2.1, 2.3, and 3.1 are special cases of the general formula (5.18). The proof of this theorem given below is based on the spectral projection Theorem 2.7 in Chapter 3 and repeats in a more abstract form a part of the argument in Sections 2 and 3.

Proof. Since for any $x \in X$ we have $B_x \supset \hat{A}_x$, it follows from Lemma 5.2 that the quasinuclear chain (5.13) is connected with the family B in the standard way. It satisfies the conditions of Theorem 2.7 in Chapter 3 as far as B is concerned and, therefore, by virtue of this theorem (more precisely, by virtue of Remark 4 in Subsection 2.10 of Chapter 3), we can write the representation

$$O^+ O = \int_{\pi} P(\lambda(\cdot))\, d\rho_\pi(\lambda(\cdot)). \tag{5.19}$$

Here O and O^+ are the imbedding operators connected with the chain (5.8); $\pi \subset \mathbb{R}^X$ is a set which appears in the formulation of this theorem; $P(\lambda(\cdot))$: $H_{+,K} \to H_{-,K}$ is a generalized projector, $\| P(\lambda(\cdot)) \| \le 1$ $(\lambda(\cdot) \in \pi)$, and the integral in (5.19) converges in the Hilbert–Schmidt norm.

Let us apply (5.19) to $\hat{\varphi} \in H_{+,K}$ and multiply by $\hat{\psi} \in H_{+,K}$ $(\varphi, \psi \in H_+ \subset G_+)$ in H_K. Then, by virtue of Lemma 5.3, we obtain

$$(K, \psi \otimes \varphi^*)_{G_0 \otimes G_0} = \langle \varphi, \psi \rangle = (\hat{\varphi}, \hat{\psi})_{H_K}$$

$$= (O^+ O \hat{\varphi}, \hat{\psi})_{H_K} = \int_{\pi} (P(\lambda(\cdot))\hat{\varphi}, \hat{\psi})_{H_K}\, d\rho_\pi(\lambda(\cdot))$$

$$= \int_\pi (\Omega(\lambda(\cdot)), \psi \otimes \varphi^*)_{G_0 \otimes G_0} d\rho_\pi(\lambda(\cdot))$$

$$= \int_\pi \left(\Omega(\lambda(\cdot))\, d\rho_\pi(\lambda(\cdot)), \psi \otimes \varphi^*{}_{G_0 \otimes G_0} \right) \quad (\varphi, \psi \in H_+). \quad (5.20)$$

Here $\Omega(\lambda(\cdot)) \in H_- \otimes H_-$ is a kernel which corresponds to the operator $P(\lambda(\cdot))$ according to Lemma 5.3; it satisfies the estimate

$$\| \Omega(\lambda(\cdot)) \|_{H_- \otimes H_-} \le k \| P(\lambda(\cdot)) \| \le k \quad (\lambda(\cdot) \in \pi).$$

Hence, the last integral in (5.20) may be regarded as an integral in the norm of the space $H_- \otimes H_-$.

Further, according to Theorem 2.7 in Chapter 3, the measure ρ_π is defined on the σ-algebra

$$\mathfrak{C}_\sigma(\pi) = \{\alpha' = \alpha \cap \pi \,|\, \alpha \in \mathfrak{C}_\sigma(\mathbb{R}^X)\}.$$

We extend it as zero to the whole of $\mathbb{R}^X$ and denote the resulting measure by ρ. This means that we set

$$\mathfrak{C}_\sigma(\mathbb{R}^X) \ni \alpha \mapsto \rho(\alpha) = \rho_\pi(\alpha \cap \pi),$$

and extend (for example, by zero) the function $\Omega(\lambda(\cdot))$ for $\lambda(\cdot) \in \mathbb{R}^X \setminus \pi$. As a result, we get

$$\int_\pi \Omega(\lambda(\cdot))\, d\rho_\pi(\lambda(\cdot)) = \int_{\mathbb{R}^X} \Omega(\lambda(\cdot))\, d\rho_\pi(\lambda(\cdot)).$$

Inserting this expression in (5.20) and using the arbitrariness of $\varphi, \psi \in H_+$, we obtain (5.18).

Let us show that the kernels $\Omega(\lambda(\cdot))$ are elementary for A. Let $\lambda(\cdot) \in \pi$, then $\Omega(\lambda(\cdot))$ is positive definite, because it is associated with the nonnegative operator $P(\lambda(\cdot))$. Further, according to Lemma 5.3, we have

$$(\Omega(\lambda(\cdot)), \psi \otimes \varphi^*)_{G_0 \otimes G_0} = (P(\lambda(\cdot))\hat{\varphi}, \hat{\psi})_{H_K},$$

hence, substituting the vector $A_x\chi \in H_+$, where $\chi \in D$ and $x \in X$, for ψ in this equality, we obtain

$$(\Omega(\lambda(\cdot)), (A_x\chi) \otimes \varphi^*)_{G_0 \otimes G_0} = (P(\lambda(\cdot))\hat{\varphi}, \hat{A}_x\hat{\chi})_{H_K}$$

$$= (P(\lambda(\cdot))\hat{\varphi}, B_x \hat{\chi})_{H_K} = \lambda(x)(P(\lambda(\cdot))\hat{\varphi}, \hat{\chi})_{H_K}$$

$$= \lambda(x)(\Omega(\lambda(\cdot)), \chi \otimes \varphi^*)_{G_0 \otimes G_0}. \tag{5.21}$$

Here we have used the fact that $\mathfrak{R}(P(\lambda(\cdot)))$ consists of generalized joint eigenvectors of the family of operators B.

The relation (5.21) means that the last equality in (5.17) holds. The first equality in (5.17) follows from the fact that the generalized kernel $\Omega(\lambda(\cdot))$ is Hermitian, and this, in turn, is a consequence of its positive definiteness.

■

We must make several remarks about this theorem.

Remark 1. Each integral of the form (5.18) is an element of the space $H_- \otimes H_-$; it is clearly positive definite, because the kernel $\Omega(\lambda(\cdot))$ is positive definite for any $\lambda(\cdot) \in \mathbb{R}^X$, and satisfies the relation (5.16) for $\varphi, \psi \in D$ (by virtue of (5.17)).

Note that in formula (5.18), we have a certain "gap in smoothness", namely, the kernel $K \in G_- \otimes G_-$ can be expanded in the integral in "less smooth" kernels

$$\Omega(\lambda(\cdot)) \in H_- \otimes H_- \quad (G_- \subset H_-)$$

(the smoothness increases when we integrate). A similar situation has already occurred in the moment problem (see Theorem 2.2) and, in fact, the same reason generates the requirement of *J*-continuity in Theorem 4.1. One can show that in the general situation, this gap is inevitable.

■

Remark 2. Clearly, Theorem 5.1 remains valid if we do not demand the selfadjointness of the closures of $\hat{A}_x$ and their commutativity but only the fact that the operators $\hat{A}_x$ can be extended in H_K to selfadjoint commuting operators B_x $(x \in X)$.

We have already encountered a similar situation in Theorem 2.1 in the indefinite case and in Theorem 3.2. In these theorems, it has been necessary to extend one of the operators $\hat{A}_x$ to obtain a family of commuting selfadjoint operators, and this has been always possible. We stress that it is, generally speaking, impossible to extend even two operators in this way. Therefore, if among $\hat{A}_x$ there are at least two operators which are not essentially selfadjoint, then the formulation of general theorems is quite difficult. In connection with this, see Remark 2 in Subsection 3.3 and the books of Berezansky [5, Chapter 8, Sections 2, 4] and Gelfand and Vilenkin [1, Chapter 2, Section 7, Subsection 2].

When proving the essential selfadjointness and commutativity of the operators $\hat{A}_x$ in Theorems 2.1, 2.3, and 3.1, we have used Theorem 1.15; in analogous situations, we can also employ the other results presented in Subsection 1.8.

■

Remark 3. In the case of a single operator A_1 ($X = \{1\}$), the differential $\Omega(\lambda)\,d\rho(\lambda)$ ($\lambda \in \mathbb{R}^1$) in formula (5.18) is uniquely determined for a given K if and only if the operator $\hat{A}_1$ is maximal in H_{K}.

The proof can be found in the book of Berezansky [5, Chapter 8, Theorem 1.1]. Some results of this type also hold in the case of more than one operator, but we shall not formulate them. In the examples, it is more convenient to prove the uniqueness of the representation directly (if this uniqueness takes place). We have already done this in Theorems 2.1 and 3.1.

■

Remark 4. If X is a topological space and the operators A_x are such that, for any $\varphi \in D$, the vector function $X \ni x \mapsto A_x\varphi \in H_+$ is weakly continuous, then the integral in (5.18) is "continual": It can be regarded as taken only over the collection $C(X) \subset \mathbb{R}^X$ of continuous functions on X.

Indeed, in this case, the vector function $X \ni x \mapsto \hat{A}_x\hat{\varphi} \in H_{+,\mathrm{K}}$ is also weakly continuous for any $\hat{\varphi} \in D_{\mathrm{K}}$. To prove this, we fix $\varphi \in \hat{\varphi} \cap D$. Then, by virtue of (5.7), we have

$$(\hat{A}_x\hat{\varphi},\, \hat{\psi})_{H_{+,\mathrm{K}}} = ((A_x\varphi)^\wedge, \psi^\wedge)_{H_{+,\mathrm{K}}} = ((A_x\varphi)_N, \psi_N)_{H_+}$$

for any $\hat{\psi} \in H_{+,\mathrm{K}}$. Since $A_x\varphi$ depends on $x \in X$ weakly continuously in H_+, the same is true for the projection $(A_x\varphi)_N$, and this yields the required continuity of $\hat{A}_x\hat{\varphi}$. It now remains to apply the result given in Subsection 2.9 of Chapter 3 and to replace the integration over $\mathbb{R}^X$ by the integration over $C(X) \supset \pi$.

■

Remark 5. One can also formulate the result of the type of Theorem 5.1 for a kernel $\mathrm{K} \in \Phi' \otimes \Phi'$ (see Remark 1 in Subsection 5.1) taking (5.11) instead of (5.13) as a chain connected with the family B in the standard way.

■

Remark 6. The results of the type of Theorem 5.1 (and the remarks to it) are valid also in the case when the operators A_x are not Hermitian ((5.16) does not hold), but it is known that they can be lowered to the operators $\hat{A}_x$ on H_{K} (for example, they are isometric with respect to $\langle \cdot, \cdot \rangle$).

■

Example 5.5. In fact, we have earlier studied three examples of the realization of the general Theorem 5.1 and Remark 2. These are Theorems 2.1–2.3, 3.1, and 3.2. Clearly, the number of examples may be increased. Thus, we can consider functions $\mathbb{R}^\infty \ni x \mapsto$

$k(x) \in \mathbb{C}^1$ for which $k(y-x)$ is replaced by $k(y+x)$ or by $1/2(k(y+x)+k(y-x))$ in the definition (3.7). We shall not dwell upon examples of this sort in our book.

5.4. The Case of Smooth Positively Definite Kernels

Assume that a positive definite kernel is an element of the positive space of the chain (5.2), i.e.,

$$\mathrm{K} = \mathrm{K} \in G_+ \otimes G_+ \subset G_- \otimes G_- .$$

Clearly, the assertions given in Subsections 5.1–5.3 are valid in this case, too. However, they can be greatly improved if we prove that elementary kernels are elements of a certain, generally speaking, positive space. The scheme presented below is adjoint in a certain sense to the scheme in Subsection 5.3.

The space H_{K} can be now constructed as follows. A bilinear form

$$G_+ \otimes G_+ \ni (\varphi, \psi) \mapsto (\varphi, \psi) = (\mathrm{K}, \psi \otimes \varphi^*)_{G_0 \otimes G_0} \in \mathbb{C}^1$$

is extended to a form

$$G_- \otimes G_- \ni (\xi, \eta) \mapsto (\xi, \eta) = (\mathrm{K}, \eta \otimes \xi^*)_{G_0 \otimes G_0} \in \mathbb{C}^1$$

by continuity. It is obvious that

$$(\mathrm{K}, \xi \otimes \xi^*)_{G_0 \otimes G_0} \geq 0, \quad |(\mathrm{K}, \eta \otimes \xi^*)_{G_0 \otimes G_0}|$$

$$\leq \|\mathrm{K}\|_{G_+ \otimes G_+} \|\xi\|_{G_-} \|\eta\|_{G_-} \quad (\xi, \eta \in G_-). \tag{5.22}$$

We set

$$\langle \xi, \eta \rangle = (\mathrm{K}, \eta \otimes \xi^*)_{G_0 \otimes G_0} \quad (\xi, \eta \in G_-).$$

Consider a factor space G_- / N with respect to the subspace

$$N = \{\xi \in G_- \mid \langle \xi, \xi \rangle = 0\},$$

and complete it with respect to $\langle \cdot, \cdot \rangle$. The space H_{K} obtained as a result coincides with H_{K} constructed by the procedure described in Subsection 5.1.

Instead of (5.6), we introduce a chain with involution "*"

$$G_- \supset H_- \supset G_0 \supset H_+ \supset G_+ . \tag{5.23}$$

Denote by S an operator from G_- to H_K which establishes the correspondence between each $\xi \in G_-$ and the class $\hat{\xi}$ which contains it. By virtue of (5.22), this operator is continuous. Let $O_- : H_- \to G_-$ be an imbedding operator, $O = S O_- : H_- \to H_K$, and let $H_{-,K}$ be an orthogonal complement to the subspace

$$N \cap H_- = \{\xi \in H_- \mid O\xi = 0\}$$

in H_-. As in Subsection 5.1, it is easy to see that the space $H_{-,K}$ can be identified with the collection of all classes $\hat{\xi}$, where $\xi \in H_-$. This identification enables us to write the imbedding $H_{-,K} \subset H_K$; the operator O is an imbedding operator.

It follows from (5.22) that

$$\| \hat{\xi} \|^2_{H_K} \le \| K \|_{G_+ \otimes G_+} \|\xi\|^2_{H_{-,K}} \quad (\xi \in H_{-,K})$$

as above, and it suffices to assume that $\| K \|_{G_+ \otimes G_+} \le 1$. Therefore, one can construct a chain

$$H_{--,K} \supset H_K \supset H_{-,K}, \tag{5.24}$$

which replaces (5.8), where $H_{--,K}$ is the corresponding negative space. If the imbedding $G_+ \to H_+$ is quasinuclear, then the same is true for the imbedding $H_- \to G_-$ and, hence, for $O: H_{-,K} \to H_K$.

Consider an operator A acting in G_- with a dense domain of definition (if such an operator acts in G_+, then we are to interpret it as an operator in G_-). Let A be Hermitian with respect to

$$\langle \xi, \eta \rangle = (K, \eta \otimes \xi^*)_{G_0 \otimes G_0},$$

i.e.,

$$\langle A\xi, \eta \rangle = \langle \xi, A\eta \rangle \quad (\xi, \eta \in \mathfrak{D}(A)). \tag{5.25}$$

Then it can be lowered to a densely defined Hermitian operator $\hat{A}$ acting in H_K ($\hat{A}\,\hat{\xi} = (A\xi)^\wedge$, $\mathfrak{D}(\hat{A}) = (\mathfrak{D}(A))^\wedge$).

Let $(A_x)_{x \in X}$ be a family of operators in G_- which are Hermitian with respect to $\langle \cdot, \cdot \rangle$ and admit a rigging in the following sense: There exists a chain

$$G_- \supset H_- \supset G_0 \supset D \tag{5.26}$$

(H_- is a HIlbert space, D is a projective limit of Hilbert spaces; and the imbeddings in

(5.26) are topological) such that for any $x \in X$, we have $D \subset \mathfrak{D}(A_x)$, and $A_x \upharpoonright D$ acts continuously from D to H_-.

Constructing a chain for given zero and negative spaces (see Lemma 1.1 in Chapter 1), we can suppose that the Hilbert spaces which appear in (5.26) are taken from the chain (5.23). Denote by D_K a collection of classes $\hat{\varphi}$, where $\varphi \in D$. It follows from Lemma 5.1 that D_K may be regarded as a projective limit of Hilbert spaces (instead of the pair $G_+ \supset \Phi$ with the subsequent factorization connected with K, one should start from the pair $H_- \supset D$ and the factorization connected with K).

Therefore, the chain (5.26) enables us to continue (5.24) and obtain a chain

$$H_{--,\mathrm{K}} \supset H_{\mathrm{K}} \supset H_{-,\mathrm{K}} \supset D_{\mathrm{K}}, \tag{5.27}$$

which replaces (5.13). By using an argument similar to that in Lemma 5.2, it is easy to understand that (5.27) is connected with the family of Hermitian operators $(A_x)_{x \in X}$ in the standard way. If the imbedding $H_- \subset G_-$ is quasinuclear, then the chain (5.27) is also quasinuclear.

By analogy, we can also construct a nuclear rigging of H_K connected with $(\hat{A}_x)_{x \in X}$ in the standard way; one should only start from the chain $G_- \supset G_0 \supset \Phi$ instead of (5.26); in this chain, Φ is nuclear and such that we have $\Phi \subset \mathfrak{D}(A_x)$, and $A_x \upharpoonright \Phi \in \mathfrak{L}(\Phi, \Phi)$ for any $x \in X$.

The role of Lemma 5.3 is now played by the following lemma.

Lemma 5.4. *Assume that the chain (5.23) is given. Each quasinuclear operator $C: H_{-,\mathrm{K}} \to H_{--,\mathrm{K}}$ is associated with the kernel $\Psi \in H_+ \otimes H_+$ such that*

$$(C\hat{\xi}, \hat{\eta})_{H_{\mathrm{K}}} = (\Psi, \eta \otimes \xi^*)_{G_0 \otimes G_0} \quad (\xi, \eta \in H_-),$$

and

$$\|\Psi\|_{H_+ \otimes H_+} \le k |C| \quad (k < \infty).$$

Proof. The proof almost repeats the proof of Lemma 5.3. Let I_1 be the operator I constructed from the chain (5.27). For $\xi, \eta \in H_-$, we have

$$(C\hat{\xi}, \hat{\eta})_{H_{\mathrm{K}}} = (I_1 CS\xi, S\eta)_{H_{-,\mathrm{K}}} = ((S \upharpoonright H_-)^* I_1 CS\xi, \eta)_{H_-} = (A\xi, \eta)_{H_-},$$

where the operator

$$A = (S \upharpoonright H_-)^* I_1 CS : H_- \to H_-$$

is quasinuclear due to the quasinuclearity of C. Therefore, by virtue of the assertion which has been employed in the proof of Lemma 5.3, there exists $\Xi \in H_- \otimes H_-$ such that

$$(A\xi, \eta)_{H_-} = (\Xi, \eta \otimes \xi^*)_{H_- \otimes H_-} \quad (\xi, \eta \in H_-).$$

It remains to set $\Psi \in I_2 \Xi \in H_+ \otimes H_+$, where I_2 is an operator I connected with the chain

$$H_- \otimes H_- \supset G_0 \otimes G_0 \supset H_+ \otimes H_+.$$

■

Let us now repeat the definitions from Subsection 5.3 in the smooth situation and formulate an analogue of Theorem 5.1. Thus, consider a chain (5.1) with involution "*". Let $K \in G_+ \otimes G_+$ be a positive definite kernel. Assume that a family of operators $A = (A_x)_{x \in X}$ is given and that the operators $A_x \colon G_- \to G_-$ have dense domains of definitions $\mathfrak{D}(A_x)$ and are Hermitian in the sense (5.25), i.e., relations (5.16) hold, where $\mathrm{K} = K \in G_+ \otimes G_+$, $\varphi = \xi$ and $\psi = \eta \in \mathfrak{D}(A_x)$ $(x \in X)$. Suppose that the family A admits a rigging in the above-mentioned sense, i.e., there exists a chain (5.26) such that, for any $x \in X$, we have $D \subset \mathfrak{D}(A_x)$ and $A_x \restriction D \in \mathfrak{L}(D, H_-)$. Suppose also that the space H_- (and, hence, H_+) is invariant with respect to the involution "*".

A positive definite kernel

$$\Omega \in H_+ \otimes H_+ \subset H_- \otimes H_-$$

is called an elementary kernel of the family A corresponding to an eigenvalue $\lambda(\cdot)$ if the equality (5.17) holds with $\varphi = \xi$ and $\psi = \eta \in D \subset H_-$.

Theorem 5.2. *Assume that the imbedding $H_- \subset G_-$ in the chain* (5.26) *is quasinuclear and $A = (A_x)_{x \in X}$ is an introduced family of operators. Then the results of Theorem* 5.1 *remain valid with the kernel* K *being replaced by* K *and the spaces H_{K} and $H_- \otimes H_-$ being replaced by H_K and $H_+ \otimes H_+$, respectively.*

Proof. The proof of this theorem is analogous to that of Theorem 5.1. We employ the constructions presented in this subsection, spectral Theorem 2.7 in Chapter 3, and Lemma 5.4 instead of Lemma 5.3.

■

Here we could also make some remarks similar to Remarks 1–6 in Subsection 5.3. We only note that the gap between the smoothness of the kernel K and the elementary kernels $\Omega(\lambda(\cdot))$ is preserved. In our case, the representation (5.18) has the form

$$G_+ \otimes G_+ \ni \mathrm{K} = \int_{\mathbb{R}^X} \Omega(\lambda(\cdot))\, d\rho(\lambda(\cdot)), \ \Omega(\lambda(\cdot)) \in H_+ \otimes H_+ \supset G_+ \otimes G_+. \quad (5.28)$$

Let us apply the scheme given in this subsection to an example, namely, to the smooth symmetrical moment problem. Consider a Hilbert space $H = W_2^l(\mathbb{R}^d, p(x)dx)$, where $d, l \in \mathbb{N}$, $l > d/2$; the weight $p \in C^l(\mathbb{R}^d)$ is positive (see Example 1.5 in Chapter 1); the involution in this space is given by an ordinary complex overbar. According to definition (2.1), a sequence of symmetric functions $s = (s_n(x_1, \dots, x_n))_{n=0}^{\infty}$ $(s_n \in H^{\otimes n},\ n \in \mathbb{N};\ s_0 \in \mathbb{C}^1)$ is called a moment sequence if, for an arbitrary finite sequence $\varphi = (\varphi_n)_{n=0}^{\infty}$, $(\varphi_n \in C_0(\mathbb{R}^{dn}), n \in \mathbb{N};\ \varphi_0 \in \mathbb{C}^1)$, the inequality

$$\sum_{j,k=0}^{\infty} \int_{\mathbb{R}^{d(j+k)}} s_{j+k}(x_1, \dots, x_{j+k}) \overline{\varphi_j(x_1, \dots, x_j)}\, \varphi_k(x_1, \dots, x_k)\, dx_1 \dots dx_{j+k} \geq 0 \quad (5.29)$$

holds (for $j + k = 0$, the integral denotes a product of numbers). It follows from (5.29) that $s_n \in (H_{\mathrm{Re}})^{\otimes n}$ (see Subsection 2.1). Assume that the condition of definiteness holds, namely, the class $C\{\| s_{2n} \|_{H^{\otimes 2n}}^{1/2}\}$ is quasianalytic.

Theorem 5.3. *Each moment sequence of functions satisfying the condition of definiteness admits a representation in the form of the scalar continual integral, namely,*

$$s_n(x_1, \dots, x_n) = \int_{C_{\mathrm{Re}}(\mathbb{R}^d)} \lambda(x_1) \dots, \lambda(x_n) d\sigma(\lambda(\cdot)) \quad (x_1, \dots, x_n \in \mathbb{R}^d;$$

$$s_0 = \sigma(C_{\mathrm{Re}}(\mathbb{R}^d))). \quad (5.30)$$

Here σ is a finite nonnegative measure given on a certain σ-algebra of the space $C_{\mathrm{Re}}(\mathbb{R}^d)$.

Proof. The proof is similar to that of Theorem 2.1, but here we use Theorem 5.2. We only outline the proof.

We set $F = L_2(\mathbb{R}^d, q(x)\, dx)$, where $q \in C^l(\mathbb{R}^d)$ is a positive weight. We take the space of sequences $f = (f_n)_{n=0}^{\infty}$ $(f_n \in F^{\otimes n})$ with the scalar product

$$(f, g)_{G_0} = \sum_{n=0}^{\infty} a_n (f_n, g_n)_{F^{\otimes n}}$$

$(a = (a_n)_{n=0}^{\infty}$, $a_n > 0$ is a weight) to play the role of G_0. The space G_+ is defined analogously, one should only substitute H and $b = (b_n)_{n=0}^{\infty}$ $(b_n > 0)$ for F and a, respectively. The weight q is chosen so that the quasinuclear imbedding $H \subset F$ takes place (this is possible by virtue of what has been said in Example 1.5 of Chapter 1). But then, for any given b, we can choose a weight a so that $G_+ \subset G_0$ quasinuclearly (it is necessary that $a_n / b_n \to 0$ as $n \to \infty$ sufficiently fast).

Let us form a matrix $K = (K_{jk})_{j,k=0}^{\infty}$, where $K_{jk} = s_{j+k}$. It can be interpreted as an element from $G_+ \otimes G_+$ provided that $b_n \to 0$ sufficiently fast (see Step IV in the proof of Theorem 2.1). We fix such a b. For the given b, we choose a so that $G_+ \subset G_0$ quasinuclearly. We assume that $H_+ = G_0$.

The space G_- now consists of the sequences $\xi = (\xi_n)_{n=0}^{\infty}$ (with appropriate behavior in n), where $\xi_n \in H_-^{\otimes n}$, and H_- is a negative space of the chain $H_- \supset F \supset H$. The space H_- contains δ-functions δ_x $(x \in \mathbb{R}^d)$, because, locally, it coincides with $W_2^{-l}(G)$ (G is a bounded region in $\mathbb{R}^d$, see Example 1.2 in Chapter 1). Hence, on finite sequences from G_-, the operation of shift

$$\mathcal{A}_{\delta_x}\xi = \mathcal{A}_{\delta_x}(\xi_0, \xi_1, \xi_2, \dots) = (0, \delta_x \otimes \xi_0, \delta_x \otimes \xi_1, \delta_x \otimes \xi_2, \dots) \quad (x \in \mathbb{R}^d) \tag{5.31}$$

is defined.

The operation (5.31) generates (in the natural way) an operator A_x acting in G_- with the domain of definition consisting of finite $\xi \in G_-$. This operator is Hermitian with respect to $\langle\cdot,\cdot\rangle$, i.e., (5.25) holds for it, because K_{jk} depends on the sum of $j + k$ indices. Therefore, for the family $A = (A_x)_{x \in \mathbb{R}^d}$, one can repeat all what has been said above concerning the existence of the quasinuclear chain (5.27) connected with $(\hat{A}_x)_{x \in \mathbb{R}^d}$ in the standard way.

Denote by B_x the closure of $\hat{A}_x$ in H_K; the family $B = (B_x)_{x \in \mathbb{R}^d}$ consists of self-adjoint commuting operators. This is proved just as in the case of Theorem 2.1 (Steps I and II of the proof; here the symmetry of s_n is essential). Then we apply Theorem 5.2 and obtain the representation (5.28) for the kernel K with the integration over $\mathbb{R}^{\mathbb{R}^d}$. The required continuity of the operators A_x with respect to $x \in \mathbb{R}^d$ follows from the form of operation (5.31). According to the remark of type 4 in Subsection 5.3, the integration in (5.28) can be carried out over $C_{\mathrm{Re}}(\mathbb{R}^d) \subset \mathbb{R}^{\mathbb{R}^d}$. Equations of type (5.17) for

$$\Omega(\lambda(\cdot)) \in H_+ \otimes H_+ = G_0 \otimes G_0 \qquad (\lambda(\cdot) \in C_{\mathrm{Re}}(\mathbb{R}^d))$$

now take the form (2.20), (2.21). If we write the solution of these equations and substitute it to (5.28), we then get (5.30).

■

Note that the σ-algebra on which the measure σ in (5.30) is given consists of the intersections of sets from $\mathfrak{C}_\sigma(\mathbb{R}^{\mathbb{R}^d})$ with $C_{\mathrm{Re}}(\mathbb{R}^d)$.

Similar "smooth" versions can be obtained also for the other results in Sections 2 and 3.

6. INFINITE-DIMENSIONAL ELLIPTIC DIFFERENTIAL OPERATORS OF THE SECOND ORDER

Infinite-dimensional differential operators appear in different branches of mathematics and their applications. Thus, in mathematical physics, they are used (though often on a formal level) as operators of energy of systems with infinitely many degrees of freedom; in the theory of random processes where diffusion processes with infinite-dimensional phase spaces are constructed with the help of operators of this sort; and, finally, the investigation of infinite-dimensional differential operators is of independent interest for infinite-dimensional analysis. These directions are not isolated, on the contrary, they interact intensively. However, each of these branches has its own specific circle of problems and, thus, if we choose this or another field of applications, then we automatically draw attention to some concrete aspects of the theory of infinite-dimensional differential operators which must be studied. For example, applications to theoretical physics are mainly based on operator problems whose contents have many similar features for different model situations and can be characterized by close connection with spectral theory. An essential place among these problems is occupied by the problems of selfadjointness and the theory of singular perturbations. These problems play a key role in both Hamiltonian and Euclidean approaches to the construction and investigation of the dynamics of many important model systems in contemporary mathematical physics. The necessity of the investigation of these problems largely determines the scope of problems in the theory of infinite-dimensional differential operators that are presented in this chapter.

In Section 1, we introduce an important class of these operators, namely, the second quantization operators in a Schrödinger representation. The second quantization operators are first defined in the Fock space $\mathfrak{F}(H_0)$ (see Chapter 2) for the given selfadjoint operator $A > 0$ acting in H_0, i.e., for every $n \in \mathbb{N}$, we construct an operator $A^{(n)}$ in $\mathfrak{F}_n(H_0) \subset H_{0,c}^{\otimes n}$ just as the Laplace operator Δ is constructed in $L_2(\mathbb{R}^n, dx_1 \ldots dx_n)$ from the operator $-d^2/dx^2$ in $L_2(\mathbb{R}^1, dx)$. We complete the collection of operators obtained by setting $A^{(0)} = 0$ in $\mathfrak{F}_0(H_0) = \mathbb{C}^1$, and introduce an operator $d\Gamma(A)$ in

$$\mathfrak{F}_0(H_0) = \bigoplus_{n=0}^{\infty} \mathfrak{F}_n(H_0),$$

given on a natural domain of definition by the restrictions $d\Gamma(A)\upharpoonright\mathfrak{F}_n(H_0) = A^{(n)}$. The operator $d\Gamma(A)$ is called a second quantization of A; the same name being applied to its image L_A under a functional realization of the Fock space (given by the Segal isomorphism described in Subsection 2.3 of Chapter 2) as $L_2(\Phi', \gamma_1)$ (γ_1 is a canonical Gaussian measure corresponding to H_0, the nuclear rigging $\Phi' \supset H_0 \supset \Phi$ is chosen so that $A \in \mathfrak{L}(\Phi, \Phi)$). We show that L_A is an elliptic differential operator of the second order associated with the Dirichlet form (or energetic form) of the measure[6] γ_1:

$$(L_A u, v)_{L_2(\Phi',\gamma_1)} = \frac{1}{2}\int_{\Phi'} (\nabla u, A\overline{\nabla v})_{H_0}\, d\gamma_1, \tag{0.1}$$

$$(L_A u)(x) = -\frac{1}{2}\operatorname{Tr}_{H_0}(Au''(x)) + (A\nabla u(x), x)_{H_0}$$

$$(u, v \in \mathcal{P}(\Phi');\ x \in \Phi').$$

Relation (0.1) shows that the operator L_A is constructed for a given Gaussian measure γ_1, just as the Laplace operator is constructed in the case of the Lebesgue measure. Its closure, which is also denoted by L_A, is a selfadjoint nonnegative operator in $L_2(\Phi', \gamma_1)$. In this section, we also establish the properties of a semigroup whose generator is L_A and give the direct construction of a functional integral connected with this semigroup. In other words, we construct a measure $\nu_{A,0}$ on a space of trajectories

$$\Omega_0 = \{\omega(\cdot) : [0, +\infty) \to \Phi' \mid \omega(0) = 0\},$$

corresponding to a diffusion process (with phase space Φ' and generating operator L_A) which starts from the point $0 \in \Phi'$. This construction is analogous to the construction of the Wiener measure from the Laplace operator.

In Section 2, we study potential perturbations of the second quantization operators. The first question which arises is whether the sum $L_A + V$ is essentially selfadjoint ($V = \overline{V}$ is a measurable function on Φ' which determines the potential). We give the conditions which should be imposed on the potential to guarantee the essential selfadjointness of the sum and show that if these conditions are satisfied, then the following analogue of the Feynman–Kac formula for a semigroup $e^{-t(L_A+V)}$ $(t \geq 0)$ holds

$$(e^{-t(L_A+V)}f)(x) = \int_{\Omega_0} f(\omega(t) + e^{-tA}x)\, e^{-\int_0^t V(\omega(\tau)+e^{-\tau A}x)d\tau}\, d\nu_{A,0}(\omega(\cdot)) \tag{0.2}$$

for any $f \in L_2(\Phi', \gamma_1)$ and for γ_1-almost all $x \in \Phi'$ (the classical Feynman–Kac for-

[6] Henceforth, in order to distinguish the Greek letter ν from the Latin v, we use the alternative version of the latter, υ, to denote the same objects as before.

mula corresponds to the case of a potential perturbation of the Laplace operator and has a form similar to (0.2) but with $\nu_{A,0}$ replaced by the Wiener measure). The representation (0.2) enables us to investigate the properties of the operator $L_A + V$ in detail.

Section 3 is devoted to the study of differential operators $L_{\mu, A}$ generated by Dirichlet forms of the form (0.1) but with an arbitrary probability measure μ on Φ' instead of the Gaussian measure γ_1, i.e.,

$$(L_{\mu,A}u, v)_{L_2(\Phi',\mu)} = \frac{1}{2}\int_{\Phi'} (\nabla u, A\overline{\nabla v})_{H_0}\, d\mu, \tag{0.3}$$

here $u, v \in C^{\infty}_{b,\text{cyl}}(\Phi')$ are smooth cylindrical functions on Φ' bounded with their derivatives. In order that the Dirichlet form may be associated (by integrating by parts in (0.3)) with the differential operator $L_{\mu, A}$ (the so-called Dirichlet operator of the measure μ), it suffices to demand that the measure μ should have the logarithmic derivative

$$\beta_{\mu}(\varphi, x) = d/dt\left(\frac{d\mu(x + t\varphi)}{d\mu(x)}\right)\bigg|_{t=0} = (\beta_{\mu}(x), \varphi)_{H_0} \in L_2(\Phi', \mu)$$

in all directions $\varphi \in \Phi$. In this case,

$$(L_{\mu, A}u)(x) = -\frac{1}{2}\,\mathrm{Tr}_{H_0}(Au''(x)) - \frac{1}{2}\,(A\nabla u(x), \beta_{\mu}(x))_{H_0} \tag{0.4}$$

$$(u \in C^{\infty}_{b,\text{cyl}}(\Phi');\ x \in \Phi').$$

Further, in Section 3, we establish the important connection between the potential perturbations of the second quantization operators $L_A + V$ and the Dirichlet operators $L_{\mu^V, A}$, where the measure $\mu^V = \phi_V^2\gamma_1$, and the density $\phi_V > 0$ is canonically determined by the potential. The transition from $L_A + V$ to $L_{\mu^V, A}$ is called a renormalization of the operator $L_A + V$; this terminology is justified by the direct analogy with the corresponding procedure in the constructive field theory. In Section 3, we also obtain conditions of essential selfadjointness of the Dirichlet operators $L_{\mu, A}$ in the case of measures μ concentrated on a certain Hilbert subspace in Φ' and $A = 1$. These conditions can be reduced to the requirements of a certain smoothness and summability of the logarithmic derivative β_{μ} of the measure μ.

In Section 4, we investigate infinite-dimensional equations connected with the Dirichlet operators. We study the smoothness properties of the generalized solutions of the corresponding elliptic equations. In the case of hyperbolic equations, we prove the energetic inequalities which give the estimate of the solution of the Cauchy problem in terms of the initial data. In particular, these estimates yield the fact that the rate of propagation of perturbations described by this hyperbolic equation is finite. This enables us to apply the hyperbolic criteria of selfadjointness presented in Chapter 5 to the investigation of the conditions of selfadjointness of the potential perturbations of the Dirichlet operators.

§1. Second Quantization Operators in the Schrödinger Representation

In this section, we study the properties of a class of elliptic differential operators of the second order acting in the space of functions of infinitely many variables. These operators appear under the functional realization of the Fock space and include many special cases important for applications among which one should mention the Hamiltonian of a free boson field.

1.1. The Second Quantization Operators

As in Subsection 2.2 of Chapter 2, we associate a real separable Hilbert space H_0 with the Fock space

$$\mathfrak{F}(H_0) = \bigoplus_{n=0}^{\infty} \mathfrak{F}_n(H_0).$$

Each operator A acting in H_0 with the domain of definition $\mathrm{D}(A) \subset H_0$ can be extended in a natural way to an operator in the complexification $H_{0,c} = \mathfrak{F}_1(H_0)$; henceforth, we shall not specify this extension separately.

For a given selfadjoint operator $A \geq 0$ with the domain of essential selfadjointness $D \subset H_0$, we define an operator

$$A^{(n)} = A \otimes 1 \otimes \ldots \otimes 1 + 1 \otimes A \otimes 1 \otimes \ldots \otimes 1 + \ldots + 1 \otimes \ldots \otimes 1 \otimes A \tag{1.1}$$

on each n-particle subspace $\mathfrak{F}_n(H_0)$, $n \in \mathbb{N}$, with the domain of definition $D^{(n)} = \mathring{\mathfrak{F}}_n(D)$ (recall that $\mathring{\mathfrak{F}}_n(D)$ is the nth algebraic symmetric tensor power of the complexification $D_c = D + iD$ (see Subsection 2.2 in Chapter 2). We set $A^{(0)} = 0$ in $\mathfrak{F}_0(H_0)$. An operator $d\Gamma(A)$ in $\mathfrak{F}(H_0)$ with the domain of definition $\mathring{\mathfrak{F}}_{\mathrm{fin}}(D)$ such that

$$d\Gamma(A)\restriction \mathfrak{F}_n(H_0) = A^{(n)}, \quad n \in \mathbb{Z}_+$$

is called a second quantization of the operator A. Since $\mathfrak{F}^{\circ}_{\text{fin}}(D)$ contains by definition only the finite sequences $f = (f_0, f_1, \ldots, f_k, 0, 0, \ldots)$, $f_j \in \mathfrak{F}_j^{\circ}(D)$, $j = 0, \ldots, k$, the operator $d\Gamma(A)$ is correctly defined on the dense linear manifold $\mathfrak{F}^{\circ}_{\text{fin}}(D) \subset \mathfrak{F}(H_0)$ and is symmetric.

It is easy to show that *the operator* $d\Gamma(A)$ *is essentially selfadjoint.*

Indeed, it follows from the general theory of operators with separated variables (see, for example, Berezansky [5, Chapter 6, Section 4, Subsection 1]) that the operator (1.1) in $H_{0,c}^{\otimes n}$ is essentially selfadjoint on the domain $\text{a.}\overset{n}{\underset{1}{\otimes}} D_c$. Since this operator commutes with the projector P_n onto $\mathfrak{F}_n(H_0)$ in $H_{0,c}^{\otimes n}$, the operator $A^{(n)}$ in $\mathfrak{F}_n(H_0)$ is essentially selfadjoint on

$$P_n\left(\text{a.}\overset{n}{\underset{1}{\otimes}} D_c\right)\mathfrak{F}_n^{\circ}(D).$$

Therefore, the range of values $\mathsf{R}(A^{(n)} + i\lambda 1)$ is dense in $\mathfrak{F}_n(H_0)$ for all $\lambda \in \mathbb{R}^1$, $\lambda \neq 0$. This implies that $\mathsf{R}(d\Gamma(A) \pm i1)$ is dense in $\mathfrak{F}(H_0)$, i.e., that the closure of the operator $d\Gamma(A)$ is selfadjoint.

■

We preserve the same notation for the closure of $d\Gamma(A)$ in $\mathfrak{F}(H_0)$.

We can always assume that the operator $A \geq 0$ is nondegenerate, i.e., that it has the kernel zero. Indeed, if this is not true, then we pass to the decomposition $H_0 = H_0' \oplus H_0''$, where $A \restriction H_0'' = 0$, and $A' = A \restriction H_0'$ has the kernel zero. Then $A = A' \oplus 0$, and employing the isomorphism between $\mathfrak{F}(H_0' \oplus H_0'')$ and $\mathfrak{F}(H_0') \oplus \mathfrak{F}(H_0'')$ under which

$$d\Gamma(A' \oplus 0) = d\Gamma(A') \otimes 1,$$

we pass to the study of the operator $d\Gamma(A')$ in $\mathfrak{F}(H_0')$. Hence, below we always assume that $A > 0$, i.e., that $\operatorname{Ker} A = \{0\}$.

Assume that B is a contracting operator in H_0, i.e., $\|B\| \leq 1$. We introduce an operator $\Gamma(B)$ in $\mathfrak{F}(H_0)$ by setting

$$\Gamma(B)\restriction \mathfrak{F}_n(H_0) = B \otimes \ldots \otimes B \ (n \text{ times}), \quad n \in \mathbb{N},$$

and $\Gamma(B)\restriction \mathfrak{F}_0(H_0) = 1$. It is clear that

$$\| \Gamma(B) \upharpoonright \mathfrak{F}_n(H_0) \| \le \| B \|^n, \quad n \in \mathbb{Z}_+;$$

therefore, $\Gamma(B)$ is a contracting operator in $\mathfrak{F}(H_0)$, i.e.,

$$\| \Gamma(B) \| = \sup_{n \in \mathbb{Z}_+} \| \Gamma(B) \upharpoonright \mathfrak{F}_n(H_0) \| = 1. \tag{1.2}$$

For contracting operators $B_1, B_2 \in \mathfrak{L}(H_0, H_0)$, we have

$$\Gamma(B_1, B_2) = \Gamma(B_1)\Gamma(B_2).$$

In particular, for the given selfadjoint operator A in H_0, the family of operators $\Gamma(e^{-tA})$, $t \ge 0$, forms a semigroup and, moreover, $\Gamma(e^{-tA}) = e^{-td\Gamma(A)}$.

Example 1.1. The operator of a number of particles N in quantum field theory is defined as an operator in $\mathfrak{F}(H_0)$ such that

$$N \upharpoonright \mathfrak{F}_n(H_0) = n1, \quad n \in \mathbb{Z}_+,$$

i.e., $N = d\Gamma(1)$. The corresponding semigroup $e^{-tN} = \Gamma(e^{-t}1)$ in $\mathfrak{F}_n(H_0)$ acts as a multiplication by e^{-tn}, $n \in \mathbb{Z}_+$, $t \ge 0$.

We fix an operator $A > 0$ selfadjoint in H_0. Let us study the image of the operator $d\Gamma(A)$ under the functional realization of the Fock space. As described in Subsection 2.3 of Chapter 1, to construct this realization one must introduce a rigging of the space H_0. As we shall see below, the choice of rigging is now not so arbitrary as above, because it should possess a series of special properties connected with the operator A. The presence of these properties guarantees (in the functional realization) the possibility of a detailed investigation of the second quantization operators.

Let $\Phi \subset H_0$ be a nuclear space densely topologically imbedded in H_0 and such that $A \in \mathfrak{L}(\Phi, \Phi)$ and $e^{-tA} \in \mathfrak{L}(\Phi, \Phi)$ for any $t \ge 0$. Moreover, Φ is a domain of essential selfadjointness of A. The existence of this Φ is established in Example 1.1 of Chapter 4. We have $\Phi = \text{pr} \lim_{\tau \in \mathrm{T}} H_\tau$, and the Hilbert spaces H_τ, $\tau \in \mathrm{T}$, are densely topologically imbedded in H_0. Denote by Φ' a space dual to Φ with respect to H_0, i.e., $\Phi' = \bigcup_{\tau \in \mathrm{T}} H_{-\tau}$ (see Lemma 1.2 in Chapter 1).

Therefore, we have the following rigging

$$\Phi' \supset H_0 \supset \Phi, \tag{1.3}$$

associated with the rigging of the Fock space

$$\mathfrak{F}'_{\text{fin}}(\Phi) \supset \mathfrak{F}(H_0) \supset \mathfrak{F}_{\text{fin}}(\Phi) \tag{1.4}$$

according to the construction in Subsection 5.1 in Chapter 2.

Due to the choice of Φ, for any $n \in \mathbb{Z}_+$, we have $A^{(n)} \in \mathcal{L}(\mathfrak{F}_n(\Phi), \mathfrak{F}_n(\Phi))$ and, therefore, $d\Gamma(A) \in \mathcal{L}(\mathfrak{F}_{\text{fin}}(\Phi), \mathfrak{F}_{\text{fin}}(\Phi))$, and $\mathfrak{F}_{\text{fin}}(\Phi)$ is a domain of essential selfadjointness of $d\Gamma(A)$. Similarly,

$$e^{-td\Gamma(A)} = \Gamma(e^{-tA}) \in \mathcal{L}(\mathfrak{F}_{\text{fin}}(\Phi), \mathfrak{F}_{\text{fin}}(\Phi)).$$

Hence, the nuclear space $\mathfrak{F}_{\text{fin}}(\Phi)$ possesses the same properties with respect to the operator $d\Gamma(A)$ as the space Φ with respect to A. According to Subsection 2.3 in Chapter 2, the Segal isomorphism realizes $\mathfrak{F}(H_0)$ as the space $L_2(\Phi', \gamma_1)$. According to the corollary of Theorem 5.2 in Chapter 2, under this isomorphism the rigging (1.4) turns into

$$\mathcal{P}'(\Phi') \supset L_2(\Phi', \gamma_1) \supset \mathcal{P}(\Phi'). \tag{1.5}$$

Denote by L_A the image of $d\Gamma(A)$ under the Segal isomorphism. As above, the operator L_A is called the second quantization of the operator A. As is already known, $L_A \in \mathcal{L}(\mathcal{P}(\Phi'), \mathcal{P}(\Phi'))$, $e^{-tL_A} \in \mathcal{L}(\mathcal{P}(\Phi'), \mathcal{P}(\Phi'))$, $t \geq 0$, and L_A is essentially selfadjoint on $\mathcal{P}(\Phi')$.

The next example gives the representation of the semigroup e^{-tL_A}, $t \geq 0$, in the simplest situation.

Example 1.2. In Example 2.2 of Chapter 2, we described the Segal isomorphism between $\mathfrak{F}(\mathbb{R}^1) = l_2(\mathbb{C}^1)$ and $L_2(\mathbb{R}^1, \gamma_1)$ for $H_0 = \mathbb{R}^1$. Recall that in order to define this isomorphism, we choose the orthonormal basis of Hermite polynomials $h_n(\cdot)$ $(n \in \mathbb{Z}_+)$ in $L_2(\mathbb{R}^1, \gamma_1)$ and each vector $f = (f_n)_{n=0}^{\infty} \in l_2(\mathbb{C}^1)$ is associated with the function

$$(I_1 f)(\cdot) = \sum_{n=0}^{\infty} f_n h_n(\cdot) \in L_2(\mathbb{R}^1, \gamma_1),$$

and vice versa, each function from $L_2(\mathbb{R}^1, \gamma_1)$ is associated with a vector from $l_2(\mathbb{C}^1)$ composed of its Fourier–Hermite coefficients. Under this isomorphism, the subspace $\mathfrak{F}_n(\mathbb{R}^1) = \mathbb{C}^1$ turns into a one-dimensional subspace of $L_2(\mathbb{R}^1, \gamma_1)$ spanned by $h_n(\cdot)$ $(n \in \mathbb{Z}_+)$.

An operator $A > 0$ in $\mathbb{R}^1$ is defined by the multiplication by a number $a > 0$ and, by definition, we have

$$\Gamma(e^{-tA})f = (e^{-tan}f_n)_{n=0}^{\infty} \in l_2(\mathbb{C}^1).$$

Therefore,

$$(e^{-tL_A}I_1 f)(\cdot) = \sum_{n=0}^{\infty} e^{-tan}f_n h_n(\cdot) = I_1(\Gamma(e^{-tA})f)(\cdot).$$

Inserting the equality

$$f_n = \int (I_1 f)(\cdot) h_n(\cdot) d\gamma_1(\cdot)$$

in the last relation and assuming that $I_1 f \in L_\infty(\mathbb{R}^1, \gamma_1)$, we obtain

$$(e^{-tL_A}f)(x) = \int_{\mathbb{R}^1} \left[\sum_{n=0}^{\infty} (e^{-ta})^n h_n(x) h_n(y)\right] (I_1 f)(y)\, d\gamma_1(y)$$

$$= \int_{\mathbb{R}^1} \frac{(I_1 f)}{\sqrt{1-e^{-2tA}}} \exp\left\{-\frac{x^2e^{-2ta} + y^2e^{-2ta} - 2xye^{-ta}}{1-e^{-2ta}}\right\} \frac{1}{\sqrt{\pi}} e^{-y^2} dy$$

(we have employed the bilinear generating formula for the Hermite polynomials to sum up the series under the integral sign). Hence,

$$(e^{-tL_A}g)(x) = \int_{\mathbb{R}^1} g(y)\, d\mu_{t,x}^A(y) \quad (x \in \mathbb{R}^1,\ t>0)$$

for $g \in L_\infty(\mathbb{R}^1, \gamma_1)$, where $\mu_{t,x}^A (t>0,\ x \in \mathbb{R}^1)$ is a family of Gaussian measures on $\mathbb{R}^1$, i.e.,

$$d\mu_{t,x}^A(y) = (\pi(1-e^{-2ta}))^{-1/2} \exp\left(-\frac{(y-e^{-ta}x)^2}{1-e^{-2ta}}\right) dy.$$

The representation obtained is known as Mehler's formula. Below, we prove an infinite-dimensional version of it which enables us to study the properties of the semigroup e^{-tL_A}, $t>0$, and the objects associated with it.

1.2. Mehler's Formula in the Infinite Case

For any $t>0$, the function

$$\exp[-1/4(\varphi, (1-e^{-2tA})\varphi)_{H_0}] \quad (\varphi \in \Phi)$$

is continuous on Φ, positive definite, and normalized. According to Theorem 1.11 of Chapter 2, it is the Fourier transform of the Gaussian measure γ_{A_t} on $\mathfrak{C}_\sigma(\Phi')$ with the correlation operator

$$A_t = 1 - e^{-2tA} > 0$$

and mean value 0. By virtue of Theorem 1.10 in Chapter 2, each Hilbert space $H_{-\tau}$, $\tau \in \mathrm{T}$, with a quasinuclear imbedding of H_0 in $H_{-\tau}$ has the full γ_{A_t}-measure, because the corresponding Fourier transform is continuous on H_0. We choose $\tau_1 \in \mathrm{T}$ such that $H_0 \subset H_{-\tau_1}$ is quasinuclear and fix it. For any $t > 0$, by assumption we have $e^{-tA} \in \mathcal{L}(\Phi, \Phi)$; therefore, one can find $H_{\tau_2} \subset H_{\tau_1}$, $\tau_2 \in \mathrm{T}$, for which $e^{-tA} \in \mathcal{L}(H_{\tau_2}, H_{\tau_1})$. Given $t > 0$, we choose this $\tau_2 \in \mathrm{T}$ and fix it, too. By using the selfadjointness of e^{-tA} in H_0, we can extend this operator to the operator $e^{-tA} \in \mathcal{L}(H_{-\tau_1}, H_{-\tau_2})$.

For every $x \in H_{-\tau_1}$, we introduce a measure $\gamma_{A_t,x}$ on $\mathfrak{C}_\sigma(\Phi')$ obtained as a result of the shift of the measure γ_{A_t} by the vector $e^{-tA}x \in H_{-\tau_2}$, namely,

$$\gamma_{A_t,x}(\alpha) = \gamma_{A_t}(\alpha - e^{-tA}x) \quad (\alpha \in \mathfrak{C}_\sigma(\Phi')). \tag{1.6}$$

In terms of the Fourier transform, the measures $\gamma_{A_t,x}$ are given by the expressions

$$\int_{\Phi'} e^{i(\varphi,y)_{H_0}} d\gamma_{A_t,x}(y) = \exp\left\{ i(\varphi, e^{-tA}x)_{H_0} - \frac{1}{4}(\varphi, A_t\varphi)_{H_0} \right\}. \tag{1.7}$$

It follows immediately from (1.6) that, according to Theorem 1.10 in Chapter 2, the measures $\gamma_{A_t,x}$ are concentrated on the Hilbert space $H_{-\tau_2}$ for all $x \in H_{-\tau_1}$. In the next lemma, we establish an even stronger result.

Lemma 1.1. *For sufficiently small* $\delta > 0$, *the functions* $\exp(\delta \|y\|^2_{H_{-\tau_2}})$ $(y \in H_{-\tau_2})$ *are summable with respect to each measure* $\gamma_{A_t,x}$, $x \in H_{-\tau}$.

Proof. Let $x = 0$. For a centered measure, the moments of the even order are given by the formula

$$\int_{\Phi'} \prod_{k=1}^{2n} (\varphi_k, y)_{H_0} d\gamma_{A_t}(y) = 2^{-n} \sum \prod_{k=1}^{n} (\varphi_{l_k}, A\varphi_{m_k})_{H_0} \tag{1.8}$$

$$(\varphi_k \in \Phi;\ k = 1, \ldots, 2n;\ n \in \mathbb{N})$$

(see (2.5) in Chapter 2), where summation is carried out over all $(2n)!/2^n n!$ possible decompositions of the numbers $1, 2, \dots, 2n$ into the pairs $(l_1, m_1), \dots, (l_n, m_n)$. We choose an orthonormal basis $(e_k)_{k=1}^{\infty} \subset \Phi$ in H_{τ_2}. Then

$$\|y\|^2_{H_{-\tau_2}} = \sum_{k=1}^{\infty} (y, e_k)^2_{H_0},$$

and by using (1.8), we get

$$\int_{\Phi'} \|y\|^2_{H_{-\tau_2}} d\gamma_{A_t}(y) = \int_{\Phi'} \sum_{n_1+\dots+n_\nu=n} \prod_{p=1}^{\nu} (y, e_p)^{2n_p}_{H_0} d\gamma_{A_t}(y)$$

$$\le \sum_{n_1+\dots+n_\nu=n} \frac{(2n)!}{4n \cdot n!} \prod_{p=1}^{\nu} \|e_p\|^{2n_p}_{H_0} \|A_t\|^n_{\mathfrak{L}(H_0, H_0)}$$

$$\le n!\left(\sum_{p=1}^{\infty} \|e_p\|^2_{H_0}\right)^n = n!\,|O_{\tau_{2,0}}|^{2n}, \tag{1.9}$$

because the imbedding $O_{\tau_{2,0}}: H_{\tau_2} \to H_0$ is quasinuclear. Independently of what has been said above, it follows from (1.9) that $\gamma_{A_t}(H_{-\tau_2}) = 1$ provided that $H_{\tau_2} \subset H_0$ is quasinuclear. Further,

$$\int_{\Phi'} e^{\delta\|y\|^2_{H_{-\tau_2}}} d\gamma_{A_t}(y) = \sum_{n=0}^{\infty} \frac{\delta^n}{n!} \int_{\Phi'} \|y\|^2_{H_{-\tau_2}} d\gamma_{A_t}(y) \le \sum_{n=0}^{\infty} \delta^n |O_{\tau_{2,0}}|^{2n} < \infty,$$

if $\delta\,|O_{\tau_{2,0}}|^2 < 1$.

For an arbitrary $x \in H_{-\tau_1}$, it suffices to employ the estimate

$$\int_{\Phi'} \exp(\delta\|y\|^2_{H_{-\tau_2}})\, d\gamma_{A_t}(y) = \int_{\Phi'} \exp(\delta\|y + e^{-tA}x\|^2_{H_{-\tau_2}})\, d\gamma_{A_t}(y)$$

$$\le \exp(2\delta\|e^{-tA}x\|^2_{H_{-\tau_2}}) \int_{\Phi'} \exp(2\delta\|y\|^2_{H_{-\tau_2}})\, d\gamma_{A_t}(y).$$

■

We are now ready to formulate the theorem which gives a representation of the semigroup e^{-tL_A}, $t > 0$

Theorem 1.1. *For any* $t > 0$ *and* $f \in L_2(\Phi', \gamma_1)$, *the following equality holds*

$$(e^{-tL_A}f)(x) = \int_{\Phi'} f(y)d\gamma_{A_t,x}(y) \tag{1.10}$$

for γ_1-almost all $x \in \Phi'$.

Proof. Let us establish (1.10) for functions of the form (see Subsection 2.3 in Chapter 2)

$$f_{s,\varphi}(y) = :e^{s(\varphi,y)_{H_0}}: = e^{s(\varphi,y)_{H_0} - s^2/4\|\varphi\|^2_{H_0}} \quad (y \in \Phi')$$

(here $s \in \mathbb{R}^1$ and $\varphi \in \Phi$ are fixed). On the one hand,

$$e^{-tL_A} = I_1\Gamma(e^{-tL_A})I_1^{-1}, \quad I_1^{-1}f_{s,\varphi} = (s^k/2^{k/2}\sqrt{k!}\,\varphi^{\otimes k})_{k=0}^{\infty} \in \mathfrak{F}(H_0)$$

and, consequently,

$$(e^{-tL_A}f_{s,\varphi})(x) = f_{s,e^{-tA}\varphi}(x). \tag{1.11}$$

On the other hand, it follows from (1.10) that

$$\int_{\Phi'} f_{s,\varphi}(y)d\gamma_{A_t,x}(y) = e^{-s^2/4\|\varphi\|^2_{H_0}}e^{s(\varphi,e^{-tA}x)_{H_0}}$$

$$\times \int_{\Phi'} e^{s(\varphi,y)_{H_0}}d\gamma_{A_t}(y) = \exp[-s^2/4\|\varphi\|^2_{H_0} + s(\varphi, e^{-tA}x)_{H_0}$$

$$+ s^2/4(\varphi, (1-e^{-2tA})\varphi)_{H_0}] = :\exp(s(e^{-tA}\varphi, x)_{H_0}):. \tag{1.12}$$

Therefore, (1.10) is established on the functions $f_{s,\varphi}(\cdot)$ ($s \in \mathbb{R}^1$, $\varphi \in \Phi$). Using the expansion

$$(f_{s,\varphi})(x) = \sum_{k=0}^{\infty}\frac{s^k}{k!}:(\varphi, x)^k_{H_0}:, \quad (e^{-tL_A}f_{s,\varphi})(x) = \sum_{k=0}^{\infty}\frac{s^k}{k!}:(e^{-tA}\varphi, x)^k_{H_0}:,$$

we obtain (1.10) for the functions of the form $:(\varphi,x)^k_{H_0}:$, $x \in \Phi'$, $\varphi \in \Phi$, $k \in \mathbb{Z}_+$, and their linear span gives the whole of the set $\mathcal{P}_{\text{cyl}}(\Phi')$.

In the next step, we establish the required equality for $f \in C_b(H_{-\tau_2})$, where $C_b(H_{-\tau_2})$ is a set of continuous bounded functions on $H_{-\tau_2}$. For $f \in C_b(H_{-\tau_2})$, one can find a sequence $(u_n)_{n=1}^{\infty} \subset \mathcal{P}_{\text{cyl}}(\Phi')$ such that

$$\sup_{y \in H_{-\tau_2}} \{|f(y) - u_n(y)| e^{-\delta \|y\|^2_{H_{-\tau_2}}}\} \xrightarrow[n \to \infty]{} 0. \tag{1.13}$$

for any $\delta > 0$. Since (1.13) implies the convergence $e^{-tL_A} u_n \to e^{-tL_A} f$ as $n \to \infty$ in $L_2(\Phi', \gamma_1)$, one can extract a subsequence $(u_{n_k})_{k=1}^{\infty}$ such that for γ_1-almost all $x \in H_{-\tau_1}$, we have

$$(e^{-tL_A} u_{n_k})(x) = \int_{\Phi'} u_{n_k} d\gamma_{A_t, x} \xrightarrow[k \to \infty]{} (e^{-tL_A} f)(x). \tag{1.14}$$

On the other hand, using the estimate

$$|u_n(y)| \le |f(y)| + C\delta e^{\delta \|y\|^2_{H_{-\tau_2}}} \quad (y \in H_{-\tau_2};\ \delta > 0;\ n \in \mathbb{N}),$$

and taking into account the Lebesgue theorem on majorized convergence, we get that

$$\int_{\Phi'} u_n(y) d\gamma_{A_t, x}(y) \xrightarrow[n \to \infty]{} \int_{\Phi'} f(y) d\gamma_{A_t, x}(y) \tag{1.15}$$

for every $x \in H_{-\tau_1}$. Combining (1.14) and (1.15), we obtain the equality

$$(e^{-tL_A} f)(x) = \int_{\Phi'} f(y) d\gamma_{A_t, x}(y)$$

for γ_1-almost all $x \in H_{-\tau_1}$.

Let us show that (1.10) holds for the characteristic functions $\kappa_\alpha(\cdot)$, $\alpha \in \mathcal{B}(H_{-\tau_2})$. If $\alpha \subset H_{-\tau_2}$ is closed, then

$$\kappa_\alpha(y) = \lim_{n \to \infty} \exp(-n \rho_{\tau_2}(y, \alpha)) \quad (y \in H_{-\tau_2}), \tag{1.16}$$

where

$$\rho_{\tau_2}(y, \alpha) = \inf_{z \in \alpha} \|y - z\|_{H_{-\tau_2}}$$

is a function continuous on $H_{-\tau_2}$. Using the previous step and the Lebesgue theorem, we get from (1.16) that

$$(e^{-tL_A} \kappa_\alpha)(x) = \int_{\Phi'} \kappa_\alpha(y) d\gamma_{A_t, x}(y) \tag{1.17}$$

for γ_1-almost all $x \in H_{-\tau_1}$.

Passing to completions and employing the equality $e^{-tL_A}1 = 1$ which follows from the definition of L_A, we establish (1.17) for the open set in $H_{-\tau_2}$. Since the σ-algebra $\mathfrak{B}(H_{-\tau_2})$ coincides with the monotone span of an algebra generated by the open and closed sets in $H_{-\tau_2}$, the equality (1.17) is valid for all $\alpha \in \mathfrak{B}(H_{-\tau_2})$. If $\alpha \in \mathfrak{C}_\sigma(\Phi')$, then $\alpha \cap H_{-\tau_2} \in \mathfrak{B}(H_{-\tau_2})$ and $\kappa_\alpha(y) = \kappa_{\alpha \cap H_{-\tau_2}}(y)$ ($\gamma_{A_t,x}$-almost everywhere) for any $x \in H_{-\tau_1}$. Hence, (1.17) holds for all $\alpha \in \mathfrak{C}_\sigma(\Phi')$.

The inequality (1.17) yields the following lemma important for what follows.

Lemma 1.2 *1. For all* $\alpha \in \mathfrak{C}_\sigma(\Phi')$, $t > 0$, *we have*

$$\int_{\Phi'} \gamma_{A_t,x}(\alpha)\, d\gamma_1(x) = \gamma_1(\alpha). \tag{1.18}$$

2. Assume that $\alpha \in \mathfrak{C}_\sigma(\Phi')$ *is such that* $\gamma_1(\alpha) = 0$. *Then* $\gamma_{A_t,x}(\alpha) = 0$ *for* γ_1-*almost all* $x \in \Phi'$.

Proof. **1.** By virtue of (1.17),

$$\int_{\Phi'} \gamma_{A_t,x}(\alpha)\, d\gamma_1(x) = \int_{\Phi'} (e^{-tL_A}\kappa_\alpha)(x)\, d\gamma_1(x)$$

$$= (e^{-tL_A}\kappa_\alpha, 1)_{L_2(\Phi',\gamma_1)} = (\kappa_\alpha, 1)_{L_2(\Phi',\gamma_1)} = \gamma_1(\alpha).$$

2. According to (1.18),

$$\gamma_1(\alpha) = \int_{\Phi'} \gamma_{A_t,x}(\alpha)\, d\gamma_1(x)$$

and $\gamma_{A_t,x}(\alpha) \geq 0$, the equality $\gamma_1(\alpha) = 0$ involves $\gamma_{A_t,x}(\alpha) = 0$ for γ_1-almost all $x \in \Phi'$.

■

Remark 1. The second assertion in Lemma 1.2 seems to be unexpected, because the measures $\gamma_{A_t,x}$ and γ_1 are orthogonal for γ_1-almost all $x \in \Phi'$. We note that this assertion holds for each fixed $\alpha \in \mathfrak{C}_\sigma(\Phi')$ or, in other words, for any $\alpha \in \mathfrak{C}_\sigma(\Phi')$ and $\gamma_1(\alpha) = 0$, there exist $\beta = \beta(\alpha) \in \mathfrak{C}_\sigma(\Phi')$ and $\gamma_1(\beta) = 1$ such that $\gamma_{A_t,x}(\alpha) = 0$ for any $x \in \beta$.

■

Remark 2. If the second assertion does not hold, then it is impossible to say that the right-hand side of (1.10) defines the space $L_2(\Phi', \gamma_1)$, because we must be sure that the functions which coincide γ_1-almost everywhere have only one image (again in the sense of coincidence almost everywhere).

■

To continue the proof of Theorem 1.1, we now take an arbitrary function

$$f \in L_\infty(\Phi', \gamma_1) = L_\infty(H_{-\tau_2}, \gamma_1).$$

Given f, we choose a sequence of simple functions $(f_n)_{n=1}^{\infty}$ uniformly convergent to f on Φ', where the functions f_n are such that

$$|f_n| \le \|f\|_{L_\infty(\Phi', \gamma_1)}$$

γ_1-almost everywhere. By virtue of Lemma 1.2, the last inequality holds also $\gamma_{A_t,x}$-almost everywhere for γ_1-almost all $x \in \Phi'$. Taking into account that

$$(e^{-tL_A} f_n)(x) = \int_{\Phi'} f_n(y)\, d\gamma_{A_t,x}(y) \tag{1.19}$$

(γ_1-almost everywhere), that $f_n \to f$ in $L_2(\Phi', \gamma_1)$, and that the limit transition in the last integral is possible (according to the Lebesgue theorem), we obtain the required inequality for all $f \in L_\infty(\Phi', \gamma_1)$.

Finally, let $f \in L_2(\Phi', \gamma_1)$, $f \ge 0$. Consider a sequence of "truncations" $(f_n)_{n=1}^{\infty}$ such that $0 \le f_n \le f_{n+1} \le n+1$, $n \in \mathbb{N}$, and $f_n \to f$ as $n \to \infty$ in $L_2(\Phi', \gamma_1)$. According to the above argument, we have the equality

$$(e^{-tL_A} f_n)(x) = \int_{\Phi'} f_n\, d\gamma_{A_t,x}$$

(γ_1-almost everywhere). Extracting (if necessary) a subsequence, we obtain the convergence of the left-hand side of the last equality to $(e^{-tL_A} f)(x)$ for γ_1-almost all $x \in \Phi'$. On the right-hand side, we apply the B. Levi theorem on the limiting transition under the integral sign and obtain the following convergence

$$\int_{\Phi'} f_n\, d\gamma_{A_t,x} \xrightarrow[n\to\infty]{} \int_{\Phi'} f\, d\gamma_{A_t,x}$$

γ_1-almost all $x \in \Phi'$. Equating the values of the limits on the right- and left-hand sides of (1.19), we get

$$(e^{-tL_A}f)(x) = \int_{\Phi'} f(y)d\gamma_{A_t,x}(y)$$

γ_1-almost all $x \in \Phi'$.

For an arbitrary function $f \in L_2(\Phi', \gamma_1)$, one should study the real and imaginary parts separately. Each part is, in turn, decomposed into a difference of two nonnegative functions from $L_2(\Phi', \gamma_1)$. Then one can apply the assertion of the theorem which is already proved for these functions.

■

1.3. Definition of the Second Quantization Operators by Using a Differential Expression

For the derivatives of functions $u \in \mathcal{P}(\Phi')$ (in the sense of H_0, see Subsection 3.2 in Chapter 2), we have

$$u'(\cdot) = \nabla u(\cdot) \in \mathcal{P}(\Phi') \otimes \Phi_c;$$

therefore, we have

$$(\nabla u(\cdot), A\overline{\nabla v}(\cdot))_{H_0} \in \mathcal{P}(\Phi')$$

for $u, v \in \mathcal{P}(\Phi')$, and the following bilinear form is defined

$$d_{\gamma_1,A}(u,v) = \frac{1}{2}\int_{\Phi'} (\nabla u, A\overline{\nabla v})_{H_0} d\gamma_1. \tag{1.20}$$

The form (1.20) is called a Dirichlet form corresponding to a measure γ_1 and a coefficient operator A. First, we note its connection with the operator L_A.

Theorem 1.2. *For $u, v \in \mathcal{P}(\Phi')$, the equality*

$$d_{\gamma_1,A}(u,v) = (L_A u, v)_{L_2(\Phi',\gamma_1)} \tag{1.21}$$

holds.

Proof. It suffices to establish the equality (1.21) on the functions from $\mathcal{P}_{\text{cyl}}(\Phi')$. Indeed, for arbitrary $u, v \in \mathcal{P}(\Phi')$, one can choose sequences $(u_n)_{n=1}^{\infty} \subset \mathcal{P}_{\text{cyl}}(\Phi')$ and $(v_n)_{n=1}^{\infty} \subset \mathcal{P}_{\text{cyl}}(\Phi')$ such that $u_n \to u$ as $n \to \infty$ in $\mathcal{P}(\Phi')$ and $v_n \to v$ as $n \to \infty$ in

$\mathcal{P}(\Phi')$ (see Subsection 5.2 in Chapter 2). Since $L_A \in \mathcal{L}(\mathcal{P}(\Phi'), \mathcal{P}(\Phi'))$, we have

$$(L_A u_n, v_n)_{L_2(\Phi',\gamma_1)} \xrightarrow[n\to\infty]{} (L_A u, v)_{L_2(\Phi',\gamma_1)},$$

and

$$(\nabla u_n, A\overline{\nabla v_n})_{H_0} \xrightarrow[n\to\infty]{} (\nabla u, A\overline{\nabla v})_{H_0}$$

in $\mathcal{P}(\Phi')$ and, moreover, in $L_1(\Phi', \gamma_1)$. In turn, the validity of the equality (1.21) on $\mathcal{P}_{\text{cyl}}(\Phi')$ follows from the fact that it holds for functions of the form

$$\prod_{k=1}^{n} (\varphi_k, x)_{H_0} : (x \in \Phi',\ \varphi_k \in \Phi,\ k = 1, \dots, n;\ n \in \mathbb{N}).$$

However, by virtue of the polarization identity, we can restrict ourselves to the case

$$u(x) = v(x) = :(\varphi, x)_{H_0}^n : (x \in \Phi',\ \varphi \in \Phi,\ n \in \mathbb{N}).$$

On one hand, by using the properties of the Segal isomorphism, we get

$$L_A : (\varphi, x)_{H_0}^n := 2^{-n/2}\sqrt{n!}\, I_1(d\Gamma(A)\varphi^{\otimes n})$$

$$= 2^{-n/2}\sqrt{n!}\, I_1(A\varphi \otimes \varphi \otimes \dots \otimes \varphi + \dots + \varphi \otimes \dots \otimes \varphi \otimes A\varphi)$$

$$= 2^{-n/2}\sqrt{n!}\, I_1\left(\frac{n!}{(n-1)!} Pn(A\varphi \otimes \varphi^{\otimes(n-1)})\right) = n : (\varphi, x)_{H_0}^{n-1} (A\varphi, x)_{H_0} : \quad (1.22)$$

whence

$$(L_A u, v)_{L_2(\Phi',\gamma_1)} = \frac{1}{2} \int_{\Phi'} n : (\varphi, x)_{H_0}^{n-1} (A\varphi, x)_{H_0} :: (\varphi, x)_{H_0}^n : d\gamma_1(x)$$

$$= n \cdot 2^{-n} n! \|\varphi\|_{H_0}^{2n-2} (A\varphi, \varphi)_{H_0}.$$

Here we have used the equality (2.26) in Chapter 2.

On the other hand, in Theorem 2.1 in Chapter 2, we established that

$$u(x) = :(\varphi, x)_{H_0}^n := (-1)^n \cdot 2^{-n} \|\varphi\|_{H_0}^n H_n(\|\varphi\|_{H_0}^{-1} (\varphi, x)_{H_0}),$$

therefore,

$$\nabla u(x) = 2n \cdot 2^{-n} (-1)^n H_{n-1}(\|\varphi\|_{H_0}^{-1} (\varphi, x)_{H_0}) \varphi / \|\varphi\|_{H_0}^{1-n}$$

$$= -n : (\varphi, x)_{H_0}^{n-1} : \varphi \in \mathcal{P}(\Phi') \otimes \Phi. \tag{1.23}$$

Then, by using the same equality as in (1.23), we obtain

$$\frac{1}{2}\int_{\Phi'} (\nabla u(x), A\overline{\nabla v}(x))_{H_0} d\gamma_1(x) = \frac{1}{2}(A\varphi, \varphi)_{H_0} n^2 \int_{\Phi'} [: (\varphi, x)_{H_0}^{n-1} :]^2 d\gamma_1(x)$$

$$= \frac{1}{2}(A\varphi, \varphi)_{H_0} n^2 2^{-(n-1)}(n-1)! \, \|\varphi\|_{H_0}^{2n-2}. \tag{1.24}$$

Comparing (1.23) and (1.24), we get the required result.

■

When discussing the differential expressions given on functions of infinitely many variables in Section 3 of Chapter 2, we have noted that the expression $\mathrm{Tr}_{H_0} u''$ is defined if the kernel of the operator $u'' \in \mathcal{L}(H_0, H_0)$ belongs to $\Phi_c \otimes \Phi_c$ (see Example 3.4). In particular, this takes place for $u \in \mathcal{P}(\Phi')$. Furthermore, the expression $\mathrm{Tr}_{H_0} Au''$ is meaningful, because the kernel of the operator Au'' has the form $(A \otimes 1)\, u'' \in \Phi_c \otimes \Phi_c$ due to the inclusion $A \in \mathcal{L}(\Phi, \Phi)$. Hence, $\mathrm{Tr}_{H_0} Au'' \in \mathcal{P}(\Phi')$ for any $u \in \mathcal{P}(\Phi')$. The expression of the first order $(A\nabla u(x), x)_{H_0}$ is also defined on the same functions as a function from $\mathcal{P}(\Phi')$; here we take into account that $\nabla u(\cdot) \in \mathcal{P}(\Phi') \otimes \Phi_c$ and, thus, $A\nabla u(\cdot) \in \mathcal{P}(\Phi') \otimes \Phi_c$. Denote

$$\mathcal{L}_A u = -\frac{1}{2}[\mathrm{Tr}_{H_0} Au''(\cdot) - 2(A\nabla u(\cdot), \cdot)_{H_0}] \in \mathcal{P}(\Phi') \quad u \in \mathcal{P}(\Phi').$$

Theorem 1.3. *For $u \in \mathcal{P}(\Phi')$, we have the representation*

$$(L_A u)(x) = -\frac{1}{2}\mathrm{Tr}_{H_0} Au''(x) + (A\nabla u(x), x)_{H_0} \quad (x \in \Phi'). \tag{1.25}$$

Proof. As in the proof of the previous theorem, we establish that it suffices to prove the representation (1.25) for the functions $u(x) = (\varphi, x)_{H_0}^n$ $(x \in \Phi')$ with arbitrary $\varphi \in \Phi$. Inserting u in $\mathcal{L}_A$, we obtain

$$(\mathcal{L}_A u)(x) = -\frac{n(n-1)}{2}(\varphi, x)_{H_0}^{n-2}(A\varphi, \varphi)_{H_0} + n(A\varphi, x)_{H_0}(\varphi, x)_{H_0}^{n-1},$$

hence,

$$(\mathcal{L}_A u, u)_{L_2(\Phi', \gamma_1)} = -\frac{n(n-1)}{2}\int_{\Phi'} (A\varphi, \varphi)_{H_0}(\varphi, x)_{H_0}^{2n-2} d\gamma_1(x)$$

$$+ n \int_{\Phi'} (A\varphi, x)_{H_0} (\varphi, x)_{H_0}^{2n-1} d\gamma_1(x). \tag{1.26}$$

Using the formula (2.5) in Chapter 2 for the moments of the Gaussian measure γ_1, we can rewrite (1.26) in the form

$$(\mathfrak{L}_A u, u)_{L_2(\Phi', \gamma_1)} = n^2 \frac{(2(n-1))!}{(n-1)!} 2^{-n} (A\varphi, \varphi)_{H_0} \|\varphi\|_{H_0}^{2n-2}. \tag{1.27}$$

It follows from (1.21) that

$$(L_A u, u)_{L_2(\Phi', \gamma_1)} = \frac{1}{2} \int_{\Phi'} (\nabla u, A\overline{\nabla u})_{H_0} d\gamma_1(x)$$

$$= \frac{1}{2} n^2 (A\varphi, \varphi)_{H_0} \int_{\Phi'} (\varphi, x)_{H_0}^{2n-2} d\gamma_1(x) = (\mathfrak{L}_A u, u)_{L_2(\Phi', \gamma_1)}$$

(here we have used the formula for moments once again).

■

Remark 1. The equality (1.25) shows that the operator A plays the role of a "matrix of coefficients" in the elliptic differential expression. Since it acts in the infinite-dimensional space, condition $A > 0$ does not yield uniform ellipticity as in the case of finitely many variables. In order to guarantee uniform ellipticity, one should demand that condition $A \geq a1$, $a > 0$, hold; henceforth, it is called a condition of uniform ellipticity of the operator L_A.

■

In the definition of the second quantization operators L_A, the differential expression (1.25) makes it natural to ask questions about the other domains of essential selfadjointness of these operators which are more traditional from the point of view of the theory of differential operators than the spaces of polynomials. It is often convenient to use a set $C_{b,\text{cyl}}^{\infty}(\Phi')$ of cylindrical functions on Φ' bounded with all their derivatives as a domain of this sort. Recall that $C_{b,\text{cyl}}^{k}(\Phi')$ $(k \in \mathbb{N} \cup \{\infty\})$ consists of functions of the form

$$u(x) = g((\varphi_1, x)_{H_0}, \dots, (\varphi_n, x)_{H_0}),$$

where $g \in C_b^k(\mathbb{R}^n)$ and $\varphi_k \in \Phi$, $k = 1, \dots, n$.

Theorem 1.4. *The operator L_A is essentially selfadjoint on $C_{b,\text{cyl}}^{\infty}(\Phi')$.*

Proof. It is necessary to show that for each polynomial $p \in \mathcal{P}_{\text{cyl}}(\Phi')$, one can construct a sequence $(u_j)_{j=1}^{\infty} \subset C^{\infty}_{b,\text{cyl}}(\Phi')$ such that $u_j \to p$ as $j \to \infty$ in $L_2(\Phi', \gamma_1)$ and $L_A u_j \to L_A p$ as $j \to \infty$ in $L_2(\Phi', \gamma_1)$. For $p \in \mathcal{P}_{\text{cyl}}(\Phi')$, the following representation holds

$$p(x) = \tilde{p}((\varphi_1, x)_{H_0}, \dots, (\varphi_n, x)_{H_0}),$$

where $\varphi_k \in \Phi$, $k = 1, \dots, n$, and $\tilde{p}(\cdot)$ is a polynomial on $\mathbb{R}^n$. Consider a sequence of functions

$$g_j(t_1, \dots, t_n) = \tilde{p}(t_1, \dots, t_n)\exp(-1/j(t_1^2 + \dots + t_n^2)).$$

The corresponding sequence of cylindrical functions

$$u_j(x) = g_j((\varphi_1, x)_{H_0}, \dots, (\varphi_n, x)_{H_0}) \quad (x \in \Phi')$$

is just the one required. Indeed,

$$|u_j(x) - p(x)| = |p(x)|\{1 - e^{-1/j((\varphi_1, x)^2_{H_0} + \dots + (\varphi_n, x)^2_{H_0})}\}$$

$$\le 1/j\,|p(x)|\,\{(\varphi_1, x)^2_{H_0} + \dots + (\varphi_n, x)^2_{H_0}\},$$

hence,

$$\|u_j - p\|_{L_2(\Phi', \gamma_1)} = c/j \to 0, \quad j \to \infty.$$

Further, by using (1.25), we get

$$(L_A u_j)(x) = (L_A l_j)(x)p(x) + (L_A p)(x)l_j(x) - 2(\nabla l_j(x), A\nabla p(x))_{H_0}, \tag{1.28}$$

where

$$l_j(x) = \exp\left[-1/j \sum_{k=1}^{n} (\varphi_k, x)^2_{H_0}\right] \quad (x \in \Phi').$$

Consider each term in (1.28) separately. It is easy to see that

$$|\mathrm{Tr}_{H_0} A l_j''(x)| \le j^{-1}|p_1(x)|, \quad p_1 \in \mathcal{P}(\Phi'),$$

$$|(A\nabla l_j(x), x)_{H_0}| \le j^{-1}|p_2(x)|, \quad p_2 \in \mathcal{P}(\Phi'),$$

therefore, $L_A l_j \to 0$ as $j \to \infty$ in $L_q(\Phi', \gamma_1)$ for any $q \ge 1$ and, thus, $p \cdot L_A l_j \to 0$ as

$j \to \infty$ in $L_2(\Phi', \gamma_1)$. Further, $l_j \to 1$ as $j \to \infty$ in $L_q(\Phi', \gamma_1)$ for any $q \geq 1$ and again $l_j L_A p \to L_A p$ as $j \to \infty$ in $L_2(\Phi', \gamma_1)$. Similarly, the last term converges to 0 in $L_2(\Phi', \gamma_1)$, because

$$\| (A \nabla l_j(\cdot) \|_{H_0} \leq j^{-1} | p_3(\cdot) |, \quad p_3 \in \mathcal{P}(\Phi').$$

As a result, we obtain the convergence $L_A u_j \to L_A p$ as $j \to \infty$ in $L_2(\Phi', \gamma_1)$ (and even in all $L_q(\Phi', \gamma_1)$, $q \geq 1$).

■

Example 1.3 (The operator of a number of particles in the Schrödinger representation). The characteristic feature of the operator of a number of particles $N = d\Gamma(1)$ in Example 1.1 is that it commutes with each operator $d\Gamma(A)$ and any nuclear rigging of H_0 can be used to study it. The image of the operator of a number of particles under the Segal isomorphism is denoted by $N = L_1$ as before. Let us find the form of it in the case of a functional realization of the Fock space.

The choice of an orthonormal basis in H_0 defines an isomorphism between H_0 and $l_2(\mathbb{R}^1)$. Consider the nuclear rigging $\mathbb{R}^\infty \supset l_2(\mathbb{R}^1) \supset \mathbb{R}_0^\infty$ and the corresponding rigging of the Fock space (1.4)

$$\mathcal{F}'_{\text{fin}}(\mathbb{R}^\infty) \supset \mathcal{F}(l_2(\mathbb{R}^1)) \supset \mathcal{F}_{\text{fin}}(\mathbb{R}^\infty),$$

which transforms into the following rigging under the Segal isomorphism

$$\mathcal{P}'(\mathbb{R}^\infty) \supset L_2(\mathbb{R}^\infty, \gamma_1) \supset \mathcal{P}(\mathbb{R}^\infty),$$

where

$$d\gamma_1(x) = \mathop{\times}_{k=1}^{\infty} \frac{1}{\sqrt{\pi}} e^{-x_k^2} dx_k$$

is a canonical Gaussian measure corresponding to the Hilbert space $l_2(\mathbb{R}^1)$, $\mathcal{P}(\mathbb{R}^\infty) = \mathcal{P}_{\text{cyl}}(\mathbb{R}^\infty)$ is a nuclear space of cylindrical polynomials on $\mathbb{R}^\infty$, i.e., of polynomials depending on finitely many variables, and $\mathcal{P}'(\mathbb{R}^\infty)$ is a space of generalized functions (formal polynomials) on $\mathbb{R}^\infty$ (see Example 2.4 in Chapter 2 and Subsection 4.3). Taking into account that here $A = 1$, we obtain the following representation for $u \in \mathcal{P}(\mathbb{R}^\infty)$ from (1.25)

$$(Nu)(x) = -\frac{1}{2} \sum_{k=1}^{\infty} \left(\frac{\partial^2 u(x)}{\partial x_k^2} - 2x_k \frac{\partial u(x)}{\partial x_k} \right) \quad (x \in \mathbb{R}^\infty) \tag{1.29}$$

the series terminates, because the polynomial u is cylindrical. If the operator A is associated with the matrix $(a_k \delta_{kj})_{k,j=1}^{\infty}$ diagonal in the chosen basis, then (1.25) yields

$$(L_A u)(x) = -\frac{1}{2} \sum_{k=1}^{\infty} a_k \left(\frac{\partial^2 u(x)}{\partial x_k^2} - 2x_k \frac{\partial u(x)}{\partial x_k} \right) \tag{1.30}$$

for $u \in \mathcal{P}(\mathbb{R}^{\infty})$.

Note that in quantum mechanics the operator

$$N_k = -\frac{1}{2} \left(\frac{\partial^2}{\partial x_k^2} - 2x_k \frac{\partial}{\partial x_k} \right)$$

is an operator of energy of one-dimensional harmonic oscillator (with unit mass and frequency) in the space of states $L_2(\mathbb{R}^1, \gamma_1(x_k))$. Thus, the operator (1.30) describes a system consisting of infinitely many noninteracting oscillators with frequencies $a_j \geq 0$, $j \in \mathbb{N}$. We shall encounter this interpretation of the second quantization operators in the study of models in quantum statistical physics.

Consider the case when the duality between Φ and Φ' is given with the help of a scalar product in a certain Hilbert space $\mathcal{H}_0 \supset \Phi$ and, moreover,

$$(\varphi_1, \varphi_2)_{\mathcal{H}_0} = (J\varphi_1, J\varphi_2)_{H_0}, \quad \varphi_1, \varphi_2 \in \Phi,$$

where $J \in \mathcal{L}(\Phi, \Phi)$. In this case, the derivatives are regarded in the sense of the space $\mathcal{H}_0$ and, acting as above, one can establish the equality

$$(L_A u)(x) = -\frac{1}{2} \operatorname{Tr}_{\mathcal{H}_0} JAJ^* u''(x) + (A\nabla u(x), x)_{\mathcal{H}_0}$$

$$(u \in \mathcal{P}(\Phi'), x \in \Phi'). \tag{1.31}$$

The necessity of the investigation of this special situation is clarified by the following important example.

Example 1.4. (the Hamiltonian of a free boson field with positive mass).

Assume that $m_0 \geq 0$ is given. We take a Hilbert space $\mathfrak{H}_{m_0}$ which is a completion of a real Hilbert space $\mathcal{H}_0 = L_{2,\mathrm{Re}}(\mathbb{R}^d)$ with respect to the scalar product

$$(\varphi, \psi)_{H_0} = ((-\Delta + m_0^2 1)^{-1/2} \varphi, \psi)_{L_2(\mathbb{R}^d)}$$

to play the role of H_0 (see Example 2.5 in Chapter 2). It is clear that H_0 is isomorphic to the Sobolev space $W_{2,\mathrm{Re}}^{-1/2}(\mathbb{R}^d)$. We regard the Schwartz space $\mathcal{S}_{\mathrm{Re}}(\mathbb{R}^d)$ of real test functions as Φ. As usual, the duality between $\mathcal{S}_{\mathrm{Re}}(\mathbb{R}^d)$ and $\mathcal{S}'_{\mathrm{Re}}(\mathbb{R}^d)$ is established in terms of the scalar product in $L_{2,\mathrm{Re}}(\mathbb{R}^d)$.

The Segal isomorphism which exploits the canonical duality of $\mathcal{S}\ (\mathbb{R}^d)$ and $\mathcal{S}'(\mathbb{R}^d)$ maps $\mathcal{F}(H_0)$ onto $L_2(\mathcal{S}'_{\mathrm{Re}}(\mathbb{R}^d), \mu_{m_0})$, where μ_{m_0} is a Gaussian measure on $\mathcal{S}'(\mathbb{R}^d)$ given by the Fourier transform ((2.36) in Chapter 2)

$$\int_{\mathcal{S}'_{\mathrm{Re}}(\mathbb{R}^d)} e^{i(\varphi, x)_{L_2(\mathbb{R}^d)}} d\mu_{m_0}(x) = e^{-1/4\|\varphi\|^2_{H_0}} \qquad (\varphi \in \mathcal{S}_{\mathrm{Re}}(\mathbb{R}^d)).$$

The measure μ_{m_0} on $\mathcal{S}'_{\mathrm{Re}}(\mathbb{R}^d)$ which appears in this way is called, in the indicated example, a measure corresponding to a free boson field with mass m_0 in the space with dimensionality d at time 0. The space $L_2(\mathcal{S}'_{\mathrm{Re}}(\mathbb{R}^d), \mu_{m_0})$ is a space of physical states of a free field and

$$L_{A_{m_0}} \equiv L_{m_0}(A_{m_0} = (-\Delta + m_0^2 1)^{1/2})$$

is its operator of energy (free Hamiltonian). Since now $J = (-\Delta + m_0^2 1)^{-1/4}$, it follows from (1.31) that

$$(L_{m_0} f)(x) = -\frac{1}{2}\,[\mathrm{Tr}_{L_2(\mathbb{R}^d)} f''(x) - 2((-\Delta + m_0^2 1)^{1/2} \nabla f(x), x)_{L_2(\mathbb{R}^d)}] \qquad (1.32)$$

for $f \in \mathcal{P}(\mathcal{S}'_{\mathrm{Re}}(\mathbb{R}^d))$, $x \in \mathcal{S}'_{\mathrm{Re}}(\mathbb{R})$, where the derivatives are regarded in the sense of $L_2(\mathbb{R}^d)$.

Using the variational derivatives, we can rewrite expression (1.32) in the form

$$(L_{m_0} f)(x) = -\frac{1}{2}\int_{\mathbb{R}^d} \frac{\delta^2 f(x)}{\delta x(s)^2} ds + \int_{\mathbb{R}^d} \frac{\delta f(x)}{\delta x(s)} (-\Delta + m_0^2 1)^{1/2} x(s)\, ds.$$

In this example, we introduced an additional Hilbert space $\mathcal{H}_0$, because we want to remain within the framework of the canonical duality between $\mathcal{S}\ (\mathbb{R}^d)$ and $\mathcal{S}'(\mathbb{R}^d)$ which is established with the help of $L_2(\mathbb{R}^d)$. Clearly, we can use the duality established by H_0 equally well; in this case, we simply obtain the isomorphic situation.

1.4. Properties of the Second Quantization Operators

In the finite-dimensional case, elliptic operators of the second order possess an important property of preserving positivity, namely, the semigroups that correspond to these operators transform nonnegative functions into nonnegative. The next theorem establishes the fact that the operator L_A has a similar property. This is not surprising, because it is also an elliptic operator of the second order (but with infinitely many variables).

Theorem 1.5. *The semigroup* e^{-tL_A}, $t \geq 0$, *preserves positivity.*

Proof. It follows from formula (1.10) that

$$(e^{-tL_A}f)(x) = \int_{\Phi'} f(y)d\gamma_{A_t, x}(y) \geq 0 \tag{1.33}$$

for γ_1-almost all $x \in \Phi'$ provided that $f \geq 0$ γ_1-almost everywhere.

■

The property of preserving positivity and the equality $e^{-tL_A}1 = 1$, $t \geq 0$, imply the L_p-contractibility of e^{-tL_A} in all $L_p(\Phi', \gamma_1)$, $p \geq 1$.

Theorem 1.6. *For every* $p \geq 1$ *and any* $f \in L_2(\Phi', \gamma_1) \cap L_p(\Phi', \gamma_1)$, *we have*

$$\| e^{-tL_A}f \|_{L_p(\Phi',\gamma_1)} \leq \|f\|_{L_p(\Phi',\gamma_1)} \quad (t \geq 0) \tag{1.34}$$

and, thus, the semigroup e^{-tL_A}, $t \geq 0$, *is* L_p*-contracting for all* $p \geq 1$ *(after restriction or extension by continuity).*

Proof. It follows from Mehler's formula (1.10) that the inequality

$$|e^{-tL_A}f| \leq e^{-tL_A}|f|$$

holds γ_1-almost everywhere for all $f \in L_2(\Phi', \gamma_1)$. Therefore,

$$\|e^{-tL_A}f\|_{L_1(\Phi',\gamma_1)} = \int_{\Phi'} |e^{-tL_A}f| d\gamma_1 \leq \int_{\Phi'} e^{-tL_A}|f| d\gamma_1$$

$$= (1, e^{-tL_A}|f|)_{L_2(\Phi',\gamma_1)} = (1, |f|)_{L_2(\Phi',\gamma_1)} = \|f\|_{L_1(\Phi',\gamma_1)}$$

(note that, for $f \geq 0$, this yields the isometricity of e^{-tL_A} in $L_1(\Phi', \gamma_1)$). For $f \in L_\infty(\Phi', \gamma_1)$, we have

$$\|e^{-tL_A}f\|_{L_\infty(\Phi',\gamma_1)} = \sup_{\|g\|_{L_1(\Phi',\gamma_1)}=1} |(e^{-tL_A}f, g)_{L_2(\Phi',\gamma_1)}|$$

$$= \sup_{\|g\|_{L_1(\Phi',\gamma_1)}=1} |(f, e^{-tL_A}g)_{L_2(\Phi',\gamma_1)}$$

$$\le \|f\|_{L_\infty(\Phi',\gamma_1)} \sup_{\|g\|_{L_1(\Phi',\gamma_1)}=1} \|(e^{-tL_A}g)\|_{L_1(\Phi',\gamma_1)} \le \|f\|_{L_\infty(\Phi',\gamma_1)}.$$

Hence, we have proved the contractibility of e^{-tL_A}, $t \ge 0$, in $L_1(\Phi', \gamma_1)$ and $L_\infty(\Phi', \gamma_1)$. According to the interpolation theorem, this implies the contractibility in all $L_p(\Phi',\gamma_1)$, $p \ge 1$.

■

In the case of the uniform ellipticity of L_A, we have a stronger result which states the hypercontractibility of the semigroup under consideration.

Theorem 1.7. *Let $A \ge 1$, $a > 0$. Then the semigroup e^{-tL_A}, $t \ge 0$, possesses the property of hypercontractibility. Namely, for $p, q > 1$ such that*

$$e^{-2tA} \le \frac{p-1}{q-1}, \tag{1.35}$$

the operator e^{-tL_A} is a contracting operator acting from $L_p(\Phi',\gamma_1)$ to $L_q(\Phi',\gamma_1)$. In particular, $e^{-tL_A} \in \mathfrak{L}(L_2(\Phi',\gamma_1), L_q(\Phi',\gamma_1))$ for $q \le 1 + e^{2ta}$. Furthermore, the estimate (1.35) *is unimprovable.*

Proof. We do not give a detailed proof of this theorem because, for this purpose, it would be necessary to present first some cumbersome technical constructions which are of no use in what follows. In addition, several versions of the proof exist and these are presented in various monographs (the corresponding references can be found in the Bibliographical Notes). We have a more restricted aim, namely, to describe the scheme of the proof, to clarify the origin of the estimate (1.35), and to show that it is unimprovable.

First, we note that it suffices to restrict ourselves to the case $A = a1$. Indeed, if we represent $A = a1 + A_1$, $A_1 \ge 0$, then we obtain

$$e^{-tL_A} = e^{-taN}e^{-tLA_1}.$$

According to Theorem 1.6, the operator $e^{-tL_{A_1}}$ is L_p-contracting and, thus, in order to prove the hypercontractibility of e^{-tL_A}, it suffices to establish the same property for e^{-taN}. We consider the operator aN in the space $L_2(\mathbb{R}^\infty, \gamma_1)$ (see Example 1.3). Then Mehler's formula (1.10) takes the form

$$(e^{-taN}f)(x) = \int_{\mathbb{R}^\infty} f(y)\,d\gamma_{A_t,x}(y),$$

where

$$d\gamma_{A_t,x}(y) = \bigtimes_{k=1}^{\infty} \frac{1}{\sqrt{\pi(1 - e^{-2ta})}} e^{-\frac{(y_k - e^{-ta}x_k)^2}{1-e^{-2ta}}} dy_k,$$

i.e., the operator e^{-taN} can be represented in the form of an infinite tensor product $\bigotimes_{k=1}^{\infty} e^{-taN_k}$, where each operator e^{-taN_k} acts in $L_2(\mathbb{R}^1, \gamma_1(y_k))$ and is given by the one-dimensional Mehler's formula in Example 1.2. Clearly, this is a consequence of the fact that aN is an operator with separated variables (in the considered functional realization). If we show that each operator e^{-taN_k} is a contracting operator from $L_p(\mathbb{R}^1, d\gamma_1)$ to $L_q(\mathbb{R}^1, d\gamma_1)$, then by virtue of Lemma 2.3 in Chapter 1 concerning the tensor product of contracting operators, we find that e^{-taN} is a contracting operator from $L_p(\mathbb{R}^\infty, d\gamma_1)$ to $L_q(\mathbb{R}^\infty, d\gamma_1)$.

Hence, the proof of the theorem is reduced to an analysis of the one-dimensional case. Let us show that if the strict inequality holds in (1.35), then the integral operator

$$(e^{-taN}f)(x) = \int_{\mathbb{R}^1} f(y)[\pi(1-e^{-2ta})]^{-1/2} e^{-\frac{(y-e^{-ta}x)^2}{1-e^{-2ta}}} dy$$

acts continuously from $L_p(\mathbb{R}^1, d\gamma_1)$ to $L_q(\mathbb{R}^1, d\gamma_1)$. For convenience, we denote $e^{-ta} = b$ and rewrite it in the form

$$(e^{-taN}f)(x) = \int_{\mathbb{R}^1} f(y)K_b(x,y)d\gamma_1(y),$$

where

$$K_b(x,y) = (1-b^2)^{-1/2} \exp\left(-\frac{y^2b^2 + x^2b^2 - 2bxy}{1-b^2}\right).$$

Denoting by p' an index conjugated to p $(1/p + 1/p' = 1)$ and taking the Hölder inequality into account, we obtain

$$\|e^{-taN}f\|_{L_q(\mathbb{R}^1,\gamma_1)} = \left[\int_{\mathbb{R}^1}\left|\int_{\mathbb{R}^1} K_b(x,y)f(y)d\gamma_1(y)\right|^q d\gamma_1(y)\right]^{1/q}$$

$$\le \|f\|_{L_p(\mathbb{R}^1,\gamma_1)}\left\{\int_{\mathbb{R}^1}\left[\int_{\mathbb{R}^1} K_b^{p'}(x,y)d\gamma_1(y)\right]^{q/p'} d\gamma_1(y)\right\}^{1/q}. \quad (1.36)$$

Computing the internal integral in (1.36) by the formula

$$\int_{\mathbb{R}^1} e^{-\alpha x^2-\beta x}\,dx = \sqrt{\frac{\pi}{\alpha}}\,e^{\beta^2/4a} \quad (\alpha>0),$$

we get

$$\|e^{-taN}f\|_{L_q(\mathbb{R}^1,\gamma_1)} \le \|f\|_{L_p(\mathbb{R}^1,\gamma_1)}\left\{c_1\int_{\mathbb{R}^1} e^{-x^2\frac{1-b^2(p'-1)(q-1)}{b^2(p'-1)+1}}\,dx\right\}^{1/q} \le c_2\|f\|_{L_p(\mathbb{R}^1,\gamma_1)}$$

provided that the condition which guarantees the convergence of the last integral holds, i.e., $b^2(p'-1)(q-1)<1$ or $b^2<(p-1)/(q-1)$. Hence, for $e^{-2ta}<(p-1)/(q-1)$, we have

$$\|e^{-taN}f\|_{L_q(\mathbb{R}^1,\gamma_1)} \le c_2\|f\|_{L_p(\mathbb{R}^1,\gamma_1)}.$$

It remains to show that it is possible to set $c_2=1$ in this estimate even in the case of the equality in (1.35). In fact, this part of the proof from the technical point of view is the most difficult one and should be analyzed separately.

Let us show that condition (1.35) cannot be made weaker even in the one-dimensional case. For $\lambda\in\mathbb{R}^1$, we consider the function

$$f_\lambda(x) = e^{\lambda x-\lambda^2/4} \quad (x\in\mathbb{R}^1);\ f_\lambda\in L_p(\mathbb{R}^1,d\gamma_1)$$

for any $p\ge 1$. Applying Mehler's formula (see Example 1.2), we obtain $(b=e^{-ta})$

$$e^{-taN}f_\lambda(x) = e^{\lambda bx-\lambda^2b^2/4},$$

therefore,

$$\|f_\lambda\|_{L_p(\mathbb{R}^1,\gamma_1)} = e^{\frac{\lambda^2(p-1)}{4}},\quad \|e^{-taN}f_\lambda\|_{L_q(\mathbb{R}^1,\gamma_1)} = e^{\frac{\lambda^2(q-1)b^2}{4}}.$$

If condition (1.35) does not hold, i.e.,

$$e^{-2ta} = b^2 > \frac{p-1}{q-1}$$

or $(q-1)b^2>p-1$, then for any $c>0$ there exists λ_0 such that

$$\|e^{-taN}f_\lambda\|_{L_q(\mathbb{R}^1,\gamma_1)} \ge c\|f_\lambda\|_{L_p(\mathbb{R}^1,\gamma_1)}$$

for any $\lambda > \lambda_0$. Hence, e^{-taN} is even not bounded as an operator from $L_p(\mathbb{R}^1, d\gamma_1)$ to $L_q(\mathbb{R}^1, d\gamma_1)$.

■

***Corollary* 1.** *For $A \geq a1$, $a > 0$, we have*
1. *for $p, q > 1$ such that $e^{-2ta} \leq (p-1)(q-1)$, we have*

$$\left| \int_{\Phi'} f(x)(e^{-tL_A} g)(x)\, d\gamma_1(x) \right| \leq \|f\|_{L_p(\Phi',\gamma_1)} \| g \|_{L_q(\Phi',\gamma_1)}; \tag{1.37}$$

2. *for $p, q > r > 1$ such that $e^{-2ta} \leq (p/r - 1)(q/r - 1)$, we have*

$$\|f(e^{-tL_A} g)\|_{L_r(\Phi',\gamma_1)} \leq \|f\|_{L_p(\Phi',\gamma_1)} \| g \|_{L_q(\Phi',\gamma_1)}. \tag{1.38}$$

Proof. **1.** Assume that $q > 1$ is chosen and that q' is an adjoint index. By the Hölder inequality, we get

$$\left| \int_{\Phi'} f(x)(e^{-tL_A} g)\, d\gamma_1 \right| \leq \|f\|_{L_q(\Phi',\gamma_1)} \| e^{-tL_A} g \|_{L_{q'}(\Phi',\gamma_1)} \leq \|f\|_{L_q(\Phi',\gamma_1)} \| g \|_{L_p(\Phi',\gamma_1)},$$

because e^{-tL_A} is a contracting operator from $L_p(\Phi', \gamma_1)$ to $L_{q'}(\Phi', \gamma_1)$ according to Theorem 1.7; here

$$e^{-2ta} \leq (p-1)(q'-1)^{-1} = (p-1)(q-1).$$

2. Using the Hölder inequality once again, we obtain

$$\|f e^{-tL_A} g \|_{L_r(\Phi',\gamma_1)} \leq \|f\|_{L_q(\Phi',\gamma_1)} \| e^{-tL_A} g \|_{L_{q'}(\Phi',\gamma_1)},$$

where

$$\frac{1}{r} = \frac{1}{q} + \frac{1}{q'} \quad \text{or} \quad 1 = \frac{r}{q} + \frac{r}{q'}.$$

Theorem 1.7 shows that e^{-tL_A} is a contracting operator from $L_p(\Phi', \gamma_1)$ to $L_{q'}(\Phi', \gamma_1)$, where $e^{-2ta} \leq (p/r - 1)(q/r - 1)$.

■

***Corollary* 2.** *For each measurable polynomial $f \in \mathcal{P}_{\gamma_{1,n}}$ whose power is not higher than n ($n \in \mathbb{N}$) and for all $p > 2$, the following inequality holds*

$$\|f\|_{L_p(\Phi',\gamma_1)} \leq (p-1)^{-n/2} \|f\|_{L_2(\Phi',\gamma_1)}.$$

Proof. We set $A \geq a1$, where $a = (\ln(p-1))^{1/2}$. Then $e^{-2a} = (p-1)^{-1}$ and, by virtue of Theorem 1.7, e^{-tL_A} is a contracting operator from $L_2(\Phi', \gamma_1)$ to $L_p(\Phi', \gamma_1)$. The polynomial f can be expanded into a sum of the Wick monomials

$$f = \sum_{k=0}^{n} f_k, \quad f_k \in \Gamma_k(\Phi', \gamma_1), \quad k = 0, \dots, n$$

(see Subsection 2.2 in Chapter 2). Moreover, the equality $\Gamma_k(\Phi', \gamma_1) = I_1(\mathfrak{F}_k(H_0))$ and the definition of L_A give

$$e^{-L_A} f_k = e^{-ka} f_k, \quad k = 0, \dots, n.$$

Thus, for $g = \sum_{k=0}^{n} e^{ka} f_k$, we have $e^{-L_A} g = f$ and, therefore,

$$\|f\|_{L_p(\Phi',\gamma_1)} = \|e^{-L_A} g\|_{L_p(\Phi',\gamma_1)} \leq \|g\|_{L_2(\Phi',\gamma_1)}$$

$$= \left(\sum_{k=0}^{n} e^{2ka} \|f_k\|^2_{L_2(\Phi',\gamma_1)}\right)^{1/2} \leq e^{na} \left(\sum_{k=0}^{n} \|f_k\|^2_{L_2(\Phi',\gamma_1)}\right)^{1/2}$$

$$= e^{na} \|f\|_{L_2(\Phi',\gamma_1)} = (p-1)^{-n/2} \|f\|_{L_2(\Phi',\gamma_1)}.$$

■

If the coefficient operator A ensures the uniform ellipticity of L_A, then the following strong version of Theorem 1.5 holds.

Theorem 1.8. *Let $A \geq a1$, $a > 0$. Then the semigroup e^{-tL_A}, $t > 0$, amplifies positivity, i.e., for any $f \in L_2(\Phi', \gamma_1)$ such that $f \geq 0$ γ_1-almost everywhere and $\gamma_1(\{x \in \Phi' \mid f(x) > 0\}) > 0$, we have $e^{-tL_A} f > 0$ γ_1-almost everywhere.*

Proof. It suffices to show that, for any $g \in L_2(\Phi', \gamma_1)$ which are nonnegative and not identically zero, the following inequality holds

$$(f, e^{-tL_A} g)_{L_2(\Phi',\gamma_1)} > 0.$$

The function

$$h(z) = (f, e^{-zL_A} g)_{L_2(\Phi',\gamma_1)}$$

is analytical on the half-plane $\operatorname{Re} z \geq 0$ and is not identically zero, because

$$h(t) \xrightarrow[t \to +\infty]{} (f, 1)_{L_2(\Phi', \gamma_1)} (g, 1)_{L_2(\Phi', \gamma_1)}.$$

Hence, for any $t > 0$, there exists $t_0 \in (0, t)$ such that

$$h(t_0) = \left(e^{-(t_0/2)L_A} f, e^{-(t_0/2)L_A} g\right)_{L_2(\Phi', \gamma_1)} > 0. \tag{1.39}$$

Denote $f_0 = e^{-(t_0/2)L_A} f$ and $g_0 = e^{-(t_0/2)L_A} g$. The functions f_0 and g_0 are nonnegative and their product is not identically zero according to (1.39). Therefore, $u = \min (f_0, g_0)$ is not identically zero. Thus, by using Theorem 1.5 for the operator $e^{-(t-t_0)L_A}$, we get

$$\begin{aligned}
(f, e^{-tL_A} g)_{L_2(\Phi', \gamma_1)} &= \left(f_0, e^{-(t-t_0)L_A} g_0\right)_{L_2(\Phi', \gamma_1)} \\
&\geq \left(u, e^{-(t-t_0)L_A} g\right)_{L_2(\Phi', \gamma_1)} \geq \left(u, e^{-(t-t_0)L_A} u\right)_{L_2(\Phi', \gamma_1)} \\
&> \left\| e^{-\frac{t-t_0}{2} L_A} u \right\|^2_{L_2(\Phi', \gamma_1)} \geq \left\| e^{-\frac{t-t_0}{2} L_A} u \right\|^2_{L_1(\Phi', \gamma_1)} \geq \| u \|^2_{L_1(\Phi', \gamma_1)} > 0.
\end{aligned}$$

Here we have used the isometricity of e^{-tL_A}, $t \geq 0$, on nonnegative functions in $L_1(\Phi', \gamma_1)$ which is mentioned in the proof of Theorem 1.6.

■

1.5. Functional Integrals Corresponding to the Second Quantization Operators

Denote by

$$\Omega(\mathbb{R}^1, \Phi') = (\Phi')^{\mathbb{R}^1} = \{\omega(\cdot) \mid \mathbb{R}^1 \ni t \mapsto \omega(t) \in \Phi'\}$$

a set of all mappings of $\mathbb{R}^1$ to Φ' which will be also called a set of trajectories with values in Φ'. In the cases when this does not lead to misunderstanding, we denote $\Omega = \Omega(\mathbb{R}^1, \Phi')$ for brevity. For fixed $t_1 < t_2 < \ldots < t_n$, $n \in \mathbb{N}$, we introduce a σ-algebra $\mathfrak{C}^{t_1, \ldots, t_n}(\Omega)$ consisting of cylindrical subsets of Ω of the form

$$\{\omega(\cdot) \in \Omega \mid \omega(t_1), \ldots, \omega(t_n) \in \alpha\},$$

where

$$\alpha \in \mathfrak{C}_\sigma(\Phi') \times \ldots \times \mathfrak{C}_\sigma(\Phi') = \mathfrak{C}_\sigma(\Phi' \times \ldots \times \Phi').$$

A family of sets

$$\mathfrak{C}(\Omega) = \bigcup \mathfrak{C}^{t_1, \dots, t_n}(\Omega)$$

(the union is taken over all possible collections $t_1, \dots, t_n$ of the indicated form) forms an algebra; its σ-span is denoted by $\mathfrak{C}_\sigma(\Omega)$ and is called a σ-algebra generated by cylindrical sets.

Each measure $\gamma_{A_t, x}$ $(t > 0, x \in \Phi')$ which takes part in the representation of the semigroup e^{-tL_A} by the formula (1.10) possesses the properties of a stochastic kernel on $(\Phi', \mathfrak{C}(\Phi'))$:

(i) for any $t > 0$ and $x \in \Phi'$, $\gamma_{A_t, x}$ is a measure on $\mathfrak{C}_\sigma(\Phi')$;

(ii) for any $\alpha \in \mathfrak{C}_\sigma(\Phi')$,

$$\gamma_{A_t, x}(\alpha) = (e^{-tL_A}\kappa_\alpha)(x) \quad (x \in \Phi') \tag{1.40}$$

is a measurable function on Φ';

(iii) for any $t > 0$ and $x \in \Phi'$, we have $\gamma_{A_t, x}(\Phi') = 1$.

The facts concerning stochastic kernels and the methods of constructing Markov processes with the help of these kernels, which are frequently used below, can be found in the standard courses on the theory of random processes (for example, see Gikhman and Skorokhod [1, Chapter 3, Section 5; Chapter 7, Section 1]).

For the collection $t_0 < t_1 < \dots < t_n$, we define a stochastic kernel $\nu_{\omega(t_0)}^{t_1, \dots, t_n}$ on $(\Phi', \mathfrak{C}^{t_1, \dots, t_n}(\Omega))$ which is called a direct product of the kernels $\gamma_{A_{t_{k+1}-t_k}, \omega(t_k)}(\omega(t_{k+1}))_{k=0}^{n-1}$. Namely, for the set

$$\beta = \{\omega(\cdot) \in \Omega \,|\, \omega(t_k) \in \alpha_k, \quad \alpha_k \in \mathfrak{C}_\sigma(\Phi'), \quad k = 1, \dots, n\},$$

we put

$$\nu_{\omega(t_0)}^{t_1, \dots, t_n}(\beta) = \int_{\alpha_1} d\gamma_{A_{t_1 - t_0}, \omega(t_0)}(\omega(t_1)) \int_{\alpha_2} \dots \int_{\alpha_n} d\gamma_{A_{t_n - t_{n-1}}, \omega(t_n - t_{n-1})}(\omega(t_n)) \tag{1.41}$$

(the integration is carried out from right to left). The standard fact in the theory of Markov processes states that $\nu_{\omega(t_0)}^{t_1, \dots, t_n}$ is a stochastic kernel on $(\Phi', \mathfrak{C}^{t_1, \dots, t_n}(\Omega))$, i.e., that the properties (1.40) hold for it with $\mathfrak{C}_\sigma(\Phi')$ being replaced by $\mathfrak{C}^{t_1, \dots, t_n}(\Omega)$.

Given stochastic kernel $\nu_{\omega(t_0)}^{t_1, \dots, t_n}$, we construct a probability measure $\nu^{t_0, \dots, t_n}$ on $\mathfrak{C}^{t_0, \dots, t_n}(\Omega)$ by setting

$$\nu^{t_0,\dots,t_n}(\delta) = \int_{\beta_0} \nu_{\omega(t_0)}^{t_1,\dots,t_n}(\beta)\, d\gamma_1(\omega(t_0)) \tag{1.42}$$

on a set $\delta = \beta_0 \times \beta$, $\beta_0 \in \mathfrak{C}^{t_0}(\Omega)$, $\beta \in \mathfrak{C}^{t_1,\dots,t_n}(\Omega)$. Hence, for any $t_0 < t_1 < \dots < t_n$, the measures $\nu^{t_0,\dots,t_n}$ are defined on the σ-algebra $\mathfrak{C}^{t_0,\dots,t_n}(\Omega)$. By construction, these measures form a consistent family of finite-dimensional distributions; this immediately follows from the representation

$$\nu^{t_0,\dots,t_n}\left(\prod_{k=0}^{n} \alpha_k\right) = \int_{\Phi'} \kappa_{\alpha_0}(e^{-(t_1-t_0)L_A}(\kappa_{\alpha_1} \dots (e^{-(t_n-t_{n-1})L_A}\kappa_{\alpha_n}) \dots)\, d\gamma_1 \tag{1.43}$$

$$(\alpha_k \in \mathfrak{C}_\sigma(\Phi'), \quad k = 0, 1, \dots, n, \quad n \in \mathbb{N}),$$

which, in turn, can be obtained from (1.41) and (1.43) if we take into account that $(e^{-tL_A}1)(\cdot) = 1(\cdot)$. In order to show that the Kolmogorov theorem on the extension of a consistent family of finite-dimensional distributions to a measure can be applied in this case, we consider the one-dimensional distributions ν^t, $t \in \mathbb{R}^1$. According to the definition and Lemma 1.2 ($t > t_0$, $\alpha \in \mathfrak{C}_\sigma(\Phi')$), we have

$$\nu^t(\alpha) = \nu^{t_0,t}(\Phi' \times \alpha) = \int_{\Phi'} \gamma_{A_{t-t_0},\,\omega(t_0)}(\alpha)\, d\gamma_1(\omega(s_0)) = \gamma_1(\alpha),$$

i.e., $\nu^t = \gamma_1$. The measure γ_1 on Φ' is concentrated on a certain Hilbert space $H_{-\tau}$, $\tau \in T$ and, consequently, it can be approximated by the values on compacts in $H_{-\tau}$. By virtue of the Kolmogorov theorem, there exists a unique measure ν_A on $(\Omega, \mathfrak{C}_\sigma(\Omega))$ which extends the given family of finite-dimensional distributions. The connection of this measure with the semigroup e^{-tL_A}, $t \geq 0$ follows from (1.41), (1.43), and formula (1.10).

Lemma 1.3. *For* $t_0 \in \mathbb{R}^1$, $t_1, \dots, t_n \geq 0$, *and*

$$f_0, \dots, f_n \in \bigcap_{p=1}^{\infty} L_p(\Phi', \gamma_1),$$

we have

$$\int_{\Phi'} f_0(x) e^{-t_1 L_A} f_1(x) \dots e^{-t_n L_A} f_n(x)\, d\gamma_1(x)$$

$$= \int_{\Omega} f_0(\omega(t_0)) f_1(\omega(t_0 + t_1)) \dots f_n(\omega(t_0 + t_1 + \dots + t_n))\, d\nu_A(\omega(\cdot)). \tag{1.44}$$

■

***Corollary* 1.** *The measure* ν_A *is invariant with respect to shifts the along the trajectories.*

Proof. For every $t \in \mathbb{R}^1$, a shift along the trajectories $\omega(\cdot) \mapsto \omega(\cdot + t)$ determines a shift of the set $T_t\beta = \{\omega(\cdot + t), \omega \in \beta\}$, $\beta \in \mathfrak{C}_\sigma(\Omega)$. It follows directly from (1.44) that $\gamma_A(T_t\beta) = \nu_A(\beta)$ for

$$\beta = \{\omega \in \Omega \,|\, \omega(t_k) \in \alpha_k, \quad \alpha_k \in \mathfrak{C}_\sigma(\Phi'), \quad k = 1, \ldots, n\} \in \mathfrak{C}^{t_1, \ldots, t_n}(\Omega)$$

and, hence, $\nu_A(T_t\beta) = \nu_A(\beta)$ for any $\beta \in \beta \in \mathfrak{C}_\sigma(\Omega)$.

■

Note that, in the special case, (1.44) yields the formula

$$(f_0, e^{-t_1 L_A} f_1)_{L_2(\Phi', \gamma_1)} = \int_\Omega f_0(\omega(t_0)) f_1(\omega(t_0 + t_1)) \, d\nu_A(\omega(\cdot)), \tag{1.45}$$

which gives the representation of the semigroup e^{-tL_A}, $t \geq 0$ in a weak form. This is why the corresponding functional integral is called weak.

Let us establish the connection between the functions on Ω and Φ'. For an arbitrary $t \in \mathbb{R}^1$ in $L_2(\Omega, \mathfrak{C}_\sigma(\Omega), \nu_A)$, we consider a closed subspace consisting of functions on Ω which are measurable with respect to the σ-algebra $\mathfrak{C}^t(\Omega) \subset \mathfrak{C}_\sigma(\Omega)$; this subspace coincides with $L_2(\Omega, \mathfrak{C}^t(\Omega), \nu_A \upharpoonright \mathfrak{C}^t(\Omega))$.

The operator of projection onto $L_2(\Omega, \mathfrak{C}^t(\Omega), \nu_A)$ in $L_2(\Omega, \mathfrak{C}_\sigma(\Omega), \nu_A)$ is denoted by E_t. This operator is called an operator of conditional mathematical expectation with respect to the σ-algebra $\mathfrak{C}^t(\Omega)$. As mentioned above, $\nu_A \upharpoonright \mathfrak{C}^t(\Omega) \equiv \nu^t = \gamma_1$ (the σ-algebra $\mathfrak{C}^t(\Omega)$ is isomorphic to $\mathfrak{C}_\sigma(\Phi')$). Therefore, we can define a natural isometry

$$I_t : L_2(\Phi', \gamma_1) \to L_2(\Omega, \mathfrak{C}^t(\Omega), \nu_A) \subset L_2(\Omega, \mathfrak{C}_\sigma(\Omega), \nu_A):$$

$$L_2(\Phi', \gamma_1) \ni f(x) \mapsto (I_t f)(\omega(\cdot)) = f(\omega(t)) \in L_2(\Omega, \mathfrak{C}^t(\Omega), \nu_A). \tag{1.46}$$

Due to the invariance of the measure ν_A with respect to shifts along the trajectories, the operators

$$U_t : f(\omega(\cdot)) \mapsto f(\omega(\cdot + t))$$

form a strongly continuous unitary group in $L_2(\Omega, \mathfrak{C}_\sigma(\Omega), \nu_A) = L_2(\Omega, \nu_A)$. It follows from (1.45) that

$$(e^{-tL_A}f)(\omega(0)) = E_0 f(\omega(t)) = E_0 U_t f(\omega(0)), \tag{1.47}$$

i.e., $e^{-tL_A} = E_0 U_t E_0$ if e^{-tL_A} is regarded in the space

$$L_2(\Omega, \mathfrak{C}^0(\Omega), \nu_A) = I_0 L_2(\Phi', \gamma_1).$$

Henceforth, we identify $L_2(\Phi', \gamma_1)$ with $L_2(\Omega, \mathfrak{C}^0\Omega), \nu_A)$ without special remarks.

Theorem 1.9. *Let $f \in L_2(\Phi', \gamma_1)$. Then for any $t < s$, a function of the form $\int_t^s f(\omega(\tau))\,d\tau$ is defined on Ω. Moreover, the integral is understood as a limit of the Riemann sums*

$$\frac{s-t}{n}\sum_{k=1}^{n} f\left(\omega\left(\frac{s-t}{n}k + t\right)\right)$$

in the sense of the strong convergence in $L_2(\Omega, \nu_A)$.

Proof. For $f \in L_2(\Phi', \gamma_1)$, we have

$$f(\omega(t)) = U_t f = U_t f(\omega, (0));$$

therefore, $f(\omega(t)) \in L_2(\Omega, \nu_A)$ and this function is strongly continuous in $t \in \mathbb{R}^1$. The statement of the theorem now follows from the properties of the Bochner integral.

■

The lemmas presented below are often used in what follows in estimates of the functional integrals.

Lemma 1.4. *Let $f \in L_2(\Phi', \gamma_1)$ and $\exp(-f) \in L_p(\Phi', \gamma_1)$ for any $p \geq 1$. Then, for any $t > 0$, we have*

(i) $$\exp\left(-\int_0^t f(\omega(\tau))\,d\tau\right) \in L_p(\Omega, \nu_A) \quad \textit{for any} \quad p \geq 1;$$

(ii) $$\left\|\exp\left(-\int_0^t f(\omega(\tau))\,d\tau\right)\right\|_{L_1(\Omega,\nu_A)} \leq \|\exp(-tf)\|_{L_1(\Phi',\gamma_1)}. \tag{1.48}$$

Proof. First, let $L_\infty(\Phi', \gamma_1)$. The Hölder inequality yields

$$\| e^{-t/n\sum_{k=1}^{n} f(\omega(t_k/n))} \|_{L_1(\Omega,\nu_A)} \leq \prod_{k=1}^{n} \| e^{-tf(\omega(t_k/n))} \|_{L_1(\Omega,\nu_A)}^{1/n}. \tag{1.49}$$

According to the definition of I_τ and (1.45), this mapping is an isometry between $L_1(\Phi', \gamma_1)$ and $L_1(\Omega, \nu_A)$ on nonnegative functions. Therefore, the right-hand side in (1.49) is equal to

$$\prod_{k=1}^{n} \| e^{-tf} \|_{L_1(\Omega,\nu_A)}^{1/n} = \| e^{-tf} \|_{L_1(\Phi',\gamma_1)}.$$

The general case is obtained from that considered above by approximation.

■

Lemma 1.5. *If* $A \geq a1$, $a > 0$, *then*

$$\| f(\omega(0)) g(\omega(t)) \|_{L_r(\Omega,\nu_A)} \leq \| f \|_{L_p(\Phi',\gamma_1)} \| g \|_{L_q(\Phi',\gamma_1)} \tag{1.50}$$

for any $t > 0$, *where* $e^{-2ta} \leq (p/r - 1)(q/r - 1)$.

Proof. For $1/p_1 + 1/p_1' = 1$ $(p_1 > 1)$, taking into account (1.45) and the hypercontractibility estimates (Theorem 1.7), where $s > 1$ is such that

$$e^{-2ta} \leq \frac{s-1}{p_1' - 1} (s-1)(p_1 - 1),$$

we obtain

$$\| f(\omega(0)) g(\omega(t)) \|_{L_r(\Omega,\nu_A)} = \left\{ \int_{\Phi'} |f|^r e^{-tL_A} |g|^r d\gamma_1 \right\}^{1/r}$$

$$\leq \left\{ \int_{\Phi'} |f|^{p_1 r} d\gamma_1 \right\}^{1/p_1 r} \left\{ \int_{\Phi'} | e^{-tL_A} |g|^r |^{p_1'} d\gamma_1 \right\}^{1/p_1' r}$$

$$= \| f \|_{L_{p_1 r}(\Phi',\gamma_1)} \| e^{-tL_A} |g|^r \|_{L_{p_1'}(\Phi',\gamma_1)}^{1/r} \leq \| f \|_{L_{p_1 r}(\Phi',\gamma_1)} \| |g|^r \|_{L_s(\Phi',\gamma_1)}^{1/r}$$

$$= \| f \|_{L_{p_1 r}(\Phi',\gamma_1)} \| g \|_{L_{sr}(\Phi',\gamma_1)}.$$

The last condition for $p = p_1 r$ and $q = sr$ takes the form $e^{-2ta} \leq (p/r - 1)(q/r - 1)$.

■

To investigate the L_p-properties of potential perturbations of the second quantization operators, it suffices to have the weak functional integral constructed above. However, to study more complicated problems, we need a representation of the action of semi-group, which is obtained by using the strong functional integral constructed below. In order to realize the corresponding construction, it is necessary to show that the family of stochastic kernels $\gamma_{A_t,x}$, $x \in \Phi'$, $t > 0$, possesses two additional properties.

Lemma 1.6. *In addition to properties (i)–(iii) formulated above, the kernels* $\gamma_{A_t,x}$, $x \in \Phi'$, $t > 0$, *satisfy the following conditions:*

(iv) for any $x \in \Phi'$, *we have*

$$\lim_{t \to 0} \gamma_{A_t,x} = \delta_x \tag{1.51}$$

in the sense of weak convergence;

(v) the Kolmogorov–Chapman condition holds, i.e.,

$$\gamma_{A_{t+s},x}(\alpha) = \int_{\Phi'} \gamma_{A_t,y}(\alpha)\, d\gamma_{A_s,x}(y) \tag{1.52}$$

for any $\alpha \in \mathcal{C}_\sigma(\Phi')$, $t, s > 0$, $x \in \Phi'$.

Proof. Property (iv) follows immediately from the fact that the Fourier transform of the measure $\gamma_{A_t,x}$ is equal to

$$\exp\{i(e^{-tA}\varphi, x)_{H_0} - 1/4(\varphi, (1 - e^{-2tA})\varphi)_{H_0}\} \xrightarrow[t \to 0]{} \exp\{i(\varphi, x)_{H_0}\} \quad (\varphi \in \Phi)$$

(see (1.7)), and this is just the Fourier transform of the δ-measure on Φ' concentrated at the point $x \in \Phi'$, and the family of Fourier transforms of the measures $\gamma_{A_t,x}$ is equi-continuous at the origin of the space Φ.

To prove the equality (1.52), we introduce a set function

$$\rho(\alpha) = \int_{\Phi'} \gamma_{A_t,y}(\alpha)\, d\gamma_{A_s,x}(y) \quad (\alpha \in \mathcal{C}_\sigma(\Phi'))$$

on $\mathcal{C}_\sigma(\Phi')$ (for fixed $t, s > 0$ and $x \in \Phi'$). It is easy to see that ρ is a probability measure on $\mathcal{C}_\sigma(\Phi')$. Indeed, the properties of finite additivity and normalization are evident. Let $\alpha_1 \supset \alpha_2 \supset \ldots \supset \alpha_n \supset \ldots$ be a decreasing sequence of sets from $\mathcal{C}_\sigma(\Phi')$ and $\bigcap_{n=1}^{\infty} \alpha_n = \varnothing$. Then, for any $y \in \Phi'$, we have $\gamma_{A_t,y}(\alpha_n) \to 0$ as $n \to \infty$, because the measure $\gamma_{A_t,y}$ is upper semi-continuous. By virtue of the Lebesgue theorem,

$$\rho(\alpha_n) = \int_{\Phi'} \gamma_{A_t,y}(\alpha_n) d\gamma_{A_s,x}(y) \to 0 \quad n \to \infty,$$

i.e., ρ is upper semi-continuous on the empty set and, consequently, it is σ-additive.

Let us show that $\rho = \gamma_{A_{t+s},x}$. First, it follows from (1.6) that

$$\rho(\alpha) = \int_{\Phi'} \gamma_{A_t}(\alpha - e^{-tA}y)\,d\gamma_{A_s,x}(y) \quad (\alpha \in \mathfrak{C}_\sigma(\Phi')),$$

and this yields the equality

$$\int_{\Phi'} f(z)\,d\rho(z) = \int_{\Phi'\times\Phi'} f(z + e^{-tA}y)\,d\gamma_{A_t}(z)\,d\gamma_{A_s,x}(y) \tag{1.53}$$

for any bounded measurable f. In particular, for $f(z) = \exp\{i(\varphi, z)_{H_0}\}$ $(\varphi \in \Phi, z \in \Phi')$, it follows from (1.53) that

$$\int_{\Phi'} e^{i(\varphi,z)_{H_0}}\,d\rho(z) = \int_{\Phi'\times\Phi'} e^{i(\varphi,z)_{H_0}} e^{i(e^{-tA}\varphi,y)_{H_0}}\,d\gamma_{A_t}(z)\,d\gamma_{A_s,x}(y)$$

$$= \int_{\Phi'} e^{i(\varphi,z)_{H_0}}\,d\gamma_{A_t}(z) \int_{\Phi'} e^{i(e^{-tA}\varphi,y)_{H_0}}\,d\gamma_{A_s,x}(y).$$

Computing the integrals by use of (1.7), we obtain the Fourier transform of the measure ρ

$$\int_{\Phi'} e^{i(\varphi,z)_{H_0}}\,d\rho(z) = \exp\{i(e^{-(t+s)A}\varphi, x)_{H_0} - 1/4(\varphi, (1 - e^{-2(t+s)A})\varphi)_{H_0}\}$$

$$= \int_{\Phi'} e^{i(\varphi,z)_{H_0}}\,d\gamma_{A_{t+s},x}(z).$$

Since the Fourier transforms of the measures $\gamma_{A_{t+s},x}$ and ρ coincide everywhere on Φ, the measures also coincide on $\mathfrak{C}_\sigma(\Phi')$, i.e., $\gamma_{A_{t+s},x} = \rho$. Therefore, we have a family of stochastic kernels $\gamma_{A_t,x}$, $x \in \Phi'$, $t > 0$, satisfying conditions (i)–(v). Let us complete this family by the kernels $\gamma_{A_0,x} = \delta_x$, $x \in \Phi'$, for $t = 0$. Then the following condition holds:

(vi) $\gamma_{A_0,x}(\Phi' \setminus \{x\}) = 0.$

Denote by

$$\Omega^+ = \Omega(\mathbb{R}^1_+, \Phi') = (\Phi')^{\mathbb{R}^1_+}$$

a set of all mappings $\mathbb{R}^1_+ = [0, +\infty)$ in Φ':

$$\Omega^+ = \{\omega(\cdot) : \mathbb{R}^1_+ \ni t \mapsto \omega(t) \in \Phi'\}.$$

For $0 \le t_1 < t_2 < \ldots < t_n$, we define the σ-algebras of cylindrical sets $\mathfrak{C}^{t_1,\ldots,t_n}(\Omega^+)$ and the σ-algebra $\mathfrak{C}_\sigma(\Omega^+)$ generated by these σ-algebras. For fixed $0 \le t_1 < \ldots < t_n$, we denote by $\nu_x^{t_1,\ldots,t_n}$ a stochastic kernel on $(\Phi', \mathfrak{C}^{t_1,\ldots,t_n}(\Omega^+))$ which is a product of the kernels $\gamma_{A_{t_1},x}(\omega(t_1))$ and $\gamma_{A_{t_{k+1}-t_k},\omega(t_k)}(\omega(t_{k+1}))$, $k = 1, \ldots, n-1$, and is given by relation (1.41).

The next theorem is taken from the theory of Markov processes. Under restrictions of the topological character imposed on the phase space, this theory establishes the possibility of constructing a Markov process (conservative and normal) for a given family of stochastic kernels with properties (i)–(v) (for example, see Gikhman and Skorokhod [1, Chapter 7, Section 1]).

Theorem 1.10. *For any $x \in \Phi'$, a probability measure $\nu_{A,x}$ for which $\nu_x^{t_1,\ldots,t_n}$ are finite-dimensional distributions is defined on $\mathfrak{C}_\sigma(\Omega^+)$.*

Proof. We present only a sketch of the proof.

For every $x \in \Phi'$, $\nu_x^{t_1,\ldots,t_n}$ is a measure on $\mathfrak{C}_\sigma(\Omega^+)$. The consistency of this family follows from the Kolmogorov–Chapman equation (1.52).

The only difference in the proof is that one must verify whether the Kolmogorov theorem on extensions is applicable in this case. To do this, we establish that the one-dimensional distributions ν_x^t on the σ-algebra $\mathfrak{C}^t(\Omega^+)$ which is isomorphic to $\mathfrak{C}_\sigma(\Phi')$ admit an approximation of values on the sets $\alpha \in \mathfrak{C}_\sigma(\Phi')$ by values on compacts. However,

$$\nu_x^t(\beta) = \int_\beta d\gamma_{A_t,x}(\omega(t)) = \gamma_{A_t,x}(\beta)$$

for any $t > 0$ and $\beta \in \mathfrak{C}^t(\Omega^+)$, i.e., $\nu_x^t = \gamma_{A_t,x}$, and this is a Gaussian measure concentrated on a certain $H_{-\tau}$, $\tau \in T$ and, consequently, it admits the required approximation.

■

Remark 1. It follows from the construction of the measure $\nu_{A,x}$ that the set $\Omega_x = \{\omega \in \Omega^+ \mid \omega(0) = x\}$ of trajectories which begin at the point $x \in \Phi'$ has the full measure

$$\nu_{A,x}(\Omega_x) = \nu_x^0\{\omega \in \Omega^+ \mid \omega(0) = x\} = \gamma_{A_0,x}(\{x\}) = 1.$$

■

Lemma 1.7. *For all*

$$f_1, \dots, f_n \in \bigcap_{p=1}^{\infty} L_p(\Phi', \gamma_1)$$

and $0 < t_1 < \dots < t_n$, *the equality*

$$\int_{\Omega_x} f_1(\omega(t_1)) \dots f_n(\omega(t_n))\, d\nu_{A,x}(\omega(\cdot))$$

$$= e^{-t_1 L_A}\big(f_1 e^{-(t_2-t_1)L_A}(\dots (e^{-(t_n-t_{n-1})L_A} f_n) \dots)(x) \qquad (1.54)$$

holds for γ_1*-almost all* $x \in \Phi'$.

Proof. Due to the conditions imposed on f_k and the contracting properties of the semigroup, the right-hand side of (1.54) has the meaning of a function from $L_2(\Phi', \gamma_1)$. If we apply formula (1.10) n times, then we can rewrite this right-hand side in the form

$$\int_{\Phi'} \dots \int_{\Phi'} f_1(y_1) \dots f_n(y_n) d\gamma_{A_t, x}(y_1) \dots d\gamma_{A_{t_n - t_{n-1}}, y_{n-1}}(y_n),$$

but according to the construction of the measure $\gamma_{A,x}$, this is just the integral of a cylindrical function on Ω_x in the left-hand side of the equality (1.54). ∎

In the special case, (1.54) yields

$$(e^{-tL_A} f)(x) = \int_{\Omega_x} f(\omega(t))\, d\nu_{A,x}(\omega(\cdot)) \quad (t > 0) \qquad (1.55)$$

for γ_1-almost all $x \in \Phi'$. Using the relation

$$\gamma_{A_t, x}(\cdot) = \gamma_{A_t}(\cdot - e^{-tA} x) \quad (t \geq 0, \;\; x \in \Phi'),$$

we easily obtain

$$\nu_{A,x}(\beta) = \nu_{A,0}(\{\omega(\cdot) \mid \tilde{\omega} \in \beta, \;\; \omega(t) + e^{-tA} x = \tilde{\omega}(t)\}) \qquad (1.56)$$

for any $x \in \Phi'$ and $\beta \in \mathfrak{C}_\sigma(\Omega^+)$. This enables us to rewrite (1.55) in the form

$$(e^{-tL_A} f)(x) = \int_{\Omega_0} f(\omega(t) + e^{-tA} x) d\nu_{A,0}(\omega(\cdot)). \qquad (1.57)$$

This form of the strong functional integral generated by the second quantization operator L_A proves to be most efficient in applications (see Subsection 2.2).

Let us study the smoothness of the trajectories on which the measures $\nu_{A,x}$ are concentrated. Equality (1.56) enables us to reduce the investigation of the measures $\nu_{A,x}$, $x \in \Phi'$, to an investigation of the properties of $\nu_{A,0}$.

Denote by $\Omega_{-\tau} \subset \Omega^+$ a collection of continuous trajectories with values in $H_{-\tau}$, $\tau \in \mathrm{T}$, namely, $\Omega_{-\tau} = \mathfrak{C}(\mathbb{R}^1_+, H_{-\tau})$. Similarly, Ω_ω denotes a set of trajectories with values in Φ' continuous as a mapping of $\mathbb{R}^1_+$ onto Φ' with the weak topology $\sigma(\Phi', \Phi')$, see Subsection 1.2 in Chapter 1.

Theorem 1.11. *Let* $\nu^*_{A,0}$ *be an external measure generated by* $\nu_{A,0}$. *Then*

(*i*) *for* $\tau_1, \tau_2 \in \mathrm{T}$ *such that* $A \in \mathfrak{L}(H_{\tau_1}, H_0)$ *and* $H_{\tau_2} \subset H_{\tau_1}$ *with quasinuclear imbedding, we have*

$$\nu^*_{A,0}(\Omega_{-\tau_2}) = 1 \tag{1.58}$$

(*ii*) $\nu^*_{A,0}(\Omega_\omega) = 1.$ (1.59)

Proof. According to the Kolmogorov theorem on the continuity of trajectories (Gikhman and Skorokhod [2, Chapter 3, Theorem 6]), to prove (1.58), it suffices to show that the following inequality holds

$$\int\limits_{\Omega^+} \| \omega(t+h) - \omega(t) \|^{\delta}_{H_{-\tau_2}} d\nu_{A,0}(\omega(\cdot)) \le ch^{1+\varepsilon} \quad (0 \le t < t+h) \tag{1.60}$$

for some $\varepsilon > 0$ and $\delta > 0$. Using the conditions of the theorem and the fact that the measure $\nu_{A,0}$ is Gaussian, we can establish the estimate

$$\int\limits_{\Omega^+} \| \omega(t+h) - \omega(t) \|^{4}_{H_{-\tau_2}} d\nu_{A,0}(\omega(\cdot)) \le ch^2,$$

which means that (1.60) holds with $\delta = 4$ and $\varepsilon = 1$ and, consequently, (1.58) is valid. It remains only to prove this estimate.

Denote $\theta_h(t) = \omega(t+h) - \omega(t)$. It is easy to see that $\theta_h(t)$ has a Gaussian distribution with respect to $\nu_{A,0}$, i.e.,

$$\int\limits_{\Omega^+} e^{i(\varphi, \theta_h(t))_{H_0}} dd\nu_{A,0}(\omega(\cdot)) = e^{-1/4(D_t\varphi, \varphi)_{H_0}} \quad (\varphi \in \Phi),$$

where

$$D_t = (1 - e^{-hA})(2 - e^{-2tA} + e^{-(2t+h)A}) \in \mathfrak{L}(H_0, H_0)$$

(we have used (1.54) to compute the integral of cylindrical function

$$\exp\left[i(\varphi, \omega(t+h))_{H_0} - i(\varphi, \omega(t))_{H_0}\right]).$$

Therefore, the moments of the function $(\varphi, \theta_h(t))_{H_0}$, $\varphi \in \Phi$ can be immediately computed by formula (2.5) in Chapter 2, namely,

$$\int_{\Omega^+} (\varphi_1, \theta_h(t))_{H_0} \cdots (\varphi_{2k}, \theta_h(t))_{H_0} d\nu_{A,0}(\omega(\cdot))$$

$$= \frac{1}{2^k} \sum \prod_{p=1}^{k} (D_t \varphi_{k_p}, \varphi_{l_p})_0 \leq \frac{1}{2^k} \sum \prod_{p=1}^{k} \| D_t \varphi_{k_p} \|_{H_0} \| \varphi_{l_p} \|_{H_0}, \quad (1.61)$$

for any $\varphi_1, \ldots, \varphi_{2k} \in \Phi$, $k \in \mathbb{N}$, where summation is carried over all pairings of the numbers $(1, 2, \ldots, 2k)$. On the other hand, by the spectral theorem

$$\| D_t \varphi \|_{H_0}^2 = \int_0^{+\infty} (1 - e^{-h\lambda})^2 (2 - e^{-2\lambda t} + e^{-(2t+h)\lambda})^2 d(E_A(\lambda)\varphi, \varphi)_0$$

$$\leq c_1 h^2 \| A\varphi \|_{H_0}^2 \quad (\varphi \in \Phi),$$

where c_1 is an absolute constant and E_A is an RI of the operator A. This shows that the right-hand side of (1.61) is not greater than

$$c_2 \| A \|_{\mathfrak{L}(H_{\tau_1}, H_0)}^k \| A \|_{H_{\tau_1}} \cdots \| \varphi_{2k} \|_{H_{\tau_1}}$$

(c_2 is a constant dependent on k). Let $(\omega)_{n=1}^{\infty} \subset \Phi$ be an orthonormal basis in H_{τ_2}. Then

$$\int_{\Omega^+} \| \theta_h(t) \|_{H_{-\tau_2}}^{2k} d\nu_{A,0}(\omega(\cdot)) = \int_{\Omega^+} \sum_{k_1 + \ldots + k_\nu = k} (\varphi_1, \theta_h(t))_{H_0}^{2k_1}$$

$$\ldots (\varphi_\nu, \theta_h(t))_{H_0}^{2k_\nu} d\nu_{A,0}(\omega(\cdot))$$

$$\leq c_2 h^k \| A \|_{\mathfrak{L}(H_{\tau_1}, H_0)}^k \left(\sum_{k_1 + \ldots + k_\nu = k} \| \varphi_1 \|_{H_{\tau_1}}^{2k_1} \cdots \| \varphi_\nu \|_{H_{\tau_1}}^{2k_\nu} \right)$$

$$\leq c_2 h^k \| A \|^k_{\mathcal{L}(H_{\tau_1}, H_0)} \left(\sum_{n=1}^{\infty} \| \varphi_n \|^2_{H_{\tau_1}} \right)^k = c_3(k) | O_{\tau_2, \tau_1} |^{2k} h^k = c(k) h^k,$$

where $c > 0$ is independent of h. In particular, for $k = 2$, we get the required estimate.

Since $A \in \mathcal{L}(\Phi, \Phi)$ and Φ is nuclear, one can always find $\tau_1, \tau_2 \in \mathrm{T}$ satisfying conditions (1.58). Thus, for some $\tau_2 \in \mathrm{T}$, we have $\nu^*_{A,0}(\Omega_{-\tau_2}) = 1$ and, moreover, Ω_ω has full outer measure, because $\Omega_\omega \supset \Omega_{-\tau}$, $\tau \in \mathrm{T}$.

■

Exploiting the connection between the measures $\nu_{A,x}$ and $\nu_{A,0}$, we obtain the following assertions from the proved theorem.

Corollary 1. 1. *Given $x \in \Phi'$, let the mapping $\mathbb{R}^1_+ \ni t \mapsto e^{-tA}x \in \Phi'$ be continuous in the weak topology $\sigma(\Phi', \Phi')$ on Φ'. Then $\nu^*_{A,x}(\Omega_\omega) = 1$.*

2. *Assume that, for the given $x \in \Phi'$, one can find $\tau \in \mathrm{T}$ such that the mapping $\mathbb{R}^1_+ \ni t \mapsto e^{-tA}x \in H_{-\tau}$ is continuous. Then $\nu^*_{A,x}(\Omega_{-\tau'}) = 1$ for some $\tau' \in \mathrm{T}$.*

■

Remark 2. One can always guarantee that propositions (i) and (ii) hold for a given A by the proper choice of a nuclear space Φ.

■

Remark 3. For the restriction of the measure ν_A (corresponding to a weak functional integral) to Ω^+, we have the following obvious representation

$$\nu_A(\beta) = \int_{\Phi'} \nu_{A,x}(\beta)\, d\gamma_1(x) \quad (\beta \in \mathcal{C}_\sigma(\Omega^+)), \tag{1.62}$$

which yields an analogy of Theorem 1.11 for the measure ν_A.

■

By virtue of Remark 3, ν_A may be regarded as a measure on certain subsets in Ω consisting of smooth trajectories (in one sense or another). Conversely, sometimes it is convenient to interpret ν_A as a measure on trajectories which are generalized functions with respect to time. This can be illustrated by the following important example which describes the functional integral corresponding to the Hamiltonian of the free boson field (see Example 1.4).

Example 1.5. (free Euclidean field)

For fixed $m_0 > 0$, we consider the function

$$\exp\{-1/2((-\Delta + m_0^2)^{-1}\psi, \psi)_{L_2(\mathbb{R}^{d+1})}\} \quad (\psi \in \mathfrak{S}_{\mathrm{Re}}(\mathbb{R}^{d+1})) \tag{1.63}$$

on a linear Schwartz space $\mathfrak{S}_{\mathrm{Re}}(\mathbb{R}^{d+1})$; here Δ is a Laplace operator on $L_2(\mathbb{R}^{d+1})$. It is easy to see that this function is positive definite, continuous on $\mathfrak{S}'(\mathbb{R}^{d+1})$, and normalized. By virtue of the Minlos theorem (Theorem 1.11 in Chapter 2), a Gaussian measure ν_{m_0} is defined on $\mathfrak{S}'_{\mathrm{Re}}(\mathbb{R}^{d+1})$. For this measure, (1.63) is a Fourier transform, i.e., this is a Gaussian measure with the correlation operator $2(\Delta + m_0^2)^{-1}$ and mean value 0. According to (1.63), the measure ν_{m_0} may be regarded as a canonical Gaussian measure on $\mathfrak{S}'_{\mathrm{Re}}(\mathbb{R}^{d+1})$ corresponding to the Hilbert space $\mathcal{H}_{m_0}$ with the scalar product

$$(\psi_1, \psi_2)_{\mathcal{H}_{m_0}} = 2((-\Delta + m_0^2)^{-1}\psi_1, \psi_2)_{L_2(\mathbb{R}^{d+1})}. \tag{1.64}$$

Thus, $\mathcal{H}_{m_0}$ is isomorphic to $W^{-1}_{2,\mathrm{Re}}(\mathbb{R}^{d+1})$. As usual, here the duality of $\mathfrak{S}_{\mathrm{Re}}(\mathbb{R}^{d+1})$ and $\mathfrak{S}'_{\mathrm{Re}}(\mathbb{R}^{d+1})$ is established by the use of $L_{2,\mathrm{Re}}(\mathbb{R}^{d+1})$.

The measure ν_{m_0} on $\mathfrak{S}'_{\mathrm{Re}}(\mathbb{R}^{d+1})$ is associated with a generalized random process Ξ which establishes the correspondence between each test function $\psi \in \mathfrak{S}(\mathbb{R}^{d+1})$ and a linear measurable function

$$\Xi(\psi) = 2(\psi(\cdot), \xi(\cdot))_{L_2(\mathbb{R}^{d+1})} \quad (\xi \in \mathfrak{S}'_{\mathrm{Re}}(\mathbb{R}^{d+1}))$$

on the measurable space $(\mathfrak{S}'_{\mathrm{Re}}(\mathbb{R}^{d+1}), \mathfrak{B}(\mathfrak{S}'_{\mathrm{Re}}(\mathbb{R}^{d+1})))$. This generalized random process is called a free Euclidean field. Since the measure ν_{m_0} is Gaussian, it is uniquely determined by the relation which defines its correlation operator, namely,

$$\int_{\mathfrak{S}'_{\mathrm{Re}}} (\psi, \xi)^2_{L_2(\mathbb{R}^{d+1})} d\nu_{m_0}(\xi) = ((-\Delta + m_0^2)^{-1}\psi, \psi)_{L_2(\mathbb{R}^{d+1})};$$

this relation yields that $\Xi(\psi)$ can be extended to the whole of $\psi \in W_2^{-1}(\mathbb{R}^{d+1})$ as a linear measurable function. In particular, by setting $s = (s_0, \mathbf{s}) \in \mathbb{R}^1 \times \mathbb{R}^d$, we define $\Xi(\psi)$ for $\psi(s) = \delta_t(s_0)\varphi(\mathbf{s})$; $\varphi \in \mathfrak{S}(\mathbb{R}^d)$, $t \in \mathbb{R}^1$.

Let us introduce the mapping

$$j : \Omega(\mathbb{R}^1, \mathfrak{S}'_{\mathrm{Re}}(\mathbb{R}^d)) \to \mathfrak{S}'_{\mathrm{Re}}(\mathbb{R}^{d+1})$$

under which

$$(\omega(t), \varphi)_{L_2(\mathbb{R}^d)} \mapsto \Xi(\delta_t \otimes \varphi) = (\delta_t \otimes \varphi, \xi)_{L_2(\mathbb{R}^{d+1})},$$

$$(\varphi \in \mathfrak{S}(\mathbb{R}^d);\ \ t \in \mathbb{R}^1;\ \ \omega(\cdot) \in \Omega(\mathbb{R}^1, \mathfrak{S}'_{\mathrm{Re}}(\mathbb{R}^d))).$$

This mapping induces a measure Ω on $\nu^j_{m_0}$. Let us show that the measure $\nu^j_{m_0}$ (or, equivalently, ν_{m_0}) may be regarded as a measure corresponding to the Hamiltonian L_{m_0} of the free boson field with mass m_0. For this purpose, one should verify the validity of the relation (1.44) with $L_{A_{m_0}} \equiv L_{m_0}$ (as in Example 1.4, here A_{m_0} is a one-particle operator $\sqrt{-\Delta + m_0^2}$ in the space $L_2(\mathbb{R}^d)$. By virtue of the fact that the set of polynomials $\mathcal{P}_{\mathrm{cyl}}(\mathfrak{S}'_{\mathrm{Re}}(\mathbb{R}^d))$ is dense in $L_2(\mathfrak{S}'_{\mathrm{Re}}(\mathbb{R}^d), \mu_{m_0})$, it suffices to check (1.44) for $f_k(x) = (\varphi_k, x)_{L_2(\mathbb{R}^d)}$, $\varphi_k \in \mathfrak{S}_{\mathrm{Re}}(\mathbb{R}^d)$, $k = 0, \ldots, n$, and taking into account that the measures μ_{m_0} and ν_{m_0} are Gaussian, we can restrict ourselves to the case $n = 1$.

On the one side,

$$\int\limits_{\mathfrak{S}'_{\mathrm{Re}}(\mathbb{R}^d)} (\varphi_0, x)_{L_2(\mathbb{R}^d)} e^{-tL_{m_0}} (\varphi_1, x)_{L_2(\mathbb{R}^d)} d\mu_{m_0}(x)$$

$$= \int\limits_{\mathfrak{S}'_{\mathrm{Re}}(\mathbb{R}^d)} (\varphi_0, x)_{L_2(\mathbb{R}^d)} (e^{-tA_{m_0}}\varphi_1, x)_{L_2(\mathbb{R}^d)} d\mu_{m_0}(x)$$

$$= \frac{1}{2} (A_{m_0}^{-1} \varphi_0, e^{-tA_{m_0}}\varphi_1)_{L_2(\mathbb{R}^d)} = \frac{1}{2} \int\limits_{\mathbb{R}^d} \frac{\tilde{\varphi}_0(\boldsymbol{k})\tilde{\varphi}_1(\boldsymbol{k}) e^{-\sqrt{\boldsymbol{k}^2 + m_0^2}}}{\sqrt{\boldsymbol{k}^2 + m_0^2}} d\boldsymbol{k} \qquad (1.65)$$

for any $t > 0$ and $\varphi_0, \varphi_1 \in \mathfrak{S}_{\mathrm{Re}}(\mathbb{R}^d)$.. On the other side,

$$\int\limits_{\mathfrak{S}'_{\mathrm{Re}}(\mathbb{R}^{d+1})} (\xi, \delta_0 \otimes \varphi_0)_{L_2(\mathbb{R}^{d+1})} (\xi, \delta_t \otimes \varphi_1)_{L_2(\mathbb{R}^{d+1})} d\nu_{m_0}(\xi)$$

$$= \frac{1}{2} (\delta_0 \otimes \varphi_0, \delta_t \otimes \varphi_1)_{\mathcal{H}_{m_0}} = \frac{1}{2\pi} \int\limits_{\mathbb{R}^{d+1}} \frac{\tilde{\varphi}_0(\boldsymbol{k})\tilde{\varphi}_1(\boldsymbol{k}) e^{-itk_0}}{\boldsymbol{k}^2 + k_0^2 + m_0^2} dk_0 d\boldsymbol{k}. \qquad (1.66)$$

In the last expression, we integrate over the variable $k_0 \in \mathbb{R}^1$, taking into account the equality

$$\int\limits_{\mathbb{R}^1} \frac{e^{-itk_0}}{k_0^2 + a^2} dk_0 = \frac{\pi e^{-at}}{a} \qquad (a > 0).$$

As a result, we obtain (1.65) from (1.66), and this completes the proof.

■

§2. Perturbations of the Second Quantization Operators by Potentials

In this section, we study potential perturbations of the second quantization operators by potentials given by measurable functions with certain L_p-properties and establish the properties of the perturbed operators. Everywhere below we assume that the nuclear space Φ, which satisfies the conditions formulated in Subsection 1.1, is chosen for the coefficient operator $A > 0$.

Each real measurable function V on Φ' is associated with a selfadjoint operator of multiplication by V in $L_2(\Phi', \gamma_1)$ which will be denoted by the same symbol. In order that the sum $L_A + V$ be densely defined, it suffices to demand the validity of the inclusion $V \in L_2(\Phi', \gamma_1)$. In this case, $L_A + V$ is well defined, for example, on a dense set of smooth cylindrical functions $C^{\infty}_{b,\mathrm{cyl}}(\Phi')$. In Section 1, we have explained that the semigroup e^{-tL_A}, $t \geq 0$, is contracting in all $L_p(\Phi', \gamma_1)$, $p \geq 1$, and possesses the property of hypercontractibility for $A \geq a1$, $a > 0$. Therefore, we can apply to the operator L_A the general criteria of selfadjointness for potential perturbations of the generators of contracting semigroups in Subsection 1.5 of Chapter 5. As a result, we obtain the following theorem (see also Reed and Simon [2, Chapter 10, Section 9]).

Theorem 2.1. (i) *Assume that* $V \in L_2(\Phi', \gamma_1)$ *and is semibounded from below, i.e.,* ess inf $V > -\infty$. *Then* $L_A + V$ *is essentially selfadjoint on* $\mathfrak{D}(L_A) \cap L_\infty(\Phi', \gamma_1)$.

(ii) *Assume that* $V \in L_p(\Phi', \gamma_1)$ *with some* $p \geq 1$ *and is bounded from below. Then* $L_A + V$ *is essentially selfadjoint on* $\mathfrak{D}(L_A) \cap L_q(\Phi', \gamma_1)$, *where* $1/p + 1/q = 1/2$.

(iii) *Under the condition of the uniform ellipticity of* L_A $(A \geq a1, a > 0)$, *the presence of the inclusions* $V \in L_p(\Phi', \gamma_1)$ $(p > 2)$ *and* $e^{-V} \in L_q(\Phi', \gamma_1)$ *for any* $q \geq 1$, *yields the essential selfadjointness of* $L_A + V$ *on the region* $\mathfrak{D}(L_A) \cap \mathfrak{D}(V)$ *and semiboundedness from below.*

Remark 1. The closure (in $L_2(\Phi', \gamma_1)$) of the sum $L_A + V$ given on the domain of its essential selfadjointness is also denoted by $L_A + V$; this will not create uncertainties.

■

Remark 2. It follows from the approximation Theorem 1.14 in Chapter 5 that, under conditions (i) or (ii), the operators $L_A + V_k$, $V_k(\cdot) = \min(V_k(\cdot), k)$ $(k \in \mathbb{N})$, converge to $L_A + V$ in the strong resolvent sense.

■

Remark 3. In the case when condition (iii) holds, we set $V = V_+ - V_-$; $V_\pm \geq 0$, and

$$V_{\pm,k}(\cdot) = \min(V_{\pm,k}(\cdot), k) \quad (k \in \mathbb{N}).$$

Then

$$\exp[-t(L_A + V)] = \lim_{k\to\infty} \lim_{n\to\infty} \exp[-t(L_A + V_{+,k} - V_{-,n})] \quad (t \geq 0)$$

in the sense of the strong convergence in $L_2(\Phi', \gamma_1)$, as follows from the same approximation theorem.

■

Since L_A is a differential operator, it is interesting to clarify the conditions of the essential selfadjointness of $L_A + V$ on the domains $C^\infty_{b,\mathrm{cyl}}(\Phi')$ or $\mathcal{P}_{\mathrm{cyl}}(\Phi')$, because they consist of smooth cylindrical functions, and the operator L_A is given on them by the differential expression (1.25). In a certain sense, these domains are minimal dense domains for the definition of differential operators on Φ', and it is important to note that they are independent of the concrete potential under consideration. Below, we show that the essential selfadjointness of $L_A + V$ can be established on the indicated sets of smooth functions on Φ'.

2.1. Kato's Inequality and Its Consequences

The well-known Kato's inequality states that if a function f on $\mathbb{R}^d$ and its Laplacian Δf, in the sense of generalized functions, belong locally to $L_1(\mathbb{R}^d)$, then the following inequality for generalized functions takes place

$$-\Delta|f| \leq \mathrm{Re}[(\mathrm{sgn}\,\bar{f})(-\Delta f)], \tag{2.1}$$

where, in the ordinary sense,

$$(\mathrm{sgn}\,\bar{f})(x) = \begin{cases} \bar{f}(x)/|f(x)|, & f(x) \neq 0 \\ 0, & f(x) = 0. \end{cases} \tag{2.2}$$

Inequality (2.1) can be also established for infinite-dimensional elliptic operators L_A.

To do this, we first define the action of L_A in the sense of the proper space of generalized functions.

Consider the rigging

$$\mathcal{P}'(\Phi') \supset L_2(\Phi', \gamma_1) \supset \mathcal{P}(\Phi'),$$

in which $\mathcal{P}(\Phi')$ is a nuclear space of continuous polynomials on Φ' (the space of test functions), and $\mathcal{P}'(\Phi')$ is its dual space with respect to $L_2(\Phi', \gamma_1)$ (the space of generalized functions) (see Subsection 5.3 in Chapter 2). Recall that according to the choice of Φ, we have

$$L_A \in \mathcal{L}(\mathcal{P}(\Phi'), \mathcal{P}(\Phi')), \quad e^{-tL_A} \in \mathcal{L}(\mathcal{P}(\Phi'), \mathcal{P}(\Phi')), \quad t \geq 0. \tag{2.3}$$

Each function $f \in L_p(\Phi', \gamma_1)$, $p > 1$, determines a (regular) generalized function from $\mathcal{P}'(\Phi')$ given by its action upon the test functions, namely, $(f, u)_{L_2(\Phi', \gamma_1)}$ $(u \in \mathcal{P}(\Phi'))$. The continuity of this functional on $\mathcal{P}(\Phi')$ follows from the fact that convergence in $\mathcal{P}(\Phi')$ implies convergence in all $L_q(\Phi', \gamma_1)$, $q \geq 1$. Further, for $f \in L_p(\Phi', \gamma_1)$, $p > 1$, we can define $L_A f \in \mathcal{P}'(\Phi')$ by setting

$$(L_A f, u)_{L_2(\Phi', \gamma_1)} = (f, L_A u)_{L_2(\Phi', \gamma_1)} \quad (u \in \mathcal{P}(\Phi')). \tag{2.4}$$

We should check only the continuity of the functional defined by formula (2.4). Let $(u_j)_{j=1}^{\infty} \subset \mathcal{P}(\Phi')$ and $u_j \xrightarrow[j\to\infty]{} u$ in $\mathcal{P}(\Phi')$. Then, according to (2.3), $L_A u_j \to L_A u$ as $j \to \infty$ in $\mathcal{P}(\Phi')$ and, consequently, in all $L_q(\Phi', \gamma_1)$, $q \geq 1$. Therefore,

$$(L_A f, u_j)_{L_2(\Phi', \gamma_1)} = (f, L_A u_j)_{L_2(\Phi', \gamma_1)} \xrightarrow[j\to\infty]{} (f, L_A u)_{L_2(\Phi', \gamma_1)} = (L_A f, u)_{L_2(\Phi', \gamma_1)}.$$

Hence, for every $p > 1$, L_A can be regarded as a mapping

$$\mathcal{L}_{A,p} : L_p(\Phi', \gamma_1) \to \mathcal{P}'(\Phi') \tag{2.5}$$

which is continuous by virtue of (2.4).

Denote by $L_{A,p}^{\min}$ the minimal operator corresponding to the differential expression $\mathcal{L}_A$ in $L_p(\Phi', \gamma_1)$, i.e., the closure of $\mathcal{L}_A$ in $L_p(\Phi', \gamma_1)$ from the set $\mathcal{P}(\Phi')$. The maximal operator is defined as a restriction of (2.5) to the set $\mathfrak{D}(L_{A,p}^{\max})$ of those $f \in L_p(\Phi', \gamma_1)$ for which L_A belongs to $L_p(\Phi', \gamma_1)$ (in the sense of generalized functions).

Lemma 2.1. *For every* $p > 1$, *we have*

$$L_{A,p}^{\min} = L_{A,p}^{\max}$$

Proof. Let $p > 1$ be given. The semigroup e^{-tL_A}, $t \geq 0$, is L_p-contracting and, consequently, after being restricted or extended by continuity, it generates a semigroup of contractions $e^{-tL_{A,p}}$, $t \geq 0$, in $L_p(\Phi', \gamma_1)$ with the generator $L_{A,p}$. The operator $L_{A,p}$ is closed and coincides with $L_{A,p}^{\min}$ on $\mathcal{P}(\Phi')$; therefore, $L_{A,p}^{\min} \subset L_{A,p}$. But

$$e^{-tL_{A,p}} \in \mathcal{L}(\mathcal{P}(\Phi'), \mathcal{P}(\Phi'))$$

and $\mathcal{P}(\Phi') \subset \mathfrak{D}(L_{A,p}^{\min})$; according to the general theorem in the theory of semigroups, this yields the equality $L_{A,p}^{\min} = L_{A,p}$.

We now assume that $f \in \mathfrak{D}(L_{A,p}^{\max})$. Let us show that $f \in \mathfrak{D}(L_{A,p'}^{*})$, where p' is an index dual to p, i.e., $1/p + 1/p' = 1$. Indeed, according to what has been said above, $L_{A,p'}^{\min} = L_{A,p'}$ and, therefore, for $g \in \mathfrak{D}(L_{A,p'})$, one can choose $(g_j)_{j=1}^{\infty} \subset \mathcal{P}(\Phi')$ such that

$$g_j \to g \quad \text{and} \quad L_{A,p'} g_j \to L_{A,p'} g \quad \text{as} \quad j \to \infty$$

in $L_p(\Phi', \gamma_1)$. Denote $h = L_{A,p}^{\max} f \in L_p(\Phi', \gamma_1)$. Then

$$(h, g)_{L_2(\Phi', \gamma_1)} = \lim_{j \to \infty} (L_A, f, g_j)_{L_2(\Phi', \gamma_1)} = (f, L_{A,p'} g)_{L_2(\Phi', \gamma_1)}. \tag{2.6}$$

Equality (2.6) shows that $f \in \mathfrak{D}(L_{A,p'}^{*})$. But the relation

$$[\exp(-tL_{A,p'})]^* = \exp(-tL_{A,p})$$

for semigroups implies the equality $L_{A,p'}^{*} = L_{A,p}$ for their generators. Thus,

$$f \in \mathfrak{D}(L_{A,p}) = \mathfrak{D}(L_{A,p}^{\min}).$$

■

The next theorem shows that the inequality analogous to (2.1) holds for the operator L_A. We immediately note that, unlike the finite-dimensional case, the absence of smooth finite functions on the space Φ' (see Subsection 4.3 in Chapter 2) leads to the non-local requirements in the formulation of this theorem.

Theorem 2.2. *Assume that $f \in L_p(\Phi', \gamma_1)$ for some $p > 1$ and $L_A f \in L_p(\Phi', \gamma_1)$, where the action of L_A is understood in the sense of generalized functions. Then the following inequality for generalized functions from $\mathcal{P}'(\Phi')$ holds*

$$L_A|f| \leq \operatorname{Re}[(\operatorname{sgn}\bar{f}) L_A f]. \tag{2.7}$$

Proof. First, we prove (2.7) for $u \in \mathcal{P}(\Phi')$. For $\varepsilon > 0$, we introduce a function

$$u_\varepsilon(x) = (|u(x)|^2 + \varepsilon^2)^{1/2} \quad (x \in \Phi').$$

Differentiating the equality

$$u_\varepsilon^2(x) = |u(x)|^2 + \varepsilon^2,$$

we obtain

$$u_\varepsilon \nabla u_\varepsilon = \mathrm{Re}\,(\bar{u}\,\nabla u),$$

whence

$$(A\nabla u_\varepsilon, \nabla u_\varepsilon)_{H_0} u_\varepsilon^2 = (\mathrm{Re}\,\bar{u}\,\nabla u, A\,\mathrm{Re}\,\bar{u}\,\nabla u)_{H_0},$$

i.e.,

$$(A\nabla u_\varepsilon, \nabla u_\varepsilon)_{H_0} = \left|\frac{\bar{u}}{u_\varepsilon}\right|^2 (A\nabla u, \nabla \bar{u})_{H_0} \le (A\nabla u, \nabla \bar{u})_{H_0}, \tag{2.8}$$

because $|u| \le u_\varepsilon$. Applying L_A to both parts of the same equality and taking into account the explicit form of this operator on smooth functions, we obtain

$$u_\varepsilon L_A u_\varepsilon - (A\nabla u_\varepsilon, \nabla u_\varepsilon)_{H_0} = 2\,\mathrm{Re}\,\bar{u}\,L_A u - (A\nabla u, \nabla \bar{u})_{H_0}.$$

By virtue of (2.8), this yields $u_\varepsilon L_A u_\varepsilon \le \mathrm{Re}\,(\bar{u}\,L_A u)$ or

$$L_A u_\varepsilon \le \mathrm{Re}\,[(\bar{u}\,/\,u_\varepsilon) L_A u]. \tag{2.9}$$

Let $\varepsilon \to \infty$. Taking into account that $u_\varepsilon \to |u|$ in all $L_p(\Phi', \gamma_1)$, $p \ge 1$, that $\bar{u}\,/\,u_\varepsilon \to$ sgn $\bar{u}$ pointwise, and $|\bar{u}\,/\,u_\varepsilon| \le 1$, we get $L_A u_\varepsilon \to L_A u$ in $\mathcal{P}'(\Phi')$ and

$$(\bar{u}\,/\,u_\varepsilon) L_A u \to (\mathrm{sgn}\,\bar{u}) L_A u$$

in $\mathcal{P}'(\Phi')$; hence, (2.7) is proved for all $u \in \mathcal{P}(\Phi')$.

According to the conditions of the theorem, we have $f \in \mathfrak{D}(L_{A,p}^{\max})$. Applying Lemma 2.1, we choose $(u_j)_{j=1}^\infty \subset \mathcal{P}(\Phi')$ such that $u_j \to f$ as $j \to \infty$ in $L_p(\Phi', \gamma_1)$ and $L_A u_j \to L_{A,p} u = L_A f$ as $j \to \infty$ in $L_p(\Phi', \gamma_1)$. For $j \in \mathbb{N}$, $\varepsilon > 0$, we have the inequality

$$L_A u_{j,\varepsilon} \le \mathrm{Re}\,[(\bar{u}_j / u_{j,\varepsilon}) L_A u_j]. \tag{2.10}$$

Choosing a subsequence $(u_{j_k})_{k=1}^{\infty}$ such that $u_{j_k} \xrightarrow[k\to\infty]{} f$ γ_1-almost everywhere, we obtain $u_{j_k,\varepsilon} \xrightarrow[k\to\infty]{} f_\varepsilon$ γ_1-almost everywhere and, thus, taking into account that $\left| \bar{u}_{j_k} u_{j_k,\varepsilon}^{-1} \right| \le 1$, we get

$$\bar{u}_{j_k} / u_{j_k,\varepsilon} \to \bar{f} / f_\varepsilon, \quad k \to \infty$$

in $L_\infty(\Phi', \gamma_1)$ and, consequently,

$$(\bar{u}_{j_k} / u_{j_k,\varepsilon}) L_A u_{j_k} \to (\bar{f} / f_\varepsilon) L_A f$$

in $L_p(\Phi', \gamma_1)$. Further, $u_{j_k,\varepsilon} \xrightarrow[k\to\infty]{} f_\varepsilon$ in $L_p(\Phi', \gamma_1)$, therefore, $L_A u_{j_k,\varepsilon} \to L_A f_\varepsilon$ in $\mathcal{P}'(\Phi')$ and, passing to the limit in (2.10), we obtain

$$L_A f_\varepsilon \le \operatorname{Re}[(\bar{f} / f_\varepsilon) L_A f]. \tag{2.11}$$

Finally, taking into account that $f_\varepsilon \to |f|$ as $\varepsilon \to 0$ in $L_p(\Phi', \gamma_1)$ and $|f/f_\varepsilon| \le 1$, we obtain from (2.11) the required inequality $L_A |f| \le \operatorname{Re}[(\operatorname{sgn} \bar{f}) L_A f]$ in the sense of $\mathcal{P}'(\Phi')$.

■

As in the finite-dimensional case, inequality (2.7) may be employed in the study of the conditions of essential selfadjointness of the perturbations of operators L_A by potentials. Unlike Theorem 2.1, here it is possible to establish essential selfadjointness on smooth functions.

Theorem 2.3. *Assume that the potential V belongs to $L_{2+\varepsilon}(\Phi', \gamma_1)$ with some $\varepsilon \ge 0$ and is semibounded from below. Then the operator $L_A + V$ is essentially selfadjoint both on the set of cylindrical polynomials $\mathcal{P}_{\text{cyl}}(\Phi')$ and on the set of smooth cylindrical functions $C_{b,\text{cyl}}^{\infty}(\Phi')$.*

Proof. First, we show that $L_A + V$ is essentially selfadjoint on $\mathcal{P}(\Phi')$. We choose $c > 0$ such that $V + c \ge 1$. Then the operator $L_A + V + c1$ is a strictly positive symmetric operator on $\mathcal{P}(\Phi')$. Consequently, to establish its essential selfadjointness, it suffices to show that

$$(L_A + V + c1)^* f = 0$$

yields $f = 0$. The latter means that $(L_A + V + c1) f = 0$ in the sense of $\mathcal{P}'(\Phi')$. But then

$$L_A f = -(V+c1)f \in L_p(\Phi', \gamma_1), \quad p > 1,$$

since $f \in L_2(\Phi', \gamma_1)$ and $V \in L_{2+\varepsilon}(\Phi', \gamma_1)$. Hence, we can apply Theorem 2.2 to the function f, and this gives

$$L_A|f| \le \operatorname{Re}[(\operatorname{sgn}\bar{f})L_A f] = -\operatorname{Re}[(V+c1)f \operatorname{sgn}\bar{f}] = -(V+c1)|f|.$$

Therefore,

$$(L_A+1)|f| \le -(V+(c-1)1)|f| \le 0$$

in the sense of $\mathcal{P}'(\Phi')$. Thus, for all $0 < v \in \mathcal{P}(\Phi')$, we have

$$((L_A+1)|f|, v)_{L_2(\Phi', \gamma_1)} = (|f|, (L_A+1)v)_{L_2(\Phi', \gamma_1)} \le 0. \tag{2.12}$$

In particular, for $v(\cdot) \equiv 1 \in \mathcal{P}(\Phi')$, taking into account (2.12) and the fact that $L_A 1 = 0$, we obtain

$$(|f|, 1)_{L_2(\Phi', \gamma_1)} \le \|f\|_{L_1(\Phi', \gamma_1)} \le 0,$$

i.e., $f = 0$.

By using the denseness of $\mathcal{P}_{\text{cyl}}(\Phi')$ in $\mathcal{P}(\Phi')$, for an arbitrary continuous polynomial $u \in \mathcal{P}(\Phi')$, we can choose a sequence $(u_j)_{j=1}^{\infty} \subset \mathcal{P}_{\text{cyl}}(\Phi')$ such that $u_j \to u$ as $j \to \infty$ in $\mathcal{P}(\Phi')$. Taking into account the fact that convergence in $\mathcal{P}(\Phi')$ involves convergence in $L_p(\Phi', \gamma_1)$, $p \ge 1$, and (2.3), we get

$$L_A u_j \to L_A u, \quad V u_j \to V u, \quad j \to \infty, \tag{2.13}$$

in $L_2(\Phi', \gamma_1)$, i.e.,

$$(L_A + V)u_j \xrightarrow[j\to\infty]{} (L_A+V)u$$

in $L_2(\Phi', \gamma_1)$, but this yields the essential selfadjointness on $\mathcal{P}_{\text{cyl}}(\Phi')$.

Let us establish the last statement of the theorem. As in the proof of Theorem 1.4, for $u \in \mathcal{P}_{\text{cyl}}(\Phi')$, we choose $(u_j)_{j=1}^{\infty} \subset C_{b,\text{cyl}}^{\infty}(\Phi')$ such that $u_j \to u$ as $j \to \infty$ in all $L_p(\Phi', \gamma_1)$, $p \ge 1$, and $L_A u_j \to L_A u$ as $j \to \infty$ in $L_2(\Phi', \gamma_1)$. Then $(L_A + V)u_j \to (L_A + V)u$ as $j \to \infty$ in $L_2(\Phi', \gamma_1)$.

■

***Corollary* 1.** *Assume that $V = V_1 + V_2$ and $0 < V_1 \in L_{2+\varepsilon}(\Phi', \gamma_1)$ with some $\varepsilon > 0$ and that V_2 satisfies the estimate*

$$\| V_2 f \|_{L_2(\Phi', \gamma_1)} \le a \| L_A f \|_{L_2(\Phi', \gamma_1)} + b \| f \|_{L_2(\Phi', \gamma_1)} \tag{2.14}$$

for all $f \in C^{\infty}_{b,\mathrm{cyl}}(\Phi')$ and $a < 1$. Then $L_A + V$ is essentially selfadjoint on $C^{\infty}_{b,\mathrm{cyl}}(\Phi')$.

Proof. By virtue of Theorem 2.3, the operator $L_A + V_1$ is essentially selfadjoint on $C^{\infty}_{b,\mathrm{cyl}}(\Phi')$. According to the Davies–Faris theorem (Reed and Simon [2, Chapter 10, Section 4, Theorem 10.31]), inequality (2.14) implies that

$$\| V_2 f \|_{L_2(\Phi', \gamma_1)} \le a \| (L_A + V_1) f \|_{L_2(\Phi', \gamma_1)} + b' \| f \|_{L_2(\Phi', \gamma_1)} \tag{2.15}$$

By virtue of the Kato–Rellich theorem (Reed and Simon [2, Chapter 10, Section 2, Theorem 10.12]), this yields the essential selfadjointness of $L_A + V_1 + V_2$ on any essential domain of the operator $L_A + V_1$, in particular, on $C^{\infty}_{b,\mathrm{cyl}}(\Phi')$.

■

2.2. The Feynman–Kac Formula

It is convenient to investigate the properties of the operators $L_A + V$ using of the functional integrals constructed in Section 1. The main technical tool for this investigation is given by the following analogy of the Feynman–Kac formula.

Theorem 2.4. *Assume that one of conditions (i)–(iii) in Theorem 2.1 guaranteeing the essential selfadjointness of the operator $L_A + V$ holds. Then*

(i) *for any $f, g \in L_2(\Phi', \gamma_1)$, $t > 0$, we have*

$$(f, e^{-t(L_A + V)} g)_{L_2(\Phi', \gamma_1)} = \int\limits_{\Omega(\mathbb{R}^1, \Phi')} f(\omega(0)) \overline{g}(\omega(t)) e^{-\int_0^t V(\omega(\tau)) d\tau} d\nu_A(\omega(\cdot)); \tag{2.16}$$

(ii) *for any $f_j \in \bigcap_{p \ge 1} L_p(\Phi', \gamma_1)$, $j = 0, 1, \ldots, n$, $n \in \mathbf{N}$, and $s_0 < s_1 < \ldots < s_n$, we have*

$$\int\limits_{\Phi'} f_0(x) e^{-t_1 (L_A + V)} f_1(x) \ldots e^{-t_n (L_A + V)} f_n(x) d\gamma_1(x)$$

$$= \int\limits_{\Omega(\mathbb{R}^1, \Phi')} f_0(\omega(s_0)) f_1(\omega(s_1)) \ldots f_n(\omega(s_n))\, e^{-\int_{s_0}^{s_n} V(\omega(\tau))d\tau} d\nu_A(\omega(\cdot)), \qquad (2.17)$$

where $t_k = s_k - s_{k-1}$, $k = 1, \ldots, n$.

Proof. Let us prove assertion (i). The proof of assertion (ii) is quite similar. For $N \in \mathbb{N}$, we have the cutoffs

$$V_N(x) = \begin{cases} V(x), & V(x) \geq -N, \\ -N, & V(x) < -N \end{cases} \qquad (x \in \Phi').$$

According to Theorem 2.1, the operator $L_A + V_N$ is essentially selfadjoint on $\mathfrak{D}(L_A) \cap \mathfrak{D}(V_N)$ and, by virtue of the fact that L_A and V_N are bounded from below, the following multiplicative formula is valid

$$e^{-t(L_A + V_N)} = s - \lim_{n \to \infty} \left(e^{-t/n L_A} e^{-t/n V_N}\right)^n \qquad (2.18)$$

(Reed and Simon [1, Chapter 8, Section 9, Theorem 8.31], see also formula (3.28) in Chapter 3). By construction of the measure ν_A, we have the representation

$$(f, (e^{-t/n L_A} e^{-t/n V_N})^n g)_{L_2(\Phi', \gamma_1)} = \int\limits_{\Omega} f(\omega(0)) \overline{g(\omega(t))}\, e^{-t/n \sum_{k=1}^{n} V_N(\omega(kt/n))} d\nu_A(\omega(\cdot)) \qquad (2.19)$$

(see (1.44). The left-hand side of (2.19) converges to $(f, e^{-t(L_A + V_N)} g)_{L_2(\Phi', \gamma_1)}$ as $n \to \infty$. By virtue of Theorem 1.9, the sums

$$\frac{t}{n} \sum_{k=1}^{n} V_N\left(\omega\left(\frac{tk}{n}\right)\right)$$

converge strongly in $L_2(\Omega, \gamma_A)$ to $\int_0^t V_N(\omega(\tau))\, d\tau$. Passing to a subsequence convergent ν_A-almost everywhere and using the Lebesgue theorem on majorized convergence, on the right-hand side of (2.19), we obtain a value of the limit equal to

$$\int\limits_{\Omega} f(\omega(0))\, \overline{g}(\omega(t))\, e^{-\int_0^t V_N(\omega(\tau))d\tau} d\nu_A(\omega(\cdot)). \qquad (2.20)$$

We have established relation (2.16) for each potential V_N truncated from below. In the remarks to Theorem 2.1, we mentioned the strong operator convergence of $e^{-t(L_A+V_N)}$ to $e^{-t(L_A+V)}$ as $N \to \infty$, whence

$$(f, e^{-t(L_A+V_N)}g)_{L_2(\Phi', \gamma_1)} \xrightarrow[N\to\infty]{} (f, e^{-t(L_A+V)}g)_{L_2(\Phi', \gamma_1)}.$$

On the other hand, the sequence

$$\exp\left(-\int_0^t V_N(\omega(\tau))\, d\tau\right), \quad N \in \mathbb{N},$$

is monotonically increasing and pointwise convergent to

$$\exp\left(-\int_0^t V(\omega(\tau))\, d\tau\right).$$

According to the theorem on monotonic convergence, expression (2.20) has the limit

$$\int_\Omega f(\omega(0))\, \overline{g(\omega(t))}\, e^{-\int_0^t V(\omega(\tau))d\tau} d\nu_A(\omega(\cdot))$$

as $N \to \infty$.

■

Equality (2.16) is not completely analogous to the classical Feynman–Kac formula, because it gives only the matrix elements of a semigroup and not a pointwise representation of its action. In order to obtain the required representation in the infinite-dimensional case, we suppose that the rigging $\Phi' \supset H_0 \supset \Phi$ is chosen so that for every $x \in \Phi'$, the mapping

$$[0, +\infty) \ni \mathbb{R}^1_+ \; t \mapsto e^{-tA}x \in \Phi'$$

is continuous in the weak topology $\sigma(\Phi', \Phi)$ on Φ'. We stress that this choice of Φ is always possible for the given coefficient A (see Example 1.1 in Chapter 4). According to Corollary 1 to Theorem 1.11, for every $x \in \Phi'$, the measure $\nu_{A,x}$ may be regarded as a measure on the space of trajectories $\Omega_\omega(\mathbb{R}^1_+, \Phi')$ continuous in the weak topology. We need the following lemma which shows that, for γ_1-almost all $x \in \Phi'$, the trajectories originating from the point $x \in \Phi'$ "rarely" hit the set of γ_1-measure 0.

Lemma 2.2. *Let* $\alpha \in \mathfrak{C}_\sigma(\Phi')$, $\gamma_1(\alpha) = 0$, *and* Ω_α *be a set of trajectories* *on* Ω_ω *such that*

$$m(\{t \in \mathbb{R}^1_+ \mid \omega(t) \in \alpha\}) = 0,$$

where m *is the Lebesgue measure. Then* $\nu_{A,x}(\Omega_\alpha) = 1$ *for* γ_1*-almost all* $x \in \Phi'$.

Proof. On the measurable space $\mathbb{R}^1_+ \times \Omega_\omega$, we introduce a mapping $X(t, \omega(\cdot)) = \omega(t)$ with values in Φ'. It is easy to see that this mapping is measurable, because due to the continuity of the trajectories, it can be represented as a limit (in the weak topology on Φ') of a sequence of mappings

$$X_n(t, \omega(\cdot)) = \omega\left(\frac{m}{n}\right), \quad t \in \left[\frac{m}{n}, \frac{m+1}{n}\right), \quad m \in \mathbb{Z}_+; \quad n \in \mathbb{N},$$

for which

$$X_n^{-1}(\beta) = \bigcup_{m \in \mathbb{Z}_+} \left[\frac{m}{n}, \frac{m+1}{n}\right) \times \left\{ \omega \in \Omega_\omega \mid \omega\left(\frac{m}{n}\right) \in \beta \right\} \in \mathcal{B}(\mathbb{R}^1_+) \times \mathcal{C}_\sigma(\Omega_\omega)$$

for $\beta \in \mathcal{C}_\sigma(\Phi')$, i.e., $X_n \colon \mathbb{R}^1_+ \times \Omega_\omega \to \Phi'$ is measurable, $n \in \mathbb{N}$. The measurability of X involves the measurability of the set

$$\hat{\Omega}_\alpha = \{(t, \omega) \mid \omega(t) \in \alpha\} \subset \mathbb{R}^1_+ \times \Omega_\omega.$$

We recall the construction of the measure $\nu_{A,x}$ and consider the expression

$$\int_{\Phi'} (m \times \nu_{A,x})(\hat{\Omega}_\alpha)\, d\gamma_1(x) = \int_{\Phi'} \left[\int_{\mathbb{R}^1_+} \int_{\Phi'} \kappa_\alpha(y)\, d\gamma_{A_t,x}(y)\, dt \right] d\gamma_1(x)$$

$$= \int_{\mathbb{R}^1_+} \left(\int_{\Phi'} \gamma_{A_t,x}(\alpha)\, d\gamma_1(x) \right) dt = \int_{\mathbb{R}^1_+} \gamma_1(\alpha)\, dt = 0$$

(we have employed statement (i) in Lemma 1.2, according to which, $\gamma_{A_t,x} = 0$ for γ_1-almost all $x \in \Phi'$). The possibility of changing the order of integration is guaranteed by the measurability of the function $\gamma_{A_t,x}$ with respect to the collection of its variables $(t, x) \in \mathbb{R}^1_+ \times \Phi'$. The proved equality immediately implies that

$$m(\{t \in \mathbb{R}^1_+ \mid \omega(t) \in \alpha\}) = 0$$

for γ_1-almost all $x \in \Phi'$ and for $\nu_{A,x}$-almost all trajectories $\omega \in \Omega_\omega$.

■

Let us establish the representation of a perturbed semigroup with the help of a strong functional integral.

Theorem 2.5. *Assume that one of the conditions* (i)–(iii) *of Theorem* 2.1 (*which guarantee the essential selfadjointness of* $L_A + V$) *holds. Then for any* $t > 0$ *and all* $f \in L_2(\Phi', \gamma_1)$, *we have*

$$(e^{-t(L_A+V)}f)(x) = \int_{\Omega_\omega} f(\omega(t))e^{-\int_0^t V(\omega(\tau))d\tau} d\nu_{A,x}(\omega(\cdot)) \tag{2.21}$$

for γ_1*-almost all* $x \in \Phi'$.

Proof. We divide the proof of (2.21) into several steps and make the conditions imposed on the potential weaker from step to step. First, assume that $V \in C_{b,\text{cyl}}(\Phi')$. As in the proof of Theorem 2.4, by using the multiplicative formula, we get that

$$(e^{-t/nL_A}e^{-t/nV})^n f \xrightarrow[n\to\infty]{} e^{-t(L_A+V)}f$$

in $L_2(\Phi', \gamma_1)$, and so there exists convergence γ_1-almost everywhere for a certain subsequence of this. On the other hand, by Lemma 1.7,

$$(e^{-t/nL_A}e^{-t/nV})^n f(x) = \int_{\Omega_\omega} f(\omega(t))e^{-t/n\sum_{k=1}^n V(\omega(kt/n))} d\nu_{A,x}(\omega(\cdot)) \tag{2.22}$$

for γ_1-almost all $x \in \Phi'$. Since V is a continuous cylindrical function, the function $V(\omega(\cdot))$ is a continuous function on $\mathbb{R}^1_+$ for $\omega(\cdot) \in \Omega_\omega$ and, consequently,

$$\frac{t}{n}\sum_{k=1}^{n} V\left(\omega\left(\frac{tk}{n}\right)\right) \to \int_0^t V(\omega(\tau))\,d\tau, \quad n \to \infty,$$

pointwise on Ω_ω. The estimate

$$\int_{\Omega_\omega} |f(\omega(t))|e^{-t/n\sum_{k=1}^n V(\omega(tk/n))} d\nu_{A,x}(\omega(\cdot))$$

$$\leq e^{t\sup|V|}\int_{\Omega_\omega} |f(\omega(t))|\,d\nu_{A,x}(\omega(\cdot)) \leq e^{t\sup|V|}(e^{-tL_A}|f|)\,(x),$$

for γ_1-almost all $x \in \Phi'$, enables us to apply the theorem on majorized convergence according to which the right-hand side of (2.22) converges to

$$\int_{\Omega_\omega} f(\omega(t)) e^{-\int_0^t V(\omega(\tau))d\tau} d\nu_{A,x}(\omega(\cdot))$$

as $n \to \infty$, for γ_1-almost all $x \in \Phi'$. The last statement proves the validity of (2.21) for all $V \in C_{b,\mathrm{cyl}}(\Phi')$.

We now assume that $V \in L_\infty(\Phi', \gamma_1)$. We choose a sequence $(V_j)_{j=1}^{\infty} \subset C_{b,\mathrm{cyl}}(\Phi')$ such that $V_j \to V$ as $j \to \infty$ γ_1-almost everywhere and

$$\| V_j \|_{L_\infty(\Phi', \gamma_1)} \le \| V \|_{L_\infty(\Phi', \gamma_1)}, \quad j \in \mathbb{N}.$$

Then the strong convergence of the semigroups $e^{-t(L_A + V_j)}$ to $e^{-t(L_A+V)}$ takes place in $L_2(\Phi', \gamma_1)$ and the following equality holds

$$(e^{-t(L_A+V_j)}f)(x) = \int_{\Omega_\omega} f(\omega(t)) e^{-\int_0^t V_j(\omega(\tau))d\tau} d\nu_{A,x} \tag{2.23}$$

for γ_1-almost all $x \in \Phi'$ and all $j \in \mathbb{N}$. Since $V_j \to V$ as $j \to \infty$ γ_1-almost everywhere on Φ', Lemma 2.2 implies the convergence $V_j(\omega(\tau)) \to V(\omega(\tau))$ as $j \to \infty$ pointwise, for almost all τ with respect to the Lebesgue measure, and this convergence takes place for γ_1-almost all $x \in \Phi'$ and $\nu_{A,x}$-almost all $\omega \in \Omega_\omega$. By virtue of the uniform boundedness of V_j, $j \in \mathbb{N}$, for γ_1-almost all $x \in \Phi'$, we have

$$\int_0^t V_j(\omega(\tau))\, d\tau \to \int_0^t V(\omega(\tau))\, d\tau, \quad j \to \infty,$$

pointwise $\nu_{A,x}$-almost everywhere on Ω_ω. By virtue of the theorem on majorized convergence, this enables us to pass to the limit on the right-hand side of (2.23). By extracting a subsequence convergent γ_1-almost everywhere on Φ' from the left-hand side of (2.23), we prove the theorem for $V \in L_\infty(\Phi', \gamma_1)$.

Finally, let V be of the most general type which satisfies the conditions of Theorem 2.5. As in Remark 3 to Theorem 2.1, we represent V in the form of a limit of the potentials $V_{k,n} = V_{+,k} - V_{-,n}$, $k, n \in \mathbb{N}$. For each pair $k, n \in \mathbb{N}$, the Feynman–Kac formula (2.21)

$$(e^{-t(L_A+V_{k,n})}f)(x) = \int_{\Omega_\omega} f(\omega(t))\, e^{-\int_0^t V_{k,n}(\omega(\tau))d\tau} d\nu_{A,x}(\omega(\cdot))$$

holds for γ_1-almost all $x \in \Phi'$. For fixed $k \in \mathbb{N}$, the left-hand side converges to $\exp(-t(L_A + V_{+,k} - V_-))f$ as $n \to \infty$ in $L_2(\Phi', \gamma_1)$ and, thus, a subsequence exists which converges γ_1-almost everywhere. On the right-hand side, we have

$$\int_0^t V_{k,n}(\omega(\tau))\,d\tau \to \int_0^t (V_{+,k} - V_-)(\omega(\tau))\,d\tau$$

$\nu_{A,x}$-almost everywhere, by the theorem on monotonic convergence and Lemma 2.2. Another application of this theorem is given by the equality

$$(e^{-t(L_A+V_{+,k}-V_-)}f)(x) = \int_{\Omega_\omega} f(\omega(t))e^{-\int_0^t [V_{+,k}(\omega(\tau))-V_-(\omega(\tau))]d\tau}\,d\nu_{A,x},$$

γ_1-almost all $x \in \Phi'$. In this equality, the limit transition as $k \to \infty$ is realized analogously. We take into account the convergence of the left-hand side in $L_2(\Phi', \gamma_1)$ and use the theorem on monotonic convergence to the right-hand side.

■

***Corollary* 1.** *Under the same restrictions on the potential and taking into account the connection between the measures* $\nu_{A,x}$ *and* $\nu_{A,0}$ *(see* (1.56) *in Section 1), we can rewrite the Feynman–Kac formula* (2.21) *in the form*

$$(e^{-t(L_A+V)}f)(x) = \int_{\Omega_\omega} f(\omega(t)+e^{-tA}x)e^{-\int_0^t V(\omega(\tau)+e^{-\tau A}x)d\tau}\,d\nu_{A,0}(\omega(\cdot)) \qquad (2.24)$$

for γ_1*-almost all* $x \in \Phi'$ *and* $f \in L_2(\Phi', \gamma_1)$, $t \geq 0$.

Written in this form, this formula becomes very convenient for applications, because dependence on $x \in \Phi'$ is given explicitly.

■

Example 2.1. The Feynman–Kac formula (2.24) enables us to investigate the properties of the semigroup $e^{-t(L_A+V)}$, $t \geq 0$, on the class of smooth functions. First, we note that unlike the finite-dimensional case, even for $V = 0$, this semigroup does not increase the smoothness. Indeed, for $h \in H_0$, a measurable linear function $l_h(x) = (h, x)_{H_0}$ $(x \in \Phi')$ belongs to all $L_p(\Phi', \gamma_1)$, $p \geq 1$ (see Subsection 2.1 in Chapter 2). By virtue of (1.10), we have

$$(e^{-tL_A}l_h)(x) = (e^{-tA}h, x)_{H_0},$$

but this function is continuous on a set of full γ_1-measure only under the condition that

$e^{-tA}h \in H_+$, where H_+ is quasinuclearly imbedded in H_0. However, generally speaking, this condition does not hold (we set $A = 1$, then $e^{-tL_A}l_h = e^{-t}l_h$). Nevertheless, under certain restrictions imposed on the potential, the semigroup under consideration preserves the smoothness of functions.

Assume that, for a given coefficient operator A, the rigging $\Phi' \supset H_0 \supset \Phi$ is chosen so that, for any $\tau_1 \in \mathrm{T}$, there exist $\tau_2 \in \mathrm{T}$ such that $e^{-\tau A} \in \mathfrak{L}(H_{\tau_2}, H_{\tau_1})$ for any $t \in \mathbb{R}^1_+$; the vector function $\mathbb{R}^1_+ \ni t \mapsto e^{-tA}x \in H_{-\tau_2}$ is continuous for any $x \in H_{-\tau_1}$. As in Subsection 3.3 in Chapter 2, we denote by $C^k_b(\Phi')$ a set of functions on Φ' which belong to $C^k_b(H_{-\tau})$ for all $\tau \in \mathrm{T}$.

By using representation (2.24), we can easily establish the following fact:

The function $(e^{-t(L_A+V)}f)(x)$ $(x \in \Phi')$ *belongs to* $C^k_b(\Phi')$ *for* $V, f \in C^k_b(\Phi')$ *and* $t > 0$.

Taking into account the equality

$$(L_A + V - \lambda 1)^{-1} = \int_0^{+\infty} e^{t\lambda} e^{-t(L_A+V)}\, dt,$$

in which $\lambda \in \mathbb{C}^1$ and $-\mathrm{Re}\,\lambda$ is sufficiently large, we can also obtain an analogous statement for the resolvent $R_\lambda(L_A + V)$ of the operator $L_A + V$ for $\mathrm{Re}\,\lambda < -\|V\|_{L_\infty(\Phi',\gamma_1)}$.

2.3. Properties of Perturbed Operators

The following theorem is a typical illustration of the application of the Feynman–Kac formula to the investigation of the operators $L_A + V$.

Theorem 2.7. *Let* $A \geq a1$, $a > 0$. *Assume that there exists* $\varepsilon > 0$ *such that* $V \in L_{2+\varepsilon}(\Phi', \gamma_1)$ *and* $e^{-V} \in L_p(\Phi', \gamma_1)$ *for any* $p \geq 1$. *Then the following facts take place:*

(i) *For* $q \geq q(t) = 2(1 - e^{-at})^{-1}$, *we have*

$$\|\exp[-t(L_A + V)]\|_{\mathfrak{L}(L_2(\Phi',\gamma_1), L_2(\Phi',\gamma_1))} \leq \left\| \exp\left(-\int_0^t V(\omega(\tau))d\tau \right) \right\|_{L_q(\Omega, \nu_A)}.$$

(ii) *For* $s < 1 + e^{2at}$, *we have* $e^{-t(L_A+V)} \in \mathfrak{L}(L_2(\Phi', \gamma_1), L_s(\Phi', \gamma_1))$ *and, moreover,*

$$\|e^{-t(L_A+V)}\|_{\mathfrak{L}(L_2(\Phi',\gamma_1), L_s(\Phi',\gamma_1))} \leq \|e^{-\int_0^t (\omega(\tau))d\tau}\|_{L_r(\Omega,\nu_A)}$$

for some $r > 1$.

(iii) *For all* $t > 0$, *the operators* $e^{-t(L_A + V)}$ *strengthen the positivity.*

Proof. (i) For $f, g \in L_\infty(\Phi', \gamma_1)$, according to (2.16), we have

$$| (f, e^{-t(L_A + V)} g)_{L_2(\Phi', \gamma_1)} |$$

$$\leq \left\| \exp\left(- \int_0^t V(\omega(\tau)) d\tau \right) \right\|_{L_{q(t)}(\Omega, \nu_A)} \| f(\omega(0))\, g(\omega(t)) \|_{L_{q'(t)}(\Omega, \nu_A)}, \quad (2.25)$$

where $1/q + 1/q'(t) = 1$, i.e., $q'(t) = 2(1 + e^{-at})^{-1}$. Let us now employ Lemma 1.5, setting $r = q'(t)$ and $p = 2$ in (1.50). Then we obtain the following condition for the index q in (1.50)

$$e^{-2ta} \leq e^{-at} \left[\frac{q(1 + e^{-at})}{2} - 1 \right],$$

which holds for $q = 2$. Therefore,

$$\| f(\omega(0))\, g(\omega(t)) \|_{L_{q'(t)}(\Omega, \mu_A)} \leq \| f \|_{L_2(\Phi', \gamma_1)} \| g \|_{L_2(\Phi', \gamma_1)}$$

and, thus, (2.25) yields the required estimate.

(ii) We set $s(t) = 1 + e^{2at}$. Then the dual index is given by $s'(t) = 1 + e^{-2at} > 1$. Hence, for $s < s(t)$, one can choose $1 < r < s'$ such that $e^{-2at} \leq (s'/r - 1)\,(2/r - 1)$. For the chosen r, by use of (1.50) we obtain

$$| (f, e^{-t(L_A + V)} g)_{L_2(\Phi', \gamma_1)} | \leq \left\| \exp\left(- \int_0^t V(\omega(\tau)) d\tau \right) \right\|_{L_{r'}(\Omega, \nu_A)} \| f \|_{L_2(\Phi', \gamma_1)} \| g \|_{L_{s'}(\Phi', \gamma_1)},$$

i.e.,

$$e^{-t(L_A + V)} \in \mathcal{L}(L_{s'}(\Phi', \gamma_1), L_2(\Phi', \gamma_1)),$$

whence (by conjugation)

$$e^{-t(L_A + V)} \in \mathcal{L}(L_2(\Phi', \gamma_1), L_s(\Phi', \gamma_1)).$$

(iii) According to Theorem 1.8, e^{-tL_A} strengthens the positivity for $t > 0$. Hence, if

$$\gamma_1(\{x \in \Phi' | f(x)g(x) > 0\}) > 0$$

for $f, g \geq 0$ γ_1-almost everywhere, then

$$(f, e^{-t(L_A+V)}g)_{L_2(\Phi',\gamma_1)} = \int_\Omega f(\omega(0))\, g(\omega(t))\, d\nu_A(\omega(\cdot)) > 0.$$

Since

$$\int_0^t V(\omega(\tau))\, d\tau \in L_2(\Omega, \nu_A),$$

this function is ν_A-almost everywhere finite and, thus,

$$\exp\left(-\int_0^t V(\omega(\tau))d\tau\right) > 0$$

ν_A-almost everywhere. Consequently,

$$(f, e^{-t(L_A+V)}g)_{L_2(\Phi',\gamma_1)} = \int_\Omega f(\omega(0))\, g(\omega(t))\, e^{-\int_0^t V(\omega(\tau))d\tau}\, d\nu_{A,x}(\omega(\cdot)) > 0,$$

and this proves that the operator $e^{-t(L_A+V)}$ strengthens the positivity.

■

The next theorem describes certain spectral properties of the operator $L_A + V$.

Theorem 2.8. *Assume that* A *and* V *satisfy the conditions of Theorem* 2.7. *Then the following facts take place:*

(i) *For* $E_V = \inf s\,(L_A + V)$, *the following estimates hold:*

$$-E_V \leq \frac{\ln(1, e^{-tq(t)V})_{L_2(\Phi',\gamma_1)}}{tq(t)},$$

$$-E_V \leq a/2 \ln \| e^{-2/aV} \|_{L_1(\Phi',\gamma_1)}. \tag{2.26}$$

(ii) E_V *is a simple eigenvalue of* $L_A + V$, *and the corresponding eigenfunction* ϕ_V *may be chosen to be strictly positive.*

(iii)

$$-E_V = \lim_{t\to+\infty} \frac{1}{t} \ln \| e^{-\int_0^t V(\omega(\tau))d\tau} \|_{L_1(\Omega,\nu_A)} \tag{2.27}$$

or, equivalently,

$$-E_V = \lim_{t\to+\infty} \ln(1, e^{-t(L_A+V)}1)_{L_2(\Phi',\gamma_1)}.$$

Proof. (i) According to assertion (i) in Theorem 2.7, we have

$$\|e^{-t(L_A+V)}\| \le \left\| \exp\left(-\int_0^t V(\omega(\tau))d\tau\right)\right\|_{L_{q(t)}(\Omega,\nu_A)}$$

$$= \left\{\int_\Omega e^{-\int_0^t V(\omega(\tau))q(t)d\tau} d\nu_A(\omega(\cdot))\right\}^{1/q(t)}$$

$$\le \|e^{-tq(t)V}\|^{1/q(t)}_{L_1(\Phi',\gamma_1)} = (1, e^{-tq(t)V}1)^{1/q(t)}_{L_2(\Phi',\gamma_1)}$$

(we have applied Lemma 1.4 to estimate the integral over Ω). Therefore,

$$e^{-tE_V} \le (1, e^{-tq(t)V}1)^{1/q(t)}_{L_2(\Phi',\gamma_1)}$$

and this yields the first inequality in (2.26). If we let $t\to 0+$ in this inequality, then $tq(t)\to 2/a$, i.e., we obtain the second estimate in (2.26).

(ii) The fact that E_V is an eigenvalue of the operator $L_A + V$ follows from the generalization of the Perron–Frobenius theorem given below (see Reed and Simon [4, Chapter 13, Section 12, Theorem 13.49]).

Lemma 2.3. *Assume that the measurable space* $(R, \mathfrak{R}, \mu)$ *is given and* $\mu(R)=1$. *If* S *is a bounded operator from* $L_2(\Phi', \gamma_1)$ *to* $L_p(\Phi', \gamma_1)$ *for some* $p > 2$ *and preserves the positivity, then* $\|S\|$ *is an eigenvalue of* S.

■

Taking into account Lemma 2.3, we set $S = e^{-t(L_A+V)}$ and exploit assertion (ii) in Theorem 2.7.

The uniqueness of the corresponding eigenfunction ϕ_V follows from the fact that $e^{-t(L_A+V)}$, $t>0$, strengthens the positivity (assertion (iii) in Theorem 2.7). Indeed, we first show that ϕ_V is a strictly positive function (to within a constant factor). Certainly, we can suppose that ϕ_V is a real function, because $e^{-t(L_A+V)}$ transforms real functions into real functions. It follows from the preservation of positivity (or directly from the Feynman–Kac formula) that

$$|e^{-t(L_A+V)}f| \le e^{-t(L_A+V)}|f| \quad (f\in L_2(\Phi',\gamma_1)).$$

Consequently,

$$(|\phi_V|, e^{-t(L_A+V)}|\phi_V|)_{L_2(\Phi',\gamma_1)} \geq (|\phi_V|, |e^{-t(L_A+V)}\phi_V|)_{L_2(\Phi',\gamma_1)} = e^{-tE_V}\|\phi_V\|^2_{L_2(\Phi',\gamma_1)}$$

hence,

$$e^{-t(L_A+V)}|\phi_V| = e^{-tE_V}|\phi_V|,$$

i.e.,

$$(L_A+V)|\phi_V| = E_V|\phi_V|.$$

If ϕ_V is not identically 0, then one of the functions $|\phi_V|+\phi_V$ or $|\phi_V|-\phi_V$ is non-negative and is not identically 0. Suppose that this is $|\phi_V|+\phi_V$. Then it follows from the strengthening of the positivity that

$$e^{-tE_V}(|\phi_V|+\phi_V) = e^{-t(L_A-V)}(|\phi_V|+\phi_V) > 0$$

γ_1-almost everywhere, i.e., $|\phi_V|+\phi_V>0$ γ_1-almost everywhere and, hence, $\phi_V>0$ γ_1-almost everywhere. In the case when $|\phi_V|-\phi_V$ is not 0, by analogy, we obtain that $\phi_V<0$ γ_1-almost everywhere; after this, it remains only to multiply ϕ_V by -1.

Let ϕ_V and $\hat{\phi}_V$ be eigenfunctions corresponding to the eigenvalue E_V (they can be regarded as strictly positive according to the argument presented above). Then $(\phi_V, \hat{\phi}_V)_{L_2(\Phi',\gamma_1)}>0$ and, thus, $\phi_V=\hat{\phi}_V$, since L_A+V is selfadjoint.

(iii) Consider the expression

$$e^{-tE_V}(1,\phi_V)^2_{L_2(\Phi',\gamma_1)} = (1, e^{-t/2(L_A+V)}\phi_V)^2_{L_2(\Phi',\gamma_1)}$$

$$\leq \|e^{-t/2(L_A+V)}1\|^2_{L_2(\Phi',\gamma_1)}\|\phi_V\|^2_{L_2(\Phi',\gamma_1)} \leq \|e^{-t(L_A+V)}\|\,\|\phi_V\|^2_{L_2(\Phi',\gamma_1)}.$$

By virtue of the fact that $(1,\phi_V)_{L_2(\Phi',\gamma_1)}>0$, after taking the logarithm, we obtain

$$-tE_V+\ln(1,\phi_V)^2_{L_2(\Phi',\gamma_1)} \leq \ln(1, e^{-t(L_A+V)}1)_{L_2(\Phi',\gamma_1)} \leq -tE_V+\ln\|\phi_V\|^2_{L_2(\Phi',\gamma_1)}$$

and this yields (2.27).

■

The estimate for the lower bound of of the spectrum of the operator L_A+V obtained in Theorem 2.8 implies the important consequence for the unperturbed operator L_A. Let us rewrite (2.26) in the form

$$-(Vf,f)_{L_2(\Phi',\gamma_1)} \le (L_A f,f)_{L_2(\Phi',\gamma_1)} + \frac{a}{2}\ln \| e^{-2V/a}\|_{L_1(\Phi',\gamma_1)} \|f\|^2_{L_2(\Phi',\gamma_1)} \quad (2.28)$$

for $f \in C^{\infty}_{b,\mathrm{cyl}}(\Phi')$. By setting $V = -a \log|f|$ for a proper dense set $f \in C^{\infty}_{b,\mathrm{cyl}}(\Phi')$, we get

$$\int_{\Phi'} |f|^2 \ln|f|\, d\gamma_1 \le \frac{1}{a}(L_A f,f)_{L_2(\Phi',\gamma_1)} + \|f\|^2_{L_2(\Phi',\gamma_1)} \ln\|f\|_{L_2(\Phi',\gamma_1)} \quad (2.29)$$

or

$$\int_{\Phi'} |f|^2 \ln|f|\, d\gamma_1 \le \frac{1}{a}\int_{\Phi'} (\nabla f, A\nabla f)_{H_0} d\gamma_1 + \|f\|^2_{L_2(\Phi',\gamma_1)} \ln\|f\|_{L_2(\Phi',\gamma_1)}. \quad (2.30)$$

By the limit transition, inequalities (2.29) and (2.30) are extended to all f from the domain of definition of the quadratic form of the operator L_A. When realizing this extension, we can use the following estimate

$$0 \le \int_{\Phi'} |f|^2 \ln|f|\, d\gamma_1 - \|f\|^2_{L_2(\Phi',\gamma_1)} \ln\|f\|_{L_2(\Phi',\gamma_1)} \le \frac{1}{a}(L_A f,f)_{L_2(\Phi',\gamma_1)},$$

the left-hand side of which follows from the Jensen inequality because the function $t\log t$, $t \ge 0$, is convex. In the case when $A = 1$, i.e., for the operator of the number of particles $N = L_1$, inequality (2.30) transforms into

$$\int_{\Phi'} |f|^2 \ln|f|\, d\gamma_1 \le \int_{\Phi'} \|\nabla f\|^2_{H_{0,c}} d\gamma_1 + \|f\|^2_{L_2(\Phi',\gamma_1)} \ln\|f\|_{L_2(\Phi',\gamma_1)}.$$

This is an (infinite-dimensional) analogy of the classical Sobolev inequality for the case of the Gaussian measure: Under the condition that $\|\nabla f\|_{H_{0,c}} \in L_2(\Phi',\gamma_1)$, the function f belongs to the Orlicz space $L_2(\log L)$. Recall that the ordinary Sobolev inequality states that the inclusion $f \in L_p(\mathbb{R}^d)$ holds for $\nabla f \in L_2(\mathbb{R}^d)$, where $1/p = 1/2 - 1/d$, and even formally it is impossible to expect something more than $f \in L_2$ as $d \to \infty$.

By using interpolation in the Orlicz spaces, we can establish the following estimate for $f \in \mathrm{D}(L_A)$

$$\int_{\Phi'} |f|^2 (\ln|f|)^2 d\gamma_1 \le c_1 \|L_A f\|^2_{L_2(\Phi',\gamma_1)} + c_2 \|f\|^2_{L_2(\Phi',\gamma_1)} \ln(\|f\|_{L_2(\Phi',\gamma_1)})^2 \quad (2.31)$$

with some $c_1, c_2 > 0$. For $f \in \mathrm{D}(L_A)$, (2.31) yields the following inclusion $f \in L_2(\log L)^2$. Similarly, we can establish the estimates for higher powers of L_A which give the inclusion for $f \in \mathfrak{D}(L_A^m)$ in the space $L_2(\log L)^{2m}$, $m \in \mathbb{N}$ (see Feissner [1]).

By using the logarithmic Sobolev inequalities (2.29) and (2.31), we can obtain the conditions of subordination in the sense of forms and operator subordination, respectively, for potential perturbations.

Theorem 2.9. *Assume that* V *is such that* $e^{|V|} \in L_p(\Phi', \gamma_1)$ *for any* $p \ge 1$. *Then, for any uniformly elliptic operator* L_A, *we have* $\mathfrak{D}(L_A) \subset \mathfrak{D}(V)$ *and*

$$\| Vf \|^2_{L_2(\Phi',\gamma_1)} \le c\| L_A f \|^2_{L_2(\Phi',\gamma_1)} + d\| f \|^2_{L_2(\Phi',\gamma_1)} \tag{2.32}$$

for all $f \in \mathfrak{D}(L_A)$. *Here,* $c > 0$ *can be chosen as arbitrarily small, i.e.,* V *is infinitely small with respect to* L_A.

Proof. It suffices to establish the estimate (2.32) on the essential domain of the operator L_A, for example, on $C^\infty_{b,\mathrm{cyl}}(\Phi')$; in this situation, due to the homogeneity, we can restrict ourselves to the case $\| f \|_{L_2(\Phi',\gamma_1)} = 1$. Using the Young inequality

$$ab \le a \ln a - a + e^b \quad (a \ge 0,\ b \in \mathbb{R}^1),$$

we estimate the product

$$|Vf|^2 = \{1/\varepsilon |V| \varepsilon |f|^2\} \le \{[\varepsilon|f| \ln(\varepsilon|f|) - \varepsilon|f|] + e^{|V|/\varepsilon}\}^2$$

$$\le 4\{\varepsilon^2 |f|^2 (\ln \varepsilon |f|)^2 + \varepsilon^2 |f|^2\} + 2e^{2|V|/\varepsilon}$$

(we have applied the Cauchy inequality $2|a||b| \le |a|^2 + |b|^2$ two times). But

$$(\ln \varepsilon |f|)^2 \le 2[(\ln \varepsilon)^2 + (\ln |f|)^2]$$

and $\varepsilon^2 \log \varepsilon^2 \le 1$ $(\varepsilon \in (0, 1))$; therefore,

$$|Vf|^2 \le 8\varepsilon^2 |f|^2 (\ln|f|)^2 + 4(2 + \varepsilon^2)|f|^2\} + 2e^{2|V|/\varepsilon}. \tag{2.33}$$

If we integrate (2.33) and employ (2.31), then we get the following inequality

$$\| Vf \|^2_{L_2(\Phi',\gamma_1)} \le 8\varepsilon^2 c_1 \| L_A f \|^2_{L_2(\Phi',\gamma_1)} + 4(2 + \varepsilon^2) + 2\| e^{|V|} \|^{2/\varepsilon}_{L_{2/\varepsilon}(\Phi',\gamma_1)};$$

by proper choice of $\varepsilon > 0$, for each $c > 0$, we obtain

$$\| Vf \|^2_{L_2(\Phi',\gamma_1)} \le c\| L_A f \|^2_{L_2(\Phi',\gamma_1)} + d(c)$$

for all $f \in C^{\infty}_{b,\mathrm{cyl}}(\Phi')$, $\|f\|_{L_2(\Phi',\gamma_1)} = 1$.

■

Corollary 1. *Let $A \geq a1$, $a > 0$, $V = V_+ - V_-$, and $V_\pm \geq 0$. Assume that there exists $\varepsilon > 0$ such that $V_+ \in L_{2+\varepsilon}(\Phi', \gamma_1)$ and $e^{-V_-} \in L_p(\Phi', \gamma_1)$ for any $p \geq 1$. Then $L_A + V$ is essentially selfadjoint on each of the domains $C^{\infty}_{b,\mathrm{cyl}}(\Phi')$ or $\mathcal{P}_{\mathrm{cyl}}(\Phi')$.*

Proof. According to what has been proved, the potential V satisfies the conditions of the corollary from Theorem 2.3.

■

2.4. More Singular Potential Perturbations. Interactions in Constructive Quantum Field Theory

Let us consider the conditions which should be imposed on a potential V in order to make the construction of a selfadjoint realization of $L_A + V$ possible. These conditions can be made somewhat weaker by passing to form sums. According to approximation theorems for form sums, the Feynman–Kac formula is valid in this case, too.

Theorem 2.1. *Let $A \geq a1$, $a > 0$, $V = V_+ - V_-$, where $V_\pm \geq 0$. Assume that $V_+ \in L_1(\Phi', \gamma_1)$, that there exists $p > 2/a$ such that $e^{-V_-} \in L_p(\Phi', \gamma_1)$, and that $L = L_A + V$ is a form sum of the operators L_A and V. Then, for all $f \in L_2(\Phi', \gamma_1)$, we have*

$$(e^{-tH}f)(x) = \int_{\Omega_\omega} f(\omega(t))e^{-\int_0^t V(\omega(t))d\tau} d\nu_{A,x}(\omega(\cdot)) \tag{2.34}$$

for γ_1-almost all $x \in \Phi'$.

Proof. First, we note that the condition imposed on V_- implies subordination in the sense of the forms

$$V_- \leq cL_A + d \tag{2.35}$$

with constant $c < 1$. Indeed, according to the Young inequality,

$$V_-|f|^2 \leq 2\tilde{c}|f|^2 \ln|f| - \tilde{c}|f|^2 + e^{-V_-/\tilde{c}}$$

for any $\tilde{c} > 0$. If we set $c = p^{-1}$ and $\|f\|_{L_2(\Phi',\gamma_1)} = 1$ and employ the inequality (2.29), then we get

$$(V_-f,f)_{L_2(\Phi',\gamma_1)} = \int_{\Phi'} V_-|f|^2\,d\gamma_1 \le \frac{2}{pa}(L_Af,f)_{L_2(\Phi',\gamma_1)} + d,$$

where $2 < pa$ by the condition of the theorem. By virtue of homogeneity, this yields (2.35).

We are now in a situation when the conditions of approximation Theorem 1.14 (see Chapter 5) are satisfied. According to this theorem, we have

$$L_A + V_{+,n} - V_{-,k} \to L_A + V_{+,n} - V_-, \quad k \to \infty,$$

$$L_A + V_{+,n} - V_- \to L_A + V, \quad n \to \infty,$$

in the sense of strong resolvent convergence ($V_{\pm,n}$, $n \in \mathbb{N}$, are defined in the remark to Theorem 2.1). For $V_{n,k} = V_{+,n} - V_{-,k} \in L_\infty(\Phi', \gamma_1)$, the Feynman–Kac formula holds

$$(e^{-t(L_A+V_{n,k})}f)(x) = \int_{\Omega_\omega} f(\omega(t))\, e^{-\int_0^t V_{n,k}(\omega(t))d\tau}\, d\nu_{A,x}(\omega(\cdot)). \tag{2.36}$$

As in the proof of Theorem 2.5, applying (twice) the theorem on monotonic convergence to the right-hand side of (2.36) and extracting a subsequence convergent γ_1-almost everywhere (to the left), we obtain (2.34).

■

If we want to study the problem of constructing dynamics for physical systems with infinitely many degrees of freedom, then we must consider many more singular potentials which never belong to the classes investigated above. To clarify the difficulties which may appear in this case, let us consider an important example.

Example 2.2. (Polynomial interactions in field theory). In Example 2.5 in Chapter 2, the physical Hilbert space of a free boson field with mass m_0 was described as $L_2(\mathfrak{S}'_{\mathrm{Re}}(\mathbb{R}^d),\ \mu_{m_0})$. The operator of energy L_{m_0} was given in Example 1.4 in Section 1 as the second quantization of the operator $A_{m_0} = (-\Delta + m_0^2)^{1/2}$. For $N \in \mathbb{N}$ and $\varphi \in L_\infty(\mathbb{R}^d)$, the interaction $:\varphi^N(\psi):$ is formally defined as a function on $\mathfrak{S}'_{\mathrm{Re}}(\mathbb{R}^d)$ of the form

$$\mathfrak{S}'_{\mathrm{Re}}(\mathbb{R}^d) \ni x \mapsto \int_{\mathbb{R}^d} :x(s)^N:\psi(s)ds. \tag{2.37}$$

The expression (2.37) can be also made meaningful in the case $d = 1$ for finite ψ, because in this case, μ_{m_0} is concentrated on regular enough elements of $\mathfrak{S}'_{\mathrm{Re}}(\mathbb{R}^d)$. However, even for $d=1$, it is impossible to take $\psi \equiv 1$ to guarantee the required invariance of interaction. For the dimensionalities $d \geq 2$, the measure μ_{m_0} is concentrated on essentially generalized functions and, thus, (2.37) is not defined even for $\psi \in C_0^\infty(\mathbb{R}^d)$, because it is impossible to make $:x^N(s):$ meaningful as a generalized function. Nevertheless, it is easy to see that (2.37) defines a generalized function $:\varphi^N(\psi):$ from $\mathcal{P}'(\mathcal{S}'_{\mathrm{Re}}(\mathbb{R}^d))$ Indeed, (2.37) can be formally written as

$$:\varphi^N(\psi):(x) = \int_{\mathbb{R}^d} :(x, \delta_s)^N_{L_2(\mathbb{R}^d)} :\psi(s)ds = I_{m_0}\xi^\psi_N, \tag{2.38}$$

where

$$I_{m_0} : \mathfrak{F}(\mathfrak{H}_{m_0}) \to L_2(\mathcal{S}'_{\mathrm{Re}}(\mathbb{R}^d), \gamma_{m_0})$$

is a Segal isomorphism (see Subsection 2.2 in Chapter 2) and

$$\xi^\psi_N = \sqrt{N!}\, 2^{N/2} \int_{\mathbb{R}^d} \delta_s^{\otimes N} \psi(s)ds \in \mathfrak{F}_N(\mathcal{S}'_{\mathrm{Re}}(\mathbb{R}^d)).$$

Similarly, for the polynomial

$$P(t) = \sum_{k=0}^{N} a_k t^k \, (a_k \in \mathbb{R}^1, \;\; k = 0, \ldots, N; \;\; N \in \mathbb{N}, \;\; \psi \in L_\infty(\mathbb{R}^d)),$$

a generalized function

$$:P(\psi): = \sum_{k=0}^{N} a_k :\varphi^k(\psi): \in \mathcal{P}'(\mathcal{S}'_{\mathrm{Re}}(\mathbb{R}^d))$$

is defined. A model in the field theory with a given polynomial interaction $P(\cdot)$ is formally described by the Hamiltonian $L = L_{m_0} + :P(1):$, which may be regarded as a mapping of $\mathcal{P}(\mathcal{S}'_{\mathrm{Re}}(\mathbb{R}^d))$ into $\mathcal{P}'(\mathcal{S}'_{\mathrm{Re}}(\mathbb{R}^d))$, but not as an operator in $L_2(\mathcal{S}'_{\mathrm{Re}}(\mathbb{R}^d), \mu_{m_0})$ (by virtue of the above argument).

Thus, the problem of constructing the operator realization for a given formal Hamiltonian L arises. The solution of this problem usually employs the approximation of the potential $:P(1):$ given as a generalized function by more regular functions. To introduce this approximation, we choose a sequence $(d_n(\cdot))_{n=1}^\infty \subset \mathcal{S}_{\mathrm{Re}}(\mathbb{R}^d)$ such that $d_n \to \delta_0$ as $n \to \infty$ in $\mathcal{S}'(\mathbb{R}^d)$. For every $s \in \mathbb{R}^d$, we consider the polynomial

$$\sum_{k=0}^{N} a_k :(d_n(\cdot - s), x(\cdot))^k_{L_2(\mathbb{R}^d)} : (x \in \mathcal{S}'_{\mathrm{Re}}(\mathbb{R}^d))$$

which clearly belongs to $\mathcal{P}(\mathcal{S}'_{\mathrm{Re}}(\mathbb{R}^d))$ and, furthermore, continuously depends on $s \in \mathbb{R}^d$ in the topology of the nuclear space $\mathcal{P}(\mathcal{S}'_{\mathrm{Re}}(\mathbb{R}^d))$. Therefore, for $0 \le \psi \le 1$, $\psi \in C_0^\infty(\mathbb{R}^d)$, we can introduce the potentials

$$V_{n,\psi}(x) = \sum_{k=0}^{N} a_k \int_{\mathbb{R}^d} :(d_n(\cdot - s), x(\cdot))^k_{L_2(\mathbb{R}^d)} : \psi(s)ds : \in \mathcal{P}(\mathcal{S}'_{\mathrm{Re}}(\mathbb{R}^d)).$$

Formula (2.26) in Chapter 2 yields the convergence $V_{n,\psi} \to :P(1):$ in $\mathcal{P}'(\mathcal{S}'_{\mathrm{Re}}(\mathbb{R}^d))$ as $n \to \infty$ and $\psi \to 1$ pointwise. For $N = 2l$, $l \in N$, and $a_N > 0$, the following inclusions hold for the potentials $V_{n,\psi}$ $(n \in \mathbb{N}, 0 \le \psi \in C_0^\infty(\mathbb{R}^d))$:

$$V_{\psi,n} \in \bigcap_{p \ge 1} L_p(\mathcal{S}'_{\mathrm{Re}}(\mathbb{R}^d), \mu_{m_0})$$

(as for each continuous polynomial) and $e^{-V_{n,\psi}} \in L_\infty(\mathcal{S}'_{\mathrm{Re}}(\mathbb{R}^d), \mu_{m_0})$ (since $V_{n,\psi}$ is semi-bounded from below, the last inclusion follows, for example, from the estimate (5.15) in Chapter 2)). According to Theorem 2.3, each Hamiltonian $L_{m_0} + V_{n,\psi}$ is essentially selfadjoint on $\mathcal{P}_{\mathrm{cyl}}(\mathcal{S}'_{\mathrm{Re}}(\mathbb{R}^d))$ As a result, we have defined a family of selfadjoint approximating Hamiltonians $L_{m_0} + V_{n,\psi}$, $n \in \mathbb{N}$, $0 \le \psi \le 1$, $0 \le \psi \in C_0^\infty(\mathbb{R}^d)$, in the Hilbert space of states of a free field $L_2(\mathcal{S}'_{\mathrm{Re}}(\mathbb{R}^d), \mu_{m_0})$ for which

$$((L_{m_0} + V_{\psi,n})u, v)_{L_2(\mathcal{S}'_{\mathrm{Re}}(\mathbb{R}^d), \mu_{m_0})} \to ((L_{m_0} + :P(1):)u, v)_{L_2(\mathcal{S}'_{\mathrm{Re}}(\mathbb{R}^d), \mu_{m_0})}$$

$$(u, v \in \mathcal{P}(\mathcal{S}'_{\mathrm{Re}}(\mathbb{R}^d))$$

as $n \to \infty$ and $\psi \to 1$. Clearly, one could hardly expect to get more than the indicated convergence. To construct the operator realization of L_{m_0} + $:P(1):$ for a given family of approximating Hamiltonians, it is necessary to use the methods of the renormalization theory in one of its possible interpretations. We shall become acquainted with one of these methods in Section 1 of Chapter 7. In Section 2 of Chapter 7, we consider its application to the construction of the operator realization of the formal Hamiltonians corresponding to the simplest models in quantum field theory.

§3. Dirichlet Forms and Differential Operators Corresponding to These Forms

The second quantization operators L_A are generated by the Dirichlet forms of a Gaussian measure γ_1 of the form (1.20) (see Theorem 1.2). Below, we establish the connection between the operator $L_A + V$ and the operator generated by the analogous form in which γ_1 is replaced by the probability measure which is a perturbation of the original Gaussian measure canonically connected with potential. In Chapter 7, this connection serves as a basis for the construction of the operator realization of formal Hamiltonians in the models of quantum field theory and quantum statistical physics. Moreover, we obtain the conditions of the essential selfadjointness of the operators associated with the Dirichlet forms.

3.1. Dirichlet Forms of Probability Measures and Dirichlet Operators

Assume that the nuclear rigging $\Phi' \supset H_0 \supset \Phi$ of a real separable Hilbert space H_0 and a selfadjoint operator $A \geq 0$, $A \in \mathcal{L}(\Phi, \Phi)$, are given. For an arbitrary probability measure μ on $\mathcal{C}_\sigma(\Phi')$, we consider the following bilinear form

$$d_{\mu,A}(u, v) = \frac{1}{2} \int_{\Phi'} (\nabla u, A\nabla \bar{v})_{H_0} d\mu \quad (u, v \in C^1_{b,\text{cyl}}(\Phi')), \tag{3.1}$$

which is defined correctly due to the fact that the function $(\nabla u, A\nabla \bar{v})_{H_0}$ belongs to $C_{b,\text{cyl}}(\Phi')$ for $u, v \in C^1_{b,\text{cyl}}(\Phi')$. We call it the Dirichlet form corresponding to the measure μ and the coefficient operator A. The form $d_{\mu,A}$ is nonnegative but, generally speaking, may be nonclosable. Hence, it is impossible to associate with it an operator acting in the space $L_2(\Phi', \mu) = L_2(\Phi', \mathcal{C}_\sigma(\Phi'))$ without some additional assumptions concerning the properties of μ. The simplest condition which should be imposed on μ to guarantee the closability of the Dirichlet form is the possibility of realizing "integration by parts" in (3.1). Thus, it is necessary to introduce some additional restrictions for the measure μ. First, we assume that μ is quasi-invariant with respect to shifts by the vectors from Φ (see Subsection 2.3 in Chapter 2). This means that for all $\varphi \in \Phi$ the measures μ and $\mu_\varphi = \mu(\cdot + \varphi)$ are mutually absolutely continuous and, therefore, the Radon–Nikodym derivative is defined

$$\rho(\varphi, x) = \frac{d\mu(x + \varphi)}{d\mu(x)} \in L_1(\Phi', \mu). \tag{3.2}$$

Further, assume the existence of the function $\beta_\mu(\varphi, x) : \Phi' \to \mathbb{R}^1$ defined as a limit in $L_2(\Phi', \mu)$ (and, consequently, defined with an accuracy to within the subsets of Φ' of μ-measure 0):

$$\beta_\mu(\varphi, x) = \lim_{t\to 0} \frac{\rho(t\varphi, x) - 1}{t} = \frac{d}{dt}\rho(t\varphi, x)\big|_{t=0} \in L_2(\Phi', \mu). \tag{3.3}$$

This function linearly depends on $\varphi \in \Phi$. Given the finite-dimensional linear manifold $K \subset \Phi$, we choose a basis $\{e_1, \dots, e_n\} \subset K$ $(n = \dim K)$ in it which is orthonormal in H_0 and define the vector

$$\beta_\mu^{(K)}(x) = \sum_{j=1}^{n} \beta_\mu(e_j, x)e_j \in \Phi \quad (x \in \Phi')$$

for which $\beta_\mu(\varphi, x) = (\beta_\mu^{(K)}(x), \varphi)_0$ for $\varphi \in K$. The value of $\beta_\mu^{(K)}(\cdot)$ is independent of the choice of the basis in K. We regard it as a mapping $\beta^{(K)}(\cdot)\colon \Phi' \to \Phi$ defined μ-almost everywhere. Thus, we have a system of mappings

$$\{\, \beta_\mu^{(K)}(\cdot) \mid K \subset \Phi, \ \dim K < \infty\}$$

consistent in the sense that

$$(\beta_\mu^{(K_1)}(\cdot), \varphi)_{H_0} = (\beta_\mu^{(K_2)}(\cdot), \varphi)_{H_0}$$

for $K_1 \subset K_2$ and any $\varphi \in K_1$.

We assume that there exists a mapping $\beta_\mu\colon \Phi' \to \Phi'$ defined μ-almost everywhere and such that $\beta_\mu^{(K)}(\cdot) = P_K\beta_\mu(\cdot)$, where P_K is a natural extension (to Φ') of the projector of Φ onto K which is orthogonal in H_0, i.e.,

$$P_K\xi = \sum_{j=1}^{n} (\xi, e_j)_{H_0} e_j \quad (\xi \in \Phi').$$

Here, $(e_j)_{j=1}^n$ $(n = \dim K)$ is an arbitrary basis in K of the form indicated above. In the case when Φ is a nuclear Fréchet space, this assumption automatically holds (see the Bibliographical Notes). The function $\beta_\mu(\varphi, \cdot)$ is called a logarithmic derivative of the measure μ in the direction φ, and the mapping β_μ is called a logarithmic derivative of the measure.

Example 3.1. Let $H_0 = \Phi = \Phi' = \mathbb{R}^d$, $d \in \mathbb{N}$. A quasi-invariant measure μ on $\mathfrak{B}(\mathbb{R}^d)$ is equivalent to the Lebesgue measure on $\mathbb{R}^d$ and $d\mu(x) = \rho(x)dx$, where $\rho > 0$ almost everywhere on $\mathbb{R}^d$ and $\rho \in L_{1,\mathrm{loc}}(\mathbb{R}^d)$ (see Gelfand and Vilenkin [1, Chapter 4, Section 5, Subsection 1, Theorem 2]). Since we now consider probability measures (i.e.,

$\mu(\mathbb{R}^d) = 1)$, the inclusion $\rho \in L_1(\mathbb{R}^d)$ holds. In what follows, it will be also convenient to use the function $\phi = \rho^{1/2} \in L_2(\mathbb{R}^d)$. For $\varphi = (\varphi_1, \dots \varphi_d) \in \mathbb{R}^d$, we have

$$\frac{d\mu(x+\varphi)}{d\mu(x)} = \rho_\mu(\varphi, x) = \frac{\rho(x+\varphi)}{\rho(x)} \quad (x \in \mathbb{R}^d)$$

and, consequently, the logarithmic derivative $\beta_\varphi(x)$ in the direction φ has the form

$$\beta_\mu(\varphi, x) = \frac{d}{dt}\frac{\rho(x+t\varphi)}{\rho(x)}\bigg|_{t=0} = \sum_{k=1}^{d} \varphi_k \frac{\partial\rho(x)}{\rho\partial x_k} = \left(\frac{\nabla\rho}{\rho}(x), \varphi\right)_{\mathbb{R}^d}$$

$$= (\nabla \ln \rho(x), \varphi)_{\mathbb{R}^d} = 2\left(\frac{\nabla\phi}{\phi}(x), \varphi\right)_{\mathbb{R}^d} \tag{3.4}$$

provided that the partial derivatives $\partial/\partial x_k(\phi(x)) \in L_2(\mathbb{R}^d)$, $k = 1, \dots, d$ exist. Note that the inclusion $\partial\phi/\partial x_k \in L_2(\mathbb{R}^d)$ is equivalent to $\frac{1}{\phi}\partial\phi/\partial x_k \in L_2(\mathbb{R}^d, \mu)$, $k = 1, \dots, d$ or to

$$\| \nabla\phi/\phi \|_{\mathbb{R}^d} \in L_2(\mathbb{R}^d, \mu).$$

It follows from (3.4) that the logarithmic derivative $\beta_\mu : \mathbb{R}^d \to \mathbb{R}^d$ of the measure μ coincides with the logarithmic derivative of the density of this measure with respect to the Lebesgue measure, i.e.,

$$\beta_\mu(x) = \nabla\rho(x)/\rho(x) = 2\nabla\phi(x)/\phi(x) \quad (x \in \mathbb{R}^d);$$

this clarifies its name. We explain that it is possible not to include the requirement $\beta_\mu(\varphi, \cdot) \in L_2(\mathbb{R}^d, \mu)$ in the definition of the logarithmic derivative if we regard in (3.3) the convergence μ-almost everywhere or in the measure μ instead of the limit in $L_2(\mathbb{R}^d, \mu)$. As far as this example is concerned, this means that we refuse the inclusion $\partial\phi/\partial x_k \in L_2(\mathbb{R}^d)$. However, for the purposes of our presentation, this requirement is necessary, and we include it in the definition of the logarithmic derivative.

Example 3.2 (Logarithmic derivative of the Gaussian measure). Let S be a positive operator in H_0 such that $S \in \mathcal{L}_0(\Phi, \Phi')$. Then, as shown in Subsection 1.4 of Chapter 2, a Gaussian measure γ_S with correlation operator S is defined on $\mathfrak{C}_\sigma(\Phi')$. It follows from the investigations in Subsection 2.3 of Chapter 2 that γ_S is quasi-invariant with respect to shifts by the vectors from Φ if $\Phi \subset \mathfrak{D}(S^{-1/2})$. Under this condition, Lemma 2.4 in Chapter 2 yields

$$\rho_{\gamma_S}(\varphi, x) = \exp\left(-\| S^{-1/2}\varphi\|^2_{H_0} - 2(S^{-1}\varphi, x)_{H_0}\right) \quad (x \in \Phi')$$

for $\varphi \in \Phi$. Using definition (3.3), we obtain

$$\beta_{\gamma_S}(\varphi, x) = -2(S^{-1}\varphi, x)_{H_0} \quad (\varphi \in \Phi,\ x \in \Phi') \tag{3.5}$$

We stress that the right-hand side of (3.5) is now defined as a linear measurable function belonging to all $L_p(\Phi', \gamma_S)$, $p \geq 1$; this follows from the inclusion $S^{-1}\varphi \in H_S$, where H_S is a completion of Φ with respect to the scalar product

$$(\varphi, \psi)_{H_S} = (\varphi, S\psi)_{H_0}$$

(see Subsection 2.1 in Chapter 2). The mapping

$$\Phi \ni \varphi \mapsto \beta_{\gamma_S}(\varphi, \cdot) = -2(S^{-1}\varphi, x)_{H_0} \in L_2(\Phi', \gamma_S)$$

is linear and continuous. Indeed, according to (2.5) in Chapter 2, we have

$$\| \beta_{\gamma_S}(\varphi, \cdot) \|^2_{L_2(\Phi', \gamma_S)} = 4 \int_{\Phi'} (S^{-1}\varphi, x)^2_{H_0}\, d\gamma_S(x) = 2 \| S^{-1/2}\varphi \|^2_{H_0}.$$

The logarithmic derivative $\beta_{\gamma_S}(x) = -2S^{-1}x$ is defined as a linear weakly measurable mapping of Φ' into Φ', i.e. for every $\varphi \in \Phi$, the function $(\beta_{\gamma_S}(x), \varphi)_{H_0}$ $(x \in \Phi')$ is a measurable linear function on Φ' (in the sense of Subsection 2.1 in Chapter 2). If $S^{-1} \in \mathfrak{L}(\Phi, \Phi)$, then $\beta_{\gamma_S}(x)$ is a continuous linear mapping of Φ' onto Φ'. In particular, this is true for $S = 1$, i.e., for the canonical Gaussian measure γ_1, we have $\beta_{\gamma_1}(x) = -2x$ $(x \in \Phi')$.

If $u \in C^2_{b,\mathrm{cyl}}(\Phi')$, then

$$u(x) = f((\varphi_1, x)_{H_0}, \dots, (\varphi_n, x)_{H_0}) \quad (x \in \Phi')$$

by definition, where $\varphi_k \in \Phi$, $k = 1, \dots, n$, $f \in C^2_b(\mathbb{R}^n)$, $n \in \mathbb{N}$. Therefore,

$$\nabla u(x) = \sum_{k=1}^{n} \frac{\partial f}{\partial s_k}((\varphi_1, x)_{H_0}, \dots, (\varphi_n, x)_{H_0})\varphi_k \in K = \text{e.s. } \{\varphi_1, \dots, \varphi_n\} \subset \Phi,$$

$$A\nabla u(\cdot) \in AK = \text{e.s. } \{A\varphi_1, \dots, A\varphi_n\}.$$

Consequently, for these functions, the following differential expression is defined

$$L_{\mu, A} u = -\frac{1}{2}\,[\mathrm{Tr}_{H_0} A u'' + (A\nabla u, \beta_\mu)_{H_0}]. \tag{3.6}$$

This expression gives the operator $L_{\mu, A}$ in $L_2(\Phi', \mu)$ with the domain of definition $C^2_{b,\mathrm{cyl}}(\Phi')$. This operator is called the Dirichlet operator corresponding to the measure μ and the coefficient operator A. Note that for each function $u \in C^2_{b,\mathrm{cyl}}(\Phi')$ in (3.6), in fact, we use only the value $\beta^{(K)}_{\mu}$ with $K \subset \Phi$, $\dim K < \infty$, depending on u and A. The connection between the operator $L_{\mu, A}$ and the Dirichlet form $d_{\mu, A}$ is established by the following theorem.

Theorem 3.1. *For $u, v \in C^2_{b,\mathrm{cyl}}(\Phi')$, the following equality holds*

$$d_{\mu, A}(u, v) = (L_{\mu, A} u, v)_{L_2(\Phi', \mu)}. \tag{3.7}$$

Proof. First, assume that $A = 1$. For given $u, v \in C^2_{b,\mathrm{cyl}}(\Phi')$, we choose an orthonormal in H_0 set $\{e_1, \ldots, e_n\} \subset \Phi$ such that $u(x) = u(P_K x)$ and $v(x) = v(P_K x)$, $x \in \Phi'$, where $K = \text{e.s.}\,\{e_1, \ldots, e_n\}$. Denoting by $\nabla_k u(x)$ the derivative of $u(x)$ in the direction e_k, $k = 1, \ldots, n$, we find that

$$\nabla u(x) = \sum_{k=1}^{n} \nabla_k u(x) e_k = \sum_{k=1}^{n} \frac{\partial u(x)}{\partial x_k} e_k, \quad x_k = (x, e_k)_{H_0}$$

and, consequently,

$$d_{\mu, 1}(u, v) = \frac{1}{2} \sum_{k=1}^{n} \int_{\Phi'} \nabla_k u \cdot \nabla_k \bar{v}\, d\mu. \tag{3.8}$$

Since

$$\nabla_k u(x) = \lim_{t \to 0} \frac{u(x + te_k) - u(x)}{t},$$

we obtain the equality

$$\int_{\Phi'} \nabla_k u(x) \nabla_k \overline{v(x)}\, d\mu(x) = \lim_{t \to 0} \int_{\Phi'} \frac{u(x + te_k) - u(x)}{t} \frac{\partial \overline{v(x)}}{\partial x_k} d\mu(x)$$

$$= \lim_{t \to 0} \frac{1}{t} \left[\int_{\Phi'} u(x) \frac{\partial \bar{v}(x - te_k)}{\partial x_k} \frac{\partial \mu(x - te_k)}{\partial \mu(x)} d\mu(x) - \int_{\Phi'} u(x) \frac{\partial \bar{v}(x)}{\partial x_k} d\mu(x) \right]$$

$$= \lim_{t \to 0} \int_{\Phi'} u(x) \frac{1}{t} \left[\frac{\partial \bar{v}(x - te_k)}{\partial x_k} \rho_{\mu}(-te_k, x) - \frac{\partial \bar{v}(x)}{\partial x_k} \right] d\mu(x)$$

$$= \int_{\Phi'} u(x)\left[-\frac{\partial^2 \bar{v}(x)}{\partial x_k^2} - \beta_\mu(e_k, x)\frac{\partial \bar{v}(x)}{\partial x_k}\right] d\mu(x).$$

Taking the equality obtained into account in each term of (3.8), we get

$$d_{\mu,1}(u, v) = \frac{1}{2}\int_{\Phi'} u(x)\left[-\sum_{k=1}^{n}\frac{\partial^2 \bar{v}(x)}{\partial x_k^2} - \sum_{k=1}^{n}\beta_\mu(e_k, x)\frac{\partial \bar{v}(x)}{\partial x_k}\right] d\mu(x)$$

$$= \frac{1}{2}\int_{\Phi'} u(x)\,\overline{[-\mathrm{Tr}_{H_0} v''(x) - (\nabla v(x), \beta_\mu(x))_{H_0}]}\, d\mu(x) = \int_{\Phi'} u(x)\,\overline{L_{\mu,1} v(x)}\, d\mu(x),$$

i.e., (3.7) in the case $A = 1$.

For the case of a general coefficient operator, one should make an analogous computation in each term of the equality

$$d_{\mu,A}(u, v) = \frac{1}{2}\sum_{k,j=1}^{n}\int_{\Phi'} \nabla_k u(x)\nabla_j \bar{v}(x)(Ae_k, e_j)_{H_0}\, d\mu(x).$$

As a result, we obtain the representation

$$d_{\mu,A}(u, v) = \frac{1}{2}\sum_{k,j=1}^{n}\int_{\Phi'} u\left[-\frac{\partial^2 \bar{v}}{\partial x_k \partial x_j}(Ae_k, e_j)_{H_0} - \frac{\partial \bar{v}}{\partial x_j}\beta_\mu(e_k, \cdot)(Ae_k, e_j)_{H_0}\right] d\mu$$

$$= \frac{1}{2}\int_{\Phi'} u\left[-\sum_{k,j=1}^{n}(\bar{v}'' e_k, e_j)_{H_0}(Ae_k, e_j)_{H_0} - \left(\nabla \bar{v}, \sum_{j=1}^{n}\beta_\mu(e_j, \cdot)Ae_j\right)_{H_0}\right] d\mu$$

$$= \frac{1}{2}\int_{\Phi'} u(x)\,\overline{[-\mathrm{Tr}_{H_0} Av''(x) - (A\nabla v(x), \beta_\mu(x))_{H_0}]}\, d\mu(x) = (u, L_{\mu,A} v)_{L_2(\Phi', \mu)}.$$

■

Remark 1. If the duality between Φ and Φ' is established by use of the Hilbert space $\mathcal{H}_0$, and $H_0 \supset \Phi$ densely topologically and $\|\varphi\|_{H_0} = \|J\varphi\|_{\mathcal{H}_0}$, $\varphi \in \Phi$, then the form

$$\frac{1}{2}\int_{\Phi'} (\nabla u, A\nabla \bar{v})_{H_0}\, d\mu = \frac{1}{2}\int_{\Phi'} (\nabla u, J^*A\nabla \bar{v})_{\mathcal{H}_0}\, d\mu,$$

in which the derivatives are understood in the sense of the space, is associated with the differential expression

$$-\frac{1}{2}[\mathrm{Tr}_{\mathcal{H}_0} J^*AJu''(x) + (A\nabla u(x), \beta_\mu(x))_{H_0}].$$

■

Remark 2. The existence of a logarithmic derivative of the measure μ enables us to associate with the Dirichlet form a symmetric nonnegative operator $L_{\mu,A}$ given densely in $L_2(\Phi', \mu)$ and, therefore, implies its closability. The operator associated with the closure $d_{\mu,A}$ (Friedrichs' extension of $L_{\mu,A}$) is denoted by $L^F_{\mu,A}$; $L^F_{\mu,A}$ is a nonnegative selfadjoint operator in $L_2(\Phi', \mu)$.

One can obtain somewhat weaker conditions sufficient for the Dirichlet form to be closable, but if we want to associate a differential operator with this form, we have to carry out integration by parts in (3.1) and this necessarily yields the requirement of the existence of a logarithmic derivative of the measure μ.

■

Remark 3. We do not now dwell upon the important problem of essential selfadjointness of the operator $L_{\mu,A}$; even in the finite-dimensional case, it is necessary to impose strong restrictions on the measure μ (we mean the requirements of smoothness of density of the measure μ and the inclusion of its derivatives in the spaces L_p with proper indices; see the Bibliographical Notes). All the more, this problem is difficult in the infinite-dimensional case when we cannot speak at all about the density of the measure μ with respect to some standard measure. It is more convenient to discuss the corresponding conditions somewhat later. Now we proceed to investigate the connection between potential perturbations of the second quantization operators and the Dirichlet operators which correspond to multiplicative perturbations of the standard Gaussian measure γ_1 on Φ'.

3.2. Renormalization of Potential Perturbations of the Second Quantization Operators by Use of the Dirichlet Forms

Assume that the coefficient operator A satisfies the condition of uniform ellipticity $A \geq a1$, $a > 0$, and the potential V is given by the real function from $L_{2+\varepsilon}(\Phi', \gamma_1)$, $\varepsilon > 0$, for which $e^{-V} \in L_p(\Phi', \gamma_1)$ for any $p \geq 1$. Then, according to the corollary to Theorem 2.9, the operator $L_A + V$ is essentially selfadjoint on $C^\infty_{b,\mathrm{cyl}}(\Phi')$ and semibounded from below. Moreover, Theorem 2.7 states that $\inf s\,(L_A + V)$ is a simple eigenvalue and the corresponding eigenfunction ϕ_V may be chosen to be strictly positive γ_1-almost everywhere. We introduce a probability measure $\mu^V = \phi_V^2 \gamma_1$ on Φ' assuming that ϕ_V

is normalized in $L_2(\Phi', \gamma_1)$. First, we show that the measure μ^V may be associated with a Dirichlet operator.

Lemma 3.1. *Under the indicated conditions imposed on* A *and* V, *the measure* μ^V *is associated with a Dirichlet operator of the form*

$$L_{\mu^V, A}u = L_A u - \phi_V^{-1}(A\nabla u, \nabla\phi_V)_{H_0} \quad (u \in C^\infty_{b,\mathrm{cyl}}(\Phi')). \tag{3.6'}$$

Proof. We now establish the existence of the logarithmic derivative of the measure μ^V and find its explicit form. Taking into account the hypercontractibility of the semigroup $\exp(-t(L_A + V))$, $t \geq 0$, which was established in Section 2, we can conclude that $\phi_V \in L_p(\Phi', \gamma_1)$ for any $p \geq 1$ and, consequently, $\phi_V \in \mathfrak{D}(V) \cap \mathfrak{D}(L_A)$. Moreover, ϕ_V belongs to the domain of definition of the corresponding quadratic form and, thus,

$$\| (A\nabla\phi_V, \nabla\phi_V)_{H_0} \|_{L_1(\Phi', \gamma_1)} < \infty.$$

But $(A\nabla\phi_V, \nabla\phi_V)_{H_0} \geq a\|\nabla\phi_V\|^2_{H_0}$. This yields the existence of the derivative $\nabla\phi_V \in H_0 \otimes L_2(\Phi', \gamma_1)$ and the derivatives in the directions $\varphi \in \Phi$, i. e., $(\nabla\phi_V, \varphi)_{H_0} \in L_2(\Phi', \gamma_1)$. Taking into account the form of the Radon–Nikodym derivative for the shifted measure γ_1 (see Example 3.2), we get

$$\frac{d\mu^V(x + t\varphi)}{d\mu^V(x)} = \frac{\phi_V^2(x + t\varphi)}{\phi_V^2(x)} \exp\{-2t(\varphi, x)_{H_0} - t^2\|\varphi\|^2_{H_0}\} \quad (\varphi \in \Phi,\ t \in \mathbb{R}^1)$$

and this yields the existence of the logarithmic derivative $\beta_{\mu^V}(\varphi, x)$ of the measure μ^V in the direction $\varphi \in \Phi$ and the equality

$$\beta_{\mu^V}(\varphi, x) = \frac{d}{dt}\left(\frac{d\mu^V(x + t\varphi)}{d\mu^V(x)}\right)\bigg|_{t=0} = 2\phi_V^{-1}(\nabla\phi_V(x), \varphi)_{H_0}$$

$$- 2(\varphi, x)_{H_0} \in L_2(\Phi', \mu^V). \tag{3.9}$$

It follows from (3.9) that $\beta_{\mu^V} : \Phi' \to \Phi'$ has the following explicit form

$$\beta_{\mu^V}(x) = 2\nabla\phi_V(x)/\phi_V(x) - 2x. \tag{3.10}$$

In view of this, (3.6) yields the following representation for the operator $L_{\mu^V, A}$

$$(L_{\mu^V, A}u)(x) = -\frac{1}{2}\mathrm{Tr}_{H_0} Au''(x) + (A\nabla u(x), x)_{H_0}$$

$$- \phi_V^{-1}(x)(A\nabla u(x), \nabla\phi_V(x))_{H_0} = (L_A u)(x) - \phi_V^{-1}(x)(A\nabla u(x), \nabla\phi_V(x))_{H_0},$$

and this proves the validity of the lemma.

■

The operator of multiplication by the function $\phi_V^{-1}(\bullet)$ realizes the unitary isomorphism of $L_2(\Phi', \gamma_1)$ and $L_2(\Phi', \mu^V)$, i. e.,

$$L_2(\Phi', \dot{\gamma}_1) \ni f \mapsto \phi_V^{-1} \cdot f \in L_2(\Phi', \mu^V).$$

Denote by $L^V_{A,\text{ren}} = \phi_V^{-1}(L_A + V - E_V)\phi_V$ the operator in $L_2(\Phi', \mu^V)$ which is unitary equivalent to $L_A + V - E_V$ under this isomorphism. We call $L^V_{A,\text{ren}}$ a renormalized operator corresponding to the potential perturbation V; the transition from $L_2(\Phi', \gamma_1)$ and $L_A + V = L_A^V$ to the new Hilbert space $L_2(\Phi', \mu^V)$ and the operator $L^V_{A,\text{ren}}$ is called a renormalization procedure or, simply, renormalization. This terminology is justified by direct analogy with a similar procedure in the constructive quantum field theory. The next theorem reveals the canonical connection between the renormalized operator and the Dirichlet form of the perturbed measure μ^V.

Theorem 3.2. *Assume that A and V satisfy the requirements imposed above. Then $C^2_{b,\text{cyl}}(\Phi') \subset \mathfrak{D}(L^V_{A,\text{ren}})$ and*

$$L^V_{A,\text{ren}} \restriction C^2_{b,\text{cyl}}(\Phi') = L_{\mu^V, A}.$$

Proof. The inclusion $C^2_{b,\text{cyl}}(\Phi') \subset \mathfrak{D}(L^V_{A,\text{ren}})$ is equivalent to

$$\{u\phi_V \mid u \in C^2_{b,\text{cyl}}(\Phi')\} \subset \mathfrak{D}(L_A + V).$$

Due to the hypercontracting properties of the semigroup $\exp(-t(L_A + V))$, $t \geq 0$, we have $\phi_V \in L_p(\Phi', \gamma_1)$ for any $p \geq 1$ and, therefore

$$\{u\phi_V \mid u \in C^2_{b,\text{cyl}}(\Phi')\} \subset \mathfrak{D}(V).$$

Thus, it suffices to establish the inclusion of the set under consideration in $\mathfrak{D}(L_A)$. Note that the explicit form of L_A yields the equality

$$L_A(u \cdot v) = (L_A u) \cdot v + (L_A v)u - (A\nabla u, \nabla v)_{H_0} \tag{3.11}$$

for $u, v \in C^2_{b,\text{cyl}}(\Phi')$. According to Theorem 1.4, $C^2_{b,\text{cyl}}(\Phi')$ is a domain of essential selfadjointness for L_A. Hence, one can choose $(v_n)_{n=1}^{\infty} \subset C^2_{b,\text{cyl}}(\Phi')$ such that $\bar{v}_n = v_n \to \phi_V$ and $L_A v_n \to L_A \phi_V$ as $n \to \infty$ strongly in $L_2(\Phi', \gamma_1)$. Substituting v_n for v in (3.11) and taking into account the fact that $uL_A v_n \to uL_A \phi_V$ and $v_n L_A u \to \phi_V L_A u$ as $n \to \infty$ strongly in $L_2(\Phi', \gamma_1)$ and the strong convergence

$$(A\nabla u, \nabla v_n)_{H_0} \to (A\nabla u, \nabla \phi_V)_{H_0}$$

which follows from the inequality

$$|(A\nabla u, \nabla v_n)_{H_0}| \le (A\nabla u, \nabla \bar{u})^{1/2}_{H_0} (A\nabla v_n, \nabla v_n)^{1/2}_{H_0}$$

and the convergence

$$(L_A v_n, v_n)_{L_2(\Phi', \gamma_1)} = \frac{1}{2} \int_{\Phi'} (A\nabla v_n, \nabla v_n)_{H_0} d\gamma_1 \xrightarrow[n \to \infty]{} (L_A \phi_V, \phi_V)_{L_2(\Phi', \gamma_1)}$$

$$= \frac{1}{2} \int_{\Phi'} (A\nabla \phi_V, \nabla \phi_V)_{H_0} d\gamma_1,$$

we obtain the inclusion $u\phi_V \in \mathfrak{D}(L_A)$ and the equality

$$L_A(u\phi_V) = \phi_V L_A u + uL_A \phi_V - (A\nabla u, \nabla \phi_V)_{H_0}.$$

The second statement of the theorem means that

$$(L_A + V - E_V)f = \phi_V L_{\mu^V, A} \phi_V^{-1} f \tag{3.12}$$

for all $f = u\phi_V$ and $u \in C^2_{b,\text{cyl}}(\Phi')$. To prove (3.12), it suffices to consider the case when $f = \bar{f}$. We have

$$((L_A + V - E_V)f, f)_{L_2(\Phi', \gamma_1)}$$

$$= (L_A f, f)_{L_2(\Phi', \gamma_1)} + (Vf, f)_{L_2(\Phi', \gamma_1)} - E_V \|f\|^2_{L_2(\Phi', \gamma_1)}. \tag{3.13}$$

But

$$(L_A f, f)_{L_2(\Phi', \gamma_1)} = \frac{1}{2} \int_{\Phi'} (\nabla(u\phi_V), A\nabla(u\phi_V))_{H_0} d\gamma_1$$

$$= \frac{1}{2} \int_{\Phi'} |u|^2 (\nabla \phi_V, A\nabla \phi_V)_{H_0} d\gamma_1 + \frac{1}{2} \int_{\Phi'} (\nabla u, A\nabla u) \phi_V^2 \, d\gamma_1$$

$$+\int_{\Phi'}(A\nabla u, \nabla\phi_V)_{H_0}d\gamma_1 = \frac{1}{2}\int_{\Phi'}(\nabla\phi_V, A\nabla(\phi_V u^2))_{H_0}d\gamma_1$$

$$+\frac{1}{2}\int_{\Phi'}(\nabla u, A\nabla u)_{H_0}d\mu^V = (L_A\phi_V, \phi_V u^2)_{L_2(\Phi',\gamma_1)} + (L_{\mu^V,A}u, u)_{L_2(\Phi',\mu^V)}.$$

If we substitute this equality in (3.13) taking into account that $L_A\phi_V = E_V\phi_V - V\phi_V$, we get

$$((L_A + V - E_V)f, f)_{L_2(\Phi',\gamma_1)} = (L_{\mu^V,A}u, u)_{L_2(\Phi',\mu^V)} = (\phi_V L_{\mu^V,A}\phi_V^{-1}f, f)_{L_2(\Phi',\gamma_1)}$$

and this proves (3.12).

■

An analysis of the calculation in the proof of Theorem 3.2 shows that its statement remains valid on the formal level also for the potential perturbations of general Dirichlet operators $L_{\mu,A}$, provided that $L_{\mu,A}+V$ possesses a ground state $\phi_V > 0$ μ-almost everywhere. In particular, this is true for the important case of the Laplace operator $-\frac{1}{2}\Delta$ in $L_2(\mathbb{R}^d)$ generated by the Dirichlet form of the Lebesgue measure on $\mathbb{R}^d$, $d \in N$. Many papers exist which are devoted to the investigation of potential perturbations of the Laplace operator from this point of view (in connection with singular problems in quantum mechanics); the reader can find references to some of these in the Bibliographical Notes. The following examples contain the minimum amount of information concerning the connection between the Dirichlet operators and potential perturbations of the Laplace operator which will be necessary in Chapter 7.

Example 3.3 (From the perturbation of the Laplace operator to the Dirichlet operator). Let $V \in L_{2,\mathrm{loc}}(\mathbb{R}^d)$, let the operator $-\frac{1}{2}\Delta + V$ defined on $C_0^\infty(\mathbb{R}^d)$ be essentially selfadjoint and semibounded from below, and let $E_V = \inf s(-\frac{1}{2}\Delta + V)$ be a simple eigenvalue with the corresponding eigenfunction (the ground state) $\phi_V > 0$. In particular, this situation is realized for $V \geq 0$ and $\lim_{|x|\to\infty} V(x) = \infty$.

Let us introduce a probability measure $d\mu^V = \phi_V^2 dx$ on $\mathfrak{B}(\mathbb{R}^d)$ and denote by

$$\left(-\frac{1}{2}\Delta + V\right)_{\mathrm{ren}} = \phi_V^{-1}(-1/2 + V - E_V)\phi_V$$

a unitary image of the operator $(-\Delta/2) + V - E_V$ under the transition from $L_2(\mathbb{R}^d)$ to $L_2(\mathbb{R}^d, \mu^V)$. As above, in the case of the second quantization operators, we call $\left(-\frac{1}{2}\Delta + V\right)_{\mathrm{ren}}$ a renormalized operator corresponding to the potential perturbation V. If $\phi_V \in \mathfrak{D}(-\frac{1}{2}\Delta)$, then by direct repetition of the transformations which follow (3.13), we find, for the case under consideration, that

$$\left(-\frac{1}{2}\Delta + V\right)_{\text{ren}} \upharpoonright C_b^\infty(\mathbb{R}^d) = L_{\mu^V}. \tag{3.14}$$

Therefore, the renormalized operator on $C_0^\infty(\mathbb{R}^d)$ coincides with the Dirichlet operator of the vacuum measure μ^V. According to Example 3.1 and the equality (3.6), this gives the representation

$$\left(-\frac{1}{2}\Delta + V\right)_{\text{ren}} u = L_{\mu^V} u = -\frac{1}{2}\Delta u - (\nabla\phi_V, \nabla u)_{\mathbb{R}^d}\phi_V^{-1} \tag{3.15}$$

$$(u \in C_b^\infty(\mathbb{R}^d)).$$

Conversely, if we start from the Dirichlet operator of the probability measure on $\mathbb{R}^d$, then (under certain conditions of smoothness of its density with respect to the Lebesgue measure) we can reconstruct the potential perturbation of the Laplace operator which leads to this operator.

Example 3.4 (From the Dirichlet operator to the perturbation of the Laplace operator). Let $\phi > 0$ almost everywhere on $\mathbb{R}^d$, $\phi \in L_2(\mathbb{R}^d)$, $\|\phi\|_{L_2(\mathbb{R}^d)} = 1$, and let $d\mu(x) = \phi^2(x)dx$ be a probability measure on $\mathbb{R}^d$ with density ϕ^2 with respect to the Lebesgue measure. Under the condition that $\phi \in W_1^2(\mathbb{R}^d)$, the Dirichlet form

$$d_\mu(u, v) = \frac{1}{2}\int_{\mathbb{R}^d} (\nabla u, \nabla \bar{v})_{\mathbb{R}^d}\, d\mu \quad (u, v \in C_b^\infty(\mathbb{R}^d))$$

is associated with the operator L_μ which generates this form, has a domain of definition $\mathfrak{D}(L_\mu) = C_b^2(\mathbb{R}^d)$, and is given on $u = C_b^2(\mathbb{R}^d)$ by the equality (cf. (3.15))

$$L_\mu u = -\frac{1}{2}\Delta u - \frac{1}{\phi}(\nabla u, \nabla\phi)_{\mathbb{R}^d}. \tag{3.16}$$

Let us find the image of the operator L_μ under the natural unitary isomorphism between $L_2(\mathbb{R}^d, \mu)$ and $L_2(\mathbb{R}^d)$ given by the correspondence

$$L_2(\mathbb{R}^d, \mu) \ni f \mapsto \phi f \in L_2(\mathbb{R}^d).$$

This operator is given by the equality $\phi L_\mu \phi^{-1}$, and below we show that

$$\phi L_\mu \phi^{-1} \upharpoonright C_0^\infty(\mathbb{R}^d) = -\frac{1}{2}\Delta + V_\mu, \quad V_\mu = \frac{\Delta\phi}{\phi}, \tag{3.17}$$

provided that the inclusions $\phi \in W^2_{2,\mathrm{loc}}(\mathbb{R}^d)$ and $\phi^{-1} \in L_{\infty,\mathrm{loc}}(\mathbb{R}^d)$ hold.

For $u = \bar{u} \in C_0^\infty(\mathbb{R}^d)$, we have

$$(\phi L_\mu \phi^{-1} u, u)_{L_2(\mathbb{R}^d)} = (L_\mu(\phi^{-1}u), (\phi^{-1}u))_{L_2(\mathbb{R}^d,\mu)} = \frac{1}{2}\int_{\mathbb{R}^d} (\nabla(\phi^{-1}u), \nabla(\phi^{-1}u))_{\mathbb{R}^d} d\mu$$

$$= \frac{1}{2}\int_{\mathbb{R}^d} \left\| \frac{\nabla u}{\phi} - \frac{u\nabla\phi}{\phi^2}\right\|^2_{\mathbb{R}^d} \phi^2 dx = \frac{1}{2}\int_{\mathbb{R}^d} (\nabla u, \nabla u)_{\mathbb{R}^d} dx$$

$$+ \frac{1}{2}\int_{\mathbb{R}^d} u^2 \frac{\|\nabla\phi\|^2_{\mathbb{R}^d}}{\phi^2} dx - \frac{1}{2}\int_{\mathbb{R}^d}\left(\nabla(u^2), \frac{\nabla\phi}{\phi}\right)_{\mathbb{R}^d} dx = \int_{\mathbb{R}^d}\left(-\frac{1}{2}\Delta u\right) u\,dx$$

$$+ \frac{1}{2}\int_{\mathbb{R}^d} u^2\left[\frac{\|\nabla\phi\|^2_{\mathbb{R}^d}}{\phi^2} + \operatorname{div}\frac{\nabla\phi}{\phi}\right] dx = \int_{\mathbb{R}^d}\left[-\frac{1}{2}\Delta u + \frac{1}{2}\frac{\Delta\phi}{\phi}\right] u\,dx.$$

Since

$$V_\mu = \frac{1}{2}\frac{\Delta\phi}{\phi} \in L_{2,\mathrm{loc}}(\mathbb{R}^d)$$

(due to the conditions imposed on ϕ), the equality obtained proves (3.14).

It is useful to rewrite the representation for the potential V_μ in terms of the logarithmic derivative of the measure μ. Taking into account the equality

$$\operatorname{div}\beta_\mu = \operatorname{div}\left(2\frac{\nabla\phi}{\phi}\right) = 2\,\frac{\phi\Delta\phi - \|\nabla\phi\|^2_{\mathbb{R}^d}}{\phi^2} = 4V_\mu - \frac{1}{2}\|\beta_\mu\|^2_{\mathbb{R}^d}$$

and the fact that Example 3.1 yields $\beta_\mu = 2\nabla\phi/\phi$, we conclude that

$$V_\mu(x) = \frac{1}{2}\|\beta_\mu(x)\|^2_{\mathbb{R}^d} + \frac{1}{8}\operatorname{div}\beta_\mu(x). \tag{3.18}$$

***Remark* 1.** Acting just as in Example 3.4, we can establish that in the case of a measure $\mu = \phi^2\gamma_1$ on $\mathcal{C}_\sigma(\Phi')$ with smooth density $\phi > 0$, $\phi \in L_2(\Phi', \gamma_1)$, under the unitary transition from $L_2(\Phi', \mu)$ to $L_2(\Phi', \gamma_1)$, the Dirichlet operator $L_{\mu,A}$ is associated with the potential perturbation V_μ of a second quantization operator L_A of the form

$$V_\mu = \frac{1}{2}[(\nabla\psi, A\nabla\psi)_{H_0} - 2L_A\psi]$$

where $\psi = \ln \phi$. Here one can easily connect the presence of the inclusion $V_\mu \in L_p(\Phi', \gamma_1)$ with the denseness properties of ϕ.

■

Relying on representation (3.18), we now discuss the connection between the Dirichlet operators of Gaussian measures on $\mathbb{R}^d$ and the potential perturbations of the Laplace operator.

Example 3.5. For given positive linear operator S on $\mathbb{R}^d$, we consider a Gaussian measure γ_S on $\mathfrak{B}(\mathbb{R}^d)$ with the correlation operator S (see Subsection 1.5 in Chapter 2) and the corresponding Dirichlet operator L_{γ_S} in $L_2(\mathbb{R}^d, \gamma_S)$. Since we now have $\beta_{\gamma_S}(x) = -2S^{-1}x$ $(x \in \mathbb{R}^d)$ (see Example 3.2), this operator has the form

$$(L_{\gamma_S} u)(x) = -\frac{1}{2}\Delta u(x) + (S^{-1}x, \nabla u(x))_{\mathbb{R}^d}$$

$$= -\frac{1}{2}\sum_{k=1}^{d}\frac{\partial^2 u(x)}{\partial x_k^2} + \sum_{k,j=1}^{d} r_{kj}x_k\frac{\partial u(x)}{\partial x_j} \quad (u \in C_b^2(\mathbb{R}^d)),$$

where $r_{k,j}$, $k, j = 1, \dots, d$, are matrix elements of the operator $R = S^{-1}$. Passing from $L_2(\mathbb{R}^d, \gamma_S)$ to $L_2(\mathbb{R}^d)$ and using the result of Example 3.4, we conclude that the operator L_{γ_S} is a renormalized operator corresponding to $-\frac{1}{2}\Delta + V_{\gamma_S}$, where

$$V_{\gamma_S}(x) = \frac{1}{2}\|\beta_{\gamma_S}(x)\|^2_{\mathbb{R}^d} + \frac{1}{8}\operatorname{div}\beta_{\gamma_S}(x) = 2\|Rx\|^2_{\mathbb{R}^d} - \frac{1}{2}\operatorname{Tr} R$$

$$= \frac{1}{2}(R^2x, x)_{\mathbb{R}^d} - \frac{1}{2}\operatorname{Tr} R = \frac{1}{2}[(R^2x, x)_{\mathbb{R}^d} - \operatorname{Tr} R] \quad (x \in \mathbb{R}^d). \tag{3.19}$$

Indeed, since the potential V_{γ_S} belongs to $L_{\infty,\mathrm{loc}}(\mathbb{R}^d)$ and is semibounded from below, the operator $-\frac{1}{2}\Delta + V_{\gamma_S}$ determined first on $C_0^\infty(\mathbb{R}^d)$, is essentially selfadjoint. Hence, the renormalized operator is essentially selfadjoint on the region

$$\{u \cdot \phi_{V_{\gamma_S}}^{-1},\ u \in C_0^\infty(\mathbb{R}^d)\} = \{u(x)e^{1/2(Rx,x)_{\mathbb{R}^d}};\ u \in C_0^\infty(\mathbb{R}^d)\} = C_0^\infty(\mathbb{R}^d) \subset C_b^2(\mathbb{R}^d).$$

This establishes the coincidence of $\left(-\frac{1}{2}\Delta + V_{\gamma_S}\right)_{\mathrm{ren}}$ and the closure of L_{γ_S} because

$$L_{\gamma_S} = \left(-\frac{1}{2}\Delta + V_{\gamma_S}\right)_{\mathrm{ren}} \restriction C_b^2(\mathbb{R}^d)$$

Conversely, the given quadratic form $(Dx, x)_{\mathbb{R}^d}$, $x \in \mathbb{R}^d$, generated by the linear operator $D > 0$ in $\mathbb{R}^d$ is associated with the selfadjoint operator $-\frac{1}{2}\Delta + (D\cdot, \cdot)_{\mathbb{R}^d}$ in $L_2(\mathbb{R}^d)$ for which the Dirichlet operator L_{γ_S} with $S = D^{-1/2}$ serves as a renormalized operator.

3.3. Dirichlet Operators in the Case of Measures on a Hilbert Space

As has been shown in Subsection 3.1, under certain conditions the probability measure μ on Φ' is associated with the Dirichlet operator on the region $C^2_{b,\text{cyl}}(\Phi')$. However, in the general situation, the problem of the essential selfadjointness of this operator remains open. The affirmative answer to this question can be obtained in the case of a Gaussian measure μ. In this situation, the Dirichlet operator can be reduced to the second quantization operator. For measures on Φ' which are multiplicative perturbations of Gaussian measures, the Dirichlet operators are unitary equivalent to the potential perturbations of the second quantization operators (see the remark after Example 3.4) and, thus, we can obtain conditions of essential selfadjointness for this case, too. Finally, for product measures, the indicated problem can be investigated by use of the general theory of operators with infinitely many separating variables (Berezansky [18, Chapter 3, Section 3] and [26, Chapter 3, Section 3]).

For measures which do not belong to the above-mentioned classes, the investigation of the selfadjointness of Dirichlet operators is a complicated problem, even in the finite-dimensional case. Below, we establish the conditions on measures under which it is possible to prove the essential selfadjointness of the Dirichlet operators corresponding to these measures. We restrict ourselves to the study of measures concentrated on Hilbert subspaces of Φ' because our method of investigation of selfadjointness exploits, in fact, the presence of a metric in the space of arguments.

Let H_0 be a real separable Hilbert space, and let $H_- \supset H_0 \supset H_+$ be its Hilbert rigging with quasinuclear imbeddings. Along with the spaces in this chain, we shall use their complexifications; this will be always clear from the context. The scalar products in the spaces of the chain (or in their complexifications) are denoted as everywhere above. Recall that for any function $f\colon H_- \to \mathbb{C}^1$, the first and second differentials are determined by the Taylor formula

$$f(x+h) = f(x) + df(x)h + \frac{1}{2}\, d^2 f(x)(h,h) + o(\|h\|^2_{H_-}), \quad \|h\|_{H_-} \to 0,$$

in which df and $d^2 f$ are continuous linear and bilinear forms on H_-, respectively. The derivatives $f' = \nabla f$ and f'' are given by the realizations of these differentials in the scalar product of the space H_0

$$df(\cdot)\,(h) = (f'(\cdot), h)_{H_0}, \quad f'(\cdot) \in H_{+,c},$$

$$d^2 f(\cdot)\,(h, h) = (f''(\cdot)h, h)_{H_0}, \quad f''(\cdot) \in \mathfrak{L}(H_-, H_{+,c}).$$

Similarly, we define the derivatives of the mapping $f\colon H_- \to H_{-,c}$ (see Subsections 3.1 and 3.2 in Chapter 2). In particular, in this case $f'(\cdot) \in \mathfrak{L}(H_+, H_{+,c})$ and $f''(\cdot) \in \mathfrak{L}(H_+, \mathfrak{L}(H_-, H_{+,c}))$. By definition, the Banach space $C_b^2(H_-, \mathbb{C}^1) = C_b^2(H_-)$ consists of all twice continuously differentiable functions on H_- bounded with their derivatives. The norm in $C_b^2(H_-)$ is given by the equality

$$\|u\|_{C_b^2(H_-)} = \sup_{x \in H_-} \{|u(x)| + \|u'(x)\|_{H_{+,c}} + \|u''(x)\|_{\mathfrak{L}(H_-, H_{+,c})}\}. \tag{3.20}$$

Similarly, we introduce a space $C_b^2(H_-, H_-)$ of twice differentiable mappings $f\colon H_- \to H_-$ with the norm

$$\|f\|_{C_b^2(H_-, H_-)} = \sup_{x \in H_-} \{\|f(x)\|_{H_-} + \|f'(x)\|_{\mathfrak{L}(H_-, H_+)}$$

$$+ \|f''(x)\|_{\mathfrak{L}(H_+, \mathfrak{L}(H_-, H_+))}\}. \tag{3.21}$$

The same set of functions (mappings), but with the condition of local boundedness on H_-, is denoted by $C_{b,\mathrm{loc}}^2(H_-)$ $C_{b,\mathrm{loc}}^2((H_-, H_-))$.

Assume that a probability measure μ quasi-invariant with respect to shifts by elements $\varphi \in H_+$ is given on the σ-algebra $\mathfrak{B}(H_-)$ of Borel subsets of H_-. As in Subsection 3.1, we assume that for all $\varphi \in H_+$ there exists a logarithmic derivative of the measure μ in the direction φ, i. e.,

$$\beta_\mu(\varphi, x) = \lim_{t \to 0} \frac{\rho_\mu(t\varphi, x) - 1}{t} \in L_2(H_-, \mu), \tag{3.22}$$

where

$$\rho_\mu(\varphi, x) = d\mu(x + \varphi)/d\mu(x)$$

as above. We demand the existence of the representation

$$\beta_\mu(\varphi, x) = (\beta_\mu(x), \varphi)_{H_0},$$

where $\beta_\mu : H_- \to H_-$ is a measurable mapping. *Below, we always assume that* $\| \beta_\mu(\cdot) \|_{H_-} \in L_2(H_-, \mu)$.

For simplicity, we consider the Dirichlet form of the measure μ with the identity coefficient operator

$$d_\mu(u, v) = \frac{1}{2} \int_{H_-} (\nabla u(x), \nabla \bar{v}(x))_{H_0} d_\mu(x) \quad (u, v \in C_b^2(H_-)). \tag{3.23}$$

"Integrating by parts" in (3.23) once again and proceeding as in the proof of Theorem 3.1, we obtain the differential operator L_μ (the Dirichlet operator of the measure μ) with the domain of definition $C_b^2(H_-)$ given by the formula

$$(L_\mu u)(x) = -\frac{1}{2} \Delta u(x) - \frac{1}{2} (\nabla u(x), \beta_\mu(x))_{H_0}. \tag{3.24}$$

Here $\Delta u = \mathrm{Tr}_{H_0} u''$ is a trace of the operator u'' in H_0 defined by virtue of the inclusion $u'' \in \mathfrak{L}(H_-, H_{+,c})$; the function

$$(\nabla u(\cdot), \beta_\mu(\cdot))_{H_0} \in L_2(H_-, \mu)$$

is defined by virtue of the estimate

$$|(\nabla u(x), \beta_\mu(x))_{H_0}| \le \|u\|_{C_b^2(H_-)} \|\beta_\mu(x)\|_{H_-}$$

and assumptions concerning β_μ. We have the evident dense topological imbedding $C_b^2(H_-) \subset L_2(H_-, \mu)$ under which the functions from $C_b^2(H_-)$ do not "paste together" (due to the quasi-invariance of the measure μ).

Lemma 3.2. *Assume that* μ *is a probability measure on* $\mathfrak{B}(H_-)$ *quasi-invariant with respect to shifts by the elements from* H_+. *Then, for every open ball*

$$B_R(x_0) = \{x \in H_- \mid \|x - x_0\|_{H_-} < R\}, \quad R > 0, \quad x_0 \in H_-,$$

we have $\mu(B_R(x_0)) > 0$.

Proof. Assume the contrary, i.e., that $\mu(B_R(x_0)) = 0$ for some $x_0 \in H_-$ and $R > 0$. Due to the denseness of H_+ in H_-, for every $x \in H_-$ one can choose $\varphi \in H_+$ such that $B_{R/2}(x) \subset B_R(x_0 + \varphi)$. By assumption of the lemma, the measures $\mu(\cdot)$ and $\mu(\cdot +$

φ) are equivalent and, consequently, $\mu(B_{R/2}(x)) = 0$. The fact that H_- is separable enables us to cover H_- by countably many balls with radius $R/2$, and this implies that $\mu(H_-) = 0$, but this, in turn, contradicts the assumption that $\mu(H_-) = 1$. Note that the proof of the lemma remains valid for any finite measure on $\mathfrak{B}(H_-)$ with a dense set of admissible shifts.

■

Lemma 3.2 shows that different functions $u_1, u_2 \in C_b^2(H_-)$ are associated with different elements of $L_2(H_-, \mu)$. Indeed, by virtue of the continuity, we can choose $\varepsilon > 0$, $R > 0$, and $x_0 \in H_-$ so that $|u_1(x) - u_2(x)| \geq \varepsilon$ for any $x \in B_R(x_0)$. Then

$$\| u_1 - u_2 \|^2_{L_2(H_-,\mu)} = \int_{H_-} | u_1 - u_2 |^2 d\mu \geq \varepsilon^2 \mu(B_R(x_0)) > 0.$$

It follows from this discussion that the space $C_b^2(H_-)$ may be regarded as a space of test functions with respect to $L_2(H_-, \mu)$.

Let us introduce the generalized functions on H_- as antilinear continuous functionals on $C_b^2(H_-)$. Their action (and the imbedding of ordinary functions in generalized functions) is interpreted in terms of the scalar product in $L_2(H_-, \mu)$. Here, we have a complete analogy with the finite-dimensional case concerning the substitution of the measure μ for the Lebesgue measure. Therefore, we have obtained the following rigging of the space $L_2(H_-, \mu)$

$$[C_b^2(H_-)]' \supset L_2(H_-, \mu) \supset C_b^2(H_-).$$

Lemma 3.3. *The operator L_μ continuously maps $C_b^2(H_-)$ in $L_2(H_-, \mu)$.*

Proof. Let us show that each mapping

$$C_b^2(H_-) \ni u \mapsto \Delta u \in L_2(H_-, \mu),$$

$$C_b^2(H_-) \ni u \mapsto (\nabla u, \beta_\mu)_{H_0} \in L_2(H_-, \mu)$$

is continuous. According to (3.24), this will imply the required result. We have

$$|\Delta u| = |\mathrm{Tr}_{H_0} u''| = \left| \sum_{k=1}^{\infty} (u''(\cdot)e_k, e_k)_{H_0} \right| \leq \sum_{k=1}^{\infty} \| u''(\cdot)e_k \|_{H_{+,c}} \| e_k \|_{H_-}$$

$$\leq \sum_{k=1}^{\infty} \| u''(\cdot) \|_{\mathfrak{L}(H_-, H_{+,c})} \| e_k \|_{H_-}^2 \leq \| u \|_{C_b^2(H_-)} | O_{H_0, H_-} |^2,$$

where $(e_k)_{k=1}^{\infty}$ is an orthonormal basis in H_0, and $O_{H_0, H_-}: H_0 \to H_-$ is an imbedding operator which is a Hilbert–Schmidt operator by the condition imposed on the chain. Consequently,

$$\| \Delta u \|_{L_2(H_-, \mu)} \leq c_1 \| u \|_{C_b^2(H_-)}$$

for any $u \in C_b^2(H_-)$. Further,

$$| (\nabla u(\cdot), \beta_\mu(\cdot))_{H_0} | \leq \| \nabla u(\cdot) \|_{H_{+,c}} \| \beta_\mu(\cdot) \|_{H_-} \leq \| u \|_{C_b^2(H_-)} \| \beta_\mu(\cdot) \|_{H_-},$$

therefore,

$$\| (\nabla u, \beta_\mu) \|_{L_2(H_-, \mu)} \leq c_2 \| u \|_{C_b^2(H_-)}$$

for any $u \in C_b^2(H_-)$.

■

By virtue of the already proved lemma, for every function $f \in L_2(H_-, \mu)$, the generalized function $L_\mu f \in [C_b^2(H_-)]'$ is defined; it is given by the equality

$$(L_\mu f, u)_{L_2(H_-, \mu)} = (f, L_\mu u)_{L_2(H_-, \mu)} \quad (u \in C_b^2(H_-)).$$

We can now formulate the following theorem which extends Kato's inequality proved in Section 2 for the second quantization operators.

Theorem 3.3. *Let* $\beta_\mu \in C_{b,\mathrm{loc}}^2(H_-, H_-)$. *If* $f \in L_2(H_-, \mu)$ *and* $L_\mu f \in L_2(H_-, \mu)$ *(the action of the operator is understood in the sense of generalized functions), then we have the following inequality for generalized functions*

$$L_\mu |f| \leq \mathrm{Re}\, [(\mathrm{sgn} \bar{f}) L_\mu f]. \tag{3.25}$$

The proof of this theorem is based on the special approximation of L_μ. We now proceed to a discussion of it. For every $r > 0$, we introduce a function $s_r \in C_b^2(\mathbb{R}^1)$ possessing the following properties:

(a) $s_r(t) = 1$ for any $t \in \mathbb{R}^1$ such that $|t| < r$;

(b) $s_r(t) = 0$ for any $t \in \mathbb{R}^1$ such that $|t| > 2r$;

(c) $s_r(t) \geq 0$ and $|s_r^{(k)}(t)| \leq cr^{-k}$ for any $t \in \mathbb{R}^1$.

Let us define a differential operator $L_r = a_r(\cdot)L_\mu$, where $a_r(\cdot) = s_r(\|\cdot\|_{H_-}^2)$, on $C_b^2(H_-)$. The formal adjoint operator L_r^+ is defined by the equality

$$L_r^+ u = L_\mu(a_r, u) = L_\mu u - \frac{1}{2}(\nabla u, \nabla a_r)_{H_0} + V_r u \quad (u \in C_b^2(H_-)), \tag{3.26}$$

in which

$$V_r = L_\mu a_r - \frac{1}{2}\Delta a_r - \frac{1}{2}(\nabla a_r, \beta_\mu)_{H_0}.$$

Let $L_r^\#$ denote one of the operators L_r or L_r^+. The next lemma establishes the fact of solvability of the Cauchy problems

$$\begin{cases} \dfrac{\partial u(t, x)}{\partial t} + L_r u(t, x) = 0, \\ u(0, x) = u_0(x) \in C_b^2(H_-), \end{cases}$$

$$\begin{cases} \dfrac{\partial v(t, x)}{\partial t} + L_r^+ v(t, x) = 0, \\ v(0, x) = v_0(x) \in C_b^2(H_-) \end{cases} \tag{3.27}$$

for equations on vector functions

$$\mathbb{R}_+^1 = [0, +\infty) \ni t \mapsto u(t, \cdot) \in C_b^2(H_-).$$

The proof of this lemma is, in fact, based on the methods of the theory of stochastic differential equations. Therefore, we give only the formulation of it (see Daletsky and Fomin [1, Chapter 4, Section 4, Subsection 1, Theorem 4.1]).

Lemma 3.4. *Let $\beta_\mu \in C_{b,\mathrm{loc}}^2(H_-, H_-)$. Then each Cauchy problem (3.27) has a unique solution in $C_b^2(H_-)$ continuously differentiable with respect to $t \in \mathbb{R}_+^1$. These solutions are given by the semigroups $T_{t,r}$ and $T_{t,r}^+$ ($t \in \mathbb{R}_+^1$) acting in $C_b^2(H_-)$ and preserving positivity, i.e.,*

$$u(t, \cdot) = T_{t,r} u_0(\cdot); \quad v(t, \cdot) = T_{t,r}^+ v_0(\cdot) \tag{3.28}$$

moreover, $T^{\#}_{t,r} \in \mathfrak{L}(C^2_b(H_-), C^2_b(H_-))$.

■

We now want to extend these semigroups $T^{\#}_{t,r}$ to $L_2(H_-, \mu)$. A technical difficulty exists connected with the fact that they are not contracting in the metric of $L_2(H_-, \mu)$ and, consequently, the possibility of this extension is not obvious.

Lemma 3.5. *Each operator of the semigroup* $T^{\#}_{t,r}$ *is bounded in* $L_2(H_-, \mu)$.

Proof. Let us establish the following estimate

$$\| T_{t,r}u\|_{L_2(H_-,\mu)} \le c_{r,t}\|u\|_{L_2(H_-,\mu)} \quad (u \in C^2_b(H_-)). \tag{3.29}$$

Since the semigroup $T_{t,r}$ preserves positivity of the functions from $C^2_b(H_-)$ (and, hence, the property of being real), it suffices to consider the case of $u = \bar{u}$. We set $u_t = T_{t,r}u$. By virtue of Lemma 3.4,

$$\frac{\partial}{\partial s}(e^{-\alpha s}u_s) = e^{-\alpha s}[-L_r u_s - \alpha u_s]$$

for all $\alpha > 0$. Let us multiply both parts of this equality by $e^{-\alpha s}u_s$ and integrate it over the set $[0, t] \times H_-$ with respect to the measure $ds \times d\mu$. We then obtain

$$\frac{1}{2}\int_{H_-}(e^{-2\alpha t}|u_t|^2 - |u|^2)\,d\mu = -\int_0^t \Big[e^{-2\alpha s}\int_{H_-}(u_s L_r u_s + \alpha u_s^2)\,d\mu\Big]ds$$

$$= -\int_0^t e^{-2\alpha s}\Big\{\int_{H_-}\Big[\frac{1}{2}a_r\|\nabla u_s\|^2_{H_0} + \frac{1}{4}(\nabla a_r, \nabla(u_s^2))_{H_0} + \alpha u_s^2\Big]d\mu\Big\}ds$$

$$\le -\int_0^t e^{-2\alpha s}\Big\{\int_{H_-}\Big[\frac{1}{2}(L_\mu a_r)u_s^2 + \alpha u_s^2\Big]d\mu\Big\}ds.$$

Let us show that one can choose $c_r > 0$ for which $|L_\mu a_r| \le c_r$. Indeed,

$$L_\mu a_r \equiv V_r = -\frac{1}{2}\Delta a_r - \frac{1}{2}(\nabla a_r, \beta_\mu)_{H_0}.$$

But

$$a_r(x) = s_r(\|x\|^2_-) = s_r(\|Jx\|^2_{H_0}), \quad x \in H_-,$$

where $J\colon H_- \to H_0$ is an operator of isometric imbedding and, moreover, $J \upharpoonright H_0$ is a Hilbert–Schmidt operator in H_0. Hence,

$$\nabla a_r(x) = 2s_r'(\|x\|_-^2)J^2 x \in H_+,$$

$$a_r''(x) = 4s_r''(\|x\|_-^2)(J^2x \otimes J^2x)^\wedge + 2s_r'(\|x\|_-^2)J^2, \quad x \in H_-,$$

where $(J^2(x) \otimes J^2(x))^\wedge$ is an operator in H_0 with the kernel $J^2(x) \otimes J^2(x) \in H_+ \otimes H_+$. Taking into account the properties of $s_r(\cdot)$, we obtain

$$|\Delta a_r(x)| = |\mathrm{Tr}_{H_0} a_r''(x)| \le 4|s_r''(\|x\|_-^2)| \cdot \|J^2 x\|_{H_0}^2$$

$$+ 2|s_r'(\|x\|_-^2)| \,\mathrm{Tr}_{H_0} J^2 \le c_1 / r,$$

$$|(\nabla a_r(x), \beta_\mu(x))_{H_0}| \le \Big(\sup_{\|x\|_-^2 \le 2r} \|\beta_\mu(x)\|_{H_-} \Big) 2|s_r'(\|x\|_-^2)| \, \|J^2 x\|_{H_+} \le c_2/\sqrt{r}.$$

Let us choose $\alpha > 0$ such that $\alpha - c_r > 0$. Then $\frac{1}{2} L_\mu a_r + \alpha \ge 0$, i.e.,

$$\int_{H_-} (e^{-2\alpha t}|u_t|^2 - |u|^2)\, d\mu \le 0$$

or

$$\|T_{t,r} u\|_{L_2(H_-,\mu)} = e^{\alpha t}\|u\|_{L_2(H_-,\mu)}.$$

Due to the denseness of $C_b^2(H_-)$ in $L_2(H_-,\mu)$, this proves the boundedness of $T_{t,r}$ in $L_2(H_-,\mu)$. By using a completely analogous argument, we can also establish the boundedness of $T_{t,r}^+$, $t \ge 0$, in $L_2(H_-,\mu)$.

■

Lemma 3.6. *Each semigroup* $T_{t,r}^\#$, $t \ge 0$, *is a semigroup of the class* C_0 *in* $L_2(H_-,\mu)$.

Proof. It suffices to verify the continuity of $T_{t,r}^\# u$ for $u \in C_b^2(H_-)$. Taking into account the inclusion $L_r^\# u \in C_b(H_-)$ and Lemma 3.4, we obtain the following equality in $C_b(H_-)$

$$T^{\#}_{t,r}u(x) - u(x) = \int_0^t T^{\#}_{s,r} L^{\#}_r u(x) ds,$$

whence

$$\sup_{0 \le s \le t} \| T^{\#}_{s,r}u - u \|_{L_2(H_-,\mu)} = te^{\alpha t} \| L^{\#}_r u \|_{L_2(H_-,\mu)} \to 0, \quad t \to 0,$$

by Lemma 3.5.

■

Denote by G_r and G_r^+ the generators of the C_0-semigroups $T_{t,r}$ and $T^+_{t,r}$ in $L_2(H_-, \mu)$, respectively. For the operator L_r, the corresponding minimal operator $L_{r,\min}$ is introduced (as usual) as a closure in $L_2(H_-, \mu)$; for the maximal operator, we have $L_{r,\max} = (L_r^+)^*$. Similarly, we define the operators $L^+_{r,\min}$ and $L^+_{r,\max}$.

Lemma 3.7. *For all $r > 0$, the equality*

$$L_{r,\min} = L_{r,\max}$$

holds.

Proof. First, we show that the inclusion $L^{\#}_{r,\min} \subset G^{\#}_r$ holds. For $u \in C_b^2(H_-)$, we have

$$\frac{1}{t}(T^{\#}_{t,r}u - u) - L^{\#}_r u = -\frac{1}{t}\int_0^t (T^{\#}_{s,r} - 1) L^{\#}_r u ds$$

and, thus,

$$\left\| \frac{T^{\#}_{t,r}u - u}{t} + L^{\#}_r u \right\|_{L_2(H_-,\mu)} \le \frac{1}{t}\int_0^t \| (T^{\#}_{s,r} - 1) L^{\#}_r u \|_{L_2(H_-,\mu)} ds$$

$$\le \sup_{0 \le s \le t} \| (T^{\#}_{s,r} - 1) L^{\#}_r u \|_{L_2(H_-,\mu)} \to 0, \quad t \to 0,$$

due to the (C_0)-property of $T^{\#}_{s,r}$. Hence, $L^{\#}_r \subset G^{\#}_r$, but the fact that $G^{\#}_r$ is closed involves $L^{\#}_{r,\min} \subset G^{\#}_r$. According to Lemma 3.4, the semigroup $T^{\#}_{t,r}$ preserves invariance of the dense subset $C_b^2(H_-)$ in $L_2(H_-, \mu)$. By virtue of the general theorem of semigroups, $C_b^2(H_-)$ is the essential domain for $G^{\#}_r$. Taking into account that $G^{\#}_r \restriction C_b^2(H_-) = L^{\#}_r$, we now obtain the equality $L^{\#}_{r,\min} = G^{\#}_r$. The equality $(G_r^+)^* = (L^+_{r,\min})^* =$

$L^+_{r,\max}$ implies that the operator $(G_r^+)^*$ is densely defined and that it is a generator of the semigroup $(T^+_{t,r})^*$, being adjoint to the generator of the C_0-semigroup $T^+_{t,r}$. Therefore, $L_{r,\max}$ is a generator of the semigroup and, thus, the inclusion $L_{r,\min} = G_r \subset L_{r,\max}$ yields $L_{r,\min} = G_r = L_{r,\max}$.

■

Proof of Theorem 3.3. We first assume that $f \in C_b^2(H_-)$. For $\varepsilon > 0$, we introduce a function $f_\varepsilon \in C_b^2(H_-)$ defined by the equality

$$f_\varepsilon^2 = |f|^2 + \varepsilon^2. \tag{3.30}$$

Differentiating (3.30), we get

$$f_\varepsilon \nabla f_\varepsilon = \operatorname{Re}(\bar f \nabla f),$$

whence

$$(\nabla f_\varepsilon, \nabla f_\varepsilon)_{H_0} f_\varepsilon^2 = (\nabla f, \nabla \bar f)_{H_0} |f|^2,$$

$$(\nabla f_\varepsilon, \nabla f_\varepsilon)_{H_0} = (\nabla f, \nabla \bar f)_{H_0}, \tag{3.31}$$

because $|f/f_\varepsilon| \le 1$. If we apply the operator L_r to (3.30), we obtain the equality

$$f_\varepsilon L_r f_\varepsilon - a_r(\nabla f_\varepsilon, \nabla f_\varepsilon)_{H_0} = \operatorname{Re}(\bar f L_r f) - a_r(\nabla f, \nabla \bar f)_{H_0}.$$

Together with (3.31), this yields

$$L_r f_\varepsilon \le \operatorname{Re}[(\bar f / f_\varepsilon) L_r f]. \tag{3.32}$$

We have $L_r f_\varepsilon \to L_r |f|$ as $\varepsilon \to 0$ in $[C_b^2(H_-)]'$. This follows from the convergence $f_\varepsilon \to |f|$ as $\varepsilon \to 0$ in $L_2(H_-, \mu)$ and from the fact that $L_r^+ \colon C_b^2(H_-) \to L_2(H_-, \mu)$ continuously. Moreover, $\bar f / f_\varepsilon \to \operatorname{sgn} \bar f$ in $L_\infty(H_-, \mu)$; therefore,

$$\operatorname{Re}[(\bar f / f_\varepsilon) L_r f] \to \operatorname{Re}[(\operatorname{sgn} \bar f) L_r f]$$

in $L_2(H_-, \mu)$ and, furthermore, the convergence in $[C_b^2(H_-)]'$ takes place. Passing to the limit in (3.32), we obtain

$$L_r |f| \le \operatorname{Re}[(\operatorname{sgn} \bar f) L_r f] \quad (f \in C_b^2(H_-)).$$

We now assume that f is an arbitrary function satisfying conditions of the theorem. Then, by definition, there exists $g \in L_2(H_-, \mu)$ such that

$$(f, L_\mu u)_{L_2(H_-, \mu)} = (g, u)_{L_2(H_-, \mu)}$$

for any $u \in C_b^2(H_-)$, i. e., $f \in \mathfrak{D}(L_{\mu,\max}) = \mathfrak{D}(L_\mu^*)$ and $L_{\mu,\max} f = g$. Let us show that in this case $f \in \mathfrak{D}(L_{r,\min})$ for any $r > 0$. Indeed, according to the definition of the action of L_r in the sense of generalized functions, we have

$$(L_r f, u)_{L_2(H_-, \mu)} = (f, L_r^+ u)_{L_2(H_-, \mu)} = (L_\mu f, a_r u)_{L_2(H_-, \mu)} = (a_r L_\mu f, u)_{L_2(H_-, \mu)},$$

i.e., $L_r f = a_r L_\mu f \in L_2(H_-, \mu)$. Hence, by virtue of Lemma 3.7, $f \in \mathfrak{D}(L_{r,\max}) = \mathfrak{D}(L_{r,\min})$. Let us choose a sequence

$$(f_j)_{j=1}^\infty \subset C_b^2(H_-) : f_j \to f, \quad L_{r,\min} f_j = L_r f_j \to L_{r,\min} f$$

in $L_2(H_-, \mu)$. According to what has been proved above, for any $\varepsilon > 0$ and $j \in \mathbb{N}$, we have

$$L_r f_{j,\varepsilon} \le \operatorname{Re}\,[(\bar f_j / f_{j,\varepsilon}) L_r f_j]. \tag{3.33}$$

Let us extract a subsequence $f_{j_k} \to f$ μ-almost everywhere as $k \to \infty$. Then $f_{j_k,\varepsilon} \to f_\varepsilon$ as $k \to \infty$ in $L_2(H_-, \mu)$, $\bar f_{j_k} / f_{j_k,\varepsilon} \to \bar f / f_\varepsilon$ μ-almost everywhere,

$$(L_r f_{j_k}) \cdot (\bar f_{j_k} / f_{j_k,\varepsilon}) \to (L_r f) \cdot (\bar f / f_\varepsilon), \quad k \to \infty,$$

as $k \to \infty$ in $L_2(H_-, \mu)$, and the limit transition in (3.33) gives

$$L_r f_\varepsilon \le \operatorname{Re}\,[(\bar f / f_\varepsilon) L_r f].$$

If we tend $\varepsilon \to 0$ in the last inequality, taking into account that $f_\varepsilon \to |f|$ in $L_2(H_-, \mu)$, $\bar f / f_\varepsilon \to \operatorname{sgn} \bar f$ μ-almost everywhere, and $|\bar f / f_\varepsilon| \le 1$, then we obtain the inequality for generalized functions from $[C_b^2(H_-)]'$

$$L_r |f| \le \operatorname{Re}\,[(\operatorname{sgn} \bar f) L_r f] \tag{3.34}$$

for any $r > 0$. It remains only to pass to the limit as $r \to \infty$ in (3.34).

Let us show that $L_r|f| \to L_\mu|f|$ as $r \to \infty$ in $[C_b^2(H_-)]'$. To do this, it is necessary to prove that

$$(L_r|f|, u)_{L_2(H_-,\mu)} = (|f|, L_r^+ u)_{L_2(H_-,\mu)} = (|f|, L_\mu u)_{L_2(H_-,\mu)}$$

for any $u \in C_b^2(H_-)$. But according to (3.26),

$$L_r^+ u = a_r L_\mu u - \frac{1}{2}(\nabla u, \nabla a_r)_{H_0} + V_r u,$$

where

$$V_r = -\frac{1}{2}\Delta a_r - \frac{1}{2}(\nabla a_r, \beta_\mu)_{H_0}.$$

In the proof of Lemma 3.5, we have established the estimates $|\Delta a_r| \le c_1/r$ and

$$(\nabla a_r, \beta_\mu)_{H_0}| \le c_2 \|\beta_\mu\|_{H_-} r^{-1/2},$$

which imply that $V_r \xrightarrow[r\to\infty]{} 0$ in $L_2(H_-, \mu)$. Further, taking into account that

$$\nabla a_r(x) = 2s_r'(\|x\|_{H_-}^2) J^2 x \in H_+, \quad x \in H_-,$$

we obtain the inequality

$$|(\nabla a_r, \nabla u)_{H_0}| \le 2\|\nabla u(x)\|_{H_0} \|J^2 x\|_{H_0} |s_r'(\|x\|_{H_-}^2|$$

$$\le c_3 \|u\|_{C_b^2(H_-)} \|x\|_{H_-} |s_r'(\|x\|_{H_-}^2| \le c_3 \|u\|_{C_b^2(H_-)} r^{-1/2},$$

which ensures the convergence $(\nabla u, \nabla a_r)_{H_0} \to 0$ as $r \to \infty$ in $L_2(H_-, \mu)$. Finally, the convergence $a_r \to 1$ as $r \to \infty$ in $L_2(H_-, \mu)$ yields $a_r L_\mu u \to L_\mu u$ as $r \to \infty$ in $L_2(H_-, \mu)$. Consequently, $L_r^+ u \to L_\mu u$ in $L_2(H_-, \mu)$, and this implies that $L_r|f| \to L_\mu|f|$ as $r \to \infty$ in $[C_b^2(H_-)]'$. On the other hand,

$$L_r f = a_r L_\mu f \xrightarrow[r\to\infty]{} L_\mu f$$

in $L_2(H_-, \mu)$. Therefore, when $r \to \infty$, the inequality (3.34) turns into the inequality

$$L_\mu|f| \le \operatorname{Re}[(\operatorname{sgn} \bar{f}) L_\mu f], \tag{3.35}$$

and this proves the theorem.

■

The selfadjointness of the operator L_μ is a direct consequence of the inequality (3.35).

Theorem 3.4. *Assume that the measure* μ *on* $\mathfrak{B}(H_-)$ *is such that the corresponding logarithmic derivative* β_μ *belongs to* $C^2_{b,\mathrm{loc}}(H_-, H_-)$ *and* $\| \beta_\mu(\cdot) \|_{H_-} \in L_2(H_-, \mu)$. *Then the operator* L_μ *is essentially selfadjoint.*

Proof. It suffices to show that the equality $(L_\mu + 1)^* f = 0$ implies that $f = 0$, because this is equivalent to the assertion: The equality $(L_\mu + 1) f = 0$ (in the sense of generalized functions) for $f \in L_2(H_-, \mu)$ yields $f = 0$. The equality $L_\mu f = -f$ gives $L_\mu f \in L_2(H_-, \mu)$ and, therefore, Theorem 3.3 can be applied. By virtue of this theorem,

$$L_\mu |f| \le \mathrm{Re}\,[(\mathrm{sgn}\,\bar{f}) L_\mu f] = -\mathrm{Re}\,[(\mathrm{sgn}\,\bar{f}) f] = -|f|,$$

whence, $(L_\mu + 1)|f| \le 0$. However,

$$((L_\mu + 1)|f|, 1)_{L_2(H_-,\mu)} = (|f|, 1)_{L_2(H_-,\mu)} = \|f\|_{L_1(H_-,\mu)}$$

and, consequently, $f = 0$.

Example 3.6. Assume that the measure μ on $\mathfrak{B}(H_-)$ is a multiplicative perturbation of the standard Gaussian measure γ_1 corresponding to H_0. In other words, $\mu = \phi^2 \gamma_1$, $\phi > 0$ γ_1-almost everywhere, and $\| \phi \|_{L_2(H_-, \gamma_1)} = 1$. Assume that ϕ is a differentiable function on H_-. It follows from (3.10) that the equality

$$\beta_\mu(x) = 2\,\frac{\nabla \phi(x)}{\phi(x)} - 2x \tag{3.36}$$

holds; this equality yields the inclusion $\| \beta_\mu(\cdot) \|_{H_-} \in L_2(H_-, \mu)$ as long as $\| \nabla \phi \|_{H_-} \in L_2(H_-, \gamma_1)$ and $\phi \in L_{2+\varepsilon}(H_-, \gamma_1)$ for some $\varepsilon > 0$. Here, one should take into account that for any $p \ge 1$, $\| \cdot \|_{H_-} \in L_p(H_-, \gamma_1)$ (see Lemma 1.1). In order to satisfy the conditions of Theorem 3.4 and, hence, to establish the essential selfadjointness of the operator L_μ, we should only demand that

$$\frac{\nabla \phi}{\phi} \in C^2_{b,\mathrm{loc}}(H_-, H_-).$$

In the particular case of cylindrical density

$$\phi(x) = f((x, \varphi_1)_{H_0}, \dots, (x, \varphi_n)_{H_0}), \quad \varphi_n \in H_+, \quad k = 1, \dots, n, \quad n \in \mathbb{N}; \quad x \in H_-,$$

it suffices, for example, for f to belong to $C^3(\mathbb{R}^n)$ and to be exponentially bounded (i.e.,

$$|f(s_1, \dots, s_n)| \le c_1 \exp c_2 \sum_{k=1}^{n} |s_k| \quad ((s_1, \dots, s_n) \in \mathbb{R}^n)$$

with some $c_1, c_2 > 0$) and strictly positive (i.e., $f(s_1, \dots, s_n) > 0$ for $(s_1, \dots, s_n) \in \mathbb{R}^n$).

§4. Infinite-Dimensional Differential Operators

Below, we consider elliptic and hyperbolic equations corresponding to the Dirichlet operators of measures on Hilbert spaces. We again restrict ourselves to the case of the identity coefficient operator; the general case, including the case of alternating coefficients, is presented in detail in the papers which are referred to in the Bibliographical Notes. The finiteness of the rate of propagation of perturbations established in Subsection 4.2 for infinite-dimensional hyperbolic equations enables us to apply the evolutionary method described in Subsections 1.1 and 1.2 of Chapter 5 to the investigation of the selfadjointness of potential perturbations of the Dirichlet operators. This application is presented in Subsection 4.3.

4.1. Coercivity Inequalities and Their Consequences

In this subsection, we consider elliptic equations corresponding to the Dirichlet operators L_μ in the case of measures on H_- (see Subsection 3.3)

$$\Delta f + (\nabla f, \beta_\mu)_{H_0} = g, \quad g \in L_2(H_-, \mu) = L_2. \tag{4.1}$$

In what follows, as in Subsection 3.3, we assume that the condition $\| \beta_\mu \|_{H_-} \in L_2(H_-, \mu)$ holds; this condition guarantee the definiteness of L_μ on $C_b^2(H_-)$. A function $f \in L_2(H_-, \mu)$ is called a generalized solution of Equation (4.1) if

$$(f, L_\mu u)_{L_2(H_-, \mu)} = (g, u)_{L_2(H_-, \mu)}$$

for any $u \in C_b^2(H_-)$. In order to study the smoothness of generalized solutions, we need an analogy of a Sobolev space consisting of the functions defined on H_-.

Denote by $W_2^1(H_-, \mu)$ a Hilbert space obtained as a result of the completion of $C_b^2(H_-)$ in norm generated by the scalar product

$$(u, v)_{W_2^1(H_-,\mu)} = \int_{H_-} [u\bar{v} + (\nabla u, \nabla \bar{v})_{H_0}]d\mu. \tag{4.2}$$

The space $W_2^2(H_-, \mu)$ is obtained by completion of $C_b^2(H_-)$ in norm corresponding to the scalar product

$$(u, v)_{W_2^2(H_-,\mu)} = \int_{H_-} [u\bar{v} + (\nabla u, \nabla \bar{v})_{H_0} + \mathrm{Tr}_{H_0}(u''\bar{v}'')]d\mu. \tag{4.3}$$

We say that $f \in W_{2,\,\mathrm{loc}}^k(H_-, \mu)$, $k = 1, 2$, if for any function $u \in C_b^2(H_-)$ with bounded support, we have $f \cdot u \in W_2^k(H_-, \mu)$.

Lemma 4.1. *Assume that the measure μ on H_- is such that for the corresponding logarithmic derivatives, we have $\beta_\mu \in C_{b,\,\mathrm{loc}}^2(H_-, H_-)$ and $\| \beta_\mu \|_{H_-} \in L_2(H_-, \mu)$. Then every generalized solution of Equation* (4.1) *belongs to $W_2^1(H_-, \mu)$.*

Proof. It follows from the definition of the generalized solution that $f \in \mathfrak{D}(L_{\mu,\max})$. But according to Theorem 3.4, we have $L_{\mu,\,\max} = L_{\mu,\,\min} = L_\mu^F$, where L_μ^F is the Friedrichs extension of L_μ. Therefore, f belongs to the domain of definition of the closure of the Dirichlet form

$$d_\mu(f,f) = \frac{1}{2} \int_{H_-} (\nabla f, \nabla \bar{f})d\mu,$$

and this yields the required result if we take into account the inclusion $f \in L_2(H_-, \mu)$. ■

As in the finite-dimensional case, in order to study completely the properties of smoothness for the generalized solution, we should employ the coercivity inequality. The validity of this inequality is established by the following theorem.

Theorem 4.1. *Assume that $\beta_\mu'(x) \in \mathfrak{L}(H_0, H_0)$, for any $x \in H_-$ and that $\| \beta_\mu'(\cdot) \|_{\mathfrak{L}(H_0,H_0)} \in L_1(H_-, \mu)$ and is locally bounded on H_-. Then for any $r > 0$,* *there exists $\lambda_0(r)$ such that the inequality*

$$c''_{\lambda,r}\|u\|_{W_2^2(H_-,\mu)} \le \|(L_\mu+\lambda)u\|_{L_2(H_-,\mu)} \le c'_{\lambda,r}\|u\|_{W_2^2(H_-,\mu)}, \tag{4.4}$$

where $c'_{\lambda,r}, c''_{\lambda,r} > 0$ depend only on λ and r, holds for any $\lambda \ge \lambda_0(r)$ and $u \in C_b^2(H_-)$ such that supp $u \subset \{x \in H_- \mid \|x\|_{H_-} \le r\}$. *In addition, if $\sup_{H_-}\|\beta'_\mu(\cdot)\|_{\mathcal{L}(H_0,H_0)} > 0$, then it is possible to choose $\lambda_0(r)$, $c'_{\lambda,r}$, and $c''_{\lambda,r}$ which are independent of r.*

Proof. We start from the equality

$$4\int_{H_-}|L_\mu u|^2 d\mu = \int_{H_-}\mathrm{Tr}_{H_0}(u''\bar{u}'')d\mu - \int_{H_-}(\nabla u, \beta'_\mu \nabla\bar{u})_{H_0}d\mu, \tag{4.5}$$

the validity of which, for the functions $u \in C_b^2(H_-)$, will be established somewhat later. Assume that $\lambda > 0$ and u satisfies conditions of the theorem. Taking into account (4.5), we obtain

$$\|(2L_\mu+\lambda 1)u\|^2_{L_2(H_-,\mu)} = 4\|L_\mu u\|^2_{L_2(H_-,\mu)} + 4\lambda(L_\mu u, u)_{L_2(H_-,\mu)}$$

$$+\lambda^2\|u\|^2_{L_2(H_-,\mu)} = \int_{H_-}\mathrm{Tr}_{H_0}(u''\bar{u}'')d\mu + 2\lambda\int_{H_-}\|\nabla u\|^2_{H_{0,c}}d\mu$$

$$+\lambda^2\|u\|^2_{L_2(H_-,\mu)} - \int_{H_-}(\nabla u, \beta'_\mu\nabla\bar{u})_{H_0}d\mu.$$

By virtue of the conditions imposed on β_μ, the following estimate holds

$$|\nabla u, \beta'_\mu\nabla\bar{u})_{H_0}| \le \|\nabla u\|_{H_{0,c}}\sup_{\|x\|_-\le r}\|\beta'_\mu(x)\|_{\mathcal{L}(H_0,H_0)} \le c_r\|\nabla u\|^2_{H_{0,c}}$$

Using this estimate, we get

$$\int_{H_-}\mathrm{Tr}_{H_0}(u''\bar{u}'')d\mu + \lambda^2\|u\|^2_{L_2(H_-,\mu)} + (2\lambda - c_r)\int_{H_-}(\nabla u, \nabla\bar{u})_{H_0}d\mu$$

$$\le \|(2L_\mu+\lambda)u\|^2_{L_2(H_-,\mu)} \le \int_{H_-}\mathrm{Tr}_{H_0}(u''\bar{u}'')d\mu$$

$$+\lambda^2\|u\|^2_{L_2(H_-,\mu)} + (2\lambda + c_r)\int_{H_-}(\nabla u, \nabla\bar{u})_{H_0}d\mu. \tag{4.6}$$

If we choose $\lambda: 2\lambda - c_r > 0$, then (4.6) can be transformed into the inequality (4.4) in which

$$4c'_{\lambda,r} = \max(1, \lambda^2, 2\lambda + c_r), \quad 4c''_{\lambda,r} = \min(1, \lambda^2, 2\lambda - c_r).$$

In the case when $\| \beta'_\mu(\cdot) \|_{\mathcal{L}(H_0,H_0)}$ is bounded on H_-, the constant c_r and, hence, $c'_{\lambda,r}$ and $c''_{\lambda,r}$ are independent of r.

It remains to establish the validity of the equality (4.5). Let $(e_k)_{k=1}^\infty \subset H_+$ be an orthonormal basis in H_0, and let $u = \bar{u} \in C_b^\infty(H_-)$ be a cylindrical function (with respect to this basis), i.e.,

$$u(x) = f((e_1, x)_{H_0}, \dots, (e_N, x)_{H_0}), \quad f \in C_b^\infty(\mathbb{R}^N), \; x \in H_-, \; N \in \mathbb{N}.$$

Recall that each cylindrical function on H_- can be represented in this form by use of a certain basis $(e_k)_{k=1}^\infty \subset H_+$ (generally speaking, this is its own basis). If we denote $x_k = (e_k, x)_{H_0}$, and take into account (3.3), then we get

$$2L_\mu u = -\sum_{k=1}^N \frac{\partial^2 u}{\partial x_k^2} - \sum_{k=1}^N \beta_k \frac{\partial u}{\partial x_k} = \sum_{k=1}^N \nabla_k^* \nabla_k u,$$

where ∇_k is an operator of differentiation in the direction of e_k in $L_2(H_-, \mu)$ and $\beta_k = \beta_\mu(e_k, \cdot)$. Then

$$4 \int_{H_-} |L_\mu u|^2 d\mu = \sum_{k,j=1}^N \int_{H_-} (\nabla_k^* \nabla_k u) \cdot (\nabla_j^* \nabla_j \bar{u}) d\mu$$

$$= \sum_{k,j=1}^N \int_{H_-} (\nabla_k u) \cdot (\nabla_k \nabla_j^* \nabla_j u) d\mu. \tag{4.7}$$

The integration by parts realized in the proof of Theorem 3.1 shows that

$$\nabla_j^* u = -\nabla_j u - \beta_j u \quad (u \in C_b^2(H_-)).$$

By direct calculation, this yields

$$(\nabla_k \nabla_j^* - \nabla_j^* \nabla_k) u = -\frac{\partial \beta_j}{\partial x_k} u = -(\nabla_k \beta_j) u. \tag{4.8}$$

Taking into account (4.8), we transform (4.7) as follows

$$4\| L_\mu u \|^2_{L_2(H_-,\mu)} = \sum_{k,j=1}^{N} \int_{H_-} \nabla_k u \cdot \nabla_j^* \nabla_k \nabla_j u d\mu$$

$$- \sum_{k,j=1}^{N} \int_{H_-} \nabla_k u \cdot \nabla_k \beta_j \cdot \nabla_j u d\mu = \sum_{k,j=1}^{N} \int_{H_-} (\nabla_j \nabla_k u)(\nabla_k \nabla_j u) d\mu$$

$$- \sum_{k,j=1}^{N} \int_{H_-} \nabla_k u \cdot \nabla_k \beta_j \cdot \nabla_j u d\mu = \int_{H_-} [\mathrm{Tr}_{H_0}(u''u'') - (\nabla u, \beta'_\mu \nabla u)_{H_0}] d\mu.$$

For an arbitrary real $u \in C_b^2(H\_)$, we choose a sequence $(u_n)_{n=1}^\infty \subset C^\infty_{b,\,\mathrm{cyl}}(H\_)$ such that $u_n \to u$ as $n \to \infty$ in $W_2^2(H_-, \mu)$. Then the required equality is obtained from (4.5) by limit transition with the help of the Lebesgue theorem on majorized convergence (taking into account the condition of summability $\|\beta'_\mu(\cdot)\|_{\mathfrak{L}(H_0, H_0)} \in L_1(H_-, \mu)$). Finally, the transition (in (4.5)) to the functions which are not real is trivial because L_μ preserves the property of being real for the functions in $C_b^2(H\_)$. Indeed, let $u = u_1 + iu_2$, $u_k = \bar{u}_k \in C_b^2(H\_)$, $k = 1, 2$. Then

$$|L_\mu u|^2 = |L_\mu u_1|^2 + |L_\mu u_2|^2$$

and, according to what has been proved,

$$4\| L_\mu u \|^2_{L_2(H_-,\mu)} = \int_{H_-} [\mathrm{Tr}_{H_0}(u_1'')^2 + \mathrm{Tr}_{H_0}(u_2'')^2] d\mu - \int_{H_-} [(\nabla u_1, \beta'_\mu \nabla u_1)_{H_0}$$

$$+ (\nabla u_2, \beta'_\mu \nabla u_2)_{H_0}] d\mu = \int_{H_-} [\mathrm{Tr}_{H_0}(u''\bar{u}'') - (\nabla u, \beta'_\mu \nabla u)_{H_0}] d\mu.$$

■

Theorem 4.2. *Assume that* $\beta_\mu \in C^2_{b,\,\mathrm{loc}}(H_-, H_-)$, $\beta'_\mu(x) \in \mathfrak{L}(H_0, H_0)$, $x \in H_-$, *and that* $\|\beta'_\mu(\cdot)\|_{\mathfrak{L}(H_0, H_0)} \in L_1(H_-, \mu)$ *and is locally bounded on* H_-. *Then the generalized solution of Equation* (4.1) *belongs to* $W^2_{2,\,\mathrm{loc}}(H_-, \mu)$. *In addition, if* $\|\beta'_\mu(\cdot)\|_{\mathfrak{L}(H_0, H_0)}$ *is bounded on* H_-, *then the solution belongs to* $W_2^2(H_-, \mu)$.

Proof. Let f be a generalized solution of Equation (4.1). Theorem 3.4 shows that $f \in \mathfrak{D}(L_{\mu,\min})$ and, by virtue of Lemma 4.1, $f \in W_2^1(H_-, \mu)$. It is easy to see that for any finite function $u \in C_b^2(H\_)$, we have $fu \in \mathfrak{D}(L_{\mu,\min})$. Indeed, let

$$(f_j)_{j=1}^\infty \subset C_b^2(H\_) : f_j \to f, \quad L_\mu f_j \to L_\mu f, \quad j \to \infty$$

in $L_2(H_-, \mu)$. Then

$$L_\mu(f_j u) = uL_\mu f_j + f_j L_\mu u - (\nabla f_j, \nabla u)_{H_0}.$$

Since u is finite and β_μ is locally bounded, we have

$$L_\mu(f_j u) \xrightarrow[j\to\infty]{} uL_\mu f + fL_\mu u - (\nabla f, \nabla u)_{H_0} \in L_2(H_-, \mu)$$

(in the last inclusion, we have used the inclusions $f \in W_2^1(H_-, \mu)$ and $u \in C_b^2(H_-)$). For every $j \in \mathbb{N}$, the following inequality holds

$$\| (L_\mu + \lambda 1) f_j u \|_{L_2(H_-,\mu)} \geq c_\lambda \| f_j u \|_{W_2^2(H_-,\mu)}$$

for $\lambda \geq \lambda_0$ (see (4.4)); here λ_0 depends only on the size of the support of u. The limit transition as $j \to \infty$ shows that $uf \in W_2^2(H_-, \mu)$, i.e., $f \in W_{2,\,\mathrm{loc}}^2(H_-, \mu)$ due to the arbitrariness of u. In the case when $\sup_{H_-} \| \beta_\mu'(\cdot) \|_{\mathfrak{L}(H_0, H_0)} < \infty$, we employ the limit transition in the inequality

$$\| (L_\mu + \lambda 1) f_j \|_{L_2(H_-,\mu)} \geq c_\lambda \| f_j \|_{W_2^2(H_-,\mu)}$$

and conclude that $f \in W_2^2(H_-, \mu)$.

■

4.2. Hyperbolic Equations Corresponding to the Dirichlet Operators

In addition to the requirement $\| \beta_\mu(\cdot) \|_{H_-} \in L_2(H_-, \mu)$, everywhere in this subsection we assume that the inclusion $\beta_\mu \in C_{b,\,\mathrm{loc}}^2(H_-, H_-)$ holds. Thus, according to Theorem 3.4, the operator L_μ is essentially selfadjoint.

Consider the Cauchy problem

$$\begin{cases} \dfrac{d^2u}{dt^2}(t) + (L_\mu + V)u(t) = 0, \\ u(0) = u_0, \quad \dfrac{du}{dt}(0) = u_1, \quad V = \overline{V} \in L_\infty(H_-, \mu). \end{cases} \tag{4.9}$$

It is known that for $u_0 \in \mathfrak{D}(L_\mu + V) = \mathfrak{D}(L_\mu)$ and

$$u_1 \in \mathfrak{D}((L_\mu + V)^{1/2}) = \mathfrak{D}(L_\mu^{1/2}),$$

the strong solution of problem (4.9) exists and is unique, i.e., there exists a twice continuously differentiable vector function

$$\mathbb{R}^1_+ = [0, +\infty) \ni t \mapsto u(t) \in L_2(H_-, \mu), \quad u(t) \in \mathfrak{D}(L_\mu), \quad t \in \mathbb{R}^1_+,$$

which satisfies the equation and the initial conditions. This solution can be represented in the form

$$u(t) = \int_{E_V}^{\infty} \cos\sqrt{\lambda}\, t dE(\lambda) u_0 + \int_{E_V}^{\infty} \lambda^{-1/2} \sin\sqrt{\lambda}\, t dE(\lambda) u_1 \quad (t \in \mathbb{R}^1_+), \tag{4.10}$$

where $E_V = \inf s\,(L_\mu + V)$ is a lower bound of the operator $L_\mu + V$, and E is its RI (see the proof of Theorem 1.2 in Chapter 5).

Let us establish *a priori* estimates in term of the initial data for the strong solutions of (4.9). We act by analogy with the finite-dimensional case.

For fixed $x_0 \in H_-$ and $c > 0$, we construct a cone

$$K_{c, x_0} = \{(t, x) \in \Pi_T \,|\, \|x - x_0\|_{H_-} \le c(T - t)\} \tag{4.11}$$

on the layer $\Pi_T = [0, T] \times H_-$ $(T > 0)$ (it is spherical and contracting in the direction of the axis t).

Let D_t be a frustrum of a cone $K_{c, x_0} \cap \Pi_t$ $(0 \le t \le T)$, let

$$S_t = \{(\tau, x) \in \Pi_t \,|\, \|x - x_0\|_{H_-} \le c(T - \tau)\}$$

be its lateral surface, and let

$$\sigma_{t_1} = \{(\tau, x) \,|\, \tau = t_1; \|x - x_0\|_{H_-} \le c(T - t_1)\} \quad (0 \le t_1 \le t)$$

be a section of D_t by the plane $\tau = t_1$. At all points (t, x) of the lateral surface of a cone K_{c, x_0}, there exists a normal in the sense of the Hilbert space $\mathbb{R}^1 \times H_0$, and it is collinear to the vector

$$n(t, x) = (n_0, n_1(t, x)) = \left(c, \frac{I(x - x_0)}{\|x - x_0\|_{H_-}}\right) \in \mathbb{R}^1 \times H_+, \tag{4.12}$$

where $I : H_- \to H_+$ is an isometry connected with the chain $H_- \supset H_0 \supset H_+$.

Theorem 4.3. *Let $u(t)$, $t \in \mathrm{R}^1_+$ be a strong solution of the problem* (4.9) *with $u_0 \in \mathfrak{D}(L_\mu)$ and $u_1 \in \mathfrak{D}(L_\mu^{1/2})$, and let K_{c,x_0} be an arbitrary cone* (4.11) *with $c \geq \sqrt{2}/2$. Then, for $t \in [0, T]$, $T \geq 0$, the energetic inequality*

$$E(t) = \sup_{0 \leq \tau \leq t} \left\{ \int_{\sigma_\tau} \left[|u(\tau)|^2 + \left|\frac{du}{d\tau}(\tau)\right|^2 + \frac{1}{2}(\nabla u(\tau), \nabla \bar{u}(\tau))_{H_0} \right] d\mu \right\} \leq c(t)E(0) \tag{4.13}$$

holds.

Remark 1. The integral

$$\int_{\sigma_\tau} (\nabla u(\tau), \nabla \bar{u}(\tau))_{H_0} d\mu$$

should be regarded as

$$\lim_{n \to \infty} \int_{\sigma_\tau} (\nabla u_n(\tau), \nabla \bar{u}_n(\tau))_{H_0} d\mu,$$

where $(u_n(\tau))_{n=1}^\infty$ is an arbitrary sequence from $C_b^2(H_-)$ convergent in norm of the graph of the operator $L_\mu^{1/2}$ to $u(\tau) \in \mathfrak{D}(L_\mu)$, $\tau \in [0, t]$.

■

The proof of the theorem is based on the following auxiliary statements.

Lemma 4.2. *For all $c \geq \sqrt{2}/2$, the cone K_{c_1, x_0} it is spatially characteristically oriented. In other words, the following inequality holds on its lateral surface*

$$n_0^2 - \frac{1}{2}(n_1, n_1)_{H_0} \geq 0. \tag{4.14}$$

Proof. According to the choice of $c \geq \sqrt{2}/2$, we have

$$c^2 - \frac{1}{2}\left\| I \frac{x - x_0}{\|x - x_0\|_{H_-}} \right\|_{H_0}^2 \geq c^2 - \frac{\|x - x_0\|_{H_+}^2}{2\|x - x_0\|_{H_-}^2} \geq c^2 - \frac{1}{2} \geq 0,$$

because $\|\cdot\|_{H_-} \leq \|\cdot\|_{H_+}$.

■

The next proposition may be regarded as a special case of the Gauss–Ostrogradsky formula for conic domains $D_t \times \mathfrak{B}([0, T] \times H_-)$ and the measure $d\tau \times d\mu$ on these domains, because it gives the connection between the integrals over D_t and the values of functions on the lateral surfaces S_t.

Lemma 4.3. *The following formula*

$$\int_{D_t} [(\nabla_x v(\tau, x), \varphi)_{H_0} + (\beta_\mu(x), \varphi)_{H_0} v(\tau, x)] d\mu(x) d\tau$$

$$= \int_{c(T-t) \le \| x - x_0 \|_{H_-} \le cT} v(T - c^{-1} \| x - x_0 \|_{H_-}, x) \left(\frac{n_1(x)}{n_0}, \varphi \right)_{H_0} d\mu(x)$$

$$((v \in C_b^1([0, T] \times H_-, \mathbb{R}^1), \ \varphi \in H_+) \tag{4.15}$$

holds.

Proof. It suffices to prove (4.15) on a set of functions of the form $v(t, x) = f(t) g(x)$, where $f \in C_b^1([0, T], \mathbb{R}^1)$ and $g \in C_b^1(H_-, \mathbb{R}^1)$, which is total in $C_b^1([0, T] \times H_-, \mathbb{R}^1)$. By virtue of the conditions $(\beta_\mu, \varphi)_{H_0} \in L_1(H_-, \mu)$ and $\| n_1(x) \|_{H_+} \le 1$, the equality (4.15) can be extended by linearity and continuity to the other functions from $C_b^2([0, T] \times H_-, \mathbb{R}^1)$.

If we apply the Fubini theorem, then the left-hand side of (4.15) becomes equivalent to the expression

$$\int_{\| x - x_0 \|_{H_-} \le cT} \left(\int_0^{T - c^{-1} \| x - x_0 \|_{H_-}} f(\tau) d\tau \right) [(\nabla g(x), \varphi)_{H_0} + (\beta_\mu(x), \varphi)_{H_0} g(x)] d\mu(x)$$

$$- \int_{\| x - x_0 \|_{H_-} \le c(T-t)} \left(\int_t^{T - c^{-1} \| x - x_0 \|_{H_-}} f(\tau) d\tau \right) [(\nabla g(x), \varphi)_{H_0} + (\beta_\mu(x), \varphi)_{H_0} g(x)] d\mu(x)$$

$$= \int_{H_-} [s_1(x) - s_2(x)] [(\nabla g(x), \varphi)_{H_0} + (\beta_\mu(x), \varphi)_{H_0} g(x)] d\mu(x),$$

where we have introduced the functions

$$s_1(x) = \begin{cases} \int_0^{T-c^{-1}\|x-x_0\|_{H_-}} f(\tau)d\tau, & \|x-x_0\|_{H_-} \le cT, \\ 0, & \|x-x_0\|_{H_-} > cT, \end{cases} \tag{4.16}$$

$$s_2(x) = \begin{cases} \int_t^{T-c^{-1}\|x-x_0\|_{H_-}} f(\tau)d\tau, & \|x-x_0\|_{H_-} \le c(T-t), \\ 0, & \|x-x_0\|_{H_-} > c(T-t). \end{cases} \tag{4.17}$$

According to the definition of the logarithmic derivative (3.3), we have

$$\int_{H_-} (s_1 - s_2)(x)[(\nabla g(x), \varphi)_{H_0} + (\beta_\mu(x), \varphi)_{H_0} g(x)]d\mu(x)$$

$$= \lim_{\lambda\to 0} \int_{H_-} \frac{(s_1 - s_2)(x+\lambda\varphi) - (s_1 - s_2)(x)}{\lambda} g(x)d\mu(x). \tag{4.18}$$

By differentiating the Riemann integrals with variable limits (4.16) and (4.17), we obtain

$$\lim_{\lambda\to 0} \frac{s_1(x+\lambda\varphi) - s_1(x)}{\lambda}$$

$$= \begin{cases} f\left(T - c^{-1}\|x-x_0\|_{H_-}\right) \dfrac{(x-x_0, \varphi)_{H_-}}{c\|x-x_0\|_{H_-}}, & 0 \le \|x-x_0\|_{H_-} < cT, \\ 0, & \|x-x_0\|_{H_-} > cT, \end{cases}$$

$$\lim_{\lambda\to 0} \frac{s_2(x+\lambda\varphi) - s_2(x)}{\lambda}$$

$$= \begin{cases} f\left(T - c^{-1}\|x-x_0\|_{H_-}\right) \dfrac{(x-x_0, \varphi)_{H_-}}{c\|x-x_0\|_{H_-}}, & 0 \le \|x-x_0\|_{H_-} < c(T-t), \\ 0, & \|x-x_0\|_{H_-} > c(T-t). \end{cases}$$

By virtue of the fact that the measure μ is quasi-invariant, we have

$$\mu(\{x \in H_- \mid \|x - x_0\|_{H_-} = r\}) = 0.$$

for any $r > 0$ and $x_0 \in H_0$. Therefore, the integrands in (4.19) converge as $\lambda \to 0$ almost everywhere; they are bounded by the constant $c_\varphi \|g\|_{L_\infty(H_-,\, \mu)} \|f\|_{L_\infty([0,\, T])}$ uniformly in λ for all λ. The reference to the Lebesgue theorem completes the proof, i.e.,

$$\int_{D_t} [(\nabla_x v(\tau, x), \varphi)_{H_0} + (\beta_\mu(x), \varphi)_{H_0} v(\tau, x)] d\mu(x)\, d\tau$$

$$= \int_{D_t} [(\nabla g(x), \varphi)_{H_0} + (\beta_\mu(x), \varphi)_{H_0} g(x)] f(\tau)\, d\mu(x)\, d\tau$$

$$= \int_{c(T-t)\le\| x-x_0 \|_{H_-} \le cT} g(x) f(T - c^{-1}\| x - x_0 \|_{H_-}) \frac{(x - x_0, \varphi)_{H_-}}{c\| x - x_0 \|_{H_-}} d\mu(x)$$

$$= \int_{c(T-t)\le\| x-x_0 \|_{H_-} \le cT} v(T - c^{-1}\| x - x_0 \|_{H_-}, x)\left(\frac{n_1(x)}{n_0}, \varphi\right)_{H_0} d\mu(x).$$

■

Lemma 4.4. *The estimate*

$$2\int_{D_t}\left[\frac{\partial^2 u}{\partial\tau^2}\frac{\partial u}{\partial\tau} + (L_\mu + V)u\frac{\partial u}{\partial\tau}\right](\tau, x)\, d\mu(x)\, d\tau$$

$$\ge \int_{\sigma_\tau}\left\{\left(\frac{\partial u}{\partial\tau}\right)^2 + \frac{1}{2}(\nabla u(\tau), \nabla u(\tau))_{H_0}\right\} d\mu\Big|_{\tau=0}^{\tau=t}$$

$$+ 2\int_{D_t}\left(Vu\frac{\partial u}{\partial\tau}\right)(\tau, x)\, d\mu(x)\, d\tau \tag{4.19}$$

holds on the class of functions $u \in C_b^2([0, T] \times H_-, \mathbb{R}^1)$.

Proof. Using the formula (4.15) and the Fubini theorem, we get

$$2\int_{D_t}\left(\frac{\partial^2 u}{\partial\tau^2}\frac{\partial u}{\partial\tau} + (L_\mu + V)u\frac{\partial u}{\partial\tau}\right)(\tau, x)\, d\mu(x) d\tau$$

$$= \int_{c(T-t)\le\| x-x_0 \|_{H_-} \le cT}\left[\left(\frac{\partial u}{\partial\tau}\right)^2 + \frac{1}{2}(\nabla u, \nabla u)_{H_0}\right.$$

$$\left. - \left(\nabla u, \frac{n_1(\cdot)}{n_0}\right)_{H_0}\frac{\partial u}{\partial\tau}\right]\left(\tau = T - \frac{\| x - x_0 \|_{H_-}}{c}, x\right) d\mu(x)$$

$$+ \int_{\sigma_\tau}\left\{\left(\frac{\partial u}{\partial\tau}\right)^2 + \frac{1}{2}(\nabla u, \nabla u)_{H_0}\right\}(\tau, x) d\mu(x)\Big|_{\tau=0}^{\tau=t} + 2\int_{D_t}\left(Vu\frac{\partial u}{\partial\tau}\right)(\tau, x)\, d\mu(x) d\tau.$$

But

$$\int_{c(T-t)\le\|x-x_0\|_{H_-}\le cT}\left[\left(\frac{\partial u}{\partial\tau}\right)^2+\frac{1}{2}(\nabla u,\nabla u)_{H_0}\right.$$

$$\left.-\left(\nabla u,\frac{n_1(\cdot)}{n_0}\right)_{H_0}\frac{\partial u}{\partial\tau}\right]\left(\tau=T-\frac{\|x-x_0\|_{H_-}}{c},x\right)d\mu(x)\ge 0,$$

since, for any $(\tau, x)\in S_t$, we have

$$\left(\frac{\partial u}{\partial t}\right)^2+\frac{1}{2}(\nabla u,\nabla u)_{H_0}-\left(\nabla u,\frac{n_1(\cdot)}{n_0}\right)_{H_0}\frac{\partial u}{\partial\tau}$$

$$=n_0^{-2}\left\{\frac{1}{2}\left(n_0\nabla u-\frac{\partial u}{\partial\tau}n_1,\ n_0\nabla u-\frac{\partial u}{\partial\tau}n_1\right)_{H_0}+\left(\frac{\partial u}{\partial\tau}\right)^2\left(n_0^2-\frac{1}{2}\|n_1\|_{H_0}^2\right)\right\}\ge 0,$$

as long as $n_0^2-\frac{1}{2}\left\|n_1(\cdot)\right\|_{H_0}^2\ge 0$; according to Lemma 4.2, the last inequality is guaranteed by the choice $c\ge\sqrt{2}/2$.

■

Lemma 4.5. *Under the conditions of the theorem, the inequality*

$$\int_{\sigma_\tau}\left[\left|\frac{du}{dt}(t)\right|^2+\frac{1}{2}(\nabla u(t),\nabla\bar{u}(t))_{H_0}\right]d\mu\le\int_{\sigma_0}\left[\left|\frac{du}{dt}(0)\right|^2\right.$$

$$\left.+\frac{1}{2}(\nabla u(0),\nabla\bar{u}(0))_{H_0}\right]d\mu-2\int_{D_t}Vu(\tau)\frac{\partial u}{\partial\tau}(\tau)d\mu\,d\tau. \qquad (4.20)$$

holds for the strong solution $u(t),\ t\in\mathbb{R}_+^1$ *of the problem* (4.9).

Proof. First, we clarify the correctness of some notations in formula (4.20). The vector functions

$$u(t),\quad\frac{du}{dt}(t),\quad\frac{d^2u}{dt^2}(t),\quad(L_\mu+V)u(t),\quad t\in[0,T],$$

defined by Equation (4.9) are continuous in mean square; consequently, they have modifications measurable with respect to $\mathfrak{B}([0,T]\times H_-)$ (Gikhman and Skorokhod [2, Chapter 3 Section 3, Theorem 1]). In other words, for every fixed $t\in[0,T]$, we can change the values of these vector functions on the sets of μ-measure 0 in H_- in order to obtain measurable functions of two variables as a result. The integrals over the regions D_t which appear in the proof will be regarded just in this sense.

The second point which should be clarified is the definition of the function

$$(\nabla u(t), \nabla \bar{u}(t))_{H_0} \in L_1(H_-, \mu), \quad t \in [0, T].$$

Let us take an arbitrary sequence $(u_n(t))_{n=1}^{\infty} \subset C_b^2(H_-)$ convergent to $u(t)$ in norm of the graph of the operator $L_\mu^{1/2}$ (these sequences exist automatically, because $u(t) \in \mathfrak{D}(L_\mu)$, and $C_b^2(H_-)$ is the domain of essential selfadjointness for L_μ). Then, clearly, $(\nabla u_n(t), \nabla \bar{u}_n(t))_{H_0}$, $n \in \mathbb{N}$ is a fundamental sequence in $L_1(H_-, \mu)$. We define $(\nabla u(t), \nabla \bar{u}(t))_{H_0}$ to be its limit in $L_1(H_-, \mu)$. Since the operator $L_\mu^{1/2}$ is symmetric, this limit is independent of the choice of a sequence $(u_n(t))_{n=1}^{\infty} \subset C_b^2(H_-)$ which gives the required approximation, and

$$\int_{\sigma_t} (\nabla u(t), \nabla \bar{u}(t))_{H_0} d\mu = \lim_{n\to\infty} \int_{\sigma_t} (\nabla u_n(t), \nabla \bar{u}_n(t))_{H_0} d\mu. \tag{4.21}$$

To check the validity of the inequality (4.20), we construct a special approximation of the solution $u(t), t \in [0, T]$ by the smooth functions $u_n \in C_b^2([0, T] \times H_-)$, $n \in \mathbb{N}$, such that

$$\int_0^T \int_{H_-} \left\{ |u - u_n|^2 + \left| \frac{d}{dt}(u - u_n) \right|^2 + \left| \frac{d^2}{dt^2}(u - u_n) \right|^2 + |(L_\mu + V)(u - u_n)|^2 \right\} d\mu\, dt \to 0,$$

and

$$\int_{H_-} \left\{ |u - u_n|^2 + \left| \frac{d}{dt}(u - u_n) \right|^2 + |(L_\mu + V)(u - u_n)|^2 \right\} d\mu \to 0, \quad t \in [0, T]$$

as $n \to \infty$, $t \in [0, T]$. To do this, we employ the representation for $u(t)$, $t \in [0, T]$, in terms of the RI of the operator $L_\mu + V$ (see (4.10)). We have

$$u(t) = \int_{E_V}^{+\infty} \cos\sqrt{\lambda}\, t\, dE(\lambda) u_0 + \int_{E_V}^{+\infty} \lambda^{-1/2} \sin\sqrt{\lambda}\, t\, dE(\lambda) u_1$$

$$= \lim_{N\to\infty} \lim_{K\to\infty} \left\{ \sum_{k=0}^{K} t^{2k} \int_{E_V}^{N} \frac{(-\lambda)^k}{(2k)!} dE(\lambda) u_0 + \sum_{k=0}^{K} t^{2k+1} \int_{E_V}^{N} \frac{(-\lambda)^k}{(2k+1)!} dE(\lambda) u_1 \right\},$$

and convergence also takes place in the sense of (4.21). Further, let us approximate the vectors

$$\varphi_{k,N} = \int\limits_{E_V}^{N} \frac{(-\lambda)^k}{(2k)!} dE(\lambda)u_0 \in \mathfrak{D}(L_\mu),$$

$$\psi_{k,N} = \int\limits_{E_V}^{N} \frac{(-\lambda)^k}{(2k+1)!} dE(\lambda)u_1 \in \mathfrak{D}(L_\mu),$$

by the functions from $C_b^2(H_-)$ in the norm of the graph of the operator L_μ. Relation (4.21) enables us to pass to the limit as $n \to \infty$ in the inequalities (4.19) written for $\operatorname{Re} u_n(t, x)$ and $\operatorname{Im} u_n(t, x)$, $n \in \mathbb{N}$. Together with the condition

$$\frac{d^2u}{dt^2}(t) + (L_\mu + V)u(t) = 0,$$

this completes the proof. ■

Denote

$$y(t) = \int\limits_{\sigma_t} \left\{ \left| \frac{du}{dt}(t) \right|^2 + \frac{1}{2}(\nabla u(t), \nabla \bar{u}(t))_{H_0} \right\} d\mu, \quad t \in [0, T].$$

As follows from Lemma 4.5,

$$y(t) \le y(0) - 2 \int\limits_{D_t} \left(Vu \frac{\partial u}{\partial \tau} \right)(\tau, x)\, d\tau\, d\mu(x)$$

$$\le y(0) + \|V\|_{L_\infty(H_-,\mu)} \int\limits_{D_t} \left(|u|^2 + \left| \frac{du}{d\tau} \right|^2 \right)(\tau, x)\, d\tau\, d\mu(x)$$

$$\le y(0) + \|V\|_{L_\infty(H_-,\mu)} \cdot t \sup_{0 \le \tau \le t} \left[\int\limits_{\sigma_\tau} |u(\tau, x)|^2\, d\mu(x) + y(\tau) \right]. \tag{4.22}$$

Let us estimate the integral

$$\int\limits_{\sigma_t} |u(t, x)|^2 d\mu(x), \quad t \in [0, T]$$

by using the representation

$$u(t) = u(0) + \int_0^t \frac{du}{d\tau}(\tau)d\tau$$

in $L_2(H_-, \mu)$. By virtue of the continuity of the mapping

$$[0, T] \ni t \mapsto \frac{du}{dt}(t) \in L_2(H_-, \mu),$$

the integral $\int_0^1 \frac{du(\tau)}{d\tau}\, d\tau$ in this representation can be understood as a Riemann integral. We have

$$\int_{\sigma_t} |u(t, x)|^2 d\mu(x) \leq 2\int_{\sigma_0} |u(0)|^2 d\mu + 2\int_{\sigma_t} \left| \int_0^t \frac{du}{d\tau}(\tau)d\tau \right|^2 d\mu$$

$$\leq 2\int_{\sigma_0} |u(0)|^2 d\mu + 2t^2 \sup_{0 \leq \tau \leq t} \int_{\sigma_\tau} \left| \frac{du}{d\tau}(\tau) \right|^2 d\mu \leq 2\int_{\sigma_0} |u(0)|^2 d\mu + 2t^2 \sup_{0 \leq \tau \leq t} y(\tau).$$

Let us take this inequality into account in (4.22). The result can be written in the form

$$E(t) = \sup_{0 \leq \tau \leq t} \left\{ \int_{\sigma_\tau} \left[|u(\tau)|^2 + \left| \frac{du}{d\tau}(\tau) \right|^2 + \frac{1}{2}(\nabla u(\tau), \nabla \bar{u}(\tau))_{H_0} \right] d\mu \right\}$$

$$\leq 2E(0) + \| V \|_{L_\infty(H_-, \mu)} tE(t) + 2t^2 E(t) \quad (t \in [0, T])$$

or

$$E(t)(1 - \| V \|_{L_\infty(H_-, \mu)} t - 2t^2) \leq 2E(0), \quad t \in [0, T]. \qquad (4.23)$$

Let us choose t sufficiently small to guarantee that

$$1 - \| V \|_{L_\infty(H_-, \mu)} t - 2t^2 \geq \frac{1}{2},$$

i.e., $0 \leq t \leq \min\{T, \tau^*\}$, where τ^* is determined from the equality

$$2(\tau^*)^2 + \| V \|_{L_\infty(H_-, \mu)} \tau^* = \frac{1}{2}.$$

For indicated t, it follows from (4.23) that $E(t) \leq 4E(0)$. If $\tau^* < T$, then we take $t = \tau^*$ as the initial time and, by virtue of the above arguments, we can state that the inequality

$$E(t) \leq 4E(\tau^*) \leq 16E(0)$$

holds for $0 \leq t \leq \min \{T, 2\tau^*\}$.

If we repeat this procedure again and again, then we finally obtain the estimate

$$E(t) \leq c(t)E(0),$$

valid for all $t \in [0, T]$, i.e., the required result.

■

Corollary 1 (Finite rate of propagation of perturbations). *The hyperbolic Equation* (4.9) *describes the propagation of perturbations whose rate does not exceed* $\sqrt{2}/2$ *in the norm of* H_-.

Proof. Assume that the initial perturbations, i.e., the functions $u_0, u_1 \in L_2(H_-, \mu)$, have their supports in a ball

$$\tilde{B}_R(x_0) = \{x \in H_- \,|\, \|x - x_0\|_{H_-} \leq R\} \quad (x_0 \in H_-, R > 0).$$

Let us show that the solution $u(t)$, $t \in \mathbb{R}^1_+$, is equal to zero outside the ball $\tilde{B}_{R(t)}(x_0)$, $R(t) = R + \sqrt{2}/2t$, at each time $t \in \mathbb{R}^1_+$.

Indeed, assume that the point $x_1 \in H_-$ lies at a distance $d > 0$ from $\tilde{B}_{R(t)}(x_0)$, i.e.,

$$\|x_1 - x_0\|_{H_-} = R + \sqrt{2}/2t + d.$$

Let us construct the frustrum of a cone

$$D_t = \{(\tau, x) \in \Pi_t \,|\, \|x - x_1\|_{H_-} \leq \sqrt{2}/2(\hat{T} - \tau)\},$$

where $\hat{T} = t + \sqrt{2}r$ and $r \in (0, d)$. The projection

$$\{x \in H_- \,|\, \|x - x_1\|_{H_-} \leq r + \sqrt{2}/2t\}$$

of its upper base σ_t onto H_- does not intersect with $\tilde{B}_{R(t)}(x_0)$, and the projection

$$\{x \in H_- \,|\, \|x - x_1\|_{H_-} \leq r + \sqrt{2}/2t\}$$

of its lower base σ_0 does not intersect with $\tilde{B}_R(x_0)$. If we now apply the inequality (4.13) to the solution $u(t)$, $t \in \mathbb{R}^1_+$, and the domain D_t, then we obtain

$$\int_{\sigma_t} |u(t)|^2 d\mu \le c(t) \int_{\sigma_0} [|u_0|^2 + \frac{1}{2}(\nabla u_0, \nabla \bar{u}_0)_{H_0} + |u_1|^2] d\mu = 0. \tag{4.24}$$

Inequality (4.24) means that the solution $u(t)$, $t \in \mathbb{R}^1_+$, is equal to 0 at each time $t \in \mathbb{R}^1_+$ almost everywhere on each ball $B_r(x_1)$ which does not intersect with $\tilde{B}_{R(t)}(x_0)$. If we construct a countable covering of the set $H_- \setminus \tilde{B}_{R(t)}(x_0)$ by these balls, then we find that outside $H_- \setminus \tilde{B}_{R(t)}(x_0)$ we have $u(t) = 0$ μ-almost everywhere. In other words, a local perturbation in the evolutionary process described by the Cauchy problem (4.9) will spread over a distance which is not greater than $\sqrt{2}\,/\,2t$ during the time t.

■

Corollary 2 (The law of conservation of energy). *Under the conditions of the theorem, the integrals of energy are equal, i.e.,*

$$\int_{H_-} \left[\left| \frac{du}{dt}(t) \right|^2 + \frac{1}{2}(\nabla u(t), \nabla \bar{u}(t))_{H_0} + V|u(t)|^2 \right] d\mu$$

$$= \int_{H_-} [|u_1|^2 + \frac{1}{2}(\nabla u_0, \nabla \bar{u}_0)_{H_0} + V|u_0|^2] d\mu. \tag{4.25}$$

Indeed, the proof of this assertion is a simplified repetition of the proof of the theorem. Let us multiply Equation (4.9) by $d\bar{u}/dt$ and integrate the result of this multiplication over $[0, T] \times H_-$. We obtain

$$\int_0^T \int_{H_-} \left[\frac{\partial^2 u}{\partial t^2} \frac{\partial \bar{u}}{\partial t} + (L_\mu + V) u \frac{\partial \bar{u}}{\partial t} \right] d\mu\, dt = 0.$$

Let $u_n(t, x) \in C_b^2([0, T] \times H_-)$, $n \in \mathbb{N}$, be an approximation of the solution $u(t)$, $t \in [0, T]$, in the sense of (4.21). Then we have the following chain of equalities

$$\int_{H_-} \left| \frac{du}{dt}(t) \right|^2 d\mu \Big|_{t=0}^{t=T} = \lim_{n\to\infty} \int_{H_-} \left| \frac{du_n}{dt}(t) \right|^2 d\mu \Big|_{t=0}^{t=T}$$

$$= \lim_{n\to\infty} 2\mathrm{Re} \int_0^T \int_{H_-} \frac{d^2 u_n}{dt^2} \frac{d\bar{u}_n}{dt} d\mu\, dt = 2\mathrm{Re} \int_0^T \int_{H_-} \frac{d^2 u}{dt^2} \frac{d\bar{u}}{dt} d\mu\, dt$$

$$= -2\mathrm{Re} \int_0^T \int_{H_-} (L_\mu + V) u \frac{d\bar{u}}{dt} d\mu\, dt$$

$$= -\lim_{n\to\infty} 2\mathrm{Re}\int_0^T \int_{H_-} (L_\mu + V)u_n \frac{d\bar{u}_n}{dt}\, d\mu\, dt$$

$$= -\lim_{n\to\infty} \int_{H_-} \int_0^T \frac{d}{dt}\Big[\frac{1}{2}(\nabla u_n, \nabla \bar{u}_n)_{H_0} + V|u_n|^2\Big]\, dt\, d\mu$$

$$= -\lim_{n\to\infty} \int_{H_-} \Big[\frac{1}{2}(\nabla u_n(t), \nabla \bar{u}_n(t))_{H_0} + V|u_n(t)|^2\Big]\, d\mu \Big|_{t=0}^{t=T}$$

$$= -\int_{H_-} \Big[\frac{1}{2}(\nabla u(t), \nabla \bar{u}(t))_{H_0} + V|u(t)|^2\Big]\, d\mu \Big|_{t=0}^{t=T}.$$

■

4.3. Potential Perturbations of the Dirichlet Operators

As stated in Theorem 3.4, for $\|\beta_\mu\|_{H_-} \in L_2(H_-, \mu)$ and $\beta_\mu \in C^2_{b,\,\mathrm{loc}}(H_-, H_-)$, the Dirichlet operator L_μ is essentially selfadjoint on $C_b^2(H_-)$. This enables us to prove (by the standard approximation) the essential selfadjointness on a subset of the finite function $C_0^2(H_-) \subset C_b^2(H_-)$. For the potential $V = \bar{V} \in L_{\infty,\mathrm{loc}}(H_-, \mu)$, the operator $L_\mu + V$ is defined on $C_0^\infty(H_-)$. By using the hyperbolic evolutionary criteria presented in Section 1 of Chapter 5, in the next theorem, we establish the conditions of essential selfadjointness of the operator $L_\mu + V$.

Theorem 4.4. *Let* $\beta_\mu \in C^2_{b,\,\mathrm{loc}}(H_-, H_-)$, $\|\beta_\mu\|_{H_-} \in L_2(H_-, \mu)$, *and* $V = \bar{V} \in L_{\infty,\mathrm{loc}}(H_-, \mu)$. *Then the operator* $L_\mu + V$ *is essentially selfadjoint on* $C_0^2(H_-)$ *provided that one of the following conditions holds*

(i) the operator $L_\mu + V$ *is semibounded from below, i.e., there exists* $c \geq 0$ *such that*

$$((L_\mu + V)u, u)_{L_2(H_-,\mu)} \geq -c\|u\|^2_{L_2(H_-,\mu)}$$

for any $u \in C_0^2(H_-)$;

(ii) there exists a nonincreasing function $\mathbb{R}^1_+ \ni s \mapsto Q(s) \in (-\infty, 0]$ *for which* $V(x) \geq Q(\|x\|_{H_-})$ *and* $|Q(s)| = O(s^2)$, *as* $s \to \infty$.

Proof. Assume that condition (i) is satisfied. For $n \in \mathbb{N}$, we define a sequence of potentials

$$V_n(x) = \begin{cases} V(x), & \|x\|_{H_-} \le n, \\ 0, & \|x\|_{H_-} > n. \end{cases} \tag{4.26}$$

Since $V_n \in L_\infty(H_-, \mu)$, the operator $L_\mu + V$ is essentially selfadjoint on $C_0^2(H_-)$. We preserve the same notation $L_\mu + V$ for its closure $(L_\mu + V)_{\min}$. According to Theorem 1.5 in Chapter 5, to prove the theorem it suffices to establish that, for every $T > 0$, the strong solutions $\mathbb{R}_+^1 \ni t \mapsto u_n(t) \in L_2(H_-, \mu)$ of the Cauchy problems

$$\begin{cases} \dfrac{d^2 u_n}{dt^2}(t) + (L_\mu + V_n)u_n(t) = 0, \\ u_n(0) = u_0, \ \dfrac{du_n}{dt}(0) = u_1\big(u_0, u_1 \in C_0^2(H_-)\big), \end{cases} \tag{4.27}$$

satisfy the relations

(a) $$u_n(t) \in \mathfrak{D}((L_\mu + V)_{\min}) \quad (t \in [0, T], \ n \in \mathbb{N}); \tag{4.28}$$

(b) $$\int_0^T (u_n(t), ((L_\mu + V)_{\min} - (L_\mu + V_n)u_n(t))_{L_2(H_-,\mu)} dt \to 0, \ n \to \infty, \tag{4.29}$$

where $[0, T] \ni t \mapsto u(t) \in L_2(H_-, \mu)$ is an arbitrary strong solution of the Cauchy problem

$$\begin{cases} \dfrac{d^2 u}{dt^2}(t) + (L_\mu + V)^* u(t) = 0, \\ u(0) = 0, \ \dfrac{du}{dt}(0) = 0, \ (t \in [0, T]). \end{cases} \tag{4.30}$$

Assume that the initial functions u_0 and u_1 are equal to 0 outside the ball

$$\tilde{B}_R(0) = \{x \in H_- \mid \|x\|_{H_-} \le R\}, \quad R > 0.$$

According to Corollary 1 from Theorem 4.3, we have

$$\operatorname{supp} u_n(t) \subset \tilde{B}_{R(T)}(0), \quad t \in [0, T], \quad R(T) = R + \sqrt{2}/2T \tag{4.31}$$

for any $n \in \mathbb{N}$.

Lemma 4.6. *If for $f \in L_2(H_-, \mu)$ and some $n \in \mathbb{N}$, we have $f \in \mathfrak{D}(L_\mu + V)$ and* $\operatorname{supp} f \subset \tilde{B}_{R'}(0)$, *where $R' < n$, then $f \in \mathfrak{D}((L_\mu + V)_{\min})$ and*

$$(L_\mu + V)_{\min} f = (L_\mu + V_n)f.$$

Proof. Let us show that there exists a sequence

$$f_m \in C_0^2(H_-), \quad \operatorname{supp} f_m \subset \tilde{B}_n(0), \quad m \in \mathbb{N},$$

such that $f_m \to f$ as $m \to \infty$ in $L_2(H_-, \mu)$, and

$$((L_\mu + V)f_m)_{m=1}^\infty = ((L_\mu + V_n)f_m)_{m=1}^\infty$$

is fundamental. Let $(g_m)_{m=1}^\infty \subset C_0^2(H_-)$ be an arbitrary sequence convergent to $f \in \mathfrak{D}(L_\mu + V_n)$ in the norm of the graph of the operator $A_n = L_\mu + V_n$. We introduce an infinitely differentiable function $\mathbb{R}_+^1 \ni s \mapsto \chi(s) \in [0, 1]$ equal to 1 for $s \in [0, R']$ and equal to 0 for $s \in [n, +\infty)$ and set

$$f_m(x) = \chi(\|x\|_{H_-})g_m(x), \quad x \in H_-, \quad m \in \mathbb{N}. \tag{4.32}$$

Among the conditions imposed above on the sequence $(f_m)_{m=1}^\infty \subset C_0^2(H_-)$, we must verify only the fact that it is fundamental.

We start from the equality which can be established by direct calculation

$$A_n(f_m - f_l)(x) = \chi(\|x\|_{H_-})A_n(g_m - g_l)(x)$$
$$+ (g_m - g_l)(x)(L_\mu \chi(\|x\|_{H_-})) - (\nabla(g_m - g_l)(x), \nabla\chi(\|x\|_{H_-}))_{H_0}. \tag{4.33}$$

Then

$$\|A_n(f_m - f_l)\|_{L_2(H_-,\mu)} \le \|\chi\|_{L_\infty}\|A_n(g_m - g_l)\|_{L_2(H_-,\mu)}$$
$$+ \|L_\mu\chi\|_{L_\infty(H_-,\mu)}\|g_m - g_l\|_{L_2(H_-,\mu)}$$
$$+ \sup_{H_-}\|\nabla\chi\|_{H_0}\|(\nabla(g_m - g_l), \nabla\overline{(g_m - g_l)})_{H_0}\|_{L_1(H_-,\mu)}^{1/2}. \tag{4.34}$$

According to the choice of $(g_m)_{m=1}^\infty$, the first two terms on the right-hand side of (4.34) tend to 0 as $m, l \to \infty$. A similar property for the third term follows from the inequality

$$((A_n - c)u, u)_{L_2(H_-,\mu)} = \int_{H_-} [(\nabla u, \nabla\bar{u})_{H_0} + (V_n - c)|u|^2]d\mu \ge \int_{H_-} (\nabla u, \nabla\bar{u})_{H_0} d\mu \tag{4.35}$$

for any $u \in C_0^2(H_-)$ and $c \le \operatorname{ess\,inf} V_n$. Inserting $u = g_m - g_l$ in (4.35), we find that

$$\| (\nabla(g_m - g_l), \overline{\nabla(g_m - g_l)})_{H_0} \|_{L_1(H_-,\mu)}$$

$$\le \| A_n(g_m - g_l) \|_{L_2(H_-,\mu)} \| g_m - g_l \|_{L_2(H_-,\mu)} - c \| g_m - g_l \|^2_{L_2(H_-,\mu)} \to 0, \ m, l \to \infty.$$

This completes the proof of the fact that the sequence is fundamental and, thus, Lemma 4.6 is proved.

■

By virtue of Lemma 4.6, we obtain

$$u_n(t) \in \mathfrak{D}((L_\mu + V)_{\min}), \quad (L_\mu + V)_{\min} u_n(t) = (L_\mu + V_n) u_n(t). \tag{4.36}$$

for any $t \in [0, T]$, and $n > R(T) = R + \sqrt{2} / 2T$.

Therefore, the relations (4.28) and (4.29) hold, and the first part of Theorem 4.4 is thus proved.

We proceed to the proof of the second assertion of the theorem. To investigate operators $L_\mu + V$ which are not semibounded, we need the following lemma.

Lemma 4.7. *Under the conditions of Theorem* 4.4 *(we do not assume that the operator* $L_\mu + V$ *is semibounded from below), the Cauchy problem*

$$\begin{cases} \dfrac{d^2u}{dt^2}(t) + (L_\mu + V)_{\min} u(t) = 0 \quad (t \in \mathbb{R}^1_+), \\ u(0) = u_0; \ \dfrac{du}{dt}(0) = u_1 \quad \left(u_0, u_1 \in C_0^2(H_-)\right), \end{cases} \tag{4.37}$$

has a strong solution $u(t)$, $t \in \mathbb{R}^1_+$, *and this solution is unique in the following sense. Let* $\operatorname{supp} u_0$, $\operatorname{supp} u_1 \subset \tilde{B}_R(0)$ $(R > 0)$, *then* $u(t)$ *coincides with the solution* $u_n(t)$ *of the problem* (4.27), *where* $n \in \mathbb{N}$ *and* $n > R(T) = R + \sqrt{2} / 2T$ *on each segment* $[0, T]$ $(T > 0)$.

Proof. It can be easily seen from (4.36) that the Cauchy problem (4.37) is locally solvable on each finite interval $[0, T]$, and the functions $u_n(t)$, $t \in [0, T]$ are its solutions. The problem of the existence and uniqueness of the solution $u(t)$ on the whole of the semi-axis $\mathbb{R}^1_+$ is reduced to the possibility of pasting together the solutions of problem (4.27) for different $T > 0$ and $n \in \mathbb{N}$. More accurately, it is necessary to establish the validity of the following statement:

Assume that $0 < T_1 < T_2$ *and*

$$[0, T_1] \ni t \mapsto u_{n_1}(t) \in L_2(H_-, \mu)$$

and

$$[0, T_2] \ni t \mapsto u_{n_2}(t) \in L_2(H_-, \mu)$$

are solutions of the problem (4.27) *on* $[0, T_1]$ *and* $[0, T_2]$, *respectively; also let* $n_1 > R(T_1)$ *and* $n_2 > R(T_2)$. *It is necessary to prove that* $u_{n_1}(t) = u_{n_2}(t)$, $t \in [0, T]$.

By virtue of the continuity of $u_{n_1}(t)$ and $u_{n_2}(t)$ in t, it suffices to check this equality for $t \in [0, T_0)$ with some $T_0 > 0$. The vector functions $u_{n_1}(t)$ and $u_{n_2}(t)$, $t \in [0, T_0)$, may be regarded as two solutions of the same Cauchy problem (4.37) on $[0, T_0)$ with the same initial data. On the other hand, taking into account the general argument which, in our situation, reflects the Holmgren principle and the solvability of problem (4.37) on $[0, T_0)$ for any $u_0, u_1 \in C_0^2(H_-)$, we conclude that, for the dual problem with the operator $(L_\mu + V)^* \supset (L_\mu + V)_{\min}$, the uniqueness of strong solutions takes place. This implies that $u_{n_1}(t) = u_{n_2}(t)$ for any $t \in [0, T_0)$.

■

In order to establish the essential selfadjointness of the operator $L_\mu + V$ (under the assumption of the second part of Theorem 4.4), it suffices to show that for any $u_0 \in C_0^2(H_-)$, the solution $u(t)$, $t \in \mathbb{R}^1_+$, of the problem (4.37), where $u_1 = 0$, is such that

$$(u(t), u_0)_{L_2(H_-,\mu)} = O(e^{ct^2}), \quad t \to +\infty, \tag{4.38}$$

with some $c = c(u_0) \geq 0$ (see Theorem 1.4 in Chapter 5). According to Lemma 4.7, $u(t) = u_n(t)$ for $t \in [0, T]$ and $n > R(T)$, where $u_n(t)$ is a solution of the problem (4.27) with the selfadjoint operator $A_n = L_\mu + V_n$. Without loss of generality, we can suppose that $n < 2R(T)$. Clearly, since $\mathbb{R}^1_+ \ni s \mapsto Q(s) \in (-\infty, 0]$ does not increase, we have

$$\begin{aligned} l_n &\geq \operatorname{ess\,inf} V_n = \min\{0, \operatorname*{ess\,inf}_{\|x\|_H \leq n} V(x)\} \\ &\geq \operatorname*{ess\,inf}_{\|x\|_{H_-} \leq 2R(T)} Q(\|x\|_{H_-}) \geq Q(2R + \sqrt{2}\,T). \end{aligned} \tag{4.39}$$

Using formula (4.10) for $u_n(t)$ in terms of the RI E_n of the operator A_n, we obtain

$$(u(T), u_0)_{L_2(H_-,\mu)} = (u_n(T), u_0)_{L_2(H_-,\mu)} = \int_{l_n}^{+\infty} \cos\sqrt{\lambda}\, t d(E_n(\lambda)u_0, u_0)_{L_2(H_-,\mu)}$$

at the point $t = T$. Taking into account (4.39), we find

$$|(u(T), u_0)_{L_2(H_-,\mu)}| \leq \sup_{Q(2R(T))\leq\lambda\leq 0} \sum_{k=0}^{\infty} \frac{|\lambda|^k T^{2k}}{(2k)!} \| u_0 \|^2_{L_2(H_-,\mu)}$$

$$\leq \sup_{Q(2R(T))\leq\lambda\leq 0} e^{\sqrt{|\lambda|}T} \| u_0 \|^2_{L_2(H_-,\mu)}$$

$$= e^{|Q(2R+\sqrt{2}T)|^{1/2} T} \| u_0 \|^2_{L_2(H_-,\mu)}. \tag{4.40}$$

By virtue of condition of the theorem, we have $Q(s) = Q(s^2)$, as $s \to +\infty$; therefore, (4.40) yields the required estimate (4.38).

■

7. INFINITE-DIMENSIONAL DIFFERENTIAL OPERATORS IN THE MODELS OF QUANTUM STATISTICAL PHYSICS AND FIELD THEORY

We have already seen in the examples of Chapter 6 that the operators of energy of the simplest physical systems with infinitely many degrees of freedom (e.g., a free boson field or a collection of noninteracting quantum oscillators) are given by elliptic differential operators of the second order acting in the spaces of the functions of infinitely many variables. At least on the formal level, the transition to more complicated systems can be often described by a perturbation of the original operator of energy by a certain potential. The following situation is typical in this case. The potential of perturbation is given by the expression which has no operator meaning in the space of states of a unperturbed system and, therefore, it is impossible to construct the perturbed Hamiltonian as a selfadjoint operator in the original Hilbert space and, hence, to define the dynamics of a perturbed system. In Example 2.2 in Chapter 6, we have shown that the polynomial models in constructive field theory lead us to just a situation of this sort: The interaction is given by potentials defined only as generalized functions.

In Section 1, we will discuss a general approach to overcoming this difficulty in the case of singular potential perturbations of the second quantization operators L_A (see Chapter 6) acting in the space $L_2(\Phi', \gamma_1)$. The idea of this approach can be formulated as follows. Under the assumption that a potential $V = \overline{V}$ possesses sufficiently good L_p-properties, we have proved that the sum $L_A + V$ is selfadjoint and has a normalized eigenfunction (ground state) $\phi_V > 0$ which corresponds to the lower bound of the spectrum $E_V = \inf s\,(L_A + V)$ γ_1-almost everywhere. Introducing a vacuum measure $\mu^V = \phi_V^2\,\gamma_1$ on Φ' and passing from $L_2(\Phi', \gamma_1)$ to $L_2(\Phi', \mu^V)$ in a unitary manner, we define a renormalized operator

$$L_{V,\mathrm{ren}} = \phi_V^{-1}(L_A + V - E_V)\phi_V$$

in $L_2(\Phi', \mu^V)$. Moreover,

$$(L_{V,\mathrm{ren}}u, v)_{L_2(\varphi', \mu^V)} = \frac{1}{2}\int_{\Phi'} (\nabla u, A\nabla v)_{H_0}\, d\mu^V, \qquad (0.1)$$

for smooth cylindrical functions u and v, i.e., the renormalized operator is connected with the Dirichlet form of the vacuum measure μ^V and acts in a new space of states $L_2(\Phi', \mu^V)$.

Now assume that the potential is given by a generalized function $V \in \mathcal{P}'(\Phi')$, and $(V_n)_{n=1}^{\infty}$ $(V_n = \overline{V}_n,\ n \in \mathbb{N})$ is a sequence of regular potentials which approximates V (in the sense of generalized functions). In this case, we cannot expect that the operators $L_A + V_n - E_{V_n}$ would converge as $n \to \infty$ in $L_2(\Phi', \gamma_1)$. This is so because in most cases for renormalized energy, we have $E_{V_n} \to \infty$ as $n \to \infty$. Nevertheless, the vacuum measures may converge to the limiting measure μ on Φ'. Therefore, by virtue of (0.1), it is natural to regard an operator associated with the Dirichlet form of the measure μ as a renormalized Hamiltonian $(L_A + V)_{\text{ren}}$ corresponding to the formal sum $L_A + V$. Hence, in our scheme of renormalization, both the Hamiltonian and the Hilbert space on which it acts are limiting objects.

In applications to the investigation of concrete models, an important role is played by the interpretation of the described renormalization procedure in terms of a functional integral corresponding to the second quantization operators. Formally, a measure on the space of trajectories $\Omega = \{\omega(\cdot) : \mathbb{R}^1 \to \Phi'\}$, corresponding to the diffusion process with generating operator $(L_A + V)_{\text{ren}}$ and stationary initial distribution μ, has the form

$$d\nu(\omega(\cdot)) = \frac{1}{N} e^{-\int_{-\infty}^{\infty} V(\omega(\tau))d\tau} d\nu_A(\omega(\cdot)), \tag{0.2}$$

where ν_A is an analogous measure corresponding to L_A and γ_1 constructed in Section 1 of Chapter 6; N^{-1} is a normalizing factor. Clearly, in each concrete case, one should attach a rigorous meaning to the representation (0.2) by introducing approximations.

In Section 2, the scheme of renormalization is applied to the construction of linear and quadratic models in quantum field theory. Using these models as examples, we illustrate the possibility of employing both aspects of the renormalization scheme, namely, in the case of the linear model (or the Van Hove model), it is convenient to investigate directly the convergence of vacuum measures, and the construction of a model with quadratic interaction is simpler in terms of a functional integral.

To be used in quantum statistical physics of lattice systems, the renormalization scheme needs a certain modification in order to take into account specific features of the limit transition in a number of interacting particles. This modification is realized in Section 3, where the case of systems in the ground state, or at temperature 0, is discussed. We investigate in detail the case of the so-called harmonic models on integral lattice $\mathbb{Z}^d$ $(d \in \mathbb{N})$ described by a formal Hamiltonian of the form

$$L = \frac{1}{2} \sum_{k \in \mathbb{Z}^d} p_k^2 + \frac{1}{2} \sum_{j,k \in \mathbb{Z}^d} d_{jk} q_j q_k. \tag{0.3}$$

Here

$$(p_k f)(x_k) = i df(x_k)/dx_k \quad \text{and} \quad (q_k f)(x_k) = (x_k f)(x_k)$$

are operators of momentum and coordinates of the particle at the vertex $k \in \mathbb{Z}^d$, respectively (they act in the space $L_2(\mathbb{R}^1, dx_k)$ (on natural domains of definition)), and $(d_{jk})_{j,k \in \mathbb{Z}^d}$ is a matrix of the operator $D > 0$, $D \in \mathfrak{L}(l_2(\mathbb{Z}^d), l_2(\mathbb{Z}^d))$. For the Hamiltonian (0.3), the renormalization scheme gives a Gaussian measure γ_S on

$$\mathbb{R}^{\mathbb{Z}^d} = \underset{k \in \mathbb{Z}^d}{\times} \mathbb{R}^1 \ni x = (x_k)_{k \in \mathbb{Z}^d}$$

with mean value zero and the correlation operator $S = D^{-1/2}$ as a limiting measure. The renormalized Hamiltonian L_{ren} coincides with the Dirichlet operator L_{γ_S} of this measure and is defined as an essential selfadjoint operator on a set of smooth cylindrical functions $u \in C^{\infty}_{b,\text{cyl}}(\mathbb{R}^{\mathbb{Z}^d})$ by the differential expression

$$(L_{\text{ren}} u)(x) = (L_{\gamma_S} u)(x) = -\frac{1}{2} \sum_{k \in \mathbb{Z}^d} \partial^2 u(x)/\partial x_k^2 + \sum_{j,k \in \mathbb{Z}^d} x_k r_{jk} \partial u(x)/\partial x_j,$$

where $(r_{jk})_{j,k \in \mathbb{Z}^d}$ is a matrix of the operator $R = D^{1/2}$. It is important to note that, by a change of variables, the operator L_{γ_S} can be reduced to the second quantization operator $L_R = L_{S^{-1}}$. In this section, we also consider the Hamiltonians (0.3) with translation invariant interactions (with respect to the lattice shifts $d_{jk} = a(j-k)$ $(a(\cdot): \mathbb{Z}^d \to \mathbb{R}^1)$) and find an explicit form for the harmonic temporal evolution of the operators of renormalized representation of canonical commutation relations constructed for the given measure γ_S.

The last section contains results concerning spectral properties of Hamiltonians of some infinite particle systems obtained by the methods of nonstationary scattering theory. Namely, we consider potential perturbations of translation invariant harmonic systems $L_{\gamma_S} + V$ (the potential $V = \overline{V}$, and the inclusion in $L_p(\mathbb{R}^{\mathbb{Z}^d}, \gamma_S)$, necessary in order that the sum be selfadjoint, holds for it). The situation under consideration has the following specific property: The wave operators for the pair of operators $L_{\gamma_S} + V$ and L_{γ_S} (defined in the nonstationary scattering theory with the help of the strong limit of the expression

$$\exp(-it(L_{\gamma_S} + V)) \exp(itL_{\gamma_S})$$

as $t \to \pm\infty$) evidently do not exist. In Section 4, we propose to pass from $L_{\gamma_S} + V$ to a renormalized operator $L_{V,\,\mathrm{ren}}$ acting in another space $L_2(\mathbb{R}^{\mathbb{Z}^d}, \mu^V)$ (as above, μ^V is a vacuum measure corresponding to $L_{\gamma_S} + V$) and consider the wave operators for the pair $L_{V,\,\mathrm{ren}}$ and L_{γ_S} (in this case, the definition of wave operators should be properly modified to take into account the fact that the groups $e^{itL_{V\mathrm{ren}}}$ and $e^{itL_{\gamma_S}}$ now act in different Hilbert spaces). Since $L_{V,\,\mathrm{ren}}$ is connected with the Dirichlet operator of the measure μ^V, it is convenient to discuss the statement of this scattering problem for a general pair of Dirichlet operators, and then make the situation more concrete to prove the criteria of the existence of wave operators in the case of perturbations of harmonic systems. As in the abstract scattering theory, the fact that the wave operators exist enables us to obtain information about the spectrum of the perturbed operator. In this section, we also consider the examples of linear and quadratic potentials. For these potentials, it is possible to find the wave operators explicitly, and this allows us to estimate the limits of applicability of the criteria of the existence of these operators proved above.

§1. The General Scheme of Constructing Renormalized Operators

When studying potential perturbations of the second quantization operators L_A (see Section 3 in Chapter 6), we introduced the procedure of renormalization which associates the operator $L_A + V$ with the Dirichlet operator L_{μ^V} corresponding to the measure μ^V connected canonically with the potential V. Under certain conditions of regularity imposed on the potential, the procedure of renormalization is trivial in the sense that it leads to an operator unitary equivalent (to within a shift) to $L_A + V$, which is itself well defined in the original Hilbert space. However, in many applications, one should use potentials which have no meaning of measurable functions and are given either by genera lized functions or by formal expressions to which even this meaning is not attached. Generally speaking, such singularity of V makes it impossible to define $L_A + V$ as an operator in $L_2(\Phi', \gamma_1)$. Nevertheless, in many cases, the formal sum $L_A + V$ may be associated with an operator in a new Hilbert space (by means of the standard procedure of renormalization which extends the described one in the natural way). Below, we discuss the general scheme of this renormalization and its interpretation in terms of the functional integral.

1.1. Operator Realization of Formal Hamiltonians

Let us first describe the construction of a special limit of a sequence of selfadjoint operators connected with the "dressing" method in the constructive field theory. Assume that a sequence $(L_n)_{n=1}^{\infty}$ of selfadjoint operators with domains of definition $\mathfrak{D}(L_n)$ is given in a separable Hilbert space $\mathcal{H}$. Also let $\mathfrak{D}$ be a region dense in $\mathcal{H}$ and let J_n, $n \in \mathbb{N}$, be invertible operators (dressing operators) with the following properties

(i) $$\mathfrak{D} \subset \mathfrak{D}(L_n), \quad \mathfrak{D} \subset \mathfrak{D}(J_n), \quad n \in \mathbb{N};$$

(ii) $$J_n \mathfrak{D} \subset \mathfrak{D}(L_n), \quad L_n \mathfrak{D} \subset \mathfrak{D}(J_n^{-1}), \quad n \in \mathbb{N};$$

(iii) for any $\varphi \in \mathfrak{D}$, there exists $\lim_{n\to\infty}(J_n\varphi, J_n\varphi)_{\mathcal{H}}$;

(iv) for any $\varphi \in \mathfrak{D}$, there exists

$$\lim_{n\to\infty} (J_n\varphi, L_n J_n\varphi)_{\mathcal{H}} = a_{\infty}(\varphi, \varphi). \tag{1.1}$$

If J_n are bounded and there exists $\lim_{n\to\infty} J_n = J_{\infty}$, then the form a_{∞} may be connected with a symmetric operator L_{∞}, namely,

$$L_{\infty} : a_{\infty}(\varphi, \varphi) = (L_{\infty} J_{\infty}\varphi, L_{\infty}\varphi_{\infty})_{\mathcal{H}}.$$

In the situation which we are interested in, we cannot expect that this condition would hold, and we *a priori* reject the possibility to obtain a limiting operator in $\mathcal{H}$. To introduce a limiting operator, we define Hilbert spaces $\mathcal{H}_n$ (for $n \in \mathbb{N}$) as a completion of $\mathfrak{D}$ with respect to the scalar product

$$(\varphi, \psi)_{\mathcal{H}_n} = (J_n\varphi, J_n\psi)_{\mathcal{H}},$$

and the space $\mathcal{H}_{\infty}$ generated by the scalar product

$$(\varphi, \psi)_{\mathcal{H}_{\infty}} = \lim_{n\to\infty} (J_n\varphi, J_n\psi)_{\mathcal{H}} = \lim_{n\to\infty} (\varphi, \psi)_{\mathcal{H}_n} \quad (\varphi, \psi \in \mathfrak{D})$$

under the assumption that it is nondegenerate (otherwise, we apply the ordinary procedure of factorization). In each space $\mathcal{H}_n$, $n \in \mathbb{N}$, we introduce operators $L_{n,\mathrm{ren}} = J_n^{-1} L_n J_n$ defined on a dense region $\mathfrak{D} \subset \mathcal{H}_n$. Then

$$(L_{n,\text{ren}}\varphi, \psi)_{\mathcal{H}_n} = (J_n\varphi, L_n J_n\psi)_{\mathcal{H}} \tag{1.2}$$

and condition (iv) in (1.1) takes the form:

For any $\varphi \in \mathfrak{D}$, there exists $\lim_{n\to\infty} (L_{n,\text{ren}}\varphi, \varphi)_{\mathcal{H}_n} = a_\infty(\varphi, \varphi)$.

If the form a_∞ is closable in $\mathcal{H}_\infty$, then it can be associated (may be, not uniquely) with a selfadjoint operator L_{ren} in $\mathcal{H}_\infty$ such that

$$(L_{\text{ren}}\varphi, \psi)_{\mathcal{H}_\infty} = a_\infty(\varphi, \psi) = \lim_{n\to\infty} (L_{n,\text{ren}}\varphi, \psi)_{\mathcal{H}_n} \tag{1.3}$$

for $\varphi, \psi \in \mathfrak{D} \subset \mathcal{H}_\infty$.

Thus, we proceed to a scheme of constructing the limiting operator in a new space using of the dressing operators. In this scheme, both the operator itself and the Hilbert space in which it acts are limiting objects.

The described "dressing" scheme takes a more concrete form when we consider singular potential perturbations of the second quantization operators. This is connected with the fact that the renormalization procedure considered in Subsection 3.2 of Chapter 6 suggests the concrete choice of dressing operators which simplifies the verification of conditions (1.1). Assume that a singular potential V is given; and let $(V_n)_{n=1}^\infty \subset L_{2+\varepsilon}(\Phi', \gamma_1)$, $\varepsilon > 0$ be a sequence such that $\exp(-V_n) \in L_p(\Phi', \gamma_1)$ and $V_n = \overline{V_n}$ for any $p \geq 1$ and $n \in \mathbf{N}$. Suppose that this sequence approximates V in a certain sense. For example, $V \in \mathcal{P}'(\Phi')$ and $V_n \to V$ as $n \to \infty$ in the sense of generalized functions. The choice of approximating sequence usually depends on the physical nature of the problem under consideration. For example, in the field theory, we consider ultraviolet and volume cutoffs; in the quantum statistical physics of lattice system we consider the transition to the situation when only the interactions of a finite number of particles are taken into account, etc. Let the unperturbed operator L_A be uniformly elliptic, i.e., $A \geq a\,1$, $a > 0$. Then, according to Subsection 2.3 in Chapter 6, for every $n \in \mathbf{N}$, we have an operator $L_A + V_n \geq E_n 1$ essentially selfadjoint on $C^2_{b,\text{cyl}}(\Phi')$ and the ground state $\phi_{V_n} > 0$ γ_1-almost everywhere. Let us pass to the operators

$$L_n = L_A + V_n - E_n 1 \geq 0$$

and choose the dressing operators J_n to be equal to the operators of multiplication by ϕ_{V_n}. We take $C^2_{b,\text{cyl}}(\Phi')$ to play the role of the domain $\mathfrak{D}$ which takes part in the dressing scheme. According to Subsection 2.3 in Chapter 6, conditions (i) and (ii) in (1.1) are satisfied. Here $\mathcal{H}_n = L(\Phi', \mu^{V_n})$, where $\mu^{V_n} = \phi^2_{V_n}\gamma_1$. Assume that the sequence of measures $(\mu^{V_n})_{n=1}^\infty$ converges to a limiting measure μ in the sense of the convergence of the integrals of the functions from $C_{b,\mathrm{cyl}}(\Phi')$. This assumption implies not only the validity of condition (iii) in (1.1) but also the validity of (iv). Indeed, according to (1.2) and Subsection 3.2 in Chapter 6, we have $L_{n,\,\mathrm{ren}} = L_{\mu^{V_n},A}$ on $C^2_{b,\mathrm{cyl}}(\Phi')$. Therefore,

$$(L_{n,\mathrm{ren}}u,\, v)_{L_2(\mu^{V_n})} = \frac{1}{2}\int_{\Phi'} (\nabla u,\, A\nabla\bar{v})_{H_0}\, d\mu^{V_n}$$

for $u, v \in C^2_{b,\mathrm{cyl}}(\Phi')$. Since

$$(\nabla u,\, A\nabla\bar{v})_{H_0} \in C_{b,\mathrm{cyl}}(\Phi'),$$

and the measures $\mu^{V_n} \underset{n\to\infty}{\longrightarrow} \mu$ in the sense indicated above, we also have

$$(L_{n,\mathrm{ren}}u,\, v)_{\mathcal{H}_n} \xrightarrow[n\to\infty]{} \frac{1}{2}\int_{\Phi'} (\nabla u,\, A\nabla\bar{v})_{H_0}\, d\mu = a_\infty(u, v).$$

Hence, the Dirichlet form of the limiting measure is just the limiting form a_∞ which takes part in the construction of a renormalized operator in the general scheme. The last question whether a_∞ is closable in $L_2(\Phi', \gamma_1)$ is reduced to the analysis of the properties of the measure μ. In particular, the answer to this question is affirmative when the logarithmic derivative of the measure μ exists (see Subsection 3.1 in Chapter 6). Here we must take into account that a_∞ may be associated with different extensions of the Dirichlet operator $L_{\mu,A}$ (but not only with the Friedrichs extension $L^F_{\mu,A}$). However, ambiguities of this sort are even encountered in the problems of quantum mechanics in the case of singular potential interactions. Some conditions which should be imposed on μ to guarantee the uniqueness of a selfadjoint extension of the Dirichlet operator were obtained in Subsection 3.3 of Chapter 6.

As above, the version of the general scheme of dressing under consideration is called a renormalization of the formal Hamiltonian $L_A + V$ because, unlike the case of a regular potential V; here, we cannot establish a unitary isomorphism of the original Hilbert space $L_2(\Phi', \gamma_1)$ and limiting Hilbert space $L_2(\Phi', \mu)$ which would connect $L_A + V$ and $L_{\mu,A}$ (however, this isomorphism exists at each approximation step).

Remark 1. It is easy to see that the renormalization scheme is suitable for the investigation of potential perturbations not only of the second quantization operators, but also of Dirichlet operators of the general type. The only essential distinction is connected with the fact that, in the case of the second quantization operator conditions, which should be imposed on the potential to guarantee the existence of the ground state ϕ_V, are known, while in the case of the general Dirichlet operators, this problem should be investigated separately.

■

To sum up what has been said above, we now outline the principal steps, the realization of which is demanded by the scheme of renormalization of a formal Hamiltonian $L_A + V$.

1. The choice of a sequence of approximating potentials. As a rule, it depends on the physical nature of the problem.

2. The proof of the convergence of the sequence of measures $(\mu^{V_n})_{n=1}^{\infty}$. When realizing this step, one should sometimes stick to the proof of the compactness of $(\mu^{V_n})_{n=1}^{\infty}$ or consider other types of convergence of measures instead of the indicated one.

3. The study of the properties of the limiting measure. In this step, the main problem is to establish the closability of its Dirichlet form.

In a sufficiently general situation, each of these steps may be quite complicated. Below, we discuss one approach to the constructive realization of the renormalization scheme. In the next sections, we consider many examples in which this scheme is applied to the construction of the dynamics of systems with infinitely many degrees of freedom.

We now restrict ourselves to the investigation of two illustrative examples. The first of these deals with the system of noninteracting quantum oscillators. The mathematical description of this system has been already presented in Example 1.3 of Chapter 6. The operator of energy of this system was defined as a second quantization operator in the Schrödinger representation. Let us show how we can obtain this operator starting from the formal Hamiltonian and applying the renormalization scheme given above.

Example 1.1 (Noninteracting oscilators).
Assume that a one-dimensional quantum harmonic oscilator with unit mass and frequency $a_k > 0$ is given for every $k \in \mathbb{N}$. It is associated with the Hilbert space of states $L_2(\mathbb{R}^1, dx_k)$ and with the operator of energy

$$L_k = -\frac{1}{2}\frac{\partial^2}{\partial x_k^2} + \frac{1}{2}a_k^2 x_k^2$$

acting in $L_2(\mathbb{R}^1, dx_k)$. Since the potential $\frac{1}{2}a_k^2 x_k^2$ belongs to $L_{\infty,\mathrm{loc}}(\mathbb{R}^1, dx_k)$, the operator L_k is essentially selfadjoint on $C_0^{\infty}(\mathbb{R}^1)$. The following formal Hamiltonian corresponds to the infinite system of noninteracting oscilators

$$L = -\frac{1}{2}\sum_{k=1}^{\infty}\frac{\partial^2}{\partial x_k^2} + \frac{1}{2}\sum_{k=1}^{\infty}a_k^2 x_k^2. \tag{1.4}$$

To construct the operator realization of the Hamiltonian (1.4) with the help of the renormalization procedure considered above, we introduce, for any $n \in \mathbb{N}$, an operator

$$L_n = -\frac{1}{2}\sum_{k=1}^{n}\frac{\partial^2}{\partial x_k^2} + \frac{1}{2}\sum_{k=1}^{n} a_k^2 x_k^2$$

acting in the Hilbert space $\mathcal{H}_n = L_2(\mathbb{R}^n)$. As shown in Example 3.5 of Chapter 6, for the operator L_n, we can construct the corresponding ground state

$$\phi_n(x_1, \ldots, x_n) = \prod_{k=1}^{n} \sqrt{\frac{a_k}{\pi}}\, e^{-a_k x_k^2/2}$$

and the renormalized operator $L_{n,\text{ren}} = L_{\gamma_{S_n}}$ generated by the Dirichlet form of the Gaussian measure γ_{S_n} with the correlation operator S_n acting in $\mathbb{R}^n$: $S_n = (a_k^{-1}\delta_{kj})_{k,\,j=1}^{n}$, i.e.,

$$d_{\gamma_{S_n}}(x) = \mathop{\times}_{k=1}^{n} \sqrt{\frac{a_k}{\pi}}\, e^{-a_k x_k^2}\, dx_k,$$

$$(L_{n,\text{ren}}u,\, v)_{L_2(\mathbb{R}^n,\gamma_{S_n})} = \frac{1}{2}\int_{\mathbb{R}^n}\sum_{k=1}^{n}\frac{\partial u}{\partial x_k}\frac{\partial \overline{v}}{\partial x_k}\, d\gamma_{S_n} \tag{1.5}$$

$$(u, v \in C_b^1(\mathbb{R}^n)).$$

As $n \to \infty$, the measures γ_{S_n} converge to a Gaussian product measure γ_S on $(\mathbb{R}^\infty, \mathcal{C}_\sigma(\mathbb{R}^\infty))$ with the correlation operator $S = (a_k^{-1}\delta_{kj})_{k,\,j=1}^{\infty}$ $S: \mathbb{R}_0^\infty \to \mathbb{R}_0^\infty$ and

$$d\gamma_S(x) = \mathop{\times}_{k=1}^{\infty} \sqrt{\frac{a_k}{\pi}}\, e^{-a_k x_k^2}\, dx_k,$$

This convergence takes place in the sense of the convergence of integrals of continuous bounded cylindrical functions, namely, for any $u \in C_{b,\text{cyl}}(\mathbb{R}^\infty)$, $u = u(x_1, \ldots, x_m)$, we have

$$\int_{\mathbb{R}^n} u(x_1, \ldots, x_m)\, d\gamma_{S_n}(x) \to \int_{\mathbb{R}^\infty} u(x_1, \ldots, x_m)\, d\gamma_S(x)$$

as $n \geq m, n \to \infty$. This implies that

$$(L_{n,\text{ren}}u,\, v)_{H_{n,\text{ren}}} \xrightarrow[n\to\infty]{} \frac{1}{2}\int_{\mathbb{R}^\infty}\sum_{k=1}^{n}\frac{\partial u}{\partial x_k}\frac{\partial \overline{v}}{\partial x_k}\, d\gamma_S(x)$$

$$= d_{\gamma_S}(u, v) = \frac{1}{2}\int_{\mathbb{R}^\infty} (\nabla u, \nabla \overline{v})_{l_2}\, d\gamma_S, \quad \mathcal{H}_{n,\text{ren}} = L_2(\mathbb{R}^n, \gamma_{S_n}) \tag{1.6}$$

for any $u, v \in C^1_{b,\mathrm{cyl}}(\mathbb{R}^\infty)$. According to the renormalization scheme, the operator $L_{\infty,\mathrm{loc}}$ corresponding to the formal Hamiltonian (1.4) acts in the Hilbert space of states $\mathcal{H}_{\mathrm{ren}} = L_2(\mathbb{R}^\infty, \gamma_S)$; it is the Dirichlet operator L_{γ_S} of a Gaussian measure γ_S. In Example 3.2 of Chapter 6, the following expression for it is presented

$$(L_{\gamma_S}u)(x) = -\frac{1}{2}\Delta u - \frac{1}{2}(\beta_{\gamma_S}(x), \nabla u)_{l_2} = -\frac{1}{2}\sum_{k=1}^{\infty}\frac{\partial^2 u}{\partial x_k^2} + \sum_{k=1}^{\infty} a_k \frac{\partial u}{\partial x_k}, \tag{1.7}$$

where $u \in C^2_{b,\mathrm{cyl}}(\mathbb{R}^\infty)$. Moreover, L_{γ_S} is essentially selfadjoint on $C^2_{b,\mathrm{cyl}}(\mathbb{R}^\infty)$. The relation (1.6) shows that L_{ren} is generated by the Dirichlet form in which the derivatives are regarded in the sense of the Hilbert space $H_0 = l_2$ with respect to which γ_S is not a canonical Gaussian measure. To proceed to this situation, we introduce a change of variables in $\mathbb{R}^\infty$, namely,

$$\mathbb{R}^\infty \ni x = (x_k)_{k=1}^\infty \mapsto y = (\sqrt{a_k}\, x_k)_{k=1}^\infty \in \mathbb{R}^\infty.$$

As a result of this change of variables, $d\gamma_S(u,v)$ is transformed to the form

$$\frac{1}{2}\int_{\mathbb{R}^\infty}\sum_{k=1}^{\infty} a_k \frac{\partial u}{\partial y_k}\frac{\partial \bar{v}}{\partial y_k}\, d\gamma_1(y) = \frac{1}{2}\int_{\mathbb{R}^\infty}(\nabla u, \nabla \bar{v})_{l_2} d\gamma_1(y) = d_{\gamma_1, A}(u, v), \tag{1.8}$$

where A is an operator in l_2 with a diagonal matrix $(a_k \delta_{kj})_{k,j=1}^\infty$ in the natural basis of l_2, $\mathrm{D}(A) \supset \mathbb{R}_0^\infty$, and $A : \mathbb{R}_0^\infty \to \mathbb{R}_0^\infty$. Taking into account Theorem 1.2 in Chapter 6 and (1.8), we conclude that if we pass unitarily from $L_2(\mathbb{R}^\infty, \gamma_S)$ to $L_2(\mathbb{R}^\infty, \gamma_1)$ with the help of the indicated change of variables, then the operator L_{ren} is associated with the second quantization operator $L_A \equiv L_{\gamma_1, A}$ with one-number operator $A = (a_k \delta_{kj})_{k,j=1}^\infty > 0$, and we are again in the situation of Example 1.3 in Chapter 6.

Let us discuss the case of coinciding frequencies separately. Assume that $a_k = \omega > 0$, $k \in \mathbb{N}$. Here $S = \omega^{-1}1$, therefore, according to the Kakutani criterion (see Example 1.1 in Chapter 2), the measures $\gamma_{\omega_1^{-1}1}$ and $\gamma_{\omega_2^{-1}1}$ corresponding to different frequencies $\omega_1 \neq \omega_2$ are orthogonal. This shows that even simple infinite systems such as the collection of noninteracting oscilators cannot be described in terms of the common space of states and require the proper choice of it for each system.

The next example also deals with mutually noninteracting oscillations; however, in this case they are influenced by external field.

Example 1.2 (Oscillators in the external field).
Let us consider a system of noninteracting oscilators with equal frequencies $a_k = \omega > 0$, $k \in \mathbb{N}$, considered in Example 1.1. After renormalization, it can be associated with the

space of states $L_2(\mathbb{R}^\infty, \gamma_{\omega^{-1}1})$ and the operator of energy $L_\omega = L_{\gamma_{\omega^{-1}1}}$. The influence of a classical external field on this system may be formally described by the potential

$$V_\lambda = \lambda \sum_{k=1}^{\infty} x_k,$$

where $\lambda \in \mathbb{R}^1$ characterizes the intensity of the field. Clearly, this potential has no meaning of a measurable function on $\mathbb{R}^\infty$ by virtue of the ergodic properties of the product measure $\gamma_{\omega^{-1}1}$. Indeed, a measurable function on $\mathbb{R}^\infty$ invariant with respect to arbitrary finite permutations of coordinates is constant on $\mathbb{R}^\infty$ $\gamma_{\omega^{-1}}$-almost everywhere (see Corollary to Theorem 1.2 in Chapter 2). We introduce a sequence of potentials

$$V_{\lambda,n}(x) = \lambda \sum_{k=1}^{n} x_k, \quad x \in \mathbb{R}^\infty, \quad n \in \mathbb{N}.$$

Each potential $V_{\lambda,n}$ satisfies the following conditions: $V_{\lambda,n} \in L_p(\mathbb{R}^\infty, \gamma_{\omega^{-1}1})$ and $e^{-V_{\lambda,n}} \in L_p(\mathbb{R}^\infty, \gamma_{\omega^{-1}1})$ for any $p \geq 1$. Therefore, the operator $L_\omega + V_{\lambda,n}$ is essentially selfadjoint on $C^2_{b,\,\mathrm{cyl}}(\mathbb{R}^\infty)$, semibounded from below, and has a ground state $\phi_n > 0$ $\gamma_{\omega^{-1}1}$-everywhere. Let us show that the corresponding measure $\mu_n = \phi_n^2\, \gamma_{\omega^{-1}1}$ has the form

$$\mu_n = \gamma_{\omega^{-1}1}(\,\cdot + a_{\lambda,n}),$$

where $a_{\lambda,n} = (a^{(j)}_{\lambda,n})_{j=1}^{\infty}$ is a vector from $\mathbb{R}^\infty$ such that $a^{(j)}_{\lambda,n} = 0$ for $j > n$ and $a^{(j)}_{\lambda,n} = \lambda\omega^{-2}$ for $j = 1, \ldots, n$.

Indeed, in this case

$$\frac{d\mu_n}{d\gamma_{\omega^{-1}1}}(x) = \prod_{j=1}^{n} e^{-\omega[\,(x_j + a^{(j)}_{\lambda,n})^2 - x_j^2\,]} = \prod_{j=1}^{n} e^{-2x_j a^{(j)}_{\lambda,n} - [a^{(j)}_{\lambda,n}]^2},$$

therefore,

$$d\mu_n(\cdot) = e^{-2\psi_n} d\gamma_{\omega^{-1}1}(\,\cdot\,),$$

where

$$\psi_n(x) = \sum_{j=1}^{n} \omega\Big(a^{(j)}_{\lambda,n} x_j + \frac{1}{2}[a^{(j)}_{\lambda,n}]^2\Big).$$

According to Remark 1 in Subsection 3.2 of Chapter 6, the potential perturbation which leads to the measure μ_n has the form (here $A = 1$)

$$V_{\psi_n} = \frac{1}{2}\left[\sum_{k=1}^{\infty}\left(\frac{\partial\psi_n}{\partial x_k}\right)^2 + 2L_\omega\psi_n\right]$$

$$= \frac{1}{2}\left[(2n+1)\lambda\omega^{-2} + 2\omega^2\sum_{j=1}^{n} a_{\lambda,n}^{(j)} x_j\right] = \lambda\sum_{j=1}^{n} x_j + E_n = V_{\lambda,n} + E_n,$$

where $E_n = (2n+1)\,\lambda\omega^{-2}$ is a shift which takes part in renormalization. Obviously, the measures μ_n converge to the measure $\mu_{\lambda,\omega} = \gamma_{\omega^{-1}1}(\cdot + a_\lambda)$, where $a_\lambda = (a_\lambda^{(j)})_{j=1}^{\infty} \in \mathbb{R}^\infty$ and $a_\lambda^{(j)} = \lambda\omega^{-2}, j \in \mathbb{N}$. According to the construction, the formal Hamiltonian $L_\omega + V_\lambda$ is associated with the renormalized operator $L_{\lambda,\,\text{ren}}$ in $L_2(\mathbb{R}^\infty, \mu_{\lambda,\omega})$ generated by the Dirichlet form

$$\frac{1}{2}\int_{\mathbb{R}^\infty}\sum_{k=1}^{\infty}\frac{\partial u}{\partial x_k}\frac{\partial \bar{v}}{\partial x_k}d\mu_{\lambda,\omega} \quad (u, v \in C_{b,\,\text{cyl}}^2(\mathbb{R}^n))$$

and defined by the differential expression

$$L_{\lambda,\text{ren}}u = -\frac{1}{2}\sum_{k=1}^{\infty}\frac{\partial^2 u}{\partial x_k^2} + \omega\sum_{k=1}^{\infty} x_k\frac{\partial u}{\partial x_k} + \frac{\lambda}{\omega}\sum_{k=1}^{\infty}\frac{\partial u}{\partial x_k} \tag{1.9}$$

for $u \in C_{b,\,\text{cyl}}^2(\mathbb{R}^\infty)$.

As is known (see Theorem 2.3 in Chapter 1), the measures $\gamma_{\omega^{-1}1}$ and $\gamma_{\omega^{-1}1}(\cdot + a)$ are equivalent if $a \in l_2$, or orthogonal if $a \,\overline{\in}\, l_2$. Here $a_\lambda \,\overline{\in}\, l_2$ for any $\lambda \neq 0$; hence, $\gamma_{\omega^{-1}1} \perp \mu_{\lambda,\omega}$. We again encounter the situation when it is impossible to describe the perturbed system in terms of the original one. The physical Hilbert spaces $L_2(\mathbb{R}^\infty, \gamma_{\omega^{-1}1})$ and $L_2(\mathbb{R}^\infty, \mu_{\lambda,\omega})$ in which the free Hamiltonian L_ω and the perturbed Hamiltonian $L_{\omega,\,\text{ren}}$ act are constructed from the orthogonal measures on $\mathbb{R}^\infty$ and, consequently, the states of perturbed system cannot be expressed in terms of the states of a free system.

1.2. The Functional Integral For a Renormalized Operator

The absence of a direct connection between the potential and the corresponding ground state ϕ_V proves to be a serious difficulty in the construction of the renormalized operator. The explicit form of ϕ_V can be found only for a series of simple model problems.

In the general case, we can speak only about the existence of ϕ_V, about its L_p-properties, and about a certain smoothness of ϕ_V. However, this information is insufficient to realize the important step of the renormalization scheme, namely, to prove the convergence of a sequence of measures $\mu^{V_n} = \phi_{V_n}^2 \gamma_1, n \in \mathbb{N}$, which corresponds to the given sequence of potentials $V_n, n \in \mathbb{N}$. It seems that the indicated difficulty in the study of the renormalized operator $L_{V,\text{ren}}$ corresponding to $L_A + V$ should become more serious when we construct the functional integral corresponding to it. However, it is possible to describe this integral in terms of L_A and V without using ϕ_V. This fact is a basis of the constructive approach to the construction of a renormalized operator with the help of measures on the space of trajectories. This approach is presented below.

Let L_A be a uniformly elliptic operator in $L_2(\Phi', \gamma_1)$ $(A \geq a\, 1, a > 0)$, and let the potential $V = \overline{V}$ satisfy conditions of Theorem 2.9 in Chapter 6 and its Corollary which state the essential selfadjointness of the sum $L_A + V$, i.e., there exists $\varepsilon > 0$ such that $V \in L_{2+\varepsilon}(\Phi', \gamma_1)$ and $e^{-V} \in L_p(\Phi', \gamma_1)$ for any $p > 1$. As above, we denote by ϕ_V the corresponding ground state; $\mu^V = \phi_V^2\, \mu_0$, and $L_{V,\text{ren}} = \phi_V^{-1} \hat{L}_V \phi_V$ is a renormalized operator in $L_2(\mu^V)$, where $\hat{L}_V = L_A + V - E_V$. By virtue of the positiveness of ϕ_V and the corresponding property of $e^{-t\hat{L}_V}$, $t \geq 0$, a semigroup

$$e^{-tL_{V,\text{ren}}} = \phi_V^{-1} e^{-t\hat{L}_V} \phi_V, \quad t \geq 0, \tag{1.10}$$

preserves the positiveness of functions from $L_2(\Phi', \mu^V)$. Moreover, $e^{-tL_{V,\text{ren}}}\, 1 = 1$ and, thus, the semigroup (1.10) is contracting in all $L_p(\Phi', \mu^V)$, $p \in [1, \infty]$.

As in Subsection 1.5 of Chapter 6, for the given semigroup $e^{-tL_{V,\text{ren}}}$, $t \geq 0$, and initial distribution μ^V, we can construct a weak functional integral given by the measure ν_A^V on the space of trajectories

$$\Omega = (\Phi')^{\mathbb{R}} = \Omega(\mathbb{R}^1, \Phi'),$$

which is invariant with respect to shifts along the trajectories. Recall that the principal relation which determines ν_A^V gives the equality

$$\int_\Omega f_0(\omega(t_0)) \ldots f_n(\omega(t_n)) d\nu_A^V(\omega(\cdot))$$

$$= \int_{\Phi'} f_0(x) e^{-s_1 L_{V,\text{ren}}} f_1(x) \ldots e^{-s_n L_{V,\text{ren}}} f_n(x)\, d\mu^V(x), \tag{1.11}$$

where $t_0 \leq t_1 \leq \ldots \leq t_n$; $t_k \in \mathbb{R}$, and

$$f_j \in L_\infty(\Phi', \mu^V) = L_\infty(\Phi', \gamma_1)$$

$s_k = t_k - t_{k-1}$. We can also take

$$f_j \in \bigcap_{p\geq 1} L_p(\Phi', \gamma_1) \subset \bigcap_{p\geq 1} L_p(\Phi', \mu^V)$$

in (1.11); in this case, the right-hand side is correctly defined due to the contracting properties of $e^{-tL_{V,\text{ren}}}$. If ν_A^V is known, then we can reconstruct ϕ_V because, by construction

$$\nu_A^V \restriction \mathfrak{C}^0(\Omega) = \phi_V^2(\omega(0))\gamma_1(\omega(0)).$$

Furthermore, the indicated connection between ν_A^V and μ^V may serve as a basis for the study of the properties of the measure μ^V. Therefore, it is desirable to have a method of construction ν_A^V in which ϕ_V is not employed.

Denote by $\nu_A^{V,t}$ a measure on Ω defined by the formula

$$d\nu_A^{V,t} = Z_t^{-1}\exp\left(-\int_{-t}^{t} V(\omega(\tau))d\tau\right)d\nu_A(\omega(\cdot)),$$

where ν_A is a measure Ω corresponding to the weak functional integral associated with L_A (see Subsection 1.5 in Chapter 6), and Z_t is a normalizing factor: $\nu_A^{V,t}(\Omega) = 1$.

Theorem 1.1. *The measure ν_A^V is a limit as $t \to +\infty$ of the measures $\nu_A^{V,t}$ in the sense of the convergence of integrals of cylindrical functions from $L_\infty(\Omega, \nu_A)$. In particular, the finite-dimensional distributions converge.*

Proof. Let $f_0, \ldots, f_n \in L_\infty(\Phi', \gamma_1)$, $t_0 \leq \ldots \leq t_n$, and let $t > 0$ be so large that $-t < t_0 \leq \ldots \leq t_n < t$. Then, according to the Feynman – Kac formula (2.16) in Chapter 6, we have

$$\int_\Omega 1(\omega(-t))f_0(\omega(t_0)) \ldots f_n(\omega(t_n))1(\omega(t))d\nu_A^{V,t}(\omega(\cdot))$$

$$= \frac{(e^{-(t+t_0)(L_A+V)} f_0 e^{-s_1(L_A+V)} f_1 \ldots f_n e^{-(t-t_n)(L_A+V)}1, 1)_{L_2(\Phi',\gamma_1)}}{(1, e^{-2t(L_A+V)}1)_{L_2(\Phi',\gamma_1)}}$$

$$= \frac{e^{-2tE_V}(e^{-(t+t_0)\hat{L}_V} f_0 \ldots e^{-s_n\hat{L}_V} f_n e^{-(t-t_n)\hat{L}_V}1, 1)_{L_2(\Phi',\gamma_1)}}{e^{-2tE_V}(1, e^{-2t\hat{L}_V}1)_{L_2(\Phi',\gamma_1)}}. \quad (1.12)$$

Here $s_k = t_k - t_{k-1}$, $1 \le k \le n$, and we have taken into account that

$$Z_t = \int_\Omega e^{-\int_{-t}^{t} V(\omega(\tau))d\tau} d\gamma_A(\omega(\cdot)) = (1, e^{-2t(L_A+V)}1)_{L_2(\Phi',\gamma_1)}.$$

If we pass to the limit as $t \to +\infty$ and note that

$$e^{-tL_A}1 \xrightarrow[t\to+\infty]{} (\phi_V, 1)_{L_2(\Phi',\gamma_1)}\phi_V,$$

by virtue of the spectral theorem, then we obtain the value of the right-hand side of (1.12):

$$(f_0 e^{-S_1 \hat{L}_V} \ldots e^{-S_n \hat{L}_V} f_n \phi_V, \phi_V)_{L_2(\Phi',\gamma_1)} = \int_{\Phi'} \phi_V^{-1} f_0 e^{-S_1 \hat{L}_V} \ldots f_n \phi_V d\mu^V.$$

Taking into account the relation

$$\phi_V^{-1} e^{-t\hat{L}_V} \phi_V = e^{-tL_{V,\mathrm{ren}}},$$

we can rewrite the last expression as

$$(f_0 e^{-S_1 L_{V,\mathrm{ren}}} f_1 \ldots e^{-S_n L_{V,\mathrm{ren}}} f_n 1, 1)_{L_2(\Phi',\mu^V)}$$

$$= \int_\Omega f_0(\omega(t_0)) f_1(\omega(t_1)) \ldots f_n(\omega(t_n)) d\nu_A^V(\omega(\cdot)). \quad (1.13)$$

The equality in (1.13) follows from the definition of the measure ν_A^V (see (1.11)). The statement of the theorem can be now extended by approximation to arbitrary cylindrical functions from $L_\infty(\Omega, \nu_A)$ which are not factorized into a product of the form $f_0(\omega(t_0)) \ldots f_n(\omega(t_n))$.

■

The theorem proved enables us to develop the constructive approach to the construction of a renormalized operator in the case of a singular perturbation. Let us associate a family of measures $\nu_A^{V_n,t}$, $n \in \mathbf{N}$, $t > 0$, on Ω with a sequence of approximating potentials $(V_n)_{n=1}^\infty$ possessing appropriate L_p-properties (which guarantees the existence of the ground state of $L_A + V_n$). If we prove the existence of the limit (in some sense) of this family or at least its compactness with respect to some convergence of measures, then the measure ν obtained (or the limiting points of the family) may be regarded as a measure on Ω generated by the renormalized operator. Under the condition of invariance of ν with respect to shifts along the trajectories, we have a unitary group U_t in

$L_2(\Omega, \nu)$ generated by these shifts

$$U_t f(\omega(\cdot)) = f(\omega(\cdot + t)), \quad t \in \mathbb{R}. \tag{1.14}$$

If we denote by P_0 the projector (in $L_2(\Omega, \nu)$) onto the subspace of functions from $L_2(\Omega, \nu)$ which are measurable with respect to $\mathfrak{C}^0(\Omega)$, then we get (just as in Subsection 1.5 of Chapter 6) a semigroup

$$T_t = P_0 U_t P_0 = e^{-|t| L_{\text{ren}}} \tag{1.15}$$

in $L(\Phi', \mu)$, where $\mu = \nu \upharpoonright \mathfrak{C}^0(\Omega)$. By definition, its generator L_{ren} is a renormalized operator. In the case when the measure μ has a logarithmic derivative, the operator L_{ren} is connected with $L_{\mu, A}$. Clearly, in each concrete case, some specific technical difficulties appear but we ignore them in the discussion of the general scheme. The method of overcoming these difficulties depends on the nature of each concrete problem. One of the merits of the considered version of renormalization is that the measures $\nu_A^{V_n, t}$ are explicitly connected with potentials. Note also that the transition to the measures on trajectories is completely analogous with the transition to Euclidean fields (and to the corresponding measures) in the constructive field theory.

§2. Several Field Theory Models

The models in the constructive quantum field theory which are investigated below are of little interest from the physical point of view, because they can be reduced to free or generalized free fields. However, even these models the exhibit principal mathematical difficulties of the quantum field theory connected with the polarization of vacuum, the existence of unitary nonequivalent representations of canonical commutation relations, etc. From the point of view of the renormalization scheme described above, these models are explicitly solvable, and this enables us to illustrate constructions presented in Section 1 by the simplest field situations.

Recall that the Schrödinger representation of a free field with mass $m_0 > 0$ at fixed time (considered in Example 2.5 of Chapter 2) was realized in the space of stated $L_2(\mathfrak{S}'_{\text{Re}}(\mathbb{R}^d), \mu_{m_0})$, where d is a dimensionality of spatial arguments and μ_{m_0} is a Gaussian measure on $\mathfrak{S}'_{\text{Re}}(\mathbb{R}^d)$ canonically connected with the Hilbert space $\mathfrak{h}_{m_0}$ with the scalar product

$$(\varphi, \psi)_{\mathfrak{h}_{m_0}} = (A_{m_0}^{-1} \varphi, \psi)_{L_2(\mathbb{R}^d)} \quad (A_{m_0} = (-\Delta + m_0^2)^{1/2}).$$

The duality of $\mathfrak{S}_{\text{Re}}(\mathbb{R}^d)$ and $\mathfrak{S}'_{\text{Re}}(\mathbb{R}^d)$ is established here by use of the standard extension of the scalar product in $L_2(\mathbb{R}^d)$ (below it is convenient to denote by $\langle\cdot,\cdot\rangle$ both the scalar product in $L_2(\mathbb{R}^d)$ and its extension).

The operator of energy $d\Gamma(A_{m_0})$ in the Fock space $\mathfrak{F}(\mathfrak{h}_{m_0})$ is associated with the selfadjoint operator l_{m_0} in $L_2(\mathfrak{S}'_{\text{Re}}(\mathbb{R}^d), \mu_{m_0})$ given on the functions $f \in C^2_{b,\text{cyl}}(\mathfrak{S}'_{\text{Re}}(\mathbb{R}^d))$ by the formula

$$(L_{m_0}f)(x) = -\frac{1}{2}[\operatorname{Tr}_{L_2(\mathbb{R}^d)}f''(x) - 2(\nabla f(x), A_{m_0}x)_{L_2(\mathbb{R}^d)}] \quad (x \in \mathfrak{S}'_{\text{Re}}(\mathbb{R}^d)) \tag{2.1}$$

(see Example 1.4 in Chapter 6); it can be also written in terms of the variational derivatives

$$(L_{m_0}f)(x) = -\frac{1}{2}\int_{\mathbb{R}^d}\frac{\delta^2 f(x)}{\delta x(s)^2}ds + \int_{\mathbb{R}^d}\frac{\delta f(x)}{\delta x(s)}\sqrt{-\Delta + m_0^2}\,x(s)ds. \tag{2.2}$$

2.1. Linear Interaction (The Van Hove Model)

We now start to study the Van Hove model. From the physical point of view, it describes the interaction of a boson field with external sources. This model is given by the formal Hamiltonian

$$L_\psi = L_{m_0} + V_\psi, \tag{2.3}$$

where

$$V_\psi(x) = (x, \psi)_{L_2(\mathbb{R}^d)}$$

is a linear function on $\mathfrak{S}'_{\text{Re}}(\mathbb{R}^d)$. The potential V_ψ has the meaning of a linear measurable function on $\mathfrak{S}'_{\text{Re}}(\mathbb{R}^d)$ only for

$$\psi \in \mathfrak{h}_{m_0} \simeq W_2^{1/2}(\mathbb{R}^d).$$

If we want to guarantee the required covariance of the field with the Hamiltonian $L_{m_0} + V_\psi$, then we must demand the invariance of the potential V_ψ with respect to the natural action of a nonuniform Euclidean group in $\mathfrak{S}'_{\text{Re}}(\mathbb{R}^d)$, i.e.,

$$V_\psi(x(A\cdot+a)) = V_\psi(x) \quad (A\in SO(d),\ a\in\mathbb{R}^d).$$

This yields the equality $\psi(\cdot)=\text{const}$. Since $1\,\overline{\in}\,W_2^{1/2}(\mathbb{R}^d)$, this potential V_ψ may be defined only as a generalized function from $\mathcal{P}'(\mathcal{S}'_{\mathrm{Re}}(\mathbb{R}^d))$, and this makes it impossible to determine the Hamiltonian $L_{m_0}+V_\psi$ directly. We shall show below that the renormalization scheme given in Section 1 enables us to construct an operator realization of (2.3) not only for $\psi=\text{const}$ but also for all $\psi\in\mathcal{S}'_{\mathrm{Re}}(\mathbb{R}^d)$.

Theorem 2.1. *Let* $\psi\in\mathcal{S}'_{\mathrm{Re}}(\mathbb{R}^d)$ *be fixed. The renormalized Hamiltonian* $L_{\psi,\mathrm{ren}}$ *corresponding to* (2.3) *is generated by the Dirichlet form of the measure* μ_ψ *on* $\mathcal{S}'_{\mathrm{Re}}(\mathbb{R}^d)$ *given by the Fourier transform*

$$\tilde{\mu}_\psi(x) = \int_{\mathcal{S}'_{\mathrm{Re}}} e^{i\langle x,\varphi\rangle}d\mu_\psi(x) = \exp\left(-\frac{1}{4}(\varphi,\varphi)_{\mathcal{H}_{m_0}} - i\langle A_{m_0}^{-2}\varphi,\psi\rangle\right) \quad (\varphi\in\mathcal{S}_{\mathrm{Re}}(\mathbb{R}^d)).$$

This operator is given by the expression

$$(L_{\psi,\mathrm{ren}}f)(x) = (L_{m_0}f)(x) + \langle\nabla f(x),\ A_{m_0}^{-1}\psi(\cdot)\rangle \tag{2.4}$$

as an essentially selfadjoint operator on $C^\infty_{b,\mathrm{cyl}}(\mathcal{S}'_{\mathrm{Re}}(\mathbb{R}^d))$.

Proof. Let us choose an arbitrary sequence $(\psi_n)_{n=1}^\infty\subset\mathcal{S}_{\mathrm{Re}}(\mathbb{R}^d)$ such that $\psi_n\to\psi$ as $n\to\infty$ in $\mathcal{S}'_{\mathrm{Re}}(\mathbb{R}^d)$. For every $n\in\mathbb{N}$, we consider an operator $L_{m_0}+V_{\psi_n}$. Since

$$V_{\psi_n}(x) = \langle\psi_n, x\rangle,\quad x\in\mathcal{S}'_{\mathrm{Re}}(\mathbb{R}^d)$$

is a continuous linear function on $\mathcal{S}'_{\mathrm{Re}}(\mathbb{R}^d)$, we have the inclusions

$$V_{\psi_n},\ e^{-V_{\psi_n}}\in L_p(\mathcal{S}'_{\mathrm{Re}}(\mathbb{R}^d),\mu_{m_0})$$

for any $p\ge 1$; therefore, all statements of the theorems in Section 2 of Chapter 6 are valid for the operator $L_{m_0}+V_{\psi_n}$. In particular, for the lower bound of its spectrum $E_{V_{\psi_n}}$, Theorem 2.8 in Chapter 6 gives the equality

$$-E_{V_{\psi_n}} = \lim_{t\to+\infty}\frac{1}{t}\ln\left\|\exp\left(-\int_0^t V_{\psi_n}(\omega(\tau))d\tau\right)\right\|_{L_1(\Omega,\nu_{m_0})},$$

where ν_{m_0} is a measure on the space of trajectories with values in $\mathcal{S}'_{\mathrm{Re}}(\mathbb{R}^d)$ generated by the semigroup $e^{-tL_{m_0}}$ and the initial distribution μ_{m_0} (see Example 1.5 in Chapter 6). By using the construction of functional integral given in Subsection 1.5 of Chapter 6 and imbedding the space of trajectories $\Omega = (\mathcal{S}'_{\mathrm{Re}}(\mathbb{R}^d))^{\mathbb{R}}$ in $\mathcal{S}'_{\mathrm{Re}}(\mathbb{R}^{d+1})$, it has been shown in Example 1.5 of Chapter 6 that ν_{m_0} is a Gaussian measure on $\mathcal{S}'_{\mathrm{Re}}(\mathbb{R}^{d+1})$ with the characteristic functional

$$\tilde{\nu}_{m_0}(u) = \exp\Big(-\frac{1}{2}((-\Delta + m_0^2)^{-1} u, u)_{L_2(\mathbb{R}^{d+1})}\Big)$$

$$\equiv \exp\Big(-\frac{1}{4}|u|^2_{\mathcal{H}_{m_0}}\Big) \quad \Big(u \in \mathcal{S}_{\mathrm{Re}}(\mathbb{R}^{d+1})\Big). \tag{2.5}$$

Therefore,

$$\left\| \exp\Big(-\int_0^t V_{\psi_n}(\omega(\tau))d\tau\Big)\right\|_{L_1(\mathcal{S}'_{\mathrm{Re}}(\mathbb{R}^{d+1}),\nu_{m_0})}$$

$$= \int_{\mathcal{S}'_{\mathrm{Re}}} e^{-(\omega(\cdot),\, (\psi_n \otimes \kappa_{[0,t]})(\cdot))_{L_2(\mathbb{R}^{d+1})}} d\nu_{m_0} \equiv e^{(1/4)\| \psi_n \otimes \kappa_{[0,t]} \|^2_{\mathcal{H}_{m_0}}}.$$

But

$$\| \psi_n \otimes \kappa_{[0,t]} \|^2_{\mathcal{H}_{m_0}} = 2 \int_{\mathbb{R}^{d+1}} \frac{|\tilde{\psi}_n(k)|^2 |\tilde{\kappa}_{[0,t]}(k_0)|^2}{k_0^2 + \|k\|^2 + m_0^2} dk_0\, dk$$

$$(k_0 \in \mathbb{R}^1,\ k \in \mathbb{R}^d)$$

(see (1.66) in Chapter 6). Taking into account that

$$\tilde{\kappa}_{[0,t]}(k_0) = \frac{1}{\sqrt{2\pi}}(1 - e^{ik_0 t}),$$

we obtain

$$\frac{1}{4}\| \psi_n \otimes \kappa_{[0,t]} \|^2_{\mathcal{H}_{m_0}} = \frac{1}{4\pi} \int_{\mathbb{R}^{d+1}} \frac{|\tilde{\psi}_n(k)|^2 (2 - 2\cos k_0 t)}{k_0^2 (k_0^2 + \|k\|^2 + m_0^2)} dk_0\, dk$$

$$= \frac{1}{\pi} \int_{\mathbb{R}^{d+1}} \frac{|\tilde{\psi}_n(k)|^2 \sin \frac{k_0 t}{2}}{k_0^2 (k_0^2 + \|k\|^2 + m_0^2)} dk_0\, dk$$

$$= \frac{1}{\pi}\int\limits_{\mathbb{R}^d} |\tilde{\psi}_n^{(k)}|^2 \Big[\int\limits_{\mathbb{R}^1} \frac{1}{\|k\|^2 + m_0^2}\left(\frac{\sin^2\frac{k_0 t}{2}}{k_0^2} - \frac{\sin^2\frac{k_0 t}{2}}{k_0^2 + \|k\|^2 + m_0^2}\right) dk_0\Big] dk.$$

But

$$\int\limits_{\mathbb{R}^1} \frac{\sin^2\frac{k_0 t}{2}}{k_0^2} dk_0 = \frac{\pi t}{2},$$

$$\int\limits_{\mathbb{R}^1} \frac{\sin^2 k_0 t}{k_0^2 + \|k\|^2 + m_0^2} dk_0 = \frac{\pi}{2\sqrt{\|k\|^2 + m_0^2}}(1 - e^{-t\sqrt{\|k\|^2 + m_0}}) \le \frac{\pi}{2\sqrt{\|k\|^2 + m_0^2}}$$

Therefore,

$$\lim_{t\to+\infty} \frac{1}{4t}\|\psi_n \otimes \kappa_{[0,t]}\|^2_{\mathcal{H}_{m_0}} = \lim_{t\to+\infty} \frac{1}{\pi t}\int\limits_{\mathbb{R}^d} \frac{|\tilde{\psi}_n(k)|^2 \pi t}{2(\|k\|^2 + m_0^2)} dk$$

$$= \int\limits_{\mathbb{R}^d} \frac{|\tilde{\psi}_n(k)|^2}{2(\|k\|^2 + m_0^2)} dk = \frac{1}{2}\|A_{m_0}^{-1}\psi_n\|^2_{L_2(\mathbb{R}^d)}.$$

Hence,

$$E_{V_{\psi_n}} = -\frac{1}{2}\|A_{m_0}^{-1}\psi_n\|^2_{L_2(\mathbb{R}^d)} = -\frac{1}{2}\langle \psi_n, A_{m_0}^{-2}\psi_n\rangle.$$

The first step of renormalization consists in the transition from the operator L_{ψ_n} to the operator

$$L_{m_0} + V_{\psi_n} + \frac{1}{2}\|A_{m_0}^{-1}\psi_n\|_{L_2(\mathbb{R}^d)}.$$

At the next step we are to find the ground state. Let us show that the function

$$\hat{\phi}_{\psi_n}(x) = \exp[-\langle A_{m_0}^{-1}\psi_n, x\rangle] \quad (x \in \mathcal{S}'_{\mathrm{Re}}(\mathbb{R}^d))$$

is an eigenfunction of $L_{m_0} + V_{\psi_n}$ with an eigenvalue $E_{V_{\psi_n}}$.

Indeed,

$$\nabla\phi_{\psi_n} = -(A_{m_0}^{-1}\psi_n)(\cdot)\hat{\phi}_{\psi_n},$$

$$\mathrm{Tr}_{L_2(\mathbb{R}^d)}(\hat{\phi}_{\psi_n})'' = \Big(\int\limits_{\mathbb{R}^d} |A_{m_0}^{-1}\psi_n(s)|^2 ds\Big)\hat{\phi}_{\psi_n} = \|A_{m_0}^{-1}\psi_n\|^2_{L_2(\mathbb{R}^d)}\hat{\phi}_{\psi_n},$$

therefore,

$$\left(L_{m_0} + V_{\psi_n} + \frac{1}{2}\| A_{m_0}^{-1}\psi_n \|_{L_2(\mathbb{R}^d)}^2 \right) \hat{\phi}_{\psi_n} = -\frac{1}{2}\| A_{m_0}^{-1}\psi_n \|_{L_2(\mathbb{R}^d)}^2 \hat{\phi}_{\psi_n}$$

$$- \langle A_{m_0}^{-1}\psi_n, A_{m_0}^{-1}x \rangle \hat{\phi}_{\psi_n} + \langle \psi_n, x \rangle + \frac{1}{2}\| A_{m_0}^{-1}\psi_n \|_{L_2(\mathbb{R}^d)}^2 \hat{\phi}_{\psi_n} = 0.$$

Since

$$\| \hat{\phi}_{\psi_n} \|_{L_2(\mathcal{S}'_{\mathrm{Re}}(\mathbb{R}^d),\mu_{m_0})}^2 = \int_{\mathcal{S}'_{\mathrm{Re}}} e^{-2\langle A_{m_0}^{-1}\psi_n, x\rangle} d\mu_{m_0}(x) = \exp(\| A_{m_0}^{-1}\psi_n \|_{L_2(\mathbb{R}^d)}^2),$$

the function

$$\phi_{\psi_n}(x) = \exp\left[- \langle A_{m_0}^{-1}\psi_n, x\rangle - \frac{1}{2}\| A_{m_0}^{-1}\psi_n \|_{L_2(\mathbb{R}^d)}^2 \right] \tag{2.6}$$

is a normalized ground state of the operator L_{ψ_n}. It follows from the definition of the measure μ_{m_0} (see (2.43) in Chapter 2) that

$$\frac{d\mu_{m_0}(x+\varphi)}{d\mu_{m_0}(x)} = e^{-2\langle A_{n_0}\varphi, x\rangle - \langle \varphi, A_{n_0}\varphi\rangle} \quad (x \in \mathcal{S}'_{\mathrm{Re}}(\mathbb{R}^d))$$

for $\varphi \in \mathcal{S}'_{\mathrm{Re}}(\mathbb{R}^d)$. Comparing the last expression with (2.6), we obtain the following representation

$$\phi_{\psi_n}(x) = \left[\frac{d\mu_{m_0}(x + A_{m_0}^{-2}\psi_n)}{d\mu_{m_0}(x)} \right]^{1/2}. \tag{2.7}$$

Thus, the renormalized operator $L_{\psi_n,\,\mathrm{ren}}$ is generated by the Dirichlet form of the measure

$$\mu_{\psi_n} = \phi_{\psi_n}^2 \mu_{m_0} = \mu_{m_0}(\cdot + A_{m_0}^{-2}\psi_n).$$

The measures μ_{ψ_n} converge as $n \to \infty$ to a measure μ_ψ of the form $\mu_\psi(\cdot) = \mu_{m_0}(\cdot + A_{m_0}^{-2}\psi)$ in the sense of the convergence of integrals of continuous bounded cylindrical functions. In fact, to prove this, it suffices to establish the convergence of their Fourier transforms, but

$$\tilde{\mu}_{\psi_n}(\varphi) = \int_{\mathcal{S}'_{\mathrm{Re}}} e^{i\langle\varphi,x\rangle} d\mu_{\psi_n}(x) = \int_{\mathcal{S}'_{\mathrm{Re}}} e^{i\langle\varphi,x\rangle} d\mu_{m_0}(x + A_{m_0}^{-2}\psi_n)$$

$$= e^{i\langle A_{m_0}^{-2}\psi_n, \varphi\rangle} e^{-(1/4)\|\varphi\|^2_{L_2(\mathbb{R}^d)}} \xrightarrow[n\to\infty]{} e^{i\langle A_{m_0}^{-2}\psi, \varphi\rangle} e^{-(1/4)\|\varphi\|^2_{L_2(\mathbb{R}^d)}}$$

$$= \tilde{\mu}_\psi(\varphi) \quad (\varphi \in \mathfrak{S}_{\mathrm{Re}}(\mathbb{R}^d)).$$

By definition, the renormalized Hamiltonian $L_{\mu,\,\mathrm{ren}}$ which corresponds to (2.3) is generated by the bilinear form

$$(L_{\psi,\mathrm{ren}}f, g)_{L_2(\mathfrak{S}'_{\mathrm{Re}}(\mathbb{R}^d), \mu_\psi)} = \frac{1}{2}\int_{\mathfrak{S}'_{\mathrm{Re}}(\mathbb{R}^d)} \langle \nabla f, \nabla \bar{g}\rangle d\mu_\psi \quad (f, g \in C^2_{b,\mathrm{cyl}}(\mathfrak{S}'_{\mathrm{Re}}(\mathbb{R}^d))).$$

In order to find the explicit form of $L_{\mu_\psi} = L_{\psi,\,\mathrm{ren}}$ on smooth functions, we must compute the logarithmic derivative of the measure μ_ψ. If we note that

$$\rho_{\mu_\psi}(\varphi, x) = \frac{d\mu_\psi(x+\varphi)}{d\mu_\psi(x)} = \frac{d\mu_{m_0}(x + A_{m_0}^{-2}\psi + \varphi)}{d\mu_{m_0}(x + A_{m_0}^{-2}\psi)} = \rho_{\mu_{m_0}}(\varphi, x + A_{m_0}^{-2}\psi)$$

for $\varphi \in \mathfrak{S}_{\mathrm{Re}}(\mathbb{R}^d)$, then we get

$$\beta_{\mu_\psi}(x) = \beta_{\mu_{m_0}}(x + A_{m_0}^{-2}\psi) = -2A_{m_0}x - 2A_{m_0}^{-1}\psi.$$

Therefore,

$$(L_{\psi,\mathrm{ren}}f)(x) = -\frac{1}{2}[\mathrm{Tr}_{L_2(\mathbb{R}^d)}f''(x) - 2(A_{m_0}x, \nabla f(x))$$

$$-2\langle A_{m_0}^{-1}\psi, \nabla f(x)\rangle] = (L_{m_0}f)(x) + \langle \nabla f, A_{m_0}^{-1}\psi\rangle \tag{2.8}$$

for $f \in C^2_{b,\mathrm{cyl}}(\mathfrak{S}'_{\mathrm{Re}}(\mathbb{R}^d))$.

The proof of essential selfadjointness can be reduced to the verification of conditions of Theorem 3.4 in Chapter 6. To do this, one should only choose a Hilbert space $H_- \subset \mathfrak{S}'_{\mathrm{Re}}(\mathbb{R}^d)$ of full μ_ψ-measure on which the logarithmic derivative β_{μ_ψ} is continuously differentiable. This is possible due to the fact that

$$\nabla\beta_{\mu_\psi}(x) = -2A_{m_0} - 2A_{m_0}^{-1} \in \mathfrak{L}(\mathfrak{S}'_{\mathrm{Re}}(\mathbb{R}^d), \mathfrak{S}'_{\mathrm{Re}}(\mathbb{R}^d)).$$

However, in this case, we can take another (simpler) way: Let us introduce a mapping

$$j_\psi : \mathfrak{S}'_{\mathrm{Re}}(\mathbb{R}^d) \ni x \mapsto j_\psi x = x - A_{m_0}^{-2}\psi \in \mathfrak{S}'_{\mathrm{Re}}(\mathbb{R}^d),$$

under which the measure μ_ψ transforms into the measure μ_{m_0}. Then the operator

$$L_2(\mathfrak{S}'_{\mathrm{Re}}(\mathbb{R}^d), \mu_\psi) \ni f(x) \mapsto (J_\psi f)(x) = f(J_\psi x) \in L_2(\mathfrak{S}'_{\mathrm{Re}}(\mathbb{R}^d), \mu_{m_0})$$

is unitary. It follows from (2.8) that, for $f \in C^2_{b,\mathrm{cyl}}(\mathfrak{S}'_{\mathrm{Re}}(\mathbb{R}^d))$,

$$\begin{aligned} & J_\psi L_{\psi,\mathrm{ren}}\, J_\psi^{-1} f(x) \\ & \quad = J_\psi L_{\psi,\mathrm{ren}}\, J_\psi^{-1} f(x + A_{m_0}^{-2}\psi) \\ & \quad = \Big[-\frac{1}{2}(\mathrm{Tr}_{L_2(\mathbb{R}^d)} f''(x + A_{m_0}^{-2}\psi) - 2\langle A_{m_0}(x + A_{m_0}^{-2}\psi), \nabla f(x + A_{m_0}^{-2}\psi)\rangle) \Big] = L_{m_0} f. \end{aligned}$$

But the operator L_{m_0} is essentially selfadjoint on $f \in C^\infty_{b,\mathrm{cyl}}(\mathfrak{S}'_{\mathrm{Re}}(\mathbb{R}^d))$ and, hence, L_{μ_ψ} is essentially selfadjoint on the region $J_\psi^{-1}(C^\infty_{b,\mathrm{cyl}}(\mathfrak{S}'_{\mathrm{Re}}(\mathbb{R}^d)))$. It remains to note the evident inclusion $J_\psi^{-1}(C^\infty_{b,\mathrm{cyl}}(\mathfrak{S}'_{\mathrm{Re}}(\mathbb{R}^d))) \subset C^\infty_{b,\mathrm{cyl}}(\mathfrak{S}'_{\mathrm{Re}}(\mathbb{R}^d))$.

■

2.2. Quadratic Interaction

Let us consider another model in constructive field theory which is also explicitly solvable and can be investigated by use of the renormalization scheme intensively exploited above. We mean the so-called $:\Phi^2:$ model described by the formal Hamiltonian

$$L_\lambda = L_{m_0} + \lambda V, \tag{2.9}$$

where $\lambda \geq 0$ is a constant (a coupling constant), and the potential V is given by the equality

$$V(x) = \int_{\mathbb{R}^d} : x^2(s) : ds \equiv \int_{\mathbb{R}^d} : \langle x(\cdot), \delta_s(\cdot)\rangle^2 : ds \tag{2.10}$$

as a generalized function from $\mathcal{P}'(\mathfrak{S}'_{\mathrm{Re}}(\mathbb{R}^d))$ (see Example 2.2 in Chapter 6). Before we begin to discuss the construction of the renormalized operator which corresponds to (2.9), we give an heuristic argument prompting the answer.

Let us introduce a free field with mass $m_0 > 0$ in the spacetime with dimensionality

$d+1$ as an operator-valued (generalized) function in $L_2((\mathfrak{S}'_{\mathrm{Re}}(\mathbb{R}^d), \mu_{m_0})$ given by the equality

$$\Phi_0(t, s) = e^{itL_{m_0}} x(s)\, e^{-itL_{m_0}} \quad (t \in \mathbb{R}^1,\ s \in \mathbb{R}^d). \tag{2.11}$$

The last equality should be understood as follows: For $\psi \in \mathfrak{S}'_{\mathrm{Re}}(\mathbb{R}^d)$, we consider an operator $A_t(\psi)$ of multiplication by a linear function $\langle x(\cdot), \psi(t, \cdot)\rangle$ in $L_2((\mathfrak{S}'_{\mathrm{Re}}(\mathbb{R}^d), \mu_{m_0})$ and define an operator

$$\Phi_0(\psi) = \int_{\mathbb{R}^{d+1}} \Phi_0(t, s)\psi(t, s)dtds = \int_{\mathbb{R}^d} e^{itL_{m_0}} A_t(\psi) e^{-itL_{m_0}} dt.$$

One can show that the field Φ_0 defines so satisfies the equation

$$(\Box + m_0^2)\Phi_0(t, s) = 0.$$

If we now perturb L_{m_0} by the potential

$$\lambda V(x) = \lambda \int_{\mathbb{R}^d} V(x(s))ds$$

and introduce a new field $\Phi(t, s)$ by replacing L_{m_0} by $L_{m_0} + \lambda V$ in (2.11), then (on the formal level) this field satisfies the equation

$$(\Box + m_0^2)\Phi(t, s) = -\lambda V'(\Phi(t, s)).$$

In our case of quadratic interaction

$$V(x) = \int_{\mathbb{R}^d} : x(s) : ds \equiv \int_{\mathbb{R}^d} x^2(s)ds - c,$$

where

$$c = \frac{1}{2} \int_{\mathbb{R}^d} \| \delta_s \|^2_{\mathfrak{H}_{m_0}} ds = +\infty$$

(see Example 2.2 in Chapter 6), this equation has the form

$$(\Box + m_0^2)\Phi = -2\lambda\Phi,$$

i.e.,

$$(\Box + m_0^2 + 2\lambda)\Phi = 0.$$

The free field with mass

$$\sqrt{m_0^2 + 2\lambda} = m_\lambda,$$

satisfies the equation obtained and, therefore, one should expect that the renormalized operator which corresponds to (2.9) will be again an operator of energy of the free field, but with another mass.

Theorem 2.2. *The operator $L_{\lambda,\,\mathrm{ren}}$ coincides with the Hamiltonian L_{m_λ} of the free field with mass*

$$m_\lambda = \sqrt{m_0^2 + 2\lambda}.$$

Proof. First, let us introduce an approximation of the generalized function

$$V(x) = \int_{\mathbb{R}^d} : x(s)^2 : ds$$

by smooth ones. Let $e_k(\cdot)$, $k \in \mathbb{Z}_+^d$, $\mathbb{Z}_+^d = (\mathbb{N} \cup \{0\})^d$, be an orthonormal basis of Hermite functions in $L_2(\mathbb{R}^d)$. For $d = 1$, this basis is described in Example 4.1, and for arbitrary $d \in \mathbb{N}$, as usual, we set

$$\psi_k(x) = \psi_{k_1}(x_1) \times \ldots \times \psi_{k_d}(x_d), \quad x \in \mathbb{Z}_+^d, \quad x \in \mathbb{R}^d.$$

For $n \in \mathbb{N}$, we define a kernel

$$G_n(s_1, s_2) = \sum_{|k| \le n} e_k(s_1) e_k(s_2).$$

It is clear that $G_n(s_1, s_2) \underset{n\to\infty}{\to} \delta(s_1 - s_2)$ in the sense of $\mathcal{S}'_{\mathrm{Re}}(\mathbb{R}^d)$. Denote by G_n the operator in $L_2(\mathbb{R}^d)$ which corresponds to the kernel $G_n(\cdot, \cdot)$. Then $G_n\colon \mathcal{S}'_{\mathrm{Re}}(\mathbb{R}^d) \to \mathcal{S}_{\mathrm{Re}}(\mathbb{R}^d)$ and, therefore, the following function is defined

$$V_n(x) = :\langle G_n x, x\rangle: = \int_{\mathbb{R}^{2d}} : x(s_1) G_n(s_1, s_2) x(s_2) : ds_1 ds_2 \in \mathcal{P}'(\mathcal{S}'_{\mathrm{Re}}(\mathbb{R}^d)).$$

Note that

$$:\langle G_n x, x\rangle : = \sum_{|k|\le n} :\langle x, e_k\rangle^2 : = \sum_{|k|\le n} \langle x, e_k\rangle^2 - \frac{1}{2}\sum_{|k|\le n} \| e_k \|^2_{\mathfrak{H}_{m_0}}$$

$$= \langle G_n x, x\rangle - \frac{1}{2}\| A_{m_0}^{-1} G_n \|^2_{L_2(\mathbb{R}^d)} \in \mathcal{P}(\mathcal{S}'_{\mathrm{Re}}(\mathbb{R}^d)).$$

The operator $L_\lambda^{(n)} = I_{m_0} + \lambda V_n$ is essentially selfadjoint on $\mathcal{P}(\mathcal{S}'_{\mathrm{Re}}(\mathbb{R}^d))$ and has the ground state $\phi_n > 0$ μ_{m_0}-almost everywhere because $V_m \ge c_n > -\infty$ and $V_n \in L_p(\mathcal{S}'_{\mathrm{Re}}(\mathbb{R}^d), \mu_{m_0})$ for any $p \ge 1$ (see Theorem 2.8 in Chapter 6). The renormalized operator

$$L_{\lambda,\,\mathrm{ren}} = \phi_V^{-1}(L_{m_0} + \lambda V_n - E_{V_n})\phi_n$$

acts in the space $L_p(\mathcal{S}'_{\mathrm{Re}}(\mathbb{R}^d), \phi_n^2\mu_{m_0})$.

In the considered model, it is convenient to realize the limit transition as $n \to \infty$ in terms of measures on the space of trajectories. Recall that the operator L_{m_0} and the initial distribution μ_{m_0} are associated with the Gaussian measure ν_{m_0} on the space $\mathcal{S}'_{\mathrm{Re}}(\mathbb{R}^{d+1})$ with the characteristic functional (2.6). It is useful to note that this form yields the following representation

$$d\nu_{m_0}(\omega(\cdot)) = \frac{1}{N}\exp\left\{-\frac{1}{2}\int_{\mathbb{R}^{d+1}} [(-\Delta + m_0^2)\omega](t, s)\omega(t, s)dtds\right\}d\omega(\cdot), \qquad (2.12)$$

which obviously has no rigorous meaning, for example, since the set of trajectories of full ν_{m_0}-measure consists of the functions which can be chosen to be continuous only with respect to time variable and, necessarily, they are generalized functions with respect to the remaining variables. However, this representation is often useful as an heuristic consideration in the study of the measure ν_{m_0} and its perturbations. According to Theorem 1.1, the operator $L_{\lambda,\,\mathrm{ren}}^{(n)}$ and initial distribution $\phi_n^2\mu_m$ are associated with the probability measure $\nu_{m_0}^{\lambda V_n}$ on $\mathcal{S}'_{\mathrm{Re}}(\mathbb{R}^{d+1})$ which can be defined as a limit of the measures

$$d\nu_n^{(t)}(\omega(\cdot)) = \frac{1}{N_{t,n}}\exp\left\{-\lambda\int_{-t}^{t} V_n(\omega(\tau))d\tau\right\}d\nu_{m_0}(\omega(\cdot))$$

$$= \frac{1}{N_{t,n}}\exp\left\{-\lambda\int_{-t}^{t} \langle G_n\omega(\tau), \omega(\tau)\rangle d\tau\right\}d\nu_{m_0}(\omega(\cdot))$$

$$= \frac{1}{N_{t,n}} \exp\{ -\lambda(G_n^{(t)} \omega(\cdot), \omega(\cdot))_{L_2(\mathbb{R}^{d+1})} \} d\nu_{m_0}(\omega(\cdot))$$

as $t \to +\infty$, where

$$(G_n^{(t)} \omega)(\tau, s) = (G_n \omega)(\tau, s) \kappa_{[-t,t]}(\tau)$$

(we note that the constant of the Wick ordering disappears by virtue of normalization). Taking (2.12) into account, we get the formal representation

$$d\nu_n^{(t)}(\omega(\cdot)) = \frac{1}{N_{t,n}} \exp\left\{ -\frac{1}{2} \int_{\mathbb{R}^{d+1}} [(-\Delta + m_0^2 + 2\lambda G_n^{(t)})] \omega(t, s) dt ds \right\} d\omega(\cdot).$$

To justify the form of the measures $\nu_n^{(t)}$ rigorously, we consider its Fourier transform

$$\tilde{\nu}_n^{(t)}(\varphi) = \int_{\mathcal{S}'_{\mathrm{Re}}} e^{i(\omega(\cdot), \varphi(\cdot))_{L_2(\mathbb{R}^{d+1})}} \frac{e^{-\lambda(G_n^{(t)} \omega, \omega)_{L_2(\mathbb{R}^{d+1})}}}{N_{t,n}} d\nu_{m_0}(\omega)$$

$$= \exp\left\{ -\frac{1}{2} (\varphi, (-\Delta + m_0^2 + 2\lambda G_n^{(t)})^{-1} \varphi)_{L_2(\mathbb{R}^{d+1})}] \right\} \quad (\varphi \in \mathcal{S}_{\mathrm{Re}}(\mathbb{R}^{d+1})).$$

Here we have used the formulas from Subsection 2.1 in Chapter 2 to compute the integral with respect to the Gaussian measure. Since the operators $G_n^{(t)}: L_2(\mathbb{R}^{d+1}) \to L_2(\mathbb{R}^{d+1})$ converge as $t \to \infty$ to $G_n \otimes 1$, we have

$$\tilde{\nu}_n^{(t)}(\varphi) \xrightarrow[t \to +\infty]{} \tilde{\nu}_n(\varphi) = \exp\left\{ -\frac{1}{2} (\varphi, (-\Delta + m_0^2 + 2\lambda(G_n \otimes 1))^{-1} \varphi)_{L_2(\mathbb{R}^{d+1})}] \right\},$$

which yields the convergence of integrals of bounded cylindrical functions. Further, we note that $G_n \to 1$ as $n \to 1$ strongly in $L_2(\mathbb{R}^d)$; consequently, the characteristic functionals converge

$$\tilde{\nu}_{m_0}^{\lambda V_n}(\varphi) \xrightarrow[n \to \infty]{} \exp\left\{ -\frac{1}{2} (\varphi, (-\Delta + m_0^2 + 2\lambda)^{-1} \varphi)_{L_2(\mathbb{R}^{d+1})}] \right\}.$$

But according to (2.5), the last expression coincides with the Fourier transform of the measure corresponding to the free field with mass $\sqrt{m_0^2 + 2\lambda} = m_\lambda$. Therefore, the restriction of it to the σ-algebra which is generated by cylindrical functions $\langle \omega(0, \cdot), \varphi(\cdot) \rangle$, $\varphi \in \mathcal{S}_{\mathrm{Re}}(\mathbb{R}^d)$, and is isomorphic to $\mathfrak{C}_\sigma(\mathcal{S}'_{\mathrm{Re}}(\mathbb{R}^d))$ leads to the measure μ_{m_λ} on $\mathcal{S}'_{\mathrm{Re}}(\mathbb{R}^d)$ which gives the representation of the free field mass $m_\lambda =$

$\sqrt{m_0^2 + 2\lambda}$. As in (1.15), here

$$P_0 U_t P_0 = e^{-|t| L_{\lambda,\mathrm{ren}}},$$

where $L_{\lambda,\mathrm{ren}} = L_{m_\lambda}$ is a Hamiltonian of the free field with mass m_λ acting in the space $L_2(\mathcal{S}'_{\mathrm{Re}}(\mathbb{R}^d), \mu_{m_\lambda})$.

We note that, in fact, the concrete choice of the operators G_n, $n \in \mathbb{N}$, has not been used yet. For any sequence of operators

$$G_n : \mathcal{S}'_{\mathrm{Re}}(\mathbb{R}^d) \to \mathcal{S}_{\mathrm{Re}}(\mathbb{R}^d)$$

such that $A_{m_0}^{-1} G_n$ is a Hilbert – Schmidt operator in $L_2(\mathbb{R}^d)$, $G_n \ge 0$, and $G_n \to 1$ as $n \to \infty$ weakly in $L_2(\mathbb{R}^d)$, the renormalization procedure gives the same answer. This follows from the convergence of the Fourier transforms $\tilde{\nu}_{m_0}^{\lambda V_n}$ which takes place under these conditions.

§3. The Dirichlet Operators in Quantum Statistical Physics

The general renormalization scheme developed in Section 1 can be applied to the construction of the dynamics of various physical systems with infinitely many degrees of freedom. In this section, we employ this scheme to investigate some models in the quantum statistical physics of lattice systems in the ground state, i.e., at temperature zero.

3.1. Construction of the Dynamics of Quantum Lattice Systems

In the case of quantum lattice systems, the renormalization procedure can be made more detailed and helpful. We describe the corresponding construction in the situation which is not the most general one, but shows the essence of the considered approach to the construction of operator realizations of formal Hamiltonians of quantum lattice systems and is sufficient for the application to the concrete models investigated below. Here we shall study the systems of interacting oscillators situated at the vertices of integral lattice $\mathbb{Z}^d$ ($d \in \mathbb{N}$). We assume that each oscillator possesses one internal degree of freedom, i.e., each vertex $k \in \mathbb{Z}^d$ is associated with the space of states $\mathcal{H}_k = L_2(\mathbb{R}^1, dx_k)$, with the canonical selfadjoint operators of momentum p_k and coordinate q_k acting in $\mathcal{H}_k$ (on natural domains of definition) according to the formulas

$$(p_k f)(x_k) = \frac{1}{i}\frac{df(x_k)}{dx_k}, \quad (q_k f)x_k = x_k f(x_k), \quad [p_k, q_k] = i1 \tag{3.1}$$

and with the operator of energy

$$\frac{p_k^2}{2} + \frac{\omega^2 q_k^2}{2}$$

($\omega > 0$ is natural frequency of oscillator, see Example 1.1). The formal Hamiltonian which describes an infinite system of interacting oscillators (under the assumption that only pair interactions exist) has the form

$$L = \frac{1}{2}\sum_{k \in \mathbb{Z}^d} (p_k^2 + \omega^2 q_k^2) + \sum_{j, k \in \mathbb{Z}^d} V_{jk}(q_j, q_k), \tag{3.2}$$

where V_{jk} is the potential of the interaction of the particles situated at vertices $j \in \mathbb{Z}^d$ and $k \in \mathbb{Z}^d$.

The construction of the dynamics of a system is realized on the basis of the operator realization (3.2) with the help of the modification of the renormalization procedure described above: Each subset of the lattice $\Lambda \subset \mathbb{Z}^d$, $|\Lambda| < \infty$ ($|\Lambda|$ is a cardinality of Λ) is associated with the system in Λ with the space of states $\mathcal{H}_\Lambda = L_2(\mathbb{R}^\Lambda, dx_\Lambda)$ (dx_Λ is a Lebesgue measure on

$$\mathbb{R}^\Lambda = \underset{k \in \Lambda}{\times} \mathbb{R}^1 \ni x_\Lambda = (x_k)_{k \in \Lambda}.$$

Denote by $\mathcal{A}_\Lambda$ a C^*-algebra of operators in $\mathcal{H}_\Lambda$ generated by the Weyl operators $U_k^\Lambda(t) = e^{itq_k}$ and $V_k^\Lambda(t) = e^{itq_k}$ ($t \in \mathbb{R}^1$, $k \in \Lambda$), which form a representation of canonical commutation relations (CCR) in Weyl's form (see Example 3.2 and 3.8 in Section 3 of Chapter 4)

$$V_k^\Lambda(t) U_j^\Lambda(s) = e^{its\delta_{jk}} U_j^\Lambda(s) V_k^\Lambda(t), \tag{3.3}$$

$$[V_k^\Lambda(t), V_j^\Lambda(s)] = [U_k^\Lambda(t), U_j^\Lambda(s)] = 0$$

$$(t, s \in \mathbb{R}^1, \ j, k \in \Lambda).$$

The dynamics of the system contained in the region Λ is determined by the local Hamiltonian

$$L_\Lambda = \frac{1}{2}\sum_{k\in\Lambda}(p_k^2+\omega^2 q_k^2) + \sum_{j,k\in\Lambda} V_{jk}(q_j, q_k). \tag{3.4}$$

We assume that the following inclusion holds for the interaction potentials

$$\exists\varepsilon > 0 : V_{jk} \in L_{2+\varepsilon}(\mathbb{R}^2,\ e^{-\omega(x_k^2+x_j^2)}dx_k dx_j),$$

$$\forall p \geq 1\, e^{-V_{jk}} \in L_p(\mathbb{R}^2,\ e^{-\omega(x_k^2+x_j^2)}dx_k dx_j). \tag{3.5}$$

Then, taking into account Examples 3.3–3.5 in Chapter 6, we can conclude from the results presented in Subsection 2.2 of Chapter 6, that L_Λ is essentially selfadjoint on $C_0^\infty(\mathrm{R}^\Lambda)$. Moreover, L_Λ is semibounded from below, and $E_\Lambda = \inf s\, L_\Lambda$ is a simple eigenvalue and the corresponding normalized eigenfunction $\phi_\Lambda > 0$ almost everywhere on $\mathbb{R}^\Lambda$ with respect to the Lebesgue measure. The Heisenberg dynamics of an observable $A \in \mathcal{A}_\Lambda$ is given by a group of unitary automorphisms

$$\tau_t^\Lambda(A) = e^{itL_\Lambda} A e^{-itL_\Lambda} \quad (t \in \mathbb{R}^1).$$

Realizing the renormalization scheme given in Section 1, we introduce a probability measure

$$d\mu_\Lambda(x_\Lambda) = \phi_\Lambda^2(x_\Lambda)dx_\Lambda$$

on $\mathbb{R}^\Lambda$ and a Hilbert space $\mathcal{H}_{\Lambda,\,\mathrm{ren}} = L_2(\mathbb{R}^\Lambda, \mu_\Lambda)$. As in Example 3.3 of Chapter 6, under the natural unitary isomorphism of $\mathcal{H}_\Lambda$ and $\mathcal{H}_{\Lambda,\,\mathrm{ren}}$, the operator $L_\Lambda - E_\Lambda$ turns into

$$L_{\Lambda,\,\mathrm{ren}} = \phi_\Lambda^{-1}(L_\Lambda - E_\Lambda)\phi_\Lambda,$$

which is called a renormalized local Hamiltonian in the region Λ. Furthermore, the same Example 3.3 shows that the operator $L_{\Lambda,\,\mathrm{ren}}$ is associated with the Dirichlet form of the measure μ_Λ

$$(L_{\Lambda,\mathrm{ren}} f, g)_{L_2(\mathbb{R}^\Lambda,\mu_\Lambda)} = \frac{1}{2}\int_{\mathbb{R}^\Lambda}\sum_{k\in\Lambda}\frac{\partial f}{\partial x_k}\frac{\overline{\partial g}}{\partial x_k}d\mu_\Lambda \tag{3.6}$$

$$(f, g \in C_b^\infty(\mathbb{R}^\Lambda)).$$

Hence, we have obtained a family $(L_{\Lambda,\,\mathrm{ren}})_{\Lambda\subset\mathbb{Z}^d,\,|\Lambda|<\infty}$ of local Hamiltonians acting on different spaces $L_2(\mathbb{R}^\Lambda, \mu_\Lambda)$. However, in contrast to Section 1, we have a family of probability measures $(\mu_\Lambda)_{\Lambda\subset\mathbb{Z}^d,\,|\Lambda|<\infty}$ given on different linear spaces $\mathbb{R}^d$ $(\Lambda \subset \mathbb{Z}^d, |\Lambda| < \infty)$.

Under renormalization, the Weyl representation of CCR in $L_2(\mathbb{R}^\Lambda, dx_\Lambda)$ is associated with a unitary equivalent representation

$$U^\Lambda_{k,\mathrm{ren}}(t) = \phi_\Lambda^{-1} U^\Lambda_k(t)\phi_\Lambda; \quad V^\Lambda_{k,\mathrm{ren}}(t) = \phi_\Lambda^{-1} V^\Lambda_k(t)\phi_\Lambda$$

$$(k \in \mathbb{Z}^d;\ t \in \mathbb{R}^1)$$

in $L_2(\mathbb{R}^\Lambda, d\mu_\Lambda(x_\Lambda))$ and a C^*-algebra of observables $\mathcal{A}_{\Lambda,\,\mathrm{ren}}$ generated by this representation. The generators of these groups are denoted by $q^\Lambda_{k,\,\mathrm{ren}}$ and $p^\Lambda_{k,\,\mathrm{ren}}$, respectively.

The next step is the limit transition to the system given on the whole of the infinite lattice $\mathbb{Z}^d$. First, we determine the sense of the limit transition $\Lambda \to \mathbb{Z}^d$. We say that the sequence $\Lambda_n \to \mathbb{Z}^d$ as $n \to \infty$ if, for any cube $K \subset \mathbb{Z}^d$, there exists $n_0 = n_0(K)$ such that $K \subset \Lambda_n$ for any $n \geq n_0(K)$. Further, we say that a convergence as $\Lambda \to \mathbb{Z}^d$ takes place if this convergence takes place for any sequence of domains $\Lambda_n \to \mathbb{Z}^d$ as $n \to \infty$ in the above-mentioned sense. Assume that the family of measures $(\mu_\Lambda)_{\Lambda\subset\mathbb{Z}^d,\,|\Lambda|<\infty}$ converges to a probability measure μ on

$$\mathcal{C}_\sigma(\mathbb{R}^{\mathbb{Z}^d}) = \mathcal{B}(\mathbb{R}^{\mathbb{Z}^d}) \quad \Big(\mathbb{R}^{\mathbb{Z}^d} = \mathop{\times}_{k\in\mathbb{Z}^d} \mathbb{R}^1 \ni x = (x_k)_{k\in\mathbb{Z}^d}\Big)$$

in the sense that for every function

$$f \in C_{b,\mathrm{cyl}}(\mathbb{R}^{\mathbb{Z}^d}) = \bigcup_\Lambda C_b(\mathbb{R}^\Lambda),$$

we have

$$\int_{\mathbb{R}^\Lambda} f\,d\mu_\Lambda \to \int_{\mathbb{R}^{\mathbb{Z}^d}} f\,d\mu, \quad \Lambda \to \mathbb{Z}^d.$$

This definition is clearly correct because, by definition, for any function $f \in C_{b,\mathrm{cyl}}(\mathbb{R}^{\mathbb{Z}^d})$, we have the representation $f(x) = g(x_{k_1}, \ldots, x_{k_p})$, $g \in C_b(\mathbb{R}^p)$ $(p \in \mathbb{N})$, and each sequence of regions $\Lambda_n \to \mathbb{Z}^d$ as $n \to \infty$ possesses the following property: There exists n_0 such that $\{k_1, \ldots, k_p\} \subset \Lambda_n$ for any $n \geq n_0$. A space $\mathcal{H} = L_2(\mathbb{R}^{\mathbb{Z}^d}, \mu)$ is called a physical Hilbert space of an infinite system. Note that (3.6) and the definition of the convergence of measures μ_Λ imply that

$$(L_{\Lambda,\text{ren}} f, g)_{L_2(\mathbb{R}^\Lambda, \mu_\Lambda)} \xrightarrow[\Lambda \to \mathbb{Z}^d]{} \frac{1}{2} \int_{\mathbb{R}^{\mathbb{Z}^d}} \sum_{k \in \mathbb{Z}^d} \frac{\partial f(x)}{\partial x_k} \overline{\frac{\partial g(x)}{\partial x_k}} d\mu(x)$$

$$(f, g \in C_b^\infty(\mathbb{R}^{\mathbb{Z}^d})).$$

Let us introduce a Hilbert space

$$l_2(\mathbb{Z}^d) = \left\{ x = (x_k)_{k \in \mathbb{Z}^d} \in \mathbb{R}^{\mathbb{Z}^d} \,\middle|\, \sum_{k \in \mathbb{Z}^d} x_k^2 < \infty \right\}$$

with a natural scalar product which is denoted in this section by $\langle \cdot, \cdot \rangle = (\cdot, \cdot)_{l_2(\mathbb{Z}^d)}$. Furthermore, we shall use a nuclear space of finite vectors $\mathbb{R}_0^{\mathbb{Z}^d} \subset l_2(\mathbb{Z}^d)$ topologized by analogy with $\mathbb{R}_0^\infty$ (see Example 1.6 in Chapter 1). The space $\mathbb{R}^{\mathbb{Z}^d}$ is dual to this with respect to $l_2(\mathbb{Z}^d)$. For $f \in C_{b,\,\text{cyl}}^1(\mathbb{R}^{\mathbb{Z}^d})$, the derivative in the sense of $l_2(\mathbb{Z}^d)$

$$\nabla f(x) = (\partial f(x) / \partial x_k)_{k \in \mathbb{Z}^d} \in \mathbb{R}^{\mathbb{Z}^d} \quad (x \in \mathbb{R}^{\mathbb{Z}^d})$$

and the indicated convergence can be rewritten as

$$(L_{\Lambda,\text{ren}} f, g)_{L_2(\mathbb{R}^\Lambda, \mu_\Lambda)} \xrightarrow[\Lambda \to \mathbb{Z}^d]{} \frac{1}{2} \int_{\mathbb{R}^{\mathbb{Z}^d}} \langle \nabla f, \overline{\nabla g} \rangle d\mu = d_\mu(f, g)$$

$$(f, g \in C_{b,\text{cyl}}^1(\mathbb{R}^{\mathbb{Z}^d})).$$

An operator associated with the Dirichlet form d_μ (under the assumption of its closability) is called a renormalized operator of energy corresponding to the formal Hamiltonian (3.2). As already mentioned in Subsection 3.1 of Chapter 6, the existence of a logarithmic derivative of μ is the simplest condition of closability of d_μ. More precisely, let us assume that, for every unit vector $e_k = (\delta_{jk})_{j \in \mathbb{Z}^d} \in l_2(\mathbb{Z}^d)$ $(k \in \mathbb{Z}^d)$, the logarithmic derivative

$$\beta_\mu(e_k, x) = \beta_k(x) = \frac{d}{dt}\left[\frac{d\mu(x + te_k)}{d\mu(x)}\right]\bigg|_{t=0} \in L_2(\mathbb{R}^{\mathbb{Z}^d}, \mu)$$

is defined (see Subsection 3.1 in Chapter 6). Then d_μ is closable; it is associated with a nonnegative selfadjoint operator L_{ren} which is the Fourier extension of the differential operator

$$(L_\mu f)(x) = -\frac{1}{2}\sum_{k\in\mathbb{Z}^d}\frac{\partial^2 f(x)}{\partial x_k^2} - \frac{1}{2}\sum_{k\in\mathbb{Z}^d}(\beta_k)(x)\frac{\partial f(x)}{\partial x_k} \tag{3.7}$$

$$(f\in C^2_{b,\mathrm{cyl}}(\mathbb{R}^{\mathbb{Z}^d})).$$

Thus, we have constructed a physical Hilbert space of $L_2(\mathbb{R}^{\mathbb{Z}^d}, \mu)$ of an infinite system and its operator of energy L_{ren}. It remains to check what representation of the canonical commutation relations is a result of the renormalization procedure. The standard representation of CCR in Weyl's form

$$(U_{\mu,k}(t)f)(x) = e^{itx_k}f(x),$$

$$(V_{\mu,k}(t)f)(x) = f(x+te_k)\sqrt{\frac{d\mu(x+te_k)}{d\mu(x)}} \tag{3.8}$$

$$(k\in\mathbb{Z}^d;\ t\in\mathbb{R}^1;\ f\in L_2(\mathbb{R}^{\mathbb{Z}^d},\mu))$$

(see Example 3.8 in Chapter 4) is connected with each $\mathbb{R}_0^{\mathbb{Z}^d}$-quasi-invariant measure μ on $\mathfrak{B}(\mathbb{R}^{\mathbb{Z}^d})$ or, in other words, with each measure for which the shift by the vectors from $\mathbb{R}_0^{\mathbb{Z}^d}$ are admissible.

The equalities (3.3) for the operators $U_{\mu,k}(t)$ and $V_{\mu,k}(t)$ are immediately verified. The generators of these unitary groups are denoted by φ_k and π_k, respectively. Under the assumption that the moments

$$\int x_k^2 d\mu(x) < \infty \quad (k\in\mathbb{Z}^d)$$

exist, the inclusion $C^1_{b,\mathrm{cyl}}(\mathbb{R}^{\mathbb{Z}^d})\subset\mathfrak{D}(\varphi_k)$ $(k\in\mathbb{Z}^d)$ and the representation

$$(\varphi_k f)(x) = x_k f(x), \quad (\pi_k f)(x) = \frac{1}{i}\left(\frac{\partial f(x)}{\partial x_k} + \beta_k(x)f(x)\right)$$

$$(k\in\mathbb{Z}^d,\ f\in C^1_{b,\mathrm{cyl}}(\mathbb{R}^{\mathbb{Z}^d}),\ x\in\mathbb{R}^d) \tag{3.9}$$

hold. This representation follows from (3.8) and the definition of β_k, $k\in\mathbb{Z}^d$. Note that for any $\Lambda\subset\mathbb{Z}^d$, $|\Lambda|<\infty$ the renormalized Weyl operators $U^\Lambda_{k,\,\mathrm{ren}}$ and $V^\Lambda_{k,\,\mathrm{ren}}$ are constructed for the given μ_Λ by analogy with (3.8). Indeed, this is obvious for $U^\Lambda_{k,\,\mathrm{ren}}$

by virtue of the fact that U_k^Λ commutes with the operator of multiplication by ϕ_Λ; in the case of $V_{k,\,\mathrm{ren}}^\Lambda$, this follows from the equality

$$(V_{k,\mathrm{ren}}^\Lambda(t)f)(x) = \phi_\Lambda^{-1}(x)(V_k^\Lambda(t)f\phi_\Lambda)(x) = \phi_\Lambda^{-1}(x)(f\phi_\Lambda)(x+te_k)$$

$$= f(x+te_k)\frac{\phi_\Lambda(x+te_k)}{\phi_\Lambda(x)} = f(x+te_k)\sqrt{\frac{d\mu_\Lambda(x+te_k)}{d\mu_\Lambda(x)}}$$

$$(f\in C_b^1(\mathbb{R}^{\mathbb{Z}^d});\ k\in\mathbb{Z}^d;\ t\in\mathbb{R}^1).$$

Consequently, the renormalized representation of the CCR in the region Λ is a standard Weyl's representation of the form (3.8) associated with the measure μ_Λ. Therefore, it is natural to call the representation (3.8) given by the limiting measure μ a renormalized Weyl's representation of CCR corresponding to the given infinite system. A C^*-algebra of operators generated by $U_{\mu,k}(t)$ and $V_{\mu,k}(t)$ $(k\in\mathbb{Z}^d;\, t\in\mathbb{R}^1)$ is denoted by $\mathcal{A}_\mu$.

Let us consider an example in which the realization of the indicated scheme of constructing the dynamics is quite simple.

Example 3.1 (A system of noninteracting anharmonic oscillators). A formal Hamiltonian of this system has the form

$$L = \frac{1}{2}\sum_{k\in\mathbb{Z}^d}(p_k^2+\omega^2 q_k^2)+\sum_{k\in\mathbb{Z}^d}W(q_k), \tag{3.10}$$

where W is a potential defined so that there exists $\varepsilon>0$ such that $W\in L_{2+\varepsilon}(\mathbb{R}^1,\ \varepsilon^{-\omega s^2}ds)$ and that $e^{-W}\in L_p(\mathbb{R}^1,\varepsilon^{-\omega s^2}ds)$ for any $p\ge 1$. The operator

$$\frac{p_k^2}{2}+\frac{\omega^2 q_k^2}{2}+W(q_k)$$

has, in this case, the ground state $\phi_\omega(x_k)>0$ $(k\in\mathbb{Z}^d)$, and the ground state of the local Hamiltonian L_Λ has the form

$$\phi_\Lambda(x_\Lambda) = \prod_{k\in\Lambda}\phi_\omega(x_k)$$

(due to the fact that L_Λ is an operator with separated variable). The measures

$$d\mu_\Lambda(x_\Lambda) = \phi_\Lambda^2(x_\Lambda)dx_\Lambda$$

converge as $\Lambda\to\mathbb{Z}^d$ to the product measure

$$d\mu_{\omega}^{W}(x) = \underset{k \in \mathbb{Z}^d}{\times} \phi_{\omega}^{2}(x_k)dx_k \quad \text{on} \quad \mathcal{B}(\mathbb{R}^{\mathbb{Z}^d})$$

(in the above-mentioned sense). Therefore, $L_2(\mathbb{R}^{\mathbb{Z}^d}, \mu_{\omega}^{W})$ is a physical Hilbert space of the system given by the Hamiltonian (3.10). The renormalized Hamiltonian $L_{\text{ren}} = L_{\mu_{\omega}^{W}}$ is given by the equality

$$(L_{\mu_{\omega}^{W}} f)(x) = -\frac{1}{2} \sum_{k \in \mathbb{Z}^d} \frac{\partial^2 f(x)}{\partial x_k^2} - \sum_{k \in \mathbb{Z}^d} \frac{\nabla \phi_{\omega}(x_k)}{\phi_{\omega}(x_k)} \frac{\partial f(x)}{\partial x_k}$$

as an essentially selfadjoint operator on the region $C_{b,\,\text{cyl}}^{\infty}(\mathbb{R}^{\mathbb{Z}^d})$. This fact follows from the general theorem concerning the essential selfadjointness of the operators with infinitely many separated variables (Berezansky [18, Chapter 3, Theorem 2.7]).

Note that, for $\omega_1 \neq \omega_2$, the measures $\mu_{\omega_1}^{W}$ and $\mu_{\omega_2}^{W}$ are necessarily mutually orthogonal. This follows from the fact that $\phi_{\omega_1} \neq \phi_{\omega_2}$ on a set of positive measures and from the Kakutani theorem (Example 1.1 in Chapter 2). Therefore, the physical Hilbert spaces corresponding to different natural frequencies are essentially distinct, and the corresponding representations of CCR are unitary nonequivalent.

3.2. Gaussian Measures

Consider an important class of models describing the system of interacting oscillators on a lattice whose formal Hamiltonian has the form

$$L = \frac{1}{2} \sum_{k \in \mathbb{Z}^d} p_k^2 + \frac{1}{2} \sum_{j,k \in \mathbb{Z}^d} d_{kj} q_k q_j \tag{3.11}$$

(conditions imposed on the coefficients $d_{kj} \in \mathbb{R}^1$ $(j, k \in \mathbb{Z}^d)$ are given below; note that, unlike (3.4), the term of the form $\omega^2 q_k^2 (k \in \mathbb{Z}^d)$ is included in the general interaction potential). These Hamiltonians appear, in particular, in the theory of solid state in the study of the dynamics of a crystal lattice in harmonic approximation. In theoretical physics, these systems are often called harmonic. They were investigated from different points of view both in the classical and quantum cases in many papers (see the Bibliographical Notes). The main reason why these systems are interesting for us is that they lead to explicitly solvable models of dynamics. The renormalization procedure gives here an explicit form of limiting measures on $\mathbb{R}^{\mathbb{Z}^d}$ generating the physical Hilbert space of the system and the renormalized Hamiltonian. Moreover, these measures turn out to be Gaussian, and this is why the models under consideration are called Gaussian.

Let us describe the Gaussian models rigorously. Assume that a linear operator

$$D : l_2(\mathbb{Z}^d) \to l_2(\mathbb{Z}^d), \quad D \geq 0,$$

is given which is strictly positive in the sense that the equality $\langle Dh, h\rangle = 0$ for $h \in l_2(\mathbb{Z}^d)$ implies $h = 0$ (as already mentioned, here and below $\langle\cdot, \cdot\rangle$ denotes the scalar product in $H_0 = l_2(\mathbb{Z}^d)$). Denote by $d_{jk} = \langle De_j, e_k\rangle$ the matrix elements of D in the natural basis $(e_k)_{k\in\mathbb{Z}^d}$ of the space $l_2(\mathbb{Z}^d)$.

The principal restriction imposed on D, which is necessary for the renormalization, can be formulated as

$$\forall k \in \mathbb{Z}^d \quad e_k \in \mathfrak{D}(D^{-1/4}) \tag{3.12}$$

or, in other words, we demand that the inclusion $\mathbb{R}_0^{\mathbb{Z}^d} \subset \mathfrak{D}(D^{-1/4})$ holds.

Theorem 3.1. *Let*

$$\mathbb{R}_0^{\mathbb{Z}^d} \subset \mathfrak{D}(D^{-1/4}).$$

The renormalization procedure applied to the Hamiltonian (3.11) *gives a Gaussian measure* γ_S *on* $\mathbb{R}^{\mathbb{Z}^d}$ *given by the Fourier transform*

$$\tilde{\gamma}_S(y) = \int_{\mathbb{R}^{\mathbb{Z}^d}} e^{i\langle y,x\rangle} d\gamma_S(x) = e^{-1/4\langle D^{-1/2}y,y\rangle} \quad (y \in \mathbb{R}_0^{\mathbb{Z}^d}), \tag{3.13}$$

(i.e., a Gaussian measure with the correlation operator $S = D^{-1/2}$*) as a limiting measure. The renormalized Hamiltonian* L_{γ_S} *is defined by the differential expression*

$$(L_{\gamma_S} f)(x) = -\frac{1}{2} \sum_{k\in\mathbb{Z}^d} \frac{\partial^2 f(x)}{\partial x_k^2} + \langle D^{1/2}\nabla f(x), x\rangle \tag{3.14}$$

as an essentially selfadjoint operator in $C^\infty_{b,\,\mathrm{cyl}}(\mathbb{R}^{\mathbb{Z}^d})$.

Proof. Denote $R = D^{-1/2} > 0$ and $R \in \mathfrak{L}(l_2(\mathbb{Z}^d), l_2(\mathbb{Z}^d))$. Then the formal Hamiltonian (3.11) can be rewritten in the form

$$L = -\frac{1}{2} \sum_{k\in\mathbb{Z}^d} \frac{\partial^2}{\partial x_k^2} + \frac{1}{2}\langle Rx, Rx\rangle.$$

For $\Lambda \subset \mathbb{Z}^d$, $|\Lambda| < \infty$, we introduce a space

$$l_2(\Lambda) = \{x \in l_2(\mathbb{Z}^d) \mid \forall k \in \Lambda \quad x_k = 0\} \subset l_2(\mathbb{Z}^d),$$

isometric to $\mathbb{R}^\Lambda$. We identify the vectors in $\mathbb{R}^\Lambda$ and $l_2(\Lambda)$ and write the scalar product in $\mathbb{R}_\Lambda$ in terms of $l_2(\Lambda)$, i.e.,

$$(x_\Lambda, y_\Lambda)\mathbb{R}^\Lambda = \langle x_\Lambda, y_\Lambda \rangle.$$

for any $x_\Lambda, y_\Lambda \in \mathbb{R}^\Lambda$. Denote by P_Λ the projector onto $l_2(\Lambda)$ and put $R_\Lambda = P_\Lambda R \,\#\, l_2(\Lambda)$. The operator R_Λ is a nonnegative operator in $l_2(\Lambda)$. The local Hamiltonian L_Λ can be now written in the form

$$(L_\Lambda f)(x_\Lambda) = -\frac{1}{2} \sum_{k \in \Lambda} \frac{\partial^2 f(x_\Lambda)}{\partial x_k} + \frac{1}{2} \langle R_\Lambda x_\Lambda, R_\Lambda x_\Lambda \rangle f(x_\Lambda)$$

$$(x_\Lambda = (x_k)_{k \in \Lambda} \in \mathbb{R}^\Lambda;\ f \in C_b^2(\mathbb{R}^\Lambda)). \tag{3.15}$$

It is easy to see that the ground state of L_Λ is given by the function

$$\phi_\Lambda(x_\Lambda) = \frac{1}{N_\Lambda} e^{-1/2 \langle R_\Lambda x_\Lambda, x_\Lambda \rangle} \quad (x_\Lambda \in \mathbb{R}^\Lambda),$$

where N_Λ is a normalizing constant (equal to $(\pi^{|\Lambda|} \det R_\Lambda^{-1})^{1/4}$). To check this, one must diagonalize the operator R_Λ and employ either the fact that the ground state of the operator of harmonic oscillator $-1/2\left(d^2/ds^2\right) + 1/2\,\omega^2 t^2$ is given by the function $\pi^{-1/4} e^{-\omega s^2/2}$, or Example 3.5 in Chapter 6. Therefore, the Gaussian measure

$$d\mu_\Lambda(x_\Lambda) = d\gamma_{R_\Lambda^{-1}}(x_\Lambda) = \frac{1}{N_\Lambda^2} e^{-\langle R_\Lambda x_\Lambda, x_\Lambda \rangle} dx_\Lambda \tag{3.16}$$

with the Fourier transform

$$\tilde{\mu}_\Lambda(y_\Lambda) = \tilde{\gamma}_{R_\Lambda^{-1}}(y_\Lambda) = e^{-1/4 \langle R_\Lambda^{-1} y_\Lambda, y_\Lambda \rangle} \quad (y_\Lambda \in \mathbb{R}^\Lambda) \tag{3.17}$$

corresponds to the renormalization in the region Λ.

To establish the convergence $\mu_\Lambda \to \gamma_S$ as $\Lambda \to \mathbb{Z}^d$ in the required sense, it suffices to show that $\tilde{\mu}_\Lambda(y) \to \tilde{\gamma}_S(y)$ as $\Lambda \to \mathbb{Z}^d$ $(y \in \mathbb{R}_0^{\mathbb{Z}^d})$. The last convergence is equivalent to

$$\langle R_\Lambda^{-1}y, y\rangle \to \| R^{-1/2}y\|^2_{l_2(\mathbb{Z}^d)}, \quad \Lambda \to \mathbb{Z}^d \quad (y \in \mathbb{R}_0^{\mathbb{Z}^d}).$$

Note that the required assertion is not obvious because $R_\Lambda^{-1} \neq (R^{-1})_\Lambda$. To verify it, we use the following lemma.

Lemma 3.1. *Let* $y \in l_2(\Lambda)$, $|\Lambda| < \infty$. *Then*

(i) $\langle (R + \lambda\, 1)^{-1} y, y\rangle$ *increases towards* $\left\| R^{-1/2}\, y\right\|^2_{l_2(\mathbb{Z}^d)}$ *monotonically as* $\lambda \to 0$ *monotonically decreasing;*

(ii) $\langle (R + \lambda\, 1)^{-1} y, y\rangle$ *increases towards* $\langle R_\Lambda^{-1}y, y\rangle$ *monotonically as* $\lambda \to 0$ *monotonically decreasing;*

(iii) if $\Lambda \subset \Lambda_1 \subset \mathbb{Z}^d$, $|\Lambda_1| < \infty$, *then* $\langle R_\Lambda^{-1}y, y\rangle \leq \langle R_{\Lambda_1}^{-1}y, y\rangle$;

(iv) for all $\lambda > 0$, $\langle (R_\Lambda + \lambda\, 1)^{-1} y, y\rangle$ *increases towards* $\langle (R + \lambda\, 1)^{-1} y, y\rangle$ *monotonically as* Λ *expands monotonically towards* $\mathbb{Z}^d$.

Proof. Statements (i) and (ii) are the general facts in the theory of selfadjoint operators in Hilbert space and follow directly from the spectral theory. To prove statement (iii), we consider the operators

$$(R_{\Lambda_1} + \lambda P_{\Lambda_1\setminus\Lambda})^{-1} \quad (\lambda \geq 0)$$

in $l_2(\Lambda_1)$. Since the operator function

$$R_{\Lambda_1} + \lambda P_{\Lambda_1\setminus\Lambda} > 0$$

is monotonically increasing with λ, the scalar function

$$\varphi(\lambda) = \langle (R_{\Lambda_1} + \lambda P_{\Lambda_1\setminus\Lambda})^{-1} y, y\rangle \quad (\lambda \geq 0)$$

is monotonically decreasing. Statement (iii) now follows from the equality $\varphi(0) = \langle R_{\Lambda_1}^{-1} y, y\rangle$ and the relation

$$\lim_{\lambda\to\infty} \varphi(\lambda) = \langle R_\Lambda^{-1}y, y\rangle. \tag{3.18}$$

To establish (3.18), we introduce a Gaussian measure

$$\begin{aligned} d\mu_{\Lambda_1}^{(\lambda)}(x_{\Lambda_1}) &= N_\lambda^{-1}\exp\Big\{-\sum_{j,k\in\Lambda_1}\langle R_{\Lambda_1}e_j, e_k\rangle x_j x_k - \lambda\sum_{k\in\Lambda_1\setminus\Lambda} x_k^2\Big\}dx_{\Lambda_1} \\ &= N_\lambda^{-1}\exp\{-\langle (R_{\Lambda_1} + \lambda P_{\Lambda_1\setminus\Lambda})^{-1} x_{\Lambda_1}, x_{\Lambda_1}\rangle\}dx_{\Lambda_1} \end{aligned}$$

on $\mathbb{R}^{\Lambda_1}$ (here N_λ is a normalizing factor). By using the formula for the moments of the Gaussian measure (see (2.5) in Chapter 2), we obtain the equality

$$\varphi(\lambda) = 2 \int_{\mathbb{R}^{\Lambda_1}} \langle y, x_{\Lambda_1}\rangle^2 d\mu_{\Lambda_1}^{(\lambda)}(x_{\Lambda_1}) \quad (\lambda \geq 0). \tag{3.19}$$

For $\lambda \to \infty$, the measures $\mu_{\Lambda_1}^{(\lambda)}$ converge to the measure $\mu_\Lambda(x_\Lambda) \times \delta_0(x_{\Lambda_1 \setminus \Lambda})$ because the dispersion of the variables x_k $(k \in \Lambda_1 \setminus \Lambda)$ converge to zero. The limit transition as $\lambda \to \infty$ in (3.19) gives

$$2 \int_{\mathbb{R}^{\Lambda_1}} \langle x_{\Lambda_1}, y\rangle^2 d\mu_{\Lambda_1}^{(\lambda)}(x_{\Lambda_1}) = 2 \int_{\mathbb{R}^{\Lambda_1}} \langle x_\Lambda, y\rangle^2 1(x_{\Lambda_1 \setminus \Lambda}) d\mu_{\Lambda_1}^{(\lambda)}(x_{\Lambda_1})$$

$$\to 2 \int_{\mathbb{R}^{\Lambda_1}} \langle x_\Lambda, y\rangle^2 1(x_{\Lambda_1 \setminus \Lambda}) d\mu_\Lambda(x_\Lambda) d\delta_0(x_{\Lambda_1 \setminus \Lambda})$$

$$= 2 \int_{\mathbb{R}^{\Lambda}} \langle x_\Lambda, y\rangle^2 d\mu_\Lambda(x_\Lambda) = \langle R_\Lambda^{-1} y, y\rangle.$$

Let us prove Statement (iv). It follows from (iii) that

$$\psi_\Lambda(\lambda) = \langle (R_\Lambda + \lambda 1)^{-1} y, y\rangle \quad (\lambda > 0)$$

increases monotonically as Λ grows. Therefore, there exists $\lim_{\Lambda \to \mathbb{Z}^d} \psi_\Lambda(\lambda)$, and one should only prove that its value is equal to $\langle (R + \lambda\, 1)^{-1} y, y\rangle$.

The function

$$\psi_\Lambda(z) = \langle (R_\Lambda + z 1)^{-1} y, y\rangle$$

is analytical on the completion to the spectrum of the operator $-R_\Lambda$, i.e., it is automatically analytical in the region $\mathbb{C}^1 \setminus [-\| R \|, 0]$. Expanding $(R + \lambda\, 1)^{-1}$ in the von Neumann series for sufficiently large $|z|$, we obtain

$$\lim_{\Lambda \to \mathbb{Z}^d} \psi_\Lambda(z) = \langle (R + z 1)^{-1} y, y\rangle.$$

Application of the Vitali theorem gives

$$\lim_{\Lambda \to \mathbb{Z}^d} \psi_\Lambda(\lambda) = \langle (R + \lambda 1)^{-1} y, y \rangle$$

for all $\lambda > 0$.

■

We continue the proof of the theorem. By using statements (i) and (iv), we get

$$\langle (R_\Lambda + \lambda 1)^{-1} y, y \rangle \le \| R^{-1/2} y \|^2_{l_2(\mathbb{Z}^d)}, \quad (\lambda \ge 0;\ \Lambda \subset \mathbb{Z}^d,\ |\Lambda| < \infty).$$

Therefore, statement (ii) yields

$$\langle R_\Lambda^{-1} y, y \rangle \le \langle R^{-1/2} y, R^{-1/2} y \rangle \quad (\Lambda \subset \mathbb{Z}^d,\ |\Lambda| < \infty).$$

By virtue of (iii), this guarantees the existence of

$$\lim_{\Lambda \to \mathbb{Z}^d} \langle R_\Lambda^{-1} y, y \rangle \le \| R^{-1/2} y \|^2_{l_2(\mathbb{Z}^d)}.$$

On the other hand, according to Statement (i), for any $\varepsilon > 0$, one can choose $\lambda > 0$ such that

$$\langle (R + \lambda 1)^{-1} y, y \rangle \ge \| R^{-1/2} y \|^2_{l_2(\mathbb{Z}^d)} - \varepsilon / 2$$

and, according to Statement (iv), we can find $\Lambda \subset \mathbb{Z}^d$ such that

$$\langle (R_\Lambda + \lambda 1)^{-1} y, y \rangle \ge \| R^{-1/2} y \|^2_{l_2(\mathbb{Z}^d)} - \varepsilon .$$

Hence,

$$\langle R_\Lambda^{-1} y, y \rangle \ge \langle (R_\Lambda + \lambda 1)^{-1} y, y \rangle \ge \| R^{-1/2} y \|^2_{l_2(\mathbb{Z}^d)} - \varepsilon .$$

Since $\varepsilon > 0$ is arbitrary, we can conclude that

$$\lim_{\Lambda \to \mathbb{Z}^d} \langle R_\Lambda^{-1} y, y \rangle = \| R^{-1/2} y \|_{l_2(\mathbb{Z}^d)}.$$

We have shown that the physical Hilbert space of an infinite system with the Hamiltonian (3.11) is given by the measure γ_S. Let us prove that the corresponding Dirichlet form which generates the renormalized Hamiltonian is closable. For this purpose, we now establish the $\mathbb{R}_0^{\mathbb{Z}^d}$-quasi-invariance of the measure γ_S and compute its logarithmic derivative. According to Theorem 2.3 in Chapter 2 concerning a set of admissible shifts of a Gaussian measure, the measures γ_S and $\gamma_S(\cdot + h)$ are equivalent if and only if $h \in \mathfrak{D}(R^{-1/2})$ and, moreover,

$$\frac{d\gamma_S(x+h)}{d\gamma_S(x)} = \exp[-2\langle Rh, x\rangle - \langle Rh, x\rangle] \tag{3.20}$$

$$(x \in \mathbb{R}^{\mathbb{Z}^d};\ h \in \mathfrak{D}(R^{-1/2})).$$

According to the condition of the theorem, $e_k \in \mathfrak{D}(R^{-1/2})$ for any $k \in \mathbb{Z}^d$ and, thus, the Radon–Nikodym derivative

$$\frac{d\gamma_S(x+te_k)}{d\gamma_S(x)} = \exp\{-2t\langle Re_k, x\rangle - t^2\langle Re_k, e_k\rangle\} \tag{3.21}$$

is defined for all $t \in \mathbb{R}^1$. According to the definition of the logarithmic derivative β_{γ_S} of the measure μ_S, we have

$$\langle \beta_{\gamma_S}(x), e_k\rangle = \frac{d}{dt}\left(\frac{d\gamma_S(x+te_k)}{d\gamma_S(x)}\right)\Big|_{t=0} = -2\langle Re_k, x\rangle$$

$$(x \in \mathbb{R}^{\mathbb{Z}^d};\ k \in \mathbb{Z}^d)$$

and, consequently, the Dirichlet form

$$d_{\gamma_S}(f, g) = \frac{1}{2}\int_{\mathbb{R}^{\mathbb{Z}^d}} \langle \nabla f(x), \overline{\nabla g(x)}\rangle d_{\gamma_S}(x) \quad (f, g \in C^2_{b,\mathrm{cyl}}(\mathbb{R}^{\mathbb{Z}^d}))$$

is associated with the differential expression

$$-\frac{1}{2}\sum_{k\in\mathbb{Z}^d}\frac{\partial^2 f(x)}{\partial x_k^2} + \langle R\nabla f(x), x\rangle \quad (f \in C^2_{b,\mathrm{cyl}}(\mathbb{R}^{\mathbb{Z}^d})) \tag{3.22}$$

(see Subsection 3.1 of Chapter 6). To prove the essential selfadjointness of the operator defined by (3.22), we apply Theorem 3.4 in Chapter 6. According to Corollary 2 of Theorem 1.8 in Chapter 2, the measure γ_S can be regarded as a measure on the Hilbert space H_- which is the completion of $l_2(\mathbb{Z}^d)$ with respect to the scalar product

$$(x, y)_{H_-} = \langle JRx, JRy\rangle \quad (x, y \in l_2(\mathbb{Z}^d)), \tag{3.23}$$

where J is a Hilbert–Schmidt operator in $l_2(\mathbb{Z}^d)$. Here

$$H_- \ni x \mapsto \beta_{\gamma_S}(x) = -2Rx \in H_-$$

is a continuous linear mapping of H_- into h_- satisfying all conditions of Theorem 3.4 in Chapter 6. Therefore, (3.22) defines an essentially selfadjoint operator L_{γ_S} in

$$L_2(H_-, \gamma_S) = L_2(\mathbb{R}^{\mathbb{Z}^d}, \gamma_S)$$

on the region $C_b^2(H_-)$. The transition to $C^2_{b,\,\mathrm{cyl}}(\mathbb{R}^{\mathbb{Z}^d})$ is realized by use of the standard cylindrical approximation.

■

3.3. Time Evolution in Harmonic Systems

Let us introduce Weyl's representation of CCR generated by the measure γ_S according to formulas (3.8). By virtue of (3.9), each vertex $k \in \mathbb{Z}^d$ is associated with the operators (generated from this representation)

$$(\varphi_k f)(x) = \langle e_k, x\rangle f(x),$$

$$(\pi_k f)(x) = \frac{1}{t}\left[\frac{\partial f(x)}{\partial x_k} - \langle Re_k, x\rangle f(x)\right]$$

$$(f \in C^2_{b,\mathrm{cyl}}(\mathbb{R}^{\mathbb{Z}^d});\ x \in \mathbb{R}^{\mathbb{Z}^d};\ k \in \mathbb{Z}^d), \tag{3.24}$$

where φ_k is an operator of the coordinate and π_k is an operator of the momentum of a particle situated at the vertex $k \in \mathbb{Z}^d$. The fact that these operators are correctly defined follows from the relations

$$\int_{\mathbb{R}^{\mathbb{Z}^d}} \langle x, e_k\rangle^2 d\gamma_S(x) = \frac{1}{2}\| R^{-1/2} e_k \|^2_{l_2(\mathbb{Z}^d)},$$

$$\int_{\mathbb{R}^{\mathbb{Z}^d}} \langle x, Re_k\rangle^2 d\gamma_S(x) = \frac{1}{2}\| R^{1/2} e_k \|^2_{l_2(\mathbb{Z}^d)}, \tag{3.25}$$

(see (2.5) in Chapter 2). We associate a vector $h \in \mathbb{R}_0^{\mathbb{Z}^d}$ with the operators $\varphi(h)$ and $\pi(h)$ on $C^\infty_{b,\,\mathrm{cyl}}(\mathbb{R}^{\mathbb{Z}^d})$ by setting

$$\varphi(h) = \sum_{k \in \mathbb{Z}^d} \varphi_k h_k, \quad \pi(h) = \sum_{k \in \mathbb{Z}^d} \pi_k h_k.$$

It follows from (3.25) that

$$\| \varphi(h) f \|_{L_2(\mathbb{R}^{\mathbb{Z}^d}, \gamma_S)} \le c'_f \| R^{-1/2} h \|_{l_2(\mathbb{Z}^d)},$$

$$\| \pi(h) f \|_{L_2(\mathbb{R}^{\mathbb{Z}^d}, \gamma_S)} \le c''_f [\| h \|_{l_2(\mathbb{Z}^d)} + \| R^{1/2} h \|_{l_2(\mathbb{Z}^d)}],$$

for $f \in C^\infty_{b,\,\mathrm{cyl}}(\mathbb{R}^{\mathbb{Z}^d})$. The latter inequality enables us to extend the definition of the operator $\varphi(t)$ to the case $h \in \mathfrak{D}(R^{-1/2})$, and the definition of $\pi(t)$ can be extended to all $h \in l_2(\mathbb{Z}^d)$. For $h \in \mathbb{R}_0^{\mathbb{Z}^d}$, we consider the time evolution of the operators π and φ defined by the renormalized Hamiltonian L_{γ_S} of the system with the help of the Heisenberg operators

$$\pi_t(h) = e^{itL_{\gamma_S}} \pi(h) e^{-itL_{\gamma_S}}; \quad \varphi_t(h) = e^{itL_{\gamma_S}} \varphi(h) e^{-itL_{\gamma_S}}$$

$$(t \in \mathbb{R}^1; \ h \in \mathfrak{D}(R^{-1/2})).$$

Theorem 3.2. *In a Gaussian model, the time evolution of the operators* π *and* φ *is given by the linear canonical Bogolyubov transformations*

$$\pi_t(h) = -\varphi(R \sin(R\,t)h) + \pi(\cos(R\,t)h),$$

$$\varphi_t(h) = \varphi(\cos(R\,t)h) + \pi(R^{-1} \sin(R\,t)h) \tag{3.26}$$

$$(h \in \mathbb{R}_0^{\mathbb{Z}^d}, \ t \in \mathbb{R}^1).$$

Proof. For $h \in \mathbb{R}_0^{\mathbb{Z}^d}$, we have

$$\cos(Rt)h,\ R^{-1} \sin(Rt)h,\ R \sin(Rt)h \in \mathfrak{D}(R^{-1/2}),$$

because the operator $\cos R\,t$ and $R^{\pm 1} \sin R\,t$ are bounded and commute with $R^{-1/2}$. By virtue of what has been said above, this ensures the definiteness of the operators on the right-hand side of (3.26) on $C^2_{b,\,\mathrm{cyl}}(\mathbb{R}^{\mathbb{Z}^d})$. For our further discussion, it is convenient to introduce the following operator

$$a(h) = \varphi(R^{1/2} h) + i\pi(R^{-1/2} h),$$

$$a^*(h) = \varphi(R^{1/2}h) - i\pi(R^{-1/2}h) \quad (h \in \mathbb{R}_0^{\mathbb{Z}^d}),$$

which are analogous to the operators of creation and annihilation in the quantum field theory. For $R = 1$, i.e., in the case of the canonical Gaussian measure γ_1, these operators were, in fact, used in Subsection 3.8 of Chapter 3. Let us find the commutators $[L_{\gamma_S}, \varphi(h)]$ and $[L_{\gamma_S}, \pi(h)]$ on a set of functions $f \in C^2_{b,\,\mathrm{cyl}}(\mathbb{R}^{\mathbb{Z}^d})$. According to (3.14), we have

$$(L_{\gamma_S}\varphi(h)f)(x) = -\frac{1}{2}[\mathrm{Tr}_{l_2}((\langle h, x\rangle f(x))'' - 2\langle(\langle h, x\rangle f(x))', Rx\rangle]$$

$$= -\frac{1}{2}[\langle h, x\rangle \mathrm{Tr}_{l_2} f''(x) + 2\langle \nabla f(x), h\rangle - 2\langle h, x\rangle\langle R\nabla f(x), x\rangle - 2f(x)\langle Rh, x\rangle],$$

$$(\varphi(h)L_{\gamma_S}f)(x) = -\frac{1}{2}[\langle h, x\rangle \mathrm{Tr}_{l_2} f''(x) - 2\langle h, x\rangle\langle R\nabla f(x), x\rangle],$$

consequently,

$$([L_{\gamma_S}, \varphi(h)]f)(x) = -\frac{1}{2}[2\langle \nabla f(x), h\rangle - 2\langle Rh, x\rangle f(x)]$$

$$= -(\langle \nabla f(x), h\rangle - \langle Rh, x\rangle f(x)) = -i(\pi(h)f)(x),$$

i.e.,

$$i\,[L_{\gamma_S}, \varphi(h)] = \pi(h). \tag{3.27}$$

Similarly, we establish the equality

$$i\,[L_{\gamma_S}, \pi(h)] = -\varphi(R^2h). \tag{3.28}$$

Taking into account (3.27) and (3.28), we obtain after evident transformations

$$[L_{\gamma_S}, a(h)] = -a(Rh), \quad [L_{\gamma_S}, a^*(h)] = -a^*(Rh). \tag{3.29}$$

It follows from (3.29) that

$$a_t^*(h) = e^{itL_{\gamma_S}} a^*(h) e^{-itL_{\gamma_S}} = a^*(e^{itR}h), \tag{3.30}$$

$$a_t(h) = e^{itL_{\gamma_S}} a(h) e^{-itL_{\gamma_S}} = a(e^{-itR}h) \quad (h \in \mathbb{R}_0^{\mathbb{Z}^d},\ t \in \mathbb{R}^1),$$

therefore,

$$\varphi_t(R^{1/2}h) = \frac{1}{2}(a_t(h) + a_t^*(h)) = a\left(\frac{e^{-itR}h}{2}\right) + a^*\left(\frac{e^{itR}h}{2}\right),$$

$$\varphi_t(h) = a\left(\frac{1}{2}R^{-1/2}e^{-itR}h\right) + a^*\left(\frac{1}{2}R^{-1/2}e^{itR}h\right)$$

$$= \varphi\left(\frac{1}{2}e^{-itR}h\right) + i\pi\left(\frac{1}{2}R^{-1}e^{-itR}h\right) + \varphi\left(\frac{1}{2}e^{itR}h\right)$$

$$- i\pi\left(\frac{1}{2}R^{-1}e^{itR}h\right) = \varphi(\cos(Rt)h) + \pi(R^{-1}\sin(Rt)h).$$

Similarly, we can derive the following equality from (3.30):

$$\pi_t(h) = \pi(\cos(Rt)h) + \varphi(R\sin(Rt)h).$$

Remark 1. The equalities (3.30) are analogous to the well-known relations which determine the evolution of the creation and annihilation operators in the case of the free boson field. And this analogy is not accidental. Let us introduce the Fock space $\mathcal{F}(H_S)$, where H_S is the completion of $\mathbb{R}_0^{\mathbb{Z}^d} \subset l_2(\mathbb{Z}^d) = H_0$ with respect to the scalar product

$$(x, y)_{H_S} = \langle x, Sy\rangle \quad (x, y \in \mathbb{R}_0^{\mathbb{Z}^d},\ S = D^{-1/2} = R^{-1})$$

and pass, as in Subsection 2.3 of Chapter 2, to a functional realization of $\mathcal{F}(H_S)$ as a space $L_2(\mathbb{R}^{\mathbb{Z}^d}, \gamma_S)$ by use of the standard duality of $\Phi = \mathbb{R}_0^{\mathbb{Z}^d}$ and $\Phi' = \mathbb{R}^{\mathbb{Z}^d}$ established by $\langle\cdot,\cdot\rangle$. It is easy to see that under this transition, the operators of creation and annihilation in the Fock space (defined in Subsection 3.8 of Chapter 3) turn into the operators a and a^* introduced above, and *the second quantization operator* $d\Gamma(R)$ *in* $\mathcal{F}(H_S)$ *is associated with the operator* L_{γ_S} (see Subsection 1.1 in Chapter 6). To establish the last assertion, one should employ the representation (1.31) in Chapter 6 for the image of the second quantization operator $d\Gamma(A)$ under the Segal isomorphism; namely,

$$(L_A u)(x) = -\frac{1}{2}\mathrm{Tr}_{\mathcal{H}_0}(JAJ^*u''(x)) + (A\nabla u(x), x)_{\mathcal{H}_0} \quad (u \in \mathcal{P}(\Phi')).$$

In the notations of Section 1 in Chapter 6, we now have $H_0 = H_S$, $\mathcal{H}_0 = l_1(\mathbb{Z}^d)$, $\Phi = \mathbb{R}_0^{\mathbb{Z}^d}$, $\Phi' = \mathbb{R}^{\mathbb{Z}^d}$, $A = R = S^{-1}$, and $J = R^{-1/2}$ and, therefore,

$$(L_R u)(x) = -\frac{1}{2}\mathrm{Tr}_{l_2} u''(x) + \langle R\nabla u(x), x\rangle \quad (u \in \mathcal{P}_{\mathrm{cyl}}(\mathbb{R}^{\mathbb{Z}^d})),$$

but according to (3.32), this expression coincides with $L_{\gamma_S} u$.

Remark 2. We have established above that the renormalized Hamiltonian of a harmonic system is a second quantization operator in the Schrödinger representation. This fact allows us to apply the results concerning potential perturbations of the second quantization operators (which were obtained in Sections 2 and 3 of Chapter 6) to the study of systems with anharmonicity. We shall encounter one of the aspects of this application in Section 4.

Remark 3. The transition from $L_2(\mathbb{R}^{\mathbb{Z}^d}, \gamma_S)$ and the Hamiltonian L_{γ_S} to the Fock space $\mathfrak{F}(H_S)$ and the second quantization operator $d\Gamma(S^{-1/2})$ given on it is called a transition to the quasiparticle description of a harmonic system. The space H_S is interpreted as a Hilbert space of states of a single quasiparticle, and the operator $S^{-1/2} = R = D^{1/2}$ on H_S is its operator of energy. The interpretation described is very popular in the physics literature, where quasiparticles obtained in this way are called phonons.

3.4. Transition Invariant Harmonic Systems

Consider the special case of a Gaussian model in which the formal Hamiltonian (3.11) is invariant with respect to the group of shifts of the lattice $\mathbb{Z}^d$. For example, this situation is realized when we investigate ideal crystals in harmonic approximation, where the forces of interaction between atoms of different cells depend only on the relative arrangement of elementary cells. In the translation invariant case, the verification of condition (3.12), which guarantees the convergence of the renormalization procedure, seems to be especially simple.

Assume that the operator D which gives the interaction between oscillators situated at different vertices commutes with a group of shifts of the lattice T_k, $k \in \mathbb{Z}^d$, acting on the basis vectors $e_k \in l_2(\mathbb{Z}^d)$, $e_k = (\delta_{jk})_{j\in\mathbb{Z}^d}$ $(k \in \mathbb{Z}^d)$ as follows

$$T_k e_j = e_{j+k} \quad (j, k \in \mathbb{Z}^d).$$

Then the matrix elements of the operator D have the form

$$d_{kj} = \langle De_j, e_k\rangle = a(k-j),$$

where $a(\cdot)\colon \mathbb{Z}^d \to \mathbb{R}^1$ and $a(-k) = a(k)$ for any $k \in \mathbb{Z}^d$.

Note that it follows from the condition of the boundedness of the operator D that

$$\sum_{j\in\mathbb{Z}^d} d_{kj}^2 = \sum_{j\in\mathbb{Z}^d} a^2(k-j) = \sum_{j\in\mathbb{Z}^d} a^2(j) < \infty. \tag{3.31}$$

for any $k\in\mathbb{Z}^d$.

Consider a unitary isomorphism F between the complexification $l_{2,c}(\mathbb{Z}^d)$ and $L_2([-\pi,\pi)^d)$ under which a vector $\xi = (\xi_k)_{k\in\mathbb{Z}^d} \in l_{2,c}(\mathbb{Z}^d)$ is associated with the function

$$\tilde{\xi}(p) = \sum_{k\in\mathbb{Z}^d} \xi_k \frac{e^{ikp}}{(2\pi)^{d/2}} = (F\xi)(p),$$

$$F: l_{2,c}(\mathbb{Z}^d) \to L_2([-\pi,\pi)^d)$$

$$\left(kp = \sum_{\nu=1}^{d} k_\nu p_\nu;\ \ p = (p_1,\ldots,p_d)\in[-\pi,\pi)^d;\ \ k = (k_1,\ldots,k_d)\in\mathbb{Z}^d \right).$$

Conversely, a function $\tilde{\xi}\in L_2([-\pi,\pi)^d)$ is associated with the vector

$$F^{-1}\tilde{\xi} = (\xi_k)_{k\in\mathbb{Z}^d} : \xi_k = (2\pi)^{-d/2} \int_{[-\pi,\pi)^d} \tilde{\xi}(p) e^{-ikp}\,dp.$$

The operator D in $l_{2,c}(\mathbb{Z}^d)$ is given by the equality

$$\langle D\xi, e_k\rangle = \sum_{j\in\mathbb{Z}^d} a(k-j)\xi_j \quad (\xi\in l_{2,c}(\mathbb{Z}^d),\ k\in\mathbb{Z}^d)$$

and, hence,

$$F(D\xi)(p) = \tilde{a}(p)\tilde{\xi}(p), \quad \tilde{a}(p) = \sum_{k\in\mathbb{Z}^d} e^{ikp}a(k) \quad (p\in[-\pi,\pi)^d). \tag{3.32}$$

The function $\tilde{a}(\cdot)$ belongs to $L_2([-\pi,\pi)^d)$ by virtue of (3.31), it is even, i.e., $\tilde{a}(p) = \tilde{a}(-p)$, $p\in[-\pi,\pi)^d$, and nonnegative due to the nonnegativity of the operator D. The condition $\langle D\xi,\xi\rangle > 0$, $\xi\in l_2(\mathbb{Z}^d)$, $\xi\neq 0$, implies the inequality $\tilde{a}(\cdot) > 0$ almost everywhere with respect to the Lebesgue measure on $[-\pi,\pi)$. In terms of the function $\tilde{a}$, condition (3.12) can be rewritten in the form

$$e_k \in \mathrm{D}(D^{-1/4}) \ \ (k\in\mathbb{Z}^d) \Leftrightarrow \int_{[-\pi,\pi)^d} \frac{dp}{\sqrt{\tilde{a}(p)}} < \infty. \tag{3.33}$$

Note that in many of the cases which are interesting for application, the function $\tilde{a}(\cdot)$ is continuous and $\tilde{a}(0) = 0$ (see Example 3.2). Since $\tilde{a}(\cdot)$ is an even function, under this condition we have $\tilde{a}(p) \le c\,|p|^2$ as $p \to 0$ ($|p| = \|p\|_{\mathbb{R}^d}$) and, thus, (3.33) may hold only for $d \ge 2$.

Consider an important example of a Gaussian translation invariant model.

Example 3.2. Assume that the formal Hamiltonian of the system has the form

$$L = \frac{1}{2}\sum_{k\in\mathbb{Z}^d} p_k^2 + \frac{1}{2}\sum_{k,k'\in\mathbb{Z}^d} (q_k - q_{k'})^2, \tag{3.34}$$

where summation is carried out over all pairs of nearest-neighbors on the lattice, i.e., $k, k' \in \mathbb{Z}^d : |k-k'| = 1$ ($|k| = \|k\|_{\mathbb{R}^d}$). Let us rewrite the potential in (3.34) in the form

$$\sum_{|k-k'|=1} (q_k - q_{k'})^2 = \sum_{k\in\mathbb{Z}^d} a(0)\, q_k^2 + \sum_{|k-k'|=1} a(1) q_k q_{k'}.$$

In this case, $a(0) = 2d$ and $a(k) = -1$ for $|k| = 1$, and $a(k) = 0$ if $|k| > 1$ ($k \in \mathbb{Z}^d$), i.e.,

$$\tilde{a}(p) = 2d - \sum_{\substack{k\in\mathbb{Z}^d\\ |k|=1}} e^{ikp} = 2 \sum_{\substack{k\in\mathbb{Z}^d;\, k_\nu \ge 0\, (\nu=1,\ldots,d)\\ |k|=1}} (1-\cos(kp))$$

$$= 2\sum_{\nu=1}^{d} (1-\cos p_\nu) \sim \sum_{\nu=1}^{d} p_\nu^2 = |p|^2, \quad p \to 0 \quad (p \in [-\pi,\pi)^d).$$

Therefore, the renormalized Hamiltonian which corresponds to (3.34) can be constructed only for the dimensionalities $d \ge 2$, since

$$\int_{[-\pi,\pi)^d} \frac{dp}{|p|} = c_d \int_0^1 t^{d-2}dt < \infty \quad \text{for } d \ge 2.$$

By analogy, we can investigate Hamiltonians of the form

$$\frac{1}{2}\sum_{k\in\mathbb{Z}^d} p_k^2 + \frac{1}{2}\sum_{j,k\in\mathbb{Z}^d} \lambda(j-k)(q_j - q_k)^2, \tag{3.35}$$

where $\lambda(k) > 0$, $k \in \mathbb{Z}^d$. The sequence $(\lambda(k))_{k\in\mathbb{Z}^d}$ must satisfy the condition

$$\sum_{k\in\mathbb{Z}^d} \lambda(k) < \infty. \tag{3.36}$$

For example, in the case where $\lambda(k) = |k|^{-d}$ ($k \in \mathbb{Z}^d, k \ne 0$), $\lambda(0) > 0$, one should demand that $q > d$. An argument similar to that presented above shows that, in this case,

$$\tilde{a}(p) \sim \sum_{k \in \mathbb{Z}^d} \lambda(k) \frac{(kp)^2}{2} \sim c|p|^2, \quad p \to 0 \quad (p \in [-\pi, \pi)^d),$$

Consequently, for $q \geq 2$, we can construct the renormalized dynamics of the model corresponding to the formal Hamiltonian (3.35) provided that (3.36) holds.

3.5. Some Remarks Concerning the Construction of Time Evolution in Models with Anharmonicity

If the formal Hamiltonian of a quantum lattice system contains anharmonic terms, then there is no explicit expression for the ground state of finite subsystems in terms of the variables which appear in the Hamiltonian. This makes the construction of the operator realization of this Hamiltonian much more difficult, because it is necessary to employ the renormalization scheme in terms of the functional integrals described in Subsection 1.2. On order to realize this scheme, it is necessary to investigate the convergence of probability measures on the space of trajectories which take values on an infinite-dimensional linear space. However, this problem may be quite complicated, and its solution may be based on the use of technical methods specific for each class of models under consideration.

Let us clarify the situation which takes place in this case with an example of the system with the formal Hamiltonian

$$L = \frac{1}{2} \sum_{k \in \mathbb{Z}^d} p_k^2 + \frac{1}{2} \sum_{k,j \in \mathbb{Z}^d} d_{kj} q_k q_j + \sum_{k,j \in \mathbb{Z}^d} V_{kj}(q_k, q_j). \tag{3.37}$$

As far as the harmonic part of the Hamiltonian L is concerned, we assume that the conditions in Subsection 3.2 which guarantee the existence of the dynamics of a Gaussian model with dynamic matrix $D = (d_{kj})_{k,j} \in \mathbb{Z}^d$, are satisfied. As above, we denote by γ_S ($S = D^{-1/2}$) a Gaussian measure on $\mathbb{R}^{\mathbb{Z}^d}$ corresponding to this system; by L_{γ_S} we denote its renormalized Hamiltonian acting in the space of states $L_2(\mathbb{R}^{\mathbb{Z}^d}, \gamma_S)$. We have no intention of clarifying the conditions imposed on the potentials V_{kj}; it proves to be reasonable to do this for the concrete models under consideration. By using the indicated representation of the harmonic part of L, we proceed to the problem of constructing the operator realization of the Hamiltonian

$$L_{\gamma_S} + \sum_{k,j \in \mathbb{Z}^d} V_{kj}. \tag{3.38}$$

Acting in accordance with the general scheme in Subsection 1.1, we introduce a family of potentials

$$V_\Lambda = \sum_{k,j\in\Lambda} V_{kj} \quad (\Lambda \subset \mathbb{Z}^d,\ |\Lambda|<\infty).$$

If the potentials V_{kj} possess certain $L_p(\mathbb{R}^{\mathbb{Z}^d}, \gamma_S)$-properties, then the operator $L_{\gamma_S} + V_\Lambda$ is essentially selfadjoint and has the ground state $\phi_{V_\Lambda} > 0$ γ_S-almost everywhere (see Section 2 in Chapter 6). Hence, we can introduce a vacuum measure $\mu_{V_\Lambda} = \phi^2_{V_\Lambda}\gamma_S$ and the renormalized Hamiltonian

$$L_{\Lambda,\mathrm{ren}} = \phi^{-1}_{V_\Lambda}(L_{\gamma_S} + V_\Lambda - E_\Lambda)\phi_\Lambda = L\mu^{V_\Lambda}$$

($E_\Lambda = \inf s\,(L_{\gamma_S} + V_\Lambda)$, the equality is on $C^2_{b,\mathrm{cyl}}(\mathbb{R}^{\mathbb{Z}^d})$.

Denote by $\Omega = \mathbb{R}^{\mathbb{R}\times\mathbb{Z}^d} \ni \omega(\cdot)$ a set of trajectories

$$\mathbb{R}^1 \ni t \mapsto \omega(t) = (\omega_k(t))_{k\in\mathbb{Z}^d} \in \mathbb{R}^{\mathbb{Z}^d}$$

and by ν_S a Gaussian measure on Ω corresponding to the stationary Markov process generated by the transient semigroup $e^{-tL\gamma_S}$, $t\geq 0$ and the initial distribution γ_S. In the terminology of Section 1 in Chapter 6, the measure ν_S corresponds to a weak functional integral associated with L_{γ_S}. Let $\nu_S^{V_\Lambda}$ be an analogous measure corresponding to the operator $L_{\Lambda,\,\mathrm{ren}}$. Theorem 1.1 states that

$$d\nu_S^{V_\Lambda}(\omega(\cdot)) = \lim_{T\to+\infty} d\nu_S^{V_\Lambda,T}(\omega(\cdot)),$$

$$d\nu_S^{V_\Lambda,T}(\omega(\cdot)) = \frac{1}{N_{\Lambda,T}}\exp\Big[-\sum_{k,j\in\Lambda}\int_{-T}^{T} V_{kj}(\omega_k(t),\omega_j(t))dt\Big]d\nu_S(\omega(\cdot)) \tag{3.39}$$

in the sense of the convergence of finite-dimensional distributions. Formally, we have

$$d\nu_S^{V_\Lambda}(\omega(\cdot)) = \frac{1}{N_\Lambda}\exp\Big[-\sum_{k,j\in\Lambda}\int_{\mathbb{R}^1} V_{kj}(\omega_k(t),\omega_j(t))dt\Big]d\nu_S(\omega(\cdot)). \tag{3.40}$$

The equality (3.40) shows that the problem of investigating the limit of a family of measures $\nu_S^{V_\Lambda}$ ($\Lambda\subset\mathbb{Z}^d$, $|\Lambda|<\infty$) as $\Lambda\to\mathbb{Z}^d$ reduces to the problem of the existence of the so-called Gibbsian reconstruction of the measure ν_S with the help of the interaction determined by the potentials V_{kj}. The last fact illustrates once again the connection between the quantum statistical physics in terms of functional integrals and the statistical physics of classical systems. However, in contrast to the well-known situation studied in many papers (see Malyshev and Minlos [1] and the Bibliographical Notes in this book), we encounter a serious difficulty connected with the fact that the space of values of "spin" variables is infinite-dimensional: In the case considered here, we have $\omega_k(\cdot) \in C(\mathbb{R}^1)$ $(k \in \mathbb{Z}^d)$. This circumstance makes the application of the technique of cluster expansions (a powerful tool for the investigation of Gibbsian random fields) more difficult or, at least, makes it necessary to significantly modify it.

Here we do not dwell upon the concrete examples of quantum lattice systems with anharmonicity in which it is possible to check the convergence of measures (3.40) as $\Lambda \to \mathbb{Z}^d$ which means, according to the constructions in Subsection 1.2, that we can construct a renormalized Hamiltonian. Each example of this sort requires rather cumbersome investigation. The references to the papers devoted to the construction of renormalized Hamiltonians in the models of quantum physics of systems with infinitely many particles can be found in the Bibliographical Notes.

§4. Investigation of the Spectral Properties of Systems with Infinitely Many Particles by the Methods of Scattering Theory

The nonstationary scattering theory enables us to obtain information about the spectrum of the Hamiltonian of a system with interaction by comparing its dynamics with the dynamics of a certain "free" system. However, this comparison may have a different meaning, depending on the class of models under consideration, and the situation may be quite different from the standard situation in the formal scattering theory. This is true, in particular, for the situation in constructive field theory, where even under the perturbation of a free Hamiltonian L_{m_0} by a smoothed polynomial interaction V (admitting the definition of $L_{m_0} + V$ as an operator in the orthogonal space), the limits

$$\lim_{t \to \pm\infty} e^{-it(L_{m_0} + V)} e^{itL_{m_0}}, \tag{4.1}$$

(in the strong sense) do not exist, i.e., the wave operators cannot be defined by the ordinary relations of nonstationary scattering theory. The last fact makes it necessary to essentially modify the definition of the wave operators and the scattering operator for models in field theory. This modification means that we use one or another form of dressing operators G for which it is possible to prove the existence of the limits

$$\lim_{t \to \pm\infty} e^{-it(L_{m_0} + V)} G e^{itL_{m_0}} . \tag{4.2}$$

Another method of overcoming this difficulty is connected with the formulation of the scattering problem on the language of C^*-algebras. The spectral properties of the Hamiltonians of the quantum field theory with cutoff were studied using this method. However, this approach proves to be particularly effective in the case of Fermi systems (see the Bibliographical Notes).

In this section, we consider the statement of the scattering problem in the case of perturbations of the Dirichlet operators. We study the potential perturbations of harmonic systems in more detail. The facts from the abstract scattering theory employed in this section can be found, for example, in the book by Reed and Simon [3, Chapter 1, Section 3].

4.1. Scattering Problem for Dirichlet Operators.

We start from the formal Hamiltonian of a harmonic system and assume that it is translation invariant with respect to shifts along the lattice (see Subsection 3.4)

$$L = \frac{1}{2} \sum_{k \in \mathbb{Z}^d} p_k^2 + \frac{1}{2} \sum_{j,\, k \in \mathbb{Z}^d} a(k-j) q_k q_j , \tag{4.3}$$

where $D > 0$ is a bounded operator in $l_2(\mathbb{Z}^d)$, and

$$a(k-j) = d_{kj} = \langle De_k, e_j \rangle (\langle \cdot, \cdot \rangle = (\cdot, \cdot)_{l_2(\mathbb{Z}^d)}, \quad k, j \in \mathbb{Z}^d),$$

are its matrix elements in the natural basis $(e_k)_{k \in \mathbb{Z}^d}$ of the space $l_2(\mathbb{Z}^d)$. In this section, we suppose that $D \geq m^2 1$, $m > 0$, and that the function $\tilde{a}(p)$, $p \in [-\pi, \pi)^d$ connected with D by formula (3.31) satisfies the conditions of sufficient smoothness and assertions that $\nabla \tilde{a}(p) \neq 0$ and let $\det \tilde{a}''(p) \neq 0$ for almost all $p \in [-\pi, \pi)^d$ with respect to the Lebesgue measure (these assertions are necessary in what follows to guarantee the applicability of the stationary-phase method). In the estimates presented below, we choose the constant m^2 more accurately, i.e., $m^2 = \min_{[-\pi,\, \pi)^d} \tilde{a}(p)$.

As in Subsection 3.2, we introduce an operator realization of the Hamiltonian (4.3) in the ground state given by the physical Hilbert space of the system $L_2(\mathbb{R}^{\mathbb{Z}^d}, \gamma_S)$, where the Gaussian measure γ_S has the correlation operator $S = D^{-1/2}$. In this case, the role of the Hamiltonian (4.3) is played by the Dirichlet operator of the measure γ_S

$$(L_{\gamma_S}u)(x) = -\frac{1}{2}\sum_{k\in\mathbb{Z}^d}\frac{\partial^2 u(x)}{\partial x_k^2} + \sum_{k\in\mathbb{Z}^d}\langle D^{1/2}e_k, x\rangle\frac{\partial u(x)}{\partial x_k} \tag{4.4}$$

$$(u\in C^2_{b,\,cyl}(\mathbb{R}^{\mathbb{Z}^d}),\quad x\in\mathbb{R}^{\mathbb{Z}^d})$$

The spectrum of the operator L_{γ_S} can be characterized with the help of the quasinuclear description of a harmonic system given in the remarks after Theorem 3.2. According to Remark 1, the operator L_{γ_S} is unitary equivalent to the second quantization operator $d\Gamma(D^{1/2})$ in the Fock space $\mathfrak{F}(H_S)$, where H_S is a completion of $\mathbb{R}_0^{\mathbb{Z}^d}$ in the scalar product

$$(x, y)_{H_S} = \langle x, Sy\rangle = \langle x, D^{-1/2}y\rangle \quad (x, y\in\mathbb{R}_0^{\mathbb{Z}^d})$$

(by assumption, $D\geq m^2 1$, $m>0$; therefore,

$$D^{-1/2} = S\in\mathfrak{L}(l_2(\mathbb{Z}^d), l_2(\mathbb{Z}^d))$$

and, thus, H_S considered as a set coincides with $l_2(\mathbb{Z}^d)$. The operator $R = D^{1/2}$ is unitary equivalent to the operator of multiplication by a smooth function

$$\tilde{r}(p) = (\tilde{a}(p))^{1/2}(p\in[-\pi,\pi)^d)$$

(see Subsection 3.4) and, hence, its spectrum is absolutely continuous, i.e., $s(R) = s_{\rm ac}(R)$ $= [m, M]$, where m is defined above and $M = \max_{[-\pi,\pi)^d}\tilde{r}(p)$. The restriction $d\Gamma(R)\upharpoonright\mathfrak{F}_n(H_S)$ is given by formula (1.1) in Chapter 6, and according to the general fact of the theory of operators with separated variables, the spectrum $R^{(n)}$ consists of points of the form $\lambda_1+\ldots+\lambda_n$ $(\lambda_k\in[m, M],\ k = 1,\ldots,n)$, i.e.,

$$(R^{(n)}) = s_{ac}(R^{(n)}) = [nm, nM] \quad (n\in N)$$

(see Berezansky [5, Chapter 6, Section 4, Subsection 1]). Finally, since $d\Gamma(R)\upharpoonright\mathfrak{F}_0(H_S) = 0$, we have $0\in s\,d\Gamma(R)$. Therefore, the operator $d\Gamma(R)$ (together with l_{γ_S}) has the spectrum in the form of a set $\{0\}\cup\left(\bigcup_{n=1}^{\infty}[nm, nM]\right)$. Moreover, 0 is an eigenvalue (corresponding for L_{γ_S} to the eigenfunction $1\in L_2(\mathbb{R}^{\mathbb{Z}}, \gamma_S)$, and on $\bigcup_{n=1}^{\infty}[nm, nM]$ the spectrum is absolutely continuous. Note that $s_{\rm ac}L_{\gamma_S} = [m, +\infty)$, provided that $M\geq 2m$.

Assume that the potential $V = \overline{V}$ is given. Suppose that it satisfies the conditions formulated in Section 2 of Chapter 6 which guarantee the essential selfadjointness of $L_{\gamma_S} + V$, namely,

$$\exists \varepsilon > 0: V \in L_{2+\varepsilon}(\mathbb{R}^{\mathbb{Z}^d}, \gamma_S); \quad \forall p \geq 1 \quad e^{-V} \in L_p(\mathbb{R}^{\mathbb{Z}^d}, \gamma_S). \tag{4.5}$$

In this case, we can try to define the wave operators $W^{\pm}(L_{\gamma_S} + V, L_{\gamma_S})$ in the ordinary way by setting (cf. (4.1))

$$W^{\pm}(L_{\gamma_S} + V, L_{\gamma_S}) = s - \lim_{t \to \pm\infty} e^{-it(L_{\gamma_S} + V)} e^{itL_{\gamma_S}} P^0_{ac}, \tag{4.6}$$

where P^0_{ac} is a projector onto an absolutely continuous subspace for L_{γ_S}, i.e., onto the orthogonal complement to 1 in $L_2(\mathbb{R}^{\mathbb{Z}^d}, \gamma_S)$. The next lemma shows that this definition of wave operators cannot be accepted in the case under consideration.

Lemma 4.1. *The strong limits in* (4.6) *do not exist and, consequently, the wave operators cannot be defined directly by the relations of nonstationary scattering theory.*

Proof. For every $h \in H_S = l_2(\mathbb{Z}^d)$, a measurable linear function $u_h(x) = \langle h, x \rangle$ ($x \in \mathbb{R}^{\mathbb{Z}^d}$) belongs to $P^0_{\text{ac}}(L_2(\mathbb{R}^{\mathbb{Z}^d}, \gamma_S))$, since $u_h \perp 1$ in $L_2(\mathbb{R}^{\mathbb{Z}^d}, \gamma_S)$ and $u_h \in L_p(\mathbb{R}^{\mathbb{Z}^d}, \gamma_S)$ for any $p \geq 1$ (see Chapter 2, Section 2, Subsection 1). As is known, the existence of the limits (4.6) for the function u_h implies that

$$\left\| V e^{itL_{\gamma_S}} u_h \right\|_{L_2(\mathbb{R}^{\mathbb{Z}^d}, \gamma_S)} \to 0, \; t \to \pm\infty \tag{4.7}$$

Let us show that (4.7) does not hold in this case.

According to formula (1.22) in Chapter 6, we have

$$(e^{itL_{\gamma_S}} u_h)(x) = \langle e^{itR} h, x \rangle = \langle h(t), x \rangle, \, h(t) = e^{itR} h (t \in \mathbb{R}^1).$$

This yields

$$\left\| V e^{itL_{\gamma_S}} u_h \right\|^2_{L_2(\mathbb{R}^{\mathbb{Z}^d}, \gamma_S)} = \int_{\mathbb{R}^{\mathbb{Z}^d}} V^2(x) \langle h(t), x \rangle \langle \overline{h(t)}, x \rangle d\gamma_S(x)$$

$$= \int_{\mathbb{R}^{\mathbb{Z}^d}} V^2(x) : \langle h(t) \otimes \overline{h(t)}, x^{\otimes 2} \rangle : d\gamma_S(x) + \frac{1}{2} \| h \|^2_{H_S} \| V \|^2_{L_2(\mathbb{R}^{\mathbb{Z}^d}, \gamma_S)}.$$

Here we have used the relation

$$:\langle h_1, x\rangle\langle h_2, x\rangle: \;=\; \langle h_1, x\rangle\langle h_2, x\rangle - \frac{1}{2}(h_1, h_2)_{H_S}$$

$$(h_1, h_2 \in H_S; \quad x \in \mathbb{R}^{\mathbb{Z}^d})$$

for Wick monomials which follows from (2.24) and (2.25) in Chapter 2. According to Remark 2 after Theorem 2.2 in Chapter 2, for any $n \in \mathbb{N}$, the orthogonal projector

$$P_{\Gamma_n}: L_2(\mathbb{R}^{\mathbb{Z}^d}, \gamma_S) \to \Gamma_n(\mathbb{R}^{\mathbb{Z}^n}, \gamma_S)$$

possesses the following property:

$$\forall\, p \geq 2, \qquad f \in L_p(\mathbb{R}^{\mathbb{Z}^d}, \gamma_S)\, \| P_{\Gamma_n} f \|_{L_p(\mathbb{Z}^d, \gamma_S)}$$

$$\leq (p-1)^{n/2} \| P_{\Gamma_n} f \|_{L_p(\mathbb{R}^{\mathbb{Z}^d}, \gamma_S)} \leq (p-1)^{n/2} \| f \|_{L_p(\mathbb{R}^{\mathbb{Z}^d}, \gamma_S)}$$

for any $p \geq 2$ and $f \in L_p(\mathbb{R}^{\mathbb{Z}^d}, \gamma_S)$. Therefore, its extension by continuity to $L_{1+\varepsilon}(\mathbb{R}^{\mathbb{Z}^d}, \gamma_S)$ $(\varepsilon > 0)$ is defined and

$$P_{\Gamma_n}(L_{1+\varepsilon}(\mathbb{R}^{\mathbb{Z}^d}, \gamma_S)) = \Gamma_n(\mathbb{R}^{\mathbb{Z}^d}, \gamma_S).$$

In particular,

$$P_{\Gamma_2}(V^2) \in \Gamma_2(\mathbb{R}^{\mathbb{Z}^d}, \gamma_S)$$

for $V^2 \in L_{1+\varepsilon}(\mathbb{R}^{\mathbb{Z}^d}, \gamma_S)$ $(\varepsilon > 0)$ and, consequently,

$$P_{\Gamma_2}(V^2) =: (a_2, x^{\otimes 2}),$$

where $a_2 \in \mathfrak{F}_2(H_S)$ (see Subsection 2.3 in Chapter 2). Hence, using Corollary 1 to Theorem 2.2 in Chapter 2 to compute the integral of the product of Wick monomials, we obtain

$$\int_{\mathbb{R}^{\mathbb{Z}^d}} V^2(x) :\langle h(t) \otimes \overline{h(t)}, x^{\otimes 2}\rangle: d\gamma_S(x) = \int_{\mathbb{R}^{\mathbb{Z}^d}} :\langle a_2, x^{\otimes 2}\rangle::\langle h(t) \otimes \overline{h(t)}, x^{\otimes 2}\rangle: d\gamma_S(x)$$

$$= c(a_2, h(t) \otimes \overline{h(t)})_{H_S^{\otimes 2}} \to 0, \quad t \to \pm\infty$$

by virtue of the continuity of the spectrum of the operator D, which implies that $h(t) \to 0$ as $t \to \pm\infty$ weakly in H_S. This yields

$$\left\| V e^{itL_{\gamma_S}} u_h \right\|_{L_2(\mathbb{R}^{\mathbb{Z}^d}, \gamma_S)} \xrightarrow[t \to \pm\infty]{} \frac{1}{2} \| h \|_{H_S} \| V \|_{L_2(\mathbb{R}^{\mathbb{Z}^d}, \gamma_S)} \neq 0$$

for $h \neq 0$ and $V \neq 0$. The analogous fact can be established in a similar way for functions of the form $:\langle \cdot, h \rangle^n:$ $(n \in \mathbb{N})$ which also belong to $P^0_{ac}(L_2(\mathbb{R}^{\mathbb{Z}^d}, \gamma_S))$. These functions form a total set in $P^0_{ac}(L_2(\mathbb{R}^{\mathbb{Z}^d}, \gamma_S))$.

■

Hence, it is now clear that it is necessary to modify the definition of wave operators to make it applicable to the case under consideration. The idea of our approach is to compare with free dynamics, not the perturbed dynamics itself, but its renormalized version. Namely, according to Theorem 2.8 in Chapter 6, under conditions (4.5) imposed on the potential V, the operator $L_V = L_{\gamma_S} + V$ has the normalized ground state $\phi_V > 0$ γ_S-almost everywhere with eigenvalue E_V. As in Subsection 3.2 of Chapter 6, we pass unitarily from the space $L_2(\mathbb{R}^{\mathbb{Z}^d}, \gamma_S)$ to $L_2(\mathbb{R}^{\mathbb{Z}^d}, \mu^V)$, where $\mu^V = \phi_V^2 \gamma_S$, and to the renormalized operator

$$L_{V,\,\text{ren}} = (L_{\gamma_S} + V)_{\text{ren}} = \phi_V^{-1}(L_{\gamma_S} + V - E_V)\phi_V$$

in $L_2(\mathbb{R}^{\mathbb{Z}^d}, \mu^V)$ which is connected with the Dirichlet form of the measure

$$\mu^V : L_{V,\,\text{ren}} \upharpoonright C^2_{b,\text{cyl}}(\mathbb{R}^{\mathbb{Z}^d}) = L_{\mu^V}$$

according to Theorem 3.1 in Chapter 6. A renormalized dynamics given by the group

$$e^{itL_{V,\,\text{ren}}} = \phi_V^{-1} e^{it(L_{\gamma_S} + V - E_V)} \phi_V \quad (t \in \mathbb{R}^1)$$

is defined in the space $L_2(\mathbb{R}^{\mathbb{Z}^d}, \mu^V)$. This equality establishes its connection with the perturbed dynamics. Let us compare the asymptotic behavior of the groups $e^{itL_{V,\,\text{ren}}}$ and $e^{itL_{\gamma_S}}$ $(t \in \mathbb{R}^1)$ on the linear manifold of trigonometric polynomials

$$\mathfrak{T} = \text{l.s.}\{ e^{i\langle h, \cdot \rangle} \mid h \in H_S = l_2(\mathbb{Z}^d) \}$$

dense in each space $L_2(\mathbb{R}^{\mathbb{Z}^d}, \gamma_S)$ and $L_2(\mathbb{R}^{\mathbb{Z}^d}, \gamma^V)$. Note that by virtue of (1.11) in Chapter 6, we have

$$(itL_{\gamma_S}) \exp(i \langle h, x\rangle) = \exp(i \langle e^{itR}h, x\rangle) \quad (x \in \mathbb{R}^{\mathbb{Z}^d}).$$

Therefore, $e^{itL_{\gamma_S}}: \mathfrak{T} \to \mathfrak{T}$ and, consequently, the following operator is defined for every $t \in \mathbb{R}^1$:

$$L_2(\mathbb{R}^{\mathbb{Z}^d}, \gamma_S) \supset \mathfrak{T} \ni u \mapsto e^{-itL_{V,\mathrm{ren}}} e^{itL_{\gamma_S}} u \in L_2(\mathbb{R}^{\mathbb{Z}^d}, \mu^V).$$

For $u \in \mathfrak{T}$, we set

$$W^{\pm}(L_{V,\mathrm{ren}}, L_{\gamma_S})u = \lim_{t \to \pm\infty} e^{-itL_{V,\mathrm{ren}}} e^{itL_{\gamma_S}} u \tag{4.8}$$

under the assumption that limits exist. The operators

$$W^{\pm}(L_{V,\mathrm{ren}}, L_{\gamma_S})u : L_2(\mathbb{R}^{\mathbb{Z}^d}, \gamma_S) \to L_2(\mathbb{R}^{\mathbb{Z}^d}, \mu^V), \quad \mathfrak{D}(W^{\pm}(L_{V,\mathrm{ren}}, L_{\gamma_S})) = \mathfrak{T}$$

obtained in this way are called the renormalized wave operators corresponding to the pair $L_V = L_{\gamma_S} + V$ and L_{γ_S}. Note that, in contrast to (4.6), it is not necessary to use the projector P^0_{ac} in (4.6), because we have $L_{V,\mathrm{ren}}\,1 = L_{\gamma_S}\,1 = 0$ for the single eigenvector 1 of the operator L_{γ_S}. In order that the wave operators defined in this way allow us to investigate the spectral structure of the operator $L_{V,\mathrm{ren}}$ (and, hence, of the unitary equivalent operator $L_{\gamma_S} + V - E_V$), it is necessary to learn more about their properties. In fact, we employ the scattering theory for the pair of spaces $L_2(\mathbb{R}^{\mathbb{Z}^d}, \gamma_S)$ and $L_2(\mathbb{R}^{\mathbb{Z}^d}, \gamma^V)$ with the operator of identification

$$J : L_2(\mathbb{R}^{\mathbb{Z}^d}, \gamma_S) \supset \mathfrak{T} \ni u \mapsto Ju = u \in L_2(\mathbb{R}^{\mathbb{Z}^d}, \mu^V)$$

which may be unbounded. If $\phi_V \in L_\infty(\mathbb{R}^{\mathbb{Z}^d}, \gamma_S)$, then J is a bounded operator and, in this case, we can apply the standard scattering theory to a pair of spaces (Reed and Simon [3, Chapter 11, Section 3]); otherwise, one should use the generalization of it for unbounded operators of identification (see Bibliographical Notes).

Taking into account the connection between perturbed and renormalized dynamics, we find that the existence of $W^{\pm}(L_{V,\mathrm{ren}}, L_{\gamma_S})$ is equivalent to the existence of the following operators :

$$W^{\pm}(L_{\gamma_S} + V - E_V, L_{\gamma_S}; J_{\phi_V}) = \lim_{t \to \pm\infty} e^{-it(L_{\gamma_S} + V - E_V)} \phi_V e^{itL_{\gamma_S}}, \tag{4.9}$$

defined on the same domain $\mathfrak{T} \subset L_2(\mathbb{R}^{\mathbb{Z}^d}, \gamma_S)$, but as operators on the single space $L_\infty(\mathbb{R}^{\mathbb{Z}^d}, \gamma_S)$. Definition (4.9) exploits the identification operator

$$J_{\phi_V} : \mathfrak{T} \ni u \mapsto J_{\phi_V} u = \phi_V u,$$

which is again bounded or unbounded depending on whether $\phi_V \in L_\infty(\mathbb{R}^{\mathbb{Z}^d}, \gamma_S)$ or $\phi_V \notin L_\infty(\mathbb{R}^{\mathbb{Z}^d}, \gamma_S)$. Note that due to the hypercontractivity of the semigroup $\exp(-t\, L_{\gamma_S})$ $(t \geq 0)$ which takes place by virtue of the fact that $S \geq m^2\, 1$, $m > 0$, we have $\phi_V \in L_p(\mathbb{R}^{\mathbb{Z}^d}, \gamma_S)$, for any $p > 1$ (see Theorem 2.7 in Chapter 6). Therefore, again by virtue of the hypercontractibility of $\exp(-t\, L_{\gamma_S})$, the operator $J'_{\phi_V} = J_{\phi_V} e^{-L_{\gamma S}}$ is bounded on $L_2(\mathbb{R}^{\mathbb{Z}^d}, \gamma_S)$, and this enables us to use a bounded operator $J'_{\phi_V} = J_{\phi_V} e^{-L_{\gamma S}}$ as an identification operator. Moreover, for the wave operators which correspond to it, we have

$$W^{\pm}(L_{\gamma_S} + V - E_V, L_{\gamma_S};\ J'_{\phi_V}) = W^{\pm}(L_{\gamma_S} + V - E_V, L_{\gamma_S};\ J_{\phi_V}) e^{-L_{\gamma_S}}. \tag{4.10}$$

Finally, the connection between the wave operators defined by (4.8) and (4.9) is given by the equality

$$W^{\pm}(L_{\gamma_S} + V - E_V, L_{\gamma_S};\ J_{\phi_V}) = \phi_V W^{\pm}(L_{V,\,\mathrm{ren}}, L_{\gamma_S}) \tag{4.11}$$

and in the case when ϕ_V is unbounded, we can pass (as in (4.10)) to the operators $W^{\pm}(L_{V,\mathrm{ren}}, L_{\gamma_S})\, e^{-L_{\gamma S}}$ which are bounded as operators from $L_2(\mathbb{R}^{\mathbb{Z}^d}, \gamma_S)$ to $L_2(\mathbb{R}^{\mathbb{Z}^d}, \mu^V)$.

The described construction of the wave operators allows natural generalization to the case of a pair of Dirichlet operators. Assume that a nuclear representation $\Phi' \supset H_0 \supset \Phi$ and measures μ_k $(k = 1, 2)$ on $\mathfrak{C}_\sigma(\Phi')$ are given and that these measures allow us to associate the selfadjoint Dirichlet operator L_{μ_k} with the Dirichlet forms

$$d\mu_k(u, v) = \frac{1}{2} \int_{\Phi'} (\nabla u, \overline{\nabla v})_{H_0}\, d\mu_k \quad (u, v \in C^2_{b,\mathrm{cyl}}(\Phi'))$$

(see Subsection 3.1 in Chapter 6). Suppose that a linear manifold $\mathcal{G}$ of functions on Φ' is given and that it is densely imbedded in the spaces $L_2(\Phi', \mu_k)$ by the imbedding operator J_k $(k = 1, 2)$ and, moreover, $e^{itL_{\mu_1}} J_1\, \mathcal{G} \subset J_1\, \mathcal{G}$ for any $t \in \mathbb{R}^1$, i.e.,

$$e^{itL_{\mu_1}} J_1 u = J_1 u(t), \quad u(t) \in \mathcal{G}, \quad u(t) \equiv e^{itL_{\mu_1}} u$$

for any $u \in \mathcal{G}$ and $t \in \mathbb{R}^1$, and the equality $\|J_1 u\|_{L_2(\Phi', \mu_1)} = 0$ yields $u = 0$. Then we can compare the asymptotics of the groups $e^{itL_{\mu_1}}$ and $e^{itL_{\mu_2}}$ on $\mathcal{G}$ by considering the mappings

$$L_2(\Phi', \mu_1) \ni J_1 u \mapsto e^{-itL_{\mu_2}} J_2 e^{itL_{\mu_1}} u \in L_2(\Phi', \mu_2) \quad (u \in \mathcal{G})$$

for $t \in \mathbb{R}^1$ and introducing the wave operators

$$W^{\pm}(L_{\mu_2}, L_{\mu_1}) J_1 u = \lim_{t \to \pm\infty} e^{-itL_{\mu_2}} J_2 e^{itL_{\mu_1}} u \quad (u \in \mathcal{G}) \tag{4.12}$$

under the assumption of the existence of limits in (4.12). In order that the wave operators be bounded, it is necessary to additionally demand that

$$\forall\, u \in \mathcal{G} \| W^{\pm}(L_{\mu_2}, L_{\mu_1}) J_1 u \|_{L_2(\Phi', \mu_2)} \geq c \| J_1 u \|_{L_2(\Phi', \mu_1)} \tag{4.13}$$

for any $u \in \mathcal{G}$ and some $c > 0$. This definition cannot be reduced to the ordinary definition of wave operators in a pair of spaces, because the operator

$$J : L_2(\Phi', \mu_1) \ni J_1 u \mapsto J_1 u \in L_2(\Phi', \mu_2) \quad (u \in \mathcal{G})$$

can be not only unbounded but also nonclosable. Nevertheless, it preserves the standard properties of the wave operators and enables us to reduce the problem of the existence of these operators to verifying an analogue of the Cook criterion.

4.2. Perturbation of a Measure in the Gaussian Model and the Wave Operators Corresponding to It

We now return to the analysis of perturbations of harmonic systems and consider the case when the perturbation is given in the form of a density $\phi > 0$ γ_S-almost everywhere, $\|\phi\|_{L_2(\Phi', \gamma_S)} = 1$. Assume that the density ϕ has weak partial derivatives, i.e.,

$$\forall\, k \in \mathbb{Z}^d\ \exists\ \nabla_k \phi = \frac{\partial \phi}{\partial x_k} = \lim_{t \to 0} \frac{\phi(\cdot + te_k) - \phi(\cdot)}{t} \in L_2(\mathbb{R}^{\mathbb{Z}^d}, \gamma_S) \tag{4.14}$$

exist for any $k \in \mathbb{R}^d$ (the limit is in $L_2(\mathbb{R}^{\mathbb{Z}^d}, \gamma_S)$). Consider a measure $\mu = \phi^2 \gamma_S$ on $\mathfrak{B}(\mathbb{R}^{\mathbb{Z}^d})$ associated with the Dirichlet operator L_μ in $L_2(\mathbb{R}^{\mathbb{Z}^d}, \gamma_S)$. According to (3.6) in Chapter 6, it has the form

$$L_\mu u = L_{\gamma_S} u - \phi^{-1}\langle \nabla\phi, \nabla u\rangle = L_{\gamma_\Sigma} u - \phi^{-1} \sum_{k \in \mathbb{Z}^d} \nabla_k \phi \nabla_k u$$

$$(u \in C^2_{b,\,\mathrm{cyl}}(\mathbb{R}^{\mathbb{Z}^d})). \tag{4.15}$$

Generally speaking, the operator L_μ may have more than one selfadjoint extension (one of these is L^F_μ, i.e., the Friedrichs extension). Denote by $\hat{L}_\mu$ an arbitrary selfadjoint extension of L_μ. The next theorem gives the conditions for the existence of wave operators of the form (4.8) (in the case where the operator $\hat{L}_\mu$ is used instead of $L_{V,\,\mathrm{ren}}$) in terms of the density ϕ.

Theorem 4.1. *Assume that* ϕ *is such that*

$$\sum_{k \in \mathbb{Z}^d} \| \nabla_k \phi \|_{L_2(\mathbb{R}^{\mathbb{Z}^d},\, \gamma_S)} (1 + |k|^{d/2-1})^{-1} < \infty. \tag{4.16}$$

Then there exist the wave operators $W^{\pm}(\hat{L}_\mu, L_{\gamma_S})$.

Proof. It suffices to establish the existence of the limits (4.8) for

$$u_\varphi(x) = \exp(i\langle\varphi, x\rangle) \quad (x \in \mathbb{R}^{\mathbb{Z}^d})$$

for all $\varphi \in \mathbb{R}^{\mathbb{Z}^d}_0$. To do this, we verify the validity of the Cook criterion (see Reed and Simon [3, Chapter 11, Theorem 11.4]):

$$\int_1^{+\infty} \left\| (\hat{L}_\mu - L_{\gamma_S}) e^{itL_{\gamma_S}} u_\varphi \right\|_{L_2(\mathbb{R}^{\mathbb{Z}^d},\, \mu)} dt < \infty. \tag{4.17}$$

Note that by virtue of the relation (1.11) in Chapter 6, we have

$$(e^{itL_{\gamma_S}} u_\varphi)(x) = e^{i\langle\varphi(t), x\rangle} = u_{\varphi(t)}(x) \quad (x \in \mathbb{R}^{\mathbb{Z}^d}),$$

$$(\varphi(t) = e^{itR}\varphi; \quad R = D^{1/2} = S^{-1}; \quad t \in \mathbb{R}^1).$$

For $\varphi \in \mathbb{R}^{\mathbb{Z}^d}_0$, we have $u_\varphi \in C^\infty_{b,\mathrm{cyl}}(\mathbb{R}^{\mathbb{Z}^d})$; therefore, by (4.15), we obtain

$$(\hat{L}_\mu - L_{\gamma_S})u_\varphi = (L_\mu - L_{\gamma_S})u_\varphi = -\phi^{-1}\langle\nabla\phi, \nabla u_\varphi\rangle = -\phi^{-1} \sum_{k \in \mathbb{Z}^d} \nabla_k\phi \nabla_k u_\varphi.$$

Approximating $\varphi \in l_2(\mathbb{Z}^d)$ by the elements of $\mathbb{R}_0^{\mathbb{Z}^d}$, we can extend this equality to those u_φ for which the series in this relation is convergent in $L_2(\mathbb{R}^{\mathbb{Z}^d}, \mu)$ or, which is the same, $\sum_{k \in \mathbb{Z}^d} \nabla_k \phi \, \nabla_k u_\varphi$ is convergent in $L_2(\mathbb{R}^{\mathbb{Z}^d}, \gamma_S)$. It can be seen from the calculation presented below that this condition holds for $\varphi(t) = e^{itR}\varphi$ $(\varphi \in \mathbb{R}_0^{\mathbb{Z}^d})$; hence,

$$(\hat{L}_\mu - L_{\gamma_S})u_{\varphi(t)} = -\phi^{-1}\langle \nabla\phi, \nabla u_{\varphi(t)}\rangle$$

Therefore, for the integral in (4.17), we have the representation

$$\begin{aligned}\int_1^\infty \left\| \left\langle \frac{\nabla\phi}{\phi}, \nabla u_{\varphi(t)} \right\rangle \right\|_{L_2(\mathbb{R}^{\mathbb{Z}^d}, \mu)} dt &= \int_1^\infty \left\| \langle \nabla\phi, \nabla u_{\varphi(t)} \rangle \right\|_{L_2(\mathbb{R}^{\mathbb{Z}^d}, \gamma_S)} dt \\ &= \int_1^{+\infty} \left\| \sum_{k \in \mathbb{Z}^d} \nabla_k \phi \nabla_k u_{\varphi(t)} \right\|_{L_2(\mathbb{R}^{\mathbb{Z}^d}, \gamma_S)} dt \\ &= \int_1^{+\infty} \left\| \sum_{k \in \mathbb{Z}^d} \nabla_k \phi \langle \varphi(t), e_k \rangle u_{\varphi(t)} \right\|_{L_2(\mathbb{R}^{\mathbb{Z}^d}, \gamma_S)} dt \\ &\le \sum_{k \in \mathbb{Z}^d} \| \nabla_k \phi \|_{L_2(\mathbb{R}^{\mathbb{Z}^d}, \gamma_S)} \left(\int_1^{+\infty} |\langle \varphi(t), e_k \rangle| dt \right). \end{aligned} \tag{4.18}$$

Let us apply the stationary-phase method to estimate

$$\int_1^{+\infty} |\langle \varphi(t), e_k \rangle| dt \quad (k \in \mathbb{Z}^d).$$

By virtue of (3.32), we have the equality

$$\langle \varphi(t), e_k \rangle = \langle e^{itR}\varphi, e_k \rangle = \int_{[-\pi, \pi)^d} e^{ikp} e^{it\tilde{r}(p)} \tilde{\varphi}(p) \, dp \quad (t \in \mathbb{R}^1;\ k \in \mathbb{Z}^d), \tag{4.19}$$

in which $\tilde{r}(p) = (\tilde{a}(p))^{1/2}$ $(p \in [-\pi, \pi)^d)$ and

$$\tilde{\varphi}(p) = \sum_{k \in \mathbb{Z}^d} \varphi_k e^{ikp}/(2\pi)^{d/2} (p \in [-\pi, \pi)^d); \quad \tilde{\varphi} \in C^\infty([-\pi, \pi)^d),$$

since $\varphi \in \mathbb{R}_0^{\mathbb{Z}^d}$. According to the condition imposed on D, for almost all $p \in [-\pi, \pi)^d$, we have $\nabla \tilde{r}(p) \neq 0$ and $\det \tilde{r}''(p) \neq 0$. Consequently, the standard theorems of the stationary-phase method (see, for example, Reed and Simon [3, Chapter 11, Theorems 11.14 and 11.15]) applied to (4.19) enables us to conclude that, for any $n > d/2$, there exists $c_{n,\varphi}$ such that

$$|\langle \varphi(t), e_k \rangle| \leq c_{n,\varphi}(1 + t + |k|)^{-m} \tag{4.20}$$

for any $t \geq 1$ and $k \in \mathbb{Z}^d$, $|k| \geq Mt$, and there exists $c_\varphi > 0$ such that

$$|\langle \varphi(t), e_k \rangle| \leq c_\varphi |t|^{-d/2}, \tag{4.21}$$

for any $t \geq 1$ and $k \in \mathbb{Z}^d$, $|k| \leq Mt$.

Using these estimates, we get

$$\int_1^{+\infty} |\langle \varphi(t), e_k \rangle| \, dt = \int_{\{t \geq 1/|k| \leq Mt\}} |\langle \varphi(t), e_k \rangle| \, dt + \int_{\{t \geq 1/|k| > Mt\}} |\langle \varphi(t), e_k \rangle| \, dt$$

$$\leq \int_{\left\{t \geq 1/t \geq \frac{|k|}{M}\right\}} c_\varphi t^{-d/2} dt + \int_{\left\{t \geq 1/t < \frac{|k|}{M}\right\}} c_{n,\varphi}(1 + |k| + t)^{-n} dt \leq \hat{c}_\varphi(1 + \|k\|)^{-d/2+1}.$$

Together with (4.18) and (4.16), this proves that the Cook criterion (4.17) holds and, consequently, that the assertion of the theorem is valid.

■

4.3. Potential Perturbations of Harmonic Systems

For applying Theorem 4.1 to investigate the operators $L_{\gamma_S} + V$, it is necessary to verify condition (4.16) for the ground state ϕ_V. This condition means that ϕ_V is weakly dependent on the variables corresponding to "far" vertices of the lattice $\mathbb{Z}^d$. Under conditions (4.5) imposed on the potential V and taking into account the result of Theorem 4.2 and the inclusion $\phi_V \in \mathcal{D}(L_{\gamma_S})$ in Chapter 6, we can conclude that $\phi_V \in W_2^2(\mathbb{R}^{\mathbb{Z}^d}, \gamma_S)$, i.e., $\|\nabla_k \phi_V\|_{L_2(\mathbb{R}^{\mathbb{Z}^d}, \gamma_S)} \to 0$ as $|k| \to \infty$. If we want to clarify the dependence of the variables $\| \nabla_k \phi_V \|_{L_2(\mathbb{R}^{\mathbb{Z}^d}, \gamma_S)}$ on $k \in \mathbb{Z}^d$, then we need some additional information about the potential. The next theorem gives the conditions on the potential which guarantee the sufficiently large rate of decreasing of these variables and, consequently, the existence of the wave operators.

Theorem 4.2. *Assume that conditions* (4.5) *on the potential are satisfied and, moreover, there exist* $\sigma > d/2$, $\delta > 0$, *and* $c > 0$ *such that*

$$\| \nabla_k V \|_{L_{2+\delta}(\mathbb{R}^{\mathbb{Z}^d}, \gamma_S)} \leq c(1 + |k|)^{-\sigma} \tag{4.22}$$

for any $k \in \mathbb{Z}^d$. *Then there exist the wave operators* $W^{\pm}(L_{V, \mathrm{ren}}, L_{\gamma_S})$.

Proof. We set

$$\hat{L}_V = L_{\gamma_S} + V - E_V = L_V - E_V$$

and

$$\phi_V^{(t)}(x) = (e^{-tL_V} 1)(x) \quad (x \in \mathbb{R}^{\mathbb{Z}^d}).$$

According to the spectral theorem,

$$\phi_V^{(t)} \to (\phi_V, 1)_{L_2(\mathbb{R}^{\mathbb{Z}^d}, \gamma_S)} \, \phi_V, \; t \to +\infty$$

in $L_2(\mathbb{R}^{\mathbb{Z}^d}, \gamma_S)$ and, moreover, $(\phi_V, 1)_{L_2(\mathbb{R}^{\mathbb{Z}^d}, \gamma_S)} > 0$, because $\phi_V > 0$ γ_S-almost everywhere. Denote by ν_0 a measure on the space of the trajectories

$$\Omega_0(\mathbb{R}^1_+, \mathbb{R}^{\mathbb{Z}^d}) = \{\omega(\cdot) \colon \mathbb{R}^1_+, \to \mathbb{R}^{\mathbb{Z}^d}\},$$

corresponding to the Markov process with the generator L_{γ_S} originating from zero. This measure is constructed in Subsection 1.5 of Chapter 6. As is shown in Theorem 2.5 of Chapter 6, the Feynman–Kac formula holds for the potentials (4.5), and this formula yields the representation

$$\phi_V^{(t)}(x) = \int_{\Omega_0} \exp\left(-\int_0^t \hat{V}(\omega(\tau) + e^{-\tau R} x) d\tau \right) d\nu_0(\omega(\cdot)) \quad \text{(for } \gamma_S\text{-almost all } x\text{)}, \tag{4.23}$$

where $\hat{V} = V - E_V$ and $R = D^{1/2}$ as above. For $k \in \mathbb{Z}^d$, it follows from (4.23) that

$$(\nabla_k \phi_V^{(t)})(x) = -\int\limits_{\Omega_0} \left[\int\limits_0^t \left\langle \nabla V(\omega(\tau) + e^{-\tau R} x),\ e^{-\tau R} e_k \right\rangle d\tau \right]$$

$$\times \exp\left(-\int\limits_0^t \hat{V}(\omega(\tau) + e^{-\tau R} x) d\tau \right) d\nu_0(\omega(\cdot)) = -\sum_{j \in \mathbb{Z}^d} \int\limits_0^t \langle e_j, e^{-\tau R} e_k \rangle$$

$$\times \left[\int\limits_{\Omega_0} (\nabla_j V)(\omega(\tau) + e^{-\tau R} x)\ \exp\left(-\int\limits_0^t \hat{V}(\omega(\tau) + e^{-\tau R} x) d\tau \right)\ d\nu_0(\omega(\cdot)) \right] dt.$$

The same Feynman–Kac formula implies that

$$\int\limits_{\Omega_0} f(\omega(\tau) + e^{-\tau R} x)\, 1\, (\omega(\tau) + e^{-\tau R} x) \exp\left(-\int\limits_0^t \hat{V}(\omega(s) + e^{-sR} x) ds \right) d\nu_0(\omega(\cdot))$$

$$= \int\limits_{\Omega_x} f(\omega(\tau))\, 1\, (\omega(\tau)) \exp\left(-\int\limits_0^t \hat{V}(\omega(s)) ds \right) d\nu_x(\omega(\cdot)) = (e^{-\tau \hat{L}_V} f e^{-(t-\tau)\hat{L}_V} 1)(x),$$

where ν_x is a measure corresponding to the process with the generator L_{γ_S} originating from the point $x \in \mathbb{R}^{\mathbb{Z}^d}$ (see Subsection 1.5 in Chapter 6). This yields the representation

$$(\nabla_k \phi_V^{(t)})(x) = -\sum_{j \in \mathbb{Z}^d} \int\limits_0^t c_{kj}(\tau) (e^{-\tau \hat{L}_V} \nabla_j V e^{-(t-\tau)\hat{L}_V} 1)(x)\, d\tau$$

$$(c_{kj}(\tau) = \langle e_k, e^{-\tau R} e_j \rangle; \quad \tau \in [0, t]; \quad k, j \in \mathbb{Z}^d).$$

After a limit transition (taking into account the hypercontractibility of the semigroup $e^{-t \hat{L}_V}$ $(t \geq 0)$ which takes place, according to Theorem 2.7 in Chapter 6), this representation implies the equality

$$(\nabla_k \phi_V) = -\sum_{j \in \mathbb{Z}^d} (\phi_V, 1)_{L_2(\mathbb{R}^{\mathbb{Z}^d}, \gamma_S)} \int\limits_0^\infty c_{kj}(\tau) (e^{-\tau \hat{L}_V} (\nabla_j V \phi_V) d\tau. \tag{4.24}$$

Using (4.24), we obtain

$$\| \nabla_k \phi_V \|_{L_2(\mathbb{R}^{\mathbb{Z}^d}, \gamma_S)} \le c \sum_{j \in \mathbb{Z}^d} \left(\int_0^{+\infty} | c_{kj}(\tau) | \, d\tau \right) \| \nabla_j V \phi_V \|_{L_2(\mathbb{R}^{\mathbb{Z}^d}, \gamma_S)}$$

$$\le c \sum_{j \in \mathbb{Z}^d} \| \phi_V \|_{L_{p(\delta)}(\mathbb{R}^{\mathbb{Z}^d}, \gamma_S)} \| \nabla_j V \|_{L_{2+\delta}(\mathbb{R}^{\mathbb{Z}^d}, \gamma_S)} \int_0^{+\infty} | c_{kj}(\tau) | \, d\tau, \quad (4.25)$$

where $p(\delta)$ is such that

$$(p(\delta))^{-1} + (2 + \delta)^{-1} = 2^{-1}$$

and $\| \phi_V \|_{L_{p(\delta)}(\mathbb{R}^{\mathbb{Z}^d}, \gamma_S)} < \infty$, because $\phi_V \in L_p(\mathbb{R}^{\mathbb{Z}^d}, \gamma_S)$ for any $p \ge 1$ (see Theorem 2.8 in Chapter 6). It remains to establish the estimate for

$$\int_0^{+\infty} | c_{kj}(\tau) | \, d\tau \quad (k, j \in \mathbb{Z}^d),$$

which has the form

$$\int_0^{+\infty} | \gamma_{kj}(\tau) | \, d\tau = \int_0^{+\infty} \left| \int_{[-\pi, \pi)^d} e^{i(k-j)p} \, e^{-\tau \tilde{r}(p)} \, dp \right| d\tau$$

$$= \int_0^{+\infty} e^{-\tau m/2} \left| \int_{[-\pi, \pi)^d} e^{i(k-j)p} \, e^{-\tau[\tilde{r}(p) - m/2]} \, dp \right| d\tau$$

$$\le c_1(\sigma)(1 + |k - j|)^{-\sigma} \quad (k, j \in \mathbb{Z}^d),$$

where we have used the inequality $\tilde{r}(p) - m/2 \ge m/2 > 0$ and the smoothness of $\tilde{r}(p), p \in [-\pi, \pi)^d$. Combining the estimate obtained with condition (4.22) in (4.25), we get the estimate

$$\| \nabla_k \phi_V \|_{L_2(\mathbb{R}^{\mathbb{Z}^d}, \gamma_S)} \le c_2 \sum_{j \in \mathbb{Z}^d} (1 + |j|)^{-\sigma} (1 + |k - j|)^{-\sigma}$$

$$\le c_3 (1 + |k|)^{-\sigma} \quad (k \in \mathbb{Z}^d),$$

which implies that condition (4.16) of Theorem 4.1 holds, because here

$$\sum_{j \in \mathbb{Z}^d} (1 + |k|)^{-\sigma} (1 + |k|^{d/2-1})^{-1} < \infty,$$

since $d/2 + \sigma - 1 > d$ by assumption about σ. The application of Theorem 4.1 completes the proof.

■

Corollary 1. *Under the conditions of Theorem 4.2, the inclusion* $s_{ac}(L_{\gamma_S} + V - E_V) \supset s_{ac} L_{\gamma_S}$ *holds, where* $s_{ac} A$ *is an absolutely continuous part of the spectrum of the operator A.*

Proof. We use the wave operators $W^{\pm}(\hat{L}_V, L_{\gamma_S}, L_{\phi_V})$ acting in $L_2(\mathbb{R}^{\mathbb{Z}^d}, \gamma_S)$ which exist simultaneously with $W^{\pm}(L_{V,\text{ren}}, L_{\gamma_S})$ (see (4.11)). Without loss of generality, we can assume that $\phi_V \in L_\infty(\mathbb{R}^{\mathbb{Z}^d}, \gamma_S)$; otherwise, we introduce the operator of identification $J'_{\phi_V} = J_{\phi_V} e^{-L_{\gamma_S}}$ as mentioned in Subsection 4.1. We set

$$\mathcal{H}_{\text{in}} = (\operatorname{Ker} W^+ (\hat{L}_V, L_{\gamma_S}; J_{\phi_V}))^{\perp}$$

and

$$\hat{\mathcal{H}}_{\text{in}} = \mathcal{R} (W^+ (\hat{L}_V, L_{\gamma_S}; J_{\phi_V}))^{\sim}.$$

It follows from the abstract scattering theory that $\mathcal{H}_{\text{in}}$ and $\hat{\mathcal{H}}_{\text{in}}$ are invariant subspaces of L_{γ_S} and $\hat{L}_V$, respectively. Furthermore, $L_{\gamma_S} \restriction \mathcal{H}_{\text{in}}$ is unitary equivalent to $\hat{L}_V \restriction \hat{\mathcal{H}}_{\text{in}}$ (see Reed and Simon [3, Chapter 11, Section 3, proposition 4]). It remains to note that the equality $\mathcal{H}_{\text{in}} = L_2(\mathbb{R}^{\mathbb{Z}^d}, \gamma_S)$ is equivalent to $\operatorname{Ker} W^+(\hat{L}_V, L_{\gamma_S}; J_{\phi_V})$. The latter means that

$$\| \phi_V e^{itL_{\gamma_S}} u \|_{L_2(\mathbb{R}^{\mathbb{Z}^d}, \gamma_S)} \xrightarrow[t \to +\infty]{} 0 \Rightarrow u = 0. \tag{4.26}$$

However, as in the proof of Lemma 4.1, we can obtain the following representation for a function $u_t = e^{itL_{\gamma_S}} u$, $u \in L_2(\mathbb{R}^{\mathbb{Z}^d}, \gamma_S)$

$$|u_t|^2 = \|u\|^2_{L_2(\mathbb{R}^{\mathbb{Z}^d}, \gamma_S)} + v_t,$$

where $v_t \xrightarrow[t \to +\infty]{} 0$ weakly in $L_1(\mathbb{R}^{\mathbb{Z}^d}, \gamma_S)$. This yields (4.26).

■

Example 4.1. Consider a potential V of the form

$$V(x) = \sum_{k \in \mathbb{Z}^d} W(x_k) \quad (x \in \mathbb{R}^{\mathbb{Z}^d}),$$

which simulates a localized impurity in the ideal crystal. Let $W \geq 0$. There exists $\varepsilon > 0$ such that $W, W' \in L_{2+\varepsilon}(\mathbb{R}^{\mathbb{Z}^d}, \gamma_S)$ $\lambda_k \geq 0$ $(k \in \mathbb{Z}^d)$. Then the validity of condition (4.5) is ensured by the requirement $\sum_{k \in \mathbb{Z}^d} \lambda_k < \infty$ which implies that λ_k decreases as $|k| \to \infty$ no slower than $|k|^{-d}$. Here $\nabla_k V(x) = \lambda_k W'(x_k)$ and

$$\| \nabla_k V \|_{L_{2+\delta}(\mathbb{R}^{\mathbb{Z}^d}, \gamma_S)} \leq c_1 \lambda_k \leq c_2 |k|^{-d}$$

and, hence, the condition of the existence of wave operators (4.22) holds for $d > d/2 + 1$, i.e. $d \geq 3$.

4.4. Scattering Problem in the Case of Exactly Solvable Models

The models of potential perturbations of harmonic systems considered below admit explicit solutions both in the sense of the construction of the renormalized dynamics and in the sense of the realization of our approach to the scattering problem for the Dirichlet operators. These models are rather simple; nevertheless, they convincingly show that our statement of the scattering problem for the Dirichlet operators with the use of the renormalization procedure and the technique of a pair of spaces, is quite natural. Furthermore, the analysis of these models shows that the requirements of Theorem 4.2 are unimprovable (for a given class of interactions).

Assume that a potential is given by

$$V_A(x) = \sum_{k \in \mathbb{Z}^d} a_k x_k^2 \quad (x \in \mathbb{R}^{\mathbb{Z}^d}), \tag{4.27}$$

where $a_k \geq 0$ is a diagonal matrix in $l_2(\mathbb{Z}^d)$. We consider an operator $L_{\gamma_S} + V_A$ which corresponds to a formal Hamiltonian of the form (4.3) with the dynamical matrix $D + A$. Hence, the renormalized operator $(L_{\gamma_S} + V_A)_{\text{ren}}$ is defined in a Hilbert space $L_2(\mathbb{R}^{\mathbb{Z}^d}, \gamma_{S_A})$ $(S_A = (D + A)^{-1/2})$ and coincides with $L_{\gamma_{S_A}}$. Therefore, we have a pair of spaces $L_2(\mathbb{R}^{\mathbb{Z}^d}, \gamma_S)$ and $L_2(\mathbb{R}^{\mathbb{Z}^d}, \gamma_{S_A})$ and a pair of Dirichlet operators L_{γ_S} and $L_{\gamma_{S_A}}$ to which we can apply the method of construction of wave operators described in Subsection 4.1.

We can now take a set of trigonometric polynomials

$$\mathcal{T} = \text{l. s.}\{\exp(i\langle\varphi, x\rangle) \mid \varphi \in \mathbb{R}_0^{\mathbb{Z}^d},\ \ x \in \mathbb{R}^{\mathbb{Z}^d}\},$$

to play the role of a set $\mathcal{G}$; then in order to construct the wave operators, it is necessary to investigate the limits

$$\lim_{t\to\pm\infty} e^{-itL_{\gamma_{S_A}}} e^{-itL_{\gamma_S}} e^{i\langle\varphi, x\rangle} = \lim_{t\to\pm\infty} \exp(i\langle e^{-it(D+A)^{1/2}}\varphi, x\rangle), \tag{4.28}$$

because

$$\exp(itL_{\gamma_S}) \exp(i\langle\varphi, x\rangle) = \exp(i\langle e^{itD^{1/2}}\varphi, x\rangle),$$

(see the proof of Theorem 4.1). If A is a nuclear operator in $l_2(\mathbb{Z}^d)$, then by virtue of the Kato–Rosenblum theorem and the Birman invariance principle, the wave operators

$$W^{\pm}(D + A, D) = \lim_{t\to\pm\infty} e^{-it(D+A)^{1/2}} e^{itD^{1/2}}$$

exist and are complete in the space $l_2(\mathbb{Z}^d)$ (Reed and Simon [3, Chapter 13, Theorem 11.8 and 11.11]), i.e., the operators $D + A$ and D are unitary equivalent. However, we must interpret these as operators acting in the pair of single-particle spaces H_S and H_{S_A}, i.e.,

$$W^{\pm}(D + A, D)\colon\ H_S \to H_{S_A},$$

and corresponding to the quasiparticle description of the harmonic systems under consideration (see the remarks after Theorem 3.2)

$$L_2(\mathbb{R}^{\mathbb{Z}^d}, \gamma_S) \simeq \mathfrak{F}(H_S),\ \ L_2(\mathbb{R}^{\mathbb{Z}^d}, \gamma_{S_A}) \simeq \mathfrak{F}(H_{S_A}).$$

Then (4.28) yields

$$W^{\pm}(L_{\gamma_{S_A}}, L_{\gamma_S})\, e^{i\langle\varphi, x\rangle} = e^{i\langle W^{\pm}(D+A,\, D)\varphi,\, x\rangle} \quad (\varphi \in \mathbb{R}_0^{\mathbb{Z}^d}).$$

Hence, the operators

$$W^{\pm}(L_{\gamma_{S_A}}, L_{\gamma_S})\colon\ L_2(\mathbb{R}^{\mathbb{Z}^d}, \gamma_S) \to L_2(\mathbb{R}^{\mathbb{Z}^d}, \gamma_{S_A})$$

coincide with the unitary image of the secondly quantized wave operators

$$\Gamma(W^{\pm}(D+A, D))\colon\ \mathfrak{F}(H_S) \to \mathfrak{F}(H_{S_A})$$

(see Subsection 1.1 in Chapter 6). Since

$$\|\varphi\|_{H_{S_A}} \le \|\varphi\|_{H_S} \quad (\varphi \in l_2(\mathbb{Z}^d)),$$

we have

$$\|W^{\pm}(D+A, D)\|_{\mathfrak{L}(H_S, H_{S_A})} \le 1$$

and, therefore,

$$\|W^{\pm}(L_{\gamma_{S_A}}, L_{\gamma_S})\| = \|\Gamma(W^{\pm}(D+A, D))\| = 1,$$

i.e., the wave operators can be extended by continuity to the operators defined on the whole of the space. In the case when A is not nuclear, the one-particle wave operators $W^{\pm}(D+A, D)$ may not exist (if A is a Hilbert–Schmidt operator, then the spectrum of $D+A$ may be purely discrete) and, consequently, the operators $W^{\pm}(L_{\gamma_{S_A}}, L_{\gamma_S})$ may not exist, too.

Let us now consider the model under investigation from the point of view of the applicability of Theorem 4.2. The validity of condition (4.5) on the potential V_A for $A \ge 0$ is equivalent to the nuclearity of A. In this case, condition (4.22) has the form $a_k \le c(1+|k|)^{-\sigma}$, $\sigma > d/2+1$ $(k \in \mathbb{Z}^d)$ and necessarily holds for $d \ge 3$. Therefore, for $d \ge 3$, conditions of Theorem 4.2 cannot be made weaker (though, possibly, they can be replaced by conditions of another nature). The case when A is a Hilbert–Schmidt operator is interesting from the following point of view: Under this condition (and only in this case), the Gaussian measures γ_{S_A} and γ_S are equivalent and, thus, the vacuum function ϕ_{V_A} is defined, i.e.,

$$\phi_{V_A}^2(x) = \frac{d\gamma_{S_A}(x)}{d\gamma_S(x)} = c\exp\left[\langle (D^{1/2} - (D+A)^{1/2})x, x\rangle\right]$$

(see Daletsky and Fomin [1, Chapter 2, Corollary 4.2]).

According to Example 2.1 in Chapter 2, for A which is not nuclear, the potential V_A has no meaning of a measurable function. However, in the same example, the potential

$$V_{A,\,\mathrm{ren}}(x) = \sum_{k\in\mathbb{Z}^d} a_k(x_k^2 - c) \in L_2(\mathbb{R}^{\mathbb{Z}^d}, \gamma_S),$$

where $c = \|x_k^2\|_{L_1(\mathbb{R}^{\mathbb{Z}^d}, \gamma_S)}$, is a measurable function (the last expression is independent of $k \in \mathbb{Z}^d$, because γ_S is translation invariant) and just $L_{\gamma_S} + V_{A,\,\mathrm{ren}}$ is associated with the vacuum measure γ_{S_A}. Formally, $V_{A,\,\text{ren}} = V_A - c\,\mathrm{Tr}_{l_2} A$, but here $\mathrm{Tr}_{l_2} A = +\infty$ and, thus, for the operator realization of $L_{\gamma_S} + V_A$, it is necessary to renormalize the energy infinitely, and the renormalized operator may be defined in the original Hilbert space $L_2(\mathbb{R}^{\mathbb{Z}^d}, \gamma_{S_A})$.

The consideration presented above show that in the case of a quadratic potential V_A, the nonsingularity of it ($V_A \in L_2(\mathbb{R}^{\mathbb{Z}^d}, \gamma_{S_A})$) involves the existence of the wave operators and, conversely, for $V_A \notin L_2(\mathbb{R}^{\mathbb{Z}^d}, \gamma_{S_A})$, the wave operators, generally speaking, do not exist. To show that this situation is not general, we consider the following Hamiltonian as an example

$$L_{\gamma_S} + V_h;\ V_h(x) = \sum_{k \in \mathbb{Z}^d} h_k x_k \quad (h = (h_k)_{k \in \mathbb{Z}^d} \in \mathbb{R}^{\mathbb{Z}^d}). \tag{4.29}$$

This Hamiltonian describes the quantum harmonic system in the classical field (see also Example 1.2). As can be easily seen by analogy with Example 1.2, the renormalization of the Hamiltonian (4.29) gives the space of states $L_2(\mathbb{R}^{\mathbb{Z}^d}, \gamma_{S_A})$, where

$$\gamma_{S,\,\psi} = \gamma_S(\cdot + \psi) \quad (\psi = D^{-1} h \in \mathbb{R}^{\mathbb{Z}^d})$$

and

$$(L_{\gamma_S} + V_h)_{\text{ren}} = L_{\gamma_{S,\,\psi}}.$$

The potential V_h has the meaning of a linear measurable function on $\mathbb{R}^{\mathbb{Z}^d}$ if and only if $h \in l_2(\mathbb{Z}^d)$; otherwise, V_h is a generalized function from $\mathcal{P}'(\mathbb{R}^{\mathbb{Z}^d})$ (see Corollary 1 to Lemma 2.1 in Chapter 2). Nevertheless, the wave operators for the pair $L_{\gamma_{S,\psi}}$ and L_{γ_S} exist independently of the inclusion $h \in l_2(\mathbb{Z}^d)$, though the case $h \notin l_2(\mathbb{Z}^d)$ leads to the infinite renormalization of energy. Indeed, let us introduce an isometric operator

$$T_\psi\colon L_2(\mathbb{R}^{\mathbb{Z}^d}, \gamma_S) \ni u \mapsto u(\cdot + \psi) \in L_2(\mathbb{R}^{\mathbb{Z}^d}, \gamma_{S,\,\psi})$$

such that, for $u \in C^\infty_{b,\text{cyl}}(\mathbb{R}^{\mathbb{Z}^d})$, we have

$$(T_\psi L_{\gamma_S} u)(x) = -\frac{1}{2}\Delta u(x+\psi) + \langle R\nabla u(x+\psi), x+\psi\rangle$$

$$= (L_{\gamma_S} u)(x+\psi) + \langle R\psi, \nabla u(x+\psi)\rangle = L_{\gamma_{S,\,\psi}} T_\psi u(x),$$

This easily yields

$$W^{+}(L_{\gamma_{S,\psi}}, L_{\gamma_S}) = W^{-}(L_{\gamma_{S,\psi}}, L_{\gamma_S}) = T_{\psi}$$

and the equality $(W^{+})^{*} W^{+} = 1$. In this model, the wave operators are defined in spite of the singular character of the interaction. The reason is that the spectral characteristics are stable with respect to the external field.

Bibliographical Notes

Volume 2

CHAPTER 5

§1. This section is, in fact, a modified presentation of the corresponding sections in the books of Berezansky [18, 26]. The idea of the method of proving selfadjointness, which is developed here, appeared first for the case of the Schrödinger operator and the wave equation in Povzner [1]. The general scheme of this approach to the problem of selfadjointness was introduced in the works of Berezansky [4], [5]. In Subsection 1.1, we improve this scheme making it more precise; the information concerning evolutionary equations can be found in the book by S. G. Krein [1]. Theorem 1.4 is, in fact, an abstract exposition of the scheme presented in the works of Levitan [1] and Orochko [1]; a result close to this was obtained by Orochko [10] without the assumption that the deficiency numbers of an operator are equal. Theorems 1.5 and 1.6 belong to Berezansky [14–16]. For an application of the given method for proving selfadjointness, see the works of Berezansky [4], [5], [14], [16], Chaus [1], [2], and Chumak [1]. In combination with other ideas, this method is used in Orochko [1], [4], [5], [12] (other works of Levitan and Orochko contain many applications of the method of hyperbolic equations to different problems in the spectral theory). A similar method of proving selfadjointness by using hyperbolic systems was employed by Chernoff [2], [4]. This idea has been developed by T. Kato [3], and Rauch and Taylor [1].

The concept of an analytical vector and the formulation of the criterion of selfadjointness in terms of these vectors belong to Nelson [1]. The concept of a quasi-analytical vector (formulated in somewhat distinct terms) was introduced by Ismagilov [1]. The formulation presented in Section 1 is due to Nussbaum [3] (after this work the concept of a quasi-analytical vector has become very popular). These authors were first to apply quasi-analytical vectors to the problems of selfadjointness; Devinatz [2] has applied this technique when proving the selfadjointness of degenerating operators. Other proofs and generalizations to Banach spaces can be found in the papers by Hasegawa [1] and Chernoff [1]. The papers of Nussbaum [4], Messon and McClary [1], and Simon [1] are devoted to Stieltjes vectors and their applications. The simple reduction of the "quasi-analytical criteria" to the "evolutionary criteria" presented in Subsection 1.3 (see Berezansky [10], [11]) is actually contained in the book of Berezansky [5, Chapter 8, § 5, Subsection 4]. We mention the review by Chernoff [3] concerning the "quasi-analytical criteria". In the book of V. I. Gorbachuk and M. L. Gorbachuk [1], the concept of an analytical vector is essentially used to construct the spaces of test and generalized functions. The joint analytical vectors for families of operators were introduced and studied in the paper of Kosyak [1] and in the book of Yu. S. Samoilenko [2, Chapter 1, Section 1].

The results concerning the parabolic criteria of selfadjointness (Subsections 1.4 and 1.5) belong to Berezansky and V. G. Samoilenko [1], [2]; Theorem 1.11 is very close to the results of Semenov [1], Faris [1], Perelmuter and Semenov [1], and Perelmuter [2]; and Theorem 1.12 is connected with the concept of hypercontractive semigroups and corresponding facts (see Reed and Simon [2, Theorem X.58]).

Theorem 1.13 and important applications of it to establish the selfadjointness of differential operators are due to Chernoff [4]. Approximation Theorem 1.14 is given, for example, in the book of Faris [2], see also the review by Kovalenko and Semenov [1].

Theorem 1.15 on the selfadjointness and commutation properties of operators in a Hilbert space goes back to Nussbaum [3]. Other results in Subsection 1.8, devoted to similar problems in the case when H is generated by a positive definite kernel, belong to Berezansky (Theorem 1.17) [4] and Kostyuchenko and Mityagin (Theorem 1.16) [1]; the book of Berezansky [5, Chapter 8, Section 2] contains other facts of this sort and more references.

Here we do not present a large list of references devoted to the application of the criteria described in this section to the proof of the selfadjointness of differential operators. Some of these references can be found in the book of Berezansky [26] (below, when speaking about Chapter 6, we give the list of papers in which the hyperbolic criterion was used).

§ 2. As mentioned in the introduction to Chapter 5, the results in Sections 2, 3, 4, and 5 have been obtained by extending the technique for spectral representations of positive definite kernels in generalized joint eigenvectors, developed by Berezansky [2–4], and [5, Chapter 8], to the case of infinitely many operators. These papers appeared, in turn, as a development of M. G. Krein's approach to the construction of the spectral representations of positive definite kernels by means of the director functional. This technique was first extended to the case of infinitely many operators by Berezansky in his papers [7], [9] devoted to the representation of functionals of the Wightman type in the form of continual integrals; an abstract version of these papers (see Berezansky [10–12]) is given in Subsection 2.4. This cycle of papers, where infinitely many field operators (i.e., operators of shift) have appeared, was inspired by the work of Gachok [1] which contains the proof of the condition of selfadjointness for a single field operator based on the moment problem technique. However, an analogy with the classical moment problem can be traced back to the paper by Borchers and Zimmerman dealing with conditions of selfadjointness for a field operator; but it was not used effectively. Later, these conditions were made more precise by using the technique of positive definite kernel representations; see the papers by Gachok [2] and Berezansky [6].

The results in Subsections 2.1–2.3 belong to Berezansky [9], [13] and to Berezansky and Shifrin [1]. The idea of the proof of Lemma 2.1 goes back to Eskin [1]. The infinite-dimensional scalar moment problem was first studied by Kostyuchenko and Mityagin [1] (by using the limiting transition from the finite-dimensional problem). We mention a recent simple example of a two-dimensional moment sequence for which the moment problem is unsolvable, i.e., the representation (2.33) does not hold (because there are no proper estimates); this example belongs to Friedrich [1]. It is connected with an example of a positive definite function on a rectangle unextendable to a positive definite function on $\mathbb{R}^2$ which has been constructed by Sakhnovich [1]. The investigation of two-dimensional moment sequences admitting this representation with a measure σ, which is based on the disintegration of σ and on the concept of a canal function, can be found in the papers of Ovcharenko [1–3]; in particular, the condition ensuring the inclusion supp $\sigma \subset G$, where G are certain regions in $\mathbb{R}^2$, has been obtained in these articles. For the necessary facts from quantum field theory, see the books by Bogolyubov, Logunov, Oksak, and Todorov [1], and Simon [2]. The Osterwalder–Schrader concept of moment sequence is given in these books and in the papers by Osterwalder and Schrader [1] and Fröhlich [1], [2].

We have already mentioned the results in Subsection 2.4 concerning the generalized power moment problem. Here we only add that an interpretation of these results for functionals of the Wightman type is contained in the paper of Berezansky [9]; a functional of the Wightman type is understood as a sequence of ordinary (not generalized) functions of an increasing number of variables satisfying some of Wightman's axioms (mainly, the condition of positive definiteness). In Subsection 2.5, we clarify the possibility of taking into account the property of group invariance. However, this is done only in the case of the symmetric moment sequences from Subsection 2.1 (see Example 2.9). The case of a generalized power moment problem is investigated analogously. The case when only the group invariance property for Wightman functionals is taken into account was investigated by Yu. S. Samoilenko [1] and Korsunsky [2].

The positive definiteness (2.1) may be regarded as a positive definiteness of a functional $s = (s_n)_{n=0}^{\infty}$ on the Borchers algebra $\mathcal{A}$ whose elements are the sequences $\varphi = (\varphi_n)_{n=0}^{\infty}$ $(\varphi_n \in \Phi^{\otimes n})$ with multiplication

$$(\varphi * \psi) = \sum_{j+k=n} \varphi_j \otimes \psi_k \quad (n \in \mathbb{Z}_+),$$

involution $\overline{\varphi} = (\overline{\varphi}_n)_{n=0}^{\infty}$, and natural operations of addition and multiplication by a scalar. The action of the functional

$$s(\varphi) = \sum_{n=0}^{\infty} s_n(\varphi_n), \quad s(\varphi * \overline{\varphi}) \geq 0 \ (\varphi \in \mathcal{A}),$$

is a condition of positive definiteness (clearly, we introduce a topology in $\mathcal{A}$). It is easy to understand that in the case of a symmetric moment sequence s, one can pass from $\mathcal{A}$ to a commutative topological algebra, and representation (2.6) may be regarded as a decomposition of a positive definite functional into multiplicative functionals (having the form $(1, \lambda, \lambda \otimes \lambda, \dots)$, where $\lambda \in \Phi'_{\mathrm{Re}}$). A similar point of view is considered in the papers by Borchers and Yngvason [1–3] containing general facts about decompositions of positive definite functionals on a commutative (and even on an arbitrary) nuclear $*$-algebra into pure states (these are positive definite functionals which cannot be decomposed further; collections of functionals of this sort contain all multiplicative functionals). In these papers, one can also find some applications to field theory. The earlier paper of Powers [1] and the recent paper by Berezansky, Lassner, and Yakovlev [1] also deal with this approach.

We mention one more delightful generalization of the classical moment problem, namely, the noncommutative moment problem. Let $\mathbb{R}^d \ni g \mapsto U(g)$ be a continuous unitary representation of an additive group $\mathbb{R}^d$ $(d \in \mathbb{N})$ by operators in a Hilbert space H, let E be an RI that corresponds to it, and let a selfadjoint operator $dU(g_k)$ acting in H be the corresponding generator in the direction of the kth coordinate axis $(k = 1, \dots, d)$. The sequence

$$s_{n_1,\dots,n_d} = ((dU(g_1))^{n_1} \dots (dU(g_d))^{n_d} \Omega, \Omega)_H \quad (n_1, \dots, n_d \in \mathbb{Z}_+), \tag{1}$$

where $\Omega \in H$ belongs to the domains of definition of all the operators in (1), is a moment sequence. Indeed, the spectral measure that corresponds to it has the form

$$\sigma(\alpha) = (E(\alpha)\,\Omega, \Omega)_H \quad (\alpha \in \mathfrak{B}(\mathbb{R}^d)).$$

It is easy to see that, conversely, every moment sequence satisfying the definiteness condition can be written in the form (1). Therefore, representation (1) is equivalent to representation (2.33).

Now let us consider a real d-dimensional Lie group G (instead of $\mathbb{R}^d$). Let $G \ni g \mapsto U(g)$ be a continuous unitary representation of it, and let $dU(g_k)$ be the corresponding generator in the direction of a one-parameter subgroup associated with the kth basis vector of its Lie algebra (the basis is fixed, $k = 1, \dots, d$). For given G, the noncommutative moment problem deals with the determination of conditions which should be imposed on the numbers $s_{n_1,\dots,n_d}$ to guarantee representation (1) (the representation with a "noncommutative measure"). This moment problem was first studied by Woronowicz [1] (see also Powers [1]) for the Heisenberg group (the "quantum moment problem"); in these papers, the Riesz method of solving the classical moment problem has been generalized (the nonnegativity of a functional on nonnegative polynomials which corresponds to a moment sequence). In the papers of Richter [1] and Schmüdgen [1], the Riesz method has been further developed and applied to the noncommutative moment problem for an arbitrary group G. In the noncommutative case, it is also possible to develop an approach such as that presented in Section 2. One must demand that the sequence of numbers $s_{n_1,\dots,n_d}$ should satisfy the condition of positive definiteness (similar to (2.32)) and a certain condition on growth which is analogous to definiteness.

The corresponding results were obtained in the papers of Yu. S. Samoilenko [2], A. Yu. Daletsky [2], [3], and A. Yu. Daletsky and Yu. S. Samoilenko [1], where the noncommutative analogue of an infinite-dimensional scalar moment problem was also considered.

§ **3.** The results in this section belong to Berezansky and Gali [1] (see also Berezansky, Gali, and Zhuk [1]). Theorem 3.2 is a generalization of M. G. Krein's classical theorem (on the extension of a positive definite function on the entire axis; see M. G. Krein [1]) for positive definite (in the integral sense) functions of infinitely many variables on a layer.

The results obtained by Kondratyev and Yu. S. Samoilenko [1] concerning the spectral representation of a generalized $\mathbb{R}_0^\infty$-quasi-invariant positive definite kernel $K \in \mathcal{A}'(\mathbb{R}^\infty) \otimes \mathcal{A}'(\mathbb{R}^\infty)$ are close to the results in Section 3. In this connection, we must also mention some constructions in the book by Yu. L. Daletsky and Fomin [1, Chapter 4, Section 4].

§ **4.** The results in Subsections 4.1–4.3, concerning the generalization of the Minlos-Sazonov theorem for a layer in the Hilbert space, belong to Berezansky and Gali [1]. For the case of positive definite functions of finitely many variables, the corresponding facts about extensions can be found in the book of Berezansky [5, Chapter 8, Sections 3, 4]. This book also contains a detailed bibliography (up to 1965). We now only note that the description of all extensions of a positive definite function of two variables defined on $(-2l, 2l) \times \mathbb{R}^1$ to a positive definite function on $\mathbb{R}^2$ was obtained by two different methods in the papers of Levin and Ovcharenko [1] and Berezansky and Gorbachuk [1]. Friedrich and Klotz [1] have recently investigated the problem of extensions (and of the description of the extensions) of a positive definite function given on $(-2l, 2l) \times G$, where G is an arbitrary commutative group, to a positive definite function on $\mathbb{R}^1 \times G$. In particular, they have obtained another proof of Theorem 4.2.

Let us dwell upon some related results obtained after 1965. Rudin [1] proved that every continuous positive definite function $k(x)$ defined on a ball centered at 0 in the space $\mathbb{R}^d$ $(d \in \mathbb{N})$. Nussbaum [7], [8] obtained a representation for radial functions defined on such a ball, positive definite in the sense that the integral condition of positive definiteness, i.e., condition of type (3.7), holds only for radial functions φ. These functions can be actually decomposed into elementary kernels connected with differential operators depending on one variable (radius $r = |x|$, cf. Section 5). These results were obtained by using the methods presented in Chapter 5 (for a single operator); they cover the above-mentioned Rudin theorem. In his paper [9], Nussbaum has generalized these results for the case where $\mathbb{R}^d$ is replaced by an arbitrary (in a certain sense) noncompact uniform Riemann space; this paper is closely connected with that of M. G. Krein [4].

For a positive definite continuous radial function $k(x) = K(\|x\|_H)$ defined on a ball centered at 0 in a real Hilbert space H, the result of Rudin concerning the extendability of it to a radial positive definite function, also remains valid. This function admits the representation

$$k(x) = \int_0^\infty e^{-\|x\|_H^2 \lambda}\, d\sigma(\lambda) \qquad (x \in H)$$

with a nonnegative finite measure σ (the Schoenberg representation; see Schoenberg [1]). A similar fact is also valid for exponentially convex radial functions on the ball in the space H; in this case, the above representation is valid with $-\|x\|_H^2$ being replaced by $\|x\|_H^2$. The results concerning extensions described in this section are due to Nussbaum [6].

As mentioned above, the Minlos-Sazonov theorem is contained in the papers of Minlos [1] and Sazonov [1]. It was also (independently) established by Gross [1]; the exposition of it (with different modifications) can be found in the books by Gelfand and Vilenkin [1, Chapter 4, Section 4], Gikhman and Skorokhod [2, Chapter 5, Section 5], Kuo [1, Chapter 1, Section 6], Yu. L. Daletsky and Fomin [1, Chapter 3, Section 1], and Vakhania, Tarieladze, and Chobanyan [1, Chapter 6]. Theorem 4.7 belongs to Tishchenko [3].

§ 5. The constructions in this section generalize the scheme of Berezansky [5, Chapter 8, Section 1], initially developed for finitely many operators, for the case of infinitely many operators. These results were presented in a less perfect form in Berezansky [8], [20]. Examples of the realization of Theorem 5.1 in the case of infinitely many operators A_x can be found in the book of Berezansky [5, Chapter 8, Section 3]; the infinite-dimensional case is also considered in Rudinsky [1]. Theorem 5.3 has been proved by Berezansky [9–11]; in connection with this theorem we also mention the paper of Shifrin [2]. For applications of the theory of representations of positive definite kernels to the spectral theory of random fields, see the book of Yadrenko [1].

CHAPTER 6

§ 1. The operators acting in the Fock space and, in particular, the second quantization operators are a traditional subject of investigation in mathematical physics. These operators appear in different models of theoretical physics (solid state physics, quantum statistical physics, and quantum field theory). From an abstract point of view the second quantization operators were studied by Cook [1]. A comprehensive investigation of the properties of these operators can be found, for example, in the books of Berezin [1], Simon [2], and Reed and Simon [1, 2]; these books also contain detailed bibliographies covering this subject. The proof of an infinite-dimensional version of the Mehler formula presented here is based on the paper of Kondratyev [5].

The Schrödinger representation for the second quantization operators is widely used in constructive field theory. However, the fact that these operators are of a differential nature, was not actually exploited until the paper of Albeverio and Høegh-Krohn [1] in which the Hamiltonian of the free boson field is described rigorously as an infinite-dimensional elliptic differential operator of the second order. The properties of the semigroup generated by a second quantization operator were discussed from different points of view in a number of papers devoted to constructive field theory (see the review by Simon and Høegh-Krohn [1] and the book of Simon [2] which contains an exposition of the history of the problem and an annotated bibliography). The hypercontraction Theorem 1.7 is due to Nelson [2]; it is one of the most important properties of this semigroup.

In our construction of functional integrals connected with the second quantization operators, we employ the construction of a Markov process from a family of stochastic kernels which can be written explicitly with the help of the Mehler formula. This construction is standard in the theory of random processes (see, for example, Gikhman and Skorokhod [2]). It is also possible to use another method of constructing functional integrals of this sort based on the methods of the theory of stochastic differential equations (Yu. L. Daletsky and Fomin [1, Chapter 7]). However, to apply these methods, one should assume that the one-particle operator is bounded, but this is inadequate for many applications. The direct construction of measures on the space of trajectories presented in this section is, in fact, a development of the approach suggested by Kondratyev [5]. An abstract scheme for constructing a functional integral from a semi-group of operators preserving positivity was developed by Klein and Landau [1]. The stochastic integral of an operator-valued function over the resolution of the identity was constructed and investigated in the paper of Berezansky, Zhernakov, and Us [1]; in the case of multiplication operators, this integral turns into an ordinary stochastic integral of a random process over a martingale.

§ 2. The Kato inequality was proved in T. Kato [2] (see also the book of Reed and Simon [2, Theorem X.27]. It was extended to the case of second quantization operators by Kondratyev [7]. In an abstract form, the Kato inequality was established in the paper of Simon [3]. However, in our case this inequality does not give the opportunity to investigate the essential selfadjointness of potential perturbations of the second quantization operators on domains consisting of smooth functions.

To construct a semi-group generated by an elliptic operator L of the second order (or, in other words, to prove the existence of a strong solution for the corresponding parabolic equation), Devinatz [3], [5] employed a stochastic equation connected with L. The fact that a stochastic equation has a solution is not affected by the degeneracy of the matrix of leading coefficients of L; due to this, Devinatz obtained results concerning the selfadjointness in the case of degeneracy, too. On the other hand, by using the technique of stochastic equations, Devinatz established (in his papers [4], [6]) the Kato inequality for the operators under consideration, and this inequality, in turn, enabled him to strengthen the results concerning selfadjointness obtained in [3], [5]. The article of Brezis and T. Kato [1] also deals with the same range of problems. Our method of proving the Kato inequality is conceptually close to that presented in the original paper of T. Kato [2], but is essentially different from the technical point of view, because it is impossible to use the standard analysis technique in the infinite-dimensional case. The Kato inequality for second quantization operators with a magnetic field has been proved by Tsykalenko [1].

Proof of the classical Feynman–Kac formula can be found, for example, in Reed and Simon [2, Chapter 10, Section 11]. The version of this formula for models of field theory with cutoff is often called the Feynman–Kac–Nelson formula (see Simon [2] and

Glimm and Jaffe [1]). Application of the Feynman–Kac formula to the investigation of properties of perturbed operators goes back to papers devoted to constructive field theory (see, for example, Simon [2]). We have extended this approach to a general class of potential perturbations of second quantization operators; here we follow the works by Kondratyev [5], [6]. The logarithmic Sobolev inequality was obtained by Gross [2]; he also noted that this inequality is connected with hypercontractibility. Our proof of this inequality (as a consequence of the estimate for a lower bound of the spectrum of perturbed operator) is based on the method proposed by Faris [2]. The statement, concerning conditions which should be imposed on a potential to guarantee its operator smallness with respect to the second quantization operator, was announced (without proof) by Faris [2, Section 9]. When proving this fact in Section 2, we use the results of Feissner [1] about interpolation in Orlicz spaces. The interpretation of polynomial interactions in field theory as potentials given by generalized functions of infinitely many variables is due to Kondratyev [2].

§ 3. A large number of works exist devoted to Dirichlet forms and operators associated with these in the finite-dimensional case; these papers were motivated both by purely mathematical considerations and by applications to quantum mechanics and theory of random processes. Since these problems remain aside from our exposition, we only refer the reader to the detailed surveys of Albeverio and Høegh-Krohn [3] and Fukushima [1]. In the infinite-dimensional case, the following problems have been mainly studied: The problem of closability of Dirichlet forms and the operators corresponding to these forms (Albeverio and Høegh-Krohn [1], [2], Paclet [1], and Kusuoka [1]. The renormalization of potential perturbations of the second quantization operators described in our book goes back conceptually to the paper of Gross [2]. The analogous method was exploited by Albeverio, Høegh-Krohn, and Streit [1], [2] when investigating singular quantum mechanical Hamiltonians and by Albeverio and Høegh-Krohn [1] when constructing models in two-dimensional field theory.

The essential selfadjointness of Dirichlet operators was studied by Takeda [1] in the case of measures which are multiplicative perturbations of a Gaussian measure by a smooth density. The investigation of this problem presented here is based on the work of Kondratyev and Tsykalenko [3].

§ 4. The coercivity inequalities for elliptic differential equations with infinitely many variables, acting in the spaces of functions square summable with respect to Gaussian measures, are due to Frolov [2]. For the case of Dirichlet operators with constant coefficients, the coercivity inequalities were proved by Kondratyev [8]; the generalization of these for the case of varying coefficients has been studied by Kondratyev and Tsykalenko [3]. In this book, we do not consider the general problems of the theory of infinitely-dimensional elliptic equations (such as the solvability of boundary-value problems, the existence of fundamental solutions of these equations, and so on) at all. There exist some well-known results in this direction due to Vishik [1], [2], Vishik and A. V. Marchenko [1], Frolov [1], Shakhbagyan [1], and many other authors. The energetic inequalities for hyperbolic equations associated with Dirichlet operators (including operators with varying coefficients) were obtained by Kondratyev and Tsykalenko [2], [3]. Statements (i) and (ii) of Theorem 4.4 are analogues of the selfadjointness criteria of Povzner [1] and Titchmarsh [1] and Sears [1], respectively, in the case of potential perturbations of Dirichlet operators.

The method of proving selfadjointness considered here is an infinite-dimensional version of a hyperbolic approach presented in Subsections 1.1 and 1.2 of Chapter 5. We now mention a series of papers in which this, or similar, technique was applied to clear up conditions of selfadjointness for a finite-dimensional elliptic operator L of the second order with regular or singular coefficients: Povzner [1], Levitan [1], Berezansky [4, 5, 14, 26], Orochko [1, 4, 5, 10, 11, 12], Chumak [1], Chernoff [2, 4], T. Kato [3], Berezansky and V. G. Samoilenko [1, 2], Rofe-Beketov [1, 2], Perelmuter and Semenov [2], Belichkova [1], and Perelmuter [3]. Some of these papers (Chumak [1], Orochko [11], Perelmuter and Semenov [2], and Rofe-Beketov [2]) are, in fact, devoted to clarifying conditions which should be imposed on the leading coefficients of L to guarantee a finite propagation velocity of perturbation for the corresponding hyperbolic equation (note that the boundedness of the leading coefficients is, in fact, assumed in Theorem 1.8 in Chapter 6 of the book by Berezansky [5]). This range of problems is studied in more detail in the book by Berezansky [26]. The hyperbolic approach is applied to prove the selfadjointness of infinite-dimensional elliptic operators in the papers by Berezansky [16, 18, 26], Berezansky and V. G. Samoilenko [1–3], and V. G. Samoilenko [2, 3]. However, in contrast to Theorem 4.4, the hyperbolic equation appears in these papers only in the case of infinitely many variables when the infinite-dimensional elliptic operator is approximated by a finite-dimensional operator.

In this book, we do not study spectral properties of certain classes of model infinite-dimensional differential operators. The corresponding results are presented in the book by Berezansky [26] and in the papers by Berezansky and V.G.Samoilenko [2, 3]. These operators have the following form.

Consider the space L_2 on $\mathbb{R}^\infty$ with respect to a certain product measure $\mu = \times_{k=1}^{\infty} \mu_k$, $d\mu_k(x_k) = p_k(x_k)\,dx_k$, where $(p_k)_{k=1}^{\infty}$ is a sequence of smooth probability weights. In this space, the "Laplacian" is constructed as an expression with infinitely many separated variables:

$$\Delta_p = -p_1^{-1/2}(x_1)\,\partial^2/\partial x_1^2\,(p_1^{1/2}(x_1)\cdot) - p_2^{-1/2}(x_2)\,\partial^2/\partial x_2^2\,(p_2^{1/2}(x_2)\cdot) - \dots$$

$$= - D_1^2 - D_1^2 - \dots . \tag{2}$$

Each term D_n^2 in (2) is an image of the second derivative which appears naturally when passing from the Lebesgue measure dx_n to the probability measure $p(x_n)\,dx_n$.

In order to correctly define the operator acting in L_2 (containing bounded functions) according to (2), it suffices to demand that $\Delta_p 1 \in L_2$; being somewhat strengthened, the last condition becomes equivalent to the requirement that the series

$$\sum_{n=1}^{\infty} \| \partial^2/\partial x_n^2 (\sqrt{p_n}) \|_{L_2(\mathbb{R}^1,\, dx_n)} < \infty \tag{3}$$

should converge. Condition (3) is a sufficiently strong restriction on the weights (for example, if

$$p_n(x_n) = (\pi\kappa_n)^{-1/2} e^{-x_n^2 \kappa_n} \qquad (\kappa_n > 0)$$

is Gaussian, then (3) is equivalent to the requirement that $\sum_{n=1}^{\infty} \kappa_n^{-1} < \infty$). Note that the operator corresponding to (2) is now a "sum with respect to distinct variables" of the operators D_n^2 with the spectrum $[0, \infty)$.

The requirement $\Delta_p 1 \in L_2$ can be also guaranteed in another way. Namely, instead of the term

$$D_n^2 = \partial^2/\partial x_n^2 + 2(D^n 1)\partial/\partial x_n + D_n^2 1,$$

one should consider $D_n^2 - D_n^2 1$, i.e., we substract a certain potential from Δ_p. For the corresponding expression

$$\Delta_{p,\,\mathrm{reg}} = -(D_1^2 - D_1^2 1) - (D_2^2 - D_2^2 1) - \dots, \tag{4}$$

it is clear that $\Delta_{p,\,\mathrm{reg}} 1 \in L_2$, and the operator in L_2 can be easily defined for (4). However, this will be another operator. Indeed, in the case of Gaussian weights, $D_n^2 - D_n^2 1$ is a harmonic oscillator which has a discrete spectrum (unlike D_n^2).

Using some general facts about operators admitting the separation of infinitely many variables, it is easy to investigate the spectral properties of the operators in L_2 which correspond to (2) and (4). However, it is much more difficult to study these properties for perturbations of these operators by a potential $V(x_1, x_2, \dots)$. Here, in addition to the classical singularities connected with the facts that the region is finite and V is not smooth, there appears a singularity "in the direction of increase in the number of variables" (we clarify that V is necessarily non-smooth, because no topology is introduced in $\mathbb{R}^\infty$ and the concept of smoothness makes no sense). Nevertheless, the properties of selfadjointness of perturbed operators can be studied; in particular, the Povzner-type and Titchmarsh–Sears-type theorems are established for these operators. This is realized by using the parabolic approaches presented in Section 1 of Chapter 5; the first approach determines the local behavior at infinity. Here, an important role is played by a cylindrical approximation of a function in L_2 and then by a smooth approximation of the corresponding function of finitely many variables.

The results described here enable us to establish some analogous facts concerning differential operators acting upon functions defined on a Hilbert space H, i.e., for the operators in "invariant form". These facts are obtained by introducing the coordinate notation for this "invariant" operator and reducing it to the perturbations of the expressions (2) or (4). As a result, we obtain the conditions of selfadjointness for elliptic operators of the second order on H with constant leading coefficients and a singular potential.

These results are presented in papers which have been referred to above and in the earlier papers of Berezansky [16, 18, 22], Berezansky and V. G. Samoilenko [1], V. G. Samoilenko [2, 3], and Perelmuter and Semenov [1]. Similar problems were studied in the case of regular or "almost regular" potentials by Yu. L. Daletsky [2, 3] and Frolov [3–5]; in the case of differential operators introduced with the help of generalized tensor products of spaces, they were investigated by A. V. Marchenko [2, 3]. By employing an approach connected with stochastic equations, Yu. L. Daletsky [1] established the selfadjointness of an operator with smooth leading coefficients but without a potential. A more detailed study of the spectral properties of these operators encounters serious difficulties; here the effect of the infinite-dimensional situation is essential (for example, in the case (2) the smallness of a potential at infinity does not guarantee the preservation of a continuous spectrum); many results in this direction were obtained by V. G. Samoilenko [1, 2, 4, 5].

These facts are closely connected with the spectral theory of operators admitting the separation of infinitely many variables which has been already mentioned in Subsections 3.6 and 3.7 of Chapter 3; this theory was developed in the works of Berezansky and Us [1, 2] and Us [2, 3] (see also the books of Berezansky [18, 26]). The enclosed derivatives $p_n^{-1/2}(x_n)\,\partial/\partial x_n\,(p_n^{1/2}(x_n)\,\cdot)$ in the space $L_2(\mathbb{R}^\infty, \mu)$ with respect to the product measure μ are classical examples of operators acting upon different variables; operators admitting the separation of variables are constructed from these derivatives. Yu. S. Samoilenko and Us [1, 2] (see also Yu. S. Samoilenko [2]) generalized these constructions; they considered a measure on $\mathbb{R}^\infty$ quasi-invariant with respect to shifts by finite sequences in $\mathbb{R}^\infty$ and a family of infinitesimal operators A_n (with respect to shifts along the nth coordinate) in the corresponding space $L_2(\mathbb{R}^\infty, \mu)$ instead of the enclosed derivatives (we clarify that, due to the existence of non-equivalent quasi-invariant measures, the families of differentiations $(A_n)_{n=1}^\infty$ which appear may be unitary non-equivalent). A series of facts (similar to those which take place in the classical case) are established in these papers for the generalized differentiations under consideration. On the other hand, these papers contain examples of functions of countable families of commuting selfadjoint operators which are "badly" approximated by cylindrical functions. Their existence and the non-triviality of the corresponding operators depend on the ergodic properties of the spectral measure of the family. These results are connected with the earlier works of Streit [1] and Reed [2] and with the paper of A. Yu. Daletsky [1] in which the spectral properties of "regular" representations of the group of shifts of a nuclear space Φ in the space $L_2(\Phi',\mu)$ are studied; here μ is a measure on Φ' quasi-invariant with respect to shifts by vectors from Φ.

CHAPTER 7

§ 1. For the method of dressing operators and applications of it in constructive field theory, see the books of Hepp [1] and A. S. Schwartz [1]. A concrete choice of dressing operators that takes into account the specific features of the second quantization operators in a Schrödinger representation was proposed by Kondratyev [5]. In the particular case of two-dimensional models in field theory, this choice was earlier employed by Albeverio and Høegh-Krohn [1]. In the case of perturbation of the Hamiltonian of a free boson field, the statement of Theorem 1.1 leads to the well-known Gell-Mann–Low formula (see, for example, Simon [2, Theorem 5.19]).

§ 2. The Van Hove model has been studied by many authors at different levels of mathematical rigorousness and with the help of different approaches (from the formal renormalization theory to the algebraic approach in field theory). The physical meaning of this model, the methods of investigation, and the important consequences of its analysis can be found in the books of Segal [2, Chapter 5] and Emch [1, Chapter 1]. In our presentation of the Van Hove model, based on the general scheme of dressing from Section 1, we follow the work of Kondratyev [5]. The quadratic model also has its own history of investigation; the corresponding list of references is given in the book by Hepp [1, Chapter 4]; this book also contains an analysis of the quadratic model with the help of renormalization theory. For the Euclidean approach to the construction of the quadratic model, see the book of Glimm and Jaffe [1, Chapter 9].

§ 3. The construction of the dynamics of quantum lattice systems based on the construction of operator realization for the corresponding formal Hamiltonians was proposed by Kondratyev [9]. Harmonic systems appear in different branches of theoretical physics. Typical problems which lead to quantum harmonic models are described in the books by Maradudin, Montroll, and Weiss [1], Lipkin [1], and Bogolyubov and Bogolyubov (jr.) [1]. The classical harmonic systems, their time evolution, and ergodic properties are studied in the work of Lanford and Labowitz [1]. For the theory of linear canonical Bogolyubov transformations, see, for example, the books by Berezin [1] and Bogolyubov and Bogolyubov (jr.) [1]. Note that using the linear character of evolution in harmonic models, Sukhov [1] studied the problem of convergence of states on the algebra of quasilocal observables which corresponds to those states. Kondratyev [5, 9] constructed the dynamics for some models with anharmonicity.

§ 4. The idea of employing dressing operators to separate the scattering effects of self-action and to construct the generalized wave operators in quantum field problems is due to Faddeev [1]. The applications of this approach to concrete models of field theory are described in the book by Hepp [1] and A. S. Schwartz [1]. The algebraic approach to the construction of wave operators in field theory models was developed by Yu. Kato and Mugibayashi [1] and Høegh-Krohn [1]. Scattering theory in terms of C^*-algebras and its applications to quantum statistical physics were investigated by Robinson [1]. In the paper of Botwich and Malyshev [1], the unitary equivalence of free and perturbed Hamiltonians is established for lattice Fermi systems with local interaction by using the algebraic approach to the scattering problem. In the remarks to the book by Reed and Simon [3], the results in the scattering theory (in the language of C^*-algebras) are discussed in detail; in this book, one can also find a detailed list of references.

The statement of the scattering problem for the case of Dirichlet operators was proposed by Kondratyev and Koshmanenko [1]; this paper also contains applications to field models with cutoff polynomial interaction. The construction of wave operators in the case of unbounded identification operators is discussed in the paper of Koshmanenko [1].

References

de Acosta, A. D.

[1] *Existence and convergence of probability measures in Banach spaces,* Trans. Amer. Math. Soc. **152** (1970), 273–298.

Akhiezer, N.I.

[1] The Classical Moment Problem and Some Related Problems in Analysis, Fizmatgiz, Moscow, 1961; ***English transl.*** Hafner, New York, 1965.

Akhiezer, N.I. and Glazman, I.M.

[1] The Theory of Linear Operators in Hilbert Space, Nauka, Moscow, 1966; ***English transl. of 1st edn.*** Ungar, New York, 1961.

Albeverio, S., Hida, T., Potthoff, J., Röckner, M., and Streit, L.

[1] *Dirichlet forms in terms of white noise analysis* **I**: *Construction and QFT examples,* Reviews in Math. Phys. **1** (1990), 291–312.

[2] *Dirichlet forms in terms of white noise analysis* **II**: *Closability and diffusion processes,* Reviews in Math. Phys. **1** (1990), 313–323.

Albeverio, S. and Høegh-Krohn, R.

[1] *Dirichlet forms and diffusion processes on rigged Hilbert spaces,* Z. Wahrsch. und verw. Geb. **40** (1977), 1–57.

[2] *Hunt processes and analytic potential theory on rigged Hilbert spaces,* Ann. Inst. H. Poincaré **A 13** (1977), 269–291.

[3] *The energy representation of Sobolev–Lie groups,* Compositio Math. **36** (1978), 37–52.

[4] *Diffusion fields, quantum fields, and fields with values in Lie groups,* Stochastic Analysis and Applications. Dekker, New York, 1985, pp. 1–98.

Albeverio, S., Høegh-Krohn, R., and Streit, L.

[1] *Energy forms, Hamiltonians, and Brownian paths,* J. Math. Phys. **18** (1977), 907–917.

[2] *Regularization of Hamiltonians and processes,* J. Math. Phys. **21** (1980), 1636–1642.

Araki, H.

[1] *Hamiltonian formalism and canonical relations in quantum field theory,* J. Math. Phys. **1** (1960), 492–504.

Araki, H. and Nakagami, Y.

[1] *A remark on an infinite tensor product of von Neumann algebras,* Publ. Res. Inst. Math. Sci. Kyoto Univ. **8** (1972), 363–374.

Arkhangelsky, A.V. and Ponomaryov, V.I.

[1] Foundations of General Topology in Problems and Exercises, Nauka, Moscow, 1974. (*Russian*)

Bargman, V.

[1] *Hilbert spaces of analytic functions,* Commun. Pure and Appl. Math. **14** (1961), 187–214.

Barut, A. and Rączka, R.

[1] Theory of Group Representations and Applications, PWN, Warsaw, 1977.

Bateman, H. and Erdeyi, A.

[1] Higher Transcendental Functions, vol. **2**, Nauka, Moscow, 1966. (*Russian translation*)

Belavkin, V.P.,

[1] *A reconstruction theorem for a random quantum field,* Uspekhi Mat. Nauk **39** (1984), No. 2 (236), 137–138; ***English transl.*** in Russian Math. Surveys **39** (1984).

Belichkova, N.V.

[1] *Selfadjointness of degenerating elliptic operators,* Dokl. Akad. Nauk Ukrain. SSR Ser. A, No. 1 (1984), 3–6. *(Russian)*

Bely, A.G., Kovalenko, V.F., and Semenov, Yu.A.

[1] *On the continuity of generalized eigenfunctions of Schrödinger operators,* Rep. Math. Phys. **12** (1977), 307–310.

Berezansky, Yu.M.

[1] *On expansion in eigenfunctions of general selfadjoint differential operators,* Dokl. Akad. Nauk SSSR **108** (1956), 379–382. *(Russian)*

[2] *Generalization of the Bochner theorem to expansions in eigenfunctions of partial differential equations,* Dokl. Akad. Nauk SSSR **110** (1956), 893–896. *(Russian)*

[3] *On expansion in eigenfunctions of selfadjoint operators,* Ukrain. Mat. Zh. **11** (1959), 16–24; ***English transl.*** in Amer. Math. Soc. Transl. (2) **93** (1970).

[4] *A generalization of the multi-dimensional Bochner theorem,* Dokl. Akad. Nauk SSSR **136** (1961), 1011–1014; ***English transl.*:** Soviet Math. Dokl. **2** (1961), 143–147.

[5] Expansions in eigenfunctions of selfadjoint operators, Naukova Dumka, Kiev, 1965; ***English transl.*:** Amer. Math. Soc. Transl. vol. **17**, Providence, R.I., 1968.

[6] *On selfadjointness of field operators and on integral representations of Wightman type functionals,* Ukrain. Mat. Zh. **18** (1966), No. 3, 3–12. *(Russian)*

[7] *Integral representations of positive definite functionals of Wightman type,* Ukrain. Mat. Zh. **19** (1967), 89–95; ***English transl.*** in Ukrain. Math. J. **19** (1967).

[8] *Expansion in generalized eigenvectors and integral representation of positive definite kernels in the form of a functional integral,* Sibirsk. Mat. Zh. **9** (1968), 998–1013; ***English transl.*** in Siberian Math. J. **9** (1968).

[9] *Representation of functionals of Wightman type by functional integrals,* Funk. Anal. i Prilozh. **3** (1969), No. 2, 1–13; ***English transl.*** in Funct. Anal. Appl. **3** (1969).

[10] *The generalized power moment problem,* Trudy Mosk. Mat. Obshch. **21** (1970), 47–102; ***English transl.*** in Trans. Mosc. Math. Soc. **21** (1970).

[11] *On the generalized power moment problem,* Ukrain. Mat. Zh. **22** (1970), 435–460; ***English transl.*** in Ukrain. Math. J. **22** (1970).

[12] *A general theorem on moment matrices,* Dokl. Akad. Nauk Ukrain. SSR Ser. A (**1971**), No. 4, 291–294. *(Russian)*

[13] *Generalized power moment problems,* Hilbert Space Operators and Operator Algebras (Tihany, Hungary, 1970), North-Holland, Amsterdam, 1972, pp. 11–20.

[14] *Selfadjointness of elliptic operators with a singular potential,* Ukrain. Mat. Zh. **26** (1974), 579–590; ***English transl.*** in Ukrain. Math. J. **26** (1974).

[15] *A remark about essential selfadjointness of powers of an operator,* Ukrain. Mat. Zh. **26** (1974), 790–793; ***English transl.*** in Ukrain. Math. J. **26** (1974).

[16] *Selfadjointness of elliptic operators with infinitely many variables,* Ukrain. Mat. Zh. **27** (1975), 729–742; ***English transl.*** in Ukrain. Math. J. **27** (1975).

[17] *On expansion in joint generalized eigenvectors of an arbitrary family of commuting normal operators,* Dokl. Akad. Nauk SSSR **229** (1976), 531–533; ***English transl.*** in Soviet Math. Dokl. **17** (1976).

[18] Selfadjoint Operators in Spaces of Functions of Infinitely Many Variables, Naukova Dumka, Kiev, 1978; ***English transl.*:** see Berezansky [26].

[19] *Spectral representations of solutions of certain classes of functional and differential equations,* Dokl. Akad. Nauk Ukrain. SSR Ser. A (**1978**), No. 7, 579–583. *(Russian)*

[20] *Representations of positive definite kernels by functional integrals,* Operators in Mathematical Physics and Infinite-Dimensional Analysis, Inst. Mat. Akad. Nauk Ukrain. SSR, Kiev, 1979, pp. 5–35. *(Russian)*

[21] *Bilinear forms and Hilbert riggings,* Spectral Analysis of Differential Operators, Inst. Mat. Akad. Nauk Ukrain. SSR, Kiev, 1980, pp. 83–106; ***English transl.*** : Selecta Math. Sovietica **9** (1990), 235–253.

[22] *On selfadjointness of infinite-dimensional elliptic differential operators with singular potentials,* Funk. Anal. i Prilozh. **16** (1982), No. 4, 55–56; ***English transl.*** in Funct. Anal. Appl. **16** (1982).

[23] *The spectral theory of commuting normal operators and applications of it,* Uspekhi Mat. Nauk **38** (1983), No. 5 (233), 138. *(Russian)*

[24] *The spectral projection theorem,* Uspekhi Mat. Nauk **39** (1984), No. 4 (238), 3–52; ***English transl.*** in Russian Math. Surveys **39** (1984).

[25] *On the spectral projection theorem,* Ukrain. Mat. Zh. **37** (1985), 146–154. *(Russian)*

[26] Selfadjoint Operators in Spaces of Functions of Infinitely Many Variables, Amer. Math. Soc. Transl. vol. **63**, Providence, R.I., 1986.

Berezansky, Yu.M. and Gali, I.M.

[1] *Positive definite functions of infinitely many variables on a layer,* Ukrain. Mat. Zh. **24** (1972), 435–464; ***English transl.*** in Ukrain. Math. J. **24** (1972).

Berezansky, Yu.M., Gali, I.M., and Kondratyev, Yu.G.

[1] *Stone's theorem for the additive group of a Hilbert space,* Funk. Anal. i Prilozh. **11** (1977), No. 4, 68–69; ***English transl.*** in Funct. Anal. Appl. **11** (1977).

Berezansky, Yu.M., Gali, I.M., and Zhuk, V.A.

[1] *On positive definite functions of infinitely many variables,* Dokl. Akad. Nauk SSSR **203** (1972), 13–15; ***English transl.*** in Soviet Math. Dokl. **13** (1972).

Berezansky, Yu.M. and Gorbachuk, M.L.

[1] *On extension of positive definite functions of two variables,* Ukrain. Mat. Zh. **17** (1965), No. 5, 96–102. *(Russian)*

Berezansky, Yu.M. and Kalyuzhny, A.A.

[1] *Nuclear spaces of functions on the basis of a hypercomplex system,* Ukrain. Mat. Zh. **35** (1983), 9–17; ***English transl.*** in Ukrain. Math. J. **35** (1983).

[2] *Representations of hypercomplex systems with locally compact basis,* Ukrain. Mat. Zh. **36** (1984), 417–421; ***English transl.*** in Ukrain. Math. J. **36** (1984).

[3] *Spectral decompositions of representations of hypercomplex systems,* Spectral Theory of Operators and Infinite-Dimensional Analysis, Inst. Mat. Akad. Nauk Ukrain. SSR, Kiev, 1984, pp. 4–19. *(Russian)*

[4] *Hypercomplex systems with locally compact bases,* Selecta Math. Sovietica **4** (1985), 151–200.

[5] *Hypercomplex systems constructed from orthogonal polynomials,* Ukrain. Mat. Zh. **38** (1986), 275–284. *(Russian)*

Berezansky, Yu.M., Kondratyev, Yu.G., and Samoilenko, Yu.S.

[1] *Generalized functions of infinitely many variables and their applications in spectral theory,* Generalized Functions and Their Applications in Mathematical Physics (Proc. Internat. Conf., Moscow, 1980), Vychisl. Tsentr Akad. Nauk SSSR, Moscow, 1981, pp. 50–70. *(Russian)*

Berezansky, Yu.M. and Krein, S.G.

[1] *Hypercomplex systems with a continuous basis,* Uspekhi Mat. Nauk **12** (1957), No. 1 (73) 147–152; ***English transl.*** in Amer. Math. Soc. Transl. (2) **16** (1960).

Berezansky, Yu.M., Lassner, G., and Yakovlev, V.S.

[1] *On decomposition of positive functionals on commutative nuclear $*$-algebras,* Ukrain. Mat. Zh. **39** (1987), 638–641. *(Russian)*

Berezansky, Yu.M. and Orochko, Yu.B.

[1] *A remark concerning the growth of eigenfunctions of selfadjoint operators,* Ukrain. Mat. Zh. **14** (1962), No. 2, 180–184. *(Russian)*

Berezansky, Yu.M., Ostrovsky, V.L., and Samoilenko, Yu.S.

[1] *Expansions in eigenfunctions of the families of commuting operators and the representations of commutation relations,* Ukrain. Mat. Zh. **40** (1988), 106–109. *(Russian)*

Berezansky, Yu.M. and Samoilenko, V.G.

[1] *On selfadjointness of differential operators with finitely or infinitely many variables,* Dokl. Akad. Nauk Ukrain. SSR Ser. A (**1979**), No. 9, 691–695. *(Russian)*

[2] *Selfadjointness of differential operators with finitely or infinitely many variables and evolution equations,* Uspekhi Mat. Nauk **36** (1981), No. 5 (221), 3–56; ***English transl.*** in Russian Math. Surveys **36** (1981).

[3] *Selfadjointness of infinite-dimensional elliptic differential operators,* Trudy Mosk. Mat. Obshch. **50** (1987), 3–54. *(Russian)*

Berezansky, Yu.M. and Samoilenko, Yu.S.

[1] *Nuclear spaces of functions of infinitely many variables,* Ukrain. Mat. Zh. **25** (1973), 723–737; ***English transl.*** in Ukrain. Math. J. **25** (1973).

Berezansky, Yu.M. and Shifrin, S.N.

[1] *The generalized symmetric power moment problem,* Ukrain. Mat. Zh. **23** (1971), 291–306; ***English transl.*** in Ukrain. Math. J. **23** (1971).

Berezansky, Yu.M. and Us, G.F.

[1] *On expansions in eigenfunctions of selfadjoint operators admitting separation of infinitely many variables,* Dokl. Akad. Nauk SSSR **213** (1973), 1005–1008; ***English transl.*** in Soviet Math. Dokl. **14** (1973).

[2] *Eigenfunction expansions of operators admitting separation of an infinite number of variables,* Rep. Math. Phys. **7** (1975), 103–126.

Berezansky, Yu.M., Zhernakov, N.V., and Us, G.F.

[1] *Operator stochastic integrals,* Ukrain. Mat. Zh. **39** (1987), 144–149. *(Russian)*

Berezin, F.A.

[1] The Second Quantization Method, Nauka, Moscow, 1986. *(Russian)*

Berezin, F.A. and Shubin, M.A.

[1] The Schrödinger Equation, Izdat. Moskov. Univ., Moscow, 1983. *(Russian)*

Billingsley, P.

[1] Convergence of Probability Measures, Wiley & Sons, New York, 1968.

Birman, M.Sh. and Solomyak, M.Z.

[1] The Spectral Theory of Selfadjoint Operators in Hilbert Spaces, Izdat. Leningrad. Univ., Leningrad, 1980. *(Russian)*

Birman, M.Sh., Vershik, A.M., and Solomyak, M.Z.

[1] *A product of commutative spectral measures can fail to be countably additive,* Funk. Anal. i Prilozh. **13** (1979), No. 1, 61–62; ***English transl.*** in Funct. Anal. Appl. **13** (1979).

Bogolyubov, N.N. and Bogolyubov, N.N. (Jr.)

[1] Introduction to Quantum Statistical Mechanics, Nauka, Moscow, 1984. *(Russian)*

Bogolyubov, N.N., Logunov, A.A., Oksak, A.I., and Todorov, I.T.

[1] General Principles of Quantum Field Theory, Nauka, Moscow, 1987. *(Russian)*

Bogolyubov, N.N. (Jr.) and Prikarpatsky, A.K.

[1] *Quantum method of N.N. Bogolyubov's generating functionals in statistical physics: the Lie algebra of currents, its representations and functional equations,* Ukrain. Mat. Zh. **38** (1986), 284–289. *(Russian)*

Borchers, H.J. and Yngvason, J.

[1] *On the algebra of field operators. The weak commutant and integral decomposition of states,* Comm. Math. Phys **42** (1975), 231–252.

[2] *Integral representations for Schwinger functionals and the moment problem over nuclear spaces,* Comm. Math. Phys **43** (1975), 255–271.

[3] *Necessary and sufficient conditions for integral representations of Wightman functionals at Schwinger points,* Comm. Math. Phys **47** (1976), 197– 213.

Borchers, H.J. and Zimmerman, W.

[1] *On the self-adjointness of field operators,* Nuovo Cim. **31** (1964), 1047–1059.

Botwich, D.D. and Malyshev, V.A.

[1] *Unitary equivalence of temperature dynamics for ideal and locally perturbed Fermi-gas,* Comm. Math. Phys **91** (1983), 301–312.

Bourbaki, N.

[1] General Topology. Real Numbers in General Topology. Functional Spaces, Nauka, Moscow, 1975. *(Russian translation)*

[2] Integration. Measures on Locally-Compact Spaces. Extension of a Measure. Integration of Measures. Measures on Separable Spaces, Nauka, Moscow, 1977 *(Russian translation)*

Brezis, H. and Kato, T.

[1] *Remarks on the Schrödinger operators with singular complex potentials,* J. Math. Pures Appl. **58** (1979), 137–151.

Brüning, E.

[1] *On the characterization of relativistic quantum field theories in terms of finitely many vacuum expectation values* **I**, Comm. Math. Phys **58** (1978), 139–166; **II**, Comm. Math. Phys **58** (1978), 167–194.

Bukhvalov, V.B.

[1] *Applications of the methods in the theory of order-bounded operators to the theory of operators in the spaces* L^p, Uspekhi Mat. Nauk **38** (1983), No. 6 (234), 37–83; ***English transl.*** in Russian Math. Surveys **38** (1983).

Cameron, R. H. and Martin, W.T.

[1] *Fourier–Wiener transforms of analytic functionals,* Duke Math. J. **12** (1945), 489–507.

Chaus, N. N.

[1] *On classes of uniqueness for solutions of the Cauchy problem and representations of positive definite kernels,* Dokl. Akad. Nauk SSSR **163** (1965), 36–39; ***English transl.*** in Soviet Math. Dokl. **6** (1965).

[2] *Classes of uniqueness for solutions of the Cauchy problem and representations of positive definite kernels,* Proc. Sem. on Functional Analysis, vol. 1, Inst. Mat. Akad. Nauk Ukrain. SSR, Kiev, 1968, pp. 176–270. *(Russian)*

Chernoff, P. D.

[1] *Some remarks on quasi-analytic vectors,* Trans. Amer. Math. Soc. **167** (1972), 105–113.

[2] *Essential self-adjointness of powers of generators of hyperbolic equations,* J. Funct. Anal. **12** (1973), 401–414.

[3] *Quasi-analytic vectors and quasi-analytic functions,* Bull. Amer. Math. Soc. **81** (1975), 637–646.

[4] *Schrödinger and Dirac operators with singular potentials and hyperbolic equations,* Pacific J. Math. **72** (1977), 361–382.

Chumak, A.A.

[1] *Selfadjointness of the Beltrami–Laplace operator on a complete paracompact Riemannian manifold without boundary,* Ukrain. Mat. Zh. **25** (1973), 784–791; ***English transl.*** in Ukrain. Math. J. **25** (1973).

Cook, J. M.

[1] *The mathematics of second quantization,* Trans. Amer. Math. Soc. **81** (1956), 106–134.

Daletsky, A. Yu.

[1] *Representations of translation groups of a nuclear space,* Dokl. Akad. Nauk Ukrain SSR Ser. A (1982), No. 12, 9–11. *(Russian)*

[2] *The moment problem on finite-dimensional Lie algebras,* Spectral Theory of Operators in the Problems of Mathematical Physics, Inst. Mat. Akad. Nauk Ukrain. SSR, Kiev, 1983, pp. 93–97. *(Russian)*

[3] *Integration of representations of nuclear Lie algebras of smooth currents,* Spectral Theory of Operators and Infinite-Dimensional Analysis, Inst. Mat. Akad. Nauk Ukrain. SSR, Kiev, 1984, pp. 77–92. *(Russian)*

Daletsky, A.Yu. and Samoilenko, Yu.S.

[1] *The noncommutative moment problem,* Funk. Anal. i Prilozh. **21** (1987), No. 2, 72–73. *(Russian)*

Daletsky, Yu.L.

[1] *Infinite-dimensional elliptic operators and the parabolic equations connected with them,* Uspekhi Mat. Nauk **22** (1967), No. 4 (136), 3–54; ***English transl.*** in Russian Math. Surveys **22** (1967).

[2] *On selfadjointness and maximal dissipativeness of differential operators for functions of an infinite-dimensional argument,* Dokl. Akad. Nauk SSSR **227** (1976), 784–787; ***English transl.*** in Soviet Math. Dokl. **17** (1976).

[3] *Probability methods in some problems of infinite-dimensional analysis,* Random Operators and Distributions in Banach Spaces, Inst. Mat. Akad. Nauk Ukrain. SSR, Kiev, 1977, pp. 108–124. *(Russian)*

Daletsky, Yu.L. and Fomin, S.V.

[1] Measures and Differential Equations in Infinite-Dimensional Spaces, Nauka, Moscow, 1983; ***English transl.:*** Kluwer AP, Dordrecht, 1991.

Devinatz, A.

[1] *A note on semi-groups of unbounded self-adjoint operators,* Proc. Amer. Math. Soc. **5** (1954), 101–102.

[2] *Essential selfadjointness of certain partial differential operators on R^n,* Proc. Amer. Math. Soc. **60** (1976), 235–242.

[3] *Selfadjointness of second-order elliptic and degenerate elliptic differential operators,* Differential Equations (Proc. Internat. Conf., Uppsala, 1977), Almqvist & Wiksell, Stockholm, 1977, pp. 37–51.

[4] *Essential self-adjointness of Schrödinger-type operators,* J. Funct. Anal. **25** (1977), 58–69.

[5] *Selfadjointness of second order degenerate-elliptic operators,* Indiana Univ. Math. J. **27** (1978), 255–266.

[6] *On an inequality of Tosio Kato for degenerate-elliptic operators,* J. Funct. Anal. **32** (1979), 312–335.

Devinatz, A. and Nussbaum, A.E.

[1] *On the permutability of normal operators,* Ann. of Math. **65** (1957), 144–152.

[2] *Real characters of certain semi-groups with applications,* Duke Math. J. **28** (1961), 221–238.

Dieudonné, J.

[1] Fundamentals of Modern Analysis, Mir, Moscow, 1964. *(Russian translation)*

Dobrushin, R.L. and Minlos, R.A.

[1] *Polynomials of linear random functions,* Uspekhi Mat. Nauk **32** (1977), No. 2 (194), 67–122. *(Russian)*

Dunford, N. and Pettis, B.J.

[1] *Linear operations on summable functions,* Trans. Amer. Math. Soc. **47** (1940), 323–392.

Dunford, N. and Schwartz, J.T.

[1] Linear Operators, vol. **I**: General Theory, Interscience, New York and London, 1958.

[2] Linear Operators, vol. **II**: Spectral Theory. Selfadjoint Operators in Hilbert Spaces, Interscience, New York and London, 1963.

Dunkl, C.F.

[1] *The measure algebra of a locally compact hypergroup,* Trans. Amer. Math. Soc. **179** (1973), 331–348.

[2] *Structure hypergroups for measure algebras,* Pacific J. Math. **47** (1973), 413–425.

Emch, G.G.

[1] Algebraic Methods in Statistical Mechanics and Quantum Field Theory, Wiley–Interscience, New York, London, 1972.

Eskin, G.I.

[1] *A sufficient condition of solvability of the multi-dimensional moment problem,* Dokl. Akad. Nauk SSSR **133** (1960), 540–543. *(Russian)*

Faddeev, L.D.

[1] *On the separation of self-action and scattering effects, according to the perturbation theory,* Dokl. Akad. Nauk SSSR **152** (1963), 573–580. *(Russian)*

Faris, W.G.

[1] *Quadratic forms and essential self-adjointness,* Helv. Phys. Acta **45** (1972), 1074–1088.

[2] *Self-adjoint operators,* Lecture Notes in Math., vol. **437**, Springer-Verlag, Berlin, 1975.

Feissner, G.

[1] *Hypercontractive semigroups and Sobolev's inequality,* Trans, Amer. Math. Soc. **210** (1975), 51–62.

Feller, W.

[1] An Introduction to Probability Theory and Its Applications, 3rd edn., vol. **2**, Wiley & Sons, New York, London, Sydney, Toronto, 1971.

Friedrich, J.

[1] *A note on the two-dimensional moment problem,* Math. Nachr. **121** (1985), 285–286.

Friedrich, J. and Klotz, L.

[1] *On extensions of positive definite operator-valued functions,* Preprint, Karl-Marx-Univ., Leipzig, 1986.

Fröhlich, J.

[1] *Schwinger functions and their generating functionals.* **I**, Helv. Phys. Acta **47** (1974), 265–306.

[2] *Verification of axioms for Euclidean and relativistic fields and Haag's theorem in a class of $P(\varphi)_2$ models,* Ann. Inst. H. Poincaré **A 21** (1974), 271–317

Frolov, N.N.

[1] *Imbedding theorems for functions of countably many variables and their application to the Dirichlet problem,* Dokl. Akad. Nauk SSSR **203** (1972), 39–42; ***English transl.*** in Soviet Math. Dokl. **13** (1972).

[2] *On a coercivity inequality for an elliptic operator with infinitely many independent variables,* Mat. Sbornik **90** (**132**) (1973), 403–414; ***English transl.*** in Math. USSR Sb. **19** (1973).

[3] *On essential selfadjointness of an infinite-dimensional differential operator,* Mat. Zametki **24** (1978), 241–248; ***English transl.*** in Math. Notes **24** (1978).

[4] *Selfadjointness of an elliptic operator with infinitely many variables,* Funk. Anal. i Prilozh. **14** (1980), No. 1, 85–86; ***English transl.*** in Funct. Anal. Appl. **14** (1980).

[5] *On selfadjointness of a Schrödinger operator with infinitely many variables,* Sibirsk. Mat. Zh. **22** (1981), No. 1, 198–204; ***English transl.*** in Siberian Math. J. **22** (1981).

Fukushima, M.

[1] *Energy forms and diffusion processes,* Mathematics + Physics. Lectures on Recent Results, vol. **1**, World Scientific, Singapore, 1985, pp. 65–98.

Gachok, V. P.

[1] *Selfadjointness of field operators and the moment problem,* Ukrain. Mat. Zh. **17** (1965), No. 5, 3–13. *(Russian)*

[2] *Quasi-analytical functionals in the quantum field theory,* Dokl. Akad. Nauk SSSR **168** (1966), 1030–1032. *(Russian)*

Gårding, L.

[1] *Notes on continuous representations of Lie groups,* Proc. Nat. Acad. Sci. U.S.A. **33** (1947), 331–332.

[2] *Eigenfunction expansions,* Supplement **I** to **L. Bers, F. John, and M. Schechter,** Partial Differential Equations (Summer Sem., Boulder, Colo., 1957), Interscience, 1964; reprint, Amer. Math. Soc., Providence, R.I., 1974, pp. 301–325.

Gårding, L. and Wightman, A.S.

[1] *Representations of the anticommutation relations,* Proc. Nat. Acad. Sci. U.S.A. **40** (1954), 617–621.

[2] *Representations of the commutation relations,* Proc. Nat. Acad. Sci. U.S.A. **40** (1954), 622–625.

Geidarov, A.G.

[1] *On expansion in eigenfunctions of selfadjoint elliptic operators with singular potential,* Proc. All-Union Conf. Partial Differential Equations, Moscow, 1976; Izdat. Moskov. Univ., Moscow, 1978, pp. 287–288. *(Russian)*

Gelfand, I.M. and Kostyuchenko, A.G.

[1] *On expansion in eigenfunctions of differential and other operators,* Dokl. Akad. Nauk SSSR **103** (1955), 349–352. *(Russian)*

Gelfand, I.M., Raikov, D.A., and Shilov, G.E.

[1] Commutative Normed Rings, Fizmatgiz, Moscow, 1960; ***English transl.*:** Chelsea, New York, 1964.

Gelfand, I.M. and Shilov, G.E.

[1] Generalized Functions and Operations on Them, 2nd edn., Fizmatgiz, Moscow, 1959; ***English transl. of 1st edn.*:** Academic Press, New York, 1964.

[2] Spaces of Test and Generalized Functions, Fizmatgiz, Moscow, 1958; ***English transls.*:** Academic Press; Gordon and Breach, New York, 1968.

Gelfand, I.M. and Vilenkin, N.Ya.

[1] Some Applications of Harmonic Analysis. Rigged Hilbert Spaces, Fizmatgiz, Moscow, 1961; ***English transl.*:** Academic Press, New York, 1964.

Gestrin, G.P.

[1] *On expansion in eigenfunctions of the Schrödinger operator with singular potential,* Mat. Zametki **15** (1974), 458–465; ***English transl.*** in Math. Notes **15** (1974).

Getoor, R.K.

[1] *On semi-groups of unbounded normal operators,* Proc. Amer. Math. Soc. **7** (1956), 387–391.

Gikhman, I.I. and Skorokhod, A.V.

[1] Introduction to the Theory of Random Processes, Nauka, Moscow, 1965; ***English transl.*:** Saunders, Philadelphia, 1969.

[2] The Theory of Stochastic Processes, vol. **1**, Nauka, Moscow, 1971; ***English transl.*** **:** Springer-Verlag, Berlin, 1974.

Glazman, I.M.

[1] Direct Methods of Qualitative Spectral Analysis of Singular Differential Operators, Fizmatgiz, Moscow, 1963; ***English transls.*** **:** Israel Program Sci. Transls., Jerusalem, 1965, and Davey, New York, 1966.

Glimm, J. and Jaffe, A.

[1] Quantum Physics. A Functional Integral Point of View, Springer-Verlag, Berlin, 1981.

Golodets, V.Ya.

[1] *Description of representations of anticommutation relations,* Uspekhi Mat. Nauk **24** (1969), No. 4 (148), 3–64. *(Russian)*

Gorbachuk, V.I. and Gorbachuk, M.L.

[1] Boundary Value Problems for Operator Differential Equations, Naukova Dumka, Kiev, 1984; ***English transl.*** **:** Kluwer AP, Dordrecht, 1991.

Gross, L.

[1] *Harmonic analysis on Hilbert space,* Mem. Amer. Math. Soc. **46** (1963), 1–62.

[2] *Logarithmic Sobolev inequalities,* Amer. J. Math. **97** (1976), 1061–1083.

Guichardet, A.

[1] *Produits tensoriels continus d'espaces et d'algébres de Banach,* Comm. Math. Phys. **5** (1967), 262–287.

Guseinov, R.V.

[1] *On the theory of the Fourier–Wiener transformation,* Vestnik Moskov. Univ. Ser. I: Mat., Mekh. (1969), No. 4, 17–25; ***English transl.*** in Moscow Univ. Math. Bull. **24** (1969).

[2] *Some properties of the Fourier–Wiener transformation,* Vestnik Moskov. Univ. Ser. I: Mat., Mekh. (1970), No. 4, 39–49; ***English transl.*** in Moscow Univ. Math. Bull. **25** (1970).

Halmos, P.R.

[1] Measure Theory, Van Nostrand, Princeton, N.J., 1950.

Hasegawa, M.

[1] *On quasi-analytic vectors for dissipative operators,* Proc. Amer. Math. Soc. **29** (1971), 81–84.

Hegerfeldt, G.C.

[1] *Gårding domain and analytic vectors for quantum fields,* J. Math. Phys. **13** (1972), 821–827.

Hegerfeldt, G.C. and Melsheimer, O.

[1] *The form of representations of CCR for Bose fields and connection with finitely many degrees of freedom,* Comm. Math. Phys. **12** (1969), 304–323.

Hepp, K.

[1] Théorie de la renormalisation, Springer-Verlag, New York, 1970.

Hida, T.

[1] *Analysis of Brownian functionals,* Carleton Math. Lecture Notes, vol. **13**, 1975.

[2] Brownian Motion, Springer-Verlag, Berlin, 1980.

[3] *White noise and random fields — old and new,* Proc. of Preseminar for Internat. Conf. on Gaussian Random Fields, Nagoya, 1990, Pt. **3**, Nagoya Univ., Nagoya, 1991, pp. 1–10.

Hida, T., Potthoff, J., and Streit, L.

[1] *White noise analysis and applications,* Mathematics + Physics. Lectures on Recent Results, vol. **3**, World Scientific, Singapore, 1990, pp. 143–178.

Hille, E.

[1] *On semi-groups of transformations in Hilbert space,* Proc. Nat. Acad. Sci. U.S.A. **24** (1938), 159–161.

Hille, E. and Phillips, R.S.

[1] Functional Analysis and Semi-Groups, rev. edn., Amer. Math. Soc., Providence, R.I., 1957.

Høegh-Krohn, R.

[1] *On the spectrum of the space cutoff* $P(\phi)_2$ *Hamiltonian,* Comm. Math. Phys. **21** (1971), 256–260.

Ionescu Tulcea, C.

[1] *Spectral representations of semigroups of normal operators,* Proc. Nat. Acad. Sci. U.S.A. **44** (1958), 44–46.

[2] *Spectral representations of certain semi-groups of operators,* J. Math. and Mech. **8** (1958), 95–109.

Ionescu Tulcea, C. and Simon, A.B.

[1] *Spectral representations and unbounded convolution operators,* Proc. Nat. Acad. Sci. U.S.A. **45** (1959), 1765–1767.

Ismagilov, R.S.

[1] *Selfadjoint extensions of a system of commuting symmetric operators,* Dokl. Akad. Nauk SSSR **133** (1960), 511–514; ***English transl.*** in Soviet Math. Dokl. **1** (1960).

[2] *On unitary representation of groups of diffeomorphisms of compact manifold,* Izvestiya Akad. Nauk SSSR. Ser. Mat. **86** (1972), 180–208 (Russian).

[3] *On unitary representations of group* $C_0^\infty(X, G)$, $G = SU$, Mat. Sbornik **100**

Itô, K.

[1] *Complex multiple Wiener integral,* Japan J. Math. **22** (1952), 63–86.

Jewett, R.I.

[1] *Spaces with an abstract convolution of measures,* Adv. Math. **18** (1975), 1–101.

Kalyuzhny, A.A.

[1] *On integral representation of exponentially convex functions,* Ukrain. Mat. Zh. **34** (1982), 370–373; ***English transl.*** in Ukrain. Math. J. **34** (1982).

[2] *A theorem on the existence of a multiplicative measure,* Ukrain. Mat. Zh. **35** (1983), 369–371; ***English transl.*** in Ukrain. Math. J. **35** (1983).

Kato, T.

[1] Perturbation Theory for Linear Operators, Springer-Verlag, Berlin, 1966.

[2] *Schrödinger operators with singular potentials,* Israel J. Math. **13** (1972), 135–148.

[3] *A remark to the preceding paper by Chernoff,* J. Funct. Anal. **12** (1973), 415–417.

Kato, Yu. and Mugibayashi, M.

[1] *Regular perturbations in quantum field theory,* Progr. Theoret. Phys. **30** (1963), 103–133.

Kats, G.I.

[1] *On expansions in eigenfunctions of selfadjoint operators,* Dokl. Akad. Nauk SSSR **119** (1958), 19–22. *(Russian)*

[2] *Generalized elements of a Hilbert space,* Ukrain. Mat. Zh. **12** (1960), No. 1, 13–24. *(Russian)*

[3] *Spectral decompositions of selfadjoint operators in generalized elements of a Hilbert space,* Ukrain. Mat. Zh. **13** (1961), No. 4, 13–33. *(Russian)*

[4] *Generalized functions on a locally compact group and decompositions of unitary representations,* Trudy Mosk. Mat. Obshch. **10** (1961), 3–41. *(Russian)*

Kirillov, A.A.

[1] *Dynamical systems, factors, and representations of groups,* Uspekhi Mat. Nauk **22** (1967), No. 5 (137), 67–80. *(Russian)*

[2] *Representations of certain infinite-dimensional Lie algebras,* Vestnik Moskov. Univ. Ser. I: Mat., Mekh. (1974), No. 1, 75–83; ***English transl.*** in Moscow Univ. Math. Bull. **29** (1974).

Klein, A. and Landau, L.J.

[1] *Singular perturbations of positivity preserving semigroups via path space techniques,* J. Funct. Anal. **20** (1975), 44–82.

Kolmogorov, A.N.

[1] Fundamental Concepts of Probability Theory, ONTI, Moscow–Leningrad, 1936; ***in German*: Kolmogoroff, A.,** Grundbegriffe der Wahrscheinlichkeitsrechnung, Springer-Verlag, Berlin, 1933; ***English transl.*:** Chelsea, New York, 1950.

Kolmogorov, A.N. and Khinchin, A.Ya. [Kolmogoroff, A. and Khintchine, A.]

[1] *Über Konvergenz von Reihen, deren Glieder durch den Zufalle bestimmt werden,* Mat. Sbornik **32** (1925), 668–677.

Kolomytsev, V.I. and Samoilenko, Yu. S.

[1] *On a countable set of commuting selfadjoint operators and the algebra of local observables,* Ukrain. Mat. Zh. **31** (1979), 365–371; ***English transl.*** in Ukrain. Math. J. **31** (1979).

Kondratyev, Yu.G.

[1] *Positive definite functions on certain spaces of sequences,* Ukrain. Mat. Zh. **28** (1976), 27–35; ***English transl.*** in Ukrain. Math. J. **28** (1976).

[2] *The Wick powers of generalized Gaussian random processes,* Methods of Functional Analysis in the Problems of Mathematical Physics, Inst. Mat. Akad. Nauk Ukrain. SSR, Kiev, 1978, pp. 129–158. *(Russian)*

[3] *Nuclear spaces of entire functions in the problems of infinite-dimensional analysis,* Dokl. Akad. Nauk SSSR **254** (1980), 1325–1329; ***English transl.*** in Soviet Math. Dokl. **22** (1981).

[4] *Spaces of entire functions of infinitely many variables associated with the riggings of the Fock spaces,* Spectral Analysis of Differential Operators, Inst. Mat. Akad. Nauk Ukrain. SSR, Kiev, 1980, pp. 18–37. *(Russian)*

[5] *The second quantization operators and perturbations of them,* Preprint No. 82.28, Inst. Mat. Akad. Nauk Ukrain SSR, Kiev, 1982. *(Russian)*

[6] *Perturbations of the second quantization operators,* Funk. Anal. i Prilozh. **16** (1982), No. 4, 76–77; ***English transl.*** in Funct. Anal. Appl. **16** (1982).

[7] *The Kato inequality for the second quantization operators,* Ukrain. Mat. Zh. **35** (1983), 753–756; ***English transl.*** in Ukrain. Math. J. **35** (1983).

[8] *Dirichlet operators and smoothness of solutions of infinite-dimensional elliptic equations,* Dokl. Akad. Nauk SSSR **282** (1985), 269–273. *(Russian)*

[9] *Construction of dynamics for quantum lattice systems,* Dokl. Akad. Nauk SSSR **283** (1985), 82–85. *(Russian)*

Kondratyev, Yu.G. and Koshmanenko, V.D.

[1] *The scattering problem for operators associated with Dirichlet forms,* Dokl. Akad. Nauk SSSR **267** (1982), 285–288; ***English transl.*** in Soviet Math. Dokl. **26** (1983).

Kondratyev, Yu.G. and Samoilenko, Yu.S.

[1] *The integral representation of generalized positive definite kernels of infinitely many variables,* Dokl. Akad. Nauk SSSR **227** (1976), 800–803; ***English transl.*** in Soviet Math. Dokl. **17** (1976).

[2] *The spaces of trial and generalized functions of infinite number of variables,* Rep. Math. Phys. **14** (1978), 325–350.

[3] *Generalized derivatives of probability measures on* $\mathbb{R}^\infty$, Methods of Functional Analysis in the Problems of Mathematical Physics, Inst. Mat. Akad. Nauk Ukrain. SSR, Kiev, 1978, pp. 159–176. *(Russian)*

Kondratyev, Yu.G. and Tsykalenko, T.V.

[1] *The spaces of test and generalized functions of infinitely many variables,* Problems in the Theory of Probability Distributions, Inst. Mat. Akad. Nauk Ukrain. SSR, Kiev, 1983, pp. 123–133. *(Russian)*

[2] *Infinite-dimensional hyperbolic equations associated with Dirichlet operators,* Dokl. Akad. Nauk SSSR **288** (1986), 814–817. *(Russian)*

[3] *Dirichlet operators and differential equations associated with them,* Preprint No. 86.37, Inst. Mat. Akad. Nauk Ukrain. SSR, Kiev, 1986. *(Russian)*

Korotkov, V.B.

[1] *On integral operators with Carleman kernels,* Dokl. Akad. Nauk SSSR **165** (1965), 748–751; ***English transl.*** in Soviet Math. Dokl. **6** (1965).

Korsunsky, L.M.

[1] *Expansions in generalized eigenvectors and integral representations of invariant positive-definite kernels,* Ukrain. Mat. Zh. **24** (1972), 341–351; ***English transl.*** in Ukrain. Math. J. **24** (1972).

[2] *Integral representations of invariant positive-definite kernels,* Spectral Theory of Operators and Infinite-Dimensional Analysis, Inst. Mat. Akad. Nauk Ukrain. SSR, Kiev, 1976, pp. 26–87. *(Russian)*

Koshmanenko, V.D.

[1] *Operator representations for nonclosable quadratic forms and the scattering theory,* Dokl. Akad. Nauk SSSR **245** (1979), 295–298. *(Russian)*

[2] *Classification of singular perturbations of selfadjoint operators,* Preprint No. 82.34, Inst. Mat. Akad. Nauk Ukrain. SSR, Kiev, 1982. *(Russian)*

Koshmanenko, V.D. and Samoilenko, Yu.S.

[1] *On an isomorphism between the Fock space and a space of functions of infinitely many variables,* Ukrain. Mat. Zh. **27** (1975), 669–674; ***English transl.*** in Ukrain. Math. J. **27** (1975).

Kostyuchenko, A.G. and Mityagin, B.S.

[1] *Positive definite functionals on nuclear spaces,* Trudy Mosk. Mat. Obshch. **9** (1960), 283–316. *(Russian)*

Kosyak, A.V.

[1] *Analytic and entire vectors for families of operators,* Spectral Analysis of Differential Operators, Inst. Mat. Akad. Nauk Ukrain. SSR, Kiev, 1980, pp. 3–18. *(Russian)*

[2] *The Gårding domain for representations of canonical commutation relations,* Ukrain. Mat. Zh. **36** (1984), 709–715; ***English transl.*** in Ukrain. Math. J. **36** (1984).

Kosyak, A.V. and Samoilenko, Yu.S.

[1] *On families of commuting selfadjoint operators,* Ukrain. Mat. Zh. **31** (1979), 555–558; ***English transl.*** in Ukrain. Math. J. **31** (1979).

[2] *The Gårding domain and entire vectors for inductive limits of commuting locally compact groups,* Ukrain. Mat. Zh. **35** (1983), 427–434; ***English transl.*** in Ukrain. Math. J. **35** (1983).

Kovalenko, V.F., Perelmuter, M.A., and Semenov, Yu.A.

[1] *Schrödinger operators with* $L_{\omega}^{1/2}(\mathbb{R}^1)$*-potentials,* J. Math. Phys. **22** (1981), 1033–1044.

Kovalenko, V.F. and Semenov, Yu.A.

[1] *Some questions about expansions in generalized eigenfunctions of the Schrödinger operator with strongly singular potentials,* Uspekhi Mat. Nauk **33** (1978), No. 4 (202), 107–140; ***English transl.*** in Russian Math. Surveys **33** (1978).

Krée, P.

[1] *Prodistributions,* Séminaire Paul Krée 1974/1975, Inst. H. Poincaré, Paris, 1975, pp. 1–14.

[2] *Calcul d'integrales et de dérivées en dimension infinie,* J. Funct. Anal. **31** (1979), 150–186.

Krée, P. and Rączka, R.

[1] *Kernels and symbols of operators in QFT,* Ann. Inst. H. Poincaré **A 28** (1978), 41–73.

Krein, M.G.

[1] *On the problem of extension of Hermitian-positive continuous functions,* Dokl. Akad. Nauk SSSR **26** (1940), 17–21. *(Russian)*

[2] *On a general method of decomposing positive definite kernels into elementary products,* Dokl. Akad. Nauk SSSR **53** (1946), 3–6. *(Russian)*

[3] *On Hermitian operators with directing functionals,* Zbirnyk Prats Inst. Mat. Akad. Nauk Ukrain. RSR **10** (1948), pp. 83–106. *(Ukrainian)*

[4] *Hermitian positive kernels on homogeneous spaces,* **I:** Ukrain. Mat. Zh. **1** (1949), No. 4, 64–98; **II:** Ukrain. Mat. Zh. **2** (1950), No. 1, 10–59; ***English transl. of Part*** **I:** Amer. Math. Soc. Transl. (2) **34** (1963), 69–108.

Krein, M.G. and Krasnoselsky, M.A.

[1] *Fundamental theorems on the extension of Hermitian operators and some applications of these to the theory of orthogonal polynomials and to the moment problem,* Uspekhi Mat. Nauk, **2** (1947), No. 3 (19), 60–106. *(Russian)*

Krein, S.G.

[1] Linear Differential Equations in Banach Space, Nauka, Moscow, 1967; ***English transl.*:** Amer. Math. Soc., Providence, R.I., 1971.

[2] (editor) Functional Analysis, 2nd rev. edn., Nauka, Moscow, 1972; ***English transl. of 1st edn.*** **:** Noordhoff, 1972.

Kristensen, P., Mejlbo, L., and Poulsen, E.T.

[1] *Tempered distributions in infinitely many dimensions.* 1. *Canonical field operators,* Comm. Math. Phys. **1** (1965), 175–214.

Kubo, I. and Takenaka, S.

[1] *Calculus on Gaussian white noise.* **I**, Proc. Japan Acad. **56(A)** (1980), 376–380.
[2] *Calculus on Gaussian white noise.* **II**, Proc. Japan Acad. **56(A)** (1980), 411–416.
[3] *Calculus on Gaussian white noise.* **III**, Proc. Japan Acad. **57(A)** (1981), 433–437.
[4] *Calculus on Gaussian white noise.* **IV**, Proc. Japan Acad. **58(A)** (1982), 186–189.

Kuo, Hui-Hsiung

[1] *Gaussian measures in Banach spaces,* Lecture Notes in Math., vol. 463, Springer-Verlag, Berlin, 1975.
[2] *Lectures on white noise analysis,* Proc. of Preseminar for Internat. Conf. on Gaussian Random Fields, Nagoya, 1990, Pt. **1**, Nagoya Univ., Nagoya, 1991, pp. 1–65.

Kurepa, S.

[1] *A cosine functional equation in Hilbert space,* Canad. J. Math. **12** (1960), 45–50.

Kusuoka, S.

[1] *Dirichlet forms and diffusion processes on Banach spaces,* J. Fac. Sci. Univ. Tokyo Sec. 1A **29** (1982), 79–95.

Lanford, O. and Lebowitz, J.

[1] *Time evolution of harmonic systems,* Lecture Notes in Phys., vol. 38, Springer-Verlag, Berlin, 1975, pp. 144–177.

Lassner, G.

[1] *Topological algebras of operators,* Rep. Math. Phys. **3** (1972), 279–293.

Lax, P.D.

[1] *On Cauchy's problem for hyperbolic equations and the differentiability of solutions of elliptic equations,* Comm. Pure Appl. Math. **8** (1955), 615–633.

Lee, Y.-J.

[1] *Analytic version of test functionals, Fourier transform, and a characterization of measures in white noise calculus,* Preprint, National Univ. Tainan, Taiwan, 1989.

Leray, J.

[1] Lectures on Hyperbolic Equations with Variable Coefficients, Inst. Advanced Study, Princeton, N.J., 1952.

Levin, B.Ya. and Ovcharenko, I.E.

[1] *Extension of Hermitian-positive functions given on a band,* Teor. Funktsij Funk. Anal. i Prilozh., Issue 5, (1967), 68–83. *(Russian)*

Levitan, B.M.

[1] *On a theorem of Titchmarsh and Sears,* Uspekhi Mat. Nauk **16** (1961), No. 4 (100), 175–178. *(Russian)*
[2] Theory of Operators of Generalized Shift, Nauka, Moscow, 1973. *(Russian)*
[3] The Inverse Sturm–Liouville Problems, Nauka, Moscow, 1984. *(Russian)*

Lipkin, H.

[1] Quantum Mechanics, Mir, Moscow, 1977. *(Russian translation)*

Litvinov, G.L.

[1] *Hypergroups and hypergroup algebras,* VINITI Series on Contemporary Problems in Mathematics. The Latest Achievements, vol. **26**, VINITI, Moscow, 1985, pp. 57–106. *(Russian)*

Mackey, G.W.

[1] *Imprimitivity for representations of locally compact groups.* **I**, Proc. Nat. Acad. Sci. U.S.A. **35** (1949), 537–545.
[2] *Induced representations of locally compact groups.* **I**, Ann. Math. **55** (1952), 101–139.

Maltese, G.

[1] *Spectral representations for solutions of certain abstract functional equations,* Compositio Math. **15** (1962), 1–22.

[2] *Spectral representations for some unbounded normal operators,* Trans. Amer. Math. Soc. **110** (1964), 79–87.

Malyshev, V.A. and Minlos, R.A.

[1] Gibbsian Random Fields, Nauka, Moscow, 1985. *(Russian)*

Mandelbrojt, S.

[1] Séries Adhérentes, Régularisation des Suites, Applications, Gauthier-Villars, Paris, 1952.

Maradudin, A., Montroll, E., and Weiss, G.

[1] Dynamical Theory of Crystal Lattice in Harmonic Approximation, Mir, Moscow, 1965. *(Russian translation)*

Marchenko, A.V.

[1] *On inductive limits of linear spaces and operators, and their applications,* Vestnik Moskov. Univ. Ser. I: Mat., Mekh. (1974), No. 2, 26–33; ***English transl.*** in Moscow Univ. Math. Bull. **29** (1974).

[2] *Selfadjoint differential operators with infinitely many independent variables,* Mat. Sbornik **96** (**138**) (1975), 276–293; ***English transl.*** in Math. USSR Sb. **25** (1975).

[3] *Spaces of functions of infinitely many variables as inductive limits of locally convex spaces of functions,* Teor. Funktsij Funk. Anal. i Prilozh., Issue 24, (1975), 86–98. *(Russian)*

Marchenko, V.A.

[1] The Sturm–Liouville Operators and Applications of Them, Nauk. Dumka, Kiev, 1977. *(Russian)*

Markushevich, A.I.

[1] Theory of Analytical Functions, vol. **2**, Nauka, Moscow, 1968. *(Russian)*

Maurin, K.

[1] *Eine Bemerkung zur allgemeinen Eigenfunktionsentwicklungen für vertauschbare Operatorensysteme beliebiger Mächtigkeit,* Bull, Acad. Polon. Sci. Sér.: Sci. Math. Astr. Phys. **8** (1960), 381–384.

[2] Methods of Hilbert Spaces, PWN, Warsaw, 1959; ***English transl.,*** 1967; ***Russian transl.*** Mir, Moscow, 1965.

[3] General Eigenfunction Expansions and Unitary Representations of Topological Groups, PWN, Warsaw, 1968.

[4] *A remark on Berezanski version of spectral theorem,* Studia Math. **34** (1970), 165–167.

Menicoff, R. and Sharp, D.H.

[1] *Representations of a local current algebras: their dynamical determination,* J. Math. Phys. **16** (1975), 2341–2360.

Messon, D. and McClary, W.

[1] *Classes of C^∞ vectors and essential self-adjointness,* J. Funct. Anal. **10** (1972), 19–32.

Minlos, R.A.

[1] *Generalized random processes and their extensions to measures,* Trudy Moskov. Mat. Obshch. **8** (1959), 497–518; ***English transl.*** in Selected Transl. Math. Statist. and Probab., vol. **3**, Amer. Math. Soc., Providence, R.I., 1963.

Nachbin, L.

[1] Topology on Spaces of Holomorphic Mappings, Springer-Verlag, Berlin, 1969.

Naimark, M.A.

[1] Normed Rings, Nauka, Moscow, 1968; ***English transl.:*** Normed Algebras, Noordhoff, 1972.

Naimark, M.A. and Fomin, S.V.

[1] *Continuous direct sums of Hilbert spaces and some applications of them,* Uspekhi Mat. Nauk **10** (1955), No. 2 (64), 111–142; ***English transl.*** in Amer. Math. Soc. Transl. (2) **5** (1957).

Nakagami, Yo.

[1] *Infinite tensor product of von Neumann algebras.* **I**, Kodai Math. Semin. Rep. **22** (1970), 341–354; **II**, Publ. Res. Inst. Math. Sci. Kyoto Univ. **6** (1970), 257–292.

[2] *Infinite tensor product of operators,* Publ. Res. Inst. Math. Sci. Kyoto Univ. **10** (1974), 111–145.

Napiórkowski, K.

[1] *Continuous tensor products of Hilbert spaces and product operators,* Studia Math. **39** (1971), 307–327.

Nelson, E.

[1] *Analytic vectors,* Ann. of Math. (2) **70** (1959), 572–615.
[2] *The free Markoff fields,* J. Funct. Anal. **12** (1973), 97–112.

von Neumann, J.

[1] *On infinite direct products,* Compositio Math. **6** (1939), 1–77.

Nussbaum, A.E.

[1] *The Hausdorff–Bernstein–Widder theorem for semi-groups in locally compact Abelian groups,* Duke Math. J. **22** (1955), 573–582.
[2] *Integral representation of semi-groups of unbounded selfadjoint operators,* Ann. of Math. (2) **69** (1959), 133–141.
[3] *Quasi-analytic vectors,* Ark. Mat. **6** (1965), 179–191.
[4] *A note on quasi-analytic vectors,* Studia Math. **33** (1969), 305–309.
[5] *Spectral representation of certain one-parametric families of symmetric operators in Hilbert space,* Trans. Amer. Math. Soc. **152** (1970), 419–429.
[6] *Radial exponentially convex functions,* J. d'Analyse Math. **25** (1972), 277–288.
[7] *Integral representation of functions and distributions positive definite relative to the orthogonal group,* Trans. Amer. Math. Soc. **175** (1973), 355–387.
[8] *On functions positive definite relative to the orthogonal group and the representation of functions as Hankel–Stieltjes transforms,* Trans. Amer. Math. Soc. **175** (1973), 389–408.
[9] *Extension of positive definite functions and representation of functions in terms of spherical functions in symmetric spaces of noncompact type of rank 1,* Math. Ann. **215** (1975), 97–116.

Orochko, Yu.B.

[1] *A uniqueness theorem for a selfadjoint extension of a Schrödinger operator,* Dopovidi. Akad. Nauk. Ukrain. RSR (**1966**), No. 11, 1391–1394. *(Ukrainian)*

[2] *Some estimates at infinity for eigenfunctions of the Schrödinger operator,* Ukrain. Mat. Zh. **19** (1967), No. 3, 39–52; ***English transl.*** in Ukrain. Math. J. **19** (1967).

[3] *An estimate of the growth of averaged eigenfunctions of selfadjoint operators,* Ukrain. Mat. Zh. **20** (1968), 400–405; ***English transl.*** in Ukrain. Math. J. **20** (1968).

[4] *A remark on the essential selfadjointness of the Schrödinger operator with a singular potential,* Mat. Zametki **20** (1976), 571–580; ***English transl.*** in Math. Notes **20** (1976).

[5] *A sufficient condition for the essential selfadjointness of polynomials in the Schrödinger operator,* Mat. Sbornik **99(141)** (1976), 192–210; ***English transl.*** in Math. USSR Sb. **28** (1976).

[6] *The Carleman estimates for the Schrödinger operator with a locally semibounded strongly singular potential,* Mat. Sbornik **104(146)** (1977), 162–174; ***English transl.*** in Math. USSR Sb. **33** (1977).

[7] *The hyperbolic equation method in the spectral theory of a Schrödinger operator with a real singular potential,* Proc. All-Union Conf. Partial Differential Equations, Moscow, 1976; Izdat. Moskov. Univ., Moscow, 1978, pp. 403–404. *(Russian)*

[8] *On the theory of expansions in generalized eigenfunctions of selfadjoint operators,* Ukrain. Mat. Zh. **30** (1978), 63–69; ***English transl.*** in Ukrain. Math. J. **30** (1978).

[9] *Some properties of expansions in generalized eigenfunctions of selfadjoint Cárleman operators,* Sibirsk. Mat. Zh. **21** (1980), No. 6, 96–107. *(Russian)*

[10] *The operator cosine method in the problem of essential selfadjointness of a symmetric operator which is not semibounded,* Ukrain. Mat. Zh. **33** (1981), 348–355; ***English transl.*** in Ukrain. Math. J. **33** (1981).

[11] *On the property of a finite global rate of propagation of a second-order elliptic differential expression,* Diff. Uravneniya **18** (1982), 1764–1772; ***English transl.*** in Diff. Equations **18** (1982).

[12] *Finite local rate of propagation of a hyperbolic equation in the problem of selfadjointness of powers of a second-order elliptic differential operator,* Izvestiya Akad. Nauk SSSR Ser. Mat. **47** (1983), 298–314. *(Russian)*

[13] *On the Carleman resolvent estimate of Schrödinger type operator with arbitrary potential,* J. Oper. Theory **9** (1983), 27–52.

Osterwalder, K. and Schrader, R.

[1] *Axioms for Euclidean Green's functions.* **II**, Comm. Math. Phys. **42** (1975), 281–305.

Ostrovsky, V.L.

[1] *An analogue of the Nelson theorem for nuclear nilpotent Lie algebras of currents,* Spectral Theory of Operators and Infinite-Dimensional Analysis, Inst. Mat. Akad. Nauk Ukrain. SSR, Kiev, 1984, pp. 120–131. *(Russian)*

[2] *Irreducible representations of the group of infinite upper triangular matrices,* Ukrain. Mat. Zh. **38** (1986), 255–259. *(Russian)*

Ostrovsky, V.L. and Samoilenko, Yu.S.

[1] *An application of the spectral projection theorem to noncommutative families of operators,* Ukrain. Mat. Zh. **40** (1988), 478–493. *(Russian)*

Ovcharenko, I.E.

[1] *Integral representations of semi-definite many-dimensional power moment sequences,* Dokl. Akad. Nauk Ukrain. SSR Ser. A (**1983**), No. 3, 17–20. *(Russian)*

[2] *Moment sequences with representing measures free of masses in certain domains,* Dokl. Akad. Nauk SSSR **273** (1983), 41–45. *(Russian)*

[3] *On two-dimensional power moment sequences,* Ukrain. Mat. Zh. **36** (1984), 51–56. *(Russian)*

Paclet, Ph.

[1] *Espaces de Dirichlet en dimension infinie,* Compt. Rend. Acad. Sci. A **288** (1979), 981–983.

Perelmuter, M.A.

[1] *Perturbations of operators with integral resolvent,* Funk. Anal. i Prilozh. **14** (1980), No. 2, 75–76; ***English transl.*** in Funct. Anal. Appl. **14** (1980).

[2] *Positivity preserving operators and one criterion of essential self-adjointness,* J. Math. Anal. Appl. **82** (1981), 406–419.

[3] *The hyperbolic equation method in L^p-spaces,* Ukrain. Mat. Zh. **40** (1988), 494–500. *(Russian)*

Perelmuter, M.A. and Semenov, Yu.A.

[1] *Selfadjointness of elliptic operators with finitely and infinitely many variables,* Funk. Anal. i Prilozh. **14** (1980), No. 1, 81–82; ***English transl.*** in Funct. Anal. Appl. **14** (1980).

[2] *On finiteness of the rate of propagation of a perturbation for hyperbolic equations,* Ukrain. Mat. Zh. **36** (1984), 56–63; ***English transl.*** in Ukrain. Math. J. **36** (1984).

Phillips, R.S.

[1] *Spectral theory for semi-groups of linear operators,* Trans. Amer. Math. Soc. **71** (1951), 393–415.

Plesner, A.I.

[1] Spectral Theory of Linear Operators, Nauka, Moscow, 1965; ***English transl.*** **:** vols. **I**, **II**, Ungar, New York, 1969.

Plesner, A.I. and Rokhlin, V.A.

[1] *Spectral theory of linear operators,* Uspekhi Mat. Nauk **1** (1946), No. 1 (11), 71–191; ***English transl.*** in Amer. Math. Soc. Transl. (2) **62** (1967).

Pontryagin, L.S.

[1] Continuous Groups, Gostekhizdat., Moscow, 1954. *(Russian)*

Potthoff, J. and Streit, L.

[1] *A characterization of Hida distributions,* Preprint BiBoS, 1989.

Povzner, A.Ya.

[1] *On expansions of arbitrary functions in eigenfunctions of the operator* $-\Delta u + cu$, Mat. Sbornik **32** (**74**) (1953), 109–156; ***English transl.*** in Amer. Math. Soc. Transl. (2) **60** (1967).

Powers, R.T.

[1] *Self-adjoint algebras of unbounded operators.* **I**, Comm. Math. Phys. **21** (1971), 85–124; **II**, Trans. Amer. Math. Soc. **187** (1974), 261–293.

Prokhorov, Yu.V.

[1] *Convergence of random processes and limit theorems in probability theory,* Teor. Veroyatn. i Primen. **1** (1956), 177–238; ***English transl.*** in Theor. Probab. Appl. **1** (1956).

Pytlik, T.

[1] *Nuclear spaces on a locally compact group,* Studia Math. **50** (1975), 225–243.

Rauch, J. and Taylor, M.

[1] *Essential selfadjointness of powers of generators of hyperbolic mixed problem,* J. Funct. Anal. **12** (1973), 491–493.

Reed, M.

[1] *A Gårding domain for quantum fields,* Comm. Math. Phys. **14** (1969), 336–346.

[2] *On self-adjointness in infinite tensor product spaces,* J. Funct. Anal. **5** (1970), 94–124.

Reed, M. and Simon, B.

[1] Methods of Modern Mathematical Physics, vol. **1**, Academic Press, New York, San Francisco, London, 1972.

[2] Methods of Modern Mathematical Physics, vol. **2**, Academic Press, New York, San Francisco, London, 1975.

[3] Methods of Modern Mathematical Physics, vol. **3**, Academic Press, New York, San Francisco, London, 1979.

[4] Methods of Modern Mathematical Physics, vol. **4**, Academic Press, New York, San Francisco, London, 1978.

Richter, P.

[1] *Zur Extremalzerlegung von Zustanden,* Wiss. Z. Karl-Marx-Univ. Leipzig. Math.-Naturwiss. Reihe **27** (1978), 293–297.

[2] *Zerlegung einer Familie symmetrischer Operatoren nach gemeinsamen verallgemeinerten Eigenvektoren,* Preprint, Karl-Marx-Univ., Leipzig, 1979.

[3] *Zerlegung positiv definiter Kerne und Entwicklung nach gemeinsamen verallgemeinerten Eigenfunktionen für Familien streng kommutierender symmetrischer Operatoren,* Wiss. Z. Karl-Marx-Univ. Leipzig. Math. Naturwiss. Reihe **31** (1982), 63–68.

Riesz, F. and Szökefalvi-Nagy, B.

[1] Leçons d'Analyse Fonctionelle, 2nd edn., Akad. Kiadó, Budapest, 1952; ***English. transl.*:** Ungar, New York, 1955.

Robinson, D.

[1] *Return to equilibrium,* Comm. Math. Phys. **31** (1973), 171–189.

Rofe-Beketov, F.S.

[1] *On positive differential operators (deficiency indices, factorization, and perturbations),* Proc. Roy. Soc. Edinburgh Sect. **A 97** (1984), 247–257.
[2] *Necessary and sufficient conditions of finite rate of propagation for elliptic operators,* Ukrain. Mat. Zh. **37** (1985), 668–670. *(Russian)*

Ross, K.A.

[1] *Hypergroups and centers of measure algebras,* Ist. Naz. Alta Mat. (Symposia Math.) **22** (1977), 189–203.

Rudin, W.

[1] *An extension theorem for positive-definite functions,* Duke Math. J. **37** (1970), 49–53.

Rudinsky, I.I.

[1] *On integral representations of even positive definite functions on nuclear spaces,* Ukrain. Mat. Zh. **36** (1984), 522–524. *(Russian)*

Sakhnovich, L.A.

[1] *Effective construction of inextensible Hermitian-positive functions of several variables,* Funk. Anal. i Prilozh. **14** (1980), No. 1, 55–60; ***English transl.*** in Funct. Anal. Appl. **14** (1980).

Samoilenko, V.G.

[1] *On wave operators for a Schrödinger-type operator in the space of functions of infinitely many variables,* Operators in Mathematical Physics and Infinite-Dimensional Analysis, Inst. Mat. Akad. Nauk Ukrain. SSR, Kiev, 1979, pp. 103–109. *(Russian)*
[2] *On second-order elliptic operators with infinitely many variables,* Ukrain. Mat. Zh. **32** (1980), 405–409. *(Russian)*
[3] *On selfadjointness of a second-order elliptic operator with infinitely many variables,* Ukrain. Mat. Zh. **34** (1982), 647–650. *(Russian)*
[4] *On the spectrum of a differential operator with infinitely many variables,* Sibirsk. Mat. Zh. **23** (1982), 147–159. *(Russian)*

[5] *On the boundedness of a multiplication operator with respect to a second quantization operator,* Dokl. Akad. Nauk SSSR **281** (1985), 1316–1319. *(Russian)*

Samoilenko, Yu.S.

[1] *Matrix-valued kernels of the Wightman functional type,* Methods of Functional Analysis in Problems of Mathematical Physics, Inst. Mat. Akad. Nauk Ukrain. SSR, Kiev, 1971, pp. 201–254. *(Russian)*

[2] Spectral Theory of Families of Self-Adjoint Operators, Naukova Dumka, Kiev, 1984; ***English transl.*:** Kluwer AP, Dordrecht, 1991.

Samoilenko, Yu.S. and Us, G.F.

[1] *On differential operators with constant coefficients in spaces of functions of countably many variables,* Spectral Analysis of Differential Operators, Inst. Mat. Akad. Nauk Ukrain. SSR, Kiev, 1980, pp. 49–67. *(Russian)*

[2] *Differential operators with constant coefficients on functions of countably many variables,* Uspekhi Mat. Nauk **39** (1984), No. 1 (235), 155–156; ***English transl.*** in Russian Math. Surveys **39** (1984).

Sazonov, V.V.

[1] *A remark concerning characteristic functionals,* Teor. Veroyatn. i Primen. **3** (1958), 201–205; ***English transl.*** in Theor. Probab. Appl. **3** (1958).

Schaefer, H.

[1] Topological Vector Spaces, Macmillan, London, 1966.

Schechter, M.

[1] Principles of Functional Analysis, Academic Press, New York, London, 1971.

Schmüdgen, K.

[1] *Positive cones in enveloping algebras,* Rep. Math. Phys. **14** (1978), 385–404.

Schoenberg, I.J.

[1] *Metric spaces and completely monotone functions,* Ann. Math. **39** (1938), 811–841.

Schwartz, L.

[1] Théorie des Distributions, vols. **1**, **2**, Hermann, Paris, 1950–51.
[2] *Théorie des noyaux,* Proc. Internat. Congr. Math. (Cambridge, Mass., 1950), vol. **1**, Amer. Math. Soc., Providence, R.I., 1952, pp. 220–230.

Sears, D.B.

[1] *Note on the uniqueness of the Green's functions associated with certain differential equations,* Canad. J. Math. **2** (1950), 314–325.

Segal, I.

[1] *Tensor algebras over Hilbert spaces.* **I**, Trans. Amer. Math. Soc. **81** (1956), 106–134.
[2] Mathematical Problems of Relativistic Physics, American Mathematical Society, Providence, R.I., 1963.
[3] *Construction of non-linear local quantum process,* Ann. Math. **92** (1970), 462–481.

Semenov, Yu.A.

[1] *On the Lie–Trotter theorems in L_p-spaces,* Lett. Math. Phys. **1** (1977), 379–385.
[2] *On the theory of eigenfunctions expansion of the Schrödinger operator,* Lett. Math. Phys. **1** (1977), 463–465.

Shakhbagyan, R.L.

[1] *The general boundary value problem for high-order elliptic operators on infinite-dimensional manifolds with boundary,* Izvestiya Akad. Nauk Armyan. SSR Ser. Mat. **19** (1984), No. 1, 3–18. *(Russian)*

Shifrin, S.N.

[1] *On infinite-dimensional symmetric analogues of the Stieltjes moment problem,* Ukrain. Mat. Zh. **26** (1974), 696–700. *(Russian)*
[2] *The infinite-dimensional smooth symmetric power moment problem on nuclear spaces,* Ukrain. Mat. Zh. **28** (1976), 793–802. *(Russian)*

Shilov, G.E.

[1] Mathematical Analysis. Functions of Several Real Variables, Nauka, Moscow, 1972. *(Russian)*

Shilov, G.E. and Tin', Fan Dyk

[1] Integral, Measure, and Derivative on Linear Spaces, Nauka, Moscow, 1967. *(Russian)*

Shiryaev, A.N.

[1] Probability, Nauka, Moscow, 1980. *(Russian)*

Shnol, E.E.

[1] *On the behavior of eigenfunctions of the Schrödinger equation,* Mat. Sbornik **42** (**84**) (1957), 273–286. *(Russian)*

Shohat, J.A. and Tamarkin, J.D.

[1] The Problem of Moments, Amer. Math. Soc., vol. 14, Providence, R.I., 1943.

Shwartz, A.S.

[1] Mathematical Foundations of Quantum Field Theory, Atomizdat., Moscow, 1975. *(Russian)*

Simon, B.

[1] *The theory of semi-analytic vectors: A new proof of a theorem of Masson and McClary,* Indiana Univ. Math. J. **20** (1970/71), 1145–1151.

[2] The $P(\phi)_2$ Euclidean (Quantum) Field Theory, Princeton University Press, Princeton, N.J., 1974.

[3] *An abstract Kato's inequality for generators of positivity preserving semigroups,* Indiana Univ. Math. J. **26** (1977), 1067–1073.

[4] *A canonical decomposition for quadratic forms with applications to monotone convergence theorem,* J. Funct. Anal. **28** (1978), 377–385.

[5] *Spectrum and continuum eigenfunctions of Schrödinger operators,* J. Funct. Anal. **42** (1981), 347–355.

[6] *Schrödinger semigroups,* Bull. Amer. Math. Soc. (N.S.) **7** (1982), 447–526.

Simon, B. and Høegh-Krohn, R.

[1] *Hypercontractive semigroups and two-dimensional self-coupled Bose fields,* J. Funct. Anal. **9** (1972), 121–180.

Skorokhod, A.V.

[1] Integration in Hilbert Space, Nauka, Moscow, 1975; ***English transl.***: Springer-Verlag, Berlin, 1974.

Smolyanov, O.G. and Fomin, S.V.

[1] *Measures on linear topological spaces,* Uspekhi Mat. Nauk **31** (1976), No. 4 (190), 3–56; ***English transl.*** in Russian Math. Surveys **31** (1976).

Sobolev, S.L.

[1] Introduction to the Theory of Cubic Formulas, Nauka, Moscow, 1974. *(Russian)*

Spector, R.

[1] *Apercu de la théorie des hypergroups,* Lecture Notes in Math., vol. **497**, Springer-Verlag, Berlin, 1975, pp. 643–673.

Strătilă, S. and Voiculescu, D.

[1] *Representations of AF-algebras and of the group $U(\infty)$*, Lecture Notes in Math., vol. **486**, Springer-Verlag, Berlin, 1975.

Streater, R.F. and Wulfsohn, A.

[1] *Continuous tensor products of Hilbert spaces and generalized random fields,* Nuovo Cim. **57** (1968), 330–339.

Streit, L.

[1] *Test function spaces for direct product representations of the canonical commutation relations,* Comm. Math. Phys. **4** (1967), 22–31.

Sukhov, Yu.M.

[1] *Linear boson models of time evolution in quantum statistical mechanics,* Izvestiya Akad. Nauk SSSR Ser. Mat. **48** (1981), 155–191. *(Russian)*

Szökefalvi-Nagy, B.

[1] *Über messbare Darstellungen Liescher Gruppen,* Math. Ann. **112** (1936), 286–296.

[2] *On semi-groups of selfadjoint transformations in Hilbert space,* Proc. Nat. Acad. Sci. U.S.A. **24** (1938), 559–560.

Takeda, M.

[1] *On the uniqueness of Markovian self-adjoint extension of diffusion operators on infinite-dimensional spaces,* Osaka J. Math. **22** (1984), 733–742.

Tishchenko, S.V.

[1] *Spectral representation of an infinite-dimensional contractive semigroup of selfadjoint operators,* Operators in Mathematical Physics and Infinite-Dimensional Analysis, Inst. Mat. Akad. Nauk Ukrain. SSR, Kiev, 1979, pp. 115–120. *(Russian)*

[2] *Spectral representations of countably parametric semi-groups,* Ukrain. Mat. Zh. **34** (1982), 532–535; ***English transl.*** in Ukrain. Math. J. **34** (1982).

[3] *On the connection of the Minlos–Sazonov theorem with Stone's theorem for an additive group in Hilbert space,* Abstracts of the School on Theory of Operators in Functional Spaces, Minsk, 1982, p. 188. *(Russian)*

Titchmarsh, E.C.

[1] *On the uniqueness of the Green's functions associated with a second-order differential equation,* Can. J. Math. **1** (1949), 191–198.

Tsykalenko, T.V.

[1] *The infinite-dimensional Schrödinger operator and potential perturbations of it,* Ukrain. Mat. Zh. **38** (1986), 534–538. *(Russian)*

Tykhonov, A.N.

[1] *Sur les espaces abstraits,* Compt. Rend. Acad. Sci. **182** (1926), 1519–1520.

Us, G.F.

[1] *The truncated generalized symmetric power moment problem,* Ukrain. Mat. Zh. **26** (1974), 348–358; ***English transl.*** in Ukrain. Math. J. **26** (1974).

[2] *Spectral decomposition of selfadjoint operators which are functions of operators acting upon different variables,* Dokl. Akad. Nauk SSSR **229** (1976), 812–815; ***English transl.*** in Soviet Math. Dokl. **17** (1976).

[3] *On functions of operators acting upon different variables,* Operators in Mathematical Physics and Infinite-Dimensional Analysis, Inst. Mat. Akad. Nauk Ukrain. SSR, Kiev, 1979, pp. 121–138. *(Russian)*

Vainerman, L.I.

[1] *The duality of algebras with involution and the operators of generalized shift,* VINITI Series on Mathematical Analysis, vol. **24**, VINITI, Moscow, 1986, pp. 165–205. *(Russian)*

Vakhaniya, N.N., Tarieladze, V.I., and Chobanyan, S. A.

[1] Probability Distributions on Banach Spaces, Nauka, Moscow, 1985.

Ventzel, A.D.

[1] A Course of the Theory of Random Processes, Nauka, Moscow, 1975; ***German transl.*: A.D.Wentzell,** Theorie zufällige Prozesse, Akademie-Verlag, Berlin and Birkhäuser, Basel and Boston, Mass., 1979.

Vershik, A.M.

[1] *Measurable realizations of groups of automorphisms and integral representations of positive operators,* Sibirsk. Mat. Zh. **28** (1987), 52–60 *(Russian)*

Vershik, A.M., Gelfand, I.M., and Graev, M.I.

[1] *Representations of the group* $SL(2, R)$, *where* R *is a ring of functions,* Uspekhi Mat. Nauk **28** (1973), No. 5 (173), 83–128. *(Russian)*

[2] *Remarks on the representation of the groups of functions with values in a compact Lie group* **I**, Compositio Math. **35** (1977), 299–334; **II**, Compositio Math. **42** (1981), 217–243.

[3] *A commutative model of the representation of the group of currents* $SL(2,\mathbb{R})^X$ *connected with a unipotent subgroup,* Funk. Anal. i Prilozh. **17** (1983), No. 2, 70–72. *(Russian)*

Vershik, A.M. and Sudakov, V.N.

[1] *Probability measures on infinite-dimensional spaces,* Zapiski Nauch. Seminarov Lenigrad. Otdel. Mat. Inst. Akad. Nauk SSSR **12** (1969), 7–67. *(Russian)*

Vishik, M.I.

[1] *Parametrix of elliptic operators with an infinite number of independent variables,* Uspekhi Mat. Nauk **26** (1971), No. 2 (158), 155–174. *(Russian)*

[2] *Fundamental solutions of infinite-dimensional elliptic operators of arbitrary order with constant coefficients,* Dokl. Akad. Nauk SSSR **208** (1973), 764–767. *(Russian)*

Vishik, M.I. and Marchenko, A.V.

[1] *Boundary value problems for elliptic and parabolic operators of the second order on infinite-dimensional manifolds with an edge,* Mat. Sbornik **90** (132) (1973), 331–371. ***English transl.*** in Math. USSR Sb. **19** (1973).

Vladimirov, V.S.

[1] Generalized Functions in Mathematical Physics, Nauka, Moscow, 1976; ***English transl. of 2nd edn.*** **:** Mir, Moscow, 1979.

Vul, E.B.

[1] *On uniqueness theorems for a class of integral representations,* Dokl. Akad. Nauk SSSR **129** (1959), 722–725. *(Russian)*

Wiener, N.

[1] Nonlinear Problems in the Theory of Random Processes, Izdat. Inostr. Lit., Moscow, 1961. *(Russian translation)*

Woronowicz, S.L.

[1] *The quantum problem of moments.* **I**, Rep. Math. Phys. **1** (1970), 135–145; **II**, Rep. Math. Phys. **1** (1970), 175–183.

Yadrenko, M.I.

[1] Spectral Theory of Random Fields, Vyshcha Shkola, Kiev, 1980. *(Russian)*

Yakovlev, V.S.

[1] *On measurability of functions with respect to spaces with resolutions of the identity,* Spectral Theory of Operators and Infinite-Dimensional Analysis, Inst. Mat. Akad. Nauk Ukrain. SSR, Kiev, 1984, pp. 138–143. *(Russian)*

[2] *On representations of some commutative nuclear $*$-algebras,* Ukrain. Mat. Zh. **39** (1987), 235–243. *(Russian)*

[3] *On representations of some $*$-algebras of smooth functions,* Ukrain. Mat. Zh. **39** (1987), 400–403. *(Russian)*

Yokoi, Y.

[1] *Positive generalized white noise functionals,* Hiroshima Math. J. **20** (1990), 137–157.

[2] *Properties of Gelfand triplet in white noise analysis and a characterization of positive Hida distributions,* Proc. of Preseminar for International Conf. on Gaussian Random Fields (Nagoya, 1990), Pt. **2**, Nagoya Univ., Nagoya, 1991, pp. 34–48.

Yosida, K.,

[1] Functional Analysis, Mir, Moscow, 1967. *(Russian translation)*

Subject Index
Volume 2

Notation

Volume 2

Contents

Volume 1

GPSR Compliance

The European Union's (EU) General Product Safety Regulation (GPSR) is a set of rules that requires consumer products to be safe and our obligations to ensure this.

If you have any concerns about our products, you can contact us on ProductSafety@springernature.com

In case Publisher is established outside the EU, the EU authorized representative is:

Springer Nature Customer Service Center GmbH
Europaplatz 3
69115 Heidelberg, Germany

Batch number: 10823775

Printed by Printforce, the Netherlands